T0362421

FRONTIERS OF ENERGY AND ENVIRONMENTAL ENGINEERING

PROCEEDINGS OF THE 2012 INTERNATIONAL CONFERENCE ON FRONTIERS OF ENERGY AND ENVIRONMENTAL ENGINEERING (ICFEEE 2012), HONG KONG, CHINA, 11–13 DECEMBER 2012

Frontiers of Energy and Environmental Engineering

Editors

Wen-Pei Sung
National Chin-Yi University of Technology, Taiwan

Jimmy C.M. Kao
National Sun Yat-Sen University, Taiwan

Ran Chen
Chongqing University of Technology, China

CRC Press
Taylor & Francis Group
Boca Raton London New York

CRC Press is an imprint of the
Taylor & Francis Group, an **informa** business

CRC Press/Balkema is an imprint of the Taylor & Francis Group, an informa business

© 2013 Taylor & Francis Group, London, UK

Typeset by V Publishing Solutions Pvt Ltd., Chennai, India
Printed and bound in Great Britain by CPI Group (UK) Ltd, Croydon, CR0 4YY

Published by: CRC Press/Balkema
P.O. Box 447, 2300 AK Leiden, The Netherlands
e-mail: Pub.NL@taylorandfrancis.com
www.crcpress.com – www.taylorandfrancis.com

ISBN: 978-0-415-66159-1 (Hbk)
ISBN: 978-0-203-38610-1 (eBook)

Frontiers of Energy and Environmental Engineering – Sung, Kao & Chen (eds)
© 2013 Taylor & Francis Group, London, ISBN 978-0-415-66159-1

Table of contents

Intelligent environments

Energy conservation and environmental protection

Preface

2012 International Conference on Frontiers of Energy and Environment Engineering (ICFEEE 2012) will be held in HongKong during December 11–13, 2012. The aim is to provide a platform for researchers, engineers, academicians as well as industrial professionals from all over the world to present their research results and development activities in Energy and Environment Engineering.

In this conference, we received more than 500 submissions from email and electronic submission system, which were reviewed by international experts, and about 200 papers have been selected for presentation, representing 7 national and international organizations. I think that ICFEEE 2012 will be the most comprehensive Conference focused on the Energy and Environment Engineering. The conference will promote the development of Energy and Environment Engineering, strengthening the international academic cooperation and communications, and exchanging research ideas.

We would like to thank the conference chairs, organization staff, the authors and the members of International Technological Committees for their hard work. Thanks are also given to Alistair Bright.

We hope that ICFEEE 2012 will be successful and enjoyable to all participants. We look forward to seeing all of you next year at the ICFEEE 2013.

September, 2012

Wen-Pei Sung
National Chin-Yi University of Technology, Taiwan

Jimmy C.M. Kao
National Sun Yat-Sen University, Taiwan

Ran Chen
Chongqing University of Technology, China

ICFEEE 2012 committee

CONFERENCE CHAIRMAN

Prof. Wen-Pei Sung, *National Chin-Yi University of Technology, Taiwan*
Prof. Jimmy C.M. Kao, *National Sun Yat-Sen University, Taiwan*
Prof. Gintaris Kaklauskas, *Vilnius Gediminas Technical University, Lithuania*
Dr. Chen Ran, *Control Engineering and Information Science Research Association (CEIS)*

PROGRAM COMMITTEE CHAIRS

Yan Wang, *The University of Nottingham, U.K.*
Yu-Kuang Zhao, *National Chin-Yi University of Technology, Taiwan*
Yi-Ying Chang, *National Chin-Yi University of Technology, Taiwan*
Darius Bacinskas, *Vilnius Gediminas Technical University, Lithuania*
Viranjay M.Srivastava, *Jaypee University of Information Technology, Solan, H.P. India*
Ming-Ju Wu, *Taichung Veterans General Hospital, Taiwan*
Wang Liying, *Institute of Water Conservancy and Hydroelectric Power, China*
Chenggui Zhao, *Yunnan University of Finance and Economics, China*
Rahim Jamian, *Universiti Kuala Lumpur Malaysian Spanish Institute, Malaysia*
Li-Xin GUO, *Northeastern University, China*
Mostafa Shokshok, *National University of Malaysia, Malaysia*
Ramezan ali Mahdavinejad, *University of Tehran, Iran*
Anita Kovač Kralj, *University of Maribor, Slovenia*
Tjamme Wiegers, *Delft University of Technology, Netherlands*
Gang Shi, *Inha University, South Korea*
Bhagavathi Tarigoppula, *Bradley University, USA*
Viranjay M.Srivastava, *Jaypee University of Information Technology, Solan, H.P. India*
Shyr-Shen Yu, *National Chung Hsing University, Taiwan*
Yen-Chieh Ouyang, *National Chung Hsing University, Taiwan*
Shen-Chuan Tai, *National Cheng Kung University, Taiwan*
Jzau-Sheng Lin, *National Chin-Yi University of Technology, Taiwan*
Chi-Jen Huang, *Kun Shan University, Taiwan*
Yean-Der Kuan, *National Chin-Yi University of Technology, Taiwan*
Qing He, *University of north China electric power*
JianHui Yang, *Henan university of technology*
JiQing Tan, *Zhejiang university*
MeiYan Hang, *Inner Mongolia university of science and technology*
XingFang Jiang, *Nanjing university*
Yi Wang, *Guizhou normal university*
ZhenYing Zhang, *Zhejiang Sci-Tech University*
LiXin Guo, *Northeastern university*
Zhong Li, *Zhejiang Sci-Tech University*
QingLong Zhan, *Tianjin vocational technology normal university*
Xin Wang, *Henan polytechnic University*
JingCheng Liu, *The Institute of*
YanHong Qin, *Chongqing jiaotong university*
LiQuan Chen, *Southeast university*

Wang Chun Huy, *Nan Jeon Institute of Technology*
JiuHe Wang, *Beijing Information Science and Technology University*
Chi-Hua Chen, *Chiao Tung University*
FuYang Chen, *Nanjing university of aeronautics*
HuanSong Yang, *Hangzhou normal university*
Ching-Yen Ho, *Hwa Hsia College of Technology and Commerce*
LiMin Wang, *Jilin university*
ZhangLi Lan, *Chongqing jiaotong university*
XuYang Gong, *National pingtung university of science and technology*
YiMin Tian, *Beijing printing college*
KeGao Liu, *Shandong construction university*
QingLi Meng, *China seismological bureau*
Wei Fan, *Hunan normal university*
ZiQiang Wang, *Henan university of technology*
AiJun Li, *Huazhong university of science and technology*
Wen-I Liao, *Taipei university of science and technology*
BaiLin Yang, *Zhejiang university of industry and commerce*
Juan Fang, *Beijing University of Technology*
LiYing Yang, *Xian university of electronic science and technology*
NengMin Wang, *Xi'an jiaotong university*
Yin Liu, *Zhongyuan University of Technology*
MingHui Deng, *Northeast China agricultural university*
GuangYuan Li, *Guangxi normal university*
YiHua Liu, *Ningbo polytechnic institute, zhejiang university*
HongQuan Sun, *Heilongjiang university*

CO-SPONSOR

International Frontiers of science and technology Research Association
HongKong Control Engineering and Information Science Research Association

Energy saving buildings and equipment

Frontiers of Energy and Environmental Engineering – Sung, Kao & Chen (eds)
© *2013 Taylor & Francis Group, London, ISBN 978-0-415-66159-1*

Reason analysis and management plan on ground collapse of a teaching building

H. Wei & J. Yi
College of Civil engineering and Architecture, Hainan University, Haikou, China

ABSTRACT: As its special geographical situation and large rainfall climate characteristics in Haikou, which often affect the project and construction safety. This paper take the ground collapse of a teaching building in Haikou as an example, through field investigation and detection, analysis the reason and put forward a reasonable management plan.

Keywords: ground collapse; reason analysis; management plan

1 INSTRUCTION

The rainfall intensity in Haikou is big and rainfall is many from July to October 2010, which not only seriously influence the construction and safety of the project under construction, also caused some completed buildings' potential disease outbreaks. This paper take the ground collapse of a teaching building in Haikou as an example, through field investigation and detection, analysis the reason and put forward a reasonable management plan.

2 THE PROCESS OF THE ACCIDENT AND THE EXTENT OF THE INJURY

On October 11, 2010 early morning, the school teacher discovered that steps next to a teaching building staircase present cracks, in the afternoon it has formed a pit which diameter is about 1.0 m (seen in Fig. 1). At this point, the teaching building was in normal use, primary school students curiously went to the side of the hole and probe to the pit. If they were not carefully, they may fell into it. In order to

Figure 1. The pit of the ground collapse.

ensure the safety of students, the school immediately cover board on the pit, and closed down the stairs.

Next day, school leaders carefully check the pit, find that the development is so fast, just one night, the pithead has reached 1.5 m in diameter, the depth of the pit has reached 2~3.0 m, the biggest diameter can be 3.2 m. The school leader is very anxious and immediately organized the staff to completely close it with materials in any possible collapse range.

3 ANALYSIS OF THE ACCIDENT CAUSE

In the spot investigation, found that the pit is located in loose or slightly dense sand, cement soil surface has built on stilts. The inner collapse depth are not the same, and the deepest position is not directly below the pithead, but gradually become deeper by the north west pithed direction (classroom building). The deepest part in the pit is on the north side of the bottom, where there are some downward motion traces of water and sand, and sand has obvious flow characteristics that flow to North West bottom.

According to the above phenomenon, initially thought that there is a a underground cavern concealed under the teaching building. Long term strong rainfall make the concealed cavern partial collapse, overlying sand collapse, peripheral sand body move to collapse parts with the groundwater, leading the ground collapse. Moreover, the collapse process did not stop and have the tendency that expands further which must make management in order to guarantee the safety of teachers and students.

However, the type, strike and the size of the cavern is not clear. Therefore, school visits to the surrounding areas. The investigation found that the nearby units had discovered many such concealed caverns in their construction which mostly buries below the ground 3.0~5.0 m, wide approximately 1.0~2.0 m, high about 1.5 m and the strike is not explicit, most caved, the interior is frequently filled with water.

It should by means of accurate detect to sure whether the concealed cavern under this teaching building is the same with them that we found in surrounding.

4 DETECTION AND MANAGEMENT PLAN

In order to manage ground collapse of this teaching economic and effective we should first find out the location, size, stable state and water filling conditions of the concealed cavern which causes ground collapse, and then take appropriate management measures. Therefore, the management work can be divided into two stages: detection and management.

4.1 Detection plan

Based on the above investigation, drilling can accurately ensure the position and status of the concealed cavern. However, the cost of drilling is high and time is long, and has big influence on surrounding residents, especially influence the school normal teaching. Therefore, the ground radon measuring technology was proposed. The technique is relatively cheap, having no influence on the environment, and the investigation time is short. Comparison of the two reconnaissance method can be seen in Table 1.

Although radon measurement method has such advantages shown in the above table, but it can only ensure flat position of the concealed cavern, cannot ensure the depth, current situation and water filling conditions. Combine two methods to detect can have double results with the half work. First of all, using cheap ground radon measuring technology to detect flat position of the concealed cavern; and then, using drilling method to ensure the depth, current situation and water filling conditions in the certain flat position.

Specific detection arrangement is as follows: In each of the four sides of teaching building layout two radon detection line, dot pitch in line is 3.0 m (according to excavation width of concealed cavern is about 1~2.0 m). When radon is abnormal in some dots, then make more measure points around the abnormal dots. We should weave radon abnormal map, and explain abnormal, qualitative determination flat position of the cavern after the completion of all radon measurement work.

Based on the cavern's flat position determined by radon method, we should arrange corresponding holes according to the actual situation, so as to be able to accurately determine the cavern's depth, shape, size, ponding and damage situation, providing a sufficient basis for management work.

4.2 Management plan

There are many methods to dispose such underground concealed cavern, like shed roof, grouting, pile foundation, cut-and-fill and so on. Considering the teaching building's use and actual situation such as layer and ground water, proposed two ends intercepted, intermediate filled or grouted.

Intercept both ends means using grout-stone to intercept at teaching building both ends which in the same cavern to prevent groundwater flowing into the cavern which will dive sand erosion, eventually cause by subsidence of the ground. In the process of the construction, cut off wall in downstream should be set up wee page equipment and make filter layer to prevent sand flow.

When filling or grouting the middle part, small scale field experiments should be first conduct to know grouting pressure and appropriate slurry ratio to ensure the quality of grouting.

REFERENCES

Hong Wei & Lansheng Wang, Characteristics of radon anomalies in unloading rock of slope.
Yayang Liu & Hong Wei, Date analysis of deep foundation pit monitor.

Table 1. Comparison table of drilling and radon measurement method.

Contrast project	A drilling	B radon measurement	C = A/B
Funds	In a 100 m exploration line is calculated, every 2 m a hole, hole depth is 12 m, 130 yuan every meter, the exploration line needs 51 holes, require 79560 yuan.	3 meters a point, each 50 yuan, 100 m exploration line needs 34 measuring points, require 1700 yuan.	46.8
Working condition	Certain area, height, water, power act.	No condition requirement.	A is far greater than B
Time	A rig for 10 days.	An apparatus for 1 day.	10
Influence on surrounding residents	Noise, mud, effect traffic act.	No effect.	A is far greater than B

Frontiers of Energy and Environmental Engineering – Sung, Kao & Chen (eds)
© 2013 Taylor & Francis Group, London, ISBN 978-0-415-66159-1

Application of quasi-heat-moisture-ratio line approximation on the design of freezing dehumidification-based radiant air-conditioning system

C.Y. Tan
Hunan University of Technology, Zhuzhou, Hunan Province, PRC

H. Zhu
Central South University, Changsha, Hunan Province, PRC

S.Y. Liu
School of Civil engineering, Hunan University of Technology, Zhuzhou, Hunan Province, PRC

ABSTRACT: For the purpose of realizing non-dewfall condition of the radiant panels, low energy consumption and nice comfort of the freezing dehumidification-based radiant air-conditioning system, the design temperature and humidity of the room were figured out according to the mean comfort index of human body during the design. And in order to lower the energy consumption, the necessity of the reasonable load allocation between the dehumidification unit and the radiant unit was discussed by means of the calculation with the quasi-heat-moisture-ratio line. As a result, much more precise design parameters were acquired through the approximation calculation. And it was proven to be a good design method by applying these parameters to the design of the reading rooms of the library. In the design example of the reading room, the load ratio between the dehumidification unit and the radiant unit was 48 to 52, and the recirculation air flow rate of the dehumidification unit was not less than 36.8% of the total air rate of the room.

Keywords: freezing dehumidification; radiant air-conditioning system; quasi-heat-moisture-ratio line; assistant design

1   INTRODUCTION

The energy consumption of the air-conditioning system accounts for 35% of the building energy consumption in China [1], as a result of which, an energy-saving air-conditioning system is in great need. Radiant air-conditioning system is such a system which can lower the energy consumption by 20% compared to the traditional air-conditioning system [2]. In addition to the heat exchange methods that are similar to the traditional air-conditioning system, the radiant air-conditioning system can provide healthy and comfortable indoor environment with the method of radiation by means of installing a radiant ceiling [3]. However, there are always dewing problems in cooling conditions, which can never be avoided by improving the materials of the radiant panels [4]. In order to avoid the dewing problems, the heat treatment process and the humidity treatment process were performed respectively. As a result, the energy-saving performance of the radiant air-conditioning system depends not only on the terminal devices but also on the energy consumption of the dehumidification unit. In this research, the load carried by each unit of the system was discussed according to energy-saving performance of the radiant air-conditioning system.

2   DESIGN AND CALCULATION OF FREEZING DEHUMIDIFICATION-BASED RADIANT AIR-CONDITIONING SYSTEM EMPLOYING THE QUASI-HEAT-MOISTURE-RATIO LINE APPROXIMATION

The freezing dehumidification-based radiant air-conditioning system usually consists of the radiation unit, dehumidification unit and fresh air unit. In the system, it is the reasonable load allocation between the dehumidification unit and the radiation unit that can realize the purpose of lowering the energy consumption. That is to say, only if the mixing temperature of the frozen, dehumidified air and the air that doesn't flow through the dehumidifier (usually with a higher temperature) equals the design temperature of the indoor air is the system an energy-saving one. Therefore, the key point of the research is the reasonable allocation of the load between the dehumidification unit and the radiation unit.

Four assumptions were presented when studying the freezing dehumidification-based radiant air-conditioning system. First, the load of the fresh air, carried only by the fresh air unit, has nothing to do with the dehumidification unit and the radiation unit. Usually the fresh air is treated to the design state by the fresh air unit. Second, the moisture load is carried by the dehumidification unit. Third, the rest of the cooling load, except the cooling load of the fresh air, is carried by both the dehumidification unit and the radiation unit. Finally, the air in the room is divided into two parts. One part of the air whose mass is m_1 varies from the design state point N to the dew point L, treated by the dehumidification unit, the other part whose mass is m_2 varies from the initial point of the indoor air to point 2, along the moisture line, treated by the radiant panel. Afterward, two parts of the air mix together and the condition of the mixed air is just at the design point N. According to these assumptions, the process diagram of combined effects is presented in Figure 1.

The load allocation of the freezing dehumidification-based radiant air-conditioning system was conducted according to the following steps.

Firstly, prepare the approximate heat-humidity-ratio line and make it go through the apparatus dew point, L. When drawing the approximate heat-humidity-ratio line, make sure that the horizontal distance between point N and point L equals the unit-time humidity load, so that the approximately calculated quasi-heat-moisture-ratio line and the actual heat-humidity-ratio line can match well. Point 2 is an endpoint in the extended line of the quasi-heat-moisture-ratio line, which is the intersection point of the consistent humidity line of the initial point (point O) of the air and the approximate heat-humidity-ratio line. Point N is the mixing point of he air whose state point is L (whose mass is m_1) and the air whose state point is 2 (whose mass is m_2).

According to the thermal balance analysis, the equation (1) to (3) could be presented.

$$m_1 i_L + m_2 i_2 = m i_N \tag{1}$$

$$m_1 d_L + m_2 d_2 = m d_N \tag{2}$$

$$m_1 + m_2 = m = \rho V \tag{3}$$

Solving the three equations simultaneously, the values of m_1 and m_2 could be obtained. After the acquisition of m_1 and m_2, the load of each unit of the freezing dehumidification-based radiant air-conditioning system could be estimated according to equation (1) to (3).

The load of the fresh air unit could be obtained according to equation (4).

$$q_1 = 0.1\, m(i_0 - i_N) + 0.001 \times 0.1\, m(d - d_N)r \tag{4}$$

The load of the dehumidification unit could be calculated as equation (5).

$$q_2 = m_1(i_N - i_L) + 0.001\, m_1(d_N - d_L)r \tag{5}$$

The load of the radiant panels could be acquired as equation (6).

$$q_3 = m_2(i_0 - i_2) + q_4 \tag{6}$$

The load of the radiant panel could be expressed as equation (7).

$$q_3 = m_2(i_0 - i_2) + q_4 S \tag{7}$$

The symbol q_4 represents the cooling load of the radiant panels, which can be found in reference [5].

$$q_4 = C_b \frac{\overset{-4}{t_r} - t_p^4}{R_d} = 4.73 \times 10^{-8}\left(\overset{-4}{t_r} - t_p^4\right) \tag{8}$$

Therefore, the total cooling load of the air-conditioning system is the sum of the load of the three units above. That's:

$$Q` = q_1 + q_2 + q_3 \tag{9}$$

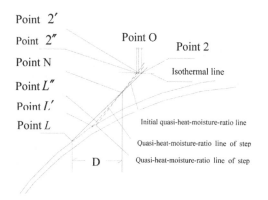

Figure 1. Process diagram of the combined effects of the dehumidifier and the radiant panel.

In the equations above, the symbol m, m_1 and m_2 means the mass of the air, and the symbol i_L, i_N, i_2 and i_0 is the enthalpy values, and d_L, d_N, d_2 represents the absolute humidity of the air. In addition, r means the latent heat of vaporization whose value is 2439 KJ/kg. And the symbols q_1, q_2, q_3, q_4 and Q mean the cooling load. The symbol S means the area of the radiant panel. The symbols t_r and t_p represent the mean radiant temperature and the mean temperature of the radiant panel respectively. Finally, C_b is the blackbody radiation constant whose value is 5.67×10^{-8} $W/(m \cdot K)$, and R_d means the radiate heat transfer coefficient that can be expressed as $R_d = 1 - \varepsilon_1/\varepsilon_1 + 1/X_{1-2} + 1 - \varepsilon_2/\varepsilon_2 (A_1/A_2)$, where ε_1 means the surface blackness of the ceiling, ε_2 is the surface blackness of the surface that is not for the cooling purpose. A_1 represents the area of the lower surface of the ceiling while A_2 represents the area of the rest of ceiling except the radiant panels. X_{1-2} means the angle factor of radiant heat transfer of the surface that is not for the cooling purpose.

Judging from the psychometric chart, the initial quasi-heat-moisture-ratio line is determined by the state of point A and the dehumidifying capacity D. Therefore, if the dehumidifying capacity is too large, the quasi-heat-moisture-ratio line will not meet the saturation line at point L, which explains one question why the humidity load can not be too large. If the humidity load is too large, it will cost so much power to remove the humidity that the energy-saving performance of the radiant air-conditioning system will not be embodied. In conclusion, it is the magnitude of the moisture load that determines whether or not to adopt the radiant air-conditioning system.

What worth attention is that the results figured out according to the quasi-heat-moisture-ratio line is inaccurate. In order to diminish the inaccuracy, approximate calculation step by step was required by increasing the air mass m_1, till the cooling load met the relationship in equation (10).

$$\frac{Q-Q}{Q} \leq 5\% \qquad (10)$$

Secondly, replace the quasi-heat-moisture-ratio line obtained from step 1 with the one determined by relationship $\varepsilon = i_N - i_L/D$, and make the new line go through point N, thus new apparatus dew-points L' and $2'$ could be acquired. Then substitute these new parameters into equation (1) to (9) and conduct the calculation, the $m_1`$, $m_2`$, $d_1`$, $d_2`$, $i_L`$, $i`_2$ could be obtained.

Thirdly, replace the quasi-heat-moisture-ratio line obtained from step 2 with the one determined by relationship $\varepsilon'' = i_N - i_L/D - \Delta d$ where Δd is the difference of moisture content between L and

L', and make the line go through point N, thus the new apparatus dew-points L''' and $2''$ could be obtained. Substitute these new parameters into equation (1) to (9) and conduct the calculation again, then the m_1'', m_2''', d_1''', d_2'', i''_L, i_2'' could be obtained. After that, conduct the approximate calculation again till the results meet the demand of the relationship in equation (10).

Finally, by allocating the load according to the relationships $q_2/q_2 + q_3$ and $q_3/q_2 + q_3$, the radiant unit and freezing dehumidification unit (whose circulating air rate is determined by m_1) could be designed.

There is an engineering example introducing the design of the reading rooms of the library of Hunan University of technology. The area of the reading rooms is 520 m^2, with a storey height of 3.4 meters. The reading rooms have a cooling load of 72.8 kw and a humidity load of 8.04 g/kg. When conducting the approximate calculation for the first time, made i_L = 29 KJ/kg, d_L = 7.5 g/kg, i_N = 68 KJ/kg, d_N = 15.5 g/kg. Then the values of i_0, i_2 and d_0 were assumed to be 82 KJ/kg, 80 KJ/kg and 18 g/kg respectively. Provided that the mass of the air was 2121 kg, the results of the calculation were m_1 = 499 kg, m_2 = 1622 kg and d_2 = 17.96 g/kg, therefore $Q - Q/Q$ = 72.8 - 67.94/72.8 = 6.7%. The results could not meet the demand of equation (10), therefore another approximate calculation was needed. Provided ε = $68 - 29/8.04 \times 10^{-3}$ = 4850, as a result of which, the values of i_L, d_L, i_2 and d_2 were acquired again and whose values turned out to be 42 KJ/kg, 10.5 g/kg, 77 KJ/kg and 17.5 g/kg respectively, therefore the results of the 2nd calculation turned out to be m_1 = 546 kg, m_2 = 1575 kg, and $Q - Q/Q$ = 72.8 - 64.64/72.8 = 11.2%. The results could not meet the demand of equation (10) either, so one more calculation was needed. Provided that the value of ε'' could be expressed as ε'' = $68 - 41/8.04 \times 10^{-3} - (10.8 - 7.5) \times 10^{-3}$ = 5485, the value of i_L, d_L, i_2 and d_2 could be obtained and these values were 43 KJ/kg, 11 g/kg, 76 KJ/kg and 17 g/kg respectively, therefore the results turned out to be m_1 = 475 kg, m_2 = 1646 kg and $Q - Q/Q$ = 72.8 - 66/72.8=9.34%, which did not meet the demand of equation (10) again. As a result, more calculations were needed. Increase the value of m_1 and conduct the approximate calculation, the results finally turned out to be m_1 = 780 kg, m_2 = 1341 kg and $Q - Q/Q$ = 72.8 - 72.4/72.8 = 0.55%, which met the relationship in equation (10).

Given that the load of the fresh air was undertaken by the fresh air unit, and the load of the dehumidification unit and the radiant panel was allocated according to a certain proportion, the results shows that q_2 equals 29.79 kw and q_3 equals 32.23 kw, which means that the dehumidification unit accounts for 48.03% of the load while the radiant panel accounts for 51.97%.

In conclusion, the cooling load of the dehumidification unit should be not less than 48% with a circulating air rate not less than 36.8% of the total air flow rate, and the cooling load of the radiant panel should be around 52%.

3 CONCLUSIONS

A series of useful information was acquired during the research, which provided a valuable reference for the engineers in this field. The results were listed below.

Applying the quasi-heat-moisture-ratio line, which is determined by the indoor air state point N and humidity load D, to the design of freezing dehumidification-based radiant air-conditioning system, more precise design parameters were acquired. So it was proven to be a good design method.

The moisture load could not be too high or the energy-saving performance of the radiant air-conditioning system would never be exhibited. And the magnitude of the moisture load was the key factor that determined whether or not to adopt the radiation air-conditioning system.

As for the freezing dehumidification-based radiant air-conditioning system, the load ratio of the dehumidification unit and the radiant panel turned out to be 48 to 52, and the recirculation air flow rate of the dehumidifier was not less than 36.8% of the total air flow rate of the room.

ACKNOWLEDGEMENTS

This study was financially supported by the Supported by Nation Key Technology R&D Program (2011BAJ03B07). Especially, it is highly appreciated that Dr Wang Han-qing, head of Hunan University of Technology, pays much attention to this study.

REFERENCES

[1] Meng Zhang, Li Ma. Summer experimental research of new ceiling radiant cooling panel. *Building energy and environment*, 29:27–31.
[2] Yang Wang. Energy Saving of Radiation Cooling System. *Building energy and environment*, 29:71–74.
[3] Baolian Niu, Qianqian Song, Xueying Xia, et al. Energy-saving refrigeration and dehumidification unit on radiant cooling and its energy-consumption analysis. *Journal of Nanjing Normal University*, 9:39–41.
[4] Qing He, Jinming Shen, Qingyun Shou, et al. Discussion on condensation of radiant cooling system combined with DOAS. *HV&AC*, 38:159–162.
[5] Xuelai Liu. Dynamic models and experimental study on capillary grating plane air-conditioning system. *China University of petroleum*, 2011.

Frontiers of Energy and Environmental Engineering – Sung, Kao & Chen (eds)
© 2013 Taylor & Francis Group, London, ISBN 978-0-415-66159-1

System research on directional cracking control blasting design in rock drift tunneling

Y.D. Bian, H.K. Pan & Y.G. Zhang
School of Civil Engineering & Architecture, Zhongyuan University of Technology, Zhengzhou, China

ABSTRACT: This paper studies the major factors influencing on the directional cracking control blasting design in rock drift tunneling systematically, and the blasting design system based on expert system technique has been developed. The system is based on the rule knowledge and blasting examples, which can improve the reliability of the results to a certain extent since the two modes can verify each other. The system has been tested and applied in Shandong Xinwen mining industry group, which can be very effective, reduce the tunneling cost and improve the speed of tunneling in engineering practice.

Keywords: expert system; rule knowledge; blasting; tunneling

1 INTRODUCTION

The technique of directional cracking control blasting is widely applied and can get the better blasting effect in different rock masses such as hard, medium hard, and soft breaking rock. The technique can control the tunnel shaping accurately, reduce the overbreak and the damage to adjacent rock, and keep the stability of the rock itself (Yang et al. 1996, Li & Zhuang. 1999). Meanwhile, the technique is advanced in reducing project cost and improving the drift tunneling speed than other general blasting techniques.

Because of many factors influencing on directional cracking control blasting design in rock drift tunneling, the project designers often intend to design with their past intuitive working experience only, the design precision just relies on the individual knowledge and experience. For the purpose of blasting rock with the scheduled cross-section in high speed and high quality, this paper studies on the blasting parameters systematically, and applies the expert system technique to the design of directional cracking control blasting.

2 EXPERT KNOWLEDGE OF DIRECTIONAL CRACKING CONTROL BLASTING IN ROCK DRIFT TUNNELING

The technique of directional cracking control blasting does not need to change the existing equipment configuration and basic drift tunneling process, controls precisely the directional cracking mainly by adjusting the blasting parameters properly. Therefore, the system research on blasting design is changed to that of blasting parameters (Gao & Yang. 1999).

2.1 *System research on blasting design*

2.1.1 *The selection of blasting plan*
The blasting plan mainly refers to the type of the blasting around the rock roadway, the depth of drilling and blasting, and the diameter of the borehole. The fracture around rock roadway can be controlled by the method of grooving the borehole or cumulating the pipe joints. It is the depth of each course of drilling and blasting which decides the amount of the drilling and muck loading, the cycle footage and cycle index per shift in every drifting cycle. According to different depth of each course of drilling and blasting, the perimeter blasting is classified into short hole, medium-length hole and deep hole blasting. It can improve the drift tunneling speed and reduce the support cost by using the same standard of drilling bolt hole and drifting borehole in rock drift tunneling.

2.1.2 *The selection of slotting mode*
Both cylinder cut and bevel cut have advantages and disadvantages respectively. The plan should be selected comprehensively by the compatibility of the geological conditions, the practicability of the construction technology, the reliability of the blasting effect and the rationality of the economic benefits in the blasting construction of rock drift tunneling.

2.1.3 The selection of blasting parameters

In addition to slotting mode, the main blasting parameters include unit explosive consumption, the diameter, the depth and the numbers of the borehole. The selection of these parameters should be determined comprehensively by the geological conditions, rock mass property, construction implement, blasting equipment, the application of the blasting technique and so on.

2.2 The application of the expert system technology

Based on a great deal of knowledge in specific areas, the expert system establishes the rule model by categorizing and inducing the knowledge. The problem can be matched reasoning according to the model, and then get the complex problems' answer that can be solved only by the human expert knowledge. In general, the expert system includes knowledge base, inference mechanism, database, man-machine interaction and explanation mechanism.

2.2.1 The establishment of expert knowledge base

The expert knowledge base of directional cracking control blasting is used storing specialized knowledge (definition, norm, specification and standard, etc.), blasting examples and so on, which provide by colliery engineers. The establishment of knowledge base is key to achieve the blasting design expert system. With the form of the rule, the process shows the relationship between the blasting design and the main factors affecting the design. These main factors include qualitative and quantitative factors, such as the rock brittle-ductility, rock mass blastability rank, rock joint development level, mash gas and coal dust status, weak interlayer, the protodyaknove's number, the tunnel cross section, and so on (Gao & Yang. 1999, Liu. 2009, M. & H. 2008).

2.2.2 Inference mechanism

The role of inference mechanism is to make use of the knowledge of base sufficiently, rationally, and efficiently. According to the classification of the expert knowledge, the inference mechanism includes two kinds, one is based on rule knowledge and the other is based on blasting examples.

1. Inference based on rule knowledge
   After entry of original data, the inference based on rule knowledge applies the corresponding rule knowledge base in turn according to the design subsequence including blasting plan, drilling and blasting equipment, slotting mode, and borehole layout. Comparing the original data and intermediate result with the knowledge of rule knowledge base, the inference is accomplished. The specific inference process is shown in Figure 1.

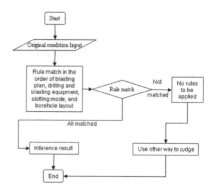

Figure 1. The process chart of inference based on rule knowledge.

Table 1. The selection of slotting mode.

Slotting method	Sectional area (M) m²	Rock mass blastability	Weak layer	Slotting mode
Bevel	$M < 8$	Grade III	Exist	Scallop
Bevel	$M < 8$	Grade IV	Exist	Scallop
Bevel	$M < 8$	Grade V	Exist	Scallop
Cylinder	$8 \leq M < 14$	Grade I	Not exist	Diamond
Cylinder	$8 \leq M < 14$	Grade II	Not exist	Diamond

Table 2. The calculation of the borehole depth.

Sectional area (M) m²	Protodyaknove number (f)	Joint develop degree	Drilling equip.	Borehole depth m
$0 < M \leq 12$	$0 < f < 4$	Most	Pusher leg drill	3.0
$0 < M \leq 12$	$0 < f < 4$	More	Pusher leg drill	2.5
$0 < M \leq 12$	$0 < f < 4$	No	Pusher leg drill	2.0

The inference process adopts production rule knowledge expression form, it can be expressed as: if A then B, which express if A is feasible, then B is feasible too 6. The partial inference mode about slotting mode is shown in Table 1. And the partial inference mode about the borehole depth is shown in Table 2. Other rules in blasting design, such as the layout of cut hole, surrounding hole and adjective hole, it cannot be discussed here for lack of space.

2. Inference based on blasting examples

The inference based on blasting examples is a method of reasoning by analogy. Its core lies in resolving new problems by using the past examples and experience. In the inference process, the question is offered firstly, the system extracts similar examples from the example library according to the initial condition of the question. And then the most similar example is found from the similar examples, or the current requirements are satisfied by modifying the object project through multi-example assembling. Finally, the final answer of the design issues is added into example library as a new example, it will be used in the following design as the new original example.

The matching degree indicates the compatibility in the inference process. The matching degree is defined as the closeness of the initial condition between the original blasting practice and the certain blasting examples. It is related to not only the numbers of the match conditions, but also every factor's weighting coefficient of the blasting design in the original conditions (Yang et al. 2004).

3 THE APPLICATION OF BLASTING DESIGN SYSTEM

The directional cracking control blasting design system in rock drift tunneling adopts modularization programming mode by expert system technique. Every module holds the maximum independence, and every knowledge base is also mutually independent. The system uses the drop down menu, the popup menu and the dialog box. The whole system is an open system because of its extensible characteristics.

The system has been tested using in affiliated coal industry of Shandong Xinwen mining industry group since the system is developed. The inference results are similar to that of based on rule knowledge and blasting examples, which shows the reliability of the system.

4 CONCLUSION

The blasting design system based on expert system technique is established by systematical research on directional cracking control blasting in rock drift tunneling. The system is accord with the engineering designers, and easy to be promoted. The inference mode is based on the rule knowledge and blasting examples, that can improve the reliability of the results to a certain extent since the two modes can verify each other. It has a profound and far reaching significance for coal mining safety and efficient production with the wide application of the system.

ACKNOWLEDGMENTS

This work was supported by the Program for Young Leading Teachers in Universities of Henan Province, China (2009GGJS-088) and the Natural Science Foundation of Henan Department of Education, China (2009A130004 and 12A130003).

REFERENCES

Gao, E.X. & Yang R.S. 1999. *Blasting engineering.* Xuzhou: China university of mining and technology press.

Li, X.L. & Zhuang J.Z. 1999. New technology of directional fracture controlled blasting and its field application. *Blasting.* 16(3): 11–14.

Liu, H.D. 2009. Study of the design of rock tunnel excavation blasting. *Sci-tech information development & economy.* 19(35): 225–227.

Monjezi M. & H. Dehghani. 2008. Evaluation of effect of blasting pattern parameters on back break using neural networks. *International Journal of Rock Mechanics and Mining Sciences.* 45(8):1446–1453.

Wang, S.L. & Yuan Z.H. 1991. *The Design Principles of Expert System.* Beijing: Science press.

Yang, R.S., Cao, H.Y., Wang, W. & et al. 2004. Key technologies of directional cracking blasting expert system for mine roadway. *Coal science and technology.* 32(2): 56–58.

Yang, Y.Q., Zhang, Q., Yu, M.S. & et al. 1996. Technology of directional split blasting. *Journal of china coal society.* 21(3): 56–60.

Frontiers of Energy and Environmental Engineering – Sung, Kao & Chen (eds)
© *2013 Taylor & Francis Group, London, ISBN 978-0-415-66159-1*

The amount estimation of interception in the thorny bamboo plantation

C. Tang & I.C. Jeng
Department of Soil and Water Conservation, National Pingtung University of Science and Technology, Pingtung, Taiwan

Y.M. Hong
Department of Design for Sustainable Environment, MingDao University, Changhua, Taiwan

W.P. Sung
Department of Landscape Architecture, National Chin-Yi University of Technology, Taichung, Taiwan

ABSTRACT: This study estimated the interception amount of Thorny Bamboo by a series of field experiment. Although the interception amount cannot be measured directly, this study measured the amount of the inner rainfall, the throughfall, and the gross rainfall to estimate the interception amount by the water budget equation. Experimental outcomes show that thorny bamboo plantation can intercept more rainfall than other species mentioned in the previous studies, so as to storage more water in the watershed. This study also established the regression equations for the amount estimation of inner rainfall, throughfall and interception by a known gross rainfall amount. Experimental outcomes showed that the canopy of thorny bamboo plantation can intercept all of the raindrops when rainfall depth is smaller than 5 mm. However, the plenty of antecedent rainfall will reduce the interception amount and increase the throughfall amount.

1 INTRODUCTION

Taiwan located in the sub-tropical region, which is warm and rainy. In addition, the fragile geology and steep terrain of Taiwan usually induces a rapid and torrential flow in streams. In general, forest can reduce the volume of surface runoff, so as to decrease the disaster occurrence such as a landslide or debris flow. According the statistics result of Forestry Bureau, Council of Agriculture, Executive Yuan (2005), the forest occupies the 58.5% of Taiwan area, a suitable method of forest management will be helpful for the disaster prevent in hillslope. Specially, the longtime drought and the mud rock geology induce many exposed areas in the southwestern area of Taiwan. i.e. a large area of forestry denudation usually triggered serious flood or debris disasters in recent years. In conclusion, forestation is an urgent mission in southwest Taiwan. Of course, an appropriate selection of species can improve the forestation performance. Thorny bamboo (Bambusa Stenostachy Hackel) can survive in the drought and infertile area. The nodes of the branch and root can densely sprout to keep the plantation canopy with well closing pattern (Tang et al. 2002). Therefore, the thorny bamboo is a suitable species for forestation in

southwestern Taiwan, and now is widely used in the protection of stream bank, the conservation of water resource, the stability of hillslope, and the economic forestation of folk. In fact, according the statistics of the Forestry Bureau of Taiwan (2005), the thorny bamboo occupied the 80% area of forestation in the southwest Taiwan.

When raindrops reached the canopy, the canopy will stop some raindrops (interception), in which part of them may slowly flow along the stem to the ground (stemflow), and the other may stop in the canopy to evaporate into the air gradually, respectively. The remaining raindrops penetrate the canopy to the ground (throughfall). In short, interception of canopy lets the forest like a natural reservoir. Arnell (2002) pointed that rainfall interception was recognized as a hydrological process of considerable importance in water resource management. Specially, interception influences the analysis results of hydrological model (Lu & Tang 1995). Therefore, the measurement of interception amount is important for the performance estimation of water resource conservation. Interception depends strongly on the rainfall duration/intensity, the vegetation structure and the meteorological conditions controlling evaporation during and after rainfall (Rutter et al. 1971). In general, the influence factors of interception include the climate, region, season, vegetation type and crown density. For instance, Gash (1979) found that a small storm may immediately wet the canopy, and the water will evaporate into the air from the saturating canopy surface. In addition, interception may be changed by the region. Lu & Tang (1995) and Lin et al. (1996) obtained that interception occupied 4.3% of the total rainfall volume in central Taiwan, and 10.5% in northeastern Taiwan, respectively. The drizzles tend to produce more interception loss than intensive showers with the same the total rainfall (Zeng et al. 2000).

Interception is not easy to obtain by direct measurement. However, interception can be estimated by the difference value between gross rainfall (rainfall outside the forest district) and stemflow. Assuming the gross rainfall equals to 100 mm, there are several studies to estimation interception amount. Pan (1964) and Pan (1965) found that the interception amount of Taiwania cryptomerioides and Taiwan Zelkova are 4.45 mm and 5.71 mm, respectively; Chiang (1964) obtained that interception amount of Taiwan giant bamboo, thorny bamboo, green bamboo and makino bamboo were 9.69 mm, 8.52 mm, 4.72 mm and 6.13 mm; Chiang (1971) estimated that interception amount of Taiwan acacia, Taiwania horsetail pine, Taiwan giant bamboo, thorny bamboo, green bamboo and makino bamboo were 12.25 mm, 12.99 mm, 19.07 mm, 12.4 mm, 11.5 mm, 9.69 mm and 0.96 mm. Especially, the aforementioned studies obtained by the same author (Chiang 1964 and Chiang 1971) but different years exist an apparent difference in the interception amount of thorny bamboo. Moreover, The interception amount of natural forest canopy in central and northeastern Taiwan was 8.30 mm and 10.74 mm, respectively (Lu & Tang 1995; Lin et al. 1996).

However, a long duration and high intensity rainfall will reduce the interception capacity. For example, the saturated interception amount of Picea-Morrisonicola Hay forest at Tatachia alpine ecosystem, was about 1.5 to 8 mm (Wey et al. 2009), which is smaller than that obtained by other studies. On the other hand, the water retained in branches and stems also influence the estimation of interception amount. For instance, the specific water retention capacity of branches and stems obtained by Llorens et al. (1997) was 0.62 mm. Moreover, high temperature can increase the evaporation, and a high speed wind will disperse the water vapor into the air. Therefore, an accurate estimated interception needs to avoid the significant vary by climate and canopy structure.

In conclusion, a suitable species with steady canopy structure and a closed topography with stable climate are the basis of accurate estimation for the interception of the canopy. Therefore, a closed topography watershed with a single species forest of thorny bamboo in southwestern Taiwan is selected as the experimental area. The inner forest rainfall, gross rainfall and throughfall are measured to estimate the interception of canopy by the water budget method, so as to give the basis of management for thorny bamboo plantations.

2 EXPERIMENTAL DESIGN

2.1 Water budget method

This study adopted the water budget equation to describe the movement of rainfall through the canopy as follows.

$$P_i = T_f + S + I \tag{1}$$

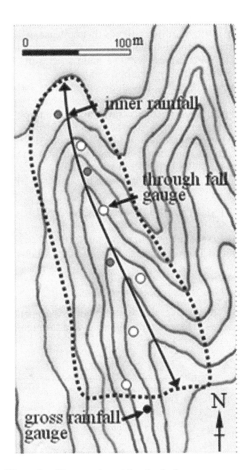

Figure 1. The experimental watershed.

where T_f is the throughfall; Pi is the inner rainfall (rainfall in the forest region); S is the stemflow; and I is the interception. Tobón et al. [15] estimated that the stemflow occupied about 1.1% of gross rainfall (P_o). The ratio between stemflow and gross rainfall in thorny bamboo is about 4.2% (Lu & Tang 1995), which is adopted in this study. The equation (1) can be revised as follows.

$$I = 0.958Pi - T_f \qquad (2)$$

This study used Equation (2) to calculate the interception.

2.2 Field experiment

The experiment watershed (Fig. 1), geographic coordinates of 22°38'20.6" N latitude and 120°36'48.4" E longitude, is located in the soil and water conservation outdoor classroom of National Pingtung University of Science and Technology. The watershed area is 3.33ha, with 310 m long, 110 m wide, elevation range between 90–50 m, and average slope 7.4°. A valley passes through the watershed from north northwest to south-southeast, where the slope of the right hand hillside is 21.8°, and left hand 18.3°, respectively. The main species of the watershed is thorny bamboo.

Along the valley, this study set up five 1×1 m^2 rainfall holders under the canopy of thorny bamboo plantation to collect travel, and three tipping-bucket rain gauges (Davis, Vantage Pro, 6152, USA) in the bare region to obtain inner rainfall. In addition, one rain gauge was set up 40 m away from the valley to collect gross rainfall.

3 RESULTS AND DISCUSSION

During September 2010 to August 2011, this study obtained 57 groups of data to process the following analysis.

3.1 Interception amount estimation

Table 1 displays the statistics of experimental outcomes. The gross/inner rainfall and throughfall

are obtained by the field measurement, and the interception amount is obtained by Equation (2). According the average value of measured outcomes, we can calculate the ratio for P_i/P_o, I/P_o, and T_f/P_o are 93.2%, 39.4% and 45.9%, respectively. However, T_f/P_o, measured in the forest is within the range of 82 to 87% (Tobón Marin et al. 2000). I/P_o, obtained in Pinus sylvestris is 30% (Llorens & Gallart 2000). Moreover, Carlyle-Moses (2004) obtained the I/P_o and T_f/P_o in a semi-arid Sierra Madre Oriental matorral community are $8.2 \pm 2.7\%$ and $83.3 \pm 1.9\%$, respectively. It is obvious that the interception amount for this study is larger than others. Consequently, thorny bamboo plantation can intercept more rainfall than other species, and reduces the throughfall amount, so as to storage more water in the watershed.

3.2 Application of regression analysis

This study used the gross rainfall as the independent variable to establish the regression equations of inner rainfall, throughfall, and interception as follows.

a. Inner rainfall

Generally speaking, it is not easy to collect P_i in the thorny bamboo plantation due to the dense canopy, is is useful to build the estimation equation of inner rainfall. This study used a linear relationship to estimate P_i by P_o as follows.

$$P_i = aP_o + b \quad (r = 0.98^{**}, \text{d.f.} = 26) \qquad (3)$$

where a and b are constant values, which can be obtained by the regression analysis of experimental outcomes. Figure 2 shows the relationship of P_i and P_o, where a = 0.44 and b = 0.95. The correlation coefficient (r) is 0.98 and the significant level reaches 1%, Pi can be accurately estimated by Po. Due to few canopy blocks the raindrop into the rain gauge in the thorny bamboo plantation, the inner rainfall is usually smaller than the gross rainfall (a < 1). Moreover, the canopy can intercept most of the raindrop for small rainfall, this study found that $P_i = 0$ when $P_o < 5$ mm.

Table 1. Experimental outcome.

Measured item	Maximum (mm)	Minimum (mm)	Average (mm)	Average/P_o (%)
Gross rainfall P_o	82.4	5.9	29.2	100
Inner rainfall P_i	80.5	1.5	27.2	93.2
Throughfall I	68.2	0.2	11.5	39.4
Interception T_f	67.6	0.1	13.4	45.9

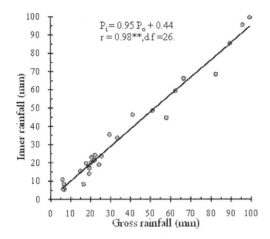

Figure 2. Relationship between P_i and P_o.

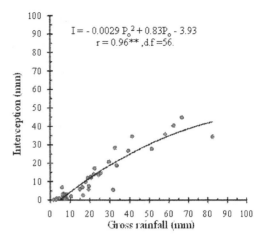

Figure 4. Relationship between I and P_o.

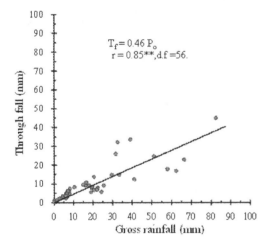

Figure 3. Relationship between T_f and P_o.

c. Interception
    The Relationship between I and P_o is displayed in Figure 4. According the regression analysis, the relationship between P_o and I can be written as follows:

$$I = -0.0029 P_o^2 + 0.83 P_o - 3.93$$
$$(r = 0.96^{**}, \text{d.f.} = 56) \qquad (5)$$

    Equation (5) is a quadratic regression equation, where correlation coefficient is 0.96 and the significance level achieved 1%.
    In conclusion, if the gross rainfall is known, we can estimate the inner rainfall, the throughfall and the interception by the regression equations developed in this study.

b. Throughfall
    The relationship between the throughfall and the gross rainfall is shown in Figure 3. A linear regression equation was built as follows:

$$T_f = 0.39 P_o + 2.3 \quad (r = 0.82^{**}, \text{d.f.} = 56) \qquad (4)$$

The correlation coefficient and significant level of Equation (4) are 0.82, and 1%, respectively. Basically, the increasement of P_o will lead to the increasement of T_f. However, the plenty of antecedent rainfall may fill the canopy with water, and then the most of raindrops may fall into the ground directly, so as to reduce the interception and to increase the throughfall. Therefore, equation (4) underestimates T_f within the P_o range of 30–40 mm due to several days of rainfall.

4 SUMMARY

This study adopted the usually used species of forestation—thorny bamboo, to estimate the interception amount by a series of field experiment, which measured the inner rainfall, the throughfall, and the gross rainfall for 57 groups of data. The water budget equation was used to calculate the interception amount by the known amount of inner rainfall and throughfall. According the comparison between the experimental outcomes and previous literatures, thorny bamboo plantation can intercept more rainfall than other species mentioned in the previous studies, and reduces the throughfall amount, so as to storage more water in the watershed.
    Moreover, this study adopted the gross rainfall as the independent variable to establish the estimation equations of inner rainfall, throughfall, and

interception. Both of the correlation coefficient and significance level are high enough to prove the availability of the regression equations. Using the regression equation developed in this study, we can calculate the amount of inner rainfall, throughfall, and interception by a known amount of gross rainfall.

Finally, experimental outcomes showed that the canopy of thorny bamboo plantation can intercept all of the raindrops when rainfall depth is smaller than 5 mm. However, the plenty of antecedent rainfall will reduce the interception amount and increase the throughfall amount.

ACKNOWLEDGEMENT

The authors would like to appreciate the research support from National Science Council of the Republic of China, Taiwan, with the project no. 100-2221-E-451-009.

REFERENCES

Arnell N. 2002. *Hydrology and Global Environmental Change*. Pearson Education, Harlow. pp. 346.

Carlylemoses D. 2004. Throughfall, stemflow, and canopy interception loss fluxes in a semi-arid Sierra Madre Oriental matorral community, *Journal of Arid Environments*, 58:181–202.

Chiang Y.C. 1964. Rainfall distribution of rainfall measured in different bamboo, *Journal of Agriculture and Forestry*, National Chung Hsing University, 13, pp. 203–218. (in Chinese).

Chiang Y.C. 1971. Interception of precipitation on the determination of forest issues, *Quarterly J. Chinese For.*, 4(1), pp. 123–129. (in Chinese).

Forestry Bureau, Council of Agriculture, Executive Yuan, Taiwan, 2005. 3rd forest resources and land use survey. (in Chinese).

Gash J.H.C. 1979. An analytical model of rainfall interception by forests, *Quarterly J. Royal Meteol.*, 105, pp. 43–55.

Lin T.C., Hsia Y.J. & King H.B. 1996. A study on rainfall interception of a natural hardwood forest in northeastern Taiwan, *Taiwan J. Forest Science*, 11(4), pp. 393–440. (in Chinese).

Llorens P., Poch R., Latron J. & Gallart F. 1997. Rainfall interception by a Pinus sylvestris forest patch overgrown in a Mediterranean mountainous abandoned area I. Monitoring design and results down to the event scale, *Journal of Hydrology*, 199:331–345.

Llorens P. & Gallart F. 2000. A simplified method for forest water storage capacity measurement, *Journal of Hydrology*, 240:131–144.

Lu S.Y & Tang K.J. 1995. *Study on rainfall interception characteristics of natural hardwood forest in central Taiwan*, Bull. Taiwan Forestry Research Institute, 10(4), pp. 447–457. (in Chinese).

Pan C.S. 1964. A study on the interception of rainfall in a chinese fir (Cunninghamia Lanceolata (Lamb.) hook) plantation, Bulletin of Taiwan Forestry Research Institute, pp.15. (in Chinese).

Pan C.S. 1965. *A study on the interception of rainfall through Taiwan zelkova plantation*, Bulletin of Taiwan Forestry Research Institute, pp. 18. (in Chinese).

Rutter A.J., Robins P.C., Morton A.J. & Kershaw K.A. 1971. A predictive model of rainfall interception in forests. 1.Derivation of the model from observations in a plantation of Corsican pine. *Agricultural Meteorology* 9, pp. 367–384.

Tang C., Yu F.C., Hsu S.H. & Lee C.Y. 2002. Primary measurement on interception of rainfall in machilus zuihoensis woods, *Quarterly J. Chinese For.*, 35(1), pp. 21–29. (in Chinese).

Tobón Marin C., Boutena W. & Sevink J. 2000. Gross rainfall and its partitioning into throughfall, stemflow and evaporation of intercepted water in four forest ecosystems in western Amazonia, *Journal of Hydrology*, 237:40–57.

Wey T.H., Chen H.H & Lin W. 2009. Canopy interception of a natural stand of hemlock specy in Tatachia alpine ecosystem, central Taiwan, *J. Chinese Soil and Water Cons.*, 40(1), pp. 105–111.

Zeng N., Shuttleworth J.W. & Gash J.H.C. 2000. Influence of temporal variability of rainfall on interception loss: Part I. Point analysis. *Journal of Hydrology*, 228, pp. 228–241.

Frontiers of Energy and Environmental Engineering – Sung, Kao & Chen (eds)
© 2013 Taylor & Francis Group, London, ISBN 978-0-415-66159-1

The research and development of steady and continuously wave power generating

D.T. Li, Y.H. Xie & D.T. Li
School of Naval Architecture and Civil Engineering, Zhejiang Ocean University, Zhoushan, Zhejiang, China
Zhengjiang (jiuhe) Advanced Marine Technology R&D Center, Zhejiang Provincial Key Laboratory of Ship
Engineering, Zhoushan, Zhejiang, China

ABSTRACT: According to the complex ocean environment and the wave randomness, main research contents of the item involve the adaptability of wave power-generator to the ocean environment and the controlling mechanism development to the wave energy accumulation. On this basis, we aim at put forward the controlling mechanism of wave energy accumulation and an effective wave impact resisting mode for the continuous and steady power-generating, and then develop a steady and continuously wave power generating set. There are three innovations in the item, including: Inventing a oriented wave energy conversion buoy, inventing a system which wave energy could be accumulated and continuously transferred with steadily and innovating a hydraulic cylinder for wave energy transferring, which could insulate sea water. The water pool test indicates that the wave energy could be continuously exported by controlling and adjusting reasonably.

Keywords: wave energy; continuous; steady; power generating

1 INSTRUCTION

With the development of economic, and the shortage of coal and petroleum, energy problem is becoming worldwide (Jebaraj et al. 2008, Meiqin Lin et al. 2010). In China, exponential increase of energy demand that we have ever experienced and far higher than the increasing rate of GDP, leads to overall tension of coal, electricity, oil and transport, and causes markedly growth of petroleum import. For protecting environment and implementing the sustainable development policy of energy, a series of actions has been taken in China. In contrast to developed countries, however, the technology of producing and utilizing energy is still laggard, which remain the significant factor of environment pollution. Continuous growth of acid rain and respiratory diseases will bring serious challenge to energy development and environmental protection. According to the comprehensive prediction, energy problems in China will emerge after 2020. The problem of energy supply is a bottleneck of social development. In order to solve this problem, people turn eyes to the ocean. The ocean has the area of approximately 70% of the earth's, and stores tremendous amounts of energy. However, compare with the advanced technology of utilizing solar energy and wind energy, the technology of utilizing wave energy is still laggard. That can be attributed to the wave randomness

and the complex ocean environment. But from the perspective of energy density, wave energy is 4~30 times higher than wind energy (Falnes J. 2007, Yongliang Zhang & Zheng Lin 2011). Therefore, development of wave energy has great potential economic benefits. Speeding up the development of wave energy has great social significance to energy security and sustainable development in our country.

2 RESEARCH STATUS AT HOME AND ABROAD OF WAVE POWER-GENERATING

2.1 Classification of wave power-generator

The most fundamental principle of wave power-generating can be described as follow: Through multi-level transformation, wave motion can convert the kinetic and the potential energy of wave into electricity. Usually, there are three levels have to pass through when wave energy transforms into electricity. In the first transformation level, wave energy will be absorbed by wave-acceptor; The second level will produce stable energy through optimizing the first level by intermediate conversion device; The third level will transform the stable energy into electricity by generating set (Dahai Zhang et al. 2009, Joao Cruz & Garrard Hassan 2008). At present wave power-generator can be classified in four different ways, shown as Table 1 (Lizhen Zhang et al. 2011).

2.2 Abroad research status

The first country in the world to invent wave energy transformation device was France (Yuan Gong 2008), then followed by Britain, Norway, India, Japan, America, Portugal and etc. Various high technologies for wave power-generating are being actively developed by countries. Right now, Japan and Britain are on the leading point of these countries (Bailin Xu et al. 2000).

2.2.1 Research status in Japan
There are more than 1500 wave power-generators have been built at present. Four fixed type and breakwater type wave power stations with the unit capacity of 40~125 kW, have been built since the midterm of 1980s. The most famous one is called "Haiming", which was built in the early 1980s, has a total installed capacity of 1250 kW. "Haiming" has been taken as the first scheme of "off-island power supply" in Japan, and is being further researched and improved (Lizhen Zhang et al. 2011).

2.2.2 Research in Britain
In 1980s, Britain has already become the research center of wave energy in the world. 75 kW oscillating water column type and 20 mW fixed wave power stations were built in Islay and Osprey in the early 1990s. The first wave power station in the world was built in vicinity of Islay and started commercial operation in November 2000. It's target is to reach the total capacity of 2 GW of wave power-generating sets (Bailin Xu et al. 2000).

2.2.3 Research in Norway
It's a prior country on developing wave energy as well. Last century, the world's largest wave power station, which has the capacity of 500 kW, was built in 1985 in Toftestallen where a 350 kW power station was built Later. Multi-resonant oscillating water column and deceleration channel flow, which

Table 1. The sorts of wave power-generating.

Classification foundation	Classical types
Distribution style	Fixed type, floating type, anchor mooring type.
Energy transfer style	Air pressure transmission type, hydraulic transmission type, mechanical transfer type.
Energy transformation style	One-level energy utilization type, multi-level energy utilization type.
Structure form	Systolic waterway type, oscillating water column type, array oscillating buoy type.

has been widely applied, were invented in Norway. The wave power station that Norway built for Tapchan, has the capacity of 350 kW, using unusual water energy transformation type, the water surface of the water tank after dyke is 3~8 meters higher than the sea surface, the design head of hydraulic turbine is 3 m, the flow rate is 4~16 m³/s. Norway has assisted Indonesia in building a 1500 kW wave power station in Bali since 1988, and planned to build hundreds of wave power stations for the purpose of incorporation.

2.3 Domestic research status

In recent 10 years, China has obtained some achievements in researching wave power-generating set. The pharos-use wave power-generating set was successfully invented, and has come into practical stage. Pharos is one of the key aid navigational facilities that ensure navigation safety. The pharos-use wave power-generating set needs not to be provided power, and is reliable as well. Therefore, it can not only improve aid navigational conditions, but also save the management cost. China's BD-type wave power-generating set has approached advanced international levels, and can compete with Japanese and British similar products. The 8 kW oscillating-water-column test wave power station built in Dawan Shan Island, Guangdong, has put into operation in 1990. However, as its scale is too small, and per kilowatt cost near 10,000 yuan, the power station can not come into practical stage in economics. In 2000, a 100 kW shore type wave power station was built in Shanwei, Guangdong, which is the first shore-type industrial demonstration wave power station in China. Through transmission and transformation circle, this power station can incorporate into the 100 kV power network for use. That marks the technology of wave power-generating in China has reached the practical levels and has meet the popularized conditions (Bing Sun & Ying Li 2005).

2.4 The key technologies of wave generating

Because it's irregularity, wave energy is not easy for being utilized. Actually, it is the most instable renewable energy. It has the following disadvantages: First the energy is hard to focus, second the development cost is too high, third the total conversion efficiency is too low, and in addition, the generating equipment is instable and unreliable. Therefore, if we want to make full use of wave energy for generating, the following key technologies have to be solved.

– *Stability*. Because of the limitations of technology, the wave energy can only be converted into instable hydraulic energy. In this way, the output electricity will still be instable as well. Even the European countries such as Britain and Portugal use the costly generating facilities, they still fail to obtain the stable electricity.

– *Control*. The controlling problem is caused by irregularity and aperiodicity of the wave. When wave swells, energy will be surplus, on the contrary, energy will be insufficient.
– The generating equipment should not have underwater moving parts.
– The overall structure should be strong enough to endure stormy waves.
– Less or no site construction.

3 THE RESEARCH CONTENT OF THE ITEM

According to the complex ocean environment and the wave randomness, main research contents of the item involve the adaptability of wave power-generator to the ocean environment and the controlling mechanism development to the wave energy accumulation. On this basis, we aim at put forward the controlling mechanism of wave energy accumulation and an effective wave impact resisting mode for the continuous and steady power-generating, and then develop a steady and continuously wave power generating set.

3.1 *The wave buoy testing in the water pool*

In order to realize application, we have to ensure the wave power-generating set can endure stormy waves. Therefore, structure of the buoy must be optimized. In this way, the buoy can not only endure horizontal impact force of waves, but also fully absorb kinetic and potential energy, so that hydraulic oil cylinder that connected with the buoy can avoid horizontal impact of waves. Through wave maker, which can produce a series of different wave for test, we can measure horizontal and vertical forces the buoy model enduring. Compare the simulation results with the maximum and minimum values in a wave circle, and make an analysis, then modify the structure design until it optimal. Figure 1 shows the wave buoy testing in the water pool, and Figure 2 is a sketch map of the wave buoy with guiding pole.

Figure 1. The wave bouy testing in the water pool.

The making wave experiment in the water pole demonstrates that this kind of buoy can work steadily, and completely meet the requirement of energy transformation, so that it can take the place of the traditional anchorage system.

3.2 *Research of the accumulation and the control of wave energy*

Because of continuously changing of the wave, wave energy sometimes big, sometimes small, sometimes occur, sometimes no exist. As the measurement results that provided by Japan's wave power ship "Haiming" during 1978~1979 reveal, no matter inter-annual, inter-seasonal or inter-monthly, average wave height changes greatly. It is small in summer while is big in winter. Even in a month, the maximum value of daily average height is several times bigger than the minimum. Also, the wave height can change in the range of 0.2~2.5 m within 4 minutes. Therefore, in order to achieve steady and continuously generating, instability and randomness of wave energy must be well solved. In other words, we have to convert the instable wave energy input into the stable output, namely, the generating set should accumulate energy when wave swells and release it when wave ebbs. In addition, when energy is fully loaded, the surplus should be unloaded automatically to protect the equipment components of the system. In order to test the storage effect, we designed an experiment. The following Figures 3 and 4

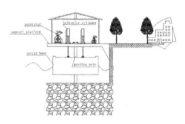

Figure 2. The wave buoy sketch map with guiding pole.

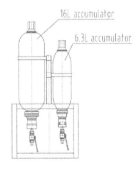

Figure 3. The test plan of storing wave energy in the water pool.

Figure 4. The test picture of storing wave energy in the water pool.

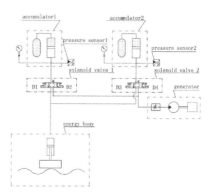

Figure 5. The wave energy storing and releasing drawing.

respectively reveal the design structure and the site picture of two different size accumulators in the laboratory. The capacity of the big accumulator is 16 L, and the small one is 6.3 L. Through the wave making experiment and the pressure setting, energy storage time of different accumulator can be determined.

4 THE CONTROLLING OF CONTINUITY AND STABILITY OF WAVE ENERGY

4.1 The working principle of the steady and continuously wave power-generating set

As shown in the Figures 2 and 5, the working principle of the steady and continuously wave power-generating set can be described as follow: when wave swells, it will push the buoy moving upward along the guiding pole, making the pressure oil of the hydraulic cylinder entering into the accumulator, then after speed regulating control, the hydraulic motor will be driven, and finally steady and continuously electricity will be obtained. When wave falls, the buoy drops down along the guiding pole. At this moment, hydraulic cylinder will produce negative pressure, makes it full of hydraulic oil, and next generating period will be ready at the same moment.

4.2 The control of accumulating and releasing of wave energy

In order to realize continuously accumulating and releasing of wave energy, two accumulators were set in the wave power-generating set, as shown in Figure 5. Accumulator 2 is prefilled under the initial state. After entering generating state, accumulator 1 will store energy, and accumulator 2 will release energy. When pressure of accumulator 1 reaching the set point, accumulator 1 and 2 will transit the state with each other, namely, accumulator 2 will

store energy and accumulator 1 will release energy. Through pressure sensor 1 and 2, the values of the accumulators will be transferred to the intelligent controller. In this way, commuting program control of the solenoid valve 1 and 2 will be realized, so that the hydraulic oil of the accumulators can be continuously conveyed to the generating set, driving the hydraulic motor generating power. Table 2 is the action control program of the 4 electromagnets of the solenoid valves, and Figure 6 shows that the researcher was debugging of the intelligent controller at the experiment water pool. Through setting the accumulators' pressure values, energy can be continuously output from the two accumulators. Figure 7 demonstrates the electric power export of wave energy.

5 INNOVATIONS OF THE ITEM

There are three innovations of the item for dealing with the complex ocean environment and realizing steady and continuously wave power-generating.

5.1 A new kind of buoy with guiding pole was invented

This kind of buoy can be efficiently adapted the complex ocean environment. There is a firm guiding column setting in its center, which can both ensure the steady up and down movement of the buoy, and reduce the investment cost.

5.2 A steady and continuous system that can accumulate and transform wave energy was invented

Through speed regulating control, this system can steady output the accumulated wave energy. Therefore, the shutdown problem that cause by overload can be avoided, the ability of continuously working, the safety, and the ability to generating in

Table 2. The electromagnetism wave controlling process to wave energy.

Equipment component	Initial state	Accumulator 1 storage/ accumulator 2 release	Accumulator 1 release/ accumulator 2 storage
D1	–	+	–
D2	–	–	+
D3	+	–	+
D4	–	+	–
Pressure sensor 1		Lower than the setting value	Reach the setting value
Pressure sensor 2		Reach the setting value	Lower than the setting value

Figure 6. The wave energy storing and releasing drawing.

Figure 7. The electric power export of wave energy.

the high sea state of the whole system can also be enhanced.

5.3 A seawater-insulated hydraulic cylinder for wave energy transferring was invented

Considering the complex ocean environment, this kind of hydraulic cylinder can efficiently protect the generating set from corrosion by marine organism, seawater, salt spray and etc. Therefore, it can reduce cost and is convenient for maintenance.

6 CONCLUSIONS

The water pool test indicates that the wave energy could be continuously exported by controlling and adjusting reasonably. That will be a new path to develop and utilize marine renewable energy in China, and will be breakthrough of the bottleneck of utilize wave energy for generating, also will be a lay foundation of industrialization of wave power-generating.

ACKNOWLEDGEMENT

We would like to express our sincere gratitude to China Oceanic Administration for it provided us the renewable energy special fund (ZJME2011BL04).

REFERENCES

Bailin Xu, Yong Ma & Yinglan Jin. 2000. Current Trends in the Development of Oceanic Power Generation. *[J]. Power Equipment* 2000(1):37–39, 42. China.

Bing Sun & Ying Li. 2005. Ocean Economics. *[M]. Harbin: Harbin Engineer University Press* 2005:188–190. China.

Dahai Zhang, Wei Li & Yonggang Lin. 2009. Wave energy in china: Current status and perspectives. *Renewable Energy* 34:2089–2092.

Falnes J. 2007. A review of wave-energy extraction. *[J]. Marine Structures* 20:185–201.

Jebaraj S., Iniyan S., Suganthi L., et al. 2008. An optimal electricity allocation model for the effective utilization of energy sources in India with focus on biofuels. *[J]. Management of Environmental Quality* 19(4):480–486.

Joao Cruz & Garrard Hassan. 2008. Ocean Wave Energy: Current Status and Future Perspectives. ST Vincent's Works.

Lizhen Zhang, Xiaosheng Yang, Shiming Wang & Yongcheng Liang. 2011. Research Status and Developing Prospect of Ocean Wave Power Generation Device. *[J]. Hubei Agricultural Sciences* 50(1):161–164. China.

Meiqin Liu, Yuan Zheng, Zhenyu Zhao, et al. 2010. The development prospect of wave energy. *[J]. Ocean Development and Management* 27(3):80. China.

Yongliang Zhang & Zheng Lin. 2011. Advances in ocean wave energy converters using piezoelectric materials. *[J]. Jounal of Hydroelectric Engineering* 30(5): 145. China.

Yuan Gong. 2008 Development trend of wave power generation technology in the world. *[J]. Power Demand Side Management* 10(6):71–72. China.

Frontiers of Energy and Environmental Engineering – Sung, Kao & Chen (eds)
© 2013 Taylor & Francis Group, London, ISBN 978-0-415-66159-1

Discussion on evaluation methods for the environmental pressure of energy consumption in buildings' life-cycle

Y. Song
College of Resources and Civil Engineering, Northeastern University, Shenyang, Liaoning, China

H. Liu
Campus Construction and Management Department, Shenyang University, Shenyang, Liaoning, China

ABSTRACT: On the basis of previous research gains, the paper constructs, by introducing quantitative evaluation method in ecological economics, ecological footprint calculating models which objectively measures environmental pressures of energy consumption in different stages in buildings' life-cycle as raw material exploitation and building materials production, construction, usage and maintenance, and the handling of dismantled and abandoned building materials, thus environmental pressures by various types of energy consumption are all presented as ecological footprint and the overall quantization of environmental pressure of energy consumption in buildings' life-cycle is realized. According to the features of resource environmental bearing capacity, indirect load strength, the evaluating indicator for environmental pressure of energy consumption, is advanced as well to thoroughly evaluate energy consumption in buildings' life-cycle and its environmental pressure. Theses quantified computational models disclose that if the energy consumption in buildings' life-cycle and its environmental pressure are within the natural bearing capacity, and may measure to what extent buildings are compatible with the ecological environment they exist in.

Keywords: building; life-cycle; energy consumption; environmental pressure; Ecological Footprint (EF); indirect load strength

1 GENERAL INTRODUCTION

Building industry is an industry that consumes energy sources and natural resources massively, and an industry that has negative influences on environment more outstandingly and more obviously. The building area newly increased every year in China is as high as 1.8–2 billion m². With the continuous increase of building area, the constraints in resources and environment are also intensified further. Therefore, it has vital practical significance to make quantitative evaluation of environmental influence by the energy consumption in buildings' life-cycle for the measure of environmental pressure, and to effectively save resources and keep its environmental pressure within the naturally bearing capacity.

At present, researches on environmental pressure of energy consumption in building's life cycle are basically analysis and researches on building's energy consumption as part of the environmental pressure of building's whole life cycle, which are made by research scholars in the field of environment by using the ideology of life cycle. The above researches have two aspects that are worth discussing. One is, in order to resolve differences on dimensionality, some subjective dimensionless weighting factors are introduced to quantify the environmental effects by using the traditional life cycle evaluation method, so the evaluation results are different from person to person, and the objectivity is inevitably affected; The other is, these quantitative computation models have not revealed whether the energy consumption in building's life cycle and its environmental pressure are within the naturally bearing capacity, and it cannot measure to what extent the building can get along harmoniously with the ecological environment where it exists. The "ecological footprint evaluation of energy saving policy" presented at Top Energy Green Building Forum has provided a new thinking for research on environmental pressure of building's energy consumption, however, the research has only presented the technique for conversion of energy consumption and ecological footprint, it has not made a deep research on the environmental pressure in each stage of building's life cycle. This paper is to research the environmental pressure of generalized energy consumption through introduction of quantitative evaluation method of ecological economics, computation model to objectively measure the ecological footprint of environmental pressure of energy consumption in each stage of building's life cycle has been constructed by introducing quantitative evaluation method of ecological economics so that the environmental pressures of different types of energy consumption are expressed with ecological footprint, which has realized the quantification of environmental pressure of energy consumption in the life cycle on the whole and shall provide beneficial references for economic development to step on the orbit of resource minimization and environment depressurization.

According to internationally accepted classification, building energy consumption usually refers to the energy consumption used in the usage process of building, including the energy consumption in such aspects as heating, air-conditioning, ventilation, hot water supply, lighting, cooking, domestic appliances and elevator. The building energy consumption dealt with in this paper is the generalized energy consumption of building, i.e. the energy consumption in building's life cycle, including the energy consumption related to the building itself in all stages of the life cycle, namely, raw material exploitation and building materials production stage, construction stage, use and maintenance stage, and dismantled and abandoned building material handling stage.

2 COMPUTATION MODEL FOR MODEL ENVIRONMENTAL PRRESSURE OF ENERGY CONSUMPTION IN BUILDING'S LIFE CYCLE

This research introduces method for measure of environmental pressure—ecological footprint method—into the research of energy consumption and its environmental pressure, to establish ecological footprint computation model for measuring energy consumption in building's life cycle and its environmental pressure. The specific formula is as follows:

$$EF_{LCEC} = EF_{em} + EF_{ec} + EF_{eu} + EF_{ed} \qquad (1)$$

where, EF_{LCEC} is total footprint of energy consumption in building's life cycle; EF_{em}, EF_{ec}, EF_{eu} and EF_{ed} are respective footprints of raw material exploitation and building materials production stage, construction stage, use and maintenance stage, and dismantled and abandoned building material handling stage.

As an indirect footprint, energy consumption footprint does not occupy any land in reality. It is assumed from the angle of ecological compensation to offset the ecological resource required for energy consumption. Indirect footprint of building's life cycle $\overline{EF_{ind}}$ is equal to the footprint of annual average energy consumption of the unit building area, which can be calculated with the formula below:

$$\overline{EF_{ind}} = \overline{EF_{LCEC}} = \frac{EF_{LCEC}}{N \times m} \qquad (2)$$

where, $\overline{EF_{LCEC}}$ is the footprint of unit building area energy consumption in building's life cycle (10^4 hm²/m²); N is building area of single building (10^4 m²); m is service life of building (a).

2.1 Raw material exploitation and building materials production stage

2.1.1 Environmental pressure computation model of energy consumption for raw material exploitation and building material production required for construction

work quantity unit is expressed in weight (t, kg), volume (m³) and area (m²). When Building material i expressed in a different unit, convert the unit and substitute it into Formula (3) to obtain the energy consumption footprint of this subitem. The computation formula is as follows:

$$EF_{emc} = \gamma_e \times \frac{\sum_{i=1}^{n} m_i \times (1 + \delta_i/100) \times M_i}{Y_c} \qquad (3)$$

where, γ_e is equivalent factor of fossil energy source land, the value is 1.35 is used; n is number of kinds of building materials used; m_i is usage amount (m², m³ or t, kg) of building material; δ_i is ratio of building material i that is discarded due to construction technology loss or poor management during construction, see Table 1 for parameters used; M_i is energy consumption (kJ) of production unit building material, which is also called contained energy or source generated energy, including energy consumptions in the whole production process of raw material exploitation, transportation, product processing, etc., see Table 1 for parameters used; Y_c shows world average footprint of standard coal (GJ/hm²).

2.1.2 Environmental pressure computation model of energy consumption for raw material exploitation and building material production required for repair and maintenance

Work quantity unit of building material i is expressed in weight (t, kg), volume (m³) and

23

Table 1. Frequently-used data of building materials.

Name of building material	Embodied energy	Apparent density of building materials (kg/m³)	Discarded ratio
Commercial concrete/m³	1 750 000 kj/m³	1 900–2 500	5
Cement/kg	5 500 kj/kg	3 100	5
Timber/m³	1 592 300 kj/m³	440–766	7
Hollow block/m³	1 600 kj/kg	901–1 100	10
Clay brick/m³	4 180 000 kj/m³	1 600–1 800	10
Hollow birck/m³	1 760 000 kj/m³	901–1 100	10
Reinforcement/kg	29 000 kj/kg	7 850	5
High polymer water-proof rolled materials/m²	108 214 kj/m²		5
Profiled steel plate/m³	29 000 kj/kg	7 850	5
Iron pieces/kg	27 000 kj/kg	7 000	5
Sand/m³	600 kj/kg	2 550–2 750	5
Aluminum product/m³	180 000 kj/kg	2 750	5
Glass/m³	16 000 kj/kg	2 500	5
Steel beam/kg	29 000 kj/kg	7 850	5
Gravel/m³	900 kj/kg	2 650–2 750	5
Granite/m²	1 058 974 kj/m²	2 630–2 750	5
Marble/m²	250 807 kj/m²	2 700	5
Floor tile/m³	4 500 kj/kg	2 431	10
Extrusion molded wall/m³	112 200 kj/kg	1 700–1 800	10
Gypsum/kg	3 800 kj/kg		10
Painting/kg	25 314 kj/kg		5
All types of non-metallic pipes/kg	22 160 kj/kg		5

area (m²). When expressed in a different unit, convert the unit and substitute it into Formula (4) to obtain the energy consumption footprint of this subitem. The computation formula is as follows:

$$EF_{emu} = \gamma_e \times \frac{\sum_{i=1}^{n} m_i \times (1 + \delta_i/100) \times M_i \times \left[\frac{m}{B_i} - 1\right]}{Y_c} \quad (4)$$

where, B_i is service life (a) of different, see Table 2 for the values used; the computed result of $[m/B-1]$ is rounded up, which shows the reused number of times of material i in the maintenance and repair stage of the building after the building materials consumed before the building is put into service have been deducted.

2.2 Construction stage

2.2.1 Computation model for environmental pressure of energy consumption of different construction technologies

When construction technology energy consumption footprint of construction technology is computed, by using unit energy consumption and world average footprint of used energy, the energy consumption of construction technology can be

Table 2. Life of common building materials or components.

Building materials	Service life
Structure part, feed pipe, cables and cords, ventilation duct	Life span of civil buildings
Doors and windows, decorative panels, external tiles, non-metal tubing	30 a
Waterproof rolling material	20 a
Coating, wallpaper	10 a

converted into ecological footprint area index with the following computation model:

$$EF_{ecp} = \gamma_e \times \sum_{j=1}^{m} \frac{p_j \times P_j}{Y_j} \quad (5)$$

where, p_j is total construction quantity (m², m³ or t, kg) when Item j construction technology is used; P_j is unit energy consumption (MJ/Unit) of Item j construction technology, see Table 3 for unit energy consumption of different construction technologies; Y_j is world average footprint (MJ/hm²) of fossil energy used for Item j construction technology.

Table 3. Unit energy consumption of various types of erections.

Construction technology	Unit energy consumption/ (MJ/unit)	Construction technology	Unit energy consumption/ (MJ/unit)
Job mix concrete	158.40 MJ/t	Crane operation	39.75 MJ/t
Cubic meter of excavated/ removed earth	115.20 MJ/m³	Cleaning of site	10.00 MJ/m²
Hygiene and heating	4.54 MJ/m²	Site arrangement	52.24 MJ/m²
Temporary heat supply	484.96 MJ/m²	Staff transportation	459.93 MJ/m²
Placing of material	5.22 MJ/m²	Roofing	1.14 MJ/m²
Temporary power supply	22.65 MJ/m²	Foundation excavation	27.26 MJ/m²

2.2.2 Computation model for environmental pressure of energy consumption of transportation of raw materials required for construction

When building material i expressed in a different unit, convert the unit and substitute it into Formula (6) to obtain the energy consumption footprint of this subitem. The computation model is as follows:

$$EF_{ect} = \gamma_e \times \frac{E_{mL} \times \sum_{i=1}^{n} m_i \times (1 + \delta_i/100) \times L_i}{Y_e} \quad (6)$$

where, E_{mL} is energy consumption (MJ/(t·km)) of unit weight of building material transported in different mode of transportation. According to reality of construction, the main mode of transportation of building materials is road transportation, and the fuel is mainly diesel oil, so the transportation energy consumption 2.055 MJ/(t·km); L_i is average distance (km) of building material i when it is transported from the supplier to the construction site, the parameters used shall be subject to the construction budget, and the data in Table 4 shall be used if the parameter is not listed in the construction budget; Y_e is world average footprint (MJ/hm²) of fossil energy used.

2.3 Service and maintenance stage

2.3.1 Computation model of environmental pressure of energy consumption of power utilization

As far as building is concerned, air-conditioning, elevator and lighting consume large quantities of energy. In our country, electricity generation mainly depends upon coal, therefore, the computation model of energy consumption footprint of power utilization in service and maintenance stage of building is as follows:

$$EF_{eue} = m \times \gamma_e \times \frac{C_e \times \rho_e}{Y_c} \quad (7)$$

Table 4. Average carrying distance of major building materials.

Name of building materials	Average carrying distance/km
Gravel	28.99
Concrete	65.38
Bricks and other wall body materials	57.75
Glass	98.84
Timber	59.01
Other cement products	65.38
Cement	65.57
Steels	122.72
Ceramic building materials	106.38
Aluminum doors and windows	128.08
Painting	103.81
Other non-metallic mineral materials	57.75

where, C_e is annual power consumption (kW·h), the parameters used are listed in Table 5; ρ_e is energy density of coal electricity (GJ/(kW·h)).

2.3.2 Computation model of environmental pressure of energy consumption of water supply and wastewater treatment

According to the research made by pertinent scholars, the electricity consumed for in taking and cleaning of each ton of tapped water is approximately 0.25–0.33 kW·h, the electricity consumed for treatment of each ton of domestic wastewater is approximately 0.15–0.20 kW·h. In our country, electricity generation mainly depends upon coal, therefore, the computation model of energy consumption footprint of water supply and wastewater treatment in service and maintenance stage of building is as follows:

$$EF_{euw} = m \times \gamma_e \times \frac{(C_w \times \psi_w + C_{ww} \times \psi_{ww}) \times \rho_e}{Y_c} \quad (8)$$

where, C_w is annual water consumption (t/a); ψ_w is electricity consumption of each ton of water

25

Table 5. The status of non-heating energy in building in cities and towns.

Types of building	Area/ 10^8 m^2	Status of energy consumption/ kW·h	Energy consumption per unit area/ (kW·h/m^2)
Residential house	100	2.0×10^{11}	10–30
Common public building	55	1.6×10^{11}	20–60
Large public building	5	1.0×10^{11}	70–300
Total quantity	160	4.6×10^{11}	29

Table 6. Energy consumption of boiler plants of 10 cities.

City	Energy consumption in heating period/ (kg/m^2)	Daily energy consumption/ (kg/m^2)
Harbin	36.30	0.20
Jilin	32.10	0.19
Dalian	22.70	0.17
Beijing	17.70	0.13
Taiyuan	46.40	0.31
Changchun	37.00	0.22
Shenyang	27.80	0.17
Baotou	40.90	0.23
Tianjin	32.40	0.27
Urumchi	35.80	0.20
Average	28.90	0.20

used (kW·h/t); C_{ww} is annual wastewater treatment quantity (t/a); ψ_{ww} is electricity consumption of each ton of wastewater treated (kW·h/t).

2.3.3 Computation model of environmental pressure of heating energy consumption

At present, the heating fuel is mainly coal in our country; therefore, the computation model of energy consumption footprint of building heating is as follows:

$$EF_{euh} = m \times \gamma_e \times \frac{N \times C_c \times Th \times \rho_c}{Y_c} \quad (9)$$

where, C_c is daily coal consumption of each square meter of building (t/(d·m^2)), the values taken are listed in Table 6; Th is the heating duration of each year (d/a); ρ_c is energy density of coal (GJ/t).

2.3.4 Computation model of environmental pressure of energy consumption when natural gas is used

$$EF_{eug} = m \times \gamma_e \times \frac{C_g \times \rho_g}{Y_g} \quad (10)$$

where, C_g is annual natural gas consumption (m^3); ρ_g is energy density of natural gas (GJ/m^3); Y_g is world average footprint of natural gas (GJ/hm^2).

2.3.5 Computation model of environmental pressure of energy consumption of transportation of building material required for repair and maintenance

In computation of energy consumption footprint of transportation of building materials for repair and maintenance, the transportation energy consumption footprint of building materials carried from the production site to the building location for repair and maintenance must be considered, the energy consumption footprint of recyclable replaced waste building materials transported to the recycling site during repair and maintenance and the energy consumption footprint of unrecyclable waste building materials transported to the final disposal yard must also be considered. The computation model is as follows:

$$EF_{eut} = \frac{\gamma_e \times E_{mL} \times \sum_{i=1}^{n} m_i \times (1 + \delta_i/100) \times \left[\frac{m}{B_i} - 1\right] \times \left[L_i + R_i \times D_i + (1 - R_i) \times l_i\right]}{Y_e}$$

(11)

where, D_i is distance (km) of recyclable waste building material carried to the recycling site; R_i is recycling ratio of waste building material, the values taken are listed in Table 7; l_i is the distance of unrecyclable waste building material carried to the final disposal yard (km).

2.4 Dismantling and waste building material disposal stage

The computation model of energy consumption footprint in this stage is as follows:

$$EF_{ed} = EF_{edd} + EF_{edrt} + EF_{edt} - EF_{edr} \quad (12)$$

where, EF_{edd} is energy consumption footprint of building dismantling,; EF_{edrt} is energy consumption footprint of transportation of recyclable waste building materials; EF_{edt} is energy consumption footprint of transportation of unrecyclable waste building materials to the disposal site; EF_{ed} is

Table 7. Recycling ratio of common wasted building materials.

Types of building materials	Recycling ratio	Types of building materials	Recycling ratio
Steel product	0.95	Concrete and cement	0.10
Materials for wall	0.60	Non-iron metal	0.90
Glass	0.80	Timber	0.10

energy consumption footprint of recycling of all waste building materials after the energy consumptions involved in transportation and processing in reusing, recycling and burning processes are deducted.

2.4.1 Computation model of environmental pressure of dismantling energy consumption

Dismantling energy consumption is mainly related to the machinery to perform dismantling operation. According to related references, dismantling energy consumption may be usually computed as per 90% of the energy consumption of the construction process. The following computation model is adopted:

$$EF_{edd} = 90\% \times \gamma_e \times \sum_{j=1}^{m} \frac{p_j \times P_j}{Y_j} \quad (13)$$

2.4.2 Computation model of environmental pressure of energy consumption for transportation of recyclable waste building materials

According to different work quantity unit i of recyclable waste building material, Computation model of transportation energy consumption footprint is as follows:

$$EF_{edrt} = \gamma_e \times \frac{E_{mL} \times \sum_{i=1}^{n} m_i \times (1 + \delta_i/100) \times R_i \times D_i}{Y_e} \quad (14)$$

2.4.3 Computation model of environmental pressure of energy consumption for transportation of unrecyclable waste building material to disposal site

According to different work quantity unit i of unrecyclable waste building material, Computation model of transportation energy consumption footprint is as follows:

$$EF_{edt} = \gamma_e \times \frac{E_{mL} \times \sum_{i=1}^{n} m_i \times (1 + \delta_i/100) \times (1 - R_i) \times l_i}{Y_e} \quad (15)$$

2.4.4 Computation model of environmental pressure of energy consumption for recyclable waste building material

It is assumed in this paper that the energy consumption footprint of all waste building materials is 20% of the energy consumption footprint of this building material. When expressed in a different unit, the computation model of energy consumption footprint of recyclable waste building material is as follows:

$$EF_{edr} = 20\% \times \gamma_e \times \frac{\sum_{i=1}^{n} m_i \times (1 + \delta_i/100) \times M_i \times \left[\dfrac{m}{B_i}\right] \times R_i}{Y_c} \quad (16)$$

3 COMPUTATION MODEL AND EVALUATION INDEX OF RESOURCE ENVIRONMENTAL BEARING CAPACITY

3.1 Computation model of resource environmental bearing capacity

According to the research results achieved by Resource and Ecological Research Center of Northeast University, indirect footprint is "carbon equivalent forest land area", which is equivalent to the carbon absorbing capacity of the forest, and which is required for measure of carbon absorbing capacity of the forest. Therefore, in computation of indirect bearing capacity of bearing energy consumption footprint, carbon absorption equivalent coefficient is introduced. Carbon absorption equivalent coefficient τ_k of the land with Class k vegetation form in the research regions is the ration of the average CO_2 absorbing capacity P_k of this class of land to the average CO_2 absorbing capacity P_c of the forests around the globe, which is specifically as follows:

$$\tau_k = \frac{P_k}{P_c} \quad (17)$$

So the indirect bearing capacity may be computed according to the ownership of lands with different vegetation forms, i.e. the carbon equivalent forest land area owned in the unit building area of the research regions, which is recorded as EC_{ind}, and the computation model is as follows:

$$\overline{EC_{ind}} = \frac{\gamma_f \times \sum_{k=1}^{N_L} S_k \times \tau_k}{N_c} \qquad (18)$$

where: N_L is the number of classes of land (vegetation form) that have CO_2 absorbing capacity; S_k is the actual area ($10^4 \, hm^2$) of land with Class k vegetation form owned by the research regions; γ_f is the equivalent factor of forest land around the globe; N_c is the annual in-process construction area of the research region ($10^4 \, m^2$).

3.2 Evaluation index of energy consumption environmental pressure—indirect load strength

The ecological footprint of unit load bearing area is similar to the intensity of pressure in physics, which is called "load strength", and μ_{ind} is used to express "indirect load strength". The computation model is as follows.

$$\mu_{ind} = \overline{EF_{ind}} / \overline{EC_{ind}} \qquad (19)$$

Indirect load strength is a relative quantity, If the indirect load strength is larger than 1, it indicates that the resource environmental bearing capacity of the research region is not enough to bear the environmental pressure of energy consumption in building's life cycle, but it needs to use foreign resource to meet the demand, as a result, the research region shall be unable to provide resource demand for other industries; if the indirect load strength is smaller than 1, the proportion of the environmental pressure of energy consumption in building's life cycle in the research regions and the resource space owned by the research region in other industries may be obtained.

4 CONCLUSION AND DISCUSSION

In this paper, it is the first time to put forward and construct the ecological footprint models of all stages of building's life cycle in accordance with the characteristics of energy consumption I the four stages of building's life cycle; based on the features of resource environmental bearing capacity, it is also for the first time to come up with the evaluation index of environmental pressure of energy consumption—indirect load strength, which is used to make overall evaluation of energy consumption in building's life cycle and the status of its environmental pressure. Such quantitative computation models

enable different types of environmental pressures of energy consumption to be expressed with ecological footprint, thus, overall quantification of environmental pressure of energy consumption in building's life cycle has been realized, whether the energy consumption and the environmental pressure of building's life cycle are within the naturally bearing capacity has been disclosed, and beneficial examples have been provided to promote economic development to walk up onto the orbit of resource minimization and environment depressurization step by step.

Thanks: Liaoning Province Science & Technology Program (No. 2011230008); Shenyang Municipality Science & Technology Program (No. F11-264-1-14).

REFERENCES

Compiled by the exam books writing group of national certified public value. 2005. Architectural Engineering Evaluation Foundation. Beijing: China Financial & Economic Publishing House.
Compiled by the training books editorial board of national cost engineer qualification exam. 2006 *Construction Engineering and Measurement Technology (civil engineering part)*. Beijing: China Planning Press.
Gao Changming. 2002. Environmental evaluation for life cycle of cement and concrete, Cement (11):11–15.
Gu Daojin, Zhu Yingxin, Gu Lijing. 2006. *Life cycle assessment for China building environment impacts*, Journal of Tsinghua University (Science and Technology) 46(12): 1953–1956.
Gu Xiaowei. 2008. Research on the Environmental Pressure for Sustainable Development of Shenyang, Graduate Dissertation Database of NEU.
Huang Boyu. Pi Xinxi.1988. Building materials. Beijing: China Architecture & Building Press.
Jiang Yi. Yang Xiu. 2006. Status of architecture energy consumption and problems in energy saving in China, Chinese Construction 13(2):12–18.
Lang Siwei. 2004. Progress in energy-efficiency standards for residential buildings in China, Energy and Buildings.
Li Hai, Sun Ruizheng. Chen Zhenxuan. 2002. Technologies and Project Cases Study on Treatment of Municipal Sewage. Beijing: Chemical Industry Press.
Li Xianrui. Lang Siwei. 2001. Status and Problems Analysis of Architecture Heating in China, District Heating (1):11–22.
Peng Wenzheng. 2003. Study on Life Cycle Evaluation Technology Applied in Architecture Energy Consumption, Taiwan Chaoyang University of Technology.
Press Conference about *Regulations on Civil Building Energy Conservation* addressed by the principal of Legislative Affairs Office of the State Council (2008). Information on http://news.qq.com/a/20080811/001953.htm.
Sustainable Building Association of Japan. CASBEE. 2005. Valuation system for comprehensive environmental performance of buildings. Beijing: China Architecture & Building Press.

Tao Zaipu. 2003. *Eco-rucksack and Eco-footprint: Beijing*, Economic Science Press.

Top Energy Green Building BBS. 2007. Green Building Assessment, Beijing: China Building Industry Press.

Tu Fengxiang, Wang Qingyi. 2004. Energy conservation: inevitable choice of Chinese strategy on energy saving (Part 1), Energy Conservation and Environmental Protection.

Wang Songqing. 2007. Life Cycle Assessment of Residential Building Energy Consumption in Severe Cold Region, Harbin: Harbin Institute of Technology.

Wei Xiaoqing. 2010. Analysis and Valuation of Energy Consumption Based on LCA for Large-scale Public Buildings.

World Bank. 2001. China-Opportunities to Improve Energy Efficiency in Buildings. World Bank.

Zhong ping. 2005. Study of Building Life-cycle Energy Use and Relevant Environmental Impacts, Chengdu: Sichuan University.

Frontiers of Energy and Environmental Engineering – Sung, Kao & Chen (eds)
© 2013 Taylor & Francis Group, London, ISBN 978-0-415-66159-1

Study on the environmental pressure assessment of water resources consumption in the life-cycle of buildings based on the method of ecological economics

H. Liu

Campus Construction and Management Department, Shenyang University, Shenyang, Liaoning, China

Y. Song

College of Resources and Civil Engineering, Northeastern University, Shenyang, Liaoning, China

ABSTRACT: By adopting the quantitative evaluation method in ecological economics, this study established an objective computation module to measure the environmental pressure of four stages of water resources consumption. Compared to the carrying capacity of water resources, the evaluation standard of water resources consumption—water loading strength was put forward to completely assess the environmental pressure of water resources consumption in the life-cycle of buildings.

Keywords: building; life-cycle; water resources consumption; environmental pressure; Ecological Footprint (EF); water load strength

1 GENERAL INTRODUCTION

In the process of building the life cycle, are accompanied by a large amount of water consumption and the generation of wastewater, the production of building materials and construction need water and wastewater, construction and use process, inevitably produce mud, use of water and sewage, waste water and so on. China's annual new construction area of up to 1.8–2 billion m². With the constant growth of building area, water consumption will be as demand increased and more and more, the water resources because of excessive consumption and become the bottleneck of restricting China's economic development. Therefore, the construction of water and waste water discharge life cycle condition, quantitative analysis of water resource consumption environment and bring pressure of the relationship between the capacity of regional water resources, and take effective measures to save water and water consumption environment will bring pressure control in the natural bear ability range has the important practical significance.

Scholars both at home and abroad based on different view, use all kinds of different analysis methods, water consumption and water resources carrying capacity around a lot quantitative study. Water consumption of quantitative research focus on the area, industry, enterprise on level. Quantitative research of the bearing capacity of

water resources is mainly by using systems, modern evolutionary algorithm, system dynamics method, the multi-objective decision analysis method, from national, provincial, region, industrial zone and other aspects of research. At present, the water consumption and water resources carrying capacity of quantitative research mainly focus on the macro and micro level, micro level of also only stay in enterprise and industrial research. For water resources consumption of micro individuals bring pressure on the environment and the study of the relationship of the water resources carrying capacity is few. TopEnergy green building BBS put forward "the ecological footprint of water-saving strategy evaluation" is the consumption of water resources for building environment pressure research provides a new concept, but the study focuses on the different ecological features is the ecological footprint of evaluation method, put forward just water resources consumption with ecological footprint conversion method, failed to further research building each stage of life cycle of water consumption environment pressure and water resources carrying capacity of the relationship. In this paper, ecological economics by introducing the quantitative evaluation method to establish the objective measure of each stage of life cycle of building water consumption environment pressure calculation model of the ecological footprint, and the capacity of regional water resources research and comparison, realize the building life cycle of water resources consumption environment pressure comprehensive quantitative, for economic development gradually took to the resources and environment of the reduction pressure track provide the beneficial reference.

2 MODEL AND EVALUATION INDEX

2.1 Water consumption environment pressure model building and decomposition

Water consumption environment pressure of two parts, one is brought by the water environmental stress (EF_{rw}), another is produced by the waste water discharge of environmental stress (EF_{ww}). According to the outdoor drainage design rules (GB50014-2006) 3.1.2 regulation, usually wastewater quantity for water consumption of 80 percent to 90 percent, this paper takes 85 percent. Therefore, the consumption of water environmental stress (EF_w) is equal to the consumption of water environmental pressure the coefficient of 1.85 times, namely:

$$EF_w = 1.85 \times EF_{rw} \tag{1}$$

2.1.1 Water consumption environment pressure calculation model

Building life cycle usually includes four stages, Therefore, the construction of water consumption of life cycle environmental pressure model also by four stage composition, namely:

$$EF_{rw} = EF_{rwm} + EF_{rwc} + EF_{rwu} + EF_{rwd} \tag{2}$$

where, EF_{rwm}, EF_{rwc}, EF_{rwu}, EF_{rwd} respectively were said raw material production stage, stage of construction, use and maintenance stages, dismantling and waste disposal water consumption of building materials stage of ecological footprint.

Water consumption that is occupied land resources is different from the water resources of the occupier, need certain water area to carry water and waste water discharge produce because of the ecological footprint. Therefore, this article will water and waste water discharge of the ecological footprint of the ecological footprint from traditional calculation separately, to say EF_w, calculation model are as follows:

$$\overline{EF_w} = \frac{1.85 \times EF_{rw}}{N \times m} \tag{3}$$

where, $\overline{EF_w}$ mean building life cycle unit building area water consumption footprint 10^4 hm²/m²; N mean the building area of individual buildings

(10^4 m²); m mean the building use fixed number of year (a).

2.1.2 Raw materials mining and building materials production stage

This stage of the water environmental stress (EF_{rwm}) for construction materials and building materials needed for mining production water environment pressure (EF_{rwmc}) and repairing maintenance ingredients needed for building materials and mining production water environment pressure (EF_{rwmu}) the sum, in the process of computation can be through the production unit of construction materials and water consumption per hectare of China have water quantity, building materials and materials will be mining production stage of water environmental pressure for corresponding conversion of water area.

1. The ingredients needed for construction and building materials production of mining water environmental pressure calculation model.

   Building materials in quantities unit weight (t, kg), volume (m³), area (m²) says, when in different units mean, after conversion unit, into the Eq.4, can be obtained for the breakdown of water consumption footprint, calculation model are as follows:

$$EF_{rwmc} = \gamma_w \times \frac{\displaystyle\sum_{i=1}^{n} m_i \times (1 + \delta_i/100) \times W_i}{P_w} \tag{4}$$

where, EF_{rwmc} mean the ingredients needed for construction and building materials mining production water consumption footprint; γ_w refers to the waters of the equal factor, take 0.35; n mean the use of the species number of building materials; mi mean the usage of the building materials (m², m³ or t, kg); δi mean during the construction process in the construction process loss or construction reasons such as poor management of building materials abandoned ratio, value see Table 1; W_i mean the water consumption of building materials during the production unit, according to the industry in Liaoning province water rules (DB21/T1237-2003) value (hereinafter referred to as the "standard"), and the specific see Table 1; P_w mean every hectare of China have water quantity, take 2.922×10^3 m³.

2. Repair maintain ingredients needed for mining production water consumption of building materials and environmental pressure calculation model.

$$EF_{rwmu} = \gamma_w \times \frac{\displaystyle\sum_{i=1}^{n} m_i \times (1 + \delta_i/100) \times \left[\dfrac{m}{B_i} - 1\right] \times W_i}{P_w} \tag{5}$$

Table 1. Frequently-used data of building materials.

Name of building material	Product unit building materials water consumption	Discarded ratio
Commercial concrete/m³	1.282 m³/m³	5
Timber/m³	12.6 m³/m³	7
Clay brick/m³	1.5 m³/ten thousand	10
Various non-metal tubing/kg	0.043 m³/kg	5
Profiled steel sheet/m³	0.01 m³/kg	5
Sand/m³	/	5
Glass/m³	0.0038–0.006 m³/kg	5
Sheet steel/m³	0.0055 m³/kg	5
Granite/m²	/	5
Ground tile/m³	/	10
Paints/kg	0.035 m³/kg	10
Reinforce steel/kg	0.0082 m³/kg	5
Cement/kg	0.00372 m³/kg	5
Cavity block/m³	1.08 m³/m³	10
Cavity brick/m³	0.47 m³/ten thousand	10
High polymer waterproof rolling material/m²	0.2 m³/m²	5
Steel piece/kg	0.0066 m³/kg	5
Aluminum product/m³	0.02 m³/kg	5
Girder steel/kg	0.0082 m³/kg	5
Rubble/m³	/	5
Marble/m²	/	5
Extrusion wall/m³	0.0067 m³/kg	10
Coating/kg	0.0093 m³/kg	5

Table 2. Life of common building materials or components.

Building materials	Service life
Structure part, feed pipe, cables and cords, ventilation duct	Life span of civil buildings
Doors and windows, decorative panels, external tiles, non-metal tubing	30 a
Waterproof rolling material	20 a
Coating, wallpaper	10 a

where, EF_{rwmu} mean the ingredients needed for the repair and maintenance of building materials production of mining water consumption footprint; γ_w, n, m_i, δ_i, W_i, P_W mean significance with the formula (4); m mean significance with formula (3); B_i mean the service life of different building materials (a), value Table 2; $[m/B_i - 1]$ operation result rounded up, mean building before use deduct the consumption of building materials, construction in the use and maintenance for the maintenance and repair stage again using the number of materials.

2.1.3 *Construction stage*

This stage water is all sorts of main construction technology and construction process of the water, the environmental pressure calculation model are as follows:

$$EF_{rwc} = \gamma_w \times \frac{N \times W_n}{P_W} \qquad (6)$$

where, EF_{rwc} mean water resources construction stage occupy footprint; γ_w, P_W mean significance with the Eq.4; N mean significance with Eq.1; W_n mean every square meter building area of construction of the calculated, according to the standard of value is 2.5 m³/m².

2.1.4 *The use and maintenance stages*

This stage water main consideration of environmental pressure brought the water for living, calculation model are as follows:

$$EF_{rwu} = m \times \gamma_w \times \frac{C_w/\rho_w}{P_W} \qquad (7)$$

where, EF_{rwu} mean the use and maintenance phase water resources occupied footprint; m mean significance with Eq.1; γ_w with meaning Eq.4; C_w mean the water consumption per year (t); ρ_w mean the density of water.

Table 3. Recycling ratio of common wasted building materials.

Types of building materials	Recycling ratio	Types of building materials	Recycling ratio
Steel product	0.95	Concrete and cement	0.10
Materials for wall	0.60	Non-iron metal	0.90
Glass	0.80	Timber	0.10

2.1.5 Demolition and abandoned building materials disposal stage

This stage is considering building materials recycling waste water consumption environment of cut pressure. This paper assumes that all the water consumption recycling waste building materials for the building materials water footprint consumption 20 percent of the footprint. Therefore, the water cut occupy footprint calculation model are as follows:

$$EF_{rwd} = 20\% \times \gamma_w$$
$$\times \frac{\sum_{i=1}^{n} m_i \times (1 + \delta_i/100) \times [m/B_i] \times R_i \times W_i}{P_w} \quad (8)$$

where, EF_{rwd} mean demolition and waste disposal said building materials of recycle of waste materials stage to cut water resources occupied footprint; γ_w, n, m_i, δ_i, W_i, P_w mean significance with the Eq.4; m mean significance with Eq.1; B_i mean with meaning Eq.5, $[m/B_i]$ operation result round up and mean building in each stage of life cycle of building materials used times; Said the recovery of waste materials ratio, value Table 3.

2.2 Water resources carrying capacity calculation model

According to Xie Gaodi etc such as high bearing capacity of water resources, the definition of a particular area that is their own water quantity can maintain and the load bearing a development the size of the components. This paper will carry water and waste water discharge of the ecological footprint in water area, said calculation model are as follows:

$$\overline{EC_w} = \frac{S_w \times \gamma_w \times \lambda_w}{N_c} \quad (9)$$

where, S_w mean the actual water area, the unit is through the amount factor and yield coefficient of the converted standardization area (10^4 hm²), value, see Table 4; γ_w mean significance with the Eq.4; λ_w say that research area of water area yield factor, take 1.00; N_c say that research area under

Table 4. The water areas and under construction per year.

Year	Water areas (10^4 hm²)	Under construction (10^4 m²)
2003	0.49	3 036.3
2004	0.49	4 158.6
2005	0.49	5 488.6
2006	0.49	6 106.2
2007	0.49	9 166.0
2008	0.49	9 824.0

construction area every year (10^4 m²), value see Table 4.

2.3 Water consumption environment pressure evaluation index-water load strength

In physics, pressure is the unit load area force. This paper is the ecological footprint of the measure of environmental pressure, although the unit is a unit of area, but its connotation is equivalent to the forces of physics, the study area have environmental bearing capacity is equivalent to the physics carrying area. So the unit load in the area of the ecological footprint is similar to the physics pressure, called "load strength", and μ_w is used to express "water load strength", the calculation model is as follows:

$$\mu_w = \overline{EF_w} / \overline{EC_w} \quad (10)$$

Water load strength is relative amounts, objectively reflect the building water consumption of the pressure on the environment and water resources of the area, the relationship between the bearing capacity, and reveals the strength of the pressure on resources and environment, and comparable, easy to different types of building the consumption of water resources life cycle pressure on the environment more. And in between different types of building comparable. If the water load strength greater than 1, of the area, that not only enough to carry water resources carrying capacity of water resources of the life cycle of building cost pressure on the environment, need to use water resources meet its needs foreign, so study area will not be able to provide water demand for other industries; If the water load strength less than 1, can get the life cycle of water consumption building environmental pressure capacity of regional water resources in the research of the proportion of the study area and other industries have water space.

3 THE MAIN CONCLUSIONS

This paper puts forward the quantitative calculation model of water resources consumption environment that pressure to ecological footprint said, realize the building life cycle of water resources consumption environment pressure comprehensive quantitative, reveal the life cycle cost and its water resources building environmental pressure is in natural bear ability range, and in order to promote economic development gradually took to the resources and environment of the reduction pressure track provide the beneficial reference. The study for both building life cycle cost of environmental pressure of water resources research method provides a new concept, and broaden the ecological footprint method field of study, and to take effective measures to save water and reduce the construction of water resources of the life cycle cost pressure on the environment to provide the scientific basis.

Thanks: Liaoning Province Science & Technology Program (No. 2011230008); Shenyang Municipality Science & Technology Program (No. F11-264-1-14).

REFERENCES

Chen Dongjing. 2008. *Structure Share and Efficiency Share of Industrial Water Consumption Intensity Change in China*. China Population, Resources and Environment 18(3):211–214.

Compiled by Bureau of Statistics of Shenyang: Shenyang Statistical Yearbook 2003–2008. China Statistics Press.

Compiled by the exam books writing group of national certified public value. 2005. Architectural Engineering Evaluation Foundation. China Financial & Economic Publishing House.

Compiled by National Bureau of Statistics of China. 2006. China Statistical Yearbook 2006. China Statistics Press.

Compiled by the training books editorial board of national cost engineer qualification exam. 2006. Construction Engineering and Measurement Technology (civil engineering part), China Planning Press.

Duan Chunqing, Liu Changming, Chen Xiaonan, et al. 2010. *Preliminary Research on Regional Water Resources Carrying Capacity Conception and Method*. ACTA Geographica Sinica 65(1):82–90.

Gu Xiaowei. 2008. *Research on the Environmental Pressure for Sustainable Development of Shenyang*. Graduate Dissertation Database of NEU.

Information on http://news.qq.com/a/20080811/001953.htm.

Liu Jiajun, Dong Shuocheng, Li Zehong. 2011. *Comprehensive Evaluation of China's Water Resources Carrying Capacity*, Journal of Nature Resources 26(2):258–269.

Niu Xiaogeng, Lian Jiting. 2010. *Dynamic Empirical Analysis of Relative Water Resources Carrying Capacity of Hebei Province*, China Population, Resources and Environment 20(3):131–134.

Peng Wenzheng. 2003. *Study on Life Cycle Evaluation Technology Applied in Architecture Energy Consumption*, Chaoyang University of Technology.

Top Energy Green Building BBS. 2007. Green Building Assessment, China Building Industry Press.

Tao Zaipu. 2003. *Eco-rucksack and Eco-footprint*, Economic Science Press.

Wang Songqing. 2007. *Life Cycle Assessment of Residential Building Energy Consumption in Severe Cold Region*, Harbin Institute of Technology.

Xie Gaodi, Zhou Hailin, Lu Chunxia, et al. 2005. *Carrying Capacity of Natural Resources in China*, China Population, Resources and Environment 15(5): 93–98.

Zhang Hongwei, He Xiabing, Wang Yuan. 2006. *Analysis of Water Consumption of China's Industries Based on the Input-Output Method*, Rescource Science 33(7):1218–1224.

Zhang Qingzhi, He feng, Zhao Xiao. 2010. *Study on the Productive Efficiency and Energy, Water Consumption of Iron and Steel Enterprises Based on DEA*, Soft Science 24(10):46–50.

Zhao Ao, Wu Chunyou. 2010. *Grey Correlation Degree and Niche Measurement of Allocation of China's Water Resources Consumption*, China Population, Resources and Environment.

Zheng Yi, Wei Wenshou, Cui Caixia. 2010. *Water Resource Carrying Capacity in Yanqi Basin Based on Multi-objective Analysis*. China Population, Resources and Environment 20(11):60–65.

Zhong ping. 2005. *Study of Building Life-cycle Energy Use and Relevant Environmental Impacts*, Sichuan University.

Frontiers of Energy and Environmental Engineering – Sung, Kao & Chen (eds)
© *2013 Taylor & Francis Group, London, ISBN 978-0-415-66159-1*

Study on the resource-environmental pressure caused by solid waste discharge from civil building life-cycle: A case study of Shenyang Library

H. Liu
Campus Construction and Management Department, Shenyang University, Shenyang, Liaoning, China

Y. Song
College of Resources and Civil Engineering, Northeastern University, Shenyang, Liaoning, China

ABSTRACT: It is a common social task to reduce the resource—environmental pressure caused by solid waste discharge from civil building life-cycle and to promote the economic development onto track of resource reduction and environmental decompression. By methods of ecological footprint and emergy analysis and taking Shenyang Library as example, resource-environmental pressure calculating model for solid waste discharge from each period of life-cycle is built for case analysis. The result shows that resource-environmental pressure caused by waste cement discharge accounts for 48.56% of ecological footprint in constructing period, 82.78% in use and maintenance period and 56.73% in dismantled and discarded building material disposal period; while resource-environmental pressure caused by waste ready-mixed concrete discharge accounts for 31.90% of ecological footprint in constructing period and 37.26% in dismantled and discarded building material disposal period. In periods of life-cycle, resource-occupied footprint of solid waste discharge in dismantled and discarded building material disposal period accounts for the largest proportion with 92.91% of the whole and in use and maintenance period for the smallest proportion with 1.06%. Thus the emphasis of cutting down the solid waste discharge amount in periods of life-cycle of civil building is to reduce the application amount of cement and ready-mixed concrete, and to develop more civil buildings with steel structure which is substitutable for concrete structure or with combined structure.

Keywords: civil building; life-cycle; solid waste; resource-environmental pressure; Ecological Footprint (EF)

1 GENERAL INTRODUCTION

Construction waste city solid waste is the main component of our country building the amount of rubbish to have accounted for 30% of the total urban waste-40%. These buildings, garbage emissions more species, using value is very low and degraded, most can only be shipped to the suburbs or according to open air are common garbage disposal, not only takes up a lot of land, change the soil characteristics and pollution about the environment, but also diverted to the life of rubbish landfill space, shorten service life of the landfill, if not treated properly, the waste will be a serious nuisance.

This paper is closely related with People's Daily life of civil building life cycle all kinds of solid waste emissions classification, with the aid of the ecological footprint method and the energy analysis method to establish the discharge of solid waste measure the pressure on resources and environment calculation model, and to shenyang library for example

empirically. To take effective measures will discharge of solid waste to bring pressure on resources and environment control in the natural bear ability range, promote economic development gradually took to the resources and environment of the reduction pressure track provide the beneficial reference.

2 THE RESEARCH OBJECT AND DATA

2.1 *Solid waste discharge from civil building life-cycle*

This paper will civil life cycle is divided into four stages, namely raw materials and building materials production stage mining, construction stage, use and maintenance stages, dismantling and abandoned building materials disposal stage, this process with all kinds of resources occupation and all kinds of waste disposal and produce huge pressure on the environment. Therefore, the calculation of civil life cycle of solid waste discharge the pressure

Table 1. Consumption inventory of main building materials of Shenyang Municipal Library.

Name of building material	Actual amount	Name of building material	Actual amount
Commercial concrete/m³	24 687.74	Cement/kg	1 427 491.70
Cavity block/m³	5 628.18	Clay brick/m³	421.42
Cavity brick/m³	172.32	Profiled steel sheet/m²	9 630.06
Sand/m³	1 213.33	Aluminum product/m²	2 958.05
Girder steel/m³	55.80	Girder steel/kg	114 238.95
Granite/m²	7 450.07	Marble/m²	5 449.19
Glass curtain wall/m²	4 210.92	Various non-metal tubing/kg	269 067.75
Timber/m³	252.40	Rubble/m³	504.57
Reinforce steel/kg	3 291 540.00	Ground tile/m²	14 836.25
Steel piece/kg	27 727.35	Sheet steel/kg	46 840.50
Glass/m²	1 003.91		

on resources and environment that is to calculate the abandoned building materials of the pressure on resources and environment.

2.2 Date

According to the research purpose, this paper selects Shenyang library for evaluation objects. The building occupies an area of 13,380 m², a total construction area of 40,269 m², and the main building height of 33.0 m, underground, the ground nine layer, volume rate 0.84. According to the code for fire protection design of high civil buildings, this project to a class of high buildings, for I fireproof grade level. Building use the new type composite insulation energy-saving wall designed, reduce the dosage of clay brick. The structure of reinforced concrete composite beams, pressed steel and concrete composite floor slab. Seismic fortification categories for c class, safety level II level, design use fixed number of year 50 years.

In this paper, it selects the concrete products, cement, wall ground material, steel, sandstone, glass and products, stone, other metal, coating and all kinds of nonmetallic pipes of the total of 10 nearly 20 species of main and common building materials, various raw materials consumption such as shown in Table 1.

3 RESEARCH METHODS

Civil building in the life cycle of the solid waste discharge the pressure on resources and environment, the main consideration stage of construction and use and maintenance stages in the construction process loss or construction, maintenance and repair bad management reason building materials abandoned the pressure on resources and environment, demolition and abandoned building materials disposal stage of construction waste discharge

the pressure on resources and environment. And building materials in the raw materials and building materials production stage mining of the discharge of solid waste the pressure on resources and environment for lack of data, this paper has not been considered. Therefore, the civil construction life cycle solid waste discharge the pressure on resources and environment of the calculation model consists of three parts, specific as follows:

$$EF_{sw} = EF_{swc} + EF_{swu} + EF_{swd} \qquad (1)$$

where, EF_{sw} mean the life cycle of the solid waste discharge resources occupation footprint; EF_{swc} mean construction stage of solid waste discharge resources occupation footprint; EF_{swu} mean the use and maintenance stages of solid waste discharge resources occupation footprint; EF_{swd} mean demolition and waste disposal stage of resources occupation footprint.

3.1 Construction stage

Work quantity unit is expressed in weight (t, kg), volume (m³) and area (m²). When Building material i express in a different unit and convert the unit. The computation formula is as follows:

$$EF_{swc} = \gamma_k \times \frac{\sum_{i=1}^{n} m_{ki} \times \delta_{ki}/100 \times STE_{ki}}{EPA} \quad k = 1,2,3 \dots 6 \qquad (2)$$

Work quantity unit is expressed in price (¥). The computation formula is as follows:

$$EF_{swc} = \gamma_k \times \frac{ERR \times \sum_{i=1}^{n} m_{ki} \times \delta_{ki}/100}{EPA} \quad k = 1,2,3 \dots 6 \qquad (3)$$

where, k mean the type of land; γ_k mean type i release factor of land; specific value Table 2; ERR(Emergy/¥ radio) mean the energy/currency ratio, the unit is sej/¥, according to the regional civil building construction time take in Table 3 numerical; EPA(Emergy per area) mean the energy density of the area, the unit is sej/hm^2, according to the regional civil construction completion time take in Table 3 numerical; n mean use building materials species; m_{ki} mean the usage of the building materials or price, usually in the area (m^2), volume (m^3), weight (t or kg) or price (¥) is said; δ_{ki} mean during the construction process in the construction process loss or construction reasons such as poor management of building materials abandoned ratio, the parameters, see Table 4; STE_{ki} (Solar transformity of Emergy) mean the Solar energy value construction material conversion, the unit is sej/kg, see Table 4.

3.2 Use and maintenance stage

Work quantity unit is expressed in weight (t, kg), volume (m^3) and area (m^2). When Building material i express in a different unit and convert the unit. The computation formula is as follows:

$$EF_{swu} = \gamma_k \times \frac{\sum_{i=1}^{n} m_{ki} \times \delta_{ki}/100 \times \left[\frac{m}{B_{ki}} - 1\right] \times STE_{ki}}{EPA} \quad (4)$$
$$k = 1,2,3 \ldots 6$$

Work quantity unit is expressed in price (¥). The computation formula is as follows:

$$EF_{swu} = \gamma_k \times \frac{ERR \times \sum_{i=1}^{n} m_{ki} \times \delta_{ki}/100 \times \left[\frac{m}{B_{ki}} - 1\right]}{EPA} \quad (5)$$
$$k = 1,2,3 \ldots 6$$

where, B_{ki} is service life (a) of different, see Table 5 for the values used; the computed result of $[m/B_i - 1]$ is rounded up, which shows the reused number of times of material i in the maintenance and repair stage of the building after the building materials consumed before the building is put into service have been deducted.

3.3 Dismantling and waste building material disposal stage

Work quantity unit is expressed in weight (t, kg), volume (m^3) and area (m^2). When Building material i express in a different unit and convert the unit. The computation formula is as follows:

$$EF_{swd} = \gamma_k \times \frac{\sum_{i=1}^{n} m_{ki} \times (1 - R_{ki}) \times STE_{ki}}{EPA} \quad (6)$$
$$k = 1,2,3 \ldots 6$$

Work quantity unit is expressed in price (¥). The computation formula is as follows:

$$EF_{swd} = \gamma_k \times \frac{ERR \times \sum_{i=1}^{n} m_{ki} \times (1 - R_{ki})}{EPA} \quad (7)$$
$$k = 1,2,3 \ldots 6$$

where, R_{ki} is recycling ratio of waste building material, the values taken are listed in Table 6.

4 RESULTS AND ANALYSIS

4.1 Construction stage

In all kinds of building materials solid waste of resources occupation in emissions footprint, cement, concrete products, steel and wall ground material is this phase solid waste discharge of the main components of the resources occupation, emissions were 82.07 hm^2, 53.91 hm^2, 14.49 hm^2 and 13.69 hm^2.

Table 2. Equivalent factor.

Land types	Equivalent factor	Land types	Equivalent factor
Cultivated land	2.11	Built land	2.11
Fossil energy land	1.35	Grassland	0.47
Forest land	1.35	Water	0.35

Table 3. Emergy per area and Emergy/¥ radio in Liaoning province.

Year	EPA ($\times 10^{16}$ sej/hm^2)	ERR ($\times 10^{11}$ sej/¥)	Year	EPA ($\times 10^{16}$ sej/hm^2)	ERR ($\times 10^{11}$ sej/¥)
2000	2.53	8.37	2001	2.67	8.18
2002	2.83	8.01	2003	3.08	7.96
2004	3.39	7.93	2005	3.69	7.25

Table 4. Frequently-used data of building materials.

Name of building material	STE_{ki} (sej/kg)	Discarded ratio
Commercial concrete/m³	5.08×10^{11}	5
Timber/m³	6.9×10^{11}	7
Clay brick/m³	2.0×10^{12}	10
Cavity brick/m³	2.0×10^{12}	10
Profiled steel sheet/m²	1.4×10^{12}	5
Sand/m³	1.0×10^{7}	5
Girder steel (kg or m³)	1.4×10^{12}	5
Rubble/m³	1.0×10^{7}	5
Ground tile/m²	2.0×10^{12}	10
Paints/¥	7.93×10^{11} sej/¥	10
High polymer waterproof rolling material/m²	4.3×10^{12}	5
Cement/kg	3.3×10^{13}	5
Cavity block/m³	5.08×10^{11}	10
Reinforce steel/kg	1.4×10^{12}	5
Steel piece/kg	8.6×10^{11}	5
Aluminum product/m²	1.6×10^{13}	5
Glass/m²	8.4×10^{11}	5
Granite/m²	5.0×10^{5}	5
Marble/m²	1.0×10^{12}	5
Extrusion wall/m³	3.8×10^{11}	10
Various non-metal tubing/¥	7.93×10^{11} sej/¥	5

Table 5. Life of common building materials or components.

Building materials	Service life
Structure part, feed pipe, cables and cords, ventilation duct	Life span of civil buildings
Doors and windows, decorative panels, external tiles, non-metal tubing	30 a
Waterproof rolling material	20 a
Coating, wallpaper	10 a

Table 6. Recycling ratio of common wasted building materials.

Types of building materials	Recycling ratio	Types of building materials	Recycling ratio
Steel product	0.95	Concrete and cement	0.10
Materials for wall	0.60	Non-iron metal	0.90
Glass	0.80	Timber	0.10

4.2 Use and maintenance stage

The use and maintenance of solid waste disposal of stage resources occupation total footprint is 29.74 hm². Among them, the abandoned cement emissions of 24.62 hm², accounting for this stage of the discharge solid waste resources occupation footprint of 82.78 percent. Therefore, control maintenance dosage of cement needed to repair is to reduce this phase of solid waste discharge the pressure on resources and environment, the main measures.

4.3 Dismantling and waste building material disposal stage

Demolition and abandoned building materials stage and the cause of the construction waste disposal is civil life cycle solid waste discharge the pressure on resources and environment of the main component, the calculated according to this stage the discharge of solid waste resources occupation footprint for 2 604.16 hm². Among them, the biggest cement emissions footprint, for 1 477.23 hm², accounting for 56.73 percent of this stage emissions footprint; Followed by concrete products emissions footprint is 970.43 hm², accounting for 37.26 percent of this stage emissions footprint; Wall ground material for 111.15 hm² emissions footprint, accounting for 4.27 percent of this stage emissions footprint; The rest of the building materials for 45.35 hm² sum emissions footprint. Therefore, to cut emissions of the construction waste the key is to reduce the dosage of cement and concrete products.

4.4 Every stage of the life cycle of the pressure on resources and environment

According to the three stages of solid waste discharge resources occupation footprint summary, we get civil building life cycle of solid waste discharge resources occupation footprint constitution. Among them, the demolition and abandoned building materials disposal of solid wastes of emissions stage resources occupation footprint is the largest, construction stage is the second smallest, and use and maintenance stages the smallest.

5 THE MAIN CONCLUSIONS

For more comprehensive to reflect all kinds of solid waste disposal to ecological environment caused by the comprehensive pressure, this paper put the ecological footprint method and the energy analysis method, and the combination of the civil building construction measure each stage of life cycle of solid waste discharge the pressure on resources and environment of the calculation model, make solid waste discharge the pressure on resources and environment of the measure have a common calculation basis and comparability, realize the discharge of solid waste of the pressure on resources and environment comprehensive quantitative.

According to the empirical study, abandoned cement emissions and abandoned concrete products emissions constitute the each stage of life cycle of solid waste discharge key component, also cause solid waste discharge the pressure on resources and environment of the source. Therefore, civil building in each stage of life cycle of solid waste emissions cut the key is to reduce the dosage of cement and concrete products, many developing alternative concrete structure steel structure or combination of civil building structure form.

Thanks: Liaoning Province Science & Technology Program (No. 2011230008); Shenyang Municipality Science & Technology Program (No. F11-264-1-14).

REFERENCES

Lan Shengfang, Qin Pei, Lu Hongfang. 2002. *Emergy Evaluation of Ecological Economic Systems*, Chemical Industry Press.

Li guangjun. 2008. *Resources Occupancy of Chinese Cities and Applied Research,* Northeasten University.

Liu Hao, Song Yang. 2008. *Emergy Evaluation on the Development of Circular Economy and Empirical Study, China Population Resources and Environment* 18 (1):79–83.

Qian Feng, Wang Weidong. 2007. *EMT and Assessment of Building Environmental Efficiency: Case Study of the Beijing University Gymnasium*, Architectural Journal.

Tao Zaipu. 2003. *Eco-rucksack and Eco-footprint*, Economic Science Press.

Wang Songqing. 2007. *Life Cycle Assessment of Residential Building Energy Consumption in Severe Cold Region,* Harbin Institute of Technology.

Zhong ping. 2005. *Study of Building Life-cycle Energy Use and Relevant Environmental Impacts,* Sichuan University.

Zhou shu. 2008. *Construction of the Environmental Impact Analysis and Evaluation of Research—to the Western Zone of Chengdu High-tech Zone Group Project as an Example of Technological innovation,* Chengdu University of Technology: 2–3.

Zhu Yanyan. 2002. *Coordinative Development Assessment of Beijing Eco-economic System Based on Energy—Emergy Analysis*, The Graduate School of Chinese Academy of Sciences.

Frontiers of Energy and Environmental Engineering – Sung, Kao & Chen (eds)
© 2013 Taylor & Francis Group, London, ISBN 978-0-415-66159-1

On the internalization of ecological costs in the energy exploitation in China

B. Xiang & H.H. Chen
Jiangxi University of Science and Technology, Jiangxi, China

ABSTRACT: Energy plays an indispensable role in the development of human beings, however, in recent years; the energy development is caught in the trap of "unsustainable development" resulting from a series of problems including irrational energy structure, high ratio of traditional energy, severe pollution, and difficult exploitation of new energy as well as troubled further development. This thesis, based on the discussion about the necessity of internalization of ecological costs, proposes the measure of a combination of direct control and indirect control to accelerate the internalization of ecological costs.

Keywords: energy, ecological, costs internalization

1 PROPOSING THE ISSUE

Energy as the important material foundation of people's daily life, its importance is self-evident. But in recent years, every country confronts various severe challenges concerning energy. As in china, it faces with the problems in energy exploitation such as: irrational production and consumption structure, intense supply and demand, low utilization ratio, outstanding energy problems in rural areas, the threats on the security of energy and so forth. Taking energy structure as an example, the consumption rate of coal has always been high in energy consumption, which reaches 69.4% in 2006, and the ratio in those developed countries in Europe and America is only 21%. China is the largest production and consumption country of coal and the amount of consumption equals to the sum total of all other countries.[i] In the consumption structure of primary energy, the share of coal in China is 41% higher than global average, the ratio of oil gas is 36% lower, and the ratio of hydroelectricity and nuclear power is 5% lower.[ii]

The renewable energy sources are abundant in China, which, however, only takes a small part except for a few technologies like solar water heater have been adopted in daily use and they are capable in market competition. The development of most industries concerning renewable resources is relatively slow and the market is limited. The main reason that lead the the condition is the defects in market which results in the higher costs of renewable sources than traditional ones and the weakening of their competitive power. Most consumers will firstly choose those "cheap" products.

Taking power generation as an example, the costs of small hydropower station is 1.2 times higher than coal electricity, biomass energy power generation 1.5 times higher, wind power generation 1.7 times higher and photovoltaic power generation is even 11–18 times higher. In fact, the costs of electricity generation by coal is much higher than that by biomass energy, wind power and photovoltaic, the ecological costs not reflected in the price of electricity has been neglected by people. The defects of market, the integrity and the public-service determine that the private laws based on market exchange has congenital shortage, and the externalization of ecological costs makes renewable energy sources not able to contend with traditional energy sources. However, the developing of renewable energy sources has important strategic significance in energy security, diversification of energy supply, especially in environmental protection. Therefore, we should keep a foothold in long term development, formulate forceful policies and rules, perfect energy costs mechanism, internalize ecological costs and support and guide the development of renewable energy sources industry.[iii]

2 THE IMPLICATION OF ECOLOGICAL COSTS

Ecological cost, also called ecological and environmental cost, Different organizations hold different understandings towards ecological costs. The Environment Management Committee of America proposes that it includes the following four aspects: environment loss costs, environment protection costs, environment work costs and the costs on abatement of pollution.[iv] UNSD in 1993 defined ecological costs as the actual costs in economy and environmental protection resulted from the quantity and quality consumption of natural resources. In china, many scholars think that ecological costs is the money paid for controlling pollution, plus money for social damage brought by pollution, plus money for the protection of natural resources in one accounting cycle, and the load is shared by the government, residents as well as enterprises.[v]

Though opinions on ecological costs vary, there is one thing in common, i.e. its definition is closely connected with its extension. This thesis holds that in order to give it's a rational definition of its extension; we should firstly discuss its appearance. Before 1970s, the concept of ecological costs has been long neglected. The reason that leads to the condition is that the ecological system in balance has the ability of self-adjusting. In the past, when the pollution of ecology and environment is under the self-restoration of the environment, people may not suffer from any loss in economy and will not be asked to pay for the discharge of wastes. However, with the fast development of industries and the sabotage on environment by human becomes more severe, it has surmounted the limit of its load, while people have become accustomed to the endless self-cleaning capability of the environment. Then the concept of ecological costs is proposed. Since the proposition of ecological costs appeared when human's sabotage is over the self-cleaning capability of environment, in the definition of the extension factors of ecological costs, we should at least remove the part that is included in the self-cleaning and self-restoration capability of environment. As for energy resources, its ecological costs means the costs on the damage of environment that is over the self-cleaning and self-restoration ability of environment in the process of energy exploitation. It includes: compensation costs in environment pollution, costs on environment management, costs on environmental protection and carrying and costs on ecological restoration.

3 AN ANALYSIS ON THE NECESSITY OF THE INTERNALIZATION OF ECOLOGICAL COSTS

The problem of ecological costs is an unavoidable one in the exploitation and development of energy sources. Environmental economics, starting from revising the original prices of price error, puts up with the concept of the internalization of the environmental costs. The internalization of ecological costs is add ecological costs as a part in the total costs of a product, so that the price not only reflects production costs, transaction costs, but also its ecological costs, and this process is known as the internalization of ecological costs. The core of ecological costs is "polluters pay principle", i.e. the polluters (in this thesis it mainly refers to the exploiters) should pay for their deeds. For a long time, the externalization of ecological costs in China has made the deficient energy resources be caught in the trap of unsustainable development, which damage the healthy move of modern economy and the sustainable development severely. This developing mode, under the promotion of quick economic development, has led to environmental pollution, degenerated ecology and large economic loss. According to the estimation of World Bank, Chinese Academy of Science and SEPA, 2/3 of Chinese economic growth is based on the overdraw of the environment, and the loss costs by environmental pollution reaches 10% of the GDP. This mode of economic development, which is based on withdrawing the environment ecological deficit, has reached a critical state of the load of Chinese ecological system. Some resources and the volume of the environment have reached the extremity.[vi] In the backdrop of building a conservation-minded society, the importance and urgency of internalizing ecological costss becomes increasingly outstanding.

3.1 The unecomomic of external elements in energy exploitation is the base of the internalization of ecological costs

The uneconomic of external elements in energy exploitation is the basic reason that leads to environmental pollution and destruction. The externalization is the inherent factor that makes the subjects in energy exploitation neglect environmental protection or be unwilling to invest in environmental protection. "The tragedy of public land" indicates that some people consume environmental resources initiatively and willingly, while some others have to bear the result of environmental consumption, which is an extremely irrational phenomenon. In order to get rid of these irrational phenomena, protect the environmental resources from excessive utilization and manage environmental resources radically, in the area of energy exploitation, we can only achieve it through the internalization of energy costs to let the exploiters bear the ecological costss; the costs should be reflected in the market price of the enterprises so that environmental resources take part in the process of resource allocation in the market to avoid the abuse of environmental resources, ensure environmental quality as well as embody the actual costs of energy exploitation objectively to stimulate the sustainable development of energy resources.

3.2 The internalization of ecological costs is the inevitable requirement of social and economic sustainable development

The mode of "pollute first and clean up later" has been used from traditional industrialization to modernization both in developed and developing countries. This mode of "exterior management" is the developing mode of the externalization of ecological costs. Just as Daley pointed out that, "the current leading mode totally excludes ecological costs"[vii], which will undoubtedly lead to the unsustainability of economic development and the dilemma faced by the ecological environment. The practices are telling us that we cannot follow the mode of "exterior management" and that we should choose a new sustainable develop mode internalizing the ecological costs. In order to build up such a mode, we must "let the market reflect all the costs of the goods and services we buy."[viii] That is to say we should let the market prices reflect the ecological costs of energy resources, or we will pay more costs.[ix]

3.3 The internalization of ecological costs is the inherent requirement embodying environmental justice

The appearance of environmental righteous follows the appearance of environmental unjustice, the nature of the environment as a public goods makes it difficult to divide the duty and compulsory on energy resources, and the market is incapable of allocate effectively on energy resources, which lead to the inequality in benefit-sharing and duty-bearing among countries, urban and rural areas, regions as well as different classes. Under the severe environmental injustice, some people even put up that facing with the environmental crisis, we have no common future.[x] Because there exists the phenomenon that "the rich pollute and the poor suffer" in environmental problems and "the powerful who trigger the problems are the leaders gaining extra benefits, while the weak who suffer losses have to be responsible," which is a structure of elitism in a broad sense.[xi] Therefore, environmental problem is not simply the imbalance between men and nature, it is more a problem of how to allocate fairly environmental benefits the assume the responsibility of the environment. Environmental justice is a environmental view combining environmental protection and social justice, which emphasize the equal allocation of environmental benefits and burden and the equal allocation is the core of this new environmental view.[xii]

In the area of energy exploitation and utilization, the phenomenon of environmental injustice is especially severe. The energy resources, as the main force in environmental destruction, lead to severe environmental injustice during it exploitation. On the one had, the exploiters seize the most part of the benefits, on the other, the environmental pollution caused by the exploitation has to be borne by the public, especially those poor people who are incapable of choose their living environment and face with the risks brought by environmental pollution. Hence, the internalization of ecological costs in the exploitation and utilization of energy resources will let the exploiters shoulder the compulsories equal to their rights, which is the inner requirement of environmental justice.

3.4 The internalization of ecological costs is the requirement of giving equal development rights to all kinds of resources, which will stimulate the exploitation and utilization of clean energy

The biggest resistance China faces with in energy exploitation and utilization is price. The irrational market cost system leads to the great price disparity in wind power, solar power and other new resources with coal, gasoline and other traditional resources, which hinder the development of new energy resources. The exploitation of new energy cannot be admitted by the market so that energy security can never be secured and environmental problems cannot be solved radically. The new energy like wind power and solar power, possessing the advantage of clean and renewable, cannot have equal developing chances as other traditional resources because of the man-made hurdles. If we can include ecological costs into the costs of exploitation and utilization, the advantages of new energy will become outstanding gradually.

4 AN ANALYSIS ON THE WAY TO REALIZE THE INTERNALIZATION OF ECOLOGICAL COSTS IN ENERGY EXPLOITATION AND UTILIZATION IN CHINA

4.1 To promote internalization through direct control

Direct control is a way combining order and control, in accordance with the requirements of laws and regulations of the country's administrative departments, to formulate and carry out unified rules and standard towards the counterparts of administrations. It requires the administrative counterparts to act or not act, and to give compulsory penalization to those who violate the rules. It is the current environmental management method in China, which owns the characteristics of technical, strict, compulsory and effective, and generally speaking, this measure can rectify the market failure in environment to some extent. With regard to resources, it requires the government to formulate and perfect the standard concerning with manufacturing and pollution discharging, attach importance to those preventive environmental protection measures center on "whole—course control" and "reduction at the source". We should stimulate the self-conscious internalization of ecological costs by the enterprises through direct control.

4.2 *To stimulate internalization through indirect control*

Although direct control owns many advantages, it reveals many disadvantages, such as the enterprises only pay attention to the end management of the pollution, which leads to the non cyclical economic mode featuring pollute first and clean up later, and it cannot solve the problem of the external ecological costs. Therefore, we must combine direct control with indirect control to accelerate the internalization of ecological costs into the enterprises. Indirect control is to adopt economic stimulus methods in accordance with the rule of compensated use of environmental resources and the "polluters pay principle", to apply the market system, they can add the external ecological costs during the exploitation into the total costs, which will make the exploiters choose a mode of production, operation and consumption that is favorable to the environment while considering their own economic condition. In this way they can gather money to be allocated by the government according to the needs to support those areas need special treatment, which is avail to the realization of the goals in sustainable development.

REFERENCES

i. Zhou Shengxian, 2012. To Optimize Economic Development through Environmental Protection—Written on "6·5" World Environment Day (eds). *People's daily.*

ii. Jiang Zheming, 2008 Reflections on the Energy Issues in China, *Science Edition of Shanghai Jiaotong University Journal*, (3):345–359.

iii. Chen Jianmin & Tian Hongyan, 2012. Our country's energy utilization status to a low carbon economy development constraints and countermeasures. *Reform and strategy.*

iv. Li Lianhua & Ding Tingxuan, 2000. The measurement and confirmation of environmental cost. *Economic obtain.*

v. Lin Wanxiang & Xiao Xu, 2002. the research on cost issues of Social environment. *Accounting Research.*

vi. Liu Sihua, 2004. Several Problems on the Scientific Outlook on Development. *Inner Mongolia journal of finance and economics.*

vii. Dai Li, 2001. Surpassing increase—Economics on Sustainable Development, Shanghai: *Shanghai Translation Publishing House.*

viii. Bu Lang, 2002. Ecological Economy: Economic Conception Favorable to the Earth, Beijing, Eastern Trade Media Pte Ltd, 24.

ix. Bu Lang, 2002. Ecological Economy: Economic Conception Favorable to the Earth, Beijing, *Eastern Trade Media Pte Ltd*, 28.

x. Ji Junjie, 1998. We Have No Common Future: The Political Economy Concerned by Western Mainstream "Environmental Protection". *Taiwan A Radical Quarterly in Social Studies*, 141.

xi. Hu Tianqing, 2011 Pursuing Environmental Justice quote from the legal expression of environmental justice. *Science Press,* 9.

xii. Liang Jianqin, 2011. The Legal Expression of Environmental Protection. *Science Press.*

Frontiers of Energy and Environmental Engineering – Sung, Kao & Chen (eds)
© 2013 Taylor & Francis Group, London, ISBN 978-0-415-66159-1

Numerical simulation on heat transfer of a rotating jet impinging on a protruding pedestal

L.M. Huang & J.W. Zhou
College of Metrology & Measurement Engineering, China Jiliang University, Hangzhou, China

ABSTRACT: *RNG κ-ε* model was used to compute the heat transfer and rotational flow field of a semi-confined turbulent axisymmetric jet, impinging onto a pedestal mounted on a flat plate. Heat transfer and rotational flow field characteristics were investigated on the pedestal top surface and side face, as well as on the plate. The influence of rotational flow intensity (rotation number *R*), Reynolds number *Re* and nozzle-to-pedestal distance *H/D* on heat transfer and flow was studied. It showed that impinging on the stagnation point was weakened by the rotating jet under different Reynolds number, and the Nusselt number on the stagnation point decreased.

Keywords: rotating jet; pedestal; *RNG κ-ε* model; heat transfer characteristics

1 INTRODUCTION

Jet impingement is a very effective measure to strengthen the heat transfer. At present, it finds extensive applications in many industrial processes of high demand for high heat and mass transfer, such as electronic components cooling, turbine blade cooling, textile drying and deicing the ice on aircraft wings.[1]

Heat transfer characteristics are effected by many parameters: Reynolds number, nozzle diameter(*D*), nozzle-to-plate distance (*H/D*), nozzle shape, jet medium and so on. At present, most impingement target surfaces have been focusing on flat surfaces, however in actual applications, such as electronic equipment and aeronautical engines, many objects are three-dimensional structures, which need to analyze several surfaces. From this point, studying the impinging jet on a protruding pedestal has an important practical significance. Previous studies on protruding pedestals were very limited. Merci[2] studied the heat transfer and flow field characteristics around a cylindrical pedestal under different Reynolds numbers by experim- ent and numerical simulation. Kim[3] studied impinging jet on a rectangular pedestal by experiment.

In the references, some rotating jets were generated through changing the nozzle structure,[1,4] some through rotating the nozzle,[5] others through adding an rotating mechanism inside the nozzle to generate a rotating jet.[6,7] Most of the studies show: rotating jet makes the heat transfer on the impingement plate more uniform.

This study is trying to combine a rotating jet and a protruding pedestal to do a numerical simulation with the aid of *Fluent* to compute its heat transfer and flow field characteristics. Previous studies about rotating jet on pedestal have not yet been found, therefore the study and research on this subject will have great significance theoretically and in practical applications.

2 ESTABLISHMENT OF MODEL

2.1 *Physical model*

Jet from a round nozzle with a 10 mm diameter (*D*) impinges on the target surface. The nozzle-to-pedestal distance is *H*; the height and diameter of pedestal is *D;* the length of semi-confined board and plate is big enough relative to nozzle diameter; the vertical distance between semi-confined board and plate is adjustable. Due to the fact of the whole flow field being axial symmetric, only the flow field within the half domain is solved in the numerical simulation.

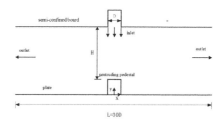

Figure 1.   Scheme of physical model.

2.2 Governing equation

Standard κ-ε model used in a strong rotating jet has distortion to some extent, however, through modifying the turbulent viscosity and time average strain rate it can better handle flow with a high strain rate and bigger streamline curvature.[8,9] RNG κ-ε equation as described below:

$$\frac{\partial(\rho\kappa)}{\partial t}+\frac{\partial(\rho\kappa u_i)}{\partial x_i}=\frac{\partial}{\partial x_j}\left[\alpha_\kappa(\mu+\mu_t)\frac{\partial\kappa}{\partial x_j}\right]+G_\kappa-\rho\varepsilon$$

$$\frac{\partial(\rho\varepsilon)}{\partial t}+\frac{\partial(\rho\varepsilon u_i)}{\partial x_i}=\frac{\partial}{\partial x_j}\left[\alpha_\varepsilon(\mu+\mu_t)\frac{\partial\varepsilon}{\partial x_j}\right]$$
$$+\frac{C_{1\varepsilon}^*\varepsilon}{\kappa}G_\kappa-C_{2\varepsilon}\rho\frac{\varepsilon^2}{\kappa}$$

$$C_{1\varepsilon}^*=C_{1\varepsilon}-\frac{\eta(1-\eta/\eta_0)}{1+\beta\eta^3}$$

$$\eta=\left(2\mathrm{E}_{ij}\bullet\mathrm{E}_{ij}\right)^{1/2}\frac{\varepsilon}{\kappa}$$

$$\mathrm{E}_{ij}=\frac{1}{2}\left(\frac{\partial u_i}{\partial x_j}+\frac{\partial u_j}{\partial x_i}\right)$$

where κ is the turbulent kinetic energy, ε the turbulent kinetic energy dissipation, μ the kinetic viscosity, $\mu_t=\rho C_\mu\varepsilon^2/\kappa$ the turbulent kinetic viscosity; $G_\kappa=\mu_t\left(\partial u_i/\partial x_j+\partial u_j/\partial x_i\right)\partial u_i/\partial x_j$ is the production term of the turbulent kinetic energy caused by average velocity gradient; constants in the model: $\alpha_\kappa=\alpha_\varepsilon=1.39$, $C_{1\varepsilon}=1.42$, $C_{2\varepsilon}=1.68$, $\eta_0=4.377$, $\beta=0.012$.

2.3 Nomenclature

Nu defined as: $Nu=qD/\lambda\left(T_w-T_f\right)$, where q is the heat flux on the pedestal and the plate, D the inner diameter of the nozzle, λ the thermal conductivity of air, T_w the temperature of the pedestal and the plate, T_f the jet temperature. Reynolds number defined as: $Re=u_iD/\nu$, where ν is the kinematic viscosity of air, u_i the axial velocity of jet at the inlet. Rotation number defined as: $R=\omega D/u_i$, where ω is the rotation angular velocity.

2.4 Boundary conditions

1. inlet(velocity inlet): the jet temperature is 293.15 K, at a uniform velocity at the inlet.
2. semi-confined board(wall): fixed wall, heat insulating; the initial temperature is 293.15 K.
3. heat plate and pedestal(wall): fixed wall, no penetration, no slip; the heat flux is 435 W/m².

4. outlet (outflow): the flow field being fully developed and running steady; the relative normal pressure gradient is zero.

3 RESULTS AND ANALYSIS

3.1 Nu number distribution under different rotation numbers

Figure 2 shows the Nusselt number distribution on the top of the pedestal at the same impinging distance ($H/D=4$) based on two different Reynolds numbers and different rotation numbers. It can be seen that the heat transfer performances are different under two different Reynolds numbers. Under a high Reynolds number (e.g when Re = 25000), different rotation jets flow down along the radial direction and the Nusselt numbers are rising, and due to the boundary conditions, it drops sharply on the edge of the pedestal. Whereas under a low Reynolds number (when Re = 5000), low rotation jets have a decreasing Nusselt number in the radial direction and the downtrend is accelerating near the edge of the pedestal. Under a bigger rotation number, the distribution of Nusselt number along the radial direction doesn't see much difference.

Comparing Figure 2(a) and (b), it can be found that the Nusselt number of the rotating jet is

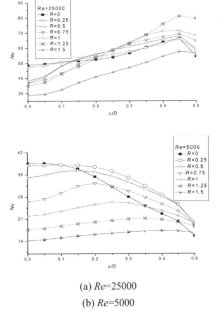

(a) Re=25000

(b) Re=5000

Figure 2. Local Nusselt number distribution on the top of protruding pedestal when $H/D=4$.

always smaller than that of the steady jet at the stagnation point, and with the increase of the rotation number, the deviation is getting bigger. This is because the rotating jet weakens the impingement at the stagnation point. Impinging jet spreads out because of rotation. Thus when there is a suitable rotating jet impinging on the pedestal, the heat transfer on the outer area of the pedestal top along the radial direction will be enhanced. If the rotating jet is over-strong ($Re = 25000$ and $R \geq 1.5$, or $Re = 5000$ and $R \geq 1.25$), the impinging jet will deflect sharply. At this moment the axial velocity will be very small, and there is almost no effective impingement on the pedestal, resulting in a very small Nusselt number on the whole top surface.

Figure 3 shows the Nusselt number distributions on the side of the pedestal under two different Reynolds numbers based on different rotation numbers. Rotating jet enhances the heat transfer on the side of the pedestal. With the increase of the rotation number, the Nusselt number tends to be bigger. When $Re = 25000$ and $R = 1.5$, due to the vertical jet being greatly affected by the rotation, the impinging jet spreads outward seriously to the extent that it can't impinge on the pedestal directly, forming a small reflux near the side, therefore resulting in the decrease of the Nusselt number on the side.

Figure 4 shows the Nusselt number distributions on the plate under two Reynolds numbers based on different rotation numbers.

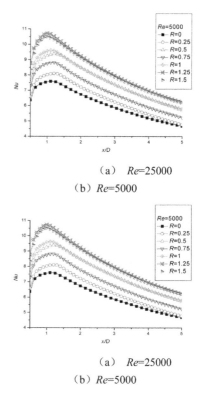

（a） Re=25000

（b） Re=5000

Figure 4. Local Nusselt number distribution on the plate when $H/D = 4$.

From Figure 4(a), we can see that when Re + 2500, the heat transfer on the plate is enhanced by rotating jet except for when $R = 1.5$. The Nusselt number distribution goes up first until it reaches the highest point, and then it starts to go down slowly.

The velocity vector diagram shows: the wall jet on the top edge will interact with the outer flow, and then impinge on the plate at a certain angle. Its heat transfer is mainly affected by the vortex near the side. In the right area, the jet develops into a wall jet after impinging the plate. Through momentum and energy exchange with the plate, the momentum of the jet will decrease accompanied by the temperature rise the jet, resulting in the slow decreasing of the Nusselt number. Figure 4 (b) shows when Re = 5000, the rotating jet enhances the heat transfer of the plate, but since the Reynolds number get smaller, the Nusselt number also decreases.

3.2 *Nu number distribution at different heights*

Figure 5 shows the Nusselt number distribution on the protruding pedestal and the plate at different heights when Re = 25000 and 4 = 0.5. Figure 5(a)

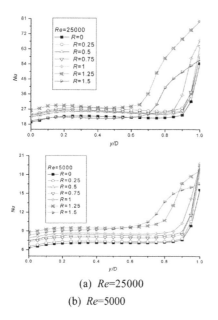

(a) Re=25000

(b) Re=5000

Figure 3. Local Nusselt number distribution on the side of the protruding pedestal when $H/D = 4$.

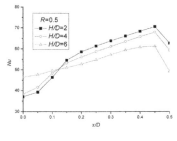

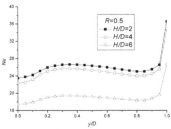

(a) *Nu* number distribution on the top

(b) *Nu* number distribution on the side

Figure 5. Local Nusselt number distribution on the top and side of the protruding pedestal when $R = 0.5$.

shows: the Nusselt number on the top surface is going up continuously at the beginning and then going down abruptly along the edge of the pedestal, At the stagnation point ($x/D = 0$) and when $H/D = 6$, the biggest Nu number appears probably because under this Reynolds number and when $H/D = 6$, the impinging jet has the maximized potential flow core. From Figure 5(b), we can see that with the increase of the height, the Nusselt number on the pedestal side and the plate are decreasing. This is because the bigger the height is, the more energy exchange between the impinging jet and the external medium, and at the same time the smaller average velocity, and thus the momentum of the impinging jet on the plate becomes smaller, resulting in a smaller Nusselt number.

4 CONCLUSIONS

Through studying the numerical simulation of the rotating jet on the protruding pedestal, the following conclusion can be drawn:

1. When a suitable rotating jet impinges on the pedestal, under the influence of the Centrifugal force and Coriolis force, the impinging near the stagnation point is weakened with the increase of the intensity of the rotating jet, but the outward turbulence of the rotating jet enhances the heat transfer in the radial direction on the top of the protruding pedestal. Under different Reynolds numbers, the influence is slightly different.

2. When the rotating jet impinges on the pedestal, due to the outward turbulence of the rotating jet, the heat transfer on the side of the pedestal and the heated plate can be enhanced.

3. The numerical simulation at three different heights combined with Re = 25000 and R = 0.5 shows: the heat transfer on the stagnation point reaches its maximum when H/D = 6 and there is the maximum potential flow core at this height. However, the heat transfer on the whole surface of the top, the side and the plate will decrease with the increase of the height.

REFERENCES

[1] Jingwei Zhou, Liping Geng, Yugang Wang. Convective heat transfer characteristics of impingement cooling for a self-excited precessing jet [J]. Journal of Aerospace Power, 2009, 24 (6): 1197–1203.

[2] Bart Merci, Masood P.E, James W. Baughn. Experimental and numerical study of turbulent heat transfer on a cylindrical pedestal [J]. International Journal of Heat and Fluid Flow, 2005, 26:233–243.

[3] Nam-Shin Kim, Andre Giovanini. Experimental study of turbulent round jet flow imgining on a square cylinder laid on a flat plate [J]. International Journal of Heat and Fluid Flow, 2007, 28:1327–1339.

[4] Rong Tao, Liping Geng, Jingwei Zhou. Large eddy simulation of a precessing jet flow [J]. Journal of China University of Metrology, 2008, 19 (2):124–128.

[5] S.Z. Shuja, B.S. Yilbas, M. Rashid. Confined swirling jet impingement onto an adiabatic wall [J]. International Journal of Heat and Mass Transfer, 2003, 46:2947–2955.

[6] Yifeng Fu, Jimin Zhou, Ying Yang. Experimental research on compulsive cooling of swirling jet impinging [J]. Conference on High Density Microsystem Design and Packaging and Component Failure Analysis, 2006, 27(30):137–140.

[7] Dae Hee Lee, Se Youl Won, Yun Taek Kim, Yong Suk Chung. Turbulent heat trsansfer from a flat surface to a swirling round impinging jet [J]. International Journal of Heat and Mass Transfer, 2002, 45:223–227.

[8] Ramezapour A, Shirvani H, Mirzaee I.A numerical study on the heat transfer characteristics of two-dimensional inclided impinging jet [C]. Electronics Packing Tecnologgy Conference, 2003, 626–632.

[9] Qingguang Chen, Zhong Xu, Yongjian Zhang. Application of RNG turbulence models on numerical computations of an impinging jet flow [C]. Mechanics and Engineering, 2002, 6 (24):21–24.

Frontiers of Energy and Environmental Engineering – Sung, Kao & Chen (eds)
© 2013 Taylor & Francis Group, London, ISBN 978-0-415-66159-1

Microcalorimetric characterization of a marine bioflocculant-producing bacterium *Halomonas* sp. V3a

F. Wang, J. Yao, H.L. Chen & Y.T. Lu

Civil & Environment Engineering School, University of Science & Technology Beijing, Beijing, China

ABSTRACT: A microcalorimeter was used to measure metabolic activity by analyzing the thermodynamic parameters (growth rate constant k, total heat released Q_T and heat released during the log phase Q_{Log}) of *Halomonas* sp. V3a growth under different conditions, that was found to yield a bioflocculant with high flocculating activity for Kaolin suspension. The results show that strain V3a had the most metabolic activity for glucose, the lowest for maltose with fixed nitrogen source (NH_4Cl). The metabolic activity of strain V3a did not depend on the nitrogen content of selected compounds. Biomass concentration and k value exhibited the similar change trend, but not consistent with Q_{Log}. Another phenomenon could also be found that flocculating activity is higher, biomass concentration or k is lower contrarily, for example, tryptone as carbon source. So, it elucidated that higher biomass can not represent the higher production of bioflocculant.

Keywords: *Halomonas* sp V3a; microcalorimetry; thermodynamic parameter; flocculating activity

1 INSTRUCTIONS

Various kinds of flocculants, such as inorganic flocculants (polyaluminum chloride, aluminum salts etc.), organic synthetic flocculants (polyacrylamide derivative etc.) and naturally occurring flocculants (chitosan, sodium alginate and microbial flocculants etc.) have been widely used in industrial processes, including water and wastewater treatments, downstream processing and food and fermentation (Wang et al., 2007). However, most of inorganic flocculants and organic synthetic flocculants have been associated with environmental and health problems, in that some of them are not readily biodegradable, neurotoxic and even as strong human carcinogens. Because of these concerns, the use of microbial flocculants is expected to increase. So bioflocculants have been investigated extensively due to their biodegradability and that their degradation products are harmless to the ecosystem [6].

Bioflocculation is a dynamic process resulting from the synthesis of extracellular polymer by living cells. Generally, soil and activated sludge samples are regarded as the best sources for isolating bioflocculant-producing microorganisms, because of sampling conveniently, which include bacteria, fungi and yeast. So far, more than 50 strains were reported (Salehizadeh et al., 2001), such as *Alcaligenes latus* B-16, *Aspergillus* sp. JS-42 and *Enterobacter* sp. from soil samples. Meanwhile, few of the bioflocculant-producing microorganisms isolated from the marine environments were reported, just reporting *Gyrodinium impudicum* Nannocystis sp. NU-2

(Zhang et al., 2002) and KG03 (Yim et al., 2007). But, marine environments are purported to be good sources of interesting bacteria.

The studies of the bioflocculants usually were emphasized on the characterization of the bioflocculants through optimizing the fermentation conditions, such as carbon and nitrogen sources, molar ratio of Carbon to Nitrogen (C/N) and initial pH and so on, rather than the activities of bioflocculant-producing microorganisms. These factors could influence the production of bioflocculants and the flocculating activities also can affect the metabolic activities of the strains.

To our knowledge, we studied the microbial activities of bioflocculant-producing bacteria under different fermentation conditions. This work is purposedly to better represent and understand microbial activities of bioflocculant-producing bacteria.

2 MATERIALS AND METHODS

2.1 *Reagent and bacteria*

Flocculant-producing bacterium, *Halomonas* sp. V3a (*Taxonomy ID*: 516958, NCBI database) was offered by Key Laboratory of Marine Biotechnology, The Third Institute of Oceanography, China. The bacterium was isolated from a deep-sea sediment sample of the west Pacific Ocean.

The compositions of the culture medium were following: glucose 10.0 g, NH_4Cl 1.0 g, KH_2PO_4 2.0 g, K_2HPO_4 5.0 g, NaCl 24.0 g, yeast extract 0.6 g, $MgSO_4 \cdot 7H_2O$ 0.5 g in 1.0 l double distilled water.

In addition, the MgSO$_4$ solution was prepared and sterilized separately. The culture temperature and initial pH were 296 K and 7.6 respectively.

All the chemicals were analytical reagents.

2.2 *Microcalorimetric determination of strain V3a at the different incubation conditions*

Four temperature points (291, 296, 301 and 306 K) were selected to assess the effect of temperature on the metabolic activity of strain V3a.

The effect of different carbon sources on the metabolic activities of V3a was determined by replacing the same amount of the carbon sources (10.0 g l^{-1}) in the medium, glucose, sucrose, starch, malic acid and maltose.

The effect of various nitrogen sources on the metabolic activities of V3a were carried out by replacing the same amount of NH$_4$Cl (1.0 g l^{-1}) in the medium with other nitrogen sources, (NH$_4$)$_2$SO$_4$, NH$_4$NO$_3$, tryptone and urea.

2.3 *Preparation of the samples for microcalorimetric determination*

In the beginning of experiment, stock cultures of strain V3a was activated at 296 K with above culture medium in a rotary shaker (160 rpm) to get enough inoculums. One loop of strain V3a was inoculated in the ampoules containing 2.0 ml of the above prepared medium with the initial concentration of 1.0×10^7 cells ml^{-1}. The cell concentration was examined with a hemocytometer.

2.4 *Determination of flocculating activity*

5.0 ml 1.0% CaCl$_2$ solution was added into 95.0 ml double distilled water in 100.0 ml test tube, and the pH was adjusted to 6.0. Then, 0.5 g kaolin clay was added into the solution. 1.0 ml fermentation medium was centrifuged at 13,000 g for 10 minutes, then 0.5 ml cell-free culture supernatant, that was the liquid bioflocculant, was added into the 100.0 ml kaolin solution. The mixture was vigorously stirred and left to stand for 5 min. The Optical Density (OD) of the upper phase was measured with a spectrophotometer at 550 nm. The flocculating activity was calculated according to the equation.

Flocculating acticity = $(B - A)/A \times 100\%$

where, A is the optical density of the sample experiment at 550 nm with the liquid bioflocculant; B is the optical density of control experiment at 550 nm with the fresh culture medium.

2.5 *Determination of biomass concentration and pH*

Strain V3a was incubated in 300 ml flask containing 150 ml culture medium. The biomass concentration was immediately measured via optical density at 2 h interval in a spectrophotometer (Spectrumlab 752s, Lingguang Ltd, Shanghai, China) at λ_{600} nm. At the same time, the flocculating activity and pH of cultural medium was measured as well.

For the calorimetric test, each experiment was repeated six times. There into, three tests were regarded as reference to find the stationary phase of microbial growth, and use to calculate thermodynamic parameters. Other tests were stopped and the ampoules were removed to measure the flocculating activity and optical density immediately, according to the reference test.

3 RESULTS AND DISCCUSSION

3.1 *Change of flocculating activity, biomass concentration and pH with incubation time*

To find the growth phase at which the flocculating activity reached the maximum value, biomass concentration (optical density) was measured. Figure 1 shows that the flocculating activity increased with incubation time until 18 h, and then reached the highest level. The turbidity kept similar trend before the 18 h with flocculating activity. The time spot (18 h) was regarded as the critical point between log phase and stationary phase. Over the period 4–18 h, strain V3a was in its exponential growth phase and the flocculation also rapidly increased from 15–91%. Beyond 18 h cultivation, the bacteria reached the stationary phase, and the highest flocculation of kaolin clay suspension attained was around 97%. Then, the flocculation decreased with the cultivation time slightly. The same phenomenon was reported in the incubation of *A. parasiticus* (Deng et al., 2005). The flocculating activity was observed at the end of exponential growth phase and the beginning of the stationary phase. It is evident that the production of bioflocculant occur during each growth phase of the bacteria.

The pH value of the culture media deceased gradually from 7.1 to 6.0 with incubation time in Figure 1 because of acidic polysaccharides yielded

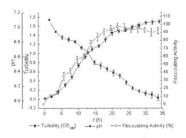

Figure 1. The strain V3a growth in flask cultures was measured at 296 K by turbidity. Simultaneously, the pH of the flask cultures and flocculating activity of the strain also were measured.

by strain V3a. The characteristics and available numbers of electric charges of functional groups on the bioflocculant also decide the flocculating activity. Electrostatic interaction plays an important role in the adsorption of kaolin clay on the bioflocculant. Decreasing pH increased the electrostatic attraction, and the bioflocculant molecules would show positive properties, and the surfaces of kaolinite carry a constant structural negative charge. So, electrostatic attraction would promote the kaolin clay to approach the bioflocculant, thus enhancing the increase in flocculation activity and accumulation of kaolin clay on cells.

3.2 Determination of the thermodynamic parameters

To test the metabolic activity of bacteria under different conditions, microcalorimetric method was applied. Thermodynamic parameters, heat flux $P(W)$, maximum specific growth rate k_{max} (h^{-1}) were calculated. The expression describing the time dependence of biomass concentrations for Monod kinetics (Eq. 1) was obtained.

$$X = X^0 \exp(k_{max}t) \tag{1}$$

here X, k_{max}, t stand for biomass concentration, maximum specific growth rate, and time, respectively. The superscript 0 indicates the situation at time zero.

Based on Eq. 1, description of the heat fluxes for the models of Monod (Eq. 2) was obtained, as follows:

$$P = A \exp(k_{max}t) \text{ with } A = \Delta^R H k_{max} X^0 \tag{2}$$

where, $\Delta^R H$ represents the reaction enthalpy. The maximum specific growth rate in Eq. 2 is the slope of $\ln P$ over t. In addition, the heat released on the log phase (Q_{Log}) was calculated through the integration of power time curve from initial point to maximum heat flux point [34], and total released heat (Q_T) was integrated the whole area under the power time curve.

3.3 Determination of the thermodynamic parameters of stain V3a under various incubation conditions

Thermogenic curves of the bacterial growth under various incubation conditions (different temperature, carbon and nitrogen sources) were obtained by microcalorimeter showing in Figure 2. The thermodynamic parameters were listed in Table 1.

For the different temperature, thermogenic curves proved that strain V3a had strong activities at 296 K with highest k and Q value (k, 1.22 h^{-1}; Q_T, 0.48 J ml^{-1}). Thermodynamic parameters increased initially and then decreased with increasing temperature, generally. In addition, the values are similar at 296 K and 301 K. It could be elucidated that

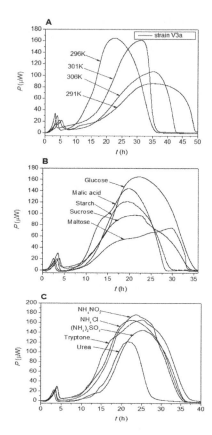

Figure 2. The power-time curves of strain V3a at 291, 296, 301 and 396 K at 5 K interval (A); different carbon sources (B) and Nitrogen sources (C) at 296 K.

Table 1. The main thermodynamic parameters of strain V3a obtained from the microcalorimetric method under various conditions. The tests using different carbon source and nitrogen source were performed at 296 K, and that under different temperatures were carried out with fixed amount of glucose and NH₄Cl.

	Experimental condition	k h^{-1}	Q_{Log} J ($\times 10$)	Q_T J ml^{-1}
Carbon source (10.0 mg l^{-1})	Maltose	1.12	0.62	0.22
	Sucrose	1.13	0.5	0.29
	Glucose	1.22	0.76	0.48
	Malic acid	1.19	0.6	0.32
	Starch	1.17	0.43	0.28
Nitrogen source (1.0 g l^{-1})	NH_4NO_3	0.90	0.79	0.46
	NH_4Cl	1.22	0.76	0.48
	$(NH_4)_2SO_4$	1.28	0.63	0.36
	Tryptone	1.18	1.0	0.52
	Urea	1.35	0.36	0.19
Incubation temp.	291 K	0.65	0.075	0.39
	296 K	1.22	0.076	0.48
	301 K	1.24	0.11	0.43
	306 K	0.83	0.086	0.37

microbial metabolism is depended on the environmental temperature. The metabolism was inhibited at higher or lower temperature, the peak of thermogenic curves delaying.

Comparing various carbon sources with fixed nitrogen source (NH_4Cl), strain V3a had the strongest metabolic activity with glucose (k, 1.22 h^{-1}; Q_T, 0.48 J mL^{-1}), the lowest with maltose (k, 1.12 h^{-1}; Q_T, 0.22 J mL^{-1}). The gap of the k is slight for all the carbon sources while that of the Q_T is relatively larger. It is a fact that monosaccharide are easier to promote microbial growth than polysaccharides and intermediate product.

For various nitrogen sources with fixed glucose dose, the metabolic activity of strain V3a did not depend on the nitrogen content of these compounds at all according to results. The growth rate constant also did not coincide with the total released metabolic heat, for tryptone (k, 1.18 h^{-1}; Q_T, 0.52 J mL^{-1}) and for urea (k, 1.35 h^{-1}; Q_T, 0.19 J mL^{-1}). Though the nitrogen content of inorganic compounds is higher than that of organic compounds; inorganic nitrogen source just provides moderate cell growth with the lower thermal parameters relatively. Considering the cost, NH_4Cl can be decided as single nitrogen source for bioflocculant-producing bacteria incubation.

All thermogenic curves had two peaks, which may be due to the decreasing of the Dissolved Oxygen Tension (DOT) in ampoule with total volume of 4.0 ml and zymotic fluid of 2.0 ml. The sealed ampoules just could offer 2.0 ml air to microbial growth. It is of importance that oxygen is a hydrophobic gas with a consequentially low solubility in liquid phase (2×10^{-4} mol l^{-1}). It is even less soluble in physiological solutions because the salting out effect reduces solubility considerably and sometimes by more than 10%, depending on the complexity of the medium. This is particularly important in closed calorimetric vessels. Without stirring, cells sediment to form a 'crowded' pile or 'heap' (the Uriah effect) with limited access to the dissolved oxygen and this phenomenon can result in increased glycolysis due to the Pasteur Effect. The dispersion of oxygen also is difficult to keep the aerobic condition of the medium. In this case, the physiological conditions in the interstices between the cells are so poor in terms of low oxygen tension, causing a decrease in the metabolic activity, when oxygen reduced gradually. Microorganisms adapted to aerobic metabolism initially for the first peak, then adjusted and converted to anaerobic metabolism with decreasing DOT for the second peak. However, the change of metabolism modes needed a transient-state exhibiting the shape of "saddle" in the thermogenic curve. Moreover, changes in the heat flux instantaneously reflect the metabolic changes (Duboc et al., 1998).

When the cultural conditions (aerobic to anaerobic) became adverse to microbial growth, the substrate is more oxidized than the product biomass. The cells would release more metabolic heat to adapt to the surrounding environment.

4 CONCLUSION

The results showed that *Halomonas* sp. V3a had ideal metabolic activity at the combination of 296 k, glucose and NH_4Cl. Another phenomenon could be found, that flocculating activity is higher, while OD_{600} or k is lower, tryptone as carbon source. So, it can be elucidated that higher biomass can not mean higher production of bioflocculant.

ACKNOWLEDGEMENTS

This work was supported in part by grants from the National Outstanding Youth Research Foundation of China (40925010), International Joint Key Project from National Natural Science Foundation of China (40920134003), International Joint Key Project from Chinese Ministry of Science and Technology (2011DFA00120, 2009DFA92830 and 2010DFA12780), and National Natural Science Foundation of China (41103060, 41103007, 41103058, 40873060), the Fundamental Research Funds for the Central Universities (FRF-TP-12-005A).

REFERENCES

Deng, S.B., Yu, G., Ting, Y.P., 2005. Production of a bioflocculant by Aspergillus parasiticus and its application in dye removal. *Colloids Surfaces B: Biointerfaces* 44: 179–186.

Duboc, P., Cascao, P.L.G., von Stockar, U. 1998. Identification and control of oxidative metabolism in Saccharomyces cerevisiae during transient growth using calorimetric measurements. *Biotechnology & Bioengineering* 57: 610–619.

Salehizadeh, H., Shojaosadati, S.A. 2001. Extracellular biopolymeric flocculants: Recent trends and biotechnological importance. *Biotechnology Advances* 19: 371–385.

Wang, S.G., Gong, W.X., Liu, X.W., Tian, L., Yue, Q.Y., Gao, B.Y. 2007. Production of a novel bioflocculant by culture of *Klebsiella mobilis* using dairy wastewater. *Biochemical Engineering Journal* 36: 81–86.

Yim, J.H., Kim, S.J., Ahn, S.H., Lee, H.K. 2007. Characterization of a novel bioflocculant, p-KG03, from a marine dinoflagellate, *Gyrodinium impudicum* KG03. *Bioresource Technology* 98: 361–367.

Zhang, J., Liu, Z., Wang, S., Jiang, P. 2002. Characterization of a bioflocculant produced by the marine myxobacterium *Nannocystis* sp. NU-2. Applied Microbiology & Biotechnology 59: 517–522.

Frontiers of Energy and Environmental Engineering – Sung, Kao & Chen (eds)
© 2013 Taylor & Francis Group, London, ISBN 978-0-415-66159-1

The finite element analysis and optimal design of aerator's blade

P. Xing & B.B. Qiu

Nanchang Hangkong University, Nanchang, China

ABSTRACT: The inverted umbrella aerator is the key equipment of biological wastewater treatment technology, which have the effects of oxygenation, mixing and plug-flow. the design and optimization of the mixing impeller is most important to the efficiency of oxygenation. In this paper, a new aerator with curve blades was presented, and more, the working process of the blades have been studied through solid modeling, mesh generation and finite element analysis process, and the stress-strain finite element analysis was offered, which identified the weak links in the leaves and put forward the effective optimal measures.

Keywords: curve blade; finite element analysis; optimization

1 INSTRUCTION

Oxidation ditch process for sewage and industrial wastewater treatment of city with secondly biological treatment technology of strong competitiveness has been widely adopted around the world. The aeration device is the foremost mechanical equipment in the oxidation ditch process, which is the one point aspect to impact treatment efficiency, energy consumption and stability of oxidation ditch. Besides the device not only has the oxygenation, mixing and plug-flow and other functions, but also determines the covering area and capital construction investment of oxidation ditch. With the increasing request oxidation ditch process relays on the aeration equipment and the energy shortage, the researches of new high efficient and low energy aeration equipment have become the important factors of impelling oxidation ditch technology development and energy saving and consumption reducing.

The present study designed a reducing curved blade aerator, whose blade is made according to logarithmic spiral rule. So that it has better ability of pushing and mixing flow. Due to the fluid characteristics of blade, the aerator can form carrier fluid's vortex and injection, which accordingly possesses better oxygenation and aeration performance.

2 THE DESIGN OF LOGARITHMIC SPIRAL BLADE

The establishing mixing blade's generatrix equation:

$$\begin{cases} x(\theta) = \rho_0 e^{\theta/\tan\beta} \cdot \cos\theta \\ y(\theta) = \rho_0 e^{\theta/\tan\beta} \cdot \sin\theta \end{cases} \tag{1}$$

where, ρ_0 is initial polar radius, θ is the spiral rotation angle, β is the spiral angle.

In here, the parameters of aerator mixing blade were selected as the research object are as follows:

$$\rho_0 = 250 \text{ mm}, \theta = 38.285°, \beta = \frac{\pi}{2}.$$

The calculation of the polar radius ρ was shown in Table 1.

And the major parameters of the aerator are as followings: 8 arc-shaped blades, impeller diameter of 3658 mm, one connecting disk, motor power 150 kW, which structure diagram shown in Figure 1.

Table 1. The calculation of the polar radius.

Points	θ	β	ρ
1	$\pi/20$	38.285	305.05
2	$\pi/10$	38.285	372.21
3	$3\pi/20$	38.285	454.17
4	$\pi/5$	38.285	554.17
5	$\pi/4$	38.285	676.20
6	$3\pi/10$	38.285	825.08
7	$7\pi/20$	38.285	1006.76
8	$2\pi/5$	38.285	1228.43
9	$9\pi/20$	38.285	1498.92
10	$\pi/2$	38.285	1828.96

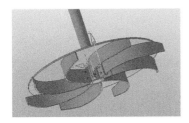

Figure 1. Curved surface impeller of inverted umbrella aerator.

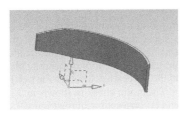

Figure 2. Blade unit.

Figure 3. Blade mapping grid.

Figure 4. The finite element model.

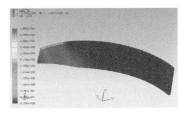

Figure 5. Blade displacement cloud.

3 THE BLADE FINITE ELEMENT ANALYSIS

3.1 The modeling

Because the inverted umbrella aeration impeller comprises 8 identical vane, so the study selects one of these leaves as a modeling unit. Firstly three-dimensional parametric model of stirring blade unit was built in the UG as shown in Figure 2, and then generated mesh, forming finite element model.

3.2 The grid model of blade

Blade material 1Cr18Ni9, density $7.9 \times 10^3 \text{kg/m}^3$, elastic modulus 206 GPa, Poisson's ratio 0.29. Considering the model of symmetry, the economics of computing and grid quality and other issues, the model was mesh-mapped with three-dimensional tetrahedral units in the system. The mesh adopted 3D unit, each unit has 4 nodes, each node has 6 degrees of freedom, which is suitable for linear, large rotation analysis. After meshing, the stirring blade is divided 16 780 solid elements. Each of these entities has 4 nodes, each node has 6 degrees of freedom, which can be used to simulate the mixing blade stress and deformation. According to the blade structure and load characteristics, the research selects the 1/8 of the structure for analysis model, calculating by axisymmetric structure. The stirring blade finite element model was divided into 20 307 units, 44 605 nodes, which grid map shown in Figure 3.

3.3 Loads processing and the boundary conditions

Assuming the blades are stressed by uniform load, the finite element model shown in Figure 4, in which, the arrows of the figure represents the vector forces.

4 FINITE ELEMENT ANALYSIS

4.1 Deformation analysis

The deformation of blade under load displayed in Figure 5. and the displacement of deformation In X, Y, Z direction are shown in Figures 6–8. It can be seen from Figures 5 to 8, the maximum displacement of blades under loading is about 182.1 mm. The blade tip was the largest deformation part of the blade, which subject to greater torque. The reason why deformation value is so large is that the design aim is to make the blade can produce a certain bending deformation as the increase of the load, thereby slowing the flow impact, and assuring the aeration rotor can keep outflow velocity stably.

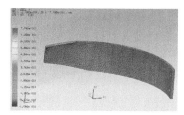

Figure 6.   Blade displacement cloud in X direction.

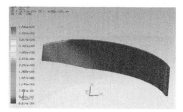

Figure 7.   Blade displacement cloud in Y direction.

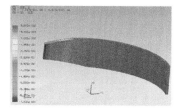

Figure 8.   Blade displacement cloud in Z direction.

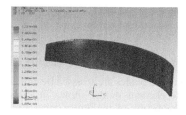

Figure 9.   The basic stress cloud.

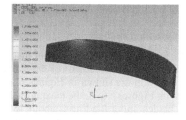

Figure 10.   Element node stress cloud.

4.2 Stress analysis

The blade stress intensity cloud under load shown in Figure 9, from which, it can be seen that the blade

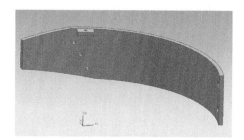

Figure 11.   The optimized blade.

overall stress is more evenly distributed. As the figure of blade stress distribution shows: the maximum stress about 77.37 MPa occurs in the arc area of the lower front part and the root of bending plate of the blade, and the lower stress locate in the web plate and blade tip. The node stress is shown in Figure 10, which trend is consistent with the Figure 9.

5 CONCLUSION

Through the finite element analysis on the aerator blade, the value of stress and deformation of blades was obtained. from the stress cloud chart, it shown that the maximum stress occurs in the blade closed to the end of the circular arc. The following methods can be used to optimize the designing:

- Blades hems here, fixed with bolts and connecting disk, in order to increase the intensity.
- Blades subjected to maximum stress need to open pore appropriately as shown in Figure 11, so that it can cut down the stress and deformation.

ACKNOWLEDGEMENTS

The project is supported by the Science Fund of Jiangxi Provincial Department of Education Science and Technology No. GJJ11510.

REFERENCES

[1] Inverted umbrella surface aerator [S]. JB/T10670-2006 industry standard.
[2] Pu XING, Jingyun ZHAO and Xiuijie YIN. The Finite Element Analysis and Optimal Design to Inverted Umbrella Aerator. Advanced Materials Research Vols. 199–200 (2011) pp 187–192.
[3] Pu XING, Jingyun ZHAO. Inverted umbrella aerator finite element and optimal design. Computer simulation 2010. 6: pp 372–376.
[4] Chwen-Jeng Tzeng, Reza Iranpour, etc. Modeling and Control of Oxygen Transfer in High Purity Oxygen Activated Sludge Process. Journal of Environmental Engineering. 3, (2003). pp 402–411.

Frontiers of Energy and Environmental Engineering – Sung, Kao & Chen (eds)
© 2013 Taylor & Francis Group, London, ISBN 978-0-415-66159-1

Using diatom assemblage to assess the water quality in Heilongjiang river (Heihe part)

Y. Liu, H.K. Hui, C.M. Sun & Y.W. Fan
College of Life Science and Technology, Harbin Normal University, Harbin, China

ABSTRACT: Heilongjiang river is the third largest river in China, it flow through Heihe city, belonging to the up and middle streams of Heilongjiang river. This study aims to be the first step to (i) identify diatom type assemblages in this part of Heilongjiang river, and (ii) find which ecological factors explain most of the variation, (iii) find the most effective indicate species. To achieve this, we collected physical, chemical and diatom community data from six sites in Heilongjiang river (Heihe part) from 2006 to 2007, for a total of 260 samples. 106 taxa were identified as diatom. Species such as Cocconeis placentula var. euglypta (Ehr.) Cl., Cymbella tumida (Gerg.) Cl., Melosira distans var. alpigena Grun. and Melosira varians Ag. are the most common species in all sites in the three seasons. The diatom assemblages changed in different sites and seasons. Species composition in site III are different variously with other sites, provide a good chance for us to study the tolerance of the species, compared with other sites, several species which are tolerant of the pollution are observed. Relationship between sample sites and environmental factors were examined using PCA, all the sites can be divided into three group, with site I, II and IV have lower BOD and higher DO, indicate good water quality, site III have higher TN, TP, BOD, COD indicate the poor water quality. Relationships between diatom communities and environmental factors were examined using Canonical Correspondence Analysis (CCA). There were strong correlations between diatoms and environmental factors, especially BOD, TP and water temperature. Heilongjiang river was characterized by mostly oligosaprobic/ß-mesosaprobic taxa. Species which can be indicator for the TP, TN et al. were also found.

1 INTRODUCTION

Monitoring of conditions in rivers has been based on either chemical or biological analyses. Biological monitoring relies on assessment of the flora or fauna at points or along stretches of rivers and continuous surveillance is necessary to reveal changes in the biota and ultimately to establish the relationships between the biological features and the overall chemical status and the reaction of the individual species to specific nutrients, pollutants etc. (Round, 1991).

Due to their sensitivity, wide distribution and relative ease of identification, diatoms are widely used for the biological assessment of running waters. Diatoms quickly respond to changes of environmental variables, thus reflecting overall ecological quality and the effects of different stressors. Many indices have been proposed to relate diatom assemblage composition to water quality. (Bona et al. 2007) Diatoms grow abundantly in rivers, live in almost all suitable habitats, they are also easily sampled and identifiable and have been used extensively in studies related to the monitoring of water quality and have been widely used in many

countries. A very detailed study of the diatoms in rivers has been completed by Descy and Coste (Descy & Coste, 1990), who consider that their diatom indices provide excellent indicators of water quality, especially with widespread increasing eutrophication. (Round, 1991).

The Heilongjiang River locate at 42°45′–53°33′N, and 108°20′–141°20′E, it is wriggled the boundaries of Northeast in China, which is the thirdly largest water area and an important national river. It is almost 4370 km long, occupy 1843000 km². Totally have 91 branches. It flow through Heihe city, belong to up and middle reaches of Heilongjiang River, with relatively low water temperature, mountainous reach, there are a plenty of hydrophytic plant and animal resources.

In years 2006 and 2007, from May to October, the samples were collected from six sites in Heilongjiang river (Heihe part) to access the water conditions. Although this river is valuable in all natural aspects, relatively few information regard to diatom communities in it and the environment change have been presented. Basic data for algae communities are poor. It is a big problem for using algae communities to assess the environmental change in such an important river. The aims of this research are, first, to explore the diatom community in this big river; second, to determine the relationship between diatom and environmental parameters; and third, to find useful indicators to the environmental change in this big river. In this paper, species composition in different years and different seasons are reported, and impact of environmental conditions on diatom community elucidated. By exploring the temporal and spatial patterns in river diatom assemblages, we will be able to provide guidance on using diatom as an indicator to environmental change in Heilongjiang river and in addition, to provide basic ecological information for the future research in Heilongjiang river.

2 MATERIALS AND METHODS

Heilongjiang River located in Heilongjiang Province, China, covers 1843000 km² and spans between 42°45′–53°33′N, and 108°20′–141°20′E. The rivers almost 4370 km long, covered by ice from November to next April. Six sampled sites (I–VI) were studied from six transverse section in Heilongjiang river (Heihe part) in 2006 and 2007 from May to October. Site I is in the entrance of the city, the water is clear and fast-flowing, less disturbance from the people; site II locate near the tap water district and the swimming place, water flowing slow, there are boats near the shore, get a lot of disturbance from the swimming people in summer; site III locate near the outfall of municipal sewage; site IV locate at the Daheihe wharf, near the outfall of waste water, the river are wide in this part and water is deep; site V is in the estuary of Jieya river, it is a cut-off point of the upsteam and downsteam of Heilongjiang river; site VI is in Changfa village, many small island in this part of the river, far away from the city, with beautiful view for both sides of the river.

Samples were collected at a water depth of 0.5 m with a plankton net (mesh size 50 μm) and were preserved using Lugol's solution. Measurements of altitude, temperature, and pH were carried out in the field using portable instruments (pH B-4, China), along with a handheld GPS device to precisely determine the geographic coordinates of the collection sites. River neighboring agricultural fields or other anthropogenic influences were not included in this study.

In laboratory, Biochemical Oxygen Demand (BOD), Chemical Oxygen Demand (COD), Total Nitrogen (TN), Total Phosphorus (TP), Dissolved Oxygen (DO) were measured using standard methods within 24 h. Diatom samples were treated with HNO_3 to remove all organic material. The cleaned materials were mounted by Naphrax mounting medium. Diatoms were identified and counted at 1000 × magnification using Olympus BH × 2. Five-hundred valves were counted from each slide. Identification of taxa followed the volumes by Krammer and Lange-Bertalot (Krammer & Lange-Bertalot,1986),(Krammer&Lange-Bertalot, 1988), (Krammer & Lange-Bertalot, 1991a), (Krammer & Lange-Bertalot, 1991b). Diatom species were expressed as relative abundances (in percentage of total diatoms).

The diatom data set was analyzed using multivariate statistics. Relationships between diatoms and environmental variables were examined using Canonical Correspondence Analysis (CCA). All ordinations were conducted using the CANOCO application for Windows 4.0

3 RESULT

3.1 *Water chemistry*

During sample collection in all sites, pH values ranged from 6.4 to 7.4, the difference between these two years and different months are not obvious. Water temperature ranged from 4 to 25°C, the highest temperature was in July, the lowest was in November. DO ranged from 4.29 mg/L to 12.5 mg/L, the average value was 6.50 mg/L and 6.32 mg/L in 2006 and 2007 respectively. BOD ranged from 1 to 5 mg/L, no obvious difference between different months. COD ranged from 5.26 to 10.78 mg/L, have obvious difference between the months and sites, site III had the highest value. TN range from 0.85 mg/L to 1.35 mg/L, TP ranged from 0.068 mg/L to 0.165 mg/L, these two parameters have obvious difference between the sites, site III have the highest values.

3.2 *Diatom diversity*

Two hundred and sixty samples were collected in 2006 and 2007 from May to October. For all the algal, a total of 146 taxa were identified, among them 33 genera 106 taxa belong to diatom, account to 72.6%. The most common genera were *Gomphonema* (11), *Fragilaria* (11), *Navicula* (9), *Cymbella* (7) and *Nitzschia* (7).

The species diversity and composition are various between different sites (Table 1), but the species

Table 1. Species number in six sites.

Site/year	I	II	III	IV	V	VI
2006	52	43	27	55	31	52
2007	50	38	23	46	32	48

composition were similar in these two years. The species diversity is lower in 2007 than it is in 2006; diatom account to 60–69% of the total phytoplankton. The total phytoplankton of diatom were various between different seasons, the highest value was 6.37×10^6 ind./L in October, the lowest value was 4.06×10^6 ind./L in July.

Dominant species were changed in different sites and seasons (Table 2). Species such as *Cocconeis placentula* var. *euglypta* (Ehr.) Cl., *Cymbella tumida* (Gerg.) Cl., *Melosira distans* Grun. and *Melosira varians* Ag. are the most common species in all sites in the three seasons. The dominant species in site I is *Ceratoneis arcus*, in spring (May and June), and it is change to *Diatoma vulgaris* var. *producta* and *Diatoma moniliformis* in summer (July and August), in autumn, *Amphora ovalis* and *Synedra acus* become dominance. In site II, the dominant species changed from *Melosira granulata* and *Cymbella excise* in spring to *Gomphonema angustatum* var. *producta* and *Achnanthes minutissima* var. *scotic*, in autumn, the dominant species is *Navicula viridula*. Site III was dominant by some Blue algae such as *Oscillatoria tenuis* and *Oscillatoria princes*. In these three seasons the dominant diatom at here are *Asterione formosa*, *Melosira varians* and *Cyclotella stelligera*. Site IV was dominant by diatom species such as *Melosira granulate*, *Cocconeis placentula* var. *euglypta* and *Cyclotella meneghiniana* in three seasons. In site V, green algae become the dominant group, *Cosmarium obtusatum* and *Ankistrodesmus spiralis* are dominant species in spring, *Ankistrodesmus sprialis* is still dominant in summer. In site VI, *Diatoma vulgaris* and *Cosmarium obtusatum* are dominant in spring, the dominant assemblage changed to *Diatoma moniliformis* and *Ankistrodesmus falcatus* in summer, and changed to *Gomphonema olivaceum* in autumn.

The diatom communities were similar in the two years with clear seasonal changes. Some species only observed in spring and autumn in 2006, but can be observed in all the three seasons in 2007, they are *Amphora ovalis* var. *affinis* Kuetz., *Asterione Formosa* Hassall, *Cocconeis pediculus* Ehr., *Craticula acidoclinata* Lange-Bertalot & Metzeltin, *Gyrosigma acuminatum* (Kuetz.) Rabh., *Diatoma vulgaris* var. *distorta* Grun., *Gomphonema pseudoaugur* Lange-Bertalot and *Synedra acus* var. *radians* (Kuetz.) Hust. *Cymbella turgidula* var. *venezolana* Lange-Bertalot and *Epithemia adant* (Kuetz.) Bréb. are observed in all the three seasons in 2006, but only were observed in spring and autumn in 2007. *Gomphonema augustatum* (Kuetz.) Grun. and *Nitzschia acicularis* (Kuetz.) Smith are observed in spring and summer in 2006, but they occurred in three seasons in 2007. *Fragilaria vaucheriae* (Kuetz.) Perterson only occurred in autumn in 2006, but it was observed in three seasons in 2007.

3.2.1 Relationship between diatom assemblage and environmental parameters

The result of PCA (principal component analysis) showed that axis 1 clearly separated the sampling sites from oligotrophic and mesotrophic (Fig. 1). TP concentration in the left side of the first axis and BOD were generally low, while DO was high, indicating site I (1–6, 37–42), site II (7–12, 43–48) and site VI (30–36, 67–72) had good water quality, consistent with the result of physical and chemical data. Higher TN, TP, BOD and COD were located on the right in the axis 1, Mainly in site III (13–18, 49–54), which was located near the sewage outfall, had poor water quality.

The DCA (detrended correspondence analysis) ordination of the 6 samples showed that the first axis eigenvalue (0.421) was higher than the second axis eigenvalue (0.092), capturing 72% of the variance in diatom data. All outliers were identified in the DCA (graph not shown); therefore, all of the 6 samples were used in the subsequent

Table 2. The dominant species in different seasons.

S	Y	Spring	Summer	Autumn
I	2006	*C. arcus*	*D. moniliformis*	*A. ovalis*
	2007	*C. arcus*	*D. vulgaris* var. *producta*	*S. arcus*
II	2006	*C. excise*	*A. minutissima*	*N. viridula*
	2007	*M. granulata*	*G. angustatum*	*N. viridula*
III	2006	*M. varians*	*A. formos*	*O. tenuis*
	2007	*O. princeps*	*M. varians*	*C. stelligera*
IV	2006	*C. placentula*	*N. radiosa*	*P. grunowii*
	2007	*M. granulata*	*C. meneghinianaa*	*C.* var. *euglypta*
V	2006	*C. obtusatum*	*A. spiralis*	*C. excise*
	2007	*D. vulgaris*	*A. minutissima*	*D. vulgaris*
VI	2006	*D. vulgaris*	*A. falcatus*	*D. tenue*
	2007	*C. obtusatum*	*D. moniliformis*	*G. olivaceum*

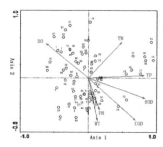

Figure 1. The results of canonical correspondence analysis(sites and environment variables, 1–6 means site I–VI in May, 7–12 means site I–VI in June et al.).

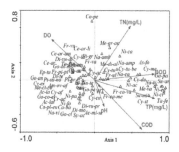

Figure 2. The results of canonical correspondence analysis (species and environment variables).

CCA (canonical correspondence analysis) analyses (including pH, Water temperature, DO, TN, TP, BOD, and COD).

The results of the CCA showed that correlation coefficient of the two axis was 0.879 and 0.725, and the total inertia (2.853) in the species data-set were explained by the chemical and physical environmental variables, and that the percentage of cumulative variance of the first two axes were 14.9% and 18%, respectively. Accounting for 25% of the relationship between diatom assemblages and environmental variables were included in the analysis (Fig. 2).

Species were closely around the axis around, and most species showed a partial distribution of small clumps and were in accordance with the rules of nutrition gradient from the Dual-axis graph (Fig. 1). The first axis of CCA was clearly divided into two parts according to the changes of BOD, increase from left to right, gradually. The samples were located on the left with relatively lower BOD, as site I, site VI, whereas site III, site IV located on the right part with high BOD value. TP and TN were closely related to axis 1, indicating a nutrient gradient along the first axis. Most species in site III were to mesotrophic closely correlated with TN and TP, such as *Cyclotella meneghiniana, Surirella*

minuta, Gomphonema parvulum. WT correlated closely with BOD and pH values, and had an inverse relationship with DO. Lower temperature were located on the top side of the second axis, mainly in Spring and Autumn, and were found high on lower side.

4 DISCUSSION

The first serious attempt to use diatoms as indicators of water quality was the study by Kolkwitz and Marsson (Kolkwitz & Marsson, 1908), who showed that water conditions determined the composition of the algal flora. They set the scene by using the presence of indicator species for waters of differing chemical composition and essentially this has been the basis of most subsequent methods. (Round, 1991) The use of diatoms as indicators or for construction of indices of water quality has been criticised due to the fact that some workers record hundreds of species (e.g. 357 in Leclercq & Depiereux (Leclercq, 1987)).

The studied diatom flora is typical for oligosaprobic/ß -mesosaprobic rivers. Many species prefer cold water and usually occurred in mountains indicate the climate characters, they are *Cymbella excise, Ceratoneis arcus, Fragilaria pinnata, Gomphonema olivaceum* var. *alcareum, Ceratoneis arcus* var. *amphioxys, Melosira distans* var. *alpigena*.

Site III is a special place in this study, this site near the outfall of municipal sewage, the diatom community showed a different pattern compared with other sites. The values of chemical parameters are relative higher than other sites. For the influence of the TN, TP and organism, species which prefer clear water disappeared, such as *Ceratonies arcus* et al.

species with wide tolerance of pollution are occur and become common in this site, such as *Melosira varians, Gomphonema parvulum* et al. A total of 53 taxa never occur in site III in both 2006 and 2007. *Achnanthes lanceolata, Achnanthes laterostrate, Caloneis bacillum, Craticula acidoclinata,*

Cymbella turgidula var. *venezolana, Gyrosigma spencerii, Cymatopleura solea, Eunotia bilunaris* and *Fragilaria capucina* var. *distans* were observed in site III in 2006, but they are gone in 2007. Compared with these species, some other species showed a totally different distribution patterns. *Fragilaria ulna, Navicula capitatoradiata, Navicula trivialis, Nitzschia acicularis, Nitzschia capitellata, Nitzschia clausiii, Nitzschia eglei, Surirella bifrons, Surirella minuta* and *Tabellaria fenestrate* are observed in site III commonly, but most of them are not occur in site I, II and VI this result indicate that these species prefer pollute habitats.

Pinnularia grunowii is the dominant species in site IV in autumn of 2006. The genus *Pinnularia* usually live in oligotrophic water with low conductivity in mountains (Krammer, 2000). It is dominant at here means the water quality are in good condition, but it become rare in this site in 2007, indicate this site is disturbed by human activity a lot in 2007, the water quality getting worse.

The CCA analysis showed there were strong correlations between diatoms and environmental factors (Fig. 2). The distribution of diatoms have strong relationship with physical-chemical parameters such as water temperature, pH, DO, TN, TP et al. In Figure 2, *Asterione formosa*, *Cyclostephanos dubius*, *Cyclotella meneghiniana*, *Surirella minuta*, *Gomphonema parvulum* are good indicator of TP, these species mainly occurred in site III. *Cyclostephanos dubius* was reported as potential indicator for the algae bloom; while *Cyclotella meneghiniana* is a eutrophic species, also have wide tolerance of temperature. *Melosira granulata* var. *angustissima* fo. *spiralis*, *Nitzschia capitellata*, *Melosira distans* var. *alpigena*, *Nitzschia acicularis* are good indicator of TN, mainly distributed in the sites with high concentration of TN.

In our study, the physical-chemical parameters as pH, DO, BOD and COD were typical of the national standard II-III. The diatom community also showed the same result. Our results also corresponding with the research of algae community in Heilongjiang river in Sun et al. [12, 13].

In summary, diatom-based assessment suggests that Heilongjiang river (Heihe part) are in a relatively good condition overall. The high natural variability in river condition and human disturbance gradients in this region might obscure the patterns between diatoms and environmental variables. Diatom can be a good indicator in the study of use plankton to access the water quality in Heilongjiang river. The physical-chemical parameters and diatom assemblages analysis give the same results that water quality in Heilongjiang river (Heihe part) belonging to mesotrophic.

ACKNOWLEDGEMENTS

This work was supported by Startup Foundation for Doctors of Harbin Normal University (10XBKQ07) and National science foundation of China (Grant No. 30870157 & 31100153).

REFERENCES

Bona, F., Falasco, E., Fassina, S., Griselli, B. and Badino, G. 2007. Characterization of diatom assemblages in mid-altitude streams of NW Italy. Hydro. vol. 583: 265–274.

Descy, J. P. and Coste, M. 1990. Utilisation des diatomées benthiques pour l'evaluation de la quality des eaux courantes. Rapport Final Contract CEE B-71-23.

Krammer, K. and Lange-Bertalot, H. 1986. Bacillariophyceae. 1. Teil: Naviculaceae. *In*: Süßwasserflora von Mitteleuropa. Band 2/1. Heidelberg: Spektrum Akademischer Verlag, 876p.

Krammer, K. and Lange-Bertalot, H. 1988. Bacillariophyceae. 2. Teil: Bacillariaceae, Epithemiaceae, Surirellaceae. In: Süßwasserflora von Mitteleuropa. Band 2/2. Heidelberg: Spektrum Akademisc—her Verlag, 611p.

Krammer, K. and Lange-Bertalot, H. 1991a. Bacillariophyceae. 3. Teil: Centrales, Fragilariaceae, Eunotiaceae. *In*: Süßwasserflora von Mitteleuropa. Band 2/3. Heidelberg: Spektrum Akademischer Verlag, 599p.

Krammer, K. and Lange-Bertalot, H. 1991b. Bacillariophyceae. 4. Teil: Achnanthaceae, Kritische Ergänzungen zu Achnanthes s. 1., Navicula s. str., Gomphonema. Gesamtliteraturverzeichnis. Teil 1–4. *In*: Süßwasserflora von Mitteleuropa. Band 2/4. Heidelberg: Spektrum Akademischer Verlag, 468p.

Kolkwitz, R. and Marsson, M. 1908. Okologie der pflanzlichen Saprobien. Ber. Deutsche Bot. Ges. 26a: 505–519.

Krammer, K. 2000. Diatoms of the European Inland Waters and comparable Habitats, Vol.1, A.R.G. Gantner Verlag K.G.

Leclercq, L. 1987. Vegétation et caratériques physico-chimique de deux rivieres de haute Ardenne (Belgique): La Helle et la Roer supérieuse. Lejeunia N.S. 88: 1–62.

Round, F.E. 1991. Diatoms in river water-monitoring studies. J. Appl. Phy. vol 3, pp. 129–145.

Frontiers of Energy and Environmental Engineering – Sung, Kao & Chen (eds)
© 2013 Taylor & Francis Group, London, ISBN 978-0-415-66159-1

Governance mechanisms to enhance the knowledge service ability of industrial cluster

Z.S. Wang, Y. Liu & J. Sun
Beijing Information Science and Technology University, Beijing, China

ABSTRACT: In this paper, fuzzy comprehensive evaluation method has been used to evaluate the knowledge service ability of industrial cluster from the staff-level, corporate-level and the overall level of industrial cluster. At last, in order to improve knowledge service ability of industrial cluster, the governance mechanism proper to enhance industrial cluster's knowledge service ability has been proposed based on the evaluation results of cluster analysis.

Keywords: knowledge service ability; industrial cluster; governance mechanism

1 GENERAL INSTRUCTIONS

21st century, economy of knowledge is developing rapidly; knowledge service ability has become an important factor which has a great effect on organizations' competitiveness. Industry cluster is a special group of organizations. In order to survive and develop, industrial clusters have to improve the capacity of knowledge service. In this paper, effective governance mechanism which is proper to enhance knowledge service ability and good for the development of industrial cluster has been explored based on the unique properties of industrial clusters.

2 THE RELEVANT THEORY SUMMARISE

In China, knowledge service has been studied in lots of fields, such as: library management, supply chain optimization, community services, agriculture and so on. For examples, Meng Wang, Kaiying Xu proposed an ecological system of library based on knowledge service, and introduce its purpose, model, structure, management, etc. [1]; Daoping Wang, Minmin Zhang researched on concept and characteristics of knowledge service architecture of the smart supply chain, and proposed four elements of the service architecture: subject of knowledge service, object of knowledge service, process of knowledge service and technical support of knowledge service and also discussed the interactive mechanism among the subjects [2]; Yilin Tian, Guangqing Teng researched community users' knowledge services model based on concept of lattice theory and explored the

establishment of "push" mode of community user knowledge service model [3]; Cui-Ping Tan, Huaiguo Zheng studied the origin and development of Chinese agricultural knowledge service and summarized the typical cases [4]. In other countries, lots of scholars focused on applications of knowledge service such as Cornell University in US developed the "My Library" system, fully reflects the personalized library service and facilitated the reader's knowledge retrieval and access; in order to improve health care service efficiency, the British National Health Service System created the site National Knowledge Service. In addition, there are some other scholars studied on knowledge clusters, such as Stephan Manning focused on the knowledge service clusters, and studied cluster development and introduction of foreign capital [5].

3 EVELUATION OF INDUSTRIAL CLUSTER'S KNOWLEDGE SERVICE ABILITY

3.1 *Calculation of knowledge service ability*

Hierarchical fuzzy comprehensive evaluation method has been used and three levels have been used to calculate the knowledge of service capabilities of industrial clusters: knowledge service ability of staff, knowledge service ability of companies within the cluster, knowledge service ability of the whole cluster level. Knowledge service ability of staff reflects in four areas: the participation of employees, the ability to meet customer needs, knowledge service efficiency, staff knowledge standard; enterprise service ability can be measured from average investment and income in knowledge services, business participation, the average stock of knowledge, innovative ability of knowledge; as to the cluster level, the total profit and input-output ratio of knowledge service, the stock of knowledge of the whole cluster, innovative ability of the cluster are the main factors to measure.

1. According to the characteristics of the cluster three different level shown in Table1.

As the Table 1 shown, the second-level index has been derived from first-level index. We use expert scoring method to determine the weight value of first-level index and second-level index. The first-level index written as $U = (U_1, U_2, U_3)$ and its weight matrix $P(U) = (P(U_1), P(U_2), P(U_3))$; the second-level index written as $U_{1j} = (U_{11}, U_{12}, U_{13})$ and its weight matrix $P(U_{1j}) = (P(U_{11}), P(U_{12}), P(U_{13}))$, $U_{2j} = (U_{21}, U_{22}, U_{23})$ and its weight matrix $P(U_{2j}) = (P(U_{21}), P(U_{22}), P(U_{23}))$, $U_{3j} = (U_{31}, U_{32}, U_{33})$ and its weight matrix $P(U_{3j}) = (P(U_{31}), P(U_{32}), P(U_{33}))$. In addition, as equation (1)

$$\sum_{i=1}^{3} P(U_i) = 1 \qquad (1)$$

2. Every index has been divided into five level based on its characteristics, $V = $ (very good, good, normal, bad, worse), $v_i (i = 1, 2 \dots n)$ is the representation of effective remark of experts.

3. By sending questionnaires to corporate officers, employees, industry experts, customers, governmental departments and other related person to get their remarks to the first-level and the second-level index, and then construct fuzzy evaluation matrix R (r_{ij} means $U_j (j = 1, 2 \dots m)$ corresponding to the evaluation of the V_i membership degree of fuzzy sets).

$$R = \begin{bmatrix} r_{11} & r_{12} & \cdots & r_{1m} \\ r_{21} & r_{22} & \cdots & r_{2m} \\ \vdots & \vdots & & \vdots \\ r_{n1} & r_{n2} & \cdots & r_{nm} \end{bmatrix}$$

4. Obtaining fuzzy evaluation results of the industrial cluster's knowledge service ability based on the equation (2)

$$K = R * P(U_j). \qquad (2)$$

3.2 The assessment after calculation of industrial cluster's knowledge service ability

Through analysis, it shows that worker' capabilities of knowledge service and companies' capabilities of knowledge service are closely related to industrial cluster's knowledge service ability.

1. The knowledge service ability of stuff has positive effect on the industrial cluster's knowledge service ability. When knowledge service ability of staff within the cluster is strong, the entire

Table 1. Industrial cluster knowledge service capacity rating system table.

	First-level index	Second-level index	Second-level index weight	First-level index weight
	Knowledge service ability of staff level (U_1)	Participation of employees (U_{11})	$P_{(U11)}$	$P_{(U1)}$
		Ability to meet customer needs (U_{12})	$P_{(U12)}$	
		Knowledge service efficiency (U_{13})	$P_{(U13)}$	
		Knowledge level (U_{14})	$P_{(U14)}$	
	Knowledge service ability of companies level (U_2)	Average investment of knowledge service (U_{21})	$P_{(U21)}$	$P_{(U2)}$
Knowledge service ability of industrial cluster		Average income of knowledge service (U_{21})	$P_{(U22)}$	
		Business participation (U_{23})	$P_{(U23)}$	
		Average stock of knowledge (U_{24})	$P_{(U24)}$	
		Innovative ability (U_{25})	$P_{(U25)}$	
	Knowledge service ability of cluster level (U_3)	Total profit of knowledge service (U_{31})	$P_{(U31)}$	$P_{(U3)}$
		Input-output ratio of knowledge service (U_{32})	$P_{(U32)}$	
		Stock of knowledge of the whole cluster (U_{33})	$P_{(U33)}$	
		Innovative ability of the cluster (U_{34})	$P_{(U34)}$	

cluster has a high efficiency of knowledge service activities, and the output of knowledge service is low as well as customers' satisfaction is also high; when the employees' knowledge service ability is poor, although the cluster's output of knowledge service is high, the cluster usually gets lower satisfaction of customers and also the income is not satisfactory.

2. The companies' knowledge service affects the knowledge service ability of industrial cluster positively. When the knowledge service ability of companies is strong, the cluster's knowledge service ability is also great. When the companies get higher income, the cluster's income is also perfect; however, if most of the companies within the cluster have poor knowledge service ability, the cluster's knowledge service ability is as well as low.

In short, stuffs' and companies' knowledge service ability have a positive relationship with the cluster's knowledge service ability. Improving the cluster's knowledge service ability must concentrate on the stuff-level and the company-level.

4 CONSTRUCTION OF INDUSTRIAL CLUSTER'S GOVERNANCE MECHANISMS

Industrial clusters Knowledge service governance mechanisms can be built from staff-level, company-level and cluster-level.

4.1 Governance mechanisms of staff-level

Employees are the basic elements of industrial clusters and also the main body provides knowledge service to customers. For employees, strengthening their skills of knowledge service to increase their stock of knowledge; establishing an effective motivation system and giving material and spiritual rewards to excellent employees; recruiting and selecting knowledgeable and innovative employees who can bring new ideas into the cluster; creating the system that different companies can exchange their employees aim to increase their stock of knowledge and the awareness of the whole cluster.

4.2 Governance mechanisms of company-level

Companies are the main skeletons of the cluster, the relationships among companies are bridges to maintain stability and development of the cluster and the companies' knowledge service ability has direct influence on the level of the industrial cluster's knowledge service. So much more attention should be paid to the construction of governance mechanisms. First, establishing an effective mechanism for knowledge sharing and cooperation, and through the cooperation to achieve stability and development of cluster; secondly, mechanisms for equitable distribution of benefits are needed, the purpose of cooperation is increasing the profits of knowledge service, so governance mechanism of benefits distribution is the most important part; the third, enhancing the flow of knowledge among companies which is good for knowledge sharing and innovation; fourthly, in order to protect the legal rights of companies, intellectual property rights protection mechanisms should be established.

4.3 Governance mechanisms of cluster-level

Cluster-level governance is the highest level, therefore, it must be overall-objective and long-term. First of all, setting the goal of sustainable development based on the whole cluster' interest; and then, creating perfect environment for companies' cooperation and knowledge-sharing; the last, improving the cluster's ability of knowledge innovation to promote the development of cluster.

5 CONCLUSION

Governance mechanisms are the security to improve industrial cluster's knowledge service ability. Good governance will improve employees' motivation, companies' cooperative sense, and increase knowledge service revenue of the whole cluster and the satisfaction of customers. Constructing governance mechanisms which is various factors should be taken into consideration, and then cluster will get stable and sustainable development.

ACKNOWLEDGEMENT

This research is supported by "National Natural Science Foundation of China" (Project No: 71073012, 71171021) and "National Technology Support Program of China-Manufacturing Industry Cluster Collaborative Spatial Information Collection and Database Optimization" (Project No: 2011BAF10B01) and "Beijing Information Science and Technology University Postgraduate Education-Key Disciplines-Management Science and Engineering Project (Project No: PXM2012-014224-000039). The authors also would like to thank the anonymous referees for their helpful comment on the early draft of the paper.

REFERENCES

Cuiping Tan, Guohuai Zheng, etc. 2011. Research of Origin and Development of Agriculture Knowledge Service in China. *Journal of Anhui Agricultural Sciences* (39):7440–7441.

Daoping Wang, Min Zhang, etc. 2009. Research on the Component Elements and Interactive Mechanism of Knowledge Service Architecture for Smart Supply Chain. *Information Studies: Theory & Application* (32):45–47.

Ferguson, J., M. Huysman. 2010. Knowledge Management in Practice: Pitfalls and Potentials for Development. *World Development* (38):1797–1810.

Hanlei Li, Wen Sun. 2011. The Research of Industrial Cluster Networks and Ability of Knowledge Integration Impact on Innovation Performance. *Commercial Times* (20):88–90.

Meng Wang, Kaiying Xu. 2011. The Construction of Library Information Ecosystem Based on Knowledge Service. *Research on Library Science* (9):43–47.

Ming Yi, Shuwang Yang. 2009. Governance and Performance Evaluation of Industrial Cluster. *Statistics and Decision* (23):50–52.

Miozzo, M., D. Grimshaw. 2005. Modularity and innovation in knowledge-intensive business services: IT outsourcing in Germany and the UK. *Research Policy* 34(9):1419–1439.

Morosini, P. 2004. Industrial Clusters Knowledge Integration and Performance. *World Development* 32(2):305–326.

Muller, E., D. Doloreux. 2009. What we should know about knowledge-intensive business services. *Technology in Society* 31(1):64–72.

Stephan Manning, Joan E. Ricart. 2010. From Blind Spots to Hotspots: How Knowledge Services Clusters Develop and Attract Foreign Investment. *International Management* (16):369–382.

Talia, D., P. Trunfio. 2010. How distributed data mining tasks can thrive as knowledge services. *Communications of the ACM* 53(7):132–138.

Yan Zeng. 2011. Research on The Relationship of High-tech Industry Cluster Characteristics and Its Performance. *Decision Making & Consultancy* (4):43–48.

Yilin Tian, Guangqing Teng, etc. 2011. Study on Service Model of Knowledge about Community Users Based on Concept Lattice. *Information Science* (8):1201–1204.

Ying Zhong, Shuping Mai, etc. 2010. Comparative Analysis of Knowledge Service Research at Home and Abroad. *Information Studies: Theory & Application* (5):110–115.

Frontiers of Energy and Environmental Engineering – Sung, Kao & Chen (eds)
© 2013 Taylor & Francis Group, London, ISBN 978-0-415-66159-1

Pyrolysis kinetics study of pine cone, corn cob & peanut shell

H.Q. Xie, Q.B. Yu, H.T. Zhang & P. Li
School of Materials and Metallurgy, Northeastern University, Shenyang, Liaoning, China

ABSTRACT: With thermogravimetric analyzer, the pyrolysis characteristics of three kinds of biomass (pine cone, corn cob and peanut shells) were studied under three heating rates (10, 20 and 40 K/min). The similar pyrolysis trends were obtained for the three samples. The reaction rate was found to rise with the increase of heating rate. On the basis of thermogravimetric analysis, the kinetic analysis was performed. In this paper, the novel two-step method of model sifting was put forward to select the most reasonable mechanism model. Shrinking core model (R2, spherical symmetry) was sifted for pine cone and corn cob, random successive nucleation growth model (A1, n = 1) for peanut shells. With the increase of heating rate, the activation energy was found to rise. The kinetic model expression was verified to agree well with the experimental result.

Keywords: biomass; pyrolysis; kinetic; mechanism model

1 INTRODUCTION

The concern with environmental problems is increasing with the rising emissions of CO_2, due to the consumption of fossil fuels. As a kind of renewable energy, biomass is an attractive alternative to fossil fuels, because of its zero CO_2 impact (Zhang 2006). In addition, the content of nitrogen and sulfur of biomass is less than fossil fuels, to produce less nitrous oxide and sulfur oxide. The types of the thermal chemical conversion of biomass are manifold, including combustion, pyrolysis, gasification, liquefaction and so on. Among them, pyrolysis is involved into the initial reaction of combustion and gasification and accompanied with liquefaction. So the kinetic analysis of pyrolysis process is relevant for process control and parameter optimization (Li & Suzuki 2009).

In the literature (Várhegyi et al. 1997, Antal et al. 1998, Hu et al. 2007, Hu et al. 2008, Saddawi et al. 2009, Lu et al. 2009), the thermogravimetric analysis technology was widely applied to elucidate the pyrolysis process of biomass. With the thermogravimetric data, the kinetic study could be done. In kinetic study, one reasonable mechanism model can well express the reaction process, and therefore how to obtain such a mechanism model is particularly important. In more cases, most researchers had an assumption that the model is one-order reaction one (Várhegyi et al. 1997, Hu et al. 2008, Saddawi et al. 2009). However, there are more other model existing and whether they meet the pyrolysis process remains unproved. So, in this paper, the novel two-step method of model sifting from various mechanism functions was put forward and applied.

2 EXPERIMENT

2.1 *Biomass feedstock*

This experiment chose Pine Cone, Corn Cob and Peanut Shell (abbreviated as PC, CC and PS, respectively), produced from the suburbs of Shenyang in northeastern China, as feedstock. Before TG analysis, all feedstock were grinded into 60~100 mesh and well mixed. The ultimate analysis and proximate analysis of biomass samples were showed in Tables 1 and 2, respectively.

Table 1. Ultimate analysis of biomass samples (%).

Sample	C	H	O*	N	S
PC	48.18	5.92	44.99	0.66	0.25
CC	45.28	5.89	47.79	0.83	0.21
PS	41.97	5.52	50.84	1.43	0.24

*Calculated by difference.

Table 2. Proximate analysis of biomass samples (%).

Sample	Volatile	Fixed carbon	Ash	Moisture
PC	74.22	17.63	2.03	6.12
CC	74.67	16.66	2.38	6.29
PS	62.97	17.29	11.83	7.19

2.2 Apparatus and experimental procedures

A Netzsch STA 409 C thermogravimetric analyzer was employed in this paper. High purity nitrogen was used as the load gas with a 30 mL/min flow. Each experiment, about 10 mg sample was heated from 293 K to 1373 K with three heating rates of 10, 20 and 40 K/min.

3 RESULTS AND DISCUSSION

3.1 TG and DTG curve of different samples

Figure 1 illustrates the TG and DTG curves of the feedstock. For different feedstock, these curves are similar in shape, with some difference in the start, end and peak point, and the residue ratio, mostly own to different content in the main components of the feedstock (see in Table 2). The whole process of biomass pyrolysis can be divided into four stages. The first stage, before 400 K, mainly is the process of surface water removal. The second stage, from 400 K to about 500 K, is the preheating process, in which the weight of sample barely changes, with some depolymerization and vitrification conversion occurred. The third stage, from 500 K to 650–800 K, is the main pyrolysis region with 40–60% weight lost and maximum weight loss velocity occurred. Under this temperature region, the weight lost order of the three samples is consistent with the volatile order (see Table 2). Usually, kinetic study of biomass pyrolysis mainly focuses on this stage. In last stage, the weight loss is minor with some carbonation reactions and the curve tends to flatten. At the end, the residual mainly contain fixed carbon and ash.

3.2 Effect of heating rates on feedstock pyrolysis

The effect of heating rate on the pyrolysis process can be reflected from the TG and DTG curves

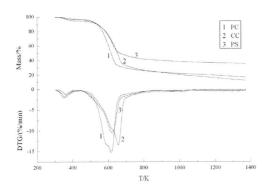

Figure 1. TG and DTG curves of biomass feedstock (heating rate $\beta = 20$ K/min).

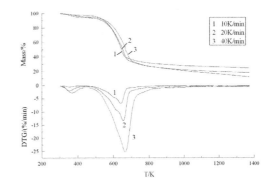

Figure 2. TG curves of PC under different heating rates.

of biomass under different heating rates (see Figure 2). From the TG curves, with the increase of heating rate the start point and end point of the pyrolysis move backward. This phenomenon is likely to be due to thermal lag effect. The DTG curves can show that the pyrolysis velocity is distinctly growing with the increase of heating rate, which is the main reason why most trials of biomass pyrolysis require high heating rate. From the DTG curves, the effect of thermal lag also is showed.

4 KINETIC ANALYSIS OF PYROLYSIS REACTION

4.1 The choice of mechanism model

In this paper, kinetic analysis focused on the largest weight loss region, i.e. the third stage of biomass pyrolysis. In this region, the temperature ranges of the sample under different heating rates were showed in Table 3.

The reaction rate equation can be written as

$$\frac{dx}{dt} = kf(x) \tag{1}$$

The conversion ratio x is defined as $x = (m_0 - m_t)/(m_0 - m_e)$, in which m_0, m_e, m_t are the sample weight at the start and the end of the largest weight loss region, and at time t, respectively.

The reaction rate constant k can be written as

$$k = A\exp(-E/RT) \tag{2}$$

where E is Arrhenius activation energy, A the frequency factor, R the gas constant, and T the reaction temperature. The heating rate $\beta = dT/dt$, thus with combinations of Eq. (1) and Eq. (2), we can get

Table 3. The temperature range of the main pyrolysis region (K).

Sample	10 K/min	20 K/min	40 K/min
PC	546.42~672.16	548.16~686.36	553.33~716.18
CC	520.41~638.87	533.52~651.02	537.73~666.63
PS	558.16~652.87	559.04~665.98	571.88~685.21

$$\frac{dx}{f(x)} = \frac{A}{\beta}\exp(-E/RT)dT \tag{3}$$

Then, by integrating Eq. (3), we can get

$$G(x) = \int_0^x \frac{dx}{f(x)} = \frac{A}{\beta}\int_{T_0}^{T}\exp(-E/RT)dT \tag{4}$$

Mechanism function, $f(x)$, is one of kinetic triplet (two other factors are E, A). Every mechanism model has a unique function expression. Obviously, to select correct mechanism function is of crucial importance to kinetic analysis. In this paper, 12 kinds of mechanism models (see Table 4) were to be selected to verify which one best fits the feedstock pyrolysis. They are one-dimensional diffusion model (D1), two-dimensional diffusion model (D2), three-dimensional diffusion model (D3, cylindrical symmetry; D4, spherical symmetry), random successive nucleation growth model (A1, n = 1; A2, n = 2; A3, n = 3), shrinking core model (R1, cylindrical symmetry; R2, spherical symmetry), chemical reaction model (C1, n = 3/2; C2, n = 2; C3, n = 3), respectively. This paper employed a new method of selecting mechanism model called two-step selecting method.

In the first step, Li Chung-Hsiung method was used to select the models with better linear correlation. Eq. (4) can be converted to

$$\ln\left[\frac{G(x)}{T^2}\right] = \ln\frac{AR}{\beta E} - \frac{E}{RT} \tag{5}$$

To a mechanism model which can correctly describe pyrolysis reaction process, $\ln[G(x)/T^2]$ should have a linear relation with $1/T$, with $(-E/R)$ as the slope and $\ln(AR/\beta E)$ as the intercept. By linear fitting of 12 mechanism functions, we can get the linear correlation coefficient R and the residual sum of squares $RESS$ of each function (see Table 5). Then, the ones with the bigger R and the less $RESS$ are selected to the second stage. For PC, D4, A1, A2, A3, R1 and R2 were selected; A1, A2, A3, R1 and R2 for CC; A1, A2, A3, R1, R2 and C2 for PS.

In the second step, using Malek method, the most reasonable mechanism function can be

Table 4. The differential and integral forms of 12 kinds of mechanism function.

Code	$f(x)$	$G(x)$
D1	$1/2x$	x^2
D2	$[-\ln(1-x)]^{-1}$	$x+(1-x)\ln(1-x)$
D3	$(3/2)[(1-x)^{-1/3}-1]^{-1}$	$[1-(2/3)x]-(1-x)^{2/3}$
D4	$(3/2)(1-x)^{2/3}[1-(1-x)^{2/3}]^{-1}$	$[1-(1-x)^{1/3}]^2$
A1	$1-x$	$-\ln(1-x)$
A2	$2(1-x[-\ln(1-x)]^{1/2}$	$[-\ln(1-x)]^{1/2}$
A3	$3(1-x)[-\ln(1-x)]^{2/3}$	$[-\ln(1-x)]^{1/3}$
R1	$2(1-x)^{1/2}$	$1-(1-x)^{1/2}$
R2	$3(1-x)^{2/3}$	$1-(1-x)^{1/3}$
C1	$2(1-x)^{3/2}$	$(1-x)^{-1/2}$
C2	$(1-x)^2$	$(1-x)^{-1}-1$
C3	$(1-x)^3$	$(1/2)[(1-x)^{-2}-1]$

Table 5. The linear correlation coefficient and the sum of squared residual of mechanism models for the biomass pyrolsyis under 20 K/min.

	PC		CC		PC	
Code	R	RESS	R	RESS	R	RESS
D1	0.9871	30.80	0.9760	48.63	0.9726	50.33
D2	0.9928	20.50	0.9848	37.09	0.9824	39.41
D3	0.9946	16.62	0.9879	31.76	0.9859	34.04
D4	0.9968	11.24	0.9926	22.38	0.9914	24.21
A1	0.9971	2.76	0.9964	3.13	0.9962	3.13
A2	0.9964	0.69	0.9955	0.81	0.9953	0.80
A3	0.9953	0.31	0.9943	0.37	0.9942	0.37
R1	0.9946	3.78	0.9881	7.32	0.9863	7.85
R2	0.9963	2.88	0.9917	5.67	0.9905	6.12
C1	0.7819	15.02	0.8548	10.75	0.8860	9.44
C2	0.9817	34.78	0.9908	15.94	0.9936	10.97
C3	0.9555	158.85	0.9715	94.32	0.9772	76.45

further selected from the ones selected from the first stage.

According to Coats-Redfern equation

$$G(x) = \int_0^x \frac{dx}{f(x)} = \frac{ART^2}{\beta E}\exp(-E/RT) \tag{6}$$

and Eq. (3), we can get

$$G(x) = \frac{RT^2}{E\beta}\frac{dx}{dt}\frac{1}{f(x)} \tag{7}$$

So, when $x = 0.5$,

$$G(0.5) = \frac{RT_{0.5}^2}{E\beta}\left(\frac{dx}{dt}\right)_{0.5}\frac{1}{f(0.5)} \tag{8}$$

where $T_{0.5}$ and $(dx/dt)_{0.5}$ are the temperature and reaction rate when $x = 0.5$. Let Eq. (7) divide by Eq. (8), and we can get

$$y(x) = \frac{f(x) * G(x)}{f(0.5) * G(0.5)} \quad (9\text{-}1)$$

$$y(x) = \frac{T^2}{T_{0.5}^2} \frac{dx/dt}{(dx/dt)_{0.5}} \quad (9\text{-}2)$$

According to Eq. (9-1), we can draw the $y(x)$–x theoretical curve of each mechanism model. Taking the thermogravimetric experiment data, i.e. x_i, $y(x_i)$, $i = 1, 2, 3\ldots$ and $x = 0.5$, $T_{0.5}$, $(dx/dt)_{0.5}$ into Eq. (9-2), the experimental $y(x)$–x curve can be drawn. Compared with the experimental curve, we can select the most similar one from the theoretical curves, and the mechanism model of the curve will be seen as the one of the sample pyrolysis process. By the least square method, R2 model was selected as the mechanism function of PC, R2 and A1 as the one of CC and PS, respectively (see Table 6).

4.2 Kinetic analysis

According to the selected mechanism function, the kinetic parameters can be gotten. The relationship between $\ln[G(x)/T^2]$ and $1/T$ for the selected mechanism function R2 of PC under different heating rates was showed in Figure 3. According to the slopes and the intercepts of the fitting straight lines, E and A can be solved (the results showed in Table 7). From Table 7, with the increase of the heating rate, the activation energy is reducing. That is, the higher heating rate can accelerate the pyrolysis reaction, i.e. increase the reaction rate (see the DTG curves).

Then, the kinetic model was to be verified by the comparison with the experimental data. Taking PC under 20 K/min for an example, the expression of the kinetic model can be erected as

$$\frac{dx}{dT} = 4.77 \times 10^4 \exp\left(-\frac{1.01 \times 10^4}{T}\right) * 3(1-x)^{2/3} \quad (10)$$

According to Eq. (10), the model relationship of the conversion ratio x and the pyrolysis temperature T was presented in Figure 4. By comparing

Table 6. The comparison result between the model and experimental curves by the error sum of square.

Sample	D4	A*	R1	R2	C2
PC	9.728	19.195	9.997	9.908	–
CC	–	11.917	12.445	9.334	–
PS	–	3.736	19.491	10.017	12.161

*A represents the A1, A2 and A3 models. In the the $y(x)$–x standard curves, the three model curves are overlapping.

Table 7. Kinetic parameters of PC pyrolysis.

Heating rate	Activation energy kJ/mol	Frequency factor min^{-1}
10 K/min (PC)	91.81	3.09×10^6
20 K/min (PC)	84.16	9.53×10^5
40 K/min (PC)	79.62	4.87×10^5
20 K/min (CC)	98.28	4.68×10^7
20 K/min (PS)	123.06	1.70×10^{10}

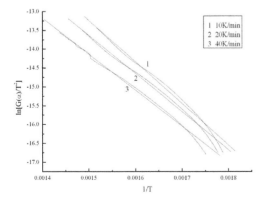

Figure 3. The relationship between $\ln[G(x)/T^2]$ and $1/T$ for R2 model of PC under different heating rates. The straight lines are fitted against the three model curves, respectively.

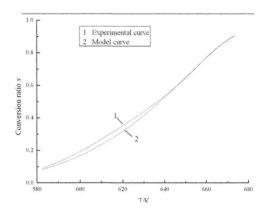

Figure 4. The comparison between the model and experimental conversion ratio curves of the pyrolysis process of PC under the heating rate of 20 K/min.

model and experimental curves in Figure 4, it was found that R2 model is reasonable to express the pyrolysis process of PC.

5 CONCLUSIONS

From the result of the thermogravimetric experiment, the different biomass samples have similar pyrolysis process with four stages (the water removal stage, the preheating stage, the main pyrolysis stage and the carbonation reaction stage, respectively). The reaction rate of the same feedstock increases with the heating rate. With the new two-step selecting method of mechanism function, R2, R2, A1 were selected as the most reasonable models of PC, CC and PS, respectively. With the increase of the heating rate, the activation energy E reduces gradually, correspond to the increase of the reaction velocity of pyrolysis process.

REFERENCES

Antal, M.J., G. Várhegyi, et al. (1998). "Cellulose Pyrolysis Kinetics: Revisited." Industrial & Engineering Chemistry Research 37(4): 1267–1275.

Hu, S., A. Jess, et al. (2007). "Kinetic study of Chinese biomass slow pyrolysis: Comparison of different kinetic models." Fuel 86(17–18): 2778–2788.

Hu, S., L. Sun, et al. (2008). "New kinetic analysis method to study biomass pyrolysis process." Taiyangneng Xuebao/Acta Energiae Solaris Sinica 29(8): 1038–1044.

Lu, C., W. Song, et al. (2009). "Kinetics of biomass catalytic pyrolysis." Biotechnology Advances 27(5): 583–587.

Saddawi, A., J.M. Jones, et al. (2009). "Kinetics of the Thermal Decomposition of Biomass." Energy & Fuels 24(2): 1274–1282.

Várhegyi, G., M.J. Antal Jr, et al. (1997). "Kinetic modeling of biomass pyrolysis." Journal of analytical and applied pyrolysis 42(1): 73–87.

Zhang, X., M. Xu, et al. (2006). "Study on biomass pyrolysis kinetics." Journal of Engineering for Gas Turbines and Power 128(3): 493–496.

Frontiers of Energy and Environmental Engineering – Sung, Kao & Chen (eds)
© *2013 Taylor & Francis Group, London, ISBN 978-0-415-66159-1*

Application status and potential of new energy resources in China

X.J. Shao, C.F. Wang, S.C. Li, S. Liu, J.W. Li & X.Q. Wang
Petroleum Engineering Department of Yanshan University, Qinhuangdao, China

ABSTRACT: China's new energy resources are abundant and have a huge development potential. CBM's recoverable reserve is 10.87 trillion cubic meters, shale gas's recoverable reserve is 26 trillion cubic meters, solar energy's reserve equates 17 trillion tons of standard coal, wind energy's total reserve is about 3.226 billion kilowatts, geothermal energy equates the heat 250 billion tons of standard coal contains, and tidal energy is 110 million kilowatts. Currently, the utilization of new energy resources is low. New energy resources account for less than 8 percent of China's total energy consumption. The main problems are following: The energy market is not open and most of the resources are monopolized by some state-owned enterprises. It is difficult to private capitals to get into it. Therefore the development speed of new energy resources is hindered. Immature and imperfect technologies influence the economy benefit of new energy resources development. Support strength of national policies is limited, and financial investments, manpower and technical strength are insufficient. The development of new energy resources needs some measures: Innovate theories and techniques. Make long-term reasonable energy strategy plan. Break resources monopoly of the state-owned enterprises. Encourage and guide the private capitals to step into new energy resources industry and promote new energy development and utilization. New energy resources will run on scaled and industrialization road as soon as possible.

Keywords: new energy resources; CBM; shale gas; solar energy; tidal energy

1 INTRODUCTION

Energy resources are important substantial resources for human-existing, and also are the motive power for society to advance continuously. Human energy society has successively experienced "firewood period", "coal period" and "petroleum period". And every energy revolution came from the contradiction between energy supply and economic society. Since the first and the second industrial revolutions happened, oil, coal and other fossil fuels have been playing more and more remarkable positions in economy production.

China is a major energy resources consumer. Especially in recent 20 years, China economy has been developing fast, and the energy resources' demand has been keeping growing rapidly. In existing energy structure of China, traditional energy production and consumption have absolute superiority. For example, from 2003 to 2008, coal had been accounting for above 70% in energy production structure while it had been stabilizing about 67% in structure of consumption. Oil is following. It accounted for 13% in production structure and 20% in consumption structure. Gas was less than 5% both in production and consumption structures. New energy resources were less than 8% (Zhou et al., 2010; Luo et al., 2010). Obviously, the characteristic of China energy structure is that

traditional one-off energy resources have been in first place in consumption, and coal takes prior status in one-off energy resources. The rise in consumption of one-off energy resources leads to seriously environmental pollution.

New energy resources generally have advantages in less pollution and bigger reserves relative to one-off energy. It is the key to solve serious environmental pollution and fossil energy exhaustion in the world. At the same time, because many new energy resources distribute uniform, it is also extremely important to reduce energy initiation's wars and maintain world peace and stability. This paper investigates and summarizes a great deal data, and analyses new energy resources application status, potential and problems in China. It can provide technical and decision-making basis for decision-making departments and related companies.

2 APPLICATION STATUS OF NEW ENERGY RESOURCES IN CHINA

New energy resources called unconventional resources contain all kinds of energy resources which are in the beginning of developing and utilizing or in process of positive study, except conventional ones. New energy resources come form the sun's or the earth's interior thermal energy directly or indirectly. Solar energy, wind energy, biological energy, geothermal energy, nuclear energy, water energy, tidal energy and energies of hydrogen and biological fuels which are derived from renewable energy sources are included. Considering actual situation and breakthrough which may be got recently in China, authors analyze and discuss CBM, shale gas, solar energy, wind energy, geothermal energy and tidal energy which can be used for society development.

2.1 CBM

CBM commonly known as mashgas is produced from the process of coalification. It is a self-generated and self-stored unconventional natural gas which keeps adsorbent status in coalbed. CBM is a mixture of gases in which methane is major. Its component is mainly same to conventional gas, so it is called the most realistic supplement source of conventional oil and gas. CBM rose in the past 20 years in the world, and it is high quality and clean energy resource and chemical industry materials.

Extensive use CBM can reduce methane emission, and protect environment at some extent. Methane is greenhouse gas following carbon dioxide. In process of mining, CBM is discharged into the air through ventilation, and leads to greenhouse effect. Because it produces carbon dioxide much less than oil and coal, developing and utilizing CBM can improve energy resources structure, and ease China's energy resources shortage situation.

China started developing and utilizing CBM a little late. The process could be divided into three stages. The first stage was pumping and releasing period from the 1950s to the late 1970s. In order to reduce mashgas damage, people pumped and released it to the air, and seldom utilized it. The second stage was pumping and utilizing period from the 1980s to the early 1990s which was development prime of CBM exploration test. From the 1990s to the present is the time to roundly test and explore CBM, and pump, utilize it in scale (Jia et al., 2007). Since 1990, China has drilled CBM exploration wells in more than 30 coal areas. Until August 2010, there were 4,657 CBM wells, and daily production had been 3,850,000 cube meter per day. It had already been developed to a certain scale. Qinshui basin and the eastern margin in Ordos are CBM commercial production areas at present in China.

2.2 Shale gas

Shale gas is natural gas that mainly accumulates in dark clay shale or high carbon clay shale. It exists in two ways: adsorption and in free status. It usually distributes in large thick shale in hydrocarbon source rocks of the basin (Zhang, 2010). Compared with conventional natural gas, shale gas development has more tremendous potential, longer mining life and cycle. The mining life of shale gas field can be up to 30 to 50 years commonly.

China explored and developed shale gas a little late. And it still remains in primary stage now. In 2004, Ministry of Land and Resources Oil and Gas Resources Strategic Research Center and China University of geosciences (Beijing) jointly carried out shale gas resources researches (Zhang, 2010). It was a prelude to China to study shale gas resource. CNPC (China National Petroleum Corporation) also started researching shale gas in 2005. CNPC signed "Shale gas united research in Weiyuan area" with the US Newfield Company in October 2007. In November 2009, it signed "Shale gas joint evaluation agreements of Fushun-Yongshun block in Sichuan basin" with Shell (Gao, 2010). China carried out the project "Shale gas resources potential analysis and favorable regions optimization in China key areas" in the special item of oil and gas resources strategic selection nationwide in 2009. On August 17th, 2009, the first Chinese shale gas development project in Qijiang of Chongqing started. It marked that China officially started shale gas exploration and development. The first Chinese special research institute of shale gas development-"National Energy Shale Gas R&D (test) Center"-was established in Langfang Branch of PetroChina Research Institute of Petroleum E&D on August 20th, 2010. Four months later, the first shale gas well-Yang 101-which CNPC and Shell cooperated started being drilled successfully. On April 1st, 2011, the first Chinese shale gas horizontal well was successfully completed.

China has already launched policies to support shale gas development. In December 2010, the State Council officially approved the document which jointly issued by four ministries—Shanxi "Comprehensively Matching Reform Testing District on Resources Economic Transformation", Commerce Department, National Development and Reform Commission etc-that newly increased CNPC, SINOPEC and CBM development Co, Ltd in Henan province to achieve foreign cooperative right in CBM and shale gas field to break exclusive monopoly of CUCBM. In addition, the enterprises who participated in shale gas bidding in 2011 not only included State-owned oil companies, but also private enterprises such as Xinjiang Guanghui groups etc. This was the first time for private enterprises to be allowed into energy development market (Hai, 2011).

2.3 Solar energy

Solar energy generally refers to the sun radiant energy. Two ways to utilize solar energy are photoelectric conversion and passive utilization (photothermal conversion). At present, the solar energy technologies can be divided into two kinds called solar Photovoltaic (PV) and solar thermal. The energy that the sun delivers to the earth each year is ten thousand times of the quantity of human demand (Xu et al., 2010). So it is greatly significant that developing and utilizing solar energy effectively to ease the energy tense situation all around the world.

1. Solar Photovoltaic (PV)

PV panel is an electricity generation device which generates dc when exposed in the sun. It consists of thin solid photovoltaic cells which are almost all made of semiconductor materials (such as silicon). China started studying photovoltaic generation in the 1970s. After more than 30 years' struggle, photovoltaic generation enters a brand new period. The "Transmit Power to Townships" project in 2002 and 2003 which was invested 5 billion in western of China holp construct nearly one thousand independent photovoltaic power stations, and solve basic life problems in the areas without electricity well. At the same time, it greatly boosted the development of domestic photovoltaic industry. Now there are more than 10 photovoltaic enterprises going public in overseas markets such as New York, London, and some domestic security markets (Fu and Cai, 2009). China is behind developed countries in solar photovoltaic power generation field, but China's photovoltaic output is growing rapidly. In 2005, China's solar cell output was 139 megawatts in total. In 2007 it was up to 1,200 megawatts, and the productive power was 2,900 megawatts. In 2008, domestic photovoltaic cell production was more than 2,000 megawatts, and the practically productive power ranked No.1 in the world with more than 3,000 megawatts (Fu and Cai, 2009). The proportion of photovoltaic cell production rose from 1% in 2001 to 15% in 2008 in the world production. The number of domestic solar cell manufacturing enterprises had reached more than 50 by 2008.

2. Solar thermal

Modern solar thermal technology is to polymerize sunlight, and use the energy to produce hot water, steam and power. There is complete industrial system of solar water heater in China and it develops steadily and rapidly in recent years. In 2008, the output value reached 43 billion dollars and the exports was $100 million. Solar water heater annual yield exceeded 40,000,000 square meters, and inventory reached 145,000,000 square meters in 2009. China became the superpower in solar water heater production in the world. According to the data, the quantity of solar heat utilization device was 75,000,000 square meters in 2005, and by 2007 it had become about 120,000,000 square meters with growth rate 30% (Fu and Cai, 2009).

The Chinese government introduced related support policies when solar energy industry developed rapidly. "Financial Subsidy Management Interim Measures of Solar PV Building Application" and "Implementation Opinions concerning Acceleration of Solar PV Building Application" established that in 2009 subsidy standard of photoelectric building was $20 per watt. "Notice about Implementing 'Golden Sun' Demonstrative Project" proposed that giving PV grid-connected and off-grid PV generation in no power areas financial subsidy were respectively 50% and 70%. "Notice about Completing 'Golden Sun' Demonstrative Project" requested to accelerate this project (Fu and Cai, 2009). A series of policies that were issued by the government have been promoting China's solar energy industry entering into another round of rapid development.

2.4 *Wind energy*

Wind energy resource is a kinetic energy formed on the earth's surface when the air flows. Wind energy resource depends on wind energy density and annual accumulation time of available wind power. Wind energy density is the power which is gained by per unit windward area. It is proportional to the air density and the cube of wind speed. Wind energy resource has many characteristics. It is renewable, clean, abundant and easy to utilize. Therefore making good use of wind energy resource is greatly significant to reduce CO_2 emissions, protect the ecological environment, relieve the energy crisis, adjust the energy structure and promote the economy sustainable development. Main ways of utilizing wind energy resource are wind dynamic and wind power generateon, and the latter leads role position.

In recent 10 years, with the rapid development of global wind power generation industry, in China, wind power generation industry has come into a high-speed development period with Chinese government's support. In 2005, in mainland, wind power generation sets were 1,864 accumulatively and installed capacity was 1.266 million kilowatts with a year-on-year growth of 65.6%. In 2006, wind power generation sets and installed capacity were respectively 3,311 and 2.599 million kilowatts with year-on-year growth of 105%. In 2007, they became 6,469 and 5.906 million kilowatts with year-on-year growth of 127.2%.

From 2008 to 2010, cumulative figures of wind power sets were respectively 16,000, 21,581 and 34,485, installed capacities were 12.169 million kilowatts, 25,805.3 megawatts and 44,733.29 megawatts. The year-on-year growth figures were 106%, 114% and 73.3% respectively. For five consecutive years, they grew with a double speed (Yang, 2010).

2.5 Geothermal energy

Geothermal energy is thermal energy stored in the earth. It has three characteristics (Ma and Tian, 2006).

1. Geothermal energy is a kind of clean energy resources. Compared with other new energy resources it is more suitable for environmental protection and sustainable development. Only in geothermal energy heating, if heating area is 10 million square meters, it can save standard coal 590,000 tons, reduce CO_2 1.6 million tons, SO_2 5,000 tons and oxynitride 4,900 tons each year.
2. Geothermal energy is a kind of extremely rich energy resources. 99% of matter in the globe is in over 1,000 degrees. Less than 1% is under 100 degrees. The available part is so small at the moment, but the reserves of geothermal energy resource which could be developed only by existing technologies is more than 30 times of all fossil energy's reserves. In short, geothermal energy is an extremely rich alternative energy.
3. Geothermal energy's applicability is much stronger. Because geothermal energy is under ground, it is difficultly influenced by external natural environment. So geothermal energy can be exploited continuously. The mining process is much easier to control. They are prior conditions to supply energy resource on a grand scale.

The utilizations of geothermal energy can be divided into two kinds. One is to use the heat directly. It includes direct utilization of geothermal water (such as geothermal heating, washing and breeding etc.), geothermal pump heating and refrigeration. The other one is geothermal power generating. Yangba well in Tibet is a model of geothermal power in high temperature. This power station's installed capacity is 25.18 megawatts, and actual power generation is about 15 megawatts which accounts for 40% in yearly supply of power grid in Lhasa and even exceeds 60% in winter. In the way of direct utilization of geothermic water in medium-low temperature in China, heating accounts for 18%, medical treatment and entertainment account for 65.2%, planting and breeding are 9.1% and the other is 7.7% (Zhao, 2007). In industry, the geothermal water's utilizations mainly concentrate on dyeing, heating oil pipes and drying and so on. In recent years, with the geothermal heat pump technology becoming better and better, geothermal energy utilizations develop rapidly in China. In 2005, geothermal fluid resource volume which could be used directly had become No.1 in the world with 445.7 million square meters (Ma and Tian, 2006). By the end of 2006, many areas had applied and popularized this technology in varying degrees except minority provinces such as Qinghai, Yunnan and Guizhou etc.

2.6 Tidal energy

Tidal energy is a kind of ocean energy resources. It is water potential energy which forms with tide rising and falling which caused by the earth's rotation and its attraction by the moon. Tidal energy is pollution-free and sustainable. Using tidal energy resource reasonably can not only reduce environmental pollution but also ease energy tense situation. The level of development and utilization of this resource is still low in the world (Xie et al., 2009). China is also in primary stage. It began developing and utilizing tidal energy in the 1950's, and the level is still low either. From building the first tidal power station in Shandong in 1957 to launching Bachimen tidal power project officially in Fujian in 2008, Sanmen tidal power station project of 20,000 kilowatts started in Zhejiang in 2009. China's technology on tidal power industry is gradually mature, and it makes great progress in reducing cost and increasing economic benefit. As a whole, tidal energy development and utilization have obtained a substantial advance. China is the country with the most tidal power stations in the world. In the past five decades, China built 76 small tidal power stations which included eight stations in the long run. Total installed capacity was 6,120 kilowatts and power generation was 10 million kilowatt-hour per year. At present, 2 power stations are still generating electricity in Zhejiang province. The bigger one is Jiangxia tidal test power station whose total installed capacity is 3,900 kilowatts and power generation is 10.7 million kilowatts-hour per year. And it is also the third largest power station in the world. China's tidal power generation capacity ranks third in the world after France and Canada now (Xie et al., 2009).

China's tidal power generation technology, as a whole, has reached advanced level in the world. In recent years, the government has promulgated and carried out a series of laws and policies such as "Renewable Energy Law" etc. to encourage development of renewable energy resources. It creates a good surrounding for ocean development and utilization.

3 NEW ENERGY RESOURCES POTENTIAL

China's new energy resources are very rich, especially in the northwestern and coastal areas. It is greatly important for China to fully develop and utilize these resources to ease the dependence on foreign energy and high environmental pressure.

3.1 *CBM*

CBM resource is rich in China. According to CBM resource evaluation report of 2006, the total CBM resource which buried less 2,000 m ranks No.2 in the world with 36.81 trillion cube meters-about 13% of the world's CBM total resource (Liu, 2008). CBM resource is as much as land conventional natural gas. It equals to 5.2 million tons of standard coal. And it mainly distributes in North China with 57% of the country's total CBM reserves and in the Northwest with 28% of the total (Jia et al., 2007). CBM recoverable volume which buried between 300~1,500 m is 10.87 trillion cube meters. And there are 15 gather gas zones which contain CBM more than 10 million cube meters. According to the size of exploitable resource, in descending order, they are respectively Erlian, Erdos, east of Yunnan and west of Guizhou, Qinshui, Junggar, Tarim and Tianshan etc. (Chi and Wu, 2000).

3.2 *Shale gas*

Shale gas exploration became popular all over the world with the US-led North America obtained great success on it. And China also began doing shale gas exploration and development work energetically. On study, China's shale gas resource is rich and has great development potential. Many basins and regions have geological conditions are conducive to large-scale accumulation of shale gas. Preliminary estimates suggest that China's shale gas recoverable resource is 26 trillion cube meters. They mainly distribute in the south zone and North China zone (Zhang and Tan, 2009; Yan et al., 2009). Shale gas resource's amount of the Paleozoic is twice of the Mesozoic's. The shale in Lower Cambrian and Lower Silurian in the southern and eastern of Sichuan basin are major exploration targets currently. Based on geological history and its changing characteristics, China's shale gas developed zones can also be divided into four areas which roughly corresponding with plates. They are respectively the South, the North China-Northeast, the Northwest and Qingzang, and every one of these zones has several sets of shale (Liu et al., 2010).

3.3 *Solar energy*

China is rich in solar energy resource, and there is great potential for development. The richest solar energy regions in China are Tuha Basin, Qaidam Basin, Erlian Basin and Yin'e Basin. The richer areas are Tarim Basin, Junggar Basin, Ordos Basin, Songliao Basin and Bohai Bay Basin. Sunshine time in two-thirds of China is more than 2,200 hours, and the theoretical reserves equates to 17 trillion of standard coal-about the total power generation of tens of thousands of Three Gorges Project. So it is of great significance to make full use of solar energy to ease Chinese energy crisis, improve energy structure, reduce CO_2 emissions, protect ecological environment and promote sustainable development (Fu and Cai, 2009).

3.4 *Wind energy*

Because of vast territory and long coastline, China has rich wind energy resource. Nonetheless, influenced by various factors such as topography wind energy resource distribution is nonuniform (Ding, 2009). Wind power is one of the ripest and the largest scale technologies in new energy electricity generated. According to estimations on more than 900 weather stations by Chinese Academy of Meteorological Sciences, average wind power density all over the country is 100 watts per square meter, and wind energy reserves is 3.226 billion kilowatts in total. Wind energy reserves which can be developed and utilized is 253 million kilowatts onshore and 750 million kilowatts offshore, about 1 billion kilowatts in total (Countryside Electrification Editorial Department, 2004). If wind power online onshore is calculated by full load 2,000 hours per year, it can provide 500 billion kilowatt-hour per year. And If wind power online offshore is calculated by full load 2,500 hour per year, it can provide 1.8 trillion kilowatt-hour per year. The total electricity is 2.3 trillion kilowatt-hour per year and equals to 81% of generation in 2006 (Yang, 2010). Onshore wind energy resource mainly centralizes in north area, including Inner Mongolla, Gansu, Xinjiang, Heilongjiang, Jilin, Liaoning, Qinghai, Tibet and Hebei and so on (Yang, 2010). Wind resource inshore and on islands is mainly in Liaoning, Hebei, Shandong, Jiangsu, Shanghai, Zhejiang, Fujian, Guangdong, Guangxi and Hainan etc.

3.5 *Geothermal energy*

According to scientists forecast, the total heat inside the earth is about 170,000,000 times as much as all the world's coal contains. Every year, the lost heat from the earth surface equals to the burning heat of 100,000,000,000 billion bbl oil. Geothermal energy has a broad development prospect. China has abundant geothermal energy resource. According to the preliminary estimates, the amount of geothermal energy resource in China is very impressive. The geothermal energy reserves in main basins within 2,000 m from the surface is equivalent to the heat of 250 billion tons of standard coal (Ma and Tian, 2006). Only based on the 2007 survey data, the total geothermic water that can be developed and used is 6.845 billion cubic meters a year—equivalent to the heat of 32.848 million tons of standard coal (Zhao, 2007).

3.6 Tidal energy

China is well known to be very rich in tidal resource. China's sea is vast. Continental coastline is 18,000 km. The total coastline is more than 32,000 km including 6,500 km of islands' coastline. According to the survey data about 424 bays whose installed capacity are more than 200 kilowatts, tidal energy reserves in China is 110,000,000 kilowatts, and total installed capacity is 21,790,000 kilowatts, annual total power generation capacity is 6,240,000 kilowatt-hours and there are 191 stations whose capacities are above 500 kilowatts (Xie et al., 2009). They mainly distribute in Zhejiang, Guangdong, Fujian and Liaoning etc. Many dams with high density energy and superior natural surrounding are on the southeast coast. Their average tidal range is 3.5~4.3 m and the longest tidal range is 7~8 m. The dams that power stations whose capacities above 10,000 kilowatts can be build are Hangzhou Bay, Leqing Bay and Xiangshangang in Zhejiang, north branch of Yangtze River, and Daguanban in Fujian etc. (Geng and Pan, 2005). Hangzhou Bay has the longest tidal range which could reach 8.9 m. Its tidal energy reserves is No.1 in China.

4 PROBLEMS

Due to the limit on technologies, fund and level of awareness, there are some problems in new energy resources development and utilization.

1. In recent years, with CBM's development policies' introduction and improvement, domestic enterprises make advantage of CBM to produce electricity more and more actively. Installed capacity of CBM power generation becomes bigger and bigger. Technology research and equipment manufacturing level are continuously improved. However, the most CBM that China exploits is low density, and domestic extraction scale is small and dispersed currently. All of these conditions hinder China making full use of CBM resource in a certain degree.

   CBM development is still in primary stage. There are no targeted and mature evaluation system, no perfect development technologies, no

enhancing production and transportation technology aiming at Chinese coal reservoir characteristics. The percentage of low production wells is high and economic benefit is bad.

Now, a few big state-owned enterprises monopolize most of favorable mining areas. But restricted by economy and technology, investment and technology are insufficient. Private enterprises want to step in, but they have no exploration rights. This management system restrains the development of CBM industry seriously in China. Thus, the monopoly should be broken and private enterprises should be encouraged to step in the industry to form diversified CBM pumping and transporting companies. These measures will improve CBM development and utilization greatly.

2. As a fresh and unconventional resource, shale gas resource's development needs a lot of technologies, fund and manpower. Because China's shale gas resource development is just at the beginning, experience and technologies are poor. These conditions constrain China's shale gas development. At the moment, there are only small tests, and becoming big scale development will take a long time. Moreover, shale gas development requires high-level technologies. Compared with conventional natural gas, shale gas development is more difficult and needs policies support.

3. Intensity of solar radiation is affected by climate, day and night, latitude, season and elevation etc. so it usually needs energy storages. It is hard to get stable output power only depending on this energy. Besides, solar energy flux density is low and the applied equipment must be large. Large area and great investment are needed. Photovoltaic system market in China, at present, primarily satisfies requirements in special conditions and remote farming areas without electricity. It has market potential as well as major limitations.

4. Wind has much effect on wind energy resource. So factors such as location that affect wind will affect wind energy collection and utilization indirectly. In addition, wind energy is a new kind of energy resources, and its equipments are not very mature. Consequently, the conversion rate of wind energy is low and the output is not stable because of different kinds of factors. How to innovate and improve technologies is very important for wind energy resource. The nation should provide preferential policies to wind power online.

5. Now, the chief problem on geothermal development is complex underground conditions, and it's hard to ensure well bores' quality. At the same time, because thermal energy resource is mainly in low-and medium temperature zones in China,

how deeply geothermal energy development does affect the Earth's crust, and how much we can exploit and so on. These problems lack enough study. It needs the government to put technologies strength and technical support to develop geothermal resource on a massive scale.

6. Tidal energy utilization is not without cost. The places where tidal energy is rich are mostly in medium-to-large size river estuary. Tidal energy stations built in these places will make influence on local ecological environment. And it is uneconomical to built power stations considering each factor comprehensively. This problem should be considered combined with environment and economy.

5 CONCLUSIONS

The development and utilization of new energy resources have become the same concept of common pursuit in the world. New energy resources are the unique chose to ease globe energy tensions situation and protect ecologic environment. It is an irresistible trend that new energy resources partly replace conventional fossil resources and become major resources in the world in the future.

In China, new energy resources are abundant and have great development potential. Innovating theory and technology, perfecting management and supervision system, enhancing investment power, researching and exploring new energy resources and making long-term reasonable strategy are needed to solve the chief problems existing at present.

The government needs to establish corresponding policies to support new energy resources and break the monopoly of state-owned business. It also should encourage and guide private capital into new energy resources industry to promote new energy development and utilization. Those can make new energy resources run the scale and industrial road as soon as possible.

ACKNOWLEDGEMENTS

This research was funded by the National Basic Research Program of China (No.2009CB219604) and National Important Science and technology Research Program of China (No.2011ZX05038-001). Corresponding author:Xianjie Shao, E-mail: shaoxjy@yahoo.com.cn.

REFERENCES

Chi W. & Wu S. 2000. A Rediscussion on The Exploration Prospect of Coalbed Gas in China. *Experimental Petroleum Geology* 22:131–135.

Countryside Electrification Editorial Department. 2004. Status and Prospects of Wind Generation in China. *NO.9 of Electrification of the Countryside*: 7–8.

Ding J. 2009. Wind Resource Reserves and Distribution in China. *3rd edition 2009 of China Meteorological Report in Jan 15, 2009.*

Fu Y. & Cai H. 2009. Current Situation and Development Solar Power Generation. *NO.9 of Rural Electrification*: 57–59.

Gao H. 2010. Recall Shale Gas in Sleep. *Shandong Land and Resource* 26: 60–61.

Geng Z. & Pan C. 2005. Development Status and Prospects of Tidal Energy Resources in China. *Proceedings of the 12th Chinese Coastal Engineering Seminar*: 637–641.

Hai X. 2011. Opening Market to Shale Gas maybe Change The Energy Shortage Situation, *B0 3rd edition of China Economy Guiding in May 5, 2011.*

Jia C. 2007. Evaluation Methods for Coalbed Methane Resources and Reserves. Beijing: Petroleum Industry press.

Liu H. & Wang H., et al. 2010. China Shale Gas Resources and Prospect Potential. *Acta Geologica Sinica* 84: 1374–1378.

Lu Y. 2008. The Geological Problem and Technical Challenge of Developing China Coalbed Methane Industry. *2008 CBM Academic Seminar Assays.* Beijing: Petroleum Industry Press.

Luo F. & Luo Y. 2010. Optimization of Energy Consumption Structure in China. *Problems and Countermeasures* 36: 21–25.

Ma L. & Tian S. 2006. The Present Situation of Geothermal Energy Exploitation and Utilization and its Development Trend in China. *Natural Resource Economics of China* 9: 19–21.

Xu H. & Li H. 2010. New Energy—Solar Energy. *Zhongguo Baozhuang Keji Bolan* 36: 184.

Xie Q. & Liao X., et al. 2009. Summary of Tidal Energy Utilization at Home and Abroad. *Water Conservancy Science and Technology and Economy* 15: 670–671.

Yang X. 2010. Study on China's Legal System for The Wind Energy Resources Development and Utilization. *Renewable Energy Resources* 28:7–10.

Yan C. & Huang Y., et al. 2009. Shale gas: enormous potential of unconventional natural gas resources. *Natural Gas Industry* 29: 1–6.

Zhang H. 2010. Shale Gas: New Bright Point of The Exploitation of The Global Oil-gas Resources—the Present Status and Key Problems of The Exploitation of Shale Gas. *Strategy & Policy Decision Research* 25: 406–410.

Zhao F. 2007. Geothermy Energy is Abundant Underground in China. *1st edition of Chinese Territory Resource News in Jan 30, 2007.*

Zhang K. & Tan Y. 2009. The Resources Potential and Development Status of Global Shale Gas and The Development Prospect of The Shale Gas in China. *Petroleum & Petrochemical Today* 17: 9–12.

Zhou Z., Zhao S. & Zhuang Y. 2010. Development Trend of New Energy in China. *Heilingjiang Science and Technology Information* 22: 18.

Frontiers of Energy and Environmental Engineering – Sung, Kao & Chen (eds)
© 2013 Taylor & Francis Group, London, ISBN 978-0-415-66159-1

Decolorization of C.I. Direct Green 6 dye in aqueous solution by electrocoagulation using iron anode

Q.P. Nguyen
Northeastern University, Shenyang, P.R. China
Water Resources University, Hanoi, Vietnam

Z.N. Sun & X.M. Hu
Northeastern University, Shenyang, P.R. China

ABSTRACT: The present study applied iron anode electrochemical method to remove color from Direct Green 6 dye wastewater. Decolorization efficiency and electrical energy consumption were also calculated to compare optimal values. The effects of factors consist of current density, initial pH, initial dye concentration and Na_2SO_4 electrolyte solution concentration on the color removal efficiency and energy consumption have been also investigated. The experimental results show that the decolorrization of Direct Green 6 in the aqueous phase was very effective with this method. The current density of $3.333 \, mA/cm^2$, initial pH value of 6.65, initial dye concentration of 50 mg/L, Na_2SO_4 concentration of 0.1 mol/L, stirring speed of 600 r/min, temperature of 20°C, interelectrode distance of 16 mm and electrolyte time of 60 minutes were optimal conditions for Direct Green 6 dye decolorization, the decolorization efficiency reached 90.3%, energy consumption for decolorization in conditions mentioned above was 4.073 kWh/kg-dye.

Keywords: electrochemical; decolorization efficiency; energy consumption; dye wastewater; direct green 6

1 INTRODUCTION

Recently, the development of textile and dyeing industry has brought huge benefits to many countries, but caused serious problems to environment by large quantities of wastewater containing pollutants and color turbidity discharged. These problems have affected not only water quality degradation but also the wastewater treatment process (Dai et al. 2000; Liu. 2007; Ding et al. 2010 and Chu et al. 2007).

For several years now, a lot of methods have been applied to decolorizing dye wastewaters or degrading dye molecules in wastewaters in order to reduce their environmental impacts such as, adsorption, coagulation, Advanced Oxidation Processes (AOPs), and microbiological treatments, chemical degradation, photodegradation (Marco & Giacomo. 2004; Apostolos et al. 2004; Lin et al. 2004; Mahmut et al. 2007). Electrochemical technique, among these methods has most advantages. This technique uses a direct electric source between metal electrodes immersed in polluted water. The electrical current causes the dissolution of metal electrodes into wastewater. Mohammad et al (2004) showed that the metal ions at an appropriate pH can form metal hydroxides that destabilize and aggregate the suspended particles or precipitate and adsorb dissolved contaminants. In addition, according to Daneshvar et al (2006) and Chen (2004), electrochemical method also reduces the amount of sludge. The most widely used electrode materials in electrochemical process are aluminum and iron (Mehmet et al. 2003).

The electrochemical technology is simple, easily automated for wastewater treatment without any need for additional chemicals; hence mitigation of secondary pollution is invented. It also reduces the amount of sludge which needs to be disposed (Daniel. 1997 and Mohammad et al. 2001). Although this is friendly environment technique, high operational costs has inhibited practical application capacity (Wang et al. 2005; Guan & Yang. 1999) that has gained increasing concerning recent years (Gutierrez & Crespi. 1999; Lorimer et al. 2001).

The previous studies on dye wastewater decolorization applied electrochemical process have revealed some disadvantages. Ayhan & Mahmut (2009), studying on Dyeing Wastewater treatment by decolorization of C.I. Reactive Black 5 in aqueous solution by electrocoagulation using sacrificial iron electrodes, the color removal efficiency reached 98.8% and electrical energy consumption was 4.96 kWh/kg dye at conditions of an initial dye concentration of 100 mg/L, initial pH of 5, current density of 4.575 mA/cm², salt concentration of 3000 mg/L, temperature of 20°C, and interelectrode distance of 2.5 cm. The removal rates of COD and chrominance from wastewater were 87.5%–90% and 99%–100% respectively. The easy operation and low consumed energy were indicated when the electrolysis voltage of 10 V, electric current of 0.1 A and electrolysis time for 1.5 h of electrolysis method of complex catalysis were applied (Yang et al. 1998). Eriochrome Black T wastewater decolorization efficiency reached to 98% was indicated by the study of Liu et al (2009), but energy consumption for decolorization required 2,76 kWh/kg-dye which performed problems of energy safety of practical applications. Song (2008) studied decolourization of red dye solution wastewater by electrochemical process with self-prepared columnar device which is consumptive iron electrode. With the conditions of initial wastewater pH 5–6, PAM 0.3 g/kg, rotating speed 300 r/min, NaCl 3 g/kg, current intensity 0.04 A/cm², voltage < 30 V, time 8 min, the decolourization rate achieved above 93%. The ion anode electrochemical method of Liang et al (2007) for reactive black KN-B wastewater decolorization gave high decolorization efficiency. The efficiency of 97% reached once NaCl electrolyte was used while it was around 93% with Na_2SO_4 electrolyte using. The method was also applied to reactive red M-3BE, reactive yellow M-3RE wastewater decolorization.

The present study focuses on decolorization of the Direct Green 6 dye by iron anode situ electrocoagulation process. The experiments of effects of current density, initial pH, initial dye concentration and electrolyte solution concentration on color removal efficiency are investigated. The study aims also to consider application ability of this technique to dye wastewater treatment in Vietnam where textile and dyeing industries have rapidly been developed.

2 MATERIALS AND METHODS

2.1 Materials and instruments

The commercial dye used in this study was Direct Green 6 (DG6). This product was supplied from Tianjin Dyestuff Chemical Co (China). The characteristic schematic structure and general characteristics of DG6 was shown in Figure 1.

Dye solution was prepared by dissolving the dye in distilled water. The electrochemical experiments were carried out in a two electrode electrochemical cell system with a steel cathode. The anode electrodes were iron with dimensions of 60 mm × 40 mm × 2 mm. The distance between iron anode and steel cathode of pieces is 16 mm. The electrodes were connected to a DC power supply

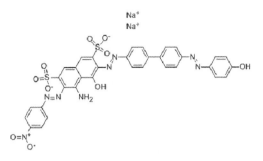

Figure 1. Molecular structure of direct green 6; molecular formula: $C_{34}H_{22}N_8Na_2O_{10}S_2$; molecular weight: 812.7 g/mol.

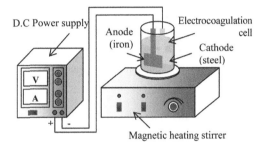

Figure 2. The diagram of electrochemical experiment.

(Matrix MPS—3030l—1) with galvanostatic operational option. Electrolytic reactor column is in diameter of 80 mm and effective volume of 400 mL.

The electrochemical experiments were set up as Figure 2. The EC unit consists of an electrochemical reactor, a D.C. power supply and electrodes. The stirrer (Magnetic heating stirrer 7–1) was used in the electrochemical cell to maintain an unchanged composition and avoid the association of the flocs in the solution.

2.2 Experimental methods

All the experiments were performed at room temperature. In each experiment, 400 mL of the dye solution was placed into the 500 mL cell. The dye concentration of 50 mg/L is used for experiments, except experiments of effect of the initial dye concentration ranging from 25–250 mg/L. Density was adjusted to a desired value and the electrolysis was started. The electrolysis concentration was adjusted to different values by addition of Na_2SO_4. The pH of the solution adjusted by adding appropriate amount of NaOH or H_2SO_4 solutions and measured by pH meter (Sartorius PB—10). Magnetic stirrer system was operated until stable temperature and constant current reached. Electrolysis time was controlled for 60 minutes; experiments were conducted at 20°C. Solution after electrolysis was measured at the wavelength corresponding to maximum absorbance λ_{max}, wich for DG6 is 621 nm using absorption spectro meter (Spectro Flex 6600).

The decolorization efficiency was calculated based on the absorbance changes

$$\eta = \frac{A_0 - A}{A_0} \times 100\% \qquad (1)$$

where η is the decolorization efficiency of dye; A_0 is the absorbance of initial solution, and A is the absorbance after decolorizing.

Operation costs cover electrode material consumption, energy, labor costs, equipment maintenance and other costs (Parag et al. 2004). Electrical energy consumption is very important economical parameters in electrochemical process (Ayhan & Mahmut. 2009). This study focused on consumed electrical energy to remove the dye. Electrical energy consumption was calculated using the equation (2):

$$E = \frac{UIt_{EC}}{1000M_{EC}} (kWh/kg\text{-}dye) \qquad (2)$$

where E is the electrical energy in kWh/kg-dye; U is the cell voltage in volt (V), I is the current in ampere (A), t_{EC} is electrolysis time in hours (h) and M_{EC} is the dye removed in the electrolytic time in kilograms (kg).

The linear equation of absorbance vs concentration (25–250 mg/L) of the dye is following

$$Abs_{621} = 0.013989C + 0.000266 \qquad (3)$$

where Abs_{621} is absorbance in abs; C is dye concentration (mg/L); correlation coefficient, $R^2 = 0.9998$.

3 RESULT AND DISCUSSION

3.1 Effect of current density on the decolorization efficiency

The dye solution with different current densities in range of 2.083–4.583 mA/cm^2 was treated in optimized dye concentration of 50 mg/L, initial pH value of 6.65, Na$_2$SO$_4$ concentration of 0.01 mol/L, and stirring speed of 600 r/min. Electrolysis time set up to 60 minutes and conducted at temperature of 20°C. The decolorization efficiency graph was plotted against relative initial dye concentration (Fig. 3).

Figure 3 shows the relationships of applied current density with the decolorization efficiency, and energy consumption in decolorization. The current density ≤ 3.333 mA/cm^2 was run and as the current

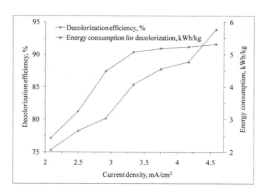

Figure 3. Effect of current density on the decolorization efficiency and energy consumption for decolorization.

density increased, decolorization efficiency rapidly increased. As can be seen from the Figure 3, the increase in current density from 2.083 mA/cm^2 to 3.333 mA/cm^2, the decolorization efficiency also significantly increased from 77.1% to 90.3%. The light increase in decolorization efficiency was determined when the current density passed 3.333 mA/cm^2 and reached 91.6% once current density was set to 4.583 mA/cm^2

The main reasons to explain this phenomenon are that the increase in current density dissolved more iron ions from anode in solution. As consequently they enhanced electrochemical reaction leading increase in color removal efficiency of the dye. When the iron ions in the electrolyte solution reached a certain concentration, dye and metal ion hydrolysis products of the condensation reaction were controlled, therefore, the increase in current density was no longer able to improve the removal efficiency of the dye.

It was notice that, energy consumption increased significantly from 2.074 to 5.765 kWh/kg-dye due to increase in current density (2.083–5.483 mA/cm^2). In general, the increase in current density is conducive to electrochemical process by ion dissolved improvement, but energy consumption is a quite problem. Therefore, the current density at 3.333 mA/cm^2 could be appropriate for decolorization of dye.

3.2 Effect of initial pH on the decolorization efficiency

In order to investigate the effect of pH value on the efficiency of color removal, electrochemical process was carried out by using different initial pH value in range of 3.12–11.56. The dye concentration of 50 mg/L, current density of 3.333 mA/cm^2, Na$_2$SO$_4$ concentration of 0.01 mol/l, stirring speed 600 r/min and temperature of 20°C were tested. The experiment results are shown in Figure 4.

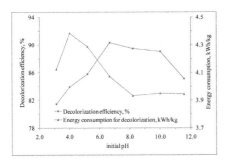

Figure 4. Effect of initial pH on the decolorization efficiency and energy consumption for decolorization.

Figure 4 shows that, pH has great influence on removal color efficiency of the dye solution. When pH of the dye solution was between 3.12 and 6.65, the decolorization efficiency increased. It increased rapidly up to 90.3% once initial pH value reached 6.65, then decreased to 85.2% when pH reached to 11.56.

Energy consumption increased with increasing in pH, energy consumption rose up to 4.384 kWh/kg dye when pH value was 4.0 and then energy consumption decreases; when decolorization efficiency reached the highest, the energy consumption was 4.073 kWh/kg and when pH value was 8.20, energy consumption dropped to 3.932 kWh/kg dye. Basically, energy consumption was maintained between 3.932–3.950 kWh/kg-dye with changes of pH.

The experimental results revealed that pH values from 6.65–10 in wastewaters are effective in removing dye color. Because high efficiency and low energy consumption for decolorization achieved in this pH condition.

3.3 Effect of initial dye concentration on the decolorization efficiency

The dye solution with different initial concentrations in range of 25–250 mg/L was treated in optimized current density of 3.333 mA/cm^2, electrolysis time set up to 60 minutes, initial pH value of 6.65, Na$_2$SO$_4$ concentration of 0.01 mol/L, stirring speed 600 r/min and temperature of 20°C. The decolorization efficiency graph was plotted against relative initial dye concentration (Fig. 5).

From Figure 5, with the increase in initial dye concentration, the decolorization efficicency and energy consumption reduced. The light decrease was indicated as the dye concentration was under 150 mg/L (91.3% to 88.7%), then dramatically dropped to 78.0% when the dye concentration increased to 250 mg/L. It was notice that, energy consumption continuously went down. It was 7.958 kWh/kg-dye as the dye concentration was 25 mg/L, then decreased to 1.307 kWh/kg-dye and 0.841 kWh/kg-dye at the

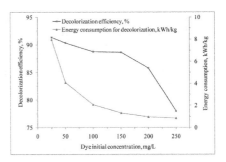

Figure 5. Effect of dye initial concentration on the decolorization efficiency and energy consumption for decolorization.

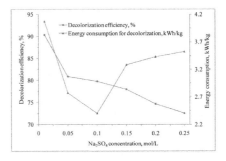

Figure 6. Effect of Na$_2$SO$_4$ concentration on the decolorization efficiency and energy consumption for decolorization.

dye concentration of 150 mg/L and 250 mg/L, respectively. Therefore, the increase in initial dye concentration led to reducing decolorization efficiency as well as energy consumption. From this study, the initial dye concentration of 150 mg/L is recommended as the most economical treatment by the electrochemical method.

3.4 Effect of Na$_2$SO$_4$ concentration on the decolorization efficiency

The experiments were conducted with dye concentration of 50 mg/L, current density of 3.333 mA/cm^2, initial pH of 6.65, stirring speed 600 r/min and temperature of 20°C, electrolysis time was set for 60 minutes and various concentration of Na$_2$SO$_4$ (from 0.01 to 0.25 mol/L). The relation of decolorization efficiency of DG6 dye and its energy consumption against Na$_2$SO$_4$ concentration was indicated in Figure 6.

As the electrolyte concentration of Na$_2$SO$_4$ increased the decolorization efficiency decreased, this is due to exceed amount of Na$_2$SO$_4$ interacted with hydrolysis group to form precipitation. Thus ionic strength contributed not only to decolorization but also to the removal of precipitated products.

The Na$_2$SO$_4$ concentration increased from 0.01 mol/L to 0.25 mol/L, the decolorization efficiency decreased from 90.3% to 72.6%. The strong decrease reached 90.3% to 80.8% when the concentration of Na$_2$SO$_4$ in range of 0.01 to 0.05 mol/L, and then slow decrease occurred later (80.8% to 72.6%).

Noticeably, the trend of energy consumption changed depending on concentration of Na$_2$SO$_4$. Energy consumption first decreased when Na$_2$SO$_4$ concentration increased from 0.01–0.1 mol/L, and it was determined 2.450 kWh/kg-dye at the Na$_2$SO$_4$ concentration of 0.1 mol/L. As Na$_2$SO$_4$ concentration passed 0.1 mol/L, the energy consumption increased.

4 CONCLUSION

In the present study, the iron anode electrochemical method was applied to decolorize simulated DG6 dye from wastewater. The effects of various factors on color removal efficiency was investigated and optimized. The results show that the current density of 3.333 mA/cm^2 is optimal for DG6 dye treatment. The DG6 wastewater decolorization reached optimum with high decolorization efficiency and low energy consumption in the pH range of 6.65 to 10. In addition, the initial dye concentration of 150 mg/L was the most effective conditions. Use Na$_2$SO$_4$ as the electrolyte brought low energy consumption but decolorization efficiency is not high, so using electrolyte of Na$_2$SO$_4$ to remove color of DG6 dye is costly.

The dye wastewater discharged from textile industrial zones of Vietnam contains chemical compounds with similar molecular structures to the DG6. Therefore, the iron anode electrochemical process investigated by the study might be feasible to decolorize dye wastewater in Vietnam.

REFERENCES

Apostolos V, Elli M.B, Sofia M. 2004. Degradation of methylparathion in aqueous solution by electrochemical oxidation. *Environmental Science & Technology* 38 (22): 6125–6131.

Ayhan S, Mahmut O. 2009. The decolorization of C.I. Reactive Black 5 in aqueous solution by electrocoagulation using sacrificial iron electrodes. *Journal of Hazardous Materials* 161(2–3): 1369–1376.

Chen G.H. 2004. Electrochemical technologies in wastewater treatment. *Separation and Purifocation Technology* 38 (1): 11–41.

Chu J.Y, Cao K.J, Wu C.D. 2007. Review of printing and dyeing wastewater processing technique. *Journal of Anhui Agri Sci* 35(7): 2041–2042, 2060.

Dai R.C, Zhang T, Guo Q, et al,. 2000. Summary of Printing—Dyeing Wastewater Treatment Technology. *Water & Wastewater Engineering* 26 (10): 33–37.

Daneshvar, N.A. Oladegaragoze, N. Djafarzadeh. 2006. Decolorization of basic dye solutions by electrocoagulation: An investigation of the effect of operational parameters. *Journal of Hazard Materials* 129 (1–3): 116–122.

Daniel S. 1997. Electrochemistry for a clearner environment. *Chemical Society Reviews* 26 (3): 181–189.

Ding S.L, Li Z.K, Wang R. 2010. Summary of treatment of dyestuff watewater. *Water resources protectiong* 26 (3): 73–78.

Guan Y.J, Yang W.S. 1999. Bipolar packed bed electrolytic reactor on the degradation of reactive dyes. *Journal of Environmental Chemistry* 18 (3): 270–273.

Gutierrez MC, Crespi M. 1999. A review of electrochemical treatments for colour elimination. *Coloration Technology* 115 (11): 342–345.

Liang J.Y, Yang Y.Z. Wang C.Z, et al,. 2007. Dye Wastewater Electrochemistry Decolorization Research. *Chinese Journal of Environment Engineering* 1 (9): 46–49.

Lin H.B, Fei J.M, Xu H, et al. 2004. Investigation on electrocatalytic oxidation tratment of the effluent wastewater in a fertilizer plant. *Industrial Water Treatment* 24 (4): 36–38.

Liu M.H. 2007. Advances in dyeing and printing wastewater treatment technologies. *Journal of Textile Research* 28 (1): 116–119.

Liu Y, Lu X.G, Zhang P, et al,. 2009. Study on Electric flocculation—flotation method of simulated dye wastewater decolorization. *Journal of East China Jiaotong University* 26 (2): 17–21.

Lorimer JP, Mason TJ, Plattes M, et al,. 2001. Degradation of dye effluent. *Pure Appl. Chem* 73 (12): 1957–1968.

Mahmut B, Murat E, Mehmet K. 2007. Treatment of the textile wastewater by electrocoagulation: Economical evaluation. *Chemical Engineering Journal* 128 (2–3): 155–161.

Marco P, Giacomo C. 2004. Electrochemical oxidation as a final treatment of synthetic tannery wastewater. *Environmental Science & Technology* 38 (20): 5470–5475.

Mehmet K, Orhan T.C, Mahmut B. 2003. Treatment of textile wastewaters by electrocoagulation using iron and aluminum electrodes, *Journal of Hazard Materials* 100 (1–3): 163–178.

Mohammad Y.A.M, Paul M, Jewel A.G.G, et al,. 2004. Fundamentals, present and future perspectives of electrocoagulation. *Journal of Hazardous Materials* 114 (1–3): 199–210.

Mohammad Y.A.M, Robert S, Jose R.P, et al. 2001. Electrocoagulation (EC)—science and applications. *Journal of Hazardous Materials* 84 (1): 29–41.

Parag R.G, Aniruddha B.P. 2004. A review of imperative technologies for wastewater treatment II: hybrid methods. *Advances in Environmental Research* 8 (3–4): 553–597.

Song W.J. 2008. Dye Wastewater Electrochemistry Decolorization Research. *Hebei Chemical* 31 (3): 66–67, 71.

Wang B, Guan Y.J, Yang W.S 2005. The electro-oxidation reduction of dye wastewater treatment. *Printing and dyeing* 11: 45–48.

Yang L.Y, Xu X, Zhu S.Y, et al,. 1998. Treatment of dyestuffs wastewater by electrolysis method of complex catalysis. *China Environment Science* 18 (6):557–560.

Frontiers of Energy and Environmental Engineering – Sung, Kao & Chen (eds)
© 2013 Taylor & Francis Group, London, ISBN 978-0-415-66159-1

Study on the movable workstation of verification on biological treatment technologies of water pollution control

W. Zhang, Q.W. Song, C.L. Xu, H.M. Huang, H.Y. Wang & J.K. Dai
Design Center of Environmental Engineering, Chinese Research Academy of Environmental Sciences, Beijing, P.R. China

ABSTRACT: The movable workstation of verification on industrial wastewater biological treatment technology is constructed to handle insufficient and not real data problem in verification on the biological treatment technologies of water pollution control. By modifying overall 20 feet standard container with strengthening the heat preservation, adding shock absorber, water supply and drainage, ventilation and air conditioning, solar power supply system, disposing on-line water quality monitoring instrument, the first movable workstation of verification is finished. According to the set verification protocols controlling sampling pumps in industrial organic wastewater treatment technology field by programming, conveying water into online instrument across the pretreatment system on time, and data obtained according to the national standard test method analysis of water quality uploaded to the server, and then integrating other water quality parameters got by other instrument inside platform, design parameters, as well as economic operation data etc., a variety of different data analysis results are obtained utilizing data processing software, finally an objective assessment conclusion is produced to the biological treatment technology according to the scientific assessment method.

Keywords: biological treatment technology; Environmental Technology Verification (ETV); pretreatment system; solar power supply system

1 INTRODUCTION

The overall level of China's environmental protection is not high, and most of the technologies are in a "usable" stage. The gap continues to expand with developed countries' advanced technologies. The problems are increasingly prominent which effluent by pollution control facilities cannot reach the national standards stably and so on. To solve these problems, China began to conduct environmental technology management system, to provide reliable technical basis for environmental management, standards formulation (amendment) and implementation, technology assessment, environmental law enforcement and supervision, to promote environmental technology innovation and guide the development of environmental protection industry through enhancing scientific, systematic and regular of environmental technology management. As an important means of environmental technology management, Environmental Technology Verification (ETV) would assess and filter environmental technologies by scientific methods and indicators, finally for the scientific decision-making for environmental management. From practices in recent years, China's ETV system is not sound to

be perfect, especially the lack of ability to support scientific and technological innovation on the environment, which is unable to meet the development needs of the environmental protection industry in the new era.

Referring to foreign environmental technology verification experience, the movable workstation of verification on biological treatment technologies of water pollution control is constructed to solve insufficient data and not real data of assessment the biological treatment technology of water pollution prevention and control. On-site monitoring of the technical process for wastewater biological treatment is realized by container transformation, online water quality monitoring system designing and data processing system setting up, which makes objective, scientific evaluation of environmental innovative technology possible.

2 NECESSITY OF BUILDING MOVABLE WORKSTATION

In China environment management system consists of three parts: the technical guidance system (the best feasibility of technical guidelines, pollution control technology policy, and environmental engineering technical specifications), the technical evaluation system, and technology demonstration and dissemination mechanisms. Because polluting industries are lack of understanding of the importance of the work of the technical assessment, business management lags behind and other reasons, technology operating data obtained through field research and data research often is not real, self-contradictory and other issues. In the evaluation of innovative environmental technologies most of domestic methods used are that government departments take the lead to convene an expert review to do it. For the lack of common criteria and basis, unclear responsibility of the expert assessment, subjective evaluation of experts and others, there are some limitations in scientific, fairness and objectivity in the assessment results, so that it is difficult to objectively and quantitatively reflect technology performance. When the technology-user makes the decision whether to adopt innovative technology, there is often a lack of accurate and reliable information support to innovative technologies, which greatly affects the promotion of innovative technology and the process of market-oriented industrialization.

Now ETV system is being constructed in our state to create a favorable market environment for development of environmental innovative technologies. ETV can accelerate environmental innovative technologies' achievements and foster new technology market. By verifying pollution control technology a lot of performance information of the technology is provided to users to shorten the communication time for users and developers. Especially for innovative technology, after verification the credibility of the technology is improved, users' confidence is enhanced, which can effectively promote the industrialization of innovative technology engineering application. The other hand, the verification can mobilize the innovation enthusiasm of enterprises, and create a virtuous circle, gradually improve the overall technological level of the national environmental protection industry.

With the continuous development of innovative environmental technologies, environmental technology assessment system is constantly improved, it is urgent to build movable workstation of verification on biological treatment of industrial organic wastewater to verify the water pollution control innovative environmental technologies in the framework of ETV system, especially to complex industrial wastewater treatment technology, which evaluate the effectiveness and practicality of the processing technology objectively and scientifically, to accelerate the application and promotion of innovative environmental technology industry.

3 MOVABLE WORKSTATION CONSTRUCTION

3.1 Container transformation

Through technical comparison, the 20 ft. container is chosen as movable workstation, with high-strength carbon steel pipe, forgings welded structure, surface sandblasting and painting. The internal area of workstation is about 13 m^2. The corrosion seal, waterproof, dustproof are taken full account of when the movable workstation is modified take, to adapt to the conditions of use. The cabinet walls, insulation material, roof, floors, etc. are to meet the requirements of the material of the chemical laboratory and emergency pollution incidents.

The movable workstation is separated laboratory module and force compartment. Laboratory module includes the experimental zone, tools areas, instrument lockers and data processing areas. Force cabin is used to place board generator, solar power control system, and air conditioning compressor. There is wall of noise-proof and thermal insulation between the two compartments, in which install the passage doors of sound and thermal insulation with double-layer vacuum glass window to facilitate the observation. There are 50 mm thick insulation layers in tops and sidewalls, the bilateral door, and ground, to reduce the influence of the external ambient temperature and noise to monitoring cabin.

3.2 Multi-point automatic continuous sampling and analysis test system

Unfamiliar with the traditional environmental monitoring car, movable workstation of verification and assessment can sample automatically, pre-treat samples, monitor synchronously online a variety of indicators of the process of verification technology, analyze and process data, through the integration of analytical instruments such as COD, total nitrogen, total phosphorus, ammonia and pH, DO, water temperature, ORP, conductivity and so on.

According to biological treatment technology verification and assessment index system and test procedures, movable workstation is allocated to online COD analyzer, online total phosphorus/total nitrogen analyzer, online ammonia analyzer and five-parameter instrument to achieve conventional indicators online monitoring of wastewater of the different processes of the craft scene, and to upload it to the monitor, for use in the verification and assessment. In addition, the integration of portable heavy metal analyzer, microscopes, ultraviolet spectrophotometer, portable BOD analyzer and other car instrument can achieve a variety of conventional indicators, animal and vegetable oils, heavy metals, toxic and hazardous pollutants, microbial phase and other special indicators which are more than 30 pieces.

Multi-point automatic continuous sampling and analysis test system mainly consists of three parts: sampling control, process monitoring, and data processing. Figure 1 shows the system structure of movable workstation of verification and workflow. By PLC programming sampling control system controls sewage pumps at sampling points to achieve order of sampling of the multiple sampling points in the entire process. After pretreatment system, output of the water quality analysis is transferred to monitoring host through the RS-485 communication interface. Process monitoring system monitors the operational status of the entire treatment process. Data processing system real-time display data curve in the different sampling sites water quality, the curve of the historical data of the different sampling points on the host. By the programming, it can display in real time on the screen effluent compliance rate, the processing efficiency of each step of the process, process stability and abnormal indicators of alarms. The data processing system analyses online water quality data, equipment operation data and economic data, in accordance with scientific processing method to arrive at a series of results, such as the compliance status of water quality parameters at different time points, rate of processing of water quality parameters within the process, the treatment effect of

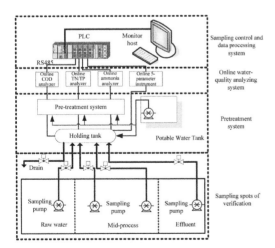

Figure 1. The system structure of movable workstation of verification and workflow. Multi-point automatic continuous sampling and analysis test system mainly includes sampling control and data processing system, online water-quality analyzing system and pretreatment system.

stability of this process in a certain period in order to verify and assess the sewage treatment process more accurately.

Pretreatment system contains two and more filtration devices. Samples first pass through a filter, filtering large particles of sediment, which the supernatant enters two or three precision filter to filter small particles in order to reach the sample requirements into instruments, and the lower containing the sediment in water samples drains from the pipe.

3.3 Multi-way power supply system

The system includes mains, solar power system and gasoline power system. Analysis test system online can be started quickly in the field without power supply. The gasoline power system generates power by gasoline, which supply of fixed equipment, lighting, equipment, etc. in movable workstations mutual backup with mains. The solar power system consists of solar panels, controllers, batteries and inverter. Double solar bracket is installed at the top of the workstation, through the cylinder to drive the rotation to the appropriate angle and to pull the top layer into a stretched state, saving space and ensuring the radiation area of solar panels. Solar panels solar energy received by the controller charge the battery, then the inverter inverse into 220 V AC, for the online monitoring instruments.

In addition, movable workstation of verification is also equipped with a 4-point support structure, security system, exhaust system, air-conditioning systems, fluctuation water system, local communication, video monitoring system and damping system to meet the basic conditions of the long-distance transport, field work, and experiments analysis, which a variety of equipment will be in the normal operating state within 20 minutes after reaching the field, to ensure the security, reliable running of the workstation, and to form the fast, reliable on-site verification and assessment capacity on industrial wastewater biological treatment technology of key industries.

4 CONCLUSIONS

The trial operation is started in one municipal wastewater treatment plant after the completion of movable workstation of verification on an innovative technology—the tail water advanced treatment technology. The verification results proved that, movable workstation of verification can monitor water quality of wastewater treatment process applying environmental new technology (ETV) on the site all the day through the development and integration of related technologies. By objective, impartial, scientific verification data and results of the environmental technical the credibility of innovative technology is improved, a reliable reference is provided for technology purchasers' decision-making, fair competition is promoted in the market, and environment management system is improved in China, environmental technology innovation and development is promoted, and the industrialization process of the outstanding environmental technology is accelerated.

The paper is supported by National Major Science and Technology Project for Water Pollution Control and Treatment of China (2009ZX07529-007).

REFERENCES

ETV Canada: Environmental Technology Verification Program on http://www.etvcanada.ca/overview.asp.

Information on http://websearch.mep.gov.cn/info/gw/huanfa/200710/W020071011272194709927.pdf.

Liu, M.Q. Zhu, D.S. & Zhou, J.Y. et al. 2007. Automatic control system of industrial wastewater pretreatment based on PLC and inverter. *Mechanical & Electrical Engineering Magazine* 24(11):43–46.

US EPA: Environmeatl Technology Verification Program on http://www.epa.gov/etv/.

Xu, C.L. Song, Q.W. & Huang, H.M. et al. 2011. Study on the Framework of Environmental Technology Verification (ETV) System in China. *Journal of Environmental Engineering Technology* 1(5):396–402.

Yi, B. Liu, Y. & Feng, Q.Y. 2007. Review of environmental protection technology in China. *Environmental Protection of Chemical Industry* 27(1):1–7.

Zhou, Z.M. & Ji, A.H. 2010. *Solar photovoltaic system design and application examples*. Beijing: Publishing House of Electronics Industry.

Frontiers of Energy and Environmental Engineering – Sung, Kao & Chen (eds)
© *2013 Taylor & Francis Group, London, ISBN 978-0-415-66159-1*

Correlation analysis on inducements of algal blooms in western Chaohu Lake

D. Guan
State Key Laboratory of Urban Water Resource and Environment, Harbin Institute of Technology, China
College of Aerospace and Civil Engineering, Harbin Engineering University, China

J. Li
Hefei Environmental Monitoring Central Station, China

L.N. Zhu
College of Aerospace and Civil Engineering, Harbin Engineering University, China

ABSTRACT: With monitor data in western Chaohu Lake, we inferred the dominant factors of algal blooms through correlation analysis. The results indicated that the Chla biomass was influenced by multiple effects of environmental factors, in which the dominant ones were atmospheric temperature, water level and COD_{Mn}. Under multifactor coupling environment, meteorologic and hydrologic factors were most significant related to algal blooms.

Keywords: Chaohu Lake; algal blooms; inducement; multifactor coupling; correlation analysis

1 INTRODUCTIONS

As one of the largest internal fresh lakes located in southern China, the Chaohu Lake was indicated that 70% of entire lake surface was covered by eutrophic water, and it might have excessive concentrations at total nitrogen and total phosphorus. Such water degradation was especially obvious in the western Chaohu Lake, and had caused serious algal blooms over 10 times in this century, which brought hazards to urban and rural drinking water safety (Water Resources Department of Anhui province 2010). The main branches (Nanfei River, Shiwuli River, Fengle River, etc.) recharge into Chaohu Lake as radial pattern with annual input quantities in about 3.7 billion m³. So the eutrophication of Chaohu Lake was influenced by the coupling effect of multiple environmental factors in water quality, weather and hydrology.

At present, most researches concerned with the relationship between the algae biomass and environmental factors in Chaohu Lake were focused on special factors with little consideration about coupling effect of multifactor (Zhong 2009, Wu 2009, Wang 2005, Chen 2006)[2–5]. By the monitor data (2005–2009) in western Chaohu Lake, correlation analysis among chlorophyll a and environmental factors was conducted in this paper, in order to indicate the multifactor coupling effect on dominant factors. The results of this paper support the

investigations of algal blooms in Chaohu Lake as theory and reference.

2 DATA AND RESEARCH METHODS

2.1 Data sources

The water quality data adopted in this paper were provided by Hefei Environmental Monitoring Central Station. These data were recorded from January 2005 to December 2009 in 6 monitor stations: Nanfeihe station (1#), Shiwulihe station (2#), Tangxi station (3#), Paihe station (4#), Xinhe station (5#), Middle station (6#). The meteorological data were provided by Hefei Meteorological Bureau; the hydrological data were provided by Hefei Hydrology and Water Resource Bureau. Chlorophyll a (Chla) was selected to express the algal biomass, and the environmental parameters with well correlative to Chla, including Total Nitrogen (TN), Total Phosphorous (TP), BOD, DO, pH, NH_3-N, COD_{Mn}, atmospheric Temperature (AT), Wind Speed (WS), Atmospheric Pressure (AP), Relative Humidity (RH), Relative Sunshine duration (RS), Water Level (WL), were chose as input variables with Chla for correlation analysis. Each parameter was determined by standard method (National Environment Protection Administration 2002).

2.2 Analysis methods

The software "SPSS.PASW.Statistics" was applied to conduct the correlation analysis between variables. Bivariate linear correlation was expressed by Pearson correlation coefficient, and the bivariate linear correlation without other variables' influence was expressed by partial correlation coefficient (Johnson & Wichern 2001).

3 RESULTS AND DISCUSSIONS

3.1 Correlation analysis of Chla concentration and environmental factors

The results of bivariate and partial correlation coefficients between Chla concentration and environmental factors in 6 monitor stations were illustrated in Figure 1. Most of bivariate correlation coefficients were appeared in the interval [−0.5, 0.5]. It indicates that the extent of linear correlation between Chla concentration and single environmental factor is relatively low in western Chaohu Lake, and the Chla concentration movement was influenced by multiple factors corporately. Based on the computing results, the atmospheric temperature, water level and COD_{Mn} performed higher correlativity to Chla concentration. Therefore, the dominant inducements of algal blooms in multifactor coupling environment were atmospheric temperature, water level and organic content in water.

Contrasted to bivariate correlation analysis, Chla concentration correlated to TN, COD_{Mn} and water level obviously in partial correlation analysis. Partial correlation coefficients reflect

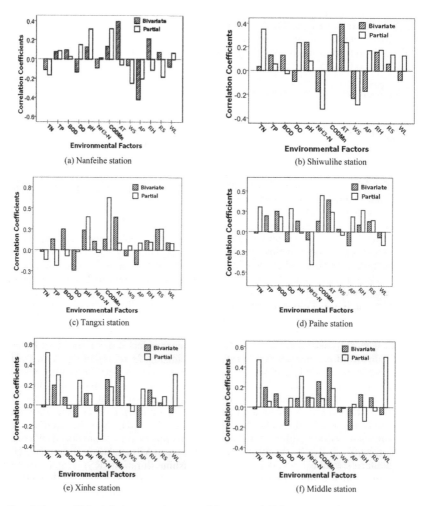

(a) Nanfeihe station

(b) Shiwulihe station

(c) Tangxi station

(d) Paihe station

(e) Xinhe station

(f) Middle station

Figure 1.   Correlation coefficients between environmental factors and Chla concentration.

the net-relationships between algal biomass and environmental factors. Because the nutrients and organic matter support the algae reproduction, the water quality factors such as TN, COD_{Mn} expressed obvious inducement function. Furthermore, water level influences the nutrition and hydrodynamic force greatly that was shown strong net-correlativity with algal blooms.

3.2 The multifactor coupling effect on dominant factors identification of algal blooms

Among the 6 monitor stations, the water quality observed in Nanfeihe River was contaminated seriously by the alongshore industrial wastewater discharge, so the nutrients nutrient and organic content of Nanfeihe inlet station were higher than others. For that background, the contaminated water could support algal growth, and weaken the effect of marked net-correlated water quality factors on algal blooms such as COD_{Mn} and pH. Besides, as the primary branch of Chaohu Lake, the Nanfeihe River provides large inlet current which may stable the water level of Nanfeihe station and reduce its influence on algal blooms. For above reasons, the correlation analysis gave the result that meteorologic factors have greater effect on Chla concentration (Fig. 1a).

The situation of Shiwulihe station was similar with Nafeihe station. The contamination content of inlet water from Shiwulihe River was relative saturation, which decreased inducement to algal blooms from water quality factors (TN, NH_3-N, COD_{Mn}) and enhanced the effect of meteorologic and hydrologic factors to algal biomass (Fig. 1b).

In the Tangxi station and Paihe station, Chla concentration was highly net-correlated to NH_3-N and pH. But the local discharge of N and P were overload in recent years, which reduced effect of aquatic nutrient loading on algal growth. As a result, the organic loading and atmospheric temperature became the obvious inducements of algal blooms (Fig. 1c and d).

For the low contamination discharge in Xinhe station and relatively strong self-purification capacity in Middle station, the water qualities of these two stations were better than others (Fig. 1e and f). Thus, the nutrient loading restricted algal biomass clearly, which led to water quality factors could influence algal blooms as similar as meteorologic and hydrologic factors.

At present, many researches pointed out that aquatic nutrient loading was the dominant inducement to algal blooms (Chapra 1997, Hai et al. 2010, Schindler 2008). However, according to the results of correlation analysis in this paper, the nutrient loading such as TN and TP preformed relatively low correlativity to Chla concentration

in western Chaohu Lake. That situation was concerned with the aquatic background in the regions surrounding monitor stations. From the year 2005 to 2009, the nutrient loading parameters in the lake could meet the requirements of algae growth, which included TN > 1.17 mg/L, TP > 0.09 mg/L, COD_{Mn} > 4.7 mg/L, detailed water quality parameters were summarized in Figures 2–4.

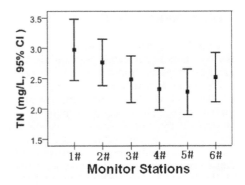

Figure 2. TN movement in various stations of western Lake Chaohu (2005–2009).

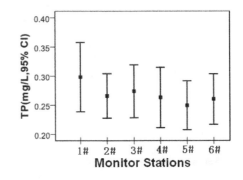

Figure 3. TP movement in various stations of western Lake Chaohu (2005–2009).

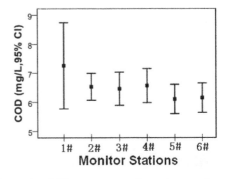

Figure 4. COD_{Mn} movement in various stations of western Lake Chaohu (2005–2009).

87

In this case, the impact of inlet nutrient loading movement was decreased. Furthermore, the local climate is warmer at summer and the water temperature changes in the range of 25~30°C that is much suitable for the blue-green algae reproduction, so the environmental factors such as organic loading and atmospheric temperature became the marked inducements of algal blooms.

4 CONCLUSION

1. The result of correlation analysis indicates that the correlation coefficients between algal biomass and single environmental factor were relatively low in western Chaohu Lake from 2005 to 2009, and the algal biomass movement was influenced by multiple factors corporately.
2. Bivariate correlation analysis pointed out that the atmospheric temperature, water level and COD_{Mn} performed higher correlativity to Chla concentration, so the dominant inducements of algal blooms were meteorologic and hydrologic factors. And partial correlation analysis showed that TN, COD_{Mn} and water level obviously netcorrelated to algal blooms.
3. Under the multifactor coupling effect in natural environment, the nutrient loading such as TN and TP preformed relatively low correlativity to Chla concentration by the varied aquatic background of monitor stations in western Chaohu Lake. On the contrary, organic loading and atmospheric temperature were dominated inducements of algal blooms.

ACKNOWLEDGMENT

This research was supported by following research grants: (1) Funded by Open Research Fund Program of State Key Laboratory of Urban Water Resource and Environment, Harbin Institute of Technology (ESK201021). (2) Fundamental research funds for the central universities (HEUCF100213). (3) Funded by Open Research Fund Program of State Key Laboratory of Water Resources and Hydropower Engineering Science (2010B077).

The author would like to thank the corresponding staff in the departments of Hefei Meteorological Bureau, Hefei Hydrology and Water Resource Bureau, Hefei environmental monitoring central station for their great help in data collection and complication during this study.

REFERENCES

Chapra, S.C. 1997. Surface Water-quality Modeling. The McGraw-Hill, New York.

Chen, Y., Yin, F. & Lu, G. 2006. The catastrophic model of water bloom: a case study on Lake Chaohu. ACTA ECOLOGICA SINICA. 26(3):878–883. (in Chinese).

Hai, X., Paerl, H.W. & Qin, B., et al. 2010. Nitrogen and phosphorus inputs control phytoplankton growth in eutrophic Lake Taihu, China. Limnology and Oceanography. 55: 420–432.

Johnson, R.A. & Wichern, D.W. 2001. Applied Multivariate Statistical Analysis. Beijing: Pearson Education North Asia Limited and Tsinghua University Press.

National Environment Protection Administration. 2002. Water and Wastewater Monitoring Analysis Method (3rd Edition). Beijing: China Environmental Science Press. (in Chinese).

Schindler, D.W. & Vallentyne, J.R. 2008. The Algal Bowl: Overfertilization of the World's Freshwaters and Estuaries. Edmonton, Alberta, Canada: University of Alberta Publications.

Wang, C., Cao, Y. & Wang, H. 2005. Analysis of Eutrophication in the West Part of Chaohu Lake. Journal of Anhui Agri.Sci., 33(8):1475–1476. (in Chinese).

Water Resources Department of Anhui province. 2010. Plan of aquatic ecosystem protection and remediation for the Hefei city. (in Chinese).

Wu, Z. 2009. Quantitiative Analysis of the Impact Factors on the Eutrophication of Western Chaohu-Lake. Anhui Agricultural University. Master Degree Thesis. (in Chinese).

Zhong, L. 2009. The Study on Nitrogen Occurrence Characteristic in Water-Sediment System and Their Relationship with Algal Bloom in Chaohu Lake. Chinese Research Academy of Environmental Sciences. Master Degree Thesis. (in Chinese).

Frontiers of Energy and Environmental Engineering – Sung, Kao & Chen (eds)
© 2013 Taylor & Francis Group, London, ISBN 978-0-415-66159-1

Influence of polycarboxylate on thermal properties of cementitious solar thermal storage materials

H.W. Yuan, Y. Shi, C.H. Lu, Z.Z. Xu, Y.R. Ni & X.H. Lan
State key Laboratory of Materials-oriented Chemical Engineering, College of Materials
Science and Engineering, Nanjing University of Technology, Nanjing, China

ABSTRACT: In the current study, influence of different contents of polycarboxylate on the mechanical and thermal properties of cementitious thermal storage materials was investigated. Compressive strength and thermal properties including thermal productivity, volume heat capacity and thermal expansion coefficient of hardened aluminate cement pastes with different contents of polycarboxylate were investigated to pursue the optimum material design for cementitious thermal energy storage system of solar parabolic trough power plant. The results show that polycarboxylate plays an important role on mechanical and thermal properties of cementitious thermal storage materials. After heat treatment at 350 °C for 6 h, compressive strength and thermal properties descended in a certain extent. XRD and FTIR were used to characterize the evolution of hydration products together. MIP was used to characterize pore distribution and porosity.

Keywords: thermal energy storage; cement; polycarboxylate; solid sensible heat

1 INTRODUCTION

In recent years, the elevated cost of fossil fuels along with the environmental concerns that they represent has fostered the development of new technologies to make use of renewable energy resources. Nowadays, solar energy thermal power generation is an attractive way to produce electricity hardly with any polluting or emissions of carbon dioxide.

Four main sections are required in the solar thermal power plant: concentrator, receiver, transport/storage media system, and power conversion device. There is no sunshine at night and limited solar energy available on cloudy days, so Thermal Energy Storage (TES) systems balancing energy supply are designed to collect more solar energy during a sunny day, which is a necessary component of four sections. Typical storage media for sensible heat consist of molten nitrate salt, rocks and pebbles, or concrete (Laing D. et al., 2010; Antoni G. et al., 2010).

The low cost of concrete makes it a desirable storage material but has a drawback of poor thermal properties and thermal stabilities (Tamme R. et al., 2004; Fernandez A.I. et al., 2010; Laing D. et al., 2006; Laing D. et al., 2008; Tamme R. et al., 2005). The problems for using concrete as an internal heat exchanger are material thermal expansion differences, excess pore vapor pressure during heating up, low thermal conductivity and low volume heat capacity. In traditional cement research, several kinds of water reducing agent are commonly introduced in cement mixtures to improve cement mechanical properties. Among them High Performance Polycarboxylate (HPP) has better performances and lower cost (Plank J. et al., 2010).

Water reducing agent plays an important role on the dispersion condition of water and cement, which greatly affect the properties of pastes (Yamada K. et al., 2001). Consequently, we here aim to provide the compressive strength and thermal properties of aluminate cement hydration products along with variation of HPP content. In addition to those properties, Mercury Intrusion Porosimetry (MIP), X-Ray powder Diffraction (XRD) and Fourier Transform Infrared spectroscopy (FTIR) were obtained to characterize the porosity, the hydration phases, and the bonds of the pastes, respectively.

2 EXPERIMENTAL

2.1 Materials and mixtures

Aluminate cement performing better corrosion resistance than Portland cement was used as cementing agent. Chemical compositions of aluminate cement are given in Table 1.

Table 1. Chemical compositions of aluminate cement (wt%).

Materials	CaO	SiO₂	Al₂O₃	Fe₂O₃	R₂O	LOI
Aluminate cement	38.79	7.17	51.68	2.07	0.29	0.30

High Performance Polycarboxylate (HPP) numbered for PCA-II was used for reducing the dosage of the water and effective in enhancing the dispersion properties of the paste. The structural formula of HPP is showed as follows:

$$
\left[-CH_2-CH- \right]\left[-CH_2-\underset{COOH}{\overset{H}{C}}- \right]\left[-CH_2-\underset{CH_3}{\overset{CH_2SOONa}{C}}- \right] \quad ---(1)
$$

with CO(OCH₂CH₂)nOCH₃

2.2 Preparation of characterization of specimens

To prepare the specimens, HPP mixed with water was added to aluminate cement. Pure paste was named as 0 J and the pastes with 0.2 wt%, 0.5 wt%, and 1 wt% HPP were named as 0.2 J, 0.5 J and 1 J, respectively. At the same time, the corresponding water/cement ratios were 0.30, 0.25, 0.23 and 0.20. The pastes cast for compressive strength, thermal conductivity and thermal expansion coefficient had moulds of 2 cm × 2 cm × 2 cm, 4.8 cm × 2 cm × 8 cm, 0.5 cm × 0.5 cm × 4 cm, respectively. After being moulded, the specimens were cured under saturated alkali solution at room temperature for 7 days and then to be ready to test. Compressive strength was obtained by automatic pressure test machine (HualongWHY-200, Hualong Ltd., China). In order to eliminate the influence of free water on pastes, the specimens should be dried at 105 °C for 24 h in the oven. Thermal conductivity and volume heat capacity were measured by thermal conductivity constant tester (TSP2500, Hot Disk Ltd., Sweden), and thermal expansion coefficient was measured by thermal expansion coefficient apparatus (PCY-II, Xiangtan xiangyu instrument Ltd., China). The specimens were then kept in muffle furnace at 350 °C for 6 h for heat treatment. After heat treatment mechanical properties and thermal properties of the pastes were tested again. Mercury Intrusion Porosimetry (MIP) (PM-60-GT, Quantachrome Ltd., America) was used to obtain pore distribution and the porosity of the pastes. XRD (Rigaku D/Max-2500, Rigaku Ltd., Japan) and FTIR Spectroscopy (Nexu s670, Nicolet Ltd., America) were used to characterize the evolution of phases and structure, respectively.

3 RESULTS AND DISCUSSION

3.1 Mechanical properties

Compressive strength of pure paste and pastes with three contents of HPP before and after heating is showed in Figure 1. It can be seen that compressive strength is improved obviously with increase of HPP content, which displays the same regularity as traditional Portland cement with HPP. In order to ensure the durability of thermal energy storage process, cementitious storage material should have better thermal shock resistance. The addition of HPP would greatly improve the dispersion of the paste and gradually reduce the water content. After heating the pastes with more HPP seemed higher compressive strength, which is favorable to mechanical properties of thermal energy storage material. When HPP content is up to 1%, compressive strength comes to 48.63 MPa, which seems much higher than pure paste without heating.

3.2 Thermal properties

3.2.1 Thermal conductivity

Thermal conductivity of pure paste and pastes with three contents of HPP before and after heating is showed in Table 2. It is observed that thermal

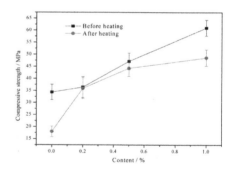

Figure 1. Compressive strength of pastes with different contents of HPP before and after heating.

Table 2. Thermal conductivity of pastes with different contents of HPP before and after heating.

Sign	Thermal conductivity ($W \cdot m^{-1} \cdot K^{-1}$)	
	Before heating	After heating
0 J	0.741	0.405
0.2 J	1.002	0.504
0.5 J	1.091	0.544
1 J	1.187	0.621

conductivity increases with the increase of the content of HPP. The addition of HPP would enhance the density of paste from the compressive strength data and hardly change the material compositions. It is known that thermal conductivity of liquid water is approximately 0.5 $W \cdot m^{-1} \cdot K^{-1}$, which is lower than prepared solid materials. Consequently, before heating the increase of the HPP content with the decrease of the water content is one of reasons for thermal conductivity increase. After heating at 350 °C, the bound water evaporated. The hydration product both changed the phases and structure. When HPP content is up to 1%, thermal conductivity comes to 0.621 $W \cdot m^{-1} \cdot K^{-1}$, which increases by 35% than pure paste after heating.

3.2.2 Volume heat capacity

Volume heat capacity of pure paste and pastes with three contents of HPP before and after heating is showed in Table 3. It can be seen that with the increase of HPP content, volume heat capacity gradually increases. When HPP content increases to 1 wt%, volume heat capacity is to 2365 $kJ \cdot m^{-3} \cdot K^{-1}$. After heating volume heat capacity of all pastes display a decline. It is known that volume heat capacity of inorganic solid materials is related to inherent materials compositions and material porosity. Decrease of volume heat capacity after heating may be due to inherent materials compositions. However, volume heat capacity still has a bit increases after adding HPP to the pastes, which is available for the optimization of the performance.

3.2.3 Thermal expansion coefficient

Thermal expansion coefficient of pastes with different contents of HPP before heating and after heating is shown in Figure 2. The results indicate that after addition of HPP thermal stability seems better than the pure paste. The pastes without preheating present unstable with the rise of temperature, while the ones suffered heat treatment can keep stable.

3.3 Characterization and analysis

3.3.1 MIP

The pore size distribution and the porosity of pure paste and pastes with three contents of HPP before and after heating are showed in Figure 3 and Table 4, respectively. It can be seen that by the addition of HPP, the pore size refines (Fig. 2), which shows that the density of cement pastes is increased and the pore structure is improved. It is shown that with increase of HPP content the porosity of pastes is decreasing (Table 4), which means that the pastes structure becomes more compact. When the HPP content is up to 1%, the porosity of the pastes comes to 23.26%. After heating at

Table 3. Volume heat capacity of pastes with different contents of HPP before and after heating.

| Sign | Volume heat capacity ($kJ \cdot m^{-3} \cdot K^{-1}$) | |
	Before heating	After heating
0 J	1785	1384
0.2 J	2140	1421
0.5 J	2295	1523
1 J	2365	1615

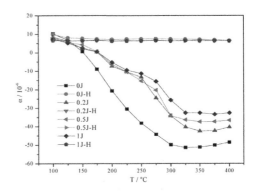

Figure 2. Thermal expansion coefficient of pastes with different contents of HPP before heating and after heating.

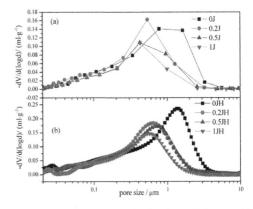

Figure 3. The pore size distribution of pastes with different contents of HPP (a) before heating (b) after heating.

350 °C the porosity of all pastes is increasing, which results in the decrease of compressive strength and the variation of thermal properties.

3.3.2 XRD

XRD patterns of pure paste and pastes with three contents of HPP before and after heating are

Table 4. Porosity of pastes with different contents of HPP before and after heating.

Sign	Porosity (%)	
	Before heating	After heating
0 J	29.80	36.43
0.2 J	26.93	29.67
0.5 J	24.80	28.98
1 J	23.26	25.40

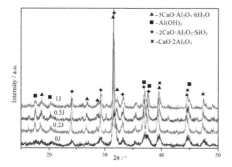

Figure 4. XRD patterns of pastes with different content of HPP before heating.

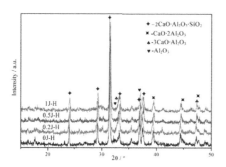

Figure 5. XRD patterns of pastes with different content of HPP after heating.

shown in Figures 4 and 5, respectively. The results indicate that after addition of HPP the degree of crystallization is improving. After heating at 350 °C for 6 h the characteristic peaks of hydration products such as C_3AH_6 and $Al(OH)_3$ disappear, which means that the decomposition reaction of the pastes with HPP still takes place.

3.3.3 FTIR

The FTIR spectra (Fig. 6) for pure paste and pastes with three contents of HPP before and after heating are discussed in this section. Combined with standard infrared spectra (Yang N.R. & Yue W.H. 2000),

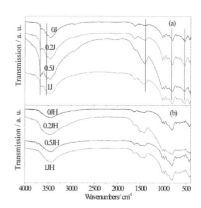

Figure 6. FTIR of pastes with different content of HPP (a) before heating (b) after heating.

for the pure paste, before heating the characteristic peaks of C_3AH_6 are 3662.3 cm⁻¹, 807.9 cm⁻¹, 525.7 cm⁻¹ and the characteristic peaks of $Al(OH)_3$ are 3526.2 cm⁻¹, 3455.5 cm⁻¹, 1019.9 cm⁻¹. After heating the intensity band of C_3AH_6 at 3662.3 cm⁻¹ and the O–H–O stretching vibration of $Al(OH)_3$ at 3526.2 cm⁻¹ both disappear. The characteristic peaks of C_3AH_6 at 525.7 cm⁻¹ and 807.9 cm⁻¹ are blue-shifted to 527.5 cm⁻¹ and 811.7 cm⁻¹, respectively, declaring expelling of the chemical bonded water of C_3AH_6. Meanwhile, the O–H–O stretching vibration and the –OH bending vibration of $Al(OH)_3$ at 3455.5 cm⁻¹ and 1019.9 cm⁻¹ are red-shifted to 3444.2 cm⁻¹ and 1016.4 cm⁻¹, respectively. Ca^{2+} in the pore liquid is easy to be carbonized. And with the increase of HPP content, the electrostatic bonding between Ca^{2+} and carboxyl groups gradually enhances, which display the wider peak at 1420 cm⁻¹. After heat treatment at 350 °C, the bonding between Ca^{2+} and carboxyl groups gradually split.

4 CONCLUSIONS

Mechanical and thermal properties of pure paste and pastes with different contents of polycarboxylate were investigated in the paper. It indicates that compressive strength increases greatly with the addition of polycarboxylate. Thermal conductivity and volume heat capacity are also improved with addition of polycarboxylate. With addition of polycarboxylate content the density of cement pastes is increased and the pore structure is improved. The properties are improved compared to pure paste. Consequently, aluminate cement pastes with polycarboxylate will be promising thermal energy storage materials. Certainly, more work about further improving mainly volume heat capacity of heat treated pastes still need to be carried out.

ACKNOWLEDGEMENT

The authors would like to express sincere thanks to Jiangsu innovation scholars climbing project (SBK200910148), Graduate Innovation Foundation of Jiangsu Province (CXLX11_0347) and Priority Academic Program Development of Jiangsu Higher Education Institutions (PAPD) for Financial Support.

REFERENCES

Antoni G. et al., 2010. State of the art on high temperature thermal energy storage for power generation. Part 1—Concepts, materials and modellization. *Renew Sust Energ Rev* 14(1): 31–55.

Fernandez A.I. et al., 2010, Selection of materials with potential in sensible thermal energy storage. *Sol Energ Mat Sol C* 94(10): 1723–1729.

Laing D. et al., 2006. Solid media thermal storage for parabolic trough power plants. *Sol Energ* 80(10): 1283–1289.

Laing D. et al., 2008. Solid media thermal storage development and analysis of modular storage operation concepts for Parabolic Trough Power Plants. *J Sol Energy Eng* 130(1): 011006 (011005).

Laing D. et al., 2010. Economic Analysis and Life Cycle Assessment of Concrete Thermal Energy Storage for Parabolic Trough Power Plants. *J Sol Energ-T Asme* 132(4): 041013.

Plank J. et al., 2010. Fundamental mechanisms for polycarboxylate intercalation into C_3A hydrate phases and the role of sulfate present in cement. *Cement Concrete Res* 40(1): 45–57.

Tamme R. et al., 2004. Advanced thermal energy storage technology for parabolic trough. *J Sol Energ Eng* 126(2): 794–800.

Tamme R. et al., 2005. Thermal energy storage technology on industrial process heat process heat applications. *Proceedings of the International Solar energy Conference* 417–422.

Yamada K. et al., 2001. Controlling of the adsorption and dispersing force of polycarboxylate-type superplasticizer by sulfate ion concentration in aqueous phase. *Cement Concrete Res* 31(3): 375–383.

Yang N.R. & Yue W.H. 2000. The handbook of Inorganic metalloid materials atlas. *Wuhan industrial university press* 11.

Frontiers of Energy and Environmental Engineering – Sung, Kao & Chen (eds)
© 2013 Taylor & Francis Group, London, ISBN 978-0-415-66159-1

The anti-negativity landscape around & under the viaducts transforming lives: Taking the Huangpu area in north of Wuhan for example

T.X. Xi, W.K. Qin & W. Wan

College of Architecture and Urban Planning, Huazhong University of Science and Technology

ABSTRACT: Combining with the public expectation, this paper summarizes the main problems of the planning and construction around & under the viaducts. Against the biggest noise pollution problem, anti-noise is the core point of view of this anti-negativity landscape design around & under the viaducts. Based on anti noise theory, it establishes an assessment model to evaluate the daily average variation of the integrated noise in the site, then to determine the efficient and economic green barrier line. Finally through the detailed design of landscape, it perfects pedestrian transport links and creates open space with different features to enhance the value of the land and achieve the anti-negative landscape around & under the viaducts.

1 BACKGROUND AND MAIN PROBLEM

Every year there is a lot of urban land with the characteristics of high efficiency space use covered by the new viaducts. For example, new viaducts occupied over 60 hectares of land in Wuhan city, 2009. Though these viaducts help alleviate traffic jams and improve the efficiency of urban travel, it brings many problems especially for the people living around the viaducts. And combining with the results of a public opinion survey on the needs of the urban construction organized by the government (Fig. 1), the negative effects of the viaducts mainly include the following points (Fig. 2).

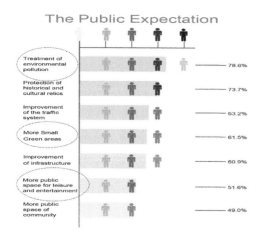

Figure 1. The public expectation.

Figure 2. The negative effects of the viaducts.

1.1 *The biggest problem is noise pollution of traffic*

On one hand, it is because the big traffic flow surrounding the elevated; on the other hand, types of native noise may include noise from the elevated roads and bridges, as well as from light rails. How to protect the city from the varied noises becomes a main problem for future urban design and planning.

1.2 *Pedestrian connections across the ground need improving*

The city viaduct as the three-dimensional vehicular carrier, the aim of its land space segmentation is to facilitate the dealers, which lead to the poor ground transport links around & under the viaducts and to the walking security risks. So it needs unified pedestrian planning to perfect pedestrian connections and effectively guarantee the walking security by using varied walking transport forms.

1.3 *Fragmentation of landscape and a lot of wasteland produced by the viaduct construction*

Landscape challenges are due to fragmentation of the landscape by viaducts. For instance, the space around and under the viaducts has the changing height and its interface is not continuous. What's more, the land around and under the viaducts becomes not suitable to plant growth, mostly because of excessive construction waste and automobile exhaust pollution.

2 CASE OVERVIEW AND DESIGN OBJECTIVE

2.1 Location and current situation

The project is located in the Huangpu area in north of Wuhan in one of the new metropolitan planning sites planned by the Wuhan Metropolitan Planning and Design Institute (Fig. 3).

The focus area is largely fragmented by various forms of transportation. A 6 lane road with emergency lanes and center median (about 40 meters wide) cuts the fabric of the area in half. An elevated rail line and light rail are planned to transverse the area and will further separate the community. Two high voltage power lines run the length of the project area from East to West, so it set up a greenbelt of 160 m to provide a measure of safety for the population (Fig. 4 and 5).

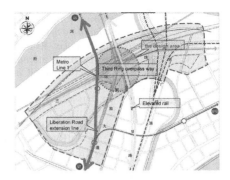

Figure 5. Site design range and analysis.

2.2 Design objective

Create a low-noise and livable space, it needs to set up the anti noise green belt controlled line with suitable plants.

Achieve convenient, safe and available connections among the different areas around the viaducts.

Change the negative space into attractive places for outdoor activity, parking or ecology to complete the city green infrastructure.

3 ANTI NOISE THEORY

3.1 Noise attenuation theory

3.1.1 Railway traffic noise

Buffer zone on both sides of the railway is generally 30 meters, the general trend of design principles to protect people's health from noise (Lili Chen, 2008).

3.1.2 Elevated road noise

The height of the elevated road is usually 8–10 meters. The influence of the traffic noise has something to do with the distance of the buildings to the road, as well as the height of the building itself (Changsheng Zhang, 2009). The noise has relatively less influence on the buildings below 4 stories. While for higher buildings, the influence is bigger, especially for the middle and higher part of the building (An-ting Zhang & Yi-ting Zhang, 2005).

3.2 Anti-noise measures

3.2.1 Anti-noise barrier designed with viaduct body

The height of the barrier should not below 1 meter, Otherwise fail to noise requirements, but not higher than 3 meters (Yan Liang, 2003), this would be a waste of materials also affects the aesthetic, appropriate length 20 h (h for height).

Figure 3. Site location.

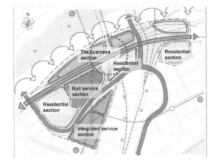

Figure 4. Site land use planning.

3.2.2 Reasonable planning around the elevated area for land use and building height

Overpass in the horizontal distance, 30 m away from the viaduct, traffic noise is greatly attenuated. At this point if more plants were planted in the buffer zone as a green sound barrier, will achieve better noise reduction effect, but also beautify the environment. For example, planting the pine forest in the median strip between 30 m and 20 m, will be able to reduce noise by 6 dB (A), coupled with natural attenuation of 3 dB (A), the reduced noise amounted to 9 dB (A), then Buildings near the road will suffer a significant reduction of noise pollution (Fig. 6) (Jianyao Wu, 1999).

In the vertical, the noise of the building the following four little influence over the noise of the four large, influenced by the location of a building in the middle and higher positions (Fig. 7) (Xiang Zhang, 2005).

3.2.3 Anti noise plant materials

Green sound barrier effect has something to do with the species of trees and its height, generally a tree is 1 m higher than the sound source.

Different plant materials have different sound deadening properties. Pine forest belt can make the 1 000 Hz frequency of the noise every ten meters to reduce 3.0 dB (A). Every 10 m of Cedar forest can reduce noise by 2.8 dB (A). Every 10 m of Pagoda forest can reduce the sound by 3.5 dB (A). Every 10 m of Grasslands (30 cm high) can reduce noise by 0.7 dB (A).

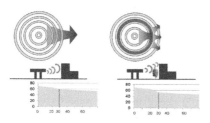

Figure 6.   The decay of the noise on the horizontal.

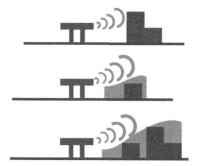

Figure 7.   The decay of the noise on the vertical.

4   ASSESSMENT MODEL OF THE ANTI NOISE GREEN BARRIER RANGE

4.1   Assessment model conditions

Considering the noise from the light rail, elevated road and high-speed railway, we could ignore the noise from the ground traffic. The assessment of noise for each noise source, the maximum amount of noise, range and laws were based on experimental data and literature.

4.2   Analysis process

Every type of noise takes longitudinal section with the center point of road noise source and noise source volume history of the average maximum amount of noise, by 10 times or 7.5 times from the external radiation, as the expression pattern of continued decay.

This evaluation is to use the data model under ideal conditions and the superposition of wave theory, reasoning the daily average of the comprehensive range of noise and trends may occur in the site (Figs. 8–11).

4.3   Green line control according to the result

The evaluation of data obtained depends on the noise attenuation theory and experimental data to determine control area of the green belt where the residence is prohibited within the building (Fig. 12).

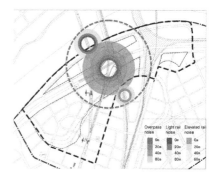

Figure 8.   The radiation range of different noises in the site.

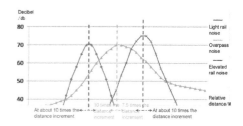

Figure 9.   The variation curve of every type of noises.

Figure 10. The variation curve by the superposition of different types of noises.

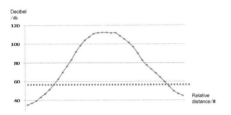

Figure 11. The daily average variation curve of the integrated noise.

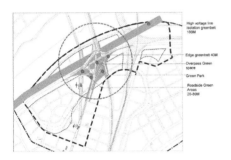

Figure 12. Anti-noise green barrier control.

5 PRODUCTS AND CONCLUSIONS

Taking Anti-noise landscape measures create a livable environment for citizens living around the elevated. A city-life pedestrian circle links the different parts around the viaducts, which are designed with the green space such as leisure square and parks. This design provides the negative areas around the viaducts with a way to realize the sustainable and economic development. It includes creating an anti-noise environment, improving the landscape and perfecting the connections of pedestrian and bicycles transport (Fig. 13).

Paying attention to the space around & under the viaducts contributes to create livable urban environment especially in our country at the present stage, the construction of the viaducts is at the climax stage, and the negative space around & under the viaducts will appear large number. The anti noise research will further strengthen the

Figure 13. The detailed landscape design.

anti-negativity landscape standpoint. It not only encourages ecological priority and economical use of land by maintaining the natural environment and designing ecological corridor, to achieve the natural environment and artificial environment symbiosis, but also advocates people-oriented, block connectivity by facilitating people to use as the basic criteria to structure the transport link between the lands separated by viaducts.

Overall, anti-negativity landscape around & under the viaducts has the concept of transformation, innovation and redevelopment of the negative area. The key of innovation is suitable land plan combined with the point of view of anti noise, so that does not affect the main transport function of viaducts, and also fit in with the needs of the society and the characteristics of land.

ACKNOWLEDGEMENTS

Supported by the National Natural Science Foundation of China (key program) under Grant No. 51078159 and Sino-US joint design, the authors are gratefully acknowledged. The authors also thank Prof. Min Wan (corresponding author) for the valuable suggestions.

REFERENCES

Changsheng Zhang. Noise control of rails and Viaduct area analysis [J]. Science & Technology Information, 2009, (29): 1037.

Jianyao Wu. Analyze and Forecast City Viaduct Traffic Noise [J]. Fujian Environment, 1999, (04): 25–28.

Lili Chen, The characteristic analysis and prediction of noise of railway station: master degree thesis, 2008.06.10.

Xiang Zhang. The traffic noise control of fly over [A]. Tenth National Conference on Noise and Vibration Control Engineering, Proceedings) [C], 2005.

Yan Liang, Research on Traffic noise of road and bridge: thesis for master degree. 2003.03.22.

ZHANG An-ting; ZHUANG Yi-ting. Study on Traffic Noise and Countermeasures of Viaducts and Overpasses [J]. Research of Environmental Sciences, 2005, 18(06) p 120–125.

Frontiers of Energy and Environmental Engineering – Sung, Kao & Chen (eds)
© 2013 Taylor & Francis Group, London, ISBN 978-0-415-66159-1

^{39}Ar-^{40}Ar dating of Gejiu tin deposit and its origin, Yunnan, China

Y.S. Li
Faculty of Land Resource Engineering, Kunming University of Science and Technology, Kunming, China

S.C. Li
Faculty of Land Resource Engineering, Kunming University of Science and Technology, Kunming, China
Bureau of Land and Resources of Jianshui, Jianshui, Yunnan, China

Y. Cai
Faculty of Application Technology, Kunming University of Science and Technology, Kunming, China

N. Chen, J.J. Chen, L. Wang, Y.K. Zhang & D.Q. He
Faculty of Land Resource Engineering, Kunming University of Science and Technology, Kunming, China

ABSTRACT: Gejiu tin ore deposit is a famous tin-polymetallic deposit throughout the world because of its old mining history and its enormous metal reserves. In addition to tin metal, there are copper, lead, zinc, silver, iron, sulphur, tungsten, bismuth, indium and rare earth elements. It was believed that there mainly are skarn-type tin deposit, stratiform tin deposit and basalt-type copper deposit in Gejiu tin poly-metallic orefield. The stratiform tin deposit are distributed in Lutangba, Malage and Huangmaoshan, which are hosted by carbonate rocks of Gejiu formation in Middle Triassic Series. In this paper, ^{40}Ar-^{39}Ar dating of cassiterite from the sratiform tin deposit in Lutangba yields plateau age of 202.18 ± 2.35 Ma and isochron age of 206.81 ± 3.23 Ma respectively. The ages are obviously older than those of the ore of the skarn type deposit of the Yanshanian epoch. The mineralization is the seabed exhalative hydrothermal sedimentary mineralization of the Indosinia epoch.

Keywords: Sn deposit; ^{40}Ar-^{39}Ar dating; exhalative hydrothermal sedimentary mineralization of the Indosinia epoch; origin; Gejiu, south-east Yunnan in China

1 INTRODUCTION

Gejiu tin ore deposit is a famous tin-polymetallic deposit throughout the world because of its old mining history and its enormous metal reserves, which is one of the largest production and export bases of tin metal in China. In addition to tin metal, there are copper, lead, zinc, silver, iron, sulphur, tungsten, bismuth, indium and rare earth elements. According to the statistics, Gejiu tin ore deposits have produced total quantity of metal of 1.35 billion tons up to 2009. Gejiu tin ore deposits were formed in the granitic mineralization of the Yanshanian epoch, whose age of mineralization was from 56 Ma to 88 Ma (Wang Zifen, 1983; Zhuang Yongqiu et al, 1996). According to tectonic evolution, mineralization and geology of mineral deposit, Qin dexian et al (2004) thought that mineralization of Gejiu tin deposits may suffer mineralization of three stage. Zhang huan et al (2007) thought that the Gejiu super-large Sn-polymetallic ore deposit is a multi-genesis deposit. In this paper, the authors study the dating of the

deposit using ^{40}Ar-^{39}Ar method of cassiterite of stratiform orebody.

2 REGIONAL GEOLOGIC SETTING OF GEJIU TIN DISTRICT

Gejiu tin deposits are located in Western Youjiang Geosyncline Fold Belt of Southern China Geosyncline Fold District. From Proterozoic to Triassic, a layer measured at 24 km was deposited in the region, and only during the Middle Triassic had there been deposited about 3 km-thick carbonate rocks, which provided a wide space for later ore deposition. There are four sets of structures which extend NE, EW, SN and NW in the mining district. The most important structures in this area are the Wuzishan anticlinorium and the Jiasha syn-clinorium. There are Triassic and Permian strata in Gejiu tin district. Gejiu tin district is divided into eastern part and western part by Gejiu Fault, while Gejiu tin deposits distribute in the eastern part (Fig. 1).

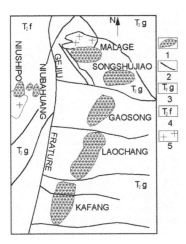

Figure 1. Schematic map for the main orefields distribution of Gejiu tin-polymetallic diggings 1-Main Orefield; 2-Fracture; 3-Gejiu Formation of Middle Triassic Series; 4-Falang Formation of Middle Triassic Series; 5-Granite.

Mineralization in the Gejiu district is found in several separate deposits, including Kafang, Laochang, Gaosong, Songshujiao and Malage (Fig. 1). The main host rocks in the mining district are limestone, dolomitic limestone and dolostone. It is found that above 90% of the total proven reserve was hosted in the layer of Gejiu formation in Middle Triassic Series ($T_2 g$) (Zhuang Yongqiu et al, 1996). There have been three magmatic cycles known as the Variscian, the Indosinian epoch and the Yanshanian epoch. The Variscian and Indosinian cycles are marked by volcanic extrusion, which gave rise to the formation of alkalic and calc-alkalic basalts and acidic volcanic rocks representing deep magmatic series.

In Gejiu eastern district, there is a strong re-melting granitic activity of the Yanshanian epoch besides basic volcanism of the Indosinian epoch. According to Wang's Rb-Sr isotopic age data (Wang Zifen, 1983), the ages of Malage-Songshujiao biotite granite and Laochang-Kafang granite are 87.83 ± 3.1 Ma and 86.3 ± 2.3 Ma respectively.

There are three types of orebody in the Gejiu Sn-polymetallic ore deposit, which are stratiform sulphide orebody, skarn sulphide orebody and vein-like orebody.

In the stratiform sulphide orebodies, the distance between the orebodies and the granite may range from tens meter to hundreds of meter. The attitude of the orebodies is controlled mainly by Strata, lithology and structure, which contain stratiform and stratoid, with a single layer or several layers, lenticular, string-of-bead-1ike, banded, tubular and irregular orebodies. Most of the orebodies have

been oxidized. The main ore minerals are pyrrhotite, pyrite, cassiterite, arsenopynte, galena, marmatite and minor chalcopyrite. The oxide ore minerals are limonite, hematite, cerusite pyrolusite, etc.

3 EXPERIMENTAL PROCEDURE AND METHODOLOGY

In terms of the results of observation for the grain sizes of casserite under microscope, the ore samples were crushed as fine as 80–100 mesh, sieved, washed by hands and dried in air. The samples were irradiated in three separate packages (M178, 181, 182) in the Isotope Laboratory of Guilin Research Institute of Geology for Mineral Resources for 48, 60 and 60 h, respectively. The background value of ^{40}Ar is 10^{-14} moles while those of ^{36}Ar, ^{37}Ar, ^{38}Ar and ^{39}Ar are 10^{-16} moles. The determining process is as follows: the samples were irradiated by the fast neutron (total flux of shines of 1.3×10^{18} n/cm^2); the cooling samples were enclosed into the purification system of stainless steel ultra high vacuum (vacuity of 1.33×10^{-6} Pa); the heating and control of temperature at every stage was done by the bombardment electron stove; the gas separated out at every stage was purified by sponge titanium and evaporating titanium; the gas enclosed into mass spectrograph and isotopic peaks of Ar (^{36}Ar, ^{37}Ar, ^{38}Ar, ^{39}Ar and ^{40}Ar) were analyzed. Normative sample adopted in this analysis was the biotite granite in Fangshan, Beijing.

4 TESTING RESULTS AND DISCUSSION

The data obtained from the step-heating experiments, age spectrum (heating steps 2–5) are presented in Table 1, shown in Figures 2 and 3.

In the temperature range of 680°C to 1300°C, cassiterite sample was measured (5 heating steps), and the ^{40}Ar-^{39}Ar spectrum is arciform. At lower and higher temperatures there appeared relatively small appearance age values. The age is smooth at 850°C and 1000°C at which the age is 189.6–194.04 ± 2.26 Ma (average age of 191.81 ± 2.26 Ma) while released ^{39}Ar quality fraction is 66.62%.

From 850°C to 1300°C (four heating steps) the appearance ages are approximate to one another, constituting a smooth age spectrum line, with the plateau age T = 202.18 ± 2.35 Ma. As can be seen on $^{40}Ar/^{36}Ar$ vs. $^{39}Ar/^{36}Ar$ diagram (Fig. 3), the age data obtained at four heating steps, which make up the plateau ages, constitute a straight line with an extremely good linearity. Correlation coefficient, R, is 0.9902, the initial $^{40}Ar/^{36}Ar$ value is 223.56, and the isochron age, T, is 206.81 ± 3.23.

Table 1. $^{40}Ar/^{39}Ar$ data table of stepwise heating analysis.

Temperature (°C)	Ar (40/36)	Ar (39/36)	Ar (37/39)	$(^{40}Ar_{Rel.}/^{39}Ar_{k})_{Cor.}$	^{39}Ar ($\times10^{-12}$ mol)	^{39}Ar (%)	$^{40}Ar_{Rel.}/^{40}Ar_{Tot.}$ (%)	Age (Ma)
680	435.6479	16.117584	0.14909	8.6953	0.092	14.13	32.16	174.28
850	540.3182	24.224264	0.13168	10.1063	0.162	24.85	45.29	197.6
1000	555.8691	25.141813	0.61938	10.356	0.232	41.77	46.82	210.06
1200	530.7112	22.913191	1.05855	10.2653	0.075	11.53	44.3	203.43
1300	526.1752	22.773429	0.42315	10.1291	0.05	7.71	43.82	198

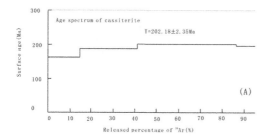

Figure 2. ^{40}Ar-^{39}Ar step-heating age spectrum of cassiterite from the Gejiu Sn polymetallic oredeposit.

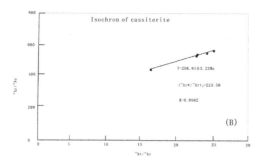

Figure 3. $^{40}Ar/^{36}Ar$-$^{39}Ar/^{36}Ar$ isochron of cassiterite from the Gejiu Sn polymetallic oredeposit.

5 DISCUSSION

The ^{40}Ar-^{39}Ar dating method is one of the newly developed techniques on the basis of the K-Ar method in the mid 1960s of the 20 century. As compared with isotope spike mass spectrometry and the volumetric method, it has some outstanding advantages. For example, the age spectrum and isochron age and initial argon isotopic value can be simultaneously obtained from the same sample with this method. So it has attracted ever increasing attention in recent years.

Jin Zude (1981, 1991) considered that there are the non-hydrothermal interlayer hematite-type tin deposits, and that is typical Triassic synsedimentary deposit; Peng Zhangxiang (1992) that gejiu tin deposit is a stratabound-type tin-polymetallic deposit of complex origin, and model of the granite mineralization has not fully consistent with the actual situation of Gejiu mining area; Zhou Jianping et al (1999) considered while investigating the mineralization of Southeastern tin-polymetallic ore genesis, by means of contrasting Gejiu, Bainiuchang, and Dulong tin-polymetallic sulfide deposits in the Southeast Yunnan, proposed the layered sulfide ore bodies are mainly sedimentary exhalative deposit and non-magmatic, and the intrusion of granitic magma of the Yanshanian epoch hydrothermally superimposed and reformed existing ore body; Zhang Huan et al (2007) deemed that carrying Sn while submarine exhalative formed sulfide, and intrusion of granite magma of Yanshanian epoch makes the enriching Sn and mineralization.

Qin Dexian et al (2004) reported that in Triassic Time, because eastsouthern Yunnan area was controlled by Mile-Shizong fracture, which was eastnorth-trending synsedimentary fracture, rift activities occurred. In the medium Indosinian epoch, because Ailaoshan-Heshuihe Benioff zone was raised and pressed, the rift valley withdrawed and descended, accompanied by seabed basic volcanism. In the medium-later Indosinian epoch, the rift valley was with drawed and descended, accompanied by seabed exhalative hydrothermal sedimentary mineralization. In the Yanshanian epoch, because the crust rose and was pressed, the rift valley closed, accompanied by granitization.

The stratiform sulphide orebodies in the Gejiu tin-polymetallic area, is mainly located in silicon-carbonate of the middle to upper part of Gejiu Formation with the distance of 400 to 800 m away from the underlying granitic body of the Yanshanian epoch. The ore body presents as layer form or lens form. The main minerals of the ore body are magnetism-pyrite, pyrite, marmatite, mispickel, chalcopyrite, cassiterite, quartz, calcite and dolomite. The ore possesses zonal and tubercular structure with the strawberry and colloidal texture of pyrite. The mineralization is the seabed exhalative hydrothermal sedimentary mineralization of the Indosinia epoch after the seabed basic volcano-sedimentary mineralization of the Indosinian epoch happened. Three isotopic ages of series II are basically identical to the real age of the geological fact (the middle-upper stratum of Gejiu Formation).

In the stratiform sulphide orebodies in the Gejiu tin-polymetallic area, the attitude of the orebodies is controlled mainly by strata and lithology, which contain stratiform and stratoi, with a single layer or several layers, lenticular, string-of-bead-1ike, banded, tubular and irregular orebodies. Most of the orebodies have been oxidized. The main ore minerals are pyrrhotite, pyrite, cassiterite, arsenopynte, galena, marmatite and minor chalcopyrite. The oxide ore minerals are limonite, hematite, cerusite, pyrolusite, etc. The main structures of ore are banded structure and nodular structure, which commonly seeing Colloidal texture, Oolitic texture and strawberry-like texture in the pyrite.

It can be seen clearly that the formation age of ore deposits of cassiterite of stratiform orebody in Gejiu tin polymetallic deposit in is from 202.18 Ma to 206.81 Ma, belonging to the Indosinian mineralization. This age represents submarine exhalative hydrothermal mineralization of the Indosinian epoch. This explain just that there is existence submarine exhalative hydrothermal mineralization of the Indosinian epoch in in Gejiu tin polymetallic area.

ACKNOWLEDGMENTS

The present study was surported by the project of tackling scientific and technical key problems of the Tenth Five-year Plan in China (2004BA615A-03). We thank Prof. Yutao Dang for his technical assistance, Dr. Chuandong Xue and Dr. Shucheng Tan for providing some samples and Prof. Guangrong Zhou, Dr. Xinqing Lee and Dr. Zhengheng Zhou for revising this manuscript.

REFERENCES

Jin Zude, 1981, Genetic discussion on earthy hematite-type tin ore deposit in Gejiu, Geology and Prospecting, Vol.17, No.7, pp 32–34.

Jin Zude, 1991, Negation of hydrothermal origin of layered hematite-type tin ore deposit in Gejiu, Geology and Prospecting, Vol.27, No.1, pp 19–20.

Peng Zhangxiang, 1992, Discussion on mineralization model of Gejiu Tin deposit, Yun Geology, Vol. 11, No.4, pp 362–368.

Qin Dexian, Tan Shucheng, Fan Zhuguo, Li Yingshu, Chen Aibing and Li Lianju, 2004, Geotectonic evolution and tin-polymetallic metallogenesis in Gejiu-dachang area, Acta mineralogica sinaica, Vol.24, No.2, pp 117–123.

Wang Zifen. 1983, Several Problems of Ore-forming of Gejiu ore deposit. Acta Geologica Sinica, Vol.57, No.2, pp 154–163.

Zhang Huan, Luo Taiyi, Gao Zhenmin, Ma Deyun and Tao Yan. Helium, 2007, lead and sulfur isotope geochemistry of the Gejiu Sn-polymetallic ore deposit and the sources of ore-forming materials, Chinese Journal of Geochemistry, Vol.26, No.4, pp 439–445.

Zhou Jianping, Xu Keqin and Hua Renmin, 1999, Discovery of sedimentary fabrics and rediscussion of origin in tin deposits such as Gejiu etc., Progress in Nature Science, Vol.9, No.5, pp 419–422.

Zhuang Yongqiu, Wang Renzhong, Yang Shupe and Yong Jinming. 1996, Gejiu Tin-polymetallic Deposit. Beijing: Seismic Publ. House. pp 108–124.

Frontiers of Energy and Environmental Engineering – Sung, Kao & Chen (eds)
© *2013 Taylor & Francis Group, London, ISBN 978-0-415-66159-1*

Application of the variogram in Dachang Sn-Zn ore deposit in Guangxi province, China

Y.S. Li
Faculty of Land Resource Engineering, Kunming University of Science and Technology, Kunming, China

S.C. Li
Faculty of Land Resource Engineering, Kunming University of Science and Technology, Kunming, China
Bureau of Land and Resources of Jianshui, Jianshui, Yunnan, China

Y. Cai
Faculty of Application Technology, Kunming University of Science and Technology, Kunming, China

J.J. Chen, N. Chen, L. Wang, Y.K. Zhang & D.Q. He
Faculty of Land Resource Engineering, Kunming University of Science and Technology, Kunming, China

ABSTRACT: The variogram is core content and basic tool of geostatistics. It can preferably research basic characteristics of regionalized variables of one ore deposit, especially it can reflect structrural change of the variables by random change of the variables. On the base of introducing to the concept of the variogram and researching on the geological characteristics of Dachang Sn-Zn deposit, Guangxi province, this paper has computed the variogram of Dachang Sn-Zn ore deposit, and analyzed those computed results.

Keywords: Sn-Zn ore deposit; application; variogram; dachang in Guangxi province, China

1 INTRODUCTION

The characteristics of the internal structure and the law of space change of external form of a ore deposit are important foundation of ensuring prospecting project, computing methods of reserves, exploiting and mining plan of mineral deposits. The geological statistics is an interdisciplinary subject in developed the last century. The existing information resources is used effectively, the structure and space is revealed, and the space information is estimated reasonably according to particular features of a ore deposit. The geological statistics have been widely used in many areas in recent years.

According to the actual situation of of Dachang Sn-Zn deposit, Guangxi province, the authors have studied geological characteristics of orebody 100 applying principle and method of variogram in the geostatistics. According to the attitude and spatial distribution characteristics of the orebody, the thickness of the orebody etc., can help us to confirm the modelling range of the mineral deposits as follows: X: From 2746778 meter to 2747967 meter, Y: from 457630 meter to 458604 and Z: from −210 meter to −784 meter.

2 BASIC PRINCIPLE AND METHOD OF VARIOGRAM

The variogram is a basic tool of the geologic statistics, which can better research basic features of the regional variables, especially can reflecting structural variables by means of random variables, and is a basis of estimating values to ore deposit with kriging method. The variogram is functions of distance, which describes relativity between variables of different position. The greater the value of variogram is, the worse relativity. Generally the variogram increases with increase of distance (h). The variogram achieves maximal value until the distance attain certain values. From then on, the variogram remains unchanged with a constant value (Qin dexian et al, 2001; Wang renduo et al, 1998).

Calculating of the experimental variogram is the basis of gaining theoretical variogram (mathematical model of ore deposit). The mathematical model of ore deposit is built by means of research on geological features of a orebody. In order to build variogram, we must summarize quantificationally general and local characteristics of a orebody, and find out space changing laws of the geological variables of the orebody. There are sill model and non-sill model in models of the variogram, of which spherical mode in the sill models is often used in the mine. Main features of the spherical mode are: (1) A linear relation can be found between variogram and distance (h) in a small range near the origin. However, this curve becomes gently sloping at a bigger distance. The variogram attains the sill value when the distance (h) get the range (a). (2) The variogram attains the sill value when a tangent of the origin attains two thirds of the range.

The formulas as follows:

$$0 \qquad\qquad h = 0$$

$$y(h) = C_0 + C(3h/2a + h^3/2a^3) \qquad 0 < h <= a$$

$$C_0 + C \qquad\qquad h > a$$

where C_0 is nugget constant, which expresses changing situation between grade of two point when the distance is very small; C is transition constant, which represents difference value between real-time estimation and nugget constant; $C + C_0$ is sill value, which represents the intensity of regional variables in the research area; h is distance; a is range, which expresses influencing bound of the variables.

Generally, the variogram value should be raised when distance among sample point increase. The space distance becomes the range when the variogram attains definite stationary value. As h is less than or equal to a, observation value between any two point is correlative, which the self correlative is decreased while h increased. As h more than a, observation value between any two point is not correlative. In fact, the range reflects changing degree of a regional variable in research object.

Definite effects of application of the variogram are gained in many the mine. For example, Gejiu Sn ore deposit(Chen aibing et al, 2003), Dahongshan Fe ore deposit (Li yingshu et al, 2005), Dulong Sn Deposit (Jia fuju et al, 2008), Lanping Pb-Zn deposit (Yan yongfeng et al, 2008), Pulang Cu deposit (Yu haijun et al, 2009) in Yunnan province, and so on.

3 GEOLOGICAL SETTING OF OREBODY 100 OF DACHANG TIN ORE DEPOSIT

There is rich mineral resources with the advantaged mineral-forming geological conditions in Dachang tin Pb-Zn area, Guangxi province (Fig. 1). Orebody 100 is one of the most important Sn ore deposits in Dachang, which is a large Sn-Zn deposit in tin leads, accompanying useful elements as lead, zinc, copper, silver, antimony, mercury, tungsten.

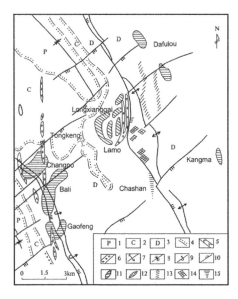

Figure 1. Distribution of deposits in Dachang, Guangxi. 1—Permian limestone, siliceous rock; 2—Carboniferous limestone; 3—Devonian limestone, shale and siliceous rock; 4—Parallel unconformity; 5—Diorite porphyrite; 6—Granite and granite porphyries; 7—anticline; 8—syncline; 9—Sn, Zn orebody; 10—Zn, Cu orebody; 11—Scheelite vein; 12—wolframite vein; 14—stibnite vein.

The main stratum from the top downwards in the mining district are the Upper Devonian and the Middle Devonian, of which organic reef limestone of the Middle Devonian Majiaao Formation ($D_2 m$) is major ore-hosting strata. Structure in the district gives priority to faults and magmatic rock gives priority to veins.

According to different characteristics of ore-formation structure and features of space distribution of the deposits, Dachang tin orefield can be divided into west ore zone, middle ore zone and east ore zone. Gaofeng tin deposit is an important Sn Pb-Zn deposit of the west ore zone in Dachang ore-formation district, Guangxi province. Orebody 100 is an important Sn-Zn-Pb orebody in Gaofeng tin deposit.

General form of the orebody 100 is irregular vein orebody in the form of reverse "S", whose strike is from NW, near EW to NE. Buried depth of the orebody is from −200 meter to 700 meter. Length of the orebody is from 200 meter to 300 meter, thickness from 1 meter to 46 meter, and depth more than 830 meter.

Originally speaking, Gaofeng Sn Pb-Zn deposit is considered as a cassiterite sulfide deposit of undergoing early biochemical sedimentary mineralization and later granitic superimposed mineralization (Su yaru et al, 2007).

4 COMPUTING OF THE VARIOGRAM

1. *Statistics and analysis of variables*

Basic statistical analysis of the original data is basis of building model. Basic statistics include the total number, mean value, standard deviation, correlative coefficient, changing coefficient, frequency distribution and grade distribution of the samples in the research district. According to changing coefficient, continuity between orebodies may be determined. According to correlative coefficient, relativity of between the elements can be determined. These analysis are basis of structural analysis of variogram and choosing kriging method.

In summary, by means of research on statistic feature of regional variables, mathematic characteristics of the regional variables and relation between the regional variables and origin of the deposits can be found out.

The length of samples with 2 meter is combined and analyzed (Table 1 and Figs. 2 and 3).

a. Transferred grades of tin, lead and zinc are characterized by lognormal normal distribution.

b. Changing coefficients of tin, lead and zinc are 116.2%, 85.2% and 58.3% respectively, which show distribution grade of tin, lead and zinc in the ore deposit is very uneven.

c. Correlative coefficient between tin and zinc is 0.005, which show relation is not closely related.

2. *Calculating of the experimental variogram of the deposit*

Calculating of the experimental variogram of the deposit is in order to acquire theoretic variogram, which is estimating grade value. According to space shape and occurrence of ore, experimental variogram along strike, dip and vertical direction of the orebody respectively have been calculated (Table 2).

3. *Fitting of the experimental variogram of the deposit*

Fitting of the experimental variogram of the deposit adopts generally weight polynomial regression method. When fitting, points whin the range must considered.

5 APPLICATION OF THE VARIOGRAM

1. Estimating value and calculating reserves of the orebody is using kriging method.

Because the data of sample are characterized by lognormal normal distribution, reserves of ore in the orebody is calculated by means of calculating and fitting of the experimental variogram and using kriging method. Calculating results are shown in Table 3. By comparison, reserves of calculating using ordinary kriging method and a traditional method are approximation with relative error within 10%.

Table 1. Statistical and analytical results of combination of the samples.

Normal statistics	Sn	Zn	Logarithm statistics	Sn	Zn
Minimum value (%)	0.004	0.010	Minimum value (%)	−5.521	−4.605
Maximum value (%)	35.110	30.170	Maximum value (%)	3.558	3.407
Mean value (%)	1.622	5.661	Mean value (%)	0.034	1.456
Standard deviation	1.885	3.298	Standard deviation	1.124	1.003
Changing coefficient	116.2%	58.3%	Total number	8605	8605

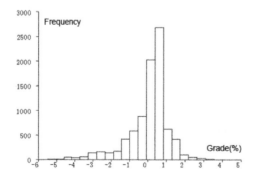

Figure 2. Statistical histogram of natural logarithm of grade of element Sn.

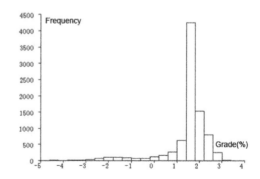

Figure 3. Statistical histogram of natural logarithm of grade of element Zn.

Table 2. Parameters of theoretic variogram of orebody 100 in Dachang.

Direction	Strike	Vergence	Vertical direction
Sn			
A	46.941	23.818	22.893
C_0	0.364	1.389	1.190
C	5.620	4.761	3.600
$C_0 + C$	5.984	6.150	4.790
C/A	0.120	0.200	0.157
Zn			
A	25.206	16.881	14.16
C_0	0.585	0.611	0.417
C	1.389	0.992	1.309
$C_0 + C$	1.974	1.603	1.726
C/A	0.055	0.059	0.092

Where a is range, C_0 nugget constant, C transition constant, $C_0 + C$ sill value, and C/A change on unit range.

Table 3. Grade and tonnage table of Sn and Zn.

Element	Data				
Sn					
Cut-off grade (%)	0.15	0.3	0.5	0.6	0.7
Average grade (%)	1.682	1.755	1.825	1.908	2.003
Amount of metal (t)	133500.3	132614.8	131137.2	128496.2	125500
Zn					
Cut-off grade (%)	1	2	3	4	5
Average grade (%)	6.213	6.67	7.194	7.735	8.924
Amount of metal (t)	481308.684	470902	450963.1	432205.7	356852.9

2. According to calculating of variogram, grade distribution, shape and occurrence of confirming orebody using ordinary kriging method and a traditional method are generally coincident.

3. According to the variogram diagrams, there are periodic fluctuations in three direction, which accounts for there is definite cavity effect in each direction and further shows that the deposit undergoes early biochemical sedimentary mineralization and later granitic superimposed mineralization.

REFERENCES

Chen Aibing, Qin Dexian and Li Yingshu. 2003, The application of the variogram in Gejiu tin deposit. Mineral Resource and Geology, Vol.17, No.5, pp 656–660.

Jia Fuju, Qin Dexian and LiYingshu. 2008, Application of variogram in the Dulong tin Pb-Zn deposit, Geology and Prospecting, Vol. 44, No. 2, pp 77–81.

Li Yingshu, Qin Dexian and Cai Yan. 2005, The application of variogram in Dahongshan iron ore deposit in Yunnan province, China Mining Magazine, Vol. 14, No. 5, pp 52–55.

Qin Dexian, Yan Yongfeng and Hong Tuo. 2001, Mathematical economic model of deposit. Kunming, China: Yunnan Science and Technology Press, pp 5–53.

Su Yaru, Li Yingshu and Qin Dexia. 2007, Foundation on model of ore-formation of orebody 100 for the Dachang orefield in Guangxi province, Nonferrous Metals, Vol. 59, No. 2, pp 21–25.

Wang Renduo and Hu Guangdao. 1998, Linear geological statistics. Beijing: Geological Publishing Press, pp 8–70.

Yan Yongfeng, Qin Dexian and Yu Yangxia. 2008, The development of mine digitization information system and its application to the Lanping Pb-Zn mine. Chin. J. Geochem., Vol. 27, pp 317–324.

Yu Haijun, Li Wenchang and Yin Guanghou. 2009, Copper grade geostatistical study of Pulang copper, Geology and Exploration, Vol. 45, No. 4, pp 437–443.

Frontiers of Energy and Environmental Engineering – Sung, Kao & Chen (eds)
© 2013 Taylor & Francis Group, London, ISBN 978-0-415-66159-1

Pollution regulation policy for SMEs under centralized pollution control

H.Y. Nie
Science and Technology Enterprise Group of Chongqing University, Chongqing, China

B. Huang
College of Economics and Business Administration, Chongqing University, Chongqing, China

Y.Y. Li
College of Computer and Information Science, Chongqing Normal University, Chongqing, China

ABSTRACT: A tripartite game model of pollution control regulation among government, Small and Medium Enterprises (SMEs) and pollution control enterprises under centralized pollution control was established and the optimal pollution discharge regulation policy was found and the main factors such as discharge coefficient and unit social cost of pollutant that influence the optimal policy were analyzed. It is shown that the government would reduce the discharge indicators of SMEs with the increase of discharge coefficient and unit social cost of pollutant and the pollution control enterprises would improve their pollution control prices; profits of SMEs with increased discharge coefficient would also be reduced and at the same time the profits of other SMEs would be increased.

Keywords: centralized pollution control; Small and Medium-sized Enterprises (SMEs); pollution discharge regulation; pollution control price

1 INTRODUCTION

Recently, SMEs in China have achieved rapid development due to the rapid responding ability to the market demands and become the market economic entities. However, SMEs have emitted large quantities of pollutants in the production process, causing great harm to the environment and social welfare due to relatively backward production process and technology and weak social responsibility sense between the entrepreneurs during the rapid development. In addition, some unique features of SMEs which are suitable for large enterprises discharge regulation strategy have no obvious effect. How to conduct the effective discharge regulation for SMEs have become the priorities and difficulties of our environmental protection and economic development (Chen & Xiao 2011, Guo 2007).

SMEs in China are small-scaled with large quantities and the entrepreneurs have weak sense of social responsibility, as most businesses are not only unaffordable to purchase pollution control equipment to conduct separate pollution control but also reluctant to make pollutant control with their own increased costs and loss of profits to increase the social welfare. In addition, the number of SMEs accounts for more than 90% of the total number of

the enterprises in China (Wang 2010), it is not only inconvenient to supervise but also will lead to diseconomies of scale, waste of social resources, and reduction of social welfare if the separate pollution control is conducted. If the SMEs are centralized for production and the pollutants are intensively managed by professional pollution control enterprises, it will not only solve the pollution control equipment investment capacity of SMEs but also the diseconomies of scale of the separate pollution control and at the same time are convenient for government supervision. Therefore, centralized pollution control mode is widely implemented at home and abroad (Godby 2002, Hintermann 2009), and domestic and foreign scholars also make in-depth research on centralized pollution control. He & He (2007) investigate how to make discharge regulation for SMEs by analyzing the characteristics and influencing factors of SMEs pollution control and the research has shown that the establishment of industrial parks to make centralized pollution control is an effective mode to achieve economic pollution control scale; Cai (2008) considers that with regard to the situation of low efficient governance and difficult environment regulation of our SEMs discharge regulation, only the introduction of professional pollution control enterprises and change of "the polluter, the governor" into "the polluter, the payer" will make it likely to improve the pollution control efficiency and the environmental quality. Htermann (2010) analyses the price formulation of EU discharge system by building the pollution control mode under the efficient market. Cui et al. (2011) establishes discharge pricing model under a centralized pollution control mode and studies the conditions of centralized pollution control mode and the discharge pricing strategy of pollution control enterprises. In fact, SEMs discharge control and governance is a key strategy of SMEs discharge regulation under a centralized control mode and too high discharge indicator or too low pollution control price will result in the damage to the environment and too low discharge indicator or too low pollution control price will also reduce the production and profits of the enterprises and then reduce the social welfare. However, little of the existing literature has studied on the SMEs discharge indicators and pollution control prices under the centralized pollution control mode. Therefore, it is necessary to make research on the optimal discharge indicator and pollution control prices and the impact on them of main factors such as discharge coefficient and pollutant unit cost to the optimal regulation and pollution control strategy.

Therefore, this paper studies the SMEs discharge indicator and pollution control prices under the centralized pollution control mode by establishing a tripartite game model of pollution control regulation among the government, SMEs and pollution control enterprises based on the centralized pollution control and obtains the optimal discharge regulation and pollution control strategy of SEMs through theoretical and simulation analysis and analyzes the influence of the main factors such as discharge coefficient and pollutant unit cost to the optimal regulation and pollution control strategy to provide the decisive basis for the SEMs discharge regulation policy in the government departments and the formulation of pollution control pricing strategy in the pollution control enterprises.

2 PROBLEM DESCRIPTIONS

Two SMEs produce the same product in the production market and they will produce some pollutants in the production process which will cause damage to the environment and social welfare and at the same time the damage to the environment and social welfare caused by the pollutant are related to the pollutant discharge, which is related to the production yield of the enterprise.

As the SMEs is small-scaled, they are unaffordable to pay the cost of purchasing pollution control equipment combined with the large number of SMEs, if each enterprise purchases pollution control equipment, it will inevitably lead to a huge waste of social resources, therefore, the government will centralize the SMEs for production and introduce professional pollution control enterprise to make pollution control of the pollutant for all the SMEs in the production process and pollution control enterprise to charge according to discharge capacity of SMEs.

As the damage of the pollutant to the environment and social welfare is borne by the whole society and the governance charges generated by pollutant control and discharge is borne by the enterprise itself, therefore, as a completely rational person, the enterprise with the maximize profit goals will naturally choose to discharge the pollutants directly. At this time, the government will regulate and supervise the discharge of the enterprises. However, if the enterprises are requested to manage all the pollutants firstly and then discharge, it will excessively increase the business costs and reduce the production and profits of the enterprise and at the same time the reduction of the yield of the products will also lead to the increase of the production price and the reduction of the consumer surplus. Therefore, the overdue control of the discharge of the enterprise will reduce the social welfare and the government with the maximum social welfare will need to regulate the pollutant discharge indicators of all the enterprises and discharge indicators in each enterprise according to factors such as regional environmental quality objectives and as well as the amounts of pollutants and the destructive power of the environment in the process of production and the market needs of each enterprise product, thus making regulations on the discharge of SEMs and achieving the maximum social welfare while reducing environmental pollution.

3 THE MODEL

Two SMEs produce the same product in the production market, the product inverse demand function of SMEs i ($i = 1, 2$) is $p = p_0 - a(q_1 + q_2)$, and $p_0 > 0$, $a > 0$; the unit production cost is c_i; certain amount of pollutants in the production process will be emerged and the discharge capacity of the enterprise are related with the product yield, i.e., the discharge capacity $e_i = \beta_i q_i$.

The pollutants will cause some damage to the environment, resulting in certain social costs (or social welfare loss), and the social costs caused by pollutants is the discharge capacity function: namely, the social costs of pollutants $C_s = \gamma(e_1 + e_2)$, γ is the unit social cost of the pollutant, that is, the loss of the social welfare caused by the environmental damage of pollutants of each unit.

In order to regulate the SMEs discharge, the government works out the total discharge indicators E for the two companies as well as the discharge indicators E_i, $i = 1, 2$ for each company, that is, SMEs i can directly discharge the pollutants within the scope of its discharge indicators and the excessive parts are required to manage and approved to discharge only after reaching the standard.

In order to achieve the economies of scale of the pollution control, the government centralizes two SMEs for production and discharge all the pollutants to the professional enterprise of pollution control and manage the SMEs excessive discharge of pollutants with the unit pollution control costs c and charge the pollution control costs by the unit pollution control price r according to the excessive amount $e_i - E_i$ of the SMEs i ($i = 1, 2$).

The unit production cost, inverse demand function, and discharge capability function of SMEs i ($i = 1, 2$), the pollutants social cost function, the discharge indicators, as well as the pollution control cost and price of the pollution control enterprise are common knowledge.

We can get that the profit of SMEs i ($i = 1, 2$) as:

$$\pi_i = (p - c_i)q_i - r(e_i - E_i), \quad i = 1, 2 \tag{1}$$

The profit of the pollution control enterprise:

$$\pi = (r - c)\sum(e_i - E_i) \tag{2}$$

The consumer surplus:

$$CS = a(q_1 + q_2)^2 / 2 \tag{3}$$

The social cost caused by untreated pollutants:

$$C_s = \gamma E \tag{4}$$

Government utility (social welfare):

$$SW = \sum \pi_i + \pi + CS - C_s \tag{5}$$

4 MODEL ANALYSIS

The order of the pollution control regulation decision-making under the centralized pollution control is as follows: first, formulate the total discharge indicators E for SMEs i ($i = 1, 2$) and the discharge indicators E_i for each enterprise with the goal of the maximum social welfare; Second, the pollution control enterprises lay down unit pollution control price r with the goal of maximizing its own profit according to the pollution control policy from the government and finally the

SMEs formulate production yield with the goal of maximizing its own profit under the discharge capacity given by the government and the unit pollution control price by the pollution control enterprise. This paper will use backward induction to solve the optimal decision of government, pollution control enterprise and SMEs i.

Firstly, SMEs i ($i = 1, 2$) determine its yield with the goal of maximum profits to solve $\partial \pi_i / \partial q_i = 0$ then the optimal product yield of the SMEs is:

$$q_i^* = \frac{p_0 - 2c_i - 2r\beta_i + c_{3-i} + r\beta_{3-i}}{3a}, \quad i = 1, 2 \tag{6}$$

Formula 6 is the reaction function of SMEs i ($i = 1, 2$), that is, given a unit pollution control price r of the pollution control enterprise, there will be a corresponding output q_i^* of the optimal product for SMEs i. As the unit production cost for the SMEs i, product inverse demand function, the volume of discharge function are all common knowledge, therefore, if the pollution control enterprises know the reaction function of the SMEs i, they will make its own profit maximized as the goal according to the reaction function and decide the optimal unit pollution control prices r^*.

Substituting formula 6 into formula 2 and solving it, then the optimal unit pollution control price of the pollution control enterprises is:

$$r^* = \frac{-3aE + \left(p_0 + \sum c_i\right)\sum \beta_i + 2\lambda c - 3\sum c_i\beta_i}{4\lambda} \tag{7}$$

where $\lambda = (\beta_1 - \beta_2)^2 + \beta_1\beta_2 > 0$.

Formula 7 is the reaction function of the pollution control enterprise, that is, giving the discharge indicators E developed by the government, there will be a corresponding optimal unit price r^* for the pollution control enterprise. As the pollution control costs and charges of the pollution control enterprise are also common knowledge, therefore, if the government knows the reaction function of pollution control enterprise i, it will make its own profit maximized as the goal according to the reaction function and decide the optimal total discharge indicators E^*.

Substituting formula 7 into formula 5 and solving $\partial SW / \partial E = 0$, we can get the optimal total discharge indicator E^* of the SEMs formulated by the government as:

$$\bar{E}^* = \frac{p_0(5\lambda + 3\beta_1\beta_2)}{3a(\beta_1 + \beta_2)}$$
$$+ \frac{2\lambda c(13\lambda + 3\beta_1\beta_2) - 48\lambda^2\gamma + \sum \delta_i c_i}{3a(\beta_1 + \beta_2)^2} \tag{8}$$

where $\delta_i = -22\beta_i^3 + 33\beta_i^2\beta_{3-i} - 36\beta_1\beta_{3-i}^2 + 17\beta_{3-i}^3$, $i = 1, 2$.

Proposition 1: the optimal unit pollution control prices charged by the pollution control enterprises are:

$$r^* = \frac{6\lambda(2\gamma - c) + \sum 9\beta_i c_i}{(\beta_1 + \beta_2)^2} - \frac{p_0 + \sum 4c_i}{(\beta_1 + \beta_2)}.$$

We can make conclusion from proposition 1 that the pollution control enterprise is in a leadership position under the centralized pollution control mode, therefore, it is accessible to work out the optimal unit pollution control price according to the reaction function of SMEs i ($i = 1, 2$) and achieve the goal of maximizing their own profits.

Proposition 2: the government will reduce SMEs discharge indicators with the increase of the pollutant unit social costs for the SMEs.

We can make conclusion from proposition 2 that with the increase of the pollutant unit social costs for the SMEs, environmental damage and loss of social welfare in the same discharge capability increase. Therefore, the government will reduce total discharge indicators of SMEs, require enterprises to enhance pollution prevention efforts and exchange higher social costs of pollutants with lower pollution control costs to better protect the environment, improve social welfare.

Proposition 3: pollution control enterprise will improve its pollution control charges with the increase of pollutant unit social costs in SMEs.

We can make conclusion from proposition 3 that the increase of SMEs pollution unit social cost will lead to the improvement of social costs in the same discharge capability and pollution control enterprise can expect that the government will reduce the total SMEs discharge indicators (which has been proved in Proposition 2), and the amount of pollutants that SMEs need to manage will be increased, that is, the pollution control service demands of the pollution control enterprises will be increased and so will the pollution control prices.

5 CONCLUSION

A tripartite game model of pollution control regulation among government, SMEs and pollution control enterprises based on the centralized pollution control was established and the optimal pollution discharge regulation and pollution control strategy of SMEs under a centralized pollution control mode was identified and the major factors such as discharge coefficient and unit social cost of pollutant that influence the optimal regulation and pollution control strategy were analyzed. It is showed that the government will reduce the discharge indicators of SMEs with the increase of discharge coefficient and social costs of pollutant unit while the pollution control enterprises will improve their pollution control prices; the profits of the SMEs with increased discharge coefficient will also be reduced, at the same time, the profits of the SMEs with reduced discharge coefficient will then be increased, therefore, enterprises should use cleaner production methods to reduce its discharge coefficient and improve enterprise profits and social welfare while protecting the environment.

ACKNOWLEDGEMENT

The corresponding author of this paper is Bo Huang. This paper is supported by National Natural Science Foundation Project of China (Granted No: 71102178), and Ministry of Education, Humanities and Social Sciences Project (Granted No: 11YJC630070).

REFERENCES

Cai, S.L. 2008. Promote the Market Operation of Pollution Control by Market System. *Contemporary Economics* (2): 66.

Chen, W. & Xiao, B. 2011. The Analysis of Environmental Regulation Policy Tools for SMEs Based on Tradable Emission Permits. *South China Journal of Economics* (10): 58–68.

Cui, Z.F. & Meng, W.D. & Liu, J.P. 2011. Pricing Model of Pollutant Emission for Small-and-Medium-Size Enterprises under Centralized Pollution Control Mode. *Industrial Engineering Journal* 14(4): 28–32.

Godby, R. 2002. Market Power in Laboratory Emission Permit Markets. *Environmental and Resource Economics* (23): 279–318.

Guo, Q. 2007. Environmental Regulations for Medium- and Small-Scale Enterprises Restrained by Pollution Control Ability. *Journal of Shandong University (Philosophy and Social Sciences)* (5): 105–110.

He, Y. & He, A.Y. 2007. The Analysis of the Impact of EMEs' Character on Pollution Control and the Solution. *China Economist* 2007, (1): 205–206.

Hintermann, B. 2009. *Market Power and Windfall Profits in Emission Permit Markets.* CEPE Working Paper No. 62, ETHZ, Zurich.

Hintermann, B. 2010. Allowance price drivers in the first phase of the EU ETS. *Journal of Environmental Economics and Management* 59(1): 43–56.

Wang, X.Y. 2010. The Analysis of the Investment and Financing Mechanism to Support the Pollution Prevention and Control in the Small and Medium-Sized Enterprises (SMEs). *Ecological Economy* (1): 99–102.

Frontiers of Energy and Environmental Engineering – Sung, Kao & Chen (eds)
© 2013 Taylor & Francis Group, London, ISBN 978-0-415-66159-1

Processing methods of heavy metal ions in beneficiation wastewater: A review

Q. Li, B.L. Ge & J. Liu

Mineral Processing Department, Faculty of Land Resource Engineering, Kunming University of Science and Technology, Kunming, Yunan Province, China

ABSTRACT: Heavy metals ions in beneficiation wastewater have caused negative effect to the environment. This study reviewed processing methods of heavy metal ions and discussed problems of part methods. Activated carbon adsorption, chemical precipitation, membrane separating method are active methods because their simple flowsheet and conspicuous effect. Some plants try new methods and flowsheet like bioadsorption, modified materials adsorption. These new methods promote beneficiation wastewater treatment technology advancing meanwhile improve utilizing percentage of cheap materials.

Keywords: beneficiation wastewater; heavy metal ions; processing methods

1 INTRODUCTION

Heavy metal ions have been excessively released in to the environment due to rapid industrialization and have created a major global concern (W.S. Wan Ngah & M.A.K.M Hanafiah, 2008). An the pollution due to the presence of harmful heavy metals from chemical process has received carefully attention of environmental organizations (OmidTavakoli et al., 2008). The most widely used methods for removing heavy metals from wastewaters include ion exchange, chemical precipitation, reverse osmosis, evaporation, membrane filtration and adsorption (M. kobya et al., 2005). Though these methods have achieved remarkable results, part of them remain have problems, one is the superficial theoretical research of some methods, the other is it's difficult to obtain the similar results due to the changeable containing of the wastewater. In addition, researchers pay attention to cheap modified materials and convenient processing methods.

Beneficiation wastewater generated enormous pollution within metal mining. Many reasons will lead to heavy metal ions exceed standard in wastewater, for example: some reagents use in ore dressing contain heavy metal ions, copper sulphate, zinc sulphate, potassium bichromate remain in wastewater or some heavy metal ions in secondary ore dissolution.

In recent years, many plant process beneficiation wastewater positively, recovery heavy metal ions or utilize backwater has brought rewards to them. This article reviewed processing methods of heavy metal ions contain in beneficiation wastewater.

2 ACTIVATED CARBON ADSORPTION

The interest in activated carbons is determined by their unique properties such as large specific surface area, strongly developed microporosity and the possibility to modify their surface and texture (PaunkaVassileva et al., 2008). Activated carbon is a non-polar adsorbent, has macroporous structure and oxygen-containing groups. Usually the adsorption character accorded with Langmuir and Freundlich adsorption model. In order to get satisfied results, conducting surface modify or use other activating reagents are viable.

3 ADSORPTION BY OTHER MATERIALS

In order to decrease cost and improve heavy metal ions adsorption rate, utilizing the cheap and easy get materials is necessary. Clays are cheap and have high specific surface areas, excellent physical and chemical stability, various helpful surface and structure properties (W.J. Chen et al., 2008). Currently, the smectite clays of benonite type are among the most investigated clays for heavy metal removal (M.G.A. Vieira et al., 2010). The main composition of bentonite is montmorillonite and it have a special superiority of adsorbing heavy metal ions such as short adsorb time and high adsorb capacity.

Attapulgite is a kind of silicate clay mineral which with exchangeable cations and relative-OH groups on it's surface (Shouyong Zhou, 2011), and there are plenty of practices that adsorbing heavy metals in wastewater by attapulgite, and pretreatment is essential to obtain better results. In beneficiation area useful pretreatment methods are: activation, modified with acid-alkaline or other ions. According practices, attapulgite is sensitive with Ph, because when Ph of wastewater increase, the negative charge also increase, apparently it's absorb ability enhanced. But higher Ph not always favorable, because heavy metal ions exist in different form under different Ph, for example in alkaline environment, Pb^{2+} in form of $Pb(OH)_2$, It's profitable for Pb^{2+} adsorption because synergetic effect of adsorbing and precipitating, while Cr^{6+} in form of CrO_4^{2-}, attapulgite show a slow absorbing rate, so control a suitable Ph is the key to get satisfying results.

Recently modified woods (Mitsuhiro Morita et al., 1987) edible fungus residues and agricultural waste (Dhiraj Sud, 2008) play an important role of process beneficiation wastes, they solved environment pollution and increase their utilizing rate. Oleic acid often use as collect when recover metals by flotation, and it always residue in beneficiation wastewater, using modified oleic acid adsorb heavy metal ions, the adsorption rate affected by temperature, adsorbing time and ion variety.

4 BIOSORPTION

Bacteria is the most common microorganism on earth, different bacteria have special role on ore dressing, wastewater treatment, food processing, and biosorption is one of the most applicable technologies in recent years (Wael M. Ibrahim 2011). It was found that various functional groups present on their cell wall offer certain forces of attractions for the metal ions and provide a high efficiency for their removal (Ashkenazy et al., 1997). Usually the cell walls have amino, alkyl, carboxyl and other functional groups, it can absorb heavy metal ions then form as complex. Due to cell wall of bacteria has penetrability, ion exchange reaction is easy for becteria. On the other hand, some microorganism rely on inorganic salt to survive, and heavy metal ions can provide such nutrition. The special cell structure and surviving condition make becteria have advantages to adsorb heavy ions.

There remain exist problems of biosorption technology: (1) most becteria need a acidic survival evrionment, but during ore dressing, plenty of alkaline reagents added such as lime, sodium carbonate, high Ph of beneficiation wastewater will hinder bacteria's adsorb ability, even worse it will lead to dying of bactria; (2) becteria have a selectivity to heavy metal ions, namely, becteria just adsorb ions which are favorable for it's survive, resulting part ions residue in wastewater; (3) Environmental temperature affect becteria

reproduce and grow, sometimes worker can't satisfy this need, so biosorption technology limited.

5 CHEMICAL PRECIPITATION

Precipitation is another method to treat beneficiation wastewater, according the reagents used, hydroxide precipitation, sulphuration process and ion exchange method belong to chemical precipitation and all of them are traditional methods used in plant. The aim of Hydroxide precipitation is let heavy metal ions unite with OH^- then formed as precipitation or complexes, for example, Mg-Al layered double hydroxide (Mg-Al LDH) behaved as a precipitation reagent for heavy metal ions such as Cu^{2+}, Pb^{2+} and Zn^{2+} in wastewater via hydroxide formation in which Mg-Al LDH acts as a hydroxide. On the other hand, Mg-Al LDH can take up cationic metals from wastewater (Tomohitokameda et al., 2008). In recent years, researchers try to use magnesium hydroxide because it can treat both acid and alkaline wastewater due to buffering ability. Sulphuration process use vulcanization reagents such as sodium sulphide, sodium sulfhydrate, iron sulfide, it's purpose is make heavy metal ions form precipitation. In order to increase incomes and utilize resources, plants try to recover the precipitation by flotation. The main technological impediment is the precitation has a fine particle size though low solubity, it always suspend in wastewater. For the sake of solving this problem, utilizing flocculant such as polyacrylamide help achieve solid-liquid separation.

Ion exchange is an alternative process for uptake of heavy metals from aqueous solutions (Nadir Dizge et al., 2009). Ion resin show a different adsorb character because it's diverse on resin materials, aperture and groups. Pretreatment is essential to remove impurity in resin, for example, soaking the resin in sodium hydroxide and sodium chloride mixed liquid. Ion exchange method affected by (1) Ph, temperature and flow velocity of wastewater; (2) regeneration times of the resin. At present, main problem of this technology are: the resin is easy poisoned then lose activity; It's difficult to remove heavy metal ions on resin; the adsorption rate decreased after the resin regenerated too many times; Due to the above problems, large scale industrial utilizing of ion resin technology still need a time.

6 PHYSICAL METHODS

6.1 *Menbrane separation method*

The major characteristics of membrane separation method are remarkable results when treat beneficiation wastewater. Compare with other materials, polyelectrolyte complex membrane won't produce secondary pollution, consume less energy, and easy to recover separation production. The usual membrane separation methods include Ultrafilltration (UF), Nanofiltration (NF), and Reverse Osmosis (RO) (H. Au Qdais & H. Mousa, 2004). Nanofiltration membrane has a large water flux and treat multivalent ions at same time, it present superiority on treat wastewater. The main technological problem is: poorly nanofiltration membrane production ability of our country, a great number of them are imported from foreign countries, so it's utilization limited. Reverse osmosis membrane method rely on pressure difference between membrane side, then solvent flow backward, namely wastewater "pass" the membrane, heavy metal ions still remain, the higher concentration liquid need further conduction like thickening.

In the purpose of obtain better results, there are many ways to improve this method, such as utilize it under a lower pressure, conduct pretreatment of the beneficiation wastewater, research new materials to synthetize membrane. A notiveable improvement of membrane UF was brought by adding a complexing water-soluble polymer that binds with the cations to be removed, forming complexes that binds with the cations to be removed, forming complexes of large size (H. Bessbousse, 2008). Add sodium hydroxide in wastewater to make heavy metal ions form precipitation is always useful, meanwhile research new membrane which contain different groups, electro-biofilm membrane or composite membrane which have different materials of film side. The membrane method will be widely used in future.

6.2 *Magnetism method*

Treat wastewater by magnetism method usually have two ways: first, magnetic polymer spherulites which are mainly made of Fe_3O_4, spherical hells adsorb heavy metal ions selectively because variety of groups on it. Compare with other materials, functional polymer spherical adsorbent's particle size and pore size can be controlled, making it an easy access to uniform size sphere (Liuqing Yang et al., 2011). The adsorption character affected by Ph and temperature of the wastewater. When the Ph ranges, guaranteeing the Ph of wastewater > isoelectric point of the polymer spherulite, because when polymer spherulite have negative charge, it will adsorb H^+ in wastewater. In order to increase the adsorption rate, modify polymer spherulite is favorable. Recently researchers attempt to utilize nano-magnetic polymer spherulite adsorb heavy metals, but consider the expensive cost, it can't massively industrial utilize. Second, recovering

heavy metal ions by magnetic field fluidized bed. The separating results affected by magnetic field intensity deeply.

7 CONCLUSIONS

Heavy metal ions in wastewater have caused immense environmental problems, activated carbon adsorption, chemical precipitation and physical methods are traditional methods to treat wastewater, biosorption is a booming method butstill have many technological impediments to solve.

For the sake of developing an environmental taste society, a variety of problems have to beat: (1) decrease the operating and repairing cost of processing wastewater thus many small-scale enterprises can bear; (2) A large number of methods still not perfect, for example, chemical precipitation will cause second pollution, some heavy metal ions may change form after processing, for example, from inorganic to organic (mercuric to salts to methylmercury), or may be oxidized or reduced to a different oxidation state (Feng Liang et al., 2011), researching ways to solve them are essential; (3) The mechanism research on some methods still superficial, intensive theory study is necessary. Processing heavy metal ions in wastewater is benefit for our environment and human health.

REFERENCES

Ashkenazy et al., 1997. Characterization of acetone-washed yeast biomass functional groups involved in lead biosorption. Biotechnol. Bioeng 55:1–10.

Au Qdais H. & H. Mousa, 2004. Removal of heavy metals from wastewater by membrane process: a comparative study. Deslination 164:105–110.

Bessbousse H. et al., 2008. Removalod heavy metal ions from aqueous solution by filtration with a novel complexing membrane containing poly(ethylenimine) in a poly(vinyl alcohol) matrix. Journal of Membrane Science 307:249–259.

Chen W.J. et al., 2008. Metal desorption from copper(II)/nickel(II)-spiked kaolin as a soil component using plant-derived saponinbiosurfant. Process Biochemistry 43: 488–498.

Dhiraj Sud et al., 2008. Agricultural waste material as potential adsorbent for sequestering heavy metal ions from aqueous solution-A review. Biothechnology 99:6017–6027.

Feng Liang et al., 2011. Discrimination of trace heavy-metal ions by filtration on Sol-Gel membrane arrays. Chemistry, vol 17:1101–1104.

Koyba M. et al., 2005. Adsorption of heavy metal ions from aqueous solution by activated carbon prepared from apricot stone. Bioresource Technology 96: 1518–1521.

Liuqing Yang et al., 2011. Preparation of novel spherical PVA/ATP composites with macroreticular structure and their adsorption behavior for methylene blue and lead in aqueous solution. Chemical Engineering Journal 173:446–455.

Mitsuhiro Morita et al., 1987. Bing of heavy metal ions by chemically modified woods. Journal of Applied Polymer Science, v34:1013–1023.

Nadir Dizege et al., 2009. Sorption of Ni (II) ions from aqueous solution by Lewatitcation-exchange resin. Journal of Hazardous Materials 167:915–926.

Omid Tavakoli & Hiroyuki Yoshida 2008. Application of sub-critical water technology for recovery of heavy metal ions from the wastes of Japanese scallopPatinopectenyessoensis. Science of the Total Environment 398:175–184.

Paunka Vassileva et al., 2008. Thiouracil modified activated carbon as a sorbent for some precious and heavy metal ions. J Porous Mater 15:593–599.

Shouyong Zhou et al., 2011. Competitive adsorption of Hg^{2+}, Pb^{2+} and Co^{2+} ions on polyacrylamide/attapulgite. Desalination 270:269–274.

Tomohitokameda et al., 2008. Uptake of heavy metal ions from aqueous solution using Mg-Al layered double hydroxides intercalated citrate, malate, and tartrate. Separation and Purification Technology 62:330–336.

Vieira M.G.A. et al., 2010. Sorptionkenetics and equilibrium for the removel of nickel ions from aqueous phase on calcinedbofebentonite clay. Journal of Hazardous Materials 177:362–371.

Wael M. Ibrahim, 2011. Biosorption of heavy metal ions from aqueous solution by red macroalgae. Journal of Hazardous Materials 192:1827–1835.

Wan Ngah W.S. & Hanafiah M.A.K.M. 2008. Removal of heavy metal ions from wastewater by chemically modified plant wastes as adsorbents: A review. Bio Resource Technology 99:3935–3948.

Frontiers of Energy and Environmental Engineering – Sung, Kao & Chen (eds)
© 2013 Taylor & Francis Group, London, ISBN 978-0-415-66159-1

Research on reasonable flow pressure of beam pumped wells based on the highest system efficiency

S.M. Dong, M.M. Xing & X.J. Wang
Mechanical Engineering College of Yanshan University, Qinhuangdao, China

ABSTRACT: Finding out the reasonable flow pressure of the beam pumped well is an important scientific subject in petroleum engineering and oil reservoir engineering. This paper presents a simulation method about finding out the reasonable flow pressure of beam pumped well with the highest system efficiency as an evaluation index. The method comprises of five simulating models. They are simulating model for beam pumped well's effective power based on multiphase pipeline flow, simulating model for the input power of the motor based on rod string's axial vibration and the polished rod indicator diagram, simulating model for the functional relationships among system efficiency, flow pressure, oil reservoir parameters and swabbing parameters, simulating model for system efficiency's extremum under certain flow pressure and simulating model for the relationships between system efficiency's extremum and flow pressure. Moreover, the author has developed the corresponding programms. The simulating results show that flow pressure has great influence on the system efficiency of the beam pumped well and the system efficiency can be improved by optimizing flow pressure; the oil reservoir parameters, such as water content ratio, the ratio of gas and oil and saturation pressure are the major factors affecting the reasonable flow pressure.

Keywords: the reasonable flow pressure; beam pumped well; the highest system efficiency; simulating models; the corresponding programms

1 INSTRUCTIONS

Flow pressure is one of the most important technical parameters of the oil well. Under the condition of certain oil reservoir characteristics, flow pressure affects oil production and the performance of the lift system. Therefore, ascertaining a beam pumped well's reasonable flow pressure is an important scientific research subject in petroleum engineering and oil reservoir engineering. Reasonable flow pressure depends on the evaluation index. So different evaluation index corresponds to different reasonable flow pressure. In these references [1–3], the author puts forward an evaluation method concerning the flow characteristic curve of beam pumped well. The fundamental principle of this method is that there are inflection points at the flow characteristic curve of the oil well. Namely, the flow pressure which corresponds to the oil well's maximum production is not zero. The flow pressure which corresponds to the oil well's maximum production is defined as the reasonable flow pressure in the reference [1], while the same flow pressure is defined as the minimum flow pressure boundary in these references [2,3]. In real fields, if not for the purpose of raising liquid to keep the oil production, the oil well's actual flow pressure is always below the flow pressure which corresponds to the maximum oil production. Therefore, the definition in these references [2,3] is much more scientific. If there is no inflection point at the oil well's flow characteristic curve, minimum flow pressure boundary is deemed as zero.

In the field of oil extraction engineering, rational submergence depth is an important design aspect in the lift system of the beam pumped well. Some optimum design methods of submergence depth taking the highest pump efficiency, the highest system efficiency or the highest economic benefits as the evaluation index are established in these references [4,5], respectively. But in the above references, the flow pressure is at a given value (the oil well has certain production), and the optimization of submergence depth is obtained by the optimization of swabbing parameters including length of stroke, pumping speed, plunger diameter and pump depth. So the flow pressure corresponding to the gained rational submergence depth is merely the rational flow pressure with a certain parameter combination but not the optimal flow pressure.

In terms of the optimization of the beam pumped well, domestic and foreign scholars extensively take the system efficiency as the evaluation index of the system optimization. In general, the oil well's flow pressure value has relations with its multiphase flow that always has a marked effect on the effective power, the input power and the system efficiency. So the evaluation model of system efficiency applied in these references [6–9] is too simple to fully reflect the reservoir parameters, production parameters and the influence of the multiphase flow on the simulation model for the system efficiency. Therefore, in this paper, based on the improved simulating model for system efficiency, a simulation optimization method taking the highest system efficiency as evaluation index for oil well's reasonable flow pressure is builded.

2 SIMULATION MODEL FOR THE RELATIONSHIPS BETWEEN THE SYSTEM EFFICIENCY THE AND FLOW PRESSURE

System efficiency of the beam pumped well can be defined as [12]:

$$\eta = \frac{P_e}{P_i} \tag{1}$$

where η is the system efficiency; P_e is the effective power of the system, kW; P_i is the input power of the system (input power of the motor.), kW.

2.1 Simulation model for relationships between effective power and flow pressure

The flow pattern of the fluid in beam pumped wells is unstable because it is multi-phase with oil, gas and water. At present, because the flow pattern is very complicated, the flow pattern with multiphase stable flow theory is widely adopted to study the oil well's flow rule [11]. As the oil well's multiphase unstable flow is simplified as stable flow, the simulation model for oil well's effective power is described as given below [12].

$$P_e = \left\{ (p_d - p_s)Q_{1s} + p_s Q_{gs} \frac{k}{k-1} \left[\left(\frac{p_d}{p_s} \right)^{\frac{k-1}{k}} - 1 \right] \right\} \tag{2}$$

where P_e is the effective power, kW; p_d is the pump's outlet pressure, Pa; p_s is the pump's inlet pressure, Pa; k is the natural gas's change process index; Q_{1s} is the average liquid flow of the pump's inlet, m³/s; Q_{gs} is the average gas flow of the pump's inlet, m³/s.

Using flow pressure to express pump's inlet pressure, Eq. (2) becomes

$$P_e = \left\{ \left[p_d - (p_f - \Delta p) \right] Q_{1s} \right.$$
$$\left. + p_s Q_{gs} \frac{k}{k-1} \left[\left(\frac{p_d}{p_f - \Delta p} \right)^{\frac{k-1}{k}} - 1 \right] \right\} \tag{3}$$

where P_f is the flow pressure, Pa; Δp is the pressure difference between the midpoint of pay zone and the pump inlet, namely, pressure difference between flow pressure and pump's inlet pressure, Pa.

2.1.1 Simulation model for pump's inlet pressure
If there is no gas in casing, the mathematical model for pump's inlet pressure is

$$p_s = p_c + \int_0^{H_d} \rho_g g dh + \int_{H_d}^{L} \rho_{og} g dh \tag{4}$$

where P_c is the surface casing pressure, Pa; h is the integral variable, m; H_d is the producing fluid level, m; L is the downhole pump depth, m; ρ_g is the density of the gas column in tubing-casing annular space, kg/m³; ρ_{og} is the density of the crude oil in tubing-casing annular space, kg/m³; g is the gravitational acceleration, m/s². ρ_g, ρ_{og} and p can be expressed as a functional relation [11].

2.1.2 Simulation model for flow pressure
The flow pattern which is multi-phase (oil, gas and water) from oil well's bottom to pump intake is unstable. Assuming the flow pattern is stable flow, taking the fluid in casing pipe below the pump intake as the research object, the distribution of the fluid pressure can be calculated by Orkiszewski method [12] as given below:

$$\begin{cases} \dfrac{dp}{dh} = \dfrac{\rho_{mc} g + \tau_{fc}}{1 - \dfrac{\rho_{mc} Q_{fc} Q_{gc}}{A_c^2 p}} \\ p|_{h=H_d} = p_s \end{cases} \tag{5}$$

where ρ_{mc} is the fluid density at point h in casing pipe, kg/m³; τ_{fc} is the friction loss gradient at point h in casing pipe, Pa/m; A_c is the flow area of casing pipe, m²; Q_{fc} is the fluid flow at point h in casing pipe, kg/s; Q_{gc} is gas flow at point h in casing pipe, m³/s.

According to Eq. (5), it is clear that the fluid density, friction loss gradient, fluid flow and gas flow have functional relations with the pressure (p)[12].

The Eq. (5) can be solved through numerical integration, so the fluid pressure value at midpoint of pay zone (Lz) can be gained as given below.

$$\begin{cases} p_f = p|_{x=L_z} \\ \Delta p = p_f - p_s \end{cases} \tag{6}$$

2.1.3 Simulation model for pump's outlet pressure

Simplifying the multi-phase and unstable flow into multi-phase and stable flow, the distributions of the fluid pressure in oil pipe can be calculated through Orkiszewski method as given below

$$\begin{cases} \dfrac{dp}{dh} = \dfrac{\rho_{mt}g + \tau_{ft}}{1 - \dfrac{\rho_{mt}Q_{ft}Q_{gt}}{A_t^2 p}} \\ p|_{h=0} = p_o \end{cases} \tag{7}$$

where P_o is the wellhead oil pressure, Pa; ρ_{mt} is the fluid density at point h in oil pipe, kg/m^3; τ_{ft} is the friction loss gradient at point h in oil pipe, Pa/m; A_t is the flow area of oil pipe, m^2; Q_{ft} is the fluid flow at point h in oil pipe, kg/s; Q_{gt} is the gas flow at point h in oil pipe, m^3/s.

The Eq. (7) can be solved by numerical method, and when h equals L, the outlet pressure (P_d)value at the midpoint of pay zone (L_z) can be gained.

2.1.4 Simulation model for average flow of the liquid-phase and the gas-phase at the pump intake

The equations for average flow of the liquid-phase and the gas-phase at the pump intake are

$$\begin{cases} Q_{ls} = [n_w + (1-n_w)B_{os}]Q \\ Q_{lg} = (1-n_w)(S_p - S_s)\dfrac{p_{st}}{T_{st}}\dfrac{T_s Z_s}{p_s}Q \end{cases} \tag{8}$$

where Q is oil well's actual output, m^3/s; n_w is water content ratio of the produced fluid in oil wells; B_{os} is volume factor of the crude oil at the pump intake; S_p is producing gas/oil ratio, m^3/m^3; S_s is dissolved gas/oil ratio at the pump intake, m^3/m^3; p_{st} is the standard pressure, Pa; T_{st} is the standard temperature, K; T_s is the temperature at the pump intake, K; Z_s is natural gas compressibility factor at the pump intake.

2.2 Simulation model for relationships between input power and flow pressure

2.2.1 Simulation model for relationships between suspended load and the flowing pressure

In terms of a vertical well, simulation model for indicator diagram of the polished rod load is composed of the wave equation which is used to describe the axial vibration of the rod string and boundary conditions as given below [10].

$$\begin{cases} \dfrac{\partial^2 u}{\partial t^2} = c^2\dfrac{\partial^2 u}{\partial x^2} - v\dfrac{\partial u}{\partial t} \\ EA\dfrac{\partial u}{\partial x}\Big|_{x=L} = P_P(t) \\ \dfrac{\partial u}{\partial t}\Big|_{x=0} = U_0(t) \end{cases} \tag{9}$$

where $P_p(t)$ is the liquid load of the plunger, N; $U_0(t)$ is the polished rod displacement (zero at top dead center, downward direction is positive direction), m; u is the rod string's displacement on any section X at time t, m; c is the propagation speed of the light wave in rod string, m/s; v is the damping coefficient, 1/s; E is the modulus of elasticity of the rod string, N/m^2; A is cross-sectional area of the rod string, m^2.

The liquid load of the plunger ($P_p(t)$) can be calculated as given below

$$P_P(t) = A_p(p_d - p) - A_{rd}p_d + F_f \tag{10}$$

where A_p is cross-sectional area of the plunger, m^2; A_{rd} is the cross-sectional area of the bottom sucker rod, m^2; p is fluid pressure at any time in pump barrel, Pa. In general, the fluid pressure which has relations with suction pressure and discharge pressure of the pump is decided by the working process of the pump; F_f is the friction between plunger and pump barrel; N.

2.2.2 Simulation model for crankshaft torque

For beam-pumped unit with crank balance, the formulation for the crankshaft torque is

$$M_N = \overline{TF}(PRL - B_W)\eta_{CL}^{k_1} - M_C\sin(\theta - \tau) \tag{11}$$

where M_N is the net torque of the crankshaft, N·m; B_w is the structural imbalanced weight, N; M_c is the maximum balanced torque of the crank, N·m; θ is crank rotation angle, rad; τ is the offset angle of the crank balance weight, rad; η_{CL} is the steering mechanism's efficiency; k_1 is the coefficient, when v_A is bigger than zero, k_1 equals—1;when v_A is no bigger than zero, k_1 equals 1.

2.2.3 Simulation model for the system's input power

The electric motor's instantaneous output power can be calculated by the equation as given below

$$N_{MO} = \frac{M_N \cdot \omega}{1000}\eta_{MB}^{k_2} \tag{12}$$

116

where N_{MO} is the electric motor's instantaneous output power, kW; η_{MB}—transmission efficiencies of the belt and the gear reducer; ω is the rotational speed of the crankshaft, 1/s; k_2 is the coefficient, if M_N is bigger than 0, then k_2 equals—1; if M_N is no bigger than 0, then k_2 equals 1.

The instantaneous input power and the average input power of the electric motor are given below respectively

$$\begin{cases} N_M = N_{MO} + P_0 + \left[\left(\dfrac{1}{\eta_N}-1\right)P_N - P_0\right]\beta^2 \\ \overline{N}_M = \dfrac{1}{T}\int_0^T N_M dt \end{cases} \quad (13)$$

where N_M is the electric motor's instantaneous input power, kW; β is utilization ratio of the electric motor's instantaneous input power, $\beta = N_{MO}/P_N$; P_N is the rated power of the electric motor, kW; P_0 is the wasted power of the electric motor, kW; η_N is the rated efficiency of the electric motor.

3 SIMULATION MODEL FOR THE REASONABLE FLOW PRESSURE

3.1 Optimization simulation model for system efficiency with given extremum of flow pressure

From the simulation model for system efficiency, the author summarizes the main factors affecting the system's efficiency of the beam pumped well as follows:

1. Oil well's inflow performance and the reservoir parameters: In certain conditions, the inflow performance can reflect that the amount of the oil production has a significant influence on the efficiency of the system. Reservoir parameters are the midpoint of pay zone, the reservoir static pressure, the water ratio, the gas/oil ratio, viscosity of the crude oil, density of the cruel oil, saturation pressure of the cruel oil, the characteristics of the natural gas and so on.
2. Device type and property parameters: The pumping equipment mainly include electric motor, pumping unit, sucker rod, sucker pump and so on. The device type and the property parameters have significant influence on the system efficiency.
3. Swabbing parameters: Swabbing parameters include pumping stroke, pumping speed, pump diameter, the depth of pump, rod string assemblage and the rated power of the electric motor (because the rated power of the electric motor is adjustable and has certain influence on the efficiency of the system, so it is classified into swabbing parameters).

Based on the given reservoir characteristics and equipment combination, the system efficiency is the function of the flow pressure and swabbing parameters. It is worth to state two points: (1) The flow pressure which is decided by the oil well's inflow performances and swabbing parameters has significant influence on the system efficiency, but not an independent variable. When the oil well's inflow performance is given, the flow pressure is the function of swabbing parameters, that is to say, the system efficiency is the function of the swabbing parameters (2) system efficiency is a multivalued function of the flow pressure. It is because that multistage assemblage or even countless swabbing parameters of the rod string can achieve the given flow pressure, and different parameter combinations have significant influence on the system efficiency. For the above reasons, the primary task is to optimize the swabbing parameters and solve the extreme value of the system efficiency under the given flow pressure.

3.1.1 Optimization of design variables
Design variables that need to be optimized are pumping stroke (S), pumping speed (n), pump diameter (D), the depth of pump (L), rated power of the electric motor (P_N) and rod string assemblage. Design variables can be expressed in the form of matrix as given below

$$\overline{X} = \left[S, n, D, L, P_N, \{d_i, L_i\}_{i=1,2,\,...,\,I_r}\right]^T \quad (14)$$

where d_i is the i—stage sucker rod diameter; L_i is the i—stage sucker rod length; I_r is the number of the stages.

3.1.2 Objective function
The optimization design objective function with the highest system efficiency is

$$\max F(\overline{X}) = \max \eta(\overline{X}) = \\ \eta\left(S, n, D, L, P_N, \{d_i, L_i\}_{i=1,2,\,...,\,I_r}\right) \quad (15)$$

3.1.3 Constraints
1. Constraints of the given flow pressure
   According to the oil well's inflow characteristic curve, if the flow pressure is given, the corresponding oil production is also definite. In this paper, the author takes Vogel function in which the hydrostatic pressure is lower than the saturation pressure as an example to illustrate a modeling approach concerning constraints of the flow pressure. The oil production at arbitrary given flow pressure is caculated as given below

$$Q_{in} = Q_{max} \left\{ 1 - 0.2\left(\frac{p_f}{p_r}\right) - 0.8\left(\frac{p_f}{p_r}\right) \right\} \qquad (16)$$

where Q_{in} is the oil production from oil layer into the bottom hole, m³/d; Q_{max} is the maximum oil production, m³/d; p_r is the oil well's hydrostatic pressure, Pa. According to the flow pressure and the oil production under the known condition, the oil well's maximum oil production can be determined.

In steady working condition, the oil production from oil layer into the bottom hole (Q_{in}) equals oil well's real production (Q). The corresponding equality constraint is found

$$H(1) = 1440 \times \frac{\pi}{4} D^2 Sn\alpha - Q_{in} = 0 \qquad (17)$$

where a is the discharge coefficient. The discharge coefficient is the function of immersible pressure and swabbing parameters[10].

2. Constraints of device's carrying capacity
   The utilization ratio of the maximum polished rod load, the utilization ratio of the crankshaft's maximum torque, the utilization ratio of the electric motor power, the strength of the rod string, constant strength and other constraints are considered comprehensively [10].

3. Boundary restrictions of the parameters
   According to the series of the basic parameters of the beam pumped unit, sucker rod, oil-well pump and the actual production needs, boundary restrictions of the parameters are

$$\begin{cases} 1\text{m} \leq S \leq 6\text{m} \\ 2\text{min}^{-1} \leq n \leq 10\text{min}^{-1} \\ 28\text{mm} \leq D \leq 95\text{mm} \\ 5kW \leq P_N \leq 75kW \\ d_i = \in \{16\text{mm},19\text{mm},22\text{mm},25\text{mm}\} \\ \quad i = 1,2,\ldots,I_r \\ L \leq L_Z \\ L = \sum_{i=1}^{I_r} L_i \end{cases} \qquad (18)$$

3.1.4 Simulation model for system efficiency's extreme

Optimal mathematical model for swabbing parameters is composed of the established objective function and corresponding constraints. Meanwhile, the system efficiency's extreme with given flow pressure and the corresponding optimal swabbing parameters can be resolved from optimization algorithm.

3.2 Computational method of system efficiency's maximum and reasonable flow pressure

Through optimizing system efficiency's extreme under different flow pressure, the curve reflecting the changing laws between system efficiency and flow pressure is obtained, and then the system efficiency's maximum and corresponding flow pressure are determined. This flow pressure can be called reasonable flow pressure.

4 COMPUTATIONAL EXAMPLE AND SIMULATION ANALYSIS OF INFLUENCING FACTORS

4.1 Computer software development

Through continuous research and collaboration with oilfield, the author has developed corresponding computer software with the name of computer simulation optimization system for highly efficient beam unit wells [10] which has been widely used in the real oil field. The actual application results show that the simulation optimization system has a high precision which can meets the practical engineering requirements, and has obtained a remarkable energy-saving effect. Because of a new simulation function of reasonable flow pressure on the basis of the highest system efficiency, this system realizes real simulation and optimization of the maximum system efficiency and its corresponding reasonable flow pressure.

4.2 Computational example for the reasonable flow pressure

Beam pumped unit's type: CYJ10-3-53HB, Motor's model: Y225S-8

Reservoir parameters: $L_z = 1400$ m, $p_s = 10.5$ MPa, $p_b = 8$ MPa, $\rho_o = 850$ kg/m³, $\mu_o = 4$ mPa·s, $S_{ng} = 0.8$, $n_w = 70\%$, $S_p = 50$ m³/m³.
Swabbing parameters: $L = 980$ m, $D = 44$ mm, $S = 3$ m, $n = 6$ min⁻¹. $p_0 = 0.5$ MPa, $p_c = 0.6$ MPa
Sucker rod assemblage: 22 mm × 500 m + 19 mm × 480 m.
Flow pressure: $p_{f0} = 5.56$ MPa.
Oil production: $Q_0 = 20.3$ m³/d.

Figure 1 shows the distribution of system efficiency with flow pressure. Through this chart, it is clear that flow pressure has a significant effect on the system efficiency which can be greatly improved by optimization of the flow pressure. The simulating results indicate that the reasonable flow pressure is 4.29 MPa and the highest system efficiency is 40.34%.

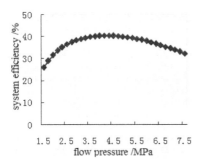

Figure 1. The highest system efficiency with flow pressure.

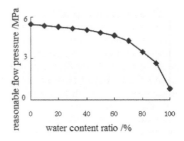

Figure 2. Reasonable flow pressure flow pressure with the water content ratio.

4.3 Effect of influencing factors on reasonable flow pressure

From this chart, it can be found that the device type merely affects the system efficiency based on the reasonable flow pressure and has little effect on the value of reasonable flow pressure while reservoir parameters have a marked effect on the value of the reasonable flow pressure. The distributions of reasonable flow pressure with water ratio, gas-oil ratio and saturation pressure are drawn in Figures 2–4, respectively pressure are drawn in Figures 2–4, respectively.

From Figures 2–4, it can be clearly demonstrated that the oil well's reasonable flow pressure decreases along with the increase of water ratio but increases along with the increase of the gas-oil ratio and the saturation pressure. Namely, the higher gas-oil ratio is, the higher the reasonable flow pressure is.

4.4 Field test of the reasonable flow pressure theory

According to the optimal design results of the reasonable flow pressure and swabbing parameters, field test is put into practice as given below:

The flow pressure and oil production corresponding to present oil well's swabbing parameters are $f_0 = 5.56$ MPa, $Q_0 = 20.3$ m³/d, respectively.

Actual measured system efficiency: 30.62%.
Optimized flow pressure: 4.29 Mpa.

Swabbing parameters corresponding to optimized flow pressure are $L = 1360$ m, $D = 56$ mm, $S = 3$ m and $n = 3.65$ min⁻¹.

The optimized result of the system efficiency is 40.34% while the actual measured system efficiency is 38.12% and the actual oil production is 23.84 t/d. Therefore, from above data, it can be seen clearly that the proportional error between actual measured system efficiency and its corresponding simulating result is just 5.82%. The actual measured

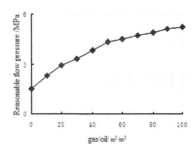

Figure 3. Reasonable flow pressure with the gas/oil ratio.

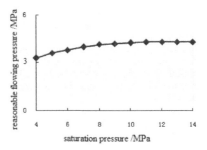

Figure 4. Reasonable flow pressure with the saturation pressure.

system efficiency increases by 24.49%, from which it can be firmly confirmed that energy-saving effect is extremely obvious.

5 CONCLUSIONS

1. Based on the dynamic simulation of the rod string, the simulation model for system efficiency of beam pumped wells and the simulation model with certain flow pressure for system efficiency's extremum are established. Through simulating different system efficiency's extremum based on certain flow pressure, a method considering reasonable flow pressure and corresponding system

efficiency's maximum is proposed. Meanwhile, the corresponding optimal simulation software based on reasonable flow pressure with the highest system efficiency is developed. Practical simulating results indicate that the optimization algorithm of established reasonable flow pressure simulation has a high precision which can meet practical requirements.

2. Flow pressure has a significant effect on the system efficiency of the beam pumped well; Water ratio, gas-oil ratio, saturation pressure and other reservoir parameters are key factors affecting reasonable flow pressure. Oil well's reasonable flow pressure decreases along with the increase of water ratio but increases along with the increase of gas-oil ratio and saturation pressure.

ACKNOWLEDGEMENTS

The Financial support for this work by the National Natural Science Foundation of China (No. 50974108, No. 51174175) is gratefully acknowledged.

REFERENCES

[1] Manping Yang. [J]. Petroleum Geology and Recovery Efficiency, 2004, 5(11):41–43.
[2] Xingli Xie, Yuxin Zhu, Jing Xia, et al. A new inflow performance equation for the IPR curves with maximum flow rate and its application [J]. Beijing: Petroleum Exploration and Development, 2005, 32(3): 113–115.
[3] Fulin Zhong, Caizhen Peng, Minhui Jia. [J]. Journal of Southwest Petroleum Institute, 2003, 25(4):30–33.
[4] Yanbin Zhang, Banglie Wan. Determination of the reasonable submergence depth of sucker rod pump [J]. Oil Drilling and Production Technology, 1999, 21(2):62–65.
[5] Riyi Lin, Maosheng Sun, Shaodong Zhang, et al. Optimization design method to determine submergence depth of sucker rod pump [J]. Journal of the University of Petroleum, China, 2005, 29(4): 87–90.
[6] Shimin Dong, Shengjie Wang, Dongfeng Lu, et al. Simulation models for optimization design of suction paramers for rod pumping system in directional wells [J]. Acta Petrolei Sinica, 2008, 29(1):120–123.
[7] J.N.McCoy, D.J. Becker, A.L. Podio. How to maintain high production efficiency in sucker rod lift Operations [A]. SPE80924, 2003.
[8] O. Lynn Rowlan, James F. Lea, and James N. McCoy. Overview of beam pump perations [A]. SPE110234, 2007.
[9] Cabor Takacs. Ways to obtain optimum power efficiency of artifical lift installations [A]. SPE126544, 2010.
[10] Shimin Dong. Computer simulation of dynamic parameters of rod pumping system optimization [M]. Beijing: Petroleum Industry Press, 2003: 87–93.
[11] Jialang Chen, Taoping Chen, Zhaosheng Wei. Gas-liquid flow in rod pumping wells [M]. Beijing: Petroleum Industry Press, 1994: 100–110.
[12] Shimin Dong, Xiuhua Huang. Calculation method of wattful power of rod-pumped wells [J]. China Petroleum Machinery, 2001, 29(7): 39–41.

Frontiers of Energy and Environmental Engineering – Sung, Kao & Chen (eds)
© 2013 Taylor & Francis Group, London, ISBN 978-0-415-66159-1

Calculation method of off-design power program for extraction and heating of turbine

Z.X. Lin & L. Fu
Centre for Energy Saving Studies, Tsinghua University, Beijing, P.R. China

Z.Y. Zhao
Huadian Distributed Energy Engineering Technology Co., Ltd, Beijing, P.R. China

ABSTRACT: The main distinguish between the heating units and the straight condensing unit is that they are appeared at the same time for the increasing of heating steam flow and the decreasing of generation power. The research object of this paper is conventional heating unit. The heating unit off-design power calculation program is developed by using Friuli Siegel formula and Excel VBA and based on the heating calculation program of straight condensing units. Finally, the generated output of units after extraction to heating can be calculated and the calculation errors of them are less than 0.5%; finally, the basic condition of 300 MW heating units can be assured through calculation.

Keywords: turbine; extraction for heating; off-design; power

1 INTRODUCTION

Currently, the off-design power calculations of steam turbine are almost used for straight condensing units, but the research datum of heating units off-design calculation is so rare. Literature [1] gets the heating units off-design power using the equivalent enthalpy drop method, the extraction parameter for heating is heat load in the calculation, not the exaction pressure or extraction flow, so the calculation accuracy is affected and the energy grade is wasted; Literature [2] programs to calculate the off-design power of heating units using the circulating function method. The program is so complex and its studying object is 50 MW non-reheat heating units, it is not applied to the large heating units. The universality of the program needs to improve. Some software can calculate the off-design power of heating units, for example, STEAM series heat balance computing software for thermal power plant developed by U.S. Thermoflow company and Ebalace performance simulation software of power plant thermal system developed U.S. Enotech, and so on [3]. The merit of these large-scale business software are powerful, flexible and simplicity, but they are all developed by the foreign companies and calculated for gas turbine. They don't fit for calculating the CHP of large-scale thermal power plant units. So, it is necessary to develop the CHP off-design power calculation program based on the previous studies.

The off-design calculation of heating units is different from that of straight condensing units. The calculation goal of straight condensing units is to obtain the main steam flow under the condition that the change of electric load is known using the sequential method. Heating units are not involved in the peak load regulation to meet the needs of heat load in the heating period. The off-design of heating units is the output of steam turbine according to the various of heating exaction parameters (heat load) under the condition that main steam flow is constant. So, its calculation method is different from that of straight condensing units. It is necessary to program for calculating heating units off-design power, which is accuracy, simple and convenient, according to the characteristics of heating units based on the previous studies.

2 THEORY AND STUDY METHOD OF HEATING UNITS OFF-DESIGN CALCULATION

2.1 Calculation theory

There is a cold source of heating-supply flow in the heating units, but it is not existence in the straight condensing units. So, one more boundary condition should be identified in the heating units calculation than in the straight condensing units. The boundary condition is ehausted steam parameters of intermediate pressure cylinder when steam turbine is running for heating-supply, that are pressures and temperatures of heating exaction of units. The boundary condition can be according to the combined characteristic of CHP system [4–6] or given parameters. Program using the Friuli Siegel formula for calculating the off-design power of heating units based on the thermal calculation program of steam turbine of straight condensing units [7,8].

Steam pressures before or after stage-group determined the through-flow capability of units. The change of through-flow leads to change of pressures before or after stage-group. Their relationship can be described in Friuli Siegel formula [9]:

$$\frac{D_1}{D_{1,0}} = \sqrt{\frac{p_1^2 - p_2^2}{p_{1,0}^2 - p_{2,0}^2}} \cdot \sqrt{\frac{T_{1,0}}{T_1}} \qquad (1)$$

where: $D_{1,0}$, D_1—through-flow before and after change of operating conditions, t/h;

$p_{1,0}$, p_1—pressures in front of stage-group before and after change of operating conditions, MPa;

$p_{2,0}$, p_2—pressures in back of stage-group before and after change of operating conditions, MPa;

$T_{1,0}$, T_1—temperatures in front of stage-group before and after change of operating conditions, K.

Initial parameters of one stage-group when conditions are changing have been known, the outlet pressure of stage-group can be gotten, then the outlet parameters of stage-group which is the inlet parameters of the next stage-group can be determined. The subsequent calculations can go on [10].

2.2 *Copying old text onto new file*

The method of calculating off-design power of heating units is that as below:

1. Calculations of parameters of each turbine part
   ① Intermediate pressure cylinder Assumption of steam flow of each turbine stage-group is reasonable (the assumption is that the steam flow is equal to the flow of each part of steam turbine under the rated extraction operating condition). The inlet initial temperature of intermediate pressure cylinder and the temperature of reheat steam are constant when the heating units are running under the off-design, the inlet parameters of intermediate pressure cylinder can be determined if the inlet pressure of intermediate pressure cylinder has been known. The each stage-group parameter of intermediate pressure

cylinder can be calculated in turn according to each stage-group relative internal efficiency of intermediate pressure cylinder, then its exhaust pressure can be obtained. Compared it with the extraction pressure for heating, the inlet pressure of intermediate pressure cylinder is calculated iteratively until the exhaust pressure of intermediate pressure cylinder is equal to extraction pressure for heating which is known before, so the parameters of each intermediate pressure cylinder stage-group can be determined.

   ② High pressure cylinder The inlet pressure of intermediate pressure cylinder is equal to the exhaust pressure of high pressure cylinder minus the pressure loss of reheat steam. So, the exhaust pressure can be obtained according to the inlet pressure of intermediate pressure cylinder. Because the flow and parameters of main steam are constant, the relative internal efficiency of governing stage is not change, the rear pressure of governing stage is assumed to be equal to the inlet pressure of high pressure cylinder, then the inlet parameters of high pressure cylinder can be obtained. Then each stage-group parameter of high pressure cylinder can be determined using the iterative calculation in turn according to the relative internal efficiency of each stage-group.

   ③ Low pressure cylinder The adiabatic iso-enthalpy process happens in the communicating pipe which joins the intermediate pressure cylinder and low pressure cylinder. So, the exhaust enthalpy of intermediate pressure cylinder is equal to inlet enthalpy of low pressure cylinder; if the back pressure of units is known, the each stage-group parameter of low pressure cylinder can be determined using the iterative calculation under the condition that the inlet pressure of low pressure cylinder is assumed.

2. Iterative calculation of each stage-group steam flow of turbine
   Each regenerative extraction steam flow can be obtained by the heat balance calculation according to the parameters of each turbine stage-group obtained from above calculations. The regenerative extraction steam flow are used to calculate for the flow of each turbine stage-group. Compared the flow calculated with the flow assumed, then iterative calculating until they are the same.

3. Calculating power with these parameters
   Calculating for power of turbine using the parameters of each turbine stage-group obtained from above calculation. The logic diagram for off-design power calculating of heating units as Figure 1 shown.

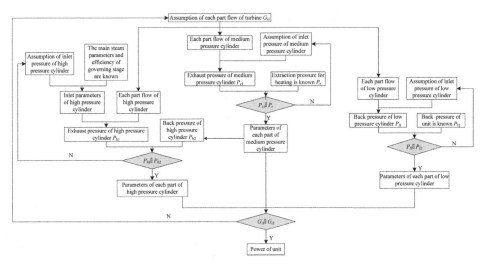

Figure 1. Off-design power calculation logic diagram of heating units.

3 INSTANCE OF OFF-DESIGN PROGRAM FOR 300 MW HEATING UNITS

3.1 *Determination of theory power*

According to the above procedures of program, calculate the power of the subcritical 300 MW heating units, which produced by Harbin Turbine Company Limited of China. 24 heating balance diagrams with different heating exaction parameters are supplied by the manufacturers. The 24 heating balance diagrams are introduced in sequence as the Table 1 shown.

If the each stage-group parameter on the every heat balance diagram has been known, the heat balance calculation for reheat system has been done firstly, and then extraction flow of regenerative heater under each condition has been checked, the third is to calculate each part through-flow of turbine and generation power, finally, the generation powers of all conditions have been checked. Compared the power obtained by the power check calculation with the power obtained from the heat balance diagram, there are some errors between them. All errors are smaller than 0.6%. So, the program can be used for the engineering. There are two reasons for the errors, one is the editions of enthalpy and entropy table used for calculations are different; the other is the significant figures retained in calculation are different.

The enthalpy and entropy table and the significant figures used for checking calculation are also used in the off-design power calculating of heating units. So, the power obtained by the checking calculation is taken as the theory value, then compared it with the datum obtained by the off-design

power calculation of heating units, the error of them is smaller than 0.5%. The program for calculating off-design power of heating units is accuracy and feasible. The relative error of power is as Figure 2 shown.

3.2 *Determination of base operating condition*

It can be known from above calculation, the calculation accuracy of program for calculating the off-design power of heating units can be assured when calculating heating units power under this condition by using the relative internal efficiency of each stage-group. But, the one purpose of this program is calculating power of heating units under different extraction conditions for heating if the only one heat balance diagram is known. So, the base operating condition for calculating the off-design power of heating units is needed.

Taking each of 24 operating conditions as the base operating condition, calculate for the other 23 conditions by using the off-design calculation program of heating units. Then compared the power obtained with the theory power, the conditions whose error are larger are deleted, finally, 8 conditions are taken as the base operating conditions to compare. The result is as the Figure 3 shown. The vertical axis is the error between the power obtained by the calculation and the theory power, the horizontal axis is the 24 conditions. The calculation power error is the smallest when taking the 8th condition in Figure 3 as the base operating condition, whose extraction parameters are 0.30 MPa and 480 t/h, its calculation result is most accurate of all. So, the base operating condition

Table 1. Serial numbers of heat balance diagrams in different working conditions (unit: MPa).

Extraction pressure/ extraction flow	No.					
	0.245	0.30	0.35	0.40	0.45	0.49
100 (t/h)	1	5	9	13	17	21
200 (t/h)	2	6	10	14	18	22
340 (t/h)	3	7	11	15	19	23
480 (t/h)	4	8	12	16	20	24

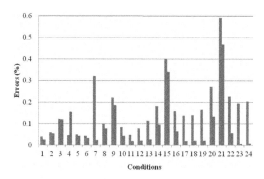

Figure 2. Relative errors of powers.

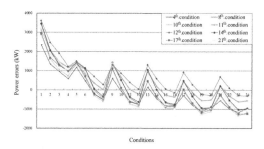

Figure 3. Power curve.

of 300 MW heating units is the condition whose extraction pressure is 0.3 MPa and the extraction flow is 480 t/h.

They can be known from the Figure 3, the eight curves showed wave shape from left to right, on the whole, their errors are decreasing gradually. At the same time, the power errors caused by the change of extraction flow are smaller; while the power errors caused by the change of extraction pressure are bigger. In the process of calculation, it can be found that the calculation power error is smaller when the extraction pressure is bigger than the extraction pressure of base operating condition; while the change of power is bigger when the extraction pressure is smaller than the extraction pressure of base operating condition.

So, it is needed to amend for the condition whose error is bigger.

3.3 *Power amendment*

It can be found through analysis that the power change of the end piece of intermediate pressure cylinder is caused by the change of relative internal efficient after extraction pressure decreased. The influence factor of relative internal efficient of the end piece of intermediate pressure cylinder is the increasing of its leaving velocity loss. The power calculation has been amending when the extraction pressure is smaller than the extraction pressure of base operating condition according to the influence of change of leaving velocity loss of intermediate pressure cylinder exhaust steam on its power. Firstly, the exhaust steam velocity of intermediate press cylinder is assumed as 100 m/s when the turbine is running at THA. The exhaust steam velocity of intermediate pressure cylinder can be obtained when its pressure is decreasing. Then, the increasing amount of leaving velocity loss caused by the increasing of exhaust steam velocity can be obtained. The influence on the power can be obtained at the same time, then the generation power of intermediate pressure cylinder can be amending. The error between the amending power and theory power is less than 0.5%.

4 CONCLUSIONS

1. The off-design power calculation of heating units is based on the thermodynamic calculation program of turbine of straight condensing units; then calculated according to the Friuli Siegel formula and eternal relative internal efficiency of stage-group. The program based on the Excel is easy to use and visual to check datum.
2. It is determined that the base operation condition of 300 MW heating unit is the condition whose extraction steam pressure is 0.3 MPa and the extraction steam flow is 480 t/h. the base operation condition is determined to clear the direct of the datum supplied by the Turbine manufacture company. The thermal power plant can obtain the power of heating unit and its heat balance diagram under the different extraction parameters condition according to the heat balance diagram of the base operating condition. The base operation condition of other heating units can also be determined using this method.
3. For the heating units, the error of the off-design calculation power produced by the change of extraction flow is less; while the error of the

off-design calculation power produced by the change of extraction pressure is bigger.

4. When the extraction pressure of operation condition is bigger than the extraction pressure of basic operation condition, the error of the off-design calculation power is less; when the extraction pressure of operation condition is less than the extraction pressure of basic operation condition, the change of the off-design calculation power is bigger. Through analysis, the reason for producing error is that the increasing of the exhaust stage residual speed loss of intermediate pressure cylinder. So, amending residual speed loss of the intermediate pressure cylinder for these operation conditions.

ACKNOWLEDGEMENT

LIN Zhen-Xian et al. thankfully acknowledge the research grant from National Key Technology R&D Program (File No. 2011BAJ07B06). The authors express their sincere thanks to Tian Jin Hua Neng Yang Liu-qing Thermal Power Plant Co., Ltd, for their kind co-operation during this work.

REFERENCES

[1] Lin Wanchao. Heating system energy-saving theory of thermal power plant [M]. Beijing: China Machine Press, 2008: 80–97.

[2] Wang Jun. Research of Performance Calculation Model in Cogeneration Units [D]. Nan Jing: Southeast University, 2004.

[3] Encotech. http://encotech.com/.

[4] Lin ZX, Yang YP. The Role and Analysis of Heat Exchanger of Steam and Water in CHP System [C]. Asia-Pacific Power and Energy Engineering Conference, Chengdu, China, 2010.

[5] Lin Zhenxian. Research on Energy Saving of Cooling Source Filed and Coupling Mechanism of Combined Heat and Power System [D]. Beijing: North China Electric Power University, 2010.

[6] Lin Zhenxian, Yang Yongping, He Jianren, et al. A new method for determining exhaust parameters of medium-pressure cylinder of heating units [J]. Journal of Central South University of Technology (Nature and Science Edition), 2011, 42(10):1–6.

[7] Chinese Society of Power Engineering. Thermal power equipment technical manuals II: steam turbine [M]. Beijing: China Machine Press, 2007, 2-9–2-14.

[8] Huang Xinyuan. Thermal Power Plant Course Design [M]. Beijing: China Power Press, 2008: 40–60.

[9] Shen Shiyi, Qing Heqing, Kang Song, et al. Theory of steam turbine [M]. Beijing: China Power Press, 2008: 40–60.

[10] Xu Damao. Technology lecture of steam turbine and its system: Introduction and application of characteristics flow area of steam turbine[R]. Beijing: North China Electric Power University.

Frontiers of Energy and Environmental Engineering – Sung, Kao & Chen (eds)
© 2013 Taylor & Francis Group, London, ISBN 978-0-415-66159-1

Study of boo-ay in coastal city planning directed by ecological security concept: A case study of Nanling district in Zhangzhou, Fujian

Q. Wang
School of Architecture, Tianjin University, Tianjin, China

K. Qi
Research Institute of Architecture Design and Urban Planning, Tianjin University, Tianjin, China

X.Y. Zang
School of Architecture, Tianjin University, Tianjin, China

W.H. Wang
Research Institute of Architecture Design and Urban Planning, Tianjin University, Tianjin, China

ABSTRACT: Ecological security has been the vital research subject that restricted city sustainable development in the process of urbanization. Studying the sustainable development of boo-ay in Coastal City, we need to analyze the condition of city itself from the view of regional ecology and social harmony to realize the complementarity and symbiosis of boo-ay and coastland. This paper applies GIS space technology analysis method to build the city planning system based on the principle of ecological security in order to provide effect guidance for the development of boo-ay.

With the acceleration of urbanization, the development of the coastal city in China is on the rising trend. The opening of the economy further increases the pace of city construction that impels the growing lack of land and resources, and makes the urban issues about traffic, environment, safety etc. getting more and more serious. Inland areas in coastal cities, including satellite towns and center towns that lack of coastline have few advantages than coastal areas in the policy, resources, and industry development in the past. And in recent years, with the rapid development of the construction of coastal areas, the land resource is becoming saturated, and the city border controlling of space growth is also gradually strictly. In this context, the inland areas in coastal city gradually get attentions of government and society. In this process, lessons under the superficial prosperity from the rapid development of coastal area have not been used for reference;rather, they speed up the pace of construction, leading to the extensive method of construction in some regions which results in the destruction of ecological environment of the city and even influence on the regional ecological security pattern of more hinterland. City problems have aggravated and presented the trend of spreading from coastal area to boo-ay. Therefore, with the ecological safety as the core of sustainable urban planning system, it has become an important research subject in the field of urban planning that establishing the strategic framework that integrated ecological security theory with city planning practice.

With the acceleration of urbanization, the development of the coastal city in China is on the rising trend. The opening of the economy further increases the pace of city construction that impels the growing lack of land and resources, and makes the urban issues about traffic, environment, safety etc. getting more and more serious. Inland areas in coastal cities, including satellite towns and center towns that lack of coastline have few advantages than coastal areas in the policy, resources, and industry development in the past. And in recent years, with the rapid development of the construction of coastal areas, the land resource is becoming saturated, and the city border controlling of space growth is also gradually strictly. In this context, the inland areas in coastal city gradually get attentions of government and society. In this process, lessons under the superficial prosperity from the rapid development of coastal area have not been used for reference;rather, they speed up the pace of construction, leading to the extensive method of construction in some regions which results in the destruction of ecological environment of the city and even influence on the regional ecological security pattern of more hinterland.

City problems have aggravated and presented the trend of spreading from coastal area to boo-ay. Therefore, with the ecological safety as the core of sustainable urban planning system, it has become an important research subject in the field of urban planning that establishing the strategic framework that integrated ecological security theory with city planning practice.

1 SIGNIFICANCE OF ECOLOGICAL SECURITY IN THE DEVELOPMENT OF BOO-AY IN COASTAL CITY

In 1989, the International Institute for Applied Systems Analysis proposed to build the optimized monitoring system of the global ecological security, and pointed out that ecological security is to protect life, health, happiness, basic rights, life security sources, necessary resources, social order, the human ability adapting the change of climate and other areas from threaten. In recent years, ecological security as an emerging study field of ecology, its connotation is gradually expanding. Most current academic works focus on analyzing its basic meaning, content, and about constructing ecological safety evaluation index system, the implementation of ecological system management. The application studies even extend to national ecological safety, regional security pattern and land use level, but the application research in combining ecological security and urban planning is few. Although a recognized complete definition of ecological security has not formed, its basic connotation can be summarized as environment resources security, biological and ecological system security, natural and social ecological security. According to the basic meaning, the elements of ecological security can be divided into a three level system. The system will clarify the impacted contents and control measures, and focus on the direct factors influencing ecological safety. The factors have been removed the general content such as population, policies, laws and regulations, the level of social productive forces, science and technology level etc. and form clear system of ecological security elements in order to provide reference basis and standard for building the organic relation between urban planning and ecological security elements system. [Table 1].

To sum up, the rapid urbanization construction boo-ay in coastal cities will face and a series of urban ecological safety issues associated show profoundly that ecological security in city planning cannot be ignored, and the attention to ecological security plays unique rule in alleviating the influence on natural ecological system of human construction activities and realizing the goal of

Table 1. System of ecological security elements based on different perspectives.

First-degree elements	Representative personage and figure	Second-degree elements	Third-degree elements
Security of environment resources	Xinfu Jiang (2000); Liuqing Chen (2002); Jingping Yang (2002); Zhangping Lin (2002)	Environmental capacity	The environmental capacity of atmospheric, water, soil, plants and animals
		Strategic natural resources	Resources of land, water, forest, grassland, marine, mineral
Security of biological and ecological system	Shirong Jia (1999); Zhiyan Xiong (2000); Guojie Chen (2002); Zesheng Tang (2002); Zhongwei Guo (2003)	Ecological system	Biosphere, biota, land ecological, human ecology, landscape ecological
		Biological species	Biodiversity, alien species invasion, the influence of the gene species
Security of natural and social ecological	Wansheng Zheng (2002); Weiyi Xu (2003); Qiaoying Zhang (2003); Wenbin Guan (2003)	Natural ecological	Natural resources and the ecological potential, biosphere self-adjustment ability
		The human society	Economic and social, national politics, environment and health

regional sustainable development. Considering the construction of the urban planning system from ecological security perspective is conducive to set explicit core objectives and principles in initial stage of planning and establish the tone of sustainable development; is conducive to ensure that the ecological security principle has been implemented into each subsystem in whole planning process and to establish a scientific eco-planning system which can afford basis for the establishment of specific ecological planning strategies and methods; is conducive to provide feedback to improve the planning system using the existing ecological safety evaluation index system.

2 BOO-AY PLANNING FACTOR ANALYSIS FROM THE PERSPECTIVE OF ECOLOGICAL SECURITY

At present, how to handle the relationship between economy development and resource, culture, ecological environment correctly, is the important guarantee to realize the sustainable development of the city. With the acceleration of urbanization, the differences between coastal and inland areas in coastal city in economic development, policy support, information acquisition, and other aspects cause the imbalance of the development of the city. From the sustainable principle, boo-ay should learn city construction experiences and lessons from the coastal areas, in order to ensure it forms stable ecological security pattern, and realize the complementation and communication between information and industry with the aid of its rapid transport links with coastal areas. Therefore, the construction of the inland areas needs to be considered from regional pattern, ecological safety, and land use, planning system construction, safeguard measures and other aspects. This paper regarding Nanling district of Zhangzhou as the study subject, attempts to conclude the analysis system of inland areas planning elements with universal significance, so it can provide direct basis for the construction of the whole planning system.

2.1 Regional factors analysis

The construction of west-straits economic zone has created unprecedented opportunity for the development of Zhangzhou. In the guidance of national policies and regional development strategy, Zhangzhou proposed the conception that Zhangzhou be the leading role in the construction of west-straits, in order to further speed up the space expansion of center city, and promote the opening and development of important strategic coastal areas. With platform of constructing

"Taiwanese investment zone" it will realize the cooperation between Taiwan and the mainland in economic, trade, culture, industry, and other aspects, strengthen regional service function of Zhangzhou, and entitle it as an important regional central city in west-straits area.

Nanling district attached to Nanjing county, Zhangzhou city, Fujian province. It is located in the western Zhangzhou plain, next to Zhangzhou city center. It is the satellite town of Zhangzhou and also an important part of west-straits economic zone. With the powerful radiation of Zhangzhou, Nanling district will have the opportunity of industrial adjustment and upgrade in the process of spatial extension in city center. So Nanling district need to take into account that integrates into high-tech industry base in main city and form a whole with the high-tech park in the north side of Jiulong River to strengthen the connection and cooperation of high-tech technology industry. At the same time, based on itself and surrounding environment, Nanling district should make full use of land resources, environment characteristics, traffic condition and forestry economy to realize the structure adjustment of compound industrial in Nanling district. It should take traditional industry and private economy like agricultural and subsidiary products processing, electronic information and furniture manufacturing etc. as basic support. Also it should develop modern logistics, business services, research and education and other high-end modern service industry, and cultivate high-tech industries and headquarters economy, in order to achieve combination patterns of high value-added industries, activate regional leaping growth, and become a livable type of ecological industry new town and one of important sub-centers of Zhangzhou.

2.2 Ecological culture inheritance

The complexity of ecological resources and the inheritability of the regional culture are the foundation of the development of inland area. Nanling district has favorable ecological environment, landscape around and natural vegetation abundant; the west rivulet of Jiulong estuary has provided the various forms of ecological runoff. The majority basin of river is plain, whose broad river valley, small rail slope, and integrated matrix-patch-corridor system provides easy-perched places for all kinds of living beings. The south Junzhai Mountain and east Round Hill are good ecological scene of Nanling district. This research uses the GIS spatial analysis technology to extract the comprehensive data of natural vegetation, water system, and human activities elements of this area, and according to present ecological status, it divides areas into ecological protection, ecological restoration and ecological construction area (Fig. 1). It makes corresponding planning principles according to different ecological zones and identifies that keeping regional security pattern is the core issues in regional development. In combination with the division of ecological zones, it determines that the development of Nanling district should focus on the protection of the capacity of land, water resources, natural legacy, green ecological network, cultural diversity and other ecological factors. And according to the characteristics of ecological factor, a targeted planning system should be built. It should pay more attention to construct the recyclable biological waste saving system. And through the planning principle as conserving water and soil reserving creature inhabits corridor, it should also protect biodiversity, constructing rivers and mountains radiation corridor.

Besides, it will prevent environmental pollution, reduce flood disaster, construct proper recreation system, coordinate the relationship between industrial development and urban construction, and realize the organic transition of the landscape pattern and ecological security in regional levels.

In addition, as the concentration of Minnan culture, Zhangzhou, in the long course of history, has formed unique natural geographical environment, costume culture, flower art and architecture style. So its development should emphasize reflecting the regional culture characteristics and integrating cultural innovative thinking into the construction of planning system and the choice of industry. By the lights of Chinese traditional urban spatial pattern, it need to extract regional architecture design vocabularies to make buildings echo landscape pattern, need to make the architectural forms, texture, color in combination with the connotation of classical Chinese space, and also need to form the regional characteristic architectural style with the new age skill.

3 PLANNING SYSTEM BUILD OF BOO-AY IN COASTAL CITY UNDER THE PRINCIPLE OF ECOLOGICAL SECURITY

3.1 Land performance optimization based on the resource intensive use

For inland areas, resource saving is one of the forward-looking planning principles. Learned from coastal cities in the experience of the construction and the insufficiency, planning of inland areas should prioritize preferring environmental sensibility analysis and ecological zoning, to protect the influence factors of ecological pattern. Land resources, as the space carrier of all kinds of construction in city, are the foundation of all kinds of factors. Therefore, we should try to improve the efficiency of land, through evaluation of land suitability, determine the compatible degree of land use, and select the land which has good compatibility and minimum damage to ecological environment as construction land preferentially (Fig. 2). According to the effects of policy and market, it need to build strong time-effective land use pattern of flexibility, in order to deal with all kinds of unknown changes.

According to the land resources distribution by characteristics, the inland area should make effective strategy of partition use, confirm that urban construction will follow the layout principle of region organic dispersed and core areas compacted, and give the top priority to public transit services. With using clean fuel, the efficient route will realize convenient contact between planned area and its hinterland. With the principle of intensive utilization of resources, it need to integrate ecological factor in the area and construct healthy

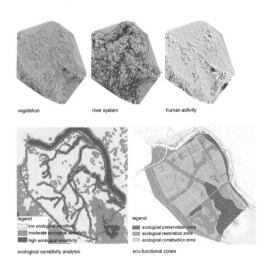

Figure 1. Ecological sensitivity analysis and ecological function regions.

Figure 2. Land suitability assessment.

efficient green transport system, to avoid the traffic interference to mountain, river, hills and forest and reduce destruction of natural terrain and vegetation, so that the traffic system and ecological factors will coexist harmoniously. By the carrier of reasonable rapid-transport network, it is recommended that the inland area develop space form in macro pattern, realize the correlation and concentration of transportation and industry between the inland and coastal areas, and integrate itself into the whole regional pattern.

3.2 Environment improvement based on the sharing of ecological culture

Combined with the GIS technology, it could integrate and optimize the matrix, patch, and corridor system internal. With the present water system as key elements of the security pattern, through the runoff and elevation analysis, and based on main flood-reserved water system we carry out the study of potential flood storage space and construct Flood-detention Lake and wetland combined with the pond, ditches and low-lying areas. The forecast and analysis of water connect system on the GIS platform are performed, and with the water system network plans, we select the best flood control ditches to connect various water in the region into network system. With keeping the dense woods, meadows and shrubs, an ecological park could be built as the basic reproduction habitat of animals and plants. Along the west rivulet of Jiulong estuary, the construction of waterfront ecological park should take good use of local characteristics vegetation, shape unique landscape pattern, and form the city's most attractive nature boundary (Fig. 3).

It needs to notice in environment improvement that landscape shaping should coordinate with current environment and pay attention to the expression of local architecture vocabulary. The design techniques need to expand to regional culture and aesthetic concept, explore culture connotation of regional culture, and realize harmony of ecological

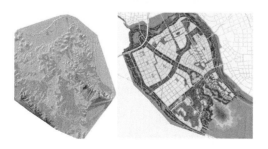

Figure 3. Land landscape value analysis and landscape pattern.

environment and green building. According to the space characteristics of the external environment, it is design that needs to realize the combination of high technology and regional culture, focus on the complexity of physique formation, appearance and space experience, and achieve natural regression of the architectural design. With the regional culture as the core, considering modern technology and materials, we will determine the style, size, color and material of architectures, in order to show the local context feature, integrate it organically with the natural environment in the hinterland, and construct architectural landscape pattern of the communication between culture and nature.

3.3 Security system based on the principle of ecological disaster prevention

To form its sustainability, the security system should have the guideline of building eco-safe living environment, pay attention to the application of disaster prevention concept in the construction of water landscape, and organically combine ecological protection, disaster prevention and resources conservation. First of all, it should establish regional flood control system, synthesize flood process simulation, and carry on historical flood disaster analysis and flood control safety pattern analysis; based on the features of terrain, set flood storage area with the elements of rivers, lakes, pond and low-lying areas in planning area. According to hydrologic process simulation, it makes catchment point of runoff as the strategic point to control water. According to the flood risk frequency, it specifies 50 years once as the flood control standard for Nanling area. Second, taking the west rivulet of Jiulong estuary as the main dredging river, and its tributaries in planning areas as subprime river, we keep the natural state of rivers, set up the flexible bank as a stable river buffer, strengthen the ecosystem of rivers and the condition of flood control, and at the same time, improve the multiformity and publicity of land use of waterfront, to keep lasting vigor and land value of waterfront and establish safe water saving model. Finally, it makes integrated ecological matrix, patch, and corridor form secure and stable ecological water system (Fig. 4).

With the objectives of livability, comfort and safety, community construction in internal planning area considers the contact between disaster-prevention space and residential public space. We design the shelter combined with the community park, square and green fields and place retaining walls and slope protections according to the geological characteristics to prevent landslides, collapse and mudslides. The use of quake-proof and local materials, such as steel structure, reinforced

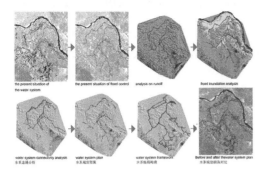

Figure 4. Water systems.

concrete mixing structure, carbon fiber composite materials will improve the disaster-prevention ability of lifeline system such as road, water, electricity and communication, avoiding the negative impact in interaction when disasters occur, and provide infrastructure support for post-disaster relief.

3.4 Dynamic feedback mechanism based on the sustainable development

After established the system of land, traffic, landscape environment, disaster prevention etc. under ecological safety concept, we should formulate the pertinence planning strategy and specific measures based on the core issues of the planning system to form a stable planning system, and establish ecological security index system to test planning system of performance. Currently, academia has been committed to the construction of the indicator system of ecological safety, and has made the preliminary research achievements, of which the environmental indicators Pressure-State-Response (PSR) concept model that Organization for Economic Co-operation and Development (OECD) and the United Nations Environment Programme (UNEP) putted forward together has strong systemic and universal significance. And then many scholars have constructed various urban ecological safety evaluation index systems that have similarity with it. Preferring ecological safety inspection by using the universal index system will form a virtuous feedback mechanism, so that it can adjust specific measures and improve planning system further.

In addition, it is important to legislate moderately and set up rewards and punishment measures to achieve the legal effect of the ecological security protection. It also needs public participation and supervision mechanism by enhancing propaganda, especially moral consciousness training of local residents to let them have the sense of responsibility that maintaining regional ecological security. All of these methods will propel the implement of ecological security concept effectively.

4 CONCLUSIONS

Ecological security is the prerequisite and basis of achieving the sustainable development of economic and social. It is an important subject of forward-looking and a complex task too that constructing the ecological security pattern of inland areas in coastal city with the ecological security idea in the field of urban planning. The case study of Nanling district, based on the analysis of the current condition of inland area, conducted the research of system and method selection and obtained the combination of ecological security and urban planning system. This study also formed planning system for inland areas that will provide effective guidelines for the construction of ecological safety livable cities.

REFERENCES

Chuanglin Fang, Xiaolei Zhang, 2001. The progress of ecological reconstruction and economic sustainable development in arid region, *Acta Ecologica Sinica* 7: 1163–1170.

Geng Wang, li Wang, Wei Wu, 2007. Recognition on regional ecological security definition and assessment system, *Acta Ecologica Sinica* 4: 1627–1637.

Ian L. McHarg, 2006. *Design with nature*. Jingwei Huang translate, Tianjin, Tianjin University Press.

Kongjian Yu, Sisi Wang, Dihua Li, Chunbo Li, 2009. The function of ecological security patterns as an urban growth framework in Beijing, *Acta Ecologica Sinica* 3: 1189–1204.

Qingyu Gong, Linchao Wang, Lin Zhu, 2007. Design guide of riparian buffers and soft banks design, *City Planning Review* 3: 51–57.

Shaolin Peng, Yanru Hao, Hongfang Lu, et al. 2004. The meaning and scales of ecological security, *Acta Scientiarum Naturalium Universitatis Sunyatseni* 6: 27–31.

Wang Jing, Zeng jian, 2007. "Complexity" Idea and Its Manifestations in Architectural Design of the Information Age, *Huazhong Architecture* 3: 66–68.

IMAGE CREDIT

Table 1 summed according to relevant data by author.

Figures 1–4 image courtesy of the Research Institute of urban planning, Tianjin University.

Frontiers of Energy and Environmental Engineering – Sung, Kao & Chen (eds)
© 2013 Taylor & Francis Group, London, ISBN 978-0-415-66159-1

The thermal performance of PCM energy storage materials in electronic thermal control

J. Yu, Q. Chen & D.K. Pan
Harbin Engineering University, Harbin, China

ABSTRACT: In electronic cooling applications, temperature stabilization requirement typically range from near ambient (20°C) up to (100°C). According this require, has a DSC test of several materials which the phase change temperature in this range. Acquire one as the best phase change material. And the simulations of absorption and cooling process of its pure material and joined graphite foam in both cases, indicate that the graphite foam greatly improves the material's absorption cooling efficiency, and reduces the impact of natural convection, which can ensure the timely cooling of electronic component and good control over its operating temperature to achieve a great energy storage effect.

Keywords: energy storage materials; thermal control; phase change; graphite foam

1 INTRODUCTION

With the extremely rapid progress in developing new micro-electronic devices and the urgent need to harvest and store energy, thermal control and the design of a new generation of energy storage have grown in importance. General, the operating temperature of electronic component range from near ambient (20°C) up to (100°C). Both higher or lower temperature are prejudicial to working. Especially higher temperature which cause failure rate increase exponentially. In recent years, The application of PCM to restrict the maximum temperature of electronic components seems very promising, especially as they act as passive elements and therefore do not require any additional source of energy. However, the traditional phase change materials have low thermal coefficient, This property reduces the thermal response of the thermal protection system and thereby may cause system overheating. graphite foam thermal conductivities are considerably high and they are strongly resistance to chemical corrosion. These good physical and chemical properties of graphite foam reveal the possibility of using them as a thermal conductivity enhancer for the PCM. We can found many research about this [1–8], In this paper one is the optimal phase change energy storage material according to the DSC tests on the materials of which the phase change temperature is in the range of the operating temperature of the electronic component. And use the commercial software Fluent simulations of absorption cooling process of its pure material and joined graphite

foam in both cases, indicate that the foamy carbon greatly improves the material's absorption cooling efficiency, and reduces the impact from natural convection, which can ensure the timely cooling of electronic component and good control over its operating temperature to achieve a great energy storage effect.

2 CHOICE PHASE CHANGE MATERIAL

Because of the operating temperature of electronic component range from near ambient (20°C) up to (100°C), we choice the following four sample materials to DSC test: record as a, b, c, and d respectively.

Figure 1 shows d have no crystallizing when the temperature down to 0°C, indicate d Melting curing amorphous, not suitable for using as energy storage materials. c will appear surfusion when solidification, however, the latent heat of c higher about 70 J/g than a and b. As a energy storage

Table 1. DSC test(NETZSCH DSC204).

Sample	Purity	Heating rate/°C/min	Cooling rate/°C/min
a	Analytically pure	5	5
b	–	5	5
c	GCS, ≥ 99.5%	5	5
d	Analytically pure, 99%	5	5

Table 2. Test results.

Sample	Melt temperature/°C	Latet heat/J/g	Crystallizing temperature/°C	Crystallizing peak/J/g
a	79.0–84.1	243.8	43.2–41.6	147.5
b	51.0–60.0	175.3	58.1–55.2	179.7
c	92.7–96.6	223.4	–	–
c	54.8–58.9	177.4	52.4–48.4	171.4

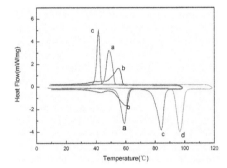

Figure 1. DSC curves of the materials.

material latent heat is the most important factor. in conclusion, we choice c as the best phase change material.

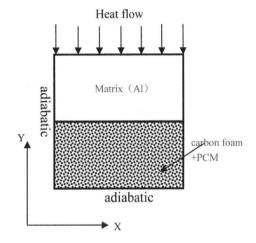

Figure 2. Physical model.

3 ENERGY STORAGE PERFORMANCE SIMULATION

3.1 *Mathematical model*

A schematic illustration of the physical problem is shown in Figure 2. The height and width of the enclosure are both 20 mm, top half of the model is Al and the other half is PCM + carbon foam. The top wall of the domain experience constant-heat flow thermal boundary conditions, while the other three walls are adiabatic. The temperature of all zone are 300 K at time t = 0.

The mathematical formulations have been done based on the following assumptions:

1. In its liquid state, the PCM is assumed to be an incompressible, Newtonian liquid;
2. The effect of buoyancy can be adequately modeled using the Boussinesq approximation.

Then the governing equations of the PCM with carbon foam can be write as:
Continuity:

$$\frac{\partial(\rho u)}{\partial x} + \frac{\partial(\rho v)}{\partial y} = 0 \qquad (1)$$

Table 3. The input parameters.

	PCM	Graphite foam matrix	Al
Melting temperature (°C)	82		
Latent heat (J/kg)	223400		
Specific heat capacity (J/kg·K)	2810 (s) 2930 (l)	710	871
Density (kg/m³)	1160	2200	2719
Thermal coefficient (W/m²·K)	0.28 (s) 0.155 (l)	1300	202.4
Expansion coefficient (K⁻¹)	0.01		
Dynamic viscosity (kg/m·s)	0.002182		

u-momentum:

$$\frac{\rho}{\varepsilon}\frac{\partial u}{\partial t} + \frac{\rho}{\varepsilon^2}\left[\frac{\partial(uu)}{\partial x} + \frac{\partial(uv)}{\partial y}\right]$$
$$= -\frac{\partial p}{\partial x} + \frac{\mu}{\varepsilon}\left(\frac{\partial^2 u}{\partial x^2} + \frac{\partial^2 u}{\partial y^2}\right) - \left(\frac{\mu}{K} + \frac{\rho F}{K^{1/2}}|u|\right)u \qquad (2)$$

133

v-momentum:

$$\frac{\rho}{\varepsilon}\frac{\partial u}{\partial t} + \frac{\rho}{\varepsilon^2}\left[\frac{\partial(vv)}{\partial y} + \frac{\partial(uv)}{\partial x}\right]$$
$$= -\frac{\partial p}{\partial y} + \frac{\mu}{\varepsilon}\left(\frac{\partial^2 v}{\partial x^2} + \frac{\partial^2 v}{\partial y^2}\right) - \left(\frac{\mu}{K} + \frac{\rho F}{K^{1/2}}\,|\,v\,|\right)v + \rho g\beta_r(T_f - T_m)$$

(3)

Energy:

$$(\rho c)_{eff}\frac{\partial T}{\partial t} = \nabla\cdot(k_{eff}\nabla T) - \rho_{pcm}\,L\delta\frac{df_l}{dt}$$

(4)

where δ is the graphite foam porosity, f_l and L are the liquid fraction and latent heat of fusion of the PCM respectively. ε is the liquid phase fraction. The value of ε is defined as the product of the PCM liquid fraction and the foam porosity. K is the permeability of the porous matrix, which is determined by the porosity and aperture. F is an empirical constant, the value of F approximate is 0.55 [9]. And $(\rho c)_{eff}$ $\delta(\rho c)_{pcm} + (1-\delta)(\rho c)_s$. The materials parameters write as follows:

3.2 *Results and discussion*

Assume that constant-heat flow of the top wall of the domain is 15 kw/m², Computing time is 400 s.

Figure 3 (b) shows the pure PCM will begin to melt about at 110 s, however, other three conditions which with carbon foam the general behavior of the melt is similar. Beginning to melt about at 190 s. As the porosity decreases, the time of beginning melt have a small advance. By comparing the case of pure PCM with other three cases, (a) shows the temperature of the top wall will rising even melt had beginning in the pure PCM conditions. However, the temperature of the top wall will maintain in the melting temperature until all of the PCM had melting. It is attributed to the high effective thermal conductivity of carbon foam enhance the PCM. The quantity of heat will transfer in time in this case. Adverse, the condition which without carbon foam enhance will accumulation quantity of heat at the matrix due to the low thermal conductivity of PCM. Lost the meaning of temperature control else. By comparing the case of different porosity. The PCM melt completely at 330 s and 280 s in the case of porosity 0.95 and 0.75 respectively. In other words, as the porosity increases, the energy storage capacity will increases.

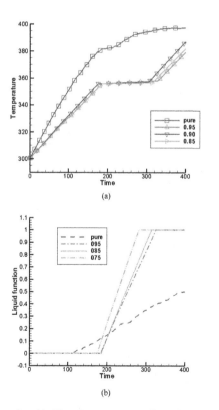

Figure 3. (a): The temperature profiles with time of the midpoint at the top wall at each cases, (b): The liquid function curves at each cases.

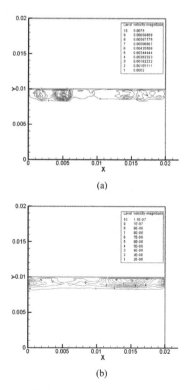

Figure 4. Velocity magnitude contours of pure PCM (a) and porosity 0.95 (b) at 200 s.

The effectiveness of natural convection to the liquid PCM is quantified by displaying the velocity magnitude. Figure 4 shows the velocity magnitude of liquid PCM can reach 10^{-3} m/s, however, the velocity order of magnitude just 10^{-8} m/s when inserting a matrix with porosity 0.95. As a energy storage material, high velocity against the stability of the materials.

As a PCM energy storage material, we should investigate the solidification of the PCM as well. In order to achieve this requirement, we change the direction of the heat flow at 400 s and continue to calculation another 400 s.

Figure 5 shows, due to the low thermal conductivity, quantity of heat can't be transferring in time, the material almost lost temperature control function in the case of pure PCM. However, after adding carbon foam to the materials, heat can be transferring quickly and controlling the zone temperature through phase change of PCM. Achieve the function of temperature control and energy storage.

To conclude, it is seen from this investigation that the improvement in the PCM storage performance depends on the carbon foams. Decreasing the porosity makes some improvement of effective thermal conductivity but this will be decreasing of the quantity of PCM and, consequently, the storage capacity of the PCM storage will decrease.

The way of using higher thermal conductivity and higher porosity seems to be a more effective way.

4 CONCLUSIONS

At first, A DSC test for four materials which the phase change temperature are in the range of the operating temperature of the electronic component. One of phase change material has been elected as the best material due to the high latent heat. Then, a numerical model is developed for an investigation of the melting and solidification of phase change materials in different cases. From the results we know that the PCM storage performance mainly depends on the carbon foams. Decreasing the porosity makes some improvement of effective thermal conductivity but this will be decreasing the storage capacity of the PCM storage.

ACKNOWLEDGEMENTS

This work was supported by the National Natural Science Foundation of China (Grant No. 51102059).

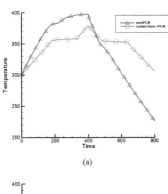

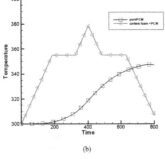

Figure 5. Temperature profiles with time of pure PCM and porosity 0.95, (a) the midpoint of the top wall; (b) the midpoint of the bottom.

REFERENCES

Beckermann C, Viskanta R. Natural. 1988. convection solid/liquid phase change in porous media. Int J Heat Mass Transfer, 31:35–46.

Fukai J, Hamada Y, Morozumi Y. 2003. Improvement of thermal characteristics of latent heat thermal energy storage units using carbon-fiber brushes: experiments and modeling [J]. International Journal of Heat and Mass Transfer, 46:4513–4525.

Hamada Y, Otsu W, Fukai J, et al. 2003. Thermal response in thermal energy storage material around heat transfer tubes: effect of additives on heat transfer rates [J]. Solar Energy, 75:317–328.

Karaipekli A, Sari A, Kaygusuz K, et al. 2007. Thermal conductivity improvement of stearic acid using expanded graphite and carbon fiber for energy storage applications[J]. Renewable Energy, 32:2201–221.

Mesalhy O, K. Lafdi, A. Elgafy. 2006. Carbon foam matrices saturated with PCM for thermal protection purposes, Carbon, 44: 2080–2088.

Sonc H, More house JH. 1991. Thermal conductivity enhancement of solid-solid phase change materials of thermal storage[J]. Thermophysics and heat transfer, 5:122–124.

Velraj R, Seeniraj RV, Hafner B. 1997. Experimental analysis and numerical modeling of inward solidification on a finned vertical tube for a latent heat storage unit[J]. Solar Energy, 60:281–290.

Zhang YW, Faghri A. 1996. Heat transfer enhancement in latent heat thermal energy storage system by using the internally finned tube [J]. International Journal of Heat Mass Transfer, 39:3165–3173.

Zhang ZG, Fang XM. 2006. Study on paraffin/expanded graphite composite phase change thermal energy storage material [J]. Energy Conversion and Management, 47:303–310.

Frontiers of Energy and Environmental Engineering – Sung, Kao & Chen (eds)
© 2013 Taylor & Francis Group, London, ISBN 978-0-415-66159-1

Graphite foam encapsulated phase change materials for thermal management of MCM

J. Yu, D.K. Pan & Q. Chen
Harbin Engineering University, Harbin, P.R. China

ABSTRACT: In this paper, the feasibility of using phase change materials with graphite foam to cooling the MCM was considered. Due to the proper melting temperature and high latent heat, erythritol as an appropriate phase change materials was chosen. Its melting temperature is 118 °C, latent heat is 339.8 kJ/kg. In order to enhance thermal conductivity of erythritol, graphite foam was considered because of high thermal conductivity. A numerical analysis was performed. The result indicates that, erythritol with graphite foam transfer heat quickly and timely, compared with the pure erythritol. It is a potential composite material to control the temperature of MCM. Different transfer coefficients and a fluctuating power are also considered to improve the feasibility.

Keywords: graphite foam; phase change materials; thermal management; MCM

1 INSTRUCTIONS

Thermal management is major issue in the design of small size electronic chips, which are subjected to high heat generation densities. The generated heat from these electronic chips may be steady or transient in MultiChip Modules (MCM) that irregularly have variable power. During transient power spikes, the temperature may reach over 100 °C. It is difficult for ordinary temperature control system to regulate well. If the heat cannot be transferred in time, it leads directly to component properties deterioration because of local overheating.

Due to the advantages: reliable performance, light weight, non energy consumption and so on, phase change temperature control technologies meet the needs of small size devices and can be applied to temperature control of MCM. Phase change temperature control technologies mainly use the Phase Change Materials (PCMs), which absorb thermal energy as latent heat of fusion during melting. Since this absorption occurs at constant temperature or within a narrow temperature range, the PCM can absorb large heat produced at the peak power operation in a short time and keep the temperature of electronic chips in a promised temperature range.

A negative aspect of PCMs that most of these materials suffer from inherent is low thermal conductivity. The lower thermal conductivity of these materials increases the charging and discharg-ing time for any PCMs energy storage. In order to improve the thermal conductivity of PCMs, extensive investigations have been carried out to improve the thermal response of PCMs through the addition of different high thermal conductivity materials, such as metal power, fins, or graphite power, carbon fibers [1–3]. Because of its lightweight and high specific surface area of heat transfer, porous matrix has gained increasing attention. Sari and Karaipekli [4] investigated thermal conductivity enhancement and storage enhancement due to the addition of expanded graphite in paraffin. D. Zhou [5] presented an experimental study on heat transfer characteristics of PCMs embedded in open cell metal foams and expanded graphite.

Some analytical studies have been carried out to investigate the potential use of PCM in the cooling applications of high power electronics. Siva [6] performed a numerical and experimentally study of the feasibility of solid-liquid PCMs for periodic power dissipating devices. Thermal enhancement has been studied with PCM enclosed inside micro channels within semiconductor devices. Evans [7] performed a thermal analysis for a package design relevant to power electronics. Lafdi [8] numerically studied the potential of using foam structures impregnated with phase change materials as heat sinks for cooling of electronic devices. Tan [9] experimentally investigated cooling of electronic devices using a heat storage unit filled with PCM inside the device.

The aim of this study is to investigate the potential of using graphite foam impregnated with PCM as heat sinks for cooling of MCM by numerical analysis. This work mainly focused on the heat transfer characteristics of pure PCM and PCM embedded with porous materials subjected to constant heat flux during charge process. In this study, erythritol was used as PCM, and graphite foam which has high thermal conductivity and high specific surface area as porous matrix. In addition, the different heat transfer coefficients were considered, and a fluctuating power was used to simulate the transient power spike.

2 MATERIALS

2.1 *Phase change materials selection*

According to literature, available PCMs used for thermal management of electronics devices should possess the following properties: (1) Suitable phase change temperature which is slightly lower than the highest temperature endured by devices, (2) large latent heat and high heat storage density to provide the minimum size of the heat sink. In this paper, suppose that the maximum temperature allowable is 125 °C, so the melting temperature of the PCMs should be below the maximum. Erythritol appears to be suitable candidates for PCMs with melting points between 100 and 125 °C, due to large latent heat and good operational safety. The thermophysical properties of sample was measured using TGA (NETZSCH TG209 F3) and DSC (TAQ200). The heating rates were 5 °C/min for the DSC and 10 °C/min for the TGA. Figure 1a shows the TGA-DTG curve for erythritol. Here, the sample mass begins to decrease at 160 °C. The mass loss suggested the decomposition of erythritol

with gas emission, and the result performed that erythritol can store thermal energy without any mass when 160 °C is the maximum operating temperature higher than 125 °C. Figure 1b shows the typical DSC pattern for erythritol, an endothermic peak appears at 118 °C with a large latent heat of 339.8 kJ/kg. The DSC and TGA results showed that erythritol could be a suitable PCM with large heat storage density and appropriate operating temperature.

2.2 *Graphite foam*

High-density electronics require more efficient and lightweight thermal management. The primary concerns in these thermal management applications are high thermal conductivity, low weight, low coefficient of thermal expansion, high specific strength and low cost. Because of the lightweight, high thermal conductivity, and high surface area, graphite foam is being evaluated as a heat sink material for cooling of power electronics. The graphitic foams exhibit a spherical morphology, and present a unique high thermal conductivity with a lower weight than metal foam. Because of the preferable orientation of large crystallites, the ligament thermal conductivity is approximately 1200–1800 W/m·K. In this paper, the ligament thermal conductivity of graphite foam is 1300 W/m·K, and the porosity is 95%.

3 NUMERICAL ANALYSIS

3.1 *Governing equations*

The governing partial differential equation in 2D describing the melting of PCM inside the porous matrix are obtained from volume averaging of the main conservation equations of mass, momentum, and energy. The property of each material is isotropic, and the thermophysical properties are constant with temperature for each material. The volume change due to the melting process is neglected.

In continuity

$$\frac{\partial(\rho u)}{\partial x} + \frac{\partial(\rho v)}{\partial y} = 0 \qquad (1)$$

In u-momentum,

$$\frac{\rho}{\varepsilon}\frac{\partial u}{\partial t} + \frac{\rho}{\varepsilon^2}\left[\frac{\partial(uu)}{\partial x} + \frac{\partial(uv)}{\partial y}\right]$$

$$= -\frac{\partial p}{\partial x} + \frac{\mu}{\varepsilon}\left(\frac{\partial^2 u}{\partial x^2} + \frac{\partial^2 u}{\partial y^2}\right) - \left(\frac{\mu}{K} + \frac{\rho F}{K^{1/2}}|u|\right)u \qquad (2)$$

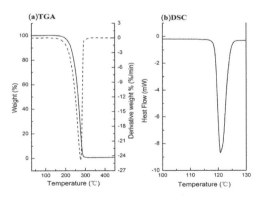

Figure 1. TGA-DTG curves and DSC melting curve of erythritol.

In v-momentum,

$$\frac{\rho}{\varepsilon}\frac{\partial v}{\partial t} + \frac{\rho}{\varepsilon^2}\left[\frac{\partial(vv)}{\partial y} + \frac{\partial(uv)}{\partial x}\right]$$
$$= -\frac{\partial p}{\partial x} + \frac{\mu}{\varepsilon}\left(\frac{\partial^2 v}{\partial x^2} + \frac{\partial^2 v}{\partial y^2}\right) - \left(\frac{\mu}{K} + \frac{\rho F}{K^{1/2}}|v|\right)v$$
$$+\rho g\beta_f\left(T_f - T_m\right) \tag{3}$$

In solid matrix energy equation,

$$(1-\delta)(\rho c)_s\frac{\partial T_s}{\partial t} = k_{se}\left(\frac{\partial^2 T_s}{\partial x^2} + \frac{\partial^2 T_s}{\partial y^2}\right) + h_{sf}a_{sf}\left(T_s - T_f\right) \tag{4}$$

In PCM phase energy equation,

$$\delta(\rho c)_f\frac{\partial T_f}{\partial t} + (\rho c)_f\left[u\frac{\partial T_f}{\partial x} + v\frac{\partial T_f}{\partial y}\right]$$
$$= k_{fe}\left(\frac{\partial^2 T_f}{\partial x^2} + \frac{\partial^2 T_f}{\partial y^2}\right) + h_{sf}a_{sf}\left(T_s - T_f\right) - \delta\rho L\frac{df_1}{dt} \tag{5}$$

where δ is the solid matrix porosity and ε is the function of liquid phase fraction. The value of $\varepsilon = \delta f_1$, where f_1 is the liquid phase fraction. K is the permeability of the porous matrix. F is the inertia resistance coefficient. k_{se} and k_{fe} are the solid matrix effective thermal conductivity of solid matrix and PCM respectively. β_f is the fluid thermal expansion coefficient and g is the acceleration of gravity. If the local thermal equilibrium is assumed, the two energy equations can be consolidated and the energy equation is given by

$$(\rho c)_{eff}\frac{\partial T}{\partial t} = \nabla\cdot\left(k_{eff}\nabla T\right) - \rho_{pcm}L\delta\frac{df_1}{dt} \tag{6}$$

where

$$(\rho c)_{eff} = \delta(\rho c)_{pcm} + (1-\delta)(\rho c)_s$$
$$k_{eff} = \delta k_{pcm} + (1-\delta)k_s$$

k_{eff} is the effective thermal conductivity of the composite material.

3.2 *Physical model*

A heat sink consisting of high thermal conductivity graphite foam encapsulated PCM (erythritol) represents the physical model. The dimensions of the heat sink are assumed to be 4 cm in height and 2 cm in width. The heat source is a MCM brazed to an aluminum substrate. The internal heat generation

of MCM is replaced by a heat flux at the aluminum substrate surface. The foam material is assumed to be in good contact with the substrate, so as to apply steady heat generation, a 136 W heat source on the left side is assumed. The top and bottom sides are adiabatic. A uniform heat transfer coefficient of 100 W/m²k with an ambient temperature of 27 °C were specified on the right wall.

4 RESULTS AND DISCUSSION

4.1 *Steady heat generation*

The pure PCM melting process is given firstly. Figure 2 presents the temperature variations with time for PCM at different locations. It is noted that the PCM temperature is higher beside the hot substrate (compare the curves: x = 6 mm, x = 11 mm, x = 16 mm) and at the top region (compare the curves: y = 5 mm, y = 20 mm, y = 35 mm). It is explained that the pure PCM transfer heat mainly by natural convection, the hot PCM liquid phase beside the heating source moved to the top region due to the buoyancy forces, it leads to uneven distribution of temperature. Figure 3 shows the tem-

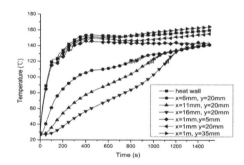

Figure 2. Temperature-time history of pure erythritol.

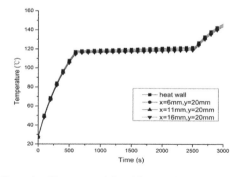

Figure 3. Temperature-time history of erythritol with graphite foam.

perature-time history of PCM with graphite foam at different location of the 5, 10, 15 mm away from the heat substrate at the axis (y = 20 mm). It is noted that the substrate temperature increases steadily with a similar rate of erythritol during the whole melting process, exhibiting quite different phenomena from pure PCM. This can be attributed to the high heat conduction through the graphite foam solid structures. Even at the solid state for PCM, the heat input at the left surface can still be quickly transferred to the whole domain of PCM through the solid matrix thermal conduction, thereby homogenizing the field temperature profile. For the pure PCM, the temperature of the heater rises sharply in the first 100s, and then turns into a moderate rise. However, the increase of the heater temperature is gently for the erythritol with graphite foam. During the melting process for the pure PCM, the temperature inside erythritol climbs slowly because of the low thermal conductivity of the solid erythritol. However, for the sample with graphite foam, the temperature inside PCM climbs quickly at the beginning, and then reaches to a plateau, about 120 °C, indicating a phase change process occurs afterwards. The appearance of the plateau implies that heat flux can be rapidly conducted to all PCM with assistance of the graphite foam skeletons. The temperature of heater can be maintain a small range about 120 °C for 20 min, it can protect the MCM from overheating for a long time, nevertheless, the pure PCM also can supply for a while, but the plateau point has reached over 150 °C. It is greater than the allowable value 125 °C. Hence, the erythritol with graphite foam is more effective method to control the temperature of MCM compared with pure erythritol.

4.2 Changing heat transfer coefficient

The heat transfer coefficient from the enclosure walls depends on the cooling system and the number of the attached fins. To investigate the effect of the heat transfer coefficient of the heat sink, three heat transfer coefficient number is considered. Figure 4 shows the results of erythritol with graphite foam with different heat transfer coefficients. It showed that the balance time increase with the heat transfer coefficient. When the heat transfer coefficient is 150 W/m² K, the balance can last over 50 min. If the power of the electronic chips is around 136 W, the heat sink consisted of PCM and graphite foam can provide a steady cooling under this heat transfer coefficient. However, the transfer coefficient cannot be enhanced infinite, because the more fins bring more weight, and the cooling system is also limiting condition. So the reasonable arrangements of

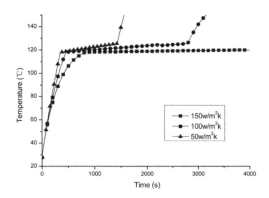

Figure 4. Heat wall temperature for different heat transfer coefficients.

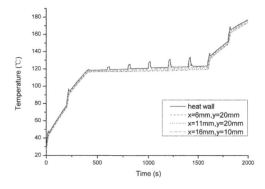

Figure 5. Temperature-time history for fluctuating heat flux.

fins and optimized regulation of cooling system are important for the heat sink.

4.3 Fluctuating power generation

The PCM played the role of a temperature damper in the heat sink, which is subjected to fluctuating heat generation. In this section, the performance of heat sink subjected to periodic spikes of heat generation has been investigated. The generated heat ranged from a minimum value of 100 W to generated energy spikes of 350 W, which lasted for 20s and repeated every 100s. To facilitate the understanding of the heat transfer performance, the heat transfer coefficient 50 W/m² K is selected. The results are shown in Figure 5. It is obvious that there are temperature peaks at the power spikes, and the peak disappears quickly. It implied that the abundance energy is transferred by graphite foam and absorbed by erythritol as latent heat. So this heat sink has a good behavior for the fluctuating power generation.

5 CONCLUSIONS

In this paper, the feasibility of using graphite foam to enhance heat transfer capability in thermal management of MCM was assessed. As a suitable PCM, erythritol is selected. Compared to the results of the pure PCM, the effect of graphite foam on solid/liquid phase change heat transfer is very significant. So the heat sink consisting of graphite foam and erythritol can maintain the temperature of MCM below the maximum allowable, and protect the MCM from overheating. It is a effective and aussichtsreich way for thermal management.

ACKNOWLEDGEMENTS

This work was sponsored by the National Natural Science Foundation of China, NO. 51102059.

REFERENCES

Bugaje, I.M. 1997. Enhancing the thermal response of latent heat storage systems. *International Journal of Energy Research* 21:759–766.

Evans, A.G. & He, M.Y. & Hutchinson, J.W. & Shaw, M. 2001. Temperature distribution in advanced power electronics systems and the effect of phase change materials on temperature suppression during power pulses. *Journal of Electronic Packaging* 123:211–217.

Fukai, J. & Hamada, Y. & Morozumi, Y. & Miyatake, O. 2002. Effect of carbon fiber brushes on conductive heat transfer in phase change materials. *International Journal of Heat and Mass Transfer* 45:4781–4792.

Lafdi, K. & Mesalhy, O. & Elgafy, A. 2008. Merits of employing foam encapsulated phase change materials for pulsed power electronics cooling applications. *Journal of Electronic Packaging* 130:021004-1-8.

Sari, A & Karaipekli, A. 2007. Thermal conductivity and latent heat thermal energy storage characteristics of paraffin/expanded graphite composite as phase change material. *Applied Thermal Engineering* 27: 1271–1277.

Siva, P.G. & Yogendra, K.J. & Jungho, K. 2002. Thermal management of high temperature pulsed electronics using metallic phase change materials. *Numerical Heat Transfer* 42:777–790.

Tan, F.L. & Tso, C.P. 2004. Cooling of mobile electronic devices using phase change materials. *Applied Thermal Engineering* 24:159–169.

Velraj, R. & Seeniraj, R.V. & Hafner, B. & Faber, C. & Schwarzer, K. 1999. Heat transfer enhancement in a latent heat storage system. *Sol Energy* 65:171–180.

Zhou, D. & C.Y. Zhao. 2011. Experimental investigations on heat transfer in phase change materials (PCMs) embedded in porous materials. *Applied Thermal Engineering* 31:970–977.

Frontiers of Energy and Environmental Engineering – Sung, Kao & Chen (eds)
© *2013 Taylor & Francis Group, London, ISBN 978-0-415-66159-1*

Analyze the influence of CO_2 from campus vehicle to the carbon emission: As a case of Guangxi University

H. Li, X. Yang, M. Tang, X.S. Lu & J.M. Chen
School of Civil Engineering and Architecture, Guangxi University, Nanning, China
Key Laboratory of Disaster Prevention and Structural Safety of Ministry of Education, Guangxi University, Nanning, China

ABSTRACT: Some researches have been completed how the modes of transportation on campus influence the carbon emission. However, the conclusions is varying because the cases done just covered one or two universities or colleges. The samples are so few that they are not representative. This paper attempts to discover the essence between the transportation mode splits and the carbon emission based on a large number of datum from 70 American schools (except McGill University). After getting something valuable, we not only try to apply this to the plan of Guangxi University but also widespread the results to universities or colleges in China.

Keywords: campus; carbon emission; auto vehicles; regression analyze

1 BACKGROUND

1.1 *Vehicle and carbon emission*

Global warming, melting in the South and North Pole have been warmly discussed by the whole world. How to control the carbon emission has also been put on schedule for many countries. The emission of vehicles takes 20 percent of the whole carbon emission on a survey of 17 large or medium-size Chinese cities [1]. Its result shows due to high rates of urbanization and the ignorance of environmental protection that the 20 percent of carbon emission was caused by transportation [1]. However, in some developing countries, air conditions are worse, where motors and vehicles will lead to high percentages constitution of the carbon emission and the rate of emission attributed by the transportation has been 50% [2].

1.2 *Motor and vehicle and carbon emission on campus*

We strongly recommend a series of measures to save energy and reduce the emission of carbon dioxide, which does not mean our campus has been or will be a polluted area but environmentally friendly elites in fact. Schools especially universities and colleges are where we make our research, which are based on three reasons.

Firstly, universities or colleges are pioneers in many fields. Their findings can be directly and efficiently applied to themselves, whose influence is positive and encourageble to the public.

Secondly, as a comparatively independent object., the elements consisted of a school are pure, simple and easily observed. Students and staff are kind and comprehensive.

Thirdly, in spite of the big difference between main roads in school and CBD (central business district), no doubt, the use of motors and vehicles have been significantly increasing in recent years, whose rate is amazing [3].

1.3 *The newest developments about controlling carbon*

New approaches to solve the contradict between the uses of motors and vehicles and the emission of carbon dioxide.

In china, many planners just put their measures on saving energy and reducing carbon emission into their new campus design [4,5] whose measures often came from their experience and their understanding. Their papers emphasized in order to deal well with the relationship between parking and campus roads they can raise the expenses of parking to limit the uses of motors and vehicles and create more buses and bicycle racks. But among a flock of ways, which one will work and how can we control the emission qualitatively? Except that Zhenlin [6] has collected the datum on the use of vehicles partly and some measures taken in built campus, the rest of the relevant articles have hardly been found.

Outside china, a number of schools have their own campus pedestrian and bicycle plans, where they mentioned the usage of motor, auto and bicycle in an elaborate and qualitative way including many datum and tables, like College of Charleston [7], University College Dublin [8]. Philippe Barla etc has analyzed the datum from the feedback of their campus transportation plans [9]. However, his article is just to pick up the datum from many school plans but readers can not get some believable conclusions in a statistical view. Some studies attempted to use statistical tools to analyze the relationship among a large number of datum but their research scopes are just limited to one or two schools.

This paper try to break up the boundary on each school, even areas. From the abundance of universities or colleges chosen and the depth of the statistical analyze, there are three goals in this paper.

1. Try to find the relationship between each items (especially for the carbon emission).
2. Analyze the contribution of each item to the emission quantitatively.
3. Prove whether the assumption has real impact on the dioxide carbon.

2 METHOD

2.1 Colecting datum

According to the representative to the statistical result, we collect the datum which come from Campus pedestrian and bicycle Plan and some websites, 70 universities and colleges (except one) in the whole America. Due to avoid the regional difference, we mostly choos American universities or colleges as our objects. The items chosen to analyze are consisted of two parts. Ones come from some objective conditions on each school such as latitude, longitude and the population of school etc. The others comes from the assumption that there are some links between parking fee and automotive vehicle, bus transit and endorsements to school and walk and bicycle and school grounds maintained organically. These 3 items we choose are practical and available. School can operate them easily.

2.2 Field research

We not only want to get a good result but also can introduce the result to our built campuses. The specific trial on Chinese school is in Guangxi University (GXU).

Like many schools in china, Guangxi University has no datum about commute split. We must collect datum by ourselves. Our group decided to get the datum by sampling randomly. The site of our survey is a part of the road beneath a bridge. The road connects eastern and western campus, which is the most significant road in Guangxi University. The information we got in our survey can be read in Table 1.

The conclusion should follow our three goals.

1. We userelevant analysis to find the relationship between each items.
2. We analyze the contribution of each item to the carbon dioxide quantitatively with stepwise regression.
3. We divide the whole into two groups whose evidences derive from the significant to the carbon emission. And this can verify whether the most significant factor has truly influence on the dependent. Our statistical tool is SPSS-15.0.

3 CONCLUSION

Though students and staff or school employees get to school with various ways or they can use blended commutes to their destinations, we just simplify this process and make some adjustments.

We summarize three kinds: automotive vehicle (high carbon), bus transit (mid carbon), Bicycle + pedestrian (low carbon) according to the carbon emission they give off (Table 2 [10], Table 3). The basic information include: the population of campus, automotive vehicle (people), bus transithe population of campus, automotive vehicle (people), bus transit (people) bicycle + pedestrian (people), endowment (million), total area (hectare), latitude (N), longitude (W), student parking fee (dollar perYear), % school grounds maintained organically, urban or not , CO_2 emission (kg CO_2 per passenger km·day (the number who use a certain transportation mode come from the sampling surveys done by their schools).

Table 3 has the names of the school where we collected the information.

In the correlation analysis (Table 4), there is a negative correlation between CO_2 emission (kg CO_2 per passenger km) and the population of school ($r = -0.390$, $p = 0.001$). There is a positive correlation between automotive vehicle and CO_2 emission(kg CO_2 per passenger km) ($r = 0.456$, $p = 0.000$). A negative correlation is existed in the parking fee and CO_2 emission (kg CO_2 per passenger km) ($r = -0.370$, $p = 0.002$). P mentioned above are all <0.01. And the correlation between longitude and CO_2 emission (kg CO_2 per passenger km) is also negative ($r = -0.294$, $p = 0.014$) whose p is <0.05.

Then we can make a stepwise regression. We choose CO_2 emission as a dependent. The rest of the items are all independents. The entry of F is more than 0.05 and the removal is fewer than 0 1.

The basic equation is $Zy = 0 + 1 \times 1 + 2 \times 2 + 3 \times 3...$ and we get the table 5.

And there are two variables entering into the equation which are the population of school and the number of people who use automotive vehicles on campus. Then, we get a useful equation which reflects the link between the two items and the emission.

$Y = 0.095 + 6.201E\text{-}006 \times$ the number of person who use automotive vehicles-2.4E-006 × the population of a school.

From the standardize coefficient, we can declare that the most powerful factor (0.780) is the number of people who use automotive vehicles.

According to the conclusion, we divide all objects into two groups (the criterion is that the objects whose percentage of auto is less than the average is defined in Group 1. The others are defined in Group 2 (Table 6). We learn that the carbon emission is indeed influenced by the number of people who use autos on campus (p = 0.000, F = 18.093). automotive vehicles-2.4E-006 × the automotive vehicles-2.4E-006 × the population of a school.

4 APPLICATION

In the correlation analysis, we learn that the emission has a negative correlation with the number of people who use automotive vehicles, the population of people and the longitude. The stepwise regression quantitively analyze the number of people who use automotive vehicles and the population of a school whose result is the same as our expectation. The further strategies to control the emission lie in these aspects. We should limit the number of autos and increase the population of the school appropriately.

With the result mentioned above, we can primarily make a plan about the carbon emission. The number of teachers and students is about 55000 in Guangxi University. We predict the number will be increased to 60000. The rate of a person who own a car achieve 26.25%. Now, there are 14441 (55000*0.2625 = 14441) cars driving in our campus. This rate include the cars just passing by. The rate is brought back to the regression equation, where there is just one unbekannte (the whole school population). Our goals contain two aspects.

Firstly, the gross of carbon emission will not be increased.

Secondly, the parking condition will not be worse than now.

The carbon emission (kg CO_2 per passenger km per day)=$0.095 + 6.201*10\string^(-6) \times 0.26256 \times 55000 - 2.4 \times 10\string^(-6) \times 55000 = 0.05255$ kg. The gross of emission = 0.05255 * 55000 = 2890.25 kg. To keep the gross constant, the designed carbon emission = 2890.25/60000 = 0.0481 kg.

The designed rate of a person who own a car = $[0.05255 - 0.095 + 2.4 \times 10\string^(-6)*60000]/6.201 \times 10\string^(-6) \times 60000 = 0.2729$. The predicted cars will be 16374 (60000 × 0.2729 = 16374). These cars include transit vehicles (cars just passing by).

Now our first measure is to limit transit vehicle to 1% of the whole cars. (0.1*16374 = 1638). But now the actual percentage is about 0.16 [11]. Except of transit cars, the rest are all owned by the teachers and students. According to this, the administration of Guangxi University can release the car permits. In order to have the same parking condition after reformation, we should define what the parking condition is. That is a rate between the cars possessed by teachers and students and the parking sites. Keep it unchanging, we should build more. Nowadays there are 2058 parking sites in Guangxi University. The rate: (1 − 0.16) × 14441/2058 = 5.894. In five years, the figure will achieve (1 − 0.1) × 16374/5.894 = 2501. Now parking sites will not be less than 443 and the parking condition will not be worse.

Every year, we should modify the figures by sampling the cars on campus. As mentioned above, parking fee is negative correlation with the number of people who use autos. The negative relationship and the development of autos on campus are significant evidences to modify the parking fee. Considering that the university has not relevant experience, we can take the local parking fee as reference and gradually coordinate the number of transit vehicles and the parking fee by sampling the possession of cars randomly.

This paper is also a reminder that we wish our campus will print some documents like campus pedestrian and bicycle plan. And it will be better that the relevant departments can provide a fixed mode for our schools which is not only advantageous to the management of the campuses but also a good case to the whole society.

REFERENCES

[1] World energy Outlook: submitted by International Energy Agency (2007).
[2] Georges Darido, Mariana Torres-Montoya and Shomik Mehndiratta: submitted to a research of The World Bank (June 2009).
[3] Informationhttp://www.weld.labs.gov.cn.

[4] Chen chengpin, Zheng chunye: On configuration of new campus construction and campus culture (Journal of zhejiang university of technology, China 2004).

[5] Hu guoxiang, Wu junmin; Study of selecting the main campus' site of the Yangze University (Optimation of Capital Constructure).

[6] zhenlinz, A dissertation submitted to Tongji University in conformity with the requirements for the degree of Master of Philosophy Research on Optimized Strategies of Chinese Campus Transportation under the Background of Mobilization.

[7] College of Charleston Campus Transportation Study (2011). http://sustainability.cofc.edu.

[8] University College Dublin, Belfield Campus Framework Commuting Strategy 2009–2012–2015. http://www.arup.ie.

[9] Philippe Barla, Nathanaël Lapierre, Ricardo Alvarez Daziano, Markus Herrmann (2012). Reducing Automobile Dependency on Campus: Evaluating the Impact TDM Using Stated. Preferences Cahier de recherche/Working Paper 2012–3.

[10] Victoria University of Wellington Travel Plan (30/09/2008).

[11] Yongshen Qian, Hailong Wan. Investigation of Urban Transit Traffic Flow and Study of Forecast Method Urban Roads Bridges & Flood Control [J], 2006, 9, 140–144.

APPENDIX

Table 1.

	Observing time	The time for class	Bus transit (people)	Automotive vehicle (people)	Bicycle + pedestrian (people)	Bus transit %	Automotive vehicle %	Bicycle + pedestrian %
1	16:55–17:25	16:55–17:25	468	925	4049	7.440	26.256	66.304
2	19:20–19:50	19:20–19:50	306	501	1577			
3	07:25–07:55	07:25–07:55	504	3084	5763			

Table 2. CO_2 conversion factor table.

Mode	kg CO_2 per passenger km	Source and assumptions
Walk	0	–
Cycle	0	–
Bus	0.0176	Used conversion factor from Environment Canada http://www.ec.gc.ca/soerree/English/Indicators/Issues/Transpo/Tables/pttb04_e.cfm
Train	0.092	Used conversion factor from Environment Canada http://www.ec.gc.ca/soerree/English/Indicators/Issues/Transpo/Tables/pttb04_e.cfm
Ferry	0.0088	Used same as buses, but divided by 2 to reflect greater numbers on board (Checking with MfE)
Car driver	0.22	Used conversion factor from Department for Environment, Food and Rural Affairs (DEFRA—UK) for a medium petrol car http://ww.defra.gov.uk/environment/business/envrp/gas/10.htm
Car passenger	0	Assume these staff are traveling with the car drivers, so their CO_2 emissions are already accounted for
Motorbike	0.085	Used conversion factor from DEFRA for a small petrol car divided by 2

Table 3.

1. Dartmouth College	36. University of California—Berkeley
2. North Carolina State University	37. Pitzer College
3. University of California—Davis	38. Pomona College
4. Catholic University of America	39. San Francisco State University
5. University of Southern California	40. Humboldt State University
6. Indiana University Bloomington	41. University of California—San Diego
7. University of California—Los Angeles	42. Western Kentucky University
8. Massachusetts Institute of Technology	43. University of Utah
9. University of Tennessee	44. University of South Carolina
10. Portland State University	45. University of Northern Iowa
11. San Jose State University	46. University of North Carolina
12. Bellevue College	at Chapel Hill
13. Arizona State University	47. University of Minnesota—Twin Cities
14. University of South Florida	48. University of Louisville
15. Smith College	49. University of Florida
16. University of California—Santa Cruz	50. University of Dayton
17. Yale University	51. Tufts University
18. Boston University	52. Southern Oregon University
19. University of Oregon	53. Princeton University
20. University of New Hampshire	54. Pomona College
21. University of Rhode Island	55. Bowdoin College
22. Harvard University	56. Ball State University
23. University of Kansas	57. Brandeis University
24. Weber State University	58. California State University, Monterey Bay
25. Radford University	59. DePaul University
26. University of California—Merced	60. Dickinson College
27. University of Kentucky	61. Drew University
28. Georgia Institute of Technology	62. Duke University
29. North Dakota State University	63. Emory University
30. Louisiana State University	64. Furman University
31. Catholic University of America	65. Grand Valley State University
32. McGill University (Canada)	67. Ithaca College
33. Michigan State University	68. Middlebury College
34. The University of Tennessee	69. Northern Arizona University
35. College of Charleston	70. Pacific Lutheran University

Table 4. Correlations.

		The population of campus	Automotive vehicle (people)	Bus transit (people)	Bicycle + pedes train (people)	Endowment (million)
The population of campus	Pearson correlation	1	.345(**)	.382(**)	.259(*)	.048
	Sig. (2-tailed)		.003	.001	.031	.698
	N	70	70	70	70	68
Automotive vehicle (people)	Pearson correlation	.345(**)	1	.406(**)	.315(**)	−.035
	Sig. (2-tailed)	.003		.000	.008	.780
	N	70	70	70	70	68
Bus transit (people)	Pearson correlation	.382(**)	.406(**)	1	.595(**)	−.016
	Sig. (2-tailed)	.001	.000		.000	.898
	N	70	70	70	70	68
Bicycle + pedes train (people)	Pearson correlation	.259(*)	.315(**)	.595(**)	1	−.023
	Sig. (2-tailed)	.031	.008	.000		.854
	N	70	70	70	70	68
Endowment (million)	Pearson correlation	.048	−.035	−.016	−.023	1
	Sig. (2-tailed)	.698	.780	.898	.854	
	N	68	68	68	68	68
Total area (hectare)	Pearson correlation	−.016	−.073	.034	.048	.039
	Sig. (2-tailed)	.904	.578	.795	.712	.768
	N	61	61	61	61	60
Latitude (N)	Pearson correlation	−.219	−.303(*)	−.059	.116	.101
	Sig. (2-tailed)	.071	.012	.632	.343	.416
	N	69	69	69	69	67
Longitude (W)	Pearson correlation	.137	−.035	−.084	−.005	−.236
	Sig. (2-tailed)	.262	.775	.494	.970	.054
	N	69	69	69	69	67
Student parking fee (dollar per year)	Pearson correlation	.195	−.071	.070	−.008	.615(**)
	Sig. (2-tailed)	.120	.577	.580	.952	.000
	N	65	65	65	65	64
% school grounds maintained organically	Pearson correlation	−.147	−.138	.020	−.045	.104
	Sig. (2-tailed)	.230	.258	.871	.715	.402
	N	69	69	69	69	67
Urban or not	Pearson correlation	.162	−.029	−.173	.083	.139
	Sig. (2-tailed)	.209	.824	.178	.523	.286
	N	62	62	62	62	61
CO_2 emission (kg CO_2 per passenger km	Pearson correlation	−.390(**)	.456(**)	−.015	.028	−.054
	Sig. (2-tailed)	.001	.000	.903	.820	.660
	N	70	70	70	70	68

Total area (hectare)	Latitude (N)	Longitude (W)	Student parking fee (dollar per year)	% school grounds maintained organically	Urban or not	CO₂ emission (kg CO₂ per passenger km)
−.016	−.219	.137	.195	−.147	.162	−.390(**)
.904	.071	.262	.120	.230	.209	.001
61	69	69	65	69	62	70
−.073	−.303(*)	−.035	−.071	−.138	−.029	.456(**)
.578	.012	.775	.577	.258	.824	.000
61	69	69	65	69	62	70
.034	−.059	−.084	.070	.020	−.173	−.015
.795	.632	.494	.580	.871	.178	.903
61	69	69	65	69	62	70
.048	.116	−.005	−.008	−.045	.083	.028
.712	.343	.970	.952	.715	.523	.820
61	69	69	65	69	62	70
.039	.101	−.236	.615(**)	.104	.139	−.054
.768	.416	.054	.000	.402	.286	.660
60	67	67	64	67	61	68
1	.093	−.072	−.129	−.024	−.220	−.056
	.479	.584	.336	.856	.098	.668
61	60	60	58	60	58	61
.093	1	−.123	.047	−.002	.096	−.035
.479		.313	.713	.988	.462	.773
60	69	69	64	68	61	69
−.072	−.123	1	.026	−.018	−.168	−.294(*)
.584	.313		.836	.884	.195	.014
60	69	69	64	68	61	69
−.129	.047	.026	1	.089	.259(*)	−.370(**)
.336	.713	.836		.483	.049	.002
58	64	64	65	64	58	65
−.024	−.002	−.018	.089	1	−.178	.019
.856	.988	.884	.483		.169	.875
60	68	68	64	69	61	69
−.220	.096	−.168	.259(*)	−.178	1	−.169
.098	.462	.195	.049	.169		.190
58	61	61	58	61	62	62
−.056	−.035	−.294(*)	−.370(**)	.019	−.169	1
.668	.773	.014	.002	.875	.190	
61	69	69	65	69	62	70

*Correlation is significant at the 0.05 level (2-tailed).
** Correlation is significant at the 0.01 level (2-tailed).

Table 5. Coefficients (a)

Model	Unstandardized coefficients		Standardized coefficients	t	Sig.	95% confidence interval for B	
	B	Std. error	Beta	Lower bound	Upper bound	B	Std. error
1							
(Constant)	.044	.010		4.585	.000	.025	.064
Automotive vehicle (people)	4.07E-006	.000	.512	4.211	.000	.000	.000
2							
(Constant)	.095	.010		9.851	.000	.076	.114
Automotive vehicle (people)	6.20E-006	.000	.780	8.457	.000	.000	.000
The population of campus	−2.41E-006	.000	−.677	−7.344	.000	.000	.000

(a) Dependent Variable: CO_2 emission (kg CO_2 per passenger km).

Table 6. Independent samples test.

		Levene's test for equality of variances		t-test for equality of means	
		F	Sig.	t	df
		Lower	Upper	Lower	Upper
The population of campus	Equal variances assumed	.351	.556	3.025	68
	Equal variances not assumed			3.143	62.065
Automotive vehicle (people)	Equal variances assumed	28.169	.000	−4.035	68
	Equal variances not assumed			−3.367	30.306
Bus transit (people)	Equal variances assumed	4.503	.037	.932	68
	Equal variances not assumed			1.061	66.608
Bicycle + pedestrian (people)	Equal variances assumed	2.188	.144	.545	68
	Equal variances not assumed			.676	47.680
Endowment (million)	Equal variances assumed	.074	.787	−.093	66
	Equal variances not assumed			−.099	63.674
Total area (hectare)	Equal variances assumed	.769	.384	−.228	59
	Equal variances not assumed			−.192	26.796
Latitude (N)	Equal variances assumed	.010	.921	.321	67
	Equal variances not assumed			.325	57.646
Longitude (W)	Equal variances assumed	2.712	.104	1.396	67
	Equal variances not assumed			1.432	60.155
Parking fee (dollar per year)	Equal variances assumed	3.139	.081	2.578	63
	Equal variances not assumed			2.978	62.795
% school grounds maintained organically	Equal variances assumed	.023	.880	−.466	67
	Equal variances not assumed			−.464	52.051
Urban or not	Equal variances assumed	3.419	.069	.985	60
	Equal variances not assumed			.963	47.408
CO_2 emission (kg CO_2 per passenger km)	Equal variances assumed	18.093	.000	−15.335	68
	Equal variances not assumed			−13.548	36.149

Sig. (2-tailed)	Mean difference	Std. error difference	95% confidence interval of the difference	
Lower	Upper	Lower	Upper	Lower
.004	13643.894	4510.869	4642.596	22645.192
.003	13643.894	4340.731	4967.076	22320.712
.000	−6430.121	1593.648	−9610.197	−3250.045
.002	−6430.121	1909.544	−10328.282	−2531.960
.355	1499.509	1609.339	−1711.877	4710.894
.293	1499.509	1413.770	−1322.694	4321.711
.587	1835.239	3366.934	−4883.374	8553.851
.502	1835.239	2715.288	−3625.162	7295.639
.926	−108.214	1166.520	−2437.246	2220.817
.921	−108.214	1088.821	−2283.599	2067.170
.820	−105.978	464.802	−1036.045	824.089
.849	−105.978	551.128	−1237.201	1025.245
.749	.3501331	1.0908787	−1.8272698	2.5275361
.747	.3501331	1.0787873	−1.8095768	2.5098430
.167	6.71369442	4.80862971	−2.88437066	16.31175950
.157	6.71369442	4.68802738	−2.66325857	16.09064741
.012	214.870	83.339	48.331	381.409
.004	214.870	72.142	70.697	359.043
.643	−.0421	.0903	−.2223	.1381
.645	−.0421	.0907	−.2241	.1400
.328	.117	.119	−.120	.354
.341	.117	.121	−.127	.361
.000	−.115952842	.007561533	−.131041647	−.100864036
.000	−.115952842	.008558924	−.133308656	−.098597027

Frontiers of Energy and Environmental Engineering – Sung, Kao & Chen (eds)
© 2013 Taylor & Francis Group, London, ISBN 978-0-415-66159-1

Study on river ecosystems restoration in Yanshan Reservoir

Y. Ji, B.H. Lu, J.Y. Li & J.Y. Shi
College of Hydrology and Water Resources, Hohai University, Nanjing, China

ABSTRACT: Yanshan Reservoir has brought many benefits for the regional social and economic development in the aspects of rational exploitation and utilization of hydropower and water resources, and flood-control, as well as disaster reduction. At the same time, the side-effect in the development and utilization has also appeared. Based on the analysis on role in social and economic development and stress that the reservoir does to river ecosystems, the study discussed the issues of the river ecosystem restoration and put forward some countermeasures to reduce the side-effect from five aspects. The research results have guiding significance for management and protection of lake and reservoir, maintaining the healthy development of ecological environment and promoting the harmony between human and water.

Keywords: ecosystem; stress; restoration; reservoir

1 ROLE OF THE RESERVOIR IN SOCIAL AND ECONOMIC DEVELOPMENT

Yanshan Reservoir has multi-function in flood-control, water supply, irrigation and hydropower generation. The reservoir is a key large flood-control reservoir in the Huaihe river basin, and plays an important role in protecting the large amount of farmlands both in Henan and Anhui Province, and the security of Beijing-Guangzhou railway and expressway routes, as well as 107 National Road.

The reservoir dam is located in Ganjianghe, the catchment area above the dam site is 1169 km². Ganjianghe is a major tributary of Lihe, east of Hongruhe, south-west of Tanghe Basin, the basin area is 1280 km², accounting for 46% of Lihe basin area and the main stream length is 98.7 km. The reservoir controls 91.3% of the Ganjiang basin area. The river systems above reservoir dam site are shown in Figure 1.

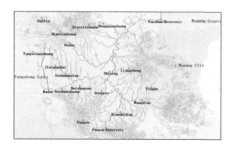

Figure 1. The river systems above Yanshan Reservoir dam site.

2 STRESS THAT THE RESERVOIR DOES TO ECOSYSTEMS WHEN COMPLETED

The reservoir is surrounded by low hills on east, west and south. Belonging to the landforms of floodplains and terraces, most of the formations in the reservoir area are composed of Quaternary loose deposits. According to the investigation results of the reservoir area made by relevant departments in July 2003, the reservoir inundation population and farmlands in different water levels are shown in Table 1.

It can be seen from the table that along with the increase of reservoir water level, the flooded area increased, on the other hand, topography, river system, streamline, soil and vegetation, groundwater, sediment and other underlying surface modalities had changed a lot.

Table 1. The relation among water level and inundation population and farmlands.

Water level (m)	Inundation population (person)	Inundation farmlands (mu)
100.0	4984	9941
102.0	6290	13756
104.0	8276	18313
106.0	10839	25502
108.0	13490	40181
110.0	20045	58052
111.0	23790	73837
113.0	36563	95022
115.0	45031	101846

2.1 Stress that reservoir construction does to river ecosystems

Construction of the dam has made the rivers non-continuous. Flowing rivers changed into relative stationary artificial lakes, not only significantly changing the geometric characteristics (such as river length, chain length, stream gradient and river cross-sectional shape) and hydrological and hydraulic characteristics (such as flow rate, water depth, water temperature and flow boundary conditions) of river system, but also changing the structure and external interference resistant capacity of food chains (networks). Mainly displays in: when the reservoir was built, the original forests, grasslands and farmlands in this area were inundated, forcing terrestrial animals to migrate; the original transformation law of river nutrient transportation changed, due to the reservoir intercepting river nutrients, higher temperature prompted the algae growing abundantly in the water surface making major plants atrophy, and the dead algae sunk to the bottom rotting at the same time consuming oxygen, water of low dissolved oxygen content could make the aquatic organisms "suffocated"; as artificial runoff regulation changed the hydrological cycles of natural rivers, the basic condition that pulse-type river corridor ecological systems formed with the hydrological cycles changed[1].

After the construction of the Yanshan Reservoir, it has made adverse effects on river ecological system of three mainstreams including Dongshahe, Xishahe and Jiahe. Due to upstream riverbed showed "V" type, midstream and downstream showed "U" type, the basin area was topographically high in the southwest and low in the northeast; basin above the dam site was wide in top and narrow in bottom, shaped like a fan, surrounded by hilly and mountainous area, with inter-hill basin in the middle, reservoir building occupied and flooded large areas of lands and cottages, vegetation was destroyed and part of the terrain was changed; because of the steep ground slope, surface lithologic resistance was poor, coupled with the low vegetation covering ratio, soil erosion was severe.

2.2 Stress that channel excavation does to river ecosystems

Reservoir construction required channel excavation or river regulation, the impacts of channel excavation or river regulation on river ecological system were mainly shown in the following respects: the natural meandering rivers transformed into linear or curved artificial rivers or river networks; as adopting the trapezoidal or rectangular or curved rules and other straight and uncluttered geometric shapes technically, the complex channel morphology of natural rivers changed; it typically used the stones or cement plates to cover in river slopes and beds. Therefore, channel excavation or river regulation changed the basic meandering form of natural rivers, diversification patterns of flows with rapid and tranquil flow, bend and shoal alternating disappeared, and river cross-section was regular in geometry, the natural form of the deep pools, shallow crisscrossing and falls and drop water varying also changed, imperceptibly making the habitat heterogeneity of river ecosystem reducing, the structure and function of river ecosystem would change, especially biological community diversity decreased, causing the degeneration of river ecosystem functions[2].

After the construction of the Yanshan Reservoir, Ganhe and tributaries of it including Dongshahe, Xishahe and Jiahe were straightened and shorten; other rivers were also cut-off and the riverbed sections expanded. The construction of various water diversion channels changed the water system and hydraulic characteristics of river systems, causing stress on river ecosystem.

3 RIVER ECOSYSTEM RESTORATION

After the construction of the Yanshan Reservoir, it played an important role in the regional social and economic development and also caused stress on river ecosystems constitute at the same time. Now and in the future, human society is inseparable from the water conservancy and hydropower engineering, how to achieve coordinated development of water conservancy and hydropower engineering construction and ecological systems is a project worth exploring.

3.1 Construct a diversity of ecological environment

Human activities, intentionally or unintentionally, made some kind of biological disappear or be introduced. These social attributes, above all, caused the destruction of the food chain of the water's edge and water. Once the food chain was destroyed, water lost the self-purification capacity.

According to the reality of social and economic development in Yanshan Reservoir region, we can do well in river ecological restoration, taking Ganjianghe as a typical river and from point to area. From the technical level, specific measures are: to return farmland to forests in water conservation and water source protection areas of the middle and upper reaches and plant some relatively large canopy trees or fruit trees around the reservoir and along the river, forming forest zones gradually; to plant shrubs or grasses on the ground, it can reduce the runoff coefficient and river erosion, on the other hand, it can play a filtering role of surface runoff and make organic matters existing in runoff in the form of suspended solids detained on shore to decomposition; to regulate river and estuary, stabilize the slope, prevent and control soil and water loss and meet the safety of flood controlling in upper and middle reaches of the river, at the same time, plant shrubs or grasses in conditional river slopes; because the river slope is a natural transition zone between the water and the land, organic combination of shrub or grass and soil can not only improve the surrounding temperature and humidity, but also create a suitable habitat for water-land transitional biology, amphibian and other organisms can multiply. In the river ecosystem restoration project, governments at all levels, especially units directly benefiting from reservoir construction, should invest some money, take engineering, management and biological measures to build natural river morphology and habitats, and to maximize ecosystem self-recovery function[3].

3.2 Adjust the industrialization structure of agriculture

Reservoir area advantageous conditions for agricultural production, has some major features that climate has both advantage of North and South, water is in good coordination with light and heat and development potential is very great. Agricultural industrialization should focus on food crops (mainly wheats and corns), economic crops (mainly fruits, vegetables, oil seed crops and cotton), many aspects of production process should linked together, promoting the transformation from traditional agriculture into modernized agriculture and from extensive pattern to intensive pattern. Since the area is relatively mountainous, dry land and paddy fields are of poor quality, dry land, mountain, plateau and the soil fertility should be transformed, to improve soil quality and realize maximize benefits in finite land resources. We should return farmlands to forests, improve ecological environment and develop ecological agriculture, to completely reverse the vicious circle that "the poorer cultivation, the cultivation and the poorer" and realize the virtuous circle of ecological environment in the reservoir area. Industrialization structure of agricultural should give priority to mega-agriculture, combine the actual conditions in different regions of the basin on basis of land and timely push agricultural economy to coordinated development of farming, forestry, animal husbandry, fishing and enterprises, and thus to promote the construction of the resource

industrialization development system, in order to stimulate social and economic development of the reservoir surrounding area.

3.3 Optimize planting structure

The growth of crops depends on soil fertility, and soil fertility maintenance is in coordination with soil biological activity, soil nutrients and water supply capacity. As a result of soil nutrients constrained by natural process, there is a special rule in the accumulation and migration process. High-intensity land use, not only destroyed the restoration and regeneration ability of desert soil, but also damaged soil nutrient balance for pursuing maximum output efficiency of lands. Climatic, soil and water conditions of the reservoir area will determine the growth of different crops. So, arranging the planting structure should be combined with local climate, soil and water conditions, optimizing planting structure is the best choice to improve the output efficiency of existing land resources[4].

Cultivating fruit trees in the areas of water and soil erosion can not only effectively control the water and soil loss, but also receive high economic benefits in short-term and make the farmer out of poverty. The development of forestry and orchard industry causes the immigrants incomes increase, at the same time, with the animal husbandry development, gradually forma benign circulation of ecological and economic benefits and raise comprehensive benefit. To develop forestry and orchard industry, primarily, agricultural and forestry departments should strengthen macro guidance and comprehensive coordination, and predict the market, clarify the existing scale and provide comprehensive information in determining which kind of fresh and dried fruits to develop the quantity to cultivate; in the areas which economic forests have developed to a certain scale, the focus should be shifted to improve the technology and management level, in order to play a better role; In remote mountainous and serious soil erosion areas, developing economic forests has not started or start late, water conservation, agriculture and forestry departments should co-organize several technology research and promotion groups, to make researches and experiments on economic forest species.

3.4 Protect ecological environment

There exist a self-renewal cycle in natural ecosystems. When utilization of the resources exceeds its update speed, it is to obtain a utility at present, and give up the more revenue opportunities in the future. Excessive utilization will cause the system population mutate, or individual species die.

Sometimes the decline of individual species will cause other dependent species decline, and emerge the chain of species declining. In the development process, the protection of environmental resources should not be ignored, consider maintaining the environment as a starting point and of coordinate development, environmental protection and environmental improvement. From the current condition, the output efficiency of planting fruit trees was significantly higher than that of planting grains, farmers have gradually transferred to the fruit growing from grain cultivation, through renting mountain fields, planting cash crops and high-quality fruits can not only obtain good economic income, but also improve and beautify the environment and realize sustainable development[5].

3.5 Analysis on impact of reservoir construction on environment

After the reservoir are built and put into operation, assessment and prejudging evaluation on the value changes of eco-environment, to reduce the adverse effects on the ecological environment to the lowest degree, and maximize the overall benefit of the reservoir in the region social economy and ecological environment. The change of river ecosystem, water resources and water environment resulting from reservoir construction is bound to affect the social and economic benefits. Therefore, the investment of river ecosystem, water resources and water environment changes, as well as the engineering measures, management measures and biological measures taken for the prevention or repairing of these changes should be evaluate as the cost of region social and economic development, to clear the value relation between input and output.

The impact of reservoir construction on ecological environment is long-term and potential, especially the structure, function and benefit changes in estuarine ecosystem, reservoir system, the aquatic ecosystem are long and slow, only after a long-term monitoring, observation, analysis and research, the inner relation and evolution can gradually be found out. Therefore, impact management units of reservoir construction on ecological environment should not only pay close attention to it and make ecosystem protection and restoration, but also plan and implement as a long-term goal.

4 SUMMARY AND CONCLUSIONS

Yanshan Reservoir construction has brought many benefits for the regional economic development. At the same time, stress or adverse effects on river ecosystem restoration has also appeared. People should follow the goal of coordinated development of reservoir construction and river ecosystems, and take appropriate engineering, management, as well as biological measures to minimize its adverse effects on river ecosystems and achieve sustainable development.

ACKNOWLEDGEMENT

This study is supported by Ministry of Water Resources' special funds for scientific research on public welfare (201201026, 200801003), and NSFC 50979023.

REFERENCES

[1] Jiang Furen, Liu Shukun, Lu Jikang: Basic connotation of sustainable development in the watershed [J], China Water Resources, 2002 (4), 20~21.
[2] Dong Zheren: Stress of water conservancy projects on the ecological system [J], Water Resources and Hydropower Engineering, 2003 (7), 37~40.
[3] Liu Changming: Water strategies of China in the 21st century [M], Beijing, Science Press, 1998, 121~156.
[4] Zhang Xiaochen: Countermeasures research on stress issue that water conservancy project do to river ecosystem [J], Water Conservancy Science and Technology and Economy, 2007 (4).
[5] Zhao Yanwei, Yang Zhifeng: Brief discussion on ecosystem restoration of urban river [J], Bulletin of Soil and Water Conservation, 2006 (1).

Frontiers of Energy and Environmental Engineering – Sung, Kao & Chen (eds)
© 2013 Taylor & Francis Group, London, ISBN 978-0-415-66159-1

Rural landscape planning & design in Jianghan Plain based on ecological views: A case of Taohuashan Town in Shishou City

N. Song

Urban Plan Department, College of Horticulture & Forestry Sciences, Huazhong Agricultural University, China

ABSTRACT: Rural urbanization is an important part of achieving urbanization in our country, so how to build rural landscape in a moderate and reasonable way has become a common topic concerned by many people. Combining the theories of landscape ecology, this paper analyses the recent problems of rural urbanization and takes the landscape planning and design of Taohuashan Town in Jianghan Plain for example to provide a kind of new idea and method for the construction of small towns.

Keywords: ecological view; rural landscape; urbanization

1 ECOLOGICAL VIEWS IN LANDSCAPE PLANNING AND DESIGN

Ecological views are human's overall understanding or views of ecological problems. Based on the basic concepts, fundamental principles and fundamental laws provided by ecological science, these views are the basic ideas which come from the summaries of philosophical worldview on the global ecosystem level of human and nature and can be used to guide human's cognition and change nature.

In the field of landscape ecology, landscape is regarded as a combination of landscape elements. Landscape elements refer to the relatively homogeneous ecological elements or units, including nature factors and human factors. There are three main types: landscape patch, corridor and matrix. In urban system, cities and towns (including villages) in different sizes and levels can be seen as one landscape unit matrix, which are composed of different types of patches and corridors (such as mountains, vegetation, rivers, farmlands, buildings, streets, parks and so on). With spatial heterogeneity, they show different spatial forms, different distribution patterns and different functions. However, in a larger regional space, they can also be seen as homogeneous landscape patches of mosaic distribution. Being interconnected and interdependent, they can form homogeneous corridors with certain functions and constitute a urban system with certain structures and functions together so as to undertake the task of operating and developing social economy.

2 RURAL LANDSCAPE PLANNING AND DESIGN

2.1 *Rural landscape*

As a kind of material expression, villages small and towns are a kind of visible landscape. Rural landscape is an whole visual image of space which is composed of buildings, structures, roads, greening, open spaces and other subjects in the village or town. Rural landscape and urban landscape have different regions and different sizes, but they have the same nature. Although the meanings of these two words town and city are different, they both are not only habitats for human but also a kind of environment where artificial conditions can dominate or control natural conditions.

The long-term backward productivity in rural areas, low educational level and weak aesthetic consciousness make residential building originally with long history and rich culture lose its own characteristics. Rural landscape construction has become dark a corner without concern. Some rural buildings in suburbs sprawl without ordered arrangement and are crumbling, so they have no sense of beauty. Even in some regions with rich culture connotations and profound historical heritage, their rural landscapes completely failed to show their futures. Especially in some scenic spots, because of having no consciousness of landscape or planning and adding the temptation of economic benefits, many beautiful natural landscapes have been seriously damaged.

There are still some problems in the development of rural urbanization in China. From the perspective of macroscopical control, these problems can be seen plainly as follows: The definition of urban function is ambiguity; The development of urban system lacks the support of scientific researches; Rural construction lacks systematic planning and usually focus on short-term benefits. From the perspective of microscopic construction, they are mainly shown as follows: similar patterns and styles, disordered land layout, indifference to the construction of infrastructure, over-exploitation and ignorance of sustainable development and ecological problems. In all, these problems can be summarized as follows:

1. Confusion of urban construction and management and low quality of urban planning
   In the process of urban construction, in order to pursue the "high, large, new and complete" features, planning is separated from practice, furthermore mass dismantlement and mass construction not only makes the construction costs extremely high, but also causes the waste of land resources and increases the pressure on the protection of basic farmland. In some areas, especially in remote mountain areas, landscape planning and design is short of instruction; Rational utilization of terrain is ignored; The division of urban internal industrial, commercial and residential function is not clear; Urban infrastructure is poor; And the urban image is bad. What's more, disorganized management mechanisms of construction, imperfect legal systems and illegal buildings are common phenomenons in rural areas.
2. Debasement of the quality of urban ecological environment
   High densities of population in time and space causes rapid deterioration of urban environment. The problem in the areas with developed township enterprises is the most serious. In these areas, township enterprises usually scatter in the village, which are in small scale and short of labour resources, physical resources and financial resources to make a unified and effective pollution government. Thus, the environmental pollution gets more and more severe, causing irreversible ecological pollution to quite a few regions. So many township enterprises along the Taihu Lake didn't take action against the pollution timely and effectively is the reason why cyanobacteria flooded Lake Taihu in 2007.
3. Degradation of natural landscapes in urban areas

In the process of urbanization, a great deal of natural landscapes in urban areas have been destroyed. Mountains were destroyed, vegetation was cut down, the loss of water and soil was serious, rivers became turbid instead of clear and the habitats of animals and plants were occupied. The shrinkage of natural landscapes directly leads to the reduction of greenery area. The reduced air quality and polluted water environment causes the deterioration of local micro-climate. The level of comfort and aesthetic values of natural environment is lower. At the same time, the normal structure and function of original natural ecosystem were also damaged, and the types of ecological environment tend to be simple and fragmentary, thus effecting the species diversity.

4. Loss of rural characteristics
   China is a multi-ethnic country where the integration of different cultures and the development of local natural environment produce many distinctive urban styles. However, the national culture characteristics in some towns are faced with the danger of losing their brilliancy and tending over comedown under the impact of the slogan appealing for economic development.

3 GENERAL SITUATION OF TOWNS IN JIANGHAN PLAIN

3.1 *Geographic situation of Jianghan Plain*

Located in south-central Hubei, Jianghan Plain is flat with no mountains. It has convenient transportation and compact cities and towns. With relatively developed commodity economy, it is one of the four fertile plains in the South of China. Including Xiajiang and Caidian District, in Wuhan, Jianghan Plain has 37 county, city and district administrative units, which is one of the main parts of the Plain of the Mid-Yangtze River.

3.2 *Current situation of towns in Jianghan Plain*

With the development of economy, road transport plays an important role in rural areas. Many new urban areas have sprung along roads, thus forming a town pattern with new and old buildings. In some original villages, some of the buildings follow spot-style layout, some of them lie on the both sides of highway, whose front part is for business while the back part is for living. However, the consequences of this layout include traffic congestion, environmental pollution, low quality of life and poor business power, and these buildings mainly constructed by red bricks, concrete and ceramic tiles, resulting in disordered rural landscape with fewer features.

4 EXAMPLES OF THE LANDSCAPE PLANNING AND DESIGN IN TAOHUASHAN TOWN

4.1 Geographic situation of Taohuashan Town and its landscape current situation

Located at the junction of Hubei and Hunan provinces, Taohuashan Town is in the southeast of Shishou City, which can be called the south gate of Shishou City. Established districts in township of Taohuashan cover 29.46 hectares. The west and south of township is surrounded by Guolao Hill and Taohua Hill. In the north-west, there is a river running through the township. The terrain of the town is rugged, which is higher in southeast and lower in northwest. The distribution of township is relatively more compact, and its development mainly takes place along the Taohuashan Road, Jiufogang Road and Hongjun Road. Roads parallel with the east-west line are open to traffic, but the links between roads are imperfect because the road system has not yet formed. There are two public green spaces in the township, one is located at the north entrance to town, the other is located at the south entrance, but there is no park there. It has better environmental quality and high quality of environmental health management. There are a dozen small furniture factories, bamboo plants and cold drinks corporations in the township, and in the west, there is a large-scale factory called Jingjiang Mineral Processing Pharmaceutical Factory.

Taohuashan has beautiful landscapes and sceneries with 54-mile meandering green peach mountains. It has 280 peaks of various shapes with unique natural landscapes. Due to the good tourist resources, it is a renowned tourism resort in Shishou City and the surrounding areas. In 2001, Taohuashan formulated a strategy for thriving the town through tourism. There is a provincial second-grade highway throughout the township level. Although a commercial street is established, its business operation is not satisfied because of the wide street and excessively quick running vehicles. The distribution of township, commercial street and buildings can not reflect the characteristics of tourist town, what's worse, the requirements of production and livelihood for different groups of people are not taken into account.

4.2 Ideas of landscape planning and design in Taohuashan Town

Based on the landscape planning of 2001, the recent landscape planning in Taohuashan is revised on the theory of landscape ecological design. It concentrates on the operating mode of rural social life, tourism and economy as well as the analysis of the relationship between a series of elements like environmental protection and landscape planning so as to make a good rural landscape planning and design which fits the local ecological environment fine. Integrate the center landscaping, group landscaping, village landscaping, field landscaping, lakeside landscaping and other kinds of landscaping together and surround it with lakes and rivers to build a modern pastoral eco-village with mountains, rivers as well as a feeling of rusticity and friendliness.

4.3 Layout of landscape planning and design in Taohuashan Town

The landscape planning and design in Taohuashan pays great attention to the strategic goal of thriving the town through tourism. So, the newly-built wetland park is regarded as the center of the township. The transit line goes through the center of this wetland park, which is separated by a 20-meter green belt so as to reduce the impact of noise and dust pollution on the town. The first level of landscape planning is to connect the Planning Road 3, Planning Road 4 with the provincial second-grade highway to build a township ring road. Commercial and residential buildings can be built along the ring road. The style of block and architecture is determined to be classical style for commercial street, and the space formed by bridge galleries and memorial arches will create a kind of traditional commercial street atmosphere. The second level of the landscape planning is to enclose distinctive pedestrian streets for each group at the external area of the ring road. Distinctive commercial pedestrian street is a commercial villages which are made up of many single-family workshops that produce small commodities. The first and second floors are retail shops, while the third and forth floors are residence of shop keeper. According to different needs, each group not only can sell pottery, bamboo and other crafts, but also can perform the function of travel reception. The shop keeper can keep the front part as yard and fix the back part as a garden or orchard, thus achieving the coexistence of business and farming. The third level of landscape planning is to construct nest building, bridge home, hanging house, the building like floating box or elevation-type house from each group to all the villages in flat places, wetlands and hillside lands. Considering residents' source of livelihood, no separate residential area will be built, so residents can be scattered in different places and support home by doing business or farming.

The distribution of villages in wetlands mainly develop the lacustrine environment without traffic interference. Building private cottages, parks, schools, resorts and other buildings is to highlight the use and fun of lakes and surrounding property to successfully excavate the value of land and water. As for architecture, give prominence to traditional styles, which must meet the requirements of reception, accommodation and entertainment. As for transport, mainly use roads, some small areas can reserve water transport. As for energy, mainly use clean energy like solar and biogas. Each household owns a methane tank, taking the way of distribution of rain and sewage to collect and process sewage together. It mainly uses livestock-raising to make money, combining with tourism industry. The available building types include nest building, bridge home and hanging house, which have the following feature: the changes of water level won't affect the safety of living. According to the features of Jianghan Plan where the buildings may be inundated with severe floods, village architectures in flat places are proposed. The available building types include overhead house, island-style building, elevation-type house and the building like floating box, which have this features: land transportation can be transformed to water transportation while suffering the floods. The distribution of the villages in hillside lands should consider restoring green space, and planting trees on the roof and using clean energy can be used to form a reasonable ecosystem.

5 SUMMARY

To apply the proper ecologic views according to the local conditions is an important basis to solve the current problems of the rural landscape construction in China. The author only takes the landscape planning and design of Taohuashan Town in Jianghan Plain for example, so the processing of many problems may be imperfect. Hoping through this discussion, people will pay more attention to rural landscape planning to avoid the occurrence of over urbanization and the destruction of beautiful rural landscape.

REFERENCES

[1] Qiang Jian: *Rural Landscape Should Be Charming* [J]: Urban and Rural Development:6(2004):p. 52–53.
[2] Xing Xueshu, Wang Hong and Jiang Shu: *Study on the Ecological Conservation of Rural Landscape in the Process of Urbanization* [J]: Journal of Guizhou University of Technology: 5,37(5) (2008): p. 207–210.
[3] Chen Shunan et al.: *Study on the Mode of Agricultural Ecological Landscape of the Wetland in Jianghan Plain (Abstract)* [C]: Design for China—The First National Environmental Arts Design Biennale; Beijing: China Architecture & Building Press; 6(2004):p. 31–35.
[4] Ye Yun, Liao Xuan and Liang Jingyun: *Transform Floods into Water Conservancy—Study on the Mode of the Lake-Wetland Village Settlement in Jianghan Plain [J]*: Technological Development of Enterprise: 2,28(2) (2009):p. 45–46.
[5] Tian Mimi and Zhao Hengyu: *Discussion of Rural Landscape Elements in the Region of Jianghan Plain* [J]: Journal of ZheJiang University of Technology: 4,36(2) (2008):p. 226–231.

Frontiers of Energy and Environmental Engineering – Sung, Kao & Chen (eds)
© 2013 Taylor & Francis Group, London, ISBN 978-0-415-66159-1

Water quality monitoring and intelligent prediction system of water-bloom in lakes

L. Sheng, X.Y. Wang, Z.W. Liu, J.P. Xu & S.Q. Dong
Beijing Technology and Business University, School of Computer and Information Engineering, Beijing, China

ABSTRACT: In view of the backward states of water quality monitoring and the difficulties of water-bloom prediction, a grey-BP neural network for water-bloom prediction and forecasting is proposed in this paper. And an intelligent system based on GPRS for water quality remote monitoring and water-bloom prediction is built, which achieves several functions including real-time water quality monitoring and water-bloom prediction. This system provides environmental protection departments with an efficient and practical method for water environment control.

Keywords: water-bloom prediction; grey-BP neural network; GPRS; water quality monitoring

1 INTRODUCTION

Water-bloom is a typical representation of Lake Eutrophication. It not only does harm to infrequent freshwater resources, but also has bad effect on water quality and ecological environment. In addition, algae toxin which is generated by water-bloom would threaten people's health by food chain. In China, this phenomenon occurs frequently. Lakes, such as Taihu Lake, Dianchi Lake, even Hankou River, the biggest tributary of Yangtze River, are inevitably included. Therefore, it is clearly that discovering a measure that applies advanced technology effectively to water-bloom prediction and water quality remote monitoring is becoming one of the most difficult focuses in the field of water environment research.

Aquatic ecosystems are very complex due the diversity and connections of the components governing the system's dynamics. The mechanism of water-bloom outbreaks is not quite clear. Currently, mathematical models for water-bloom prediction have not been set. Some intelligent prediction theories, such as neural network and supporting vector machine, are used to predict the occurrence of water-bloom, which provide several efficient ways for water-bloom prediction. However, these methods are more suitable for short-term water-bloom prediction. At present, two measures are adopted in water quality monitoring: the first one is to collect water quality data with a portable monitor. Then the data are analyzed in laboratory. Another one is to use water quality monitoring systems, which are consisted of a monitoring center and several monitoring stations. To the former measure, however, the real-time water environmental

parameters can't be monitored. There are other issues, such as longer monitoring cycle, larger labor requirement, and lower data accuracy. Although the latter method can compensate for the above-mentioned drawbacks, problems still exist, such as limited scope of monitoring and higher cost of equipment investment.

In order to improve the backward states of water quality monitoring and solve the problems of water-bloom prediction, a composite prediction model based on grey-BP neural network is proposed. It combines two advantages—higher computing speed of grey prediction model and BP neural network's superiority of nonlinear prediction. Moreover, the composite model avoids single neural network' disadvantages in information loss and reduces the randomness of prediction. The model improves the accuracy of water-bloom prediction, and extends the forecast period. In the paper, an intelligent system based on GPRS for water quality remote monitoring and water-bloom pre-warning is proposed. The system can obtain real-time water quality data collecting and send the data to the host system. Meanwhile, according to the data being collected, the host system predicts the outbreak of water bloom through grey-BP neural network model.

2 WATER-BLOOM PREDICTION MODEL BASED ON GREY-BP NEURAL NETWORK

Taking into account that the mechanism of water-bloom outbreaks is complicated and nonlinear, it is difficult to build up an ecological model through water-bloom mechanism directly. Intelligent prediction method is able to make the impossibility possible. In the paper, Principal Component Analysis is used to pick up prediction elements which have significant influence on water-bloom outbreak. And a multi-input and mono-output neural network prediction model is built.

In the process of building the water-bloom hybrid prediction model based on grey-neural network, BP neural network with multi-input and mono-output has been used. The deviation value of chlorophyll is the difference between prediction values $\hat{x}^{(0)}(l)$ obtained by GM(1,1) model and values of $x^{(0)}(l)$, for every stage, acquired every day (24 hours as a stage) in time series, that is $e^{(0)}(l) = x^{(0)}(l) - \hat{x}^{(0)}(l)$, after addressing which gets the deviation (residual difference) sequence of the n day, which is treated as the input of BP network training, combined with other prediction elements which is obtained by principal component analysis method and orthogonal experiments. The neural network was trained with a large number of measured data. The deviation of chlorophyll in different stages can be predicted. Getting the sum of all the deviation prediction sequences and previous corresponding grey prediction sequences, the prediction values of chlorophyll can be obtained.

The neural network consists of three layers: 10 nerve cells of hidden layer, transfer function tansig (tangent curve); 1 neural unit of output layer, transfer function purelin (linear). The composite model structure of neural network is shown in Figure 1.

Inputs: Temperature T, total phosphorus TP, total nitrogen TN, light value, Dissolved Oxygen DO, deviation sequence value of chlorophyll a moment ago;
Target outputs: T = prediction deviation value of chlorophyll at the next moment;
Training algorithm: Trainbpx;
Training accuracy: 0.001.

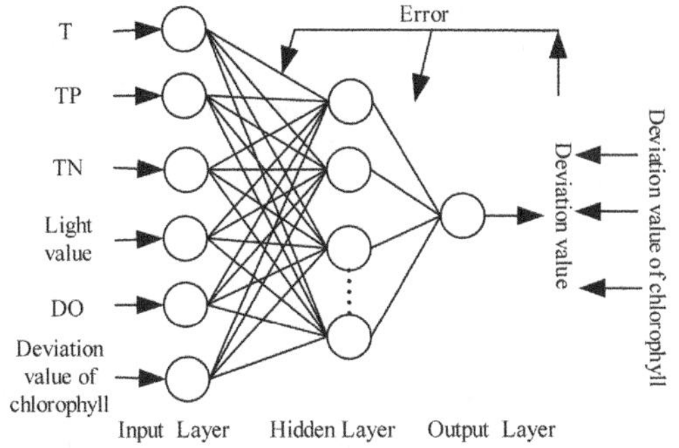

3 WATER QUALITY REMOTE MONITORING AND WATER-BLOOM PREDICTION SYSTEM IN LAKES

The structure of water quality remote monitoring and water-bloom prediction system in lakes is shown in Figure 2.

The system is divided into four layers: data collection, data transmission, data processing and results display. The main function of the system is to transfer water quality information which is obtained by sensors to monitoring center through GPRS remote transmission technology, then the data are stored in the database and analyzed by intelligent algorithm which is inserted in monitoring center to achieve water-bloom prediction. At last, monitoring and forecasting results will be published through the Internet. So that users can view the Web page to access the results in time.

3.1 *GPRS-based wireless transmission technology in data transmission module*

GPRS-based data transmission module is in data transmission layer, connecting with water quality monitoring sensors through RS232 port. The water quality data which are obtained by sensors are transferred to monitoring center through GPRS wireless data transmission module. Then the data are stored in the database. The structure of data transmission module is shown in Figure 3.

TT60 module is at the core of the data transmission module. It supports automatic dial-up, automatic reconnection, TCP/UDP protocol and heartbeat protocol. A SCM (single-chip microcomputer)—C8051F023, produced by Silicon Labs is used for TT60 module control. DS12887 chip is used for clock chip which is inserted in the battery so that the data won't be lost if power is cut. Module's supply voltage is 5 V. The realization of data transmission function of GPRS-based wireless transmission module includes establishment

Figure 1. The structure of composite neural network model.

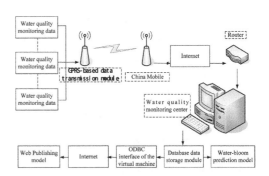

Figure 2. The structure of system.

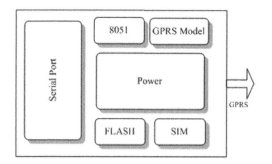

Figure 3. The structure of TT60 in data transmission module.

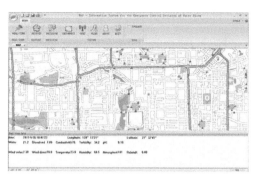

Figure 4. The real-time monitoring interface.

of the IP protocol, creating a connection, access to the network and sending messages, which are achieved by MCU programming.

3.2 Host system design

The host system—Water Quality Monitoring center is consisted of database data storage module, Web publishing module and water-bloom prediction module. The system has multiple functions including water quality monitoring function in lakes and water-bloom prediction function. Meanwhile, the prediction model can be upgraded according to the variation of water quality data in different fields of water.

The functions of Web publishing module include real-time data display, historical data and corresponding curves display, the water-bloom prediction results and corresponding curves display. By using dynamic Web development technology combined with PHP and MySQL database, the publishing module has many advantages including easy maintaining, expanding and operating the databases. The users are able to visit the website to see the water quality monitoring data and the results of water-bloom prediction with an ordinary PC which can access to the Internet. The real-time monitoring interface is shown in Figure 4.

Using Ajax (Asynchronous JavaScript and XML) technology, the real time data monitoring interface renews the data without refreshing the Web page, so that the burden of servers and databases caused by refreshing Web page frequently is avoided. Meanwhile the performance of system is improved, and it is user-friendly.

Through COM component technology, the water-bloom prediction model built by MATLAB is embedded in the host system which is designed with VC++ in Windows platform. This mixed programming technology can take the advantage of programming convenience given by MATLAB for complex algorithm and VC++'s excellence

Figure 5. Water-bloom prediction module.

characteristics: powerful programming capabilities and high speed of program execution. Moreover, it is easy for the host system to update the neural network module according to users' needs and enhance practicability and expansibility of the system. Water-bloom prediction module is shown in Figure 5.

Jpgraph technology is adopted to draw data diagrams for historical data inquiry interface and water-bloom prediction module. So that monotonous and inaccessible number can be transformed into straightforward and vivid diagrams. Diagrams strengthen the readability of data and it is convenient for users to comprehend the water quality information and water-bloom outbreak trend. While providing series of decision-making support for water environment protection departments, the system has significant effect on technical innovation in water environment protection.

4 CONCLUSION

According to the characteristics of the outbreak of water-bloom, grey-BP neural network model can simulate the complex nonlinear relationship between impacting factors and prediction results, reduce the influences on prediction values caused

160

by cycle and random components and improve the prediction accuracy. Combined with Water quality remote monitoring system and water-bloom prediction model, automatic and intelligent water quality monitoring, water quality data analysis, water-bloom prediction and results publishing are obtained, which provides a valuable and efficient solution for automatic control in water quality real-time monitoring and water-bloom prediction. Meanwhile, monitoring and prediction results provide the environment protection department with references helping to decide what treatment should be made before water-bloom outbreak.

ACKNOWLEDGMENT

This study is supported by the National Natural Science Foundation of China, Beijing Science and Technology New Star Project (51179002, 2010B007), Beijing Technology and Business University Graduate Innovation Fund in 2011. Those supports are gratefully acknowledged.

REFERENCES

Deng Julong. Grey Prediction and Grey Decision. Wuhan, Huazhong University of Science and Technology Publishing House. 2002.

Liu Zaiwen, Cui Lifeng, Wang Xiaoyi. Soft-sensing Method for Water-bloom in River-Lake based on RBF Neural Network. Beijing University of Aeronautics and Astronautics Press. 2007: 108–111.

Liu Guangzhong, Li Xiaofeng. The Improvement of BP Algorithm and Self-Adjustment of Structural Parameters. Journal of Operations research. 2001. 5 (1): 81–8.

Pei Hongping, Luo Nina. Applications of back propagation neural network for predicting the concentration of chlorophyll-a in West Lake. Acta Ecologica Sinica, 2004, 24 (2):246–251.

Qi Wenqi, Chen Guang, Sun Zongguang. The water environment monitoring technology and the development of instrument. Modern Scientific Instruments, 2003, 13(6): 8–12.

Wang Xiaoyi, Liu Zaiwen etc. Short-term prediction on water-bloom based on evidence theory neural network. The 7th World Congress on Intelligent Control and Automation (WCICA'08) 2008:8981–8984.

Xi Junqing, Wu Huaimin, Jiang Huohua. The state and advice of China's environmental monitoring capacity building. Management and Technology of Environmental Monitoring, 2001, 13(3):102–106.

Zhang Xuegong. Introduction to statistical learning theory and support vector machine. Acta Automatic Sinica, 2000, 26 (1): 32–42.

Frontiers of Energy and Environmental Engineering – Sung, Kao & Chen (eds)
© 2013 Taylor & Francis Group, London, ISBN 978-0-415-66159-1

Effects of environmental factors on phytoplankton assemblage in the Plateau lakes

J. Li

Institute of Environmental Sciences and Ecological Restoration, Yunnan University, Kunming, China
Yunnan Environment Science Institute, Kunming, China

Y. Huang

School of Resource Environment and Earth Science, Yunnan University, Kunming, China

C.Q. Duan

Institute of Environmental Sciences and Ecological Restoration, Yunnan University, Kunming, China

ABSTRACT: Data on phytoplankton specious composition and environmental factors, collected from the plateau lakes of Yunnan (China), are examined using canonical correspondence analysis. The relationship between the change of phytoplankton population and environmental factors is investigated. It shows that depth appears to be the main driving factors for phytoplankton dynamics in the lake. Dissolved oxygen is linked to depth variation in the plateau lakes. TP, TN and COD_{Mn} also play an important role in assembling the communities. The gradual increase in water nutrition could have led to the selection of more tolerant self-shading species.

Keywords: environmental factors; phytoplankton; lakes; canonical correspondence analysis

1 INTRODUCTION

The influence of environment factors on phytoplankton dynamics differs significantly. Environmental variables are likely to influence phytoplankton assemblages directly, by influencing cellular production and content (Long et al., 2001). Physical factors, such as temperature, depth, transparency are the most important. Chemical factors, such as DO, pH, salinity and nutrient level are also important (Reynolds, 1984), and indirectly influence phytoplankton species and concentrations (Chorus, 2001). However, the behavior of phytoplankton population could not be predictable, the effects of environment factors are difficult to evaluate. In this respect, multivariate statistical techniques have proved to be the valuable tool in the study of the environmental factors influencing phytoplankton community dynamics (Alvarez-Cobelas et al., 1994). Canonical Correspondence Analysis (CCA) is a multivariate method to detect the internal structure in data sets and elucidate the relationships between species assemblages and their environment (Ter Braak, 1987;1989).

The purpose of the study was to investigate the relationship between phytoplankton population and environmental factors in the plateau lakes. It will be shown that a clear picture of the effect of various environmental factors, which potentially influence assembles of the phytoplankton.

2 MATERIALS AND METHODS

This area included a total of 2 lakes from the same region of the Yunnan-Guizhou Plateau: the Fuxian Lake and the Xingyun Lake. The two lakes were connected by an artificial channel. The climate of this region is warm and wet. Annual mean air temperature around lake area is 15.6° C. Mean annual precipitation is 942.4 mm (84% occurring in May–October). The two lakes receive surface runoff from the catchment through many streamlets (Cui et al, 2008). A summary of the physical, chemical and biological characteristics and the location of the two lakes are shown in Table 1.

Ninety-nine simples were carried out seasonally from July 2010 to April 2011 (Fig. 1). Samples for chemical and phytoplankton analyses were taken simultaneously. Chemical characteristics were measured by Yunnan Institute of Environmental Science. Nitrate was measured using Ultraviolet Spectrophtometric Screening method, and phosphate was measured using Vanadomolybdophosphoric Acid Colorimetric method. Phytoplankton samples were fixed in 4%

Table 1. A Summary of the Fuxian lake and the Xingyun lake.

	Fuxian lake	Xingyun lake
Latitude	24°21′N-4°38′N	24°18′N24°23′N
Longitude	102°49′E-02°58′E	102°45′E-102°49′E
Altitude	1721 m	1720 m
Water cover area	211 km²	34.7 km²
Maximum depth	155 m	9 m
Eutrophic state	Oligotrophe	Hyper eutropher

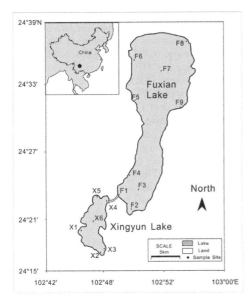

Figure 1. Sampling stations in Fuxian Lake and Xingyun Lake and its location in China.

formaldehyde immediately after collection and identified and counted using Nikon E200 microscope. All phytoplankton were identified to species or genus level. The matrices of species assemblage and environmental variables were examined to canonical correspondence analysis (Ter Braak, 1987) using the CANOCO 4.5 program (Ter Braak, 1989).

3 RESULT AND DISCUSSION

The positions of the phytoplankton species and samples are based on their relations to the environmental variables in a CCA ordination diagram. The eigenvalues of species for CCA axes 1 and 2 are 0.485 and 0.083. The results were tested by the unrestricted Monte Carlo permutation

(499 runs) (Table 2). The canonical axes both based on their p-values demonstrate with statistically significant (Table 3).

CCA analysis has proved to be an efficient method in yielding valuable information on phytoplankton-environment interactions and trends (Figs. 2 and 3). They seem to explain dynamics of smaller algal assemblages (Romo, 1991).

The environmental variables used here are depth, Dissolved Oxygen (DO), Total Nitrogen (TN), Total Phosphorus (TP) and Permanganate Index (COD_{Mn}). The correlations between

Table 2. Summary of CCA ordination diagram.

Axes	1	2
Eigenvalues	0.485	0.083
Species-environment correlations	0.996	0.993
Cumulative percentage variance of species data	46.0	53.9
Cumulative percentage variance of species-environment relation	58.8	68.9

Table 3. Summary of Monte Carlo test (499 permutations under reduced model).

Significance of first canonical axis		Significance of all canonical axes	
Eigenvalue	0.485	Trace	0.824
F-ratio	4.260	F-ratio	1.992
P-value	0.002	P-value	0.004

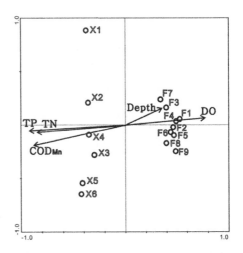

Figure 2. The positions of the surface sample sites and their relations to the environmental variables.

163

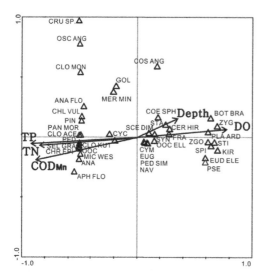

Figure 3.   The positions of the surface sample sites and their relations to the environmental variables.

the environmental variables and two axes can be explored in Finger 2. The relation of depth and DO is closer. While the correlation among the TN, TP and COD$_{Mn}$ is more positively correlated. Axis 1 from diagrams of CCA (Figs. 2 and 3) has highest eigenvalue (0.485), represented mostly phytoplankton species. Therefore, variables are highly positively correlated to the axis 1. Axis 2 (Figs. 2 and 3) has too much "noisy" (eigenvalue 0.083) too reflect phytoplankton species. Therefore, there isn't meaning to discuss axis 2.

Figure 2 shows the positions of the surface sample sites in the CCA ordination diagram and their relations to the five environmental variables. Samples from the Fuxian Lake project positive axes of depth and DO and negative axes of TN, TP and COD$_{Mn}$, suggesting that they represent the better water quality. Samples from the Xingyun Lake are placed in the positive axes of TN, TP and COD$_{Mn}$, implying that they are characterized by deteriorating environment. The diagram of variables and samples is consistent with the known facts, indicating that is credible.

The positions of the phytoplankton species are based on their relations to the environmental variables in a CCA ordination diagram (Fig. 3). Phytoplankton species could be classified four groups according to the positions of environment variables which they projected.

1. The bottom layer, excellent water
   Cyanophyta taxa, e.g. Pseudanabaena sp. (PSE SP.), Planktothrixag ardhii (PLA ARD), and Chlorophyta taxa, e.g. Eudorina elegans

(EUD ELE), Kirchneriella spp. (KIR), Zgogonium sp. (ZGO), Botryococcus braunii (BOT BRA), Zygnema spp. (ZYG), Stigeoclonium spp. (STI), Spirogyra sp. (SPI), project onto the positive axis of depth and DO. Meanwhile, they have negative value on the TN, TP and COD$_{Mn}$ axis. Assemblage of Cyanophyta taxa and Chlorophyta taxa are related the lowest level of eutrophication. All of those indicate the excellent water quality with the highest concentration of DO, and the lowest concentration of TN, TP and COD$_{Mn}$ in the bottom layer.

2. Middle and lower layer, better water
   Oocystis taxa, e.g. *Elliptica* (OOC ELL), *Scenedesmus dimorphus* (SCE DIM), *Staurastrum sp.* (STA), *Coelastrum sphasricum* (COE SPH), *Cosmarium angulosum* (COS ANG), *Golenkinia spp.* (GOL), and Euglenophyta taxa, e.g. *Euglena spp.* (EUG), *Navicula spp.* (NAV), *Fragilaria spp.* (FRA), *Synedra spp.* (SYN), *Cyclotella spp.* (CYC), project onto the middle of the depth and DO positive axis. And, they have less negative value on the TN, TP and COD$_{Mn}$ axis. This situation occurs in Bacillariophyta taxa, e.g. Cymbella spp. (CYM) and Dinophyta taxa, e.g. Ceratium hirundinella (CER HIR). These phytoplankton species related to the higher concentration of DO, and the lower concentration of TN, TP and COD$_{Mn}$, represent better water quality in the middle and lower layer.

3. Middle and upper layer, less pollution water
   Cyanophyta taxa, e.g. *Merismopedia minima* (MER MIN), Oocystis taxa, e.g. *Golenkinia spp.* (GOL), *Closterium Kutzingii* (CLO KUT), and Euglenophyta taxa, e.g. *Cyclotella spp.* (CYC), project onto the middle of the depth and DO negative axis. They have less positive value on the TN, TP and COD$_{Mn}$ axis. These phytoplankton species represent less pollution water in the middle and upper layer.

4. The surface layer, pollution water
   Cyanophyta taxa, e.g. Oscillatoria angustissima (OSC ANG), Anabaena flosaguas (ANA FLO), Chroococcus epiphyticus (CHR EPI), Microcystis wesenbergii (MIC WES), Anabaena sp. (ANA), Aphanizomenon flosaquae (APH FLO), Oocystis taxa, e.g. Crucigenia sp. (CRU), Closterium moniliferum (CLO MON), Chlorella vulgaris (CHL VUL), Pandorina morum (PAN MOR), Closterium acerosum (CLO ACE), Selenastrum gracile (SEL GRA), Oocystis sp. (OOC), Pediatrum sp. (PED), and Euglenophyta taxa, e.g. Pinnularia sp. (PIN), preferred negative axis of depth and DO. They relate to positive value on the TN, TP and COD$_{Mn}$ axis. These phytoplankton species are associated with pollution water in the surface layer.

164

Depth appears to be the main driving factors for phytoplankton dynamics in the lake. Dissolved oxygen is linked to depth variation in the plateau lakes. TP, TN and COD_{Mn} also play an important role in assembling the communities. Water depth affecting phytoplankton vertical distribution in lakes is thought to be underwater light gradients and thermal stratification which can influence access to nutrient resources (Klausmeier and Litchman, 2001).

This result indicates it is not the only factor which determines the phytoplankton assembles (Kudo and Matsunaga, 1999). Surface and bottom values are not different. The gradual increase in water nutrition could have led to the selection of more tolerant self-shading species, such as *Oscillatoria angustissima, Anabaena flosaguas, Closterium moniliferum, Chlorella vulgaris, Pinnularia sp.*. In response to the gradients of essential resources (depth or nutrients), the phytoplankton groups are predicted to show differential vertical distributions as they optimize growth conditions according to their specific physiologies and motilities (Longhi and Beisner, 2009).

4 CONCLUSIONS

The relationship between the change of phytoplankton population and environmental factors is investigated. It shows that depth appears to be the main driving factors for phytoplankton dynamics in the lake. Dissolved oxygen is linked to depth variation in the plateau lakes. Water depth affecting phytoplankton vertical distribution in lakes is thought to be underwater light gradients and thermal stratification which can influence access to TP, TN and COD_{Mn} resources. The gradual increase in water nutrition could have led to the selection of more tolerant self-shading species. Phytoplankton groups show differential vertical distributions as they optimize growth conditions according to their specific physiologies and motilities.

ACKNOWLEDGEMENT

Yue Huang, corresponding author, the School of Resource Environment and Earth Science, Yunnan University, North Cuhu Road 02, Kunming 650091, China. Tel: +86 871 4114918; E-mail: yuehuang@yun.edu.cn.

This paper was supported by Applied Basic Research fund (No. 2009CD146) and Social Development fund (No. 2001CA008), both from Yunnan Provincial Department of Science and Technology.

REFERENCES

Alvarez-Cobelas, M., J.L. Velasco, A. Rubio, C. Rojo, 1994. The time course of phytoplankton biomass and related limnological factors in shallow and deep lakes: a multivariate approach. Hydrobiologia, 94: 139–151.

Chorus, I. (Ed.), 2001, Cyanotoxins: Occurrence, Causes, Consequences. Springer, Berlin.

Cui Y.D., Liu X.Q., Wang H.Z. 2008. Macrozoobenthic community of Fuxian Lake, the deepest lake of southwest China. Limnologica 38: 116–125.

Klug, J.L. and Cottingham, K.L. (2001) Interactions among environmental drivers: community responses to changing nutrients and dissolved organic carbon. Ecology, 82, 3390–3403.

Kudo I. and Matsunaga K. 1999. Environmental Factors Affecting the Occurrence and Production of the Spring Phytoplankton Bloom in Funka Bay, Japan. Journal of Oceanography, 55, 505–513.

Long, B.M., Jones, G.J., Orr, P.T., 2001. Cellular microcystin content in N-limited Microcystis aeruginosa can be predicted from growth rate. Appl. Environ. Microbiol. 67 (1), 278–283.

Longhi M.L. and Beisner B.E. 2009. Environmental factors controlling the vertical distribution of phytoplankton in lakes. Journal of Plankton Research, 31(10):1195–1207.

Reynolds, C.S. 1984. Phytoplankton periodicity: The interactions of form, function and environment variability. Freshwater Biology, 14:111–142.

Romo, S., 1991. Estudio del fitoplancton de la Albufera de Valencia, una laguna hipertrofica y somera, entre 1980 y 1988. Ph. D Thesis, University of Valencia (Spain), 303.

Ter Braak, C.J.F., 1989. CANOCO—an extension of decorama to analyze species-environment relationships. Hydrobiologia 181:169–170.

Ter Braak, C.J.F., 1987. Data Analysis in Community and Landscape Ecology. Pudoc Wageningen 299 pp.

Frontiers of Energy and Environmental Engineering – Sung, Kao & Chen (eds)
© 2013 Taylor & Francis Group, London, ISBN 978-0-415-66159-1

Research on effect of prior probability for maximum likelihood classification accuracy in remote sensing image analysis

X. Zhu
Yunnan Environment Science Insititute, Kunming, China

H.J. Duan
Department of Physics, Faculty of Basic Science and Technology, Kunming, China

J.M. Chen
Kunming University of Science and Technology, Kunming, China

J. Li
Yunnan Environment Science Insititute, Kunming, China

ABSTRACT: Maximum likelihood classification is classic and widely used in remote sensing image analysis. The mathematical theory foundation is based on Bayesian formula for the minimizing Bayesian error. Many researches have indicated that using prior probability can make Maximum likelihood classifier perform better, so the correct setting of Prior probability does improve the accuracy of Maximum likelihood classifier. In this paper, Firstly, Based on the Bayesian formula is deduced from the Maximum likelihood discriminant function. Secondly, the method of determining prior probability for Iterative loop algorithm is shown. Finally, an experiment about Songhuaba reservoir in northwest of Kunming, Yunnan province in China, which is carried out by Iterative loop algorithm verify and explain the effect of Prior probability on Maximum likelihood classification, In comparison with the first classification, then after the sixth classification, Overall Accuracy and Kappa coefficient were improved by 2.04% and 0.0543 respectively.

Keywords: maximum likelihood classification; prior probability; iterative loop algorithm; Bayesian formula

1 INTRODUCTION

Maximum Likelihood Classification (MLC) is implemented in almost all remote sensing and image processing software. MLC is known to be optimal in the sense of minimizing Bayesian error (Fuatince. 1987). The MLC algorithm is based on Bayesian formula and Bayesian rule, Prior probabilities is the probabilities of occurrence of every class, and through the experiments prior probabilities can improve accuracy of classification results by helping to resolve confusion among classes that are poorly separable (Mather, 1985; McIver and Friedl; 2002). So it can be indeed a powerful and effective aid for improving classification accuracy, but it is a common method to set up the equal prior probabilities as reliable prior probabilities are not always easy to obtain in advance. In the past ten years, Much work has been done on the incorporation of prior probabilities into MLC (Jayantha and Siamak. 1997; Strahler, 1980; Maselli et al, 1990).

Prior probability of Maximum likelihood classification results is a certain effect. through a large number of experiments, Strahler show the prior probability properly set will increase the overall accuracy of classification, when prior probability is set up for extreme value (i.e. close to 0 or 1), the classification results will be effected (Pedroni, 2003; Zheng and Cai, 2005). When dealing with spectrum close to the classes most affected, and with good spectral separation of the class is almost no effect. When there are many classes in the image, for the distribution of large-size classes, Producer's accuracy decrease and User's accuracy increase (Li, 2005; Peng, 1991; Congalton, 1991). For the distribution of small-size classes, Producer's accuracy increase and User's accuracy decrease.

2 THE THEORY OF BAYESIAN MAXIMUM LIKELIHOOD CLASSIFICATION

Maximum likelihood classification is one of most classical method in remote sensing image classification, it is based on Bayesian criterion, and the attribution probability (posteriori-probability) of pixels is calculated by the formula (1):

$$P\left(G_k \middle| X\right) = \frac{P(G_k)P(X \middle| G_k)}{P(X)} \quad (1)$$

For the general case of n-dimensional, $P(G_k \middle| X)$ under n-band (X k-th class in the probability density function (likelihood probability)) of the expression is as follows (John et al, 2007):

$$P\left(X \middle| G_k\right) = \frac{1}{(2\pi)^{\frac{n}{2}} |\Sigma|^{1/2}} \exp\left[-\frac{1}{2}(X - \mu_k)^t \Sigma_k^{-1}(X - \mu_k)\right] \quad (2)$$

Show formula (2) into Bayesian criterion (1), then you can obtain the following formula, since $P(X)$ is only considered all image datas, and it isn't relationship with category, so transform formula can be equivalent to (3):

$$P(X \middle| G_k) = \frac{1}{(2\pi)^{\frac{n}{2}} |\Sigma|^{1/2}} \exp\left[-\frac{1}{2}(X - \mu_k)^t \Sigma_k^{-1}(X - \mu_k)\right] \quad (3)$$

Because formula (3) is not convenient in the computation, so we take the transformation of natural logarithm on both sides of the equation so that the formula is converted into (4):

$$g_k(X) = \ln(P(G_k)) - \frac{1}{2}(X - \mu_k)^T \Sigma_k^{-1}(X - \mu_k) \quad (4)$$

Parameter description: $P(G_k)$ is priori probability of the k-class features; Σ_k is covariance matrix of the k-class features in n-bands, Σ_k^{-1} is inverse matrix of the matrix Σ_k, $|\Sigma_k|$ is determinant of matrix Σ_k; μ_k is mean-vector of the k-class features; X is the feature vector of input image.

Based on above discriminant function formula, Prior probability must be obtained before the MLC in the remote sensing image, otherwise we will get bad classification accuracy.

3 METHOD FOR DETERMINING PRIOR PROBABILITIES ON MAXIMUM LIKELIHOOD CLASSIFIER

In the present paper, researcher often chose the method of Iterative loop algorithm to determine prior probabilities. This method is that one simple approach is designed to dispel the effect of prior probabilities. The first classification assumes that the prior probabilities are equal for all classes, i.e. all classes have equal area, and later the relative area proportion in the first classification result is assigned to prior probabilities in the second

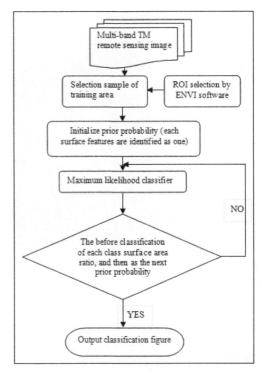

Figure 1. Iterative loop algorithm flow chart.

classification. If the second classification has higher overall accuracy and kappa coefficient than the first classification, which implies that the second classification result is closer to reality than the assumption of equal area, then this process can continue until the overall accuracy and the kappa coefficient has a stable value. The flow chart of algorithm is as follows Figure 1.

4 EXPERIMENT AND ANALYSIS

4.1 Experiment study area

Experimental study area choose Songhuaba reservoir which is located in the in the northwest of Kunming, Yunnan province in China, the altitude is 1966 meters and the size of study area is 314*261 pixels. Songhua reservoir is a medium-sized reservoir of Kunming, which main supply the Kunming city's drinking water. Based on the Figure 2, the main features of the reservoir area have five categories, such as cultivated land, forest land/grass, urban/rural land use, water (including reservoir and pool). The original image choose Kunming LANDSAT TM 7-bands remote sensing image, as shown in Figure 2: the combination of TM4-TM3-TM2 bands image.

Figure 2. The combination TM4-TM3-TM2 bands image.

Table 1. Experimental area selection the number of polygons and pixels of the training samples and test samples.

	Cultivated land	Forest land/ grass	Urban/rural land use	Water (including reservoir and pool)	Shadow
Number of training set polygons	5	7	8	4	2
Number of training set pixels	667	2062	965	330	150
Number of testing set polygons	5	8	8	4	2
Number of testing set pixels	568	1505	1009	435	80

4.2 Select the training samples and test samples

The experiment was carried out under the Matlab7.11 programming environment, and the training set sample is selected by ENVI4.7 through the ROI tools, as MLC is based on statistical properties to create a discriminant function, so the number of training set samples selection demand as many as possible, the test set samples are selected in accordance with Google's Google Earth software and at the same time in combination with the original TM remote sensing images, the numbers of training samples and test set samples as follows Table 1.

4.3 Accuracy assessment and analysis

The experiment was done under Matlab environment, based on iterative loop algorithms for six experiments, the initialize prior probabilities of each class surface features are one, then the classification of each experimental area ratio are calculated by statistics surface features area, which is used as a prior probability of the next classification, and so on to the sixth, then this process can continue until the overall accuracy and the kappa coefficient has a stable value. In the selection feature-bands, except the spectral band of TM6 and TM1, all five other bands and NDVI were used in classification, so the method of mentioned above was used to classify the image.

Table 2 and Figure 3 is comparison between the first classification and sixth classification on Producer's accuracy and User's accuracy, Table 3 is comparison between the first classification and sixth classification.

As show in Table 2, Producer's accuracy and User's accuracy are calculated by the confusion matrix of the first classification and sixth classification, at the same time, it is displayed through the graph in Figure 3. Based on Table 2 and Figure 3, Priori probability indeed affect the accuracy of Maximum likelihood classification, when dealing with spectrum close to the class most affected, and with good spectral separation of the class is almost no effect. When there are many classes in the image, for the distribution of large-size classes, Producer's accuracy decreases and User's accuracy increase. For the distribution of small-size classes, it is just the opposite.

Figure 3: Abscissa show each type surface features, Ordinate show producer's accuracy and user's accuracy of each type surface features. Through visual interpreting the original image, the size of Forest land/grass is biggest, and Water (including

Table 2. Comparison between the first classification and sixth classification on producer's accuracy and user's accuracy.

Classification number and accuracy type	Cultivated land	Forest land/ grass	Urban /rural land use	Water (including reservoir and pool)	Shadow
The first classification					
Producer's accuracy	87.15%	99.00%	91.58%	97.70%	83.93%
User's accuracy	84.76%	99.00%	92.49%	98.38%	85.45%
The sixth classification					
Producer's accuracy	90.85%	99.80%	93.40%	99.54%	89.88%
User's accuracy	88.81%	99.40%	94.61%	99.54%	96.79%

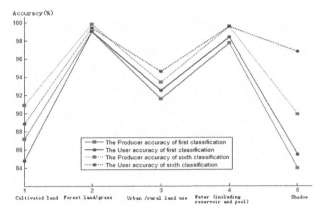

Figure 3. Comparison producer's accuracy and user's accuracy.

Table 3. Comparison between the first classification and sixth classification.

Classification number and category number	Number of each classification pixel statistics	Area ratio of each type surface features	Overall accuracy	Kappa coefficient
The first classification				
Cultivated land	24445	29.828%	94.30%	0.8934
Forest land/grass	40907	49.915%		
Urban/rural land use	14112	17.219%		
Water (including reservoir and pool)	2102	2.565%		
Shadow	388	0.473%		
The sixth classification				
Cultivated land	28984	35.366%	96.34%	0.9477
Forest land/grass	41174	50.240%		
Urban/rural land use	8477	10.423%		
Water (including reservoir and pool)	2957	3.608%		
Shadow	412	0.482%		

reservoir and pool) have better spectral separability, but Cultivated land and Urban/rural land use have bad spectral separability, So Forest land/grass is almost no effect and Water has little effect, others has more effect in Producer's accuracy User's accuracy.

As show in the Table 3, which is made up by every classification about the number pixels for each class surface features, area ratio for each class surface features, overall accuracy and kappa coefficient. In comparison with the first classification which is carried out by MLC with equal prior

| Cultivated land | Forest land/grass | Urban /rural land use | Water (including reservoir and pool) | Shadow |

Figure 4. Comparison between the first classification and sixth classification.

Table 4. Values of overall accuracy and kappa coefficient from the first to sixth classification.

The first to sixth classification	1	2	3	4	5	6
Overall accuracy	0.9430	0.9524	0.9593	0.9630	0.9632	0.9634
Kappa coefficient	0.8934	0.9326	0.9426	0.9473	0.9475	0.9477

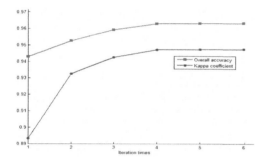

Figure 5. Comparison accuracy and kappa form first to sixth.

Figure 5. Abscissa show iteration times from first to sixth classification. Ordinate show Overall accuracy and Kappa coefficient. Among them overall accuracy is replaced by decimal not percentage, The red line display the change of overall accuracy, The blue line display the change of kappa coefficient, when iteration loop to fourth classification, Overall accuracy and Kappa coefficient all tend to be a very small change, and in the sixth it almost tend to be unchanged.

probabilities one, Overall Accuracy and Kappa coefficient were improved by 2.04% and 0.0543 respectively in the sixth classification.

As show in the Figure 4, it is the first classification and sixth classification figure. Through the Table 3 and the Figure 4, Cultivated land and Urban/rural land use become more than other surface features classes, this is that because the spectral characteristics of two types surface features is very similar and Area is also relatively large, so that prior probabilities is greater effect; because of better spectral separation and larger area for Forest land/grass, it is almost no effect; in the upstream reservoir, due to both sides of a large number mixed pixels, so also have some effect.

5 CONCLUSION

Based on above experiments, priori probabilities are indeed certified to affect the accuracy of MLC. Though using iteration loop method on MLC for remote sensing images, to ensure minimal loss of misclassification and eliminate the influence of priori probabilities, then improve the Overall accuracy and Kappa coefficient. In this paper, after the sixth iterations loop, then in comparison with the first classification, Overall Accuracy and Kappa coefficient were improved by 2.04% and 0.0543 respectively.

Prior probabilities can minimize the incorrect assignment by using a posteriori probability as classification criterion, so prior probabilities is an effective technique to improve the classification accuracy. When dealing with spectrum close to the classes most affected, and with good spectral separation of the class is almost no effect, and it is not true that all individual classes will have higher classification accuracy after prior probabilities are incorporated on MLC. For individual class, the change of Producer's accuracy and User's accuracy after setting prior probabilities are used depends on the size of the classes. In general, for the large-size classes, Maximum likelihood classification using prior probabilities tends to be higher Producer's accuracy and lower User's accuracy, and for the small-size classes, the tendency is just opposite.

ACKNOWLEDGEMENT

Jianming Chen, corresponding author, Kunming University of Science and Technology, 121 Street, Kunming 650034, China. Tel: +868714173209; Fax: +868714173209; E-mail: jmchen@inems.com.

This paper was supported by Social Development fund (No. 2001CA008) from Yunnan Provincial Department of Science and Technology.

REFERENCES

Congalton R.G. 1991. A review of assessing the accuracy of classification of remotely sensed data. Remote Sensing of Environment, 37, 35–46.

Fuatince. 1987. Maximum likelihood classification, optimal or problematic, A comparison with the nearest neighbor classification. International Journal of Remote Sensing, 12, 1829–1838.

Jayantha, Siamak. 1997. Hierarchical Maximum-Likelihood Classification for Improved Accuracy. IEEE Transactions on Geoscience and Remote sensing, 35, 810–816.

John R. Jensen (U.S.), Chen X. and other translation. 2007. Introductory of Remote Sensing Digital Image Processing, Beijing: Mechanical Industry Press, in Chinese, 359–363.

Li Q., Wang H., Li L. 2005. A Study of accuracy control of land cover classification based on MLH Algorithm. Land and Resource Technology Management, in Chinese, 4, 42–45.

Maselli F., Conese C., Zipoli G. 1990. Use of error probabilities to improve area estimates based on maximum likelihood classification. Remote Sensing of Environment, 31, 155–160.

Mather, P.M. 1985. A computationally-efficient maximum-likelihood classifier employing prior probabilities for remote-sensed data. INT. J. Remote Sensing, 62, 369–376.

McIver D.K., Friedl M.A. 2002. Using prior probabilities in decision-tree classification of remotely sensed data. Remote Sensing of Environment, 81, 253–261.

Pedroni L. 2003. Improved classification of Landsat Thematic Mapper data using modified prior probabilities in large and complex landscapes. International Journal of Remote sensing, 24, 91–113.

Peng W. 1991. The computer data processing of Remote Sensing data and Geographical Information System. Beijing: Beijing normal university in Chinese, 150–175.

Strahler H. 1980. The use of prior probabilities in Maximum Likelihood Classification of Remotely Sensed Data. Remote Sensing of Environment, 10, 135–163.

Zheng M., Cai Q. 2005. Effect of Prior Probabilities on Maximum Likelihood Classifier. IEEE conference, 3753–3756.

Frontiers of Energy and Environmental Engineering – Sung, Kao & Chen (eds)
© 2013 Taylor & Francis Group, London, ISBN 978-0-415-66159-1

Research on GHGs emission accounting method in shipbuilding industry

Q. Pan, H.Z. Yan & N. Wang
Key Laboratory Yangtze River Water Environment, Ministry of Education, College of Environmental Science and Engineering, Tongji University, Shanghai, China

R. Guo
College of Environmental Science and Engineering, UNEP-Tongji Institute of Environment for Sustainable Development, Tongji University, Shanghai, China

ABSTRACT: To investigate the Greenhouse Gases (GHGs) emission characteristics of ship manufacturing industry, a research was made on the accounting method of GHG emission. Based on the processes and technical characteristics of the ship manufacturing industry, the detailed methods and parameters of GHG Protocol and IPCC 2006, a method of material balance was adopted to analysis the GHGs emission of a ship manufacturing factory in Shanghai as a typical case. Results indicate that the direct GHGs emission intensity of the case factory was 0.055 ton CO_2 per output value of 10,000 RMB. Involving indirect emission such as electric and heating power, the total emission intensity reach 0.53 ton CO_2 per output value of 10,000 RMB. Comparison of the calculated results to the average standards of ship manufacturing industry and the overall Shanghai industry demonstrated that the carbon emission intensity of the case ship manufacturing factory was below average emission intensities. The integration of direct GHGs emission into processes specifies the major processes of direct carbon emission, and supplements the blank of carbon emission calculation in ship manufacturing industry.

Keywords: greenhouse gases; shipbuilding industry; accounting method

1 INTRODUCTION

Ship manufacturing, which assembles advanced modern industrial manufacturing technology and theory method, embodies a country's national scientific level and heavy industrial ability. In recent years, according to the demand of low carbon development, the concepts of Green Shipbuilding and Cleaner Production have been put forward and the low carbon development has become a tendency of ship construction industry. To select a scientific and reasonable technical principle for low carbon development, the Greenhouse Gases (GHGs) emission inventory and characteristics must be figured out at first. Therefore, the appropriate methods for shipbuilding GHGs emissions accounting remains an urgent problem, which needs research and discussion.

For comparison, there must be a uniform GHGs emissions accounting standard. The current main accounting standards and assessment method are as follows. At the country level, the UN's Intergovernmental Panel on Climate Change (IPCC) released in 1996, 2000 and 2006 respectively Revised 1996 IPCC Guidelines for National Greenhouse Gas Inventories (IPCC 1996), Good Practice Guidance and Uncertainty Management in National Greenhouse Gas Inventories (IPCC 2000) and IPCC guidelines for national greenhouse gas inventories (IPCC 2006). At the enterprise level, the World Business Council on Sustainable Development and World Resources Institute announced the Green House Gas Protocol (GHG Protocol) in 2002 (WBCSD/WRI 2002). At products and services level, the British Standard Institute (BSI), Carbon Trust and the Department for Environment (CTDE), Food and Rural Affairs (FRA) jointly issued PAS 2050 Specification for the Assessment of the Life Cycle Greenhouse Gas Emissions of Goods and Services (PAS 2050) in October, 2008 (BSI 2008).

In the respect of carbon accounting subjects, international studies have gradually shift from national level to region and city level (Betsill, M.M. & Bulkeley, H. 2006, Koehn, P.H. 2008, Wheeler, S. 2008). Guo J. et al. (2008) calculated the amount of GHGs emission of Liaoning Province in 2004, based on the data of Statistical

Yearbooks. Zhao M. et al. (2009) accounted the carbon emission due to the energy consumption of Shanghai in 2006. Currently, however, there are relatively few researches on a specific industry carbon emission accounting method, especially on shipbuilding industry and most studies focused on discussion of low carbon development concept and technologies (Zhao, Z.P. 2009, Li, B.Y. 2008). This study filled the gap in this aspect.

The main GHGs emission accounting methods are including material balance algorithm, measurement, emission factor, model and life cycle method (Zhang, D.Y. 2005). Nejadkoorki F. et al (2008) studied the CO_2 emission of road traffic in urbanized regions by model approach.

This study involved the method of GHGs emission accounting of the shipbuilding industry based on material balance algorithm, referring to the detailed methods and parameters of GHG Protocol and IPCC. A typical ship construction factory was selected as a case study. Finally, a comparison study was also conducted to evaluate the carbon emission intensity of the shipbuilding factory.

2 METHOD

2.1 Accounting boundary

GHG Protocol defined the GHGs accounting boundary according to the project effects, including primary effects and secondary effects. (WBCSD/WRI 2002) This study mainly accounting the primary effects, the intended change caused by a project activity in GHGs emissions, removals, or storage associated with a GHGs source or sink. (WBCSD/WRI 2002).

2.2 Process analysis

The analysis of shipbuilding process figured out the main energy consumption and carbon emission session, providing the foundation of the emission. The main process of shipbuilding and carbon emission were shown in Figure 1.

2.3 Accounting method

In IPCC 2006 (IPCC 2006), the most common method of accounting was multiplying the fuel consumption by the corresponding emission factor. So, the fundamental equation was as follows.

$$\text{Emissions} = \text{Fuel consumption} \times \text{Emission factor} \tag{1}$$

The default emission factors of each kind of fuel, recommend in IPCC 2006, were used in this study. The average lower heating value (Qnet) of each fuel was from China Energy Statistical Yearbook 2011. (China's National Bureau of Statistic & China's National Energy Administration 2011) Emission factors of some kinds of gas fuel without mentioning in IPCC 2006 were calculated based on the combustion value and mass fraction of carbon compared to the known fuel. The emission factors were listed in Table 1.

Emission factors of coating process was from EMEP/EEA air pollutant emission inventory

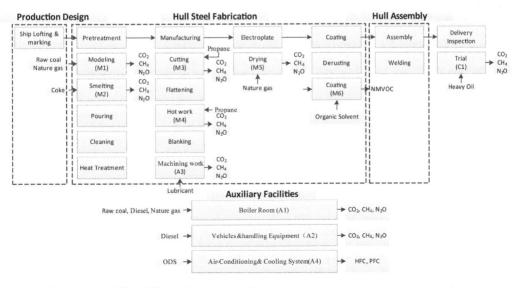

Figure 1. The process of shipbuilding and carbon emission sectors.

Table 1. Emission factors calculation table.

Fuel/other products	Emission factor (kg/TJ)			Qnet	Emission factor in mass or volume unit (kg/t)		
	CO_2	CH_4	N_2O	kJ/kg	CO_2	CH_4	N_2O
Gas/diesel oil	74100	3	0.6	42652	3160.51	0.13	0.026
Heavy oil (residual fuel oil)	77400	3	0.6	41816	3236.56	0.13	0.025
Coke	107000	1	1.5	28435	3042.55	0.028	0.043
Nature gas	56100	1	0.1	38931[a]	2184.03[b]	0.039[b]	0.0039[b]
Raw coal	96920	1	1.5	20908	2026.40	0.021	0.031
Propane	99881.6	1	0.1	2217.8[c]	5034.49	0.050	0.0050
Lubricant	75200	10	2	40000	3008	0.4	0.08

a. Unit of data: kJ/m^3. b. Unit of data: kg/m^3. c. Unit of data: kJ/mol.

Table 2. Emission factors of NMVOC in coating process (EMEP/EEAAIR 2009).

Emission factors	Unit	Default emission factor	95% confidence interval	
			Lower	Upper
Default	g/kg paint	400	100	800
Shipbuilding	g/m^2 painting area	125	100	150

guidebook 2009 (EMEP/EEAAIR 2009), shown in Table 2.

3 CASE STUDY

A typical shipbuilding factory, one of the largest and most advanced in China located in Shanghai.

Shipbuilding Base was selected as a case study. The planning manufacturing capacity of this factory was 3.32 million dead weight tons and the major products included the 760,000-ton Panamax, 2970,000-ton VLCC and 1760,000-ton Capesize.

4 RESULTS AND DISCUSSION

4.1 Carbon emission

The inventory of GHGs emissions of JN shipyard in 2009 was showed in Table 3. Since the limitation of acquiring data, this study did not calculate the emission of the alternatives for ODS (Ozone Depleting Substances) and the mobile combustion source from vehicle transportation in the fuel combustion emission source. The energy use statistics in the Table 3 are gathered from the JN shipyard's environmental assessment report and investigation. All the coating process in the JN shipyard is operated indoor, its pollution control measures are implemented and the average removal rate

of NMVOC is about 85%–90%, thus the average removal rate η is considered when calculating.

The emission of CO_2, CH_4, N_2O and NMVOC in 2009 from the whole JN shipbuilding factory is respectively 30400 t, 8500 t, 1700 t and 280.8 kg. As to the kinds of the GHGs discharged, the emissions of CH_4, N_2O and NMVOC are 4 orders of magnitude fewer than that of CO_2, even if the greenhouse effect of CH_4 and N_2O is one hundred times higher than that of CO_2, the emission is very little. The GHGs from the shipyard is mainly CO_2.

Figure 2 showed that the main processes which released CO_2 were gas cutting, hot work, trial run, heating station and boiler room, the emission of CO_2 from these processes accounting for over 88% the emission of CO_2 from the whole factory.

The main emissions of the CO_2 from the factory were caused by the usage of heavy oil, propane and diesel (Fig. 3), the emission of CO_2 from these fuel consumption accounting for more than 85% the total emission of CO_2. Propane was mainly used for shipbuilding enterprises cutting, plate bending and initiating process and Diesel for large lifting equipment and the engine of the handling vehicle as well as pretreatment and the boiler power. The key to reduce this part of emission was a reasonable arrangement of production process and the improvement of boiler equipment combustion efficiency. Heavy oil was mainly used for the whole ship trial.

Table 3. The GHGs emission accounting results of JN shipbuilding factory.

Process	Fuel/products	Fuel consumption	Unit	CO_2 t	CH_4 kg	N_2O kg	NMVOC kg
M1	Raw coal	28.36	t	5.75	0.59	0.89	–
	Nature gas	515377	m³	112.56	20.06	2.01	–
M2	Coke	392	t	119.27	11.15	16.72	–
M3	Propane	2175.2	m³	1095.1	109.64	11.49	–
M4							
M5	Nature gas	511402	m³	111.69	19.91	1.99	–
A1	Coke	10.84	t	2.2	0.23	0.34	–
	Diesel	1220	t	385.58	156.11	31.22	–
	Nature gas	351471	m³	76.76	13.68	1.37	–
A2	Gas/diesel oil	–	–	–	–	–	–
A2	Diesel	1020	t	322.37	130.52	26.1	–
C1	Heavy oil	2461	t	796.52	308.73	61.75	–
A3	Lubricant	200000	kg	12.03	80	16	–
M6	Organic solvent	4680	kg	–	–	–	280.8
A4	ODS	–	–	–	–	–	
Summation		30398.31		850.61	169.35	280.8	30398.31

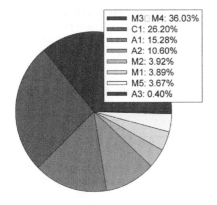

Figure 2. Proportion of CO_2 emission from each process to total emission.

- M3 M4: 36.03%
- C1: 26.20%
- A1: 15.28%
- A2: 10.60%
- M2: 3.92%
- M1: 3.89%
- M5: 3.67%
- A3: 0.40%

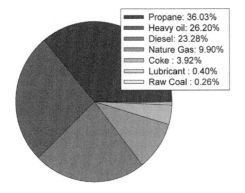

Figure 3. Proportion of CO_2 emission from each fuel to total emission.

- Propane: 36.03%
- Heavy oil: 26.20%
- Diesel: 23.28%
- Nature Gas: 9.90%
- Coke : 3.92%
- Lubricant : 0.40%
- Raw Coal : 0.26%

Table 4. Results of carbon emission intensity accounting.

CO_2 (t/10,000 RMB)	CH_4 (kg/10,000 RMB)	N_2O (kg/10,000 RMB)	NMVOC (kg/10,000 RMB)
0.055	$1.56*10^{-3}$	$5.79*10^{-4}$	$3.09*10^{-4}$

Therefore, the main sources of GHGs emissions from JN shipyard were the hull process, trial, plant handling equipment. The effective control of the GHGs emission from there processes of the ship factory emissions was very important and crucial.

4.2 Carbon emission intensity

Carbon emission intensity was evaluated by the carbon emissions per output value of 10,000 RMB, which was shown in Table 4.

In order to further analysis of shipbuilding carbon emission characteristics, the scope of accounting was expanded to the secondary effects in GHG Protocol, including the indirect emissions caused by purchasing power. (WBCSD/WRI 2002) The East China power grid emission factor was 0.8367 $kgCO_2$/MW (China's National Development and Reform Commission's Climate Change Department 2011). 349 million kWh electric power was bought by JN shipbuilding factory in 2009, resulting in 261668.41 tons CO_2 emission. Therefore, the total emissions of CO_2 form JN shipbuilding factory was 292066.71 tons when accounting the indirect emissions, the carbon intensity is

Table 5. The Results Carbon Emissions Intensity Accounting of the Chinese Industry (CI), Chinese Manufacturing Industry (CMI) And Chinese Transportation Equipment Manufacturing Industry (CTEMI) and Shanghai Industry (SI) from other studies.

Year	Idustry	Secondary effects	Source of emission factors	tCO_2/10,000 RMB	References
2007	CTEMI	Not included	References (Tan D. & Huang X.J. 2008)	0.255	Xu, D.F. 2011
2007	CTEMI	Not included	China's National Development and Reform Commission	0.02	Xu, Y.Z. & Hu, Y.S. 2011
2007	CMI	Not included	IPCC 2006	0.28	Pan, X.F. et al. 2011
2009	CMI	Included	IPCC 2006	1.34	Ai, M.Y. et al. 2012
2007	CI	Not included	References (Ma C. & Stern D.I. 2007)	1.39	Pan, J.J. & Li, L.S. 2011
2007	SI	Included	IPCC2006 &References (Dhakal S. 2009)	0.61	Shen, W et al. 2010

0.53 tCO_2 per output value of 10,000 RMB. The direct emissions of CO_2 made up to only 11.6% of total emissions of CO_2. So if considering the indirect emission, the key to reduce the carbon emissions intensity of the shipyard was the energy saving design and improvements of the main power consumption procedures.

Table 5 summarized the recent studies on the carbon emissions intensity accounting of the Chinese industry, manufacturing industry, transportation equipment manufacturing industry and Shanghai industry.

Pan et al. (2011) divided the Chinese manufacturing industry into high carbon and low carbon industry. In contrast to the carbon emission intensity of this study, shipbuilding was close to the average level of low carbon industry and lower than the average level of manufacturing industry.

Due to the difficulty in obtaining sufficient data, this study did not calculate the emissions of the ODS substitution and vehicle transportation, the actual carbon intensity value of the JN shipyard would be slightly higher than the calculated results. This study did not refine the power consumption into each process. Most of the emission factors were default value without considering specific engineering processes, and these would also skew the accounting results. Due to the lack of data, this research only studied the carbon emissions intensity in 2009 and dynamic changes from year to year had not been considered.

5 CONCLUSIONS

In general, the main kind GHGs of direct emissions was CO_2. The hull workshop cutting, hot work, ship trial and handling vehicle were the main carbon emissions processes in shipyard. The main CO_2 emissions from the shipyard was the indirect emissions caused by power consumption, and direct emission makd up a low rate to the total emission. From comparison, the shipyard carbon emission intensity is lower than the industry average level. The calculation results provided a certain guidance and reference for the low carbon development and technology for ship manufacturing. The future studies needed to refine the power consumption to each working procedure to provide a more complete scientific description of ship manufacturing carbon emission.

ACKNOWLEDGEMENTS

The authors thank to the financial support of 'Chongming low carbon economy development key technology research and integrated application demonstration' Program (No. 2009BAC62B02) and the program of Shanghai Municipal Science and Technology Commission (No. 10230712200).

REFERENCES

Ai, M.Y., Bi, K.X., Li, W.H. 2012. A study on model dynamic evolution and the industrial structure optimization of China's manufacturing industry. *Inquiry into Economic* (1):48–54.

Betsill, M.M., Bulkeley, H. 2006. Cities and the multilevel governance of global climate change. *Global Governance* 12(2):141–159.

British Standards Institution 2008. PAS 2050 Specification for the Assessment of the Life Cycle Greenhouse Gas Emissions of Goods and Services. UK: British Standards Institution.

China's National Development and Reform Commission's Department of Climate Change 2011. *Baseline emission factors for regional power grids in China.* http://cdm.ccchina.gov.cn.

China's National Bureau of Statistic, China's National Energy Administration 2011. *China Energy Statistical Yearbook 2011*. Beijing: China Statistics Press.

Dhakal, S. 2009. Urban energy use and cities CO_2 emissions in China and policy implications. *Energy Policy* 37(11): 4208–4219.

Guo, J., Zhao, L.C., Jia, H.R. et al. 2008. Current status of CO_2 emission and its controlling methods in Liaoning province. *Renewable Energy Resources* 26(2): 110–113.

IPCC 1996. Revised 1996 IPCC Guidelines for National Greenhouse Gas Inventories. Bracknell: Hadley Centre.

IPCC 2000. Good Practice Guidance and Uncertainty Management in National Greenhouse Gas Inventories. Kanagawa: Institute for Global Environmental Strategies.

IPCC 2006. *IPCC guidelines for national greenhouse gas inventories*. Kanagawa: Institute for Global Environmental Strategies.

Koehn, P.H. 2008. Under neath Kyoto: emerging sub national government initiatives and incipient issue-bundling opportunities in China and the United States. *Global Environmental polities* 8(1):53.

Li, B.Y. 2008. Study on Green Ship and it's Assessment Index System. *Shipbuilding of China* (S1): 27–35.

Ma, C., Stern, D.I. 2007. China's Carbon Emissions 1971–2003. Rensselaer Working Papers in Economics, Number 0706.

Nejadkoorki, F., Nicholson, K., Lake, I., et al. 2008. An approach for modeling CO_2 emissions from road traffic in urban areas. *The Science of the Total Environment* 406 (1–2): 269–278.

Pan, J.J., Li, L.S. 2011. Analysis of Factors Affecting Industrial Carbon Dioxide Emission in China. *Environmental Science & Technology* 34(04): 86–92.

Pan, X.F., Shu, T., Xu, D.W. 2011. On the Changes in the Carbon Emission Intensity of China's Manufacturing Industry and Its Factors Decomposition. *China Population, Resources and Environment* 21(05):101–105.

Shen, W., Zhu, D.J., Bai, Z.L. 2010. Research on Relationship Between Industrial Carbon Emissions and Carbon Productivity in Shanghai. *China Population, Resources and Environment* 20(9):24–29.

Tan, D., Huang, X.J. 2008. Correlation Analysis and Comparison of the Economic Development and Carbon Emissions in the Eastern, Central and Western Part of China. *China Population Resources and Environment* 18(3):54–57.

WBCSD/WRI 2002. *Greenhouse Gas Protocol*. http://www.ghgprotocol.org.

Wheeler, S. 2008. State and municipal climate change plans: the first generation. *Journal of the American Planning Association* 74(4):481–496.

Xu, D.F. 2011. Carbon Productivity, Industrial Correlation and Low-carbon Economic Structure Adjustment—Empirical Analysis Based on Input-output Table in China. *Soft Science* 25(03): 42–56.

Xu, Y.Z., Hu, Y.S. 2011. Analysis of the Difference of Carbon Emissions of China's Manufacturing Sector: A Decomposition Study Based on Input-output Model. *Soft Science* 25(04): 69–75.

Zhang, D.Y. 2005. A Study on Estimation Method of Carbon Emission in Industry Branch. Beijing: Beijing Forestry University.

Zhao, M., Zhang, W.G., Yu, L.Z. 2009. Carbon Emissions from Energy Consumption in Shanghai City. *Research of Environmental Sciences* 22(8): 984–989.

Zhao, Z.P. 2009. Discussion on Green Shipbuilding Technology System. *Marine Technology* (2): 42–44.

Frontiers of Energy and Environmental Engineering – Sung, Kao & Chen (eds)
© 2013 Taylor & Francis Group, London, ISBN 978-0-415-66159-1

Feasibility analysis on the phase change thermal storage heating system of solar energy

L. Pan

Energy and Power Engineering Departement of Changchun Institute of Technology, Changchun, China

ABSTRACT: The paper used solar energy for heating in the region where the solar energy was rich combined with paraffin phase change thermal energy storage technology in winter. The paper mainly discussed the technology feasibility on two schemes, using whole year solar energy after storing energy by paraffin in winter heating period and using some colder daytime solar energy in supplying heat after storing energy by paraffin. By computing, the two schemes were feasible in technology. This system would generate more value on energy saving emission reduction if was used in heating system in winter.

Keywords: phase change; thermal storage; heating; feasibility

1 CHARACTERISTICS OF SOLAR ENERGY RESOURCES AND PHASE CHANGE TECHNOLOGY

We have rich solar energy resources in our country with annual solar radiation is over 5000 MJ/m² in more than 2/3 region, annual sunshine hours are more than 2200 h. The solar energy resource is uneven distribution thought rich. The solar energy resource is distributing as the Table 1.

It's not feasible for solar energy to use as heating resource because it's lean in the north China. While, it's feasible for solar energy to be used as heating resource, because it's rich in the south China. But it's a bit difficult to use solar energy for heating immediately in the south China. In the whole year, the solar energy can is bigger but not be used because it's not necessary for heating in summer. In winter, it's not necessary for heating in daytime because of high temperature. Heating is only necessary in the nighttime. So between solar energy and heating energy there is time difference. To solve this problem, using phase change thermal energy storage is a better method.

Energy storage has become an important method. Using solar energy and phase change thermal energy storage for heating in south China. Paraffin is selected as energy storage materials for solar energy storage heating system. Paraffin phase transition temperature is lower than the temperature supplied by solar water collector. Paraffin can be used repeatedly for it's lower price, better economy, stable chemical properties, higher economic benefit.

2 SCHEME INTRODUCING

Because solar energy can is low, if using solar energy in heating system, the local heat load couldn't be so high. It is possible only make the heat load equivalent to solar energy. According to solar energy distribution zones and local building heat consumption, the scheme chooses Lasa, Luoyang and Yancheng. The index of the solar energy radiation amount in these three places can be seen in Table 2[1]. Lasa is the number one place because it's long sunshine time and rich solar energy. Luoyang and Yancheng may use local solar energy for heating system because they are in southern heating area with lower heat consumption in winter.

Scheme 1: using the whole year solar energy.

Scheme 2: using daytime solar energy phase storage heat for nighttime heating in colder days in winter.

Table 1. Index of solar energy resource regionalization in our country.

Region code	Name	Radiation amount (GJ/m²·a)
I	Rich region	≥6700
II	More rich region	5400~6700
III	General region	4200~5400
IV	Poverty region	<4200

Table 2. Index of each region solar energy radiation amount.

Area	q_a (GJ/m²·a)	q_d (GJ/m²·d)
Lasa	7.6	16000
Luoyang	4.4	8997
Yancheng	4.4	9500

Table 3. Amount of each region heating heat in heating period.

Area	q_h (W/m²)	F (m²)	T (d)	Q_a (MJ)
Lasa	20.2	300	142	74348.9
Luoyang	20.0	300	98	50803.2
Yancheng	20.0	300	90	46656.0

Analysis on technical feasibility: calculating solar energy collection, phase materials storage heat and heating heat in each scheme. Draw the technical feasibility in that area after comparing the relationship among solar energy collection, phase materials storage heat and heating heat.

Table 4. Solar phase storage energy.

Area	q_h (GJ/m²·a)	f^{*1} (m²)	η_s^{*2} (%)	η_x^{*3} (%)	η_h^{*4} (%)	Q_{ax} (MJ)
Lasa	7.6	30	60	92	94	102530
Luoyang	4.4	30	60	92	94	59328
Yancheng	4.4	30	60	92	94	59328

*1 Collector area of the villa on its roof is 30 m²; *2 efficiency of solar energy collector is 60%[3]; *3 efficiency of phase materials storage is 92%; *4 heat changing efficiency of phase storage energy is 94%[3].

3 ANALYSIS ON THE TECHNICAL FEASIBILITY OF THE SCHEME

Heating object: villa with architectural area is 300 m², the design indoor temperature is unified to 18 °C.

3.1 Scheme 1

3.1.1 Calculating amount of heating heat during the heating period in Table 3

$$Q_a = q_h \times F \times T \qquad (1)$$

where Q_a = the amount of heating heat, MJ; q_h = the index of the heat consumption, W/m²; F = the architectural area, m²; T = the days of the heating period.

3.1.2 Calculating the amount of solar phase storage energy in Table 4

$$Q_{ax} = q_a \times f \times \eta_s \times \eta_x \times \eta_h \qquad (2)$$

where Q_{ax} = annual needed phase energy, MJ; q_a = index of annual radiation quantity, $GJ/m^2·a$; f = the collector area, m^2; η_s = the collector efficiency, %; η_x = efficiency of the phase materials storage, %; η_h = heat changing efficiency of phase storage energy, %.

3.1.3 Comparison on technical feasibility

The economy feasibility of scheme1 can be seen in Table 5 after comparing solar phase storage energy and solar phase heating energy.

Analyzing Table 5, we can see solar phase storage energy is more than heating needed energy, so scheme1 is technical feasible in Lasa, Luoyang and Yancheng area.

Table 5. Scheme1 technical feasibility analysis.

Area	Q_a (MJ)	Q_{ax} (MJ)	Technical feasibility
Lasa	74349	102530	Feasible
Luoyang	50803	59328	Feasible
Yancheng	46656	59328	Feasible

3.2 Scheme 2

Though the designed heating load is unknown in each area, we can calculate by index of heat consumption.

3.2.1 Designed heat load calculating

Designed heat load[4]:

$$Q_j = q_e + q_{\text{int}} \qquad (3)$$

$$q_e = (t_i - t_w)\left(\sum_i^m \varepsilon_i F_i K_i\right)\frac{1}{A} \qquad (4)$$

$$q_{\text{int}} = (t_i - t_w)(c_p \rho N V)\frac{1}{A} \qquad (5)$$

Index of heat consumption[4]:

$$q_h = q_e' + q_{\text{int}}' \qquad (6)$$

$$q_e' = (t_i - t_p)\left(\sum_i^m \varepsilon_i F_i K_i\right)\frac{1}{A} \qquad (7)$$

$$q_{\text{int}}' = (t_i - t_p)(c_p \rho N V)\frac{1}{A} \qquad (8)$$

179

(3)/(6):

$$\frac{Q_j}{q_h} = \frac{(18 - t_w)}{(18 - t_p)} \qquad (9)$$

Figured out the designed heat load can be seen in Table 6.

3.2.2 *Calculated the energy in the coldest days in the three places can be seen in Table 7*

3.2.2.1 Calculating the solar energy collector area

Supposed solar phase storage energy equal to heating needed heat, calculated the collector area makes the scheme technical feasible. According to the day ration index, using formula 11, we can figure out the solar energy collector area, seeing Table 8.

$$f = \frac{Q_d}{\eta_x \eta_h \eta_s q_d} \qquad (11)$$

Absolutely, the collector area is not very big, so it has application prospect completely.

Table 6. Designed heat load in each area.

Area	Q_j (W/m²)
Lasa	26.4
Luoyang	26.6
Yancheng	24.6

Table 7. Amount of each region heating heat in heating period.

Area	Q_j (W/m²)	F (m²)	t* (h)	Q_d (MJ)
Lasa	26.4	300	8	228.096
Luoyang	26.6	300	8	229.824
Yancheng	24.6	300	8	212.544

*Considering there is nobody in the villa in daytime, only heating at night or intermittent at night, so the heating time is 8 hours.

Table 8. Solar phase storage energy.

Area	q_h (GJ/m²·a)	f^{*1} (m²)	η_s^{*2} (%)	η_x^{*3} (%)	η_h^{*4} (%)	Q_{ax} (MJ)
Lasa	7.6	30	60	92	94	102530
Luoyang	4.4	30	60	92	94	59328
Yancheng	4.4	30	60	92	94	59328

4 SUMMARY

In the paper, we use solar energy and phase materials in heating system in winter. Making full use of the advantages of solar, clean and rich and phase storage energy, calculate detailed on technical aspect. We make the conclusion that this technical is feasible at the region that solar energy is rich and the heating load is lower in winter. This technical possess popular value that can deduce carbon emission, strongly meaningful at energy saving and emission reduction.

REFERENCES

[1] Zheng ruicheng. Solar water heating system of civil buildings technology handbook. Chemical Industry Press, (2006). p. 52.
[2] Lu yaoqing Heating and air conditioning Design Handbook Second Edition (China Building Industry Press, (2008). p. 288.
[3] Zhang yanping. Phase energy storage theory and use University of Science & Technology China press (1996). p. 34.
[4] Jilin province construction standard management office. Energy saving design standard of residential buildings in Jilin Province. p. 12.

Frontiers of Energy and Environmental Engineering – Sung, Kao & Chen (eds)
© 2013 Taylor & Francis Group, London, ISBN 978-0-415-66159-1

Air pollutants concentrations prediction model based on adaptive artificial neural network

L. Li & P. Wang
Guangzhou Institute of Energy Conversion, Chinese Academy of Sciences, China

M. Cai & Y.H. Liu
School of Engineering, Sun Yat-sen University, China

ABSTRACT: This paper proposes an air pollutants concentrations prediction model based on adaptive artificial neural network. Two main approaches are put forward to improve the prediction accuracy. One is sample adaptive optimization, which is based on the criterion of meteorological similarity with three-tiered screening mechanism; the other is GA-ANN, combining genetic algorithm to ANN, which is built to make the prediction model more adaptive. These two approaches achieve good results as they mainly focus on predicting day by eliminating the impact of invalid situation. A case study is carried out in Guangzhou and Hong Kong to demonstrate the prediction capabilities and applicability in contrast with the conventional modeling approach. The results show that the mean prediction absolute error of SO_2, PM_{10} and NO_2 is 0.015 mg/m^3, 0.023 mg/m^3 and 0.014 mg/m^3 respectively in Guangzhou, while 0.011 mg/m^3 on average in Hong Kong.

Keywords: air pollutant concentration prediction; sample optimization; genetic algorithm; BP neural network

1 INTRODUCTION

Nowadays, people have a better sense of reducing the air pollution and protecting the nature. Urban air quality forecast has become more and more important in people's daily life. However, the prediction accuracy is hard to guarantee.

There are two urban air pollution prediction approaches widely used today, one is the numerical prediction, the other is the statistical way. Numerical prediction is popular these years, but a detail and accurate pollutant inventory is needed. With a large amount of calculation, it is hard to be widely applied. Statistical forecast is the most widely used prediction method today[1]. By means of regression analyzing and fuzzy mathematics, the recorded concentration of pollutants and meteorological conditions will be associated with future conditions, so as to establish corresponding functions to forecast the pollutant concentration, air pollution index, etc[2–3]. In recent years, neural network prediction has emerged and developed rapidly. Li et al[4]. was the first domestic researcher to employ neural network for the purpose of forecasting the air pollution, after which comes applications of neural network in predicting air quality[5–7]. However, disadvantages had been clearly shown in these predictions, such as slow convergence speed, minimum error falls onto local minimum, etc.

To guide the daily life for citizens by improving the prediction accuracy, this paper establishes an air pollutant prediction model based on a three-tiered screening optimization mechanism and GA-ANN to improve the qualities of sample and model. Besides, this adaptive prediction model was successfully applied to several monitoring sites in Guangzhou and Hong Kong, all of which achieved good results. By comparing with the traditional prediction method, the accuracy has been enhanced approximately 5% for three pollutants.

2 PREDICTION MODEL

2.1 Sample adaptive optimization

This paper proposed a sample optimization approach based on meteorological similarity. The meteorological condition of predicting day is the criterion of sample optimization. A set of samples which have a higher relation with the predicting day is selected from the historical database by three-tiered screening. This three-tiered screening, including screening of single meteorological factors, integrated meteorological factors and correlation degree, is performed to obtain the superior samples, while several thresholds and weights should be confirmed in the beginning of these screening.

After three-tiered screening, the samples used to build up the prediction model are those who have high similarity and could reflect the changing rule of pollutant concentration under a certain meteorological condition. Eliminating the inferior impact of those unrelated rules, each day's pollutants concentrations prediction will be more targeted.

2.2 Adaptive model initialization

This paper adopted the genetic algorithm to optimize the initial threshold and weighing of the neural network, taking global search on optimal solution instead of random generation. After the neural network optimization, not only may the neural network generalization ability be enhanced, but also the speed of the network convergence can be improved, amending the defect of easily falling into the local minimum. This can enhance the accuracy of air pollutants prediction model to some extent.

Figure 1 shows the implement flow of the genetic algorithm. 100 was taken as the population size; Model error was put forward as the fitness target; Selection employed the Hollan method; Crossover used the linear crossing model; Mutation adopted the uniform mutagenic operator with the mutational rate of 0.05. After specific times' iteration, the initial threshold and weighing of the neural network were acquired as the optimal solution.

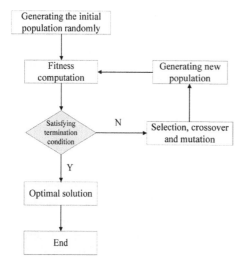

Figure 1.   Flow chart of genetic algorithm.

2.3 BP neural network

BP neural network is specialized in non-linear modeling on account of its highly nonlinear mapping ability. However, there is no common rule to specify how to determine the structure of the network for different nonlinear object, while the network topology has a great impact on the objectivity, rationality and accuracy of the prediction.

This paper defined the network topology into three layers: input layer, output layer and hidden layer, according to the practical considerations of pollutant concentration prediction. The neuron numbers of the former two were decided through the actual situation, while the last one was obtained by trial and error. By making a mean absolute error analysis between the testing result and the target output, the hidden neuron number of the best predicted effect model was chosen as the superior one.

3 APPLICATION

3.1 Site description and data

Eight air quality monitoring sites located in Guangzhou are chosen as the study objects. The air pollutant data is from these eight monitoring sites while the meteorological data is from a monitoring site located in the center part of Guangzhou.

Five years data is used in this model started from 2006. Nine parameters are applied to predict the concentration of SO_2, PM_{10} and NO_2. They are daily average temperature (°C), daily average wind speed (m/s), daily average wind direction (°), daily rainfall (mm), atmospheric pressure (hPa), relative humidity (%) and daily average concentration of SO_2, PM_{10} and NO_2 (mg/m³).

Taking almost six months (January 1st 2010 to June 30th 2010) as the prediction period, using the model built above, an effect analysis of the prediction can be present.

3.2 Results and discussion

For the time period investigated in this study, the mean absolute error of these three pollutants is low, while the PM_{10} prediction effect is not so perfect, with the error of 0.023 mg/m³, which is shown in Figure 2. However, it has already had a great improvement on the prediction accuracy, with SO_2, PM_{10} and NO_2 prediction enhancing 7.6%, 7% and 1% respectively, compared to the traditional neural network model.

To verify the applicability of this model, three air quality monitoring sites which are the sub stations in Hong Kong contained in PRD regional air quality monitoring network are chosen in this

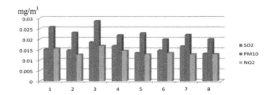

Figure 2. Eight sites mean absolute error of three air pollutants prediction.

Table 1. Absolute error of air pollutants concentration prediction in Hong Kong.

Site	SO$_2$ (mg/m^3)	RSP (mg/m^3)	NO$_2$ (mg/m^3)
Tung Chung	0.007	0.018	0.013
Tsuen Wan	0.008	0.017	0.009
Tap Mun	0.003	0.018	0.004
Average	0.006	0.017	0.009

study. They are Tap Mun, Tsuen Wan and Tung Chung, which located in the north-east, center and south-west of Hong Kong. The air pollutant data is from these three monitoring site while the meteorological data is from Hong Kong Observatory.

Nine years data is used in this model started from 2002. Taking almost six months (December 1st 2009 to May 18th 2010) as the prediction period, a good prediction effect was shown in Table 1.

Among these three sites, Tap Mun has the best prediction effect, with the average concentration error of 0.008 mg/m^3, while Tung Chung is the worst, with the average error of 0.013 mg/m^3.

In these cases, it can be seen that the prediction on different pollutants and different sites present a different effect. This might have a great relevance to the inputs of the impact factors, which appear diverse impact effects on different pollutants. Therefore, the selection of the impact factors to different pollutants should be precise, a formal approach, using to select the input factors, should be put forward.

4 CONCLUSIONS

1. This paper adopt two approaches, sample optimization and model optimization, to enhance the air pollutants concentrations prediction accuracy. It is found out that the input of the model had a great influence on it, while superior samples can highly improve the accuracy. Besides, the model structure and modeling ability also play an important role in accuracy enhancement.

2. The adaptive prediction model is applied to eleven sites located in Guangzhou and Hong Kong, with the prediction mean absolute errors of three pollutants reaching 0.011 mg/m^3 in Hong Kong, and 0.017 mg/m^3 in Guangzhou. Simultaneously, by comparing with the traditional neural network prediction method, the result shows that the adaptive prediction model has a better forecast effect.

ACKNOWLEDGMENT

This work was supported by Guangdong Natural Science Foundation (S2011040002839) and National Key Technology R&D Program (2011BAG07B00).

REFERENCES

[1] WANG Qin-geng, XIA Si-jia, WAN Yi-xue, JIN Long-shan. A New Idea for Urban Air Pollution Forecast [J]. Environmental Science & Technology, 2009, 32(3):189–190. (in Chinese).

[2] TU Li-juan, XU Zhong, ZHANG Hui-qing. A PM_(10) Pollution Forecast Simulation Model for Xi an City [J]. Computer Simulation, 2008, 25(012): 110–113. (in Chinese).

[3] QU Dan, LIU Miao, Lian Xiufeng. The Research and Development of Air Quality Forecast in Changchun City [J]. Journal of Guangdong Polytechnic Normal University, 2007, (012): 46–49. (in Chinese).

[4] LI Zuoyong, DENG Xinmin. Prediction Model of Environmental Pollution Using Artificial Neural Network [J]. Journal of chengdu institute of meteorology, 1997, 4(2):279–283. (in Chinese).

[5] FANG Li. Utilization of Artificial Nerve Cell Network in Air Pollution Prediction [J]. Environmental Protection Science, 2006, 32(3):4–6 (in Chinese).

[6] HOU Chun-hua. The Application of BP Network Mode in the Atmosphere Pollution Predication in Chaoyang Area [J]. Environmental Protection Science, 2007, 33(5):53–54. (in Chinese).

[7] BAI Xiaoping et al. Application of Artificial Neural Network to Air Pollution Prediction in Suzhou City [J]. Science & Technology Review, 2007, 25(3):45–49. (in Chinese).

Frontiers of Energy and Environmental Engineering – Sung, Kao & Chen (eds)
© 2013 Taylor & Francis Group, London, ISBN 978-0-415-66159-1

The numerical simulation and experimental research of Low-NO$_x$ burner retrofitting on a 420 T/h utility boiler

Q. Zhang & J.H. Zhou
State Key Laboratory of Clean Energy Utilization (Zhejiang University), Hangzhou, Zhejiang Province, China

C.J. Zhao
Zhejiang Pyneo Co., Ltd., Hangzhou, China

N. Ding & J.R. Xu
Hangzhou Dianzi University, Hangzhou, China

ABSTRACT: The thermal test was carried out on a 420 T/h utility boiler of Xiaoshan power plant, the boiler can reach satisfactory performance and the thermal efficiency reaches 91.07%; Then a numerical simulation of the pulverized coal combustion was carried out on the basis of the test, by means of thermal aero dynamical field analysis and temperature field analysis, as well as the research for the pollutant, the numerical simulation results are in good agreement with the experiment, which provide good guidance for the Low-NO$_x$ burner transformation of the future large-scale station.

Keywords: boiler; burner; NO$_x$; numerical simulation

1 INTRODUCTION

In response to the national energy-saving and environmental protection policy appeals, two natural circulation boilers of Xiaoshan power plant were desired to be retrofitted by low NO$_x$ combustion technology. After transformation, the coal type and boiler output required remain unchanged, but the NO$_x$ emissions have to be reduced to a certain value. The design principle to achieve low NO$_x$ combustion by retrofitting burner is forming a "less oxygen" reductive environment in main combustion zone, so the formation of NO$_x$ will be inhibited in this environment. Even though the amount of NO$_x$ slightly increased in the later period influenced by SOFA, the NO$_x$ content in the export of combustion area is the major factor to influence the final content.

According to the requirements of the programmer and the design requirements for air distribution, the retrofitted burner was divided into two groups: the SOFA burner and the main burner, the 10 spout of the retrofitted burner are lower secondary air, lower primary air, middle primary air, mid-secondary air, upper primary air, tertiary air, upper secondary air, CCOFA, lower SOFA, upper SOFA. The position of flame center is regulated by the SOFA swimming angle adjustment, so the main steam temperature, reheat temperature and the combustion efficiency of pulverized coal can adjust by it. The burner structure is shown in Figure 1.

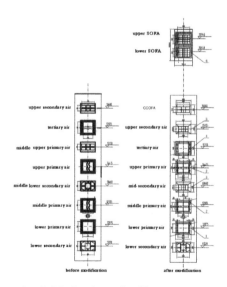

Figure 1. Original and retrofitted burner.

2 NUMERICAL MODEL

A three-dimension steady model was adopted to simulate the object. To avoid the effect of false diffusion[1], QUICK format is used in the numerical method, and the SIMPLE algorithm is adopted to solve the equations. The boundary conditions of inlets and outlets are respectively set as pressure inlet, pressure outlet, at the walls a non-slip condition is assumed for this simulation. The realizable k-ε turbulent model[2] and the Lagrange Discrete Phase Model (DPM)[3] were used to simulate and analysis the flow field characteristics. When the combustion characteristics was simulated by numerical method, the P1 radiation model and two parallel reaction model[4] were used in the simulation, a kinitics/diffusion limited combustion model for simulating surface combustion of pulverized coal particles, and a Probability Density Function (PDF) method for simulating turbulent combustion[5].

3 RESULTS AND DISCUSSION

3.1 *The flow field distribution in the furnace*

The horizontal section of middle primary air field was selected to analysis the flow field distribution.

The SOFA burner supplied parts of the total air-flow after modification, so the air-flow at the primary combustion zone reduced. The rigidity decreases and the tangent circle diameter become smaller. The flow field was mainly distributed at the center line area of the furnace.

The flow field distribution varies significantly at the vertical section due to increasing the SOFA burner, the flow field was mainly distributed at the bottom of the primary combustion zone before modification, but after the burner' retrofitted, the flow field also concentrated distribution in the area between the primary combustion zone and the SOFA burner. So the low field become long and narrow, which helps burning more completely and improve the efficiency of the furnace.

3.2 *The temperature distribution in the furnace*

From the Figure 4 we can see that the temperature rises near the primary combustion zone after

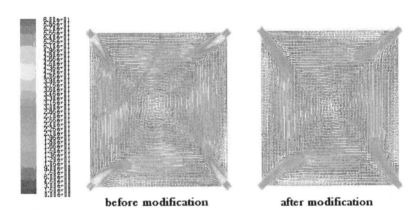

before modification   **after modification**

Figure 2.   Flow field distribution at the horizontal section of middle primary air.

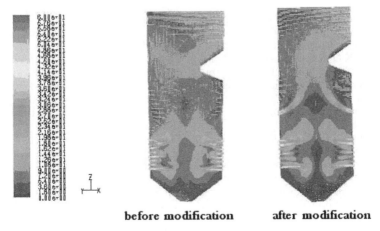

before modification   **after modification**

Figure 3.   Flow field distribution at the vertical section of diagonal.

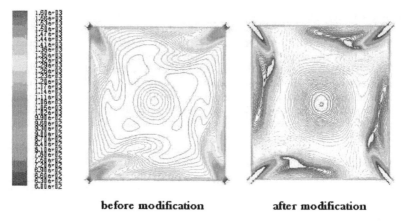

before modification   **after modification**

Figure 4.   Temperature distribution at the horizontal section of middle primary air.

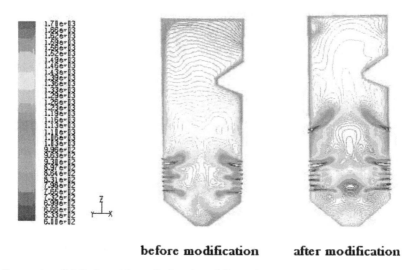

before modification   **after modification**

Figure 5.   Temperature distribution at the vertical section of diagonal.

modification, which was influenced by the reduced air flow at this area.

Figure 5 show the calculated value of the maximum is about 1700 K at the bottom of the primary combustion zone after modification, and the average temperature is about 1400 K. The experimental values of them are 1448 K and 1186 K, so the result of the numerical simulation agrees well with the experimental data.

The temperature distribution at the vertical section of diagonal shows the temperature concentrated distribution in the area between the primary combustion zone and the SOFA burner, it gives the similar result as the air flow distribution. As the air-flow reduced at the primary combustion zone, the temperature increased at the primary combustion zone. But as the SOFA increases, the temperature reduced at the top of the SOFA spout.

3.3   The distribution of NO in the furnace

The NO distributions of the middle primary air at the horizontal section were displayed in Figure 6. The distribution of NO increased at the primary combustion zone after modification, which was influenced by the reduced secondary air and the increased temperature at this area.

The NO distributions of the middle primary air at the vertical section were displayed in Figure 7. The NO content slightly higher at the primary combustion zone after modification, but it declined obviously at the furnace outlet due to the SOFA. This shows that the Low-NO$_x$ Burner can reach the target level for decreasing the NO content. The combustion experiment results suggest that the content of NO declared from 550~750 mg/Nm^3 to 310 mg/Nm^3 at the inlet of air pre-heater.

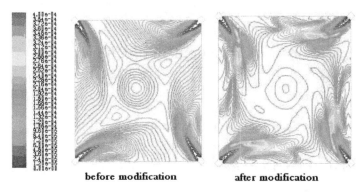

before modification　　**after modification**

Figure 6. NO distribution at the horizontal section of middle primary air.

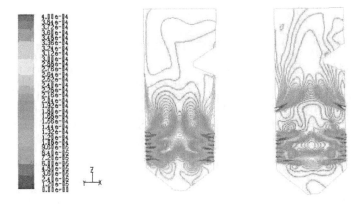

Figure 7. NO distribution at the vertical section of diagonal.

4 CONCLUSION

After retrofitting the boilers of the Xiaoshan power plant by the low-NO_x burner, the variability of air flow, temperature and NO distribution were very significant. By increasing the SOFA spout, the air-flow at the primary combustion zone reduced and the low field become long and narrow; the temperature increased at the primary combustion zone, and it concentrated distribution in the area between the primary combustion zone and the SOFA burner, it's more rational then before; The NO content slightly higher as the temperature increased, but it declined obviously at the top of the SOFA spout; The agreement between the calculated results and experiment results is very good, the results suggest that the retrofitted boiler can effectively reduce NO_x emission level.

ACKNOWLEDGMENT

The authors acknowledge the financial support of the natural science foundation of Zhejiang Province (GK110901002).

REFERENCES

[1] Zheng Changhao, Tang Qing, Xu Xuchang, et al. Numerical modeling of 3-D isothermal flow in tangentially-fired boiler with body-fitted meshes [J]. *Proceedings of the CSEE)*. 2002, 22(7): 29–34.

[2] Shih T.H., Liou W.W., Shabbir A, et al. A new k-e eddy-viscosity model for high Reynolds number turbulent flows-model development and validation [J]. *Computers Fluids*, 1995, 24(3): 227–238.

[3] Cen Ke-fa. Engineering gas-solid multiphase flow theory and calculation of [M]. Zhejiang University press, 1990.

[4] Yu Hailong, Zhao Xiang, Zhou Zhijun, et al. Numerical simulation analysis on the effects of O/C ratio and concentration in coal water slurry on gasification process [J]. *Journal of Fuel Chemistry and Technology*, 2004, 32(4): 390–394.

[5] Sivathanu, Y.R., Faeth G.M. Generalized State Relationships for Scalar Properties in Non-Premixed Hydrocarbon/Air Flames [J]. *Combust. Flame*, 1990, 82(2): 211–230.

Frontiers of Energy and Environmental Engineering – Sung, Kao & Chen (eds)
© 2013 Taylor & Francis Group, London, ISBN 978-0-415-66159-1

Policy and legal analysis of development of China's low carbon buildings

K. Zhou & W. Luo
School of Law, Renmin University of China, Beijing, P.R. China

B. Li
School of International and Public Affairs, Columbia University, NY, USA

S. Ouyang, J.Y. Fu & Y.Y. Zeng
School of Law, Renmin University of China, Beijing, P.R. China

ABSTRACT: The energy consumption of the construction industry is a major source of greenhouse gases, while the construction industry is one that have faster positive effects to save energy. To fulfill the goal of energy-saving and emission reduction, China's construction industry has follow "low carbon" and "energy conservation". Construction of low-carbon development is inseparable from strong policy support and legal protection. China's low-carbon construction policy and legal system are still in infancy at present, few specialized legislation, low hierarchy of most specifically norms, and not concentrated. It need to be further improved and integrated based on the national situation, as well as experience from the advanced foreign countries.

Keywords: low-carbon building; energy-saving and emission reduction; policy and law

1 LOW-CARBON BUILDINGS

Low-carbon building is a building development model which reduces fossil energy use, improves energy efficiency and minimizes greenhouse gas emissions throughout the whole life cycle of construction of manufacture of building materials and equipment, the construction and building use based on low power consumption, low pollution, low-emission, while at the same time to meet people's basic comfort requirements and special service needs[1].

2 CONSTRUCTION INDUSTRY— THE MOST LIKELY FIELD TO REACH THE TARGETS OF ENERGY-SAVING AND EMISSION REDUCTION

According to the global cost curve for greenhouse gas issued by McKinsey in 2007, the top five most effective ways are building insulation (thermal insulation), fuel efficiency in commercial vehicle, lighting systems, air conditioning and water heating (plumbing). It can be seen that four of them are associated with the construction industry.

In addition, according to the China's response to climate change solutions announced by the

Lawrence Berkeley National Laboratory (LBNL), reducing the buildings carbon emissions was classified as one of the most critical path for China's climate change solutions. The table[1] as shown.

The study shows that though the construction industry is one of the biggest parts of social energy consumption; it is also the most flexible industry to reduce the consumption. In order to complete the energy-saving emission reduction goals and fulfill international commitments towards low carbon development path, the construction of "low carbon" and "energy" is destined to become China's inevitable choice. The reforms of construction industry and policy making should be starting from building capacity for sustainable development and energy

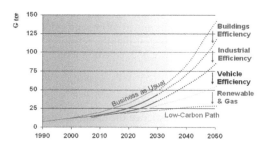

efficiency, and it is the only way to promote the transition of construction industry to a low carbon industry[2].

3 POLICY AND LAW OF LOW-CARBON BUILDING

The health development of low-carbon construction industry cannot depart from strong policy support and legal protection. The legal evolution of China's current low-carbon construction policies and systems are as follows.

3.1 Policy

The earliest policy document on sustainable development on construction, is the "China's Agenda 21—White Paper on Population, Environment and Development in China in the 21st century" promulgated by the State Council in July 1994. As a guiding document required the governments and departments at all levels to document their economic and social development plans, China's Agenda 21 is also basic principles and program of action to promote sustainable development of China's construction industry.

Chinese government enacted "the Outline of China's Action for Sustainable Development in the early 21st Century" (hereafter referred to as "Outline") in July 2003. "Outline" put forward specific requirements in its guiding principles and objectives as well as specific protection measures for the sustainable development of the construction industry.

The Fourth Session of the Tenth National People's Congress passed the National Economic and Social Development Five-Year Plan (referred to as the 11th Five-Year Plan) on March 14, 2006, and the main provisions are all related to low-carbon buildings. Subsequently, the former Ministry of Construction released "The Construction Industry Outline of Eleventh Five-Year Plan" on March 15, 2006, This Outline took rapid development of land-efficient buildings as one of the main objectives of the reform and development of construction, and put forward vigorous development of land-efficient housing and public buildings.

The National Development and Reform Commission released, the "energy saving and long-term special plan" in November 2006 by consent of the State Council. This is the most systematic and explicit low carbon policy regulation and the Plan indicated that the full implementation of energy-saving standard.

On June 4, 2007 the State Council issued the "China National Program on Climate Change".

The content involves the construction of low-carbon development and many measures will have a significant impact of mitigating greenhouse gas emissions.

The State Council issued the White Paper on "China's policies and actions for addressing climate change" on October 29, 2008. It treated the construction industry as a key energy saving areas and proposed to actively promote energy-saving and environmentally friendly construction and green buildings, strictly implement new buildings mandatory energy efficiency standards, speed up the energy conservation and transformation of existing buildings.

In order to promote the application of renewable energy construction, the government unveiled a series of documents including notice of pushing the renewable resource and management of special funds. At the same time, government should positively guide and promote model application, laws and regulations, technical standards, technical support and capacity building.

3.2 Laws

Resolution of the Standing Committee of the National People's Congress on Making Active Responses to Climate Change issued on August 27, 2009, which is the highest level of legal norms in the legal system of low-carbon construction. The legal resolution plays a great role on China to address climate change and the legislative process for the development of low-carbon economy and adaptation of existing laws.

National People's Congress Standing Committee approved the "United Nations Framework Convention on Climate Change" in 1992, and Chinese government signed the Kyoto Protocol in 1998 and ratified in 2002. According to the provision of Article 46 of the "Environmental Protection Law of the People's Republic of China", these conventions have the same effect as Chinese law.

It is prescribed in the Article 4 of the Construction Law of the People's Republic of China issued in 1998 that the State assists the development of construction industry, supports research on construction science and technology in order to raise the level of housing and construction design, encourages saving of energy and environmental protection and advocates the adoption of advanced technology, advanced equipment, advanced process, new building materials and modern managerial methods.

It is formulated in the Article 24 of the People's Republic of China on Cleaner Production Promotion Law which took effect in 2003, that the construction projects should be conducive to environmental protection and resource conservation using energy, water and other architectural design, construction and decoration materials.

The People's Republic of China Renewable Energy Law took effect in 2006, and revised in 2009, Article 17 which involves the construction industry states that the administrative department of construction of the State Council shall, in conjunction with other relevant departments of the State Council, formulate technical and economic policies and technical criteria for the combination of solar energy utilization systems with the construction of buildings.

The People's Republic of China Energy Conservation Law, amended in 2007 and implemented in 2008, determines the construction industry as a key energy-saving area, including the detailed provisions from the formulation of building energy efficiency standards to the specific implementation of building energy efficiency as well as legal responsibility, and building energy saving as a separate one chapter to be laid down.

The People's Republic of China Circular Economy Promotion Law issued in 2009, contains the provisions that are directly applicable to the construction industry .

The "Building Energy Conservation Ordinance" promulgated on October 8, 2008 is a specialized administrative regulation which summarizes the Energy conservation work practices over the years and developed on the basis of the relevant legislative experience at home and abroad. It is an important supporting regulations of "Energy Conservation Law".

"Energy Conservation Regulation for State-funded Institutions" was promulgated by the State Council in October 2008, according to the "Energy Conservation Law". "Regulation" provides supervision and management of energy management, energy-saving measures of existing buildings and new construction of public institutions in detail.

The Ministry of Construction and the State Administration of Quality Supervision, Inspection and Quarantine jointly issued a national standard Green Building Evaluation Standard on March 7, 2006. On June 1, the same year, this standard was formally implemented, and became China's first recommended national standard for green building which prescribes multi-objective, multi-level comprehensive evaluation from the perspective of whole life cycle of the residential and public buildings.

3.3 *Evaluation*

Overall, the low-carbon economic legal and policy system of the construction is still in its infancy. Policy prescriptive takes up the major part, while specialized legislations are comparatively less. With more attention being put on the energy saving of the construction industry, policy support and legal protection system of low-carbon development has being gradually improved. However, there are still some inadequacies in the low carbon buildings policy and legal system:

"Construction Law", as a national law, is with high authority. Though targeting on building energy efficiency, it lacks mandatory provisions on "support" and "encourage" and "advocate" of building energy efficiency. In addition, "Construction Law" doesn't provide green buildings and low-carbon buildings, let alone the specific terms of the constraints.

Although China has enacted and modified the "Energy Conservation Law", and done more detailed provisions of building energy efficiency, problems such as deficient observation and lax enforcement are serious. In addition, the supporting regulations are incomplete, and interoperability regulations need to be improved, and there are still articles not to be refined and implemented. Since there are no appraisal, no quantization, no audit, there is still a big loophole in energy management. Further, the process of setting efficiency standards lags behind. Despite that over 50% efficiency design standards of building energy in each climate zone has been successively formulated and promulgated, there are still a lack of implementation rules and enforcement .

"Renewable Energy Law" provides that real estate development enterprises shall create necessary conditions for the solar building design and construction. But the Construction Law of the People's Republic of China has no relevant requirement. Defects in cooperation and coordination of institutional, will no doubt severely constraint the implementation and effect of the Renewable Energy Law.

The green building rating system in China impacted greatly by LEED (Leadership in Energy and Environmental Design) of the United States, which lays more emphasis on energy conservation, but little attention to carbon emission. The EU started a number of demonstration projects as early as 1990s. In addition to the requirements of saving energy and water, those projects require control of carbon dioxide and other greenhouse gases and pollution gas emissions as well.

Lacking of incentive policies of saving building energy is another problem. Domestic and international practices show that the energy saving goes to market failure, and it needs government's macro-regulation and guidance. The tax policies on building energy-saving, energy-saving equipment exploration and applications as well as energy-saving incentives, are far from enough. China has not set up an effective building energy incentive mechanism.

4 ESTABLISH AND IMPROVE THE POLICY AND THE LEGAL SYSTEM TO PROMOTE CHINA'S LOW CARBON BUILDING DEVELOPMENT

4.1 *The advanced experience of the low carbon buildings policy and legislation of the major developed countries*

4.1.1 *United Kingdom*
Energy consumption standards of construction industry were regulated by the building codes issued in 2002. In 2006, the UK sustainable housing regulations determines the well-known standard assessment procedure rating (Standard Assessment Procedure Rating) which requested to evaluate mandatorily new buildings energy consumption, carbon dioxide emissions for all new buildings from 2008, with the objective of zero emissions of carbon dioxide by 2016[3].

4.1.2 *United States*
In January 2009, the U.S. Obama government has proposed an economic stimulus plan called "Green New Deal", including investment in building energy-saving projects, especially energy-saving facilities of government agencies and public institutions and promotion of energy saving of ordinary family homes.

4.1.3 *Japan*
As an island nation with meager resources Japan has long been carrying out a low-carbon society scenarios and action plans. In May 2008, the research team issued "Scenario Team. Japan Scenarios and Actions towards Low-carbon Societies", which put forward a low-carbon roadmap for the housing industry[4]. In April, 2009, Japan published the economic stimulus plan regarding low carbon revolution as the core.

4.1.4 *Germany*
Energy Conservation Act provides that when consumers purchase or rent housing, builders must produce an "energy certificate" and inform consumers of the annual energy consumption in the residential area. New regulations also encourage businesses and individuals to conduct energy saving for the old building, and implement measures to mandatorily scrap the old ones.

To sum up the practical experience of developed countries in the low-carbon buildings areas, we can conclude that their common practices are: (1) make feasible technology standard; ②combine mandatory legislation with economic incentives; ③cultivate public low carbon environmental awareness; ④carry out the section of low-carbon buildings in the government agencies and other public buildings; ⑤strong support for low carbon technology research. China should learn much from developed countries in promoting low-carbon buildings. Since the developed countries have experienced different stages of social development and levels of economic development, Chinese government should learn from their advanced experience, and give full attention to the functions of the government and the promotion of low carbon buildings according to China's national conditions.

4.2 *Improve the legal system of China's low carbon building policy combining with national conditions*

Recommendations from the following aspects to improve the legal system of China's low carbon building policy are:

The diverse and coordination of policy. A lot of laws, regulations and policy measures were implemented to promote building energy efficiency by a way of administrative order. The role of economic incentives should be strengthened. In the policy design process, government should not only take more diverse and different policy instruments, but also pay attention to the synergies between different policy measures.

Handle the relationship between policy and law. Although policy is flexible and timely, it is instable and variable. On the contrary, the law is stable and predictable. Therefore, how to exert a regulatory function and make full use of its advantage, and realize their complementarity and unity in the low-carbon building policy and legal security system in the future is the focal point.

Improving the relevant policies to ensure the implementation of "Building Energy Conservation Ordinance". Great change has taken place since Building Energy Conservation Ordinance was promulgated and implemented. The key change is from the entirely executive order-driven to a legal based protection to promote building energy efficiency. Such as building energy consumption statistics, energy efficiency audits and other methods to determine the usage of fixed energy; creating a standard of construction and building energy efficiency evaluation system operation management system, promoting the optimization of building energy-saving technology upgrades; establishing a building energy-saving special inspection, supervision, management and appraisal system; establishing contract energy management practices and supporting financing system; building energy efficiency market regulatory system; establishing a complete system of building energy efficiency standard. Special attention should be paid to rural housing construction standard and the current situation of building energy consumption in order to promote energy saving and environmental protection of the rural housing when improve the relevant policies.

Promote the CDM (Clean Development Mechanism) in the field of construction industry. The potential of greenhouse gas emission reduction in the field of building is enormous. But so far, there is still no registered CDM project of building[5]. Implementation of CDM creates obstacles for the construction sector. PCDM projects can integrate dispersed emission reduction projects into a project, and it is very suitable for replication and application of policies or measures, especially for construction projects. In addition CDM project pilots can be taken into implementation Pilot projects can simplify the building energy use, reduce the incompatible part of the construction sector in the CDM, advocate CDM in the whole society and improve its social impact.

Amend the laws and to add the low carbon content. To amend the Building Law, the Environmental Protection Law and other relevant laws, it is necessary to add the content of the low carbon buildings and green building, and to increase binding provisions. The new provisions must be mandatory and operable. The content of carbon emission constraints should be increased in the Green Building Evaluation Standard. Further, it is necessary to connect the departmental rules and regulations to the upper law.

To carry out the publicity of the low carbon buildings and low-carbon life. Media and public opinion should not on one hand boost carbon reduction, on the one hand, overwhelmingly boost the high-carbon energy-consumed lifestyle (such as luxury villas, big cars, etc.) Low carbon concept is supposed to be spread, with the widely accepted low-carbon lifestyle as well as the large market-based low carbon construction, via the government making policy, mass media campaigns and compulsory education.

5 CONCLUSION

The construction industry is one of the important areas towards energy conservation and low carbon development. The Concept of low carbon building is actively promoted in the world, but it has different meanings in different countries. Interest is different, so is the cost. During the meeting in Copenhagen, China under apparent international pressure is asked to assume responsibility for emissions reductions with a loud voice. Yet low-carbon development in China is not just energy-saving emission reduction and limiting their emissions. What we need is a smarter, environmentally, friendly, resource-saving way. In china, low carbon development of construction and the relevant legal system of policy should be established and perfected upon this concept.

REFERENCES

[1] LI Qiming, OU Xiaoxing. Analysis of the Definition and Development of Low Carbon Buildings [J]. Construction Economy. 2010(2).
[2] Steve Sorrell. Making the Link: Climate Policy and the Reform of the UK Construction Industry [J]. Energy Policy, 2003(9).
[3] Camilleri, M.J., A draft climate change sustainability index for house[R], study report No. 95 (2000), BRANZ.
[4] 2050 Japan Low -carbon Society Scenario Team. Japan Scenarios and Actions towards Low-carbon Societies(LCSs) [R]. 2008.
[5] HAO Bin, LIN Ze, Research on application of the clean development mechanism in building energy conservation [J]. Heating Ventilating & Air Conditioning. 2009 (11).

Analysis and discussion on intelligent induced ventilating system of underground garage

D.L. Cheng, Z.H. Zhou & X.F. Tou
College of Urban Construction and Safety Engineering, Shanghai Institute of Technology, Shanghai, China

ABSTRACT: With the development of population and economy, the numbers of small and medium-sized cars have become increasingly growth, especially in Shanghai, Beijing and other big cities. The problem of parking has become more and more difficult, even though underground parking garage are developed well. However, with the disadvantages of the growing tensions in the global energy today, the traditional underground garage ventilation modes can not meet our requirements. And induced ventilation system is paid more and more attention to its no ventilation ducts, energy-saving, saving investment, good ventilation and other features, especially intelligent induced ventilation. The paper has discussed the intelligent induced ventilation system for underground park on the basis of the survey and understanding of some underground garage ventilation systems' status, summarizes the status and characteristics of the system, offers the Intelligent design, proposes the principles of layout and been applied in the project example at last.

1 INSTRUCTIONS

Automobile parking garages can be partially open or fully enclosed. Fully enclosed parking garages are usually underground and require mechanical ventilation. Indeed, in the absence of ventilation, enclosed parking facilities present several indoor air quality problems. The most serious is the emission of high levels of Carbon Monoxide (CO) by cars within the parking garages. Other concerns related to enclosed garages are the presence of oil and gasoline fumes, and other contaminants such as oxides of nitrogen (NOx) and smoke haze from diesel engines.

The design of intelligent induced ventilation is very important to the underground garages for energy-saving, saving investment, good ventilation, or environment protection. To this reason, combines the expertise and field experience, and combined with the investigation and understanding of a number of underground garage ventilation system as well, the author discussed the induced ventilation system in the underground garages. Investigate the intelligent design of induced ventilation system in the underground garages on the basic of the summary of intelligent induced ventilation system's status and characteristics. And propose the system's layout principle, applied to the engineering example in the end as well.

2 THE PRINCIPLE AND COMPOSITION OF INTELLIGENT SYSTEM

The system conclude blower, multiple induced fan units, exhaust air (smoke) machines, the centralized controller, sending and exhaust fan controlling modules. The principle is that the induced fan unit's nozzle shot the directional high-speed air flow driven by the ambient air flow. Under the conditions of no wind pipe, formed directional air flow from the blower to the exhaust fan to achieve the purpose of diluting CO and ventilation. When the air under a certain speed spread from diameter D_0 of the vent to the space, which is not limited by the interface surface around, it is called free jet. The air flow shot by the vent can be regarded as isothermal free jet. Due to the turbulent momentum exchange between the jet boundary and the surrounding medium, the ambient air continues to be involved, and the jet range expanding. The velocity field of the jet cross section gradual decays from the jet center to the border, and continuously decreases along the range, while the flow along the range direction increases, the jet diameter also increases, but the total momentum of each section remains unchanged (Fig. 1).

Set the vent air flow as Q_0 and air flow velocity as V_0, set the air flow at the nozzle parallel to the cross section which is X away from the nozzle as

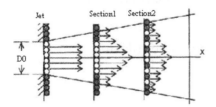

Figure 1. Induction of the jet on the ambient air.

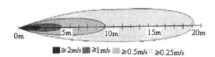

Figure 2. Ranges and velocity distribution of the jet.

QX, air flow rate is the VX, according to the law of momentum conservation:

$$M_0 = M_x \tag{1}$$

$$M_0 = Q_0 V_0 \rho \tag{2}$$

$$M_x = Q_x V_x \rho \tag{3}$$

$$Q_0 V_0 = Q_x V_x \tag{4}$$

where M is the air momentum, kg·m/s; Q is the air flow, m³/s; V is the wind speed, m/s; ρ is the air density, kg/m³.[1]

Although in theory, the width of jet has been increased infinitely, induced air flow will be increased to infinity as well and point's rate will be reduced to an infinitesimal. But reality environment always has many non-ideal conditions, for example, there are obstructions such as beams, columns and other natural airflow from all directions in the buildings. So there must to be another nozzle relay when the jet center speed decays to a certain speed to form the "air push-pull effect" to produce the flow velocity field in whole space.

Figure 2 is a distribution map of the range and speed of the jet (nozzle diameter is 80 mm; nozzle speed is 12 m/s).

3 THE CHARACTERISTICS OF INTELLIGENT SYSTEM

Researches show there are many characteristics for intelligent system as followings.[2]

1. Lower storey, smaller engine room, less investment. Such as, induced fan box is generally thinner, about 250 mm high, hanging directly under the floor, can installed between the beam and even can wear a bean laying, reducing the cross opportunities with other professional.
2. Easier calculate and shorter preliminary time. Each set of system burdens of same size. It is a

modular design. So it can reduce the hydraulic calculations, air speed accounting tedious work and so on, greatly improve the efficiency of the design work.
3. Simple loading, saving time and costs, beautiful and flexible.
4. Saving energy and having low noise.
5. The ventilation effect is ideal and rapid.
6. Intelligence and varied control methods. Because garages have different needs in the management and use, the ventilation system should provide multiple modes of operation. It is necessary to set a automatic detection of air quality equipment. It may automatic open and close system according to the carbon monoxide sensor value or user settings and protects the linkage of the system and its main exhaust fan in the same partition in order to ensure the continuity of the tubular type ventilation and the effect of induced ventilation system.
7. Low leakage trap and easy to check. The system has simple structure, few contacts and small possibility of leakage. Although leaks, we can easily use the bubble test and other methods of leak detection. And it only takes a few minutes to alright finish if use special heat-shrinkable closures duct tape and silicone.

4 DESIGN OF GOOD INDUCTION VENTILATION SYSTEM

To design an economic, security, and application of good induction ventilation system, the first is to clear ventilation volume, and then determine the design principles of ventilation and exhaust systems (such as proposing the layout principles of induced fans-FYA-type induced fans).

4.1 The determination of the amount of ventilation

It is not the same for the regulations between different countries. In Japan, for parking garage which is greater than 500 m², if open area is less 1/10 of floor area, it should use machinery ventilation system to provide more than 25 m³ of fresh air per hour. For indoor parking which open area is less than 1/10, the number of ventilators should above 10 times per hour. The recommendations of the times of ventilation for the underground garage in American are 4 to 6 times per hour or each m² 4 L/s air flow. These garages which is connect with outside should have 2.5~5% open area to offer fresh air. Finnish building codes require minimum fresh air of the office buildings' underground garage of 2.7 L/s·m².[3] According to the provisions of the relevant technical measures in china, the exhaust is not less than 6 times per hour and ventilate is not less than 5 times per hour as design basic when there has no information to calculate.

4.2 *Design principles of ventilation and exhaust system[4]*

1. Underground garage must use mechanical ventilation; ventilation system of combination building within the garage and underground garage should be set up independently and should not mixed set with other building ventilation systems.
2. The duct should use flammable materials and it should not pass through the firewall. When it must to through, we should set fire draper. Damper operating temperature should be 70°C. Duct insulation materials should be used not burning or burning hard materials.
3. An underground garage of over 2000 m² areas should set a mechanical exhaust system. Mechanical exhaust system and civil defense, health and other exhaust ventilation systems can be combined.
4. In the garage with a mechanical exhaust system, the construction area of each smoke partition should not be more than 2000 m² and the smoke partition should not cross the fire district. Smoke partitions can use smoke screen, wall or beam which not less than 0.5 m from the ceiling to divided.
5. Each smoke partition should set exhaust ports. The exhaust port should be established on the ceiling or wall near the ceiling. Exhaust ports' furthest point of the horizontal distance should be no more than 30 m away from the smoke partition.
6. Exhaust ventilation should be not less than 6 times per hour.
7. A centrifugal fan or exhaust axial fan can be used as exhaust fan. And there must to be a smoke damper in the exhaust manifold which can self-close when the flue gas temperature exceeds 280°C. The exhaust fan should ensure to work continuously for 30 minutes at 280°C. The smoke damper should be interlocked to shut down the exhaust fan.
8. The wind speed should not exceed 20 m/s in metal pipes and not exceed 15 m/s in other smooth surface pipes. The wind speed of exhaust ports should not exceed 10 m/s.
9. It should set ventilation system as well in the smoke partition which hasn't car evacuation export direct access to the outdoor to evacuate the export. And air supply should not be less than 50% of the amount of smoke.

4.3 *Layout principles of induced ventilation system*

Here we use FYA type induced fan as example to put forward the layout principle[5,6] of induced ventilation system. As follows.

Regard each fire district has one induced ventilation system to design. Use inducing fan between the air vents and air intakes to form the piston "air wall" to exhaust.

Make the lane in the garage as mainstream field to layout the induced fan in order to set up a stable piston space.

Form a certain angle between induced ventilation fans and the fans of the mainstream field as Secondary flow field.

Back to consider of the layout of the wind shaft in the underground garage, sometimes delivery and exhaust vents are very close. This requires using induced ventilation fans to virtual separate, setup process and prevent short circuit.

Each induced fan vertical distance may be not less than 0.5 m/s by the end of the control velocity, the 8–9 m as a "relay". (See Fig. 3 example of design node).

Centralized controller should be placed in the middle of the partition of their control.

The controller can output a passive on-off signal to the main exhaust fan to control the main exhaust fan.

Centralized controller for each sub-region will be paralleled to the BA host by a five twisted-pair. Use control software and the signal converter provided by the manufacturers to remote monitoring.

5 THE PROJECT APPLICATION OF GOOD INDUCTION VENTILATION SYSTEM

Here, one project application of good induction ventilation system is given. The system has an underground garage with a total construction area of 5326 m², divided into two fire district. Storey

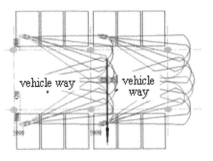

Figure 3. Design node of vehicle way.

for 3.6 m high, beam for 0.8 m and under the beam height is 2.8 m.

5.1 Equipment layout of ventilation system

Here describes the first fire district for example. The fire district has 2778 m^2 area, put it into two smoke partitions according to the regulations. Remove the housing and other facilities, it can be divided into an area of 1250 and an area of 1200 fire district. Each fire district sets up an exhaust fan.

The fire district has ramp to ground, so it can use natural ventilation. Take 6 times per hour for ventilation rate. The exhaust system and exhaust system share the same duct. And set the exhaust valve. Exhaust valve is normally open for the usual garage exhaust; the exhaust valve will be closed and the exhaust will be opened when fires. Use type FYA induced fans and layout by its principle.

Then provide the intelligent design of induced ventilation system, propose the layout principle of the system and apply in the project example at last.

By all counts, induced ventilation system is quite good. It is worth to promote in underground or enclosed space ventilation. Due to different performance between different companies they made, design principles are different as well. In addition, architectural design often has some special cases. So HVAC engineers and architects should to strengthen exchanges in the system design.

5.2 The equipment selection of ventilation system

Here, the selected major equipment of ventilation system is given below (Table 1).

6 CONCLUSIONS

The paper has discussed the intelligent induced ventilation system for underground park on the basis of the survey and understanding of some underground garage ventilation systems' status, summarizes the status and characteristics of the system, offers the Intelligent design, proposes the principles of layout and been applied in the project example at last.

ACKNOWLEDGEMENT

This paper is supported by the Shanghai excellent curriculum "Auto Control and Thermal Instrument".

BIOGRAPHY

DaoLai CHENG: born in 1965, Male, Professor, Ph.D. The vice director of College of Urban Construction and Safety Engineering, Shanghai Institute of Technology, Shanghai. His research interests include energy saving technology, signal process, mechanical fault diagnosis.

CONTACT METHOD

College of Urban Construction and Safety Engineering, Shanghai Institute of Technology A308, No. 2 Teaching Building, No. 100 Haiquan Road, Fengxian District, Shanghai City, China.
(P.C: 201418).
E-mail: daolaicheng@163.com.
Tel: +86-21-60873631, 13311998959.

REFERENCES

[1] Gao Jun Fen G. Discuss on the application of induced ventilation system with no wind duct into underground park. Refrigeration and Aircondition, 2009, 16(19):163–164.
[2] Zhang Li Xin, Meng Xiang Feng. Design and application the induced ventilation system with flat and war in duct into underground parking. China People Aerial Defence, 2009, 8(10):78–81.
[3] GB50067-9. Design and Fireproofing Criterion on Automobile Garage, Repairing Garage, Park lot [M]. China Plan Publishing Company, 1997.
[4] Moncef Krarti, Ph.D. and Arselene Ayari, Ph.D. Ventilation for Enclosed Parking Garages. ASHRAE Journal, February 2001.
[5] Analyses Reports on Induced Ventilation System Hangdling [R]. Shanghai Wind Limb Air-congdition Equipment Limited Company, 2011.
[6] Krarti, M., A. Ayari, and R.A. Grot. 1999. "Evaluation of fixed and variable rate ventilation system requirements for enclosed parking facilities." Final report for ASHRAE Project 945-RP.

Table 1. Major equipment in underground garage.

Name	Type	S
Wind/smoke exhaust fan	IMX-1000 D6-7.5	65% L = 33969 m^3/h H = 452 Pa N = 7.5 kW
Induced fan	FYA-3-Z	L = 680 m^3/h D = 80 mm H = 250 Pa N = 120 W dB(A) < 55
Centralized controller	FYK-1	U = 220 V N = 20 W
Fan control module	FYM-2	U = 220 V N = 5 W

Frontiers of Energy and Environmental Engineering – Sung, Kao & Chen (eds)
© *2013 Taylor & Francis Group, London, ISBN 978-0-415-66159-1*

Influence of the combined packing on removal efficiency of ammonia nitrogen from high concentration ammonia nitrogen wastewater

G.H. Wang
College of Chemical Engineering and Technology, Wuhan University of Science and Technology, Wuhan, Hubei, China

W. Wang
College of Resources and Environmental Engineering, Wuhan University of Science and Technology, Wuhan, Hubei, China

W.B. Li, X.L. Ren & C. Liu
Hubei Coal Conversion and New Carbon Materials Key Laboratory, Wuhan Hubei, China

ABSTRACT: Two types of packing were applied to improve ammonia nitrogen removal efficiency and study stripping condition. Through laboratory experiment of air stripping, it was found the same optimum condition with column packed with two types of packing respectively: wastewater temperature T = 80°C and pH = 11.5, gas-liquid volume ratio r = 250:1, dosage of surfactant C = 20 mg·L⁻¹. And the overall removal efficiency using pall ring is 4% higher than that of using combined packing.

Keywords: air stripping; ammonia nitrogen; combined packing; foam ceramic packing

1 INTRODUCTION

Surplus ammonia from coking plant is a high concentration of ammonia nitrogen wastewater, as the main source of ammonia in the coke plant wastewater. Ammonia nitrogen, added as a new binding target in the twelfth five-year environmental protection plan of China, is required to be reduced by 10% in 2015 (Zeng & Shen 2012). Usually high concentration of ammonia nitrogen wastewater is pretreated to significantly reduce the concentration of ammonia nitrogen, and then treated with biochemical method. At present, the common methods of pretreating high concentration of ammonia nitrogen wastewater are ammonia distillation, chemical precipitation and air stripping method, etc. And air stripping method has received wide attention for its energy-saving, economic, and environment friendly characteristics (He et al. 2008).

Air stripping method applied to remove ammonia nitrogen from simulated high concentration of ammonia nitrogen wastewater in this paper. And its optimum conditions affecting air stripping efficiency were studied in a organic glass column with two different types of packing. The foam ceramic packing, mixed with pall rings in a certain proportion, is introduced to study the influence of packing with different specific surface area to ammonia nitrogen removal.

2 EXPERIMENT

2.1 *Experiment theory*

Taken air as desorption medium and the difference of ammonia concentration in different phases as mass-transfer driving force, ammonia nitrogen is removed from liquid phase (wastewater) in form of ammonia diffusing along with the gas phase (air).

Two forms of ammonium salt, including fixed and volatile ammonium salt, exist in the surplus ammonia. Fixed ammonium salt such as NH_4Cl and $(NH_4)_2SO_4$ have to transfer into volatile ammonium (NH_3) by adding alkali to increase pH. And then with large difference between liquid and gas phase, ammonia is easy to diffuse into gas phase. According to the balance relationship referred to equation (1), the mole fraction of volatile ammonium is more than 98%.

$$NH_4^+ + OH^- = NH_3 + H_2O \qquad (1)$$

By increasing the temperature, the solubility of ammonia decreases and rate of fixed ammonia transferring into volatile ammonium becomes higher, which makes desorption of ammonia easier finally. To some extent, high specific surface area of packing can enhance the efficiency of mass-transfer. Earlier research shows that surfactant could improve mess-transfer process by changing the surface tension of wastewater (Wang et al. 2011). Relevant literatures indicate that combined packing (random packing and structured packing), by improving distribution of the liquid phase in the random packing layer, could increase the apparent height of transfer unit and improves the mass transfer efficiency (Wu et al. 2010).

2.2 Experimental apparatus and flow

The size of pall ring is 25 mm × 25 mm × 1.2 mm. And foam ceramic packing has a round plate structure with size of 120 mm (diameter) × 37.5 mm (height). Properties of packing is shown in Table 1. On the basis of preliminary experiment, firstly the organic glass column is packed with pall ring, and then with pall ring in which is evenly distributed with 4 pieces of foam ceramic packing. The inside diameter of organic glass column is 130 mm, and total height of packing layer is 1 m.

Concentration of raw simulated ammonia nitrogen wastewater prepared with ammonium chloride should be maintain at 2500 mg/L by dosing equal quality of ammonium chloride. The pH is determined by pH (Thermo Orion, 9107 WLMD, USA). Experimental procedure is as follows: a) Adjust raw wastewater to the required condition by heating and adding technical pure NaOH and surfactant. After that, sample raw wastewater and keep sealed; b) Adjust air rotameter after starting air blower. And adjust wastewater rotameter after starting pump; c) Control the water valve and keep proper height of wastewater at the bottom of the column to prevent air leaking; d) When the air stripping process is finished, sample from wastewater container.

2.3 Experimental scheme

Preliminary experiment is prepared for orthogonal experiment.

Preliminary Experiment: To realize the pressure-drop characteristic of combined packing and determinate range of gas-liquid ratio in the later orthogonal experiment, pressure-drop per meter test is conducted. Single-factor experiment of surfactant conducted to study the relationship between surfactant concentration and ammonia nitrogen

Table 1. Properties of packing.

Packing types	Porosity /%	Specific surface area/m²·m⁻³	Plate height/mm
Pall ring	91	219	–
Foam ceramic packing	≥78	2500	37.5

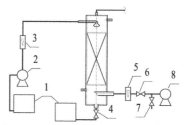

1.wastewater container; 2.pump; 3.wastewater rotameter; 4.wastewater valve; 5.air rotameter; 6. air valve; 7. vent valve; 8. air blower;

Figure 1. The experimental flowchart.

Table 2. Range of influencing factors.

Level	pH	Temperature T/°C	Surfactant dosage C/mg·L⁻¹	Gas-liquid ratio r
1	10.0	65	10	1:100
2	10.5	70	20	1:150
3	11.0	75	30	1:200
4	11.5	80	40	1:250

removal rate is for determinating the dosage range of surfactant in the later orthogonal experiment.

Orthogonal Experiment: The orthogonal experiment $L_9(4^4)$ is designed in Table 2.

3 RESULTS AND DISCUSSIONS

3.1 Pressure drop measurement

As is shown in the Figure 2, pressure drop per meter of dry combined packing is quite low and has a linear relationship with air flow. The rising air flow enhances the drag force to the liquid flow, which makes high liquid holdup and takes more interspaces. When pressure drop increases sharply and liquid holdup becomes more, the flooding phenomena happened. An appropriate liquid/gas flow can be obtained from Figure 2. Obviously the pressure drop becomes high along with liquid flow increasing when gas flow is constant. The pressure drop of combined packing is mainly caused by foam ceramic packing.

The measuring range of rotameter $(0.25~2.5 \text{ m} \cdot \text{h}^{-1})$, wastewater flow is setted at $10 \text{ L} \cdot \text{h}^{-1}$, and air flow range from $1.0~2.5 \text{ m} \cdot \text{h}^{-1}$. The pressure drop per meter is round $140~700 \text{ Pa} \cdot \text{m}^{-1}$.

3.2 Single-factor experiment of surfactant

As is shown in Figure 3, with the increase of the surfactant from 0 to 20 mg·L⁻¹, the removal

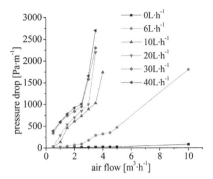

Figure 2. Pressure drop per meter.

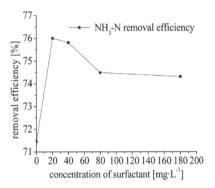

Figure 3. Relationship between removal efficiency and surfactant concentration.

Table 3. Result of orthogonal experiment.

	pH	T/°C	C/mg·L⁻¹	r	Removal efficiency/% P*	C*
1	1	1	1	1	22.47	20.87
2	1	2	2	2	33.20	30.81
3	1	3	3	3	38.15	36.31
4	1	4	4	4	48.39	36.94
5	2	1	2	3	50.40	45.07
6	2	2	1	4	58.27	52.01
7	2	3	4	1	43.49	40.00
8	2	4	3	2	55.16	51.26
9	3	1	3	4	57.19	54.37
10	3	2	4	3	60.09	55.60
11	3	3	1	2	55.59	40.51
12	3	4	2	1	51.29	43.82
13	4	1	4	2	53.04	46.36
14	4	2	3	1	41.82	39.39
15	4	3	2	4	71.29	64.16
16	4	4	1	3	68.28	61.63
k_{a1}	35.55	45.78	51.15	39.75		
k_{a2}	51.83	48.35	51.57	49.25		
k_{a3}	56.04	52.13	51.25	54.23		
k_{a4}	58.61	55.78	51.25	58.79		
R_1	23.06	10.00	3.49	19.04		
k_{a1}	31.23	41.67	43.76	36.02		
k_{a2}	47.09	44.46	45.97	42.24		
k_{a3}	48.58	45.25	45.33	49.66		
k_{a4}	52.89	48.41	44.73	51.87		
R_2	21.66	6.74	2.21	15.85		

* P and C is short for pall ring and combined packing respectively.

efficiency enhanced by 4%. And the removal efficiency decreased to 74.4% slowly with the concentration reaching 180 mg·L⁻¹ at last.

Surfactant changed the surface tension of the wastewater, and affected the uniform performance of the liquid film in the packing surface, thus it affected the mass transfer area (Zuiderweg & Harmens 1958). When surfactant concentration reaches its micelle concentration, and excess foam generated will increase the liquid film resistance and hinder the gas-liquid mass transfer process, which finally reduce ammonia nitrogen removal rate. These two aspects together make the ammonia stripping efficiency increase first and then decrease.

3.3 Orthogonal experiment

Results are shown in Table 3.

1. Analysis of Influence Factors: According to table 3.1, the same result obtained with two types of packing. The sequence of influential significance is pH > r > T > C. The optimum condition is T = 80°C, pH = 11.5, r = 250:1, C = 20 mg·L⁻¹.

2. Comparison Analysis: For different types of packing, wetting property, surface treatment and specific surface area together influence the mass transfer process (Yang et al. 2010). Structured packing with a large specific surface area has large K_La_e and K_Ga_e, making the HETP decreased. This is the main reason for the structured packing with high separation efficiency (Huang et al. 2000). Combined packing used in this experiment has a larger specific surface area than pall ring. Theoretically, the stripping efficiency should be better.

But the experimental results indicate that Pall ring has better ammonia nitrogen removal efficiency than combined packing. The reasons may be as follows:

a. Foam ceramic packing has a large specific surface area of 2500 m²·m⁻³ which limits the proportion of foam ceramic packing. According to the proportion of foam ceramic packing in the combined packing, the average specific surface area is 561 m²·m⁻³ calculated by Eq.(2).

The specific surface area is not high enough to improve removal efficiency.

$$\sigma_{ave} = (\sigma_1 V_1 + \sigma_2 V_2)/(V_1 + V_2) \qquad (2)$$

b. Flooding often happens in the layer of foam ceramic packing during pressure drop measurement. It is better to choose packing with appropriate specific surface area and uniform property in the further study. Because of overlarge specific surface area the flow path within the packing is so small that the poor gas-liquid contact makes mass transfer efficiency low.

c. Bias flow occurs when liquid passes through foam ceramic packing layer and decreases the gas liquid contact area.

4 CONCLUSIONS

The pressure drop per meter is round $140{\sim}700$ Pa$\cdot$m^{-1}, mainly caused by foam ceramic packing of combined packing. Packing with appropriate specific surface area and uniform property will be chosen in the further study.

The same result obtained with two types of packing. The sequence of influential significance is pH > r > T > C. The optimum condition is T = 80°C, pH = 11.5, r = 250:1, C = 20 mg$\cdot$L^{-1}.

Because of gas-liquid ratio limitation of stripping method is 250:1 (Liu 2012). Gas-liquid ratio can be increased after choosing packing with proper specific surface area to improve the removal efficiency.

ACKNOWLEDGEMENT

This work was financially supported by National Science Foundation (20976141), Hubei Provincial Department of Education (C2010019), fund from Coal Conversion and Open-end Fund of the Science Research Center of Green Manufacture and Energy-saving and Emission-reduction of Wuhan University of Science and Technology (A1201, B1207).

REFERENCES

Huang Jie, Zeng Bing & Zhang Xue. 2000. Comparison of Properties of Mass Transfer between Structured Packing and Random Packing. *Chemical Engineering* 28(3): 15.

He Yan, Zhao Youcai & Zhou Gongming. 2008. Research process on the denitrogenation of highly concentrated ammonia-nitrogen wastewater. *Industrial Water Treatment* 28(1): 1–4.

Liu Tiejun. 2012. Study on Effects of Surfactants on Ammonia Nitrogen Mass Transfer in Buffer System of High Concentrations of Ammonia Nitrogen Wastewater. Wuhan: *Wuhan University of Science and Technology*.

Wang Guanghua, Gong Fanjie & Li Wenbing. 2011. Ammonia removal from coal dry distillation wastewater. *China Energy Society*.

Wu Shaomin, Hu Xueqin & Liu Chunjiang. 2010. Mass transfer efficiency of a novel hybrid structured packing used in liquid-liquid extraction. *Chemical Engineering* 38(11): 15–16.

Yang Yuncai, Chen Guizhen & Yang Yinru. 2010. Research progress on mass-transfer model calculation of structured packing. *Journal of Chemical Industry & Engineering* 31(2): 31.

Zeng Qingling & Shen Chunhua. 2012. Experimental Study on Recovery of Ammonia-nitrogen by Struvite Precipitation Method. *Environmental Science & Technology* 35(1): 80–83.

Zuiderweg F.J. & A. Harmens. 1958. The influence of surface phenomena on the performance of distillation columns. *Chem Eng Sci* 9(2–3): 89–103.

Frontiers of Energy and Environmental Engineering – Sung, Kao & Chen (eds)
© 2013 Taylor & Francis Group, London, ISBN 978-0-415-66159-1

SWOT analysis of new nuilding materials industry based on circular economy

J.L. Zheng & D.W. Qin

Department of Management and Economy, Kunming University of Science and Technology, Yunnan, China

ABSTRACT: In the construction of circular economy, new building materials industry is growing rapidly, but also faced many problems. New building materials industry join the metallurgy industry, chemicals industry, coal industry, thermal power and other industries. It is the combination of high-energy consuming industries. And it is also a strategic and emerging industry to save energy in the circular economy. Simultaneously, it bears the important task of industrial innovation. To this end, from the perspective of circular economy, this paper carries out a SWOT analysis for the new building materials, and propose countermeasures.

Keywords: new building materials industry; circular economy; SWOT analysis

1 INTRODUCTION

1.1 *The concept of new building materials*

New building materials is different from the traditional building materials like brick, gray sandstone. It is the raw material to Substitute or recycle the industrial raw material. Through new technologies, it is able to save energy and protect environment during its production or use.

From the aspects of use, it can be divided into the wall materials, decorative materials, doors and windows materials; from the aspects of function, there are insulation materials, waterproofing materials, bonding and sealing materials, as well as for supporting hardware accessories, plastic parts and a variety of supporting material; from the aspects of material, it can be divided into natural materials, chemical materials, metal materials, non-metallic materials and so on.

1.2 *Overview*

The new building materials industry is rapidly developing. For one hand, energy conservation and circular economy has become an important part of the government's work, on the other hand, China's building materials technologies and energy-saving technologies are sustainable developing. The technology has been more generally applied to the building construction, such as renewable energy in the building envelope, equipment, etc. In 2010, the production of new wall materials accounted for more than 55% of total wall materials, an increase of 11 percentage points compared with 2005. The area of new construction, which use new wall materials, is 4.8 billion square meters, an increase of more than three times over five years ago. Due to industrialized construction and policies to encourage, housing construction scale is the direction of the development of construction industry and new building materials industry in China is experiencing an unprecedented development opportunity.

2 CHARACTERISTICS

2.1 *Combine other industries*

New building materials industry can absorb a wide range of material. It associated with coal, electricity, metallurgy, chemical and many other traditional industries. It can absorb the residue and by-products which generated by the chemical industry, coal industry, mining, metallurgy industry and other traditional industries, such as slag, coal gangue, fly ash, gypsum, phosphorus gypsum, slag, iron slag, copper slag, etc. The waste can reuse as hollow blocks, cement and other building materials, relying on new technology.

2.2 *Performance improvement*

The building materials recycled by waste, has good performance. For example, slag cement has good stability. It is suitable for requirements of high temperature concrete structures.

2.3 Absorptive capacity

New building materials, especially wall materials, can absorb many kinds of waste, and does not require complex chemical changes. In general, it can almost fully absorb any waste by crushing, mixing.

3 SWOT ANALYSIS

3.1 Strength

3.1.1 Rapid technological development
New building materials technology is rapidly developing. Wall materials made by new technology have been basically mature at life, fire safety, sound insulation absorption performance, and is applying wide range. Other building materials technologies, such as green glass, asphalt, wall panels and other technologies are fast development and progress. [2]

3.1.2 Easily combined with other industries, and became the co-generation mode
One of the features of the building materials industry is the link. Through the use of gypsum, fly ash, coal gangue of the thermal power industry, the slag, tailings, of coal mining industry and phosphogypsum of chemical industry and so on, it can be synthesized wall material or modified cement.

3.1.3 Effectively absorb the pollution of the waste, and benefit in environment
China's traditional industries such as coal-fired electricity, chemicals, etc. produce large amounts of waste and hazardous substances. However, due to high cost and the lack of processing capacity, large amounts of toxic industrial waste residue stack everywhere, these seriously pollute groundwater, soil environment. New building materials industry reuse waste, avoid pollution, and effectively solve the problem of industrial waste, and gradually ease the pressure on the environment, resulting in a good environmental benefits. From 2006 to 2010, the new wall materials recycle solid waste is about 1.5 billion tons, including coal gangue, fly ash and tailings.

3.1.4 Effective conservation of resources
New building materials completely replace the traditional building materials. Traditional building materials use all kinds of natural resources about 4 billion tons each year. New Building Materials significantly save resources, through an replacement of traditional building materials. For example, using new building blocks instead of the traditional clay bricks, has been able to save land about 200,000 hectares from 2006 to 2010. [3]

3.1.5 Energy conservation
The energy savings include direct saving energy in process and indirect saving energy during use building materials. By using equipment to reuse heat for power generation and heating etc. The new building materials industry effectively save energy in the production process. The energy loss in residential housing in China is: the wall about 50%, roofing about 10%, windows and doors account for about 25%, the basement and ground about 15%. Lack of thermal insulation is one of the causes of poor performance of housing insulation. New building materials such as thermal insulation coating, block, reduce energy consumption through its unique structure.

3.1.6 Other strength
New building materials' performance is improving and upgrading, and cost is gradually reducing.

3.2 Weakness

3.2.1 Development and applications is still a gap compared with developed countries
New building materials and the application is less. Types of building materials are small, and need to broaden the scope of application. At present, China's building energy consumption is high, and the energy utilization is low, which result in units of building energy consumption is 2–3 times higher than the same climate countries such as Sweden, Denmark, Finland. For example, the average energy consumption is 20.6 W/m^2 in Beijing each year, after implementation of "energy efficiency standards". The average energy consumption is 11 W/m^2, in the same climate countries, like Sweden, Denmark or Finland.

3.2.2 Business management skills need to improve
New building materials are rapidly developing, but extensive management still exists.

3.2.3 Some building materials exist defects
As the technology is not yet mature, there are some unavoidable problems in some building materials. For example, hollow bricks made of industrial waste, more or less contain radioactive elements. Some products due to these defects, to some extent limit the long-term development of the product.

3.2.4 Lack of detailed standards and norms for new products
Since a lot of new building materials is lack of standards, and production and application is out of touch. Lagging on the market and technology research, some products are not included in the engineering and construction standards, and can not be successfully enter the field of construction applications.

3.3 Opportunity

3.3.1 Policy support

The style of China's economic is growing to the "resource conserving, environment-friendly". After proposing the goal of "Emission reduction", many policies have emerged to support new building materials industry. *Building materials industry development plan* issued by the China Building Materials Federation said: building energy saving and green building has become a major demand in China's sustainable development and the promotion of external wall insulation is an effective way to reduce energy consumption in China.

3.3.2 Large quantities of industrial wastes

In China's industrial park, industrial waste is piling up too much, resulting in environmental pollution, and need to handle the waste well. Large quantities of industrial waste t provide resources for the new building materials industry.

3.3.3 Large demand

"Low-carbon building materials, recycling economy" has become a new trend of the global materials industry. With the national importance attached to the circular economy, energy saving, and the increasing in consumers awareness of energy-saving and environmental protection, the future demand for new materials is large. The main objectives of our country to develop programs are: build 1.1 billion square meters green buildings, while upgrade 570 million square meters buildings for energy saving. [5]

3.4 Threat

3.4.1 Downstream industry—The real estate industry is in the doldrums

The real estate industry as a downstream industry of building materials, is easily affected by the policy. With the harsh policies of the national real estate industry, from September 2011, the real estate industry has entered a cold period. National real estate transaction volume dropped significantly, which may lead to a decline in demand for building materials.

3.4.2 Increasingly fierce competition— Investors are easily to enter the market

Because of relatively low investment of most varieties of building materials. The huge market and higher growth rate has attracted a large number of competitors in the market, resulting in an increasingly competitive market. The market mechanism is imperfect, and it is difficult to compete orderly.

3.4.3 The policy has not yet comprehensive and detailed

Manufacture of new building materials require new technology, large initial investment, and will take some time to adapt to the market. Policy has not yet provided good protection, resulting in many premature death of some products.

4 COUNTERMEASURE

4.1 Improve the technological level

Technology development is a fundamental factor in the development of new building material, there are a big gap between developed countries in resource reuse and product performance. The adoption of new technology and equipment can make up for product defects, reduce energy consumption, and improve energy structure. New building materials industry should improve product technology, optimize its performance. And it is the key and basis of "resource-conserving and environment-friendly", as well as objectives of sustainable development. Relying on science and technology dominate the market.

4.2 Reduce costs and improve competitiveness

Improve the product technology, simultaneously, enhance the capacity of resources and energy reuse in the production process, and improve production efficiency. Meet market demand, and produce the building materials needed by the market.

4.3 Enhance the joint of building materials industry and other industries

Joint of new building materials industry and metallurgy industry, chemical industry, coal industry, mining industry can effectively absorb industrial waste. Waste reuse is able to reduce the impact and destruction of industrial waste on the environment.

4.4 Promote policies and standards

First, give the R & D convenience and tax benefits, and limit the high energy consumption industry, and closely linked energy-saving targets and business benefits. Second, the Government should pay attention to the development of new building materials industry and timely make product standards, to help building material products to successfully enter the market. Thirdly, government agencies should step up publicity to make environment and promote the use of new building materials.

5 SUMMARY

Policy support and future increase in demand of new building materials provide favorable conditions for the building materials industry. The characteristics of the new building materials industry will produce good economic and environmental benefits and ecological benefits, which comply with the policy direction. In aspects of product optimization and application, enterprises should be further research and development, and change disadvantages to advantages; At the same time, should be effective to avoid industry risk and to ensure the long-term development of the building materials industry to face the treat.

REFERENCES

[1] Tao Li, The development and application of new building materials: *globle market information guide* Vol. 24 (2011):90.

[2] Xianghua Cheng, Discussion on the regeneration of waste cement concrete aggregate technology: *construction machinery technoloyic management* Vol. 5 (2010):77.

[3] Zhibin Wu, Development of low-carbon economy leapfrog development of building materials enterprises: *Building materials technology and applications* Vol. 2 (2011):47–48.

[4] Rulai Yang, On the new building materials and building energy conservation technology development status: *Industrial Technology* Vol. 7 (2011): 97–98.

[5] Baoxing Qiu, Green energy-saving construction pregnant trillion business opportunities, new building materials promising: *Liaoning Building Material* Vol. 10 (2011):25.

Frontiers of Energy and Environmental Engineering – Sung, Kao & Chen (eds)
© *2013 Taylor & Francis Group, London, ISBN 978-0-415-66159-1*

The primary study on the role of ocean to CO_2

C.H. Ma, K. You & W.W. Ma
Ocean University of China, Qingdao, P.R. China

ABSTRACT: Over the ocean, the uptake of anthropogenic CO_2 by the global ocean induces fundamental changes in seawater chemistry that could have dramatic impacts on biological ecosystems in the upper ocean. In this paper, we analyze the mutability of carbon and its role in the ocean. It certifies the acidification around the release point may be the most important for the global warming. It is going to take a lot of time to let these processes can reach new carbon dioxide equilibrium. In view of literature analysis, chemical analysis, and bio-analysis and so on, to save energy and significantly reduce greenhouse gas emissions is the only solution.

Keywords: mutability of carbon; biological ecosystems; oceans; value

We all know the oceans cover around 70 per cent of the Earth's surface. They thus play an important role in the Earth's climate and in global warming. One important function of the oceans is to transport heat from tropics to higher latitudes. Besides heat, the Oceans absorb substantial amounts of carbon dioxide, and thereby consume a large portion of this greenhouse gas, which is released by human activity. Moreover it cannot be predicted how the marine biosphere will react to the uptake of additional CO_2. The exchange of carbon dioxide (CO_2) between the atmosphere and ocean is a critical process of the global carbon cycle and an important determinant of the future of the Earth system (*Fung et al.*, 2005; *Friedlingstein et al.*, 2006; *Denman et al.*, 2007). From 1800 until 1994, the Ocean removed about 118 ± 19 Pg C (1 Pg $= 1015$ g) from the atmosphere (*Sabine et al.*, 2004).

The Earth's climate is influenced by many factors, including solar radiation, wind, and ocean currents. But interaction among the various factors is very complex and numerous questions remain unresolved. Scientists are making extensive measurements to determine how much of the human-made CO_2 is being absorbed by the oceans.

1 THE MUTABILITY OF CARBON

Carbon is the element of life. Plants on land and algae in the ocean assimilate it in the form of carbon dioxide (CO_2) from the atmosphere or water, and transform it through photosynthesis into energy-rich molecules such as sugars and starches. Carbon constantly changes its state through the metabolism of organisms and by natural chemical processes. Carbon can be stored in and exchanges between particulate and dissolved inorganic and organic forms and exchanged with the atmosphere as CO_2. The ocean store much more carbon than the atmosphere and the terrestrial biosphere (plant and animals). Even more carbon, however, is stored in the lithosphere (calcium carbonate, $CaCO_3$).

Atmosphere, terrestrial biosphere and ocean, which are the three most important repositories within the context of anthropogenic climate change, are constantly exchanging carbon. Although the process can occur over time spans of up to centuries, yet considering that carbon remains different Earth's crust for millions of years, then the exchange between the three factors mentioned above is relatively rapid. The amount of carbon stored in the individual reservoirs can be fairly accurately estimated. The ocean, with around 38,000 gigatons (Gt) of carbon (1 gigaton = 1 billion tons), contains 16 times as much carbon as the terrestrial biosphere, and around 60 times as much as the pre-industrial atmosphere. The ocean is therefore the greatest of the carbon reservoirs, and essentially determines the atmospheric CO_2 content.

Consequently, changes in atmospheric carbon content that are induced by the oceans also occur over a time frame of centuries. In geological time that is quite fast, but from a human perspective it is too slow to extensively buffer climate change.

1.1 *The ocean as a sink for anthropogenic CO_2*

With respect of climate change, the greenhouse gas CO_2 is of primary interest in the global carbon cycle. From the early 19th to the end of the 20th century, humankind released around 400 GT C in the form of carbon dioxide. This has created a serious imbalance in today's carbon cycle. In addition to the atmosphere, the oceans and land plants permanently absorb a portion of this anthropogenic CO_2.

As soon as CO_2 migrates from the atmosphere into the water, it can react chemically with water molecules to form carbonic acid. Because carbon dioxide is thus immediately processed in the sea, the CO_2 capacity of the oceans is ten times higher than that of freshwater, and they therefore can absorb large quantities of it. This kind of assimilation of CO_2 is referred as a sink. The ocean absorbs human-made atmospheric CO_2, and this special property of seawater is primarily attributable to carbonation, which, at 10 per cent, represents a significant proportion of the dissolved inorganic carbon in the ocean. At the same time, the carbon dissolved in the form of CO_2, bicarbonate and carbonate is referred to as inorganic carbon.

1.2 *Measuring exchange between the atmosphere and ocean*

For dependable climate predictions, it is extremely important to determine exactly how much CO_2 is absorbed by the ocean sink. Researchers have therefore developed a variety of independent methods to quantify the present role of the ocean in the anthropogenic impacted carbon cycle. Two procedures in particular have played an important role: the first method (atmosphere-ocean flux) is based on the measurement of CO_2 partial-pressure differences between the ocean surface and the atmosphere. The second method attempts, with the application of rather elaborate geochemical or statistical procedures, to calculate how much CO_2 of the ocean is derived from natural sources and how much is from anthropogenic sources.

2 HOW CLIMATE CHANGE IMPACTS THE OCEANS

2.1 *Climate change impacts the marine carbon cycle*

The natural carbon cycle transports many billions of tons of carbon annually. In a physical sense, the carbon is spatially transported by ocean currents. Chemically, it changes from one state to another. The foundation for this continuous transport and conversion is made up of a great number of biological, chemical and physical processes that constitute what is also known as carbon pumps. These processes are driven by climatic factors, or at least strongly influenced by them.

Changes in the carbon cycle are also becoming apparent in another way: the increasing accumulation of carbon dioxide in the sea leads to acidification of the oceans or a decline in the PH value. This could have a detrimental impact on marine organisms and ecosystems. Carbonate—secreting organisms are particularly susceptible to this because an acidifying environment is less favorable for carbonate production. Laboratory experiments have shown that acidification.

We will have to carry out focused scientific studies to see what impact global change will have on the natural carbon cycle in the ocean. It would be naïve to assume that this is insignificant and irrelevant for the future climate of our planet. To the contrary, our limited knowledge of the relationships should motivate us to study the ocean even more intensely and to develop new methods of observation.

2.2 *How climate change acidifies the oceans*

Carbon dioxide is a determining factor for our climate and, as a greenhouse gas, it contributes considerably to the warming of the Earth's atmosphere and thus also of the ocean. The global climate has changed drastically many times through the course of Earth history. These changes, in part, were associated with natural fluctuation in the atmospheric CO_2 content. The drastic increase in atmospheric CO_2 concentrations by more than 30 per cent since the beginning of industrialization, by contrast, is of anthropogenic origin, i.e. caused by humans. The largest sources are the burning of fossil fuels, including natural gas, oil, and coal, and change in land usage: clearing of forests, draining of swamps, and expansion of agricultural areas.

There is a permanent exchange of gas between the air and the ocean. If the CO_2 levels in the atmosphere increase, then the concentrations in the near-surface layers of the ocean increase accordingly. The dissolved carbon dioxide reacts to form carbonic acid. This reaction releases protons, which leads to acidification of the seawater. It has been demonstrated that the PH value of seawater has in fact already fallen, parallel to the carbon dioxide increase in the atmosphere, by an average of 0.1 units. Depending on the future trend of carbon dioxide emissions, this value could fall by another 0.3 to 0.4 units by the end of this century. This may appear to be negligible, but in fact it is equivalent to an increased proton concentration of 100 to 150 per cent.

The total dissolved inorganic carbon of sea water is defined as:

$$C_T = [CO_2^*] + [HCO_3^-] + [CO_3^{2-}] \tag{1}$$

Where brackets represent total concentrations of these constituents in solution (in mol kg^{-1}) and [CO_2^*] represents the total concentration of all unionized carbon dioxide, whether present as H_2CO_3 or as CO_2.

2.3 *Equilibrium constants*

All the equations for the equilibrium constants presented here use concentrations expressed in moles per kilogram of solution.

2.3.1 *Solubility of carbon dioxide in sea water*

$$K_0 = [CO_2^*]/f(CO_2) \qquad (2)$$

is given by the expression (Weiss, 1974)

$$K_1 = [H^+][HCO_3^-]/[CO_2^*] \qquad (3)$$

is given by the expression (Lueker *et al.*, 2000)

$$K_2 = [H^+][CO_3^{2-}][HCO_3^-] \qquad (4)$$

is given by the expression (Lueker *et al.*, 2000).

2.3.2 *The consequences of ocean acidification*

Climate change not only leads to warming of the atmosphere and water, but also to an acidification of the oceans. It is not yet clear what the ultimate consequences of this will be for marine organisms and communities as only a few species have been studied. Extensive long-term studies on a large variety of organisms and communities are needed to understand potential consequences of ocean acidification.

On the basis of ice core data, it is therefore possible to compare man-induced and climate-induced atmospheric composition changes. We will see that in many cases, the anthropogenic effect exceeds the natural fluctuation range. For instance, the present-day atmospheric levels of two major greenhouse gases (CO_2 and CH_4) are unprecedented over the last 400 ka (Robert J. Delmas, 1998).

2.3.3 *The effect of PH on the metabolism of marine organisms*

The currently observed increased of CO_2 concentrations in the oceans is, in terms of its magnitude and rate, unparalleled in the evolutionary history of the past 20 million years. It is therefore very uncertain to what extent the marine fauna can adapt to it over extended time period. After all, the low PH values in seawater have an adverse effect on the formation of carbonate minerals, which is critical for many invertebrate marine animals with carbonate skeletons, such as mussels, corals or sea urchins. Processes similar to the dissolution of CO_2 in seawater also occur within the organic tissue

of the affected organisms. CO_2, as a gas, diffuses through cell membranes into the blood, or in some animals into the hemolymph which is analogous to blood. The organism has to compensate for this disturbance of its natural acid-base balance, and some animals are better at this than others.

Benthic invertebrates (bottom-dwelling animals without a vertebral column) with limited ability to move great distances, such as mussels, starfish or sea urchins often cannot accumulate large amounts of bicarbonate in their body fluids to compensate for acidification and the excess protons. Long-term experiments show that some of these species grow more slowly under acidic conditions. But when they are exposed to long-term CO_2 stress, this protective mechanism could become a disadvantage for the sessile animals. With the long-term increase in carbon dioxide levels in seawater, the energy-saving behavior and the suppression of metabolism inevitably leads to limited growth, lower levels of activity, and thus a reduced ability to compete within the ecosystem.

2.4 *Threat to the nutrition base in the oceans-phytoplankton and acidification*

The entire food chain in the ocean is represented by the microscopic organisms of the marine phytoplankton. These include diatoms (siliceous algae), calcareous algae, and the cyanobacteria (formerly called blue algae), which, because of their photosynthetic activity, are responsible for around half of the global primary productivity.

Ocean acidification is not the only consequence of increase CO_2. This gas is, above all, the elixir of life for plants, which take up CO_2 from the air or seawater and produce biomass. Except for the acidification problem, increasing CO_2 levels in seawater should therefore favour the growth of those species whose photosynthetic processes were formerly limited by carbon dioxide. This is also true for certain coccolithophores, such as *Emilliania huxleyi*. But even for *Emilliania* the initially beneficial rising CO_2 levels could become fatal. Species posses a calcareous shell comprised of numerous individual plates. There is now evidence that the formation of these plates in impaired by lower PH levels. In contrast, shell formation by diatoms, as well as their photosynthetic activity, seems to be hardly affected by carbon dioxide. For diatoms also, however, shifts in species composition have been reported under conditions of increased CO_2 concentration.

2.5 *Challenge for the future*

In order to develop a comprehensive understanding of the impacts of ocean acidification on life in the sea, we have to learn how and why CO_2 affects various physiological processes in marine organisms.

The ultimate critical challenge is how the combination of individual processed determines the overall CO_2 tolerance of the organism. So far, investigations have mostly been limited to short-term studies. To find out how and where an organism can grow, remain active and reproduce successfully in a more acidified ocean, long term (months) and multiple-generation studies are necessary.

The final, and most difficult step, thus is to integrate the knowledge gained from species or groups at the ecosystem level. Because of the diverse interactions among species within ecosystems, it is infinitely more difficult to predict the behavior of such a complex system under ocean acidification.

The further development of such ecosystem-based studies is a great challenge for the future. Such investigations are prerequisite to a broader understanding of future trends in the ocean. In addition, deep-sea ecosystems, which could be directly affected by the possible impacts of future CO_2 disposal under the sea floor, also have to be considered.

In addition, answers have to be found to the question of how climate change affects reproduction in various organisms in the marine environment. Up to now there have been only a few exemplary studies carried out and current science is still far from a complete understanding. Whether and how different species react to chemical changes in the ocean, whether they suffer from stress or not is, for the most part, still unknown. There is an enormous need for further research in this area.

3 CONCLUDING REMARKS

There are several important environmental factors impact on CO_2 disposed in Ocean. Above all, the acidification around the release point certified may be the most important in this paper. Impacts around the release point are inevitable. Yet, the size and severity of the impacted area will depend on the exact release technology.

In addition, some expert addresses zooplankton mortality which could result from exposure to low pH plumes associated with ocean CO_2 disposal. Mortality was shown to correlate with two factors: mass loading per unit area, and the travel time to a pH of 7.

Analyzing the CO_2 cycle reveals the principle of the CO_2 reservoirs of the atmosphere, land biomass and ocean. The oceans are buffering CO_2 concentrations of atmosphere trace gases. But it

will take millennia to reach a new equilibrium for CO_2. Natural processes therefore cannot keep up with the speed at which humans continue to discharge CO_2 and other climate-relevant trace gases into the air. The only solution is to save energy and significantly reduce greenhouse gas emissions.

ACKNOWLEDGEMENTS

This work was supported financially by projects of the National Science and Foundation of China (41106094) and Department of Science and Technology of Shandong Province (BS2010NY030).
Corresponding author: ykmch@ouc.edu.cn.

REFERENCES

Eric Adams, E.* Jennifer A. Caulfield, Howard J. Herzog and David L Auerbach, 1998. Impacts of reduced pH from ocean CO_2 disposal: sensitivity of zooplankton mortality to model parameters, waste management, Vol. 17, No. 5/6, pp. 375–380, 1997.

Gruber, N., et al., 2009. Oceanic sources, sinks, and transport of atmospheric CO_2, Global Biogeochem. Cycles, 23, GB1005, doi:10.1029/2008GB003349.

Gruber, N. (1998), Anthropogenic CO_2 in the Atlantic Ocean, Global Biogeochem. Cycles, 12(1), 165–191, doi:10.1029/97GB03,658.

Gruber, N., and C. D. Keeling (2001), An improved estimate of the isotopic air-sea disequilibrium of CO_2: Implications for the oceanic uptake of anthropogenic CO_2, Geophys. Res. Lett., 28(3), 555–558, doi:10.1029/2000GL011853.

Gruber, N., and J. L. Sarmiento (2002), Large-scale biogeochemical/physical interactions in elemental cycles, The Sea: Biological-Physical Interactions in the Oceans, edited by A. R. Robinson, J. J. McCarthy, and B. J. Rothschild, vol. 12, pp. 337–399, John Wiley, New York.

Lueker, T. J., Dickson, A. G. and Keeling, C. D. 2000. Ocean pCO_2 calculated from dissolved inorganic carbon, alkalinity, and equations for K_1 and K_2: validation based on laboratory measurements of CO_2 in gas and seawater at equilibrium. Mar. Chem. 70: 105–119.

Robert J. Delmas. Ice-core records of global climate and environment changes. Proc. Indian Acad. Sci. (Earth Planet. Sci.), 107, No. 4, December 1998, pp. 307–319.

Weiss, R. F. 1974. Carbon dioxide in water and seawater: the solubility of a non-ideal gas. Mar. Chem. 2: 203–215.

Frontiers of Energy and Environmental Engineering – Sung, Kao & Chen (eds)
© 2013 Taylor & Francis Group, London, ISBN 978-0-415-66159-1

Simulation of continuous production in horizontal planetary ball mill

J.L. Ye, H. Dong, L.J. Zhang & X.C. Ye
College of Materials Science and Engineering, Nanjing University of Technology, Nanjing, China

ABSTRACT: A self-designed horizontal planetary ball mill was employed to study the influence of grinding time to production and yield ratio, total energy consumption of equipment, and energy consumption of available powder, while the revolution speed was 300 r/min, the centrifugal acceleration was 14 G and the ratio of rotation-to-revolution speed was 3. The results showed that the yield ratio of available powder, i.e. the feeding rate of continuous production per minute, was 32.9% of filling rate; the total energy consumption of equipment in 10 min was 196.53 KJ, while the energy consumption of available powder was 14.6% of it.

Keywords: horizontal planetary ball mill; production and yield ratio; energy consumption

1 INTRODUCTION

Energy consumption in cement production is large. The grinding power consumption of raw materials, coal and clinker accounts for about 70% to 72% of the total electricity consumption of cement plant. So the key to reducing the power consumption of the cement production is to reduce the grinding power consumption. The limitation of the critical speed of ordinary ball mill makes the efficiency of cement grinding low[1]. Planetary ball mill is based on ordinary ball mill, and it is regarded as an efficient and energy-saving grinding equipment due to its unique structure and excellent grinding efficiency. As early as the 1980s, Prof. Yan worked on grinding theoretical and experimental of planetary ball mill, and put forward to the concept of production capacity[2,3], power consumption[4], materials distribution[5–7], the limit of particle size[8] and self-protection[9]. Optimum operating parameters and energy-saving mechanism of planetary ball mill were studied by Prof. Yan[3,6–10], who analyzed the critical speed, optimum ball charge, optimum rotation speed and optimum material and ball ratio. He obtained that grinding efficiency was highest when rotation speed was 80% of the critical speed and ball charge ratio was 40%~50%. Huaitao Sun[11] made a study on movement law of planetary ball mill. Shizhu Chen[12] worked on structure, kinetics and kinematics of planetary ball mill. Hiroshi Mio and Junya Kano[13–17] studied on the influence of the structure parameters of vertical planetary ball mill on grinding media, and the relationship between impact energy and magnification factor of planetary ball mill. They obtained that the increasing of the revolution speed and the revolution radius could increase the impact energy of vertical planetary ball mill.

2 EXPERIMENT

There are two processes of cement production called open-circuit and close-circuit grinding. The former is characterized by discharging product after the cement was grinded once, less required auxiliary equipment and convenience to control. The latter is characterized by meal returns the mill to be grinded after the cement powder passed through the powder concentrator, until the powder reaches the required fineness. Close-circuit grinding is more complex, whose most notable feature is reducing the over grinding phenomenon compared with open-circuit grinding. This article simplified the close-circuit grinding process and made simulation experiment in the horizontal planetary ball mill.

2.1 Experiment condition

The pot rotation direction is set opposite to the disk revolution direction. Grinding media is steel ball whose density is 7.78 g/cm³, and ball graduation of each pot is Φ20 mm × 5, Φ12 mm × 12 and Φ5 mm × 12. The total volume of the pot is 1.526 L whose effective diameter is 90 mm. The charging amount of each pot is 150 g. The revolution speed is 300 r/min while the centrifugal acceleration is accordingly 14 G, and the ratio of rotation-to-revolution is 3. The feed particle diameter of cement clinker is 10–16 mesh. Table 1 shows the chemical composition of cement clinker.

Table 1. The chemical composition of cement clinker.

Component	CaO	SiO$_2$	Al$_2$O$_3$	Fe$_2$O$_3$	MgO	K$_2$O	Na$_2$O	SO$_3$	Others
Content/%	64.8	21.31	4.98	3.51	2.34	0.724	0.232	0.34	1.764

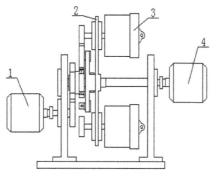

1-rotation motor; 2-large cap; 3-grinding tube;
4-revolution motor

Figure 1. Multifunctional horizontal planetary ball mill chart.

Table 2. Output and yield ratio of the finished powder.

Time (each 30s)	Output (g)	Yield ratio (%)
1	80.8	17.96
2	77.5	17.22
3	74.1	16.47
4	75.9	16.87
5	73.9	16.42
6	74.7	16.6
7	76	16.89
8	68.1	15.13
9	77	17.11
10	72.9	16.2
11	74	16.44
12	74	16.44
13	74.8	16.62
14	73.9	16.42
15	74.3	16.51
16	72.4	16.09
17	72.3	16.07
18	71.4	15.87
19	74.9	16.64
20	74.3	16.51
Average	74.02	16.45

2.2 Main equipment

Figure 1 shows the horizontal planetary ball mill which is used in the experiment. The rotation and revolution motors, respectively, control the rotation and revolution speed of the mill by the inverter.

2.3 Experiment scheme

Firstly, cement clinker was grinded for 30s, then removed and sieved for 2 min. Afterwards the 190 mesh powder was taken away and some new cement clinker (particle size is 10 to16 mesh) was added in till the total weight is 150 g, grinded for another 30s, followed by being sieved for 2 min again. Similarly, the 190 mesh powder was removed and some new cement clinker was supplemented. The process was repeated to ensure that the cumulative grinding time was 10 min. Meanwhile, the reaction force of motor was recorded in each grinding period, i.e. every 30s. Eventually, the output, yield ratio of the finished powder and the total energy consumption of the equipment were calculated.

3 RESULTS AND DISCUSSION

3.1 Output and yield analysis

Table 2 shows the output and yield ratio of the finished powder in each grinding period when the horizontal planetary ball mill carried out the close-circuit grinding. From the table, we could know that the output and yield ratio of the finished powder are fluctuating around the average value (the value of the first 30s is not included) and fluctuation range is small. In each period, the yield ratio is relatively stable and the average yield ratio of every 30s is 16.45%. Furthermore, the yield ratio of 1 min is 32.9% and this could provide a reference for the device of how to set the feeding speed in industrial production.

We could also obtain from Table 2 that the output of the finished powder almost grows linearly with time. The reason could be that the fine powder was constantly moved out so that the impact of the grinding medium would not be reduced. And new cylinder was constantly supplemented, which made the device more efficient.

3.2 Energy consumption analysis

The total energy consumption of the equipment is shown in Figure 2. From which we could know that energy consumption, including the friction energy

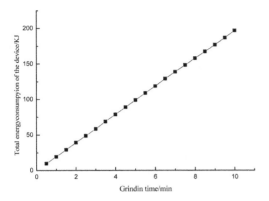

Figure 2.   Equipment total energy consumption.

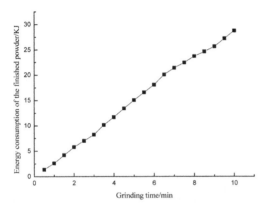

Figure 3.   Finished powder energy consumption.

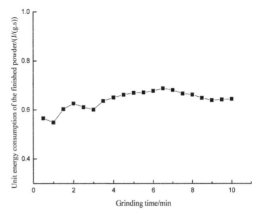

Figure 4.   Energy consumption per unit time and per unit mass.

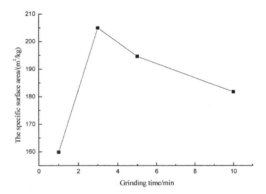

Figure 5.   The specific surface area of the finished powder.

consumption of the equipment, grinding balls' collision, friction and grinding energy consumption, is proportional to time. The energy consumption of the equipment is 196.53 kJ in 10 min, and sequently is about 1179.2 kJ in an hour, which may provide another reference in industrial production.

Figure 3 shows the relationship between the energy consumption of the finished powder which doesn't include the energy consumption of grinding medium, device friction and mesh powder less than 190 and grinding time, which is almost a linear relationship. To get the finished powder, 28.76 kJ is needed in 10 min. And combining with Figure 2, we can estimate the energy efficiency is 14.6%. The energy used for grinding clinker in deed is only about 2% of the total, and the rest are mostly turns into heat[18]. Therefore, timely shipping the finished powder out of the grinding cylinder can greatly improve the effective energy ratio and equipment ratio.

Figure 4 shows the changing curve of the finished powder energy consumption per unit time and per unit mass. It fluctuates between 0.5~0.7 J/(g·s). In 1 min, the smallest unit energy consumption is 0.548 J/(g·s). In 6 min, the largest unit energy consumption is 0.687 J/(g·s). However, the average is 0.64 J/(g·s) within 10 min. Thus, it is seen that the energy consumption per unit time per unit mass fluctuates around the average value and the fluctuation range is small, which also serves as an experimental basis.

3.3   *Specific surface area analysis*

Figure 5 shows the changing curve of the specific surface area in different grinding time. With the increasing of grinding time, the specific surface area trends to increase and then decrease. There is a largest specific surface area (205 m²/kg) when grinding time is 3 min. The reasons are listed as follows: the specific surface area is increasing as the cement clinker is constantly being grinded;

the energy consumption is increasing as well in the initial grinding time. With the increasing of grinding time, mesh powder less than 190 becomes smaller, so that the powder is increasingly difficult to further refine. The specific surface area and energy consumption trends to decrease after 3 min.

4 CONCLUSION

Through the simulation experiments of continuous production in horizontal planetary ball mill, we could obtain the yield ratio of the finished powder per unit time, the feeding speed in close-circuit grinding, energy consumption and so on. The specific conclusions are showed as follows:

1. Under the conditions of this experiment, the finished powder yield ratio is 32.9% per minute. Thus, the closed-circuit grinding feeding speed of the continuous production is 32.9% of charge amount every minute.
2. The energy consumed by finished powder is 14.6%. Timely shipping the finished powder out of the grinding cylinder can greatly improve the yield ratio, the effective energy utilization ratio and equipment ratio.

ACKNOWLEDGEMENT

This work was supported by the grant from the National Basic Research Program of China (973 Program: 2009CB623100).

REFERENCES

[1] Yao Lu zhi. 2002. Analysts on reason resulting in inefficiency of ball mill [J]. *Cement gulde for new epoch*, (2): 37–39.
[2] Yan Jing ping. 1990. Theoretical analysis of the optimal parameters in planetary ball mill [J]. *Equipment for electronic products manufacturing*, (2): 47–57.
[3] Yan Jing ping. Dang Gen mao. 1990. Study of ultrafine grinding mechanism in planetary ball mill [J]. *Equipment for electronic products manufacturing*, (2): 28–30.
[4] Zhu Mei ling. Yan Jing ping. 1995. Study of mechanism and fracture energy on preparation of ultra-fine powder [J]. *Journal of southeast university: English edition*, 11(1): 38–43.
[5] Zhou Jia chun. Yan Jing ping. 1999. Parametric analysis of feed distribution in thimble mill [J]. *West-China exploration engineering*, 11(4): 77–81.
[6] Cao Jin hua. Zhang Zhi sheng. Zhang Chao. 1999. Study of interface of feed distribution in mill [J]. *Equipment for electronic products manufacturing*, 28(4): 12–15.
[7] Chen Fang. Cao Jin hua. Zhang Zhi sheng. 1999. Theoretical research on feed distribution in mill [J]. *Equipment for electronic products manufacturing*, 28(4): 9–11.
[8] Zhou Jia chun. Yan Jing ping. 1999. Disscussion on the crushing capability of planetary ball mill [J]. *Mining and processing equipment*, (4): 22–24.
[9] Zhang Zhi sheng. Zhou Jia chun. Yan Jing ping. 1999. Research on Self Protectional Conditions of Ball Mill [J]. *Journal of southeast university: Natural science edition*, 29(4): 78–81.
[10] Yan Jing ping. Yi Hong. Shi Jin fei. 2008. Development of planetary mill and its energy-saving mechanism [J]. *Journal of southeast university (Natural science edition)*, 38(1): 27–31.
[11] Sun Huai tao. Fang Ying. Wan Yong min. 2007. Study on ball movement law in planetary ball mill [J]. *Metal mine*, (10): 104–106.
[12] Chen Shi zhu. Li Wen xian. Yin Zhi min. 1997. Research on working principle of a planetary high energy ball mill [J]. *Mining and metallurgical engineering*, 17(4): 62–65.
[13] Feng, Y.T. Han, K. Owen, D.R.J. 2004. Discrete element simulation of the dynamics of high energy planetary ball milling processes [J]. *Materials science and engineering A*, (375–377): 815–819.
[14] Mio, Hiroshi. Kano, Junya. Saito, Fumio. 2002. Effects of rotational direction and rotation-to-revolution speed ratio in planetary ball milling [J]. *Materials science and engineering A*, 332(1–2): 75–80.
[15] Mio, Hiroshi. Kano, Junya. Saito, Fumio. 2004. Optimum revolution and rotational directions and their speeds in planetary ball milling [J]. *International journal of mineral processing*, 74(Supplement 1): S85–S92.
[16] Mio, Hiroshi. Kano, Junya. Saito, Fumio. 2004. Scale-up method of planetary ball mill [J]. *Chemical engineering science*, 59(24): 5909–5916.
[17] Sato, A. Kano, J. Saito, F. 2001. Analysis of abrasion mechanism of grinding media in a planetary mill with DEM simulation [J]. *Advanced powder technology*, (2): 212–216.
[18] Zhang Shao ming. Zhai Xu dong. Liu Ya yun. 1994. *Powder engineering* [M]. Beijing: China Building Materials Industry Press.

Frontiers of Energy and Environmental Engineering – Sung, Kao & Chen (eds)
© 2013 Taylor & Francis Group, London, ISBN 978-0-415-66159-1

Design and experimental study on powering divider of the multi-function harvester

C.H. Zhao & Z.Y. Cheng
Engineering College, Gansu Agricultural University, Lanzhou, China

ABSTRACT: In order to make further study for powering divider of herbage harvester, its mechanical model was established by ANSYS. Its stress deformation and distribution were analyzed. The results showed that choosing 6 mm tooth thickness and 430 r/min rotating speed was the best effect, deformation displacement and equivalent stress were respectively reduced by 53.5% and 54.2% compared with original dates. Based on the above analysis, the field test was done. The experiments showed that the disc upright cutter with power can realize dividing the creeping tangled herbage quickly. It can effectively reduce the harvesting loss to 0.933 percent than manual harvesting, and the productivity was the 30 times than that of manual harvesting .The transmission efficiency was up to 0.88.

Keywords: harvester; powering divider; deformation and stress; experiments

1 INTRODUCTION

Coronilla varia l.cv.'l is the most common perennial leguminous plant in the forage. Its nutrition constituent is slightly higher than the *alfalfa*. It contains more protein and fat and hay crude protein is 20%.In addition, it also has powerful vitality, big coverage and high yield. As western development policy, its growing area is expanding unceasingly. However, *Coronilla varia l.cv.'l* receives a fatal flaw that its grass layer is very dense and usually crisscrosses each other,and what is more it owns a strong creeping tangled root[1]. At present, many qualified forages are almost totally harvested by people alone because of the tangled stems at the period of cutting, such as *Coronilla*, podding *alfalfa*. There is a main obstructive factor to popularize and apply forage because the artificial harvest uses intensive labor and receives low efficiency.

The powering divider can high-speedily pick creeping tangled forage open from down to up and take cutting grass apart from the others, the result was a bit ideal, but it was usually inserted grass and easily ran into hard things, such as stones. If make a little mistake, it will lead to harvester in poor condition. In view of the above reasons, the study of powering divider has certain practical significance. Reference to the researches of corn stalks stress deformation and sound scattering field distribution[2,3], finite element method was used to analyze loads on powering divider based on 4GH-120 harvester, obtained main factors affected it by comparing the simulation

results, So as to make preparations for further optimization.

2 THE TRANSMISSION SYSTEM OF TANGLE HERBAGE HARVESTER

According to working principle and structure characteristics of herbage harvester dividing, cutting and conveying system, direct type of transmission system was chosen. The transmission system was shown as Figure 1. It showed that diesel power was sent to the power input shaft by the power input pulley, on the left of the power input shaft,

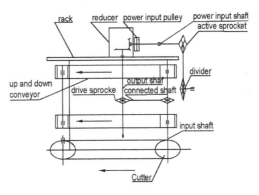

Figure 1. Schematic of the transmission system of the harvester.

bevel gear reducer which had only one level was connected, On the right of it, the active sprocket was connected, the divider were received the power. It took separating and cut grasses with branch interlocking, helped cutter work smoothly. After reduced speed and increased torsion, through output shaft of reducer, power reached drive sprocket that passed on power to input shaft by connected shaft .The input shaft drove cutter and up and down conveyor belt turn coaxially. The cut pasture was transported from left to right quickly by up and down nail conveyor belt and formed a certain thickness pasture bed on the right.

3   WORK PRINCIPLE OF THE DIVIDER

Reference to the research of Wu Mingliang et al[4] and constraint conditions and design principles of tangled herbage harvester, the divider with longitudinal tooth disk was chosen and shown in figure 2. Engine passed power to driven sprocket through chain wheel. The divider dish was connected by driven sprocket and clinch bolt and did high-speed rotary around shaft. It divided the creeping tangled herbage by powering divider, and made the stems of standing forage quickly rushing to cutting attachment, and then pushed by conveyer belt to one side, and with a reliable performance, cut-layers of herbage had a uniform thickness and well laying. According to different herbage height, the divider was kept at suitable height with the help of support stem by up-down adjustment.

4   THE MODEL FOR POWERING DIVIDER

The powering divider with longitudinal tooth disk was designed. It was made of 45 steel with 3 mm thickness, There were four tooth blades riveted at the equal distance on the edge. The blade was a small sword recreated by linear cutting on the type of standard. Its shear area was 23.8 cm², cutting angle was 40° and the pitch was 76 mm. As the divider model was simpler, it was established directly by ANSYS finite element model in its creating module[5]. In order to accurately simulate the nearside divider structure, assumed that there was no abrasion during its working process and disk was keeping completely vertical with chain wheel shaft. The physical parameters were shown in Table 1.

The Tet10 node92 unit was used to the model. It was consisted of 10 nodes and each of them had three degrees freedom. At the same time, it owned many functions, such as plastic, creep, inflation, rigidization stress, large deformation, big strain and so on. The established simulation model of divider was shown in Figure 3.

Table 1.   The physical parameters of testing for divider.

Parameter	Numerical value
Elastic modulus [N/m^2]	2.1×10^{11} N/m^2
Poisson ratio	0.3
Disc diameter [mm]	380
Tooth thickness [mm]	3
Tooth height [mm]	55
Rotating speed [r/min]	894

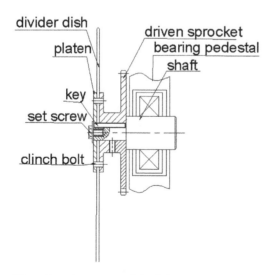

Figure 2.   The structure of the divider.

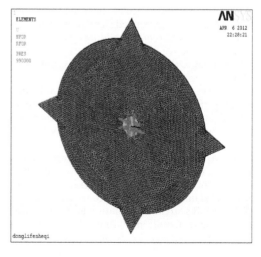

Figure 3.   The finite element model of divider.

5 THE DIVIDER SIMULATION AND RESULT ANALYSIS

Relative to tooth thickness and rotational speed, the diameter of divider dish had simper influence on it. Due to 60~150 cm pasture height and structure of divider, 380 mm diameter was selected. In order to avoid disorder that the front parts were heavier than the back ones of the whole pasture harvester, the lighter the divider weight is, the better it is. The scope of tooth thickness was chosen for 4~10 mm at last. Considered with the precision of simulation and tested efficiency, four typical values instead of the changing range, they were 4 mm, 6 mm, 8 mm and 10 mm. According to the highest cutting speed Vcmax 7.5 m/s, 430~894 r/min rotational speed was decided on the divider and took three typical values 430 r/min, 690 r/min and 894 r/min for example. On the basis of the above, simulation was done.

The result showed that displacement deformation and equivalent stress were decreased gradually as rotational speed declining in the same tooth thickness and the biggest decreased rates were 51.6% and 52.7% respectively. At the same speed, when tooth thickness increased, they decreased at first and then began to grow up. The greater thickness was good to dividing, but mechanical deformation and the whole weight imbalance became more serious. Therefore, choosing appropriate tooth thickness was very important. The high rotational speed caused shock, mechanical deformation and more serious power consumption. As a result,

the minimum was the best choice in a certain speed range. Concluded by analysis, choosing tooth thickness 6 mm and rotating speed 430 r/min was the best result. Maximum displacement deformation and equivalent stress distribution were 0.0121 mm and 13.8 MPa. It was shown as Figure 4. Compared with original date, optimization parameter deformation decreased 53.5% and 54.2% respectively. For diameter, if conditions permit, as far as possible to choose the minimum value. It yielded valuable information for divider further melioration. Besides, the maximal displacement deformation appeared on the left divider tooth, and stress value decreased from center to edge.

6 EXPERIMENTAL INVESTIGATION

On the basis of analysis and study, the field test of the divider was down. The creeping tangled

Table 2. Restrain condition on technical parameter of the forage harvester.

Parameter	Restrain	Parameter	Restrain
Swath	120 cm	Total loss rate	<1.5%
Stubble height	4~7 cm	Reliability	≥0.9
Operating speed	≥0.6 m/s	Productivity	≥0.22 hm²/h
Cutting speed	≥3 m/s	Power	≤8 kw
Oil consumption	≤15 kg/hm²	Weight	<280 kg

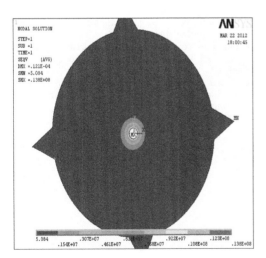

Figure 4. The displacement deformation and equivalent stress distribution when tooth thickness for 6 mm and 430r/min rotating speed.

Table 3. Detection result on technical parameters of the forage harvester.

Test items	Unit	Maximum value	Minimum value
Cutting speed	m/s	7	6.5
Operation speed	mls	1.3	0.8
Total lass rate	%	1.33	0.92
Stubble height	cm	7	4.5
Swath	cm	120	114
Oil consumption	kg/hm²	15	12
Productivity	hm²/h	0.23	0.22
Reliability	%	92	91
Cutter trouble-free quantity	hm²	24	23.3
Drive unit trouble-free quality	hm²	24	23.3
Vulnerable parts trouble-free quality	hm²	240	180

herbage harvester was used for pasture harvester operation which are less than 20° grass slope, 16000 kg/hm²~22000 kg/hm² output, 50~150 cm plant height, more than 380 plant/m² plant density, less than 85% grass stem moisture. Its main performance conformed for the project evaluation index rule. It was shown as the Table 2.

After comprehensively analyzing the experiment result of prototype. The consequence was shown as Table 3.

The operation performance of the whole machine is good. It owned high operation quality, smoothly dividing, lower stubble, tidy laying, less loss and can fully satisfy all kinds of grass agricultural demands[6].The efficiency is 30 times compared with artificial harvest. The high speed cutting and high efficiency operation have realized. The cutting speed is 7 m/s and 3 times than reciprocating cutter. The pure working time productivity is 0.23 hm²/h.

7 CONCLUSION

1. The creepy tangled herbage harvester is a kind of high speed and high efficient lawn mower which with powering divider and adapt to reap a variety of grass. The powering divider with longitudinal tooth disk was designed on the pasture harvester. It achieved high cutting and efficient delivery. Its structure is simple and the transmission efficiency is 0.88.
2. The text structured divider model and analyzed its mechanical deformation which effected by different tooth thickness, rotational speed and diameter. The results showed that choosing 430 r/min rotational speed and 6 mm tooth thickness was the best. Compared with original dates result, displacement deformation and equivalent stress decreased rate were 53.5% and 54.2% respectively. For diameter, if conditions permit, as far as possible to choose the minimum value. It yielded valuable information for divider further melioration.

3. The powering divider realized high speed dividing, stable operation, smooth dividing, no jam and met the demand of clear the way harvest divider. The unit cutting length productively is improved 1.5 times. The optimization creep tangled pasture harvester which with powering divider, the total loss rate reaches 0.92%, fuel consumption is 12 kg/hm², using reliability reached attains 92%, owns good cost performance and the main technical performance indexes has reached the leading domestic level.

ACKNOWLEDGEMENTS

The mission has been funded by The National Natural Science Foundation of China (50965001); 2011 innovation talent plan in Gansu province; The special research of public welfare profession (201203024) and The Gansu province agricultural accomplishment of science and technology transforms (1105NCNA095).

REFERENCES

[1] Chunhua Zhao. etal. Prostrate tangle stem biomechanical characteristics test at harvest time [J]. Journal of Agricultural Machinery, 2010,41(6):65–69.
[2] Junlin He, et al. The movement simulation of the helped hetian stem in the unlined corn [J]. Transaction of The CSAE, 2007,(6):125–129.
[3] Apaydin, G. On the use of physical spline finite element method for acoustic scattering [J]. Applied mathematics and computation, 2010,215 (10).
[4] Mingliang Wu. et al The test ofrape stem cutting force influence foctor [J]. Transaction of the CSAE, 2009,25(6):141–144.
[5] Liu Wei. et al. A valuable book about ANSYS 12.0 [M]. Publishing house of electronics industry, 2010,7.
[6] Hanna H Mar K, kohl Kr is D, Haden David A. Machine losses convetional versus narrow row cornharvest [J]. Applied Engineering Agrichiture, 2002,8(4):405–409.

Frontiers of Energy and Environmental Engineering – Sung, Kao & Chen (eds)
© 2013 Taylor & Francis Group, London, ISBN 978-0-415-66159-1

Research on inclined shaft deformation control technology in soft surrounding rock

Z.H. Liu
Mechanics and Construction Engineering, China University of Mining and Technology, Beijing, China
Huaibei Mining Shares and Limited Liability Company, Huaibei, Anhui, China

L.W. Jing, P.W. Hao, L.P. Ye & K.X. Zhao
Supporting Technology Research Institute, Anhui University of Science and Technology, Huainan, Anhui, China

ABSTRACT: This paper is to reveal soft rock deformation mechanism and corresponding control measures of the inclined roadway. Taking Huaibei Mining circulated air inclined shaft in Xutuan coal mine 83 inferior Mining area as an example, the paper analyzes the main factors of inclined roadway deformation in detail, introduces the roadway surrounding rock deformation mechanism, gives the specific supporting program, and proves the feasibility of program and mechanism analysis by field test. As a successful design case, the article emphasizes importance to control the floor heave of roadway stability. Research on achievement ion for a variety of the inclined shaft, down the supporting scheme research has important reference significance. The background of this paper is a successful cooperation in scientific research between the coal enterprises and universities, its remarkable achievements in scientific research is not only reflected in the huge economic benefits, but also in high academic value.

Keywords: soft surrounding rock; subinclined shaft; deformation control; floor heave

1 INTRODUCTION

Circulated air inclined shaft in Xutuan coal mine 83 inferior mining area is between faults F5-1 and F15. The faults are all reverse fault. The entire subinclined shaft is divided into two section construction, the construction starting point is auxiliary parking lots associated Lane at −497.169 m. −497.169 m—−432.921 m range for uphill construction, −497.169 m—−640 m range for downhill construction. The construction of −455 m—−460 m will cross the 32 coal mine, downhill construction in −595 m–600 m will cross 4_2 coal mine. And roadway cross the area, composition is mostly mudstone, the nature of the rock and soil is very poor.

As a result of the special location design in the 83 mining area which under air inclined shaft, inclined shaft will be influenced by high pressure effect caused by the stress redistribution and the mining area tectonic stress. In addition, the inclined shaft will cross the long loose coal and soft rock, so is a high degree of difficulty of roadway. The supporting difficult reason is mainly reflected in the following aspects: (1) Because rock diagenetic is late, so cementation degree is not high, recirculated air rubinclined shaft in 83 inferior

mining area roof and floor rocks are very soft and broken, easily weathered, fear of water, fear of earthquakes. (2) The strength of rock is low; the lithology through the rock in circulated air inclined shaft 83 in inferior mining area is mudstone, aluminum mudstone and sandy mudstone etc. Therefore the single axis compressive strength is relatively low, the supporting is very difficult. (3) A high level of surrounding rock stress; It is shown in three aspects: ①large roadway buried depth; ②A certain amount of tectonic stress; ③with coal rock mined, the surrounding rock will produce asymmetric secondary high pressure.

Based on the present field observation of Xutuan mine rock roadway, general supporting way is very difficult to guarantee the stability of roadway, such as the southern track roadway, after many repair, but deformation of the roadway will continue. It is still in repair. In addition, according to our country a lot of mining area on the mountain tunnel and inclined shaft research [1–5], most repair rate is high, some roadways are almost repaired once a year, and the cost of maintenance is very large.

This project is based on the background of this research project.

2 ENGINEERING GEOLOGICAL PROFILES

83 mining area is located in the middle of the mine, it is generally a monoclinal structure, strata dip is 4°~12°. the zone has 4 folds, 74 faults, the account that drop is greater than or equal to 5 meters of the fault is 45, less than 5 meters of the fault is 29, the main strike of the fault is NE, NW is the secondary. Small fault orientation is controlled by large and medium-sized construction, the distribution and characteristics of them is nearly identical, both NE strike of fault is the most development; Small faults often lead to seam sliding between layers and traction constructing. Top pressure, bottom convex, puncture deformation will generate, so that the coal seam thin, thicken, roof and floor rock crush, fault is often associated with fold circulated and air inclined shaft services in 83 inferior mining area, design engineering quantity 854 m, predict duration 300 days, expected service life of 30 years. Net size of circulated and air inclined shaft roadway is that W × H = 4.6 × 4.1 m, Inclination = 16°, circulated air inclined shaft engineering uphill section is about 262 m, downhill section is about 592 m, the entire roadway is divided into two section to construct, the construction starting point is auxiliary parking lots associated Lane at −497.2 m. The tunnel during the construction process cross the F5-1 and F15 faults as well as the 32 and 42 coal beds. F5-1 is a reverse fault, the inclination is 40°~70°, the drop is 0~50 m; F15 is also reverse fault, the inclination is 40°~50°, drop is 0~35 m. water filling source of the roadway is mainly sandstone fissure water. Sandstone fracture aquifer is mainly to static reserve. There is phenomenon such as roof drip and sprinkling in the area with large fault tectonic fissures. It has certain influence on tunneling construction. Predicting the normal water yield of drifting tunnel is 4 m³/h, and the maximum water emission of drifting tunnel can reach 8 m³/h.

The surrounding rocks strength of return air dark oblique lane is weak, then it is specific shown in Table 1.

3 TUNNEL DEFORMATION FACTORS AND MECHANISM ANALYSIS

3.1 Surrounding rocks have rheological characteristics

From Figure 1, there have two kinds of rock creep, an unstable creep (the creep under the effect of σ_A, σ_B), a stable creep (the creep under the effect of σ_C). Experimental results show [6–10], a rock producing stable or unstable creep depends on the size of rock stress. When stress is above a certain critical stress, creep tends to be unstable creep; and when it is less than this critical stress, creep is stable. The critical stress is the long-term strength of rocks. The long-term strength of rocks is a very valuable time indicator. Under a long-term constant load, the rock will be destruction than the much smaller instantaneous intensity, and the ratio between long-term strength an instantaneous strength is usually 0.4–0.8, then the soft rock and secondary solid rock is 0.4–0.6, solid rock 0.7–0.8.

Area of Xutuan 83 is smaller, therefore, the corresponding long intensity is low, and under the high pressure and the centralized stress, surrounding rock instability creep will occur.

3.2 Stress state of rock mass displacement change trigger

The excavation of tunnel leads to the changing of surrounding rock stress state, the first consequence of stress state change is particle shape elastic changing and shape elastic changing, second one is long-term creep, and special calculation formula is seen in Figure 2.

Shape elastic changes:

$$\begin{cases} \varepsilon_1 = \dfrac{1}{E}\left\{\left(\sigma_1' - \sigma_1\right) - u\left[\left(\sigma_2' - \sigma_2\right) + \left(\sigma_3' - \sigma_3\right)\right]\right\} \\ \varepsilon_2 = \dfrac{1}{E}\left\{\left(\sigma_2' - \sigma_2\right) - u\left[\left(\sigma_3' - \sigma_3\right) + \left(\sigma_1' - \sigma_1\right)\right]\right\} \\ \varepsilon_3 = \dfrac{1}{E}\left\{\left(\sigma_3' - \sigma_3\right) - u\left[\left(\sigma_1' - \sigma_1\right) + \left(\sigma_2' - \sigma_2\right)\right]\right\} \end{cases} \quad (1)$$

Table 1. The surrounding rocks strength of return air dark oblique lane.

Number of rock	Moisture condition	Geometry size (mm)		Uniaxial compressive strength/Mpa	Modulus of elasticity/ Gpa	Modulus of deformation/ Gpa	Poisson's ratio/
		Height	Diameter				
1	Natural	84.39	50.14	34.398	7.254	3.798	0.324
2	Natural	106.76	52.88	26.373	13.803	5.693	0.481
3	Natural	79.15	52.43	49.051	15.328	8.767	0.157

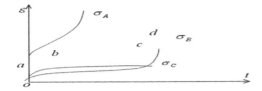

Figure 1. Rock creep curve.

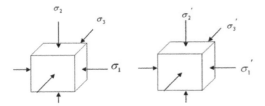

Figure 2. Rock mass stress state change diagram.

Shape of the volume change:

$$\theta = \frac{1-2u}{E}\left[\left(\sigma_1' - \sigma_1\right) + \left(\sigma_2' - \sigma_2\right) + \left(\sigma_3' - \sigma_3\right)\right] \quad (2)$$

In addition to the elastic shape change and volume change, the particle occur long creep. It not only lead to changes in particle shape and size, but also lead to the movement of adjacent particles which forming long-term slow displacement or flow of the surrounding rock.

3.3 The floor with no support

The failure of roadway and 90 percent of deformation in our country is caused by floor-heave, The mass stress state of within a certain range of rock in the backplane have more substantial changes than the tunnel excavation, since there is no support in the floor. It provides excellent conditions for long-term occurrence of the creep. The two sides to close are caused by floor-heave. It will lead to support failure of the roof, and finally lead to the overall collapse of the roadway.

These three areas are the main factors of the inclined destruction of the coal mine.

4 THE DESIGN OF SUPPORT

4.1 The basic idea

As shown Figure 3, a bearing arch in the surrounding rock, the arch must be balanced by force in order to withstand the higher load. There are three anchor cables in the top of the figure and

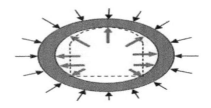

Figure 3. Control of deformation force balance thought.

backplane without the support reaction force. In order to compensate for this shortcoming, So in the lower part and points of two side each role on the repugnant forces. And it can improve the carrying capacity of the bottom of the arch in a large extent.

4.2 Design

According to the above mechanism and thought, the designs of the dark inclined roadway back to the wind are as follows.

The dark inclined roadway back to the wind use the joint support of anchors, nets, rope, spray plus grouting. Select GM22/2400 m type high strength resin bolt, each bolt use two Z2550 resin anchoring agent. Anchor lines, row spacing is 700×700 mm, longer anchoring, rectangular layout. The anchor cable arrange at the top of the roadway. The anchor cable specifications for $\phi17.8$ mm $\times$ L6300 mm, pitch is 1600 mm $\times$ 1600 mm. full-face arrangement 3 are arranged symmetrically along the roadway centerline. Two rows GM24/3000 mm high strength bolt are arranged in the each two sides of roadway. Row spacing is 700 mm $\times$ 700 mm intervals and layout with the basic support bolt. A row of anchor layout at the end of the junction, specifications GM24/2500 mm high strength, row spacing 700 mm and a 450 angle with the horizontal.

5 INDUSTRIAL TEST AT THE SITE

5.1 Measuring-point arrangement

The monitoring were set up 10 stations, about 5 stations in testing of side reinforcing anchor, another 5 stations in reinforce testing of without side anchor, two testing are apart from 30 m. Stations separation distance 5 m (as shown in Fig. 4 shows), the station 1 distance assist depot combined ways about 70 m, and in stations of setting areas. A roadway supporting design scheme is anchor nets cable spray combined support (no floor anchor). Each monitoring stations in two sides respectively from orbit face 300 mm and

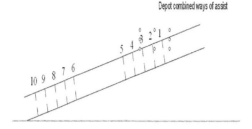

Figure 4.   The station layout diagram.

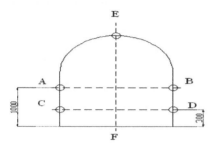

Figure 5.   Double cross layout diagram.

1000 mm position set four monitoring point, and separately set 2 point in the midline of roof and floor, As shown in Figure 5 shows, the between two points about AB, CD and EF respectively show surface displacement of two sides arch Angle, side bottom and top bottom.

This subject adopts double tenth glyph decorate point to determine of roof and floor, and two sides of the relative amount for move.

5.2   Monitoring data

Figure 6a, b, and c are the sidewall reinforcement and without sidewall reinforcement monitoring contrast. Figure 6d, e, and f are the bottom corner reinforcing and without bottom corner reinforcement monitoring contrast.

5.3   Data analysis

Up to May 4, 2010, the end of the tunnel top, arch angle and the bottom of roadway side of the sidewall reinforcing bolts, the maximum deformation are up to 0.114 mm/d, 0.074 mm/d and 0.085 mm/d, while the average deformation of without the sidewall reinforcing bolts are still up to 0.3958 mm/d, 0.20414 mm/d and 0.3422 mm/d, which is respectively 3.472 times, 2.76 times, 4.03 times than reinforced segments.

Up to July 28th 2010, the end of the tunnel top, arch angle and the bottom of roadway side of the

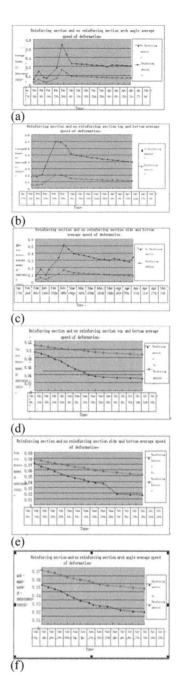

(a)

(b)

(c)

(d)

(e)

(f)

Figure 6.   Monitoring data.

floor anchor reinforcement segment, the average deformation are respectively up to 0.03002 mm/d, 0.01736 mm/d, 0.01548 mm/d. while the average deformation of without floor anchor reinforcement segment are still up to 0.08748 mm/d,

0.05208 mm/d, 0.04344 mm/d, which is respectively 2.91 times, 3.0 times and 2.8 times than reinforced segments. Obviously, the floor anchor reinforcement on the roadway reinforcement has a very significant results, this technique can be extended to the support of the Rock Lane and coal Lane.

6 CONCLUSION

From the monitoring data, it can be identified that the bottom corner bolt has a very good control of floor heave and roadway side, and control the role of the roof. In combination with the first stage test results, without the sidewall reinforcing bolts, the end of the tunnel top, arch angle and the bottom of roadway side of the maximum deformation are 3.472 times, 2.76 times, 4.03 times than reinforcing segments. Based on the measured results, without floor anchor test section were 2.91 times, 3.0 times and 2.8 times than reinforced segments. It could be projected that in two cases whether has the sidewall reinforcing bolts and the bottom corner reinforcing bolts, the end of the tunnel top, arch angle and the bottom of roadway side of the maximum deformation velocity ratio are 3.472 times, 2.76 times, 4.03 times. Therefore we can conclude that, sidewall and bottom corner reinforcing bolt have an important role in the control of the overall stability of surrounding rock.

ACKNOWLEDGEMENT

This work is supported by Natural Science Foundation of Anhui Province (11040606M101).

REFERENCES

[1] Keyu Gong, Wei Niu, Yunde Zou. Strata behavior and controlling in ventilation district dip [J]. *Ground Pressure and Strata Control*. 2001, (3):53–55.

[2] Qinghe Zhang, Zhengquan Wu. Project practices on combined support technology in high stressed and large cross section mine blind inclined shaft [J]. *Coal Engineering*. 2008, (11):30–31.

[3] Jianguo Yang, Xianchao Qu. Application of prestressed anchor and grouting combined support technology under complicated geological conditions [J]. *Coal Science and Technology*. 2007, 35(7):58–60.

[4] Genshen Dong, Gang Shao. 3_2 Seam Leather Belt Inside Slope Uncovers Coal to Support and Protect Engineering Research [J]. *Coal Technology*. 2007, 26(1):122–124.

[5] Xinxian Zhai, Xiaolin Yang, Xiantao Zeng. Relationship between Deformation Failure of Blind Incline Shafts and Their Safety Pillars [J]. *Mining and Metallurgical Engineeping*. 2003, 23(2):20–22.

[6] Guangzhe Deng, Weishen Zhu. An Experiment Research on the Crack Propagation in Rockmass[J]. *Journal of Experimental Mechanics*. 2002, 17(2): 177–183.

[7] O da M. An equivalent continuum model for coup led stress and flint of low analysis in jointed rock masses [J]. *Water Resources Research*, 1986, 22(13): 1845–1856.

[8] Napier JAL, Malan DF. A viscop last ic discont inuum model of time dependent fracture and seismicity effects in brittle rock [J]. *Int . J. Rock Mech. Min . Sci . Geomech*, 1997, 34:1075–1089.

[9] Song D. Non2linear visco2 plastic creep of rock surrounding an underground cavation [J]. *Int. J.Rock Mech. Min. Sci. Geomech. Abst r*. 1993, 30: 653–658.

[10] Guangzhe Deng, Yik Zhang. A analysis model of the mechanics of jointed rock mass [J]. *Journal of coal science and engineering*, 2000, 6(1):30–36.

Frontiers of Energy and Environmental Engineering – Sung, Kao & Chen (eds)
© 2013 Taylor & Francis Group, London, ISBN 978-0-415-66159-1

The impact of temperature changing on the force of mine shaft

C.X. Lei

Mechanics and Construction Engineering, China University of Mining and Technology, Beijing, China
Huainan Mining Shares and Limited Liability Company, Huainan, Anhui, China

L.W. Jing, K.X. Zhao & L.P. Ye

Supporting Technology Research Institute, Anhui University of Science and Technology, Huainan, Anhui, China

ABSTRACT: The purpose of the article is to reveal the presence of temperature stress in mine shaft sidewall and the related solving method. It detailed analyzes temperature stress caused by axial expansion of mine shaft due to temperature changes with modern mechanical theory. It lists some measured data of typical mine shaft wall stress changes due to temperature changing with concrete examples. At the same time it informs the majority wellbore rupture occurred in the summer in our country. Finally, for example, it specifically solves the temperature stress of the air shaft sidewall in Anhui coal mines. The research results of this article have a very important reference value on the structure design of mine shaft wall and the sidewall rupture prediction.

Keywords: temperature stress, sidewall, rupture

1 INTRODUCTION

The temperature of the wellbore outer wall contacted topsoil is almost constant, while the temperature of the air flowing through the wellbore is constantly changing with the seasons. Therefore, the temperature difference between the within wall and outer wall of the wellbore is always exist (when Ta = To = Tb, excluded), and this temperature difference will lead to the vertical temperature stress and also lead to the circumferential and radial temperature stress. The mine shaft sidewall rupture characteristics inform us in References [1]: the rupture of the shaft sidewall occurs mainly in the 6–8 months each year and the time of the rupture of the shaft sidewall is sufficient proof of the influence of temperature in the rupture of the shaft sidewall. In the following, there is a brief introduction to some of the measured data in this regard.

2 THE INFLUENCE OF TEMPERATURE CHANGING WITHIN WELLBORE ON THE FORCE OF SIDEWALL IN THE WELLBORE

In many mining area of our country, air temperature difference is generally up to more than 30°C [2] in the air shaft, and the highest temperature difference is up to 44°C in some places (The minimum temperatures can reach −6°C in winter, while the highest summer temperature is up to 38°C in Huainan mining area). Because there is such a large temperature difference, the set of the Surface Soil Segment (or bedrock section) and the constraints of topsoil friction, the temperature stress of sidewall is inevitable in the annual summer. The vast majority of our country is mainly concentrated in 4–10 months, the majority of which takes place in 6–8 months [1]. It fully demonstrates the impact of temperature changing on the sidewall rupture that can not be ignored.

In addition, the time characteristics from the sidewall rupture may be able to explain the impact of temperature changing on the sidewall rupture, but also from the much on-site measured data to be confirmed. So far, many experts, professors and coal businesses have engaged in this research, such as, Professor Renhe Wang, Huainan Mining Institute, monitors the sidewall stress of the main shaft in Linhuan miners(see Figures 1 and 2) [3]. In China University of Mining and Technology, Henglin Lv, Guangxin Cui and others monitor the temperature and deformation of sidewall in Baodian south air shaft, and Shuanglou Zhang monitors the deformation of sidewall in air shaft. For details, please see Table 2.3.

From Professor Renhe Wang monitoring results, the vertical temperature stress difference is 4.22 MPa (8.63 MPa–4.41 MPa) and the circumferential stress difference is 2.41 MPa (4.65 MPa–2.24 MPa) on the −240 mlevel in Linhuan main shaft.

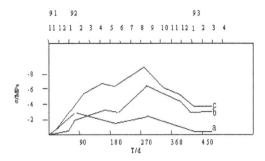

Figure 1. Sidewall vertical stress increment versus time curve a, b, c represent −100 m, −200 m, −240 m level curve.

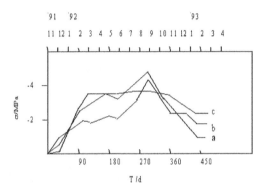

Figure 2. Sidewall circumferential stress increment versus time curve a, b, c represent −100 m, −200 m, −240 m level curve.

From the temperature variation within a year and a half of the vertical depth of 139.5 m the Baodian south air shaft, at the borehole wall temperature variation of about 3°C (Fig. 3), the maximum temperature in July and September, the District is about 23°C, the lowest temperature zone at the end of January, about 20°C.

According to data [4] documented that level at the strain—time variation can be obtained within one year, the temperature differential stress is 0.9 MPa, the topsoil and bedrock near the interface of the sidewall at the highest temperature period additional temperature compressive stress of 0.9 MPa.

It is worth mentioning here that the two wellbore stress calculation depending on the strain tests exist special circumstances. Namely, Linhuan main well strain tests (−240 m level) occur after the main shaft wall reinforcement, while Baodian south air shaft strain tests occur after the completion of construction of the relief groove. Therefore, the temperature stress value of untreated sidewall should be a more large number than the above-mentioned

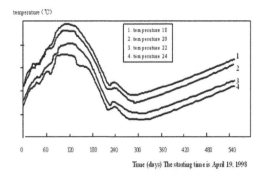

Figure 3. The third test level of the air shaft in Baodian south test level (vertical depth 139.5 m) temperature—time curve.

corresponding stress value. Under normal circumstances, the range of the sidewall vertical thermal stress generated by the temperature changing is about 2–8 MPa [6], but because the wellbore air of ventilation shaft is received the surrounding rock temperature regulation, the annual range of temperature is smaller than that, the value of the sidewall temperature stress is relatively smaller.

3 SIDEWALL TEMPERATURE STRESS ANALYSIS

Due to temperature rising, the vertical shaft usually occur vertical and horizontal direction size expansion, the topsoil on the wellbore occurring the vertical additional force is caused by the vertical expansion, and resulted the vertical temperature stress in the wellbore cross section. The horizontal expansion in addition to causes the circumferential temperature stress. According to Hooke's law under complex stress state, at this point, it also produces vertical stress. Two cases were discussed in the following.

3.1 *The determination of the elastic-plastic junction*

As we all know, the higher temperatures will cause the wellbore elongate along the axial. Due to the constraints of the lower part of the topsoil segment wall socket or bedrock, the elongation behavior can only be up along the axial direction, so that it must exist shear between the wellbore and topsoil, and make the wellbore subjected the temperature additional force. According to the analysis of the plastic shear force and the elastic shear in reference [7], it is not difficult to infer that in the up expansion process of the wellbore, within the soil depth of the wellbore through the entire topsoil, the shear deformation between the wellbore and the topsoil can be divided into two clearly defined scope—the plastic shear zone and the elastic shear zone. For details, please see Figure 4.

According to "Figure 4" on vertical additional force distribution law, we can know: Whether the soil and borehole which are regardless of the plastic shear in the plastic range, or the soil and the wellbore which are within the elastic range elastic shear stress, they should be along the wellbore axial curve distribution, which we can easily find the back of an example. Due to the temperature rise which caused by plastic shear, the length of OA_0 than the length of the depth of the topsoil section of the OC is much smaller, that is, the cross section of A_0 is very small. Therefore, at this time the plastic shear force distribution of the OA_0 should be close to linear distribution, and calculated according to linear processing to calculate. About elastic shear of A_0C, there also can be approximated by linear processing in order to calculate convenience. Approximate treatment of the results will inevitably cause some error, but the total error which due to thermal stress in the limited proportion of the sidewall stress, which caused by the linear processing is very small and negligible.

In order to calculate the vertical expansion more accurately, which caused by temperature stress, it is very important to determine the location of A_0 in Figure 5. Based on [7] shows that the following formula to calculate the location of the basis for A_0.

$$Z_{A_0} = \sqrt[3]{-\frac{q}{2} + \sqrt{\left(\frac{q}{2}\right)^2 + \left(\frac{p}{3}\right)^3}}$$

$$+ \sqrt[3]{-\frac{q}{2} - \sqrt{\left(\frac{q}{2}\right)^2 + \left(\frac{p}{3}\right)^3}} \quad (1)$$

Formula

$$\begin{cases} p = \dfrac{3d_1 d_3 - d_2^2}{3d_1^2}, \ q = \dfrac{2d_2^3 - 9d_1 d_2 d_3 + 27 d_1^2 d_4}{27 d_1^3} \\ d_1 = D_2, \ d_2 = D_2 L, \ d_3 = -(2D_2 L^2 + 3D_1), \\ d_4 = 3D_1 - 3u_0 \\ D_1 = \alpha \cdot \left\{ \dfrac{(T_a - T_b)}{2 \ln \frac{b}{a}} \left[1 - \dfrac{2a^2}{b^2 - a^2} \ln \frac{b}{a} \right] + T_b \right\}, \\ D_2 = \dfrac{\pi \cdot R \cdot \gamma_e \cdot g \cdot \lambda \cdot k_0}{EA} \end{cases} \quad (2)$$

Above all formulas, $(\sigma_r)_{td}$, $(\sigma_\theta)_{td}$, $(\sigma_z)_{td}$ differentiate show temperature of the sidewall radial between inside with outside, circumferential and vertical stress, MN/m²; T_a for borehole inner wall temperature and use To for the basis to calculated (°C); Tb for borehole outer wall temperature and use To for the basis to calculated. (The entire topsoil outside the borehole wall temperature can be considered approximately equal (°C); a, b for inner and outer wall radius, m; r for the distance between a dot pitch of the wellbore axis radius for the well bor, m; σ_r, σ_θ, σ_z differentiate show center r radial, ring and vertical stress about the point of the wellbore (MPa); u is the poisson ratio of the concrete shaft lining; ε_z, ε_H for vertical and ring strain which away from the wellbore center r; lA_0C for segment length of the wellbore, m.

Discussion: T_a, T_b and T_0 to determine.

For the outer wall of wellbore is always connecting with the topsoil, and the vast majority of surface soil temperature throughout the year change is very small (except near the surface and shallow soil outside). Coupled the wellbore rupture occurred in a number of years after the mine into production, so the wells temperature of the borehole for rupture the wall stress, which can be neglected and need not recorded. According to the distribution of surface temperature in China, where the T_0 temperature value approximately equal to 18°C, then $T_b = 0$°C; As for T_a can be determined based on

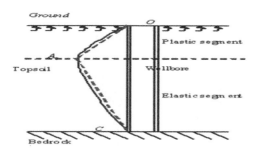

Figure 4. Vertical additional force of the elastic segment and plastic sections.

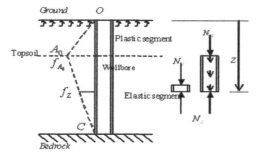

Figure 5. Shaft wall force analysis.

the difference between the local summer maximum temperature of T_0.

3.2 Calculation of thermal stress

3.2.1 Calculation of vertical expansion caused by thermal stress

For the distribution law of additional force of the vertical temperature different from OA_0 to A_0C, therefore, the corresponding temperature stress should be considered separately.

OA_0: According to Figure five, in any one cross section of OA_0 (Z-section) at the vertical additional force should be:

$$f_Z = \frac{Z}{Z_A} f_{A_0} = \frac{Z}{Z_{A_0}} \gamma_e \cdot g \cdot z_{A_0} \cdot \lambda \cdot k_0 (0 \leq Z \leq Z_A)$$

(3)

so

$$N_Z = \left(\frac{1}{2} f_Z Z\right) \cdot 2\pi R$$

$$= \left(\frac{Z^2}{Z_{A_0}}\right) \pi R \cdot \gamma_e \cdot g \cdot z_{A_0} \cdot \lambda \cdot k_0 (0 \leq Z \leq Z_{A_0}) \quad (4)$$

And thus can be obtained, this section of the wellbore cross section of the vertical thermal stress caused by the axial expansion of the distribution law:

$$(\sigma_z)_{tae} = \frac{N_Z}{A} = \frac{\left(\dfrac{Z^2}{Z_{A_0}}\right) \pi R \cdot \gamma_e \cdot g \cdot z_{A_0} \cdot \lambda \cdot k_0}{\pi R^2}$$

$$= \left(\frac{Z^2}{Z_{A_0} R}\right) \gamma_e \cdot g \cdot z_{A_0} \cdot \lambda \cdot k_0 (0 \leq Z \leq Z_{A_0}) \quad (5)$$

A_0C: This section of any cross-section (Z section) on the axial force, by the formula (4) that obtained, be divided by the cross-sectional area A_0, which can be caused by this section of the wellbore cross section of the axial expansion of the vertical thermal stress distribution law as follows

$$\sigma_{Ztae} = \frac{N_Z}{A}$$

$$= \frac{\gamma_e \cdot g \cdot z_{A_0} \cdot \lambda \cdot k_0 \cdot \left[L - \dfrac{(L-Z)^2}{L - Z_{A_0}}\right]}{R} (Z_{A_0} \leq Z \leq L)$$

(6)

3.2.2 Temperature stress which caused by temperature difference form inside and outside wall

Temperature changes in addition to cause the wellbore vertical expansion, and thus lead to the outside temperature stress. Can also cause thermal stress which due to temperature difference between inside and outside the borehole wall. Now the difference in sidewall vertical, the circumferential and radial stress components, respectively $(\sigma_r)_t$, $(\sigma_\theta)_t$, $(\sigma_z)_t$, so

$$\begin{cases} (\sigma_r)_{td} = \dfrac{\alpha_f \cdot E}{1-u} \cdot \dfrac{T_a - T_b}{2\ln\dfrac{b}{a}} \left[-\ln\dfrac{b}{r} + \dfrac{a^2}{b^2-a^2}(\dfrac{b^2}{r^2}-1)\ln\dfrac{b}{a}\right] \\[4mm] (\sigma_\theta)_{td} = \dfrac{\alpha_f \cdot E}{1-u} \cdot \dfrac{T_a - T_b}{2\ln\dfrac{b}{a}} \left[1 - \ln\dfrac{b}{r} - \dfrac{a^2}{b^2-a^2}(\dfrac{b^2}{r^2}+1)\ln\dfrac{b}{a}\right] \\[4mm] (\sigma_z)_{td} = \dfrac{\alpha_f \cdot E}{1-u} \cdot \dfrac{T_a - T_b}{2\ln\dfrac{b}{a}} \left[1 - 2\ln\dfrac{b}{r} - \dfrac{2a^2}{b^2-a^2}\ln\dfrac{b}{a}\right] \end{cases}$$

(7)

3.2.3 Temperature stress which caused by temperature changes

Temperature stress which caused by temperature changes, which should include the vertical expansion caused by temperature stress and inner or outer wall temperature which difference cause temperature stress, that is:

$$\begin{cases} (\sigma_r)_t = (\sigma_r)_{td} \\ (\sigma_\theta)_t = (\sigma_\theta)_{td} \\ (\sigma_Z)_t = (\sigma_z)_{td} - (\sigma_z)_{tae} \end{cases}$$

(8)

4 COMPUTING EXAMPLE

The basic parameters of the air shaft of Anhui mine in the Table 1.

Table 1. Basic parameters into the air shaft and the list of weeks side topsoil.

D/m	d/m	A/m²	E_1/MPa	E_2/MPa	L/m
9.1	6.5	31.8396	2×10^4	2.4×10^4	243.15
λ	α_f(°C)	ρ_c(kg/m³)	σ_r/MPa	σ_t/MPa	E_S/MPa
0.333	10×10^{-6}	2500	50	5.0	50.9
u_0/m	k_0	γ_c(kg/m³)	μ	ρ/(kg/m³)	φ/(°)
0.007	0.25	1900	0.3	2500	18.1

E_1 corresponding to deformation secant modulus of the segment BC concrete shaft lining. E_2 for the concrete secant modulus for corresponding segment AB sidewall deformation.

225

For temperature stress, Thermostat with temperature in most of the mining area of China Quaternary alluvium is basically near 18°C. When the shaft wall temperature is 18°C that does not produce thermal stress. Therefore here the reference temperature of 18°C given for the borehole wall temperature, the temperature compressive stress generated in the sidewall when the temperature is higher than in the sidewall 18°C. On the contrary, the resulting temperature tensile stress, which can be inferred $T_0 = 18°C$, then $T_b = 0°C$; The shaft wall temperature can be determined according to the local summer maximum temperatures, here take $T_a = 35°C$ $-18°C = 17°C$, So according to the formula (1), then the location of the cross section of A_0 as follows:

$$Z_{A_0} = 3.1506\ m$$

We can take the relevant data into the formula (7) within the borehole wall temperature variation of the stress as follows:

$$(\sigma_r)_{r=a} = 0, (\sigma_\theta)_{r=a} = 1858442.7\ N/m^2$$

The vertical temperature stress along the vertical variation shown in Table 2 (A_0C).

Temperature stress can be seen from the table, from top to bottom was gradually increasing trend.

5 CONCLUSION

Article given in the coal mine shaft lining temperature stress method for solving the opposing shaft wall structure design and anti-breakdown measures of great reference value

Computing instance specific results reveal that a coal mine shaft lining rupture occurs mostly in the reason of every summer, and the thermal stress induced in the shaft wall accident.

ACKNOWLEDGEMENT

This work is supported by Natural Science Foundation of Anhui Province (11040606M101).

REFERENCES

[1] Laiwang Jing. Law of shaft distribution secondary pressure and prediction of the shaft wall [Doctoral Dissertation D]. *Beijing. China University of Mining and Technology*. 2006:1–2, 10.
[2] Yanchun Xu, Dejing Xi. Stress and Deformation Feature of Pressure Releasing Slot on Mine Shaft Wall and Analysis on Results of Prevention Mine Shaft Wall Failure[J]. *Construction Technology*. 2001, 22(4):26–28.
[3] Renhe Wang. Calculation of Linhuan coal borehole wall stress monitoring and wellbore additional force[J]. *Mining*. 1997, 6(4):23–26.
[4] Mengmin Gao, Henglin Lv, Guangxin Cui. Practical study of the decompression method for dealing with the disruptive mechanism of the shaft lining[J]. *Coal Engineering*. 2002, (1): 36–39.
[5] Zhongjiang Liu. Grouting Applied to Mine Shaft Cracking Treatment[J]. *Coal Science and Technology*. 1999, 27(5):6–9.
[6] Guangxin Cui. The special geological conditions of shaft wall failure mechanism and control techniques[J]. *Construction Technology*. 1998, 19(1):28–32.
[7] Xiliang Liu, Weishen Zhu, Shucai Li. Simple Shear Test of Interface between Sand and Structure under High Stress[J]. *Rock Mechanics and Engineering*. 2004, 23(3):408–414.

Table 2. Rule of vertical temperature stress change.

Borehole depth	70	80	90	100	110	120	130	140	150
Vertical stress	2.503	2.692	2.869	3.035	3.190	3.323	3.465	3.586	3.696
Borehole depth	160	170	180	190	200	210	220	230	240
Vertical stress	3.795	3.882	3.959	4.024	4.078	4.121	4.152	4.172	4.191

Frontiers of Energy and Environmental Engineering – Sung, Kao & Chen (eds)
© 2013 Taylor & Francis Group, London, ISBN 978-0-415-66159-1

Research of thermodynamic calculation for boiler bended with blast furnace gas

L.J. Fang & S. Wu
School of Energy Power and Mechanical Engineering, North China Electric Power University, Baoding, China

ABSTRACT: In order to improve a 300 MW power plant's blending ratio of Blast Furnace Gas (BFG). Based on thermodynamic calculation of the mixed fuel combustion, we separately calculated the influences of different coals and different blending ratio of BFG on the theoretical combustion temperature, furnace outlet gas temperature, density of fly ash, exhausted gas temperature, boiler efficiency and so on. Result shows that with the blending ratio of BFG increasing the boiler efficiency decreases. Where as, the exhausted gas temperature increases. When combusting the coal NO. 1 with the blending ratio of BFG 20%, the exhausted gas temperature increases by 25°C and the boiler efficiency decreases by 2.07%; When combusting the coal NO. 2 with the blending ratio of BFG 20%, the exhausted gas temperature increases by 33°C and the boiler efficiency decreases by 2.75%.

Keywords: BFG; blending ratio; boiler efficiency; thermodynamic calculation

1   INTRODUCTION

In the productive process of the iron and steel enterprises, this will produce large amounts of secondary energy such as Blast Furnace Gas (BFG). If these secondary energy can be used fully, it will generate considerable economic benefit. However, the heat value of BFG is very low, generally about 3350 kJ/m³, 20% CO and a small amount of hydrocarbons consisted, and the rest is non-combustible material, which is difficult to be fully utilized. The work on the use of BFG began earlier in abroad. BFG has been used as power fuel since 1970s, and the area mainly concentrated on these two aspects: Firstly, BFG and the nature gas of high heat value were blended to form mixed gas to increase the heat value of BFG, constructing the gas cabinet to adjust the gas supply peak, to make the gas-fired boiler combust in a steady and efficient situation. Secondly, through the combined cycle of BFG turbine to improve the thermal efficiency of BFG. The recycling and utilization of BFG haven't come up in our country until 1995 when the average emission rate of BFG in 18 key steel enterprises was still about 12%. The main usage of BFG is main in two aspects: Firstly, blending BFG with coal to generate power and heat. Secondly, generating power through the combined cycle of BFG turbine. In terms of energy use, the combined cycle of BFG turbine should be the developing direction of the utilization of BFG, but the initial investment of gas turbine combined cycle power generation is very large. At present, how to increase the blending

ratio of BFG in coal boilers is the key element to solve the problem of utilization of the BFG.

This paper argues that the 300 MW coal-fired boiler in Jing Tang Shou Gang. Through the thermodynamic calculation of gas coals with different ash content in the condition of different blending ratio of BFG, we conclude the influence of the blending ratio of BFG on the theoretical combustion temperature, furnace outlet gas temperature, density of the fly ash, boiler efficiency, exhausted gas temperature, radiation heat transfer of the furnace and convection heat transfer and so on. The results can provide a reference to the operation of coal—fired boiler blended with BFG.

2   STUDY OBJECTS

The plant boilers features a single drum natural circulation, swing corner firing, three level spray

Table 1.   Main technology parameters of the boiler.

Main team flow (t/h)	25
Main steam pressure (MPa·g)	17.50
Main steam temperature (°C)	541
Reheat steam flow (t/h)	853.18
Reheat steam inlet/outlet pressure (MPa·g)	3.810/3.630
Reheat steam inlet/outlet temperature (°C)	324/541
Feed water (°C)	278
Air temperature (°C)	20
Hot air temperature (°C)	320

Table 2. Composition of coals and BFG.

Properties	Car (%)	Har (%)	Oar (%)	Nar (%)	Sar (%)	Aar (%)	Mar (%)	Vdaf (%)	$Q_{net,ar}$ (kJ/kg)
Coal NO. 1	65.23	3.78	8.47	0.74	0.74	11.24	9.82	33.15	24,911
Coal NO. 2	52.99	3.63	5.70	0.57	0.13	28.10	8.88	22.64	20,546
Properties	CO (%)		CO_2 (%)	H_2 (%)	N_2 (%)		CH_4 (%)	$Q_{net,ar}$ (kJ/kg)	
BFG	20.3		22.5	3	51.5		0.5	3066	

desuperheating system, solid state slag-tap, furnace membrane water wall in furnace, two Tri-Sectional tubular, air-preheater are installed in boiler tail, there is protection device for BFG either. Designed fuel is gas coal, with medium speed mill and positive-pressure direct-fired pulverizing system and the ability of blending ratio of BFG from 0 to 30% at same time. Main technology parameters of the boiler shown in Table 1, composition of coals and BFG shown in Table 2. The blending ratio are proved to be 0%, 3%, 7%, 10%, 13%, 15%, 18%, 20%, 25% in the thermodynamic calculation. In order to analysis the result comfortably we treat the excess air ratio α to a constant 1.2.

3 CALCULATION METHOD OF THE MIXED FUEL

3.1 *The co-firing of solid–fuel (liquid–fuel) and gas-fuel*

As for the co-firing of solid–fuel (liquid–fuel) and gas-fuel, combustion is not calculated according to the mixed fuel per kilogram, but the solid–fuel (liquid–fuel) per kilogram, and add the gas-fuel to 1 kg solid–fuel (liquid–fuel). The calculation method is as follows:

$$X = \frac{1-q_1}{q_1} \times \frac{Q_1}{Q_2} \qquad (1)$$

$$Q = Q_1 + XQ_2 \qquad (2)$$

where, q_1 is the heat share of the solid–fuel (liquid–fuel) (%); Q, Q_1 and Q_2 are the heating value of mixed fuel, solid–fuel (liquid–fuel) and gas-fuel (kJ/kg); X is the additional amount of gas-fuel of solid–fuel (liquid–fuel) per kilogram (m^3).

4 RESULT

4.1 *Influences on radiation heat transfer of the furnace and furnace outlet gas temperature*

When blending BFG, as the heat value of BFG (3066 kJ/m³) is very low, but the heat value of gas coal is generally 20,000 kJ/kg–30,000 kJ/kg, so the

theoretical combustion temperature of the mixed fuel will decrease, as shown in Figure 1a, the theoretical combustion temperature of the two kinds of gas coal changed with the blending ratio of BFG shows a hyperbolic decline. The theoretical combustion temperature of coal NO. 1 is higher than that of coal NO. 2 because of its higher heat value. When combusting the coal NO. 2 with the blending ratio of BFG 20%, the theoretical combustion temperature is 1758°C. This level of temperature has no impact on the ignition and burnout of coal particles, and the decrease of theoretical combustion temperature is meaningful to reducing the generation of NO_x. However, this decreasing makes the average furnace temperature level decrease, also the furnace radiation heat transfer lowered. As is shown from the calculation, when combusting the coal NO. 1 with the blending ratio of BFG 20%, the theoretical combustion temperature is lower than that with the pure coal-fired combustion condition by 220°C, and the radiant transfer heat decreased by 12.22%. When combusting the coal NO. 2 with the blending ratio of BFG 20%, the radiant transfer heat decreases by 11.50%.

As is shown in Figure 1b, the furnace exit flue gas temperature changes little with the blending ratio of BFG (0%–25%). When combusting the coal NO. 1 with the blending ratio of BFG 20%, the furnace outlet flue gas temperature is 1061°C. The temperature changes little compared with the pure coal-fired combustion condition. The reason is that with the increasing of the blending ratio of BFG, although the theoretical combustion temperature reduces, the N_2 which is inert gas that does not have the ability to radiate increased in the boiler. These inert gas will absorb a large amount of combustion heat in the combustion chamber, and reduce the temperature of the combustion products, also decrease the average furnace temperature level, lowering the furnace radiation heat transfer. In addition, with the increase of the blending ratio of BFG, the position of flame center shift, the time that flue gas in the furnace shortened, and the furnace radiant heat transfer lowered either. The heat loss due to the reducing of theoretical combustion temperature and the decreasing of radiant heat transfer are equal, so the furnace exit flue gas temperature changes little. We

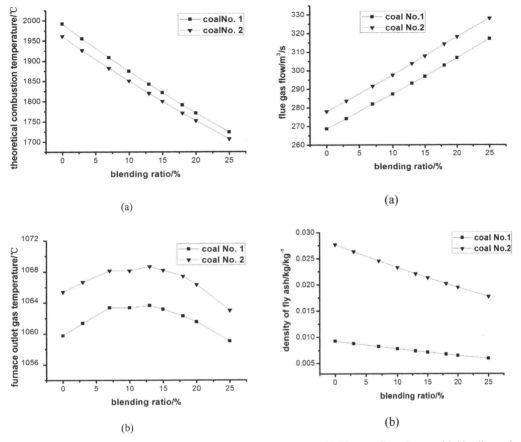

(a)

(a)

(b)

(b)

Figure 1. (a) Theoretical combustion temperature change with blending ratio of BFG; (b) Furnace outlet gas temperature change with blending ratio of BFG.

Figure 2. (a) Flue gas flow change with blending ratio of BFG; (b) Density of fly ash change with blending ratio of BFG.

can also find the furnace exit flue gas temperature of coal NO. 2 is higher than that of coal NO. 1, that is because combusting coal NO. 2 would produce less CO_2, H_2O which have the ability to radiate compared with coal NO. 1, so radiant heat transfer is reduced, leading the furnace exit flue gas temperature of coal NO. 2 higher than that of coal NO. 1.

4.2 *Influences on flue gas flow and density of fly ash*

Figure 2a shows the flue gas flow changes with the blending ratio of BFG, treating the blending ratio of BFG 0% a baseline, the flue gas flow increases with the increasing of the blending ratio of BFG. When combusting coal NO. 1 with the blending ratio of BFG 20%, the flue gas flow is 306.8 m³/s, an increase of 38.1 m³; When combusting coal NO. 2 with the blending ratio

of BFG 20%, the flue gas flow is 318.2 m³/s, an increase of 40.2 m³; The rise of the flue gas flow will increase the heat loss due to exhausted gas, the exhaust gas temperature and the flue gas flow are two main factors affecting the heat loss due to exhausted gas.

Almost free of dust in BFG. So density of fly ash decreases with the blending ratio of BFG increasing. As is shown in Figure 2b, when combusting coal NO. 1 with the blending ratio of BFG 20%, the density of fly ash is 0.00648 kg/kg, 0.00277 kg lower than that with the blending ratio of BFG 0%, we can see the blending ratio of BFG has little impact on density of fly ash. Therefore in the range of 0%–25% of the blending ratio, it has little effect on burnout of pulverized coal. So we treat heat loss due to unburned carbon in refuse q_4 a constant 1%. When combusting coal NO. 2 with the blending ratio of BFG 20%, we treat it 1.42%.

4.3 Influences on exhaust gas temperature and boiler efficiency

When combusting coal NO. 1 without blending BFG, the exhaust gas temperature is 122°C. The temperature would be 127°C when combusting coal NO. 2, the exhaust gas temperature increases with the blending ratio of BFG increasing. As is showed in Figure 3a, when combusting coal NO. 1 with the blending ratio of BFG 20%, the Exhaust gas temperature increased by 25°C, the reason is furnace outlet flue gas flow increased with the blending ratio of BFG increasing, even though the convection heat transfer increases either, the heat from increment of flue gas is more than that from increment of convection heat transfer, leading the exhaust gas temperature shows a hyperbolic increase with blending ratio of BFG. When combusting coal NO. 2 with the blending ratio of

BFG 20%, the exhaust gas temperature increased by 33°C. In the boiler operation, exhaust gas temperature plays a key limiting factor to the blending ratio of BFG. So when blending with BFG, the coal of high calorific value and low ash is proposed.

From the data that concentration of CO in flue gas changes with the blending ratio of BFG, we can see the unburned gas changes little with the blending ratio of BFG, so we treat the heat loss due to unburned gases q_3 as a constant 0.5%.

As is shown in Figure 3b, the boiler efficiency decreases with the blending ratio of BFG increasing. When combusting the coal NO. 1 with the blending ratio of BFG 20%, the boiler efficiency is 91.11%, decreased by 2.07% compared with that without BFG. When combusting the coal NO. 2 with the blending ratio of BFG 20%, the boiler efficiency is 88.72%, decreased by 2.75%.

5 CONCLUSIONS

As BFG is a kind of gas-fuel of low heat value, so the radiation heat transfer of furnace reduced after blending with BFG. When combusting the coal NO. 1 with the blending ratio of BFG 20%, the radiation heat transfer of furnace reduced by 12.22%; When combusting the coal NO. 2 with the blending ratio of BFG 20%, the radiation heat transfer of furnace reduced by 11.50%. The blending ratio of BFG makes little impact on the furnace outlet gas temperature.

The flue gas flow increases after blending with BFG, the convection heat transfer increases either, but the heat from increment of flue gas is more than that from increment of convection heat transfer, leading the exhaust gas temperature changed with blending ratio of BFG shows a hyperbolic increase. When combusting coal NO. 1 with the blending ratio of BFG 20%, the exhaust gas temperature increased by 25°C; When combusting coal NO. 2 with the blending ratio of BFG 20%, the exhaust gas temperature increased by 33°C. Exhaust gas temperature plays a key limiting factor to the blending ratio of BFG in the boiler's operation, so when blending with BFG the gas coal of low ash and high heat value is proposed.

In case of excess air ratio α constant to 1.2, with the increasing of blending ratio of BFG, the boiler efficiency declines, and the decline is more obvious when the blending ratio is high. When combusting the coal NO. 1 with the blending ratio of BFG 20%, the boiler efficiency is 91.11%, decreased by 2.07% compared with that without BFG. When combusting the coal NO. 2, with the blending ratio of BFG 20%, the boiler efficiency is 88.72%, decreased by 2.75%.

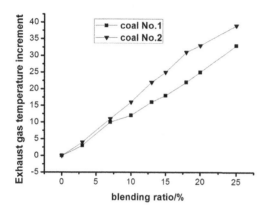

(a)

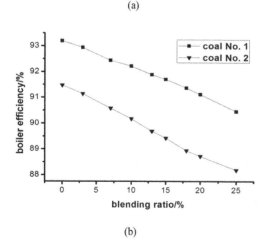

(b)

Figure 3. (a) Exhaust gas temperature increment change with blending ratio of BFG; (b) Boiler efficiency change with blending ratio of BFG.

REFERENCES

Anzhong Jiang, Zhimin Xiong & Yahui Yu. 1999. Some problems about the thermal calculation of boiler that burn mixed fuels. *Boiler Technology* 30(2):10–13.

Guanglei Han. 2010. The energy efficiency benefit and characteristics of 300 MW Pulverized-coal and Blast Furnace Gas (BFG)-fired boiler. *Energy Technology.*

Junkai Feng, Youting Shen & Ruichang Yang. 2003. *Pringciple and Operation of Boiler.* Science Press.

Quangui. Fan & Weiping Yan. 2004. *Pringciple of Boiler.* China Electric Power Press.

Shin, J.Y. Y.J. Jeon & D.J. Maeng. 2002. Analysis of the dynamic characteristics of a combined-cycle power plant. *Energy* (27):1085–1098.

Somasundaram. M & Chattopadhyay. D. 1997. Reliability improvement and life extension of mixed fuel-fired steel plant boilers. *Fuel and Energy Abstract* 29(2):618–623.

Wenzhong Zhang, Xuemin Gong & Yanguang Zang. 2003. Application of blast furnace gas in cogeneration of heat electricity and hot air. *Power Science and Engineering* 19(2):66–67.

Frontiers of Energy and Environmental Engineering – Sung, Kao & Chen (eds)
© 2013 Taylor & Francis Group, London, ISBN 978-0-415-66159-1

Influence of injection rate on coal gasification in molten blast furnace slag

P. Li, Q.B. Yu, Q. Qin & W. Lei

School of Materials and Metallurgy, Northeastern University, Shenyang, Liaoning, P.R. China

ABSTRACT: In this work, the influence of injection rate on carbon conversion, gas composition, calorific capacity of synthesis gas and the ratio of carbon to hydrogen were studied. The results show that the inject rate has a significant effect on coal gasification in molten blast furnace slag. The calorific capacity, content of CO and H_2 in synthesis gas and carbon conversion are increased firstly and then decreased as injection rate is increased. The ratio of carbon to hydrogen increases with increasing injection rate. When the gasification agent injection rate is larger than 0.18, the ratios of carbon to hydrogen increase rapidly. When the injection rate is 0.14 m^3/h, the content of CO is over 75%, the content of H_2 is over 15%, the calorific capacity is over 11000 kJ/m^3, the carbon conversion is about 0.95, and the quality of the synthesis gas is high. In this experiment, when the depth of slag pool was kept at 100 mm, the optimum gasification agent injection rate was found to be 0.14 m^3/h.

Keywords: waste heat recovery; molten slag; coal gasification; blast furnace; injection rate

1 INTRODUCTION

Steelmaking contributes greatly to the environmental problem. The environmental problem is caused by the use of fossil fuel that emits CO_2 and waste heat. The energy consumption of steelmaking is very high, and the energy efficiency is about 50%. So the steel industry releases much waste heat, which gives rise to environmental problems. Molten slag is discharged from the steelmaking industry as a byproduct at higher temperatures over 1723 K. The amount of Chinese molten blast furnace slag was about 162 million tons in 2009. The sensible heat in molten slag is currently emitted into atmosphere without any recovery. As a result, the current process consumed water and polluted environment. Therefore, how to effectively recover the sensible heat from molten slag has been extensively studied recently. Some of studies have been focused on evaluating energy requirement, proposing new systems, and designing new equipments. Some of promising methods regard dry granulation (Bisio 1997, Mizuochi 2001, Purwanto 2004 & 2005) and reforming of methane (Akiyama 2000, Kasai 1997, Shimada 2001, Maruoka 2004 & Purmanto 2006). The method of utilizing reaction heat instead of latent heat and sensible heat becomes quite attractive in terms of energy storage without heat loss and has made a connection to other industries. However, both of them are in laboratory so the sensible heat of slag cannot be efficiently recovered when the slag is at a liquid state. To resolve this problem, Yu and Li (Li 2010) proposed a new route to recover the sensible heat of liquid slag. It is to use molten BF slag as heat carrier for coal gasification.

The combined process to generate syngas by utilizing coal gasification reaction heat from high temperature waste heat in molten slag is an interesting technology. Coal gasification reaction is a highly endothermic reaction whilst molten slag processing is an exothermic process. This system consists of three parts namely a melting gasifier, a 2nd gasifier, and a boiler. The operating temperature of a melting gasifier ranges from 1573 K to 1773 K, the 2nd gasifier from 1073 K to 1573 K and the boiler from 473 K to 1073 K. With regard to this process, our research has been focused on the coal gasification in a melting gasifier. In the melting gasifier, there is a three-phase zone and the complicated reactions are involved. Molten slag bath is not only heat source, but also the main reaction district. Coal and gasification agent was injected into the molten BF slag through a high-speed nozzle and drove the molten slag rotating. Coal particle was also kept rotating due to centrifugal force. Coal particles in molten slag have two states: one is coal particles without bubble, the other is coal particles within bubble. For coal particles without bubble, coal pyrolysis takes place firstly followed by coal gasification. Pyrolysis and gasification occur at the same time for coal particles within bubble.

In this process, the gasification agent injection rate and depth of slag pool determines the residence time of coal particles and the injection rate greatly affects mass transfer. Therefore, the gasification agent injection rate is very important to the coal gasification in molten slag. In this work, we will report the influence of injection rate on coal gasification and calculated gasification indexes in the process of coal gasification with molten BF slag.

2 EXPERIMENTAL SECTION

2.1 Raw materials

In the present series of experiments, four different types of coal: Datong (DT) coal, Fuxin (FX) coal, Shennan (SN) coal, Coke (CO) and a BF slag sample were used. The proximate analyses of the coal samples in the gasification are shown in Table 1. The granularity of coal samples is about 0.125 mm. The chemical compositions of BF slag are 41.21 mass% of CaO, 34.38 mass% SiO_2, 11.05 mass% Al_2O_3, 8.22 mass% MgO, 0.35 mass% TiO_2 and some minor constituents of iron, sulphur and manganese as well as phosphor oxides. The gasification agent is CO_2 gas whose purity is 99.9%.

2.2 Apparatus and procedure

The experiments were conducted in a tube-type setup. Figure 1 shows the schematic diagram of experiment apparatus of coal gasification in molten BF slag. The reactor was made of high-purity alumina tube whose bottom was closed. The inner diameter of the reactor was 30 mm. The reactor was vertically placed in an electric furnace (about 12 KW) whose temperature could be measured by an S-type thermocouple. About 10 kg of BF slag was set at the bottom of the reactor. Another alumina tube whose inner diameter was 5 mm was inserted into the slag; through it coal particles and gasification agent can be introduced into the molten slag pool. The gasification agent came from gas cylinder, and it was as the carrier gas of the coal particles from the screw feeder. The screw feeder can be controlled by the controller which

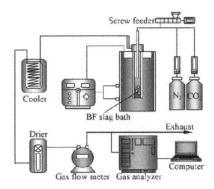

Figure 1. Schematic diagram of experiment apparatus.

can change the feeding rate from 2 g/s to 15 g/s. First, the slag was heated up to the desired temperature. After holding for 40 min and the temperature of reactor reaching in constant, the coal particles and gasification agent was introduced into the molten slag pool at constant conditions of ratio of CO_2 to coal and atmosphere pressure. Then the gasification of coal immediately took place.

For each experiment, the synthesis gas was analyzed by gas analyzer and was measured by gas flow meter. The gas analyses were conducted for the gas species of H_2, CO, CO_2, and CH_4.

2.3 Gasification indexes

The gasification indexes of gasifier consist of calorific value of synthesis gas, the yield of synthesis gas, gasification efficiency and carbon conversion rate. In this experiment, the gas analyzer analyzed the composition of synthesis gas, and then the gasification indexes were calculated.

The computing formulas of gasification indexes are shown as follows:

$$Q = \sum Q_i V_i \qquad (1)$$

where, Q is the calorific value of synthesis gas, kJ/Nm³; V_i is the volume of the content i, %; Q_i is the calorific value of content i, kJ/Nm³.

$$V = \frac{M}{G} \times 22.4 \qquad (2)$$

where, V is the yield of synthesis gas, Nm³/kg; M is the volume of synthesis gas, kmol; G is the consumption of coal, kg.

$$\eta_G = \frac{Q_G V}{Q_C} \qquad (3)$$

Table 1. Proximate analysis of the raw coal and char.

	Proximate analysis, %				Fusion point, K		
	M	A	V	FC	T_{def}	T_{hem}	T_{flow}
CO	0.07	12.5	1.59	85.84	>1773	>1773	>1773
DT	9.05	14.15	38.38	38.42	1498	1578	1653
FX	6.59	34.41	26.05	32.95	1673	1723	1773
SN	2.30	58.47	26.38	12.85	–	–	–

where, η_G is the gasification efficiency, %; Q_G is the calorific capacity of synthesis gas, kJ/Nm^3; Q_c is the calorific capacity of raw material, kJ/kg.

$$\eta_C = \frac{C_q}{C_m} \qquad (4)$$

where, η_C is the carbon conversion rate, %; C_q is the carbon in synthesis gas, %; C_m is the carbon in coal, %.

3 RESULTS AND DISCUSSION

3.1 *Effect of injection rate on product gas composition*

The reaction of gasification was tested at the temperature of 1773 K. In order to investigate the effect of injection rate on gas composition, a constant ratio of CO_2 to coal was kept and the feeding rate of feed screw and the flow rate of gasification agent (CO_2) were varied.

Figures 2 and 3 show variations of the gas composition with reaction time at 1773 K with a CO_2-Coal ratio equaling to 1, respectively, when the gasification agent injection rate were kept at 0.06 and 0.1 m^3/h. We can see that the gasification reaction process can be divided into three stages, namely, startup stage, stable stage and downtime stage.

Figure 4 shows the change of gas composition with different injection rates at 1773 K. It is evident that the content of CO and H_2 increases firstly and then decreases with increasing gasification agent injection rate. When the flow rate of gasification agent is 0.14 m^3/h, the contents of CO and H_2 have reached the highest with about 75 vol% of CO and 15 vol% of H_2. At a given depth of slag pool, the gasification reaction is dependent on the resident time of coal particles, heat transfer and mass transfer. The resident time of coal particles in molten slag pool is determined by the gasification agent injection rate and the depth of slag pool. The

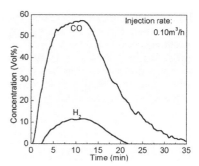

Figure 3. The gas composition vs. time.

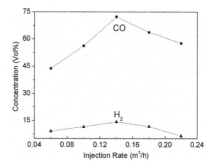

Figure 4. Effect of injection rate on gas composition.

heat transfer and mass transfer are determined by the injection rate. When the injection rate was less than 0.14 m^3/h, both heat transfer and mass transfer were enhanced as injection rate was increased, which resulted in an improved condition of gasification reaction thus the concentration of CO and H_2 increasing with increasing injection rate. However, with the injection rate larger than 0.14 m^3/h, the resident time of coal particles became shorter and parts of coal particles left the slag pool without sufficient reaction. As a result, the gasification reaction rate, the carbon conversion and the concentration of CO and H_2 were decreased. Therefore, to ensure the coal particles react fully, further studies on understanding the relationship of depth of slag pool, the injection rate and resident time are essential to acquiring the suitable depth of slag pool and injection rate for coal gasification in molten BF slag. In the present test, the optimum gasification injection rate was 0.14 m^3/h when the depth of slag pool was kept at 150 mm and CO_2 was used as the gasification agent.

3.2 *Effect of injection rate on calorific capacity of synthesis gas*

Figure 5 shows the calorific capacity of synthesis gas at various gasification agent injection rates (0.05, 0.1, 0.14, 0.18, and 0.22 m^3/h) when

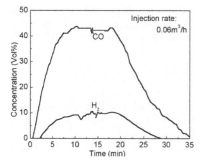

Figure 2. The gas composition vs. time.

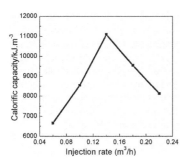

Figure 5. Effect of injection rate on calorific capacity of syngas.

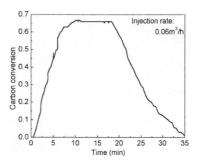

Figure 6. Carbon conversion vs. time.

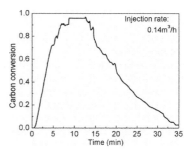

Figure 7. Carbon conversion vs. time.

the gasification temperature was 1773 K and the CO_2-Coal ratio was 1. We can see that the calorific capacity increases firstly and then decreases with increasing gasification agent injection rate. When the gasification agent injection rate equals to 0.14 m³/h, the calorific capacity of synthesis gas is about 11000 kJ/m³, whereas the calorific capacity is only 6500 kJ/m³ when the gasification agent injection rate is 0.06 m³/h. This can also be explained by the reduced resident time while increasing injection rate. As discussed earlier, When the injection rate was larger than 0.14 m³/h, the resident time of coal particles became too short to have sufficient time to completely react, which was unfavorable to coal gasification reaction. Therefore, the calorific capacity was observed to increase firstly and then decreased. This issue can be mitigated by reducing the injection rate, increasing the depth of slag pool and decreasing the ratio of gasification agent/coal.

As shown in Figure 5, under the conditions of CO_2-Coal ratio = 1 and gasification agent injection rate = 0.14 m³/h, the calorific capacity of syngas pro duced by this system can reach over 11000 kJ/m³. This indicates that the quality of the generated syngas is high, which can be used as not only the fuel for heating furnace or boiler, but also the feedstock in the chemical industry.

3.3 Effect of injection rate on carbon conversion

Figures 6 and 7 show the carbon conversion changed with time at the temperature of 1773 K, when the CO_2-Coal ratio was 1 and the gasification agent injection rate was 0.06 m³/h and 0.14 m³/h, respectively. There are also three stages: starting, stable and downtime.

We can see that the carbon conversion is about 0.65 when the gasification agent injection rate is 0.06 m³/h whilst the carbon conversion is increased to over 0.95 when the gasification agent injection rate is 0.14 m³/h. Figure 8 shows the carbon conversion changed with injection rates. It is exhibited that the carbon conversion increased firstly and

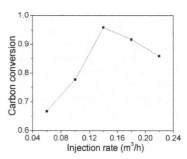

Figure 8. Effect of injection rate on carbon conversion.

then decreased with increasing gasification agent injection rate. This is also because the resident time of coal particle in molten slag became shorter as the injection rate became larger. We also can see that the carbon conversion of this system is quite high when the slag pool matching with the injection rate. In the present test, the carbon conversion reached the maximum when the gasification agent injection rate was 0.14 m³/h.

3.4 Effect of injection rate on the ratio of carbon to hydrogen

Figure 9 shows variation of the C-H ratio with different gasification agent injection rates (0.05, 0.1,

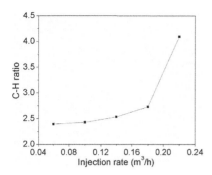

Figure 9. Effect of injection rate on C-H ratio.

0.14, 0.18 and 0.22 m³/h) at 1773 K and with the CO_2-Coal ratio being 1.

It can be seen that the C-H ratio increases with increasing gasification agent injection rate. When the gasification agent injection rate is more than 0.18, the C-H ratio increases rapidly. This may be due to the following reasons: One could be that when the resident time of coal particles in slag pool became shorter, the volatile matter and water vapor of coal did not have enough time to decompose. The other could be that with the increasing injection rate, parts of coal particles were brought out and the actual coal particles that involved in the chemical reaction were not enough, which resulted in an excessive gasification agent.

Thus, the reaction of CO_2 with carbon restrained the reaction of carbon and oxygen that decomposed from water vapor of coal. Hence the C-H ratio increased with increasing injection rate.

4 CONCLUSIONS

The influence of gasification agent injection rate on coal gasification in molten blast furnace slag is significant. Under the condition that CO_2-Coal ratio equaled to 1, the calorific capacity of syngas were vaired at different injection rates. The main results are summarized as follows:

1. The concentrations of CO and H_2 increased firstly and then decreased with increasing gasification agent injection rate. When the gasification agent was 0.14 m³/h, the concentrations of CO and H_2 reached the highest with about 75% of CO and about 15% of H_2 and the calorific capacity was over 11000 kJ/m³. The quality of the produced synthesis gas was high under these conditions.
2. The carbon conversion increased firstly and then decreased with increasing gasification agent injection rate. When the gasification agent

injection rate was 0.14 m³/h, the maximum carbon conversion of 0.95 was obtained. The C-H ratio increased with increasing injection rate.
3. In order to generate high quality syngas, it is important to have a reasonable depth of slag pool with a matched optimum injection rate. In the current tests, when the depth of slag pool was 100 mm, the optimum gasification agent injection rate was 0.14 m³/h.

ACKNOWLEDGMENTS

This research was supposed by National Natural Science Fund (51274066), Fundamental Research Funds for the Central Universities (N110602002) and Academic New Artist Ministry of Education Doctoral Post Graduate.

REFERENCES

Akiyama, T., Oikawa, K., Shimada, T., Kasai, E. and Yagi, J. 2000. Thermodynamic analysis of thermo-chemical recovery of high temperature wastes, ISIJ International 40: 286–291.

Bisio, G. 1997. Energy recovery from molten slag and exploitation of the recovered energy. Energy 22: 501–509.

Kasai, E., Kitajima, T., Akiyama, T., Yagi, J. and Saito, F. 1997. Rate of methane-steam reforming reaction on the surface of molten BF slag—for heat recovery from molten slag by using a chemical reaction. ISIJ International 37: 1031–1036.

Li, P., Qin, Q., Yu, Q. and Du, W. 2010. Feasibility Study for the System of Coal Gasification by Molten Blast Furnace Slag. Advanced Materials Research 97–101: 2347–2351.

Maruoka, N., Mizuochi, T., Purwanto, H. and Akiyama, T. 2004. Feasibility study for recovering waste heat in the steelmaking industry using a chemical recuperator. ISIJ international 44: 257–262.

Mizuochi, T., Akiyama, T., Shimada, T., Kasai, E., Yagi, J-I. 2001. Feasibility of rotary cup atomizer for slag granulation. ISIJ International 41: 1423–1428.

Purmanto, H. and Akiyama, T. 2006. Hydrogen production from biogas using hot slag. International Journal of Hydrogen energy 31: 491–495.

Purwanto, H., Mizuochi, T., Tobo, H., Takai, M. and Akiyama, T. 2001. Characteristics of glass beads from molten slag produced by Rotary Cup Atomizer. Materials Transactions 45: 3286–3290.

Purwanto, H., Mizuochi, T. and Akiyama, T. 2005. Prediction of granulated slag properties produced from spinning disk atomizer by mathematical model. Materials Transactions 46: 1324–1330.

Shimada, T., Kochura, V., Akiyama, T., Kasai, E. and Yagi, J. 2001. Effects of slag compositions on the rate of methane-steam reaction. ISIJ international 41: 111–115.

Frontiers of Energy and Environmental Engineering – Sung, Kao & Chen (eds)
© 2013 Taylor & Francis Group, London, ISBN 978-0-415-66159-1

The quantitative evaluation of environmental load in littoral mixing cement pile

J.Y. Shao, W.G. Chen, J. Zong & F.F. Cui
School of Management, Qingdao Technological, Qingdao, P.R. China

ABSTRACT: Life cycle assessment is used to establish evaluation system for the energy consumption and the pollution of mixing cement pile. It has three stages, construction material materialization, construction material transportation and construction process. The results show that construction material transportation has the largest environment burden and the sum of energy consumed for construction material materialization is the biggest. Greenhouse effect is the key environmental pollution problem in mixing cement pile. Based on the above results, corresponding improvement and precautionary measures can be put forward by concerned unit.

Keywords: mixing cement pile; LCA; environmental assessment

1 INTRODUCTION

Mixing cement pile uses cement and other materials as curing agent, through a special mixing machine, forced mixing the soft soil and the curing agent (slurry or powder) in situ, and make the soft soil to form a holistic, water stability and a certain intensity plus solid, thereby increasing the strength of the foundation soil and the modulus of deformation [1]. In China the cement mixing pile technology was introduced in the early eighties of the last century, now this technology has been widely used for Soft Foundation Consolidation in our country's railway, highway, municipal engineering, port and industrial and civil construction industry engineering and so on. Engineering Practice has proved that this method is simple, fast, efficient, and can reduce and control settlement and Residual Settlement effectively [2]. In recent years, our study about mixing cement pile is focused on the carrying capacity [3, 4] and construction quality control [5], but lack of environmental impact. Therefore, a comprehensive whole life cycle study of energy consumption and emissions about the various stages of mixing cement pile is of great significance.

Life Cycle Assessment (LCA) is a method of evaluating the environmental carries of products' whole life. Midwest Research Institute has done the study on beverage containers using LCA method since 1969 in America. From then on LCA has widely used in the system of product—development, and has became an important component part of ISO certification [6].

2 LCA OF MIXING CEMENT PILE

ISO14040~ISO14049 has been enacted by the ISO, and has carried out a detailed description of LCA. The work process of LCA can be divided into four associated phases, Goal and Scope Definition, Inventory Analysis, Impact Assessment and Interpretation [7].

2.1 Goal and scope definition

On the base of quantitative evaluating of energy consumption and environment pollution, this paper tries to find out the possibility and the key point of environmental improvement, which can provide basic database and theory support to the construction of mixing cement pile.

1. *Functional unit.* This paper uses one meter as the functional unit.
2. *Scope Definition.* (1) After the removal of mixing cement pile, materials treatment is not considered. (2) The construction process of dust and noise pollution is not considered. (3) The indirect energy consumption in the process of material production is not considered. (4) The transportation of supplementary materials and construction machinery is not considered.

2.2 Life cycle inventory analysis

The inventory analysis of mixing cement pile in life cycle is a process that set up a system of input and output service which can correspond with functional unit. This paper has done quantified research on the date of energy consumption and environmental pollution for mixing cement pile in its life cycle. Due to limited data, this paper established a life cycle inventory as shows in Figure 1.

2.3 Life cycle inventory analysis

The cement-mixture ratio 500 of mixing cement pile is usually between 12% and 20%, computational formula is shown in equation 1. The cement strength grade can select ordinary portland cement which 32.5 above magnitude. This paper selectes ordinary portland cement which has a strength of grade 42.5. We can calculate the usage of cement between 32 kg/m and 75 kg/m in one 500 mixing cement pile by equation 1. Table 1 shows the construction material materialization inventory, the data of ordinary portland cement which has a strength of grade 42.5 comes from references [6].

$$a_w(\%) = \frac{\text{Weight of cement}}{\text{Weight of soft soil}} \times 100\% \qquad (1)$$

Assuming the mode of transportation is gasoline truck, in that way energy consumption of transportation is 2.42 MJ/(t·km). The median transportation distance 67 km is used in this paper. "Emission Factor Handbook" and IPCC have done research on emission factors of motor vehicle. The caloric value of gasoline is 44 MJ/kg. Through the research all above, this paper calculated the c construction material transportation inventory as shown in Table 2.

In the construction process of mixing cement pile, cement mixers, crawler cranes, grout pump and grout mixers are widely used. "Construction Machines Pricelist in Shandong Province" (2008) records each mechanical equipment's energy consumption during eight hours. Calculating for gas emissions can use motor vehicle emission factors of fuel. This paper also considers the production of

"process energy". Table 3 shows the construction inventory.

2.4 Impact assessment

Impact Assessment is the procession of setting the sort order of the life cycle inventory data by quantitative or qualitative analysis, which contains classify, characterization and valuation.

1. *Classify*. According to life cycle inventory data, environmental impact assessment is divided into four parts: energy consumption, greenhouse effect, environment acidification and eutrophication.
2. *Characterization*. Characterization is the procession of summarizing environmental influence factors, which based on relevant science and technology. And different types of factors are normalized, which in order to increase comparability of different factors, as well as provide reference for valuation. This paper adopts the equivalent from Yang Jian-Xin [7]. Table 4 shows the environmental impact potential after factors are normalized.
3. *Valuation*. Valuation is the impact of environmental load which can determine the relative values of affecting degrees of different environmental influence factors to get the general environment influence data. This paper adopts the index weight from Yang Jian-Xin [7]. Specific results are as follows, construction material materialization: 5.764E-03~1.34E-02, construction material transportation: 2.15E-01~5.07E-01, construction process: 2.39E-04, LCA: 2.22E-01~5.20E-01.
4. *Interpretation*. (1) Construction material transportation of 500 mixing cement pile brings the strongest effect on the environment, and the next is construction material materialization and construction process. Strengthening the ability of construction material transportation

Table 1. Construction material materialization inventory.

Electricity consumption (kJ/m)	Heat consumption (kJ/m)	CO_2 (g/m)	SO_2 (g/m)	NO_X (g/m)	CO (g/m)	COD (g/m)
1.39E+04~ 3.26E+04	8.79E+04~ 2.06E+05	2.94E+04~ 6.90E+04	8.13E+00~ 1.91E+01	4.59E+01~ 1.08E+02	1.14E+01~ 2.67E+01	9.95E-01~ 2.33E+00

Table 2. Construction material transportation inventory.

Energy consumption (KJ/m)	CO_2 (g/m)	SO_2 (g/m)	NO_X (g/m)	CO (g/m)	COD (g/m)
5.19E+03~1.22E+04	3.69E+02~ 8.68E+02	4.84E+01~ 1.14E+02	3.46E+03~ 8.13E+03	2.77E+04~ 6.51E+04	–

Table 3. Construction inventory.

Energy consumption (KJ/m)	CO_2 (g/m)	SO_2 (g/m)	NO_X (g/m)	CO (g/m)	COD (g/m)
3.15E+04	5.67E+02	1.63E+00	2.50E+00	1.49E+00	1.49E-02

Table 4. Environmental impact potential of ϕ 500 of mixing cement pile in one meter.

Environmental impact	GWP	AP	NP
Construction material materialization	5.08E-03~1.18E-02	1.12E-03~2.62E-03	1.00E-03~2.35E-03
Construction material transportation	1.33E-01~3.14E-01	6.86E-02~1.61E-01	7.53E-02~1.77E-01
Construction process	1.57E-04	9.39E-05	5.45E-05
LCA	1.39E-01~3.25E-01	6.97E-02~1.64E-01	7.65E-02~1.79E-01

to manage the environment is effective in reducing the environmental impact on mixing cement pile of LCA. We should also take dust pollution into consideration in transportation environment protection. The dust pollution can reduce through the way of completing truck environment allocation, adding flushing system or many other ways. (2) Construction material materialization of 500 mixing cement pile brings largest energy consumption in these three stages. Improving construction material produces technology continuously can reduce energy consumption in mixing cement pile of LCA effectively. (3) In the stage of construction material materialization, transportation and construction, the most serious problem is the greenhouse effect in each stage. So the greenhouse effect is one of the major causes of environmental problems in mixing cement pile. Global warming brings about the greenhouse effect and the greenhouse effect brings about abnormal climate change, causing floods, droughts, tsunamis, storms, earthquakes and other problems. We need to begin curbing global greenhouse emissions right now, and our government is through the way of developing low-carbon energy, increasing the level of industrialization, improving the independent innovation capability and other ways to control the greenhouse effect.

3 SUMMARY

LCA evaluation has become an effective promotional tool of sustainable development, according to LCA evaluation model, the soil cement

mixing method can effectively predict the level of environmental pollution and energy consumption of resources, and the relevant parties can make prevention and improvement measures in advance, which is in line with the requirements of the global sustainable development. But in our country, the application of LCA is also subject to many limitations, the evaluation method is still not perfect. This paper put forward the restrictions about how to use the LCA model in mixing cement pile model, and how to combine this with circular economy can also be considered in further days.

This paper is supported by Science and Technology Development Projects in Qingdao(JK2012-6).

REFERENCES

[1] Cui jinghao, Technique of foundation treatment in soft soil. China WaterPower Press, Beijing, 2011, 36–37.
[2] Gong xiaonan, The Handbook of ground treatment four ed., Cina Architecture & Buillding Press, Beijing, 2008, 465–545.
[3] Yu wei, Wang Yong-heng, Lin Hong-lei, Gray Prediction on Ultimate Bearing Capacity of Composite Foundation Reinforced by Cement-soil Mixing Pile, Subgrade Engineering,. 2010, (3):142–144.
[4] Zhang Weili, Cai Jianl Lin Yixi, Huang Liangji. Experimental study of the load transfer mechanism of cement-soil pile composite foundation. China Civil Engineering Journal, 2010, 43(6):116–121.
[5] Gong Zhi-Qi. A Quantitative Method to the Assessment of the Life Cycle Embodied Environmental Profile of Building Materials. 2004.
[6] Yang Jian-Xin, Xu Cheng, Wang Ru-Song. Application of Life Cycle Assessment. China Meteorological Press, Beijing 2002.

Frontiers of Energy and Environmental Engineering – Sung, Kao & Chen (eds)
© 2013 Taylor & Francis Group, London, ISBN 978-0-415-66159-1

Tidal level forecast of Wenjiayan station on Qiantangjiang River based on support vector machine

J.P. Liu, Z.Z. Wu & S.X. Zheng

College of Civil Engineering and Architecture, Zhejiang University of Technology, Hangzhou, China

ABSTRACT: Support Vector Machine (SVM) is a sort of automatic computerized learning method based on statistical learning theory, which is a new theory in the field of machine learning. Statistical learning theory not only to consider the generalization ability, and the pursuit of optimal results under the conditions of the existing limited information is the best theory for the small sample statistical estimation and forecasting study. Use the support vector machine to set up a hourly non-linear relationship between the tidal based on the tidal data for grasping the regularity of tides and making short-term forecasts on tidal level. The conclusion could provide references for seawall seepage analysis, stability analysis and engineering of safe operation. Examples show that the prediction model based on the support vector machine has higher accuracy, in line with the standards of the hydrological information and forecasting.

Keywords: support vector machine; Wenjiayan station; tidal level; forecast

1 INTRODUCTION

Qiantangjiang River, named Zhejiang River in the ancient time, is the largest river in Zhejiang Province, which is known for the spectacular tidal bore at home and abroad. Qiantangjiang River's tidal bore is a tide phenomenon. Tide is a periodic regular fluctuation movement, which caused by the celestial gravitation. Tidal process includes the fluctuation of sea level affected by the gravitation of the moon and sun (astronomical tide) and exceptional fluctuation affected by the gale or drastic change of the atmospheric pressure brought by the weather system including the typhoon and cold wave (storm surge) (Chen 1980), and also affected by the seasons, submarine relief and even El Nino phenomenon.

The construction of buildings along the coast, the coastal port settings, the Estuarine engineering, land reclamation, changes in the marine ecosystem, marine resources development and utilization are all connected with tide. The downstream of Fuchunjiang power plant in Qiantangjing River is identified as tidal reach. The outside tidal waves from the estuary fluid from the southeast to the northwest of Hangzhou Bay, deforming rapidly under the effect of the trumpet-shaped shoreline. High tide level elevate, low tide level reduce and the tidal range increases. However, tide is one of the major marine natural disasters having a great impact on economic and social development of coastal areas (Wang 1999). Thus, a precise forecasting of tide is important for guaranteeing sustained and stable development.

Under normal circumstances, the tide forecast mainly refers to the astronomical tide forecast. Traditional tidal prediction model regards the ocean tide as composed of many different amplitude, period and phase tide. It works through the determination of the sub-tidal factors. For instance, Doodson uses the least squares method to determine the harmonic constants; Yen (Yen 1996) uses Kalman filtering theory to determine the harmonic constants. Compared to the traditional harmonic analysis of tidal prediction method, intelligent machine learning algorithm like neural networks and support vector machines should be widely used, which can provide highly accurate forecasts of both long-term and short-term based on a limited number of known data. For example, Lee (2002) and his colleague make forecasts of different types of tidal in both long, and short term using this method. Limingchang (2005) had put forward a cycle analysis of pre-treatment method dealing with the time lag occurs in Artificial Neural Networks tidal prediction. In recent years, many scholars carried out extensive studies about tide time series analysis, who pointed out that hourly tidal time series have the nonlinear law. Support vector machines has shown great advantages in dealing with complex nonlinear problems. It can be widely used in nonlinear system modeling, identification and prediction. As a result, we use support vector machines to carry out a research of hourly tide level of Wenjiayan Station in a short term.

2 SUPPORT VECTOR MACHINE REGRESSION ALGORITHM

Support Vector Machine (SVW) is a new machine learning methods based on statistical learning theory proposed by Vapnik (1995) and his colleagues. Recently, support vector machine has draw extensive attention from the academic community and has been widely used to solve the problem of classification and regression. Support vector machine is characterized by solving the small sample learning problems, which has changed the traditional empirical risk minimization principle to structural risk minimization principle. Consequently, it has better generalization ability. This method can avoid difficulties in determining the network structure, over learning, less learning, and local minima occurring in Artificial Neural Network. It provides a new way to solve nonlinear problems.

For regression problems, which use the function $f(x) = w \cdot x + b$ to fitted to the data (x_i, y_i) $i = 1, 2, \dots, n$, $x_i \in R^d$, $y_i \in R$, linear function is set as

$$f(x) = w \cdot x + b \qquad (1)$$

where, w is weight vector; b is deviation. Considering to the allow fitting error, we introduce the relaxation factor ξ_i and ξ_i^*, which represent respectively the upper and lower limits of the training error in the error constraints. Optimization problem is to minimize function,

$$\min\left(\frac{1}{2} <w \cdot w> + C\sum_{i=1}^{q}(\xi_i + \xi_i^*)\right) \qquad (2)$$

where, C is a constant, $C > 0$, indicating the punishment of the samples which beyond the error ε, for controlling model complexity and the compromise of approximation error. ε is used for controlling regression approximating to error pipeline. Constraint condition is

$$\begin{aligned} f(x_i) - y_i &\leq \xi_i^* + \varepsilon \\ y_i - f(x_i) &\leq \xi_i + \varepsilon \\ \xi_i, \xi_i^* &\geq 0 \end{aligned} \qquad (3)$$

As to this convex quadratic optimization problem, we introduce the Lagrange function

$$\begin{aligned} L = &\frac{1}{2} w \cdot w + C\sum_{i=1}^{n}(\xi_i + \xi_i^*) - \sum_{i=1}^{n}\alpha_i[f(x_i) - y_i + \xi_i + \varepsilon] \\ &- \sum_{i=1}^{n}\alpha_i^*[y_i - f(x_i) + \xi_i^* + \varepsilon] - \sum_{i=1}^{n}(\xi_i\gamma_i + \xi_i^*\gamma_i^*) \end{aligned} \qquad (4)$$

where, α_i and α_i^* are Lagrange factors, $\alpha_i \geq 0$, $\alpha_i^* \geq 0$, $\gamma_i \geq 0$, $\gamma_i^* \geq 0$, $i = 1, 2, \dots, n$. Take partial differential on equation (4), and to make all equal to zero,

$$\begin{cases} \dfrac{\partial L}{\partial w} = w - \sum_{i=1}^{n}(\alpha_i - \alpha_i^*)x_i = 0 \\ \dfrac{\partial L}{\partial \xi_i} = \gamma_i - C + \alpha_i = 0 \\ \dfrac{\partial L}{\partial \xi_i^*} = \gamma_i^* - C + \alpha_i^* = 0 \end{cases} \qquad (5)$$

Eq (5) into equation (4), getting dual form of the optimization problems, the maximum function

$$\begin{aligned} G(\alpha_i, \alpha_i^*) = &-\frac{1}{2}\sum_{i,j=1}^{n}(\alpha_i - \alpha_i^*)(\alpha_j - \alpha_j^*)(x \cdot x_j) \\ &+ \sum_{i=1}^{n}(\alpha_i - \alpha_i^*)y_i - \sum_{i=1}^{n}(\alpha_i + \alpha_i^*)\varepsilon \end{aligned} \qquad (6)$$

Constraint condition is

$$\begin{cases} \sum_{i=1}^{n}(\alpha_i - \alpha_i^*) = 0 \\ 0 \leq \alpha_i, \alpha_i^* \leq C \end{cases} \qquad (7)$$

w can be get from equation (5), using corresponding sample of $y_i(w \cdot x_i + b) = 1$ and $\alpha_i \in (0, C)$ to calculate b.

To take nonlinear regression firstly, we uses nonlinear mapping data to map to a high dimensional feature space. Secondly, we take regression in high dimensional feature space. The selection of the kernel function is a key issue. The optimization problem is to maximize the function under the constraint in equation (7).

$$\begin{aligned} G(\alpha_i, \alpha_i^*) = &-\frac{1}{2}\sum_{i,j=1}^{n}(\alpha_i - \alpha_i^*)(\alpha_j - \alpha_j^*)K(x_i, x_j) \\ &+ \sum_{i=1}^{n}(\alpha_i - \alpha_i^*)y_i - \sum_{i=1}^{n}(\alpha_i + \alpha_i^*)\varepsilon \end{aligned} \qquad (8)$$

In this circumstance,

$$w = \sum_{i=1}^{n}(\alpha_i - \alpha_i^*)\Phi(x_i) \qquad (9)$$

Regression function can be expressed as

$$f(x) = \sum_{i=1}^{n}(\alpha_i - \alpha_i^*)K(x, x_i) + b \qquad (10)$$

Use function $K(x_i, x_j) = (\Phi(x_i) \cdot \Phi(x_j))$ to change the inner product in the transform space into the calculation of a function in the original space. Thus indirectly solving the mapping of the input space to high dimensional feature space. However, this inner product can be achieved using a function in the original space even without know the transform form (Bian 2000). In SVM, a different inner product kernel function will form a different algorithm. The choice of kernel function must satisfy the Mercer conditions. Currently the most studied kernel functions are as follows

1. Linear kernel function

$$K(x, y) = x^T \cdot y \qquad (11)$$

2. Polynomial kernel function

$$K(x, y) = (x \cdot y + 1)^d, d = 1, 2, \dots \qquad (12)$$

3. RBF kernel function

$$K(x, y) = \exp\left|-\frac{\|x - y\|^2}{2\sigma^2}\right| \qquad (13)$$

4. Sigmoid kernel function

$$K(x, y) = \tanh[\upsilon(x \cdot y) + c] \qquad (14)$$

5. Bspline spline kernel function

$$K(x, y) = B_{2N+1}(x - y) \qquad (15)$$

With the introduction of kernel function we can map the sample space into a high-dimensional and even infinite dimensional feature space (Hibert space) through nonlinear mapping method. Consequently, in the feature space, it is available to use the linear learning machine method to solve the problems about highly nonlinear classification and regression in the sample space (Lin 2006).

3 TIDE PREDICTION BASED ON SUPPORT VECTOR MACHINES

3.1 *Sample selection*

The tide time series is a series in accordance with the order of the time line. Though there is a certain randomness on the value of every moment, numerical correlation before and after shows a trend or cyclical changes. As a result, through the analysis of historical data of tidal time series, we can reveal the inner regulation and predict the future tidal level.

The location of the stations along the Qiantang River estuary are as showed in Figure 1 (You 2010).

Taking the Wenjiayan Station as an example, we establish the short-term forecasting model based on support vector machine tide, in which we use the hourly whole point of tide level from October 1, 2011 to 2011, October 21, 2010 as foundational data. Choice of the sample use the sliding value. Tidal time series is expressed as $\{Xt\}$, in which $Xt = X(t), t = 1, 2, \dots, 504$. There are some potential relationship existing between the present and future value of time series and the previous m data, m = 48. The functional relationships is $x_{n+1} = F(x_n, x_{n-1}, \dots x_{n-m+1})$.

Normalized to the original sample data. After the time series is normalized, which is within [−1 1], it would be easier for training and learning for support vector machine method.

3.2 *Model training*

To verify the feasibility of support vector machine algorithm and its use in the tide prediction superiority, we use Qiantang River tide Wen Jiayan station measured data as raw data. Selecting the former 365 data as learning examples, build a support vector machine model to learn learning samples for determining the parameters of support vector machine model. After this, take forecasts on the later 100 tests samples based on the former training, which means the forecasting of whole point of tidal level.

After the establishment of the input and output of SVM model, we need to determine the model's parameters, among which the selection of kernel function is critical, which has direct impact on calculation accuracy of models. And Radial Basis Function (RBF) is the most widely used in support vector machine kernels.

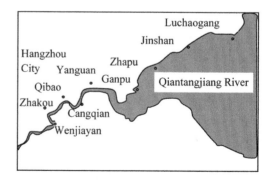

Figure 1. Tidal stations along the Qiantangjiang River Estuary.

We use 356 samples for training the training model. Parameter C, ε, and the width of the radial basis kernel function σ would have a great impact to the result of the prediction, which need to be determined specifically based on characteristics of training data. Generally, reducing the value of ε and σ will increase the accuracy of the training. However, if these two parameters are too small will decrease the performance of the generalization of the model. On the other hand, increase c also increases the accuracy of precision training. Through the test of different C, ε and σ, we choose the parameter $C = 100$, $\varepsilon = 0.001$, $\sigma = 0.1$. The average relative error and mean-square error of training samples are, respectively, 1.84% and 0.15.

3.3 Forecast results and analysis

In real-world applications, Trained models need be tested before it can be put into practical application. Consequently, we use 100 test data to take forecasting, which shows the minimum relative prediction error is 0.002% with a maximum relative prediction error 15.7%. The average relative error and root mean square error of the test samples are 1.52% and 0.10. Detailed data is shown in Figure 2.

The minimum relative prediction error is 0.002% with a maximum relative prediction error 15.7%. The average relative error and root mean square error of the test samples are 1.52% and 0.10.

In 100 test samples, there 65 samples have an absolute error in 0~5 cm, which is 65%, There is only one sample has larger errors, 71 cm, which is 1%. Detailed data is shown in Table 1.

There are 94, 5 and 1 example respectively have an absolute error of 0~5%, 5%~10% and 10~20%, which is 94%, 5% and 1%. All the tested samples have an absolute error below 20%. Detailed data is shown in Table 2.

In the specification of hydrological information and forecasting, the admitted errors of tidal level prediction (the highest level) is ±0.30 m. In the calculation examples, among the 100 prediction samples of Qiantang River WenJiayan Station, there are 98 samples have an absolute error within

Table 1. The predicting absolute error of SVM.

Absolute error range/cm	Number of samples	Proportion/%	Accumulation/%
0~5	65	65	65
5~10	22	22	87
10~20	9	9	96
20~30	2	2	98
30~40	1	1	99
40~50		0	99
50~60		0	99
60~70		0	99
70~80	1	1	100
Sum	100		

Table 2. The predicting relative error of SVM.

Relative error range/%	Number of samples	Proportion/%	Accumulation/%
0~5	94	94	94
5~10	5	5	99
10~20	1	1	100
Sum	100	100	

30 cm, which is 98% of the whole. It can be seen that tidal level forecasting model based on support vector machine prediction has high precision. And it is possible to predict the tidal level based on the former 48 hourly tidal data. This method provide a reference for the safe operation of the seawall project.

4 CONCLUSION

Statistical learning theory to a large extent, solve the model selection, over learning, the curse of dimensionality and local minima problems, providing a better theoretical framework for the establishment of small sample of machine learning. The models which established using SVM method have better generalization ability for it is based on a more stringent statistical learning theory. Tidal level prediction in short term based on SVM has a higher precision. The relative prediction error is within 20% and 98% of the absolute error of the sample are in line with the forecast regulatory requirements. Yet, how to select the type of kernel function and loss function and the perimeters are still difficult to be solved currently, which mostly rely on experience or repeated spreadsheet.

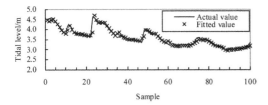

Figure 2. The predicting value of SVM model.

ACKNOWLEDGMENT

Financial support from the Key Social Development Project of Zhejiang Special Science and technology project (Grant No. 2009C13012) and Key Scientific and Technological Innovations Teamwork Project (Grant No. 2012R10035-11) is acknowledged.

REFERENCES

Bian Zhaoqi, Zhang Xuegong. 2000. Pattern Recognition. Beijing: Tsinghua University Press.
Chen Zong-yong. 1980. Tide. Beijing: Science Press.
Lee T.L, Jeng D.S. 2002. Application of artificial neural networks in tide forecasting. Ocean Engineering 29(9):1003–1022.
Li Ming-chang, Liang Shu-xiu, Sun Zhao-chen. 2007. Application of artificial neural networks to tide forecasting. Journal of Dalian University of Technology 47(1): 101–105.
Lin Jianyi, Cheng Chun-tian. 2006. Application of support vector machine method to long-term runoff forecast. Journal of Hydraulic Engineering 37(6):681–686.
Vapnik, V.N. 1995. The Nature of Statistical Learning Theory. New York: Springer.
Wang Yi-hong, Shang Si-rong. 1999. Storm tide disaster and its countermeasure in bohai bay. Journal of Catastrophology 14(3): 70–74.
Yen P.H., Jan C.D., Lee Y.P., et al. 1996. Application of Kalman filter to short-term tide level prediction. Journal of Waterway Port Coastal and Ocean Engineering, ASCE122(5):226–231.
You Ai-ju, Han Zeng-cui, He Ruo-ying. 2010. Characteristics and effecting factors of the tidal level in the Qiantangjiang River Estuary under changing environment. Journal of Marine Science 28(1):18–24.

Frontiers of Energy and Environmental Engineering – Sung, Kao & Chen (eds)
© 2013 Taylor & Francis Group, London, ISBN 978-0-415-66159-1

Study on pyrolysis and liquefaction of sludge from oil tank bottom

G.J. Li, Y.L. Ma, L.P. Bai & R.H. Zhao
*Department of Environmental Science and Engineering, Tianjin University of Science & Technology,
Tianjin, P.R. China*

ABSTRACT: The catalytic effects of two inorganic compounds (Na_2CO_3 and KOH) on pyrolytic behavior of sludge from oil tank bottom were investigated by thermal analysis experiments. The influences of catalyst categories and dosage on sludge pyrolysis from oil tank bottom were studied by Thermogravimetric and Differential Thermogravimetric (TG-DTG) analysis. The oil products obtained by direct thermochemical liquefaction under the optimal conditions were detected by GC-MS. The results showed that the catalytic effect followed Na_2CO_3 > KOH, and 2 wt% Na_2CO_3 was chosen as the optimal dosage of the catalysts. The components of oil product obtained at 2 wt% Na_2CO_3, reaction temperature of 553 K and holding time of 30 min were complicated, including benzene species and hydrocarbon species etc.

Keywords: pyrolysis; catalyst; liquefaction; sludge; oil product

1   INTRODUCTION

In the petroleum refineries, the considerable amount of sludge is accumulated from the treating process of cleaning oil storage tanks. The kind of sludge is designated as hazardous waste in Resource Conservation and Recovery Act (U.S. 1989). It can be handled via landfill, incineration, solvent extraction (Taiwo & Otolorin 2009) and microbial degradation (Jose Luis et al. 2007), but it has been found that such methods can result in secondary pollution. Pyrolysis has been generally accepted as a promising disposal method for sludge, since pyrolytic process can not only minimize solid waste but also yield valuable oil products.

The relevant researches of sludge pyrolysis had been reported earlier. The kinetic and thermal conversion behaviors of oily sludge pyrolysis were studied at different heating rates by TGA (Punnaruttanakun et al. 2003, Wu et al. 2006). The pyrolysis reaction of oily sludge started at a low temperature of about 473 K and the maximum conversion rate was observed within the temperature range of 623–773 K by using TG/MS (Wang et al. 2007). The major products obtained from the pyrolysis of oily sludge were studied by using nitrogen as carrier gas in the temperature range from 378 K to 873 K (Chang et al. 2000). However, these studies were mainly focused on the pyrolysis kinetics of sludge. At the present, the reports about the influence of catalyst on sludge pyrolysis and oil-producing process are less. Catalyst plays a vital role in achieving the net energy output. In the catalytic process, pyrolysis can be promoted because the catalyst can shorten phrolysis time, reducing phrolysis temperature and raising conversion rate. So, catalysts are necessary for improving the efficiency of sludge pyrolysis and oil yield.

In the paper, the catalytic effects of two inorganic compounds (Na_2CO_3 and KOH) on pyrolysis of sludge from oil tank bottom were investigated by thermal analysis experiments with changing the catalyst dosage. The effects of catalysts on product distribution and oil yield were studied by direct thermochemical liquefaction. Finally, the optimal conditions of sludge pyrolysis and oil-producing reaction were obtained. Thus it can supply basic data for practical application in sludge pyrolysis.

2   EXPERIMENT

2.1   *Experimental materials*

Sludge from tank bottom (Huabei Oilfield Company, China) was characteristic of black, viscidity and representative odor of petroleum. The proximate analysis of sludge was presented in Table 1, where Mad (moisture) was weight loss percentage on air dry basis at 378 K, A_d (ash) was the residue percentage on dry basis after complete combustion

Table 1.   Property analysis of oily sludge.

	Ma_d (%)	V_d (%)	A_d (%)	FC_d (%)	Na, K (‰)
Percentage	76.74	58.53	38.38	3.09	<1

at 1073 K., V_d (volatile) was weight loss percentage on dry basis after devolatilization at 973 K for 10 min under scarce oxygen atmosphere and FC_d (fixed carbon) was calculated by difference.

From Table 1, it was seen that sludge from oil tank bottom had relatively high content of 58.53% volatile, which made it a good material for oil recovery. The content of Na and K in oily sludge is less than 1‰, so the effects of Na and K in the sludge on catalysts can be ignored.

2.2 Experimental methods

The samples were pyrolyzed on a Q50 TGA thermogravimetric analyzer from US TA Company at different heating rates, terminal temperatures and holding time. 10 mg sample were heated from ambient temperature to 378 K and dried at 378 K to constant weight. Then, the experiment samples were obtained with quarter method after grained into powders of 150~200 mesh.

Experiment samples were further heated to 1173 K under the nitrogen flow of 60 cm³/min. The heating rate was controlled at 20, 30, 40 and 50 K/min. The data of Thermogravimetric and Differential Thermogravimetric (TG-DTG) analysis were used for characterizing the pyrolysis behavior of the sludge samples and estimating their kinetic parameters. Finally, the optimal catalyst category and catalyst dosage were obtained according to DTG curves of sludge pyrolysis. Liquefaction experiment of sludge was studied with the optimal pyrolytic conditions. And the components of oil product were detected by GC-MS.

3 RESULTS AND DISCUSSION

3.1 Thermogravimetric analysis of sludge from oil tank bottom

Figure 1 showed the TG-DTG curves at a heating rate 5 K/min. The profiles of TG-DTG curves at different heating rates were almost similar and TG-DTG curves at heating rate of 5 K/min were taken as the example to characterize the pyrolysis behavior of the sludge samples.

The pyrolysis process is divided into four stages by the TG-DTG curves at a fixed heating rate, as follows: ①Ambient Temperature-393 K is a dehydration stage, in which the mass loss is about 1.54%. ②The temperature range of 394 K–633 K is the volatilization stage of light component in the sludge, in which the total mass loss is 26.82%. The mass loss rate is obviously controlled by evaporation temperature of light component in the sludge. It is also the main reaction stage for oil-producing process. ③The temperature range of 634 K–923 K is the decomposition stage of intermediates, in which the total mass loss is 19.37%. ④The temperature

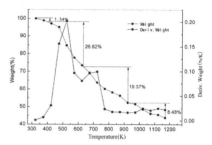

Figure 1. TG-DTG curves of sludge pyrolysis at heating rate of 5 K/min.

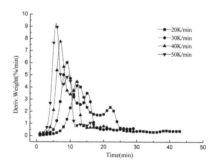

Figure 2. DTG curves of sludge pyrolysis at different heating rates.

range of 924 K–1173 K is the last stage, in which the mass loss rate begins to decrease and the total loss is about 8.48%. The minerals and the rest organics are broken down in this stage. So, the second stage (394 K–633 K) in which the main weight lose process happens is taken into research on catalytic pyrolysis.

3.2 Sludge pyrolysis at different heating rates

The DTG curves at different heating rates are shown in Figure 2. The maximum weight loss rates in the first peak are 4.179%/min, 6.033%/min, 7.751%/min and 9.559%/min, respectively, corresponding to 20 K/min, 30 K/min, 40 K/min, 50 K/min. It was observed that the maximum weight loss rate was also lower with the decrease of heating rate. Thus, the increase of heating rate can shorten the time to reach the maximum weight loss rate.

3.3 Effects of catalyst type on sludge pyrolysis

Previous work had confirmed that Na_2CO_3 and KOH was the best catalyst for sludge pyrolysis in sodium compounds and potassium compounds. Figure 3 is the TG-DTG curves of Na_2CO_3. It is seen that the weight loss before 373 K is mainly caused by water evaporation, whereas within the temperature range of 373–1023 K, the weight of Na_2CO_3 remains basically constant, indicating that Na_2CO_3 can act as a catalyst in the temperature range. Similarly, KOH exhibits the same TG-DTG curves.

From Figure 4 and Table 2, it was seen that both the two catalysts lead to pyrolyze sludge samples at lower temperatures, and also increase the maximum weight loss rates. The results showed that Na_2CO_3 made the larger conversion rate of sludge appear at the lower temperatures. The effects of catalytic performance followed the order of $Na_2CO_3 > KOH$. So, Na_2CO_3 was more efficient for sludge pyrolysis.

3.4 Effects of catalyst dosage on sludge pyrolysis

In Figure 5 and Table 3, when the dosage increased, the pyrolysis happened at lower temperature, suggesting that the increase dosage of Na_2CO_3 promoted sludge pyrolysis. The accordingly temperature reduced, but the conversion rate also decreased. Considering these two sides, so, 2 wt% Na^+ was chosen as the optimal dosage.

3.5 Catalytic effects on oil yield of sludge pyrolysis

Na_2CO_3 was chosen as the catalyst, and the proportions of Na^+ concentration in the sludge were 0%, 1%, 2%, 4%, 5%, respectively. Reaction temperature was set at 553 K. Reaction pressure was the saturated vapor pressure corresponding to the temperature, and holding time was zero minute. Figure 6 showed the relation between catalyst dosage and oil yield.

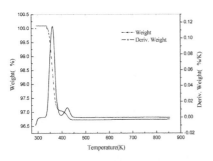

Figure 3. TG and DTG curves of Na_2CO_3.

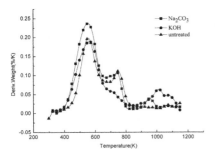

Figure 4. DTG curves of sludge pyrolysis with different catalysts.

Table 2. Experimental results of sludge pyrolysis with different catalysts.

	Untreated	KOH	Na_2CO_3
Conversion at 1023 K (%)	90.41	94.48	95.31
Temperature accordingly (K)	569	555	553
Maximum weight loss rate (%/K)	0.1890	0.1986	0.2373

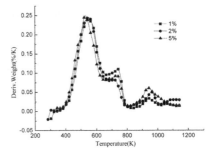

Figure 5. DTG curves of sludge pyrolysis with different proportions of Na_2CO_3.

Table 3. Experimental results of sludge pyrolysis with different proportions of Na_2CO_3.

	1 wt% Na_2CO_3	2 wt% Na_2CO_3	5 wt% Na_2CO_3
Conversion at 1023 K (%)	96.24	95.19	94.13
Temperature accordingly (K)	556	553	548
Maximum weight loss rate (%/K)	0.2399	0.2438	0.2476

From Figure 6, oil yield increased with the increase of the catalyst dosage. After Na_2CO_3 concentration was more than 2% wt, the oil yield increased indistinctively. Herein, the dosage of catalyst was set at 2 wt%. It basically matched with the results from thermogravimetric analysis.

3.6 Oil analysis

The color of the oil was puce or black, and the oil had typical petroleum odor. It was reported that multi-benzene ring compounds were detected in pyrolysis oil of oily sludge, and they were closely similar (Heuer & Reynolds 1992). Table 4 showed that the major chemical compositions of oil product were gained at the ideal reaction parameters of liquefaction: 2 wt% Na_2CO_3, reaction temperature of 553 K and holding time of 30 min, including benzene species, hydrocarbon species and so on.

Table 4. Chemical compositions of the oil product.

Peak	Residence time (min)	compound	Formula	Relative content (%)
1	2.093	Ethylenzene	C_8H_{10}	9.019
2	2.211	Stytene	C_8H_8	21.500
3	2.346	Benzene (1-methylethyl)	C_9H_{12}	5.832
4	2.657	Alpha-methylstytene	C_9H_{10}	1.994
5	3.107	Benzene, 1-methyl-3-propyl	$C_{10}H_{14}$	5.949
6	3.466	Benzene, 1-ethyl-4-ethyl	$C_{10}H_{12}$	3.660
7	3.830	Benzene, 1, 3-diethyl-5-methyl	$C_{11}H_{16}$	5.852
8	4.410	1H, Indene, 2, 3-dihydro-1, 2-dime	$C_{11}H_{14}$	6.081
9	4.510	N-methyl-9-aza-tricyclo	$C_{12}H_{11}$ NO	2.071
10	5.700	Naphthalene, 2-methyl	$C_{11}H_{10}$	5.851
11	6.137	Benzene, (2, 2-dimethy-1-methyl)	$C_{12}H_{16}$	1.161
12	6.576	Pentadecane, 7-methyl	$C_{16}H_{34}$	2.409
13	6.844	Naphthalene, 1-ethyl	$C_{12}H_{12}$	7.819
14	7.924	Naphthalene, 2, 3, 6-trimethyl	$C_{13}H_{14}$	2.932
15	8.901	Tetracosane	$C_{24}H_{50}$	3.140
16	10.009	Heptacosane	$C_{27}H_{56}$	8.234
17	11.136	Nonadecane	$C_{19}H_{40}$	6.494

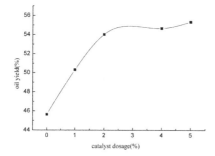

Figure 6. Relation graph of catalyst dosage and oil yield.

4 CONCLUSION

1. Thermal decomposition behaviors of oily sludge were divided into four stages. The temperature range of 394 K–633 K was the main reaction stage for oil making, in which the total mass loss was 26.82%. The increase of heating rate can shorten the time to reach the maximum weight loss rate.
2. The catalytic effects of the two compounds followed Na_2CO_3 > KOH. The larger conversion rate was obtained at lower temperature by using Na_2CO_3. With the increase of Na_2CO_3 dosage, the pyrolysis temperatures decreased and the conversion rate also decreased. The optimal dosage of the catalysts was 2 wt% Na_2CO_3.
3. The oil yield was increased with the increase of the catalyst dosage. After more than 2% wt Na_2CO_3, the oil yield increased indistinctively. Herein, the dosage of catalyst was set at 2 wt%.
4. The components of oil product obtained by direct thermochemical liquefaction under the optimal

conditions were complicated, mainly including benzene species and hydrocarbon species etc.

ACKNOWLEDGMENT

This project was supported by the Science and Technology Development Fund of Tianjin Colleges and Universities (20090519).

REFERENCES

Chang C.Y., Shie J.L., Lin J.P., et al. 2000. Major Products Obtained from the Pyrolysis of Oil Sludge. Energy & Fuels 14:1176–1183.

Heuer S.R. & Reynolds V. 1992. Process for the Recovery Oil from Waste Oil Sludge. Environment International 18:22–29.

Jose Luis R.G., Juan F.L., et al. 2007. Biodegradation of oil tank bottom sludge using microbial consortia. Biodegradation 18(3):269–281.

Punnaruttanakun P., Meeyoo V., Kalambaheti C., et al. 2003. Pyrolysis of API separator sludge. Journal of Analytical and Applied Pyrolysis 68–69:547–560.

Taiwo E.A. & Otolorin J.A. 2009. Oil recovery from petroleum sludge by solvent extraction. Petroleum Science and Technology 27(8):836–844.

U.S. Environmental Protection Agency Office of Solid Waste and Emergency Response. 1989. Guidance Manual for Hazardous Waste Permits PB 84-10057. Washington DC.

Wu R.M., Lee D.J., Chang C.Y. 2006. Fitting TGA data of oil sludge pyrolysis and oxidation by applying a model free approximation of the Arrhenius parameters. Journal of Analytical and Applied Pyrolysis 76: 132–137.

Wang Z., Guo Q., Liu X., et al. 2007. Low temperature pyrolysis characteristics of oil sludge under various heating conditions. Energy Fuels 21: 957–962.

Frontiers of Energy and Environmental Engineering – Sung, Kao & Chen (eds)
© 2013 Taylor & Francis Group, London, ISBN 978-0-415-66159-1

Integrated design of solar water heating systems for a landscape city: A case study of RunHongShuiShang residential quarter-phase II, Guilin

L. Wang & J. He

Key Laboratory of Disaster Prevention and Structural Safety, College of civil engineering and architecture, Guangxi University, Nanning, China

ABSTRACT: This paper describes integrated design of solar water heating systems in buildings of RunHongShuiShang Residential Quarter-Phase II which is located in Guilin city called the landscape city. This study aims to provide constructive suggestions and knowledge for the integration design of solar water heating systems in this region or other regions with similar climate.

Keywords: landscape city; solar energy; water heating system; integrated design

1 INSTRUCTIONS

At present, China has become the second largest countries of energy consumption in the world. The contradiction between the natural environment and energy demands is getting significant. A top priority of building a harmonious society lies in the green energy is applied comprehensively, not only energy conservation but also the matter of carbon emission reduction. After the "Eleventh Five-year" plan, related governments of China announced the "Twelfth Five-year" plan which is a comprehensive plan including energy saving and carbon emission reduction. From governments to enterprises as well as individuals, all is asked to carry out the plan deeply in the further steps of energy saving and emission reduction. In such a policy guidance and background, solar water heating systems are turned out to be one of significant environmental protections of sustainable development, because of its reasonable and moderate cost and green measures. In 2012, a solar building design competition was constructed by the Guilin city government. Real projects under construction were selected as competition targets, attempted to find solutions for the plan of solar energy utilization which are suitable for the development of Guilin, and strived to explore a new way to solve the problem between energy shortage and polluted natural environment in the current society.

As is well-known, Guilin city is famous as its unique landscapes. Landscape is the main content of the whole city, while buildings just belong to the body. Therefore, the basic principle of urban planning in Guilin is going to follow the character of the city's landscape. "Landscape city" that is the bond between natural and artificial environment. Hence, the application of solar water heating systems should be given to careful considerations in the integration design. This must be strictly checked from building facades to color collocations.

This study is aimed to discuss the integrated design of solar water heating systems that are well-matched with Guilin's landscape. And the design competition will be treated as an opportunity to provide some experience in the area and somewhere has the similar context afterwards, through analysis of buildings in Run Hong Shui Shang Residential Quarter-Phase II.

2 DESCRIPTION OF THE PROJECT

The RunHongShuiShang Community Project (hereafter as "The Project") is located at the joint of the Kaifeng Road and Chadian Road in Xiangshan District of Guilin city, and near Kaifeng Road which is the Golden Path to the scenic spots of Yangshuo country. The project covers more than 300 Mu (1 Mu is equal to about 666.7 m^2) of ground area and has more than 400,000 m^2 of total construction area. It has a central lake covering an area of 20,000 m^2 which introduces the core value of the project to be a community with the landscape of Chinese Garden Style. This paper selects building No. 29 which has been constructed in stage 2 of the project, and has similar New Chinese Garden Style as stage 1 of the project. (See Fig. 1).

Figure 1. A view of RunHongShuiShang residential quarter.

3 TECHNICAL PARAMETERS AND DESIGN OF SOLAR WATER HEATING SYSTEMS

3.1 *Thermal parameters*

Guilin is ranked as a Class III area of solar radiation with annual average temperature of 19.3°C. Its annual average daily solar radiation is 11293 kJ/(m²·d). Temperature of city water is 25°C in summer and 11°C in winter, 16°C for the transition season. The location of the project is Latitude 25°11′N and Longitude 110°17′E. The targeted building has 11 floors with 3 units, and 82 households in total. Unit No. 1 has 9 + 1 floors with two households in a staircase; Unit No. 2 has 10 + 1 floors with two households in a staircase; Unit No. 3 has 11 floors with four households in a staircase. The slope of the sloped roof is 21° with solar azimuth of 16°.

3.2 *Selection of heat collector*

Vacuum Tube Heat Collector (VTHC) and Flat Plate Heat Collector (FPHC) are the most popular in the domestic market. The latter could have a better combination with buildings and make buildings more beautiful from the viewpoint of integrative design. FPHC also has advantages of good resistance to damage, long using time, no scaling when water temperature is below 60°C, and fits for the hard water areas such as Guilin. Blue roof tiles are used for the project. Dark blue is also the most common color for FPHC, so that dark blue FPHC are included in the design for a matching purpose. And FPHC is of a uniform size of 2000 × 1000 × 80 (mm).

3.3 *Design of heat collecting system*

A sloped roof has been taken into consideration for the original design of the project. There is another fact that the roof area is insufficient for solar collection when a family of 3 people with water consumption of 50 L/d per capita, and part of FPHC should be installed at balconies above the 5th floor. Moreover, solar radiation in Guilin might be not enough and auxiliary heating energy is required. It was found the hybrid system of solar and Air Source Heating Pump (ASHP) has better performance than the system of solar and electrical heating according to calculation results. Considering climatic conditions of Guilin, the appearance design, the area of FPHCs, the auxiliary heating system, economy, etc., we then have proposed the following protocol for the application based on a precise calculation: the solar azimuth of 16°, the slope of FPHC installed on the roof is the same as 21°, the slope of FPHC installed on facades is 35°. Households below the 5th floor and 22 households in the north face of Unit No. 3 use the solar water heating system with a single tank to which water is indirectly supplied. FPHCs are installed on the sloped roof with mounting brackets. ASHPs are installed on the jacket of the sloped roof with necessary components e.g. water tanks as the auxiliary heating energy. Split type pressure-bearing water-heating systems are adopt for the households above the 5th floor, one set with electricity heating device as the auxiliary energy per household. An additional dedicated water flow meter shall be installed for each household with FPHCs installed on the roof. Corresponding fees shall be calculated and paid on the basis of the consumed hot water, which is more convenient for the property management to provide the post-service of management and maintenance for the system.

4 INTEGRATED DESIGN OF SOLAR WATER HEATING SYSTEMS IN BUILDINGS

ASHPs are installed on the jacket of the sloped roof with necessary components e.g. water tanks as the auxiliary energy, and this action shall not introduce negative impacts on the appearance of facades. Since FPHCs are installed on the roof and balconies, there shall be plenty applicable options of the integrated design and three construction schemes are designed for this project.

Scheme 1 is a combination of FPHC and decorative frame (see Fig. 2). Most building styles in Guilin are of free-of-burden and opened due to the famous beautiful scenery, which was known as "the Lijiang river looks like a green belt, and hills look like jade jewellerys". And the scheme is aimed to simulate a scenery of "reflection of the mountains and hills" by the arrangement for FPHCs installed on the sloping roof, and the combination of the

Figure 2. A view of the building for the first scheme.

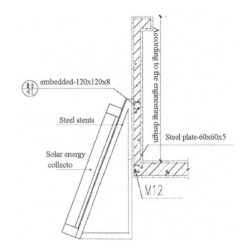

embedded-120x120x8

Steel stents

Solar energy collecto

According to the engineering design

Steel plate-60x60x5

M12

Figure 3. Details of the balcony's solar panels for the first scheme.

comparison of the REAL element "FPHCs" and the VIRTUAL element "Frame" has a suitable connection under the railing on the balcony, this design shows a light and elegant appearance of the involved building, but also has its functions such as sun-shedding, etc. for the daily living (see Fig. 3). Table 1 provides detailed configurations of the system.

Scheme 2 is a combination of FPHC and Tridimensional Virescene (see Fig. 4). FPHCs installed on the roof are well organized, and FPHCs installed on balconies also have the decorational function for the railing on the facade, the distance of cantilever between the balcony and the window shall be increased to have the additional function of flower beds, thus to introduce the Tridimensional Virescene system for the buildings, providing a combination of artificial technology and natural greening, showing a more harmonious environment for both systems (see Fig. 5). Table 2 indicates detailed configurations of the system.

Scheme 3 is a combination of the FPHC and glass railing (see Fig. 6). FPHCs installed on the sloped roof are well arranged in permutation and combination, the color of those installed FPHCs creates an image of traditional black tiles; and FPHCs installed on the facade have an combination of sloped glass railings and FPHCs, showing a picture of "the folds of mountains" to all people, also reflecting the impression of the beautiful natural scenery of Guilin as "East or West, Guilin landscape is best". Meanwhile, lights illuminate the building through those glass components and become a part of the decoration for the facade. The selection of FPHC should take a good efficiency and reasonable price of cost into considerations, and we suggest that normal size FPHCs available on the market could be procured, and the glass railings shall be used for the appearance of facade

Table 1. Detailed configurations of the system for Scheme 1.

Location	Details
Unit 1 (two households)	1~6 floors: FPHCs installed on the sloping roof. 7~9 floors: FPHCs installed on roof and balconies.
Unit 2 (two households)	1~7 floors: FPHCs installed on the sloping roof. 8~10 floors: FPHCs installed on roof and balconies.
Unit 3 (four households in a staircase)	1~7 floors of the South face and 1~11 floors of the North face: FPHCs installed on the sloping roof. 8~11 floors of the South face: FPHCs installed on roof and balconies.

Figure 4. A view of the building for second scheme.

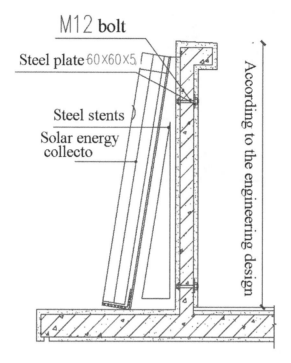

Figure 5. Structural details of the balcony's solar panels for the second scheme.

Table 2. Detailed configurations of the system for scheme 2.

Unit 1 (two households)	1~5 floors: FPHCs installed on the sloping roof. 6~9 floors: FPHCs installed on roof and balconies.
Unit 2 (two households)	1~5 floors: FPHCs installed on the sloping roof. 6~10 floors: FPHCs installed on roof and balconies.
Unit 3 (four households)	1~5 floors of the South face and 1~11 floors of the North face: FPHCs installed on the sloping roof. 6~11 floors of the south face: FPHCs installed on roof and balconies.

Figure 6. A view of the building for third scheme.

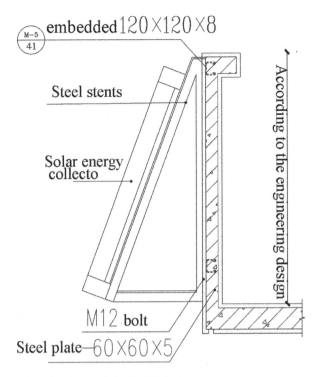

embedded 120×120×8

M-5 / 41

Steel stents

Solar energy collecto

M12 bolt

Steel plate 60×60×5

According to the engineering design

Figure 7.   Details of the balcony's solar panels for the third scheme.

Table 3.   Detailed configurations of the system for scheme 3.

Unit 1 (two households in a staircase)	1~4 floors: FPHCs installed on the sloping roof. 5~9 floors: FPHCs installed on roof and balconies.
Unit 2 (two households in a staircase)	1~4 floors: FPHCs installed on the sloping roof. 5~10 floors: FPHCs installed on roof and balconies.
Unit 3 (four households a staircase)	1~4 floors of the South face and 1~11 floors of the North face: FPHCs installed on the sloping roof. 5~11 floors of the South face: FPHCs installed on roof and balconies.

Table 4.   Comparison for three schemes of solar water heating systems.

	Scheme 1	Scheme 2	Scheme 3
Tank location	Jacket under the sloped roof	Jacket under the sloped roof	Jacket under the sloping roof
Location of FPHC	Centralized on roof; separated on roof and balconies for above 7F	Centralized on sloped roofs; separated on roofs and balconies for above 6F	Centralized on roof; separated on roof and balconies for above 5F
FPHC color	Dark blue	Dark blue	Dark blue
FPHCs on roof	FPHCs and frame in sequence distribution	FPHCs in two rows distribution	FPHCs in one row of crisscross pattern
FPHCs on facade	FPHCs and frame on balconies	FPHCs and virescene on balconies	FPHCs and glass railings on balconies
Functions of FPHCs	Sun-shading, decoration	Flower beds, decoration	Railing, decoration

(see Fig. 7). Table 3 shows detailed configurations of the system.

Different styles of buildings using FPHCs shall be reflected in these three schemes, and a detailed comparison for these three schemes is listed in Table 4.

5 CONCLUSIONS

This study has presented the integrated design of solar water heating systems in residential buildings for a landscape city like Guilin. The analyzed results described in the paper are summarized as follows:

1. Comprehensive survey and analysis on local weather conditions and surrounding environments shall be performed, and considerations shall be taken from the viewpoint of economy and aesthetics in order to select a proper solar water heating system and FPHC.
2. Installation of water tanks shall not interfere appearance of building facades, e.g. installed at the jacket of the sloped roof, the inner face of the balcony or a special position for the installation with the uniform design of façade.
3. The color of FPHCs shall be in accordance to the color of facades for the building, the installation of FPHCs shall have assurance of solar energy utilization performance, and the installation location could be roof, balcony, rain-shield, window, etc.
4. If the building roof is flat, then supporting frames could be used to simulate the "sloped roof" together with those installed FPHCs, which is also a well-known recommended approach for the system in the Guilin area. This option could enhance the performance of heat-resistance for the roof, as well as creating the "Scenery reflecting on the Fifth facade".

FPHCs installed on the facade shall be decorated with other elements, e.g. frames, glass railings, tridimentional virescene, etc. mentioned in the paper.

Integrated design of solar water heating systems in buildings is an essential part in the solar energy utilization for landscape cities, and also an approach for the integration of solar energy utilization into the general building design. Advanced technology and art shall be reflected by this design, while components of solar systems become a part of the building with appropriate combination for each other and decorate the building without the interference of traditional solar energy utilization, this design shall make great contributions to a more beautiful city with more energy saving.

ACKNOWLEDGEMENT

The corresponding author of this paper is He Jiang. The authors gratefully acknowledge the financial support from the natural science foundation of Guangxi.

REFERENCES

Deng X. et al. 2006. Simply discussion on the landscape city and architectural culture, *Shanxi Building*, (9):12–13.
Du, X.H. et al. *Exploration of integrated design of solar collecting roof of high-rise residential building. Industrial Construction. 42(2):5–9.*
Zheng, R. et al. 2006. The Technical Manual of Solar Water Heating System Engineering of the Civil Building. Beijing: Chemical Industry Press.
Zhou, M.L. 2010. The climate characteristics and its impact assessment of Guangxi in 2009. *Agriculture Newspaper of Guangxi.* (6):35–38.

Frontiers of Energy and Environmental Engineering – Sung, Kao & Chen (eds)
© *2013 Taylor & Francis Group, London, ISBN 978-0-415-66159-1*

The influence analysis of temperature and economy to household water consumption

Q. Liu, P. Zhou, J.X. Fu, W.R. Wang & X.H. Chen
Shenyang Jianzhu University, Shenyang, Liaoning, China

ABSTRACT: According to the data analysis. The passage choices canonical variable of water factors. By using the grey relational analysis method, combined with non-dimensional method, calculating the grey relational grades researches the influence degree of economic and temperature to household water consumption. The results show that the correlation of the household water consumption of district representative urban with July average temperatures is higher than with per capita GDP, also Chongqing and Shenyang household water consumption have higher correlation with July average temperatures and per capita GDP than other cities. This indicates that the temperature has a higher influence to water consumption than the economic, as well as the temperature and the economy influence Chongqing and Shenyang household water consumption greater than other cities respectively.

Keywords: household water consumption; grey relational analysis; influence degree of economic and temperature; per capita GDP

1 INTRODUCTION

Household water is refers to the water for the urban residents' daily life needs in households, including drinking, washing, flushing, bathing, etc.

Domestic and foreign scholars did a considerable amount of research about the influenced factors of household water consumption, the United States for each area research water needs in the early days, then extend to Europe. OECD study results show that the invention and the use of the water-saving instruments play a role for the water saving. Arbues has did some systematic research on water demand in terms of water price, variable selection and data collection. Martinez-Espineira researched the influence of climate, population and water price for household water consumption of northwestern Spain, and it turned out that climate is the important factor. Gao Huiyun discussed the impact of the various climate meteorological elements to water consumption in Shanghai during the summer, which showed that climate is the important factor bringing about changes of the water consumption during midsummer. Sun Yong and Xu Zuxin comprehensive analysis the various influence factors of urban water consumption forecast, and establish the weight scoring method to provide a guidance method for water forecast. Zhou Jingbo to 180 cities as the research object, by constructing a econometric model of urban life and the various factors affecting the water demands for preliminary analysis, found that with the increase of water consumption per capita GDP increased.

Based on former research results, author studies the gray related degrees for the household water consumption of district representative urban with economic and temperature by gray correlation method in order to provide a affirmance theory foundation for planning urban household water rationally.

2 GRAY CORRELATION METHOD INTRODUCTION

The correlation degree is the relevance measure between two system or two factors, which describes the relative changes of factors in the course of development of systems, such as size, speed and direction. If the relative change is consistent, the correlation is bigger, conversely it's smaller. Therefore, the grey relational grade analysis quantitatively describes and compares a system development situation.

The household water consumption and influence factors constitutes a gray system, because the influence factors are many and the relationship between them is indeterminacy, which this article uses the grey correlation degree analysis method to study it, because the calendar year relevant statistical data is limited.

3 ANALYSIS ON INFLUENCE OF ECONOMY AND TEMPERATURE ON HOUSEHOLD WATER CONSUMPTION IN DIFFERENT REGIONS

3.1 *The choice of factors index and representative city*

The relevant studies have shown that per capita GDP growth will make residents living environment comfortable, which lead to increase water demands of residents. Meanwhile The perfect family water instruments can make water consumption increased in a certain range. Some informations show that bath water consumption that is the major content in household water consumption will increase in summer. Figures 1 and 2 respectively is trend chart of water consumption with per capita GDP and July average temperature change. These data from 23 cities in 2004. The two pictures

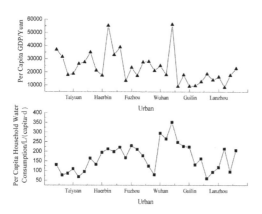

Figure 1. Trend chart of water consumption with per capita GDP change.

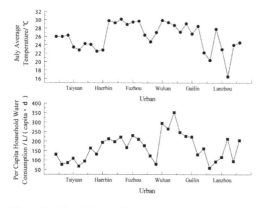

Figure 2. Trend chart of water consumption with July average temperature change.

indicate that there is some link between water consumption and two factors. So the author will do an in-depth study of economic and temperature on household water consumption by selecting the per capita GDP and July average temperature as their index respectively.

The author selected Shanghai, Guangzhou, Wuhan, Tianjin, Lanzhou, Chongqing, and Shenyang as representative city of East China, South China, Center China, North China, Northwest China, Southwest China and Northeast China respectively.

3.2 *Data base*

The article got per capita GDP data of representative cities from China's urban statistical yearbook, and got July average temperature data from China's meteorological yearbook and China's statistical yearbook from 1999 to 2008. The relevant data from China's city construction statistical yearbook is calculated to get per capita household water consumption.

3.3 *Analytic process*

Different indicators are different units, so there are no comparability among the different kinds of data. From the foregoing, the data is made into a comparable data column by mean dimensionless processing whose characteristic is that it has the only dimension of which value is greater than 0, but most close to 1.

When the distinguishing coefficient is 0.1, the relation degree index of mother and son sequence is calculated, which the mother sequence is per capita household water consumption, and the son sequence is factors date. The correlation coefficient reflects the close degree of two sequence compared in a time. Finally, the correlation degree was calculated by average correlation degree of two sequence each time.

The data was calculated in accordance with the above method to respectively obtain associate degrees July average temperature, and per capita GDP with per capita residential water consumption, shown in Table 1.

3.4 *Results analysis*

On the whole, the correlation of the average temperature and the residents consumption was greater than the per capita GDP and consumption of water in the representative city of each region in July. This shows the hottest temperatures for city residents consumption effect more than that economic effect on it.

Table 1. The link between characteristic factors and residents water consumption.

Region	Representative city	Correlation degree	
		July average temperature	Per capita GDP
East China	Shanghai	0.37242	0.19959
South China	Guangzhou	0.42922	0.22875
Center China	Wuhan	0.47452	0.2138
North China	Tianjin	0.53793	0.18638
Northwest China	Lanzhou	0.41378	0.18021
Southwest China	Chongqing	0.62825	0.33504
Northeast China	Shenyang	0.62367	0.25147

The correlation between the consumption of water and the average July temperature int he representative city of different areas is not the same degree. Chongqing and Shenyang are the most, Respectively is 0.62825 and 0.62367. Shanghai is 0.37242 which is lowest in all.

The four seasons division obvious in Chongqing, summer heat, people want to have a comfortable summer, the consumption of water will be great. And last for a long time, it may be the cause of most of time the water and temperature change trend more closely.

Shenyang is located in the northeast area, the average annual temperature of 8°C, cool climate, therefore, lead to residents of the temperature of a high degree of sensitivity, temperature increase leads to increased consumption of water. For example, in 2004 August, the temperature of Shenyang rises sharply, leading to the city's average daily water supply volume than the same period of more than 13000 m³, 2008 August 8th water volume achieves 1320000 m³, achieve the water supply peak since this year into the summer.

The climate of shanghai is monsoon climate, spring and autumn is longer than winter and summer. It may be lead to the correlation between water supply and the temperature of July is not great. Together with the Shanghai more developed, residents' water saving consciousness is strong, water content and temperature of covariation trend is not very good.

As can be seen from Table 1, the water consumption of Chongqing and Shenyang has a great relation with per capita GDP, respectively is 0.33504 and 0.25147. And Tianjing and Lanzhou has a little relation with per capita GDP, respectively is 0.18638 and 0.18021. It is not difficult to find the water consumption of Chongqing and

Shenyang also has a great correlation with average July temperature, possible causes, the two city is in general developed city over to a more developed city in the process. Household water facilities continue to improve, therefore, the correlation between consumption and per capita GDP is higher than other city. But compared with the temperature, GDP of average per capita has little effect on it.

Tianjin is a developed city, there is a very small proportion that residential with obsolete Water supply equipment in total residential volume. At the same time it is the early national water-saving city, residents' water saving consciousness is strong, may be lead to a little correlation with economic. Lanzhou is located in the northwest, water resource shortage, economic development lags slightly behind, water supply facilities are not perfect. Economic development does not quickly pull up the household water facilities, this could be the result of correlation degree is smaller.

4 CONCLUSION

In general, the influence of temperature on water consumption of city resident is greater than the influence of economic on it. Indicates that temperature is the major factor.

Temperature and economy effect on resident water consumption in Chongqing and Shenyang is more than other representative city.

The influence of economic on resident water consumption is smaller than other representative city.

ACKNOWLEDGEMENT

This work was financially supported by Water pollution controlling and management in the national science and technology major special project (2009ZX07318-008-007-001).

REFERENCES

Arbues, F. & Garcia-Valinas, M.A. 2002. Estimation of residential water demand: a state-of-the-art review. *Journal of Sociol Economics* 32(1): 81–102.

Cao, Hongren & Zheng, Yaowen 1988. *The Narrative of Grey System Theory*. Beijing: China Meteorological Press.

Cong, junRao & Jin, Peng 2009. Fuzzy group decision making model based on credibility theory and gray relative degree. *International Journal of Information Technology & Decision Making* 8(3): 515–527.

Cui, Huishan & Deng, Yiqun 2009. Factors influencing residential water consumption. *Water Resources Protection* 25(1): 83–85.

Chen, Xiaoguang & Xu, Jintao 2005. Research on determinant factors in residential water demand. *Water Economy* 23(6): 23–71.

Deng, Julong 2002. *Gray forecast and decision making*. Wuhan: Press of Huazhong University of Science and Technology.

Gao, Huiyun & Yao, Zhizhan 2001. The analysis of the mid-summer influencing Shanghai urban water and establishing the models of water consumption assessments. T*he Paper Collection of Urban Meteorological Service Symposium*.

Gan, Hongbin 2002. Simple discuss on using water in city and water-saving. *Shanxi Architecture* 28(4): 4–7.

Hu, Feng 2006. Analysis of influence factors on residential water needs in Nantong City. Hangzhou: Zhejiang University.

Liu, Jianlin & Huang, Tinglin 2007. Relational analysis of user group and city water consumption. *Yellow River* 29(2): 56–57.

Martinez-Espineira, R. 2002. Residential water demand in Northwest of Spain. *Environmental and Resource Economics* 21(2): 161–187.

Organization for Economic Cooperation and Development 2001. *Household energy &water consumption and waste generation*. Paris: OECD.

Schefter, J.E. & David, E.L. 1985. Relational analysis of influence factors on residential water consumption in Jinhua City. *Land Economics* 61(3): 272–280.

Sun, Yong & Xu, Zuxin 2008. Impact factors classification Urban water consumption forecast and priority weight analysis for forecasting target. *Water Supply & Sewage* 34: 114–117.

Wang, Yali & Feng, Lihua 2011. Relational analysis of influence factors on residential water consumption in Jinhua City. Journal of Water Resources & Water Engineering 22(3): 51–54.

Zhou, Jingbo 2005. Analysis impact factors on China residential water. *Statistics & Decision Making* (6): 75–76.

Frontiers of Energy and Environmental Engineering – Sung, Kao & Chen (eds)
© *2013 Taylor & Francis Group, London, ISBN 978-0-415-66159-1*

Forecast of the scaling tendency of gas recovery wellbore

Z.Y. Qiu
Chongqing University of Science and Technology, Chongqing, China

H.Y. Wang
Chongqing Gas Group Corporation Ltd., Chongqing, China

J. Meng & Y. Tian
Chongqing University of Science and Technology, Chongqing, China

ABSTRACT: The forecasting models of the scaling tendency and the maximum amount of scale buildup ($CaCO_3$) of gas-recovery wellbore of Southern Sichuan Gas Field were established; the analog calculation was carried out by means of the compilation of computer program, resulting in the positions appearing the scaling tendency and the maximum amount of scale build up of research well, so as to provide the basis for the targeted scaling on the production site.

Keywords: gas well; scaling; tendency; forecast

For the water-cut gas-recovery well, with the fluid flowing to the wellhead and the temperature & pressure drops, it was easy to form the scale in the wellbore. The scale formation had a significant adverse effect on the normal production of gas-recovery well and the wellbore corrosion. To date, there has been the more mature scaling forecasting methodology for the oil well, rather than the scientific forecasting theory for the gas well. Based on the working conditions of typical gas wells (i.e. Auxiliary 1# Well, H1# Well and Tong 18# Well) of Southern Sichuan Gas Field of CNPC Western Oil Company, the scaling prediction of gas well was explored in this paper.

1 DETERMINATION OF THE FORECASTING MODEL

The main component of scale formed by the gas-well produced water in Southern Sichuan Gas Field should be $CaCO_3$. The scaling of gas wellbore was forecasted by means of the common forecasting ($CaCO_3$) scaling model of oil well.

1.1 *Forecasting model of the scaling tendency*

Firstly, the forecasting model of scaling tendency of liquid single-phase system should be used; for the calculation formula on the scaling forecasting parameter SI (i.e. the saturation index), see the Formula (1):

$$SI = \log(T_{Ca} \times Alk) + pH - 2.78$$
$$+ 1.143 \times 10^{-2}T - 4.72 \times 10^{-6}T^2$$
$$- 4.37 \times 10^{-5}P - 2.05\mu^{1/2} + 0.727\mu \quad (1)$$

where: T_{Ca}—Ca^{2+} concentration, mol/L; Alk—HCO_3^- concentration, mol/L; P—total pressure, Pa; T—formation temperature, °F; μ—ionic strength, mol/L.

Secondly, due to the water-gas two-phase flow gas-well, the scaling tendency should be also forecasted as per the scaling two-phase flow formula; for the formula, see the Formula (2):

$$SI = \log\left(\frac{T_{Ca^{2-}} Alk^2}{P X_{CO_3^{2-}}}\right) + 5.89 + 1.549 \times 10^{-2}T$$
$$- 4.26 \times 10^{-6}T^2 - 7.44 \times 10^{-5}P$$
$$- 2.52\mu^{1/2} + 0.919\mu \quad (2)$$

where: T_{ca}—Ca^{2+} concentration, mol/L; Alk—HCO_3^- concentration, mol/L; P—total pressure, Pa; T—formation temperature, °F; X_{CO_2}—mole fraction of CO_2 in the gas phase; μ—ionic strength, mol/L.

The saturation index SI should represent the possibility of precipitation of $CaCO_3$, $BaSO_4$, $SrSO_4$ and $CaSO_4$, etc. in the solution. Based on the scale of *SI*, the possibility of precipitation could be forecasted, rather than the number of scaling. If *SI* < 0, the solution should not be saturated, not resulting in the scaling; if *SI* = 0, the solution should be saturated, resulting in the equilibrium state; if the $SI > 0$, the solution should be over-saturated, resulting in the scaling.

1.2 Forecasting model of the maximum amount of scale buildup

For the forecast of the maximum amount of scale buildup ($CaCO_3$), the maximum amount of scale buildup ($CaCO_3$) should be forecasted as per the prediction formula proposed by Valone and Skillern. For the calculation formula, see the Formula (3):

$$PTB = 17500 \times [C - (X^2 + 4 \times 10^{K-pH})^{0.5}] \qquad (3)$$

where: PTB—the maximum amount of scale buildup ($CaCO_3$), no dimension; C, X—ionic concentration, $C = CO_3^{2-} + HCO_3^-$; $X = CO_3^{2-} - HCO_3^-$, mol/L; pH—pH value, $pH = pH_{ground} + \Delta pH$.

$$\Delta pH = (4150 \times 10^{-3} \Delta T)$$
$$+ [4185 \times 10^{-7} \times (T_{d2} - T_{s2})]$$
$$- 3107 \times 10^{-5} \Delta p$$

$$\Delta p = p_d - p_s$$

where: p_d—subsurface pressure, MPa; p_s—surface pressure, MPa.

$$\Delta T = T_d - T_s$$

where: T_d—subsurface temperature, °C; T_s—surface temperature, °C.

Forecasting judgment standards for the maximum amount of scale buildup ($CaCO_3$): If $PTB < 0$, not resulting in the scaling; if $0 < PTB < 100$, resulting in the less scaling; if $100 < PTB < 250$, resulting in the more and hard scaling; if $PTB > 250$, resulting in the extremely-serious scaling.

1.3 Determination of the forecasting model

In order to facilitate the study on the scaling rule of water-cut gas-well wellbore, the following assumptions would be made: The wellbore temperature should be consistent with the linear changes; there should be the same temperature, pressure, pH value and ionic concentration of the fluid with the same depth in the wellbore; the water should be characterized by the fluid.

As shown in Table 1, there would be no any high pressures of several studied gas wells; there would be no any large difference between the wellhead pressure and the bottom-hole pressure. The production site data showed that, due to the less CO_2 content in the produced gas, the higher mineralization of produced water and the larger $Ca^2 +$ and HCO_3^- concentration, the pressure would less impact on the amount of scale buildup. Since the formation reaction of $CaCO_3$ was the endothermic reaction, the impact of temperature on the scaling should be mainly analyzed. By means of the two-phase flow scaling forecasting model and the single-phase flow scaling forecasting model (only considering the effects of temperature and pH value changes, and ignoring the effects of pressures), the analysis of scaling tendency should be carried out for such wells, so as to compare with the accuracy and consistency of forecasting results of both models, to determine which model used by the studied gas-well should be more appropriate, and to determine its reliability thereof.

The parameters for the Auxiliary 1# Well, H1# Well and Tong 18# Well should be shown in Table 1. By means of the compilation of computer program, it was necessary to calculate the scaling tendency and the maximum amount of scale buildup of various wells with the different depths and temperatures.

Figure 1: The curve diagram of the SI value of Auxiliary 1# Well with the temperature change as achieved by means of both scaling tendency whether forecasting models. From the SI curve analysis, the

Table 1. The table of parameters for the typical wells.

Parameters	Auxiliary 1# well	H1# well	Tong 18# well
ID, mm	ID of tubing 53 ID of casing 110	ID of tubing 78 ID of casing 161	ID of tubing 78 ID of casing 161
Well depth, m	1366.85	3570	2760
Relative density of natural gas as measured	0.587	0.591	0.586
Wellhead pressure, MPa	Tubing pressure 0.76 Casing pressure 0.82	Tubing pressure 2.92 Casing pressure 13.7	Tubing pressure 1.55 Casing pressure 7.3
Wellhead temperature, °C	40	90	60
Gas production rate, m³/d	16000	43900	4320
Ground fluid production rate, m³/d	0.1	470	180

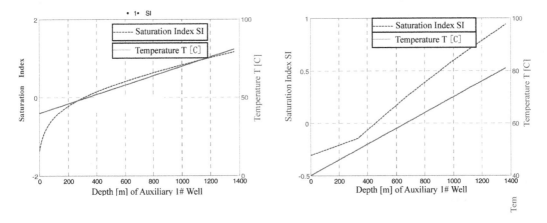

Figure 1. Forecast of the scaling tendency of auxiliary 1# well. a) scaling forecasting curve of the two-phase forecasting model of auxiliary 1# well. b) scaling forecasting curve of the single-phase forecasting model of auxiliary 1# well.

two-phase flow scaling forecasting model or the single-phase flow scaling forecasting model, there would be the consistent forecasting results. After the scaling tendencies of H1# Well and Tong 18# Well were analyzed by means of both models, there would be the consistent forecasting results as well. Therefore, the scaling tendency might be forecasted by means of the single-phase flow scaling forecasting model.

1.4 Solving of the forecasting model

Firstly, a fixed step distance of calculation should be taken; based on the parameters for average daily water production rate and well structure, etc., the wellbore temperature, fluid pressure and pH value

should be calculated; then, the changes of mass concentration of scale ions should start to be considered from the shaft bottom; the saturation index SI and the precipitation volume of $CaCO_3$ scaling at each section in the wellbore should be calculated.

2 ANALYSIS OF THE FORCASTING RESULTS

Figure 2: The curve diagram of the maximum amount of scale buildup of Auxiliary 1# Well with the temperature and well-depth changes. As shown in Figure 2a, the scaling would be started at the temperature of 55 °C; the SI value would almost

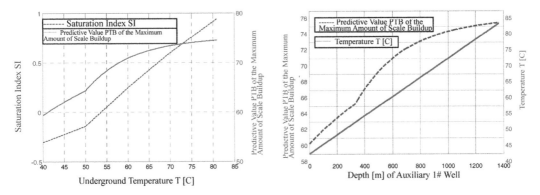

Figure 2. Forecast of the maximum amount of scale buildup of auxiliary 1# well. a) Curve graph of the relation between the scaling forecasting index, the maximum amount of scale buildup and the temperature of auxiliary 1# well. b) Curve graph of the relation between the maximum amount of scale buildup and the depth of auxiliary 1# well.

show the linear increase at the temperature of 55 °C or above; the *SI* value would be up to about 1 at the temperature of 80 °C or so; therefore, from the *SI* value, there would be the weaker scaling tendency of Auxiliary 1# Well; from the curve analysis of the maximum amount of scale build up, the maximum amount of scale build up would be stabilized at the temperature of 70 °C, while the whole *PTB* value would be 55–76, namely, the slight scaling arrange. Therefore, from the analysis of two curves, there would be the slight scaling of Auxiliary 1# Well wellbore, which should be consistent with the on-site investigated basic non-scaling phase of the said well. From the curves of the maximum amount of scale buildup and the well depth as shown in Figure 2b, in spite of the slight scaling of Auxiliary 1# Well wellbore as a whole, the amount of scale buildup would be increased rapidly at the well depth of 370 m or above, until it was stabilized at the well depth of 1200 m or so. Therefore, for the Auxiliary 1# Well, the main part of scaling should be located at 1200 m of shaft bottom, i.e. at 500 m from the shaft Bottom.

From the same analysis of H1# Well and Tong 18# Well, it indicated that, the scaling tendency of H1# Well would be very serious, and the amount of scale buildup would appear at the well depth of 3000 m or so thereof; the scaling tendency of Tong 18# Well would be weaker, and the amount of scale buildup would appear at the well depth of 2500 m or so thereof.

3 CONCLUSIONS

1. For the typical wells of Southern Sichuan Gas Field, under certain conditions, the scaling tendency might be forecasted by means of the single-phase flow scaling forecasting model, resulting in the more accurate results.
2. Through the analysis of scaling tendency of typical wells, it was necessary for the gas wells of Southern Sichuan Gas Field to appear the scaling phenomenon, except for the different scaling severity of each well.
3. From the forecast of the whole scaling position and the analysis of the amount of scale buildup, the maximum amount of scale buildup of wellbore would be located at 500 m from the shaft bottom, which should not only be consistent with the structure studied in the Reference [1], but also the on-site investigated condition.

REFERENCE

[1] Gao Bo, Wang Yong, Li Bing, et al., *Forecast of the Scaling Tendency of Tianwaitian Gas Field* [J], Natural Gas Exploration and Development, 2009, 32 (1): 71–73.

Frontiers of Energy and Environmental Engineering – Sung, Kao & Chen (eds)
© *2013 Taylor & Francis Group, London, ISBN 978-0-415-66159-1*

Study on current and future development of wind power grid integration in China

J.F. Mi
Guodian Energy Research Institute, Beijing, China

J. Sun
Guodian Power Development Co., Ltd., Beijing, China

W. Luo
North China Electric Power Research Institute Co., Ltd., Beijing, China

ABSTRACT: The wind power in China has been going through a period of rapid development in recent years. In this paper, the main data of wind power installed capacity, grid-connected capacity, and the generation during 2008 to 2011 has been collected and researched to analyze the proportion of wind power grid integration in top 10 provinces in China and the situation of wind power curtailment in 2011. Furthermore, difficulties in wind power grid integration have been concluded based on perspectives of technical, political and economical factors. In the end, some recommendations have been proposed in the issue of future wind power grid integration.

Keywords: wind power grid integration; wind power curtailment; wind power accommodation; peak regulation; wind power transmission projects

1 INTRODUCTION

China's wind power installed capacity has almost doubled in each year from 2006 to 2010. However, accompanied with its rapid development, several problems have been clearly exposing in wind power grid integration and electricity generation. In recent years, the proportion of wind power integration has remained about 70%, even some installed wind turbines cannot be allowed to generate effectively, and result in significant economic loss. The installed capacity and grid integration of wind power in recent years is researched comparatively in this paper. The construction of wind power with its deliveries and utilization in China is analyzed. In the last, the wind-power development in the future in China is also suggested in this paper.

2 CURRENT SITUATION OF WIND POWER INSTALLED CAPACITY IN CHINA

In the year of 2011, the growth of wind power installed capacity in China has reached 39.4%,

ranking first in the world. By the end of 2011, the total wind turbines have been 45894 with the increment of 11409, and the installed capacity has reached 62364 MW with the increment of 17631 MW during the year. Table 1 shows the top 10 provinces by wind power installed capacity in China from 2008 to 2011.

As is shown in Table 1,

1. The wind power installed capacity in each province above has exceeded 2000 MW in the year of 2011. Especially in Inner Mongolia, the wind power installed capacity has reached 17594 MW, which accounts for 28.2% of China's total wind power installed capacity.
2. The wind power installed capacity in top 10 provinces has almost doubled in each year from 2008 to 2011, which has multiplied 5 times in four years, from 10902 MW in 2008 to 53958 MW in 2011.
3. The sum of top 10 provinces' wind power installed capacity accounts for 86.5% of that in China. Study on the current situation and problems of wind power installed capacity in these ten provinces is instructively significant to improve the wind power integration in China.

Table 1. Top 10 provinces by wind power installed capacity in China from 2008 to 2011.

No.	Province	Installed capacity (MW)				Percentage of installed capacity in China in 2011 (%)
		2008	2009	2010	2011	
1	Inner Mongolia	3762	9203	13859	17594	28.2
2	Hebei	1104	2791	4792	6971	11.2
3	Gansu	610	1188	4936	5408	8.6
4	Liaoning	1188	2437	4073	5251	8.4
5	Shandong	582	1222	2635	4564	7.3
6	Jilin	1157	2061	2944	3557	5.7
7	Heilongjiang	829	1662	2372	3448	5.5
8	Ningxia	391	681	1181	2872	4.6
9	Xinjiang	631	998	1361	2318	3.7
10	Jiangsu	648	1097	1597	1975	3.1
Total		10902	23340	39750	53958	86.5

3 CURRENT SITUATION OF WIND POWER GRID INTEGRATION IN CHINA

3.1 *The condition of the wind power grid-connected capacity and generation in top 10 provinces in China from 2008 to 2011*

In 2011, the total wind power grid-connected capacity has reached 45050 MW in China with the new built increment of 15850 MW, and the generation has been 73200 GWh. Table 2 shows the completion of the wind power grid-connected capacity and generation in top provinces in China.

As is shown in Tables 1 and 2,

1. In 2011, the wind power installed capacity in top 10 provinces accounted for 86.5% of the total amount, and was 1.5 percentage points higher than the average. The wind power generation in these provinces was 64400 GWh, accounting for 88.0% of the total wind power generation amount in China.
2. In 2011, the proportion of wind power grid integration in top 10 provinces was 73%, and slightly higher than country's 72%.
3. Due to the characteristics of the wind turbine generating which are instability, stochasticity and intermittence, to improve the proportion of wind power grid integration will promote the wind power generation, enhance the efficiency, and cut down the cost.

3.2 *The condition of wind power grid integration in top 10 provinces in China from 2008 to 2011*

As is shown in Figure 1,

1. The proportion of wind power grid integration in China has been increasing stably, which is 63%,

Table 2. 2008–2011 total wind power grid integration and generation in China.

No.	Province	Grid-connected capacity (MW)			
		2008	2009	2010	2011
1	Inner Mongolia	2300	5030	10000	13640
2	Hebei	700	1210	3720	4320
3	Gansu	600	750	2190	5460
4	Liaoning	850	1630	3080	4020
5	Heilongjiang	620	1400	1910	2550
6	Shandong	370	700	1580	2420
7	Jilin	760	1410	2210	2850
8	Xinjiang	510	810	1360	1880
9	Jiangsu	610	990	1370	1580
10	Ningxia	170	540	720	1120
Total		7490	14470	28140	39840
Proportion of wind power grid integration		68	62	70	73

		Wind power generation (GWh)			
		2008	2009	2010	2011
1	Inner Mongolia	3740	8700	17500	22700
2	Hebei	1360	2500	5700	8800
3	Gansu	630	1200	2100	7100
4	Liaoning	1080	2700	4700	6600
5	Heilongjiang	1090	2000	3300	4400
6	Shandong	540	1100	2700	4100
7	Jilin	1370	2300	3300	4000
8	Xinjiang	780	1500	2300	2900
9	Jiangsu	770	1400	2300	2700
10	Ningxia	240	700	1200	1100
Total		11600	24100	45100	64400

69%, 72% during 2009–2011. The similar increasing situation also represented in two provinces Inner Mongolia and Jilin, especially in Inner Mongolia of which installed capacity accounts for 28.2% in China. Its proportion of wind power grid integration has increased rapidly from 55% which lower than the average level in 2009 to 78% in 2011 which becomes higher than the average.

2. The proportion of wind power grid integration remains steadily about higher than 80% in Xinjiang and Jiangsu provinces.
3. The proportion of wind power grid integration in Heilongjiang province kept above the average level in China, but reduced 10 percentage points from 84% in 2009 to 74% in 2011. The cause needs to be considered.

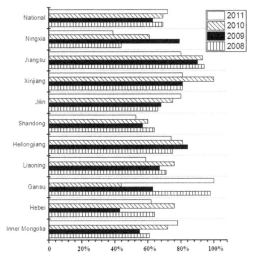

Figure 1. The condition of wind power grid integration in top 10 provinces in China during 2008 to 2011.

4. The proportion of wind power grid integration in Gansu province of which installed capacity accounted for 8.6% in 2011, enhanced from 44% in 2010 to 100% in 2011. Its growth experience should be researched by other provinces.
5. The proportion of wind power grid integration in some provinces such as Hebei, Shandong and Ningxia, kept lower than average level. Meanwhile, the proportion in Liaoning province showed unstable in recent years. This situation should be noticed.

3.3 Wind power curtailment in China in 2011

According to the research, Inner Mongolia, Gansu and other provinces as wind power base, showed evident curtailment situation. In "A primary statistics of wind power curtailment in China in 2011", the wind power curtailment has been calculated, and the sample coverage reached 18691 MW of installed capacity. The final statistical results are shown in Table 3.

As is shown in Table 3,

1. The loss of wind power generation due to the curtailment has been up to 5980 GWh in the total sample coverage in 2011. According the data of CEC, the total wind power grid-connected capacity in China has been 45050 MW, and based on this number we can estimate that more than 10000 GWh of wind power generation would be lost in China in 2011.
2. The most loss of wind power generation due to the curtailment happened in Gansu, Inner Mongolia, Jilin and Heilongjiang provinces, which have the higher proportion of wind power grid integration than average level in China.

Table 3. The loss of wind power curtailment in some provinces.

No.	Province	Capacity surveryed (MW)	Generation without loss (GWh)	Loss of generation (GWh)	Percentage of generation loss (%)
1	Gansu	2103	4333	1094	25.25
2	Inner Mongolia	7647	14225	3287	23.10
3	Jilin	1187	2260	475	21.02
4	Heilongjiang	1551	3117	448	14.39
5	Liaoning	2181	3732	390	10.45
6	Xinjiang	597	1339	70	5.20
7	Yunnan	325	490	24	4.90
8	Hebei	2370	4614	178	3.86
9	Shandong	601	871	10	1.17
10	Guangdong	128	366	4	1.00
Total		18691	35347	5980	16.92

4 THE MAIN PROBLEMS IN WIND POWER GRID INTEGRATION

Many factors lead to the difficulties in wind power grid integration, including technical factors, Political factors and economical factors.

4.1 Technical factors

At generation side, wind power output is instable, intermittent and stochastic. However, electricity grid prefers to accommodate more controllable wind power, which has anti-trouble ability, can be able to adapt system disturbance and meet the demand. In this respect, wind farm should improve the wind power's controllability, response and anti-interference ability to meet the grid's demand.

At demand side, firstly the area which is abundant in wind power resource is far away from the load center in China, and its load demand cannot accommodate the generation of local grid-connected wind power. Secondly, grid auxiliary facilities cannot absolutely adapt to wind power interconnection in some degree. Transmission bottlenecks in the electricity grid restrict the wind power to transmit in long-distance. Finally, peak regulation capacity of grid in some area not enough, so that some wind turbines need participate in peak regulation in certain period which seems impossible to comply.

4.2 Political factors

First of all, wind power development lacks comprehensive planning program. When a local government makes a plan to develop the wind power, the local development scale and construction sequence does not comply with the central government's approved scale, but follows local wind energy resources, which would bring much more wind power development than the central government's overall plan.

Secondly, the approval and constructing period of grid project is longer than that of wind farm project. It is hard to achieve the coordinate development between wind farm projects and grid auxiliary facilities for interconnection.

Finally, more and more sufficient demonstration project should be made in wind power planning in order to avoid or decrease the bottlenecks in the large-scale wind power development.

4.3 Economical factors

Nowadays, the practical on-grid electricity price of wind power is higher than that of thermal power. The rapidly development of wind power, to a large extent, attributes to the subsidy from government and grid company which leads to the lack of initiatives in wind power grid integration of grid company. Furthermore, grid dispatch agency sometimes requires that wind farm to participate in certain peak regulation, which would reduce the wind power generation equipment utilization hours and harm the wind power enterprises' benefit. In a word, rational wind power pricing mechanism considering each market entity's benefit must be established.

5 CONCLUSION

First of all, the renewable energy planning, such as wind power and photovoltaic power, should keep coordinate with conventional energy planning as well as electricity grid planning. Meanwhile, regional interconnection should be strengthened and accommodation scale for wind power should be expanded. The wind power development strategy must observe the principles of according to local conditions, overall consideration, rational distribution as well as orderly development, so that blind development and unnecessary waste or loss could be avoided.

Secondly, generation structure and distribution should be optimized and conventional power source should be rational allocated. Rationally planning and constructing of peak-shaving power source such as pumped storage power station and diesel generator should be carefully considered, in order to solve the peak-load regulating problems caused by large-scale wind power grid integration.

Thirdly, it is very important to make rational arrangements of the sequence of wind power development and construction. It can be very helpful to eliminate disordered development of wind power project, and implement coordinated and orderly development between wind power projects and grid auxiliary projects for wind power interconnection.

At last, to strengthen the research on wind power grid integration is significant issue in the future, especially the technology related to the Chinese mode of wind power development. Such technologies should include wind-hydro or wind-photovoltaic complementation, deep peak-regulation, wind power forecasting, new transmission technology for wind power delivery. And furthermore, with the consideration of the benefit of wind power. the technology of grid dispatching and power system controlling should also be researched for the whole energy system's security and stability.

REFERENCES

China Electricity Council. 2009. Statistical bulletin of national electric power industry in 2008.

China Electricity Council. 2010. Statistical bulletin of national electric power industry in 2009.

China Electricity Council. 2011. Statistical bulletin of national electric power industry in 2010.

China Electricity Council. 2012. Statistical bulletin of national electric power industry in 2011.

China Wind Energy Association. 2009. A statistics of wind power installed capacity in China in 2008. *Wind Energy* 2009(01): 25–29.

China Wind Energy Association. 2010. A statistics of wind power installed capacity in China in 2009. *Wind Energy* 2010(01): 28–33.

China Wind Energy Association. 2011. A statistics of wind power installed capacity in China in 2010. *Wind Energy* 2011(03): 34–40.

China Wind Energy Association. 2012. A statistics of wind power installed capacity in China in 2011. *Wind Energy* 2012(03): 40–48.

China Wind Energy Association. 2012. A primary statistics of wind power curtailment in China in 2011. *Wind Energy* 2012(04): 38–39.

Ming YIN, Chengshan WANG & Xubo Ge. 2011. Comparison and analysis of wind power development between China and Germany. *Transactions of China Electrotechnical Society* 25(9): 157–162, 182.

Xiaoyou JIAO, Jianfeng MI & Jiao Sun. 2010. The trend new energy development research: wind power. *The Awarding Achievement Collection of Important Project Research of Guodian Corporation in 2010*: 56–74.

Zheyi PEI, Cun DONG & Yaozhong XIN. 2010. Review of operation andmanagement of integrating wind power in China. *Electric Power* 43(11): 78–82.

Frontiers of Energy and Environmental Engineering – Sung, Kao & Chen (eds)
© *2013 Taylor & Francis Group, London, ISBN 978-0-415-66159-1*

Measurement of gas-solid flow in a square cyclone

Y.X. Su
School of Environment Science and Engineering, Donghua University, Shanghai, China

B.T. Zhao
School of Power Engineering, Shanghai University of Science and Technology, Shanghai, China

ABSTRACT: A Three-Dimensional Particle Dynamic Analyzer (3D-PDA) was used to measure the gas-solid flow in a lab-scale square cyclone separator. Glass bead was used as the particle tracer. Results showed that the flow field in the square cyclone included forced vortex of strongly swirling flow in the center region and quasi-free vortex near the wall. The suspension carried out a downward swirling flow near the wall and upward swirling flow between the bottom of the vortex finder and the top of the exit, which was similar to the dual helix flow pattern in a traditional cyclone. Local small vortex was formed at the corners. The impacts between particles or particles-to-wall at the corners enhanced the quasi-laminar fluctuating motion, leading to the local peak values of the turbulent kinetic energy and turbulent intensity. The corners were one of the major regions that caused the pressure drop. Re-circulation flow was found in the central region. The re-circulation between the bottom of the vortex finder and the top of the exit reduced the chance of particles to escape with the gas flow to the open air.

Keywords: gas-solid flow; square cyclone separator; PDA measurement

1 INTRODUCTION

Gas-solid suspension flow is a fundamental problem related to many industrial fields, especially coal-fired boilers, chemical engineering, metallurgy engineering, etc. Due to the merits of geometrical simplicity, low operation cost and high separation efficiency, cyclone separator is generally adopted as the first choice for particle separation from suspension flow. The cyclone is usually designed to have a cylindrical body with tangential inlet. Details can be referred to the monograph compiled by Hoffmann and Stein[1] and a recent review by Cortes and Gil[2]. Many authors have given their efforts to the research on traditional cyclones by different methods[3–12]. Darling[13] reported a square cyclone separator that was first designed by Ahlstrom Pyropower Company and applied to compact CFB boiler design. The square cyclone separator is formed from water-cooled tube panels with a thin refractory lining and is connected directly to the furnace walls. The lower radiation loss provided by the external insulation and lagging increases boiler efficiency and reduces fuel costs, resulting in cost effective operation and maintenance[13]. Shibagaki and Nishiyama[14], Makkonen[15] and Lu et al[16] reported the application of such a separator to commercial CFB boilers.

Since the flow pattern plays a very important role in understanding its operating principle and separation performance, many authors[17–21] tried to measure the flow field inside cyclones by different methods including Pitot tubes, hot-wire anemometers, as well as the non-intrusive techniques such as Laser Doppler Anemometry (LDA), Particle Dynamics Analyzer (PDA) and Particle Image Velocimetry (PIV). In order to understand the nature and characteristics of the suspension flow in the square cyclone separator, the authors employed a Three Dimensional Particle Dynamic Analyzer (3D-PDA) to measure the suspension flow in a lab-scale square cyclone separator.

2 EXPERIMENTAL DETAILS

The experiment setup is showed in Figure 1. Particles are fed into the riser by a screw feeder which was controlled by an electromagnetic engine. The suspension goes through the riser and a horizontal duct and enters into the separator. The inlet of the separator is 20×60 mm. The separator is of square cross-section and its edge length is 120 mm, body height 180 mm. The diameter of the vortex finder and the exhaust exit is both 60 mm. The height of the vortex finder is 90 mm. The distance between the bottom of the vortex finder and the top

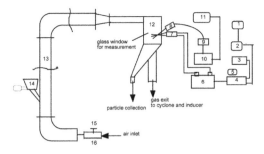

Figure 1. Experimental setup.
1: transformer 2: power source 3: water cleaner 4: laser 5: laser controller 6: beam splitter 7: laser transmitter 8: receiver 9: A/D con-verter 10: signal processors 11: computer 12: square separator 13: CFB riser 14: particle feeder 15: air control valve 16: flow meter.

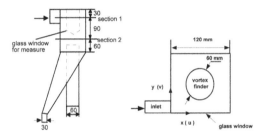

Figure 2. a) measure position; b) cross section.

of the exhaust is 60 mm. The separated particles are collected by a discharge bin. The exhaust gas goes through the downward exit to open air by an induced blower. The front side of the separator is glass window for PDA measurement. Figure 2(a) and (b) show the separator model and the measure sections.

The particle used was glass beads of mean diameter 30~40 μm and of density 2400 kg/m³. The glass bead has good physical properties, e.g., sphericity around 0.95 and refraction index 1.5. Particles of diameter 0~5 μm were selected as gas tracer. The optical arrangement of the 3D-PDA is a backward-scattered-light system supplied by Dentec, including a 5 W (max. power) Argon-Ion laser source, laser transmitting and receiving systems, signal processor and a computer. A frequency shift of 40 MHz is added to the green and blue beam of the laser. The measure point coordinate is automatically controlled by the transverse system of the PDA following the given coordinate before the running. The PDA measures the velocity of a certain number of particles and calculates the statistically-averaged value as the result. Hence the more particles measured, the better the result is. At each measure point, 5000 particles are enough

for the present study and the time limit is one minute.

3 RESULTS

The experiment was carried out for several cases at different inlet velocity and particle concentration and this paper reports two cases listed in Table 1. The coordinate origin was set at the front-left corner of the separator, x-coordinate (also U velocity) parallels to the glass plane and the positive direction is from left to right. *Y-coordinate* (also V velocity) is perpendicular to the glass plane and z-coordinate (also W velocity) is upward, as shown by Figure 2(b).

3.1 Flow field

Figure 3(a) presents the flow pattern at section 1. It can be seen that the swirling flow field was not fully developed at section 1 and the swirling intensity is still low.

Figure 3(b) shows that swirling flow is formed at section 2. Local vortex exists at the corners, e.g., $x = 0$, $y = 0$ and $x = 120$ mm, $y = 0$, where the particle motion is very disordered. The swirling flow in the separator is mostly strong swirling turbulent flow. Local vortex exists at the corners, where the quasi-laminar flow happened. The quasi-laminar flow includes two types of fluctuating motion; one is the irregular motion of particles due to the inter-impact between particles or particle and wall surfaces as well as the fluid fluctuating motion induced by the particle fluctuating motion. This kind of fluctuating flow is almost random and has

Table 1. Cases for PDA measurement.

Case	Inlet velocity (m/s)	Inlet particle load (kg/m³)	Note
1	25.3	0.231	Section 1
2	28.32	0.18	Section 2

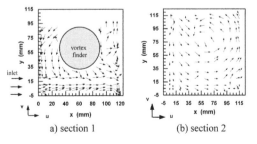

a) section 1   (b) section 2

Figure 3. Gas phase (0–5 μm) u-v vector.

no coherence structure. It's similar to the thermal motion of gas molecular, but it's not completely chaotic. The other is the fluctuating motion in the particle wakes due to the relative motion of fluid and particles. This kind of fluctuating flow has coherence characteristics, but it's of very small scale, the order of particle diameter or smaller. There is an obvious wake condition behind the particles, namely the condition for vortex shedding at the particle surface, by and large when the particle Reynolds number exceeds 100. The condition for particle impact is high particle concentration, especially for the particle flow of asymmetrical diameter. The impact between particle and wall surface mainly happens when the channel shrinks, bends or turns sharply. In general the fluctuating motion due to the above reasons exists only at local position. It's at the corner that the quasi-laminar fluctuating motion comes into being and enhances the gas or particle fluctuating velocity to form local vortex at the corners.

3.2 Mean velocity distribution

The mean velocity was statistically defined by the mean value of the detected velocities of all the particles that passed the measured point as the following:

$$U = \sum_{i=1}^{N} u_i / N \tag{1}$$

where u_i is the transient particle velocity, N is the total number of particles at the measure point. In the present experiment, N is set to be 5000. The mean velocity distribution at section 1 is showed in Figure 4.

When the suspension entered the cyclone at a high speed, the horizontal velocity component, U had a relatively higher value in the region of $y = 0\text{--}20$ mm and the maximum U velocity was found at $y = 15$ mm. The U velocity was very small near the wall, such as at $y = 2$ mm, because of the friction between the suspension and wall surface. When x increased, i.e., when the suspension traveled further into the cyclone, the U velocity decreased. In the region right to the vortex finder, the gas velocity is usually larger than that of particles, while in the region left to vortex finder, the particle velocity is usually larger than that of gas, as showed in Figure 4(a). At section 1, the suspension flowed downward, i.e., the vertical velocity component, W was negative. The largest downward velocity is not near the wall of the separator, but at the right side of the vertex finder. The velocity W near the right wall, i.e., the wall facing the inlet, is much larger than that near the left wall, as showed in Figure 4(b).

Figure 5 shows the comparison of the mean velocity U and W of gas and particles at section 2. Usually at the central region the smaller particles have a relatively higher velocity U than larger particles which means smaller particles were carried by the gas to follow the swirling flow easily. The W velocity of particle phase is usually a little higher than that of gas. Therefore the particles could be collected by the wall.

The distribution of the velocity W was different at section 2 from that at section 1. At section 2, both gas and particles have upward velocity, i.e., re-circulation flow, at the center part of the cyclone ($x = 30\text{--}110$ mm). The re-circulation flow is due to the change of the pressure distribution and is beneficial for particles collection before they flow out of the cyclone separator with a downward exhaust

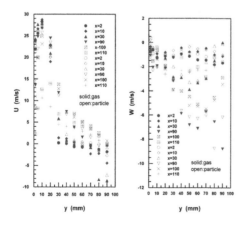

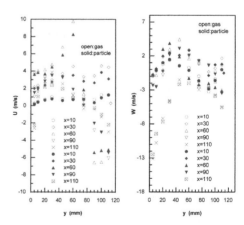

Figure 4. Gas and particles (30–40 μm) velocity at section 1—a) horizontal velocity U; b) vertical velocity W.

Figure 5. Gas and particles (30–40 μm) velocity at section 2—a) horizontal velocity U; b) vertical velocity W.

because the re-circulation flow reduces the chance for particles to escape with the gas flow into the open. The largest downward velocity W is near the right wall of the separator at section 2. The W at the left wall is relatively much smaller, as showed in Figure 5(b). This means the right wall gives more contribution for particle collection than the left wall does, especially the front right corner.

We can see from the distribution of the mean velocity W that the suspension flowed downward near the wall and upward in the core region. While at section 1, the W velocity between the wall and the vortex finder is always downward. Hence we can draw such a picture of the flow field in the square separator that the suspension carried out a downward swirling flow near the wall and an upward swirling flow between the bottom of the vortex finder and the top of the exit, which was similar to the dual helix field in a traditional cyclone or circular cross-section. Because the exhaust exit is at the lower part of the separator, the re-circulation between the bottom of the vortex finder and the top of the exit reduced the chance of particles to escape with the gas flow to the open air. A reasonable design of the vortex finder to enhance the flow field would be beneficial to the particle separation.

3.3 Fluctuating velocity

The turbulent fluctuating velocity is the root-mean value of the fluctuating velocity by statistical calculation and defined as

$$\bar{u}' = \sqrt{\sum_{i=1}^{N} (u_i - U)^2 / (N-1)} \qquad (2)$$

Figure 6 is the results of fluctuating velocity. It's seen that the fluctuating velocity is very small at $x = 0$ and that is much larger at $x = 120$ mm. The fluctuating velocity near the front wall is smaller than that near the rear wall. The difference between the fluctuating velocities along the three coordinates indicates the anisotropy of the flow inside the separator. Usually the fluctuating velocity of particles is larger than that of gas phase. It's at the corner that the quasi-laminar fluctuating motion comes into being and enhances the gas or particle fluctuating velocity.

3.4 Turbulent kinetic energy and turbulent intensity

The turbulent kinetic energy is defined as

$$k = \frac{1}{2} \left(\bar{u}'^2 + \bar{v}'^2 + \bar{w}'^2 \right) \qquad (3)$$

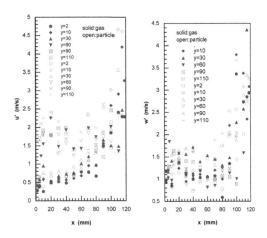

Figure 6. Fluctuating velocity for gas and particles (30–40 μm) at section 2.

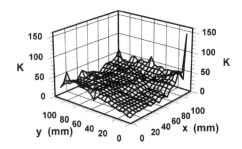

Figure 7. Particle (30–40 μm) turbulent kinetic energy at section 2.

The local turbulent intensity is defined as the ratio of the mean square value of the fluctuating velocity and the mean velocity at any local position.

$$Tu = \sqrt{\frac{1}{3} (\bar{u}'^2 + \bar{v}'^2 + \bar{w}'^2)} / \sqrt{U_{mean}^2 + V_{mean}^2 + W_{mean}^2} \qquad (4)$$

where $\bar{u}'$ and U_{mean} are the mean-square-root of the local fluctuating velocity and the mean velocity, respectively.

The turbulent kinetic energy is deduced by the distribution of the root-mean-square of the fluctuating velocity. Figure 7 shows that the turbulent kinetic energy inside the cyclone is larger at the right part than that at the left part. The peak turbulent kinetic energy was found at the front corner facing the inlet side. The motion at the four corners is different from each other mainly due to the the different intensity of the quasi-laminar fluctuating

271

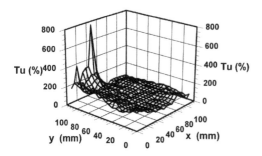

Figure 8. Particles (30–40 μm) turbulent intensity at section 2.

motion there. The local turbulent intensity was maximal at the corners as showed in Fig. 8, however the local maximum turbulent intensity was not found at the front corners, but at the rear part. The strong turbulent flow at the corners consumed much of the flow energy and caused pressure drop.

4 CONCLUSION

The flow field in the square cyclone included forced vortex of strongly swirling flow in the center region and quasi-free vortex near the wall. The suspension carried out a downward swirling flow near the wall and upward swirling flow between the bottom of the vortex finder and the top of the exit, which was similar to the dual helix flow pattern in a traditional cyclone. Local small vortex existed at the corners. The impacts between particles or particles-to-wall at the corners enhanced the quasi-laminar fluctuating motion, leading to the local peak values of the turbulent kinetic energy and turbulent intensity. The corners were one of the major regions that caused the pressure drop. Re-circulation flow was found in the central region. The re-circulation between the bottom of the vortex finder and the top of the exit reduced the chance of particles to escape with the gas flow to the open air.

ACKNOWLEDGEMENTS

The work was supported by Shanghai Natural Science Foundation (No. 11ZR1401000) and the Fundamental Research Funds for the Central Universities (No. 12D11311), which are gratefully acknowledged.

REFERENCES

[1] C. Hoffmann and L.E. Stein, Gas Cyclones and Swirl Tubes Principles, Design and Operation, Springer, New York (2004).
[2] Cortes and A. Gil, Prog. Energy Combust. Sci. 33, 409 (2007).
[3] J. Lee, H. Yang, D. Lee, Powder Technol. 165, 30 (2006).
[4] S. Bernardo, M. Mori, A Peres, R.P. Dionısio, Powder Technol. 162, 190 (2006).
[5] Qian, J. Zhang, M. Zhang, J. Hazard. Mater. B136, 822 (2006).
[6] J. Chen, X. Lu, H. Liu, C. Yang, Chem. Eng. J. 129, 85(2007).
[7] K. Fuat and I. Karagoz, Chem. Eng. J. 151, 39 (2009).
[8] J. Cui, X. Chen, X. Gong, G. Yu, Ind. Eng. Chem. Res. 49, 5450 (2010).
[9] Gimbun, T.G. Chuah, A. Fakhrul-Razi, T.S.Y. Choong, Chem. Eng. Process. 44, 7 (2005).
[10] Zhao, Chem. Eng. Process. 44, 447 (2005).
[11] B. Zhao, H. Shen, Y. Kang, Powder Technol. 145, 47 (2004).
[12] Lim, S. Kwon, K. Lee, Aerosol Sci. 34, 1085 (2003).
[13] S.L. Darling, Pyroflow Compact: The next generation CFB boiler. Proceedings of the 1995 International Joint Power Generation Conference, Part 1 (of 4), (1995) October 8–12; Minneapolis, MN, United States.
[14] G Shibagaki and N. Akio, Jpn. J. Pap. Technol. 42, 59 (1999) (in Japanese).
[15] Makkonen, VGB Power Technol. 80, 30 (2000).
[16] J. Lu, J. Zhang, H. Zhang, Q. Liu, G. Yue, Fuel Process. Technol. 88, 129 (2007).
[17] A. Patterson, R.J. Munz, Can. J. Chem. Eng. 74, 213 (1996).
[18] J. Hoekstra, J.J. Derksen, H.E.A. Van Den Akker, Chem. Eng. Sci. 54, 2055 (1999).
[19] W. Peng, A. C. Hoffmann, P.J.A.J. Boot, A. Udding, H.W.A. Dries, A. Ekker, J. Kater, Powder Technol. 127, 212 (2002).
[20] Z. Liu, Y. Zheng, L. Jia, J. Jiao, Q. Zhang, Chem. Eng. Sci. 61, 4252 (2006).
[21] Y. Su, Chem. Eng. Sci. 61, 1395 (2006).

Frontiers of Energy and Environmental Engineering – Sung, Kao & Chen (eds)
© 2013 Taylor & Francis Group, London, ISBN 978-0-415-66159-1

The treatment of newsprint deinking wastewater using electro-sorption technology

Z.S. Wang & W.J. Han
Key Laboratory of Pulp and Paper Science and Technology of Ministry of Education, Shandong Polytechnic University, Ji'nan, China

ABSTRACT: In this paper, the newsprint deinking wastewater was treated by electric adsorption technology. The optimum condition of treatment deinking waste water in dealing with voltage of 20 volts DC, water flow rate of 200 ml/minute and the appropriate processing time is 120 minutes. The conductivity of deinking waste water was lowered to 740 µs/cm (25 °C) from 3000 µs/cm (25 °C) which meet the requirements of water usage under these conditions. The chroma turbidity decreased to varying degrees with 45.4% reduction in turbidity and chromaticity decreased by 64.6%. BOD5 and CODcr were reduced by 37.4% and 41.4%, 32.6% reduction.

Keywords: newsprint deinking wastewater; electro-sorption; conductivity; turbidity and chromaticity

1 INTRODUCTION

At present, the methods of paper wastewater treatment were mainly physical, chemical and biochemical methods. This kind of paper wastewater can't be recycled due to its high conductivity. In terms of traditional reverse osmosis to reduce the conductivity, processing and equipment maintenance costs are too high (Zhang, 2011). It is important to paper industry to find a stable and reliable low-cost wastewater treatment technology in order to increase the reuse rate of waste water, reduce wastewater emissions (Robert et al., 2008).

The process of electro-sorption is a method to remove the ions in the solution, so as to achieve desalination purification of solution and reduce waste water conductivity. Electrochemical sorption electrode plate of the water treatment module has a high surface area. Electrodes and external power supply were connected. The migration of ions is moving to the electrode plate under the action of an electric field which eventually be attracted to the electrode a solution surface of the electric double layer (Zhang et al., 2005). The desalination of the solution is finished. Electro-sorption of water treatment technology is treated as effective and economical water desalination due to low pressure drop, low power consumption, no secondary pollution, less investment and long cycle life (over 100,000) and regeneration of the advantages of easy (Sun et al., 2005).

Basing on the project of a newsprint waste water treatment, the wastewater was treated using electro-sorption technology. The experimental results show that an electro-sorption wastewater treatment technology to deal with newsprint paper-making wastewater is reasonable in the respect of economic and technology.

2 EXPERIMENT

2.1 Material

Wastewater was supplied by Enso Huatai Paper Co., Ltd.

2.2 Equipments

Constant flow pump (YZ151SX) was purchased from Baoding LanGeHeng flow pump Co., Ltd. Conductivity meter and pH meter were provided by Shanghai precision & scientific instrument company. Electric adsorption module prepared by ourselves.

Adjustable DC power supply was made by Wuhan jinnuo electronic Co., Ltd. Experimental facility is consist of instrument verification system, wastewater recycle system and electric adsorption module system which is shown in Figure 1.

2.3 Methods

10 L deinking waste water was added into the deinking wastewater container and then starts the constant flow pump. The flow rate of water was controlled at 200 ml/minute meanwhile the

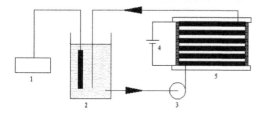

Figure 1. Schematic diagram of the CDI device. 1-Conductivity meter 2-Wastewater container 3-Constant flow pump 4-Adjustable DC power 5-Electric adsorption module.

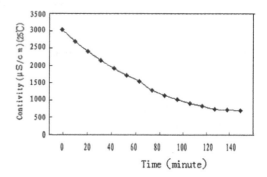

Figure 2. Effect of treating time on the conductivity of wastewater.

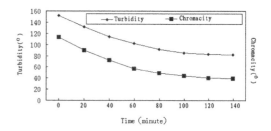

Figure 3. Effect of treatment time on turbidity and chromaticity of wastewater.

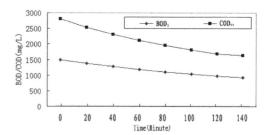

Figure 4. Effect of treating time on the BOD5 and CODcr.

electro-sorption module was opened. The voltage was 20 volts and timing. The conductivity value of processing time was recorded.

Taking the waste water with different processing time, its turbidity, chromaticity, CODcr, pH value, BOD5 and the TSS were detected.

3 RESULTS AND DISCUSSION

3.1 *Effect of treating time on conductivity*

The effect of treating time on the conductivity was shown in Figure 2. The conductivity of deinking waste water decreased with the electrosorption longer processing time increases. The change was quickly in first 100 minutes and then the conductivity decreases slowly after 120 minutes maintaining to a steady state. After with 120 minutes treatment, the electric double layer electro-sorption module has basically been saturated and can no longer continue to absorb a large number of charged particles. In other words, the further period of time to waste water treatment has no practical significance. It should be transferred to an electric adsorption cycle after the time. The final conclusion was as followings, the DC voltage of 20 volts, deinking waste water in order to deal with 2 hours for a loop.

3.2 *Effect of treatment time on turbidity and chromaticity of wastewater*

It can be seen from Figure 3, the turbidity and chromaticity curve of deinking waste water is basically consistent with the electro-sorption of treatment time and decreased gradually becomes slowly when lower after two hours after the turbidity and chroma. After 140 minutes of treatment, the turbidity decreased by 45.4%, 64.6% reduction in chroma. This indicates that waste water can effectively reduce turbidity and chroma with electro-sorption technology. The changes in trends are basically the same as changes in the conductivity trend.

It confirms that the electrosorption technology can remove the waste water Fe and Mg non-ferrous metal ions and some organic complexes.

3.3 *Effect of treating time on the BOD5 and CODcr*

It can be seen from Figure 4, the BOD5 and CODcr were all declined with the electric adsorption treatment of deinking waste water for 2 hours, BOD5 was reduced from 1485 mg/L to 930 mg/L, by 37.4%.

Meanwhile the CODcr was down from 2800 mg/L to 1640 mg/L by 41.4%. This indicates

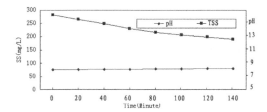

Figure 5. Effect of treating time on the pH and TSS.

that electrosorption processing technology of BOD5 of the water recycling and CODcr not concentrate. This provides the possibility of long-term system of wastewater recycling.

3.4 *Effect of treating time on the pH and TSS*

From Figure 5, the water treated with electric adsorption treatment was weakly alkaline and pH was 8.0 which had a increasing of 0.4 comparing with starting water pH 7.6. The TSS reduced from 282 mg/L to 190 mg/L by 32.6%. It demonstrates that TSS can be reduced greatly by treatment electros-orption. At the same time, it should be proper control of pH value, the pH value to meet the wastewater treatment system and back to the water requirements.

3.5 *The consumption of energy for electro-sorption treatment wastewater*

The energy consumption is below 1.5 KWh (1.2 RMB) which can reduce conductivity of inking wastewater from 3000 μs/cm (25 °C) to the 850 μs/cm (25 °C). Therefore, from an economic point of view, it is acceptable comparing with other processing methods such as reverse osmosis treatment costs about 4 to 6 RMB per ton water

greatly reduce the cost, and thus have great market potential.

4 CONCLUSION

The optimum condition of treatment deinking waste water using homemade the electrosorption processing module in dealing with voltage to 20 volts DC, water flow rate of 200 ml/minute and the appropriate processing time is 120 minutes.

The conductivity of deinking waste water was lowered to 740 μs/cm (25 °C) from 3000 μs/cm (25 °C) which meet the requirements of water usage under these conditions. The chroma turbidity decreased to varying degrees with 45.4% reduction in turbidity and chromaticity decreased by 64.6%. BOD5 and CODcr were reduced by 37.4% and 41.4%, 32.6% reduction.

Through experiment and practice of engineering, the electro-sorption treatment deinking waste water, it is acceptable in technically feasible and economically.

REFERENCES

Robert, D.V. 2008. Hybrid capacitive deionization and electro-deionization (CDI-EDI) electrochemical cell for fluid purification, US. Patent, us 008672AI.

Sun S.Q, Zhang W.F., Cheng, J. 2005, Study on progress and capacity of activated carbon electrode capacitive deionization, water treatment technology 34(3):31–34.

Zhang, H.P. Wu, L.M., Yang, Z., Guan X.H. 2005. Study on electrosorption deionization technology. Northeast journal of electricity institute 25(4):72–74.

Zhang, Z.C. 2011. Design and study on capacitive deionization module, Beijing University of Chemical Technology 21(5) 11–16.

Frontiers of Energy and Environmental Engineering – Sung, Kao & Chen (eds)
© 2013 Taylor & Francis Group, London, ISBN 978-0-415-66159-1

Regulating drinking water contaminants as group(s)

Y.N. Gao, X.G. Ma, P.P. Zhang & J.X. Fu
School of Municipal and Environmental Engineering, Shenyang Jianzhu University, Liaoning Shenyang, China

ABSTRACT: The deteriorating quality and the increasing number of detected contaminants in drinking water, brought challenges to the drinking water treatment agent and detecting system, if single to control and detect each type of pollution will not only time consuming, and increases energy consumption, but also not conducive to the protection of drinking water safety. This paper proposing a protective new strategy for health improving and drinking water safety, that is regulate contaminants as group(s) rather than one at a time. This paper is to provide publics an opportunity to understand the concept of regulating contaminants as group(s), present some background information on the current regulatory approach include its limitations, explain how those limitations might be overcome by regulating contamination as group(s), put forward possible factors for grouping contamination and regulatory mechanism for regulating contamination group(s), analyze the current research status and future research direction of regulating contaminants as group(s).

Keywords: regulate contaminants as group(s); new strategy; factors; regulatory mechanism

1 INSTRUCTIONS

Nowadays, people are increasingly concerned about the safety of their drinking water. As improvements in analytical methods allow us to detect impurities at very low concentrations in water, water supplies once considered pure are found to have contaminants. Consumers cannot expect pure water, but they want safe water. Drinking water can become contaminated for some natural causes, at the original water source, during treatment, or during distribution to the home. If provided water comes from surface water (river or lake), it can be exposed to acid rain, storm water runoff, pesticide runoff, fertilizer runoff, domestic sewage and industrial waste. If provided water comes from groundwater (private wells and some public water supplies), it generally takes longer to become contaminated but the natural cleansing process also may take much longer. Groundwater moves slowly and not exposes to sunlight, aeration, or aerobic (requiring oxygen) microorganisms. Groundwater can be contaminated by disease-producing pathogens, leachate that discharge from landfills and septic systems, hazardous household products for careless disposal, agricultural chemicals and leaking underground storage tanks. Furthermore, with herbicide and fertilizer constantly applying to agriculture, new industry emerging, the species of detected contaminants increasing in drinking water sources, some of those detected contaminants are included in drinking water quality standards, others are not included.

Those materials could be divided into materials dissolved and suspended in water [1]. The ahead group include inorganic compound (dissolved gas, metal and metalloid positive ions, negative ions, radon) and organic compound (synthetic organic chemicals, Trihalomethanes), the latter group include pathogens, viruses, bacteria, protozoan, cryptosporidium, asbestos and other suspended solids.

In order to guarantee drinking water protection can be achieved more quickly, cost-effectively and availably deal with large number of new contaminations and the continuous accumulation of new information on these contaminants. Paper comes up with the suggestion of regulating contaminants as group(s), refer to U.S. EPA Administrator Lisa Jackson's four principles to provide greater protection of drinking water, On March 22, 2010, one of which is to regulate contaminants as group(s) [2].

2 CORRELATION ANALYSIS FOR REGULATE DRINKING WATER CONTAMINANTS AS GROUP(S)

2.1 *Drinking water quality in China*

Based on the results released by the Chinese State of the environment bulletin display: In 2007 national surface water was serious polluted, seven major river systems in general moderate pollution, Pearl River, Yangtze River water quality is good overall, Songhua River is mild polluted, the Yellow River, Huai River is moderately polluted,

Liao River, Huai River is severely polluted. In 2008 national surface water pollution situation remain serious, seven major river water quality is overall moderate pollution. The Pearl River, the Yangtze River water quality overall is good, for slightly polluted the Songhua River, the Yellow River, Huai River, Liao River is moderately polluted, Haihe is severe pollution. In 2009 the seven major river overall for slightly polluted, the Pearl River, the Yangtze River water quality is good, pollution of Songhua River, Huai River as mild, Yellow River, Liao River is moderately polluted, Huai River severe pollution. In 2010 National surface water pollution is still heavier. Major pollution indicators are permanganate index and biochemical oxygen demand and ammonium—nitrogen. The Yangtze River and Pearl River water quality is good, pollution status of Songhua River, Huai River is mild, Yellow River, Liao River is moderately polluted, Huai River is severely polluted. The proportion of seven major river systems water quality comparison from 2007 to 2010 is shown in Table 1.

2.2 The current regulatory framework problem statement

The current regulatory framework for drinking water protection mainly focuses on assessing risks from exposure to individual contaminants. The water relevant authority and utilities find it not only increasingly difficult to effectively deal with large number of new contaminations and the continuous accumulation of new information on these contaminants, but also results in slow progress in regulating unregulated contaminants, in addition, it fails to take advantage of strategies for enhancing health protection cost-effectively. Evaluating and regulating contaminants as groups during the regulatory process may better protect public health; consume less time and resources; account for risks from multiple contaminants; deal more effectively with an increasing number of emerging contaminants; use advanced treatment technologies that regulate several contaminants at once; provide water systems with an opportunity to make better long-term decisions on capital investments[3].

2.3 The concept of regulating contaminants as group(s)

Nowadays, more and more new contaminations were detected at an increasingly faster rate than we are regulating them in drinking water. Some efforts have been made to define approaches to regulate contaminants as group(s) rather than one at a time so that enhancement of drinking water protection can be achieved cost-effectively. The opportunity afforded by the groups concept to protect the public from the large numbers of chemicals in the environment that are not currently regulated, the main mechanism of regulate contaminant as group is to included all eligible contaminations in the same group.

2.4 General considerations in regulating contaminants as group(s)

There are various factors should be considered to determine appropriate contaminant groups [4]. First, it is important to check and recheck several times to be sure there are no consequences. Second, if contaminations are regulated as groups then operators should be required to monitor those contaminations as groups, which significantly reduce monitoring costs. Third, it is also important to consider if there is anything can be done within drinking water strategy to keep contaminants out of the water in the first place.

3 THE CURRENT RESEARCH STATUS OF REGULATING CONTAMINANTS AS GROUP(S)

February 2, 2011, U.S. EPA announced that it would move forward with efforts to regulate, as a single group, up to 16 carcinogenic volatile organic compounds in drinking water, as shown in Table 2.

The VOCs would be the first group of drinking water contaminants to be regulated as a single group under the Agency's new Drinking Water Strategy announced in March of 2010. Under this new regulatory strategy, EPA is attempting to

Table 2. 16 carcinogenic volatile organic compounds in drinking water.

Already regulated VOCs	Non-regulated VOCs
Benzene	Aniline
Carbon tetrachloride	Benzyl chloride
1,2-dichoroethane	1,3-butadiene
1,2-dichloropropane	1,1-dichloroethane
Dichloromethane	Nitrobenzene
Tetrachloroethylene	Oxirane methyl
Trichloroethylene	1,2,3-trichloroproane
Vinyl chloride	Urehane

Table 1. The proportion of the seven major river systems water quality comparison from 2007 to 2010.

Year	Proportion of water quality comparison		
	I–III (%)	IV~V (%)	Lower V (%)
2007	49.9	26.5	23.6
2008	55.0	24.2	20.8
2009	57.3	24.3	18.4
2010	59.9	23.7	16.4

regulate drinking water contaminants in groups, rather than the traditional method of individual contaminant regulation, in an effort to speed up the regulatory process, while encouraging development of more cost-effective and available treatment technologies that can removal of multiple contaminants at the same time [5].

The first eight VOCs in this list are already regulated and the last eight are not yet regulated in the current drinking water quality standard. EPA included them all in the same group on the basis that they are all known or suspected to cause cancer. At present, EPA predicts that a proposed rule for this first VOCs group may take another 2 to 2.5 years to develop.

3.1 Factors in deciding regulate VOCs group

The Agency consider in deciding regulate VOCs group could be the first group of contaminants to be regulated under the new Drinking Water Strategy, there are several factors in evaluating which contaminants might effectively be regulated as a group, and these factors include: whether contaminants have a similar health endpoint; whether contaminants can be measured by the same analytical methods; whether contaminants can be treated using the same technology or treatment technique approach; whether contaminants have been shown to occur together(i.e., co-occur) [4].

EPA conducted extensive national outreach to solicit input from stakeholders and stakeholders generally agreed that these were some of the more important factors to consider in evaluating which contaminants would work best in a group regulation.

3.2 The reason for EPA choose carcinogenic volatile organic compounds (VOCs) as the first group to regulate in the near-term

After carefully considering input from stakeholders, EPA decided to regulate as a group up to 16 volatile organic compounds (VOCs) that may cause cancer. The Agency determined that they represent a near term opportunity [5].

- the public health goal for all is currently or would likely be set at zero because they may cause cancer
- most of this group of VOCs can be measured by the same analytical method (i.e., EPA 524.2)
- many can be treated by the same treatment (i.e., aeration and/or granular activated carbon)
- a preliminary evaluation of occurrence indicates that some of these VOCs may co-occur.

This group will include trichloroethylene (TCE) and tetrachloroethylene (PCE). EPA determined in March, 2010 that the drinking water standards for these two currently regulated contaminants need to be revised. Regulating these VOCs as a group will help reduce exposure to these contaminants. Until now, the Agency has not decided what the group approach will be for these VOCs and what level will be set. Any decisions will be based on the best available science and responsibilities under the law.

4 FUTURE RESEARCH DIRECTION OF REGULATING CONTAMINANTS AS GROUP(S)

In the near-term, EPA also will evaluate whether to regulate nitrosamine disinfection byproducts as part of the Contaminant Candidate List Regulatory Determination process. Data from the second Unregulated Contaminant Monitoring Rule indicate that these compounds are frequently being found in public water systems. In the long-term, EPA will continue to work with stakeholders to evaluate and fill the data gaps for other groups of interest for drinking water. Other potential groups are listed blow [6].

4.1 Regulating nitrosamines as a group paper title

The group nitrosamines were at the top of the list of a chemical group with similar toxicological characteristics. Unregulated Contaminant Monitoring Rule 2 (UCMR2) has already surveyed the nation's drinking water supply for 6 nitrosamines. Although N-nitrosodimethylamine (NDMA) has been the most prevalent N-nitrosamine detected in disinfected waters, it remains unclear whether NDMA is indeed the most significant N-nitrosamine or just one representative of a larger pool of N-nitrosamines. A widely used assay applied to quantify nitrite, S-nitrosothiols, and N-nitrosamines in biological samples involves their reduction to nitric oxide by acidic tri-iodide, followed by chemiluminescence detection of the evolved nitric oxide in the gas phase. NDMA is a human cancer initiator suspect. Ethanol, a cancer risk factor, may synergize with nitrosamines by suppressing hepatic clearance, to increase internal exposure [7].

Nitrosamines are potent carcinogens and have been found in latex products, food and versus water. One paper named estimation of the total daily oral intake of NDMA attributable to drinking water, which published in 2007, estimate NDMA concentrations from food and water versus NDMA formed in the body and estimate the Proportional Oral Intake (POI) for drinking water. And the bottom line is the POI for drinking water was estimated to be 0.02% compared to the other sources.

4.2 Regulating goitrogenic anions as a group

Ammonium Perchlorate (AP) and Sodium Chlorate (SC) have been detected in public drinking water supplies in many parts of the United States. These chemicals cause perturbations in pituitary-thyroid homeostasis in animals by competitively inhibiting iodide uptake, thus hindering the synthesis of thyroglobulin and reducing circulating thyroxine T(4). Little is known about the short-term exposure effects of mixtures of perchlorate and chlorate. The present study investigated the potential for the response to a mixture of these chemicals on the pituitary-thyroid axis in rats to be greater than that induced by the individual chemicals.

The implementation of grouping goitrogens becomes quite challenging in the case of oxyhalides (perchlorate, chlorate, and chlorite and bromated). Take for example, a utility that uses hypochlorite as a disinfectant. There is no question that the hypochlorite solution will contain chlorite, chlorate, and perchlorate, the concentration of the oxyhalides will change during storage.

4.3 Regulating estrogenic mimics as a group

Endocrine disruptors are recognized as problematic chemicals. Some are as old as Dichloro-Diphenyl-Tricgloroethane (DDT) and as new as contemporary pharmaceuticals. A significant number mimic estrogen or other estrogenic steroids essential to normal endocrine function for both humans and many other organisms. The mechanism of disruption is frequently at the estrogen receptor with a cell's cytoplasm. A bioassay could be used to measure the intensity of estrogen receptor binding, to which an MCL could be established. If further information about specific compounds causing the reaction is needed, more detailed analysis could follow. This approach could detect a large number of contaminants sharing a common negative impact using a relatively inexpensive and rapid technique. Natural estrogenic compounds would also likely be detected by the above. Such detection may be useful since exposure to excessive natural estrogenic compounds may be as problematic as exposure to artificial ones. This conceptual approach could be used for other endocrine disruptors causing a common cellular disruption.

4.4 Regulating Herbicides as a group

On a broader scale EPA work with other federal and state agencies to institute voluntary practices that reduce the input of groups of contaminants to drinking water resource. For example, Triazine pesticides such as Atrazine and Simazine are potential surface water contaminants particularly in the Midwest. Searching better methods of application and runoff control that could reduce the release to water resource is a main job of environmental protection. Of course,

this is not a new issue with this group of pesticides but is worth rising as an example. This would be best handled on a non-regulatory manner.

In additional to the above introduction, there are also some other proposed contaminant groups that EPA going to regulate on a follow-up work, that are Herbicides, Algaltoxins, Pharmacecals and Personal Care Products (PPCPs).

5 CONCLUSION

Regulate contaminants as group(s) rather than one at a time is a protective new strategy for health improving and drinking water safety. The concept of regulating contaminants as group(s), correlation analysis for regulate contaminants as group, the current research status and future re-search direction of regulating contaminants as group(s) is analyzed in this paper. Health goal, analytical method, treatment, occurrence are four decisive factors for the feasibility of appropriate contaminant groups. Regulate contaminants as group(s) is a strategy that not only time saving, energy conservation, but ensure to the safety of drinking water.

ACKNOWLEDGEMENTS

This work was financially supported by National Water pollution control and management of major special science and technology (NO. 2012ZX07505-003-002), Doctor Start-up Foundation of Liaoning Province (201104326) and Plan of the Ministry of Education of Liaoning Province (L2011088).

REFERENCES

[1] http://www.cybernook.com/water/contam.html#intro.
[2] US EPA. Paradigm for Regulating Drinking Water Contaminants As Groups to Enhance Public Health Protection. Draft Discussion Paper: 2010, 3.
[3] Drinking Water Strategy: A New Framework for Regulating Contaminants as Group(s). Web Dialogue Summary, 2010, 7, 28–29. Available from: http://www.webdialogues.net/epa/dwcontaminantgroups.
[4] EPA. A new approach to protect drinking water and public health. Office of Water (4607M), EPA 815F10001, 2010, 3. Available from: http://www.epa.gov/safewater.
[5] EPA's First "Group" Drinking Water Regulations-Taft Stettinius & Hollister LLP. 2011, 3, 9. Available from: www.taftlaw.com//745-epa-s-first-group-drinking-water-regulate.
[6] EPA. Basic Questions and Answers for the Drinking Water Strategy Contaminant Groups effort. Office of Water (4607M) EPA 815-F-11-002, 2011, 1.
[7] EPA. National primary drinking water regulation. Update safe drinking water act, HDR's 13th edition, 2011, 2.

Frontiers of Energy and Environmental Engineering – Sung, Kao & Chen (eds)
© 2013 Taylor & Francis Group, London, ISBN 978-0-415-66159-1

The design of pin shaping device for IC chips based on recycling

Y.L. Wang, H. Jiang & G.F. Liu
School of Mechanical & Automotive Engineering, Hefei University of Technology, Hefei, China

ABSTRACT: With the rapid development of information and electronic industry, IC chips are widely used in various kinds of electronic products. However, a large number of abandoned IC chips are engendered in the recycling and dismantling process of the electronic products and the mechanical deformations of IC chip pins are inevitably engendered. If the pins of the IC chips are shaped appropriately through proper methods and devices, the IC chips can be reused by refurbishment after being tested. In this paper, the package types of IC chips and the various mechanical deformations of IC chip pins are described. Compared with the existing pin shaping devices, a kind of pin shaping device is developed for IC chips packaged with DIP (Double In-line Package) and SOP (Small Out-Line Package). The various mechanical deformed pins of the abandoned IC chips packaged with DIP and SOP can be shaped correctly by this device which has advantages of wide adapting range, high operating efficiency and high ratio of performance to price.

Keywords: IC chip; recycle; pin deformation; pin shaping

1 INSTRUCTIONS

In recent years, with the rapid development of technology and consumption level, electronic products were widely used and a large number of wasted electronic products were abandoned because of the rapid updating of electronic products, thus the environment has been polluted inevitably.[1–6] In the meantime, a large number of abandoned IC chips have been engendered in the recycling and dismantling process of the abandoned electronic products, if the old and wasted IC chips cannot be recycled and reused properly, resources will be wasted seriously and environment will be polluted inevitably. Therefore it appears to be more important to recycle and reuse the IC chips. Except for mechanical deformations, these IC chips can be reused after being tested if the pins are shaped appropriately through proper methods and devices.

There have been many studies about the pin shaping for IC chips at home and abroad, a kind of pin shaping device was developed for IC chips in the Chinese invention patent (CN201188745[7]), the pins of the IC chips were shaped by closing mold and opening mold between the upper and lower mould, so the co-planarity of pins was improved cheaply. A kind of pin shaping device was developed for IC chips packaged with DIP in the Chinese invention patent (CN101102663A[8]), the double rows pins were set on both sides of the bracket, the cylindrical rolling-bearings were fixed on both sides of the bracket, the double rows pins

were squeezed by rolling of the rolling-bearings. Although the pins of the IC chips could be shaped by the above methods, the devices only worked with the IC chips packaged with DIP and only achieved the co-planarity of pins. A kind of pin shaping device was developed for IC chips in the Singapore invention patent (69929[9]), the adjustment device of IC chip pins was described, the irregular pins were rearranged by the programmable adjustment tool of IC chip pins, although this device had a high degree of automation, the structure was complex, the cost was high and not all kinds of the mechanical deformations of pins could be shaped. In this paper, compared with the existing pin shaping devices, a kind of pin shaping device will be described for IC chips packaged with DIP (Double In-line Package) and SOP (Small Out-Line Package), and the various mechanical deformed pins of the abandoned IC chips packaged with DIP and SOP will can be shaped correctly by this device.

2 PACKAGE TYPES OF IC CHIPS

The package types of IC chips can be divided into two types: In-Line package and Surface Mounted package, as shown in Table 1.[10] The usual surface mounted package includes chips packaged with SOJ, SOP, PLCC, TQFP and PQFP; and the chip packaged with SOP is most widely used, whose pin pitch is 1.27 mm and pin number from 8 to 44; the usual in-line package includes chips packaged with

Table 1. The usual surface mounted package and in-line package.

Surface mounted package		
SOJ	SOP	PLCC
TQFP	PQFP	TSOP

In-line package		
DIP	SIP	ZIP

Table 2. Mechanical deformations types of IC chip pins.

IC chips packaged with SOP		IC chips packaged with DIP	

DIP, SIP and ZIP, and the chip packaged with DIP is most widely used, whose pin pitch is 2.54 mm and pin number from 6 to 64.

3 MECHANICAL DEFORMATIONS TYPES OF IC CHIP PINS

The mechanical deformations of IC chip pins were inevitably engendered in the recycling and dismantling process of the IC chips, so the pins were curved into various types, as shown in Table 2. SO a kind of proper pin shaping device is needed urgently to shape the various mechanical deformations for refurbishment of the IC chips.

3.1 Pin shaping device for IC chips

Compared with the existing pin shaping devices, a kind of pin shaping device has been developed for IC chips packaged with DIP and SOP. The pins of IC chips which have been abandoned in the dismantling process of the electronic products can be shaped by this device, then the IC chips will be retread and reused, so that nature resource can be recycled.

3.2 Pin shaping method for IC chips packaged with SOJ

The pin shaping method for IC chips packaged with SOJ contains three steps:

1. Pretreatment;
2. Adjust the pin pitch;
3. Shape the pins contacting with the PCB board.

1. Pretreatment: The pins which have been bended to inside of abandoned IC chips packaged with SOJ are bended to outside with manual way by sheet tools, as shown in Figure 1.
2. Adjust the pin pitch: The whole device is shown in Figure 2; the IC chip which was treated through the step 1 is put on the convex lower mold (21), whose pins adown; the guide hole (20) of the convex lower mold which is shown in Figure 3 is contacted with a vacuum suction device to fasten the IC chip, the IC chip (15) is adsorbed on the upper surface of the convex lower mold by negative pressure of air; the toothed upper mold (18) is moved down toward by the eccentric handle (1) being pressed down, a pair of raised teeth (22) of the toothed upper mold are embed between two pins just right through the movement of the convex lower mold in the V-shaped guide groove (6), then the convex lower mold is fastened by the two screws (7) being screwed; the eccentric handle is continued to being pressed down, all the teeth of the toothed upper mold are embed between every two pins when the toothed upper mold and convex lower mold are closed, as shown in Figure 4, so the pin pitch is adjusted, as shown in Figure 5.
3. Shape the pin contacting with the PCB board: The toothed upper mold and convex lower mold of the step 2 are changed into no-toothed upper mold (23) and double concave lower mold (24), the IC chip which was treated through the step 2 is put on the double concave lower mold, the pins of IC chip are sustained by the interior panel (25) of the double concave lower mold; the guide hole (27) of the double concave lower mold is contacted with a vacuum suction device to fasten the IC chip, the IC chip (15) is adsorbed on the upper surface of the double

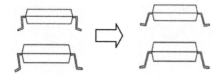

Figure 1.   Pretreatment.

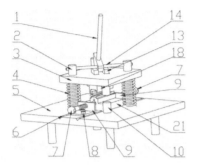

Figure 2.   The pin shaping device.

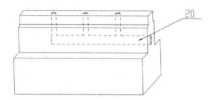

Figure 3.   The guide hole of the convex lower mold.

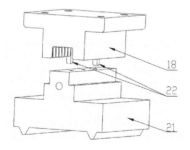

Figure 4.   The toothed upper mold and convex lower mold closed.

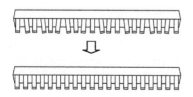

Figure 5.   Adjustment of the pin pitch.

concave lower mold by negative pressure of air; then the double concave lower mold is fasten by the two screws (7) being screwed; the no-toothed upper mold is moved down toward by the eccentric handle (1) being pressed down, as shown in Figure 6, so the ram pressure of gradient panel on both sides of the no-toothed upper mold is acted on the flank pins of IC chip and J-lead pins, as shown in Figure 7. Then the IC chip is taken out, the whole process of pin shaping finishes.

3.3   *Pin shaping method for IC chips packaged with DIP*

The pin shaping method for IC chips packaged with DIP contains five steps:

1. Pretreatment;
2. The co-planarity shaping of pins;
3. The root teeth shaping in-plane;
4. The top teeth shaping in-plane;
5. Pin finishing.

1. Pretreatment: The pins which have been bended to inside of abandoned IC chips packaged with DIP are bended to outside with manual way by sheet tools, as shown in Figure 8.

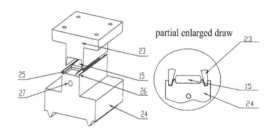

Figure 6.   Schematic diagram of shaping.

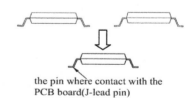

the pin where contact with the PCB board(J-lead pin)

Figure 7.   The J-lead pin shaping.

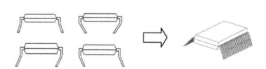

Figure 8.   Pretreatment.

2. The co-planarity shaping of pins: The whole device is shown in Figure 9; the IC chip which was treated through the step 1 is put on the concave lower mold (11), whose pins adown; the packaging part of IC chip is sustained by the sunken portion of the concave lower mold, the pins of IC chip are sustained by the both sides of panel of the concave lower mold; the guide hole of the concave lower mold is contacted with a vacuum suction device to fasten the IC chip, the IC chip (15) is adsorbed on the upper surface of the concave lower mold by negative pressure of air; then the concave lower mold is fastened by the two screws (7) being screwed; the no-toothed upper mold (12) is moved down toward by the eccentric handle (1) being pressed down, so the ram pressure of the both sides panel of the no-toothed upper mold is acted on the both flank pins, as shown in Figure 10, and the co-planarity shaping of pins is realized, an shown in Figure 11.

3 The root teeth shaping in-plane: The no-toothed upper mold and concave lower mold of the step 2 are changed into toothed upper mold (18) and convex lower mold (21); the IC chip which was treated through the step 2 is put on the convex lower mold, whose pins adown; the guide hole

of the convex lower mold is contacted with a vacuum suction device to fasten the IC chip, the IC chip (15) is adsorbed on the upper surface of the convex lower mold by negative pressure of air; the toothed upper mold is moved down toward by the eccentric handle (1) being pressed down, a pair of raised teeth (22) of the toothed upper mold are embed between two pins just right through the movement of the convex lower mold in the V-shaped guide groove (6), then the convex lower mold is fastened by the two screws (7) being screwed; the eccentric handle is continued to being pressed down, all the teeth of the toothed upper mold are embed between every two pins when the toothed upper mold and convex lower mold are closed, as shown in Figure 12, so the root teeth of pins are shaped, as shown in Figure 13.

4. The top teeth shaping in-plane: The whole device is shown in Figure 14; the IC chip which was treated through the step 3 is put on the concave lower mold (11), whose pins upwards; the guide hole of the concave lower mold (11) is contacted with a vacuum suction device to fasten the IC chip, the IC chip (15) is adsorbed on the upper surface of the concave lower mold by negative pressure of air; a raised teeth (28) of the pectinate mold (16) are embed between two pins just right through the movement of the pectinate mold in the T-shaped guide groove (17), as shown in Figure 15; then the concave lower mold is fastened by the two screws (7) being screwed; the pectinate mold is continued to move along the T-shaped guide groove, until all

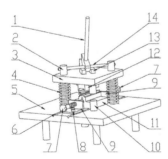

Figure 9.   The pin shaping device.

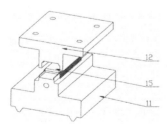

Figure 10.   The co-planarity shaping of pins.

Figure 11.   The IC chip after co-planarity shaping.

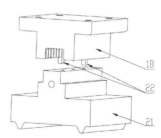

Figure 12.   The toothed upper mold and convex lower mold closed.

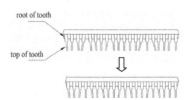

Figure 13.   The IC chip after root teeth shaping.

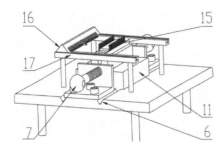

Figure 14. The pin shaping device.

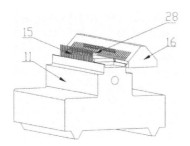

Figure 15. Schematic diagram of top teeth shaping.

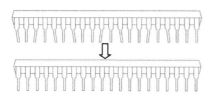

Figure 16. The IC chip after top teeth shaping.

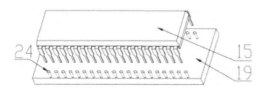

Figure 17. Schematic diagram of pin finishing.

the teeth of the pectinate mold (18) are embed between every two pins, so the top teeth of pins are shaped, as shown in Figure 16.

5. Pin finishing: the pins of IC chip which was treated through the step 4 are set into the tapered holes (24) of mold (19) from small hole entering and big hole out, as shown in Figure 17; the diameter of the small hole need to match with the diameter of PCB board and the distance of every two tapered holes need to match with the distance of every two holes of PCB board. Then

the IC chip is taken out, the whole process of pin shaping finishes.

4 SUMMARY

The whole device has advantages of wide adapting range, high operating efficiency and high ratio of performance to price. Only different of upper and lower mould are needed to change to the different IC chips packaged with SOJ and DIP, and only different sizes of upper and lower mould are needed to change to different sizes IC chips. The IC chips are tested after being shaped; then the qualified IC chips can be reused after sanding and lettering.

ACKNOWLEDGEMENT

We gratefully acknowledge the heuristic suggestions of Professor Liu Zhifeng, we also think the assistance of Zhao Kai in preparing the figures. This work was supported by the National Natural Science Foundation of China (Grant No. 51075115).

REFERENCES

[1] Goosey M, Kellner R. 2003. Recycling Technologies for the Treatment of End of Life Printed Circuit Boards (PCBs).*Circuit World* 29(3):33–37.
[2] Xin Zhao, Jiaqi Hu, Xiaoping Zhang et al. 2010. Disposal of Electronic Wastes Based on Circular Economy. *Electrical Appliances* (9):12–14.
[3] Li JZ, Shrivastava P, Gao Z, et al. 2004. Printed Circuit Board Recycling: a State-of-the-Art Survey.*IEEE Transactions on Electronics Packaging Manufacturing* 27(1):33–42.
[4] Zhou Lei. 2007. Planning Research of Abandoned Electric Appliance Recycling. Tianjin: Tianjin University.
[5] LI Li, LIU YuQiang, WANG Qi. 2009. National Plan for Recycling and Disposal of Waste Electrical and Electronic Equipment. *Research of Environmental Sciences* 22(1):119–124.
[6] Hicks C, Dietmar R, Eugster M. 2005.The recycling and disposal of electrical and electronic waste in China: legislative and market response. *Environmental Impact Assessment Review* 25 (5):459–471.
[7] Pu Guang rong, Shu Haibo, Tang Zhineng. 2008. Pin Adjustment Device, Chinese patent: CN201188745Y.
[8] Li Qingbin, Fan Yin. 2007. Pin Shaping Device for IC Chips Packaged with DIP, Chinese patent: CN101102663A.
[9] Zhu Yi, Wang YuLin, Song ShouXu. 2010. The Pin-shaping Device for IC chip packaged with DIP Based on Recycling. *Modular Machine Tool & Automatic Manufacturing Technique* 6:79–81.
[10] Zhu Yi. 2010. Research on Recycle Method of IC Chips Packaged with DIP of Abandoned Circuit Board. Hefei: Hefei university of technology.

Frontiers of Energy and Environmental Engineering – Sung, Kao & Chen (eds)
© 2013 Taylor & Francis Group, London, ISBN 978-0-415-66159-1

IOWGA operator and Markov chain to groundwater depth prediction

D.M. Zhang & X.J. Liang
College of Environment and Resources, Jilin University, Changchun, China

Q.W. Li
Jilin Provincial Electric Power Survey and Design Institute, Changchun, China

ABSTRACT: The groundwater depth change is a complex nonlinear process. Falling groundwater level will rise to a series of environmental problems. A groundwater depth combination model which based on vector cosine angle IOWGA Operator and Markov chain is proposed to solve the problems about not extensive source of information for the single model and the shortage of fixed weights for the combined model. Take the long-term observation well (16 years) groundwater depth date in Daan for example, the combination prediction method which based on index model, grey model and linear regression model is used to simulate and predict groundwater depth. The example shows that the model has a better prediction accuracy.

Keywords: groundwater depth; Markov chain; IOWGA operator; combination prediction

1 INTRODUCTION

The groundwater level is an important indicator to measure the merits of ecological environment and groundwater resources. Falling groundwater level will cause land subsidence, ground subsidence, the landing funnel, seawater intrusion and other issues. Therefore, the groundwater level prediction is very important to the ecological environment protection and groundwater resources management. There are many methods about groundwater depth prediction, such as regression analysis, the gray forecasting method, exponential smoothing, neural network prediction method, spectrum analysis and the combination prediction method (Deng et al. 2010, Wu et al. 1998, Wang et al. 2009, Yang et al. 2005, Zhu et al. 2006). The combined model based on vector included angle cosine IOWGA Operator and Markov chain is used to predict the groundwater depth.

2 PRINCIPLE AND METHOD

2.1 IOWGA operator

Induced ordered weighted geometric averaging (IOWGA) operator is defined by: set $<v_1, a_1>, <v_2, a_2>, ..., <v_m, a_m>$ for the two-dimensional arrays.

$$f_w(<v_1, a_1>, <v_2, a_2>, ..., <v_m, a_m>) = \prod_{i=1}^{m} a_{v\text{-}index(i)}^{w_i} \qquad (1)$$

where the function f_w is generated by m-dimensional induced ordered weighted geometric averaging operator, marked as IOWGA Operator, v_i is called the induced value of a_i. $v\text{-}index(i)$ is subscript of the ith number which $v_1, v_2, ..., v_m$ in descending order, $W = (w_1, w_2, ..., w_m)^T$ is weighted vector relating to the IOWGA, $\sum_{i=1}^{m} w_i = 1, w_i \geq 0, i = 1, 2, ..., m$. (Cheng et al. 2005, Zhou et al. 2010).

2.2 Markov chain (Liu et al. 2011)

Assume that there is a random process $\{Y(t), t = 0, 1, 2, ...\}$, set of states $E = \{0, 1, 2, ...\}$. For any positive integer b, c, d and any non-negative integer $j_b > \cdots > j_2 > j_1 (c > j_b)$, $i_{c+d}, i_c, i_{j_b}, ..., i_{j_2}, i_{j_1}$.

As long as

$$P\begin{cases} Y(c) = i_c, Y(j_b) = i_{j_b}, ..., Y(j_2) = i_{j_2}, \\ Y(j_1) = i_{j_1} \end{cases} > 0,$$

$$P\begin{cases} Y(c+d) = i_{c+d} \mid Y(c) = i_c, Y(j_b) = i_{j_b}, ..., \\ Y(j_2) = i_{j_2}, Y(j_1) = i_{j_1} \end{cases}$$

$$= P\{Y(c+d) = i_{c+d} \mid Y(c) = i_c\}$$

$\{Y(t)\}$ is known as Markov chain.

2.3 Model

Assume that the observations of groundwater depth sequence is $\{x_t, t = 1, 2, ..., n\}$. The combined model has m kinds of feasible individual prediction methods, Assume that predictive value of the ith prediction methods at time t is x_{it} ($i = 1, 2, ..., m$; $t = 1, 2, ..., n$), Weighting vectors of m kinds of individual forecasts in combination forecast is $W = [w_1, w_2, ... w_m]^T$, $\sum_{i=1}^{m} w_i = 1, w_i \geq 0$. ($i = 1, 2, ..., m$) Based on vector included angle cosine IOWGA Operator and Markov chain, the combined modeling steps are as follows:

1. Individual model forecasts
   M ways to predict water depth and calculate the prediction accuracy a_{it} of the ith method at time t. It can be expressed as a formula (2).

$$a_{it} = \begin{cases} 1 - |(x_t - x_{it})/x_t|, |(x_t - x_{it})/x_t| \leq 1, \\ 0, |(x_t - x_{it})/x_t| > 1 \end{cases}$$
$$i = 1, 2, ..., m; \quad t = 1, 2 ..., n \tag{2}$$

2. Calculate the IOWGA combination predictive value of m kinds of individual prediction methods at time t.
   Put the prediction accuracy as induced value of the predictive value x_{it}, A two-dimensional array $<a_{1t}, x_{1t}>, <a_{2t}, x_{2t}>, ..., <a_{mt}, x_{mt}>$ is composed of m kinds of prediction methods prediction precision at t moment and its corresponding predicted values in the sample interval. We order the prediction accuracy $a_{1t}, a_{2t}, ..., a_{mt}$ of m kinds of single prediction methods at t moment from large to small sequence. Hypothesis a-index(it) is the subscript of the ith prediction precision. IOWGA operator combination which is generated by prediction accuracy sequence $a_{1t}, a_{2t}, ..., a_{mt}$ can be predicted by formula (3).

$$x_t' = f_w(<a_{1t}, x_{1t}>, <a_{2t}, x_{2t}>, ..., <a_{mt}, x_{mt}>)$$
$$= \prod_{i=1}^{m} x_{a\text{-}index(it)}{}^{w_i} \tag{3}$$

Take the log of formula (3). On either side, get formula (4).

$$\ln x_t' = \sum_{i=1}^{m} w_i \ln x_{a\text{-}index(it)} \tag{4}$$

Make $X = (lnx_1, lnx_2, ..., lnx_n)^T$ said the logarithmic vector of the forecasted object measured, $X_i = (lnx_{i1}, lnx_{i2}, ..., lnx_{in})^T$ said the logarithm vector of the first i kinds of forecasting method prediction, $i = 1, 2, ..., m$, $X' = (\ln x_1', \ln x_2', ..., \ln x_n')$ said prediction combination logarithmic vector.

3. Set a combination forecast model which based on the vector cosine angle IOWGA Operator.

$$\begin{cases} \eta' = \dfrac{\sum_{t=1}^{n} \ln x_t \cdot \ln x_t'}{\sqrt{\sum_{t=1}^{n} (\ln x_t)^2} \cdot \sqrt{\sum_{t=1}^{n} (\ln x_t')^2}} \\ F_{ij} = \sum_{t=1}^{n} \ln x_{a\text{-}index(it)} \ln x_{a\text{-}index(jt)}, i, j = 1, 2, ..., m \end{cases}$$
$$\tag{5}$$

where η' is included angle cosine between logarithmic vector X' of the combined predictive value and the measured values vector X.

$$\sum_{t=1}^{n} (\ln x_t')^2 = \sum_{t=1}^{n} \left(\sum_{i=1}^{m} w_i \ln x_{a\text{-}index(it)} \right)^2$$
$$= \sum_{i=1}^{m} \sum_{j=1}^{m} w_i w_j \left(\sum_{t=1}^{n} \ln x_{a\text{-}index(it)} \ln x_{a\text{-}index(jt)} \right)$$
$$= \sum_{i=1}^{m} \sum_{j=1}^{m} w_i w_j F_{ij} = W^T F W \tag{6}$$

The larger angle cosine is, the higher prediction accuracy is. The optimal combination forecasting model can be expressed as a formula (7).

$$\max \eta'(w_1, w_2, ..., w_m) = \dfrac{\sum_{i=1}^{m} w_i \sum_{t=1}^{n} \ln x_t \ln x_{a\text{-}index(it)}}{\sqrt{\sum_{t=1}^{n} (\ln x_t)^2} \sqrt{W^T F W}}$$
$$s.t \begin{cases} \sum_{i=1}^{m} w_i = 1, \\ w_i \geq 0, i = 1, 2, ..., m \end{cases} \tag{7}$$

Model is a nonlinear programming problem, m kinds of single model weight coefficient can be solved by MATLAB toolbox.

4. The size of the combined model weights is determined by the order of different methods prediction accuracy. But the groundwater depth is unknown in the forecast period. Application IOWGA Operator has its limitations. In this paper, the Markov chain forecasting model is used to predict future precision state in order to determine the combination model weight, then calculate the predicted values.
   1. Prediction accuracy based on mean-standard deviation classification method is divided into the K state space.
   2. According to the above classification criteria, the state of each prediction accuracy is determined.

3. Calculate the one-step Markov chain transition probability matrix.
4. Markov test. Suppose $(f_{ij})_{m \times n}$ is one-step transition frequency matrix, $p \cdot_j = \sum_{i=1}^{m} f_{ij} / \sum_{i=1}^{m} \sum_{j=1}^{m} f_{ij}, p_{ij} = f_{ij} / \sum_{j=1}^{m} f_{ij}.$ Statistic $\chi^2 = 2 \cdot \sum_{i=1}^{m} \sum_{j=1}^{m} f_{ij} \cdot \left| \ln \frac{p_{ij}}{p \cdot_j} \right| \sim \chi^2((m-1)^2).$ Given significance level, the look-up table can score the value of sub-sites $\chi^2((m-1)^2)$. The calculated statistical measure is less than the value of sub-sites, you can use the Markov chain to predict.
5. Calculate the combination model weight coefficient and predicted values according to the size of the prediction accuracy at predicted moment.

3 APPLICATION EXAMPLES

Daan is located in western Jilin province. Water resources have become an important factor in restricting local economic and social development. The groundwater level prediction model establishment can achieve the purposes of water resources management and economic development protection The long-term observation well (number 26,631,007) groundwater depth data from 1991 to 2006 in Daan is used to predict future groundwater level changes. Model validation year is from 2007 to 2009.

3.1 *Individual model for predicting*

The index, gray model and regression method are used to forecast the water level. The groundwater depth measured value, predicted value and prediction accuracy are shown in Table 1.

3.2 *Combined model for predicting*

According to the three kinds of single prediction method, we construct the t moment predictive accuracy and predictive value two-dimensional array in the corresponding sample interval. Compute IOWGA combined predictive value according to formula (3). Process of 1992 combined predictive value calculation is as follows:

$$x_1' = f_W(<a_{11}, x_{11}>, <a_{21}, x_{21}>, <a_{31}, x_{31}>)$$
$$= 4.92^{w_1} \times 4.89^{w_2} \times 4.83^{w_3}$$

The combined model predictive values also can be obtained in the remaining years. We can substituted it into the formula (7), the optimal combination predictive model can be written as:

$$\max \eta(w_1, w_2, \ldots, w_m) = \frac{46.8170w_1 + 46.8355w_2 + 46.8490w_3}{6.8380\sqrt{W^\tau FW}}$$
$$s.t \begin{cases} \sum_{i=1}^{m} w_i = 1, \\ w_i \geq 0, i = 1, 2, \ldots, m \end{cases}$$

(8)

Calculation is $w_1 = 1$, $w_2 = 0$, $w_3 = 0$. The standard deviation S and correlation coefficient R are used to analyze predictive results. $\bar{x}$ is the actual observed average value. Evaluation on predictive volatility are shown in Table 2.

Table 1. The measured value, predictive value of single prediction methods and prediction accuracy.

Year	Measured values	Index		GM (1, 1)		Linear regression analysis	
		Predictive value	Predictive accuracy	Predictive value	Predictive accuracy	Predictive value	Predictive accuracy
1991	5.18	4.80	–	5.18	–	4.74	–
1992	5.23	4.92	0.9409	4.83	0.9242	4.89	0.9346
1993	4.84	5.04	0.9578	4.96	0.9744	5.03	0.9605
1994	4.6	5.17	0.8760	5.10	0.8918	5.17	0.8750
1995	5.07	5.30	0.9547	5.24	0.9673	5.32	0.9510
1996	5.48	5.43	0.9912	5.38	0.9812	5.46	0.9967
1997	5.83	5.57	0.9550	5.52	0.9472	5.61	0.9615
1998	5.64	5.71	0.9881	5.67	0.9944	5.75	0.9807
1999	5.78	5.85	0.9880	5.82	0.9923	5.89	0.9806
2000	5.91	6.00	0.9855	5.98	0.9878	6.04	0.9787
2001	6.32	6.15	0.9724	6.14	0.9721	6.18	0.9778
2002	6.43	6.30	0.9797	6.31	0.9812	6.32	0.9833
2003	6.34	6.46	0.9816	6.48	0.9780	6.47	0.9801
2004	6.69	6.62	0.9892	6.65	0.9947	6.61	0.9880
2005	6.84	6.78	0.9917	6.83	0.9992	6.75	0.9874
2006	6.95	6.95	0.9996	7.02	0.9901	6.90	0.9924

Table 2. The predictive validity of evaluation form.

Model	Index	GM (1, 1)	Linear regression analysis	Vector cosine angle IOWGA operator combination model	The divided weight combination model
S	0.2191	0.2126	0.2237	0.1873	0.2162
R	0.9539	0.9566	0.9518	0.9665	0.9551

Table 3. The model predictions and relative error absolute value from 2007 to 2009.

Year	Measured values	Index Predictive value	Relative error	GM (1, 1) Predictive value	Relative error	Linear regression analysis Predictive value	Relative error	Vector cosine angle IOWGA operator Predictive value	Relative error
2007	7.18	7.13	0.0070	7.21	0.0042	7.04	0.0195	7.21	0.0042
2008	7.42	7.30	0.0162	7.40	0.0027	7.18	0.0323	7.40	0.0027
2009	7.55	7.49	0.0079	7.60	0.0066	7.33	0.0291	7.60	0.0066

By comparison, the combined model accuracy based on the vector cosine angle IOWGA operator is the highest.

3.3 The prediction of the accuracy state

According to Table 1, the state space is divided by the mean standard deviation. Calculated from 1992 to 2006, the mean and standard deviation of the prediction accuracy is 0.9701, 0.0300. Prediction accuracy is divided into three states: state 1 (0, 0.956), state 2 [0.956, 0.986], state 3 (0.986, 1). State the result of division is omitted.

Then Markov inspect and calculate the transfer frequency matrix. The statistic values of the Index, gray GM (1, 1) and linear regression analysis are 9.73, 14.99 and 12.73. Take a significant level a = 0.05, the look-up table can inquire: Subsite $\chi_a^2(4) = 9.49$, $\chi^2 > \chi_a^2((m-1)^2)$, so Markov chain model is available for the prediction analysis of the accuracy state.

One-step transition probability matrix $p_1^{(1)}, p_2^{(1)}, p_3^{(1)}$ of Index, GM (1,1), Linear regression analysis are:

$$P_1^{(1)} = \begin{bmatrix} 1/4 & 1/4 & 2/4 \\ 1/5 & 3/5 & 1/5 \\ 1/5 & 1/5 & 3/5 \end{bmatrix}, P_2^{(1)} = \begin{bmatrix} 0 & 2/3 & 1/3 \\ 2/6 & 3/6 & 1/6 \\ 0 & 1/5 & 4/5 \end{bmatrix},$$

$$P_3^{(1)} = \begin{bmatrix} 1/3 & 1/3 & 1/3 \\ 1/8 & 6/8 & 1/8 \\ 0 & 1/3 & 2/3 \end{bmatrix}$$

Two-step transition probability matrix $P_1^{(2)}$, $P_2^{(2)}$, $P_3^{(2)}$ are:

$$P_1^{(2)} = \begin{bmatrix} 0.2125 & 0.3125 & 0.4750 \\ 0.2100 & 0.4500 & 0.3400 \\ 0.2100 & 0.2900 & 0.5000 \end{bmatrix}$$

$$P_2^{(2)} = \begin{bmatrix} 0.2222 & 0.4000 & 0.3778 \\ 0.1667 & 0.5056 & 0.3278 \\ 0.0667 & 0.2600 & 0.6733 \end{bmatrix}$$

$$P_3^{(2)} = \begin{bmatrix} 0.1528 & 0.4722 & 0.3750 \\ 0.1354 & 0.6458 & 0.2188 \\ 0.0417 & 0.4722 & 0.4861 \end{bmatrix}$$

Three-step transition probability matrix $P_1^{(3)}$, $P_2^{(3)}$, $P_3^{(3)}$ are:

$$P_1^{(3)} = \begin{bmatrix} 0.2106 & 0.3356 & 0.4537 \\ 0.2105 & 0.3905 & 0.3990 \\ 0.2105 & 0.3265 & 0.4630 \end{bmatrix}$$

$$P_2^{(3)} = \begin{bmatrix} 0.1333 & 0.4237 & 0.4430 \\ 0.1685 & 0.4294 & 0.4020 \\ 0.0867 & 0.3091 & 0.6042 \end{bmatrix}$$

$$P_3^{(3)} = \begin{bmatrix} 0.1100 & 0.5301 & 0.3600 \\ 0.1259 & 0.6024 & 0.2717 \\ 0.0729 & 0.5301 & 0.3970 \end{bmatrix}$$

From 2007 to 2009, the calculated highest prediction accuracy is the gray GM (1, 1). This can be determined to take over the three-year forecast, using the gray model is better.

From Table 3, the vector cosine angle IOWGA Operator combined model can be better predicted the groundwater depth change.

4 CONCLUSION

1. Vector cosine angle IOWGA Operator combined model can be fully rational taken advantage of each prediction model information and improved the prediction accuracy. It provides a new way to forecast the groundwater depth.
2. Markov chain has been widely applied in practice, but many scholars have ignored the inspection of the Markov chain. In this paper, whether the random variables sequence has "Ma" is tested, so Markov chain is more perfect.
3. The division of prediction accuracy state space has an important influence on the results. How to choose a reasonable division of space need to be further researched.

REFERENCES

Chen Hua-you, Sheng Zhao-han. A Kind of New Combination Forecasting Method Based on Induced Ordered Weighted Geometric Averaging (IOWGA) Operator. [J]. Journal of Industrial Engineering P Engineering Management, 2005, 19(4): 36–39.

Deng Hong-yan, Wang Cheng-hua. Prediction of Groundwater Level for Reservoir Slope with Non-linear-Combined Model [J]. Journal of Civil, Architectural & Environmental Engineering, 2010, 32(1): 31–35.

Liu Jia-jun, Wang Ming-jun, Xue Mei-juan, et al. A new combination forecasting model based on Theil coefficient and the Induced Ordered Weighted Averaging (IOWA) operator and Markov chain (MC) for annual electricity consumption [J]. Power System Protection and Control 2011, 39(19): 30–36.

Wang Xin-min, Cui Wei. Application of Changeable Weight Combination Forecasting Model to Groundwater Level Prediction [J]. Journal of Jilin University (Earth Science Edition), 2009, 39(6): 1101–1105.

Wu Dong-jie, Wang Jin-sheng, TENG Yan-guo. Application of wavelet decomposition and wavelet transform method to forecasting of groundwater regime., SHULI XUEBAO, 2004, 39(5): 39–44.

Yang Zhong-ping, Lu Wen-xi, LI Ping. Application of time-series model to predict groundwater regime [J]. SHULI XUEBAO, 2005, 36(12): 1475–1479.

Zhu Chun-jiang, Tang De-shan. Study on Ecological Agriculture Groundwater Level Forecast Based on Grey Theory BP Neural Network and D-S Theory [J]. Journal of Anhui Agri. Sci, 2006, 34(5): 831–832.

Zhou Li-gang, Zhao Juan, Chen Hua-you, et al. Combination forecasting model based on vectorial angle cosine and IOWGA operator [J]. Journal Of Hefei University Of Technology, 2010, 33(9): 1425–1429.

Frontiers of Energy and Environmental Engineering – Sung, Kao & Chen (eds)
© *2013 Taylor & Francis Group, London, ISBN 978-0-415-66159-1*

Emergency treatment process research of tailings reservoir failure and pollution accident

Y.L. Tang, X.Y. Cao, J.X. Fu & X. Song
Municipal and Environmental Engineering Institute of Shen Yang Jianzhu University, Shen Yang, China

ABSTRACT: Tailings reservoir waste water have a large number of pollutants, especially heavy mental ion, thus the key work in emergency treatment of tailings reservoir failure and pollution accident is treatment of heavy mental ion. With the main heavy metal ion—copper and zinc ion in metal mine tailings reservoir waste water as the object of study, this paper studied the method of active carbon adsorption and sulphide precipitation by doing experiment. These experiments were determined the related parameters and scope of application, which providing technical support for tailings reservoir failure and pollution accident.

Keywords: authors of; papers to proceedings; tailings reservoir; failure; pollution; heavy metal; emergency; experimental

1 INTRODUCTION

Recently, the sporadic environmental pollution accidents occurred frequently in China. Such as, dam-break in Tal Fibre mountain iron ore in Xinta mining industry LTD CO in 2008, in Xiangfen, Shanxi, lead to 26800 m³ tailings loss and 276 people dead (Zhengpan Shi, 2008); a tailings dam in Damiao village, WoLong town of PingQuan county, Hebei province in 2008 occurred local piping, part dam collapse, tailings flow gone to the road and then bury a running pickup car, lead to 3 people dead (Guohua Wang, et al. 2008). Almost every dam-break would trigger the sporadic environmental pollution events, furthermore threat drinking water environment safety repeatedly, these were especially worth to pay attention to. Recently, there had 10 events which were caused by tailings reservoir accidents in 56 sporadic environmental events involving in drinking safety, about 18% of the total (speaking of Lijun Zhang. 2010). China has increased emergency management and treatment dynamics of sporadic environment events, at the same time done an in-depth study in emergency treatment of tailings sporadic environment events from 2003. This paper studied the common emergency treatment craft in tailings failure and pollution water accidents by doing experiment.

2 EXPERIMENTAL MATERIALS AND METHODS

2.1 *Experimental materials*

1. Key instrument: WFX-320 atomic absorption spectrophotometer; six combination synchronization automatic lifting mixer; centrifuge; oven; colorimetric cylinder of 100 ml; precision pH test paper.
2. Key reagent: $CuSO_4 \cdot 5H_2O$, $ZnSO_4 \cdot 7H_2O$; powder activated carbon; HCl; $Na_2S \cdot 9H_2O$; NaOH.
3. Preparing for polluted water: Tap water and standard reserve liquid contained copper and zinc ion was confected to different concentrations polluted water sample 30 min before experiment. These water sample was used in simulate the water sample of tailings failure and pollution water accident and used in experimental.
4. Preparing for activated carbon pulp: put the activated carbon into the oven, and then dry it 4h at 120°C; get 2.00 g powder activated carbon into the 400 ml distilled water after it cooling, confect it to 5 g/L carbon pulp, then soak it about 24h and reserve it to use.

2.2 *Experimental methods*

1. Using the static stirring experimental.
   a. Using 500 ml measuring cylinder to get 500 ml water sample into six beaker respectively after the polluted water sample was confected in the tank;
   b. Adding the drug, then put the 1~6 beakers in the six combination synchronization automatic lifting mixer to stir: for active carbon adsorption, stirred several minutes at the speed of 110 r/min. For sulphide precipitation, stirred 30 min at the speed of 150 r/min;
   c. Letting the water sample static sank 30 min after stirring;

d. Taking the water sample at the superstratum into 100 ml centrifuge tube (attention: don't let the floating pollutants into the centrifuge tube);

e. Putting the centrifuge tube into the centrifuge according to four or two as a group, for active carbon adsorption, centrifuge 5 min at the speed of 3000 r/min, for sulphide precipitation centrifuge 15 min at the speed of 3000 r/min, so that to separate the floating substances in the water sample;

f. Using the pipette to move 10 mL supernatant liquor from the centrifuge tube for analyzing.

2. Analyzing method for Cu (II), Zn (II): use the atomic absorption spectrophotometry.

3 RESULTS AND DISCUSSION

3.1 *Emegency craft to remove heavy metal ion by using active carbon adsorption*

3.1.1 *Ascertaining the dosing quantity of active carbon*

According to experiment in 2.2, making other condition (concentration of Cu^{2+}:0.1 mg/L, concentration of Zn^{2+}:1.0 mg/L, pH:4.5, water temperature: about $18°C$) constant, then taking 2, 3, 4, 5, 6, 7 mL prepared 5 mol/L activated carbon pulp into 1~6 number beaker, the test result was in Figure 1.

The adsorption effect had relationship with the effective contact area of the adsorbent and adsorbates in the static state adsorption experiment (Meng Chao, et al. 2011). From Figure 1 we know: with the quantity of active carbon increased from 10 mg/L to 25 mg/L, the precipitation removal rate is improving gradually. When the dosing quantity of actived carbon is 25 mg/L, the removal rate of copper ion and zinc ion is 96.22% and 95.23%, and the effluent are all reach the standard. When the dosing quantity of actived carbon increased from 25 mg/L to 35 mg/L,

removal rate enhanced slowly, only improved about 1%. Comprehensive consideration all factors, the best dosing quantity of actived carbon is 25 mL/L in this experiment.

The adsorption effect had relationship with the effective contact area of the adsorbent and adsorbates in the static state adsorption experiment (Meng Chao, et al. 2011). With the increasing quantity of actived carbon, the provided adsorbent active locus increase (Guifang Wang, et al. 2004), and the effective contact area increased, then the absolute quantity of adsorbates in the adsorbent increased, thus the removal rate improved; but when the active carbon increased at a certain amount, the effective contact area of the adsorbent and adsorbates in the solution would not increase, the increasing range of the effective contact area would become little, therefore the removal rate of copper and zinc ion become slowly.

3.1.2 *The influence of pH on removal effect of copper and zinc ion*

Making other condition constant, then regulating pH become 2,3,4,5,6,7 by adding dropwise HCl into the 1~6 water sample respectively, the test result showed in Figure 2. From the chart we knew: with the increasing of pH, the active carbon adsorbance for Cu^{2+}, Zn^{2+} increased, but when pH increased continue the adsorbance reduced slowly when the pH among 2~7. It was because when pH was small, there was abundant H^+ in the solution which was combined with the functional group in the surface of active carbon, then the affinity of active carbon surface was changed because the effective active centers was occupied by H^+, so the adsorbance was less relatively. With the increasing of solution pH, the H^+ combined in the surface of active carbon was dissociated, then a mass of active centers was exposed, which was occupied rapidly by copper and zinc ion, thus can be adsorbed effectively by active carbon. At the moment, after pH increased, the functional group in the surface of active carbon was protonized, which made the

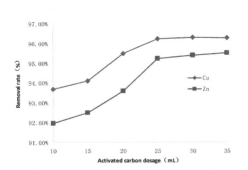

Figure 1. Different coagulant activated carbon dosage for Cu (II), Zn (II) removal.

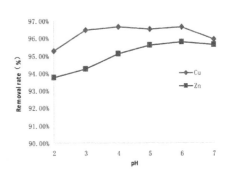

Figure 2. Effect of pH on Cu^{2+}, Zn^{2+} removal rate.

surface potential density was reduced, therefore it weakened the electrostatic repulsion of the active carbon surface and Cu^{2+}, Zn^{2+}, and increased the adsorbance (Chen J Paul, et al. 2001; Chen J Paul, et al. 2001; Yanhui Li, et al. 2003). But with the continue increased of pH, the chemistry acting force of OH^- and heavy metal ion in the solution enhanced, that resulted produce metal precipitation of hydroxide, then the adsorbance decreased relatively.

Different metal ion had different optimum adsorption pH, we knew the best pH of the test by using powder active carbon to wipe off Cu^{2+}, Zn^{2+} should be controlled among 3.0~6.0,5.0~7.0 respectively. The main reaction on metal ion in the surface of active carbon was ion exchange adsorption, taking more electric charge would be good for adsorbing, (Yanhui Li, et al. 2003). But the atomic number of copper ion was higher than zinc ion's, the adsorptive capacity of copper ion exceeded zinc ion's. Thus the adsorption effect of Cu^{2+} was better than Zn^{2+}. The pH of effluent should be regulated to among the required standards of Living and Drinking Water Health Standards in practical emergency craft process of water works.

3.1.3 *Ascertaining the adsorption time*

According to the experiment method showed in 2.2, added equivalent powder active carbon into 1~6 water samples, then stirred 10 min, 20 min, 30 min, 60 min, 90 min, 120 min respectively, detected concentration of Cu^{2+}, Zn^{2+} in the water samples at last. The test result showed in Figure 3. We knew that the powder active carbon adsorption for Cu^{2+}, Zn^{2+} experienced two process—rapidly adsorption and slowly adsorption from Figure 3. In the former 10 min of the whole reaction, the speed of adsorption was quickly and the adsorbance of Cu^{2+}, Zn^{2+} was 79.35%, 72.79% of the whole respectively. Afterwards, the adsorbance gradually reduce, the adsorbance of Cu^{2+}, Zn^{2+} was 13.29%, 16.22%

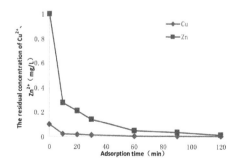

Figure 3. Adsorption time and residual Cu^{2+}, Zn^{2+} concentration.

of the whole e respectively in the 40 min among 20 min~60 min. The adsorbance of Cu^{2+}, Zn^{2+} was invariant mainly in the succedent 1 h. After a enough long time, the concentration of Cu^{2+}, Zn^{2+} in water was not changed and a dynamic balance relationship of the concentration and adsorbance of Cu^{2+}, Zn^{2+} in water was builded.

At first, the pollutants' concentration in the water were high relatively which made the probability of active carbon catching pollutants high relatively, moreover the concentration difference of interspace inside and outside was big that made the pollutants' diffusion velocity become rapidly, therefore the adsorption speed was more rapid (Guifang Wang, et al. 2004). But as time went on, pollutants concentration in interspace gradually raised which lead to the concentration difference of interspace inside and outside decreased, then the pollutants' diffusion velocity became slowly, and the rate of adsorbing pollutants became slowly (Youzhen Zhu, et al. 2001). When adsorption reached dynamic balance, the pollutants' concentration of interspace inside and outside became the same, then the pollutants' concentration in water would not change. From Figure 3 we knew among the former 60 min the adsorbance of Cu^{2+}, Zn^{2+} was all over 90% of the whole, thus the adsorption time was 60 min in the experiment.

3.1.4 *The removal rate of Cu (II), Zn (II) on different exceed standard times*

Confecting water sample respectively, which contain copper and its concentration is 0.10 mg/L, 0.20 mg/L, 0.50 mg/L, 1.00 mg/L, 1.50 mg/L, 2.00 mg/L, the corresponding zinc ion's concentration is 0.50 mg/L,1.00 mg/L, 2.50 mg/L, 5.00 mg/L, 7.50 mg/L, 10.00 mg/L. Adding different quantity powder active carbon into above water sample to do repeatedly experiment in order to ascertain the best dosing quantity of powder activated carbon. The needing dosing quantity of powder activated carbon, the concentration of Cu^{2+}, Zn^{2+} in effluent and the removal rate when the pH is 4.5 are listed in Table 1.

From Table 1, we knew, it could fully be removed effectively by using this method when the concentration of Cu^{2+}, Zn^{2+} in raw water were under 20 times of the standards. When the concentration of Cu^{2+}, Zn^{2+} in raw water were 10 times and 20 times respectively, only added 25 mg/L and 50 m mg/L powder active carbon correspondingly, the concentration of Cu^{2+}, Zn^{2+} in effluent were under 0.01 mg/L and 0.05 mg/L and all reached the standards. But it should be ensure the adsorption time of powder active carbon was about 60 min.

It was very difficult to reach A class of surface water standards by using this method when the concentration of Cu^{2+}, Zn^{2+} were over 20 times of the standards. Even the dosing quantity of powder

Table 1. Suitable dosage of activated carbon on different Cu²⁺, Zn²⁺ concentration and Cu²⁺, Zn²⁺ removal.

Concentration of raw water (mg/L)						
Cu (II)	0.1	0.2	0.5	1	1.5	2
Zn (II)	0.5	1	2.5	5	7.5	10
The times of exceed standard	10	20	50	100	150	200
Dosing quantity (mg/L)	25	50	125	250	375	500
The adsorption time (min)	60	60	120	120	120	120
Cu (II)						
Residual concentration (mg/L)	0.009	0.008	0.045	0.163	0.257	0.390
Removal rate (%)	91.58%	96.20%	91.06%	83.71%	82.86%	80.48%
Zn (II)						
Residual concentration (mg/L)	0.042	0.042	0.364	0.867	1.303	2.150
Removal rate (%)	91.56%	95.78%	85.43%	82.67%	82.62%	78.50%

active carbon was 250 mg/L, the effluent could not reach the standards when the when the concentration of Cu²⁺, Zn²⁺ were over 100 times. The adsorption method by using powder active carbon had quantities of shortcomings, such as the adsorption equilibrium time was too long which make it not suitable to use in the system of river emergency treatment, moreover it was not suitable to use in the system of water pipe emergency treatment when the water pipe was not enough long; the dosing quantity of powder active carbon was more and the cost was higher; it should regulate the pH two times when using in the system of water works emergency treatment, and so on.

3.2 Emergency craft to remove heavy metal ion by using method of sodium sulfide precipitation

3.2.1 Ascertaining the dosing quantity of sodium sulfide

According to experiment in 2.2, making other condition constant, then taking 2, 3, 4, 5, 6, 7 mL prepared 0.01 mol/L sodium sulfide solution into 1~6 number beaker, the test result was in Figure 4. From the chart we knew: with the increasing of sodium sulfide, the concentration of Cu²⁺, Zn²⁺ decreased, moreover the precipitation removal effect was enhancing, but the unit precipitation capacity decreased. When the dosing quantity of sodium sulfide increased from 2 mL to 5 mL, the removal rate of copper ion and zinc ion enhanced respectively from 96.53% to 97.1% and 94.85% to 96.96%, the removal effect was notably and the effluent all reached the standard; Continuing to increase dosing quantity of sodium sulfide, the improving of removal rate went to gentle.

Comprehensive consideration all factors, the best dosing quantity of sodium sulfide was 5 mL the 1.3 times of its theoretical calculation quantity.

The removal rate of copper ion was higher than zinc ion under the circumstance of the same dosing quantity of sodium sulfide for the same

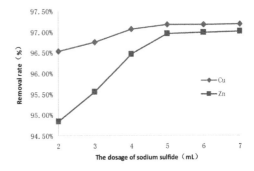

Figure 4. Cu²⁺, Zn²⁺ removal rate on different dosage of sodium sulfide.

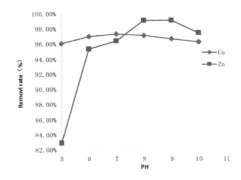

Figure 5. Effect of pH on Cu²⁺, Zn²⁺ removal rate.

solution in the certain concentration, because solubility product of copper sulfide was 4×10^{-38} and the solubility product of zinc sulfide was 8×10^{-26} (Yueyuan Zhang. 2010), comparing with zinc ion, solubility product of copper sulfide was more small, the affinity of its with sulphur ion was more stronger. So, copper ion would be prior sediment, the removal rate would higher than zinc ion.

Table 2. Suitable dosage of sodium sulfide on different Cu^{2+}, Zn^{2+} concentration and Cu^{2+}, Zn^{2+} removal.

Concentration of raw water (mg/L)						
Cu (II)	0.1	0.2	0.5	1	1.5	2
Zn (II)	0.5	1	2.5	5	7.5	10
The times of exceed standard	10	20	50	100	150	200
Dosing quantity (mg/L)	0.3×10^{-3}	0.6×10^{-3}	1.5×10^{-3}	3×10^{-3}	4.5×10^{-3}	6×10^{-3}
Cu (II)						
Residual concentration (mg/L)	0.003	0.005	0.01	0.02	0.032	0.035
Removal rate (%)	96.65%	97.63%	98.01%	97.96%	97.87%	98.23%
Zn (II)						
Residual concentration (mg/L)	0.009	0.01	0.047	0.364	0.435	0.545
Removal rate (%)	98.11%	99.00%	98.11%	92.71%	94.20%	94.55%

3.2.2 *The influence of pH for removal Cu (II), Zn (II) by the method of sodium sulfide precipitation*

Previous study showed that we could precipitate different heavy metal ion selectively by regulating pH[9], thus it could be seen the key influencing factor of sodium sulfide precipitate copper and zinc ion.

According to experiment in 2.2, Making other condition constant, then regulated pH become 5, 6, 7, 8, 9, 10 by adding dropwise HCl or NaOH into the 1~6 water sample respectively, the test result was in Figure 5.

From the chart we knew: the pH's influence for copper ion removal rate greater than zinc ion. When the pH regulated from 5 to 10, the removal rate of copper ion were all higher than 95%, even the range of variation was not higher than 1%, oppositely the removal rate of copper ion improved from 82.85% when the pH was 5 to 99.06% when the pH was 8, then decreased. Because zinc was metalloid, when the pH was overtop, it would occur opposite dissolve phenomenon; But, when pH over 7, the removal rate of zinc ion higher than copper ion; The best pH of sodium sulfide precipitate copper ion was 6~9, the zinc ion's was 8~9, comprehensive consideration the best pH of this experiment was 8.

3.2.3 *The removal rate of Cu (II), Zn (II) on different exceed standard times*

Confecting six water sample respectively, which contain copper and its concentration was 0.10 mg/L, 0.20 mg/L, 0.50 mg/L, 1.00 mg/L, 1.50 mg/L, 2.00 mg/L, the corresponding zinc ion's concentration was 0.50 mg/L, 1.00 mg/L, 2.50 mg/L, 5.00 mg/L, 7.50 mg/L, 10.00 mg/L. The dosing quantity of every sample were 1.3 times of its theoretical calculation quantity and the pH was 8. The needing dosing quantity of sodium sulfide, the concentration of Cu^{2+}, Zn^{2+} in effluent and the removal rate were listed in the Table 2.

From Table 2, we knew sodium sulfide has a preferable removal effect for Cu (II), Zn (II). The removal rate were over 96% and the effluent would all reach the standard when the concentration of Cu^{2+}, Zn^{2+} in raw water were 20 times over the standard; the effluent would mainly reach the standard when the concentration of Cu^{2+}, Zn^{2+} in raw water were below 50 times over the standard, moreover it would lower if increased the dosing quantity of sodium sulfide. Compared with power active carbon adsorption, this method had advantages: low dosing quantity, when the concentration of Cu^{2+}, Zn^{2+} in raw water were 50 times over the standard, the dosing quantity of sodium sulfide was 1.5×10^{-3} mg/L, and the effluent all reached the standard; short reaction time, method of sodium sulfide precipitation only had 30 min stirring time while the method of power active carbon adsorption had at least 60 min adsorption time; compared with the method of power active carbon adsorption, sodium sulfide precipitation had a higher removal rate for Cu^{2+}, Zn^{2+} at the same times over the standard. The removal rate of sodium sulfide precipitation were all over 96% when the concentration of Cu^{2+}, Zn^{2+} in raw water were 10 times over the standard.

4 CONCLUSIONS

1. The best dosage were 25 mg/L, the best pH rang was 5~6 and the best adsorption time was above 60 min when using method of active carbon adsorption to deal with the waste water in sudden water pollution incident which the concentration of Cu^{2+}, Zn^{2+} were 10 times and 20 times respectively. But it had an unsatisfactory removal effect when deal with the high concentration waste water, such as it could not reach the standard when the concentration of Cu^{2+}, Zn^{2+} in raw water were above 20 times even added 100 mg/L powder active carbon.

2. When using method of sodium sulfide precipitation to deal with the waste water in sudden water pollution incident which the concentration of Cu^{2+}, Zn^{2+} were 10 times and 20 times respectively, the best dosage was 1.3 times of its theoretical calculation quantity and the best pH range was 8~9. The effluent would reach the standard and the removal rate would over 98% when the concentration of Cu^{2+}, Zn^{2+} in raw water were above 50 times.
3. The method of active carbon adsorption was suitable in the system of river and water pipe and water works emergency treatment under controlled condition; The method of sodium sulfide precipitation should be used with coagulation and precipitation method in the system of river emergency treatment because of the produced particle was so small and it would produce hydrothion when the pH of waste water was acidity, so it should be careful use in the system of water pipe emergency treatment.

ACKNOWLEDGEMENTS

This research has been funded by the project "National water pollution control and management science and technology major projects" (contract no. 2009ZX07528-006-01) and by the Shenyang science and technology plan projects (F11-263-5-12). The work has benefited greatly from numerous discussions with tutors, notably Yulan Tang, Jinxiang Fu, Xinguan Ma. The constructive reviews by two anonymous referees is also acknowledged.

REFERENCES

Chen J Paul, Minsheng Lin. 2001. Equilibrium and kinetics of metal ion adsorption onto a commercial H-type granular activated carbon: experimental and modeling studies [J]. Water Research, 35(10): 2385–2394.

Chen J Paul, Minsheng Lin. 2001. Surface charge and metal ion adsorption on a H-type activated carbon: experimental observation and modeling simulation by the surface complex formation approach [J]. Carbon, (39): 1491–1504.

Guifang Wang, Mingfeng Bao, Zezhi Han. 2004. Application Study on Heavy Metal Wastewater Treatment With Active Carbon [J]. Environment Protection Science, 2004,(02).

Guohua Wang, Xixiang Duan, Yangang Miao, etc. 2008: The experience and lessons from tailings failure in domestic and overseas [J]. Science & Technology Information, 2008(01).

Meng Chao, Xiaofang Hu. 2011. The experiment study on removal effect of Hg, Pb, Cr, etc nine heavy metal by using method of powder active carbon adsorption. [J] City and Town Water Supply, 2011,(03).

Speaking of Lijun Zhang. 2010. Figure out fast and sum up experiences, advancing tailings' environment emergency management works—the undersecretary in Environment Protection Department on-the-spot meeting of tailings' environment emergency management works in nationwide. 2010,10,15.

Yanhui Li, Jun Ding, Zhaokun Luan. 2003. Competitive adsorption of Cu^{2+}, Pb^{2+} and Cd^{2+} ions from aqueous solutions by multiwalled carbon nanotubes [J]. Carbon, 2003(41): 2787–2792.

Youzhen Zhu, Youcai Zhao. 2001. Selective Separation of Lead From Alkaline Zinc Solution by Sulfide Precipitation [J]. Journal of Shanghai Institute of Technology, 2001,(01).

Yueyuan Zhang. 2010. Using Method of Precipitation to Deal with Solution which Contains Heavy Metal [J]. Guangxi Journal of Light Industry, 2010,(07).

Zhengpan Shi. 2009. Study on the Preventing and Controlling of Pollution and the Urgent Processing Measures of Accident about the Tailings Pond [D]. Changan University, 2009.

Frontiers of Energy and Environmental Engineering – Sung, Kao & Chen (eds)
© 2013 Taylor & Francis Group, London, ISBN 978-0-415-66159-1

Reconfiguration of electric vehicle access to micro-grid

X. Zhan, T.Y. Xiang & B. Zhou
School of Electric Engineering of Wuhan University, Wuhan, Hubei Province, China

ABSTRACT: Taking the characteristics of the micro-grid and the EVs into account, this paper studies the micro-grid reconfiguration considering several scenarios of EVs, such as the influence of different connection points and different charging regimes. Taking CERTS system as an example, the paper analyzes the micro-grid reconfiguration with EVs by using discrete particle swarm optimization algorithm in terms of providing the theoretical basis for future micro-grid with the additional EVs.

Keywords: electric vehicle; micro-grid; vehicle to grid; reconfiguration

1 INTRODUCTION

Electric Vehicles (EVs), which are considered as the random charging loads or the mobile storages for renewable generation expansion, show the potential to be one of the most important part of the power grid [1]. In addition, micro-grid as a dispatchable unit has proven its economic and environmental benefits comparing to the large scale power grid. The penetration of a large number of EVs will certainly influence the power flow, voltage profile and network reconfiguration. EVs transmit power energy to power grid to ensure the supply for the important loads when faults happen in the micro-grid. It is obvious that the reconfiguration will be affected by the EVs, which are a role of the source now.

This paper investigates the reconfiguration of EVs charging on the micro-grid, considering several scenarios of EVs. The paper is organized in the following way: Section II presents a brief overview of the micro-grid. Section III gives the reconfiguration problems formulation with the addition of EVs in the micro-grid. Section IV provides an overview of Discrete Particle Swarms Optimization (DPSO) algorithm and describes how it can be applied to the network reconfiguration problem. Section V, taking CERTS system as an example, proposes the different connection points of aggregation that are taken into account in order to reflect different configuration scenarios. Additionally, the cases considered for investigation of different EVs charging regimes and the simulation results for reconfiguration are following. Section VI outlines the conclusions.

2 OVERVIEW OF MICRO-GRID

The basic structure of the micro-grid contains a number of DGs that are multiple energy forms.

There are several feeders and a bus in the micro-grid. Feeders which are connected to the distribution system through the main transformer, operate between the isolated network and grid-connected. The switch is called PCC [2,3]. Under the normal operating state, the PCC is closed and the micro-grid is connected to distributed grid. The power is exchanged between the two systems. When the faults are happened in the distributed grid, the PCC is turned off so that the micro-grid switches to island operation mode. The micro-grid is still able to ensure the normal power supply for the loads.

3 PROBLEMS FORMULATION

Reconfiguration is certainly a mixed integer non-linear optimization problem to find a best configuration of radial network that gives minimum power loss while the imposed operating constraints are satisfied, which are current capacity of the feeder, voltage profile of the system, and radial structure of the micro-grid [4,5]. The objective function aimed at the minimization of real power loss is described as

$$\min P_{loss} = \sum_{i=1}^{n} X_i R_i \frac{P_i^2 + Q_i^2}{V_i^2} \tag{1}$$

Subjected to

$$AP = D \tag{2}$$

$$g_k \in G_k \tag{3}$$

$$V_{\min} \leq V_i \leq V_{\max} \quad i = 1, 2 \ldots n \tag{4}$$

$$I_{\min} \leq I_i \leq I_{\max} \quad i = 1, 2 \ldots n \tag{5}$$

where: P_{loss} is the total real power loss of micro-grid; X_i is the witch status of branch i; n is the number of lines; R_i is the resistance of branch i; P_i, Q_i are the active power and reactive power of branch i, respectively; V_i is the voltage magnitude of node i; A is the bus incidence matrix; P is the vector of feeder power flow; D is the vector of load; g_k is the current configuration of micro-grid; G_k is the set of all the radial structure; V_{min}, V_{max} are the bus minimum and maximum voltage limits, respectively; I_i is the current of branch i; I_{min}, I_{max} are the bus minimum and maximum current limits, respectively.

4 OVERVIEW OF THE ALGORITHM

Discrete Particle Swarm Optimization (DPSO) algorithm is carried out to solve it in this paper. DPSO also has a position vector $\mathbf{x} = \{x_1, x_2 \ldots, x_i, \ldots x_n\}$, $1 \leq i \leq n$ and a speed vector $\mathbf{v} = \{v_1, v_2 \ldots, v_i, \ldots v_n\}$, $1 \leq i \leq n$, which are the same as the standard particle swarm algorithm. Several rules are defined below:

Definition 1: Subtraction calculation of position vector
The position vector x_1 subtracts x_2 to get the speed vector v. When v and x_2 take a compute, we can get the position vector x_1

$$x_1 \ominus x_2 = v \qquad (6)$$

Definition 2: Addition calculation of speed vector and position vector
The addition calculation of speed vector and position vector enable the particle to move:

$$x_1 = x_2 \oplus v \qquad (7)$$

Definition 3: Addition operation (calculation) of speed vector
New speed vector is the sum of two speed vectors, which is:

$$v = v_1 \oplus v_2 \qquad (8)$$

Therefore, the basic equations of the discrete particle swarm algorithm is listed as follows

$$v_i(t+1) = c_1 \cdot v_i(t) \oplus c_2 \cdot \left(p_i(t) \ominus x_i(t)\right) \oplus c_3 \left(p_g \ominus x_i(t)\right) \qquad (9)$$

$$x_i(t+1) = x_i(t) \oplus v_i(t+1) \qquad (10)$$

where: x_i is a discrete position vector of the particle i, v_i is a discrete speed vector of the particle i, p_i is the individual optimal solution of the particle i, p_g is the global optimal solution, $\oplus, \ominus$ is

addition and subtraction calculation in the DPSO algorithm, respectively.

5 CASE STUDY AND RESULTS

In this research, using a MATLAB model of the CERTS micro-grid which is shown in Figure 1, different EVs integration scenarios have been considered. Toyota RAV4 EVs with a 3.6 kW charger are used in the simulation. Results are shown by per-unit value. Figures 2 and 3 illustrate the load duration curves and output power of micro-source, respectively.

A. Scheme 1
Analyze and compare the reconfiguration results in an assumption of EVs charging at different connection points at thirteen o'clock. The calculation results are listed in Table 1 and shown from Figures 4–8.

In Table 1, it is observed that the micro-grid power loss is increased by the additional EVs

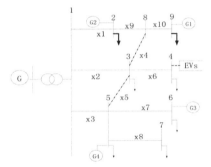

Figure 1.   CERTS system.

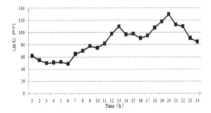

Figure 2.   Load duration curves.

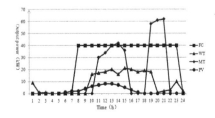

Figure 3.   Output power curves of micro-grid.

Table 1. Reconfiguration results of EVs in different connection points.

	Without EVs	Before reconfiguration		After reconfiguration		
Case	Real power loss	Connection points	Real power loss	Real power loss	Bus number of lowest voltage	Breaker reconfiguration
Case 1	0.0063	Bus 4	0.0228	0.0187	7	x2, x5
Case 2	0.0063	Bus 2	0.0132	0.0116	4	x1, x4

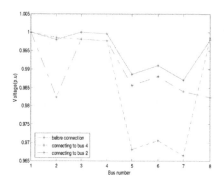

Figure 4. Voltage of nodes before and after integration.

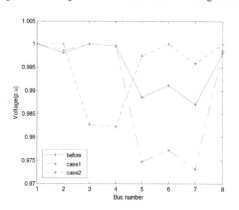

Figure 5. Voltage of nodes before and after reconfiguration in scheme 1.

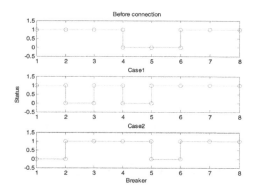

Figure 6. Reconfiguration of scheme 1.

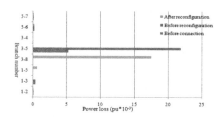

Figure 7. Power loss of case 1.

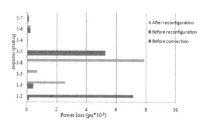

Figure 8. Power loss of case 2.

penetration, so the operators should do the reconfiguration according to the current load demand. Different connection points of EVs can cause different impacts on the micro-grid. In Figure 4, when EVs connecting to the bus 4, voltage of bus 4 is reduced rapidly due to the heavy loads connected. Besides, voltage of bus 7 after reconfiguration is relatively higher than before. Because bus 4 is at the end of the feeder, it can be observed in Figure 5. that voltage of bus 4 after reconfiguration is the lowest when EVs connecting to the bus 2 in parallel with G2 (wind generators) leading the load heavier near bus 2. The results of power loss are obviously reduced after reconfiguration in Figures 7 and 8. Figure 6 represents the optimal status of switches after reconfiguration.

B. Scheme 2

As shown in Figure 2, The peak hours of the micro-grid are happened at thirteen o'clock and twenty-one o'clock. Hence, study the impacts of uncoordinated charging and coordinated charging model at twenty-one o'clock on the assumption of EVs connecting to the bus 6 in scheme 2. The calculation results are shown in Table 2 and from Figures 9–11.

Table 2. Reconfiguration results of EVs in different charging regimes.

Models	Before reconfiguration	After reconfiguration		
	Real power loss	Real power loss	Bus number of lowest voltage	Breaker reconfiguration
Coordinated charging (V2G)	0.0078	0.0078	2	x2, x4
uncoordinated charging	0.0349	0.0247	6	x1, x4

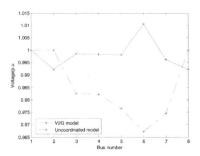

Figure 9. Voltage of nodes in scheme 2.

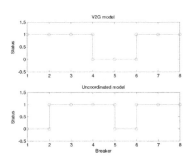

Figure 10. Reconfiguration of scheme 2.

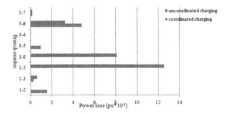

Figure 11. Power loss results of scheme 2.

The real power loss of micro-grid without EVs is 0.0119. In Table 2, it shows that coordinated charging model (V2G model) can obviously reduce the power loss. All the bus voltage under the coordinated charging model is near the per-unit value of 1 after reconfiguration in Figure 9. On the contrary, all the bus voltage under the uncoordinated charging model will reduce heavily, especially bus 6 due to the phenomenon of peaks store. In addition, the power loss under the coordinated charging model is further less than that under the uncoordinated charging model. Figure 10 represents the optimal status of switches after reconfiguration.

6 CONCLUSION

This paper introduces the reconfiguration with EVs access to the micro-grid on the research basis of the characterastics of the EVs and the micro-grid. Some of the issues that concern the impacts of EVs on micro-grid were provided and case study scenarios were built. The simulation results can be summarized as follows: (1) Loads and the bus voltage will certainly be changed when EVs are connected into different points, especially the heavy loads will have an obvious impact on the grid; (2) EVs under the V2G model can reduce the power loss efficiently, and the reconfiguration carrying out the coordinated models is able to optimize the voltage profiles and the operation of the micro-grid; however, the uncoordinated model can burden the micro-grid during the peak hours as well as the peak-valley difference, affect the load balancing and voltage quality.

REFERENCES

[1] Kristien C, Edwin H, Johan D. The impact of charging plug-in hybrid electric vehicles on a residential distribution grid [J]. IEEE Trans on Power Systems, 2010, 25(1):371–380.
[2] Wang Jian, Liu Tianqi, Li Xingyuan. Influences of Connecting Wind Farms and Energy Storage Devices to Power Grid on Reliability of Power Generation and Transmission System [J]. Power System Technology, 2011, 35(5):165–170.
[3] Robert H. Lasseter, Paolo Piagi. Control and Design of Micro-grid Components. Final Project Report, 2006, 1:34–36.
[4] Gao Ciwei, Zhang Liang. A survey of influence of electric vehicle charging on power grid [J]. Power System Technology, 2011, 35(2):127–131 (in Chinese).
[5] Cao Yijia, Tan Yi, Li Canbing, Li Jian, Tang Shengwei, Zhang Zhikun. Typical Schemes of Electric Vehicle Charging Infrastructure Connected to Grid [J]. Automation of Electric Power Systems, 2011, 35(14):48–52.

Frontiers of Energy and Environmental Engineering – Sung, Kao & Chen (eds)
© 2013 Taylor & Francis Group, London, ISBN 978-0-415-66159-1

Existence of equilibrium point for impulsive stochastic neural networks with mixed delays

X.A. Li & J.Z. Zou
School of Mathematics and Statistics, Central South University, Changsha, Hunan, P.R. China

E.W. Zhu
School of Mathematics and Computational Science, Changsha University of Science and Technology, Changsha, Hunan, P.R. China

ABSTRACT: This paper studies existence of equilibrium point for stochastic neural networks with time-varying delays and distributed delays under nonlinear impulsive perturbations. Based on definition of homeomorphism and inequality technology, the conditions of existence are derived.

Keywords: equilibrium point; stochastic neural networks; mixed delays

1 GENERAL INSTRUCTIONS

Delayed neural networks have drawn considerable attention in recent years and many results have been reported in the literatures during the past few decades. Arik S. & Tavsanoglu V. (1998) presents a sufficient condition for the uniqueness of the equilibrium point of cellular neural networks with constant delays by using M-matrix. Liu B. & Huang L.H. (2007) and Mohamad S. (2007) studies the sufficient conditions for existences of equilibrium point for neural networks by applying homeomorphism theory, but in these models, there is no delays. Hu C. (2010) investigates the dynamics behavirs of a class of neural networks with time-varying delays and the delays are boundary. In this paper, we consider the existence of equilibrium point for neural networks with time-varying delays and distributed delays under impulsive control and obtain the conditions of equilibrium point of this model.

2 MODEL DESCRIPTION AND PRELIMINARIES

In this paper, we consider the following model:

$$
\begin{cases}
dx_i(t) = \left[-a_i(x_i(t)) + \sum_{j=1}^{n} b_{ij} f_j(x_j(t)) + \sum_{j=1}^{n} c_{ij} g_j(x_j(t - \tau_j(t))) \right. \\
\left. + \sum_{j=1}^{n} d_{ij} \int_{-\infty}^{t} K_{ij}(t - s) h_j(x_j(s)) ds + I_i \right] dt \\
+ \sum_{l=1}^{m} \sigma_{il}(t, x_i(t), x_i(t - \tau_i(t))) dw_l(t), \quad t \neq t_k, \\
x_i(t_k) = p_{ik}(x(t_k^-)), \quad k \in \mathbb{Z}^+, i \in \Lambda.
\end{cases}
\tag{1}
$$

where $\Lambda = \{1, 2, \dots, n\}$, $\{t_k\}$ satisfy $0 < t_1 < t_2 < \dots < t_k < t_{k+1} \dots$, $\lim_{k \to \infty} t_k = \infty$; $x_i(t)$ corresponds to the state of the ith unit at time t; $a_i(x_i(t))$ represents the rate with which the ith unit will reset its potential to the resting states of the ith unit before and after impulse perturbation at the moment t_k; b_{ij}, c_{ij} and d_{ij} denote the constant connection weight and the constant delayed connection weight of the jth neuron on the ith neuron, respectively. $\tau_j(t)$ is the time-varying transmission delay, $j \in \Lambda$. $f_i(\cdot)$, $g_i(\cdot)$ denote the activation functions of the jth neuron; n corresponds to the numbers of units in a neural network; I_i denotes the external bias on the ith unit, and $p_{ik}(x(t_k))$ represents the abrupt change of the state $x_i(t)$ at the impulsive moment t_k. $w(t) = (w_1(t), \dots, w_m(t))^T$ is m-dimensional Brownian motion defined on a complete probability space $(\Omega, \mathcal{F}, P)$, $\sigma(t, x, y) = (\sigma_{il}(t, x_i, y_i))_{n \times m} \in R^{n \times m}$.

System (1) is supplemented the initial condition

$$
x_i(s) = \varphi_i(s), \, s \in (-\infty, 0], \, i \in \Lambda.
\tag{2}
$$

where $\varphi(s) = (\varphi_1(s), \varphi_2(s), \dots, \varphi_n(s))^T \in PCB_{\mathcal{F}_0}^b((-\infty, 0], R^n)$. Denote by $PCB_{\mathcal{F}_0}^b$ the family of all bounded $\mathcal{F}_0$-measuable, $PC((-\infty, 0], R^n) = \{\varphi : (-\infty, 0] \to R^n\}$ is continuous everywhere except at finite number of points t_k, at which $\varphi(t_k^+)$ and $\varphi(t_k^-)$ exist and $\varphi(t_k^+) = \varphi(t_k)$.

The norms are defined by the following norms:

$$
\|\varphi\|_p = \sup_{s \in (-\infty, 0]} \left(\sum_{i=1}^{n} |\varphi_i(s)|^p \right)^{1/p},
$$

$$
\|x\|_p = \left(\sum_{i=1}^{n} |x_i|^p \right)^{1/p}.
$$

Throughout this paper, the following standard hypothesis are needed:

(H1) Functions $a_i(x): \mathbb{R} \to \mathbb{R}$ are continuous and monotone increasing, that is, there exist real number $a_i > 0, for\ \forall u, v \in \mathbb{R}, u \neq v, i \in \Lambda$, such that $a_i(u) - a_i(v)/u - v \geq a_i$.

(H2) Functions $f_j(\cdot), g_j(\cdot), h_j(\cdot)$ are Lipschitz-continuous on R with Lipschitz constants $L_j^f > 0, L_j^g > 0$ and $L_j^h > 0$, respectively. That is For $\forall u, v \in R, u \neq v, i \in \Lambda$,

$$|f_j(u) - f_j(v)| \leq L_j^f |u - v|,$$

$$|g_j(u) - g_j(v)| \leq L_j^g |u - v|,$$

$$|h_j(u) - h_j(v)| \leq L_j^h |u - v|$$

(H3) The delay kernels $K_{ij}: [0, \infty) \to R$ are piecewise continuous and satisfy

$$|K_{ij}(s)| \leq \mathcal{K}(s) \forall i, j \in \Lambda, s \in [0, \infty)$$

And

$$\int_0^\infty \mathcal{K}(s) e^{\mu_0 s} ds < \infty.$$

in which $\mathcal{K}(t)$ corresponds to some nonnegative function defined on $[0, \infty)$ and the constant μ_0 denotes some positive number.

We end this section by introducing two definitions.

Definition 1 A constant vector $x^* = (x_1^*, x_2^*, ..., x_n^*)^T \in \mathbb{R}^n$ is said to be an equilibrium point of system (1) if x^* is governed by the algebraic system

$$a_i(x_i^*) = \sum_{j=1}^n b_{ij} f_j(x_j^*) + \sum_{j=1}^n c_{ij} g_j(x_j^*)$$
$$+ \sum_{j=1}^n d_{ij} \int_{-\infty}^t K_{ij}(t-s) h_j(x_j^*) ds + I_i, \qquad (3)$$

where it is assumed that impulse functions $p_{ik}(\cdot)$ satisfy $p_{ik}(x_1^*, x_2^*, ..., x_n^*) = x_i^*$ for all $i \in \Lambda$ and $\sigma(t, x_i^*, x_i^*) = 0$.

Definition 2 (Forti M. & Tesi (1995)) A map $H: \mathbb{R}^n \to \mathbb{R}^n$ is a homeomorphism of $\mathbb{R}^n$ onto itself if H is continuous and one-to-one and its inverse map H^{-1} is also continuous.

3 MAIN RESULT

For convenience, we denote that

$$\phi_i = p a_i - \sum_{j=1}^n \sum_{l=1}^{p-1} \left(|b_{ij}|^{p\alpha_{l,ij}} L_j^{f p \beta_{l,ij}} + |c_{ij}|^{p\gamma_{l,ij}} L_j^{g p \delta_{l,ij}} \right)$$
$$- \sum_{j=1}^n \sum_{l=1}^{p-1} |d_{ij}|^{p\xi_{l,ij}} L_j^{h p \eta_{l,ij}} \int_0^\infty \mathcal{K}(s) ds$$

$$- \sum_{j=1}^n \frac{\mu_j}{\mu_i} |b_{ji}|^{p\alpha_{p,ji}} L_i^{f p \beta_{p,ji}},$$

$$\psi_i = \sum_{j=1}^n \frac{\mu_j}{\mu_i} |c_{ji}|^{p\gamma_{p,ji}} L_i^{g p \delta_{p,ji}}, \quad \zeta_{ij} = \frac{\mu_i}{\mu_j} |d_{ij}|^{p\xi_{p,ij}} L_j^{h p \eta_{p,ij}}.$$

where μ_i are positive constant, $\alpha_{l,ij}, \beta_{l,ij}, \gamma_{l,ij}, \delta_{l,ij}, \xi_{l,ij}$ and $\eta_{l,ij}$ are real numbers and satisfy

$$\sum_{l=1}^p \alpha_{l,ij} = 1, \sum_{l=1}^p \beta_{l,ij} = 1, \sum_{l=1}^p \gamma_{l,ij} = 1,$$

and

$$\sum_{l=1}^p \delta_{l,ij} = 1, \sum_{l=1}^p \xi_{l,ij} = 1, \sum_{l=1}^p \eta_{l,ij} = 1.$$

Lemma 3.1 (Beckenbach EF. & Bellman R. (1965))
If $a_i (i = 1, 2, ..., p)$ denote p nonnegative real numbers, then

$$a_1 a_2 ... a_p \leq \frac{a_1^p + a_2^p + \cdots + a_p^p}{p}. \qquad (4)$$

where $p \geq 1$ denotes an integer. A particular form of (4), namely

$$a_1^{p-1} a_2 \leq \frac{(p-1)a_1^p}{p} + \frac{a_2^p}{p},$$

for $p = 1, 2, 3, \cdots$.

Lemma 3.2 (Forti M. & Tesi (1995))
If $H: \mathbb{R}^n \to \mathbb{R}^n$ is a continuous function and satisfies the following conditions:

1. $H(x)$ is injective on $\mathbb{R}^n$, that is $H(x) \neq H(y)$ for all $x \neq y$.
2. $\|H(x)\| \to \infty$ as $\|x\| \to \infty$.

Then $H(x)$ is homeomorphism of $\mathbb{R}^n$.

Theorem 3.3 System (1) exists a unique equilibrium x^* under the assumptions (H1)–(H3) if the following condition is also satisfied:

(H4) $\phi_i > \psi_i + \sum_{j=1}^n \zeta_{ji} \int_0^\infty \mathcal{K}(s) ds$.

Proof: Defining a map

$$H(x) = (h_1(x), h_2(x), ..., h_n(x))^T \in C^0(\mathbb{R}^n, \mathbb{R}^n).$$

where

$$h_i(x) = -a_i(x_i) + \sum_{j=1}^n b_{ij} f_j(x_j) + \sum_{j=1}^n c_{ij} g_j(x_j)$$
$$+ \sum_{j=1}^n d_{ij} \int_{-\infty}^t K_{ij}(t-s) h_j(x_j) ds + I_i \qquad (4)$$

301

Then by Lemma 3.2 the map H is a homeomorphism on $\mathbb{R}^n$ if it injective on $\mathbb{R}^n$ and satisfies $\|H(x)\| \to \infty$ as $\|x\| \to \infty$. In the following, we will prove that $H(x)$ is a homeomorphism.

Firstly, we claim that $H(x)$ is injective map on $\mathbb{R}^n$. namely

$H(x) \neq H(y)$ for all $x \neq y$.

There exist $x^T, y^T \in \mathbb{R}^n$ and $x^T \neq y^T$ such that

$H(x) = H(y)$, then

$$a_i(x_i) - a_i(y_i)$$
$$= \sum_{j=1}^{n} b_{ij}\left[f_j(x_j) - f_j(y_j)\right] + \sum_{j=1}^{n} c_{ij}\left[g_j(x_j) - g_j(y_j)\right]$$
$$+ \sum_{j=1}^{n} d_{ij} \int_{-\infty}^{t} K_{ij}(t-s)\left[h_j(x_j) - h_j(y_j)\right] ds.$$

$$(5)$$

It follows from (H1)–(H3) that

$$a_i |x_i - y_i|$$
$$\leq \sum_{j=1}^{n} |b_{ij}| L_j^f |x_j - y_j| + \sum_{j=1}^{n} |c_{ij}| L_j^g |x_j - y_j|$$
$$+ \sum_{j=1}^{n} |d_{ij}| L_j^h |x_j - y_j| \int_{-\infty}^{t} K(t-s) ds.$$

$$(6)$$

Therefore by lemma 3.1

$$\sum_{i=1}^{n} p\mu_i a_i |x_i - y_i|^p$$
$$\leq \sum_{i=1}^{n}\sum_{j=1}^{n} p\mu_i |b_{ij}| L_j^f |x_i - y_i|^{p-1} |x_j - y_j|$$
$$+ \sum_{i=1}^{n}\sum_{j=1}^{n} p\mu_i |c_{ij}| L_j^g |x_i - y_i|^{p-1} |x_j - y_j|$$
$$+ \sum_{i=1}^{n}\sum_{j=1}^{n} p\mu_i |d_{ij}| L_j^h |x_i - y_i|^{p-1} |x_j - y_j| \int_{-\infty}^{t} K(t-s) ds$$
$$\leq \sum_{i=1}^{n}\sum_{j=1}^{n} \mu_i \sum_{l=1}^{p-1} |b_{ij}|^{p\alpha_{1,ij}} L_j^{p\beta_{1,ij}} |x_i - y_i|^p$$
$$+ \sum_{i=1}^{n}\sum_{j=1}^{n} \mu_i |b_{ij}|^{p\alpha_{p,ij}} L_j^{p\beta_{p,ij}} |x_j - y_j|^p$$
$$+ \sum_{i=1}^{n}\sum_{j=1}^{n} \mu_i \sum_{l=1}^{p-1} |c_{ij}|^{p\gamma_{1,ij}} L_j^{g\,p\delta_{1,ij}} |x_i - y_i|^p$$
$$+ \sum_{i=1}^{n}\sum_{j=1}^{n} \mu_i |c_{ij}|^{p\gamma_{p,ij}} L_j^{g\,p\delta_{p,ij}} |x_j - y_j|^p$$
$$+ \sum_{i=1}^{n}\sum_{j=1}^{n} \mu_i \int_{-\infty}^{t} K(t-s) ds \sum_{l=1}^{p-1} |d_{ij}|^{p\xi_{1,ij}} L_j^{h\,p\eta_{1,ij}} |x_i - y_i|^p$$
$$+ \sum_{i=1}^{n}\sum_{j=1}^{n} \mu_i \int_{-\infty}^{t} K(t-s) ds |d_{ij}|^{p\xi_{p,ij}} L_j^{h\,p\eta_{p,ij}} |x_j - y_j|^p$$
$$= \sum_{i=1}^{n} \mu_i \left(pa_i - \phi_i + \psi_i + \sum_{j=1}^{n} \zeta_{ji} \int_0^{\infty} K(s) ds\right) |x_i - y_i|^p.$$

$$(7)$$

From (H4), which leads to a contradiction with our assumption. Therefore, $H(x)$ is an injective map on $\mathbb{R}^n$.

To demonstrate the property

$$\|H(x)\| \to \infty \text{ as } \|x\| \to \infty.$$

We have

$$\sum_{i=1}^{n} sgn(x_i) p\mu_i (h_i(x) - h_i(0)) |x_i|^{p-1}$$
$$= \sum_{i=1}^{n} sgn(x_i) p\mu_i |x_i|^{p-1}\left(-a_i(x_i) + \sum_{j=1}^{n} b_{ij} f_j(x_j) + \sum_{j=1}^{n} c_{ij} g_j(x_j)\right)$$
$$+ \sum_{i=1}^{n} sgn(x_i) p\mu_i |x_i|^{p-1} \sum_{j=1}^{n} d_{ij}\int_{-\infty}^{t} K_{ij}(t-s) h_j(x_j) ds$$
$$\leq -\sum_{i=1}^{n} pa_i\mu_i |x_i|^p + \sum_{i=1}^{n} \mu_i\left(pa_i - \phi_i + \psi_i + \sum_{j=1}^{n} \zeta_{ji}\int_0^{\infty} K(s) ds\right) |x_i|^p$$
$$= -\sum_{i=0}^{n} \mu_i(\phi_i - \psi_i - \sum_{j=1}^{n}\zeta_{ji}\int_0^{\infty} K(s) ds) |x_i|^p \leq -\omega \|x\|_p^p.$$

where $\omega = \min_{1\leq i\leq n}\{\mu_i(\phi_i - \psi_i - \sum_{j=1}^{n}\zeta_{ji}\int_0^{\infty} K(s) ds)\}$.

Using the *Hölder* inequality, we obtain

$$\|x\|_p^p \leq \frac{p\overline{\mu}}{\omega}\left(\sum_{i=1}^{n} |x_i|^p\right)^{1-1/p}\left(\sum_{i=1}^{n} |h_i(x) - h_i(0)|^p\right)^{1/p}.$$

$$(8)$$

which leads to

$$\|x\|_p \leq \frac{p\overline{\mu}}{\omega}(\|H(x)\|_p + \|H(0)\|_p).$$

$$(9)$$

From (9), we see that $\|H(x)\| \to \infty$ as $\|x\| \to \infty$. Thus, under the sufficient condition (H4), the map $H(x)$ is a homeomorphism on $\mathbb{R}^n$, and hence it has a unique fixed point x^*. This fixed point is the unique solution of the algebraic system (1) defining the unique equilibrium state of the impulsive network (2). The proof is now completed.

4 EXAMPLES

In this section, we will give an example to show the conditions given in the previous sections are more applicable than those given in some earlier literatures.

302

Consider the following neural networks with variable delays and distributed delays:

$$\begin{cases} dx_i(t) = \left[-a_i x_i(t) + \sum_{j=1}^{2} b_{ij} f_j(x_j(t)) \right. \\ \qquad + \sum_{j=1}^{2} c_{ij} g_j(x_j(t - \tau_j(t))) \\ \qquad \left. + \sum_{j=1}^{2} d_{ij} \int_{-\infty}^{t} K_{ij}(t-s) h_j(x_j(s)) ds + I_i \right] dt \\ \qquad + \sigma_i(t, x_i(t), x_i(t - \tau_i(t))) dw(t), t \neq t_k, \\ x_i(t_k) = p_{ik}(x(t_k^-)) = -0.2 x_i(t_k^-), k \in \mathbb{Z}^+, i \in \Lambda. \end{cases}$$

$$(10)$$

where $t_k = k, K_{ij}(s) = e^{-s}$,

$$f_j(s) = g_j(s) = h_j(s) = \frac{1}{2}(|x+1| - |x-1|)$$

$$(a_i)_{2\times 1} = \begin{pmatrix} 2 \\ 2 \end{pmatrix}, (b_{ij})_{2\times 2} = \begin{pmatrix} 0.5 & -0.3 \\ 0.4 & 0.2 \end{pmatrix}$$

$$(c_{ij})_{2\times 2} = \begin{pmatrix} 0.2 & 0.1 \\ -0.4 & 0.3 \end{pmatrix}, (I_i)_{2\times 1} = \begin{pmatrix} 0 \\ 0 \end{pmatrix},$$

$$(d_{ij})_{2\times 2} = \begin{pmatrix} 0.2 & 0.4 \\ 0.5 & 0.1 \end{pmatrix}, (\sigma_i)_{2\times 1} = \begin{pmatrix} 0.1 x_1(t) \\ 0.1 x_2(t) \end{pmatrix}.$$

In this case, we have

$$L_j^f = L_j^g = L_j^h = 1, \mathcal{K}(s) = e^{-s} \text{ for } j = 1, 2, \mu_0 = 0.9.$$

For $p = 2, \mu_1 = \mu_2 = 1,$

$$\alpha_{l,ij} = \beta_{l,ij} = \gamma_{l,ij} = \delta_{l,ij} = \xi_{l,ij} = \eta_{l,ij} = \frac{1}{2}$$

We can compute that

$$\phi_1 = 1.4, \psi_1 = 0.6, \zeta_{11} = 0.2, \zeta_{12} = 0.4,$$

$$\phi_2 = 1.6, \psi_2 = 0.4, \zeta_{21} = 0.5, \zeta_{22} = 0.1.$$

Obviously, (H4) is satisfied with this example, there is a unique point of system (10).

5 CONCLUSION

Delayed neural networks have been widely studied by many authors due to their extensive applications. In this paper, we consider the existence of unique equilibrium point for stochastic neural networks with mixed delays under nonlinear impulses. By virtue of homeomorphism, We derive the results. A numerical example illustrates the effectiveness of our results.

ACKNOWLEDGEMENTS

This work was supported in part by the National Natural Science Foundation of China under Grants no. 11101054, the Scientific Research Funds of Hunan Provincial Education Department of China under Grants no. 09C059, the Scientific Research Funds of Hunan Provincial Science and Technology Department of China under Grants no. 2010FJ6036 and the Open Fund Project of Key Research Institute of Philosophies and Social Sciences in Hunan Universities under Grants no. 11FEFM11.

REFERENCES

Arik, S., V. Tavsanoglu. 1998. Equilibrium analysis of delayed CNNs, *IEEE Trans Circuits Syst.* I. 45: 168–171.

Beckenbach EF, Bellman R. 1965. *Inequalities*, New York: Springer-Verlag.

Bingwen Liu, Lihong Huang. 2007. Existence of periodic solutions for cellular neural networks with complex deviating arguments. *Applied Mathematics Letters.* 20: 103–109.

Cheng Hu, Haijun Jiang, Zhidong Teng. Globally Exponential Stability for Delayed Neural Networks Under Impulsive Control, *Neural Process Lett.* 31:105–127.

Forti, M., Tesi A. 1995. New conditions for global stability of neural networks with application to linear and quadratic programming problems. *IEEE Transactions on Circuits and Systems I*, 42: 354–3695.

Sannay Mohamad. 2007. Exponential stability in Hopfield-type neural networks with impulses. *Chaos, Solitons and Fractals.* 32: 456–467.

303

Frontiers of Energy and Environmental Engineering – Sung, Kao & Chen (eds)
© 2013 Taylor & Francis Group, London, ISBN 978-0-415-66159-1

Modeling and simulation of variable-speed direct-driven PMSG wind turbine

S. Yang & L.D. Zhang
College of Energy & Environment, Xihua University, Chengdu, China

ABSTRACT: With the growing of electrical energy demand, wind power gains much more potential and plays a important position in energy sectors. In this paper present the dynamic model and control strategy of a direct drive permanent magnetic synchronous generator wind turbines. The PSMG model use d-q synchronous rotating reference frame and the wind turbine is pitch controllable. Three types of controllers, such as PI controller, Nonlinear PI controller for anti-windup, PID controller, present in the paper to analyze their performance in wind turbine. Matlab/Simulink used to verify the presented model and the control method. The model and the control strategy are verified by simulation results, at the same time, the Nonlinear PI controller for anti-windup shows the best performance in simulation progress which can makes up the effects of over-simplifications in modeling.

Keywords: wind turbine; variable speed; direct-drive PMSG; pitch control; optimum control

1 GENERAL INSTRUCTIONS

Wind power is rapid developing in recent years.

Variable-speed wind turbines make it could obtain maximum energy coefficient from wind of wide range. For variable speed wind turbines which can change the blade pitch angle to limit the capture efficiency of wind power. The pitch controller allows operating wind turbines at the optimum tip-speed ratio and enhance electricity production.

This technology has been widely use in Megawatt (MW) class wind turbines which are equipped with induction generator, Doubly Fed Induction Generators (DFIG) or Permanent Magnetic Synchronous Generators (PMSG). Compare with DFIG, the direct drive permanent magnetic synchronous generator wind turbines are without of gearbox, which could induce potential problems and require regular maintenance, can be more reliable [1–5].

The PI and PID control schemes are widely used in wind turbine [4–6]. Both of two kinds of control schemes are individual used in design and simulation. In order to reduce costs and improve wind turbine system reliability, a rational choice could help to analyze model in early period. This paper focuses on the dynamic modeling and control issue of a 2 MW direct drive PMSG wind turbines.

Using the Optimum method [7] to compare PI controller, Nonlinear PI controller for anti-windup, PID controller effect on wind turbines capture power from wind and operation.

This paper is structured as follows: the models of the wind PMSG wind turbine is present in section 2. Control of system will be presented in section 3. The simulation result are given in section 4, and some conclusion are draw in Section 5.

2 MODEL OF D-D PMSG WIND TURBINE

2.1 *Wind turbine rotor model*

According to Betz theory, the output of mechanical power capture from a variable speed wind turbine can be expressed as follow:

$$P_m = 0.5\rho A v^3 C_p(\lambda, \beta) \tag{1}$$

where ρ is the air density (Kg/m³), A is the swept area (m²) by the wind turbine rotor. v is the wind speed (m/s), and C_p is the power coefficient which is a nonlinear function of the pitch angle β and tip speed ratio λ is given by:

$$\lambda = \frac{R\omega_m}{v} \tag{2}$$

where R is the radius of the rotor (m) and ω_m is the mechanical angular speed velocity of the turbine rotor. For the D-D PMSG wind turbines, the angular velocity of the generator is equal to the turbine rotor.

The ratio between the power captured from the wind P_m(W), and the wind turbine rotor speed ω_m (rad/s) is aerodynamic torque T_m(N·m) can be expressed as:

$$T_m = \frac{P_m}{\omega_m} \tag{3}$$

A generic equation models power coefficient C_p. It indicates the relationship between power coefficient and tip speed ratio is illustrated in Figure 1, it based on the modeling turbine characteristics of [8], is given by:

$$C_p(\lambda, \beta) = 0.22\left(\frac{116}{\lambda_i} - 0.4\beta - 5\right)\exp\left(-\frac{12.5}{\lambda_i}\right) \tag{4}$$

with

$$\lambda_i = \left[\frac{1}{(\lambda + 0.089)} - \frac{0.035}{(\beta^3 + 1)}\right]^{-1} \tag{5}$$

The maximum value $C_{pmax} = 0.4382$, it is achieved for $\beta = 0°$ and $\lambda = 6.325$. This special value λ_{opt} is the optimal point where the maximum energy is wind turbine captured from wind.

2.2 Drive train

The wind turbine drive train dynamics can be modeled as:

$$\frac{d\omega_g}{dt} = \frac{T_e - T_w - B_f\omega_g}{J_{eq}} \tag{6}$$

where T_e is electro magnetic torque of the electric machine, B_f is viscous friction coefficient, ω_g is rotor mechanical speed, which is related to the rotor angular speed of the electric machine ω_e, through:

$$\omega_e = \omega_g P_n \tag{7}$$

where P_n is the number of poles of the PMSG, and ω_g equals to ω_m is the characteristic of direct drive

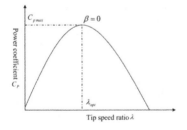

Figure 1. Power coefficient versus tip-speed ratio.

wind turbines. The model of drive train put into the Matlab/Simulink is illustrated in Figure 2.

2.3 Permanent magnetic synchronous generator

The permanent magnetic synchronous generator dynamic equation are expressed in the "d-q reference" frame as. The model of electrical dynamics in terms of voltage and current can be given as [9]:

$$U_{sd} = R_s i_{sd} + L_{sd}\frac{di_{sd}}{dt} - \omega_e L_{sq} i_{sq}$$
$$U_{sq} = R_s i_{sq} + L_{sq}\frac{di_{sq}}{dt} + \omega_e L_{sd} i_{sd} + E_s \tag{8}$$

where $E_s = \omega_e \psi_f$, ψ_f is the amplitude of flux linkages established by the permanent magnet, ω_e is the electrical angular velocity, u_{sd}, u_{sq} are d-axis and q-axis current respectively, the R_s is the generator resistance, L_{sd}, L_{sq} are d-axis and q-axis inductance respectively. Figure 3 shows the equivalent circuit of the PMSG in the the d-q synchronous rotating reference frame. The model of the PMSG put into operation in Matlab/Simulink is illustrated in Figure 4.

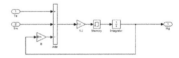

Figure 2. Drive train modeled in Matlab/Simulink.

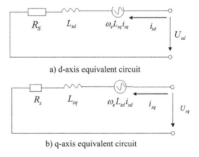

a) d-axis equivalent circuit

b) q-axis equivalent circuit

Figure 3. Equivalent circuit of the PMSG in the synchronous frame.

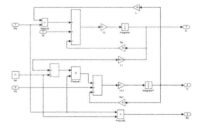

Figure 4. PMSG modeled in Matlab/Simulink.

The expression for the Electromagnetic (EM) torque in the rotor is given by [9,10]:

$$T_e = \left(\frac{3}{2}\right)\left(\frac{P_n}{2}\right)\left[\left(L_{sd} - L_{sq}\right)i_{qs}i_d - \psi_f i_{sq}\right] \quad (9)$$

If $i_{sd} = 0$, the electromagnetic torque is expressed as:

$$T_e = 1.5 P_n \psi_f i_{sq} \quad (10)$$

3 CONTROL STRATEGIES

3.1 Blade pitch angle control

The pitch angle effects wind turbines capture the power from wind. Pitch angle controller adjusts the pitch angle to changing the power coefficient. It maintains pitch angle in optimal position to adjust the rotor and the generator in optimal speed, thus, conducts wind turbine to optimum power. In the point of fact, the blade of pitch angle controller operates as same as rotational speed limiter in any wind speed conditions [1,3,11]. This model is illustrated in Figure 5.

3.2 Optimum control strategy

The optimal rotational speed of the wind turbine roto can be simply estimated as follows [7]:

$$\omega_{opt}(t) = \frac{\lambda_{opt}}{R} \cdot v(t) \quad (11)$$

By measuring wind speed, corresponding optimum rotor speed can be calculated and set as the reference rotor speed. Then, the calculated result put in to pitch controller for change the pitch angle, it is in order to reduce or improve wind turbines capture the power from wind.

3.3 Nonlinear PI controller for anti-windup

The windup phenomenon appears and result in performances degradation when the PI controller output is saturated. Compare with traditional PI controller, and the integral term of Nonlinear PI controller for anti-windup is separately controlled,

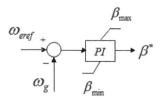

Figure 5. Pitch control diagram.

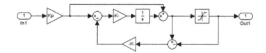

Figure 6. Nonlinear PI controller for anti-windup.

Table 1. Parameter of wind turbine and genrator.

	Symbol	Value	Units
Wind turbine			
Air density	ρ	1.225	kg/m³
Radius of blades	R	34	M
PMSG			
Rated power	P_m	2.1	MW
Rated speed	$\omega_{g\,rated}$	2.4	rad/s
Pole number	P_n	60	
Inductances	$L_{sd} = L_{sq}$	2.7	mH
Stator phase resistance	R_S	0.00665	Ω

corresponding to whether the PI controller output is saturated or not. The back-calculation gain is used to reduce the integrator wind effect at the windup phenomenon [12]. The design of Nonlinear PI controller for anti-windup is illustrated in Figure 6.

4 SIMULATION RESULTS

In order to investigated the proposed wind turbine models is available and analyze the PI controller, the Nonlinear PI controller for anti-windup and the PID controller performance in wind turbine. The model is built in Matlab/Simulink dynamic system simulation software. Table 1 shows the parameters of the turbine and the PMSG.

Simulation depicted in Figures 7 and 8 show the different controller in wind turbine, and appear different performance. The controller uses in pitch controller. The reaction of PI controller and nonlinear PI controller for anti-windup are more accurate for control of power. Because of the part of differentiation induces PID has a dip at the change of wind speed. The more changing wind speed has, the more drop appears. It can find out nonlinear PI controller for anti-windup is stablest controller in this situation. Nonlinear PI controller for anti-windup is successfully implemented and proved to be very effective and can makes up the effects of over-simplifications in modeling.

5 CONCLUSION

This paper presents the complete model of the variable speed direct drive permanent magnetic synchronous generators wind turbine. This whole

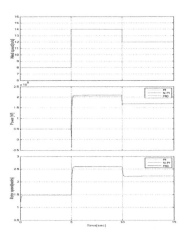

Figure 7. Results for wind speed in PMSG-wind speed, power and rotor speed.

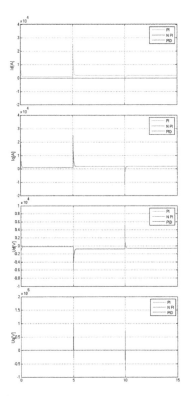

Figure 8. Result for wind speed in PMSG electric current I_d, I_q and generator voltage U_d, U_q.

model comprises of a drive train model, a pitch angle controllable wind turbine model, a permanent magnetic synchronous generator model. It also has been addressed optimum control schemes of pitch angle and generator, and built in Matlab/Simulink. At the same time, it compares with PI controller, Nonlinear PI controller for anti-windup, PID controller effect on operation of wind turbine, nonlinear PI controller for anti-windup is justified the effective and makes up the effects of over-simplifications in modeling.

Future research will concentrate on the control strategy of back to back PWM power converter used to variable speed direct drive permanent magnetic synchronous generators wind power system. It is help to improve research a wind farm with variable speed direct drive permanent magnetic synchronous generators.

REFERENCES

[1] Heier S. 2006. *Grid integration of wind energy conversion systems. 2nd ed. London:* John Wily & Sons.
[2] A. Grauers. 1996. Design of direct-permanent magnet generators for wind turbines, *in: Technical Report. 292,* Göteborg, Chalmers University of Technology.
[3] Eduard Muljadi, Butterfield C. 2001. Pitch controlled variable speed wind turbine generation, *IEEE Trans. On* IA, 1(37), pp. 240–246.
[4] Muljadi W, Pierce K, Migliore P. 2000. *A conservative control strategy for variable-speed stall-regulated wind turbines.* Technical Report NREL/CP-500-24791, Colorado, U.S.A, National Renewable Energy Laboratory.
[5] Wang Youqing. 2003(1). General PI control Strategy of PLC. Industrial control and Instrumentation, pp: 46–48.
[6] Xing Zou-xia, Zheng Qiong-lin, Yao Xing-jia, Wang Fada. 2006. PID control in adjustable pitch wind turbine system based on BP neural network, *Shenyang University* Technol, 28, pp: 681–5.
[7] Connor B, Leithead WE. 1993. Investigation of fundamental trade-off in tracking the Cpmax curve of a variable speed wind turbine, *In: Proceedings of the 12* the British Wind Energy Conference, pp. 313–319.
[8] Knight AM, Peters G.E. Simple wind energy controller for and expanded operating range. *IEEE Trans on* Energy Conversion, 20(2). pp. 459–466, 2005.
[9] Yin, M. G Li, M. Zhou and C. Zhao. 2007. Modeling of the Wind Turbine With a Permanent Magnet Synchronous Generator for Inteergration, *IEEE Power Engineering* Society General Meeting, Tampa, Florida, pp: 1–6.
[10] Keyuan Huang, Shoudao Hhuang, Feng she, Baimin Luo, Luoqiang Cai. 2008. A Control Strategy for Direct-drive Permanent-magnet Wind-power Generator Using Back-to-back PWM Converter, *ICEMS, pp. 2283–2288.*
[11] E.L van der Hooft. 2001. Dowec Blade pitch control algorithms for blade optimisation purposes, *Technical* Report ECN-CX-00-083, ECN Patten.
[12] S. Luo. 2011. Dynamic simulation of wind turbine generator system variable-pitch process, *Computer* Engineering and Applications, Vol. 47(25), pp. 14–17.

Frontiers of Energy and Environmental Engineering – Sung, Kao & Chen (eds)
© 2013 Taylor & Francis Group, London, ISBN 978-0-415-66159-1

The traditional materials in our country's modern packaging design of inheritance and application

J. Deng
Beihua University, Jilin, China

J.Y. Pang
Jilin woodiness material science and engineering key laboratories, Jilin, China

ABSTRACT: In recent years in international and domestic important packaging design competition of won awards of excellent packing design work, mostly in the design of Chinese traditional elements fusion. From traditional material to expound on. The traditional materials in the Chinese modern packaging design, the application will make the product more attractive, power and competitiveness.

Keywords: traditional elements; packaging design; aesthetic philosophy; traditional modeling; the traditional color

1 INTRODUCTION

China have their own unique and ancient historical culture, it is different from foreign economic structure and custom, traditional culture and folk art form in the national ideology of entrenched. Therefore, in modern packaging design, use of traditional elements of goods directly or indirectly displays the cultural value. That is to the modern products, consumption, marketing competition for the time background, through the prominent cultural value of the merchandise image, so that people in the spiritual resonate, and satisfy the people's aesthetic needs.

This paper in packaging design was a concept from traditional, extract the traditional materials of element; What is worth to put forward specific reference and application; Combine in recent years in the international and domestic winning good instance to packaging design analysis and explanation; Meanwhile, in the course of the study, as far as possible and modern packaging design trend linked up to find traditional elements and the prospective of packaging design, to find specific what traditional elements to the Chinese modem packaging design what significance, notice and improvement scheme, in order to enhance this topic research practical significance.

2 THE PROBLEMS

At present, there are four academic circles about the aspect, The first kind is a separate study of the traditional literature, such as a special study of the traditional philosophy, traditional culture, traditional folk, traditional aesthetics of literature such as number is the most, but are in the theoretical level, and from the perspective of the history of traditional form, no and combined with the reality, and the packaging design more no link. Such as "the Chinese tradition and modern", mainly from the economic and social customs, custom belief customs, customs and so on, entertainment, discusses the Chinese tradition and modernization of the wave collision, calls for a new customs to the Chinese nation and the new civilization.

The second is a separate study packaging design of literature, and all kinds of college textbooks or is specially provide excellent packing design yearbook. College textbook just specify the creation of the production method, packaging means and production process, and traditional culture elements from the view point of analysis are less. Such as "the modern packaging design" just use the modern production technology, craft level to illustrate how to better for packaging design, not how much is about tradition of the meaning of packing. Be like again the product packaging design case is pithy just listing the excellent packaging design picture, not for its cultural connotation.

The third kind is a traditional element extracted analysis in the design of the application of the small articles. Such as "the traditional cultural values in modern packaging design of the application", just analysis in modern packaging design, traditional culture and philosophy thoughts should be combined with age. But no one from the packaging design of traditional elements to find again.

A fourth is the traditional elements and packaging design of conscious connected, such as "the nationalization packaging design", it basically tells the story of national culture and design concept, the form feature, performance strategy, structure, design element, the relation with environment, and a large number of works. And as the visual culture research—contemporary visual culture and traditional Chinese aesthetic culture, it is mainly to contemporary image technology for the research field of vision and cultural background, based on the traditional aesthetic culture as foothold, focuses on the analysis of the modern society two different culture, the relationships between the traditional aesthetic culture and the future prospects. But these can't avoid packaging talk about packing, talk about the defects of the tradition.

2.1 The traditional aesthetic philosophy in China the significance of modern packaging design

The traditional aesthetic philosophy penetration in human culture and social life in all areas, create and rich man's whole life space, improve people's artistic taste, satisfy the human spiritual world, the healthy development of human nature, and gradually to the harmony of man and nature. Scientific comb tradition aesthetic philosophy, is to be in Chinese traditional culture, and on the basis of more comprehensive to think, explore and establish traditional Chinese modern packaging design.

Combination, exquisite aesthetic subject and object the understanding of the contents of the form harmonious, and claims that the harmony of nature and man, is China's culture in the spirit in thought and a prominent feature. Mr. Xu fuguan said: "in the ancient world cultural system, the no any system of culture, people and the nature, as happened in ancient China the same affinity with". Think nature and people, people and society spirit unified beauty is the highest state, is also the source of the artist's thinking.

"Nature and humanity" and "the green packaging" thoughts have completely place. "The green packaging" is the general trend of modern packaging design. China's modern packaging design can emphasize the harmony of human and nature and advocate natural, primitive, under the guidance of the concept of health, the use of advanced production technology, use as far as possible "small volume package", "no pollution packing", "pollution-free package" to alleviate the pressure of environment.

2.2 Traditional packaging material in China of the application of the modern packaging design

Packaging is not only the inevitable outcome of the development of human society, and the Chinese nation with a long history and the birth of. Our packing experience from the original to civilization, by the simple to prosperity of the development process. As for how the origin packing, the explanations, but at least one thing is for sure human first used for packaging materials, should belong to use local materials, is a natural, namely using bamboo, wood, grass, hemp, liu, cane, thorns, melon and fruit, hides and natural materials to packaging items. And later due to the development of productive forces, the progress of the society, human gradually enlarged packing material choice scope, natural materials manufactured goods packaging were born, such as pottery and porcelain, paper, etc. So the traditional packaging materials to see, should be a natural materials and natural reengineering material two types.

2.3 Natural material

Natural material is an important aspect of the traditional packaging materials, it means that a bamboo, wood, grass, hemp, liu, cane, thorns, melon and fruit, hides as packaging material. It can be said to be one of the longest, survival force more exuberant a packaging materials, almost the entire history of packaging with. Human picked up from natural materials used for packaging items began, has been to the packaging materials diverse today, it has been in packaging design played an important role.

In the new Stone Age before an pottery, humans have begun to use natural material packing. Engels in the family, private ownership and the origins of nation, "a book written in" pottery manufacturing is the establishment or wooden vessel paint to clay refractory and of generation." That is natural material as early as in pottery appear before play the role of packaging has .... A lot of the original pottery fragments unearthed a print table lines, rope lines, it shows people at mature compile technical and use rope skill. One rope lines in pottery decorated a large number of adoption, illustrating the rope has become the original packing of a kind of important form. The rope is strapping characteristics and simple operation, easy to help, not easy to slide etc, and is other packaging form to be, so rope as a packing a can't replace form, have been used for many years. If use soft wheat straw pole written into the rope packed in fragile ceramic products outside, played a very good protection function; Hollow-out the bandage technique on one hand save packing material, on the other hand are also more likely to convey the information about the commodities inner packing. This kind of packing form, a response to the human and the nature harmonious "green packaging" scheme, avoid to use the decomposition of the white pollution one by one earthquake bubble, and also it can be more intuitive performance goods.

Natural material in modern packaging recognition of the development and utilization of, with its relaxed, simple sense and the industrial production bring delicate, rigorous packaging design to compete. Use the natural material, can make the person can gain from the breath of nature, to today's people advocate natural state of mind completely.

2.4 *Natural recycling materials*

Natural reengineering materials are to point to natural materials after many processing happened after material qualitative change. Such as pottery and porcelain, paper, etc. Such as pottery clay after high temperature by after roasting changed its chemical structure formed. For natural reengineering materials for, traditional and modern change is the biggest is packaged. The earliest paper material is similar to the ancient egyptians in three thousand BC invent paper grass, the real paper should be Chinese in A.D. 105 by CAI lun invented. But paper as a packaging material specific when and where was the textual research has not, in the historic "volume," zhao queen preach "recorded in the paper parcel Chinese traditional medicine, has been the most widely as at least paper materials are used in packaging design.

In the tradition of the chosen commonly used wrapping paper CaiHou thick paper, now in the countryside can also see, with thick yellow straw paper will food packaging into the small DouFang type, and at the same time in the above attach a DaGongZhi, known as "anise package", it whole concise and bright, also has a strong ethnic characteristics.

3 CONCLUSIONS

In a word, the traditional material, whether natural or natural materials recycling materials, is in the use of packaging that the value of the premise condition, formed a distinctive green packaging, the connotation of simplicity, natural, the significance of environmental protection. If use of aquatic plants of eggs and wrapped with lotus leaf to wrap the meat, with leaves and flowers made of dry envelopes, bags, with gourd to install Dan pills and storage tea, the full of wits and the milk of human kindness traditional packing material, the green packaging design is in development, should get to good use and perfect.

REFERENCES

[1] Kai Wang. "Twenty-four grades of poetry" of si kong-tu to Chinese aesthetics and poetics of builds important [J]. Journal of wuhan university (humanities and science edition), 2009(5):579.

[2] Jing Huang. The classification of the tourist souvenirs packaging design and design elements of exploring positioning [J]. Packaging engineering, 2009, 30(9):173.

[3] Baudrillard. Let. Consumer society [M]. Chengfu Liu, Zhigang Jin, transcribing. Nanjing: nanjing university press, 2008.

[4] Jianjun Sun. China folk art appreciation [M]. Chongqing: southwest normal university press, 2006.

[5] Shiyou Zheng. Chinese folk belief annals-volunteer [M]. Jinan, shandong education press, 2005.

[6] Yongbin Xu. The traditional Chinese landscape painting aesthetic contemplation [J]. Journal of zhengzhou university (philosophy and social science edition), 2009(2): 174.

[7] Yueen Li, LiYan. China traditional culture of humanistic spirit reading and design application [J]. Packaging engineering, 2007, 28 (4): 152.

[8] Jing Huang. Modern packaging design of the traditional culture elements [J]. Packaging engineering, 2005, 26 (1): 181.

Intelligent environments

Frontiers of Energy and Environmental Engineering – Sung, Kao & Chen (eds)
© 2013 Taylor & Francis Group, London, ISBN 978-0-415-66159-1

Design of video indoor phone of Android building intercom system

W.B. Zhou, Y. Wu, Q.L. Guo, R. Lu, L.J. Yao & H.Z. Ni
The 718 Research Institute of CSIC, Handan, China

ABSTRACT: Android building intercom system has several advantages, such as open source, low cost, good expansibility and compatibility, great UI (User Interface), etc. It has broad application prospects in building intercom field. The paper presents an architecture for a video indoor phone of android building intercom system using Android and JNI (Java Native Interface) technology. The architecture contains two layers, upper layer and lower layer. Upper layer is signaling control layer and SIP (Session Initiation Protocol) protocol is utilized to complete signaling control. Lower layer is realization layer, RTP (Real-time Transport Protocol) transport module, media and other modules are used to do concrete actions specified from upper layer signaling.

Keywords: Android; video indoor phone; building intercom; SIP; RTP; JNI

1 INTRODUCTION

The traditional analog building intercom or mixed analog/digital intercom have many shortcomings, such as complex networking, poor compatibility, easy to busy, short transmission distance, etc. Pure IP building intercom system can overcome these limitations. It adopts standard TCP/IP protocol and Ethernet structure, all equipment including software and hardware terminal is in digital form. However, the current digital building intercom system basically based on Linux also has some problems: closed system, high cost, poor application expansibility, simple UI (User Interface), etc.

Android [1,2] is a Linux-based operating system for mobile devices and it is developed by Google. If building intercom system adopts Android platform, the system will have several advantages such as open source, low cost, good expansibility and compatibility, great UI (User Interface), etc and it can solve the existed problems of current IP building intercom.

In the paper, we present an architecture for a video indoor phone of android building intercom system using Android and JNI (Java Native Interface) technology. The architecture contains two layers, upper layer and lower layer. Upper layer is signaling control layer and SIP (Session Initiation Protocol) protocol is utilized to complete signaling control. Lower layer is realization layer, RTP (Real-time Transport Protocol) transport module, media and other modules are used to do concrete actions specified from upper layer signaling.

2 TECHNICAL PRINCIPLE

Current IP building intercom basically uses SIP (Session Initiation Protocol) protocol as the control core. SIP [3] is an IETF-defined signaling protocol widely used for controlling communication sessions over IP. The sessions may consist of message, audio intercom, video intercom, video conference, etc.

The general SIP system can be divided into four sections: user agent, proxy server, registration server and redirect server. Actually these three servers can be synthesized in one server. In the digital building intercom, the user agent can be used as the indoor terminal, outdoor terminal, and the SIP server can be run on an independent computer or run on outdoor intercom terminal. Figure 1

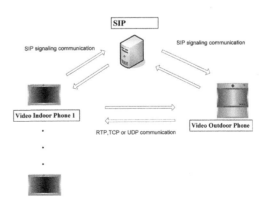

Figure 1. The basic process of SIP signaling and RTP, TCP or UDP communication.

shows the basic process of SIP signaling and RTP, TCP or UDP communication.

In Figure 1, the SIP server consists of proxy server, registration server and redirect server. The SIP server is responsible for recording, managing and transmitting the registration information, IP address and communication port of all users. Through SIP server and SIP signaling communication, video indoor phone and video outdoor phone can acquire each other's IP address and communication port. Then video indoor phone and video outdoor phone can use acquired each other's user information to complete the detail sessions, such as message, audio intercom, video intercom, etc.

3 DESIGN OF ANDROID VIDEO INDOOR PHONE

We present an architecture for Android video indoor phone using Android technology, see details in Figure 2. The architecture contains two layers, upper layer and lower layer. Upper layer is signaling control layer and SIP (Session Initiation Protocol) protocol is utilized to complete signaling control. Lower layer is realization layer, RTP (Real-time Transport Protocol) [4,5] transport module, media and other modules are used to do concrete actions specified from upper layer.

Now, we combine with video intercom to describe the running process of our Android video indoor phone briefly. If outdoor terminal want to have a video intercom with video indoor phone, the detailed work for video indoor phone is as follows:

i. Through SIP server and SIP signaling communication, video indoor phone acquires video outdoor phone's IP address and communication port. Meanwhile, video indoor phone send

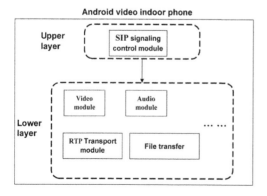

Figure 2. The architecture for Android video indoor phone.

its own IP address and communication port to video outdoor phone.

ii. With SIP signaling communication, video indoor phone knows that video outdoor phone wants to have a video intercom and the format of multimedia.

iii. Then video indoor phone starts RTP transport, video, audio modules which are necessary for video intercom to complete video talk.

In the following, we will introduce these two layers.

3.1 Upper layer

Upper layer is signaling control layer and we usually adopt SIP stack to develop signaling control. There have been many mature open source SIP stack at present, such as Osip/Exosip, OPAL, VOCAL, etc. Among these SIP stacks, Osip/Exosip [6], which is small and fast, is most suitable for portable applications.

Osip/Exosip stack is developed by C language. However, we want to develop an Android video indoor phone and Android application program adopts Java language. This means that Osip/Exosip stack can't be applied on Android directly. Fortunately, Android supports other development tools now, including Native Development Kit (NDK) [7] for applications or extensions in C or C++, etc. For concrete realization, we also need an interface programming—Java Native Interface (JNI) [8]. JNI is a programming framework that can provide interface between Java and C.

With Android NDK and Java Native Interface (JNI), Osip/Exosip can be ported to Android. Figure 3 shows the basic implementation process of SIP on Android. In the process, the SIP stack and JNI program should be compiled into dynamic link library on Linux using Android NDK so that it is convenient for upper Android application layer to use. In addition, the SIP stack and JNI program working on Linux kernel (Android) have faster run rate than Java language.

3.2 Lower layer

Lower layer is realization layer. Generally, it consists of RTP transport, video, audio, file transfer module, etc. In the following, we will describe these modules' basic content briefly.

i. RTP transport
RTP transport module adopts RTP Protocol. RTP designed for end-to-end, real-time is used for transfer of multimedia stream data. There have been two famous RTP program now: one is Jrtp (C++ language) and the other is Ortp (C language). Both Jrtp and Ortp can

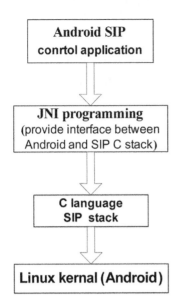

```
┌─────────────────────────┐
│     Android SIP         │
│  conrtol application    │
└─────────────────────────┘
            │
            ▼
┌─────────────────────────┐
│    JNI programming      │
│ (provide interface      │
│  between Android and    │
│  SIP C stack)           │
└─────────────────────────┘
            │
            ▼
┌─────────────────────────┐
│    C language           │
│    SIP  stack           │
└─────────────────────────┘
            │
            ▼
┌─────────────────────────┐
│  Linux kernal (Android) │
└─────────────────────────┘
```

Figure 3. The implementation process of SIP on Android.

be ported to Android to complete the transfer task of audio or video stream data. The detail ported method is similar to SIP stack.

ii. Video

Video module consists of video encoding, video decoding, video capture, video playback and it is in charge of video multimedia application. Video encoding and decoding can use the standard video formats include: H.263, H.264, etc. In android, video capture and playback programming mainly depends on Android Hardware Abstraction Layer (HAL) [9] and Linux kernel.

iii. Audio

Audio module is similar to video module and consists of audio encoding, audio decoding, audio capture, audio playback. Generally, the audio encoding and decoding formats include: G.711, G.729, MP3, etc.

iv. File transfer

This module mainly uses TCP/IP protocol to transfer the files including advertisement and information. File transfer programming in Android also utilizes Android HAL and Linux kernel.

4 SUMMARY

The paper designs an architecture for Android video indoor phone of building intercom system using Android technology. The architecture contains two layers, upper layer and lower layer. Upper layer is signaling control layer and lower layer is realization layer. And we describe the details of these two layers briefly. Android video indoor phone has several advantages, such as low cost, good expansibility and compatibility, great UI (User Interface), etc and it has broad application prospects in building intercom field.

REFERENCES

[1] Mark L Murphy, The Busy Coder's Guide to Advanced Android Development, CommonsWare, LLC, Macungie, 2011.
[2] Reto Meier, Professional Android 2 Application Development, Wrox, UK, 2010.
[3] Information on http://tools.ietf.org/html/rfc3261, "RFC 3261, SIP: Session Initiation Protocol".
[4] Colin Perkins, Rtp: Audio and Video for the Internet, Addison-Wesley Professional, Boston, 2003.
[5] Larry L. Peterson, Bruce S. Davie, Computer Networks: A Systems Approach, Morgan Kaufmann, Massachusetts, 2007.
[6] Information on http://www.gnu.org/software/osip.
[7] Sylvain Ratabouil, Android NDK Beginner's Guide, Packt Publishing, Birmingham, 2012.
[8] Sayed Hashimi, Pro Android 2, Apress, New York, 2010.
[9] Lucas Jordan, Practical Android Projects, Apress, New York, 2011.

Frontiers of Energy and Environmental Engineering – Sung, Kao & Chen (eds)
© 2013 Taylor & Francis Group, London, ISBN 978-0-415-66159-1

Experimental study on surrounding rock deformation characteristics of gateway driving along next goaf with different thickness limestone roof

S. Zhong

Shandong Province Bureau of Coal Geology, Taian Shandong, China

C.Q. Wang

Shandong Province Caozhuang Coal Mine of Tengzhou Limited Liability Company, Tengzhou Shandong, China

M.F. Guo & N.N. Zhao

C.C. Shandong University of Science and Technology, Qingdao Shandong, China

ABSTRACT: The immediate roofs of gateways in Seams No. 15U, 16U and 17 of Yanzhou Coal-field are generally limestone with different thickness. In order to maximize the self-bearing capacity of limestone, reduce the support costs of gateways and improve the production efficiency of the coal mine, research on the surrounding rock deformation of gateway 17304 under different limestone thickness driving along next goaf in Liyan Coal Mine was carried out. And then, it comes to the conclusion of the deformation laws and characteristics of surrounding rock which provide reliable basis for reasonable support designing of gateway driving along next goaf under different limestone.

Keywords: gateway driving along next goaf; limestone roof; thickness of roof; surrounding rock of gateway; deformation characteristics

1 INTRODUCTION

In coal mine production, the support of limestone roof of mining gateway without scientific theoretical guidance, generally is determined the corresponding supporting parameters or not support according to the critical safety thickness confirmed by mine production experience, and this goes against the mine production safety as well as the mine production efficiency. The immediate roofs of gateways in Seams No. 15U, 16U and 17 of Yan Zhou hypoground coal are generally limestone with different thickness. In order to maximize the high load capacity of limestone[1,2], reduce the support costs of gateways and improve the production efficiency of the mine, research and analysis on the deformation characteristics of surrounding rocks under different limestone thickness of gateway 17304 in LiYan Coal Mine driving along next goaf were carried out. The results provide the basis and reference for optimizing support parameters of mining gateway with thin limestone roof.

2 THE GEOLOGICAL AND TECHNICAL CONDITIONS OF COAL FACE AND GATEWAY

2.1 Geological conditions of gateway 17304 driving along next goaf

The gateway 17304 of Liyan Coal Mine is located in 17 coal seam, and the average coal thickness of 17 coal seam is 0.97 m, f = 2.5–3, moreover, the coal-seam thickness is steady and not changing much, but coal seam undulates much and exists local simple texture. The floor is clay rock, the average thickness of which is 1.2 m, and the roof is 11 limestone or siltstone with a higher hardness and no false roof. The lithological characters of the roof and floor in 17304 coal face are shown in Table 1.

2.2 Original support method of gateway 17304 driving along next goaf

The size of gateway 17304 section is 2.6 m × 2.3 m, with non-support when the roof integrity is good and limestone thickness is bigger than the thickness

Table 1. The lithological characters of the roof and floor of coal seam.

Roof and floor	Lithology	Thickness/m	Lithologic characteristics
Roof	11 limestone/siltstone	0.9	11 limestone seam, not pure and high shale content, siltstone, shale cementation and poor integrity
Direct floor	Clay rock	1.2	Light grey, easily crushing in water, volume without obvious expansion

Figure 1. The original support method of gateway 17304 driving along next goaf.

of 0.7 m; otherwise anchor ladder is used as a permanent support, the selection of which are $\Phi16 \times 1800$ mm steel bolt, $130 \times 130 \times 6$ mm square anchor plate, each hole with volume one of MSZ2860 type of resin-anchored bolts, $\Phi14 \times 2600$ mm steel ladder and rows between bolts are all equal to 800 mm, besides, anchor-hold of each anchor is not less than 60 kN[3]. Two sides need support protection only when lower weak rock masses exist, moreover, to control the broken surrounding rock of weak rock surface, the guard net is used to avoid the destruction of deep surrounding rock caused by the excessive side slabbing of soft rock towards to the gateway. The support method of gateway 17304 driving along next goaf is shown as Figure 1.

3 CONTENTS AND PROJECTS OF OBSERVATION ON GATEWAY DEFORMATION

3.1 Deep deformation observation in surrounding rock of gateway

The multi-point deformation indicator is used to observe different surrounding rock deformation of deep gateway, fixed in the middle of roof and the wall of gateway, meanwhile a total of three spots. Roof basis points are fixed in the rock stratum and that of gateway's sides are fixed uniformly according to the spacing of 1.2 m.

3.2 Methods of observation and setting in the surveyed area

To improve the precision of the evaluation, observation stations are respectively set in gateway 17304 driving along next goaf where the limestone thickness is around 0.5 m, 0.7 m, with two observation sections of each one. Monitoring once a day during the time when the gateway deformation rate is greater than 1 mm/d or the period of mining influence; decreases of observation frequency depend on the concrete conditions when the surrounding rock is steady and with no evident deformation.

4 ANALYSIS ON THE DEFORMATION LAWS OF SURROUNDING ROCK GATEWAYS

4.1 Deep surrounding rock deformation of roof

It comes to the conclusion that the basis point deformation was at 1.2 m depth of roof internal under different limestone thickness: when limestone thickness of direct roof is 0.5 m, the maximum deformation of basic point is 24 mm and the maximum relative velocity of convergence is 2.6 mm/d; when limestone thickness of direct roof is 0.7 m, the maximum deformation of basic point is 20 mm and the maximum relative velocity of convergence is 2.0 mm/d with an average velocity of 0.4 mm/d; when limestone thickness of direct roof is 1.0 m, the maximum deformation of basic point is 10 mm and the maximum relative velocity of convergence is 1.2 mm/d. Generally the deformations begin to increase by 28.6 m distance from working face, and significantly to 12.6 m.

Comprehensive analysis of observations on the deep surrounding rock deformation of roof reviews on the deep surrounding rock deformation of roof under different limestone thickness as shown in Figure 2.

Figure 2 shows: (1) with the increase of limestone thickness, the total amount of roof deformation decreases gradually; (2) bolt support can effectively control bed separation of anchorage zone; (3) the limestone rock mass is in good integrity without obvious separation inside.

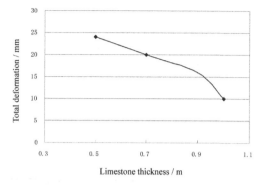

Figure 2. Total roof deformation under different limestone thickness.

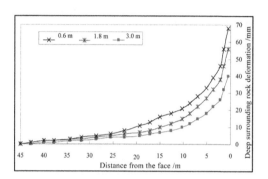

Figure 3. Deep deformation of coal side surrounding rock in face side with limestone thickness 0.7 m.

4.2 Deep surrounding rock deformation of gateway's sides

As the limestone thickness is 0.7 m, observations on deep surrounding rock deformation of gateway's sides are as shown in Figures 3 and 4.

When limestone thickness is 0.7 m, the amount of coal side deformation with working face is greater than that of entity coal side obviously, and whether in side of working face or coal mass, the two sidewalls deformation of gateway is mainly under the range of 0–1.2 m, and it also comes to the conclusion that wall rock looseness range is within 1.2 m.

When the limestone thickness is 0.5 m, observations on deep surrounding rock deformation of gateway's sides are as shown in Figures 5 and 6.

As the limestone thickness is 0.5 m, the surrounding rock deformation is mainly concentrated in the surface range of 1.2 m, what is more, the amount of coal side deformation with working face is greater than that of entity coal side obviously.

The observation results from deep deformation of coal sides under different limestone thickness show that: (1) no matter the coal side of

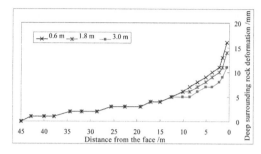

Figure 4. Deep deformation of coal side surrounding rock in entity coal side with limestone thickness 0.7 m.

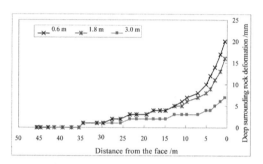

Figure 5. Deep deformation of coal side surrounding rock in entity coal side with limestone thickness 0.5 m.

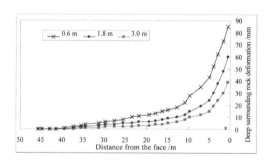

Figure 6. Deep deformation of coal side surrounding rock in face side with limestone thickness 0.5 m.

surrounding rock in entity coal side or face side, the deep surrounding rock deformation all decrease gradually with the increase of limestone thickness; (2) under the influence of the mining activities on the working faces, deformation values and velocity of surrounding rock at the side of working face are all greater than that in entity coal side obviously, moreover, the surrounding rock deformation of gateway's sides is mainly concentrated in the shallow surrounding rock area at the range of 0–1.2 m.

5 CONCLUSIONS

1. The roof thickness of limestone has great influence on the stability of surrounding rocks, therefore, deformation value and velocity of surrounding rock on the surface and depth of gateways are all going down gradually along with the increase of limestone thickness.
2. The limestone rock mass is in good integrity without obvious bed separation inside, and this is helpful to the roof support and reduce the difficulty of support.
3. In the gateway driven along next goaf, deformation values and velocity of surrounding rock at the side of working face are all greater than that in entity coal side obviously, moreover, the surrounding rock deformation of gateway's sides is mainly concentrated in the shallow surrounding rock area at the range of 0–1.2 m.

REFERENCES

[1] Li Wenfeng, SunYinghui, Yang Bo, Yan Shuai, Hu Zhongchao. Practices on surrounding rock control technology of gateway driving along goaf towards to coal mining face [J]. Coal Engineering, 2010 (2):24–26.

[2] Li Jinping. Study on the Technology of Gob-side Entry Retaining with Longwall Top Coal Caving System and Its Application in Lu'an Mining District [D]. Bingjing: China Coal Research Institute, 2005.

[3] Liu Zenghui, Gao Qian, Hua Xinzhu, Wu Yuhua, Fu Kunlan, Liu Yu. Aging Characteristics of Wall Rock Control in Roadway Driving Along Goaf, Journal of Mining & Safety Engineering 2009, 26 (4):465–469.

[4] Bai Jianbiao. Controlling surround rock of gateway driven along goaf [M]. Xu Zhou: China University of Mining and Technology Press, 2006.

[5] Yang Yongjie, Jiang Fuxing, Ning Jianguo, et al. Method of determining rational width of small coal pillar protecting roadways driving along next goaf supported by bolting and of fully mechanized sublevel face [J]. The Chinese Journal of Geological Hazard and control, 2001, 12 (4):81–84.

Frontiers of Energy and Environmental Engineering – Sung, Kao & Chen (eds)
© 2013 Taylor & Francis Group, London, ISBN 978-0-415-66159-1

Experimental study and theoretical analysis on flame characteristic of blast burner combustion with oxygen-enriched

L. Pan

Energy and Power Engineering Departement of Changchun Institute of Technology, Changchun, China

ABSTRACT: Using blast burner on the study of the LPG oxygen-enriched combustion, and composed the experimental platform with flame burning test device and oxygen-enriched supplying system. By this platform, through experiment, the temperature region, flame length, angle etc, changed as below when oxygen concentration was between 21% to 30%. The length shortened 16.7%, the angle of flame entrance increased, flame temperature gradient increased, the highest flame temperature improved by 24.8%, flame intensity increased. Analyzing by combustion theory, burning speed increased linearly with the increasing of oxygen concentration, and the combustion region decreased. Theoretical analysis and the experimental result were uniform.

Keywords: LPG; oxygen-enriched combustion; flame characteristic; test investigation; theoretical analysis

1 INTRODUCTION

Oxygen-enriched combustion can reduce fuel combustion point, quicken combustion speed, accelerate entire combustion, improve flame temperature, deduce exhaust after burning with preponderance energy saving performance.[1] Because the study on flame characteristic of oxygen-enriched combustion is seldom overseas, its significance to understand the thermal character of the flame with oxygen enriched and thus to oxygen-enriched combustion extension.

The author used LPG as the fuel, and designed and made blast burner and the measuring device of the flame. The author used different ratio with ordinary air and oxygen cylinder supplying pure oxygen to make the oxygen-enriched air with different volume fraction. The heat energy utilization benefit is highest while the volume fraction of oxygen is between 21% to 30%.[2] The paper studied the flame character parameters change of the blast burner using measuring device of the flame while the volume fraction of oxygen is between 21% to 30%.

2 EXPERIMENTAL DEVICE AND DESIGN

2.1 *Experimental device*

The experimental device is mainly composed of blast burner, measuring device of the flame and the oxygen supplying system is this paper.

The blast burner is tangential air feed swirl, and its thermal power is 15 kW. The measuring device of the flame is made of stent, back shutters, temperature measuring platform (see Fig. 1). Through the scale and the protractor of the back shutters, we can measure the length and the cone angle of the flame. The cone angle is the angle between the flame contour and the medial axis of the burner. The temperature measuring platform is a moving thermocouple.

2.2 *Experimental design*

Under changing the volume flow of the LPG and oxygen-enriched air, study the flame character

Figure 1. Flame burning test devices.

Figure 1 flame burning test devices parameters varies while only changing the volume fraction of oxygen with the flame flow characteristics parameters not changing.

3 EXPERIMENTAL RESULT

3.1 *Flame status and flame temperature field*

The length and the cone angle of the flame can be seen in Figures 2, 4 and 6, while the volume fraction of oxygen is 21%, 25% and 30%. Fitting the temperature value with origin 8.0, generate half flame temperature field with oxygen concentration: 21%, 25%, 30%. the coordinate system of the temperature field see Figures 3, 5 and 7.

Figure 2.   Flame fig when oxygen was 21%.

Figure 3.   Flame temperature field when oxygen was 21%.

Figure 4.   Flame fig when oxygen was 25%.

Figure 5.   Flame temperature field when oxygen was 25%.

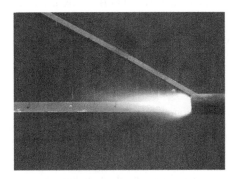

Figure 6.   Flame fig when oxygen was 30%.

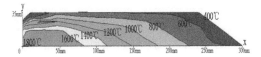

Figure 7.   Flame temperature field when oxygen was 30%.

According to the experiment, the flame changed as flow with the volume fraction of oxygen increasing.

a. Flame combustion noise increasing gradually.
b. Flame surface generating lots folds Flame shape changing, from drifting to level burning gradually upward.
c. With oxygen volume fraction increasing, the temperature of the temperature field gradient increased gradually. The highest temperature of the flame was at the angle of 12°of the axis not at the central axis. The length shortened 16.7%, the angle of flame entrance increased, flame temperature gradient increased, the highest flame temperature improved by 24.8%, flame intensity increased.

3 EXPERIMENTAL RESULT ANALYSES

3.1 *The relationship between the burning speed and the oxygen volume*

Using the combustion kinetics to analyze the changing of the Oxygen-enriched combustion speed, LPG could be treat as a single gas fuel as a mixed gas of alkanet and olefin. In qualitative analysis, it could be used as a single-component gas fuel. Supposing the LPG's chemical equation was:

$$aA + bB \xrightarrow{K} cC + dD \qquad (1)$$

In equation, A was LPG;B was oxygen containing air; C and D was combustion products; a b c d was amount of the reactants and products.

When the reaction series was Eq. 1, the formula of the reaction speed was:

$$w_2 = k_2(p/RT)y_A y_B \qquad (2)$$

In equation, w_2 was reaction speed; k_2 was reaction constant; y_A was LPG mole fraction; y_B was mole fraction of oxygen containing air; p was pressure of mixture gas.

Because the whole mixed gas volume was certain, supposed the mixed gas volume was 1, the combustible gas could be expressed as:

$$y_A + y_B(\varphi_{oxygen} + \varphi_{inert}) = 1 \qquad (3)$$

In equation, y_A was LPG mole fraction; y_B was air mole fraction; φ_{oxygen} was fraction volume of the pure oxygen in air; φ_{inert} was fraction volume of the inert gas in air.

So, when the fraction volume was φ_{oxygen}, that was the combustion aided with oxygen-enriched air, the burning reaction speed was:

$$w_2 = \varphi_{oxygen}k_2(p/RT)y_A y_B \qquad (4)$$

From analyzing Eq. 4, the speed of the LPG increased linearly with the increasing of the oxygen volume fraction, when the oxygen-enriched air aided burning with fixed pressure and temperature of the reactants.

3.2 Relationship between combustion regional volume and oxygen concentration

The flame of blast burner was large scale turbulent. The relation of large scale turbulent combustion regional volume was:

$$V \propto \frac{w'}{u_n} \qquad (5)$$

In equation, V was combustion regional volume; w' was turbulent pulsation speed; u_n was laminar flame propagation velocity.

The turbulent pulsation speed could be thought as no changing with fixed volume flow of the mixed inflammable gas. So the relationship between the combustion regional volume and the laminar flame was inverse ratio. The flume of the laminar flame propagation velocity was:

$$u_n = \left(\frac{2a}{t_m}\right)^{1/2} \propto \sqrt{\frac{a}{t_m}}^{[5]} \qquad (6)$$

In equation, a was thermal diffusivity pre mixed inflammable gas; t_m was average reaction time.

The thermal diffusivity of the pre mixed inflammable gas in this experiment a was almost invariant. Because the relationship between the combustion regional volume and the laminar flame was inverse ratio, from Eq. 4, the reaction speed was increasing while the oxygen volume fraction increased which lead to the average reaction time t_m becoming shorter. In result, the laminar flame propagation velocity u_n gradually increased when the oxygen volume fraction gradually increased. So from Eq. 6, the combustion regional volume became smaller with the oxygen content getting bigger when the oxygen-enriched air aided burning.

The flame bright regional was the main regional of the blast burner. After analyzing the combustion flame pictured and the fitting temperature field, we could conclude that the combustion regional volume was larger, light regional was wider when ordinary air aided burning. The light regional was getting smaller and concentrated to the entrance of the burner with oxygen volume fraction increasing. So the experimental phenomena and the theoretical analysis was consistent.

3.3 The relation of the combustion temperature and the oxygen content

Calculation of the combustion temperature was:

$$\theta_{products} = \frac{Q_{low}}{Vc_{products}}^{[6]} \qquad (7)$$

In equation, $\theta_{products}$ was theoretical heating temperature of combustion; Q_{low} was lower heating amount of LPG; V was combustion region volume; $c_{products}$ was specific volume of combustion products.

Because the combustion regional became smaller, namely the combustion region V decreased, when the specific heat was at constant pressure $c_{products}$ and the input power Q_{low} were not changing, and the oxygen volume fraction became higher, the temperature of the combustion regional would regularly increase, and the burning speed would fasten, the combustion region V would decrease.

4 SUMMARY

Through experimental study and theatrical analysis, the paper got the flame characteristics of oxygen-enriched combustion: With the oxygen volume fraction increasing, combustion reaction speed gradually increased, and the flame propagation speed got faster which led to reducing of the combustion regional. This resulted to that the flame highest regional gradually reduced, and temperature gradient gradually increased, the highest flame temperature gradually increased, the flame intensity gradually enhanced.

REFERENCES

[1] Kiga T, Takano S, Kimura N, et al. Characteristics of pulverized-coal combustion in the system of oxygen-recycled flue gas combustion [J]. Energy Conversion and Management, 1997,(38), Supplement 1:129–134.

[2] Kimura, Browall W.R. Membrane Oxygen Enrichment Demonstration of Membrane Oxygen Enrichment for Natural Gas Combustion [J]. Membrane Sei, 1986:62.

[3] Gu Hengxiang . Fuel and Combustion [M]. Xian Northwest Industrial University Press, 1993:44–52.

[4] Chang Hongzhe Fuel and Combustion [M]. Shanghai: Shanghai Jiaotong University Press. 1993:90.

[5] Zhang Songshou Engineering Combustion Science [M]. Shanghai: Shanghai Jiaotong University Press. 1987:174.

[6] Han Zhaocang. Fuel and Combustion [M]. Beijing Metallurgical Industry Press. 1994:39.

Frontiers of Energy and Environmental Engineering – Sung, Kao & Chen (eds)
© *2013 Taylor & Francis Group, London, ISBN 978-0-415-66159-1*

Preparation of nano alumina modified phenolic resin and wear-resisting property

H.F. Chen

Department of Materials & Chemistry, Huzhou Teachers College, Zhejiang, P.R. China

ABSTRACT: Phenolic resin, N, N-dimethyl formamide, nano-Al_2O_3 were used as raw materials, DMF as solvent, and the modified phenolic resin was prepared by adding nano-Al_2O_3 directly. The density, wear resistance, Rockwell hardness, melt flow rate, specific surface area, etc, the morphologies of the samples and composite materials in the crystal components of Al_2O_3 were observed by optical microscope and XRD respectively. Conclusion: when the mass fraction of nano-Al_2O_3 is 5%, the hardness of phenol-formaldehyde resin reaches the maximum value; the melt rate experiment indicated that the volume rate of phenolic resin decreased by adding nano-Al_2O_3; phenolic resin wearability reached the maximum when the nano-Al_2O_3 increased to 16%, then decreased, that because the phenolic resin as a binder in the case of a large number of nano-Al_2O_3 can not play an effective role in bonding; nano-Al_2O_3 affects the brittleness of phenolic resin, which makes the phenolic resin structure more stable.

Keywords: phenolic resin; nano-Al_2O_3; wear-resisting

1   INTRODUCTION

Phenolic resins were derived from phenol and formaldehyde polycondensation, and modified phenolic resin has good mechanical properties, wear resistance and heat resistance. In addition, it is well compatible with variety of organic and inorganic filler, comparing with other resin systems; the phenolic resin system has the advantage of low smoke and toxicity. Therefore, phenolic resin has been widely used in refractory, and the harsh environment at high temperatures. The thermal shock resistance, wear resistance, friction properties, mechanical strength and heat resistance presents certain requirements, carrying out a series of studies, using phenolic resin, Schwetz, K.A. & Grellner W. 1981. modified material B_4C, Byrun, H.Y. 2001. developed montmorillonite, Zhang, S.Q. 2003. researched molybdenum oxides, chlorides and acids, Lin, R.H. 2006. developed nano-copper, Li C.H. 2009. modified TiO_2 & SiO_2, Sarkar, S. & Adhikari, B. 2000. modified lignin, Hsiue, G.H. 2001. researched dicyclopentadiene, Matsumoto A. 1991. modified N-phenylmaleimide and Su, F.H., Zhang, Z.Z., Liu, W.M. 2005. judged that these inorganic nano-particles have high strength, high hardness, corrosion resistance, heat resistance and a series of outstanding features. These two above-mentioned combinations improved the properties of resin; the most difficult problem by far is how

to effectively disperse the nano-powder in a thermoplastic phenolic resin.

In order to make the phenolic resin to reach the viscous flow state, N, N-Dimethyl Formamide (DMF) was used as solvent in the experiment, nano-Al_2O_3 was added to phenol-formaldehyde resin according to certain proportion of the amount, and then the modified phenolic resin was obtained. In this article, the different content of nano-Al_2O_3 on the phenolic resin has the different influence of performance, especially the wear resistance. On self-made instruments, wearing capacity was characterized under controlled pressure conditions.

2   EXPERIMENTAL SECTION

2.1   *Preparation of samples*

Cracked the massive phenolic resin of the industrial production by hammer, then smashed the solid sample masticator for 30 seconds and collected the obtained phenolic resin powder. Five copies of the same quality of phenolic resin powder were dissolved in the same amount of DMF solution for 24 hours, respectively. Then added the different quality of nano-Al_2O_3, and stirred well, covered with plastic film to prevent volatilizing of DMF, then the sample was vacuumized for 15 minutes in the vacuum machine, then the sample was baked for 6 hours at 180 °C in dryer, then took out and cooled naturally. The process is not added the nano-Al$_2$O$_3$ phenolic resin which is the blank control samples.

2.2 *Characterization*

The nano-Al$_2$O$_3$ powder was measured by XD-6 X-ray diffractomer (Beijing Puxi Instrument Co., Ltd.), voltage of 36 KV, step width of 0.02 and Cu target. The nano-Al$_2$O$_3$ dried for two hours at 105 °C in vacuum, then put them into JW-K surface area and pore size analyzer tube (Beijing JWGB Sci & Tech Co., Ltd.), using three different helium and nitrogen flow ratio to test the surface area of nano-Al$_2$O$_3$. The hardness of the sample was measured by the XHR-150 plastic Rockwell hardness tester (Shanghai Alliance Seoul Test Equipment Co., Ltd.); the flow rate of modified resin was measured by RL-Z1B1 melt flow rate instrument (Shanghai S.R.D Scientific Instruments Co., Ltd), and Eq. 1 & 2 are as follows:

$$MFR = \frac{600\,m}{t} \qquad (1)$$

$$MVR = \frac{600\,m}{\rho \times t} \qquad (2)$$

where MRF = melt flow rate; m = the average mass of the sample, t = cutting time; MVR = melt volume flow rate; ρ = the density of the composite phenolic resin.

The FT-IR spectra were measured by The Nicolet 5700 by FT-IR infrared spectrometer.

2.3 *Relative wearability test*

The samples were re-melted, and made into diameter d (~1.5 cm) cylindrical. Placed G (~200 g) force at the top of the sample, the samples were placed in 8 cm distance from the wheel center marker position. Then the wheel began to rotate, the sample was contacted the surface of grinding wheel surface, meanwhile, the timer starts timing for the T, the mass of the sample before and after testing the friction properties was measured, and noted the change of the mass. The relative wearability Eq. 3 is calculated as follows:

$$S = \frac{T}{m_1 - m_2} \qquad (3)$$

where S = the relative wearability of sample, s/g; m_1 = the mass of the sample before the test, g; m_2 = the mass of the sample after the test, g; T = the time of the sample test.

3 RESULTS AND DISCUSSION

3.1 *XRD*

Figure 1 shows the XRD patterns of alumina, there are nine obvious peak, the positions of highest three diffraction peaks are 35, 43.5 and 57.5, respectively, and its positions are fully consistent with the standard pattern of alumina (JCPDF: 46-1212), which indicates that the measured material is alumina, and no other crystal interference, which is correspond to the experimental requirements.

3.2 *FT-IR*

The phenolic resin is made by the polycondensation of phenol and formaldehyde, from Figure 2 we can know that the infrared spectrum of the pure phenolic resin has a series of peaks in region of 500–800 cm^{-1}, 1000–1800 cm^{-1} and 2500–3600 cm^{-1} and appears peak at 900 cm^{-1} and 1880 cm^{-1} nearby, comparing with infrared absorption peaks of phenolic resin added nano-Al$_2$O$_3$. Through the comparison, except the transmittance of the composite phenolic resin is not as good as the pure phenolic resin, and the absorption peaks in 500–800 cm^{-1} and 2800–3200 cm^{-1} are not the

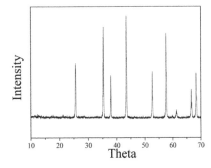

Figure 1.   X-ray of r-alumina.

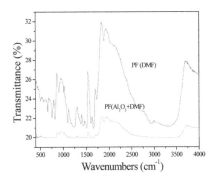

Figure 2.   The FT-IR spectra of samples.

325

same, the others absorption peaks are basically coincident. The different absorption peaks may connect with –OH.

3.3 Density of phenolic resin

According to the above preparation process and the six samples in Table 1 (DMF 8.0 mL), considering the density of pure Phenolic Resin (PR) is slightly larger than water and the density of nano-Al_2O_3 (NA) is relatively heavy, adding Nano-Al_2O_3 will change the density of the composite phenolic resin, and with the increasing of the content of nano-Al_2O_3 composite phenolic resin, the density of the composite phenolic resin also increased. As it seen from Figure 3, with the adding of nano-Al_2O_3, the composite phenolic resin density is in linear growth, at the beginning, but it will be infinitely close to the density of nano-Al_2O_3.

3.4 Hardness of phenolic resin

As it seen in Figure 4, with the nano-Al_2O_3 adding gradually to phenolic resin, the phenolic resin composite hardness increased rapidly. The hardness reaches its maximum when the mass fraction of nano-Al_2O_3 was around 5%. With the continued adding of nano-Al_2O_3, the hardness of composite phenolic resin decreased rapidly. When the mass fraction of nano-Al_2O_3 reached 10%, the hardness of the composite phenolic resin is no longer rapidly

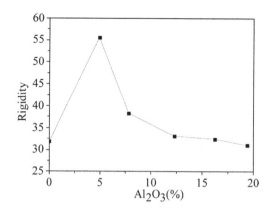

Figure 4. The hardness of samples versus nano-Al_2O_3 contents (25°C).

decline. With the continued adding of nano-Al_2O_3, the hardness decline of composite phenolic resin is not obvious, but it will still decline more or less.

3.5 The melt flow rate of phenolic resin

Figure 5 shows that the flow rate of phenolic resin melt grew slowly with the increasing of nano-alumina. When the mass fraction of nano-Al_2O_3 content is 7.8%, its growth rate declines, and then the flow mass rate increased steadily. The overall melt flow rate grew with the increasing of the mass fraction of nano-Al_2O_3 composite phenolic resin.

Figure 6 shows the volume flow rate of composite phenolic resin melt declined steadily with increasing mass fraction nano-Al_2O_3, because the density of the composite phenolic resin increased with increasing mass fraction of nano-Al_2O_3. And the density of nano-Al_2O_3 in the phenolic resin is larger than phenolic resin. However, with the added amount increased, its density is not greater than the density of the nano-Al_2O_3. Owing to some minor errors caused by the devices, which impacted the data of melt mass flow rate. However, in this figure, the error still existed. But it did not affect the steady change of the melt volume rate.

3.6 Phenolic resin relative wearability

As it known in Figure 7, the wear resistance of composite phenolic resin was significantly improved by adding the nano-Al_2O_3. When the mass fraction of nano-Al_2O_3 accounted for the total mass is around 16%, the relative wearability of composite phenolic resin reaches the maximum. Continue to increasing the amount of nano-Al_2O_3, the material itself chalking owing to the amount of nano-Al_2O_3 composite phenolic resin added too much. The degree

Table 1. The ratio of the raw samples.

No.	1	2	3	4	5	6
NA/g	0	0.464	0.766	1.261	1.739	2.167
PR/g	9.025	9.025	9.022	9.023	9.028	9.023

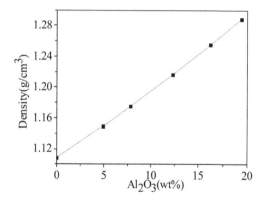

Figure 3. The density of samples versus nano-Al_2O_3 contents.

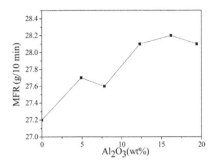

Figure 5. The melt flow rates of samples versus nano-Al_2O_3 contents.

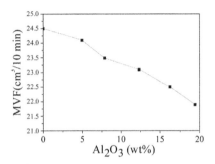

Figure 6. The volume flow rate of samples versus nano-Al_2O_3 contents.

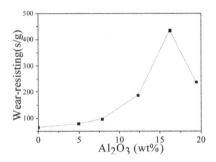

Figure 7. The relative wearability of samples versus nano-Al_2O_3 contents.

of adhesion of the nano-Al_2O_3 from the adhesive role of the phenolic resin substantially declined, which resulted in lower wearability.

4 CONCLUSIONS

In this experiment, the phenolic resin dissolved by DMF, the phenolic resin composite was obtained by adding different content of nano-Al_2O_3. The performance of the modified phenolic resin was changed. The main conclusions of this paper as follows: (1) The density of phenolic resin increased with adding the nano-Al_2O_3, but the density does not exceed the density of nano-Al_2O_3; (2) The hardness of the composite phenolic resin increased rapidly with the addition of nano-Al_2O_3 and then declined rapidly, The alumina mass fraction is 5% under the maximum hardness, (3) The impacts of nano-Al_2O_3 on the phenolic resin melt mass flow rate change is not large, but the volume rate turn into a downward trend when the content of nano-Al_2O_3 increased, which indicates that nano-Al_2O_3 caused a certain influence on the fluidity of the phenolic resin; (4) Phenolic resin wearability reached the maximum when the nano-Al_2O_3 content increased to 16%, then decreased, which because the phenolic resin as a binder in the case of a large number of nano-Al_2O_3 added can not play an effective role in bonding; (5) nano-Al_2O_3 affects the brittleness of phenolic resin, which makes the phenolic resin structure more stable.

REFERENCES

Byrun, H.Y., Chol, M.H., Chung, I.J. 2001. Synthesis and Characterization of Rosol Type Phenolic Resin/Layered Silicate Nanocomposites. *Chem. Master.* 13: 4221–4226.

Hsiue, G.H., Shiao, S.J., Wei, H.F., Kuo, W.J., Sha, Y.A. 2001. Novel phosphorus-containing dicyclopentadiene-modified phenolic resins for flame-retardancy applications. *Journal of Applied Polymer Science.* 79: 342–349.

Lin, R.H., Fang, L., Li, X.P., Xi, Y.X. 2006. Sufeng Zhang and Ping Sun, Study on Phenolic Resins Modified by Copper Nanoparticles, *Polymer Journal.* 38: 178–183.

Li C.H., Feng, Y., He, L., Yan, J.W., Xie Z.P., Liu, Y., Yang, Z.J. 2009. The Effects of Nano-Particles on Tribo-Properties of NBR Modified Phenolic Resin. *Polymer Materials Science & Engineering.* 25: 59–61.

Matsumoto A., Hasegawa, K., Fukuda, A., Otsuki, K. 1991. Study on modified phenolic resin. Modification with homopolymer prepared from p-hydroxyphenyl-maleimide. *Journal of Applied Polymer Science.* 43: 365–372.

Sarkar, S.; Adhikari, B. 2000. Lignin-modified phenolic resin: synthesis optimization, adhesive strength, and thermal stability. *Journal of Adhesion Science and Technology.* 14: 1179–1193.

Schwetz, K.A., Grellner W. 1981. The influence of carbon on the microstructure and mechanical properties of sintered boron carbide. *Journal of the Less Common Metals.* 82: 37–47.

Su, F.H., Zhang, Z.Z., Liu, W.M. 2005. Study on the friction and wear properties of glass fabric composites filled with nano- and micro-particles under different conditions. *Materials Science and Engineering: A.* 392: 359–365.

Zhang, S.Q., Qiang M., Lin H.S., Guan J., Li J.Q. 2003. Synthesis Technology for Thermosetting Molybdic Acid-Modified Phenolic Resin. *Journal of Wuhan Yejin University of Science and Technology.* 26: 370–373.

Frontiers of Energy and Environmental Engineering – Sung, Kao & Chen (eds)
© 2013 Taylor & Francis Group, London, ISBN 978-0-415-66159-1

Influence of wind farm's wind conditions on wind turbine's flicker

J.H. Zhang, Y.Q. Liu, D. Tian & K. Yang
State Key Laboratory of Alternate Electrical Power System with Renewable Energy Sources
(North China Electric Power University), Beijing, China
School of power, North China University of Water Resources and Electric Power, Zhengzhou,
Henan Province, China(s)

ABSTRACT: Power quality which hinders the development of large-scale wind power is one of the main factors. Flicker is one important indicator of power quality measurement. According to International Electrotechnical Commission standards IEC61400-21 and national standards GB/T12326-2008, wind flicker calculation of one wind unit operating was derived, the simulation was built by Bladed software, the modes were set up by taking the linear method, nonlinear method and BP neural network, the mapping of wind farm's wind conditions and flicker was established. It is proved that flicker synchronous grows with wind speed and turbulence intensity, and the error of BP neural network model is the smallest.

Keywords: wind farm's wind conditions; power quality; flicker; BP neural network

1 INTRODUCTION

Due to the random variations of wind speed, the output power of wind power units fluctuates. Frequent starting, stopping and switching the wind power units will have an effect on the power quality of regional power network, and increase the maintain cost of the management of wind farms. As the proportion of the wind power capacitance increases, the power quality will become one of the major constraints of the development of wind power. Therefore, the research on the measure of power quality is of great importance. The paper mainly studies the relations between the flicker and the wind condition of the wind farm when a wind power unit operates continuously.

References [1,2] show that wind condition has great impact on the Voltage fluctuation and flicker caused by grid-connected wind power units, especially on the average wind speed and turbulence intensity. Apart from the features of wind itself, the fluctuation of the output power of wind power units may be caused by the features or the switch of wind power units, the effect of tower shadow, the wind shear, the yawing distance error, or the network structure of the connecting system, etc. Reference [3–6] studies the flicker produced by grid-connected wind power units and the influence of various factors on the flicker, adopting International Electrotechnical Commission standards IEC61400-21 and national standards GB/T12326-2008. These references haven't carried out further research on quantifying the influence of wind condition on flicker.

According to International Electrotechnical Commission standards IEC61400-21 and national standards GB/T12326-2008, this paper studies the wind flicker calculation with one wind unit operating. It also quantifies the influence of wind condition on flicker with wind power units in continuous operation, and studies the relations between them, by applying the software GH Bladed, the liner and nonlinear approach, and BP Neural network algorithm.

2 THE CALCULATION OF FLICKER

The short-term flicker value P_{st} is measures the flicker in a short time (eg. several minutes). The basic recording cycle of the short-term flicker is 10 min. Long-term flicker value P_{lt} measures flicker in a long time (eg. several hours). The basic recording cycle of the short-term flicker is 2h, which is calculated by the short-term flicker value P_{st}.

2.1 *The calculation of the short-term flicker with wind power units in continuous operation*

According to International Electrotechnical Commission standards IEC 61400-21 and national standards GB/T 12326-2008, flicker can be calculated through voltage variation d and the frequency of voltage variation r, when the load is periodic equal interval rectangular wave (or square pulse wave). As wind speed is a random variable, wind power is considered as periodic equal interval rectangular wave (or square pulse wave).

1. The calculation of voltage variation d

   Voltage variation d is the difference of two adjacent extreme points on the root mean square curve, shown in percentage of System nominal voltage. When the given active power and reactive power of the three phase load are ΔP_i and ΔQ_i, it Can be calculated as follows.

$$d = \frac{\Delta P_i R_L + \Delta Q_i X_L}{U^2{}_N} \times 100\% \qquad (1)$$

   R_L and X_L are the resistance and Reactance component of the Grid Impedance respectively. U_N is the System nominal voltage. According to the rule that the recording cycle of the short-term flicker P_{st} is 10 min, the Mean square errors of active power and reactive power are ΔP_i and ΔQ_i.

2. The calculation of voltage variation frequency r

   The calculation of voltage variation frequency r is the times of voltage variation per unit time (Voltage in ascending or descending each count as one change). If the time interval is less than 30 ms, several changes in different directions count as one change. Voltage variation frequency r in this paper can be obtained from wind module of Bladed software.

3. The calculation of voltage short-term flicker P_{st} If the voltage variation d and voltage variation frequency r are given, short-term flicker can be calculated, using the voltage variation $d_{Lim}(P_{st} = 1)$ in Table 1.

$$P_{st} = \frac{d}{d_{Lim}} \qquad (2)$$

   As fluctuating load, wind power units meet the estimated conditions of the flicker.

2.2 The calculation of short-term flicker of multiple wind farm units

As the short-circuit capacity of the PCC (Point of Common Coupling) which wind turbine generator connected to should be consistent, total short-term flicker of multiple wind farm units $P_{st\Sigma}$ can be shown as follows.

$$P_{st\Sigma} = \sqrt{\sum_i P_{st,i}} \qquad (3)$$

$P_{st,i}$ is the real-time short-term flicker value of the corresponding wind power unit.

3 THE CALCULATION OF FLICKER OF THE WIND POWER UNIT BASED ON BLADED SOFTWARE

Based on the common wind condition generated by Bladed Software, the paper simulates the

operation of 1.5 MW single wind power units, and then fits the flicker value of the grid access point with the wind condition. The paper defines the 3D wind model in GH. Bladed the wind menu, sets the average wind speed, and generates different wind condition documents, using the basic von Kaman turbulence model.

a. Wind turbine parameters.

   ①Cut-in wind speed is 3.5 m/s, cut-out wind speed is 25 m/s, rated wind speed is 11.3 m/s;

   ②Aerofoil is SINOMA40.2, 40.25 m in length, three-blade, impeller on the wind direction arrangement;

   ③Tower height is 67.8 m, obliquity is 4°, cone angle is 0;

b. Wind condition combination.

   ①Wind speed V: Common major operation range is set as [5, 12] considering the actual situation, resolution ratio is 1 m/s.

   ②Turbulence intensity I_n: Because the average Turbulence intensity on land In is 0.12–0.15, Windward longitudinal turbulence intensity I_{nx} takes the place of I_n according to von Kaman model. Operation range is set as [0.1, 0.165], resolution ratio is 0.005;

   ③The default of X longitudinal fluctuations frequency is 6.8266 Hz (Table 1 shows that $d\%$ changes little when the frequency reaches hundreds of times. Meanwhile, to reduce the computational complexity, it is enough to simulate.) The essence of flicker is power fluctuation, and that of power fluctuation is wind speed fluctuation. Therefore, the corresponding voltage variation frequency is 409.6 times/min. According to Table 1, $d_{Lim} = 0.54\%$.

c. Power parameters are set as in Figure 1.

   The voltage of single power unit is 0.69 kV and that of PCC is 0.69/35 kV. The average line length is 3 km. The distance between PCC and Transmission booster station 35/110 kV is 10 km. The best relation between flicker value and Line impedance angle is (60°–70°). Referring to relevant data, current carrying capacity and data of load, wire LGJ-240 is used in three-phase transmission line, the Line impedance angle Φ is approximately 70.2° (R = 0.132 Ω/km, X = 0.362 Ω/km).

d. After simulation analysis of random (.wnd) document of various wind conditions, the active and reactive changes in the infinity nexus can be obtained.

4 THE RELATION BETWEEN THE WIND CONDITIONS OF WIND FARM AND THE FLICKER OF WIND POWER UNIT

Fitting relations of $V, I_n - P_{st}$ according to the data in Table 2.

Table 1. corresponding data of voltage variation with unit flicker ($P_{st} = 1$) on periodic rectangular wave (or square pulse wave).

d/%	3.0	2.9	2.8	2.7	2.6	2.5	2.4	2.3	2.2	2.1	2	1.9	1.8
r (time/min)	0.76	0.84	0.95	1.06	1.20	1.36	1.55	1.78	2.05	2.39	2.79	3.29	3.92
d/%	1.7	1.6	1.5	1.4	1.3	1.2	1.1	1.0	0.95	0.90	0.85	0.80	0.75
r (time/min)	4.71	5.72	7.04	8.79	11.16	14.44	19.10	26.6	32.0	39.0	48.7	61.8	80.5
d/%	0.70	0.65	0.60	0.55	0.50	0.45	0.40	0.35	0.29	0.3	0.35	0.40	0.45
r (time/min)	110	175	275	380	475	580	690	795	1052	1180	1400	1620	1800

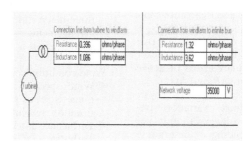

Figure 1. Access network parameters.

Table 2. Relational table of $V, I_n - P_{st}$.

$P_{st}/I_n/V$	0.100 / 0.135	0.105 / 0.140	0.110 / 0.145	0.115 / 0.150	0.120 / 0.155	0.125 / 0.160	0.130 / 0.165
5	0.007373 / 0.009464	0.006050 / 0.009856	0.006358953 / 0.010136258	0.006647 / 0.010606	0.006184 / 0.010951	0.007733 / 0.009672	0.007928 / 0.010024
6	0.005532 / 0.009161	0.005957 / 0.009821	0.006407240 / 0.015725780	0.006898 / 0.016421	0.007271 / 0.017150	0.012853 / 0.018376	0.014242 / 0.019316
7	0.011967 / 0.026251	0.012653 / 0.027713	0.013336587 / 0.028388518	0.013813 / 0.030233	0.014993 / 0.019685	0.016363 / 0.032742	0.016433 / 0.034876
8	0.019414 / 0.041329	0.020769 / 0.043438	0.022263511 / 0.032391323	0.023613 / 0.048737	0.023589 / 0.049049	0.026508 / 0.051227	0.040655 / 0.056990
9	0.028134 / 0.059081	0.030043 / 0.045238	0.031692650 / 0.064071974	0.033924 / 0.066078	0.035148 / 0.068712	0.053649 / 0.070297	0.056533 / 0.055046
10	0.039272 / 0.057811	0.041349 / 0.059851	0.044342365 / 0.062186463	0.046488 / 0.063017	0.048265 / 0.065518	0.055903 / 0.068089	0.058026 / 0.067859
11	0.032830 / 0.047776	0.034399 / 0.048001	0.036108565 / 0.047919083	0.037989 / 0.052958	0.040043 / 0.054518	0.040948 / 0.068010	0.045244 / 0.073842
12	0.012056 / 0.023539	0.013701 / 0.025775	0.015481871 / 0.041471735	0.016583 / 0.044163	0.018287 / 0.045104	0.016145 / 0.047518	0.022153 / 0.049514

4.1 Linear fitting

In the form of $z = a_1x + a_2x^2 + a_3xy + a_4x^2y + a_5xy^2 + a_6x^2y^2 + a_7$, fitting them with liner regression approach.

According to Figures 2 and 3 the result of calculation is as follows:

$a_1 = 0.0171$; $a_2 = -0.0002$; $a_3 = 0.3101$; $a_4 = -0.0305$; $a_5 = -1.0122$; $a_6 = 0.1197$; $a_7 = -0.1486$;

The relation after fitting:

$$P_{st} = 0.1197V^2 \cdot In^2 - 0.0305V^2 \cdot In \\ - 1.0122V \cdot In^2 + 0.3101V \cdot In - 0.0002V^2 \\ + 0.0171V - 0.1486 \qquad (4)$$

4.2 Nonlinear fitting

Fitting with nonlinear approach in the form of $z = a_1 \times y + a_2xy + a_3$ and marking the condition at Coordinate points, the result can be obtained as follows: $a_1 = 0.1074$; $a_2 = 0.0394$; $a_3 = -0.0263$;

330

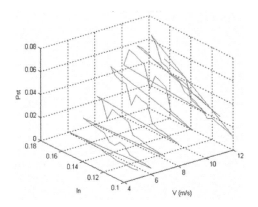

Figure 2. The tendency of the linear fitting relationships for $V, I_n - P_{st}$.

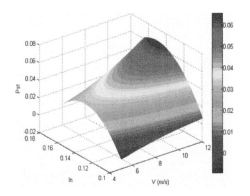

Figure 3. Surface charts of the linear fitting relationships for $V, I_n - P_{st}$.

The relation after fitting:

$$P_{st} = -0.0263 + 0.1074 I_n + 0.0394 V * I_n \qquad (5)$$

4.3 BP neural networks model

4.3.1 Introduction

Artificial Neural Networks (ANN) is a mathematical model of algorithm, which is characterized by simulating animal neural network behavior and distributed parallel information processing. This network relies on the complexity of the system, by adjusting the connection between the numbers of nodes within the relationship, so as to achieve the purpose of processing information. BP (Back Propagation) network used in this paper is put forward by the team of scientists led by Rumelhart and McCelland. It is a feed forward neural network practiced by back propagation algorithm, and one of the most

widely used neural networks models. The circuit topology of neural networks model includes input, hide layer and output layer. Its learning rule is to use the steepest descent method to adjust weights and threshold value of net through and minimize the network error square sum. Therefore, BP neural network can learn and store a lot of input—output mode mapping, without revealing in advance the mathematical equations which describe the mapping.

4.3.2 BP neural network modeling

According to Funahashi, the three-layer BP neural network with S-shaped function can approximate any continuous function, so three-layer BP neural network is established. Wind speed V and turbulence intensity In are input. Four Neurons are set in the intermediate hide layer and there is one output layer, which is the corresponding P_{st}. Neuron transfer functions are S-shaped logarithmic function logsig and LM algorithm training function trainlm.

The time of training is 10000 and the target error is 0.00001. When the calculations achieve the

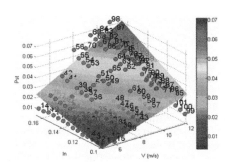

Figure 4. Surface charts of the nonlinear fitting relationships for $V, I_n - P_{st}$.

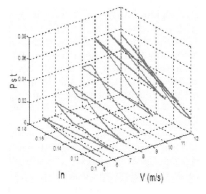

Figure 5. The tendency of the relationships of $V, I_n - P_{st}$ based on BP neural network.

Table 3. The relational table of V, $I_n - P_{st}$ based on BP neural network.

$P_{st}/I_n/V$	0.100 / 0.135	0.105 / 0.140	0.110 / 0.145	0.115 / 0.150	0.120 / 0.155	0.125 / 0.160	0.130 / 0.165
5	0.006360285	0.006582	0.006865	0.007222	0.007659	0.008084	0.007173
	0.007406254	0.007986	0.008624	0.009327	0.010109	0.01099	0.011995
6	0.007210711	0.007608	0.008112	0.00875	0.009553	0.010549	0.011761
	0.01309562	0.012342	0.013208	0.014917	0.016837	0.01896	0.021283
7	0.009856796	0.010792	0.011945	0.013368	0.01512	0.017256	0.019823
	0.022835396	0.026266	0.029867	0.028761	0.030158	0.033447	0.036706
8	0.016938021	0.018992	0.021346	0.02403	0.027071	0.030492	0.034295
	0.038441025	0.042839	0.047343	0.051777	0.05574	0.051567	0.050169
9	0.029454555	0.03223	0.035085	0.038025	0.041071	0.044258	0.047615
	0.051151218	0.054828	0.058549	0.062175	0.065554	0.068567	0.070935
10	0.038434123	0.040824	0.043129	0.045385	0.047649	0.049987	0.052465
	0.055135449	0.058012	0.061051	0.064147	0.067153	0.069925	0.072356
11	0.030863637	0.033479	0.036023	0.038514	0.040983	0.043479	0.046061
	0.048790184	0.05171	0.054827	0.05809	0.06139	0.064582	0.067526
12	0.00959328	0.011536	0.013698	0.016069	0.018639	0.021406	0.024373
	0.027555656	0.030975	0.034646	0.038557	0.042653	0.046828	0.05094

target, they will be stored as. mat file, used as application function relationships. Finally Table 3 and Figure 4 can be obtained.

5 CONCLUSIONS

1. The relationship tendency of the wind speed V, turbulence intensity In and the short time flicker P_{st} is verified through liner and nonlinear fitting, which is consistent with the theory of reference [2]. With the increase of wind speed, the voltage variation and flicker generated by wind power units increase continuously. Turbulence intensity and the voltage fluctuation and flicker almost grow in direct proportion.
2. Using the data of the given points to testify the above-mentioned functional form and considering the mutual influence of the numerical size, the paper finds that the missing non-linear fitting of equation (4) and(5) has a better accuracy. But the error of the fitting results of the two functional forms is still large, the average of which is about 30% at most. Fitting data relationships established by the neural network has good generalization ability, the error of which is less than the error of functional fitting mentioned above. The average error is about 10%, which is desirable.
3. The voltage fluctuation and flicker in the next period can be predicted through BP neural network model and the wind speed and turbulence intensity of single power unit of the wind farm predicted in advance, which is of great significance to the operation and maintenance of the wind farm.

REFERENCES

[1] Sun Tao, et al. 2003. Voltage Fluctuation and Flicker Caused by Wind Power Generation. Power System Technology, 27(12):62–70.
[2] Papadopoulos PM, et al. 1998. Investigation of the flicker emission by grid connected wind turbines. Proceedings of the 8th International Conference on Harmonics and Quality of Power. pp: 1152–1157.
[3] GAO Yujie. 2009. Power Quality Analysis with the Wind Farm Integrating in Power System. Southern Power System Technology, 3(4):68–72.
[4] Wang Haiyun & Wang Weiqing. 2008. Analysis and calculation on flicker severity for wind farm. Renewable Energy, 26(2):87–89.
[5] Nguyen Tung Linh. 2009. Power Quality Investigation of Grid Connected Wind Turbines. Industrial Electronics and Applications, ICIEA 2009 4th IEEE Conference on, pp: 2218–2222.
[6] Larsson Å. 2002. Flicker emission of wind turbines caused by switching operation. IEEE Transactions on Energy Conversion, 17(1):119–123.
[7] Liu Haitao, et al. 2011. Supervision on Power Supply Quality in Wind Farm and Analysis to Power Characteristic Curve Calibration of Wind Generator. Inner Mongolia Electric Power. 29(1):34–36.
[8] Liu Kai. 2005. Short-Term Load Forecasting Based on Improved BP Neural Network. Hohai University.
[9] Zhao Gang, et al. 2001. A Study on Calculation of Short Term Flicker Severity. Power System Technology, 25(11):15–18.
[10] Zhao Haixiang, et al. 2004. Error Analysis of Discrete Calculation Method of Flicker Severity and Its Application. Power System Technology. 28(13):84–87.
[11] GB12326-2008, Power Quality-Voltage fluctuations and flicker.
[12] EA Bossanyi. 2005. GH Bladed Principles of Manual. Garrad Hassan and Partners.
[13] Liu Weiguo. 2006. MATLAB Program Design and Application. Beijing. Higher Education Press.

The whole process of risk management research on EPC project

L. Zhao & H.B. Li
School of Management, Hebei Institute of Architectural Engineering, Zhangjiakou, China

X.M. Wang
School of Management Science and Engineering, Shenyang Jianzhu University, Shenyang, China

ABSTRACT: From the angle of general contractor, we put forward the thinking of the whole process of risk management by analyzing the risk of EPC contracting mode faced by the contractor. This thought has certain enlightenment function to EPC general contractors better doing risk management.

Keywords: EPC; the whole process; risk management

1   INSTRUCTIONS

EPC contracting mode as the mainstream mode of project contracting market, is used in construction market more and more. In this mode, the contractors take more risks. That means the contractors must prevent the risks through the effective risks management to reduce project risk and get more profit.

2   THE RISK ANALYSIS OF GENERAL CONTRACTOR UNDER EPC CONTRACTING MODE

2.1   *The risk of project content*

Although owner ought to take the responsibility of anticipated goal, functional requirement and the design basis, the general contractor has to undertake the uncertain risk of project content if some are unreasonable, omitted and the owner change the command in the process of project construction, When doing EPC project tendering work.

2.2   *The risk of blind tender offer*

In the period of EPC bidding, the general contractor usually faces many uncertain situations. In so many uncertain situations, the biding risk of general contractor is much higher when general contractor signing general contract in the way of fixed lump sum. So the general contract faces double pressure of the opportunity of losing the project and obtaining the potential of the financial risk when acquiring project.

2.3   *The risk of rashly entering market*

In different areas of the project, backgrounds often exist differences. If rashly entering market without understanding the market situation, the general contractor will face big risks. In order to obtain the general contracting project, Some companies rashly bid without insight into the factors of local politics, economy and geographic environment of project, enough analyzing terms of tender, their own conditions and risk of biding. The lost by this way is typical risk of rashly entering market.

2.4   *The risk of project design*

Although the general contractor takes the task of design in EPC mode, the owner has the power of examining and verifying design documents. The owner may repeatedly put forward advises which may increase workload and extend the planned project duration. The general contractor takes the responsibility of increased cost when optimizing, deepening design and modifying the project design in order to satisfy the functional requirement of project. All of these costs need be taken by general contractor.

2.5   *The risk of project purchasing*

In the phase of purchasing materials and equipments, the cases such as delivery delay of supplier, disqualification of equipments and materials, the goods damaged or lost in transit, may cause purchasing risk to the general contractor.

2.6   *The risk of project construction*

In the engineering construction process, the general contractor bears the loss of engineering equipment damage owing to unexpected events and the risk of casualties. In addition, the general contractor needs to undertake the unforeseeable risks by bad weather in the engineering construction process and so on.

2.7 The risk of EPC project itself

Because of all kinds of influences of uncertainty factors in the EPC project, the general contractor will face many risks. This risks can be represented as many forms such as default of employer, declining or delaying engineering payment, terminating the contract before owner completing the contract, default of subcontractor, engineering delay and technical indexes can not reaching the contract and so on.

3 THE RISK MANAGEMENT OF EPC PROJECT

The whole process risk management of EPC project should run through the whole process of each project. The project risk identification, risk analysis, risk control and risk process are a recycle process in engineering project execution.

3.1 The risk management of project bidding and discussing bid

If the risk management of project is more earlier began, the cost is lower, and the project risk control is also more effective. The contractor needs to make a full understanding of the project as much as possible before bidding. It is beneficial to make a correct bidding decision.

The contractor ought to start the process of risk management when getting the information of tender and discussing bid. Mainly includes: Deeply investigating the politics, economy, society, law and other relevant circumstances in project location. Carefully studying on the contract documents, written form for clarification if the clauses are not clear or rigorous and keeping the file of result. Investigating the payment of owner. The contractor investigates the financial position and the previous payment. Careful investigating the spot. The contractor needs to inspect the spot as far as possible, and understand the spot's condition. Providing alternative list of suppliers as more as possible, reducing the possibility of occurring risk events for the reason of suppliers.

Through the expert meeting or other risk identification technology to the project risk identification, confirming the main risk source, and using experience estimation or other risk analysis technique to the project risk analysis, determining

the probability of occurrence of all kinds of risks and the influence of contractor. The contractor decides whether to attend in project bidding and tender discussion or not according to the results of analysis. Generally speaking, the contractor need to give up the bidding existing deadly risk, and find more reasonable measures to avoid the risk for the common risk of project. For existing slightly risk of project, The contractor should carefully study on the method of risk controlling from bidding, bid negotiation, signing contract to the each phase of project.

3.2 The risk management of business negotiation of project contract and the signing process

In the signing stage of contract, the general contractor more accurately identifies the risk based on the risk analysis in the process of bidding. The contractor estimates the probability of risk occurrence and the influence to contractor, and makes more detailed risks source table of the project. The contractor analyses the different causes of risks and takes different business negotiation strategies. In the contract signing stage, the contractor eliminates or reduces some risk factors for the purpose of reducing the possibility of risk occurring in the phase of contract implementation. These risk factors include the following aspects:

3.2.1 The contracting work scope
The contractor should examine and verify work scope in the contract is clear or not, and the divided responsibility of contracting parties, and the handling of the problems among the interfaces with other contractors.

3.2.2 Contract money
It includes the constituent part of contract money, the kind of currency in payment and adjusting method of contract money and so on, and try best to avoid the lost because of the improper price adjusting method caused change by exchange rate.

3.2.3 The mode of payment
It mainly audits the ability to pay of the owner in the spot exchange payment project, and the Guarantee of payment providing by the owner in the deferred payment project.

3.2.4 The guarantee of three banks
The effective time of advanced payment guarantee, whose amount of money limited to advanced payment in full, must be the same as the receipt time of advanced payment in full. The effective time of advanced payment guarantee is the sooner the better. The ratio of amount of money in performance guarantee will strictly be controlled

and the ratio is the lower the better. The effective time of retention money guarantee must be the same as the payment time of final payment in project and in amount of money. All of the guarantees need to avoid overlapping with follow-up guarantee as far as possible. In other way, the period of validity is the shorter the better.

3.2.5 *The default fine of contractor*
It agrees accumulative maximum limit as far as possible based on the subentry fines, and pays attention to the reasonable of accumulative maximum limit.

3.2.6 *The responsibility of owner*
It mainly includes responsibility of default for owner deferring payment, providing spot condition by owner, technical parameters of design requirement and so on.

3.2.7 *Application of law clause and resolution terms of dispute*
It should use the related law clause of project country to get more project convention. The dispute needs be solved in the third country as far as possible, and the arbitral authority and the clauses are clear as far as possible and the final award results must be obeyed by both sides.

3.3 *The risk management in the process of project executing*

The risk management in execution phase of project is the key link of risk management of the whole project and plays a decisive role to the success or failure of the whole risk management. We have to identify the project risk again, and analyze and perfect the risks source table, when identifying the project risk again. According to the different risks causing of risk, we make different risk controlling strategy, and prevent risk by the way of strengthening engineering management.

3.3.1 *Making the insurance planning according to analyze the table of risks source*
According to analyze the table of risks source, we make the insurance planning, appoint special persons to be responsible for insurance management, and claim indemnity for the risk of project. We transfer the risk using the way of engineering insurance. The risk administrators must set the goal of obtaining the most ideal risk guarantee in the best engineering insurance.

3.3.2 *The risk management of signing the subcontract and equipment purchasing contract*
We seriously understand the credit standing and respectability, contractual capacity and the ability of bearing risk of subcontractors and suppliers and so on, first according to the client file. We may go to the bank or work place of user to inspect, and as far as possible reduce the possibility of risk to contractors without the right selection of subcontractors and suppliers.

At the same time, contractor needs transfer the responsibility given by owner to subcontractor and supplier by means of subcontract and purchasing contract as far as possible. In the subcontract and purchasing contract, the contractor ought to adopt the same warranty clause, limited responsibility clause, guarantee clause, undertaking the responsibility of breach of clause and so on as the main contract and maximum transfer or disperse the responsibility of the contractor. These are the main methods to the contractor transfer the risk. Signing a high quality subcontract and purchasing contract is the basis of project risk deflection.

3.3.3 *The management of cash flow*
The contractor makes cash flow statement of subcontract and equipment purchasing contract according to the payment cash flow statement of the main contract, and timely adjust the payment cash flow statement of subcontract and equipment purchasing contract based on the changes of cash flow in main contract in order to reduce the expenses of engineering financial and strengthen the ability of project preventing risk. According to the compared cash flow table with the actual cash flow table, the company timely finds out cases such as extra expenses and so on and deals with them in time.

3.3.4 *The management of network program*
According to the project network program in the main contract, the contractor makes progress network program and timely adjusts it based on specific implementation of project. According to the analyze influence of the time limit for a project caused by some works delay, the contractor timely finds out or eliminate the factors which restrict time limit finish, and takes steps in the way of controlling and supervising the subcontractor and supplier, and effectively controls the risk of delay. Controlling and managing the engineering progress by network program is the main method for the contractor to control the delay risk.

3.3.5 *The management of contract*
The contractor needs strengthen contract management and equip special contract management personnel to be responsible for the work of contract management. The contractor improves themselves the ability of fulfilling the contract and avoids the all kinds of risks caused by Contractor breach. Because of the change

factors so many in the EPC project, all the sides of contract need interpret, confirm and archive all recorded documents as a contract files except contract itself. This is extremely good measures of making sure the project progresses smoothly according to the contract requirement and avoiding the risk of project. This is also the important basis for the engineering claims after the risk occurring. Effectively contract management can prevent disagreement in both contractor sides, and restrict contracting parties to comply with the responsibility and obligation in contractual stipulations, and eliminate some potential risks.

4 CONCLUSIONS

EPC contract mode has many advantages and is developing into the mainstream mode of the project contracting. But in the practice of enterprise management model, there are many aspects need to be discussed and improved. We can avoid or reduce the risk of EPC projects from the fundamental and carry out the profit goal of EPC contract though deeply analyzing the main risk of EPC contract and managing the whole process of EPC contract.

REFERENCES

[1] CNAEC. Design procurement construction(EPC)/ Turnkey contract conditions [M]. Beijing: China machine press, 2002.
[2] Wuren WANG. EPC general contract management [M]. Beijing: China architecture and building press. 2008.5.
[3] China International Contractors Association, International engineering contracting practical manual [M]. Beijing: China railway publishing house, 2007.
[4] [U.S.] John M. Nicholas. Project Management for Business and Technology: Principles and Practice. TSINGHUA University Press, 2006, 1–59.

Frontiers of Energy and Environmental Engineering – Sung, Kao & Chen (eds)
© 2013 Taylor & Francis Group, London, ISBN 978-0-415-66159-1

Research and application of large section pillar reduction technology in thick coal seam

Z.P. Guo & Y. Wang

College of Resources and Environmental Engineering, Shandong University of Science and Technology, Qingdao, Shandong, China

ABSTRACT: In order to increase the recovery rate of coal resources and to improve the maintenance situation of section drift, a new way of roadway layout of the working face is put forward. Through theoretical analysis and engineering practice, section coal pillar is cut from original 50 m to 7.5 m. The large pillar left in the original design of roadway layout can be recovered, which achieved very good economic benefits. This research result provides theoretical basis for section drift layout design under similar geological conditions.

Keywords: thick coal seam; large coal pillar; lateral stress; coal recovery rate

1   GENERAL INSTRUCTIONS

Fully-mechanized mining technology with large height is used in a working face of a coal mine, in which the mining height is 5 m. Because of the restriction of double tunneling, the section coal pillar that reaches 20~50 m which is left between working faces in the mining district cannot be recovered, which is becoming an important factor to restrict the resources recovery rate to further improve. So far there is about 50 m-long pillar left in the gob side and the reason is that the auxiliary transportation crossheading is going to be used as the return airway of the next working face after recovery. The size of pillar is designed too big, which not only influences the production continue, but also causes great waste of coal resources. So that determining the size of pillar reasonably, cutting the width of the pillar is of great practical significance.

2   GETTING STARTED

2.1   *Situation of working face*

The dip angle of the 8# coal seam of the working face is 3~8° and the thickness is 3.19~8.84 m and the average thickness is 5.01 m. The coal seam structure is complex which has 3~9 parting layers, reaching a maximum of 0.65 m. The main rock is mudstone, followed by lime mudstone. The roof of the coal seam is a typical compound structure and the false roof is combined with coal and rock which is 2~2.8 m thick. The lithology differences significantly and the stability and cohesiveness is poor which occurs softened abscission layer when encountered water. The thickness of immediate roof is 0~10.97 m and the lithology is mudstone and sandy mudstone. The thickness of main roof is 12~34 m and the form of the main roof is coarse sandstone, which is thick-bedded.

2.2   *Original roadway layout*

When working face replaced, according to the original design, the auxiliary transportation crossheading is going to be used as the return airway of the nest working face after recovery. There are 4 crossheadings in the working face. The auxiliary transportation crossheading 1 is used to transport materials, persons and intake air. The belt crossheading 2 is on the same level of main roadway, using to transport coal. The air return crossheading 3 is used to return air. The auxiliary transportation crossheading of last working face is used as gas tail roadway to drain gas. The pillar A which is 25~30 m is left between the belt crossheading and auxiliary transportation crossheading on the bottom of the face. The pillar B which is 25~30 m is left between air return crossheading and gas tail roadway on the top of the face. The total width X is A+B = 50~60 m. The original roadway layout is shown as Figure 1.

Where, A,B,C—large pillar, about 25 m in depth, 0—link-roadway,   L1—198.5 m,   1—auxiliary transportation crossheading, to be used as gas tail roadway of next face, 2—belt crossheading, 3—air return crossheading, 4—gas tail roadway, left by auxiliary transportation crossheading of last face, 5—belt crossheading of last face.

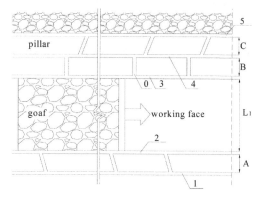

Figure 1. Original roadway layout drawing.

3 PROGRAM DESIGN OF SMALL PILLAR

3.1 Small pillar program

Given the large amount of resources waste and for the face replacement carrying out smoothly, the program design based on pillar mining and gob-side entry driving is put forward. Still use the way of double tunneling to dig the auxiliary transportation crossheading 1 and belt crossheading 2 left a large pillar A (A = 15~28 m). The roadway 1 is used as an auxiliary transportation crossheading, while the roadway 2 on the edge of the face is used as a belt crossheading. The auxiliary transportation crossheading is used to transport materials, persons and intake air. Mobile substation and belt conveyor are placed in the belt crossheading which is used to transport coal and intake air. Because the belt crossheading is lay outside the auxiliary transportation crossheading, when the new face mining begins, the large pillar A between auxiliary transportation crossheading and belt crossheading can be recovered. So that the coal pillar produced by the technology of double tunneling is recovered thoroughly. The design of new roadway layout with small pillar is shown as Figure 2.

Where, A—large pillar, 18~25 m in depth, B—small pillar, 3~8 m in depth, 1—auxiliary transportation crossheading of new working face, 2—belt crossheading of new working face, 3—air return crossheading designed by the way of gob-side entry driving with small pillar.

3.2 Theoretical calculations of small pillar size

After the recovery of last face, generally, there will be an area of internal stress field with low abutment pressure. Gob-side entry retaining and small pillar should be lay inside this stress decreasing zone. The best place of roadway driving along gob-side is place 2, which can be shown in Figure 3.

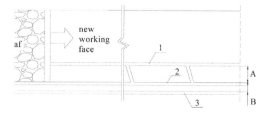

Figure 2. New roadway layout with small pillar.

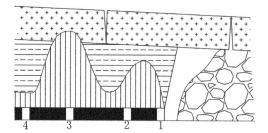

Figure 3. Side abutment pressure distribution and roadway driving place.

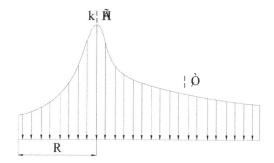

Figure 4. Calculation model of plastic area width.

It can be seen from Figure 3 that the internal stress field has an advantage of low stress, little coal losses and easy to anchor, which is the best place. The width of internal stress field can be calculated through Eq.1. The calculation model is shown as Figure 4.

The width of plastic zone in stress limit equilibrium zone of 8# coal seam in the coal mine is:

$$R = \frac{h\lambda}{2\tan\theta} \ln\left[\frac{k\gamma H + \dfrac{C}{\tan\theta}}{\dfrac{C}{\tan\theta} + \dfrac{P_x}{\lambda}}\right] \quad (1)$$

where, R—width of plastic zone, h—height of the roadway, here is 5.2 m, λ—side pressure coefficient, here is 0.4, θ—internal friction angle, here is 36.96°, k—stress concentration factor, here is 4, γ—bulk density, here is 0.025 MN/m³, H—mining depth, here is 267.6 m, C—cohesion of coal seam, here is 0.9 MPa, P_x—support strength to roadway side wall, here is 0.

Put all the parameters to Eq.1, the width of plastic zone R can be calculated to be 4.35 m.

The minimum of the pillar width should bigger than width of plastic zone plus length of anchor inside coal body, and an appropriate safety factor must be considered.

The length of anchor is 2 m. When considered with the prosperity coefficient and the size of pillar should satisfy the effect of isolate goal, the width of small pillar is 7.5 m.

4 PRACTICAL MEASURE RESEARCH OF THE RANGE OF INTERNAL STRESS FIELD

Borehole stress meters are applied to measure the abutment pressure distribution in the roadway to obtain the width of pillar when mining is in process. Survey area is set inside the section coal pillar, and observing station is set in one side. Totally, 5 drillings of Φ45 mm are set along the coal seam strike with 2 m distance from each other. The drillings are distributed on the level difference of 2 m, denoted by 1~5 measuring point to measure the pressure in different depth. The depth of each point is 2 m, 4 m, 6 m, 8 m and 10 m. Model of KS-1 borehole stress meters are set in the drillings to observe the whole process before mining, mining in process and after mining. Observing begins from the place 150 m ahead working face and ends 130 m behind. The survey area is shown as Figure 5.

Where, 1—gas tail roadway, 2—air return crossheading, 3—auxiliary transportation crossheading, 4—belt crossheading.

The observing results of observing station is shown as Figure 6.

According to the monitor results, except the 1, 2 measuring point, the pressure in other measuring

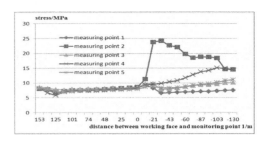

Figure 6.   Stress distribution.

points is close, which is about 7~12 Mpa. The stress distribution is relatively flat and the variation is little. The pressure of No. 3, 4, 5 measuring point is going to decrease after 15 m behind the working face passed, which illustrates that the coal body is going to enter the stage of plastic. The pressure of 4 and 5 measuring point which is 8 m and 10 m in depth is close to each other and is the least. So the range of the internal stress field is 0~8 m.

From the analysis above, under effectively control of surrounding rock, goaf gas and spontaneous combustion of coal of the current face, and also the mining safety, according to the monitor result in field measurement, the size of small pillar is designed to be 7.5 m.

5   CONCLUSIONS

A new way of working face roadway layout in thick coal seam which is the large pillar reduction program of pillar mining and gob-side entry driving is put forward. Through the way of ground pressure control and field measurement, the rational size of pillar is determined to be 7.5 m. During the test, 256 thousand tons of coal is recovered after pillar mining of the working face. The working section recovery is increased by 15.5%, which achieved very good economic benefits.

REFERENCES

Du Jiping & Wang Liquan. 2003. *Special Coal Mining Methods*. Xuzhou: China University of Mining and Technology Press, China. (In Chinese).

Lin Jian & Wu Yongzheng. 2005. Application and technology of narrow pillar drift in Ningwu coal mine. *Coal Science and Technology*. 33:5–7. (In Chinese).

Ma Qihua, Wang Yitai. Theory of narrow pillar drift and support technology of gob-side entry in deep coal mine. *Journal of Mining & Safety Engineering*. 26: 520–523. (In Chinese).

Qian Minggao & Shi Pingwu. 2003. *Ground Pressure and Strata Control*. Xuzhou: China University of Mining and Technology Press. (In Chinese).

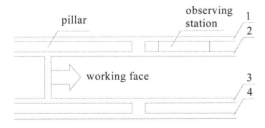

Figure 5.   Layout of the stress observing station in pillar.

Frontiers of Energy and Environmental Engineering – Sung, Kao & Chen (eds)
© 2013 Taylor & Francis Group, London, ISBN 978-0-415-66159-1

A review: Research on highway runoff pollution

L. Wang & B.L. Wei
Ocean University of China, College of Environmental Science and Engineering, Qingdao, China

ABSTRACT: Highway runoff pollution as a non-point source is the key factor that causes water quality deterioration of waters around the highway. The study of highway runoff pollution of researches abroad began in the 1970s. There has been a clear understanding about the type of pollutants, characters of highway runoff pollution, its sources, influencing factors. Through researching domestic and international literatures on the content above, this work endeavored to give an overview of the highway runoff pollution.

Keywords: highway pollution; water quality; pollutant sources; influencing factors

1 INTRODUCTION

After decades of effort, point source pollution, such as municipal sewage and industrial wastewater have been effectively controlled. However, surface water quality did not reach expected result; as a result, non-point source polluting receiving water gets more attention (Hamilton & Harrison 1991). Compared with point source pollution, non-point source pollution is more difficult to measure, evaluate and control, especially, non-point source pollution caused by highway. Lots of environmental pollution, including noise pollution, air pollution, water pollution and soil pollution are caused by road traffic, road construction and road maintenance. Highway runoff mixed pollutants causing received water pollution is important road pollution. Road runoff contains large quantities of pollutants such as heavy metals, hydrocarbons, pesticides and so on (Legret & Pagotto 1999, Kim et al. 2007, Barrett et al. 1988, Ball et al. 1998, Lee et al. 2011, Desta et al. 2007, Shen et al. 2005). During rainfall period, pollutants in road runoff flow into surrounding water, resulting in the deterioration of water quality. Highway runoff pollution is a key factor of road surrounding water quality deterioration and ecosystem damage (Hamilton & Harrison 1991, Zong et al. 2003, Perdikaki & Mason 1999, Denny et al. 1999).

Foreign scholars began to study highway runoff pollution from the seventies of last century. After forty years of accumulation, they have a comprehensive and in-depth understanding of the characteristic of runoff pollution. Combining the analysis of the results with mathematics and computer technology, researchers developed runoff pollution prediction models for certain conditions (Tamar & Eran 2009, Morrison & Rauch 2007, Kim et al. 2005, Lyn et al. 1998). The purpose of the study of highway runoff pollution is to propose effective measures to control pollution according to the characteristic of pollution (Terzakis et al. 2008, Karin et al. 1999, Ana & Thorkild 2001, Shutes & Revitt 1999).

2 CHARACTERISTIC OF RUNOFF POLLUTION

2.1 *Major pollutants in runoff*

Road surface runoff pollution is that road runoff mixing with the harmful substances which including particles spilled by the transportation projects on the road, particulate matter in vehicle exhaust settled on the road, the vehicle fuel dripped on the pavement, wear of tire and road and etc discharges into waters or farmland and causes.

There is a wide range of pollutants in road runoff, such as SS, the nutrients N and P, heavy metals (Pb, Zn, Fe, Cu, Cd, Cr, Ni and Mn), cyanide, sodium, calcium, chloride, sulfate, oil and PCB. Ellis (Ellis & Revitt 1987) found that SS is the most important pollutants in road surface runoff, the most abundant heavy metals are Pb and Zn.

2.2 *Characteristic of highway runoff*

2.2.1 *Characteristic of water quality*
Foreign researchers began to study the characteristic of highway runoff quality and quantity in an early age, and the got a more mature findings. Lee-Hyung et al. (Lee-Hyung et al. 2007) studied 7 rainfall events on the bridge deck of KONG JU City, South Korea in 2004, results showed that EMC of COD and TSS vary in a large

played a decisive role on the concentrations of heavy metals in highway: concentrations of heavy metals in rainfall events with short rainfall duration, little rainfall volume, low rainfall intensity were at a high level and fluctuated within a certain range; concentrations of pollutants in the initial runoff were higher, and were lower in the later runoff in events with long rainfall duration, high rainfall volume and heavy initial rainfall intensity; the discharge of heavy metals appeared 'twice wash' phenomenon in bimodal rainfall events with short rainfall duration, heavy rainfall volume.

Otherwise, some studies showed that there were some correlation between the discharge of road runoff and the total traffic volume during rainfall. Chui et al. (Chui et al. 1982) studied 500 rainfall events on 9 roads with different weather conditions, traffic conditions, and land use in 1982, Washington, USA, results showed that TSS load in road runoff and other pollution index load were in proportional to the cumulative traffic volume during rainfall.

5 SUMMARY

Researchers have conducted extensive research on the characteristics of pollutants in highway runoff, pollutants sources and pollution impact factors and accumulated lots of valuable experience. They pointed out that highway runoff pollution was important non-point source pollution, which caused deterioration of the waters around the road. Highway runoff pollution mainly results from runoff washing road particles. There is a wide range of pollutants in road runoff, such as SS, the nutrients N and P, heavy metals (Pb, Zn, Fe, Cu, Cd, Cr, Ni and Mn), cyanide, sodium, calcium, chloride, sulfate, oil and PCB. Concentrations of pollutants in runoff are high and change in a large range.

REFERENCES

Ana Estela Barbosa, Thorkild Hvitved-Jacobsen. 2001. Infiltration pond design for highway runoff treatment in semiarid climates. Journal of Environmental Engineering 127: 1014–1022.

Ball J.E. R. Jenks, D. Aubourg. 1998. An assessment of the availability of pollutant constituents on road surfaces. The Science of the Total Environment 209: 243–254.

Barrett, M.E., Irish, L.B. Jr., Charbeneau, R.J. 1988. Characterization of highway runoff in Austin, Texas, areas. Journal of Environmental Engineering 124: 131–137.

Brad Stephenson J. W.F. Zhou, Barry F. Beck, Tom S. Green. 1999. Highway stormwater runoff in karst areas-preliminary results of baseline monitoring and design of a treatment system for a sinkhole in Knoxville, Tennessee. Engineering Geology 52: 51–59.

Brigitte Helmreich, Rita Hilliges, Alexander Schriewer, Harald Horn. 2010. Runoff pollutants of a highly trafficked urban road—Correlation analysis and seasonal influences. Chemosphere 80: 991–997.

Brizonik PL, Stadelmann TH. 2002. Analysis and predictive models of stormwater runoff volumes, loads, and pollutant concentrations from watersheds in Twin Cities metropolitan area, Minnesota, USA. Water research 36: 1743–1757.

Chui TW, Mar BM, Horner RR. 1982. Pollutant loading model for highway runoff. Environmental Engineering 108: 1193–1210.

Dean CM, Sansalone JJ, Cartledge FK, et al. 2005. Influence of hydrology on rainfall-runoff metal element speciation [J]. Journal of Environmental Engineering 131: 632–642.

Denny R. Buckler, Gregory E. Granato. 1999. Assessing Biological Effects from Highway-Runoff Constituents. U.S. Geological Survey Information Services.

Driver NE, Troutman BM. 1989. Regression models for estimating urban storm-runoff quality and quantity in the United States. Journal of Hydrology 109: 221–236.

Dropper D, Tomlinson R, Williams P. 1999. Pollutant concentrations in road runoff; southeast Queensland case study. Environmental Engineering 126: 313–320.

Ellis JB, Revitt DM. 1987. The contribution of highway surfaces to urban stormwater sediments and metal loadings [J]. The Science of Total Environment 59: 339–349.

Gan Hua-yang, Zhuo Mu-ning, Li Ding-qiang, et al. 2007. Characteristic of heavy metal in highway rainfall runoff. Urban Environment and Urban Ecology 20(3): 34–37.

Gupta K, Saul AJ. 1996. Specific relationships for the first flush load in combined sewer flows. Water Research 30: 1244–1252.

Haejin Lee, Sim-Lin Lau, Masoud Kayhanian, Michael K. Stenstrom. 2004. Seasonal first flush phenomenon of urban stormwater discharges. Water Research 38: 4153–4163.

Huayang Gan, Muning Zhuo, Dingqiang Li. 2008. Quality characterization and impact assessment of highway runoff in urban and rural area of Guangzhou, China. Environmental Monitoring and assessment 140: 147–159.

Joo-Hyon Kanga, Masoud Kayhanianb, Micheal K. Stenstroma. 2008. Predicting the existence of stormwater first flush from the time of concentration. Water Research 42: 220–228.

Ju Young Lee, Hyoungjun Kim, Youngjin Kim, et al. 2011. Characteristics of the event mean concentration (EMC) from rainfall runoff on an urban highway. Environmental Pollution 195: 884–888.

Karin Lundberg, Maria Carling, Per Lindmark. 1999. Treatment of highway runoff: a study of three detention ponds. The Science of the Total Environment 235: 363–365.

Katerina Perdikaki, Christopher F. Mason. 1999. Impact of road run-off on receiving streams in eastern England. Water Research 33: 1627–1633.

Kayhanian M. C. Suverkropp, A. Ruby, K. Tsay. 2007. Characterization and prediction of highway runoff constituent event mean concentration. Journal of Environmental Management 85: 279–295.

Kim LH, Kayhanian M, Zoh KD, et al. 2005. Modeling of highway stormwater runoff. Science of the Total Environment 348: 1–18.

Kimberly M. Camponelli, Steven M. Lev, Joel W. Snodgrass, et al. 2010. Chemical fractionation of Cu and Zn in stormwater, roadway dust and stormwater pond sediments. Environmental Pollution 158: 2143–2149.

Lee-Hyung Kim, Seok-Oh Ko, Sangman Jeong, et al. 2007. Characteristics of washed-off pollutants and dynamic EMCs in parking lots and bridges during a storm. Science of the total Environment 376: 178–184.

Legret M. C. Pagotto. 1999. Evaluation of pollutant loadings in the runoff waters from a major rural highway. The Science of the Total Environment 235: 143–150.

Li He, Shi Jun-qing, Shen Gang, et al. 2003. Characteristics of rainfall runoff discharge rule caused by heavy metals on express highway [J]. Journal of Southeast University 39(2): 345–349.

Li He, Zhang Xue, Gao Hai-yan, et al. 2008. Characterization of contaminated runoff on free way surface. China Environmental Science 28(11): 1037–1041.

Li LQ, Yin CQ, Kong LL, et al. 2007. Effect of antecedent dry weather period on urban storm runoff pollution load. Environment Science 28: 2287–2293.

Lyn B. Irish Jr, Michael E. Barrett, Joseph F. Malina Jr, Randall J. Charbeneau. 1998. Use of regression models for analyzing highway storm-water loads [J]. Journal of Environmental Engineering 124: 987–993.

Mcleod SM, Kell JA, Putz GJ. 2006. Urban runoff quality characterization and load estimation in Saskatoon, Canada. Environmental Engineering 132: 1470–1481.

Mesfin Berhanu Desta, Michael Bruen, Neil Higginsc, et al. 2007. Highway runoff quality in Ireland. Journal of Environmental Monitoring 9: 366–371.

Morrison G.M. S. Rauch. 2007. Establishing a procedure to predict highway runoff quality in Portugal. Highway and Urban Environment 12: 371–383.

Pagotto C. M. Legret, P. Le Cloirec. 2000. Comparison of the hydraulic behavior and the quality of highway runoff water according to the type of pavement. Water Research 34: 4446–4454.

Pyeong-Koo Leea, Jean-Claude Touray, Patrick Baillif, et al. 1997. Heavy metal contamination of settling particles in a retention pond along the A-71 motorway in Sologne, France. The Science of the Total Environment 201: 1–15.

Ronald S. Hamilton, Roy M. Harrison. 1991. Highway pollution. Elsevier Science Publishers B.V.

Shen Gui-fen, Zhang Jing-dong, Yan Xiao-xuan, et al. 2005. Characteristics of runoff water quality in Wuhan and its main influencing factors. Water resources protection 21(2): 57–58, 71.

Shinya M, Tsurubo K, Konishi T, et al. 2003. Evaluation of factors influencing diffusion of pollutant loads in urban highway runoff. Water Science Technology 47: 227–232.

Shutes R.B.E. D.M. Revitt, I.M. Lagerberg, et al. 1999. The design of vegetative constructed wetlands for the treatment of highway runoff. The Science of the Total Environment 235: 189–197.

Sim-Lin Lau, Jiun-Shiu Ma, Masoud Kayhanian, Michael K. Stenstrom. 2002. First Flush of Organics in Highway Runoff. Urban Drainage 219: 1–12.

Tamar Opher, Eran Friedler. 2009. A preliminary coupled MT–GA model for the prediction of highway runoff quality. Science of the Total Environment 407: 4490–4496.

Terzakis S. M.S, Fountoulakis, I. Georgaki, et al. 2008. Constructed wetlands treating highway runoff in the central Mediterranean region. Chemosphere 72: 141–149.

Terzakis S. M.S. Fountoulakis, I. Georgaki, et al. 2008. Constructed wetlands treating highway runoff in the central Mediterranean region. Chemosphere 72: 141–149.

Westerlund, C., Viklander, M., Backstrom, M. 2003. Seasonal variations in road runoff quality in Lulea. Sweden Water Science. Technology 48: 93–101.

Xie Jian-guang, Zhang xue, Li He. 2009. Partitioning fraction of organic pollutants in expressway stormwater runoff. China Environmental Science 29(10): 1047–1051.

Yiping Zhu, Pu Liu, Haifeng Liu. 2009. Pollutant washoff characterization of expressway runoff in Shanghai. Bull Environ Contaim Toxicol 83: 398–402.

Zanders J.M. 2005. Road sediment: characterization and implications for the performance of vegetated strips for treating road run-off. Science of the Total Environment 339: 41–47.

Zong Yue-fuang, Zhou Shang-yi, Peng Ping, et al. 2003. Perspective of road ecology development. Acta Ecologica Sinica 23(11): 2396–2405.

Frontiers of Energy and Environmental Engineering – Sung, Kao & Chen (eds)
© 2013 Taylor & Francis Group, London, ISBN 978-0-415-66159-1

A comprehensive review of photovoltaic envelopes in China

L. Chen, Y.F. Zhang, L.L. Jiang & Z. Yi
Beijing Jiaotong University, Beijing, P.R. China

ABSTRACT: Photovoltaic envelopes are both building energy sources and building envelopes, and they are consist of BIPV modules which replace conventional building materials or components of building envelopes such as roofs, skylights, or facades. This paper aims to review research developments that have made by Chinese scholars during the recent decade in this field of BIPV systems, especially of photovoltaic roofs, photovoltaic walls and photovoltaic shades.

Keywords: building-integrated photovoltaic system; photovoltaic envelope; photovoltaic roof; photovoltaic wall; photovoltaic facade; photovoltaic exterior shade

1 INTRODUCTION

At present, BIPV systems which integrate photovoltaic modules with building envelopes are vigorously promoted as key technologies of renewable energy utilization in some developed countries and China. The process of Building Integrated Photovoltaic (BIPV) development is gradually accelerated in China, and more and more photovoltaic systems are applied to all kinds of constructions [1]. At the same time, lots of academic researches have been done in China.

This paper aims to review studies carried out by Chinese scholars during the last 10 years in photovoltaic envelopes such as roofs, curtain walls and shades. A quantitative summary of the literature review is shown in Figure 1. This literature review was separated into general summary, theoretical and experimental studies, design and technology studies, application and case studies (Fig. 2).

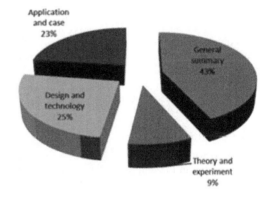

Figure 2. Specific subjects of academic papers distribution during the current decade.

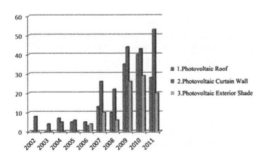

Figure 1. Academic papers distribution per year.

2 PHOTOVOLTAIC ENVELOPES

2.1 Building Integrated Photovoltaic (BIPV) system and photovoltaic envelope

BIPV system is a photovoltaic system used to replace conventional building materials or components in parts of the building envelopes such as roofs, skylights, curtain walls or exterior shades. BIPV systems are both building energy sources and building envelopes. Photovoltaic envelopes will be produced when photovoltaic modules are architecturally incorporating into buildings.

With characteristics of energy saving and carbon dioxide emission reduction, photovoltaic buildings meet the requirements of future green buildings in the background of global energy crisis. Developing

Table 1. Main forms of photovoltaic envelopes.

Forms	Schematic diagram
Photovoltaic roof	
Flat roof	
Pitched roof	
Photovoltaic curtain wall	
Photovoltaic exterior shade	

photovoltaic building envelopes will be one of the trends of future buildings [2].

2.2 *Main forms of photovoltaic envelopes*

Photovoltaic roofs, photovoltaic curtain walls, and photovoltaic shade are main forms of photovoltaic envelopes (Table 1). Photovoltaic roofs can be classified into flat photovoltaic roofs and pitched photovoltaic roofs, or into opaque photovoltaic roofs and photovoltaic skylights.

3 PHOTOVOLTAIC ROOF

3.1 *General summary*

Compared with other parts of buildings, the sunshine intensity in the unit area of roof is stronger, and the sunshine time is longer, so the generating efficiency of photovoltaic roof is higher, and photovoltaic roof has been the main choice of building-integrated photovoltaic system in the world [3].

There are two kinds of integration way for photovoltaic roof. One is substituted system which means photovoltaic modules are designed to replace the outside protection layer of roofs, and are combined with other layers of roofs, such as solar tile roofs. The other one is integrated system which means photovoltaic modules with all architecture functions completely take the place of roof components directly, such as solar skylights. The latter one has more strict requirements of photovoltaic modules [4].

3.2 *Theory and experiment*

As some defects and shortcomings exist in current theory models of building-integrated photovoltaic system, Jian-Bo Ren [5] put forward a new theory model for three typical photovoltaic roofs including the photovoltaic roof with ventilation port, the photovoltaic roof with closed ventilation port, and

the photovoltaic roof without ventilation port, and ordinary roof. He also established contrast experimental units and compared theoretical calculation results with experiment results. And then a conclusion came as that it is more suitable for such areas that the weather is severe cold, is cold, is hot in summer and cold in winter, is hot in summer and warm in winter to choose the photovoltaic roof with ventilation port in summer, and choose the photovoltaic roof with closed ventilation port in winter. It is good for temperate regions to choose the photovoltaic roof with ventilation port, but there is no difference for the photovoltaic roof with closed ventilation port and the photovoltaic roof without ventilation port in winter.

Li-Min Liu et al. [6] analyzed typical actual operation data of a 50 kWp grid-connected photovoltaic power plant and drew a conclusion that due to differences in installation angle, installation azimuth and component material, the output power of different photovoltaic modules is influenced differently.

Solar roof modules are ordinarily fixed into roofs in current China. Xiao-Jing Hu [7] does some research on the application of sun tracking technologies on solar roof by a mean of computer simulation, and proved the advantage of this intelligent solar roof quantitatively.

3.3 *Design and technology*

A variety of factors needed to be considered when photovoltaic roof is designed. Chun-Jiang Zhao et al. [8] pointed out that the main considerations were the connection between the solar cell modules, the combination between photovoltaic arrays and roof or wall around, waterproof construction and the standardization and modular of photovoltaic modules. In addition, safety, appearance and construction convenience also are needed to be considered.

Hai-liang Liu et al. [9] studied the photovoltaic roof tile components, and developed a photovoltaic tile product with clay base, a photovoltaic tile product with cement base and a photovoltaic tile product of aluminum alloy.

Feng-Yuan Li [10] pointed out that although in other countries photovoltaic grid-connected systems had been applied in many real cases, a lot of detailed studies are needed for their implementation in large cities in China.

3.4 *Application and case*

Due to shortage of natural resources and serious environmental problems, urgent needs to construct photovoltaic roof in large scale in China emerge.

Feng-Yuan Li's study [10] showed that there were less BIPV applications in large housing project in China, but there had been many applications in large-scale public buildings, for example, the photovoltaic skylight of Nanjing South Railway Station was the world's largest photovoltaic roof project which used CIS film solar cells. Photovoltaic grid-connected systems serve as roofs and curtain walls of most major stadiums of the 2008 Olympic Games in Beijing and exhibition buildings of the 2010 Shanghai World Expo [16].

4 PHOTOVOLTAIC CURTAIN WALL

4.1 General summary

Compared with photovoltaic roofs, photovoltaic walls can provide photovoltaic modules with more installation areas, especially for high-rise buildings. Photovoltaic curtain walls are superior to common glass curtain walls by adding a new function of electricity power generation to function of building envelopes, and serve multiple functions including power generation, waterproof, sound insulation, heat insulation, safeguard and decoration [12].

Owing to the policy support of Chinese government and the maturity of the photovoltaic facade technology, according to statistics of De-Lin Yang [11], photovoltaic curtain walls have become hot fields of development and research in recent ten years in China. Even so, Chinese prosperity of photoelectric curtain walls is delayed by a lot of obstacles such as high cost, weak innovation, lack of technical standards and certification systems, instability of power generation, and so on [3]. In addition, Hui-Qin Ma [12] thought that difficult to determine electricity prices generated by grid-connected systems of photoelectric curtain wall and hard to control and adjust the systems were also problems must to be faced in China.

4.2 Theory and experiment

Theory and experimental studies regarding solar photovoltaic curtain wall are insufficient in China presently. Some researchers stated some heat transfer model of photovoltaic wall. Hong-Xing Yang et al. [13] set a heat transfer model for estimating heat gains of photovoltaic wall structures. Ye Yi [14] set a planar heat transfer model. Lei Zhu [15] set a mathematical model and use CFD fluid software for numerical simulation. Wei He et al. [16] establishes theoretical models of two primary structures of photovoltaic/thermal integrated buildings (BIPV/T). All above studies concluded that photovoltaic wall can highly reduce the heat

gain of walls in summer, so as to reduce cooling load of air conditioner.

Hai Chen et al. [17] analyzed and calculated the influence of different work environment temperatures on solar cells conversion efficiency, flowing rules of hot air on the effect of the mechanical ventilation and temperature difference in double-skin photovoltaic curtain wall, and cooling performance of hot air flow inside on both buildings and photovoltaic modules. Finally, a physical model of energy-saving photovoltaic double-skin curtain wall was tested and calculated results were compared and analyzed.

4.3 Design and technology

As a new skin form of buildings, the design of photovoltaic curtain wall is different from traditional curtain wall. Shi-Tao Xie [18], Wen-Zhi Long [19], Chuan-Dong Huang et al. [20], put forward that the design of photovoltaic curtain walls need to meet requirements of safety, energy saving, convenience for construction and industrialization. According to the difference in construction, Hui-Qin Ma [12] grouped photovoltaic curtain walls into five types: exposed framing type, hidden framing type, point-supporting type, adornment type and double-skin type. Miao Li et al. [21] proposed that point-supporting type and hidden framing type ought to be chosen considering structure requirements. Chun-Tao He [22] studied the construction technique of curtain walls integrated with single crystalline silicon solar cell systems.

In addition, some scholars studied the design of such photovoltaic wall components as solar cell modules, frames, joints and so on. Ren-Qiang Bao et al. [23] made a technology innovation in developing a new type of double-skin BIPV modules. Xiao-Yang Huang et al. [24] developed a new frame structure to solve stress problems of EVA layers. Xiao-Min Huang et al. [25] designed a Ventilation Photovoltaic Curtain Wall (VPVCW).

4.4 Application and case

Overall, the application of photovoltaic curtain walls in china is still in its infancy, and a number of demonstration projects of photovoltaic curtain walls has been built in Beijing, Shanghai, Jiangsu, and Guangdong, such as the enterprise joint hall of the 2010 Shanghai World Expo, Baoding Jinjiang international hotel, Changsha Zhongjian building, Guangzhou TV tower [12].

New technologies of double-skin facade systems and photovoltaic curtain walls have been utilized in the Tsinghua University's demonstration building with super low energy consumption [18].

These successful cases accumulate the technical data and practical experiences for a large-scale promotion of photovoltaic curtain walls, so that they are very important to the development of photo-electric curtain walls in China [26].

5 PHOTOVOLTAIC EXTERIOR SHADE

Yong-Jun Huang et al. [27] put forward that photovoltaic exterior shades are outermost skin of buildings and have less shadow on them, so their power efficiency is high; moreover, they have less effect on building functions, therefore they would be leading forms of BIPV systems in the future.

5.1 General summary

Photovoltaic shades consisted primarily of photovoltaic modules achieve double functions as energy conservation and power generation of buildings. As they are outside buildings, there are almost no requirements of their thermal performance, but an appropriate integration way needs to be chosen based on consideration of both characteristics of buildings and photovoltaic systems. Tao Shi [3] grouped building photovoltaic shades to three forms as level ones, vertical ones and baffle–type ones, and he concluded that power efficiency will change with their positions and arrangements.

5.2 Theory and experiment

Few scholars conducted theoretical and experimental research of photovoltaic shade. Han-Cheng Yu et al. [28] put forward a mathematical method of optimization design of photovoltaic shades, and set up a model of heat transfer and power generation for numerical simulation.

5.3 Design and technology

Zhou Wu et al. [29] summarized that polycrystalline silicon cells and amorphous silicon cells are suitable for photovoltaic shade components, and they presented two available encapsulation forms of photovoltaic shade modules as following, one is assembling aluminum alloy laminas with photovoltaic shade modules encapsulated under a single glass plate, and the other one is encapsulating photovoltaic modules among double glass plates.

Ting Ren [30] designed four specific architectural forms of open photovoltaic shades in regions of hot and humid based on meteorological data of Guangzhou. In addition, Han-Cheng Yu et al. [28] present a design and a reasonable technology of photovoltaic shades as that 12 × 12 arrays of solar cells are pasted on aluminum alloy laminas, and then are covered by glass plates.

5.4 Application and case

Zhou Wu et al. [29] pointed out applications of photovoltaic shades have been done in some public construction projects, such as Shanghai Solar Energy Technology and Engineering Research Center which is a BIPV construction integrated most kinds of photovoltaic modules. Its photovoltaic shades system with installed capacity of about 8.3 kW is highly economical and effective which can be automatically adjusted according to different actual needs.

Yu-Ping Xu et al. [31] described that many Chinese researchers were studying the combination of photovoltaic systems and architectural shades, but the promotion of photovoltaic shades is still difficult because of the immature market in China currently.

6 SUMMARY

A comprehensive literature survey of studies carried out in china during the last 10 years in photovoltaic envelopes has been performed. The studies review was grouped into the following three systems: photovoltaic roof, photovoltaic curtain wall, and photovoltaic exterior shade.

As a new form of building skin, the development of photovoltaic envelopes is just at the beginning in China because of a series of policy problems and technical difficulties. This study showed that photovoltaic envelopes are demanded and promising in China, but more studies especially innovation studies on theory and technology are needed to achieve their optimum performance and large scale application.

ACKNOWLEDGEMENTS

The work of this paper is supported by "the Fundamental Research Funds for the Central Universities" (2011JBM317).

REFERENCES

[1] Ying Sun, Zhong-Chun Mi, Lin Luo: Analysis on Key Issues about Extension of BIPV. Building Energy Efficiency. 12(2011):41–43.
[2] Chao-Hong Wang, Hui Gao, Jian-Jun Wang: Photovoltaic integration design research of modern architecture skin. New Architecture. 5(2009):73–78.
[3] Tao Shi, Jia-Yan Shen, Jin-Yang Jiang: Current application about building-integrated photovoltaic. New Building Materials. 11(2011):38–41.

[4] Wei Wei: BIPV Study from Architect's a Dissertation. Nanjing University. (2011).

[5] Jian-Bo Ren: Experimental and Theoretical Study on Photovoltaic Roofs. Tianjin University. (2006).

[6] Li-Min Liu, Zhi-Feng Cao, Hong-Hua Xu: Design and Analysis of 50 kWp Grid-Connected PV Station. Acta Energiae Solaris Sinica. 27(2) (2006):146–151.

[7] Xiao-Jing Hu: Intelligent Solar Roof Modal Design and Realization. Computer Knowledge and Technology (Academic Exchange). 03(2011):627–629.

[8] Chun-Jiang Zhao, Rong-Qiang Cui: Technical Research and Development Of Solar Energy Building Materials: Thermal Investigation of Integrated PV Roof. Acta Energiae Solaris Sinica. 03(2003): 352–356.

[9] Hai-Liang Liu, Yu-Zhi Xue, Peng-Yuan Qi, Shun-Bo Zhang: The design and making of photovoltaic tile. Solar Energy. 4(2008):21–23.

[10] Feng-Yuan Li: Discussion on the Application of Photovoltage Generator to Public Building. Building Electricity. 23(4) (2004):3–6.

[11] De-Lin Yang: Photovoltaic Curtain Wall Development Present Situation and the Prospect. Friend of Science Amateurs. 18(2010):19–20.

[12] Hui-Qin Ma: The Development and prospect of photoelectric curtain wall in China. Door & Windows. 06(2011):11–15.

[13] Hong-Xing Yang, Jie Ji: Study on the Heat Gain of a PV-Wall. Acta Energiae Solaris Sinica. 03(1999): 270–273.

[14] Hua Yi: Theoretical and experimental research of new PV-Trombe Wall system. China Science and Technology University. (2007)

[15] Lei Zhu: The Theory analysis and Experimental study of Photovoltaic curtain wall system hot environment. (2007). 2007 China Renewable Energy Industry Forum.

[16] Wei He, Jie Ji: Theoretical study of photovoltaic/thermal integrated buildings on energy efficiency. Heating Ventilating & Air Conditioning. 06(2003):8–11.

[17] Hai Chen, Huo-Nan Mao, QiuWang, Qing-Hai-Jiang, DongPan, Zhi-Jun Long, Jin-Ji Guo, Feng Chen: Experiment Research and Calculation of the Hot Air Flow Inside the Energy-saving Photovoltaic Double Curtain Wall. Acta Scientiarum Naturalium Universitatis Sunyatseni. 05(2011):39–43.

[18] Shi-Tao Xie: The technology and Application of building-integrated photovoltaic. Door & Windows. 09(2007):42–45.

[19] Wen-Zhi Long: Building Integrated Photovoltaics. Architecture Technology. 10(2009):935–937.

[20] Chuan-Dong Huang: The Technology and Standard Analysis of BIPV Component. Construction Science and Technology. 04(2011):86–88.

[21] Miao Li, Yi-Qun Li: Analyses of Solar Photoelectric Glass Curtain Wall. Door & Windows. 12(2009):11–15.

[22] Chun-Tao He, Xiu-Qiong Deng, Zhen Wang: Four technology innovations of Green building construction. Architecture and Construction. 16(2010):74–77.

[23] Ren-Qiang Bao, Li-Qiong You, Yue-Chao Wu: The Manufacture of New BIPV Solar Cell Module. Zhejiang Construction. 10(2010):59–60+81.

[24] Xiang-Yang Huang, QingHe, FangRen, Shi-Tao Xie: Introduction on a Newly Photovoltaic Façade. Construction Conserves Energy. 02(2008):49–50.

[25] Xiao-Min Huang, Ying-Ying Zeng: Research on ventilation photovoltaic curtain wall. New Building Materials. 06(2011):92–94.

[26] Shi-Tao Xie, Xiao-Wu Zeng: Technology research of Low energy consumption demonstrational building construction curtain wall. Housing Industry. 09(2009):75–77.

[27] Yong-Jun Huang, Wei Tao: Photovoltaic building shade: the two fitting for power generation and shading. City & House. 4(2010):96–97.

[28] Han-cheng Yu, Cheng-long Luo. Study on optimizations design and characteristic of building Integrated with the photovoltaic Industrial Construction. S1(2009):31–34.

[29] Zhou Wu, Guo-QiangHao, Xiao-Tong Yu, Yong Huang: High Power PVMPPT Controller Based on Dual Boost. Modern Building Electricity. 4(2010): 50–53.

[30] Ting Ren: Guangzhou Based Study on BIPV Design with Unclosed Skin for Shading and Cooling in Hot-Humid Area. South China University of Technology. (2011).

[31] Yu-Ping Xu, Yao-Long Zhang. Energy-saving Architecture Design Viewed from shading. Modern Science. 20(2009):126.

Frontiers of Energy and Environmental Engineering – Sung, Kao & Chen (eds)
© *2013 Taylor & Francis Group, London, ISBN 978-0-415-66159-1*

Economical thought in ecological landscape design[1]

Q.L. Sun
College of Design Art, Henan University of Technology, Zhengzhou Henan, China

ABSTRACT: This article from the ecological landscape generation and concept, discusses the conservation-oriented landscape design and ecological landscape design consistency, the concept of savings contained in the ecological landscape design, and proves economical thought in the ecological landscape design works.

Keywords: ecological landscape; landscape architecture; affordable landscape design; saving

1 INTRODUCTION

In the book of "Introduction to sciences of human settlements", Wu Liang-yong academician takes the architecture, city planning, landscape architecture as three leading professional in science of human settlements, peripheral multidisciplinary group fuse along with the development of the times. In three subjects, architecture often leads the development of the times; landscape architecture often lags behind the times. But the relationship between architecture and environment are inseparable, architecture exists in the site, site of environmental design for landscape architecture content. In the second half of the twentieth Century, the influence of ecology, green architecture theory expands increasingly, traditional landscape architecture thought shifted from practical, functional gradually to a new design concept, which emphasizes the people-oriented, pays attention to create organic environmental.

2 GENERATION AND CONCEPT OF ECOLOGICAL LANDSCAPE

After the Second World War, all countries in the world focused on the post-war reconstruction, the industry developed rapidly, has promoted economic development, on the other hand, the human environment suffers serious pollution and destruction. A person with breadth of vision are keenly aware of the serious consequences of environmental damage, the United States Marine Biologist Kason published the "Silent Spring"

(1962), Rome club D. mythos put forward "the limit of growth" theory in 1970, the alarm bell of environmental crisis sounded for human.

In order to cope with the environmental crisis, some landscape architects design based on ecological principles, they show people the ecological phenomenon, ecological effect in the environment surrounding through their design, in order to arouse emotional connection between Man and nature. Ecological landscape is produced in such circumstances. In 1969, Ian McHarg (1920–2001) published a book "design with nature", which marked the landscape planning and design professional bravely had assumed design task of humans overall ecological environment in post-industrial.

Ecological landscape takes sustainable development as the guiding ideology, taking ecology as the principle of LA. It elevated LA a height to protect city ecological balance, improve the human ecological system and living environment. It is a maximally design by natural force minimum, a regeneration design based on self organic renewal ability of natural system, which create sustainable landscape.

3 ECONOMICAL THOUGHT IN ECOLOGICAL LANDSCAPE DESIGN

3.1 *Economical landscape design*

Face of increasingly enhanced resources and environmental constraints, Chinese government put forward the construction of low investment, high yield, low consumption, low emissions, energy recycling, sustainable national economic system and construct resource saving, environment friendly society, in order to implement the basic national policy of saving resources and protecting environment. Economical landscape design is

[1]Project of Education Department of Henan Province 2011B220002.

"design with minimal use of land and water, minimal funding, selection on the minimum of interference patterns to the ecological environment". Not only such, conservation-oriented landscape requires resources, energy minimization, on the other hand requires comprehensive benefits maximization, the key is to reduce energy consumption, improve resource utilization [1].

3.2 *Consistency of ecological landscape and economical thought*

Landscape ecology theory and method were introduced into theories of modern landscape planning, landscape planning gradually tend to the landscape ecological planning model. It is a maximally design by natural force minimum, reduce as far as possible the use of energy, land, water and biomass resources, improve efficiency in the use, namely reduction, it is also the most narrow definitions of saving [2]. By improving the utilization of natural resources, energy and resource recycling and regeneration, in fact emphasis of circular economy theory in the "reuse" and "recycle" principle, they create a sustainable landscape.

From the perspective of ecological environment least disturbed, conservation-oriented landscape design and ecological landscape is consistent. Some scholar thinks, conservation-oriented landscape means "ecological environment with minimal interference", emphasizes damage to the environment reduced to the minimum. This means that the design and ecological processes are in harmony, respect for diversity, maintaining wildlife habitat quality, the system energy and materials used judiciously. According to the natural process of effective adaptation and combination, the full measure of design approach to environmental impact, in order to obtain an overall consideration, so as to establish a rational structure, efficient, coordinating relations of ecological system, create a harmonious environment for the human being and nature.

4 APPROACHES OF ECOLOGICAL LANDSCAPE DESIGN

4.1 *Saving through preservation and reuse of historic elements*

There are a large number of domestic and foreign outstanding landscape design works aimed for transformation of industrial wastelands in recent years. The ecological design approach embodies the concept of savings in landscape architecture, such as Landschaftspark designed in 1991 by Latz + Partner (Peter Latz). It is a public park located in Duisburg Nord, Germany. The park closely associates itself with the past use of the site: a coal and steel production plant (abandoned in 1985, leaving the area significantly polluted) and the agricultural land it had been prior to the mid 19th century [3]. Unlike his competitors, Latz recognized the value of the site's current condition [4]. He found new uses for many of the old structures, for example, tall concrete structures used as a climbing training facilities, the original gas tank transformed into a water rescue training centre, metal framework left by factory used as a climbing plant support, abandoned elevated railway as park walk, waste iron paved "metal square". In his design, Peter Latz attempted to preserve as much of the existing site as possible [5]. A series of parks were designed by using the same method, such as Seattle Gas Works Park designed by Richard Haag, Brickworks Park in Heilbronn by Bauer, Seoul Park etc.

In China, Guangdong Zhongshan Shipyard Park was the first design of this kind. The park closely associates itself with the past use of the site, cement frame dock was retained in the original place, large gantry crane and transformers, large machines were combined in site design [2]. The retained and regenerative design method embodies the economical design idea. After this, a lot of landscape works are on retaining and reuse of the history element in domestic [2,6], they are all inexpensive and affordable.

4.2 *Saving in water treatment using*

Water as a natural resource, is the basic condition of human survival, life, production and maintenance of the good ecological environment. Chinese ancient gardens emphasized "anhydrous garden does not live", modern landscape emphasizes water reusing, landscape of water reusing reflects progress of human thinking. Chengdu living water park is the first city ecological environment protection park "water conservation" as the theme in the world, showing the international advanced "artificial wetland sewage treatment system". Living water park makes use of artificial wetland sewage treatment system, water flowing sequentially through water purification system including the water sculpture, anaerobic deposition pool, plant pond, plant bed, pond and oxidation ditch etc. which demonstrates changing process of water quality. Artificial wetland sewage treatment system has advantages over traditional two stage biochemical treatment, realizes the cyclic utilization of water, and in the process creates conditions for large animal and plant growth in favor of a benign ecological environment construction.

Hangzhou Xixi National Wetland Park is another classic case, explores the human minimal intervention of wetland protection mode. Through network configuration, or build firedamp pool and a sewage treatment device, the XiXi Wetland water quality changed, while ecological purification system established [7,8].

4.3 *Saving by making good use of nature*

Every designer should protect the natural environment not to be disturbed from human by application of ecology principle. Such as Brickworks Park of Heilbronn, landscape architect noted, when the brick and tile was idle for 7 years, base of the ecological environment had changed, plants, insects, birds came back here, the original side of yellow clay cliff into wildlife "paradise". In order to protect it, designers used the abandoned stone masonry as retaining wall, which well protect the wild environment, but also created a suitable environment for the growth of plants, promoting the natural vegetation regeneration. This is less design, which opens natural self-organization or self design process, in accordance with the saving idea.

The planting design of Duisburg Park is characteristic. The original vegetation and the growth of weeds has been preserved in factory, Peter Latz allowed the polluted soils to remain in place and be remediated through phytoremediation, and sequestered soils with high toxicity in the existing bunkers [3], that provided conditions for the field ecosystem natural regeneration.

4.4 *Saving by planting design*

Planting design is an important part of landscape design; its content includes the choice of plant species and plant configuration. The most basic principle of selection of plant species is the preferred native plants. A clear request for local species application is indicated in National Ecological Garden City Construction of interim standards. There are many advantages on application of native plants, rich in resources, low cost, grow well, easy living, drought resistance, disease and pest resistance tube etc. They can maximize the ecological benefits, highlight the local characteristics, and greatly reflect economical thought. Plant configuration required comprehensive application of trees, shrubs and herbaceous plants, Fujimoto and other material, through artistic methods, give full play to the physical plant, lines, texture, color and other natural beauty to create natural plant community landscape. Green pattern with multiple layer type community simulates the natural plant community structure, result of long-term natural competition. With a certain amount of healthy and stable native plant species can resist the invasion of alien species, protect ecological security. This community structure is relatively stable, can obtain the maximum ecological benefit in the limited green space area.

Two aspects are mainly considered, first, retention of original plant, second, application of native plants. In Hangzhou Xixi National Wetland Park, on the basis of respecting for the original topography, landforms, vegetation, a large number of native plant species were applied for vegetation restoration, original persimmon forest, bamboo, Merlin in wetland was preserved in all [6]. In order to retain the old banyan tree, landscape architects deliberately built a river, forming a banyan tree island, meet the flood control needs in Zhongshan Shipyard Park. Native plants are widely applied in many works. The designers introduced a great deal of wild native plants into Zhongshan Shipyard Park, with large areas of water, forming the aquatic—Marsh—wet plant communities, such as *Nelumbo nucifera, Zizania latifolia, Acorus calamus, Cyperus alternifolius, Sagittaria sagittifolia, Phragmites australis, Pennisetum purpureum, Imperata cylindrical* and other thatch etc. The park has become a native aquatic plants demonstration base.

5 CONCLUSIONS

Ecological landscape is a product of the development of times which centers on nature; agrees with economical design concept based on the minimum input, the least interference on the ecological environment without prior without previous consultation. On one hand, this paper focuses on the economical thought included in ecological landscape design; on the other hand, ecological landscape design pays attention to solve environmental problems while taking other aspects into account. Landscape architecture itself is comprehensive, need to solve economic, social, aesthetic and other problems, a large number of successful ecological landscape design work has to be confirmed.

REFERENCES

[1] ZHAO Yan et al. Application of Resource-efficient Concept to Urban Green Space System Planning. Journal of Anhui Agri Sci, 2009, 37(34):17182–17183, 17193.
[2] YU Kong-jian. Principles and Practices of Affordable Urban Green Space. Landscape Architecture. 2007, 1:55–64.
[3] Information on http://en.wikipedia.org/wiki/Landschaftspark_Duisburg-Nord.

[4] Diedrich, Lisa. "No Politics, No Park: The Duisburg-Nord Model." Topos: European Landscape Magazine, no. 26 (1999): 69–78.

[5] Weilacher, Udo (2008): Syntax of Landscape. The Landscape Architecture by Peter Latz and Partners. Basel Berlin Boston: Birkhauser Publisher. ISBN 978-3-7643-7615-4.

[6] WANG Ying et al. The Ecological Ideas of Economic Gardens. Journal of Anhui Agri. Sci. 2009, 37(1):119–120, 143.

[7] GAO Yi-Liang. The Practices and Improvements of the Xixi National Wetland Park Model, Hangzhou. Wetland Science&Management. 2006, 3:55–59.

[8] CHEN Jiu-he. Study on the Green Design of Ecotourism Landscape of Xixi National Wetland Park in Hangzhou. Areal Research and Development. 2006, 10:72–75.

Frontiers of Energy and Environmental Engineering – Sung, Kao & Chen (eds)
© 2013 Taylor & Francis Group, London, ISBN 978-0-415-66159-1

Analysis of optimum U-tube diameter in ground source heat pump

H.L. Zhang & Y.J. Yu

Energy and Mechanical Engineering Academy, Nanjing Normal University, Nanjing, China

ABSTRACT: In order to reduce the pump energy consumption in GSHP, pipe diameter is increased to solve the problem. When flux of pump is constant, by choosing pipes of different diameters for numerical calculation and simulation, the pump energy consumption and the heat transfer of ground heat exchanger are both analyzed. It could be concluded that an appropriate increase of pipe diameter could reduce pump energy consumption and enhance heat transfer. However, when the pipe diameter is too large, the hot short circuit of ground heat exchanger would be serious, thus, increase of pipe diameter should be based on the hot short circuit between branch pipes. It is given in the paper that the suggest pipe diameter would exist during 30 mm to 40 mm. The analysis would reduce energy consumption of pump, improve a guarantee for the selection of pipe diameter and the design of GSHP with great guide and economic benefits.

Keywords: ground source heat pump; pipe diameter; energy consumption; hot short circuit

1 INSTRUCTIONS

In recent years, with the energy shortage and the climate problem brought about by consumption of fossil energy to power homes and businesses, more and more people are concerned about ground source heat pump system. For vertical U-tube ground heat exchanger, the main factors impacting the heat exchanger capacity are fluid in the pipe, the pipe material, backfilling material, soil thermal properties, groundwater seepage, etc. Researches related to these factors are detailed. In addition, for different velocities and different diameters, the heat transfer abilities between fluid in U-tube and the soil are different. It is exactly important to analysis the various factors impacting heat transfer ability of U-tube in optimization design of ground heat exchanger, further promotion, and application of ground source heat pump system.

To ensure the exhaust timely and strengthen heat transfer of the system, In practice, the determination of pipe diameter should meet two requirements: (1) the diameter should be large enough to maintain the minimum transmission power; (2) the diameter should be small enough to make the fluid in pipe in the region of turbulent flow in order to ensure good convective heat transfer between the fluid and the inner surface of pipe. Both of them are based on the same flux. Obviously, the above two requirements are contradictory. Pipe diameter could be neither too large nor too small. However, the final selection is not only affected by these two effects but also the pump energy consumption and heat transfer performance of U-tube.

2 PRESENT ENGINEERING SITUATION

According to the existing engineering design experience, small diameter pipes are generally used in parallel loop, and trunk lines choose big diameter pipes. Diameter of ground heat exchanger is commonly used with 20 mm, 25 mm, 32 mm, 40 mm and 50 mm. Velocity of fluid in pipe should be controlled in 0.6 to 1.22 m/s, and if diameter of pipe is beyond it, velocity should be controlled in less than 2.44 m/s, or equivalent length of pressure loss in every pipe should be controlled at 0.4 kpa/m below. Compared to the antifreeze joined with $CaCl_2$ or glycol, when they are in the same diameter and velocity, water would have a larger Reynolds number. When using antifreeze, its velocity and flux should be larger than water's to ensure fluid in pipe in turbulent region. Reynolds number of fluid in pipe usually keeps form 5000 to 40000 in summer condition, while 8000 to 25000 in winter condition. Therefore, fluid in U-tube is in the turbulent region all the time, and there would be large convection heat transfer.

3 THEORETICAL ANALYSIS

The impacts of the pipe diameter are in two directions: one is for pump energy consumption, and the other is for heat transfer of ground heat exchanger caused by changed surface and velocity.

3.1 Influence on energy consumption of pump for changed diameter

First of all, we make hydraulic calculation and pressure loss calculations of ground heat exchanger.

To calculate velocity of fluid in pipe:

$$u_m = \frac{G}{3600 \times A} \tag{1}$$

where u_m = flow velocity; G = flux of fluid; A = cross area of pipe.

To calculate Re:

$$Re = \frac{\rho u_m d}{\mu} \tag{2}$$

where Re = Reynolds number; ρ = density of fluid; d = pipe diameter; μ = dynamic viscosity of fluid.

Due to the existence of viscous shear stress and eddy current, when fluid flows in pipe, loss of mechanical energy is inevitable, including on-way resistance caused by the fluid flowing along the pipe and local resistance caused by the fluid flowing at the bottom of the U-tube for changed flow direction. Factors affecting the resistance loss are density of fluid, dynamic viscosity, flow velocity, pipe diameter, pipe length, roughness of pipe's inner wall, etc. Because of polyethylene pipe, smooth inner wall, and that viscous sublayer is greater than the rough bump height in pipe, impact of roughness could be neglected.

To calculate on-way resistance P_y:

$$P_d = \frac{\lambda}{d} \cdot \frac{\rho u_m^2}{2} \tag{3}$$

where P_d = on-way resistance per unit length of pipe; λ = on-way resistance coefficient.

Because the turbulence of fluid in ground heat exchanger is in hydraulically smooth region, on-way resistance coefficient λ is only related to Reynolds number and has nothing to do with roughness of pipe's inner wall.

When Reynolds number is from 4000 to 100000:

$$\lambda = \frac{0.3164}{Re^{0.25}} \tag{4}$$

Then:

$$p_d = \frac{k}{d^{4.75}} \tag{5}$$

where k = a constant.

It is obviously that change of on-way resistance per unit length of pipe is very large because of change of pipe diameter. If pipe diameter is reduced half, the on-way resistance will be increased to 27 times of original value.

$$P_y = P_d \cdot L \tag{6}$$

where P_y = on-way resistance; L = length of pipe.

The calculation of local resistance is a complex problem when fluid flow in turbulent flow region. There are usually two ways to calculate: resistance coefficient method and equivalent length method. The resistance coefficient method is to make the value of energy loss caused by overcoming local resistance to be approximated as multiples of the kinetic energy, while the equivalent length method is to make the local resistance to be the on-way resistance of straight pipes. In this article, the resistance coefficient method is to be used.

For the local resistance:

$$P_d' = \xi \frac{\rho u_m^2}{2} \tag{7}$$

Where P_d' = local resistance per unit length of pipe; ξ = local resistance coefficient.

The local resistance coefficient only depends on components of pipes and geometrical flow channel of the equipments, usually without considering relatively roughness and Reynolds number. It is usually be measured by experiments. When the flux is certain it can be seen the velocity of fluid increased as pipe diameter reduced, and both of them would make the local resistance increased exponentially.

To calculate total resistance P_z:

$$P_z = P_y + P_j \tag{8}$$

where P_z = total resistance.

Five different pipe diameters are taken for calculation and analysis, and they are 20 mm, 25 mm, 32 mm, 40 mm and 50 mm. Ground heat exchanger used single U-tube filled with water has a depth of 20 m. In order to ensure the minimum flow velocity of 0.6 m/s, flux of water G is selected to be 4.239 m³/h based on 50 mm diameter pipe. Calculation results are shown in Table 1.

It can be seen from Table 1, the Reynolds number is reduced with the increase of the diameter. However, water in U-tube is still in turbulent region. It is desirable that the actual project set the upper limit of pipe diameter to 50 mm. As pipe diameter increased, change rate of Reynolds number is gradually reduced.

Table 1. The impact on on-way resistance of U-tube for the diameter changed.

Pipe diameter mm	Reynolds number	On-way resistance pa/m
20	75000	6721.62
25	60000	2328.90
32	46875	720.95
40	37500	249.79
50	30000	86.55

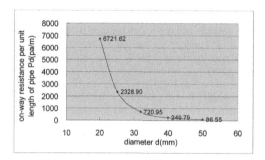

Figure 1. On-way resistance per unit length of pipe changes with the changes of diameter.

Take on-way resistance per unit length of pipe for analysis, and list its change for changed pipe diameter in Figure 1.

Figure 1 shows that with the increase of the pipe diameter, on-way resistance per unit length of pipe is reducing, especially pipe diameter in the region of 20 mm to 40 mm. the trend could be seen in the Figure 1 that when pipe diameter is larger than 50 mm, with the increase of the pipe diameter, on-way resistance per unit length of pipe keeps the same mostly. Therefore, the selection of pipe diameter is persuasive. The suggested pipe diameter is from 30 mm to 40 mm.

Rated pumping head of a water pump mainly depends on the resistance in pipe. From the analysis, it could be concluded that, with the decrease of the pipe diameter, resistance in U-tube is increasing. Finally it would make water pump selection larger, and energy consumption would be greatly increased. In order to reduce energy consumption, tube diameter should be appropriately increased.

3.2 Influence on heat transfer of ground heat exchanger for changed diameter

In the case of a certain flux of water pump, the diameter changes also bring corresponding changes on the heat transfer area of the U-tube and flow velocity of water, etc. All these changes would affect the heat transfer of ground heat exchanger. With the increase of the pipe diameter, the increased heat transfer surface would be enhanced in the whole heat transfer process.

For heat convective, according to the turbulent heat transfer theory, convective heat transfer coefficient is related to fluid's flow velocity, heat transfer coefficient, density, dynamic viscosity coefficient, Prandtl number and pipe diameter, especially fluid's flow velocity and pipe diameter. If heat transfer analysis only considers turbulent heat transfer of fluid in pipe, and ignores impact of fluid's density change for temperature change, increase of the flow velocity can improve the heat transfer coefficient. For U-tube with the same diameter, Heat transfer per unit length increases with flow velocity of fluid increased, especially in the case of low flow rate. And if the velocity increases more, growth of heat transfer per unit length will be slower. However, the paper studies the heat transfer in different diameters under the same flux of water in pipes.

For convective heat transfer coefficient on project:

$$h_c = 0.023 \left(\frac{u_m d}{v} \right)^{0.8} \left(\frac{v}{a} \right)^m \frac{\lambda}{d} = A \frac{u_m^{0.8}}{d^{0.2}} \tag{9}$$

where a = thermal diffusion coefficient; $m = a$ ratio (when the fluid is heated by pipes, $m = 0.4$, otherwise $m = 0.3$).

For water:

$$A = 1384 + 25 t_f - 0.052 t_f^2 \tag{10}$$

where t_f = average temperature of water in pipes.

Also take the five pipe diameters for simulation and analysis, and take the conditions in winter for the study object. The soil average temperature is 291 K, inlet water temperature is 283 K, heat conductivity coefficient of backfill materials is 2 W/m² · K, and the single U-tube is at a depth of 20 m. To ensure the minimum flow velocity 0.6 m/s, the flux based on 50 mm diameter is selected as 4.239 m³/h. List the form as shown in Table 2.

It can be seen from Table 2, heat transfer area is increased with the increase of the pipe diameter, while the convection heat transfer coefficient is reduced seriously. The thermal resistance of convective heat decided by these two factors together is increased.

Therefore, it could be concluded that under the condition of constant flux, convective heat transfer between circulating water in pipes and walls of pipes would be worse with the increase of pipe diameter.

Obviously, above analysis about heat transfer performance of ground heat exchanger is only

Table 2. The impact on convective heat of U-tube for the diameter changed.

Pipe diameter mm	Heat transfer coefficient W/m²·K	Thermal resistance W/K
20	75000	3.67E-05
25	60000	4.39E-05
32	46875	5.34E-05
40	37500	6.39E-05
50	30000	7.64E-05

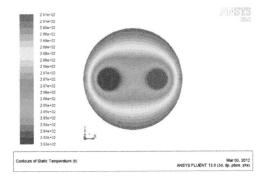

Figure 2. Temperature field (32 mm).

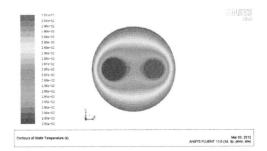

Figure 3. Temperature field (50 mm).

about convective heat transfer, while the whole heat transfer is a complex process. Under the condition of constant flux, different diameters lead to different heat exchange efficiency. Now, take these five kinds of U-tube for numerical calculation and simulation, and list 2 figures of temperature field cut from the same depth for analysis.

It can be concluded from Figures 2 and 3 that under the condition of constant flux, heat transfer between circulating water and the inner surface of pipe will be enhanced with the increase of pipe diameter. When pipe diameter is 40 mm the hot short circuit begins to appear. When pipe diameter

is 50 mm, the hot short circuit is very serious, and it is unacceptable for GSHP. Thus, the suggested pipe diameter is about 40 mm.

Obviously, before we made the choice of pipe diameters, we should think more about geological conditions and economic factors, and consider influence of water pump energy consumption and the hot short circuit. An appropriate pipe diameter would help us to reduce the operation cost.

4 CONCLUSION AND PROSPECT

The change of the pipe diameter is related to both water pump energy consumption and heat transfer of ground heat exchanger. Under the condition of constant flux, resistance in pipe and energy consumption of the pump is reducing with the increase of the pipe diameter. At the same time, heat transfer performance of U-tube would be enhanced, and then the hot short circuit should be taken into consideration. Therefore, the final trend isn't a monotonous increase. From this paper, it could be concluded that the suggest pipe diameter would exist during 30 mm to 40 mm, and the best pipe diameter should be obtained from the physical truth, a large number of experiments and simulation calculations, including economic analysis.

The set of pipe diameter is more than a simple estimation, and it should be taken into comprehensive consideration, not only making fluid in pipe in turbulent region, but also considering the pump energy consumption, the heat transfer performance of the ground heat exchanger and the hot short circuit. The paper only analyses the impact of the change in the pipe diameter to the single U-tube, without considering other U-tube. The drilling spacing and heat-flow density should be considered to determine the best pipe diameter. Further researches and analysis should be taken in the later study.

REFERENCES

Bernier, M.A. 2001. Ground-coupled heat pump system simulation, *Ashrae Trans* 107: 605–616.
Christopher, J. Wood & Liu, H. 2012. Comparative performance of 'U-tube' and 'coaxial' loop designs for use with a ground source heat pump, *Applied Thermal Engineering* 37: 190–195.
Gu, Y. & ONeal, D.L. 1998. Development of an equivalent diameter expression for vertical U-tubes used in ground-coupled heat pumps, *Ashrae Trans* 104: 347–355.
Hepbasli, A. Akdemir, O. & Hancioglu, E. 2003. Experimental study of a closed loop vertical ground source heat pump system, *Energy Conv Manage* 44: 527–548.
Song, J. Lee, K. et al. 2010. Heating performance of a ground source heat pump system installed in a school building, *Science China Technological Sciences* 53: 80–84.

Frontiers of Energy and Environmental Engineering – Sung, Kao & Chen (eds)
© *2013 Taylor & Francis Group, London, ISBN 978-0-415-66159-1*

Interrelation of pollutants in roads runoff of Zhenjiang urban area

J.G. Sheng, Y.D. Shan & C.S. Liu

College of Biology and Chemical Engineering, Jiangsu University of Science and Technology, Zhen Jiang, P.R. China

ABSTRACT: Considering the main outfall of ancient canal as studied object, to analyze the quality of road runoff from Zhenjiang urban area, to study the relations among the main pollutants in the road runoff. It is helpful to test effectively, to quantitative analysis and to make effective control measures. The results shows that COD is well related with TSS, ammonia nitrogen, TP, BOD_5 and TOC.

Keywords: road runoff; runoff water quality; pollution index

1 INTRODUCTION

With the acceleration of urbanization, the human activity ability strength increases, the impervious area of the city to rapid growth, the rain runoff also increase, rainwater runoff pollution is also more and more serious [1,2]. At present, the Zhenjiang city is combined with the comprehensive training works of ancient canal, planning to reconstruct the pipeline network which along the line of ancient canal and cutting the overflow of pollution load through the engineering measures [3–6]. Consider the main outfall of ancient canal as studied object in the paper, to study the relations among the main pollutants in the road runoff. It is helpful to test effectively, to quantitative analysis and to make effective control measures [7].

2 RESEARCH METHODS

2.1 Sampling points and sampling scheme

According to the historical data and the investigation and analysis of the drainage outlet of the ancient canal, the finally nine sample points: Zhongshan bridge, Nanmen street, Tashan bridge, HuJu bridge, LimingHe river, Siming river, Zhoujia river, Tuanjie river and Yudai river. Sampling starts at runoff formation and sampling methods according to GB (HJ 494-2009), dwell time began from the runoff formation after 0 min later in 5, 10, 15, 20, 25, 30, 40, 50 min ... respectively, and take mixed samples. Considering mixed samples as research object in the paper.

2.2 The pollution index and method of determination

Pollution indicators: Chemical Oxygen Demand (COD), Total Suspended Solids (TSS), Total Phosphorus (TP), Ammonia Nitrogen (NH_3-N), Total Organic matter Content (TOC), Biological Oxygen Demand (BOD_5) and Heavy metals (CU, Zn), etc. According to the national environmental protection standard [8], the determination methods are respectively: dichromate titration, filter paper, Molybdenum antimony anti—spectrophotometric method, Nessler's reagent spectrophotometric method, TOC analyzer, Method for determination of dissolved oxygen instrument, Atomic absorption method, etc.

3 THE RESULTS AND DISCUSSION

3.1 The pollution condition and index of each drainage outlet

The each contamination index of mixed rainwater is studied from July to October, 2011. Such as pH, TSS, ammonia nitrogen (NH_3-N) and phosphorus (TP), COD, BOD_5, TOC and heavy metals (CU, Zn, etc), the oil material. The result showed in Table 1.

Although, it often rains from July to October, the COD is rather high, the road is scoured comparatively clean, but the maximum of COD of mixed rain to 400 mg/L, and higher COD of the primary rainwater. The source of pollutants are complex, because the pollutants of the rainwater except from pavement rainwater, in addition from residential areas and residential areas. Table 1 shows, generally, COD of mixed water is about 100~400 mg/L, SS is 300~1500 mg/L or so. NH_3-N and TP respectively in 1.0~7.0 mg/L and 1.0~5.0 mg/L or so, NH_3-N and TP mainly comes from the toilet and laundry sewage. Both TOC and BOD_5 is 10~80 mg/L or so. We can see from the table, it is serious pollutants of petroleum and organic pollution. Mostly water sample of the

Table 1. Part of contamination index of road mixed rainwater.

Places	TSS (mg/L)	NH₃-N (mg/L)	TP (mg/L)	COD (mg/L)	BOD₅ (mg/L)	TOC (mg/L)	CU (μg/L)	Zn (μg/L)	Oil material (mg/L)
Zhongshan bridge	184~576	0.99~4.28	0.41~3.01	58.4~252	23.5~129	8.69~45.8	2.25~43.2	35.6~336	2.58~33.8
Nanmen street	123~405	1.23~7.32	0.23~4.34	168~389	56.7~98.5	10.7~53.1	14~30.08	42.0~245	0.12~45.6
Tashan bridge	201~414	0.69~6.01	0.42~3.25	32.4~168	11.8~83.94	3.29~28.48	2.42~40.7	13.7~64.9	0.65~2.81
HuJu bridge	98~498	1.31~3.23	0.19~1.24	52~133.6	8.7~97.14	6.26~16.34	5.58~41.3	12.4~80.8	0.86~2.34
LimingHe river	75~395	0.64~4.53	0.48~1.56	27.1~180	8.92~83.9	5.04~18.2	6.83~64.7	6.9~75.4	0.45~9.4
Siming river	165~563	1.35~5.43	0.67~1.13	20.5~85.7	7.8~39.7	6.2~13.8	5.2~34.9	28.9~133	0.23~1.95
Zhoujia river	123~453	1.51~5.34	0.31~2.31	23.1~262	4.2~33.4	11.3~49.2	12.3~38.3	21.9~86.2	0.357~1.92
Tuanjie river	23.7~234	1.98~5.88	0.56~3.2	20.2~168	9.2~54.3	4.5~88.6	4.2~16.2	51.5~117	0.16~1.49
Yudai river	141~1496	0.60~6.44	0.33~0.90	21.2~97.6	5.4~34.84	4.85~17.75	4~33.55	23.0~71.4	0.28~1.4

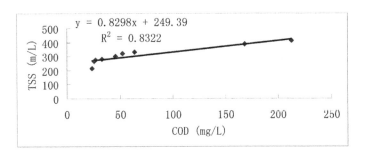

Figure 1. The relation between COD and TSS (7.4).

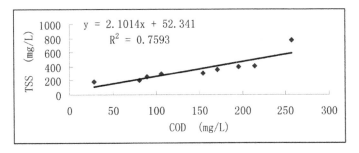

Figure 2. The relation between COD and TSS (8.2).

heavy metal exceeds bid, especially Zn, the highest up to 400 μg/L.

3.2 Interrelation of pollutants in roads runoff

Take COD and SS as the key point detection and control object, also they are the main pollutants of urban road. In order to determine the relationship between COD and SS of mixed rainwater of the main outlet along ancient canal, take 4th of July and 2nd of August rainwater as example.

Figures 1 and 2 show that that COD is well related with TSS usually, Compared them, the 4th of July has better relation. Maybe it is that continuous rain before 4th of July, so the pollutants have scoured cleanly and pollution type is simple. However, there is less rain before 4th of August, there is more pavement pollution, TSS is higher, pollution kinds complex, more influence factors, and the correlation is poorer.

4 CONCLUSIONS

1. COD and SS are the main pollution of mixed rainwater by each drainage outlet of ancient canal, generally, COD is well related with TSS. However, there is less rain, there is more pavement pollution, TSS is higher, pollution kinds complex, more influence factors, and the correlation is poorer.
2. Although the content of both NH_3-N and TP are very low, still have well relation with COD. NH_3-N and TP is key factors to cause waterbody eutrophication, and the rain NH_3-N and TP should not be neglected pollution factors. So control runoff COD quantity can carry effectively cut into the water content of nitrogen and phosphorus.
3. The same rain, both COD/TOC and COD/BOD_5 have good linear relationship, the higher content of dissolubility organism, the better linearity about COD/TOC.
4. Metal ions in mixed rainwater of each of the drainage outlet of ancient canal are very big. While some other organic matter content is higher levels such as fats, all of them have a threat to groundwater. Need to take further management plan.

ACKNOWLEDGMENTS

The project is supported by Zhenjiang Technology Bureau of Social Programmes (SH2011013), Zhenjiang Water Industry Corporation Project, Zhenjiang Water Conservancy Investment Corporation Project.

REFERENCES

[1] Yuan Mingdao. 1986. The Water Pollution Control and Development Situation of American [M]. Beijing: China Environmental Science Press: 206–232.
[2] Wang Huizhen, Li Xianfa. 2002. Beijing City Rain Runoff Pollution and Control [J]. City Environment and Urban Ecological, 15(2): 16–18.
[3] The State Environmental Protection Agency. 2002. Water and Wastewater Monitoring Method (4th edition) [M]. Beijing: China Environmental Science Press: 82–84.
[4] Zhang Yadong, Che Wu, Liu Yan etc. 2003. Interrelation of Pollutants in Road Runoff of Beijing Urban Area [J]. Urban Environment & Urban Ecology, 16(6): 182–184.
[5] Gnecco I, Berretta C, LanZa L G, et al. 2005. Storm water pollution in the urban environment of Genoa, Italy [J]. *Atmospheric Research*, 77: 60–73.
[6] Wu Chun tuk, RuMei, Huang Weieast, zhang wave. 2008. Zhenjiang stormwater runoff pollution forecast [J], *Journal of kingdom university of science and technology*. 38(4): 337–342.
[7] NieFa. 2007. Xinhui city rain ecological comprehensive utilization technology scintigraphy [J]. *Journal of east China jiaotong university*, 24(1): 27–31.
[8] Wang XingQin, Liang ShiJun. 2010. Urban fainfall runoff pollution and the best treatment plan to explore [J]. *Journal of environmental science and management. res social admi pharm*, 378(3): 50–53.

Frontiers of Energy and Environmental Engineering – Sung, Kao & Chen (eds)
© 2013 Taylor & Francis Group, London, ISBN 978-0-415-66159-1

Separate layer feature in overlapping mining overburden strata

R.H. Sun & W.P. Li

China University of Mining and Technology, Xuzhou, Jiangsu, China

ABSTRACT: Deformation-failure of overburden strata of overlapping mining is more complex than caving of monolayer. On the purpose of getting the development law, the authors work through the overburden strata geological characteristics with analysis of well-measured and drilling television, simulate the deformation-failure of roof overlapping face with discrete element model. And we can obtain the laws of deformation-failure of overburden strata of overlapping mining, which can effectively guide coal mining roof prevention and control work. Of overlapping mining.

Keywords: overlapping mining; overburden strata; discrete element model; separate layer

1 INTRODUCTION

Water bursting on roof occurs and instantly floods working face and all the nearby roadways in Huaibei Haizi coal mine of China in May 2005, causing 5 person death, with maximal water bursting volume of 3887 m³/h. It is already ascertained by checking and controlling flooding injury on working face 745 that leading cause is: overlapping mining of multiple seams(coal 10 and coal 7) leads to a great deal of seeper existing in roof conduit pipe and polymineralic caving of main roof causes water bursting.

The working face 844 in coal 8 is located at the side below working face 745. The lower working face 1048 and 10410 have been mined. The upper working face 744 has been mined. In order to prevent such flooding injury accident from occurring again, it is necessary to do research on failure law of roof bed deformation, upgrowth and development of separate layer in roof sandstone and developmental condition of roof caving zone and water-flowing fractured zone in overlapping mining process.

2 ANALYSIS FOR ENGINEERING GEOLOGY FEATURE AND ROOF SEPARATE LAYER UPGROWTH FEATURE

Occurrence of hugely thick polymineralic and thick bed sandstone in overburden strata of coal 7 in Haizi colliery forms special engineering geologic conditions of coal 7, 8, 9 or even coal 10 and generates special engineering geology problems.

2.1 Analysis for polymineralic formation's engineering geology feature and destruction influence on overburden strata

The average uniaxial compressive strength of polymineralic is 144.21 MPa. Meanwhile it is judged by borehole core that rock mass evidently assumes overall structure, with little primary structural plane, simplex lithology and unobvious ground-water porcess. Therefore it is judged as "hugely thick overall structure", which displays the following features after mining of underlying coal seam:

1. Water-physical property: With average osmotic coefficient of 1.15×10^{-5} cm/s, it is affirmed as an impermeable lithologic horizon. But when there is any faultage grown in polymineralic and containing crevice water, it would have water-resisting property lost locally.
2. The roof bed structure and engineering geology property of working face 745 are large in difference, especially polymineralic structure is very intact and compression strength and tensile strength are very large, so back production on working face 1049 under coal 7 would have the overlaying hugely thick integral polymineralic not easy to sink, thereby forming separate layer to make coal measure strata above coal 7 remain in the scope of slow subsidence zone.
3. When sandstone below polymineralic generates caving in mining process of working face 745 in coal 7, polymineralic would not generate caving because of its large intensity and intact structure, so separate layer space between polymineralic layer and lower coal measure strata is increased.

4. It can be known from engineering geology feature indices of polymineralic formation that it is large in elasticity modulus (average 28.64 GPa) and small in Poisson ratio (average 0.17), but silttrock (baking zone) below polymineralic is small in elasticity modulus and large in Poisson ratio, so it is easy to generate distortional incongruity in mining process of coal 10 and coal 7, thereby forming separate layer.

5. Along with advance of working face, hanging roof area of separate layer is gradually enlarged. When hanging separate layer develops to enough length, polymineralic bursts with intensity destabilization and generates a powerful impulse force to water body in separate layer to form super high water pressure and make coal measure strata on working face top generate instant rupture along intensity weakness zone, forming water bursting channel and generating water bursting.

6. Because there are crevices grown below polymineralic, filled by calcite vein and increased in structural planes, it would form a "stretching failure zone" under strong dynamic action.

2.2 Analysis for medium sandstone formation's engineering geology feature and destruction influence on overburden strata

The average uniaxial compressive strength of medium sandstone formation is 71.35 MPa. It is judged from borehole core that it is quite intact in rock mass structure and fewer in structural surfaces, so it is judged as "thick layer structure".

1. Water-physical property of medium sandstone: osmotic coefficient ($1.3 \times 10^{-4} \sim 4.4 \times 10^{-4}$ cm/s), belonging to poorly permeable strata.

2. On the whole, medium sandstone is quite intact in structure and large in compression strength and tensile strength, but local crevices are quite grown, low in intensity and shattered in structure. Therefore, separate layer would appear in medium sandstone after coal 7 is fully mined. Because it is a poorly permeable stratum, after

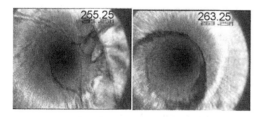

Figure 1. Television picture of bore on the stretching failure zone in polymineralic.

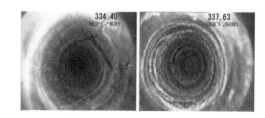

Figure 2. Television picture of bore on the separate layer in sandstone.

separate layer is formed, it is filled by water to form "the seeper separate layer". The result verifies correctness of analyzing occurrence of "separate layer" from mechanical behavior of rock formation, see Figure 2.

3. The medium sandstone is large in elasticity modulus and small in Poisson ratio, but silttrock is small in elasticity modulus (2.73 Gpa) and large in Poisson ratio, so it is easy to generate distortional incongruity in mining process of coal 10 and coal 7, thereby forming separate layer at interface between medium sandstone and silttrock.

4. Along with advance of working face, stress in medium sandstone accumulates to certain degree, and under condition of free face with separate layer and stretching cracking space, medium sandstone would also generate intense dynamic shock destruction.

3 SIMULATION RESEARCH ON DISCRETE ELEMENT MODEL OF OVERLAPPING MINING OF COAL 10 AND 7

3.1 Establishment for discrete element model of mining overlapping

If it is determined according to the correlative formula of mining subsidence theory and pilot calculation of numerical model that scope of the mining influence zone is 200 m, then horizontal size of model will be 800 m. When determining vertical size of numerical calculation zone, take model bottom as 20 m below soleplate of coal 10 (namely deepness 510 m) and model top as 30 m thick polymineralic (namely deepness 300 m). Therefore, finally decided computation model scope is 800 m × 210 m.

On boundary treatment, lower boundary and left right boundaries of computation model can be regarded as boundaries where displacement constraint is zero. The buried depth on cross-section boundary is 300 m. According to R456 boring information, thickness of soil is about 250 m and the polymineralic thickness not included in model is about 50 m. The weight density of soil layer and rock formation are respectively taken as 0.02 MN/m and 0.025 MN/m. It is derived from calculation that load on cross-section boundary is 6.25 MPa.

3.2 Dynamic variation process of overburden strata in mining overlapping simulation

It is indicated by analysis that it is feasible to adopt discrete element method to have overburden strata movement destruction feature recur. Figure 3 is the dynamic variation process of roof bed caving and overburden strata separate layer in back production process of working face 745 after completion of back production on working face 1049. The dynamic variation process for overburden strata separate layer under mining overlapping influence is as follows:

1. When coal 7 advances to 30 m, direct top falls in layering. The overlying rock gradually moves down and begins to generate separate layer at interface between packsand and siltrock about 18 m apart from roof of coal 7. When it advances to 60 m, mudstone of direct top falls and separate layer between packsand and siltrock further expands to make packsand form temporary bearing beam structure. The tiny slippage and dislocation begin to appear among various stratification planes of overlying rock.

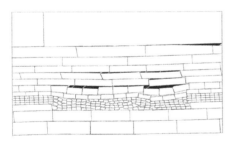

（a）Overlapping mining of 90m

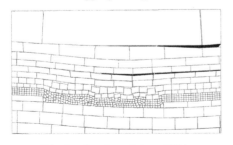

（b）Overlapping mining of 150m

Figure 3.   Calculation result of discrete element in rock overburden while overlapping mining.

2. When the working face advances to 90 m, separate layer between packsand and siltrock previously formed is gradually compacted and closed. The separate layer develops along advance direction of the working face. Meanwhile, because lower packsand and siltrock are gradually compacted and sunk, separate layer at thick bed medium sandstone layer formed by previous mining of 1049 (about 34 m apart from roof of coal 7) appears as continually increasing.

3. When working face continually advances to 150 m, mining scope of coal 7 enters into the section with the most centralized and biggest separate layer formed at polymineralic bottom on account of coal 10 mining. The separate layer between lower packsand and siltrock has been closed basically. The separate layer scale in thick medium sandstone layer reaches the maximal value. The rock formation fissure zone develops gradually to the bottom of thick bed medium sandstone separate layer. By this time, coal 7 reaches full mining and water-flowing fractured zone grows to maximum height, which is displayed as about 30~31 m in the picture. The distance between its top interface and thereon sandstone separate layer bottom boundary (seeper separate layer) is about 3~4 m. Because the working face already approaches left side of separate layer formed under prebiously mined polymineralic of coal 10 and left groundwork position of polymineralic limit equilibrium arch by this time, destroying polymineralic limiting equilibrium arch groundwork and making polymineralic generate power impact unbalance.

4   CONCLUSION

1. Through simulation result, in combination with engineering geology feature of roof overburden strata, flow measurement in bore well and borehole TV data analysis, it is feasible to adopt discrete element method to research discontinuity of overburden strata movement destruction.

2. The research analysis result is effective and intuitionistic, summarizing the deformation destruction features of roof overburden strata in mining process. It can effectively save partial exploration on the deformation failure zone of overburden strata.

3. Because overburden strata destruction height for seam top of overlapping mining working face is evidently different from single coal seam mining, it is not completely applicable to the computing formula of water-flowing fractured zone height in regulations.

4. Such method plays very good application and popularization roles in researching roof deformation destruction features in colliery of overlapping mining coal seam.

REFERENCES

Gundall. P.A Computer Model for Simulating Progressive Large Movements in Klocky Rock Systems. Proc. Symp. on Rock Fracture ISRM NANCY, (1971).

HAOYan-jin, WULi-xin, HU Jin-xing.: Mechanism research on divided layer in mining Coal Technology, 18(6): 40–41 (1999).

Tang CA, Kaiser PK. Numerical simulation of cumulative damagev and seismic energy release during brittle rock failure (part I, II): fun-damental [J]. Int. J. Rock Mech. and Min. Sci, 35(2):120–124 (1998).

WANG Shuren, WANGJin'an.: Distinct element analysis on coal movement law and failure mechanism during mechanized top-coal caving in steep thick seam. Journal of University of Science and Technology Beijing, 1:5–8 (2005).

Frontiers of Energy and Environmental Engineering – Sung, Kao & Chen (eds)
© 2013 Taylor & Francis Group, London, ISBN 978-0-415-66159-1

15000 killed ducks for 10 hours production line cold item-engine room's design

R.Z. Jia, Y.C. Wang & R. Wang
School of Mechanical and Vellicular Engineering, Beijing Institute of Technology, Beijing, China

ABSTRACT: This design is that 15000 killed ducks for 10 hours production line cold item-engine room's design in LuoYang. Through the project design, construction and operation inspection, to explore the methods of cold storage's coldly refrigeration system design and design principles. In this design, I proposed a cooling program to meet the higher latitude regions, which has a certain theoretical meaning and practical application value.

Keywords: cold database design; engine room design; thermal calculation; the refrigeration system

1 THE PROJECT OVERVIEW

1.1 *The cold storage overview*

This cold storage is a small production of cold storage, which is located in Luoyang, and its main production is meat ducks. Its main processing sectors are as follows: firstly, take the slaughtered meat ducks to the freezing room by through the normal temperature hallway, and then take it back through the hallway to the refrigerated room after the meat duck carcass temperature dropped to the required temperature.

There is a fire cistern at the 25 meters south of the ware house, this pool is two meters deep and water storage is about 2800 M^3. The meat duck storage temperature is about minus 20 degrees, owing to nearing the water and water quality is better, so the water is stable. Luoyang City is located at latitude 34° 35, the outdoor design temperature is 33 °C in the summer ventilation, the daily average outdoor design temperature is 31 °C in the Summer air-conditioning. The extreme minimum temperature is −20 °C and the extreme maximum temperature is 45 °C. Taking into account, ammonia has a good thermal performance, the Cooling capacity of per unit volume is great and the pressure is moderate, the condensing pressure is not more than 1.5 MP_a in the room temperature. As long as the evaporation temperature is not less than −33.4 °C, the evaporation pressure is always greater than 1 atm. In addition, the technology requirements are relatively low by using ammonia as the refrigerant, and comply with environmental requirements. Ammonia is cheap, readily available, so decide to select the ammonia refrigeration system. If the condensing temperature is set as 37 °C and the evaporation temperature is set

as 33 °C, then the condensing saturation pressure is $P_k = 1.514$ MP_a and the evaporative saturation pressure $P_o = 0.103$ MP_a, $P_K/P_o > 8$, so the cold storage use the style of single machine two-stage compression. The entire cold storage is divided into four districts: four freezing rooms, which are Used to freeze fresh meat ducks shipped over from the outside; one Refrigerated room, which is Used to store frozen ducks transported over from the freezing room; Another plus one engine room and one duty room. At the same time, according to the needs of frozen and refrigerated, the design temperature of freezing room is −23 °C, the relative humidity is 90%, the design temperature of Storage Room is −18 °C and the relative humidity 85%~90%. The construction plane shown in Figure 1-1.

1.2 *The engine room overview*

The machine room is single separate buildings, which size is 9 m × 19 m and area is 171 m^2. Divided In to equipment, pump house, power distribution room, duty room, etc. In the among of them, the

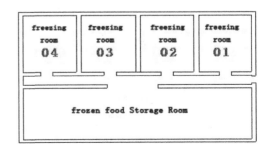

Figure 1-1. Storehouse construction plane diagram.

Compressor, oil separator, intercooler are furnished in the machine room; Condenser. High pressure liquid storage barrels, barrels of low-pressure circulation, Pai liquid barrel, oil interceptors, Total regulating stations are arranged in the equipment room; the pump is furnished in the pump room; Plus ammonia station, put air control, emergency vent ammonia is arranged in the wall close to the equipment.

2 THERMAL CALCULATION

2.1 Cold storage calculation of tonnage calculations

The formula of cold storage tonnage calculations is as follows:

$$G = \frac{\sum V_1 r_i \eta}{1000}$$

In the formula: G—the calculating tonnage of Cold storage, t;

V_1—the Nominal volume of refrigeration room, m^3;

η—the volume utilization factor of refrigeration room;

r_i—the Calculating density of food, kg/m^3.

The heat load calculation of Cold storage is shown in Table 1.

2.2 Heat flow summary

We calculate the heat flow of the various cold room by the formula $Q = Q_1 + Q_2 + Q_3 + Q_4 + Q_5$, being summarized as follows.

Table 1. Load calculation.

No.	Load types	Formula
1.	Heat load of building envelope	$Q_1 = K_w \times F \times a(t_w - t_n)(K_w = 1/R_0)$
2.	Heat load of goods	$Q_2 = Q_{2a} + Q_{2b} + Q_{2c} + Q_{2d}$ $(Q_{2a} = G'(h_1 - h_2) \times 10^3/\tau \ Q_{2b} = G'BC_b(t_1 - t_2) \times 10^3/\tau$ $Q_{2c} = G'(q_1 + q_2)/2 \ Q_{2d} = (G_n - G')q_2)$
3.	Ventilation heat load	$Q_3 = 0.0083 \, n_r \cdot r(h_w - h_n) \times 10^3$
4.	Heat load of motor running	$Q_4 = 1000 \sum N \xi \rho$
5.	Heat load of Operation and management	$Q_5 = Q_{5a} + Q_{5b} + Q_{5c}$
		$(Q_{5a} = q_d \times F \ Q_{5b} = \dfrac{V \times n \times (h_w - h_n) \times M \times r_n \times 10^3}{3600 \times T} \ Q_{5c} = \dfrac{3}{24} \times n_r \times q_r)$
6.	Heat load of each cold cooling equipment	$Q_q = Q_1 + PQ_2 + Q_3 + Q_4 + Q_5$
7.	Mechanical load of cold room	$Q_j = R\left(n_1 \sum Q_1 + n_2 \sum Q_2 + n_3 \sum Q_3 + n_4 \sum Q_4 + n_5 \sum Q_5\right)$

K_w—the heat transfer coefficient of building envelope;
a—the correction factor of building envelope's temperature difference in both sides;
t_w, t_n—the outdoor (indoor) calculated temperature °C;
Q_{2a}—the heat release of food W;
Q_{2b}—the heat release of food packaging materials and carrying tools;
Q_{2c}—the respiratory heat in the process of food cold working;
Q_{2d}—the respiratory heat in the process of food refrigeration;
G'—the daily purchase quality of the cold room kg;
ρ—the factor of motor operating time;
F—the floor area of the cold room m^2;
h_w—the enthalpy of the air outside the cold room kJ/kg;
h_n—the enthalpy of the air inside the cold room kJ/kg;
M—the correction coefficient of air curtains, being taken as 1;
r_n—air bulk density kg/m^3;
n_r—the number of operators;
q_r—the heat release of each operator per second;
P—the load coefficient of the cooling or freezing process;
n_1—the heat transfer season correction factor of the building envelope;
n_2—the techanical load reduction factor;
n_3—the concurrent number of air changes;
n_4—the same period running coefficient of the Electric device in the cold room;
n_5—the operating parameters over the same period in the cold room.

Table 2. Old calculation of cooling load summary.

Room and No.	Q_1 (W)	Q_2 (W)	Q_3 (W)	Q_4 (W)	Q_5 (W)	Q (W)
Freezing room 01	1344.125	23567.5	2445	8800	2492.24	38648.9
Freezing room 02	1095.59	23567.5	2445	8800	2492.24	38400.3
Freezing room 03	1095.59	23567.5	2445	8800	2492.24	38400.3
Freezing room 04	1344.125	23567.5	2445	8800	2492.24	38648.9
Frozen food storage room 05	6479.6	1780	2740	4400	11658.5	27058.1

Table 3. Cooling equipment in each cold load summary.

Room and No.	Q_1 (W)	P	Q_2 (W)	Q_3 (W)	Q_4 (W)	Q_5 (W)	Q_q (W)
Freezing room 01	1344.125	1.3	23567.5	2445	8800	2492.24	45719.115
Freezing room 02	1095.59	1.3	23567.5	2445	8800	2492.24	45470.58
Freezing room 03	1095.59	1.3	23567.5	2445	8800	2492.24	45470.58
Freezing room 04	1344.125	1.3	23567.5	2445	8800	2492.24	45719.115
Frozen food storage room 05	6479.6	1.0	1780	2740	4400	11658.5	27058.1

Table 4. All cold mechanical load summary.

Evaporation Temperature (°C)	Name	R	$n_1 Q_1$ (W)	$n_2 Q_2$ (W)	$n_3 Q_3$ (W)	$n_4 Q_4$ (W)	$n_5 Q_5$ (W)	Q_J (W)	Total (W)
−28	Freezing room	1.07	2293.45	94270	5868	35200	4984.48	152706	176049
−33	Frozen food storage room	1.07	3045.4	1068	1644	4400	11658.5	23343	

3 IDENTIFICATION OF THE REFRIGERATION SYSTEM PROGRAM

3.1 Identification of the cold storage refrigeration system parameters

Evaporation temperature is the temperature of refrigerant in the evaporator to vaporize, the determination of the evaporation temperature in association with refrigerant, being also associated with chilled water and air.

According to the file [2]: when air was taken as refrigerant in cold storage, the evaporation temperature is low 8 to 10 °C than the bank required air temperature. Therefore, although there are two evaporation loop in the design of the system, the temperature difference is small, So you can use an evaporation temperature $t_0 = -33$ °C loop, and the refrigerated room loop can take the temperature-controlled device to make the space temperature achieve the necessary requirements. Meanwhile, due to the ambient air temperature of the inhalation tube, compressor suction gas temperature is higher than the evaporation temperature of the refrigerant. This is called the gas superheat. Because the pump supplying liquid is superior than the other methods, it has several advantages, such as high efficiency of the refrigeration equipment,

good safety profile, management and operation easily, and easy to implement automatic control and so on. So the system uses the ammonia pump for supplying liquid. Ammonia compressor suction temperature higher than the evaporation temperature of 5~8 °C, we take 8 °C in the design, so each circuit suction temperature is −25 °C. At the same time, when using a water-cooled condenser, the condensation temperature is 5~7 °C higher than the average temperature of cooling water import and export. so they chose: $t_k = 39$ °C.

3.2 Main equipment selection

Because the calculated pressure ratio is greater than 8, when take $t_k = 39$ °C $P_k = 1.514$ MP$_a$, $t_0 = -33$ °C $P_o = 0.103$ MP$_a$ so it can be identified as the two-stage compression. Meanwhile, the S8–17 compressor is being selected according to the cooling capacity Q = 164.45 KW. Based on the performance curve of the S8–17, we investigate something as follows: each S8–17 cooling capacity is 178 KW in this condition, So we Choose one S8–17 piston compressor, and the other one is alternated. In addition, the equipped motor power is 62 KW.

Due to $Q_K = 266350$ W $q_f = 3500$ w/m², the condenser heat transfer area F = Q_k/q_f = 76.1 m², in which Q_K is condenser heat load. Cooling water $V_w = F_1V_m = 76.1*0.7 = 53.27$ m³/h, So the condenser is elected as: 1 set DWN-90 horizontal-type condenser, each cooling area is 90 m².

4 CONCLUSION

Cold storage includes freezing room and the frozen food storage room. Each r room's design temperature is −23 °C, −18 °C; An evaporation loop, the evaporation temperature is −33 °C; The calculated cooling load is 181.1 kw; The mechanical load is 176.05 kw; The loop selection is single machine two-stage refrigeration unit and it's cooling capacity is 178 kw; the selection of engine room condenser is a models for the DWG-90 horizontal condenser, which cooling area is 90 m², and the condenser heat load is 266.35 kw, fully able to meet the requirements. Some other auxiliary equipment, such as high-pressure liquid receiver, low-pressure circulation barrel and so on, all being selected on the basis of the choice of compressor and condenser, Its performance is also able to achieve to meet the cold storage requirements.

In this engineering design process, from the choice of the refrigeration compressor to the selection of cooling system auxiliary equipment and all storehouse cooling method selection, we all had done a reasonable consideration. We determined engine room's design according to the actual size of the space, which engine room's equipment of the cooling system required in the layout process, and on the basis of compliance with the principle of piping layout. Finally, in line with the principle of reasonable, suitable, compact structure, and ensure the principle of reserved space, we confirmed the refrigeration process scheme above.

REFERENCES

[1] Zhuang Youming. Refrigeration equipment design, Xiamen University Press, 2006, p.1–176.
[2] Zhao Rongyi, Qian Yiming, Fan Cunyang and so on. condensed air-conditioning Design Manual, ninth edition, Beijing, China Architectural Industry Press, 2006, p.26–59.
[3] Guo Qingtang. Practical Refrigeration Engineering Design Manual. Beijing. China Architecture Industry Press, 2006.
[4] Wu Yezheng. Refrigeration equipment. Xi'an, Xi'an Jiaotong University Press.
[5] Chen Weigang. refrigeration engineering and equipment. Shanghai Jiaotong University Press.
[6] Wang Chun. cold storage refrigeration technology. Machinery Industry Press.

Frontiers of Energy and Environmental Engineering – Sung, Kao & Chen (eds)
© 2013 Taylor & Francis Group, London, ISBN 978-0-415-66159-1

Latest development and application of FGD with organic amine in Pangang group

W. Jianshan, L. Jianming, Q. Zhengqiu, Z. Weijia & Z. Xiaolong
Pangang Group Research Institute Co., Ltd., State Key Laboratory of Vanadium and Titanium Resources Comprehensive Utilization, Panzhihua, China

ABSTRACT: The operation of FGD system for the 173 m² No. 6 sintering machine in Pangang Group is introduced. Through over two-years' process and equipment improvement, availability of the system working synchronously with the sintering machine reaches 85% with a desulfurization rate of above 88%. Following experimental research and field tests, the independently developed liquid desulfurizer for FGD has been put into industrial operation.

Keywords: flue gas desulfurization; desulphurization process; organic amine; desulphurization solution; desulfurizer

1   INTRODUCTION

With the rapid development of industry, air pollution is getting worse and becoming a worldwide issues that attract all the countries' attention. SO_2 is one of the atmospheric pollutants and its major emission sources are energy and metallurgical industry. Among those SO_2 emission plants, coal-fired power plants account for about a half of SO_2 emission, the non-ferrous metallurgy; iron and steel smelting and sulfuric acid, oil refining and other industrial enterprises account for about 1/3; the rest of SO_2 comes from the transportation and commercial and civil stoves.

It is predicted that China's total coal demand will reach 2.9 billion tons by 2020, and coal-fired generating units will increase to 660 million kilowatts with a total SO_2 emission of more than 4350 tons from burning coal. According to targets for pollutant reduction, the SO_2 emission in and after 2020 shall be controlled at 12 million tons per year. Though China overfulfill its target of SO_2 emission reduction in the 11th Five-Year Period, in the long run, there is still a long way to go. To this end, the Ministry of Environmental Protection of China requires that the sintering machines and pellet production equipments located in the built-up urban area should be equipped with flue gas desulfurization system with desulfurization efficiency of above 80%. Thus, 530 sets of new desulfurization facilities, i.e. 78,000 m² of sintering area in total (90 m² for single unit) shall be provided with new desulfurization facilities, and the flue gas desulfurization facilities for 129 sets of sintering machines in operation, i.e. 17,000 m² of sintering area, need to be transformed during the 12th Five-Year Period.

2   OVERVIEW OF FLUE GAS DESULFURIZATION WITH ORGANIC AMINE IN PANGANG GROUP

No. 6 sintering machine with a sintering area of 173 m² in Ironmaking Plant of Panzhihua Steel & Vanadium Co., Ltd., is equipped with a flue gas desulfurization system employing organic amine to absorb flue gas. The system is designed to deal with 550000 Nm³/h of flue gas with inlet SO_2 concentration of 5000 mg/Nm³ and inlet flue gas temperature of 130 °C. Its designed efficiency of desulfurization is no less than 97%, which means SO_2 concentration after treatment is no more than 150 mg/Nm³. Covering an area of approximately 1000 m², the plant realizes a maximum reduction of SO_2 of 21200 t/a and a largest output of sulfuric acid for industrial use of 32000 t/a. Designed by China Enfi Engineering Corporation and constructed by Chengdu Huaxi Industrial Gas Co., Ltd., the project is estimated to cost 91,253,400 yuan. The construction of the project started on March 31, 2008 and completed on December 23, 2008.

3   SINTERING FLUE GAS DESULFURIZATION TECHNOLOGY OF ORGANIC AMINE

3.1   *Desulfurization principle*

Desulfurization solution prepared with organic amine (ionic liquid) consists of organic cations and inorganic anions. Desulfurization reaction is as follows.

$$R_1R_2N\text{-}R_3\text{-}NR_4R_5 + HX \rightarrow R_1R_2NH^+\text{-}R_3\text{-}NR_4R_5 + X^- \tag{1}$$

$$SO_2 + H_2O \rightleftharpoons H^+ + HSO_3^- \tag{2}$$

$$R_1R_2NH^+\text{-}R_3\text{-}NR_4R_5 + SO_2 + H_2O \rightleftharpoons$$
$$R_1R_2NH^+ - R_3 - NR_4R_5H^+ + HSO_3^-. \tag{3}$$

$$R_1R_2NH^+\text{-}R_3\text{-}NR_4R_5 + X^- \rightarrow R_1R_2N\text{-}R_3\text{-}NR_4R_5$$
$$+ HX. \tag{4}$$

The organic amine (ionic liquid) in desulfurization solution reacts with strong acid according to the formula 1. X^- represented strong radical ion, such as Cl^-, NO_3^-, F^- and SO_4^{2-}. However, the strong acid hydrogen ions are adsorbed to the amine molecules in the strong alkaline base. When SO_2 in flue gas contacts with desulfurization solution, it dissolves into the water at the beginning and dissociates into hydrogen ions and sulfite ions in accordance with formula 2. If there is no organic cation, formula 2 will soon establish a balance and little SO_2 will dissolve. When the desulfurization solution contains organic cations, due to its another weak alkaline base's absorption of hydrogen ions, the equilibrium shifts to the right and the dissolution of SO_2 increases in accordance with formula 3, which greatly facilitates the dissolution of SO_2 into the desulfurization solution. The above process is carried out at around 50 °C. But when the SO_2-absorbed desulfurization solution is heated to about 110 °C, the equilibrium shifts to the left in accordance with formula 3. As a result, SO_2 is obtained reversely and organic amine is regenerated for the next cycle of SO_2 absorption.

During the absorption and desorption process, the accumulated thermal stable salts, such as sulfate, thiosulfate, nitrate, chloride and thiocyanate are removed by electrodialysis or ion exchange resin in accordance with formula 4 to ensure the balance of thermal stable salt in the system, no influence on the desulfurization capacity of desulfurization solution, and less corrosion of desulfurization solution on the equipment.

3.2 Operation

3.2.1 Properties of sintering flue gas

Flue gas obtained after the blower fan of the desulfurization system has flow rates of 350,000 Nm³/h~450,000 Nm³/h, temperatures of 90~130 °C, dust concentration of 150~300 mg/Nm³, and moisture concentration of 6%~7%. The compositions of flue gas generated in Pangang are shown in Table 1.

From Table 1, some conclusions can be drawn as follows.

Table 1. The composition of flue gas generated in Pangang.

CO (mg × ⁻³m)	NOx (mg × ⁻³m)	SO₂ (mg × ⁻³m)	SO₃ (mg × ⁻³m)	HCl (mg × ⁻³m)
14000~ 16000	150~ 160	4000~ 5000	200~ 300	50~ 100

1. Very high concentration of SO_2 in the flue gas (the highest in domestic and international plants);
2. High concentration of reducing gas CO in the flue gas with the highest up to 16,000 mg/Nm³;
3. Relatively high concentration of strong acid gases SO_3 and HCl in the flue gas;
4. High content of dust in the flue gas due to the single electric field of the electrostatic precipitator in sintering machine head.

3.2.2 Existing problems and research progress

It is the first time in the world to apply organic amine to remove SO_2 in the flue gas and then preparing sulfuric acid, so no experience can be learned. Due to insufficient knowledge about the particularity of the sintering flue gas in Pangang, there are lots of defects in system design and deviations in process and technology, which lead to unsteady operation of the system in the whole year of 2009 with a low service rate of 16.83%. The main reasons for unsatisfactory operation are as follows.

1. High dust concentration in sintering flue gas. The designed inlet dust concentration was 100 mg/Nm³, but it actually achieved more than 200 mg/Nm³ or even up to 700~800 mg/Nm³ in certain time interval. Excessive dust in the desulfurization solution could easily leads to blocking in some part of the system.
2. No sulfur melting and removal device in the design. The sulfur in desulfurization solution sublimated on heating in the resolution process, then condensed and deposited in the upper part of resolving tower and the subsequent condenser, causing blockage in these parts.
3. Incapability of the filter for suspended solids removing in the desulfurization system. Thus, suspended solids in the desulfurization solution could not be effectively removed in time, resulting in blockage in the heat exchanger, filters or even pipes.
4. The strong acidic compositions' getting into the washing water and the desulfurization solution in the process of washing and desulfurization. It led to high concentration of strong radical ions such as Cl^-, SO_4^{2-} and NO_3^-. In addition, the washing water and desulfurization solution

exhibited strong corrosion with the pH value of below 2 and 4~6 respectively. These two factors caused severe corrosion of pipes, valves, containers, tanks, pumps, padding, testing instruments etc., and often resulted in equipment damage or failure.

5. The content of acid components including SO_3 and HCl in the sintering flue gas was higher than expected in design while the ability of SO_4^{2-} and Cl^- removal for the device was smaller than desired in practice. The enrichment of SO_4^{2-} and Cl^- in the desulfurization solution affected the desulfurization performance.

6. Very complex compositions, high content of O_2, and high content and complex compositions of dust in the sintering flue gas. Once they contacted with the desulfurization solution, the latter would cease to be effective, the mechanism for which has not been understood by far.

7. Serious loss of desulfurizer. The effective concentration of the desulfurization solution could not be effectively maintained, especially when the flue gas was at top of the tower and the resin desalting position.

Since 2010, great efforts have been made by the researchers of Panzhihua Iron and Steel Research Institute. As a result, the following technical measures have been taken to address these problems:

1. Strengthen the management and maintenance of the sintering electrostatic precipitator and make sure that the dust concentration in the flue gas decreases from 300~500 mg/Nm³ to about 200 mg/Nm³ and keep in this level;

2. Install a spray in the flue dust removal device to further enhance the effect of wetting and washing for the dust in the flue gas;

3. Increase the height of packing layer in washing section to improve the performance of dust and mist washing;

4. Improve the filter of the lean solution to improve the filtering performance;

5. Install a filter for sulfur in condensate at the regeneration tower to remove the sulfur produced timely;

6. Establish a new freeze crystallization process, so as to remove the sulfate ion in the desulfurization solution;

7. Improve the demisting devices for flue gas at top of the tower, and optimize each step of the desalination process.

3.2.3 Development of sintering flue gas desulfurization technique

After the exploration test in 2008, the lab formulation development in 2009 and on-site small scale

test in 2010, from December 20th to 30th, 2010, an industrial application test for 30 tons of self-developed sintering flue gas desulphurization solution was performed in the desulfurization system for No. 6 sintering machine. During the test, the running time of the desulfurization system reached 175.8 hours in total with an average desulfurization efficiency of 92% to 95% and sulfuric acid output of 196 tons, indicating an initial success had been achieved.

The industrial test proved that the self-developed sintering flue gas desulfurization technology has good desulfurization performance. Consequently, at late April 2011, the technology was transferred to industrial applications. This system operated from April 24 to May 31 with total running time of 35 days, sulfuric acid output of 1259.5 tons (i.e. the average daily acid production of 36 tons) and the average desulfurization efficiency of above 88.1%.

The sulfuric acid production, pH, SO_2 concentration for each day during operation can be seen in Figures 1–3.

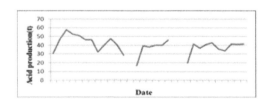

Figure 1. Daily sulfuric acid production during the period of industrial application.

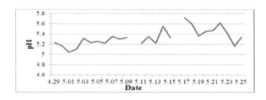

Figure 2. Daily PH of the solution during the period of industrial application.

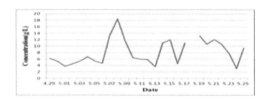

Figure 3. SO_2 concentration in the solution during the period of industrial application.

371

4 CONCLUSION

1. At present, the sintering flue gas desulfurization technology is still under continuous development. There are various types of desulfurization technologies with their own advantages and disadvantages. However, only those that can transfer SO_2 to sulfur products and don't produce secondary pollutants are worth developing and applying, because they represent the direction of technological development and can satisfy the requirements of environmental protection.
2. Developed by in Panzhihua Iron and Steel Research Institute, the technology of flue gas desulfurization with organic amine for No. 6 sintering machine in Pangang Group is the first of its kind in the world. After more than two years' research on this area, most important engineering problems that hinder the normal operation of the sintering flue gas desulfurization system have been solved. The process is now mature and the system can be operated continuously with the operating rate and desulfurization rate up to 85% and 88% respectively. Researches on prolonging the efficiency of desulfurizer and service life of equipments will be carried out at next phase.

REFERENCES

China steel news. 2011. The iron and steel enterprises will be the full implementation in sintering flue gas desulfurization. Beijing.
Jianming, L. 2011. The national sintering flue gas desulfurization technology symposium, Beijing.

Frontiers of Energy and Environmental Engineering – Sung, Kao & Chen (eds)
© *2013 Taylor & Francis Group, London, ISBN 978-0-415-66159-1*

Improved particle swarm optimization for power economic dispatch considering emission

C.L. Chiang
Nan Kai University of Technology, Taiwan, P.R. China

ABSTRACT: This paper presents an Improved Particle Swarm Optimization with Multiplier Updating (IPSO-MU) for solving Power Economic Dispatch (PED) problems considering emission. The Improved Particle Swarm Optimization (IPSO) has the ability to efficiently search and actively explore solutions. Multiplier Updating (MU) is introduced to avoid deforming the augmented Lagrange function and resulting in difficulty to solution searching. To handle the PED problems considering emission, the ε-constraint technique is employed. The proposed approach integrates the ε-constraint technique, IPSO, and the MU. The simulation using the proposed method is carried out on a 6-unit test system, and results are compared with that obtained using other different methods. Numerical results indicate that the proposed approach is superior to other methods in solution quality and computational burden.

Keywords: emission; particle swarm optimization; power economic dispatch

1 INTRODUCTION

Generally, the power economic dispatch problem involves allocation of generations to different thermal units to minimize the cost of generation, while satisfying the equality and inequality constraints of the power system and keeping pollution within limits. Some papers modeled the problem as a multi-objective problem and solved it using the constraint method (El-Keib et al. 1994), the weighting method (Heslin et al. 1989) or the separate runs method (Farag et al. 1995). In the constraint method, the multi-objective problem was reduced to a single objective problem by treating the emissions as a constraint. The weighting method linearly combined the objectives as a weighted sum. The objective function so formed may lose significance due to the incorporation of multiple non-commensurable factors into a single function. Farag et al. (1995) took two separate runs for the two objectives by Linear Programming method (LP) with bounded variables and an incorporated technique of the Section Reduction method and the Third Simplex method. This paper presents a multi-objective optimization algorithm employing the ε-constraint technique (Lin 1976) to handle the multi-objective problem.

2 FORMULATION

In the multi-objective problem formulation, two important non-commensurable objectives in an electrical thermal power system are considered. These are economy and environmental impacts.

2.1 *Economy objective F_1*

The economy objective F_1 of generator power output P_i is represented as (Farag et al. 1995).

$$F_1 = \sum_{i=1}^{N_g} a_i P_i^2 + b_i P_i + c_i \qquad (\$/h) \qquad (1)$$

where F_1 is the total cost of generation, P_i is the generation of the ith generator, a_i, b_i and c_i are coefficients of the cost curve of the ith generator, and N_g is the total number of the generators.

2.2 *Environmental objective F_2*

The emission of sulfur dioxide, nitrogen oxides, carbon monoxide gases etc., which cause atmospheric hazards, can be mathematically modeled as (Farag et al. 1995).

$$F_2 = 10^{-2}\left(\alpha_i + \beta_i P_i + \gamma_i P_i^2\right) + \xi_i e^{(\zeta_i P_i)} \qquad (2)$$

where α, β, γ, ξ, and ζ are coefficients of generator emission characteristics.

2.3 *System constraints*

To ensure a real power balance, an equality constraint is imposed:

$$\sum_{i=1}^{N_g} P_i - P_D - P_{loss} = 0 \qquad (3)$$

where P_D is the total demand, and P_{loss} is the real power loss in the transmission lines. The inequality constraint imposed on generator output is

$$P_{i\min} \leq P_i \leq P_{i\max} \qquad (4)$$

where $P_{i\min}$ and $P_{i\max}$ are the minimum and maximum limits on the loadings of the ith generator. Aggregating equations (1) to (4), the multi-objective optimization problem is formulated as

$$\underset{P_i}{minimize} \quad [F_1(P_i), F_2(P_i)]$$

$$subject\,to \quad \sum_{i=1}^{N_g} P_i - P_{loss} = P_D \qquad (5)$$
$$P_{i\min} \leq P_i \leq P_{i\max}; i = 1,2,...,N_g$$

where $F_1(P_i)$, $F_2(P_i)$ are the objective functions to be minimized over the set of admissible decision vector P_i.

3 THE PROPOSED ALGORITHM

3.1 *The ε-constraint technique*

The ε-constraint technique (Lin 1976) is used to generate pareto-optimal solutions to the multi-objective problem. To proceed, one of the objective functions constitutes the primary objective function and all other objectives act as constraints. To be more specific, this procedure is implemented by replacing one objective in the problem as defined by (5) with one constraint. Re-formulate the problem as follows:

$$\text{min} \quad F_j(P_i), j = 1 \text{ or } 2$$
$$subject\,to \quad F_k(P_i) \leq \varepsilon_k; k = 1 \text{ or } 2, \text{ and } k \neq j$$
$$\sum_{i=1}^{N_g} P_i - P_{loss} = P_D \qquad (6)$$
$$P_{i\min} \leq P_i \leq P_{i\max}; i = 1,2,...,N_g$$

where ε_k is the maximum tolerable objective level. The value of ε_k is chosen for which the objective constraints in problem (6) are binding at the optimal solution. The level of ε_k is varied parametrically to evaluate the impact on the single objective function $F_j(P_i)$.

3.2 *The IPSO*

Generally, the PSO (Wang et al. 2010) involves two critical issues (evolutionary direction and

population diversity). As the evolutionary direction is effective in searching, the strong evolutionary direction can reduce the computational burden and increase the probability of rapidly finding an (possibly local) optimum. As population diversity is increased, the genotype of the offspring differs more from the parent. Accordingly, a highly diverse population can increase the probability of exploring the global optimum and prevent a premature convergence to a local optimum. These two important factors are here balanced by both employing the accelerated operation and migration (Chiang 2005) in the proposed IPSO that can determine an efficacious direction in which to search for a solution and simultaneously maintain an appropriate diversity for a small population.

3.3 *MU technique*

Consider the nonlinear programming problem with general constraints as follows.

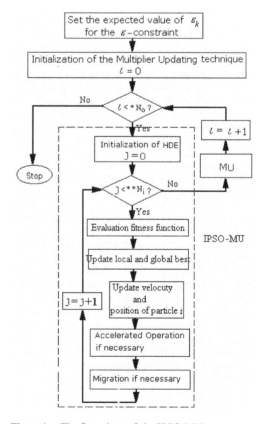

Figure 1. The flow chart of the IPSO-MU.
*N_o: maximum number of iterations of outer loop
** N_i: maximum number of iterations of inner loop.

$$\min_x F(x)$$

subject to $\quad h_k(x) = 0, \quad k = 1, ..., m_e$ $\qquad$ (7)

$$g_k(x) \le 0, \quad k = 1, ..., m_i$$

where x represents a n_c-dimensional variables, and $h_k(x)$ and $g_k(x)$ stand for equality and inequality constraints, respectively. The augmented Lagrange function (Chiang et al. 2002) is combined with the Lagrange function and penalty terms, yielding,

$$L_a(x, \nu, \upsilon) = f(x) + \sum_{k=1}^{m_e} \alpha_k \{[h_k(x) + \nu_k]^2 - \nu_k^2\}$$

$$+ \sum_{k=1}^{m_i} \beta_k \{\langle g_k(x) + \upsilon_k \rangle_+^2 - \upsilon_k^2\}$$
$\qquad$ (8)

where α_k and β_k are the positive penalty parameters, and the corresponding Lagrange multipliers $\nu = (\nu_1, ..., \nu_{m_e})$ and $\upsilon = (\upsilon_1, ..., \upsilon_{m_i}) \ge 0$ are associated with equality and inequality constraints, respectively.

3.4 The solution procedure of the proposed approach

The merit of the proposed approach is that the augmented Lagrange function can be scaled to avoid the ill condition resulting in difficulty of solution searching. The proposed algorithm has two iterative loops, and is shown as Figure 1. The augmented Lagrange function is solved for a minimum value in the inner loop with the given penalty parameters and multipliers, which are then updated in the outer loop toward obtaining an upper bound of $L_a(x, \nu, \upsilon)$. When both inner and outer iterations are sufficiently large, the augmented Lagrange function will converge to a saddle point of the dual problem (Chiang et al. 2002).

4 SYSTEM SIMULATION

This section investigates an example to illustrate the effectiveness of the proposed approach with respect to the quality of the solutions obtained. The example compares the proposed approach with two other different methods, the LP method and the PSO, in terms of the obtained results for a 6-generator test system given in Farag et al. (1995).

The computation was implemented on a personal computer (P4-2.0GHz) in FORTRAN-90 language. The population size N_p is set to 5 for the proposed approach in this example, and is set to 20 for the PSO. The iteration numbers of outer loop and inner loop, (outer, inner), is set to (50, 3000). To proceed the constrained problem (6) is converted into an unconstrained problem by augmented Lagrange function (8) and is described as follows:

$$L_a(P_i, \nu, \upsilon) = F_1(P_i) + \alpha_1 \{[h_1(x) + \nu_1]^2 - \nu_1^2\}$$

$$+ \beta_1 \{\langle g_1(P_i) + \upsilon_1 \rangle_+^2 - \upsilon_1^2\}$$

subject to $h_1: P_D - \sum_{i=1}^{N_g} P_i - P_{loss} = 0$
$\qquad$ (9)

$$g_1: F_2(P_i) - \varepsilon_2 \le 0$$

where the prime objective function is $F_1(P_i)$, $P_{i\,min} < P_i < P_{i\,max}, i = 1, ..., N_g$, h_1 stands the violation of power balance constraint (3), and g_1 stands the violation of emission objective for expected ε_2, $(\varepsilon_2 \in [F_2^{min}, F_2^{max}] = [0.1942, 0.2215])$.

The augmented Lagrange function (9) is solved by the proposed approach. Since cost and emission are of conflicting nature, the value of objective F_2 will be the maximum when the value of F_1 objective is the minimum and vice versa. So, the values of the best cost with F_2^{max} and the minimum emission with F_2^{min} are obtained by performing the augmented Lagrange function (9) separately. The best compromise indicates the minimum cost within expected ε_2. For comparison, Table 1 gives three types of solution obtained from the proposed approach, the LP method (Farag et al. 1995), and the PSO. In this table, best cost and best emission indicate the minimum cost and minimum emission taken individually. The best compromise indicates

Table 1. Computational results of the proposed approach and two other different methods.

| Items | The proposed IPSO-MU ($N_p = 5$) | | | LP | | PSO ($N_p = 20$) | | |
	Cost ($/h)	Emission index	CPU time (s)	Cost ($/h)	Emission index	Cost ($/h)	Emission index	CPU time (s)
Best cost	605.8946	0.2215	5.57	606.04	0.2215	605.8951	0.2215	12.16
Best emission	643.9285	0.1942	6.16	645.88	0.1952	643.9478	0.1942	13.25
Best compromise	620.8583	0.1979	5.84	–	–	620.8586	0.1979	11.83

the minimum cost within ε_2 with a value of 0.1979. The total cost obtained by the proposed approach is satisfactory compared with that obtained by the LP and the PSO. The proposed approach is superior in solutions quality than the LP and PSO, and has less CPU time than the PSO.

5 CONCLUSION

The proposed IPSO-MU for power economic dispatch considering emission was proposed herein. The IPSO helps the proposed method efficiently search and actively explore solutions, the MU technique helps the proposed method effectively solve the augmented Lagrange function, and the ε-constraint technique is employed to handle the multi-objective problem. The proposed method requires relatively smaller CPU time in comparison with the PSO. The numerical results of the simulations verified the advantages of the proposed approach.

ACKNOWLEDGEMENT

Financial support given to this research by the National Science Council, Taiwan, ROC under Grant No. 100-2632-E-252-001-MY3 is greatly appreciated.

REFERENCES

Chiang, C.L. 2005. Improved genetic algorithm for power economic dispatch of units with valve-point effects and multiple fuels. *IEEE Trans. on Power Systems* 20(4): 1690–1699.

Chiang, C.L., Su, C.T. & Wang, F.S. 2002. Augmented Lagrangian method for evolutionary optimization of mixed-integer nonlinear constrained problems. *International Math. Journal* 2(2): 119–154.

El-Keib, A.A., Ma, H. & Hart, J.L. 1994. Environmentally constrained economic dispatch using the LaGrangian Relaxation method. *IEEE Trans. on Power Systems* 9(4): 1723–1729.

Farag, A., Al-Bayiat, S. & Cheng, T.C. 1995. Economic load dispatch multiobjective optimization procedures using linear programming techniques. *IEEE Trans. on Power Systems* 10(2): 731–738.

Heslin, J.S. & Hobbs, F.B. 1989. Multiobjective production costing model for analyzing emissions dispatching and fuel switching. *IEEE Trans. on Power Systems* 4(3): 836–842.

Lin, J.G. 1976. Multi-objective problems: Pareto-optimal solutions by method of proper equality constraints. *IEEE Trans. on Autom. Control* 21(5): 641–650.

Wang, Y. & Yang, Y. 2010. Particle swarm with equilibrium strategy of selection for multi-objective optimization. *European Journal of Operational Research* 200(1): 187–197.

Frontiers of Energy and Environmental Engineering – Sung, Kao & Chen (eds)
© *2013 Taylor & Francis Group, London, ISBN 978-0-415-66159-1*

Hydropower project investment and decision making based on improved fuzzy probability method

L. Liu, Y. Kong & Z.J. Liu
College of Hydraulic & Environmental Engineering, China Three Gorges University, Yichang, P.R.China

ABSTRACT: This paper presents a method named infinite irrelevance method for complex correlation analysis which analyses the evaluation index about the multiple correlations, we establish the hydropower project investment and decision making model combined with the fuzzy probability theory of fuzzy mathematics to check index factors for repeatability and to simplify the evaluation of the system. The analysis shows that the modified fuzzy probability model of fuzzy mathematics not only inherits the classical fuzzy comprehensive evaluation theory and advantages but also overcomes the limitations of the evaluation index weight value in practice. So it is obvious rationality.

Keywords: hydropower project; investment and decision making; fuzzy probability; infinite irrelevance method

1 INTRODUCTION

As an important element of the national economy and the social development, the hydropower project plays an irreplaceable role in the flood safety, water supply security, food security, economic security, ecological security. But it is obviously that it lives beyond its income as a result that the hydropower engineering development lags behind and the investment channels are instability. Many factors such as the social, the economic, the nature in the hydropower investment engineering and the mutual influence should be considered, so it still have no uniform standard about the selection of evaluation factors. The paper presents infinite irrelevance method to analyze the evaluation index about the correlations; meanwhile uses the fuzzy probability model[1,2] developed in recent years to evaluate the hydropower project investment and provides a new way for hydropower investment decision making method.

2 THE SCREENING FACTORS OF THE INFINITE IRRELEVANCE METHOD

Generally, using the complex correlation from p indexes, selecting one part of indicators to fully reflect the original p variable indexes, making evaluation results be more real and effective[3]. The calculation steps can be expressed as follows:

1. Let us define an index vector $x = (x_1, x_2 \ldots x_p)$, the corresponding N group dates make up matrix and expressed by X.

$$X = \begin{pmatrix} x_{11} & x_{12} & \cdots & x_{1p} \\ x_{21} & x_{22} & \cdots & x_{2p} \\ \vdots & \vdots & \ddots & \vdots \\ x_{n1} & x_{n2} & \cdots & x_{np} \end{pmatrix} \tag{1}$$

where x_{ij} represents an observation value of one sample.

2. We can calculate the mean value $\bar{x}_i = 1/n \sum_{j=1}^{n} x_{jk}$ of the variable x_i from the data of X, the variance s_{ij} of the variable x_i and the variable x_j is

$$s_{ij} = \frac{1}{n} \sum_{k=1}^{n} (x_{ki} - \bar{x}_i)(x_{kj} - \bar{x}_j), \quad j = 1, 2, \ldots, p \quad i = 1, 2, \ldots, p \tag{2}$$

The complex correlation matrix is

$$R = (r_{ij})_{p \times p} \tag{3}$$

where $r_{ij} = s_{ij}/\sqrt{s_{ii}s_{jj}}$ reflects the linear correlation degree between x_i and x_j. The complex correlation coefficient which reflects the linear correlation degree between x_i and the rest $p - 1$ variables usually be noted as $\rho_{xi}|x_1, x_2, \ldots, x_{i-1}, x_{i+1}, \ldots, x$, Simple as ρ_i, it can be made by the below formula to calculate:

$$R(x) = \begin{pmatrix} R_{-i} & r_i \\ r_i^{\mathrm{T}} & 1 \end{pmatrix} \tag{4}$$

where $r_i = (r_{1r}, r_{2r}, \ldots, r_{r-1,r}, r_{r+1,r}, \ldots, r_{pr})$, R_{-i} is related matrix of removing x_i. Meanwhile, The correlation coefficient ρ_i between x_i and the rest $p-1$ variable indexes is

$$\rho_i^2 = r_i^T R_{-i}^{-1} r_i \quad i = 1, 2, \ldots, p \tag{5}$$

3 THE MODEL OF THE FUZZY PROBABILITY

We can get the fuzzy probability theory of fuzzy mathematics from the document ranked [5]; the main steps of evaluation model are as follows:

1. Establish the factors of evaluation $U = \{u_1, u_2, \ldots, u_n\}$ and the judge level set $V = \{v_1, v_2, \ldots, v_m\}$.
2. Make single-factor evaluation for every factor of evaluation u_i, Confirm the membership $v_j(u_i)$ based on the fuzzy relationship between u_j and v_j, then establish the fuzzy relation matrix, $Z = (z_{ij})_{m \times n}$, where the element z_{ij} is the membership named $v_j(u_i)$.
3. Evaluate the fuzzy weight π_i of the evaluation indexes u_i, namely look the weight as fuzzy number:

$$\pi_i = \frac{\beta_{i,1}}{w_{i,1}} + \cdots + \frac{\beta_{i,k}}{w_{i,k}} + \frac{1}{w_{i,0}} + \frac{\beta_{1,k+1}}{w_{i,k+1}} + \cdots + \frac{\beta_{i,2k}}{w_{i,2k}} \tag{6}$$

where the weight-values $w_{i,0}$ must be satisfied $\sum_{i=1}^{w} \lambda_{i,0} = 1$; the coefficient $\beta_{i,l}$ could take the one of the number 0.5, 0.6, 0.7, 0.8, 0.9; the value $w_{i,l}$ ($l = 0, 2 \ldots 2k$) is determined to the value of $w_{i,0}$.

4. By the theory of the fuzzy probability formula, namely:

$$(a_1\pi_1 + a_2\pi_2 + \cdots + a_n\pi_n)(P)$$
$$\triangleq \left(\bigvee_{\substack{a_1p_1 + \cdots + a_np_n = P \\ p_1 + \cdots p_n = 1}} (\pi_1(p_1) \wedge \cdots \wedge \pi_n(p_n)) \right) / u \tag{7}$$

where $\mu = \bigvee_{p_1 + \cdots p_n = 1} (\pi_1(p_1) \wedge \ldots \wedge \pi_n(p_n))$; π_i is the fuzzy weight of the evaluation index ranked u_i of i; p_i is the any value for the corresponding domain $\{\lambda_{i,1}, \ldots, \lambda_{i,k}, \lambda_{i,0}, \lambda_{i,k+1}, \ldots, \lambda_{i,2k}\}$; a_i is the membership of any evaluation ranked i, a_i is the membership $v_j(u_{ij})$ for the evaluation level ranked j.

5. Work out the fuzzy probability $P(v_j)$ of evolution level v_j, namely

$$P(v_j) = v_j(u_1)\pi_1 + v_j(u_2)\pi_2 + \cdots + v_j(u_n)\pi_n \tag{8}$$

6. We can obtain comprehensive evaluation by the principle of the information focus, the principle about information focus: if the fuzzy probability of the evolution level v_j is

$$P(v_j) = \frac{\alpha_{j1}}{x_{j1}} + \frac{\alpha_{j2}}{x_{j2}} + \cdots + \frac{\alpha_{jk}}{x_{jk}}, \quad \sigma_j = \sum_{i=1}^{k} \alpha_{jl} x_{jl},$$
$$j = 1, 2, \ldots, m \tag{9}$$

where α_{jl} is the degree of membership of y_{jl} ($l = 1, 2, \ldots, k$).

If the value of the judge level about $\bar{\alpha}_M$ and $\bar{\alpha}_N$ is required

$$\frac{\bar{\alpha}_N - \bar{\alpha}_M}{\bar{\alpha}_N} \leq 5\% \quad \bar{\alpha}_N = \max_{1 \leq j \leq m}[\bar{\alpha}_j] \tag{10}$$

Then the judge level may be set for $M - N$, or evaluation level may be set for N.

4 THE INVESTMENT EVALUATION EXAMPLE IN HYDRAULIC PROJECT

We select the economic factor, social factor and natural factors of these three aspects which include 12 factors are as investment project evaluation indexes from the document [4], Table 1 contains the original data. In this paper, according to the actual situation of each application unit, the demand situation of the hydropower engineering is ranked into {not need, haven't need, a little need, need, very need}. The rest of the evaluation index constitutes a new evaluation system, is shown in Figure 1 below: (just part of the data).

(Where A_1 is Proportion of water shortage (/%), A_2 is Immigration area ration (/%), A_3 is per capita income/(RMB/per), A_4 is statistics of food sale/(kg·per^{-1}), A_5 is cultivated land per capita/(hm^2/per), A_6 is adult illiteracy rate (/%), A_7 is sick bed rate/(bed/10^4), A_8 is slope farmland proportion (/%).

The rest of the evaluation index constitutes a new evaluation system, which is shown in Figure 2:

Combining with 10 experts' evaluation and scoring who have researched for many years in the field, getting the fuzzy relation matrix of food aid program of the voltage S1.

voltage	A_1	A_2	A_3	A_4	A_5	A_6	A_7	A_8
S_1	20.00	3.46	365.00	8.28	0.062	15.63	27.25	30.22
S_2	64.00	14.74	271.00	1.59	0.098	35.78	2.46	52.82
S_3	50.00	0	218.00	36.68	0.078	44.36	3.51	75.10
S_4	90.00	0	195.67	13.33	0.074	45.30	4.91	73.74
S_5	48.00	6.79	256.00	5.02	0.077	44.32	7.08	69.13

Figure 1. The index of food aid program.

378

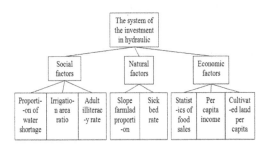

Figure 2. The investment evaluation system in hydraulic project.

$$\tilde{R}_{S1} = \begin{pmatrix} \tilde{R}^1 \\ \tilde{R}^2 \\ \tilde{R}^3 \end{pmatrix}$$

$$= \begin{bmatrix} 0.3 & 0.1 & 0.3 & 0.4 & 0.0 & 0.3 & 0.3 & 0.1 & 0.2 & 0.3 \\ 0.3 & 0.2 & 0.4 & 0.3 & 0.1 & 0.3 & 0.2 & 0.2 & 0.2 & 0.4 \\ 0.2 & 0.3 & 0.2 & 0.2 & 0.2 & 0.3 & 0.2 & 0.3 & 0.4 & 0.1 \\ 0.2 & 0.2 & 0.1 & 0.1 & 0.5 & 0.1 & 0.2 & 0.3 & 0.1 & 0.1 \\ 0.0 & 0.2 & 0.0 & 0.0 & 0.2 & 0.0 & 0.1 & 0.1 & 0.1 & 0.1 \end{bmatrix}^T$$

5 CONCLUSIONS

This article proposes a new decision method for hydraulic project investment based on the improved fuzzy mathematical method of the hydraulic project investment decision-making model. Through the validation of practical examples, the comprehensive evaluation value of this method is higher resolution, at the same time, it's not need too much mathematical operation, then can get the result of hydraulic project investment decisions. In addition, this method can also be applied to the safety evaluation of water resources, and the capacity evaluation of regional water resources, etc.

REFERENCES

[1] Z.J. Liu & L.Y. Ye, 2007, Fuzzy Probability Model and Its Application to Evaluation of Groundwater Quality, *Journal of Basic Science and Engineering 15 (3)*: 286–293.
[2] Z.J. Liu & L.Y. Ye, 2007, Evaluation Method for Water Resources Renewal Capacity Based on Fuzzy Probability Theory, *China Rural Water and Hydropower 7(2)*: 1–5.
[3] Y.H. Hu & S.H. He, 2000, *Comprehensive evaluation method*, Beijing: Science Press.
[4] G.X. Shi & Y.F. Zhang, 1999, Application of Fuzzy Comprehensive Evaluation in Water Conservancy Projects, *China Rural Water and Hydropower*, 12:14–16.
[5] L. Bao & J.S. Lei & Q. Liu, 2011, Synthetically Evaluation of Metro System Safety Based on Principal Component Analysis, *Journal of China Three Gorges University* (Natural science) 44(3):57–59.

Frontiers of Energy and Environmental Engineering – Sung, Kao & Chen (eds)
© 2013 Taylor & Francis Group, London, ISBN 978-0-415-66159-1

Distribution, enrichment, and accumulation of copper in sediments of Canon River estuary, Taiwan

C.D. Dong, C.F. Chen & C.W. Chen
Department of Marine Environmental Engineering, National Kaohsiung Marine University, Kaohsiung, Taiwan, China

M.S. Ko
Institute of Mineral Resources Engineering, National Taipei University of Technology, Taipei, Taiwan, China

ABSTRACT: This study was conducted using the data collected at the Canon River estuary, Taiwan to investigate and analyze Copper (Cu) contained in the sediments, and to evaluate the enrichment and accumulation of Cu. Results of laboratory analyses show that concentrations of Cu in the sediments are between 33 and 385 mg/kg with an average of 232 ± 117 mg/kg. The spatial distribution of Cu reveals is relatively high in the boundary of the river estuary region. This indicates that upstream industrial and municipal wastewater discharges along the river bank are major sources of pollution. Results from the enrichment factor and geo-accumulation index analyses imply that the sediments collected from the river estuary can be characterized between minor to severe degree enrichment and between none to moderately strong accumulation of Cu, respectively. Base on the comparison with sediment quality guidelines, the sediments Cu concentrations may cause acute biological damage. The results can provide regulatory valuable information to be referenced for developing future strategies to renovate and manage river estuary and harbor.

1 INTRODUCTION

Copper (Cu) is a common environmental contaminant; It is an essential trace element for the growth of most aquatic organisms however it becomes toxic to aquatic organisms at levels as low as 10 µg/g (Callender 2003). Therefore, much research effort has been directed toward the distribution of Cu in water environment. Anthropogenic activities including mining, smelting, incinerator emissions, domestic and industrial wastewaters, steam electrical production, and sewage sludge are the major source of Cu pollution (Callender 2003, Pertsemli & Voutsa 2007). Cu has low solubility in aqueous solution; it is easily adsorbed on waterborne suspended particles. After a series of natural processes, the water-borne Cu finally accumulates in the sediment, and the quantity of Cu contained in the sediment reflect the degree of pollution for the water body (Selvaraj et al. 2004).

The Canon River is approximately 5 km long, and drains a catchment of less than 11.3 km². The river flows through the downtown area of Kaohsiung City and is finally discharged into Kaohsiung Harbor (Fig. 1). Regions along Canon River have dense population with prosperous business and industrial establishments. The major pollution source includes domestic wastewater discharges, industrial wastewater discharges,

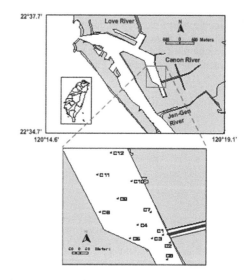

Figure 1. Map of the study area and sampling locations.

municipal surface runoff, and transportation pollution (Chen et al. 2007). All the pollutants will eventually be transported to the river mouth to deposit and accumulate in the bottom sediment. Results from recent investigation indicate that the Kaohsiung Harbor is heavily polluted, and the

Canon River is one of the major pollution sources (Chen et al. 2007). The river received untreated municipal and industrial wastewater discharges causing serious deterioration of the river water quality and the environmental quality near the river mouth to threaten the water environmental ecological system seriously. The objective of this study is to investigate the Cu distribution in the surface sediment near Canon River estuary so that the degree of Cu accumulation and potential ecological risk can be evaluated.

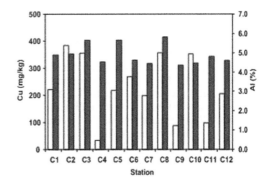

Figure 2. Distribution of Al and Cu contents in surface sediments of Canon River estuary.

2 MATERIALS AND METHODS

Surface sediment samples were collected at 12 stations near Canon River mouth (Fig. 1) with Ekman Dredge Grab aboard a fishing boat. The sampling stations, sample collection, and characteristics of the sediment (e.g. particle size and organic matter) have been reported in detail previously (Chen et al. 2012a). For Al and Cu analyses, the sediments were screened through 1 mm nylon net to remove particles with diameters larger than 1 mm. 0.5 g dry weight of the sediment sample was mixed with a mixture of ultra-pure acids (HNO_3:HCl:HF = 5:2:5), and was then heated to digest. The digested sample was filter through 0.45 μm filter paper; the filtrate was diluted with ultra-pure water to a pre-selected final volume. The Al and Cu contents were determined using a flame atomic absorption spectrophotometry (Hitachi Z-6100, Japan).

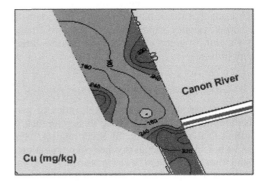

Figure 3. Contour map of surface sediments Cu contents in Canon River estuary.

3 RESULTS AND DISCUSSION

3.1 Distribution of Cu in sediments

Contents of Al in the sediment of Canon River estuary are between 4.35 and 5.8% with an average of 4.90 ± 0.52% (Fig. 2). All surface sediment samples collected at 12 monitoring stations studied contain 33–385 mg/kg of Cu with an average of 232 ± 117 mg/kg (Fig. 3). Concentration distributions of Cu in the Canon River estuary sediment shown in Figure 3 reveal that the sediment Cu content is relatively higher in the boundary of the river estuary. These observations indicate that the upstream pollutants brought over by the river are the major sources of estuary Cu pollution. The Canon River receives a great amount of industrial and domestic Cu from Kaohsiung city because about 44% domestic wastewater is discharged directly without adequate treatment. Moreover, several industrial plants discharge industrial wastewater effluents into the tributaries in or adjacent to Kaohsiung city (Chen et al. 2007), and the pollutants are transported by river flow and finally accumulate near the river estuary. Some pollutants may drift with sea current and are dispersed into open sea (Chen et al. 2007, Chen et al. 2012b).

3.2 Enrichment factor

The Enrich Factor (EF) is a useful tool for differentiating the man-made and natural sources of metal enrichment (Chen et al. 2007, Chen et al. 2012b). This evaluating technique is carried out by normalizing the metal concentration based on geological characteristics of sediment. Aluminum is a major metallic element found in the earth crust; its concentration is somewhat high in sediments and is not affected by man-made factors. Thus, Al has been widely used for normalizing the metal concentration in sediments (Chen et al. 2007, Chen et al. 2012b). EF is defined as: $EF = (X/Al)_{sediment}/(X/Al)_{crust}$, where (X/Al) is the ratio of Cu to Al. The average Cu and Al content in the earth crust were 55 mg/kg and 8.23% that excerpted from the data published by Taylor (1964). When the EF of a metal is greater than 1, the metal in the

381

sediment originates from man-made activities, and vice versa. The EF value can be classified into 7 categories (Chen et al. 2007): 1, no enrichment for EF < 1; 2, minor for 1 < EF < 3; 3, moderate for $3 \leq EF < 5$; 4, moderately severe for $5 \leq EF < 10$; 5, severe for $10 \leq EF < 25$; 6, very severe for $25 \leq EF < 50$; and 7, extremely severe for $EF \geq 50$.

Table 1 show EF values of the sediment Cu for the Canon River estuary region; the Cu concentration is consistent with the Cu EF value for all sampling stations, and all EF values are greater than 1. This indicates that the sediment Cu has enrichment phenomenon with respect to the earth crust and that all Cu originates from man-made sources. Station C4 is classified as minor enrichment, Stations C9 and C11 are classified as moderate enrichment, and the other stations are classified as either moderately severe or severe enrichment. These results point out that the sediment near the boundary of the river estuary experiences severe enrichment of Cu that originates from the upstream sources of pollution.

3.3 Geo-accumulation index

Similar to metal enrichment factor, index of geo-accumulation (I_{geo}) can be used as a reference to estimate the extent of metal accumulation. The I_{geo} values for the metals studied were calculated using the Müller's (1979) expression: $I_{geo} = log_2$ ($Cn/1.5Bn$), where Cn is the measured content of element Cu, and Bn is the background content of Cu 55 mg/kg, in the average shale. Factor 1.5 is the background matrix correction factor due to lithogenic effects. The I_{geo} value can be classified into 7 classes: 0, none for $I_{geo} < 0$; 1, none to medium for $I_{geo} = 0–1$; 2, moderate for $I_{geo} = 1–2$; 3, moderately strong for $I_{geo} = 2–3$; 4, strong for

$I_{geo} = 3–4$; 5, strong to very strong for $I_{geo} = 4–5$; and 6, very strong for $I_{geo} > 5$. Based on the I_{geo} data and Müller's (1979) geo-accumulation indexes, the accumulation levels with respect to Cu at each station are ranked in Table 2. Stations C4, C9, and C11 are classified as either none or none to medium accumulation, Stations C2, C3, C8, and C10 are classified as moderately strong accumulation, and the other stations are classified as moderate accumulation.

3.4 Comparison with sediment quality guidelines

One of the widely used sediment toxicity screening guideline of the US National Oceanic and Atmospheric Administration provides two target values to estimate potential biological effects: Effects Range Low (ERL) and Effect Range Median (ERM) (Long et al. 1995). The guideline was developed by comparing various sediment toxicity responses of marine organisms or communities with observed metals concentrations in sediments. These two values delineate three concentration ranges for each particular chemical. When the concentration is below the ERL, it indicates that the biological effect is rare. If concentration equals to or greater than the ERL but below the ERM, it indicates that a biological effect would occur occasionally. Concentrations at or above the ERM indicate that a negative biological effect would frequently occur. Figure 4 shows the measured concentrations of Cu in comparison with the ERM and ERL values. Among the 12 sediment samples collected, the Cu is between ERL (34 mg/kg) and ERM (270 mg/kg) in 7 samples (58%), and 4 samples (33%) are exceed ERM for Cu. This indicates that the sediment concentrations of Cu may cause adverse impact on aquatic lives.

Table 1. Enrichment Factor (EF) of Cd for each station studied at Jen-Gen River estuary.

Site	EF value	EF class	EF level
C1	6.7	4	Moderately severe
C2	11.7	5	Severe
C3	9.4	4	Moderately severe
C4	1.1	2	Minor
C5	5.8	4	Moderately severe
C6	8.7	4	Moderately severe
C7	6.7	4	Moderately severe
C8	9.2	4	Moderately severe
C9	3.0	3	Moderate
C10	11.8	5	Severe
C11	3.1	3	Moderate
C12	6.7	4	Moderately severe
Mean	7.1	4	Moderately severe

Table 2. Geo-accumulation index (I_{geo}) of Cd for each station studied at Jen-Gen River estuary.

Site	I_{geo} value	I_{geo} class	I_{geo} level
C1	1.4	2	Moderate
C2	2.2	3	Moderately strong
C3	2.1	3	Moderately strong
C4	−1.3	0	None
C5	1.4	2	Moderate
C6	1.7	2	Moderate
C7	1.3	2	Moderate
C8	2.1	3	Moderately strong
C9	0.1	1	None to medium
C10	2.1	3	Moderately strong
C11	0.3	1	None to medium
C12	1.3	2	Moderate
Mean	1.5	2	Moderate

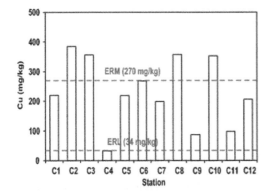

Figure 4. Distribution of Cu contents in surface sediments of Canon River estuary. Lines mark the levels or risk considered [9]. ERL (Effects Range Low) and ERM (Effects Range Median).

4 CONCLUSIONS

The sediment samples collected at Canon River estuary contain 33–385 mg/kg of Cu with an average of 232 ± 117 mg/kg. The distribution of Cu in sediment reveals that the Cu originates from the river upstream discharges of industrial and domestic wastewaters; it is transported along the river and finally deposited and accumulated near the river estuary. Results of EF analysis indicate that the Canon River estuary sediments were minor to severe enrichment with Cu. Results of I_{geo} analysis show that the Canon River estuary sediments were none to moderately strong accumulation with Cu. Base on the comparison with SQGs, the sediments Cu concentrations may cause acute biological damage. The results can provide regulatory valuable information to be referenced for developing future strategies to renovate and manage river estuary and harbor.

REFERENCES

Callender, E. 2003. Heavy metals in the environment–historical trends. In H.D. Holland & K.K. Turekian (eds), Treatise on Geochemistry: 67–105. New York: Elsevier.

Chen, C.W. Chen, C.F. & Dong, C.D. 2012b. Evaluation of zinc contamination in the sediments of Canon River mouth, Taiwan. *Adv. Mater. Res.* 468–471: 1767–1770.

Chen, C.W. Chen, C.F. Dong, C.D. & Tu, Y.T. 2012a. Composition and source apportionment of PAHs in sediments at river mouths and channel in Kaohsiung Harbor, Taiwan. *J. Environ. Monit.* 14: 105–115.

Chen, C.W. Kao, C.M. Chen, C.F. & Dong, C.D. 2007. Distribution and accumulation of heavy metals in the sediments of Kaohsiung Harbor, Taiwan. *Chemosphere* 66: 1431–1440.

Long, E.R. Macdonald, D.D. Smith S.L. & Calder F.D. 1995. Incidence of adverse biological effects within ranges of chemical concentrations in marine and estuarine sediments. *Environ. Manage.* 19: 81–97.

Müller, G. 1979. Schwermetalle in den sediments des Rheins-Veranderungen seitt 1971. *Umschan* 79: 778–783.

Pertsemli, E. & Voutsa, D. 2007. Distribution of heavy metals in Lakes Doirani and Kerkini, Northern Greece. *J. Hazard, Mater.* 148: 529–537.

Selvaraj, K. Ram-Mohan, V. & Szefer, P. 2004. Evaluation of metal contamination in coastal sediments of the Bay of Bengal, India: geochemical and statistical approaches. *Mar. Pollut. Bull.* 49: 174–185.

Taylor, S.R. 1964. Abundance of chemical elements in the continental crust: a new table. *Geochem. Cosmochim. Acta* 28: 1273–1285.

Frontiers of Energy and Environmental Engineering – Sung, Kao & Chen (eds)
© *2013 Taylor & Francis Group, London, ISBN 978-0-415-66159-1*

The research on benefits sharing model of efficiency power plant

J.M. Wang & G.J. Liu
School of Business Administration, North China Electric Power University, Baoding, Hebei, P.R. China

ABSTRACT: The development of Efficiency Power Plant is an important measure to achieve the emission reduction targets in the twelfth "Five-Year Plan". But, as a new energy-saving mode, many troubles have to be dealt with. Benefit sharing is a key premise to build EPP for a business. Paper synthesized all aspects of risk factors in the implementation of EPP and constructed benefits sharing model, and proved its feasibility and accuracy through empirical analysis.

Keywords: efficiency power plant; risk; benefits sharing model

1 INTRODUCTION

The emission reduction is a breakthrough in optimizing the economic structure and upgrade the production structure. What's more, it's the important content of China's the twelfth "Five-Year Plan". According to the actual situation of China at this stage, the implementation and development of Efficiency Power Plant is an important measure to achieve emission reduction targets. Efficiency Power Plant or EPP is virtual power plant. To implement a package of energy-saving measures encourage users to adopt energy-efficient electrical equipment and improve energy efficiency in an area. Reaching the same purpose with the new or expanded electric power supply system. compared with conventional power plants. It does not burn fuel, zero pollution emissions, as well as power generation and low cost, etc. The role in promoting its construction and social development is significantly. On the one hand, it can promote energy conservation, improve energy efficiency, low carbon economy and green sustainable development. On the other hand, it can reduce costs, help increase the competitiveness of enterprises.

2 THE IMPLEMENTATION RISK OF EPP PROJECT

The Efficiency Power Plant as a new energy-saving mode, it is not yet ripe for development in China. There are many barriers to implement related to the concept of legal, policy and institutional constraints.

1. Electricity consumers lack awareness of energy-saving technologies and equipment.

2. The cost of financing is very high. In addition to the interest and commitment fees, the project also charged to the energy audit, energy-saving effect of detection and confirmation fees, loan guarantees, fees, financial institutions, fees and project office management costs.

3. China's energy prices and the tax didn't reflect the energy used to internalize external costs. For most power users, only electricity costs enough significant impact on its balance of payments will be interest in such energy-saving projects involved in the EPP. In addition to the TOU price and the difference in price, China's current tariff level and structure does not encourage energy efficiency. It is not conducive to fully mobilize the power users to generate enthusiasm and interest in energy efficiency investments.

4. The benefits of energy-saving is the whole of society enjoy, it is an external income.

2.1 Risk identification

As the results of the risk is expected to have a lot of uncertainties exist that it is difficult to achieve precise quantification, in order to increase the rationality and applicability of the indicators, paper take the fuzzy mathematics membership assignment method for evaluation.

In order to calculated logic and thinking clearly, the existing risk of 17 to be divided into five kinds, they are policy risk, financial risk, operational risk, benefit risk and market risk. Various indicators of risk classification table below.

2.2 The determination of the risk coefficient

The risk coefficient of policy risk, financial risk, operational risk, benefit risk and market risk

Table 1. Risk index system table.

u_1 policy risk	u_{11} The complexity of the examination and approval procedures
	u_{12} Supporting policies, the law is imperfect
	u_{13} Organizational system is imperfect
u_2 financial risk	u_{21} Financing costs
	u_{22} Energy prices
	u_{23} Energy tax
	u_{24} Difficulties in funding lending
u_3 operational risk	u_{31} engineering construction
	u_{32} The turnover of project funds
	u_{33} Project operation and maintenance
	u_{34} Force majeure caused by failure
u_4 benefit risk	u_{41} The desired effect can be achieved
	u_{42} Preliminary analysis and diagnosis of the energy wasted
u_5 market risk	u_{51} Information is not smooth
	u_{52} Advanced technology
	u_{53} Changes in cost of major equipment production

abbreviated as follows: R_n, R_f, R_q, R_b, R_m. Using the fuzzy comprehensive evaluation method to measure the risk of each risk factor values.

1. Divided Rank

   Using membership assignment method of fuzzy mathematical evaluation to assess and measure each risk. To determine the judge set{v_1 (Low-risk), v_2 (Lower risk), v_3 (Medium risk), v_4 (Higher risk), v_5 (High-risk)}. Using expert scoring method to determine its membership in the evaluation rating, endow with evaluation set each element to vector-valued $V = \{0.1, 0.3, 0.5, 0.7, 0.9\}$, to show the judge to focus on the size relationship between the corresponding values of various elements and project risk.

2. AHP (Analytic Hierarchy Process) method to measure the weighing

   Hierarchical processing of the evaluation factors, there into first factors including: {policy risk, financial risk, operational risk, benefit risk, market risk} five major categories, denoted by $U = \{u_1, u_2, u_3, u_4, u_5\}$, its weights $W = \{w_1, w_2, w_3, w_4, w_5\}$. Through the AHP to determine the weights of each indicator, the key is to accurately determine the relative degree of importance from top to bottom at all levels of risk indicators. Weights to determine generally start with the contract energy management features to determine the pairwise comparison matrix, then use the mathematical methods to calculate the weights, finally judge the consistency test for the weights.

3. The calculation of the risk factor

   The weights of each level indicators can be drawn based on the AHP, starting from a single factor

to judge to determine the degree of membership of the evaluation set. Analysis the operational risk, build operational risk evaluation matrix B_3. To set up the evaluation factors u_3 which is the i factor u_{3i}, define the membership degree b_{ij} of evaluation set V which is the j factor v_j, then the evaluation results of u_{3j} can represent by fuzzy set $B_{3j} = (b_{j1}, b_{j2}, b_{j3}, b_{j4}, b_{j5})$.

B_{3j} is the single factor evaluation matrix which is single factor judge set $B_3 = (B_{31}, B_{32}, B_{33}, B_{34}, B_{35})^T$ of operational risk u_3, through the matrix B_3 to achieve fuzzy transformation from the factor set u_3 and evaluation set V. Using expert scoring method to the degree of risk to each index ($u_{31}, u_{32}, u_{33}, u_{34}$), according to evaluation sets $V = \{v_1, v_2, v_3, v_4, v_5\}$ to vote, statistics the n experts in index j for operational risk u_3. When there are m_{ij} experts choose index j that the probability of this index is $b_{ij} = m_{ij}/n$. Their selection results for probability and statistics can get a row matrix $B_{3i} = (b_{i1}, b_{i2}, b_{i3}, b_{i4}, b_{i5}) = (m_{i1}/n, m_{i2}/n, m_{i3}/n, m_{i4}/n, m_{i5}/n)$ to the i indicator evaluation of u_3, There into. $\sum_{j=1}^{n} b_{ij} = 1$ Operational risk evaluation matrix is

$$B_3 = \begin{pmatrix} B_{31} \\ B_{32} \\ B_{33} \\ B_{34} \end{pmatrix} = \begin{pmatrix} b_{11} & b_{12} & b_{13} & b_{14} \\ b_{21} & b_{22} & b_{23} & b_{24} \\ b_{31} & b_{32} & b_{33} & b_{34} \\ b_{41} & b_{42} & b_{43} & b_{44} \end{pmatrix}$$

to u_3: $u_3 = w_3^T \cdot B_3 = \{b_1, b_2, b_3, b_4, b_5\}$, normalization processing: $b_i' = b_i / \sum_{j=1}^{5} b_j (i = 1, 2, \ldots 5)$

Then use vector $u_3' = (b_1', b_2', b_3', b_4', b_5')$ to represent the expert group on operational risk indicators by the results of the fuzzy transformation in the judgment set V. At last the operational risk factor of contract energy management is $R_q = u_3 \cdot V^T$.

In a similar way, can be obtained policy risk factor R_n, financial risk factor R_f, Benefit risk factor R_b, market risk factor R_m. The total risk factor $R = 1 - (1 - R_p)(1 - R_q)(1 - R_f)(1 - R_b)(1 - R_m)$.

3 BENEFITS SHARING MODEL

Benefit sharing is to divide maintenance costs spent on costs as well as its risk-based project construction and during operation, its relation is positive correlation between benefits and risks. Defined the cost is the key, include two side. (The cost of energy-saving products and related equipment, project construction and maintenance costs incurred in the process can denote by C.)

3.1 Determine the benefit sharing amount

According to the principle which is benefits and risks of peer-to-peer, determine costs C and total risk factor R, the on the basis of $I = C \times (1 + R)$ calculate a real return that the energy service companies should get.

3.2 Sharing period

There are many sides factors when determine the sharing period. Not circumvent the risk of accidents make the input costs increase when in the project implementation process. The impact of the economic environment for investment income will produce varying degrees impact to sharing period, that make the sharing period to put out the appropriate changes. But these all belong to not easy grasp factors, not easy quantitative estimates in the theoretical study, so appear a variety of different methods to measure. Here refer to document [5] as a reference, using the way to calculate.

Assuming unit time that can take measurements and calculate the energy efficiency of the unit time period, so it through formula $I \cong \sum_{j=0}^{S-1} m \times (1+i)^j$ calculate the total time units s, there into t is unit time, m is the energy efficiency of the unit time period, i is discount rate (using bank lending rates).

Working-out S, EPP benefit-sharing period is: $T = S \times t$.

4 EMPIRICAL ANALYSIS

Supposing company R want to implement alteration of EPP, total investment is 26,300 yuan, the entire project to save the load is 77.7 kW, integrated energy-saving rate is 54.8%, saving electricity per year is 311,000 kWh (one year run-time meter by 4,000 hours), saving energy charge per year is 201,700 yuan (electrovalency is 0.65 yuan/kWh). In accordance with the model estimates that the total risk factor is 0.6, based on $I = C \times (1 + R)$ can draw the efficiency and share amount of the energy effectiveness company is $26.3 \times 1.6 = 42.08$. According to $I \cong \sum_{j=0}^{S-1} m \times (1 + i)^j$, using Matlab to

run basis, the result is $j = 374$, so benefit-sharing period is $T = S = 374$ d, recovery period of investment approximately equal 1.3 year (Here is not considering to reduce equipment maintenance costs).

5 CONCLUSION

EPP is effective way for implementation of energy saving in China, is a new energy-saving mode, its construction and implementation have a lot of questions need to be resolved. To determine the sharing of energy efficiency and earnings recovery period of EPP is the key link to ensure the smooth progress of the EPP construction. This text constructs Sharing Model to consider various factors, so the model is more reasonable, fuzzy comprehensive evaluation and AHP level analysis combined that making it more operational and higher calculation accuracy.

REFERENCES

[1] Gao Xiang-feng. Efficiency Power Plant to promote the bottleneck analysis and counter measures. China Electric Power Education. 2008, (S2):97–99.
[2] Qin Feng-hua. Efficiency Power Plant with the general promotion conditions—Fang Junshi, head of Policy Unit in National Energy Leading Group Office, analysis Efficiency Power Plant prospects. China Investment. 2007,(7): 60–63.
[3] Gao Xiang-feng. The Study of Efficiency Power Plant Theory and Implement Scheme. North China Electric Power University. 2009.
[4] Liu De-jun & LV Lin. Evaluation of risk and benefit in energy management contract project. Power Demand Side Management. 2009, 11(1): 21–22.
[5] Li Jing & He Jiang-bo & Wu Xi-ping. The energy-saving reform of HVAC system of a hotel. Power Demand Side Managemen. 2007, (1):41–43.

Frontiers of Energy and Environmental Engineering – Sung, Kao & Chen (eds)
© 2013 Taylor & Francis Group, London, ISBN 978-0-415-66159-1

Characteristics and CO_2 sequestration efficiency of mafic tailings distributed in five regions of China

L.J. Zhang, X.Y. Yuan & X.Y. Wang
College of Environment, Hohai University, Nanjing, China

L.W. Liu & J.F. Ji
Key Lab of Surficial Geochemistry, Ministry of Education, Nanjing, China

ABSTRACT: The characteristics of mafic tailings in the Donghai, Rizhao, Chicheng, Chaoyang and Chengde regions of China have been researched in this paper. We also assess the carbon sequestration efficiency of these mafic tailings by analysing the chemical and mineral compositions of these samples and their specific surface areas. The results show that there are significant differences in chemical and mineral compositions in five regions. Serpentine and clinochlore are the major minerals in Donghai and Rizhao whereas magnesium amphibole and magnetite are dominated in the Chicheng, Chaoyang and Chengde. The MgO components in Donghai and Rizhao tailings are higher compared with those other regions. There are no significant differences in the specific surface areas of tailings from five regions. The chemical and mineral compositions of Donghai tailings show the greater carbon sequestration efficiency, whereas the specific surface area of Rizhao tailings reveals a better carbon sequestration potential. The carbon sequestration efficiencies of tailings in five regions follow the order of Donghai > Rizhao > Chaoyang > Chicheng > Chengde based on a comprehensive assessment.

Keywords: mafic tailings; chemical and mineral characteristics; specific surface area; carbon sequestration efficiency; eastern China

1 GENERAL INSTRUCTIONS

Tailings generated from the mining industry have brought about a lot of environmental problems such as land occupying, soil loss and water pollution (Zhang et al. 2006). Mafic tailings are generated by exploiting and using deposits in ultrabasic and basic rocks. These tailings have high Ca and Mg contents, can react with CO_2 to form carbonate minerals. But there are few researches on using mafic tailings to store CO_2 in China at present. This paper will mainly discuss the efficiencies of sequestrating CO_2 by comparing three characteristics (chemical composition, mineral composition and specific surface area) of tailings from five regions in eastern China.

2 MATERIALS AND METHODS

2.1 Sample collection

The samples were collected in five areas including Donghai, Rizhao, Chicheng, Chaoyang and Chengde (Fig. 1), and 11, 8, 13, 8 and 6 samples were obtained in these regions respectively. The samples

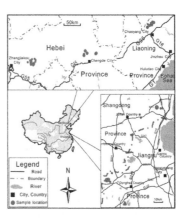

Figure 1. Location and sample distribution map of the research areas.

have a weight about 1.5 kg. The detrital materials were removed before bagging for analysis.

2.2 Experimental methods

All the samples are air dried and screened through 20 mesh sieve before experimental determining.

Major components were determined by X-ray Fluorescence spectrometer (XRF, ARL9800XP, USA) after sample digestion. Mineral compositions were determined by D/Max-B type X-ray Diffraction (XRD), with Cu target powder pressing piece, 35 KV voltage, 20 mA current, scanning speed (2 θ) 2 °/min. 10 ml 10% H_2O_2 and 10 ml 10% of hydrochloric acid were added in beaker with 1 g tailing, then 0.05 N sodium hexametaphosphate was added as dispersant. These samples were tested on the Mastersizer-2000 type laser particle size analyzer for size analysis.

The half quantitative analysis of minerals is achieved by Jade5.0 software. Data statistics and plotting are processed by Excel 2007.

3 RESULT AND DISCUSSION

3.1 *The chemical composition of the tailings*

The chemical compositions of the tailings from 5 regions have showed in Table 1. There are significant differences in these chemical compositions such as Al_2O_3, CaO, Fe_2O_3 and MgO whereas the contents of MnO and SiO_2 are steady. Compared with Chicheng, Chaoyang and Chengde tailings, Donghai and Rizhao tailings have obviously higher MgO (>30%), and lower CaO, Al_2O_3, Fe_2O_3 and Na_2O contents. On the other hand, LOI of these tailings are high as a result of serpentinization, choritization and other strong alterations.

The main material acted with CO_2 is $(Mg,Ca)_x$ $Si_yO_xH_{2z}$ in mafic tailings. Therefore, it is reasonable to utilize the Mg and Ca contents for evaluating

the sequestration efficiency. The general reaction is shown in Equation 1.

$$(Mg,Ca)_xSi_yO_{x+2y+z}H_{2z}(s) + xCO_2(g)$$
$$\rightarrow x(Mg,Ca)CO_3(s) + ySiO_2(s) + zH_2O \qquad (1)$$

The m/f value of samples collected from Donghai region is 7.63, the m/f value of samples collected from Rizhao region range from 2.93 to 6.88 with the mean value of 5.26. The m/f value of samples collected from Chicheng, Chaoyang and Chengde regions are 1.10–1.58. Bigger m/f values indicate much magnesium in tailings, which is benefit to the rate of CO_2 storage (Wu 1963).

3.2 *Mineral components of tailings*

We can observe data from Table 1 that w (MgO)/ w (SiO_2) values of tailings in Donghai and Rizhao are 0.81–1.12 and 0.33–1.07 respectively. The wide ranges of variation indicate that pyroxene-olive and olivine-pyroxenite are the dominant components in original rock (Qiu & Lin 1991).

According to Table 2, serpentine, magnetite, chlorite and phlogopite are major minerals in these samples. The mineral species listed increase from bottom to up in Table 2. The availability of Ca and Mg for antigorite, achromaite, phlogopite, kupfferite, clinochlore, magnesioferrite, chlorite are 26.09%, 23.41%, 17.31%, 14.08%, 12.44%, 12%, 11% respectively, which are effective for CO_2 sequestration efficiency.

$Mg(OH)_2$ and Mg not combined with Si in the mineral structure can react with CO_2 whereas Mg in the lattice of serpentine and silicon-rich layer

Table 1. Chemical compositions of tailings in 5 regions [%].

	Donghai		Rizhao		Chicheng		Chaoyang		Chengde	
	Range	Mean	Range	Mean	Range	Mean	Range	Mean	Range	Mean
Al_2O_3	0.53–3.70	1.99	1.14–4.93	2.11	2.88–10.94	6.14	7.32–17.02	12.79	7.17–10.29	8.96
Fe_2O_3	6.56–9.48	7.77	8.83–13.16	10.28	10.75–26.85	17.11	9.95–17.22	14.38	15.5–18.36	17.08
K_2O	0.02–0.59	0.14	0.03–1.31	0.36	0.03–2.57	0.68	0.83–1.81	1.33	0.41–1.07	0.70
MgO	32.7–39.72	34.99	31.46–38.7	31.46	8.47–18.93	13.97	6.78–15.28	9.49	9.25–13.43	11.06
MnO	0.08–0.11	0.09	0.14–0.19	0.14	0.09–0.21	0.14	0.12–0.20	0.16	0.16–0.17	0.17
Na_2O	0–0.37	0.07	0.71–2.82	0.71	0.04–3.14	0.71	0.81–3.16	2.04	0.24–1.75	1.01
CaO	0.64–6.54	2.8	2.53–7.15	2.54	10.23–19.57	16.81	4.05–9.24	7.59	14.16–20.8	16.83
P_2O_5	0.02–0.10	0.03	0.10–0.30	0.10	0.06–5.05	1.43	0.37–0.76	0.54	0.48–4.16	2.1
SiO_2	34.1–40.53	37.35	40.8–52.23	40.83	32.9–44.99	40.19	40.5–55.48	47.18	34.88–41.8	38.01
TiO_2	0.07–0.30	0.13	0.34–1.06	0.34	0.44–44.99	1.33	0.63–2.16	1.31	1.16–2.54	1.71
LOI	12.4–17.68	14.82	3.31–15.15	11.36	0.99–2.47	2.30	2.21–7.39	2.88	1.73–5.49	2.94
m/f	5.87–8.75	7.63	2.93–6.88	5.26	0.53–2.93	1.58	0.72–2.56	1.21	0.86–1.45	1.10
w (MgO + CaO)	33.3–46.26	37.79	33.99–45.8	34.00	18.7–38.5	30.78	10.83–24.5	17.08	23.41–34.2	27.89
w (MgO)/ w (SiO_2)	0.81–1.12	0.94	0.33–1.07	0.82	0.21–0.47	0.35	0.15–0.33	0.20	0.25–0.32	0.29

Table 2. Results of X-ray diffraction analysis for tailing minerals.

Region	Mineral	Molecular formula
Donghai	Antigorite	$Mg_{3-x}(Si_2O_5)(OH)_{4-2x}$
	Kupfferite	$(Ca,Na)_{2.26}(Mg,Fe,Al)_{5.15}(Si,Al)_8O_{22}(OH)_2$
	Clinochlore	$(Mg,Fe,Al)_6(Si,Al)_4O_{10}(OH)_8$
	Achromaite	$NaCa_2Mg_5Si_7AlO_{22}F_2$
	Omphacite	$NaCaMgAl(Si_2O_6)_2$
Rizhao	Antigorite	$Mg_{3-x}(Si_2O_5)(OH)_{4-2x}$
	Clinochlore	$(Mg,Fe,Al)_6(Si,Al)_4O_{10}(OH)_8$
	Kupfferite	$(Ca,Na)_{2.26}(Mg,Fe,Al)_{5.15}(Si,Al)_8O_{22}(OH)_2$
	Manganese chlorite	$(Mn,Mg,Fe)_6Si_4O_{10}(OH)_8$
Chicheng	Kupfferite	$(Na,Ca)_2(Mg,Fe)_5Si_8O_{22}(OH)_2$
	Magnetite	$Fe^{+2}Fe_2^{+3}O_4$
	Phlogopite	$KMg_3(Si_3Al)O_{10}(OH)_2$
	Riebeckite	$(Na,Ca)_2(Fe,Mn)_3Fe_2(Si,Al)_8O_{22}(OH,F)_2$
	Clinochlore	$(Mg,Fe,Al)_6(Si,Al)_4O_{10}(OH)_8$
Chaoyang	Kupfferite	$(Na,Ca)_2(Mg,Fe)_5Si_8O_{22}(OH)_2$
	Magnetite	$Fe^{+2}Fe_2^{+3}O_4$
	Achromaite	$NaCa_2Mg_5Si_7AlO_{22}F_2$
	Phlogopite	$KMg_3(Si_3Al)O_{10}(OH)_2$
	Magnesioferrite	$MgFe_2^{+3}O_4$
Chengde	Kupfferite	$(Na,Ca)_2(Mg,Fe)_5Si_8O_{22}(OH)_2$
	Magnetite	$Fe^{+2}Fe_2^{+3}O_4$
	Achromaite	$NaCa_2Mg_5Si_7AlO_{22}F_2$
	Riebeckite	$(Na,Ca)_2(Fe,Mn)_3Fe_2(Si,Al)_8O_{22}(OH,F)_2$
	Clinochlore	$(Mg,Fe,Al)_6(Si,Al)_4O_{10}(OH)_8$

can't take the reaction (Larachi et al. 2010). So Mg not bound with Si in mineral is easy to react with CO_2. From the molecular formula of the minerals we can realize that Mg in the antigorite and clinochlore is apt to release. Compared to achromaite and pyroxene, Mg and Ca in the kupfferite are easily to release, or it is easier to react with CO_2. In Donghai and Rizhao regions, antigorite, kupfferite and clinochlore are dominated in the tailings and the content of magnetite is quite high in the samples collected from Chicheng, Chaoyang and Chengde regions. As a result, the reaction efficiency with CO_2 and the availability of the tailings in latter three regions are low.

The dissolution of magniferous minerals are mainly influenced by the pH. The chemical bonds between Mg-O-Si of minerals will break down in acidic conditions and release Mg atoms (Larachi et al. 2010, Teir et al. 2007). At 70°C, some acids (H_2SO_4, HNO_3 and HCOOH) are able to leach 100% of magnesium in serpentinite in 2 h, and 94% of magnesium can be transformed to carbonates (Teir et al. 2009, Bobicki 2012). In this way, the CO_2 sequestration efficiency can increase in mafic tailings with loose Mg atom bonds. In aspect of Mg releasing, potentially, the CO_2 storage efficiency of tailings in Donghai and Rizhao areas are highest in natural conditions.

3.3 Grain size and specific surface area of tailings

Due to the direct relationship between the particulate size and the specific surface area, Sverdrup has proposed a specific surface area formula (Eq.2) from which we can obtain the specific surface area of tailing samples (Sverdeup 1996).

$$Aw = (0.08X_{clay} + 0.022X_{silt} + 0.003X_{fine\ sand} + 0.0005X_{coarse\ sand})$$ (2)

Aw is specific surface area (m^2/g), X is the mass percentage of clay, silt, fine sand and coarse sand, showing $X_{clay} + X_{silt} + X_{fine\ sand} + X_{coarse\ sand} = 100$. The sizes of clay, silt, fine sand and coarse sand are <2 μm, 2–60 μm, 60–250 μm, >250 μm respectively.

The tailing samples collected from Rizhao have the largest mean specific surface area (Fig. 2). The mean specific surface area of samples from Chaoyang, Donghai, Chengde, Chicheng areas decrease in sequence. Mineral weathering leads to the difference in particle size of mineral and is associated with conditions like geographic location, amount of rainfall, wind strength and temperature (Caillaud 2009). The weather of Donghai and Rizhao is wetter than other regions, which benefits to mineral weathering. These tailings from

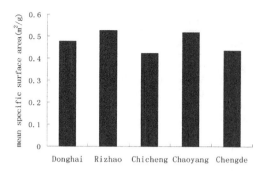

Figure 2. The average specific surface area of tailings in the five regions.

Chicheng, Chaoyang and Chengde regions have a lower availability.

In addition, the sufficient reaction rate is associated with specific surface area. Although sulfuric acid is considered to be the most efficient chemical activator which can increase the superficial area from 8 m^2/g to 330 m^2/g or even more, it's increase will depend on specific surface area of mineral (Maroto-valer et al. 2005). Therefore we could estimate the CO_2 sequestration efficiency by comparing specific surface area of the tailing, which decreases with Rizhao > Chaoyang > Donghai > Chengde > Chicheng.

the two regions show relatively high mean specific surface area.

B. Garcia proves that mineral surface area has a good linear relationship with mineral crystallization process and magnesium carbonate formation (Pronost et al. 2011, Garcia 2011). The larger specific surface area displays the faster reaction rate between tailings and CO_2. We can draw a conclusion that tailings in the Rizhao and Chaoyang regions have the faster carbon sequestration rate than other regions from the analysis above.

3.4 *Analysis of efficiency of carbon sequestration*

In carbon sequestration reaction, CO_2 reacts with MgO and CaO in the tailings directly, so the contents of MgO and CaO are very important in the process. Some researches indicate that the absorptivity of CO_2 rise with MgO increase in tailings under that same reaction condition (Larachi et al. 2010). When the concentration of MgO reaches at 40%, the absorbtivity of CO_2 can even increase from 0.5% to 60% (Information online).

The contents of MgO in tailings of Donghai and Rizhao regions are 34.99% and 31.46% respectively whereas they are lower in other regions. The total contents of MgO and CaO in Donghai, Rizhao areas are also bigger (37.79% and 34.00%) compared to other areas. In the view of total reaction rate of MgO and CaO, the carbon sequestration efficiency decreases with Donghai > Rizhao > Chicheng > Chengde > Chaoyang.

The mineral component of the tailings is the important factor for the carbon sequestration efficiency. Serpentine and chlorite containing more Mg and Ca are the main materials of carbon sequestration reaction (Vogeli et al. 2011). The major minerals in tailings of Donghai and Rizhao regions are serpentine, kupfferite and clinochlore, Mg and Ca is easy to release from these minerals. The utilizability of tailings with these minerals is high. Due to the less Mg contents, tailings in

4 CONCLUSIONS

Trailings from Donghai, Rizhao, Chicheng, Chaoyang and Chengde regions have different geochemical characteristics, which influence the CO_2 sequestration efficiency.

MgO and CaO contents of tailings are more important compared to mineral component and specific surface area. They will significantly increase or decrease the reaction rate between CO_2 and tailings.

Based on a comprehensive assessment, the carbon sequestration efficiency in tailings of these five regions ranks in order of Donghai > Rizhao > Chaoyang > Chicheng > Chengde.

REFERENCES

Bobicki, E.R. Liu, Q. Xu, Z. et al. 2012. Carbon capture and storage using alkaline industrial wastes. *Progress in Energy and Combustion Science* 38: 302–320.

Caillaud, J. Proust, D. Philippe, S. et al. 2009. Trace metals distribution from a serpentinite weathering at the scales of the weathering profile and its related weathering microsystems and clay minerals. *Geoderma* 149: 199–208.

Garcia, B. Beaumont, V. Perfetti, E. et al. 2010. Experiments and geochemical modelling of CO_2 sequestration by olivine: Potential, quantification. *Applied Geochemistry* 25: 1383–1396.

Information on http: //www.iea.org/CO_2highlights/CO_2 highlights.pdf.

Larachi, F. Daldoul, I. & Beaudoin, G. 2010. Fixation of CO_2 by chrysotile in low-pressure dry and moist carbonation: Ex-situ and in-situ characterizations. *Geochimica Et Cosmochimica Acta* 74: 3051–3075.

Maroto-valer, M.M. Fauth, D.J. Kuchta, M.E. 2005. Activation of magnesium rich minerals as carbonation feedstock materials for CO_2 sequestration. *Fuel Processing Technology* 86: 1627–1645.

Pronost, J. Beaudoin, G. Tremblay, J. 2011. Carbon Sequestration Kinetic and Storage Capacity of Ultramafic Mining Waste. *Environmental Science and Technology* 45: 9413–9420.

Qiu, J. & Lin J. 1991. *Rock chemistry*. Beijing: Geological Publishing House.

Sverdeup, H. 1996. Geochemistry, the key to understanding environmental chemistry. *Science of the Total Environment* 183: 67–87.

Teir, S. Eloneva, S. Fogelholm, C. et al. 2009. Fixation of carbon dioxide by producing hydromagnesite from serpentinite. *Applied Energy* 86: 214–218.

Teir, S. Revitzerb, H. Elonevaa, S. et al. 2007. Dissolution of natural serpentinite in mineral and organic acids. *International Journal of Mineral Process* 83: 36–46.

Vogeli, J. Reid, D.L. Becker, M. 2011. Investigation of the potential for mineral carbonation of PGM tailings in South Africa. *Minerals Engineering* 24: 1348–1356.

Wu, Liren. 1963. On metallogenic specialization of basic-ultrabasic rocks in China. *Geochimica*: 77–78.

Zhang, W. & Li Y. 2006. Research status and prospects of carbon dioxide sequestration technique. *Environmental Pollution and Control* 28(12): 950–953.

Frontiers of Energy and Environmental Engineering – Sung, Kao & Chen (eds)
© *2013 Taylor & Francis Group, London, ISBN 978-0-415-66159-1*

UASB reactor bio-hydrogen production characteristics of stable operation

Y.F. Li, C.Y. Liu, Y.J. Zhang, J.Y. Yang & X.L. Wang
School of Forestry, University of Northeast Forestry Harbin, China

ABSTRACT: This study chose brown sugar as substrate for bio-hydrogen production using an up flow anaerobic sludge bed (UASB) to prove the feasibility of biological hydrogen and the characteristics during the phases of steady operation. In this paper, brown sugar was used as the substrate to be the artificial wastewater, Temperature was controlled at about (35 ± 1) °C HRT was 8 h. During the whole operation, ORP stably remained at the range of −380 mV~−440 mV. COD removal rate was above 28%. Although, pH value in feeding box varied a lot, the pH value in effluent was always stably maintained at the range of 4.5~5.0. Fermentative gas production kept significant increasing, the maximum value of which is 27.040 L/d while the hydrogen content is about 50% at the same time. Terminal liquid production was consisting of ethanol, acetic acid, butyric acid and propionic acid. As the expected aimed production, ethanol acid and acetic acid kept increasing, the proportion of which was large and stably maintaining at the range of 80%~85%. Furthermore, the content of ethanol in terminal liquid production was significantly higher than acetic acid, propionic acid and butyric acid, which indicated that the system, in UASB reactor, successfully formed ethanol-type fermentation.

Keywords: anaerobic fermentation; UASB reactor; ethanol-type fermentation

1   INTRODUCTION

Energy demand is expected to grow in the following centuries because of the increasing world population. Although the fossil fuels have been the major choice of energy since the beginning of the 20th century, forecasts show a gradual transition from fossil fuel domination to a more balanced distribution of energy sources[1]. Among the various candidates, hydrogen gas (H_2) is regarded as the most promising future energy carrier as it has higher energy content by 2.75 times compared to hydrocarbon fuels (gasoline) and produces only water upon combustion. In addition, H_2 can be directly used in a fuel cell, generating electricity with high efficiency[2].

1966 Lewis first proposed the idea of biological hydrogen production. Until the 1970's energy crisis, the idea make the study of biological hydrogen production move a major step forward[3]. By making the use of gravity effects on different densities of material differences, Professor Lettinga invented the three-phase separator[4]. Activated sludge retention time and separation of waste water residence time, the formation of up flow anaerobic sludge blanket (UASB) reactor prototype[5]. In this study, the advantages of UASB and anaerobic bio-hydrogen production technology, to study its operation characteristics. I hope this study can provide a reference for promoting the industrialization of anaerobic hydrogen.

2   EXPERIMENTAL APPARATUS AND METHOD

2.1   *Experimental setup*

Figure 1 shows the experimental UASB reactor used in the main and auxiliary equipment. Effective volume of 18 L. In the reaction zone and the precipitation area of the outer winding resistance and the temperature control device monitoring. In order to achieve continuity of the experience, I supply a peristaltic pump with artificial wastewater to the reactor at a constant flow rate.

2.2   *Experimental methods*

Species used in this study is taken from the sewage sludge sewage treatment plant discharge ditch activated sludge secondary sedimentation tank. After washing the sludge after the treatment with aeration cultured for 20 days, by COD:N:P is 100:5:1. Add brown sugar to the sludge, NH4Cl and KH2PO4 everyday. When the color of sludge changed from black to brown, the sludge floc settling turned out

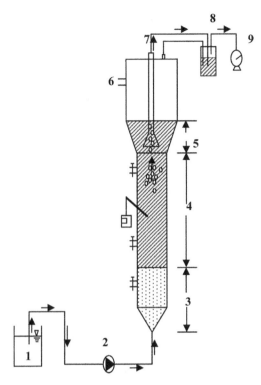

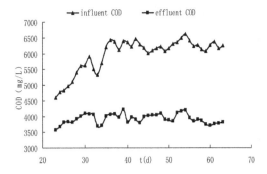

Figure 2. UASB reactor stable operation of phase change in influent COD and effluent COD.

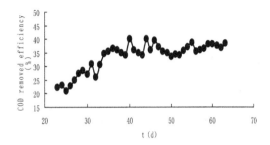

Figure 3. UASB reactor stable operation of phase change in COD removal efficiency.

Figure 1. Used in the experiment UASB reactor bio-hydrogen production plant diagram.
1. into tank 2. constant pump 3. sludge bed 4. sludge floating zone 5. three-phase separator 6. water outlet 7. gas outlet 8. filter 9. wet gas flow meter 10. temperature control device 11. the sampling port.

to be good. Concentration of suspended solids in sewage sludge 2000 mg/L or less can be inoculated into the reactor for continuous flow experiments.

The brown sugar was chosen as the bio-hydrogen substrate. Adding nitrogen and phosphorus fertilizer in the water to keep the microbes needed nutrients, including COD:N:P (concentration ratio) of 1000:5:1.

3 RESULTS AND DISCUSSION

3.1 *Stable operation phase of the regulation and the removal efficiency of COD*

In the stable operation of the first 10 days, we will COD concentration from 4500 mg/L adjusted to 6500 mg/L (Fig. 2), and to maintain the COD concentration. It can be seen from Figure 3, with the micro-organisms gradually adapt to the environment, COD removal increased from 22% to 36%, up to 40%, the formation of a stable ethanol-type fermentation. This process of change shows that

the organic load of the reactor changes the composition of microbial populations had a great impact, and because of the different microbial flora of the advantages of the different physiological metabolism and strength, making the system generates the COD removal efficiency some changes[6].

3.2 *Liquid end products and fermentation type*

The main liquid end products is ethanol and acetic acid, and there is produced a small amount of propionic acid and butyric acid. Including ethanol and acetic acid in the total amount of volatile acids in the proportion of very large, stable and maintained at 80% to 85%, The end product of ethanol in the liquid phase in the content of 1400 mg/L, significantly higher than acetic acid, propionic acid and butyric acid content. The stage of the stable operation, propionic acid of the average concentration is 180 mg/L, while the average concentration of butyric acid is 266 mg/L. Professor Nanqi Ren proposed: In bio-hydrogen fermentation reactor liquid end products, when the total concentration of ethanol and acetic acid accounted for about 70%, the dominant hydrogen-producing microorganisms is ethanol-type fermentation bacteria. This stage of the stable operation of the reactor

393

liquid at the end of the distribution of the product with this theory can be compared in this experiment to determine the anaerobic fermentation of ethanol-based hydrogen fermentation eventually formed the dominant species.

3.3 pH value of the stable operational phase

Hydrogen of anaerobic fermentation, pH is an important limiting factor. In this study, although the reactor influent pH range of 6 to 7 vary widely, but the effluent is always maintained at pH 4 to 4.5, it reached the ethanol-type fermentation bacteria required pH niche, acetic acid and ethanol accounted for liquid end products 80% of the total, so it is a ethanol-type fermentation of standard. Effluent pH is not fluctuations influent pH, the phenomenon that once the reaction within the reactor system environment to form ethanol-type fermentation, fermentation hydrogen production

reactor in the ecosystem will have a strong ability of resistance the organic load and stability.

3.4 The case of stable operation stage of the hydrogen production

Enhance the COD concentration of the stage, gas production increased significantly explain changes in organic loading on hydrogen production stability to a certain extent. In the COD concentrations in the 6500 mg/L, the maximum gas production reached 27.040 L/d, while the hydrogen content is also more than 50%. Describing the COD concentration suitable for growth of hydrogen-producing micro-organisms, and the system of gas production and hydrogen production gradually to reach a new steady state (see Fig. 5).

3.5 The ORP of stable operation phase

ORP have the main partial pressure of oxygen in the environment, both directly proportional relationship; it will be the environment of pH, pH

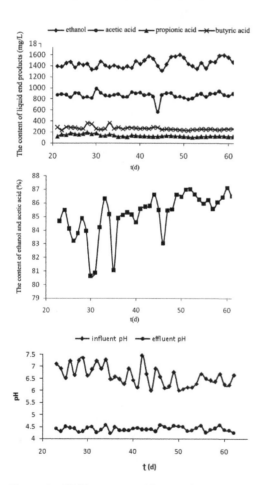

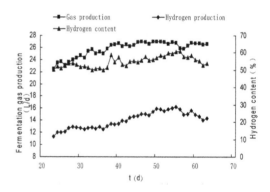

Figure 5. UASB reactor stable operation stage fermentation gas and hydrogen production and hydrogen content.

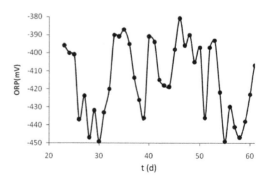

Figure 6. UASB reactor stable operation of phase changes in the ORP.

Figure 4. UASB reactor stable operation of phase changes in influent pH and effluent pH.

lower in environment will leading to the increase of ORP. Into the stage of the stable operation, due to acid fermentation bacteria domestication and environmental factors within the system is relatively stable, although there are substantial changes in COD concentration, but did not have a great impact on the ORP has been maintained at a stable between −380 mV~−440 mV, can be seen the UASB reactor system have a good organic load capacity (see Fig. 6). According to the findings of other researchers, the ORP range are basically niche of ethanol-type fermentation.

4 CONCLUSION

1. In the stable operation of the first 10 days, we will COD concentration from 4500 mg/L adjusted to 6500 mg/L, and to maintain the COD concentration in the next 30 days. Adjust the initial of COD concentration, COD removal efficiency remained at 22%, when the system of micro-organisms adapted to the organic load after impact, COD removal up to 40%.

2. The main liquid end products is ethanol and acetic acid, and there is produced a small amount of propionic acid and butyric acid. Including ethanol and acetic acid in the total amount of volatile acids in the proportion of very large, stable and maintained at 80% to 85%, but the end product of ethanol in the liquid phase were significantly higher than in acetic acid, propionic acid and butyric acid of content.

3. Although the reactor influent pH 6 to 7 has a lot changed, but the effluent is always maintained at pH 4 to 4.5, wastewater alcoholic fermentation is not affected.

4. In the stage of enhance the COD concentration, gas production increased significantly, and up to 27.040 L/d, while the hydrogen content is also more than 50%.

5. From the stable operation stage, although there are substantial changes in COD concentration, but did not have a great impact on the ORP has been maintained at a stable −380 mV~−440 mV.

6. Based on the above experimental results can be judged in the final form of anaerobic fermentation hydrogen of dominant species of ethanol-type fermentation. Once the reactor and the reaction system environment formation the ethanol-type fermentation, hydrogen in the reactor ecosystem will have a strong anti-shock organic loading capacity and stability.

REFERENCES

[1] Efe Boran, Ebru Özgür, Job van der Burg. Biological hydrogen production by Rhodobacter capsulatus in solar tubular photo bioreactor. Journal of Cleaner Production, 2010, 529~535.

[2] Dong-Hoon Kim, Mi-Sun Kim. Hydrogenases for biological hydrogen production. Bioresource Technology, 2011, 8423~8431.

[3] Buranakaral L, Fan Ch. Y, Ito K. Production of Molecular Hydrogen by Photosynthetic Bacteria with Raw Starch [J]. Agric. Biol. Chem., 1985, 49:3339~3341.

[4] Atsala TM, Raj SM, Manimaran A. A pilot-scale study of biohydrogen production from distillery effluent using defined bacterial co-culture [J]. International Journal of Hydrogen Energy, 2008, 33: 5404~5415.

[5] Ren N.Q., Wang B.Z., Huang J.C., Ethanol-type fermentation from carbohydrate in high rate acidogenic reactor [J]. Biotechnol Bioeng, 1997, 54(5):428~433.

[6] Jaime MN, Dinsdale R, Guwy A. Hydrogen production from sewage sludge using mixed microflora inoculums: Effect of pH and enzymatic pretreatment [J]. Bioresource Technology, 2008, 99: 6325~6331.

Frontiers of Energy and Environmental Engineering – Sung, Kao & Chen (eds)
© 2013 Taylor & Francis Group, London, ISBN 978-0-415-66159-1

Study on the construction of catastrophe insurance system in China

Z.S. Wang

Economics and Management School of Wuhan University, Luo-jia-shan, Wuchang, Wuhan, Hubei Province, P.R. China

ABSTRACT: The total loss amount and the insurance compensation amount of catastrophe risk events have showed a rising trend around the world in nearly 30 years. China's catastrophe risks have also brought huge losses to national economy, but the insurance compensation mechanism has played the limited role. China has not constructed catastrophe insurance system. China's collective loss-sharing mechanism is restricted to finance income amount in one hand, and it cannot effectively compensate the affected groups, on the other hand it also exists the higher opportunity costs. It hungers for structuring the catastrophe insurance system led by the government and depending on the market.

Keywords: catastrophe insurance; government intervention; collective loss-sharing

1 INTRODUCTION

Faced with huge losses caused by the catastrophe, the countries take a different management model according to their own national conditions. How to build an effective catastrophe loss compensation system has become an important task around the world. In view of the unique advantages of the insurance mechanism in compensation, governments regard it as an important means of catastrophe losses compensation system. It hungers for structuring the catastrophe insurance system leaded by the government and depending on the market.

1.1 *Worldwide catastrophe risk*

According to the statistics of the Munich Reinsurance Company from 1980 to 2010, the number of global extreme natural disasters was a significant increase in the trend (refer with Fig. 1). Extreme natural disasters in this context include the geological disasters such as earthquakes, volcanic eruptions; meteorological disasters such as storm; hydrological disasters such as floods; climatic events, such as extreme temperatures, droughts, forest fires. Since the 1990s, the economic losses and insurance losses caused by extreme natural disasters increase rapidly (refer with Fig. 2). For example, the losses caused by extreme natural disasters have exceeded $ 200 billion in 1995 and 170 billion in 2008. The sizes of insured losses are also rising. Especially since the 1990s, the United States has repeatedly encountered severe hurricanes such as Hurricane Andrew in 1992 and Hurricane Katrina in 2005. Insurance coverage in the United States is

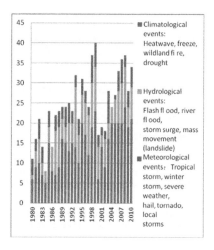

Figure 1. Frequency of occurrence of extreme natural disasters (1980–2010).
Source: http://www.munichre.com, natural catastrophes 2010 analyses, assessments, positions.

relatively high, which makes insured losses in these years much higher.

1.2 *China's catastrophe risk*

The natural disaster occurs frequency in China, it brings serious burden to the national economy and social life (refer with Fig. 3). We can see that the direct economic loss to GDP is approximately 2% except in 1998, 2008 and 2010, in which China has suffered more serious disasters.

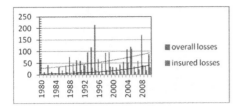

Figure 2. Overall losses and insured losses caused by extreme natural disasters: calculated at 2010 prices (1980–2010).
Source: http://www.munichre.com, natural catastrophes 2010 analyses, assessments, positions.

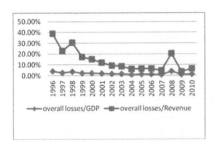

Figure 3. Direct economic losses to GDP and revenue (1996–2010) caused by Catastrophe.
Source: According to the Ministry of civil affairs statistical bulletin and the Statistical Yearbook of China (1996–2010).

China has not constructed catastrophe insurance system. The subsequent sections of this paper are organized as follows, the second part is the problem of catastrophe insurance market in China; third part research on china Catastrophe Insurance System; the last part is conclusion, which points out the future research directions.

2 THE PROBLEM OF CATASTROPHE INSURANCE MARKET IN CHINA

There are three problems in catastrophe insurance market in China.

2.1 Insufficient supply of catastrophe insurance

Even though consumers in the insurance market want to buy the catastrophe insurance such as earthquakes, the supply of these types of insurance is limited. For example, property insurance companies providing insurance of earthquake risks set compensation limits and deductibles, both of which are generally 20%. When the earthquake risks occure, the policy holder does not have access to the full payout.

2.2 Lack of demand for catastrophe insurance

The commercial property insurance premium shared 3.5% of the total premium income in property insurance market, family property insurance premium income accounted for 1%, both of less than 5% view of the real effective demand of the property insurance in 2010. Taking into account the awareness of Chinese enterprises and residents of the catastrophe risk insurance, the ratio of the actual insured catastrophe insurance will be lower.

2.3 Catastrophe insurance loss compensation playing limited role

All kinds of natural disasters caused direct economic losses of 204.2 billion Yuan in 2005, while the insurance indemnity was only 10 billion Yuan, accounting for less than 5%. In 2008 snowstorm brought the insurance payment less than 2% of the direct economic losses. The Wenchuan earthquake caused direct economic losses of 845.1 billion Yuan, yet insurance claim was no more than 1.66 billion Yuan, accounting for less than 0.2%. Worldwide catastrophe losses was $ 218 billion in 2010, and the insurance payments was $ 43 billion, accounting for 19.7%. In contrast, China's catastrophe insurance is far from its function.

3 RESEARCH ON CHINA CATASTROPHE INSURANCE SYSTEM

Accordance with relevant international experience, the government should not over-react before markets working in order to avoid "crowding out" private financing. So the initial task is clear about the government function in order to build the system.

3.1 Government function: Advance premium subsidies

China manages catastrophe risk through ex-post compensation. There are more drawbacks. In condition of compulsory insurance, providing premium subsidies to the insured can increase the government expected utility. The certificate is as follows.

Assume that the government is risk neutral, the government utility function relative to the financial expenditure is

$$U(G) = aG + b \qquad (1)$$

which G is on behalf of the government expenditure, a is the negative correlation coefficient between fiscal spending and the government utility function, a < 0, b is a constant.

Assume that there are N policyholders in the market; its wealth is W_0, W_1, ..., W_i, W_N

respectively, the probability of catastrophic risk is π, and $\pi < 1$, when the release of catastrophe risk L, each insured is total loss. Then the catastrophe occurs, the expectation loss is,

$$E(L) = a\pi \sum_{i=0}^{N} W_i \qquad (2)$$

In the post-compensation mode, the government needs to pay all the losses caused by the catastrophe. In this case, the expected utility is as follows:

$$E_P[U(G)] = \pi \sum_{i=0}^{N} W_i + b \qquad (3)$$

Under the compulsory insurance system, the total premium is P. Assume the insured pay premium αP, the government subsidies for the premium is $(1-\alpha)$P. Insurers facing of catastrophe risk L, its underwriting fair premium P are,

$$P = E(L) = \pi \sum_{i=0}^{N} W_i P = E(L) = \pi \sum_{i=0}^{N} W_i \qquad (4)$$

In this mode, the government only transfers $(1-\alpha)$P, in this case, the utility function is,

$$E_T[U(G)] = a(1-\alpha)\pi \sum_{i=0}^{N} W_i + b \qquad (5)$$

We use equation (5) minus (4), can be

$$E_P[U(G)] - E_T[U(G)] = a\alpha\pi \sum_{i=0}^{N} W_i < 0,$$

That is, $E_P[U(G)] < E_T[U(G)]$

To the government, its utility of compensation after the catastrophe losses is lower than that of the premium subsidies. Conclusion is proved.

3.2 Demand for catastrophe insurance: Compulsory insurance

In accordance with the law of large numbers, the premise of the steady operation of the insurance company is its independence of the risk exposure. Otherwise the dependent risk exposure affects its financial stability.

Assume $X_1, X_2, ..., X_N$ is a set of random variables, which denote insurance claims, and obey the normal distribution (μ_i, σ_i^2). According to the law of large numbers, we have,

$$p\left[\frac{\sum_{i=1}^{N} X_i - N\bar{\mu}}{\sigma_N} < Z_\varepsilon\right] = 1-\varepsilon$$

$$\bar{\mu} = \sum_i \frac{\mu_i}{N}, \bar{\sigma}^2 = \sum_{i=1}^{N} \sigma_i^2, \sigma_N^2$$

$$= \sum_{i=1}^{N} \sigma_i^2 + 2\sum_{j=2}^{N} \sum_{i=1}^{j-1} \sigma_{ij}, \sigma_{ij}$$

$$= \text{Cov}(X_i, X_j)$$

when $X_1, X_2, ..., X_N$ is independent,

$$\lim_{N\to\infty} \frac{Z_\varepsilon \sigma_N}{N} = \lim_{N\to\infty} \frac{Z_\varepsilon \sqrt{\bar{\sigma}^2}}{\sqrt{N}} \to 0;$$

when $X_1, X_2, ..., X_N$ is independent,

$$\lim_{N\to\infty} \frac{Z_\varepsilon \sigma_N}{N} = \lim_{N\to\infty} \frac{Z_\varepsilon \sqrt{\bar{\sigma}^2}}{\sqrt{N}}$$

$$= \lim_{N\to\infty} \frac{\sqrt{N\bar{\sigma}^2 + N(N-1)\bar{\sigma}_{ij}}}{N} \to \sqrt{\bar{\sigma}_{ij}},$$

Here $\bar{\sigma}_{ij}$ denote the average covariance.

Thus, when underwriting risks in line with the conditions of independence and in the case of a sufficient number of the subject matter insured, the capital of each of the subject matter insured of the insurance company is close to zero; When underwriting risk does not meet the conditions of independence, the average equity required by the subject matter insured is dependent on the correlation of the risk. The insurance companies do not have the desire to underwriting catastrophe risk, unless the risk can be reduced, and the implementation of compulsory insurance is a solution.

3.3 Supply of catastrophe insurance: Co-insurance

We follow the above analysis, how should the insurer provide products? Co-insurance is a good way. Cooperation between the insurance companies enhanced tolerance for risk, which can reduce the additional security costs. Therefore, in the face of catastrophe risk, co-insurance can be considered by the domestic property and casualty insurance company and the proportion of co-insurance is decided by its maximum underwriting capacity. Co-insurance can reduce premiums on the one hand, on the one hand it can reduce the risk of insolvency of the individual insurance companies involved in the catastrophe risk which affecting the stability of the industry as a whole; It can mobilize all social insurance resources to jointly cope with the catastrophe risk.

4 CONCLUSION

By analyzing the characteristics of global catastrophe risks and state of catastrophe insurance in China, we explain the necessity of establishment of catastrophe insurance system in China. Government should play to the function of relief, and completed the compensation by the market, which has the advantage of Pareto improvement of social welfare. In order to make full use

398

of resources of the insurance market, we need to introduce compulsory catastrophe insurance system from the point of view of catastrophe insurance market demand. From catastrophe insurance market supply point of view, the property insurance companies co-insure the risks according their maximum underwriting capacity. In this way, China formed the catastrophe insurance risk management system led by the government and operated based on the market. In the system, all insurance companies account catastrophe risk accounts separately, which realize the time and spatial smoothing of catastrophe risk. This article clearly has thought of construction of catastrophe insurance system in China, but how to implement it require further study.

REFERENCES

Cummins, J.D. & M.A. Weiss, Convergence of Insurance and Financial Markets Hybrid and Securitized Risk Transfer Solutions. 2008.

Joanne Linnerrooth-Bayer & Aniello Amerndola, 2000, Global Change, Natural Disasters and Loss-sharing: *Issues of Efficiency and Equity,* The Geneva Papers on Risk and Insurance, Vol. 25 (No. 2), 203–219.

Richard A. Posner, *Catastrophe: Risk and Response,* 2004, ISBN 978-0-19-530647-7.

Zeckhauser, R. 1995. *Insurance and Catastrophes.* The Geneva Papers on Risk and Insurance Theory. Vol. 20. No. 2. p157–175.

Information on http://siteresources.worldbank.org.

Information on http://www.munichre.com.

Frontiers of Energy and Environmental Engineering – Sung, Kao & Chen (eds)
© *2013 Taylor & Francis Group, London, ISBN 978-0-415-66159-1*

Research on characteristic and mechanism of shaft rupture with the seepage sedimentation in mining

L.W. Jing
Mechanics Department, Anhui University of Science and Technology, Huainan, China

K.X. Zhao & L.P. Ye
Mathematics Department, Anhui University of Science and Technology, Huainan, China

J.W. Zhou & X.G. Xue
Mechanics Department, Anhui University of Science and Technology, Huainan, China

ABSTRACT: The deformation and rupture control of shaft is one of the heavily difficult problems in coal mining. Only when the mechanism of the shaft rupture is correctly understood, the difficult problems can be solved. Based on the filed investigation and statistic analysis of the freezing shaft rupture in many mining areas, the basic characteristic of shaft rupture is discussed considering the aspects of the stratum structure, the time, the position and the form of shaft rupture. And then, the relevant rupture characteristic which can not be explained reasonably by the theory in existence is analyzed. The conception that waterpower connection exists between the upper water-bearing layer and the underlying around the shaft is brought forward. Further more, the comprehension about the mechanism of the shaft rupture comes into being. It reckons that the underlying water-bearing loses water and arise drainage in the upper. As a result, soil mass is compressed and subsides, the additional stress of shaft changes, therefore, shaft rupture occurs finally. The paper analyses the mechanism of the shaft rupture using the theory above aiming at the representative phenomenon of shaft rupture. Research indicates that the new theory can explain the phenomenon of shaft rupture and reveal the mechanism. It can be applied to guide the design of reinforcing and repairing the ruptured shaft.

Keywords: water-bearing layer; hydraulic method; the shaft rupture; surface subsidence; hydraulic channel; the mechanism of rupture of shaft lining

1 INTRODUCTION

Since 1980s, it appears varying degrees of deformation on both radial and longitudinal in the construction process and put into operation at different stages in the wall, such as more than 90 shaft mines in China's Xuzhou, Huainan, Huaibei, Datun, Yong Xia, Yanzhou mining area, Showed that wall rupture, cans tank road to the I-beam bending and so on, making the cage can not be properly upgraded, which seriously affected the mine's normal production equipment and personnel transport, reducing the efficiency of coal production, resulting in several hundred million dollars in economic losses, mine safety has long been plagued by the heavily difficult technical problems. For understanding the mechanism of rupture of shaft lining, researchers of our country have proposed a variety of hypotheses [1~3], to reveal the deformation and fracture mechanism

of shaft wall. Under the hypothesis that the lower aquifer seepage caused the settlement of the soil compression deformation, the force has led to the breakdown and destruction of wall, which generated in the radial wall down and in the vertical. In practice, this hypothesis can explain some wall damage phenomena well, but burst through the wall a lot of researches the phenomenon reflecting the series of hypothesis that, in theory is imperfect and one-sided, due to it is difficult to justify a greater number of walls rupture phenomena and can not correctly reveal the shaft wall deformation mechanism. Therefore, this article researches the rock mass deformation and changes, where contacts with water drainage in the wall under the side of the aquifer around, which through shaft broken by the phenomenon of a large number of surveys, analysis of the basic characteristics of wall rupture, and theoretical analysis, field observation and other means, from the point contacts between aquifer of the water in the around shaft, due to reveal the hydrophobic surface subsidence caused by mining shaft wall rupture mechanism, so as to provide prevention and treatment of rupture of shaft lining theoretical guidance. Research get some useful conclusions.

2 BASIC CHARACTERISTICS OF SHAFT WALL RUPTURE

According to the pit shaft's surveys and borehole deformation observation and study of its damage and fracture, such as more than 70 fields in the Xuzhou, Huainan, Datun, Yong Xia, Yanzhou mining area, the results shown in Table 1 and found that rupture with freezing shaft the following basic features:

1. Broken shaft has obvious geographical distribution of concentration, and correlation with stratum structure, which through the formation and borehole. Statistics showed that rupture shaft

are concentrated in the typical range of the alluvial plain where in eastern China along the Huang-Huai, the basic north-south to the zonal distribution; in the stratum structure, broken shaft through the thick quaternary alluvial topsoil. It is water-rich gravel in the alluvial the bottom layer. This stratum structure, started in the mine after the mining activities, including gravel will contain a series of seepage of groundwater into the mine passage, causing the hydrophobic surface subsidence; it is created the conditions to the friction where in the deep alluvium and the shaft being larger in the negative, due to the hydrophobic surface subsidence.

2. It is obvious time to the shaft fracture. Shaft breakdown is related decreased significantly under water, when it ruptures the basic position under the water head away from the wellhead down the range of about 85~95 m. For the loss of under water, however, it will only occur in the situation of underground mining. So the wall rupture usually occurs several

Table 1. The shaft rupture information of part frozen shafts in China.

Shaft name	Time of completion/ time to failure	Thickness of shock layer/m	Position with the bottom layer (Thickness)	The depth of the actual damage/m	Depth of freezing/m
Zhangshuang lou auxiliary shaft	1982.12.31/1987.07.29	243.15	233.15–243.15 (10)	225.0, 229.3~230.5	285
Haizi auxiliary shaft	1983.03/1987.08.21	247.24	214.81–247.24 (32.43)	232.8~237.5	285
Haizi main shaft	1982.08/1988.10.06	248.69	214.81–247.24 (32.43)	211.79~219.79	285
Haizi central air shaft	1980.05/1988.06	245.18	/	226.75~234.75	285
Hing Lung main shaft	1977.08.13/1997.06.23	189.31	165.91–189.31 (23.4)	150, 184	216
Hing Lung auxiliary shaft	1978.09/1997.06.26	190.41	167.61–189.31 (22.8)	154	221
Hing Lung Westerly shaft	1976.08/1995.10	183.9	163.2–183.9 (20.7)	165.6~171.6	219
Hing Lung Easterlies shaft	1977.05.31/1997.06.07	176.45	Missing	157	205
Yang village main shaft	1984.12/1997.02.29	185.42	165.92–185.42 (19.5)	176.5	207
Yang village auxiliary shaft	1985.01.23/1997.12.02	184.45	174.00–185.42 (10.45)	160, 176	212
Yang village North shaft	1984.10.31/1997.02.04	173.4	160.67–173.4 (12.73)	150~156.6	213
Baodian main shaft	1979.05.14/1995.07.12	148.69	130.31–148.69 (18.38)	136~144	256
Baodian auxiliary shaft	1979.11.26/1995.06.05	148.6	130.85–148.6 (17.75)	126.9	256
Baodian souther shaft	1979.08.01/1996.08.09	157.92	140.36–157.92 (17.56)	158.1~159.3	189
Baodian North shaft	1979.10.21/1996.08.02	202.56	175.67–202.56 (26.89)	168.4, 180, 204	234

years after the mine into production rooms. And wall temperature depends on the temperature changes. It will cause additional stress wall temperature produced and the concrete rose fission shape because of the temperature increases, so it is normally April to October each year mostly concentrated in the July to August month.

3. Rupture location in the shaft has obvious concentration in the spatial distribution. Wall rupture location mainly in two parts, one part concentrated in the deep Quaternary alluvium and the bedrock interface and its vicinity, and the other parts are concentrated in the upper aquifer within a specified range or clay layer.

4. It is very obviously for the bore rupture of the precursors and the associated phenomenon. It is typically exhibit bending bunton, cans road and narrowing the gap between wall and concrete wall localized longitudinal and radial deformation before rupture. Another occurrence of the mine shaft fracture the decline in industrial square more in the 250~500 mm, and the radial shaft increases as the center, that center wellbore to the surface showed significant bending deformation.

5. Shaft fracture deformation characteristics have obvious similarities. Wall rupture, the wellbore was wedge-shaped inner edge of spalled concrete off the block, most of the broken shaft was similar to the level of ring fracture zone, a few thick topsoil and larger hole diameter shaft of the rupture zone of the rupture was about 45° oblique, well inside the longitudinal and circumferential reinforcement both within and outside the mutant form of the well.

6. In addition to the aforementioned five bore similarities, there is also on the deformation and failure of individual characteristics to the wall rupture shaft. For example, after some broken shaft treatment by grouting, the week-side ground side of the wellhead in a performance of a continuous updraft in a long period of time. some use of "stress relief slot + grouting" in the wellbore after the treatment, longitudinal strain relief slot at the slot less than before against the corresponding parts of the longitudinal strain values; Governance seems broken shaft effect is not caused by the stress relief slot; Most broken shaft with the bottom of the aquifer, but a few shaft without the bottom of the aquifer; most of the broken shaft through the bottom of the aquifer thickness is larger but there are a few shaft through the bottom of the aquifer thickness is small. Toward these particular phenomenon, it is difficult to reasonably explain if use the hypothesis, which that wall rupture caused by dehydration aquifer under.

3 FRACTURE RESEARCH MECHANISM OF WALL BY HYDROPHOBIC SETTLEMENT

More than 20 years of research [4–6] that led to the breakdown of the main shaft force is acting on the shaft between the outer wall and the surface vertical additional force, and hydrophobic surface mining would cause subsidence and changes in the distribution of additional force. It has long been in the aquifer will be no significant change in pressure as the next head of the aquifer on the aquifer and its hydraulic no connection between the basis of free and up based on this analysis and surface characteristics of settlement by the shaft wall of the force characteristics. We proposed a series of hypothesis that the root cause of the mine shaft lining is breakdown of mine drainage in the aquifer under the surface subsidence, result in friction of the negative between the outer wall surface and the shaft. Under the hypothesis based on the understanding that although the vast majority of wall breakdown of the problem can be better explain, but then it is self-contradictory and no reasonable explanation of [7–8] when a considerable number of broken shafts in the sixth article appeared in the preceding statement of the phenomenon. The fundamental reason is to uphold the irrational point of view that "under the layer and the other with no hydraulic connection between aquifers". According to field monitoring and related simulations showed that there is a specific kind of seepage around the shaft channel, exists between each aquifer within the hydraulic connection, the formation of the corresponding shaft fracture mechanism discussed on this basis.

4 THE HYDRAULIC CONNECTION BETWEEN AQUIFERS

The formation of freezing drilling shaft, as shown in Figure 1 can be divided into two regions by

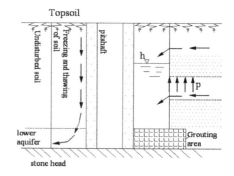

Figure 1. Relationship between aquifer hydraulic diagram.

Table 2. Comparison of the physical parameters between original and thawing soils [9].

Soil samples	Void ratio/%	Plasticity index	Liquidity index	Horizontal permeability/cms^{-1}	Vertical permeability/cms^{-1}
North-Asia clay undisturbed soil	0.99	16.5	1.10	2.23×10^{-6}	0.672×10^{-6}
North-Asia clay freezing and thawing of soil	1.06	14.7	1.28	20.7×10^{-6}	6.88×10^{-6}
South-Asia clay undisturbed soil	1.04	16.8	0.91	3.72×10^{-6}	0.722×10^{-6}
South-Asia clay freezing and thawing of soil	1.04	14.0	1.08	11.8×10^{-6}	3.23×10^{-6}

around shaft of surface: regional freeze-thaw and undisturbed soil areas. In the region of freezing and thawing, on the one hand because of permafrost melts, causing soil liquefaction [4], making the dramatic increase of permeability of clay (Table 2), formation of a seepage channel in the radial side of the shaft within the shaft week, On the other hand, with the shaft excavation process, the ice wall in the ground under pressure will have a greater circumferential stress and radial displacement, destruction of the original soil structure, but also create the conditions for the formation of seepage channels. In addition, the following factors will also affect form a seepage channel in the freezing and thawing area.

1. Substandard quality of pore filling after frozen pipes removal.
2. In the construction period shaft outer wall formation a series of small cracks due to the role of permafrost around the frost heaving force.
3. Use of cushioning material after the wall and the expansion and contraction of the radial shaft will be in the side wall and the weeks have gaps between rock and soil; Impermeable layer is not completely impermeable, Dongfeng mine Xinglongzhuang larger compartment deformation and lack of other aquifers under the shaft (such as wells and booming east wind Tongting wells) show that the rupture may occur relatively impermeable layer of hydrophobic severe deformation. It received after the disturbance and melting of frozen water permeability confining layer has undoubtedly expected. For the drilling shaft, the wall of the borehole drilling construction method is the post-filling cement and gravel (or mine Slag), Concrete construction, the two sub-cross-filling material, General rubble fill far more than the amount of slurry (surface section). Because slag is a good gravel and permeable body, and cement in the intensity of the process will increase the side for weeks, the role of soil lateral pressure becomes loose body, therefore, backfilling area became seepage channel of communication between the upper and lower aquifer hydraulic connection.

4.1 In the aquifer under hydraulic connection stability of head

Accounting for mining subsidence of mine coal caused by falling roof and bottom water which across the hydrophobic layer deposition, this will lead to the formation of a mining area of the concave surface soil filling area, it leads to making the depression, body rock aquifer region and peripheral region in the corresponding free water aquifer the formation of a potential difference between, and enhance the regional aquifer outside the free flow of water to the mining depression of the ability of, When the external supply of water seepage and the water aquifer under the equilibrium, In addition to the range of flow funnel containing water layer head will fall significantly better than the original head, the flow outside the funnel of the aquifer (the aquifer than) the head will be relatively stable. This is the reason why next compartment above the head of the aquifer (60 m away from the wellbore outside of) the process of settlement in the topsoil remained relatively stable long-term causes. Therefore, many mining areas in the aquifer is not the reason of the instability of the head in the bottom of the aquifer and the hydraulic isolation between the aquifer.

5 SHAFT FRACTURE CAUSED BY THE HYDROPHOBIC MECHANISM OF SETTLEMENT

Accounting for there are a lot of water seepage channel around the shaft, the effective stress within the aquifer will increase, and lead to aquifer compression within the rock mass deformation. Due to the above the aquifer is less than the radial hydraulic conductivity of soil water seepage channel permeability, a range of soil around the well bore the body of water is the water level under the water, with the loss of water under the upper aquifer will be reduced due to hydrophobic and then. Because of the hydrophobic upper aquifer, the settlement process due to shaft the support effect on soil and makes part of the role of soil weight transfer to the shaft above the, The week so that the shaft on the vertical side of the soil re-distribution of ground pressure, resulting in the lower aquifer is declining in the head little increase in the case of effective stress, it means that the lower side of the shaft weeks of soil moisture lost during the next will occur only slight deformation.

In addition, under the theory of seepage channel exists, in the broken shaft after grouting the implementation period of time, Seepage water level rises in the channel less consequential decrease the rates are less effective stress, which trigger the vertical expansion of the soil is also too small to restore the soil by lateral wellbore weeks reduce the water level caused by the original wellbore pressure drop in the value of, In this case the creep characteristics of concrete and the upper part of the continued settlement of the soil will continue to be subject to the shaft compressed state when the water level continues to rise and reached a certain level, the well bore by the compression state will gradually stretched into. Shaft can be compressed into the stretch besides the existence of the reasons above, the other impermeable layer of water loss from the original state of buoyancy resulting from reduced sink into the case to regain buoyancy resulting upward displacement of the situation, This is compressed into the stretched shaft is an important reason. Therefore, after implementation date of grouting, in the subsequent period of time, experience will be followed by surface tension shaft section—compression—tension—a stable process, a lot of field data have proved the existence of this process.

6 FIELD MEASUREMENT OF THE DEFORMATION OF SHAFT

The "seepage channel" theories is correct or not exist, it is clearly inadequate on the basis of theoretical analysis along. The following are measured from a variety of related data, in which we started more detailed and in-depth research and analysis.

7 THE OBSERVATION AND ANALYSIS OF REPAIRING THE DEFORMATION OF SHAFT WITH SLURRY INJECTION METHOD

The surface soil thickness of Baodian south well is 157.92 m; the surface freezing method is used in the surface soil segment. In 1996, 7 to 8 months, Wall is broken in the vertical depth of 158 m, where broken blocks of concrete were falling off,

the height of the rupture zone is about 1.5 m, the radial crack depth is 50–70 mm, steel convex, which is a typical fracturing damage. April 1~July 19, 1998, a method of" broken grouting + relief groove + wall units" is used in the implementation of the comprehensive management of broken wall. It use the method of "upward-type initial grouting, down-type re-grouting". Opening relief slot in the 153 m and 163 m deep shaft, the size is $300{\sim}350 \times 400$ mm, Sets of wall thickness of the last set of wall stress relief slot is 200 mm concrete. It established a borehole monitoring system march 1998; we collect the hoop and the vertical wall strain data at the same year April 19 to October 22, 1999. (See Table 3). Observational data in Table 3. During the grouting, three measured wall are in the tensile deformation, the maximum tensile deformation occurred in the second part of measurement level; the cumulative tensile capacity is 289.5 mm. Within a certain period (0–9 months) after the treatment, shafts are in the deformation. the level of the third test after 5 months, the maximum deformation reached 63.75 mm; later (10 months after) shafts again show tensile, the maximum tensile capacity of the second test level reach 80.3 mm. This is mainly because there is a channel seepage around the shaft as shown in Figure 1, The density of the grouting to the aquifer, which take hydraulic connection broken between the lower part and the upper part of the aquifer, making the original channel of water seepage into a closed circular groove, the upper part of the continuous flow aquifers makes the tank's water level rise, effective stress of the soil in side of the shaft is decreasing, leading to compressed soil occurred rebound. And when the water level exceeds a certain water layer interface, the aquifer head values (near the water seepage channels) will significantly increase, causing the impermeable by the upward pressure p, which significantly increasing, This result in the aquifer to produce upward movement and make the thickness of the water increase due to the rebound deformation. These two factors together make the topsoil near the seepage channel produced the overall upward movement, and displacement distribute in space, it presents the characteristics of bottom-up increases, upward additional effect force on the wellbore, causing the tensile deformation along the longitudinal shaft. As the water level continues to rise, when it reaches a stable level, the ground uplift and the shaft will be the end of the tensile deformation. The experimental results using the traditional hypothesis can not be reasonably explained, while the view of the wellbore weeks existing channel seepage can be well explained, To some extent, a direct proof of water seepage channel is an objective reality.

Table 3. The detection result of the deformation of the air Shaft in Baodian mine [10] Unit: mm.

Time quantum	The first test level, vertical depth 113.5 m		The second test level, vertical depth 125..3 m		The third test level, vertical depth 139.5 m	
	Central to the average	Vertical average	Central to the average	Vertical average	Central to the average	Vertical average
Grouting behind segment stage	+32	+21.5	−28.3	+263.5	+23.75	+115
On stage pressure relief groove	+7.5	+5.5	0	+12.5	+7.5	−6.25
Jacket wall stage	+12.5	+12.25	+18.3	+13.75	+13	+10
During the cumulative treatment	+52.5	+39.25	−10	+289.5	+44.25	+118.75
0–5 months after treatment	−49.7	−47.5	−13.33	−58.75	−25.5	−63.75
6–9 months after treatment	−4.25	−2.75	+18.33	−15	+5.5	−8.75
10–15 months after treatment	+43	+45.25	+80.3	+27.5	+66.25	+37.25
The total	+41.25	+34.25	+75.3	+243.25	+90.5	+83.5

8 STATISTICAL ANALYSIS OF OBSERVATIONS IN WALL RUPTURE LOCATION

If the lower aquifer and other aquifers do not exist the hydraulic connection, the compression of the aquifer is an absolute factor on surface subsidence. The stiffness of the shaft deformation is much larger than the deformation of soil stiffness, so the next aquifer losing water will cause the earth to produce compression deformation, it will be generated uncoordinated deformation between the stiffness, so that vertical additional force between soil and wall inevitable within the next aquifer, the largest vertical shaft wall stress is also bound in the soil and rock at the junction. If the shaft is damaged, the damage must occur in the location at the interface of Rock, the probability is very small in theory when damage takes place in other places, unless the wellbore exists quality problems. But in fact, through the broken wall in table 1, after research we found a considerable number of broken shaft position is not at the interface in the soil and rock, but in the side on the site, and with the increasing depth of topsoil the deviation from the characteristics of the distance is to increase. For thick topsoil and a thin shaft wall, often in two different locations simultaneously broken. For the broken wall, if use the view that lower aquifer and other aquifers don't exist hydraulic connection, no matter what will not work. However, if considering hydraulic connection between aquifers, combined with the redistribution of earth pressure theory of shaft, then this phenomenon can get a reasonable description.

9 THE OBSERVATION AND ANALYSIS OF MISSING AQUIFER AND THICKNESS OF BROKEN SHAFT

"Below aquifer and above aquifer don't have hydraulic connection, Surface soil subsidence depends on the below aquifer compression." This view is rejected impermeable layer of water permeability, thus contributing to people will naturally have the point of view that" so long as the bottom layer containing missing, the surface soil will not subside, shaft does not rupture". However, not all is the case, the fact that the missing below aquifer of Xinglong east well and the shaft rupture of Tongting well is very powerful evidence. In addition, the thickness of the aquifer is very thin, in poor permeability of the mine, the broken shaft also shows the same principle, such as dragon east shaft [11], the end gravel layer thickness is only about one meter, water abundance is more weak, gravel layer itself has a lower compression, reduced under the sink aquifer is the main factor that lead to wall rupture, this view is difficult to explain the mechanism of wall rupture. Such as the breakdown of shaft, if use the "seepage channel" theory to explain the existence, it is more convincing. Loss of below aquifer caused the shaft breakdown, which indirectly proves that freeze-thaw of clay have a good permeability, this also provides the basis point of view for the "leakage

Table 4. The detection result of the absolute stress of Linhuan mine yard [13].

Name	The first test level		The second test level		The third test level	
	South		South	North	South	North
Hydraulic fracturing tensile strength of concrete (Mpa)	18		18.5	21	19	18
Inner wall tangential stress (Mpa)	11.5		19	19.5	22	24
Inner wall vertical stress (Mpa)	14.5		25	26.5	32.5	34
Inner wall the projected value of the external loads inner wall tangential stress (Mpa)	1.8		3.11	3.19	3.6	3.9

channel". by the hypothesis of inference that there is no hydraulic connection between below aquifer and other aquifers, when the shock layer thickness and other factors (hole diameter, the wall, the concrete strength, etc.) in the same circumstances, the thicker end of the aquifer, the number of the rupture should be more. It is not the case, from Table 1, it is clear that topsoil at Haizi well is 4.09 m thicker than ZhangShuanglou well, the bellow aquifer is 22.43 m thicker, according to corollary the number of Haizi well rupture should be more than ZhangShuanglou well, in fact, the number of ZhangShuanglou well rupture is 5, the number of Haizi well rupture is 3. Corresponding the topsoil of Tongting is 230.5 m thick, bottom water is 12 meters thick, broken only twice [12]; similar phenomena such as the Longdong well, the topsoil is 212 meters thick, However, it destroyed in 1987.08 and 1990.07, by contrast, in Huang-Huai region, the thickness of both shaft and topsoil is greater than Longdong shaft, but no damage.

Therefore, to some of the vertical shaft, this view is not realistic: the compression of below aquifer cause shaft rupture and no hydraulic connection between aquifers, it is difficult to explain the phenomenon of a similar wall rupture, and it will come to conflicting inferences and conclusions.

10 THE OBSERVATION AND ANALYSIS OF ABSOLUTE STRESS OF SHAFT RUPTURE

At present, the additional absolute stress in the vertical shaft measured very little, so far the only absolute stress tests carried out over by Professor of Anhui University Chou Wan Xi. Professor Chou Wan Xi developed his own sleeve fracturing method, this method take five main drilling test on Linhuan well [13]. Testing conducted in June 1993; five holes are located at three test levels. The first level is 100 m vertical depth (the middle of

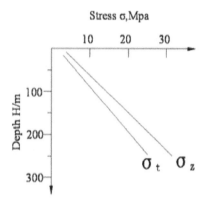

Figure 2. The change laws of shaft stress along the vertical of shaft.

the second compartment), set up a hole, which is arranged in sections of the south wall; the second level is 200 m vertical depth (three-compartment bottom), set up two holes which were arranged in wall cross-section of the south and north sides; the third level is 240 m vertical depth (including the end of the junction area and the bedrock), set two holes, the same wall section arranged in the south and north sides; the test results are in Table 4. See from Figure 2, regardless of the tangential stress or the vertical stress, they are all called to be similar to a linear relationship with the depth of the wall, This linear relationship require the vertical additional force on the wall approximately uniform distribution along the vertical depth down. (From the shaft within the upper range of a certain depth), and this is very consistent with the distribution pattern of vertical additional stress, which is corresponding with "Leakage channel". For this shaft, the hydraulic between the below aquifer and other aquifers is isolated, according to the results of literature [14], the vertical additional stress should be showed a rapid increase of the curve, obviously this is not consistent with Linhuan wall's vertical stress distribution, which also shows it really exist seepage channels around some of the shaft.

11 CONCLUSION

The engineering examples are typical shafts in China's major mining areas, the measured data's are all come from well-known experts in the field of scientific research, it is extremely convincing. Through research, we made the following useful conclusions:

1. Through research, we propose that there is leakage channels between each aquifer around shaft, it exist hydraulic connection between aquifers. Based on this wall rupture mechanism can reasonably explain well break phenomenon that many of the traditional hypothesis can not better account for: ground around shaft sustain elevate after grouting, Shaft fracture and the aquifer water level under the distance away from the wellhead is to be the corresponding relationship. After grouting the wall appears slightly stretched in vertical and compression in ring.
2. The existence of water seepage channels is confirmed by the conclusion. This is an important basic role for the right analysis of the pressure distribution for the future shaft, it not only provides a theoretical basis for the shaft wall rupture prediction theory, but also gives great important reference for the optimization of shaft structural designing in deep soil layer.

ACKNOWLEDGEMENT

This work is supported by Natural Science Foundation of Anhui Province (11040606M101).

REFERENCES

[1] Zhou Zhian, Yang Weimin. Analysis of lateral frictional resistance for shaft rupture [J]. *Coal Geology & Exploration*. 2003, 35 (5): 37–39.
[2] Ge Xiaoguang. Engineering analysis of disaster for ground and grouting wall broken rupture [J]. *Coal*. 2002, 27 (1):42–44.
[3] Zhang Shifang, Yang Xiaolin. *The mine construction technology in deep alluvium* [M]. Beijing: Coal Industry Press. 2002: 78–123.
[4] Chen Xiangsheng. Simple analysis of shaft damage in east region [J]. *Construction Technology*. 1991, 18 (6): 1–3.
[5] Wang Chang, GE Hong Zhang. Analysis of causes and Control for Yanzhou Mine Shaft Wall Rupture [J]. *China University of Mining Technology*. 1999, 28 (5): 494–498.
[6] LiWenPing. Experimental study for deep mine water loss compress deformation in Xuhuai mine [J]. *Coal*. 1999, 24 (3): 231–235.
[7] Jing Laiwang, ZhANG Tianyong, XU Huidong and so on. Mining wall topsoil settlement mechanism and relationship of shaft rupture [J]. *Coal Geology & Exploration*. 2005, 33 (3): 60–63.
[8] Jing Laiwang, Liu Fei, Zhang Tianyong. Mechanism study for shaft rupture by grouting [J]. *China's mining industry*. 2005, 14 (9): 53–56.
[9] Yang Ping, Zhang Ting. Study for artificial freezing of physical and mechanical properties [J]. *Glaciology*. 2002, 24 (5): 665–667.
[10] Gao Mengmin, Lv Henglin, Cui Guangxin and so on. The Mapping research of shaft rupture by destress [J]. *Coal*. 2002, (1): 36–39.
[11] Yu Xianxiang. Analysis and Control of shaft wall rupture in Datun mining [C]. Shaft destruction treatment technology Engineers. China Coal Society. Beijing. 2000: 192–197.
[12] Xie Hongbin. Comparison of deformation treatment and Issues discussed between Tongting auxiliary shaft fengjing shaft [J]. *Mine surveying*. 1998, (2): 39–41.
[13] Chou Wanxi, Wei Shan Bin, Zang Desheng. Sleeve fracturing measurement and analysis of Shaft stress test [J]. *Geotechnical Engineering*. 1995, 1 (17):61–65.

Frontiers of Energy and Environmental Engineering – Sung, Kao & Chen (eds)
© 2013 Taylor & Francis Group, London, ISBN 978-0-415-66159-1

Research on the thick very loose coal bed driven along the bottom of roadway supporting

J.P. Li
Mechanics and Construction Engineering, China University of Mining and Technology, Beijing, China
Huaibei Mining Shares and Limited Liability Company, Huaibei, Anhui, China

L.W. Jing
Supporting Technology Research Institute, Anhui University of Science and Technology, Huainan, Anhui

Z.H. Liu
Huaibei Mining Shares and Limited Liability Company, Huainan, Anhui, China

P.W. Hao & D.D. Liu
Supporting Technology Research Institute, Anhui University of Science and Technology, Huainan, Anhui, China

ABSTRACT: This paper using 8203 machines tunnel and wind tunnel as the research object comprehensively introduces a thick very loose coal bed driven along the bottom of roadway support technology in Anhui some coal mine. It is not only describes the specific supporting programs, but also introduces support basis in detail, field trials, monitoring reports about proving the effect of supporting. It also creates a huge effective technological change. The background of this paper is a successful scientific collaboration between coal enterprises and universities. Its significant achievements in scientific researches are not only reflecting in the huge economic benefits, but also having high academic value. It revealed "forced supporting weak pressure, weak supporting forced pressure" theory not only has the great significance to improve the supporting theory of the existing mine roadway, but also has important reference value to change the supporting difficulties of the thick very loose coal bed and the status of repeatedly repair in our country.

Keywords: the thick very loose coal bed; driven along the bottom of roadway; forced supporting weak pressure; weak supporting forced pressure

1 PREFACE

The thick very loose coal bed of roadway supporting has been one of the academic and technical problems in coal field at home and abroad [1–8]. The severe deformation of the tunnel and substantially muster of the floor not only have severely impact on the running of motorcycle, the transport of material, the operation of the conveyor belt, but also block the ventilation result within coal face and result a great security risk. The occurrences caused by gas gauge and coal bed spontaneous combustion are frequent in the years. At the same time, the repeated repair of the roadway has not only significantly increased the production costs, but also seriously affected the operations of normal production and huge indirect economic losses. In fact, the impact on the key sticking of the thick very loose coal of roadway supporting theory and technological development does not lie in the low level of technical means, but rather to have the erroneous on the recognition concept, which is mainly reflected in the following aspects: (1) So far, it is said only metal bracket can use in the very loose coal bed roadway supporting while the bolt can not show the role in the very loose coal bed roadway supporting. (2) "Relief", the key factors, should be considered in the very loose coal bed roadway supporting design, which does not understand modern mechanics theory and the related strength criterion enough, even violates the some classical principles in modern mechanics. (3) Ignoring the surrounding rock (coal) is a whole structure, and the relationship between the various parts is interaction and mutual influence, and the relationship between the roadway's sides, the bottom, the top is linkage effects; (4) Ignoring the loose coal still has a high carrying capacity under certain conditions; (5) Ignoring the bearing capacity of a metal stent can be increased dramatically by means of bolt force. It is because of the above series of awareness of factors, that the theory and technology of the very loose coal bed roadway supporting difficultly develop and the way of supporting is constant pattern in several decades.

8_1, 8_2 coal bed is very loose coal bed in Anhui some coal mine. Roadway deformation is extremely serious in this coal bed, and almost all of the roadways are equipped with a repair team to the implementation of 24-hour repair. Even so, the roadway can only maintain 30%–50% of the original section, in some places less than 20% and people can only bend over sideways through the tunnel. 500 mm the row spacing of 36 U scaffolding seems unable to support the tremendous pressure and to occur extremely severe bending and twisting.

The deformation of Wind Tunnel roadway at the beginning 200 m in 8203 face is basically no different from the same coal seam roadway: after digging then swollen, shed legs substantially adduction, arch waist bending, column legs distorted. In response to this situation, Anhui University of Science and technology cooperate with Huaibei mining shares and limited liability company, under the basis of modern mechanics, after a large number of theoretical calculation and experiments, and greatly succeed in the 8203 face roadway. Not only promotes the supporting theory further improved, but also creates huge economic benefits.

2 PROJECT OVERVIEW

8203 caving working face is the first recovery prepared face of two mining area in this mine. The face's elevation is −557.0~−655. Only return air door is under construction, and there are no other extractive activities. The face's long is 1212 m, and tilt long is 146 m, with a total area is 176 952 m^2. The main mining of 8203 face is 81 and 82 coal. The coal seam thickness is 8.78 m, and dip is 19o to 31o, with an average inclination is 26o (see Fig. 1). 81 coal's thickness is 4.28–6.84 m, and its average thickness is 5.55 mm, 82 coal's thickness is 2.46–3.84 m, and its average is 3.23 m. The two layers of coal is stability, loose structure, low intensity, very loose, and it is broken by grip. The partings

between the two layers are mostly mudstone, their thicknesses are 0.78–5.49 m, and their average is 2.30 m. Driven along the bottom of roadway of 8203 face, the directly bottom is a layer mudstone, and it thickness is 1.44 m. Because there is a 26o angle in the horizontal between the floor strata and roadway floor, so only one side of the roadway locate on the floor rock, As Figure 1 shows.

2.1 Roadway basic stability principles

Because coal bed is thick and very loose, so the roadway's loose circle is too big, and its coal is very broken. Based on deformation of the same coal bed roadway, the principle of "yielding pressure" is impossible in this situation. Based on modern mechanical principles, it can use "forced supporting weak pressure, weak supporting forced pressure" principle in here, and it Specifically analyzes as following.

2.2 The basic principles

For coal mine roadway, when we research supporting, we hope that the surrounding rock loose circle as small as possible, and can stabilize quickly. For the smaller loose circle, effective section size of the roadways will get smaller (in this case, loose circle round which is out of the loop boundary of the section is called the effective cross section, the size of the corresponding section is called effective section size), the stress of corresponding degree of concentration will be tangent to the boundary of the stress, and the value is smaller. The value of Re-distribution on the boundary of the effective cross section is relatively small. If the Coulomb-Navier criteria: $\sigma_1\left[(f^2+1)^{1/2}-f\right]-\sigma_3\left[(f^2+1)^{1/2}+f\right]=2c$, as a fragmentation of strength criterion for rock, when σ_1 is clearly smaller, but σ_3 is larger, and surrounding rock will not be fractured, loose circle will be stable down. The bracket by the force will not be further increased, the deformation of the roadways will not further occur (specifically shown in Fig. 2), if σ_3 Can't keep a certain value, then loose

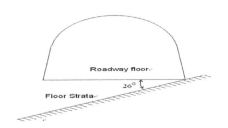

Figure 1. Roadway floor and floor strata schematic.

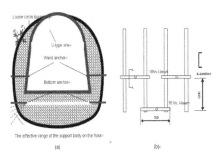

Figure 2. "Forced supporting weak pressure, weak supporting forced pressure" schematic diagram.

circle periphery will broken rock, this will lead to the following three consequence:

1. With the increase of effective cross-section size, σ_1 will further increase, if σ_3 can not be improved effectively, the loose circle will be bound to be increasing;
2. With the expansion of the loose circle, the periphery of the rock mass destruction, which will result in continue to occur, and the hulking will give more pressure to support body.
3. With the expansion of the loose circle, will also lead to the loss of anchor role. Once when loose circle is over the scope of the anchor bolt, the anchor would lose their effective anchor role.

From above, we can see that the greater the inhibition of the circle loose, which extensions needed to support with the higher strength, otherwise it is difficult to make sure the value of the σ_3. There is the necessary measures that effectively improve the strength of the supporting body, which make ensure to get roadway stability, this is well known as "forced supporting weak pressure, weak supporting forced pressure" principle.

Based on the above principle, according to the specific situation of 8203, which takes a series of measures is maintain stability.

3 THE STABILIZATION MEASURES

For 8203 roadways, the basis principle of the "forced supporting weak pressure, weak supporting forced pressure". Put forward the following four aspects of the concrete measures:

1. Through the role of the channel steel + anchor, the carry capacity of the metal stent gets increasing;
2. Increase the strength of the steel mesh, which use to avoid the steel mesh excessive deformation, the value of σ_3 gets effective controlled, thus restrain the loose circle gets further expansion;
3. As far as possible to avoid digging super when in the process of excavating roadway, which is difficult to improve the value of σ_3, then it will be resulted in the generation of large loose circle;
4. Increase double anti-cloth in the back of the reinforced net, which can prevent the outflow of pieces of coal.

4 ANALYSIS OF THE ROLE

4.1 *Analysis of effect of the bolt + channel*

Figure 3 is for simple 36 U metal stent support (row spacing for 500 mm), the distribution of the edge of the inside and outside of the stress of the supporting pure metal stents. Figure 4 the distribution of

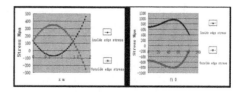

Figure 3. The distribution of the edge of the inside and outside of stress of the supporting pure metal stents.

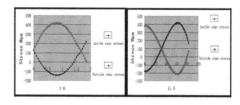

Figure 4. The distribution of the edge of inside and Outside of stress of support metal stent + bolt + channel.

the edge of inside and outside of stress of support metal stent + bolt + channel, which Contain Metal stents model and Row & Line Space unchanged. Figure 3 shows the maximum tensile stress for $-811.707\ MP_a$, and the maximum compressive stress for $962.507\ MP_a$, which be Produced in the outer edge and inner edge of the arch and angle φ 49 degrees, Figure 3 shows the maximum tensile stress for $-212.76\ MP_a$, and the maximum compressive stress for $418.12\ MP_a$, which be Produced in the outer edge and inner edge of the arch and angle φ 62 and 49 degrees, and we can easy see the Effect which is very obvious.

4.2 *Analysis the effect that increase the strength of the steel mesh and increase the role of double cloth spread*

With the increase in the strength of steel mesh, and additional the double anti-cloth, these can effectively inhibit the outflow of the pieces of coal, those can improve the value of a loose circle hulking force. This will not only effectively increase the value of σ_3 in Figure 2, but also effectively inhibit the expansion of the loose circle, and get the original carrying capacity into a carrying capacity tremendous load-bearing structures, at the same time, the anchoring force of the bolt get a very high value, and to improve the carrying capacity of metal stents.

4.3 *Effectively inhibit the occurrence of floor heave*

Due to the internal helps linkage between the roadways for department of effective control, which also makes the stability of the floor got a promotion.

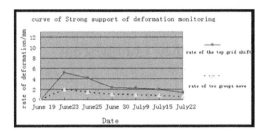

Figure 5. Detection curve about strong supporting effect.

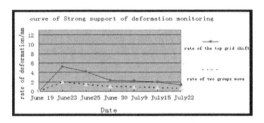

Figure 6. Deformation monitoring curve about pure metal stent support segment.

4.4 *Effective of the analysis*

Figures 5 and 6 for two kinds of support cases tunnel deformation monitoring curve, the data show that a significant effect is extremely strong, roadway deformation reduced about five times, and deformation rate is in a continuous convergence in, significant support results show the validity and rationality of the supporting method.

5 CONCLUSIONS

Mechanism analysis, the measures to research and field test, which means a more in-depth analysis of the rock support techniques and principles of deep very loose coal Lane, a series of higher-value theory and practical results, concrete can be summarized as follows:

1. Based on modern mechanical principles, it analyzes the basic idea of "forced supporting weak pressure, weak supporting forced pressure" principle.
2. It proposes an effective basic method to prevent the roadway's deformation, inhibit the floor heave in the thick very loose coal bed roadway;
3. The field test demonstrates the rationality and effectiveness of steel support reinforcement technology.

ACKNOWLEDGEMENT

This work is supported by Natural Science Foundation of Anhui Province (11040606M101).

REFERENCES

[1] Guo S. New thought of hot Zontal stress to the stability of surrounding rock of roadway [J]. *Coal Mine.* 1998, 5(4):14–17.
[2] Zhang N, Gao MS, Xu XL. Pretentioned supporting system ot roadway and Its engineering application [J]. *Ground PIessuie and Strata Control*, 2002, 19(4):1–4.
[3] Qi Ty, Lu SL, Gao B. Mechanical characteristics of rock bolts in roadways with large defOrmatron[J]. *Journal of China University of Mining and Technology*, 2002, 31(5):354–357.
[4] Bai Jianbiao, Hou Chaojiong, Du Mum_in, et al. On bolting support of roadway in extremely soft seam of coal mine with complex roof [J]. *Chinese Journal of Rock Mechanics and Engineering*, 2001, 20(1):53–56. (in Chinese).
[5] He Manchao, Xie Heping, Peng Suping, et al. Study on rock mechanics in deep mining engineering [J]. *Chinese Journal of Rock Mechanics and Engineering*, 2005, 24(16):2 803–2 813. (in Chinese).
[6] Gong Yao, Yin Jianguo. Bolt/anchor/steel mesh support technology for high inclined seam gateway with complex roof [J]. *Coal Science and Technology*, 2005, 33(7):18–19. (in Chinese).
[7] LI Bo. Analysis and control about stability of compound roof on the coal mine returns picks the tunnel [J]. *Coal Technology*, 2006, 25(2): 70–73. (in Chinese).
[8] Liu Haiyan, Wu Faqunn, Li Zengxue, et al. Study Oil stability characteristics of main seam roof in Yanzhou coalfield [J]. *Chinese Journal of Rock Mechanics and Engineering*, 2006, 25(7): 1450–1456. (in Chinese).
[9] Zhou Weiyuan. Advanced Rock Mechanics [M]. *Beijing: China Water Conservancy Hydropower Press*, 1989. 13–25. (in Chinese).

Frontiers of Energy and Environmental Engineering – Sung, Kao & Chen (eds)
© *2013 Taylor & Francis Group, London, ISBN 978-0-415-66159-1*

Synthesis and thermal stability of Sm(p-FBA) Phen Luminescent complex

J. Tao, Y.R. Ni, C.H. Lu, J. Chen & Z.Z. Xu
Materials Science & Technology, Nanjing University of Technology, Nanjing, China

ABSTRACT: Sm(p-FBA) Phen complex have been synthesized by wet chemical method. People study the thermal stability in the application. They were characterized using Fourier Transform Infrared (FT-IR) spectroscopy and fluorescence spectroscopy. The thermal stability of complex was analyzed by TG-DSC. The results shows the Sm(p-FBA) Phen complex is stable below 310°C. The photoluminescence (PL) analyses exhibit that the complex emits the characteristic red fluorescence of Sm (III) ions at 644 nm under ultraviolet light excitation. The highest peak of Sm(p-FBA) Phen fluorescence intensity only declined 41% after heat treatment 280°C. The morphology of the comlplex with diameter of 100 ± 50 nm nm is characterized by Scanning Electron Microscopy (SEM).

Keywords: samarium (III); 1,10-phenanthroline; thermal stability; luminescent complex

1 INSTRUCTIONS

Air pollution brings the level of carbon dioxide rising [1]. Both carbon dioxide rising and the ozone layer destruction make the greenhouse effect increasing, which result in a huge life-threatening to life on earth [2,3]. It can use natural photocatalytic—photosynthesis reactions in the nature to absorb large amounts of carbon dioxide to control greenhouse. For these purpose researchers prove that Rare Earth (RE) organic complex with high absorption efficiency, high optical conversion ability and long-life characteristics are now being extensive research [4,5]. Nowadays, more and more researchers focus on the preparation of ternary complexes [6]. There are two main ways to increase fluorescence intensity. The fluorescence of the lanthanides in the complexes can be further enhanced by the use of synergistic agents, such as 1,10-phenantroline, organic phosphatess and sulfoxides. These compounds provide an insulating layer around the lanthanide complex, reducing the probability of radiationless energy [7,8]. The complexes have dual effects that they can both absorb and convert harmful ultraviolet in to something useful and minimize the greenhouse pollution. In order to improve the stability of complexes, it should maintain the structure after they absorbed a lot of ultraviolet stability. In the past synthesis of Sm(TTA) Phen (Sm: Samarium, HTTA: α-thenoyltrifluoroacetone, Phen: 1,10-phenanthroline) samples has poor thermal and ultraviolet stability. The Sm(TTA) Phen complex

is stable below 245°C and the highest peak fluorescence intensity declined 61% after heat treatment 280°C [9]. So it cannot be composite to polymer materials and short for the life span.

In this paper, the properties of Sm(p-FBA) Phen (p-FBA: 4-Fluorobenzoicacid) Luminescent complex, the FT-IR absorption spectra, TG-DSC and fluorescence spectra are also discussed. It also observed that the Sm(p-FBA) Phen complex is stable below 310°C. The highest peak of Sm(p-FBA) Phen fluorescence intensity only declined 41% after heat treatment 280°C.

2 EXPERIMENT

2.1 *Apparatus and reagents*

$SmCl_3 \cdot 6H_2O$ (99.9%) was purchased from Funing Rare Earh Industrial Company Ltd. 4-Fluorobenzoicacid (p-FBA) (99.00%) and 1,10-phenanthroline (Phen) (99.00%) was obtained from Sinopharm Chemical Reagent Company Ltd.

2.2 *Charcterization*

Infrared spectra (FT-IR) were measured at room temperature in 4000–400 cm^{-1} region using a Nexus 670 FT-IR spectrophotometer with KBr pellet technique. Photoluminescence (PL) spectra were recorded in a HORIBA JOBIN YVON FL3-221 spectrofluorometer. The band pass for the excitation and emission monochromators was set at 2.0 nm. Thermogravimetric (TG) analysis of the sample was performed with STA 449C/6F analyzer with the temperature from room temperature to 900°C, at a rate of 20°C/min in air. The complex morphologies were observed on a JSM-5900 Scanning Electron Microscope (SEM) in 15 kv accelerating voltage.

2.3 Preparation and composition of the Sm(p-FBA) Phen sample

Firstly, the ethanolic solution of 6.0 mmol p-FBA was added dropwise to the 2.0 mmol $SmCl_3 \cdot 6H_2O$ ethanolic solutions. Next ethanolic solution of 2.0 mmol Phen that the molar ratio of Phen to Sm^{3+} ion being 1:1 was added to the mixture. Then the pH value of the mixture was adjusted to 5–6 by adding NaOH solution (1 mol/L), The solution was stirred for 6 h at 60°C. Finally, the suction filter taken precipitation, which precipitation washed with water and ethanol. Complexes powder stored in a silica-gel drier after drying at 50°C in vacuum drying oven.

3 RESULTS AND DISCUSSION

3.1 TG-DSC

The powder can added to the polymer material to application that need improved the stability of the powder. The TG-DSC curves of the complex were shown in Figure 3. The complex Sm(p-FBA) Phen initial decomposition temperature was 310 + 10°C. However, The Sm(TTA) Phen complex is stable below 245°C [9]. This is due to the stability of the benzene ring is higher than the thiophene ring, increasing the thermal stability of the complex. It proves that the higher rigidity of the organic ligands, the stronger thermal stability of the complex.

3.2 FT-IR

We choose Polycarbonate (PC) and powder composite, so we select 280 degrees temperature as the obsrved temperature points. Infrared spectra of FT-IR spectra of ligand Phen and Sm(p-FBA) Phen complex at 280°C before and after heat treatment are shown in Figures 1 and 2. The second ligand Phen has stretching vibration of $v_{C=N}$ in 1560 cm^{-1} in Figures 1 and 2 shown I curve represented Sm(p-FBA) Phen complex and II curve repeaented Sm(p-FBA) Phen complex after heating 280°C (holding time 0 min). The two curves peak almost no changed. After the formation of coordination bonds with samarium, the peaks move to lower frequencies to 1539 cm^{-1} [11]. The FT-IR spectra of the Sm(p-FBA) Phen evidenced intense absorption bands characteristic of the carboxylate groups at

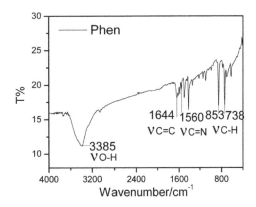

Figure 1.   FT-IR spectra of Phen.

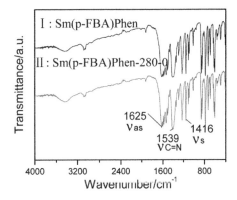

Figure 2.   FT-IR spectra of Sm(p-FBA) Phen at 280°C before and after heat treatment.

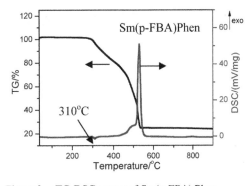

Figure 3.   TG-DSC curves of Sm(p-FBA) Phen.

1416, 1625 cm^{-1}, and are attributable to the symmetric v_s (C=O) and asymmetric v_{as} (C=O) vibrations, respectively [11]. And the vibration peak still exists when the Sm(p-FBA) Phen heated after 280°C. It proved the bonds has not breaking after heating treatment.

3.3 SEM

Figure 4 exhited the SEM image of the rare earth organic complex. It can be observed from Figure 4 that the complex sample diameters of 100 ± 50 nm. The poweder can be uniforrmly dispersed in the polymer material. The polymer material have high stabilty and luminescence propertise.

3.4 PL spectra

Figures 5 and 6 show the excitation and emission spectra of the Sm(p-FBA) Phen complex, which no heat treatment and 280°C heat treatment (holding time 0 min, 10 min, 20 min). It can be seen from the excitation spectra (Fig. 4) that there is the stronger and wider excitation band, which the highest excitation is nearly 360 nm. So the emission spectra of Sm(p-FBA) Phen complex (Fig. 5) is obtained 360 nm as excitation wavelength. We can find in Figure 5 that there are three emission peaks with different intensities at 564, 598 and 644 nm, which belong to the transition of $^4G_{5/2} \rightarrow {}^6H_{5/2}$, $^4G_{5/2} \rightarrow {}^6H_{7/2}$

Figure 4.   SEM image of the Sm(p-FBA) Phen complex.

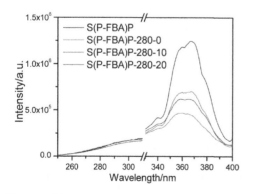

Figure 5.   Fluorescence excitation spectra of Sm(p-FBA) Phen@280°C at different time (λ_{em} = 644 nm).

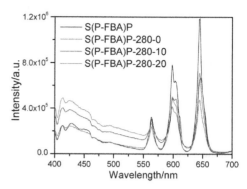

Figure 6.   Fluorescence emission spectra of Sm(p-FBA) Phen@280°C at different time (λ_{ex} = 360 nm).

and $^4G_{5/2} \rightarrow {}^6H_{9/2}$ of Sm^{3+} [12]. The strongest peak near 644 nm is strong and sharp, which the characteristic red fluorescence of Sm^{3+} resulting from $^4G_{5/2} \rightarrow {}^6H_{9/2}$ electron dipole transition. And the complex has a broad peak in the blue-violet light.

In Figure 5 the strongest peak 644 nm fluorescence intensity of Sm(p-FBA) Phen heat 280°C treatment (holding time 0 min) decreased by 41% than no heat treatment sample. The highest peak 647 nm fluorescence intensity of Sm(TTA) Phen declined 61% after heat treatment 280°C [9]. Therefore, the groups of fluorescence intensity that Sm(p-FBA) Phen sample has the smaller decrease at 280°C heat treatment. We considered that maybe due to the fluorescence quenching effect and the bond energy fact.

4   SUMMARY

Using a co-precipitating method, complexes Sm (p-FBA) Phen were synthesized. Based on the above discussion, The results shows the Sm(p-FBA) Phen complex is stable below 310°C. The highest peak of Sm(p-FBA) Phen fluorescence intensity only declined 41% after heat treatment 280°C. Also proved that the higher rigidity of the organic ligand, the stronger thermal stability of the complex.

ACKNOWLEDGEMENTS

This work is supported by the National Natural Science Foundation of China (Grant No. 20901040/B0111), the Natural Science Foundation of Jiangsu Province of China (BK2010553), the University Natural Science Grand Basic Research Project of Jiangsu Province of China

(Grant No. 10KJA430016) and the Innovation Scholars National "Climbing" Program of Jiangsu Province of China (Grant No. SBK200910148). A Project Funded by the Priority Academic Program Development of Jiangsu Higher Education Institutions.

REFERENCES

[1] Shakun, J.D. & Clark, P.U. & He, F. et al, 2012. Nature. 484: 49–54.

[2] S.L. Piao, P. Ciais, Y. Huang, et al: Nature. Vol. 467 (2010) p. 43–51.

[3] W.J. Manning, A.V Tiedemann: Environmental pollution. Vol. 88 (2) (1995) p. 219–45.

[4] H.F. Jiu, J.J. Ding, Y.Y. Sun: Journal of Non-Crystalline Solids. Vol. 352 (2006) p. 197–202.

[5] J. Wang, Z. Wang, H.S. Wang: Journal of Alloys and Compounds. Vol. 376 (2004) p. 68–72.

[6] H.F. Jiu, J.J. Ding, Y.Y. Sun, Journal of Non-Crystalline Solids. 352 (2006) 197–202.

[7] J. Wang, Z. Wang, H.S. Wang, Journal of Alloys and Compounds. 376 (2004) 68–72.

[8] A.V. Polishchuk, E.T. Kara Seva, A.T. Korpel, et al, Journal of Luminescence. 128 (2008) 1753–1757.

[9] Y.R. Ni, C. Xu, C.H. Lu, Z.Z. Xu: Acta Photonica Sinica. Vol. 39 (2010) p. 1424–1430.

[10] C.J. Xu, F. Xie, X.Z. Guo, et al: Spectrochim Acta A. Vol. 61 (2005) p. 2005–2008.

[11] S. Sivakumar, M.L.P. Reddy, A. H. Cowley, et al: Inorganic Chem. Vol. 50 (2011) p. 4882–2891.

[12] Y. Hasegawa, S. Tsuruoka, T. Yoshida, et al: Thin Solid Films. Vol. 516 (2008) p. 2704–2707.

Frontiers of Energy and Environmental Engineering – Sung, Kao & Chen (eds)
© 2013 Taylor & Francis Group, London, ISBN 978-0-415-66159-1

The study on performance of desulfuration bacteria degradation hydrogen sulfide and resistance toxicity of phosphine

S. Chen, L. Chen, B. Huang, H.Y. Zhao, X.M. Zhu, J. Deng & Z. Xiao
Faculty of Environmental Science & Engineering, Kunming University of Science & Technology, Kunming, Yunnan, China
Fuzhou Port Group Co., Ltd., Fuzhou, Fujian, China

ABSTRACT: A group desulfuration bacteria for degradation hydrogen sulfide in carbon monoxide was obtained from anaerobic pool of city sewage treatment plant by inductive domestication method with carbon monoxide contained H_2S gas. And the ability of the bacterium degradation H_2S in water, toxicity resistance of desulfuration bacterium to phosphine and another influence factors for the bacterium growth were investigated. The results showed that: the desulfuration bacteria have an ability with degradation rate of 30 mg/(L.h) and 60% degradation efficiency for H_2S in water. And the desulfuration bacterium is life propitious to weak alkaline environment of pH = 6~9; but phosphine in CO gas has a greater influence on activity of the desulfuration bacterium.

Keywords: carbon monoxide; hydrogen sulfide; phosphine; desulfuration bacteria; domestication

1 INTRODUCTION

The throughput of yellow phosphorus production equipment have exceeded 2×10^6 tons, and the production was close to 9×10^5 tons in 2010 (Tao J.F. & Yang J.Z. 2011). Yellow phosphorus production process by electric stove will produced a by-product, yellow phosphorus tail gas (Dai C.H. et al. 2009, Wu M.C. et al. 2003), which contain more than 85% (volume ratio) carbon monoxide (CO). Because of the tail gas is contained Sulfur (H_2S, COS) of 0.6–3.0 g/Nm³, Phosphorus (PH_3, P_4) of 0.4–1.0 g/Nm³, Fluoride (HF, SiF_4) of 0.2–0.5 g/Nm³, and 1–5% CO_2, 0.1–0.5% O_2 et al (Chen J.H. 2010), it made that the tail gas contained high concentration CO is difficulty to be used producing chemicals, and can only be used as raw material for drying phosphor ore or other low-grade fuel. The utilization rate is insufficient 40%, and can not satisfy the national regulation for effective utilization the tail gas. Like this direct combustion venting of the tail gas means that there are 1 million tons CO_2, sulfide of 2000–13,000 t (SO_2), phosphide of 4000–10,000 t (with phosphorus calculation), fluoride of 300–3000 t, and particulate matter of 1000–2000 t was discharge to atmosphere, it is equivalent to combustion 4.5–5 million ton coal, which pollute greatly the atmospheric environment (Liao M.D. et al. 2010).

At present, there are some methods, Water washing and alkali wash (Ren Z.D. & Chen L. 2004), Alkali washing and catalytic oxidation (Ning P. et al. 2004), Pressure swing adsorption process

(Chen Z.M. et al. 2001), Activated carbon adsorption process (Zhang Y. et al. 2009), and Liquid phase catalytic oxidation method (Tang X.L. et al. 2005) et al, which have a good removal effect to sulfur, phosphorus, fluorine and other impurities in yellow phosphorus tail gas. But there methods have some problems, high energy consumption, high cost, easy to produce intractable waste, or lead to secondary pollution. In order to achieve "energy saving and mitigation of CO_2" of the yellow phosphorus industry, to explore a purification process for yellow phosphorus tail gas with safety and practical is very important. With the advent of the patent of using soil bacteria to treat discharge gas contained hydrogen sulfide (H_2S), the study for purification H_2S emissions used microorganism has become one of the research focus at home and abroad (Oyarzun P. 2003, Luc M. 2006, Morgan-Sagastume J.M. 2006, Deng L.W. et al. 2009, Jensen H.S. et al. 2010), more mature technology is trickling biofilter purification of exhaust gases containing H_2S (Wu Y.G. et al. 2006, Qian D.S. et al. 2011). But the research about removal H_2S in CO with microorganism method and the resistance toxicity of phosphine (PH_3) for the desulfurization bacteria is no reports.

A group desulfuration bacterium with a capacity for removal H_2S in the carbon monoxide gas was obtained by inductive domestication method (Xiao Z. et al. 2011). For the particularity of impurities in the yellow phosphorus tail gas, this study focuses on the domestication proceeds desulfuration bacterium to resistance the toxicity of PH_3 and other affecting factors for the bacteria growth. The research achievement will provide a new theoretical basis and breakthrough for the method of microorganism purifying yellow phosphorus tail gas.

2 EXPERIMENTAL PART

2.1 The main raw material, the experimental setup and analysis methods

2.1.1 The main raw material
CO: cylinder gas, 99%; H_2S: made by NaS (AR) and HCl (AR) to be reaction; PH_3: made by AlP (AR) and HCl (AR) to be reaction. The domestication gas and PH_3 test gas: Home-made, H_2S and PH_3 collected in airbags to be leading into a container filled with CO gas, and mixed 30 min by the circulating pump, mensurating the concentration of H_2S and (or) of PH_3. Nutrient solution A (for domesticated of desulfuration bacterium): $NaNO_3$ (AR) 26.8 g/L, KH_2PO_3 (AR) 4.4 g/L in 1000 mL volumetric flask and add distilled water to the mark. Nutrient solution B (no phosphate, for mensuration toxicity of PH_3): $NaNO_3$ (AR) 26.8 g/L, KNO_3 (AR) 19.1 g/L to the 1000 mL volumetric flask and add distilled water to the mark.

2.1.2 Analytical methods
H_2S in the liquid phase: iodometry (State Environmental Protection Administration. 2002a), H_2S in the gas: determine by gas detection tubes, specification of H_2S test tube: 2–50 ppm, 0–200 ppm, 0–1000 ppm, 0–5000 ppm.

Phosphate in liquid phase: Molybdenum Antimony anti-spectrophotometric method (State Environmental Protection Administration. 2002b), PH_3 in the gas: 1) by PH_3 gas detection tube (for test gas without H_2S); 2) by PH_3 test tube in series H_2S detection tube (for test gas contained H_2S). Specification of PH_3 test tube: 0–50 ppm, 0–100 ppm, 0–1000 ppm.

2.1.3 Activity test of desulfurization bacteria
To take 50 mL domesticated desulfuration bacteria added into the flask prior to N_2 replacement the air, and filled with 500 mL water contained H_2S (known its concentration), control temperature by water bath. Test S^{2-} concentration in the liquid after degradation a certain period of time (2 h). Use degradation efficiency, $\eta_{S^{2-}}$, (%) (equation 1) and degradation rate, $r_{S^{2-}}$, (mg/(L·h)) (equation 2) as indicators for activity of the desulfuration bacteria.

$$\eta_{S^{2-}} = ([S^{2-}]_1 - [S^{2-}]_2)/[S^{2-}]_1 \times 100\% \qquad (1)$$

$$r_{S^{2-}} = ([S^{2-}]_1 - [S^{2-}]_2)/t \qquad (2)$$

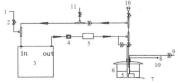

1. inlet 2. sealing clip 3. mixer (25L) 4. rotameter
5. air pump 6. domesticated device 7. submersible pumps
8. power supply 9. outfall 10. funnel 11. gas test port

Figure 1. Domestication experiment installation.

where $[S^{2-}]_1$ = S^{2-} concentration in the solution before microorganism degradation, mg/L; $[S^{2-}]_2$ = S^{2-} concentration in the solution after microorganism degradation, mg/L; T = the time of microbial degradation, h.

2.2 Domestication of desulfuration bacteria in CO atmosphere

At room temperature and ordinary pressure, to take 10 L sewage from anaerobic pool (muddy water volume ratio of about 1:9) of Kunming first sewage treatment plant placed in the incubator (Fig. 1). The desulfuration bacteria were domesticated by the domestication gas (contained H_2S only in CO), introducing the domestication gas 10–60 min very times, increasing the concentration of H_2S day by day, the flow rate of the domestication gas was controlled at 0.25–1.0 m^3/h. The domestication time and H_2S concentration were chosen according to the activity of the desulfuration bacteria last day. Replacing 1 L liquid contained bacteria body, and adding 0.9 L of fresh water and 0.1 L nutrient solution very day. The bacteria activity is test before replacing liquid.

3 RESULTS AND DISCUSSION

3.1 Domestication of desulfurization bacteria

The domestication of desulfuration bacteria is according to the method in 2.2, the results are shown in Figure 2.

It is distinct from Figure 2 that the degradation rate and degradation efficiency of the desulfuration bacteria have an increase with time quickly in the previous six days. It indicates that the domestication method is easy to domesticate desulfuration bacteria for degradation H_2S in CO, and the bacteria can be rapid growth and reproduction in CO environment. Degradation efficiency and degradation rate go into mildly from 7th days, the degradation rate and degradation efficiency of the domesticated bacteria at the 8th day and 9th day is basically equal to that of the seventh day. It can be identified that domestication is success.

417

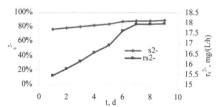

Figure 2.   Degradation rate and degradation efficiency of desulfuration bacteria.

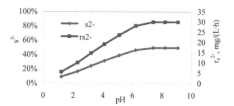

Figure 3.   Degradation rate and degradation efficiency in different pH value.

3.2   A fitting pH environment of desulfuration bacteria

The following study is for the desulfuration application in yellow phosphorus tail gas. Due to some acidic gases in the tail gas, the pH value of domestication environment of desulfuration bacteria was simulated, and the adaptability of domesticated bacteria in different pH was tested, where the temperature was controlled at 22–23 °C, the pH value of aqueous solution was adjust with dilute HCl and NaOH, The starting concentration of H_2S is 40.28 mg (S^{2-})/L in the liquid phase for test. The results were shown in Figure 3.

View the Figure 3, when the pH value less than 6, the solution was tiny acidic, the degradation rate and degradation efficiency of desulfuration bacteria are low, and have an increase with the down of acidity of the solution. The degradation of efficiency and degradation rate is greater at pH = 6.2. When the pH of the solution is 7.4 to 9.1, or, the degradation liquid is tiny alkaline, the degradation rate and degradation efficiency of the domesticated bacteria is slightly at a higher level. It indicate that the desulfuration bacteria is suitable for pH = 6–9 at purifying H_2S in CO gas.

3.3   Sulfur resistance capacity of desulfuration bacteria

The system temperature was controlled at 21–22 °C, the degradation ability of successful domesticated desulfuration bacteria at different concentrations of H_2S was tested. The results are shown in Figure 4.

To be seen from Figure 4, the bacteria have a certain of degradation ability for H_2S in water (no air) within the 22.14–74.58 mg (S^{2-})/L, it is showed that the domesticated bacteria is better, they can adapt a wide concentration range of S^{2-}. The desulfuration bacteria has a higher degradation efficiency and degradation rate in 46.28 mg (S^{2-})/L. Because S is a very important nutrient for microorganism, the too low S concentration will affect the microbial life activities, so domesticated bacteria have a lower degradation capacity on the S^{2-} at 22.14 mg (S^{2-})/L. But H_2S can also have toxic effects on biological cells.

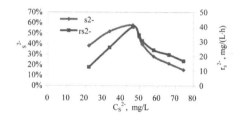

Figure 4.   Degradation rate and degradation efficiency in different concentration of S^{2-}.

Due to S^{2-} concentration is relatively low at the domestication process, the domesticated desulfuration bacteria does not yet suited the higher H_2S concentrations, So the degradation ability of the domesticated bacteria for S^{2-} in water is down, while the S^{2-} concentration is over 46.28 mg/L. When the technology is used to desulfurization for carbon monoxide contained H_2S, there are mass transfer process of low concentration H_2S transfer to liquid from gas, H_2S in liquid phase is usually not too high. It is imagined that the degradation ability on higher H_2S can be enhanced, using a method of gradually increase the H_2S concentration in domestication gas at the post—domestication.

3.4   The effect of PH_3 to the activity of desulfuration bacteria

The domestication experiment is carried out continuously for domesticated successful desulfuration bacteria with CO gas contained a certain concentration of PH_3 (60–100 ppm) and H_2S by the method in section 2.2, the no-phosphorus domesticated nutrition liquid B was used for replacing contained phosphorus domesticated nutrition liquid A for excluding the influence of P in water, the flux of domestication gas was controlled in the 0.25–1.0 m^3/h, and other conditions and actions are same to above domestication process. The ability of desulfuration bacteria degradation H_2S in water was determined for exploring the influence of the PH_3 in CO gas to domesticated bacteria activity. The results are shown in Figure 5.

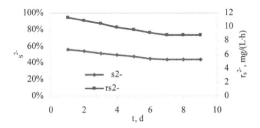

Figure 5. Effect of PH_3 to the activity of desulfuration bacteria.

Look at Figure 5, the degradation rate and degradation efficiency of desulfuration bacteria have a continuous decrease during the first six days after adding PH_3 in the domestication gas, but they are not decreased after 7th day. Which indicate that the desulfurization bacteria began to adapt the environment of the PH_3, and have a certain of desulfuration ability. However, to compare with Figure 2, we can find that there are a greater effect for activity of desulfuration bacteria, while adding PH_3 in CO gas, and the degradation rate of desulfurization bacteria on H_2S in water is 8.5 mg $(S^{2-})/(L.h)$, it is far low 30 mg/(L.h) without PH_3 in CO gas. This results show that PH_3 has a toxicity to desulfurization bacteria with the presence of PH_3 in CO gas. Otherwise, it is difficulty for PH_3 in gas substitution phosphorus in the water as a phosphorus source for microorganism growth, of course, it need further study.

4 CONCLUSION

It is possible that using inductive domestication method domesticate the bacterium from anaerobic pool of city sewage treatment plant with carbon monoxide contained H_2S, and the domesticated bacteria have an ability for degradation H_2S in water. And the desulfurization bacteria have a better activity in the neutral and tiny alkaline. However it had a greater influence to desulfurization bacteria activity when there is phosphine gas in CO gas.

Thinks Liang Chen as Corresponding Author to help for me. *Corresponding Author: Liang Chen E-mail: kmchenliang@hotmail.com.

REFERENCES

Chen Ji-shan. 2010. Comprehensive utilization of by-produce ferrophosphorus in yellow phosphorus production. *Phosphate & Compound Fertilizer* 25(6):62–64.
Chen Zhong-ming, et al. 2001. Purification and recovering of CO from yellow phosphorus tail gas by TSA and PSA. *Natral gas chemical industry*, 26(4):24–26.

Dai Chun-hao, et al. 2009. Design and verification of methanol synthesis gas by shift of purified yellow phosphorus off-gas. *Chemical Engineering*. 37(8):71–74.
Deng Liangwei et al. 2009. Process of simultaneous hydrogen sulfide removal from biogas and nitrogen removal form swine wastewater. *Bioresource Technology* 100:5600–5608.
Jensen H.S. et al. 2010. Growth kinetics of hydrogen sulfide oxidizing bacteria in corrodede concrete from sewers. *Journal of Hazardous Materials* 189:685–691.
Liao Ming-dian, et al. 2010. Experimental Study on Absorption of PH_3 and H_2S in Yellow Phosphorus Tail Gas by Composite Solvent. *Chemistry & Bioengineering*, 27(7): 84–86.
Luc M., 2006. Biological treatment process of air loaded with an ammonia and hydrogensulfide mixture. *Chemosphere* 50:145–153.
Morgan-Sagastume J M. 2006. Hydrogen sulfide removal by compost biofiltration: Effect of mixing the filter media on operational factors. *Bioresource Technology* 97:1546–1553.
Ning ping, et al. 2004. Purifying yellow phosphorus tail gas by caustic washing-catalytic oxidation, *Chemical Engineering*, 32(5): 61–65.
Oyarzun P., 2003. Biofiltration of high concentration of hydrogen sulphide using thiobacillus thioparus. *Process Biochemistry* 39:165–170.
Qian Dong-sheng, et al. 2011. Removal of hydrogen sulfide by plate type-biotrickling filter. *Environmental science*, 32(9): 2786–2793.
Ren Zhan-dong & Chen Liang. 2004. Oxidative removal of PH_3 and H_2S from yellow phosphorus tail gas by JC series catalysts. *Natral gas chemical industry*, 29(6): 19–23.
Tang Xiao-long, et al. 2005. Study of liquid phase catalytic oxidation of hydrogen sulfide in low concentration. *Techniques and equipment for environmental pollution control* 6(9): 19–23.
Tao Jun-fa & Yang Jian-zhong. 2011. Present status and developmental direction of phosphorus chemical industry in China. *Inorganic chemicals industry*, 43(1):1–3.
The State Environmental Protection Administration. 2002. *Shui he feishui jiance fenxi fangfa* (Book 4). Beijing: China Environmental Science Press, 133–136.
The State Environmental Protection Administration. 2002. *Shui he feishui jiance fenxi fangfa* (Book 4). Beijing: China Environmental Science Press. 246–248.
Wu Man-chang, et al. 2003. Approach to purification of tail gas from phosphorus production. *Phosphate & Compound Fertilizer* 18(4):41–43. 71–74.
Wu Yong gang, et al. 2006. Study on the start up of an innovative polyethylene carrier biotrickling filter treating waste gas containing hydrogen sulphide. *Environmental science*, 27(12): 2396–2400.
Xiao Zhuo, et al. 2011. The study on biological removes H_2S in simulation yellow phosphorus tail gas. In *2011 International Conference on Electrical and Control Engineering* [ICECE] 1861–1864, Sept 2011. Yichang, China.
Zhang Yong, et al. 2009. Removal of PH_3 and H_2S from yellow Phosphorus off-gas by metallic modified activated carbon. *Environmental Science & Technology*, 32(8): 57–61.

Frontiers of Energy and Environmental Engineering – Sung, Kao & Chen (eds)
© 2013 Taylor & Francis Group, London, ISBN 978-0-415-66159-1

Effects of mass flow rate and the throat diameter of diffuser on vapor absorption into aqueous LiBr solution for a liquid-gas ejector using convergent nozzle

H.T. Gao & R. Wang
Department of Marine engineering, Dalian Maritime University, Dalian, China

ABSTRACT: In order to improve the mass transfer efficiency, liquid-gas ejector is applied to the lithium bromide refrigeration system. Experimental analysis on liquid-gas ejector driven by aqueous LiBr solution is presented. Meanwhile, a model has been developed to predict the cooling capacity. The effects of mass flow rate and throat diameter of diffuser have been investigated. It has been observed that the cooling capacity increases with mass flow rate, and also increases with the throat diameter of diffuser from both the experiments and simulations; the required absorber volume per kW of cooling capacity increases as the throat diameter decreases; the empirical relationship could well predict the cooling capacity.

Keywords: lithium bromide solution; experiment; model; ejector

1 INTRODUCTION

The LiBr absorption refrigeration technique is widely used in central air-conditioning refrigeration technology because it can make use of low-grade heat, and refrigerant is safe and environmentally friendly. To develop the absorption refrigeration technology, its miniaturization and efficiency improvement is very important. It was studied widely by experts and scholars from domestic and abroad. Islam, M.A. 2009, described the development of a novel film-inverting design concept for falling-film absorbers. A conventional tubular absorber is modified by introducing film-guiding fins between tubes to produce a film inverting arrangement. The numerical simulation indicates that the vapor absorption rate can be increased by using a large number of film inverting segments in the absorber. Karami, S. 2011, focused on the numerical study of the combined heat and mass transfer process in absorption of water refrigerant vapor into a LiBr solution of incline plate absorber. The effects of plate angle and film Reynolds number on absorption process have been investigated. The average Nusselt and Sherwood numbers, which are essential parameters to design an absorber, are correlated as a function of plate angle and film Reynolds number. Yoon, J. 2005, proposed a model of simultaneous heat and mass transfer process in absorption of refrigerant vapor into a lithium bromide solution of water-cooled vertical plate absorber was developed. The absorption heat and mass fluxes, the total heat and

mass transfer rates and the heat and mass transfer coefficients get high values at the inlet region but decrease at the outside of the inlet region. Gao, D. 2003, studied the surface wave dynamics of vertical falling films under monochromatic-frequency flow rate-forcing perturbations by the direct simulation of Navier–Stokes equations using the Volume of Fluid (VOF) method to track free surfaces and the Continuum Surface Force (CSF) model to account for dynamic boundary conditions at free surfaces. At low frequency and high flow rate, the small inlet disturbance develops into large solitary waves preceded by small capillary bow waves. On the other hand, at high frequency and low Re, small-amplitude waves in nearly sinusoidal shape without forerunning capillary waves are formed on the surface. The wavy motions of the laminar wavy film flow with the Reynolds number 200–1000 are successfully found by the VOF and PLIC (Piecewise Linear Interface Calculation) method Tong, A. 2007. The numerical results, including the average film thickness, and the wave's amplitude, frequency and velocity, are compared with the experimental results. Yigit, A. 1999, describes a model of the absorption process in a falling film Lithium Bromide-Water absorber. The results showed that coolant side flow rate effects small on the values of mass absorbed, the outlet film temperature, and also the outlet mass fraction. Inlet coolant temperature influences on the mass absorbed.

In this paper the liquid-gas ejector was used in the lithium bromide absorption refrigerator in order to improve the heat and mass transfer efficiency.

In other words, LiBr solution is used as working fluid to suck the water vapor for the purpose of promoting miniaturization and efficiency improvement of lithium bromide absorption refrigerator. The goal of this work is to investigate the effects of mass flow rate and the throat diameter of diffuser on vapor absorption into aqueous LiBr solution for a Liquid-Gas ejector using convergent nozzle.

2 EXPERIMENTAL FACILITIES

The main components of the experiment set-up shown in Figure 1 are an absorber, a generator, a condenser, an evaporator, and control and measurement devices.

The absorption refrigeration system in this paper is based on single-effect absorption refrigeration cycle. The ejector-type absorber consists of an ejector and a heat exchanger. The ejector includes a nozzle (N), a mixing chamber (M) and a diffuser (D). The density and temperature of the inlet and outlet solutions can be measured by the mass flow meter. The concentrations of inlet and outlet solutions are determined, applying the water-LiBr density correlation as a function of temperature and concentration.

Table 1 shows the experiment conditions, and Table 2 shows the accuracy of measurement devices.

The heat and mass balance for the absorber are expressed respectively as

$$h_{Sin}W_{Sin} + h_{Vin}W_{ab} = Q_C + h_{Sout}(W_{Sin} + W_{ab}) \quad (1)$$

$$W_{in}x_{in} = (W_{in} + W_{ab})x_{out} \quad (2)$$

Table 1. Experiment conditions.

Experiment conditions	Parameters
Convergent nozzle	Outlet diameter 1.0 mm
Mixing chamber	Diameter 200 mm
Diffuser	Throat diameter 5 mm
Temperature of strong solution	18°C
Concentration of strong solution	50.7 wt%
Flow rate of strong solutin9	1.2~2.9 kg/min
Evaporator temperature	5°C

Table 2. Accuracy of the measurement instruments.

Measurement insutrument	Accuracy
PT100	±0.1°C
Pressure sensor	±0.01%
Densitometer	±0.01 g/cm³
Flow meter	±0.1%

where h is the enthalpy, W is the mass flow rate, W_{ab} is the mass flow rate of absorbed vapor, x is the mass fraction of LiBr, Q_C is the heat transfer rate. The subscript S and V refer to the solution and vapor, respectively, and the subscripts *in* and *out* indicate the inlet and outlet of the absorber, respectively.

3 CFD MODELING STRATEGY

In order to present a simple model, the following assumptions have been made:

1. The physical properties of the liquid solution are constant.
2. The mass flux at the interface as

$$M_{ab} = 9 \times 10^{-7}u + 6 \times 10^{-6} \quad (3)$$

where u is the nozzle outlet velocity. There is almost a linear relationship between the entrainment rate and velocity, Kandakure M.T. 2005. And our former experimental data also show that there is a linear relationship between cooling capacity and mass flow rate. Based on these experiment data and theoretical analysis, this empirical relationship (equation 3) is proposed.

In the present work, the ejector is located so that the convergent nozzle is downward directly. The entrained vapor flows around the lithium bromide solution jet in the annular space between the solution jet and the ejector. As a result, there is no bubble formation inside the ejector. Both the phases

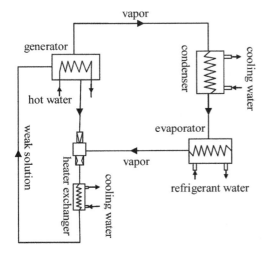

Figure 1. The schematic diagram of experiment.

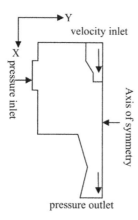

Figure 2. The geometries of model and the boundary conditions.

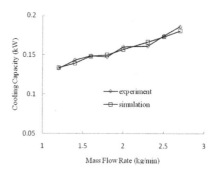

Figure 3. Comparison of the experimental chamber and the predicted cooling capacity for throat diameter of diffuser 6 mm.

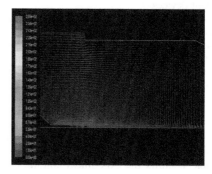

Figure 4. The velocity profiles of the mixture chamber as the throat diameter of diffuser is 6 mm.

flow co-currently. Two-dimensional axisymmetric geometry is considered with quadrilateral-meshing scheme. Lithium bromide solution as the motive fluid and water vapor as the entrained fluid are considered. Since the ejector geometry is down-flow, the gravity is taken in the positive X direction. Standard κ-ε model is used for modeling the turbulent behavior of the flow. Figure 2 shows a geometry studied and the boundary conditions used. Since the mass flow rate of lithium bromide solution through the nozzle were known, the inlet velocity is used as the boundary condition.

4 RESULTS

The predicted cooling capacity is compared with the measured values in Figure 3 with the diameter of mixing chamber of 50 mm, throat diameter of diffuser of 6 mm and 2EH of 50ppm. The predicted cooling capacity from simulation matches well with the measured cooling capacity. The curves in Figure 3 show that the cooling capacity increases with the mass flow rate and there is almost a linear relationship between the cooling capacity and mass flow rate. As the speed increases, the momentum transfer from the lithium bromide solution to water vapor increases resulting in an increase of the water vapor entrainment rate. Meanwhile, as the speed increases, the concentration gradient of LiBr aqueous solution at the gas-liquid interface increases, which make the mass transfer rate increase. Therefore, the cooling capacity increases from 0.133 kW to 0.186 kW with the improvement of mass flow rate from 1.2 kg/min to 2.7 kg/min.

Figure 4 shows the vapor velocity profiles of the mixture chamber as the throat diameter of diffuser

is 6 mm. The inlet velocity is different at the vapor inlet. At the gas-liquid interface, the velocity gradient is relatively large, it confirms the momentum transfer from LiBr aqueous solution to water vapor. There is no recirculation in the mixture chamber.

In order to clearly observe the velocity profile near the gas-liquid interface, this part is amplified and shown in Figure 5. The vapor velocity has a radial velocity component towards to liquid interface, which indicates that the gas is absorbed by LiBr solution.

The measured data and predicated results with throat diameter of diffuser of 7 mm and 5 mm are shown in Figures 6 and 7, respectively. The cooling capacity also increases with the mass flow rate, which is similar to the throat diameter of diffuser of 6 mm shown in Figure 3. The predicted cooling capacity from simulations matches also well with the experimentally measured cooling capacity.

Figure 8 shows the required absorber volume per kW of cooling capacity for different throat diameters of diffuser of 5 mm, 6 mm and 7 mm respectively for different mass flows ranging

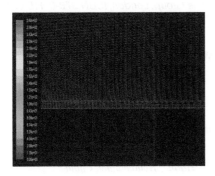

Figure 5. Part of the velocity profiles of the mixture chamber, the throat diameter of diffuser is 6 mm.

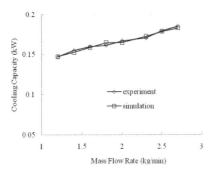

Figure 6. Comparison of the experimental and the predicted cooling capacity for the throat diameter of diffuser 7 mm.

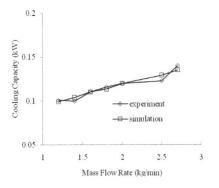

Figure 7. Comparison of the experimental and the predicted cooling capacity for the throat diameter of diffuser 5 mm.

from 1.2 kg/min to 2.7 kg/min. It shows that the required absorber volume per kW of cooling capacity increases as the throat diameter decreases. The required absorber volume per kW of cooling capacity has about 5.1% and 34.4% increases when the throat diameter is decreased from 7 mm to 6 mm and from 6 mm to 5 mm, respectively.

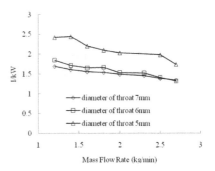

Figure 8. The influence of throat diameter of diffuser on volume per kW.

5 CONCLUSIONS

Liquid-gas ejector was applied to absorption refrigerator to enhance mass transfer efficiency, and a model was developed to analyze the experimental results. The predicted cooling capacity matched well with the experimentally measured cooling capacity. The conclusions are as follows:

1. The cooling capacity increases with the mass flow rate.
2. The cooling capacity increases with the throat diameter of diffuser.
3. The required absorber volume per kW of cooling capacity increases as the throat diameter decreases.
4. The empirical relationship could well predict the cooling capacity.

ACKNOWLEDGEMENT

The authors are grateful for the financial support from National Natural Science Foundation of China (No. 50776011) to this project.

REFERENCES

Gao D., Morley N.B., Dhir V. 2003 *Journal of Computational Physics* 192(2), p. 624–642.
Islam M.A., Miyara A., Setoguchi T. 2009. *International Journal of Refrigeration* 32(7):1597–1603.
Kandakure M.T., Gaikar V.G., Patwardhan A.W. 2005. *Chemical Engineering Science* 60(22):6391–6402.
Karami S., Farhanieh B. 2011. *Heat and Mass Transfer* 47(3), pp. 259–267.
Tong A.Y., Wang Z. 2007 *Journal of Computational Physics* 21(2):509–523.
Yigit A. 1999. International Communications in Heat and Mass Transfer 26(2):269–278.
Yoon J., T Phan T., Moon C G. 2005. *Applied Thermal Engineering* 25(14–15):2219–2235.

Frontiers of Energy and Environmental Engineering – Sung, Kao & Chen (eds)
© 2013 Taylor & Francis Group, London, ISBN 978-0-415-66159-1

Study on coal town land development law in China's Heilongjiang province based on resource exhaustion

M. Sun & W.B. Liu
School of civil engineering, Northeast Forestry University, Harbin, Heilongjiang, China

L.B. Dai
School of Civil Engineering, Heilongjiang Institute of Technology, Harbin, Heilongjiang, China

ABSTRACT: At present, the coal town land formation law is coal town planning research important one of coal town planning, representing the future development direction. This paper it is in this context, this paper discusses Heilongjiang coal town land formation law the necessity and feasibility of seeking guidance, coal town space health, sustainable development path. The article discusses and analyzes Heilongjiang coal town periodic law, space evolution rule, population growth rule and urbanization pattern of cold to coal town planning land development law provide certain reference for the study.

Keywords: coal town; resource exhaustion; spatial evolution

1 GENERAL INSTRUCTIONS

Coal town mostly take development mode of build with the mine margin, which form relatively the dispersed spatial structure of towns. Along with the weakening coal resources advantage, coal town space, create new development power which makes coal town space structure has a certain degree of change. The paper discusses Heilongjiang coal town law of development; combine the theory of exhausted resources.

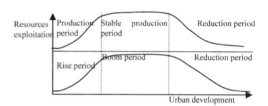

Figure 1. General process of coal city development evolution [1].

2 PERIODIC LAW OF THE COAL TOWN DEVELOPMENT

2.1 Cyclical theory

In the coal town development process, Figure 1 shows that the resources development can be divided into three stages: Increase production period, stable production period, reduction of output phase.

(1) Resources development initial Along with increasing escalation of the development enterprise funds and personnel the, resource development ability strengthens constantly, mining resources yield is rising year by year, the scale of resources zone is a increasing expansion; (2) stable stage, various production technology and management of resource development enterprise are better and better, the production have gradually stability; (3) Decline period with the reserves decline, resources mining difficulty is increasing, the cost of mining is rising, resource development enter into a recession stage. Table 1 shows that towns have different characteristics of the landscape in the various stages.

2.2 Periodic development process of coal town

Periodic development law of the coal town produces some major influence on the promotion of the coal town such as large-scale population migration, the decline of the city. Western countries completed the resources exploitation activities in mining area, if no new development opportunities, many towns can only keep a community in the state, and are hard to be developed. Many town's population would gradually reduce, until eventually become ghost town. Only a few towns will get another chance for opportunity. These towns often perform edge function of social and economic life, contact with other regions Less.

Table 1. Coal town development process & trait [2].

Development stage	Stage of development	Population growth stage	Economic development characteristics	Social development characteristics	Urban landscape facility
Production period	Rise period	Mechanical growth, rapid rise	Single enterprise, single industry, lack of service industry	Emerging risk crowd, unstable community culture, stable community	More crude facilities and the scattered settlements
Stable production period	Boom period	Natural growth, steady growth	Various supporting service industry	Cultural form, stable communities	More complete town facilities and the scale of the settlements
Reduction phase	Degenerating stage	Tends to reduce	Scattered local services	Mature culture, decline community	Broken landscape

3 COAL TOWN SPACE EVOLUTION RULE

3.1 Coal town space evolution process

According to the coal mining degree and the expansion of the urban areas, Figure 2 shows that the evolution process of coal town space structure present a round state, linear, the space of the expansion enclaves type evolution rule.

1. The forming stage. There are rich coal resources in a region, under the joint action of the reserves and mining conditions traffic condition, people start investing in building ore, this time the town has small scale, has obvious centripetal trend. With the increasing people, near mining area, workers living spots are formed, and will be coal town Initial built-up area.
2. Town extension stages. As the coal mine are put into operation and the corresponding related industry of coal washing, coking, chemical etc are developed. Heavy demand of the productive service and life service demand led to the development of the third industry, further attracted population growth, and make the town territorial scope to expand.
3. Town enclave's stage. When earlier mines decline or need to expand the scale of production, the construction of the new mines produces new heartland. Town space begins to "enclave's type" expansion, town space give priority to scattered centrifugation.
4. Town development again. When coal resources tend to dry up, other industry departments replace coal become the new pillar industry, industrial structure changes. Urban enclaves expansion stop, mainly enter into filling extension stages. At this time opposite direction of the adjacent residential areas development could eventually lead to interconnection, form a larger center, make whole town to form major

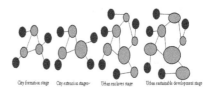

Figure 2. Coal town space stage of development.

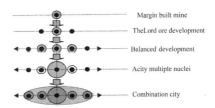

Figure 3. Coal city land space evolution.

centre. Figure 3 shows that if the resource distribution concentrated, and has the small scope, coal city space evolution stage may not apparent, and also not all of the coal town experience the same regional expanded way.

3.2 Coal town space evolution characteristics

Coal town space evolution process perform from gather to diffusion, and then to concentration [3]. (1) Spheres expansion, this is a kind of the most common way of urban spatial expansion. The formation mechanism is in accordance with the time sequence by the town center, to the different direction, commonly known as "over pie". In different economic development stage, the speed of expansion is different. Concentric expansion is also the basic way of world urban space expansion. (2) Thread ness expansion along the transit line is due to potentially high economy efficiency of transit, and also because of the coal industrial big traffic volume and the necessity of social and economic contact of the new and old mining area. High traffic accessibility is suited to transport processed product, and attract more enterprise layout along the lines.

3.3 Coal town space evolution characteristics

Coal town space evolution process perform from gather to diffusion, and then to concentration [3]. (1) Spheres expansion, this is a kind of the most common way of urban spatial expansion. The formation mechanism is in accordance with the time sequence by the town center, to the different direction, commonly known as "over pie". In different economic development stage, the speed of expansion is different. Concentric expansion is also the basic way of world urban space expansion. (2) Thread ness expansion along the transit line is due to potentially high economy efficiency of transit, and also because of the coal industrial big traffic volume and the necessity of social and economic contact of the new and old mining area. High traffic accessibility is suited to transport processed product, and attract more enterprise layout along the lines.

Middle development stage, if no other factors to promote and stimulation, urban population growth

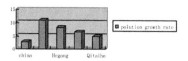

Figure 4. Average annual growth rate with Heilongjiang coal town population in 1949–2000.

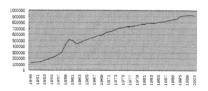

Figure 5. Jixi city total population.

will relatively stable. Due to less mining industrial labor requirements and the growth of new Labor, mechanical population growth gradually decrease, its growth mainly add up to proportion of natural growth the old and minors. Development later stage, the mining industry increasingly drop, urban population growth may be two kinds of trend. In the undeveloped processing industrial town, urban pollution will gradually reduce as the abandonment of coal mine, miners moved. Figure 5 shows that Jixi city population change, as its mining industry development, bring population increasing, but since 1975, it is obviously to slow down.

4 THE DEVELOPMENT LAW OF COAL TOWN

4.1 Periodic evolution rule

The development history of Heilongjiang coal town is short, many coal town are developed in large-scale after founding a state. Countries spend a lot of money and human, focus on developing the coal resources, coal town obtains rapid promotion. When resources gradually reduce, the state's input scaled down correspondingly, development of towns also tends to slow down. Interests, investment, production, marketing of the coal town enterprise are planed control by the state. Table 2 can be seen that the most towns of Heilongjiang have been in the big boom.

4.2 Low levels of generalized the urbanization process

The development of the coal town is a generalized urbanization process. From the economic development perspective, it is generally thought that agricultural development is the foundation and initial motive force of the urbanization establishment, industrialization is the essential force of urbanization, post-industrialization period, and the second industry gradually become the subject of urban economy and employment, and the subsequent power of urbanization. Information technology revolution further improves the quality

Table 2. Heilongjiang coal town stage of development.

Period	Large town	Middle coal town	Small coal town
Rise	–	–	Suzhou, Tiefa
boom	Pingdingshan, Zaozhuang, Huainan, Huaibei	Qitaihe, Shuangyashan, Yangquan, Shizuishan, Tong chuan, Xintai, Tengzhou, Cifeng, Liaoyuan, Changzhi, Wuhai, Fengcheng, Liupanshui	Hancheng, Jincheng, Huozhou, Yima, Shuzhou, Zixing, Laiyang, Manzhouli, Nuzhou, Gujiao
Recession	Hegang, Jixi, Fushun, Fuxin	Hebi, Beipiao, Pingxiang	Heshan

of urbanization, speed up the development of the town modernization. Figure 6 show that the economic development and the urbanization is a long-term interactive process [4].

4.3 *Gusty urbanization process*

The development of the coal town is not relying on natural population concentration, but a sudden growth process. Although population size has a certain extent, the corresponding infrastructure, industry and culture construction have quite a hysteresis. Such as the basic facilities, generally speaking, the town's infrastructure is always corresponding to its population scale. But, coal town infrastructure level is corresponding to population size; relatively higher level in the small towns of infrastructure, the infrastructure level of the big cities was far lower than the same scale of general town.

Table 3 can be seen that urban infrastructure construction lagged far behind the development needs, and higher urbanization level index can't reflect really the low quality urban life and public service. The construction of the coal town is passivity, temporary and variability. the residential areas has small scale, urban facilities only is branched passage by the relevant unit, culture education facilities is charged by the factories, public facilities and municipal construction is untouched.

5 CONCLUSIONS

Along with the construction and development of coal mine, town space also increased accordingly.

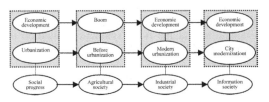

Figure 6. The interactive relationship model of urbanization and social economic development.

From building the mine centralization, city expansion of the development period, building a new mine enclaves of development, to town extension stages, it is shows that its development has the obvious periodic, this periodic is fit of the development of coal mine and periodic, Heilongjiang most coal town is in a boom, and large town gradually entered a recession, urbanization for sudden and low level of performance, etc. The paper discusses four formation rule of the urban land use of Heilongjiang, laid a theoretical foundation for the mechanism and strategy research of cold coal towns' development.

ACKNOWLEDGMENTS

This work was financially supported by the science and technology research funds for Heilongjiang provincial Department of Education, China (11541317).

REFERENCES

[1] Bradbury JH, "Living with boom and cycles: new towns on the resource frontier in Canada". Resource Communities, Australia, 1988:pp. 36–42.
[2] LiuYunGang. China resources city development mechanism and control countermeasures study. Northeast Normal University PhD thesis, pp. 9–63, 2002 (In Chinese).
[3] JiaoHuaFu. China coal city development model research [D]. Beijing university PhD thesis, PP. 5–76, 1998 (In Chinese).
[4] ZhaoJingHai. Chinese resources city space development [D]. The northeast normal university Ph.D. Thesis, pp. 7–42, 2007 (In Chinese).
[5] Sun Ming. Study on eco-planning of coal town special land based on Extenics [J]. Advanced Materials Research Vols. 450–451 (2012) pp 1108–1111.
[6] Sun Ming, Li Jie, Qin Xin. Innovation Research on Eco-Planning of Heilongjiang Coal-Exhausted Town Subsidence Area [J]. Advanced Materials Research vols. 368–373 (2012) pp 1849–1853.
[7] Sun Ming, DONG Jun. Urban Ecological Planning Opposite Problem Innovation Research Based on the Transforming Bridge Strategy [J]. Advanced Materials Research vols. 368–373 (2012) pp 1831–1834.

Table 3. Heilongjiang coal town infrastructure level compares.

Town scale	Town name	Water consumption per capita [tons]	Power consumption per capita n [KWH]	Road area of pavement per capita [M²]	Bus number per 10 thousand people	Infrastructure total amount per capita
Large town	State	81.8	315.6	5.8	7.3	850
	Hegang, Jixi	52.4	187.8	5.0	5.4	347.4
Middle town	State	79.3	213.7	4.4	4.0	496.0
	Qitaihe	67.6	260.9	5.7	4.9	338.0

Research on evaluation model of MME relationship in aviation maintenance

Y. Zhang

Department of Management for Postgraduate, Naval Aeronautics Engineering Institute, Yantai, China

L. Wang & J.J. Yang

Unit 92095, PLA, Taizhou, China

ABSTRACT: To estimate aviation maintenance man-machine-environment scientifically, evaluation index system is established and AHP-Fuzzy method is adopted. The man's suitability to machine, machine's suitability to man and influence factor of the environment on man and machine are three first-level indices. Then, the model to evaluating aviation maintenance man-machine-environment system is built. The detailed implementation and solving processes of evaluation are given. AHP-Fuzzy method reduces subjectivity in system evaluation dramatically. The faults of MME system and bottlenecks hindering aviation maintenance performance can be found through evaluation. It helps to improve aviation maintenance quality and efficiency practically.

Keywords: man-machine-environment; evaluation model; aviation maintenance

1 INTRODUCTION

AHP-Fuzzy synthetic evaluation method is the combination of AHP and fuzzy methods. Firstly, evaluation indices are hierarchized. Secondly, the weights of the indices are determined by AHP. Thirdly, fuzzy synthetic evaluation is implemented and conclusions can be drawn (Dong Du & Qinghua Pang, 2005, Modern …).

AHP-Fuzzy synthetic evaluation introduces the AHP method to evaluation model. As a result, subjectivity problem occurring in indices' weight determination is reduced dramatically. Also, shortcomings of determining the weights of indices relaying on expert to excess are overcame (Chunxiu Wang, 2005 Applied …). It filters the uncertainty factors to some extent fundamentally and errors that are caused by traditional weight determining method are eliminated. AHP supplies a reliable base for fuzzy synthetic evaluation. Fuzzy synthetic evaluation integrates suggestions of evaluators.

2 ESTABLISHING OF EVALUATION INDEX SYSTEM

In accordance with the principle of index system establishment, three main factors which are aircrew, aircraft and environment are analyzed. Finally, the evaluation index system related to Man-Machine-Environment (MME) system of aviation maintenance is established. The evaluation index system is established as shown in Table 1.

3 SYSTEM EVALUATION BY AHP-FUZZY SYNTHETIC METHODS

3.1 *Establishing of AHP-fuzzy synthetic evaluation model*

AHP-Fuzzy synthetic Evaluation model consist of two parts: one is AHP, the other is fuzzy synthetic evaluation, and the fuzzy synthetic evaluation is implemented based on AHP.

3.2 *Indices weight determination by means of AHP*

AHP decomposes complex problem to components so as to form hierarchy structure according the dominance relationship.

3.2.1 *Construction judgment matrix*
Dominance relationship is defined as the evaluation index system is established. Then relative importance of components in the same level should be determined.

Evaluation index system of MME includes 3 levels. Correspondingly, factors of evaluated object include 3 levels too. Indices set corresponding to first-level factors is $U = \{U_1, U_2, U_3\}$; Indices set corresponding to second-level factors which is contained in U_i is $V_i = \{V_1^i, V_2^i, ..., V_m^i\}$, $(i = 1, 2, 3)$;

Table 1. Evaluation index system of MME in aviation maintenance.

First-level factors	Second-level factors	Third-level factors
Man's suitability to machine	Quality	Suitability of human body size to working space
		Anti-fatigue ability
		Reaction ability
		Physiological adaptability
	Psychological quality	Moral level
		Regulation consciousness, safety consciousness
		Will quality
		Psychological adaptability
		Personality tendency
		Emotion quality
	Professional quality	Education level
		Level of knowledge mastering
		Maintenance skill level
Machine's suitability to man	Maintainability	Accessibility
		Interchangeability of components
		Working reliability
		Essential reliability
		Average workload of maintenance
		Degree of difficulty to master maintenance knowledge
		Instruction and warning signs
	Safety	Working safety
		Personnel safety
Environment influence o man and machine	Microclimate environment of space	Air quality
		Temperature
		Moisture
		Noise
		Visual surroundings
		Chemical substances
	Humanistic environment	Maintenance cost
		Maintenance management
		Team spirit
	Natural environment of airfield	Aircrafts' suitability to natural environment

Indices set corresponding to third-level factors which is contained in V_j^i is $W_{ij} = \{W_1^{ij}, W_2^{ij}, ..., W_n^{ij}\}$, $(j = 1, 2, ..., m)$. In order to quantify the importance of factors, an appropriate scaling is introduced to the judgment. As a result, a judgment matrix $A = (a_{ij})$ is formed. It is produced by multiple comparison of factor in current level. Judgment matrix often refers the values shown in Table 2.

$1/a_{ij}$ means that if the comparison of a_i to a_j is a_{ij}, then the comparison of a_j to a_i is $1/a_{ij}$. The judgment matrix corresponding to first-level is

$$A = \begin{matrix} M & U_1 & U_2 & U_3 \\ U_1 & \begin{bmatrix} 1 & U_{12} & U_{13} \\ U_{21} & 1 & U_{23} \\ U_{31} & U_{32} & 1 \end{bmatrix} \end{matrix} = \left(U_{fg}\right)_{3\times3} \left(U_{gf} = 1/U_{fg}\right) \quad (1)$$

Table 2. Scaling of judgment matrix.

a_{ij}	Meaning
1	i is as important as j
3	i is a little more important than j
5	i is apparently more important than j
7	i is significantly more important than j
9	i is extremely more important than j
2, 4, 6, 8	Intermediate value
$1/a_{ij}$	Comparison value of a_j to a_i

The value of U_{fg} represents the relative importance of U_f to U_g. The diagonal of the matrix is 1, which means that each factor is as important as itself. The judgment matrix corresponding to second-level is

$$B_i = \begin{array}{c} \\ V_1^i \\ V_2^i \\ \cdots \\ V_m^i \end{array}
\begin{array}{c} U_i \;\; V_1^i \;\; V_2^i \;\; \cdots \;\; V_m^i \end{array}
\begin{bmatrix} 1 & V_{12}^i & \cdots & V_{1m}^i \\ V_{21}^i & 1 & \cdots & V_{2m}^i \\ \cdots & \cdots & \cdots & \cdots \\ V_{m1}^i & V_{m2}^i & \cdots & 1 \end{bmatrix} = (V_{pq}^i)_{m \times m} \left(V_{qp} = 1/V_{pq} \right)$$

(2)

The value of V_{pg}^i means the relative importance of V_p^i to V_p^i. The judgment matrix corresponding to third-level is

$$C_{ij} = \begin{array}{c} \\ W_1^{ij} \\ W_2^{ij} \\ \cdots \\ W_n^{ij} \end{array}
\begin{array}{c} V_i \;\; W_1^{ij} \;\; W_2^{ij} \;\; \cdots \;\; W_n^{ij} \end{array}
\begin{bmatrix} 1 & W_{12}^{ij} & \cdots & W_{1n}^i \\ W_{21}^{ij} & 1 & \cdots & W_{2n}^{ij} \\ \cdots & \cdots & \cdots & \cdots \\ W_{n1}^{ij} & W_{n2}^{ij} & \cdots & 1 \end{bmatrix} = (W_{st}^{ij})_{n \times n} \;\; (W_{ts} = 1/W_{st})$$

(3)

The value of W_{st}^{ij} means the relative importance of W_s^{ij} to W_t^{ij}.

3.2.2 Consistency test

1. Consistency test of single matrix

Because judgment matrix obtained is sometimes not subject to consistency, the consistency test is necessary (Yimin Yang, Xubing Yang and Fengjie Jing, 2004, Judgment...). The consistency test method is shown as below.

 i. Computing consistency test index $C.I.$

$$C.I. = \frac{\lambda_{\max} - n}{n - 1}$$

(4)

$\lambda_{\max}$ is the maximum eigenvalue of the judgment matrix A.

 ii. Calculating consistency test discriminant $C.R.$

Equation (4) indicates that judgment matrix A is related to divisor n. That is to say, the acceptable critical values of $C.I.$ is different when the rank of matrix changes (Jijun Zhang, 2000, Fuzzy ...).

$$C.R. = \frac{C.I.}{RI}$$

(5)

The value of R.I. which is related to the rank of matrix can be accessed from Table 3 (Yingluo Wang, 2003, System...).

 iii. Consistency discriminating

When $C.R. < 0.1$ is tenable, the matrix is thought to be consistent. Otherwise, it is

required to be adjusted until it is satisfied to consistency constraint.

2. Hierarchically consistency test of judgment matrix

The consistency of judgment matrices above level k can be implemented by the formulas below. Where $RI_j^{(k)}$ and $CI_j^{(k)}$ are the R.I. and C.I. of the factor j in level $k-1$.

$$CI^{(k)} = \left(CI_1^{(k)}, CI_1^{(k)}, \dots CI_{n_{k-1}}^{(k)} \right) w^{(k-1)}$$

(6)

$$RI = \left(RI_1^{(k)}, RI_1^{(k)}, \dots RI_{n_{k-1}}^{(k)} \right) w^{(k-1)}$$

(7)

$$CR^{(k)} = \frac{CI^{(k)}}{RI^{(k)}}$$

(8)

when $CR^{(k)} < 0.1$ is tenable, judgment matrices above level k are thought to be consistent.

3.2.3 Index weight calculation

The judgment matrix corresponding to first-level is

$$A = \begin{bmatrix} U_{11} & U_{12} & U_{13} \\ U_{21} & U_{22} & U_{23} \\ U_{31} & U_{32} & U_{33} \end{bmatrix}$$

(9)

The weight vector $w^1 = (\bar{w}_1^1, \bar{w}_2^1, \bar{w}_3^1)$ is calculated as below.

$$w_f^1 = \left(\prod_{j=1}^{3} U_{fg} \right)^{\frac{1}{3}}$$

(10)

$$\bar{w}_f^1 = \frac{w_f^1}{\sum_{f=1}^{3} w_f^1}$$

(11)

The weight vectors of the second-level and third-level can be obtained in the same way.

It is assumed that weight vector of each matrix is obtained and amount of level is s. The number of elements in level k is n_k, k = 1, 2, ..., s, $n_1 = 1$. It is assumed that the relative weight vector of level 2 to level 1 is $w^{(2)} = \left(\varpi_1^{(2)}, \varpi_2^{(2)}, \dots, \varpi_{n_2}^{(2)} \right)^T$, the relative weight vector of level $k - 1$ to level 1 is $w^{(k-1)} = \left(\varpi_1^{(k-1)}, \varpi_2^{(k-1)}, \dots, \varpi_{n_{k-1}}^{(k-1)} \right)^T$ and the relative

Table 3. Mean random consistency index.

n	1	2	3	4	5	6	7
R.I.	0	0	0.52	0.89	1.12	1.26	1.36
n	8	9	10	11	12	13	14
R.I.	1.41	1.46	1.49	1.52	1.54	1.56	1.58

weight vector of level $k - 1$ to element j in level k is $p_j^k = \left(p_{1j}^{(k)}, p_{2j}^{(k)}, ..., p_{n_{k_j}}^{(k)} \right)^T$. The he relative weight vector of level $k - 1$ to the whole elements in level k is constructed as $p^{(k)} = \left[p_1^{(k)}, p_2^{(k)}, ..., p_{n_{k-1}}^{(k)} \right]_{n_k n_{k-1}}$.

Notice that vector $\left(p_{i1}^{(k)}, p_{i2}^{(k)}, ..., p_{m_{k-1}}^{(k)} \right)$ in the ith row of $p^{(k)}$ is the relative weight of the ith element in level k to level $k - 1$. Then relative weight of the element i to the first level is a inner produce:

$$\left(p_{i1}^{(k)}, p_{i2}^{(k)}, ..., p_{m_{k-1}}^{(k)} \right) \left(\omega_1^{(k-1)}, \omega_2^{(k-1)}, ..., \omega_{n_{k-1}}^{(k-1)} \right)^T = \sum_{j=1}^{n_{k-1}} p_{ij}^{(k)} \omega_j^{(k-1)}.$$

Then, the relative weight of elements in level k to level 1 is $w^{(k)} = p^{(k)} w^{(k-1)}$, $k = 3, 4, ..., s$.

The comprehensive weight vector is: $w^{(s)} = p^{(s)} p^{(s-1)} ... p^{(3)} w^{(2)}$.

3.3 System valuation by means of fuzzy synthetic method

The subjectivity of the evaluator and the fuzzy phenomenon in practice can be processed well by using fuzzy synthetic method.

3.3.1 Evaluation set

The evaluation set is $U = \{u_1, u_2, u_3, u_4\} = \{$excellent, good, general, poor$\}$.

3.3.2 Fuzzy evaluation matrix

What extent the element belongs to evaluation set U is described by membership degree. The fuzzy evaluation matrix of single element is

$$D_{ij} = \begin{matrix} & \begin{matrix} u_1 & u_2 & u_3 & u_4 \end{matrix} \\ \begin{matrix} W_1^{ij} \\ W_2^{ij} \\ ... \\ W_n^{ij} \end{matrix} & \begin{bmatrix} s_{11}^{ij} & s_{12}^{ij} & s_{13}^{ij} & s_{14}^{ij} \\ s_{21}^{ij} & s_{22}^{ij} & s_{23}^{ij} & s_{24}^{ij} \\ ... & ... & ... & ... \\ s_{n1}^{ij} & s_{n2}^{ij} & s_{n3}^{ij} & s_{n4}^{ij} \end{bmatrix} \end{matrix} (i = 1, 2, 3; j = 1, 2, ..., m)$$

(12)

s_{kl}^{ij} represents what extent is the kth element in level 3 of the jth element in level 2 of the ith element in level 1 belongs to first element of the evaluation set. The s_{kl}^{ij} is calculated as below.

First, the elements in level 3 are evaluated by experts. Then some remarks corresponding to W_k^{ij} are obtained. The number of remarks belonging to u_1 is W_{k1}^{ij}. The number of remarks belonging to u_2 is W_{k2}^{ij}, and so on. Then, the extent W_k^{ij} belongs to the remark of u_r is

$$s_{kr}^{ij} = \frac{W_{kr}^{ij}}{\sum_{r=1}^{4} W_{kr}^{ij}}$$

(13)

The membership degree of element W_k^{ij} in level 3 is $S_k^{ij} = (s_{k1}^{ij}, s_{k2}^{ij}, s_{k3}^{ij}, s_{k4}^{ij})$. Then, D_{ij} is obtained. Fuzzy synthetic evaluation of level 1: determining fuzzy relation matrix $R_i = \left(R_{i1}, R_{i2}, ..., R_{ij}, ..., R_{im} \right)^T$.

$$R_{ij} = (w_1^{ij}, w_1^{ij}, ..., w_n^{ij}) \begin{bmatrix} s_{11}^{ij} & s_{12}^{ij} & s_{13}^{ij} & s_{14}^{ij} \\ s_{21}^{ij} & s_{22}^{ij} & s_{23}^{ij} & s_{24}^{ij} \\ ... & ... & ... & ... \\ s_{n1}^{ij} & s_{n2}^{ij} & s_{n3}^{ij} & s_{n4}^{ij} \end{bmatrix} = (\bar{r}_{ij1}, \bar{r}_{ij2}, \bar{r}_{ij3}, \bar{r}_{ij4})$$

(14)

$(w_1^{ij}, w_1^{ij}, ..., w_n^{ij})$ is the ordered weight vector in level 3 of the jth element in level 2 of the ith element in level 1. The fuzzy synthetic evaluation of level 2 and evaluation of level 3 are calculated in the same way. The vector below is obtained: $S = (s_1, s_2, s_3, s_4)$. Evaluation results can be drawn based on the vectors.

3.3.3 Level judgment

The level of evaluated object is determined according to maximum membership degree law (Shuili Chen, Jinggong Li and Xianggong Wang, 2005, Principals ...). If $s_k = \max (s_1, s_2, s_3, s_4)$, that is s_k is the kth component of the vector, the valuated object belongs to the level k. The evaluators fall into P types, the result of synthetic evaluation is the vector $S_1, S_2, ..., S_p$ and the corresponding weight is $T_1, T_2, ..., T_k$. Then, the last evaluation result is (Yanmei Zhou & Weihua Li, 2008, Enhanced..., Li Yang & Nan Li, 2010, Application of ...): $E' = (T_1, T_2, ... T_k)(S_1, S_2, ..., S_p)^T$.

4 CONCLUSIONS

The MME system theory is applied to aviation maintenance on the basis of regarding it as a MME system which is comprised of aircrew, aircraft and environment. The research and evaluation of MME system of aviation maintenance is helpful to discover the unreasonable factors in the course of aircraft and peripheral devices design. Also, it helps to reduce the mistake of the aircrews in aviation maintenance. Similarly, adverse effect induced by environment will be found and eliminated based on analyzing the system. Therefore, the quality and efficiency of aviation maintenance can be improved.

REFERENCES

Chunxiu Wang. 2005. Applied Research on Job Appraisal and Performance Evaluation by AHP-Fuzzy Comprehensive Evaluation. Beijing: North China Electric Power University, pp. 35–40.

431

Dong Du, Qinghua Pang. 2005. *Modern synthetic evaluation method and cases selected*. Beijing: Tsinghua University Press, pp. 5–8.

Jijun Zhang. 2000. Fuzzy Analytical Hierarchy Process. *Fuzzy Systems and Mathematics*. Vol. 14, pp. 80–88.

Li Yang, Nan Li. 2010. Application of FAHP in software project risk prioritization. *Computer Engineer and Application*, Vol. 46, pp. 65–66.

Shuili Chen, Jinggong Li, Xianggong Wang. *Principals and Application of Fuzzy Set*. Beijing: Science Press, (2005), pp. 217.

Yanmei Zhou, Weihua Li. 2008. Enhanced FAHP and its application to task scheme evaluation. *Computer Engineer and Application*, Vol. 44, pp. 212–214.

Yimin Yang, Xubing Yang, Fengjie Jing. 2004. Judgment and Correction of Global Consistency in AHP. *Journal of Wuhan University*. Vol. 50, pp. 306–310.

Yingluo Wang. 2003. *System Engineering*. China Machine Press, pp. 122–125.

Frontiers of Energy and Environmental Engineering – Sung, Kao & Chen (eds)
© 2013 Taylor & Francis Group, London, ISBN 978-0-415-66159-1

Green engineering of urban river—in the case of WU Sha-river

J. Chen

NanChang Institute of Technology, Jiangxi, P.R. China

ABSTRACT: This paper analyses the problems that has existed in Wusha-river at present firstly, then it designs general scheme according to the requirements of urban master plan and flood control plan. These design methods of eco-hydraulic engineering are mainly used in ecologica 1 restoration of the River internal channel, slopes and banks protection of embankment are green, fishway of diversion dams and reengineering riverine wetlands are made, and they can provide certain experiences for the using of eco-hydraulic engineering in urban river green engineering.

Keywords: urban river; green engineering; Wusha-river channel; ecological design

1 INSTRUCTIONS

River is one important factor of urban ecological balance. it is green life line of urban. It has many works which contain flood control, waterway transportation, water supply, tourist recreation, environmental beautification, and natural ecological keeping, etc. The river also can decrease urban heat island effect. Urban residents are entertained by literary style near the river. So more flood control project need to be strengthen, and completely independent flood control project system need to be established. while we need to maintain channel landscape level; water environment and residential environment become better; water flows unblocked and the ecological wetlands are protected in the bank. In the end, we make river to be a secure healthy ecological river, which is also the needs of urban developments. Now, we make green engineering of urban river in the case of Wusha-river in Nanchang city.

2 WUSHA-RIVER CHANNEL FACING PROBLEMS

2.1 Condition of channel engineering

The Wusha-river divide into two parts by Chai bridge. Now, the channel above Chai bridge reach of wusha-river is almost nature statue. There are a lot of garbages in the channel, weeds have overgrown on the bank, water has been polluted seriously, and water also has been prevented by the bridge and culvert. The river was affected by the septations of roads and bridges, which digging or filling field arbitrarily, and discarding indiscriminately, such

as Figure 1. So the conveyance capacity of river was lower. The channel below Chai bridge reach of wusha-river has more than ten bridges. There were so many channel-control sections which prevented water most seriously. Some opening areas of these bridges were only 105 m and 155 m respectively, which made flood channel severe obstruction.

2.2 Flood control capacity

During the flood season, Wusha-river channel blocked drain, flood and watelogging were serious; while during the dry season, its riverbed exposed, waste water cross flowed. Otherwise, some faces of lakes and channels had been impropriated for other use, the flood diversion and detention capacity of basin would decrease continuously.

2.3 Flood control facilities

Current embankment of flood control capacity is low and single that could not meet the needs of city developing. Some embankments which had been excavated in the last 1960s have occupied river channels partly, it affected the flood diversion, and the wide of some channel is 20 ~ 50 m. Only a few embankments have been designed for heightening and consolidation. Others' embankment still is low tenuity. The flood control capacity is about 2~10 years encountered.

2.4 Present situation of water pollution

At present, Wusha-river channel hasn't set up water quality monitoring station. So it lacks of the special water quality monitoring data. According to water quality monitoring (annual, 2003~2004)

Figure 1. Some pictures of Wusha-river channel before regulation.

at Wusha river section of Shuangang water plant in the north zone in Nachang city, NH3-N has exceeded the Standards sometime. In addition, it had exceeded the standards in permanganate index, Mn, escherichia coli, etc. Wusha-river ecological system had been destroyed, its water self-purification capacity decreased, water quality was worse especially in dry season. The water quality couldn't meet the requirements which had affected domestic water use and industrial water use of urban residents, we must put more money into water treatment.

3 GENERAL DESIGN SCHEME

Wusha-river regulation works main contain: the internal of channel regulation works, the embankment project of consolidation and new building, diversion dam project, the sluice project of extension, rebuilding and new building, and riverine wetlands construction project. Mainly according to the river current basic situation, hydrological regime characteristics, the needs of flood and watelogging control, urban construction planning, and environment needed, with the asking of municipal, traffic and sight constructions, the project have implemented (see examples in Section 1). On the one hand, the design wants to make riverbed dredged, section extended, flood controlled by existing and reservoirs, which make the capacity of flood diversion and carrying improved, while in order to exclude watelogging and control river level we need to built engineering facilities which contain electric power station, sluice, diversion dam, and so on. On the other hand, it asks to make full use of theory method of eco-hydraulic engineering in channel design and various hydraulic buildings, which lays a foundation of channel

ecological diversity and river ecological capacity removing, while the river sight constructions of urban become brightly beautiful (see examples in Section 2).

4 GREEN ENGINEERING

4.1 *The ecological design of internal channel*

The internal channel ecological design makes vegetation systems along the river and biology underwater recovered, it makes water clean and bank green, while the channel water system is ecological which can meet the asking of residents living nearby river. We make a River plan which preserved existing channels, lakes and drainage channels. The plan made channel water surface as widen as possible, relieved the draining pressure of flood season in the river, while it enlarged the range of water infiltration, the growth of microorganisms are promoted in soil. The works don't take sludge landfill and silting.

We select compound section as river channel design, which have shown in Figures 2 and 3. These section not only meets the needs of draining flood and excluded waterlogging, but also meet the requirements of ecological environment and landscape. It contains planting concrete slope, riverine wetland, grid spring-grass concrete slope, and spring grass slope. Those can offer biology more activity space by wetland greening and channel enlarging. The channels was widen that may be best for groundwater recharged, flood flow inhibitory, and atmospheric purification.

4.2 *Greening the slopes and banks protection*

Because of the higher flood prevention, bank protections of Wusha river regulation are

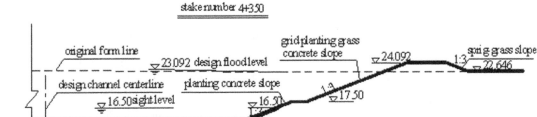

Figure 2. Channel duplex sectional drawing.

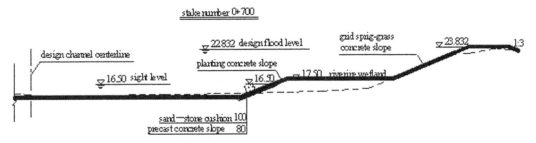

Figure 3. Channel duplex sectional drawing (contains riverine wetland).

containing of nature banks and mankind-nature banks. When making the design of slopes, we have considered two factors: water power stability design and ecological design. The kinds of slopes and banks protection projections are vertical types and stepped slope types. Materials usually adopted precast concrete, planting concrete, planting grid, etc. On the slower slopes, we take soil bank natural slope and gentle natural slope, it's protected by planting tree and grass, paving and packing rock blocks, gabion bank protection, etc. while on the higher slopes, we take stakes, wood frame add rubble, imitating wood pile revetment, planting concrete revetment, etc. These are shown as Figures 2 and 3. These designs stable riverbed, improve ecology and environment, promote self-purifying of water, make river water quality better. It has established a healthy channel ecosystem which is among sun, water, biology, soil and banks, on the natural topography and geomorphology.

4.3 Green design in lasher

Lashers are mainly applied to control sight water level of non-rainy season in Wusha-river, keeping sight water depth of dry season. When making the ecological design of lasher, we mainly consider two contents.

One is the problem how to solve river blocked problem of lasher, that is shoal's migration problem. We adopt fish way and fish weir for fish protected. The dam type of Lotus lake lasher and Huangjia lake lasher is smaller, we adopt plastic pipe or concrete pipe for fish passage, exports are at downstream or scour steady pool in the stilling basin. Fish weirs of which pore size is less than 10 mm (see examples in Section 5). which have been made in front of diversion and intake sluice to prevent from un safety. Those have shown in Figure 4.

Another is the problem of water quality decreased by lasher impoundment. Lotus lake lasher energy is dissipated by grid waterfall, that is one lay or two lays reinforced concrete level grid in downstream overflow weir. It is changed from over-flow lasher water into thin-layer face water firstly, and declined by grid crushing secondly, the polluted water has been aerated largely in the dissipation of energy, which can purify water quality by aeration. Its another feature is that floating garbage could be intercepted by grid which can be cleared by handwork or mechanical facilities on time. The grid prevents downstream water from floating garbage polluted.

4.4 Flood plain wetland engineering

The greening range of Wusha-river regulation engineering mainly includes dyke body, embankment

435

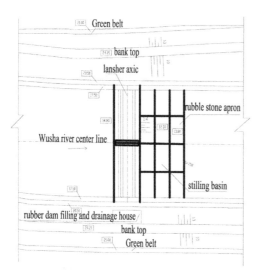

Figure 4. Lasher sectional drawing.

Figure 5. Wetland engineering effect figure.

management zone which is the lateral 50 m range of tsutsumiuchi foot, regulating storage flood area coastline riverine wetland, etc. which has shown in Figure 5. The key wetland constructions are Kong-mu-lake, Mingya Europe town, Huangjia-lake, Lianhua-lake, etc. We construct wetland waste-water treatment system: firstly, we imbed aquatic plant before waste water flowing into the channel; then, we adapt the work level of wetland system according to plant growth habits and design level, so as to promote the growth of aquatic plant and its foot growing into basement depths (see examples in Section 6).

In the stable mature stage, the system is in the dynamic balance while plants growing periodic variation with the season, and its gives full play to clean and treat, works stabled. The plants can reap and refresh to the needs in wetland. The spending of work is low. When water quality and quantity have a little change, the system which wouldn't use power equipment and any other maintenance can self-flow, The aquatic plants design of compound contributed wetland: stair oxidation plants hya-cinth, water lettuce, plant in the floating cultivation mode; secondary oxidation pond plants mainly are water dropwort and water lettuce; subsurface flow wetland plans mainly are cattail, calamus, canna, black rush, cyperus alternifolius; gravel collector drain plants mainly are cattail and canna.

5 CONCLUSIONS

Eco-environmental problems of urban river are complexly multivariable, which involved hydraulic, ecology, mathematics, physical, chemical, biology, meteorological, hydrogeology, system engineer-ing and computer scientific technologies. In this paper, firstly we build ecological river bank which can recover plant community along the bank and aquatic ecosystem in water; secondly we increase wetland and flood plain areas; the end we maintain the biological diversity, these are international inev-itable trend of river construction. In the future, We will analysis on structure and function of hydraulic ecological system comprehensively, and then well establish the evaluation index system of river eco-system functions to predict the ecological impact of water conservancy works, so as to promote eco-logical balance an improve eco-environment.

REFERENCES

[1] Zhe-ren Dong. submitted to Journal of Hydraulic Engineering, in Chinese, 2004. 10.
[2] Shiguo Xu, Yongmin Gao, et al. Planning and Construction of Modern Riverbank—creating harmonious riverside environment between human and natrue, in Chinese, M, China Water Conservancy And Hydropower Publish, 2006.
[3] VA Compeman Meng Mphil. CIWEM, 1997.
[4] Changfeng Fu. Water Resources Planning and Design, in Chinese. 2007.
[5] Vannote RL, Minshall GM, CumminsKW, et al. The River Continuum Concept, J, Canadian Journal of Fisheries and A-quatie Sciences, 1980, p:130–137.
[6] Kai-qi CHEN. submitted to Journal of Hydraulic Engineering, in Chinese, 2012. 2, p:183–187.

Frontiers of Energy and Environmental Engineering – Sung, Kao & Chen (eds)
© *2013 Taylor & Francis Group, London, ISBN 978-0-415-66159-1*

Calculation and trend analysis of China's international shipping fuel consumption/GHG emission

W.Q. Wu, Q.G. Zheng & X. Feng
Dalian Maritime University, Dalian, Liaoning, China

ABSTRACT: First, this paper introduces two main methods of calculating the GHG emission: top-down method and bottom-up method briefly, and by five different bottom-up modes, calculates the total fuel consumption and GHG emission in 2009 & 2011 of China's international shipping industry. Then, the paper contrasts and analyzes the different results of different calculation mode and of different years. The study shows that the total GHG emission of China's international shipping industry increased a lot in recent years, and the major ship fleet structure are continuously changing and large size is the trend.

Keywords: international shipping; fuel consumption; GHG emission; top-down; bottom-up method

1 GENERAL INSTRUCTIONS

As well as giving us abundant products, the industrial revolution brings up enormous amount of GHGs. Global warming caused by the increase of GHG in atmosphere has caused wide concern. According to relevant study, international shipping attributed about 2.7% of total GHG emission in recent years, and the most major GHG given out by ships is CO_2. So it is important, for the control of the global GHG emission, to study the total fuel consumption and GHG emission of international shipping[1-3].

According to different methods given by IMO, this paper calculates the total fuel consumption and GHG emission of China's shipping industry. By this study, the total emission level and the change trend could be obtained and the conclusion would be helpful for the emission control strategy.

2 CALCULATION METHODS

2.1 General description

There are two different main methods to calculate the total GHG emission of a certain industry: top-down method and bottom-up method. By top-down method, the total emission of the industry is calculated according the total energy consumed by this industry, the type of the energy resource, and the GHG converting coefficient, where the total energy consumed is given by statistics. As for the international shipping industry, the total fuel consumed is obtained from shipping company or the state statistics department and the rotation volume

of freight is obtained from the state transportation department. Similar to top-down method, bottom-up method calculates the total emission by multiplying the total energy with the GHG emission factors of the certain energy resource, the difference lies in the total energy is obtained by setting a reasonable active level and a efficiency level to the equipments, where the equipment quantity is achieved by statistics [4].

Comparatively, top-down method is easy to use; the reliability of the result depends on the accuracy of the data. Bottom-up method needs more work in statistics, and the active level and efficiency level are affected by many factors and difficult to set accurately, so the result will fluctuate in certain range. As the international shipping is concerned, it is not suitable to use top-down method to calculate a certain country's total emission because the ships may take bunkers in different country.

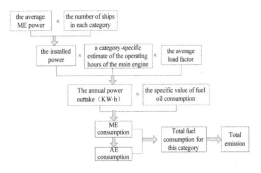

Figure 1. Flow chart of bottom-up method.

Table 1. Preferences for the five bottom-up methods.[5]

	Ship category: working time (average days)	specific fuel consumption (g/KW*h)	Average load factor of main engine
Corbett et al., 2003	Cargo ship: 229–292 (271)	206 g/KW*h (185–225 g/KW·h)	65%~70%
Eyring et al., 2005	Cargo ship: 225–275	210 g/KW*h	70%~80%
IMO expert group (2007)	All ships: 175–310 (226)	185 g/KW*h	62%~90% (average 80%)
Endresen et al., 2007	Cargo ship: (181)	221 g/KW*h	70%
MEPC 59th	All ships: 100–285 (240)	196 g/KW*h	For dry cargo ships: 65%~80%, Average: 70% All ships: 16%~80%, average: 64%

2.2 Bottom-up method

Bottom-up method is to calculate the CO_2 emission by means of, getting the number, type and the rated power of the ships through researches, setting a reasonable active level to the ships through the study of the activities of the ships. The calculating flow is shown as Figure 1.

2.3 The methods used in this paper and the parameter setting

This paper uses the flowing five popular bottom-up methods: Corbett et al. (2003), Eyring et al. (2005), IMO expert group (2007), Endresen et al. (2007) and the method given by MEPC 59th.

Corbett et al., 2003 can only be used to deal with cargo ships, including tankers and dry cargo ships. In this method, the working time of main engine varies from 229 days to 292 days per year, default setting is 271 days, average load percentage of main engine is 65%~70%, the specific fuel consumption is 206 g/KW*h, which may change from 185 g/KW*h to 225 g/KW*h. Other concerned parameters are shown in Table 1.

According to the calculating flow and the preferences, the fuel consumption of main engines can be obtained. So after, the fuel consumed by ships is obtained according to the power ratio of AE/ME given by IPCC (here the emission of boiler and other equipments is skipped). In the end, the CO_2 emission equals to the fuel consumption times 3.130 [6].

So far, to know the annual fuel consumption and the emission of the ships, the following data are needed: the installed main engine power, the working time of the main engine, the average load factor and the specific fuel consumption, among which, the working time is difficult to determine, and has a lower accuracy. For ships with unknown power, the emission is estimated according to the tonnage of the ship and the relationship between fuel consumption and the ship tonnage[7].

2.4 Study scope

This paper aims to calculate international ships registered in China, and calculates and analyzes this kind of ships in 2009 and 2011 respectively.

By statistics, on 2009, April 12, 1330 ships are registered in China, among which bulk carrier, container ship, oil tanker, general cargo ship, chemical/oil tanker and multipurpose ships count up to 661, accounting for 49.7%.

On 2011, March 15, 1136 ships are registered in China, among which bulk carrier, container ship, oil tanker, general cargo ship, chemical/oil tanker and multipurpose ships count up to 692, accounting for 60.9%.

In calculation, the ships are categorized as container ship, oil tanker, bulk carrier, general cargo ship, reefer ship, Tug, working ship, Public Service Ship, sand dredger, multi-purpose ship, dredger, chemical tanker, chemical/oil tanker, aquatic products carrier, roll-roll ship, engineering ship, heavy cargo ship, salvage ship, special purpose ship, roll-roll passenger ship, liquefied gas carrier, bulk asphalt ship, semi-submerged vessel, barge, floating platform, piling ship, wood ship, floating crane, drilling ship, sand mining, supply ship, research vessel, passenger/cargo ship, passenger/container ship. The emission of each category of ships is calculated and then summed up.

3 RESULTS

3.1 Calculation results

The calculation results of fuel consumption and emission of China's international shipping fleet are given in Tables 2 and 3.

3.2 The comparison among the results of different bottom-up modes

The comparison among the results of different calculation modes of 2009 and 2011 is shown as Figure 2.

438

Table 2. Fuel consumption (unit, million tons).

Methods	2009		2011	
	Range of fuel consumption	Default consumption	Range of fuel consumption	Default consumption
1. IMO expert group	3.952027~9.59208	6.29945	4.39811~11.3094	7.32884
2. MEPC 59th		5.681618		6.60697
3. Corbett et al., 2003	4.314~7.2051	5.904043	5.18838~8.66512	7.09989
4. Eyring et al., 2005	5.18174~7.23799		6.23176~8.70468	
5. Endresen et al., 2007		4.386768		5.2757

Table 3. Total emission (unit, million tons).

Methods	2009		2011	
	Range of emission	Default value	Range of emission	Default value
1. IMO expert group	12.35323~30.0066	19.70067	13.76609~35.39852	22.93926
2. MEPC 59th		17.9672		20.67982
3. Corbett et al., 2003	13.503329~22.551921	18.479654	16.23962~27.12182	22.22264
4. Eyring et al., 2005	16.218852~22.654899		19.50542~27.24566	
5. Endresen et al., 2007		13.62168		16.51293

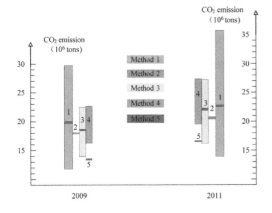

Figure 2. The comparison among five results in two years.

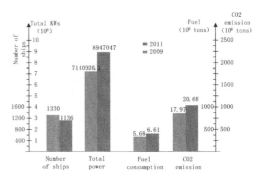

Figure 3. The comparison between 2009 and 2011.

Method 1 to method 5 in Figure 2 are in the same sequence as in Tables 2 and 3. The comparison shows that, the result regularities in each year of different modes confirm each other and the calculation method is stable.

3.3 The comparison between the results in two years

According to the statistics data, MEPC 59th mode gives the following results: 17.9672 million tons of CO_2 emission in 2009, 20.67982 million tons of CO_2 emission in 2011, the year-on-year growth is 15.1%.

To analyze the fleet data, 1330 ships were registered in China in the year of 2009, 1136 ships were registered in China in the year of 2011, the number of ships fell 14.58% year-on-year. Though the number fell, the power has increased obviously, which leads to the increase of the emission. Figure 3 shows the comparison between 2009 and 2011 about ship number, total power, calculation results of fuel consumption and total emission.

By comparing the data of the fleet in 2009 and 2011, the study also finds the fleet structure has changed. Both the number and total power of tankers has decreased. As for container ships,

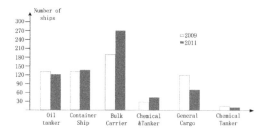

Figure 4. The number changing of main ship types.

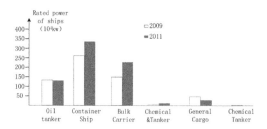

Figure 5. The total power changing of main ship types.

the number has increased slightly, and the power has increased to much more extent. This proves the China's container ship is developing toward large size. The number and power of bulk carriers has increased in same trend. Total volume of chemical & oil tanker scale is small but is rising up. Both the number and the total power of general cargo ships declined significantly. The changing trend of the number and the total power of each ship type is shown as Figures 4 and 5.

4 CONCLUSIONS

According to the analysis, the following conclusions can be summarized:

1. By the end of March 15, 2011, the CO_2 emission of international ships registered in China reached 20.68 million tons (as calculated by the MEPC 59th method), raised about 15.1% by comparing to the data of April 21, 2009.

2. The major ship fleet structure in China changed a lot and large size is the trend, especially for the container fleet. The total number of the container fleet remained unchanged but the power increased a lot. The number and power of the bulk carriers show an upward trend, which indicate that the number of new bulk carrier may increase. At the same time, the number of tanker fleet shows a steady downward trend.

3. The five popular calculation methods were studied and verified in this study. And then the CO_2 emissions were analyzed from different angles.

REFERENCES

Farrell Alex, Glick Mark. 2000. Natural Gas as a Marine Propulsion Fuel. *Transportation research record. volume number: 1738.*

ICCT. Air Pollution And Greenhouse Gas Emission From Ocean-going Ships. 2007.

IMO. Second IMO GHG Study 2009. 2009.

IMO. 2005. Interim guidelines for voluntary ship co_2 emission indexing for use in trials. *MEPC/Circ. 471.*

IMO. 2009. Guidelines for voluntary use of the ship energy efficiency operational indicator (EEOI). *MEPC. 1/Circ. 684.*

MARINTEK, ECON, DNV. Study of greenhouse gas emissions from ships. IMO Issue no. 2–31, 2000.

MARINTEK, ECON, DNV. 2000. Study of greenhouse gas emissions from ships. *IMO Issue no. 2–31.*

Yang Su. 2005. The former and existing status of Kyoto Protocol. *Ecological-economic.*

Frontiers of Energy and Environmental Engineering – Sung, Kao & Chen (eds)
© 2013 Taylor & Francis Group, London, ISBN 978-0-415-66159-1

Spatial expansion of development zone in Changchun city and its impact on urban space

R.Q. Pang & Y.P. Gao
School of Urban and Environmental Science, Northeast Normal University, Changchun, Jilin, China

X.R. Wei
Liaoning Urban and Rural Construction and Planning Design Institute, LURDI, Shenyang, China

D.X. Zhang
Department of Architecture, Northeastern University, Shenyang, China

ABSTRACT: The development zones of Changchun city have promoted changes in economic structure, spatial form and structure of urban space. This paper focuses on the research of the four development zones, their development situation and the characteristic of spatial expansion. In order to make a healthy and orderly development both in Changchun city and in its development zones, we analyze the crucial problem on the construction of development zones and its impact on urban space, also make a discussion in the selection of point to axial development model, which concerns about the quality of spatial expansion, coordinates the layout of the space structure and improves the efficiency of land use.

Keywords: urban spatial structure; spatial expansion; development zone

1 DEVELOPMENT STATUS OF DEVELOPMENT ZONES

There are three national development zones, nine provincial development zones and three provincial industrial zones in Changchun, so it has already formed a multi-level development zone construction system. Among them, there are four zones, namely high tech industrial zone, economic and technological development zone, Xixin and Jingyue economic and technological development zone, in the geographical space is connected with the downtown of Changchun (see Fig. 1).

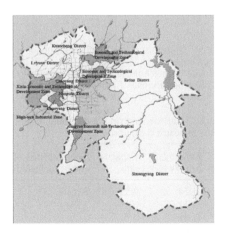

Figure 1. Location map of four development zones.
Reference: "Refer to Changchun overall urban planning and drawing by myself".

2 THE CHARACTERISTICS OF SPATIAL EXPANSION OF CHANGCHUN DEVELOPMENT ZONE

2.1 *Large scale, speed, and the accelerating expansion trend*[1]

From the city land transfer TAB we can see that the transfer of the four major land development zone in Changchun City area accounts for more than half of the total land area (see Figs. 2 and 3). In 2003, due to the starting stage, the land transfer area accounts for 30.9% of the total area. From 2004 to 2008, the development zones ahead into the rapid development stage and the annual transfer lands account for total urban land area were 69.3%, 77.8%, 53.7%, 58.8%, 68.8%. In 2009, the transfer area of the development zones is almost the same as other regional total transfer area of the city.

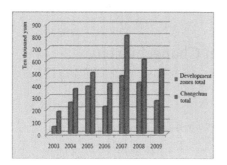

Figure 2. Land transfer situation in urban area of Changchun and four development zones.
Reference: "Changchun Bureau of Land and Resources".

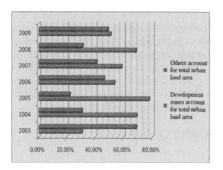

Figure 3. The proportion of Changchun's four development zones to urban area of Changchun in land transfer.
Reference: "Changchun Bureau of Land and Resources".

2.2 Industrial space leads the development of living space, becoming the agglomeration of urban new living space

The construction of the development zone makes economic growth as its goal, the policy-driven and foreign investment have become the core power of spatial expansion, and the expansion mainly based on the industrial space [1]. In recent years, the development zones try to improve its residential, commercial, education, scientific research, and other facilities; the new type of modern living community has begun to take shape.

2.3 Intensive land use

2.3.1 Low land output efficiency

Assuming the transfer of land can form a production capacity after one year, through the proportion between development zones' GDP and the transfer area, we can see, the zones annual output coefficient increased gradually, but individual years decline. Although this analysis is still not stringent

enough, but generally reflects the current zone of land and output efficiency is not satisfactory[1] (Figs. 4 and 5).

2.3.2 Low intensity of land development

Industrial land use planning condition rules: building density < 50%, capacity rate < 0.7, the height < 24 m, administrative office and living facilities < 7%, green rate < 30%. In fact, only the Xixin Zone new building density index is greater than the

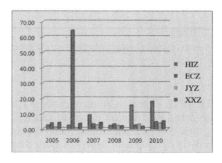

Figure 4. Output coefficient analysis of four development zones respectively.
Reference: "Changchun Statistical Yearbook (2006–2011) and Changchun Bureau of Land and Resources".

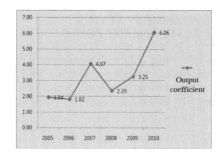

Figure 5. Output coefficient analysis of the sum of four development zones.
Reference: "Changchun Statistical Yearbook (2006–2011) and Changchun Bureau of Land and Resources".

Table 1. Four development zones land exploitation intensity indicator comparison.

	Building density (%)	Capacity rate	Administrative office and living facilities (%)
HIZ	41.54	0.59	11
ECZ	32.11	0.67	7.78
JYZ	41.48	0.72	6
XXZ	55.64	0.70	5

Reference: "The land intensive use project evaluation of development zones in changchun city (Data Time 2008)".

national standard, the other zones are less than 50%. The volume rate of high-tech zone and the development zone are lower than the national standard, building layers is 1–2 mainly. High-tech zone and the open area administrative office and living facilities ratio is higher than the national standard, especially high-tech zone is far more than the national standard (Table 1). According to the above data, there is still room for improvement.

3 THE EFFECTS OF DEVELOPMENT ZONES' CONSTRUCTION ON CHANGCHUN CITY

The agglomeration of Changchun industrial production activities inevitably lead to the change of the structure of urban space, since large-scale land development zone and building speed, lead to its acceleration in the evolution of urban space structure. Meanwhile, the construction of the development zones has been the new power of the expansion of urban land. With the development of industry growth, lead to the agglomeration of city's development and population, so that leads to the change of city's space form and its function[2].

3.1 Effects on the boundary of changchun city

The four development zones are next to Changchun city, lead to the rapid expansion of urban space, and promote the urbanization of fringe areas. Meanwhile, the function of the downtown transfer, the boundary of the city is getting more and more indistinct.

3.2 Effects on the urban space morphology

The scale of the development zone administrative area expands rapidly, and has already overcome the hundred-year old city, dramatically changed the city space form [3]. The urban form, namely "three groups and main urban areas", is transferring into "over pie", and they have already been connected by urban construction land.

3.3 Effects on changchun urban structure

3.3.1 Effects on the economic

Due to the policy of "back three binary", the enterprises in town move to the development zone, as well as the new manufacturing enterprises, thus the city main manufacturing enterprises are moving to the development zone gradually. By the comparison of GDP between Changchun city and the four development zones, we can see the growing trend of economic development in the zones (Table 2). Meanwhile, the construction of the development zones within the city has an important impact on the formation of industrial space, mainly in the urban manufacturing suburbanization and the tertiary industry agglomeration.

3.3.2 Effects on the society

With the reform of housing system, the government will release the distribution of housing management power, so the old living pattern is gone, and the difference of income and profession has become the main reason of the differentiation of urban space. Depending on the level of social status and income, the commercialization and marketization of housing also provide a residential location choice[4]. Because of the beautiful natural environment and perfect infrastructure, the development zones have become high salary groups' first choice, however, the low income groups still live in the old area and it leads to the differentiation of social classes living space.

3.3.3 Effects on the transportation

The construction of the development zone makes the city layout structure function more complex, and the growing phenomenon of job lived separation leads to a high pressure on transportation. Before constructed development zones, people seldom pass by the main roads as they live near their workplace. So, the traffic was not that heavy at that time. However, due to the construction and development of the zones, a large number of units, businesses, factories migrate to the zones, while

Table 2. The proportion of each development zone's GDP to Changchun city (Unit: billion yuan).

Years	2005	2006	2007	2008	2009
Changchun	1503	1737	2089	2501	2849
Development zones	210	263	305	365	455
ECZ	150	194	260	285	342
JYZ	131	160	192	231	277
XXZ	60	86	136	230	285
Subtotal	551	703	893	1111	1359
Proportion (%)	36.66	40.47	42.75	44.42	47.70

Reference: "Changchun Statistical Yearbook (2006–2011) and Changchun Statistics Bureau".

the workers still live in the city center. Meanwhile, some people live in the zones and work in downtown, they have to travel long through the main road, this has led to the heavy traffic.

3.3.4 *Effects on the entertainment space*

Before constructed development zones, the entertainment places are mainly located in Chongqing Road, Hong qi Street and Guilin Road. Meanwhile, with the construction of some high-grade leisure market, it leads to multi-centralization and marginalization of the city entertainment space.

4 CONSTRUCTION OF DEVELOPMENT ZONES BASED ON THE IDEA OF OPTIMIZATION OF URBAN SPACE

4.1 *Optimize the urban space form*

The designer of the development zones should change the mode "over pie" into "dot-axis". The development zones should insist a dynamic develop goal which is reflected by the intensive and interval develop concept in the process of urban outward expanding. They also stress to develop the growth axis and the growth point, and to keep ecological quarantine between the growth points[5]. In the process of urban space development, the development zone is the point of the mode "dot-axis". Therefore, when building the development zone, we should segregate it with downtown in fixed scale of ecology. At the same time, we should use the urban develop axis to perfect the aggregation of social and economic factors in development zones. Finally, we should develop the development zone to be an independent satellite city.

4.2 *The transfer of "single center" into "multi-center" of the space form*

As the "single center" of the city, the Changchun municipal people square concentrates the functions of administrative, commercial, cultural, and financial, also it has many structural difficulties, in order to solve the difficulties, we should depend on the developing zones outside the city, namely to construct a new charming city center. During the transform, the regulation effects of urban planning should be further strengthened.

4.3 *Achieve the organic mixed function*

While maintaining its main feature, we should establish a relative balance function structure of life and production. Through the coordination of industrial, residential, commercial and land distribution and the optimal allocation, thereby contributing a rationalization of the structure of land use. Optimizing the internal structure of various types of land use while enhancing the level of public services, to ensure the healthy functioning of the whole zone.

5 CONCLUSION

The building of Development zones has become an important economic growth project in Changchun city, leading the city's economic development. However, the mode of "over pie" leads to the expansion of development zones, showing a large scale of speeding up expansion of the state. Industrial space drives the development of the living space, becoming the agglomeration of new urban living space. The expansion of the Development Zones makes a dramatic effect on the boundaries, spatial form and spatial structure of Changchun City. The Development Zones in Changchun City need more attention on the quality of spatial expansion in the context of the "Changing the style of economic growth", the "over pie" development model should be replaced by the path of the dot-axis. In the aspect of optimizing the city space, we should coordinate the spatial structure of the layout to achieve the organic mix function, improve the land utilization, and achieve a healthy and orderly development between Changchun and the Development Zones.

REFERENCES

[1] Chanchun Institute Of Urban Planning & Designning, Changchun Urban District Planning (2005–2020—High-tech Industrial Zone.
[2] Changchun Statistical Yearbook 2010.
[3] Chanchun Institute Of Urban Planning & Designning, Changchun Urban District Planning (2005–2020—Economic and Technological Development Zone.
[4] Chanchun Institute Of Urban Planning & Designning, Changchun Urban District Planning (2005–2020—Automobile Industry Development Zone Xixin Economic And Technological Development Zone.
[5] Chanchun Institute Of Urban Planning & Designning, Changchun Urban District Planning (2005–2020—Jingyue Economic Development Zone.
[6] Zhang. Yan. The Spatial Expansion of Development Zones and Urban Spatial Restructuring-The Cases of Suzhou, Wuxi and Changzhou[J]. Urban Planning Forum. 2007,(1):49–54.
[7] He. Dan, Cai. Jianming, Zhou. Can. Analysis on Development Zone and Urban Spatial Structure Evolution in Tianjin[J]. Progress in Geography. 2008,(6):97–103.

Frontiers of Energy and Environmental Engineering – Sung, Kao & Chen (eds)
© 2013 Taylor & Francis Group, London, ISBN 978-0-415-66159-1

The investigation of establishing assessment index system in water environment sudden pollution emergency

X.G. Ma, B.L. Hu, J.P. Sun & W.J. Ji
Municipal and Environment Engineering, Shenyang Jianzhu University, Shenyang, Liaoning, China

ABSTRACT: The emergency assessment is the fundamental part of the emergency in water environment sudden pollution. Establishing emergency assessment index system is the important part of the emergency assessment. The assessment index system in water environment sudden pollution emergency includes 6 first grade indexes, 28 second grade indexes and 66 third grade indexes. By analyzing the first grade index weights in the water environment sudden pollution emergency and the result shows that the emergency response occupies in the most important position, and the weight is 0.3583.

Keywords: sudden pollution; emergency assessment; index system

1 INTRODUCTION

Facing all kinds of emergencies, governments are committing to improve the emergency response system to make up for the defects of the contingency, to the greatest extent possible to reduce the loss brought from emergency. In recent years, water environment pollution incidents frequent occur in a lot of place lead to that the water environment security has been a great threat (Wu 2010, Wang 2007 & Hu). Emergency of sporadic pollution of the water environment must be given great attention.

Therefore, the emergency assessment of the sporadic pollution of the water environment is of great theoretical significance and practical value for enhancing the capacity of government in deal with unexpected pollution (Pan 2007, Cui 2005 & Cao 2008).

2 MAIN CONTENT OF THE EMERGENCY ASSESSMENT OF SPORADIC POLLUTION OF THE WATER ENVIRONMENT

The emergency of sudden pollution of the water environment is based on the emergency chronological. Sudden water environment pollution response is divided into the emergency prevention, emergency support, emergency response, emergency decision, recovery after emergency and the impact followed-up emergency based on various stages of the sudden water environment pollution emergency (Zhang 2009). Therefore, the sudden pollution emergency assessment of the water environment is divided into early prevention and preparation, the emergency response, rescue after the pollution occurred, recovery after the emergency response and the ecological impact followed-up emergency, etc (Tian 2008).

3 STUDY OF THE SUDDEN POLLUTION EMERGENCY ASSESSMENT INDICATORS OF THE WATER ENVIRONMENT

Sudden water environment pollution emergency assessment involves a wide range of content, and the complete emergency assessment should include the comprehensive assessment that from prevention and prepared before the sudden pollution to the followed-up impact that may arise at the end of the emergency (Wu 2006). Determining the index of the pollution emergency assessment process of the sudden water environment is extremely important, and it relates to the results of the assessment which should be accurate and reliable.

1. Emergency prevention U_1
   Laws and regulations C_1
   Emergency training C_2
   Control of pollution sources C_3
   Promotional education C_4
   Formulation for the plans C_5
2. Emergency support U_2
   Emergency response team C_6
   Emergency expert group C_7
   Emergency supplies C_8
   Emergency funds C_9

Emergency technical C_{10}
Medical protection C_{11}
Communications protection C_{12}
Traffic protection C_{13}
3. Emergency response U_3
   Information receiving C_{14}
   Information passed C_{15}
   Accident response to deal with C_{16}
4. Emergency Decision U_4
   Emergency processes C_{17}
   Analysis of decision-making C_{18}
   Emergency response C_{19}
   The division of labor and coordination of emergency personnel C_{20}
   Emergency command and coordination C_{21}
   Materials shipped and sent to the configuration C_{22}
   Accident response C_{23}
5. Recovery after the emergency U_5
   Announced case of emergency C_{24}
   Accident investigation C_{25}
   Emergency summary and improvement C_{26}
6. The ecological impact U_6
   Ecological function C_{27}
   Ecological structure C_{28}

4 EMERGENCY ASSESSMENT INDICATORS OF SPORADIC POLLUTION OF THE WATER ENVIRONMENT AND THEIR WEIGHTS

4.1 *The evaluation index system*

Based on the current sudden water environment pollution and emergency status quo and future trend, combined with a sudden water environment pollution emergency process, determination of the 6 first grade assessment indexes, and emergency prevention corresponds to 5 indexes; emergency support corresponds to 8 indexes, emergency response corresponds to 3 indexes; emergency decision corresponds to 7 indexes; recovery after the event corresponds to 3 indexes, the followed-up ecological impact corresponds to 2 indexes; the total of 28 second grade indexes. They correspond to a total of 66 third grade indexes, which to constitute the assessment index system in water environment sudden pollution emergency, which included the first, second and third indexes as showed below:

1. Emergency prevention U_1
   Laws and regulations C_1
   The formulation of laws and regulations, the improvement of laws and regulations and the execution of the laws and regulations;
   Emergency training C_2
   The training cycle, the scope of training and the exercise cycle;

Control of pollution sources C_3
The source control, the control method and the control effect;
Promotional education C_4
The modes of publicity, the publicity coverage and the publicity cycle;
Formulation for the plans C_5
The targeted content, effective implementation and revised plan in a timely manner.
2. Emergency support U_2
   Emergency response team C_6
   The commanding officers, the force of action and the social assistance;
   Emergency expert group C_7
   The members of the experts, the composition of the specialty, the level of scientific research and their work experience;
   Emergency supplies C_8
   The supplies site, the material reserves and the deployment of goods
   Emergency funds C_9
   Whether the funds are in palace and used reasonable;
   Emergency technical C_{10}
   The scientific and reasonable and the proficiency application of the technology;
   Medical protection C_{11}
   The medical force and medical equipment;
   Communications protection C_{12}
   The wired communications, the wireless communications, the network communications and the satellite communications;
   Traffic protection C_{13}
   The capacity of transportation.
3. Emergency response U_3
   Information receiving C_{14}
   The normative of alarm processes and the analysis ability to cope with information.
   Information passed C_{15}
   The timeliness and accuracy of the information transfer;
   Accident response to deal with C_{16}
   The timeliness of analysis information, start timely early emergency system, Emergency Command timely establishment and timely released the accident information.
4. Emergency Decision U_4
   Emergency processes C_{17}
   The normative and completeness of the emergency processes;
   Analysis of decision-making C_{18}
   The capacity of analysis and decision-making of emergency;
   Emergency response C_{19}
   The agility of first reaction and the rapidity of first decisions;
   The division of labor and coordination of emergency personnel C_{20}

The rationality of the division of labor and timely communication between emergency workers;
Emergency command and coordination C_{21}
The ability of organization and command, the ability of interdepartmental coordination, the speed to start the decision making system and the coordination with neighboring areas;
Materials shipped and sent to the configuration C_{22}
The timeliness and rationality of 6te deployment of goods;
Accident response C_{23}
The organizational coordination to respond the pollution.

5. Recovery after the emergency U_5
Announced case of emergency C_{24}
The openness and transparency of the emergency situation;
Accident investigation C_{25}
The investigation of the cause of the pollution;
Emergency summary and improvement C_{26}
The experience summary and rectification after the emergency.

6. The ecological impact U_6
Ecological function C_{27}
The biological ability to reproduce and the abundance of biological quantity;
Ecological structure C_{28}
The diversity of biological species and the structure of the population age.

4.2 Determination of the weights

According to the basic principles of the Analytic Hierarchy Process, using the 6 first evaluation indicators to establish the judgment matrix derived the relative weights between the various indicators.

Build the judgment matrix constructed as shows in Table 1.

From the above we know that the maximum characteristic root of judgment matrix:

$\lambda_{max} = 6.472$
$CI = (\lambda_{max} - n)/(n - 1) = 0.0944$
Know $RI = 1.12$;
Testing indicators: $CR = CI/RI = 0.0843 < 0.1$

Table 1. Expert assignment table of the first indicators.

Judgment matrix	U_1	U_2	U_3	U_4	U_5	U_6
U_1	1	1/2	1/6	1/4	1	1
U_2	2	1	1/4	1/3	2	2
U_3	6	4	1	2	8	5
U_4	4	3	1/2	1	3	3
U_5	1	1/2	1/8	1/3	1	1
U_6	1	1/2	1/5	1/3	1	1

Table 2. Weights of the second indicators.

Second indicators C_i	C_1	C_2	C_3	C_4
Weight W_{Ci}	0.0991	0.2887	0.4108	0.0506
Second indicators C_i	C_5	C_6	C_7	C_8
Weight W_{Ci}	0.1958	0.1992	0.1927	0.1161
Second indicators C_i	C_9	C_{10}	C_{11}	C_{12}
Weight W_{Ci}	0.1129	0.1312	0.0650	0.0852
Second indicators C_i	C_{13}	C_{14}	C_{15}	C_{16}
Weight W_{Ci}	0.0977	0.2008	0.2401	0.5591
Second indicators C_i	C_{17}	C_{18}	C_{19}	C_{20}
Weight W_{Ci}	0.0860	0.2395	0.1435	0.2298
Second indicators C_i	C_{21}	C_{22}	C_{23}	C_{24}
Weight W_{Ci}	0.1006	0.0706	0.1300	0.6483
Second indicators C_i	C_{25}	C_{26}	C_{27}	C_{28}
Weight W_{Ci}	0.1220	0.2297	0.6667	0.3333

It can be concluded that the judgment matrix has relatively good consistency. The corresponded eigenvectors of the matrix respectively are 0.0782, 0.1107, 0.3583, 0.2900, 0.0782 and 0.0846. They are the weights of the first assessment indexes in water environment sudden pollution emergency. The order: $U_3 > U_4 > U_2 > U_6 > U_1 \geq U_5$. It can be seen that U_3 in the 6 first indexes occupies the most important position. It means that emergency response occupies the most important position in the process of emergency, and, secondly is the emergency decision, and then is the emergency support.

The method of determining the weights of the second indexes is same as the first indexes. The specific weights of the second indexes are shown in Table 2.

5 CONCLUSIONS

1. According to the analysis, the indexes screening in the process of establishment of Index System are critical.
2. As above study, there are 100 indexes remained in the final assessment index system in water environment sudden pollution emergency, which includes 6 first grade indexes, 28 second grade indexes and 66 third grade indexes from six aspects, emergency prevention, emergency support, emergency response, emergency decision, recovery after emergency and emergency follow-up ecological impact.

3. After the weighted analysis of the 6 first grade indexes, we know that emergency response accounts for the most important position followed by emergency decision, emergency support, emergency follow-up ecological impact, emergency prevention and recovery after emergency.

ACKNOWLEDGEMENTS

Financial support from the National Water pollution control and governance important specialized science and technology item (2011ZX07530-04) is greatly appreciated.

REFERENCES

Cao Jia, Chen Ling. Study on urban abrupt water crisis and emergency management. JianSu Environmental Science and Technology. 2008, 21(4): 60–63.

Cui Weizhong, Liu Chen. Study on establishing emergency mechanism for handling major sudden accidents of water contamination. Pearl River. 2005(5):1–3.

Hu Wangjun. Emergency response technology and monitoring methods of common toxic chemicals and environmental incidents. Chinese Environmental Science Press.

Jinruiling, Yuanhui. Study of Emergency Decision Support System. Chinese Science and Technology Press. 1990:116–118.

Pan Bo, Wang Jie. Emergency mechanisms of water pollution accidents for Yangtze valley. Water Resources Protection. 2007, 23(1):87–90.

Tian Yilin. Study of the evaluation index system model of the emergency capability on emergency. Journal of Science Engineering. 2008, 16(2): 200–208.

Wang Donghui. The control and damage of organic pollutant in Song-hua River to the ecological environment. Environmental Science and Management. 2007, 32(6):67–69.

Wu Panfeng. Statistical analysis on environmental pollution incidents in China during the year 1985 to 2005 and counter measure research. Environmental Science and Management. 2010, 35(2):18–21.

Wu Xiaogang. Study on the emergency mechanism for sudden pollution of water resources. Water Resources Protection. 2006, 22(2)76–79.

Zhang Haibo, Tong Xing. Theory framework of capability assessment for emergency management. Chinese Public Administration. 2009(4):33–37.

Frontiers of Energy and Environmental Engineering – Sung, Kao & Chen (eds)
© 2013 Taylor & Francis Group, London, ISBN 978-0-415-66159-1

Research about applying the incubator to the transformation and development of national characteristic industries in minority areas

L. Zuo & Y.H. Li

Economics and Management Faculty Dalian Nationalities University Dalian, China

ABSTRACT: The transformation and development of national characteristic industries in minority areas play an important role in the transformation and development of national economy, and the incubator theory gives a theoretical and technological support. In this paper, we introduce the basic knowledge about the characteristic industries in minority areas and the incubator, and then we analyze the necessity of the transformation and development of characteristic industries in minority areas. Following, we analyze the feasibility of applying the incubator to transform; then we give a model of applying the incubator to transform and develop the characteristic industries in minority areas; at last, we put forward several suggestions. The characteristic industries are the new focus of our national economy, and the incubator is a new theory, we believe that the combination of the two new things can produce gorgeous sparks.

Keywords: characteristic industries; minority areas; transformation; incubator

1 INTRODUCTIONS

1.1 *The concept of minority areas*

According to the data of the China Statistical Year-book 2011, until the year 2010, there are 55 ethnic minorities and 5 ethnic autonomous regions. There are 77 regional districts, including 31 prefecture-level cities and 30 autonomous prefectures. And the number at the county level division is 698, which contains 65 county-level cities and 120 autonomous counties. The total area of the ethnic minority areas is 6, 1173 million square kilometers, accounting for 63.72% of the country's total area. And the total population in the ethnic minority areas is 18484.30 million, which accounts for 13.79% of the total population 134,091.00 million. The minority population is 8851.74 million, which takes up 47.86% of the total population of the autonomous areas.

1.2 *The introduction of national characteristic industries in minority areas*

The so-called national characteristic industries in minority areas is that within the national scope, basing on the regional and unique national resources (including tangible and intangible ones), using the modern industry and agriculture production technology which have the national regional characteristics and modern management techniques, and through the operation of the market economy, unlike traditional industries, with national regional specialties, industries meet the market demand for special products and services sector or industry.

The core segment of the national characteristic industries in minority areas is that it is different from the traditional industries that it has the national regional characteristics and that it offers the special products and service to meet the demand of the market. It's based on the national region-specific tangible and intangible resources. The important conditions to form characteristic industries in minority areas is the modern industrial and agricultural production technology which have the national regional characteristics and the modern management techniques. And its implement is the operation of the market.

The characteristic industries in minority areas have the nationality character, regional character, historical character, competitive character and the sustainable development and so on. Nationality is that the basis of the characteristic industries is coming from is national culture and national resources. The regional character refers to the formation of the characteristic industries. When the industry forms, it must be subject to the geographical conditions, weather conditions and so on. It can not transfer from this area to other place, just like the saying goes, "if you plant the orange in the Huainan area, you will get orange. But if you plant the orange in the Huaibei area, what you get will not be orange". The products and technology in the minority areas is developing through several generations, they are the historical heritage, and therefore they must have historic features. Every product and every industry must have their own competition, or they can not survive from the fiercely competing market, so they must have the competitive feature. Sustainable development feature is the new demand and character the industry should have or develop, only the ecological, economic and the social sustainable development come true, can the industry really exiting for a long time.

1.3 *The introduction of incubator*

The original meaning of incubator is the specialized equipment to artificial incubation of the eggs. And now, as the introduction of the economic field, it used to refer to a centralized place. When the enterprise meet difficulties at the beginning, the incubator will offer the shared facilities using for the research, production, business venues, communications and net-working and so on, and the incubator will provide the training and consulting in system, and it also offer the support about policy, financing, legal and marketing. The incubator is aiming at incubating the high-tech achievements, science and technology enterprises and start-ups to promote cooperation and exchanges, so that the goods from the entrepreneurs' inventions and results will enter the market as soon as possible. It will also provide comprehensive service to help the emerging company become mature and to be scale, to reduce the risk and cost of start-ups and to improve the survival and success rate, and ultimately make the enterprise bigger. At the same time, develop the successful business and entrepreneurs.

According to some statistics, in 2009, China has built 674 business incubators in total, and there is 2,300 million square meters area for technology business incubator. There are nearly 930 thousand people do this work directly, and now, there are 4.4 million companies in incubating. The direct entrepreneurs, tech business incubator area, the number of incubating companies and the success rate of incubation in our country place ahead in the world.

1.4 *The relationship between the characteristic industries in minority areas and incubator*

The relationship between the national characteristic industries in minority areas and the incubator are as follows.

Under the guidance of the incubation theory and within the incubation technique, we can improve the weakness and inadequacy in the process of transformation and development of the national characteristic industries in minority areas. Under the guidance of the incubation theory and according to the local conditions and its advantages, we can achieve a reasonable transformation of

characteristic industries in minority areas, raise the level of the development and market competitiveness of characteristic industries in minority areas, and make a scientific, rational and sustainable development of national characteristic industries in minority areas come true.

By applying the incubation theory and technology to the transformation and development of national characteristic industries in minority areas successfully, it can not only confirm the correctness of the incubation theory, but also let the government, enterprises and other stakeholders see the feasibility of the theory again, and then it will improve the credibility of the incubation theory. At the same time, the success use of the incubation theory in the transformation and development of characteristic industries in minority areas can make the theory of incubator plentiful and make up for the gaps of the incubation theory of ethnic regions application, and it let the incubation theoretical system be more complete.

2 NECESSITY OF THE TRANSFORMATION AND DEVELOPMENT OF CHARACTERISTIC INDUSTRIES IN MINORITY AREAS

2.1 *The necessity of national economy*

In the post-crisis era, Chinese economy is in the stage of overall adjustment and upgrading. And as an integral part of china' economy, the development of national special economy plays a vital role. As the pillar of the national special economy, the national regional characteristic industries, whether it can get this chance to achieve the restructuring and development or not, have direct relations to the comprehensive takeoff of the national regional special economy and the comprehensive development and upgrade of our country's economy.

2.2 *In order to solve the social problems happened in the ethic areas to some degree*

The social problems happen from time to time in the minority areas. In the last analysis, one of the reasons that caused these problems is just that the economy development in the ethnic regions does not meet the demand of local people and that the economy develops imperfect. So we should, through the transformation of the national characteristic industries in minority areas, create more jobs and enhance the level of the national economy and the income of local people to ease the plight of the ethnic people in the region, to achieve industrial transformation and to make contribution to economic development. And then we could make the incidence of social problems of the ethnic minority areas lower to some degree.

2.3 The needs of the realization of common prosperity

China is a socialistic country, and it's our final aim to achieve the common prosperity of the nation. However, the development of the national regions is very slow at present, and the gap between the minority areas and the east and the middle areas is growing. In order to fill in the gap, and to achieve the common prosperity, we need to speed up the development of the minority nationality regions. The combination of the incubator and the characteristic industries in minority areas provide a quick and stable road for the development of the minority nationalities regions. And the whole country will speed up its step to achieve the final goal.

3 THE FEASIBILITY OF USING THE INCUBATOR TO TRANSFORM AND DEVELOP CHARACTERISTIC INDUSTRIES IN MINORITY AREAS

3.1 The good era

In the post-crisis era, the world economy patterns come into a new round of adjustment. And some countries, especially the developed countries, have increased the investment in science and technology innovation, and they strive to foster new industry, transform the original industry and create new economic point by using the new technologies, and then they could catch the strategic high ground in the new economic forms. The world's development is inseparable from China, and the development of China can not get away from the world. China now is also in the adjustment period, and as the important part of China's economy, the national characteristic industries should make full use of these opportunities and speed up the transformation and development of the structure.

3.2 Theory support

After several years' development, the theory of incubator has made a lot of successful results, and it has been more perfect. What's more, the better development of the incubator theory and its usage give a shining future of the national characteristic industries. The incubator theory is a totally new theory, and it has made some achievement. What's more, the characteristic industries are the new focus in the process of development in minority areas. And the combination between them will produce gorgeous sparks.

3.3 Policy support

National and local governments have different levels of policy to support the industries which are in

use of incubator theory. And the ethnic minority areas is just the focal point of the state support and help developing, and the transformation of the national characteristic industries in the minority areas can get a lot of help and support from the state and local government.

3.4 Funds from the society

It is the new economic developing point that applying the incubator theory to the characteristic industries in the minority areas. And the entrepreneurs who have the strategic vision have or will have diverted their attention to this filed, in order to achieve a long business life.

More over, a lot of funds from the society joined the transformation of characteristic industries in minority areas, the transformation will ensure less worry about the capital and the economy will have a relatively stable road to go.

4 MODELING OF USING THE INCUBATOR TO TRANSFORM AND DEVELOP NATIONAL CHARACTERISTIC INDUSTRIES IN MINORITY AREAS

According the relative theory and the basic conditions of the characteristic industries in minority areas, we give the model of using the incubator to transform and develop national characteristic industries in minority areas. The main part of this model is the government, companies, colleges, incubator and the companies/industries which have finished incubation.

The model is shown in the Figure 1.

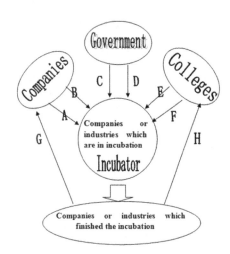

Figure 1. The model of using the incubator to transform and develop the characteristic industries in minority areas.

451

In this model, the government should put the policies and funds support into the incubator for the companies/industries which are in incubation. Just as the letter C and D stand for.

For the companies, they could provide the capital and experience of management for the companies/industries which are in incubation. And the letters A and B represent for these.

For the colleges, they have a lot of talent people and have a compacted theory foundation. Letters E and F are standing for these things and the colleges should put them into the incubator.

In the middle of the model, it is the incubator, which is the core part of the incubation. The incubator accepts the support from the government, companies and colleges, and the companies and/or industries which are in incubation choose the best choices assemble for themselves to incubate, according the conditions of themselves and the environment of incubation.

At last, when the companies and/or industries finished the incubation, they should take actions just as the letters G and H standing for. They should share the experience of incubation and be a member of the companies' parts which support the incubation activity. On the other hand, they should share the knowledge of incubation and the knowledge obtained during the incubation with the colleges.

The five parts of this model are an organic entirety. Only all of them taking part in the incubation actively and doing the things they should do can the incubation give better results. And the model could enjoy a benign loop.

5 THE SUGGESTIONS ABOUT USING THE INCUBATOR TO TRANSFORM AND DEVELOP NATIONAL CHARACTERISTIC INDUSTRIES IN MINORITY AREAS

The government should ensure that the polices formulated by the state really put into practice, and then let the transformation and development of national characteristic industries make the contribution it should make to economic development of the minority areas and nation.

More over, the government should not intervene in the operation of the incubator too much. We should let the incubator industry have less political color. The new industry should be suitable to the market economy and take part in the competition freely. Unless it is necessary, or the government should not put his hand in the markertization and free management of the incubator industry.

It is a new developing pattern to combine the characteristic industries in minority areas and the incubator. There must be some uncertainty, but it is also full of business opportunities. As the entrepreneurs who have long-term strategic vision or the companies which want to find a new developing point, the new pattern is absolutely a good point of penetration. In addition, if they take part in this model, they not only response to the policies of the state, but also realize their own development. They should have enough courage to breakthrough the traditional things and they should have the responsibility to take the social duty. The enterprise should respond to the national policies and guidelines actively, and they could seize the opportunities to achieve the success of themselves.

And the colleges and universities, as the major component of the incubator, should participate the transformation and development of national characteristic industries in minority areas actively. They should give the suggestions which can be put in to practice successfully for the development of ethnic founding on the basic conditions of national characteristic industries in minority areas, using the incubator theory, and make a contribution to the development of national economy.

6 CONCLUSION

Through the leading and using of incubation theory and under the corporation of several parts like government, companies and colleges, it make the transformation and development of the national characteristic industries in minority areas come true, and it let the economy in minority regions raise a new high level. At the same time, it also makes the incubator theory itself more perfect, and then offers a better guide for the following economic development.

REFERENCES

[1] Barbara Becker, Oliver Gasman. Corporate Incubators: Industrial R&D and What Universities Can Learn from them. The Journal of Technology Transfer, 2006–07.
[2] Daisong Li, Ruidan Wang and Xin Ma. A study on the characteristics of incubator industry and operation models of Chinese's business incubator. The science research management, May 2005: 8–11.
[3] Yunzhi Liang, Chunlin Si. Study on the incubator business model: theory structure and empirical analysis. R&D management, February 2010: 43–51.
[4] Liqing Ma, Weiwei Hu. On constructing tourist product chain with marine characteristics in Changjiang River delta area during the transitional period. Human geography, April 2009: 125–128.

Frontiers of Energy and Environmental Engineering – Sung, Kao & Chen (eds)
© 2013 Taylor & Francis Group, London, ISBN 978-0-415-66159-1

Study on vertical flow filter in dealing with rural domestic sewage

Q. Zhang
Institute of Resources and Environment Science, Agricultural University of Hebei, Baoding, China

J.W. Ma
Institute of Chemistry and Environmental Science, Hebei University, Baoding, China

Y. Li & J.L. Liu
Institute of Urban and Rural Construction, Agricultural University of Hebei, Baoding, China

ABSTRACT: The effects of different filter media when loaded in vertical flow filters on removal of main pollutants in rural domestic sewage is studied in this paper. The experimental result has shown that the comprehensive removal efficiency of main pollutants by different filter media was in sequence of slag > haydite > river sand. Under the condition of 0.35 $m^3/m^2 \cdot d^{-1}$ hydraulic loading and that filter media layer was not immersed, the COD and NH_4^+-N influent concentration of slag filter, respectively, could stay at 50 mg/L and 2.2 mg/L, and removal rate of TP was always 70%.

Keywords: rural domestic sewage; vertical flow filter; slag; haydite; river sand

1 INTRODUCTION

Since most villages have no perfect wastewater collection system and sewage treatment facilities can play a role only after organized sewage discharge and complete collection, rural domestic sewage treatment and utilization can't perform until it is collected. In view of this, the author put forward a kind of low power sewage treatment process which can guide rural sewage collection and reduce the concentration of pollutants. By applying common bio-filter principle that combined physics filtration with biological filtration of particle filter material, this process not only could intercept effectively particles and suspension but also reduce dissolved pollutants in sewage to provide appropriate material ratio and load conditions for the follow-up wastewater advanced treatment and reclamation[1].

2 MATERIAL AND METHOD

2.1 *Test material*

Considering rural living conditions, Coal slag, shale haydite and natural river sand were chosen as filter media for experiment because of their low price, easy obtainment and suitability for water treatment. Their physic properties were shown in Table 1[2].

Table 1. Physical properties of the filter media.

Material	Effective diameter (d_{10}) mm	Nonuniform coefficient (k_{80})	Porosity
Coal slag	2–4	1.99	0.55
Natural river sand	1–2	1.5	0.45
Shale haydite	3–5	1.8	0.65

Many kinds of domestic sewage from some peasant household was collected and mixed as test water, and then the mixture was stored in an interior tank to perform experiment. Concrete indexes of test wastewater were showed in Table 2.

2.2 *Test device*

Test device was shown in Figure 1. Biofilter was made up of three pump columns whose diameters was 200 mm, and its total height was 1.2 m. Content of each group of biofilter, from up to down, was detailed as follows: The first part was super high and water distribution layer with its height 0.5 m, the second part was biological filtering layer with its height 0.9 m and the third part was filter supporting bed and outflow layer with its height 0.5 m.

Table 2. Quality of test wastewater.

Water quality index	COD (mg/L)	NH₄⁺-N (mg/L)	TP (mg/L)	Temperature (°C)
Range	80–480	30–40	5–15	16–30

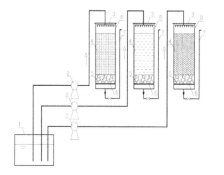

Figure 1. The diagram of the biofilter test device.
1 water regulating tank 2 peristaltic pump 3 water distributor 4 biological filter material 5 filter supporting bed 6 perforated sup-porting plate 7 adjustable outlet 8 overflow port.

The test device was mainly made up of three parts, water tank, pump set and biofilter. Test water was stored in water tank (number 1) and suction pipes of pump set independently took water from the water tank. Pump set consisted of three sets of the same type of peristaltic pump (number 2). Biofilter was composed of three pmma columns whose diameters were 200 mm, and on the top of each reactor there were hindu lotus seedpod-like water distributors (number 3) which can ensure raw water uniform distribution. When row water flowed through filter media (number 4) from top to bottom, it was adsorbed and decomposed by the filter media and microbe attached to the surface of filter media. In order to avoid the loss of filter media, there are gravel were put at the bottom of each reactor whose diameters were 3–5 mm as filter supporting bed (number 5), and perforated supporting plate (number 6) whose diameter were 2 mm was set under it. Effluent discharged from adjustable outlet (number 7), and at the same time, water sample was obtained there. Overflow port (number 8) was set to prevent influent overflowing.

2.3 *Analysis items and determination method*

Water quality analysis and test mainly referred to standard method of *Water and Exhausted water Monitoring Analysis Method (fourth edition)* written by National Environmental Protection Agency

Table 3. Main testing instruments and analytical methods.

Analysis items	Determination methods	Instruments and equipment
COD	Potassium dichromate colorimetric method	Reflux and titration device
NH₄⁺-N	Reagent colorimetric method	Spectrophotometer/ HACH DR5000
TP	Ammonium molybdate spectrophotometry	Spectrophotometer/ HACH DR5000
Water temperature	Determination of dissolved oxygen instrument	Magnetic portable oxygen meter/ JPSJ-608

and Editorial board of Water and Exhausted water Monitoring Analysis Method, and the methods of *Water Quality—Determination of Ammonium-Reagent Colorimetric Method (GB7479-87)*, and *Water Quality—Determination of Total phosphorus- Ammonium Molybdate Spectrophotometry (GB11893-89)*[3–5]. Main testing instruments and analytical methods were showed in Table 3.

3 RESULTS AND ANALYSIS

3.1 *Analysis on removal efficiency of pollutants when filter media layer was not immersed*

3.1.1 *Analysis on removal efficiency of COD*
When influent concentration changed, removal efficiency of COD by the three filters was shown in Figures 2 and 3. In the three filters, as the COD influent concentration increased, its concentration in effluent also grew. While the COD influent concentration was lower than 350 mg/L, the COD effluent concentration of the river sand filter and the haydite filter rose greatly, but that of the slag filter increased little and still stayed at 50 mg/L or so. When the COD influent concentration was 400–500 mg/L, the three filters' COD effluent concentration was relatively stable. In all, the COD removal efficiency of the slag filter was better than that of the river sand filter and haydite filter.

As shown in Figure 3, in the whole process, the COD removal rate of the slag filter always stayed about 90% and the average of this rate was 89% which was obviously higher than that of the river sand filter's COD removal rate which was 73%. When

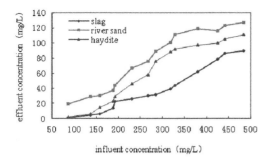

Figure 2. The COD removal efficiency by filters.

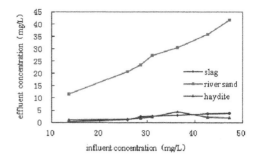

Figure 4. The NH$_4^+$-N removal efficiency by filters.

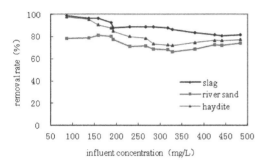

Figure 3. The COD removal rate by filters.

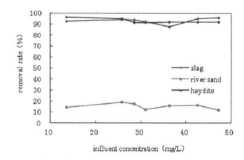

Figure 5. The NH$_4^+$-N removal rate by filters.

the COD influent concentration was not high, the haydite filter had a similar COD removal rate to the slag filter, but its COD removal rate decreased sharply with the COD influent concentration increasing.

3.1.2 Analysis on removal efficiency of NH$_4^+$-N

When influent concentration changed, removal efficiency of NH$_4^+$-N by the three filters was shown in Figures 4 and 5. As the COD influent concentration increased, the NH$_4^+$-N effluent concentration of the slag filter and the haydite filter stayed lower than 5 mg/L, and the average NH$_4^+$-N effluent concentration of the slag was 2.2 mg/L. It was concluded that both the slag filter and the haydite filter have good NH$_4^+$-N removal efficiency.

As indicated in Figure 5, the NH$_4^+$-N removal rate didn't change as the NH$_4^+$-N influent concentration changed under the operating condition. In the whole process, the NH$_4^+$-N removal rate of the river sand filter always stayed 10%–20% and the average of this rate was only 15.2% which was obviously lower than that of the slag filter and haydite filter's NH$_4^+$-N removal rate which was 92.6% and 92.9% repectively.

3.1.3 Analysis on removal efficiency of TP

When influent concentration changed, removal efficiency of TP by the three filters was shown in Figures 6 and 7. In the three filters, as the TP

influent concentration increased, its concentration in effluent also grew. And the slag fiter had a much better control on the TP effluent concentration than the other two. While the TP influent concentration was more than 8 mg/L, the growth of the TP effluent concentration in the slag filter began to get gentle. At the same time, the river sand filter and the haydite filter have a similar TP removal process, and their TP effluent concentration rose with the TP influent concentration growing and kept constant increase. But the haydite had a better TP absorbability than the river sand filter.

As shown in Figure 7, the TP removal rate by the three kinds of filters didn't change greatly as the TP influent concentration increased. The slag filter had a much better removal efficiency of TP than the other two. In the whole process, the TP removal rate of the slag filter always stayed 70% or so, while the removal rate of the river sand filter was only about 15.2% and the haydite filter's was only around 20%.

3.2 Analysis on removal efficiency of pollutants when filter media layer was immersed

3.2.1 Analysis on removal efficiency of COD

When influent concentration changed, removal efficiency of COD by the three filters is showed in Figures 8 and 9. Under the operating condition that filter media layer was immersed, as the COD

455

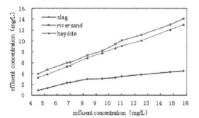

Figure 6. The TP removal efficiency by filters.

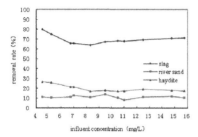

Figure 7. The TP removal rate by filters.

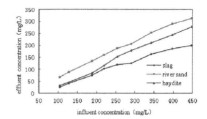

Figure 8. The COD removal efficiency under continuous influent.

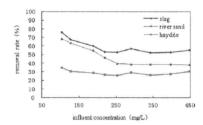

Figure 9. The COD removal rate under continuous influent.

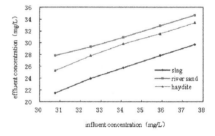

Figure 10. The NH_4^+-N removal efficiency under continuous influent.

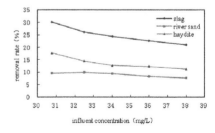

Figure 11. The NH_4^+-N removal rate under continuous influent.

influent concentration increased, in the three filters its concentration in effluent also grew. It was similar to the result under the condition that filter media layer was not immersed but the difference was that the COD effluent concentration showed no obvious fluctuation. Under the condition, the slag had a stronger COD removal ability than the other filters and the average of its COD effluent concentration was 116 mg/L, and when the COD influent concentration was above 250 mg/L, the growth of COD effluent concentration of the slag filter began to get slower. However, with the COD influent concentration growing, the COD effluent concentration of the river sand filter and the haydite filter kept a relatively stable growth.

As shown in Figure 9, under the operating condition, the COD removal rate of the three filters increasingly got stable with the COD influent concentration growing. The average of the COD removal rate by the slag filter, the river sand filter and the haydite filter was 58%, 29%, 47% respectively. Although their COD removal rates were different, they had no significant differences as mentioned above. Under this condition, the river sand filter had a more stable COD removal rate, and the breakpoint of the removal rate of the other two filters appeared when the COD influent concentration was 250 mg/L.

3.2.2 *Analysis on removal efficiency of NH_4^+-N*

When influent concentration changed, removal efficiency of NH_4^+-N by the three filters was shown in Figures 10 and 11. As the COD influent concentration increased, the NH_4^+-N effluent concentration of the three filters rose sharply. Among the three filters, the slag filter still showed a much stronger ability to absorb NH_4^+-N, and its average NH_4^+-N effluent concentration was lower than the average 5.3 mg/L of the river sand filter and the average 3.8 mg/L of the haydite filter. Only taking the effluent concentration into consideration, the river sand filter had a bad removal

456

efficiency of NH_4^+-N and its NH_4^+-N was only lower than the influent concentration 2 mg/L.

As indicated in Figure 11, it can be known the NH_4^+-N removal rate by the three filters decreased as the NH_4^+-N influent concentration increased under the operating condition, and the NH_4^+-N removal rate of the haydite filter and the slag filter decreased much more apparently. Though the river sand filter have a stable removal rate of NH_4^+-N, but its average removal rate was not high and only stayed 9% or so.

3.2.3 *Analysis on Removal efficiency of TP*

When influent concentration changed, removal efficiency of TP by the three filters was shown in Figures 12 and 13. In the three filters, with the average of the TP effluent concentration 5.2 mg/L, the slag filter had a stronger removal efficiency of TP than the other two filters, and its TP effluent concentration changed gently. As the TP influent concentration grew, the river sand filter and the haydite filter showed a similar TP absorption process which had a very great growth and their TP effluent concentration was 8.3 mg/L and 8.0 mg/L, respectively, which was relatively close to each other.

The slag had a much better removal efficiency of TP than the other two filters and its minimum of TP removal rate was close to 50%, and the average was 53.2%, all of which was shown as Figure 13. Under the operating condition, the TP removal rate of the river sand filter and the haydite filter went down quickly with the increase of the TP influent concentration, and the minimum was only 10% or so.

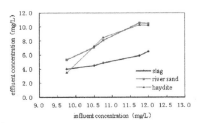

Figure 12. The TP removal efficiency under continuous influent.

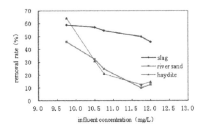

Figure 13. The TP removal rate under continuous influent.

4 CONCLUSIONS

– Under the different operating conditions, keeping the hydraulic loading 0.35 $m^3/m^2 \cdot d^{-1}$, the slag filter had a better removal efficiency of COD, NH_4^+-N and TP than the other two filters. From the view of the filters' operating stability, when filter media layer was not immersed, the slag filter had the best operating efficiency, and its effluent concentration of COD, NH_4^+-N, respectively stayed 50 mg/L, 2.2 mg/L and its removal rate of TP kept 70% or so.
– The result showed that vertical flow slag filter could achieve the effective removal of COD, NH_4^+-N and TP, and the operating mode of the process was basically identical to living habits of rural residents. Moreover, the process has the advantages of small floor area, simple operation, powerful resistance to impact load and stable effluent quality.
– The filter has relatively low cost in designing and making, low level of management technology, and low running cost, so it was suitable for popularization and application in the rural.

ACKNOWLEDGMENTS

The authors thanks the anonymous reviewers who provided extensive criticism that substantially improved the study and manuscript. The study is financially supported by the subject—the research on implementation of quality standard and safeguard mechanism of rural security drinking water project, which is included in people's livelihood special investigation of social science development research of Hebei province in 2012 (Project approval number: 201201201).

REFERENCES

[1] Cauchie HM, Salvia M, Weicherding J et al. Performance of a single-cell aerated wast stabilization pond treating domestic wastewater: a three-year study [J]. International Review of Hydrobiology, 2000, 85(231–251).
[2] Creese, E.E., Robinson, J.E. Urban Water Systems for Sustainable Development [J]. Canadian Water Resources Journal. 1996, 21(3): 209–220.
[3] National Standard of PRC. Water Quality-Determination of Total nitrogen-Alkaline Potassium Persulfate Digestion-UVSpectrophotometric Method (GB11894-89) [S]. Beijing: Public of National Standard of PRC, 1989:192~195.
[4] National Standard of PRC. Water Quality-Determination of Ammonium-Reagent Colorimetric Method (GB7479-87) [S]. Beijing: Public of National Standard of PRC, 1989:192~195.
[5] National Standard of PRC. Water Quality-Determination of Total phosphorus-Ammonium Molybdate Spectrophotometry (GB11893-89) [S]. Beijing: Public of National Standard of PRC, 1989:192~195.

Frontiers of Energy and Environmental Engineering – Sung, Kao & Chen (eds)
© 2013 Taylor & Francis Group, London, ISBN 978-0-415-66159-1

Modeling and optimization for the hydrothermal gasification process of olive mill wastewater as a biomass source in supercritical water

D.Z. Li & Q.L. Zeng

Department of Automation, North China Electric Power University, Baoding, China

ABSTRACT: The LSSVM (Least Square Supporting Vector Machine) model of hydrothermal gasification process of Olive Mill Wastewater (OMW) as a biomass source in supercritical water was established in this paper. The calculation verification has been done to the model according to the experimental data. It shows that the maximum absolute relative error of the experimental value and the predictive value is 7.82% and the average absolute relative error is about 3.5%. The model established in this paper can better simulate the OMW hydrothermal gasification process in supercritical water. Based on the model, a multi-objective optimization function was put forward, by which the Pareto-optimal solution set was gained. The optimization results show that both the content of CH_4, H_2 and the TOC transformation rate are close to the maximum experimental values.

Keywords: OMW; supercritical water; hydrothermal gasification; modeling; optimization

1 INSTRUCTION

The biomass energy is one of the most promising renewable energy in new clean energy. The effective development of biomass energy would make it the dominant of the new energy. Among the technology of biomass energy transformation, the hydrothermal gasification of biomass in supercritical water has received extensive attention and studies in recent years because the technology could gasify the wet biomass directly and get hydrogen, methane and other clean gas [1].

The hydrothermal gasification is currently being investigated for biomass and biodegradable organic wastes treatment, such as municipal solid wastes, industrial waste and residues etc [2]. These wastes have a common characteristic that the organic content is high. The organic content of municipal solid wastes is around 50% and the value of industrial wastewater is 3.5~15%. The Olive Mill Wastewater (OMW) is selected as the biomass source in this paper, according to the experimental data of hydrothermal gasification of OMW, The LSSVM model of OMW hydrothermal gasification process was established. The model can be used to predict the gas component in gasification process. The optimization researches of related reaction parameters and gasification products have been made as well. It shows that the results are satisfactory.

2 MODELING FOR THE PROCESS OF OMW HYDROTHERMAL GASIFICATION

2.1 Determination of model parameters

There are many factors that affect the product composition in the process of hydrothermal gasification, including the raw material type, temperature, pressure, residence time, catalyst, reaction environment and so on [3, 4]. The most significant affecting factors of them are temperature and residence time in non-catalytic gasification. Paper [5] used the experimental equipment of hydrothermal gasification in continuous tubular reactor system and OMW was used as a biomass source. The Total Organic Carbon (TOC) of OMW in the experiment is 6138 mg/L. The reaction pressure is 25 MPa, part of the final experimental data is shown in Table 1. In this paper, temperature and residence time are considered as the main control parameters, the gas components (CH_4, H_2, C_2H_6, CO_2) and TOC transformation rate are the evaluation indexes.

2.2 Modeling

In this paper, the LSSVM (least square supporting vector machine) method [6] is adopted to build the model of the process of OMW hydrothermal gasification in supercritical water. The input of the model were Temperature (T) and residence time (t),

Table 1. Experimental date from OMW hydrothermal gasification in supercritical water.

Temperature (°C)	Residence time (s)	CH$_4$ (mol%)	H$_2$ (mol%)	C$_2$H$_6$ (mol%)	CO$_2$ (mol%)	TOC transformation rate (%)
400	30	1.13	4.07	1.28	64.74	31.09
400	90	2.20	6.12	2.30	77.74	40.70
400	120	3.01	8.30	2.17	76.47	42.98
450	30	6.39	6.57	3.06	74.06	45.73
450	90	13.35	8.92	2.69	72.64	51.86
450	120	12.98	9.30	2.95	72.74	56.58
500	30	28.62	6.98	2.04	57.09	46.02
500	90	28.69	7.06	2.80	56.54	55.20
500	120	31.44	7.43	3.11	55.81	68.07
550	30	34.84	9.23	4.04	49.34	74.42
550	90	33.68	8.92	4.29	50.35	85.81
550	120	29.15	9.06	4.70	54.70	87.63
600	30	23.11	10.59	6.76	56.70	83.46
600	90	22.34	10.33	4.69	58.58	87.45
600	120	18.34	9.69	4.05	66.13	91.68

the gas components (CH$_4$, H$_2$, C$_2$H$_6$, CO$_2$) and TOC transformation rate are the output, which is shown in Figure 1.

The LSSVM function estimation model is:

$$y(x) = \sum_{i=1}^{d} a_i K(x_i, x) + b \qquad (1)$$

where the input vector is $x = [T, t]$ (T = temperature, t = residence time), the output vector is $y = [y_{CH_4}, y_{H_2}, y_{C_2H_6}, y_{CO_2}, \eta_{TOC}]$ ($y_{CH_4}, y_{H_2}, y_{C_2H_6}, y_{CO_2}$ are the mole fraction/mol% of CH$_4$, H$_2$, C$_2$H$_6$, CO$_2$ respectively, η_{TOC} = TOC transformation rate/%), d = number of training samples, $K(x_i, x)$ = kernel function, a_i = regression coefficient and b = bias.

The main problem of using LSSVM to build a model is to select the kernel function. The kernel function is always selected as the Radial Basis Function in practical applications, which is called RBF kernel function:

$$K(x, y) = \exp(-(x-y)^2 / 2\sigma^2) \qquad (2)$$

The two parameters, the regularization parameter γ and the width of kernel δ must be adjusted, for that the value of parameters will determine the LSSVM model's training and generalization ability directly. In this paper, the genetic algorithm is used for parameters optimization [7] about γ and δ. The ranges of parameters optimization are $\gamma \in [1,1000], \delta^2 \in [0.1,10]$, and the optimization results are $\gamma = 497.5153, \delta^2 = 0.6613$. Based on the optimal parameters γ, δ^2, 21 groups of experimental data

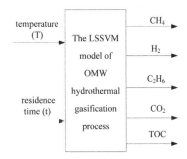

Figure 1. The LSSVM model of OMW hydrothermal gasification.

were selected as the model training data to define the model of biomass hydrothermal gasification process.

2.3 *Model verification*

Based on the trained model, the test data were input to the trained model to examine the generalization ability of the model. Four groups of data were selected from Table 1 as the test data, t = 120 s, T = 450°C, 500°C, 550°C, 600°C. The test results of the model are shown in Table 2. It can be seen in Table 2 that the maximum absolute value of the relative error of the experimental value and predictive value is 7.82% and the average value is about 3.5%. It shows that the model has good qualities in generalization and simulation and the model can be used to simulate the process of OMW hydrothermal gasification well.

Table 2. Comparison of experimental value and predictive value of OMW hydrothermal gasification.

Gasification conditions	Gasification indexes	CH$_4$ (mol%)	H$_2$ (mol%)	C$_2$H$_6$ (mol%)	CO$_2$ (mol%)	TOC transformation rate (%)
t = 120 s						
T = 450°C	Experimental value	12.98	9.30	2.95	72.74	56.58
	Predictive value	12.25	9.49	3.08	70.29	52.34
	Diff.*/%	5.64	2.09	4.36	3.37	7.47
T = 500°C	Experimental value	31.44	7.43	3.11	55.81	68.07
	Predictive value	28.98	7.55	2.90	55.43	66.67
	Diff./%	7.82	1.62	6.87	0.68	2.05
T = 550°C	Experimental value	29.15	9.06	4.70	54.70	87.63
	Predictive value	29.29	8.77	4.56	53.22	84.44
	Diff./%	0.47	3.19	2.87	2.70	3.64
T = 600°C	Experimental value	18.34	9.69	3.55	66.13	91.68
	Predictive value	19.36	9.41	3.46	63.92	87.51
	Diff./%	5.56	2.89	2.45	3.35	4.55

*The absolute value of the relative error of the experimental value and predictive value.

3 OPTIMIZATION FOR THE PROCESS OF OMW HYDROTHERMAL GASIFICATION

Based on the model, optimization for the process of OMW hydrothermal gasification is made to find out the optimization values of temperature and residence time when the gasification gas component CH$_4$, H$_2$, C$_2$H$_6$, CO$_2$ and TOC transformation rate are optimal respectively. The essence of the optimal process is a Multi-Objective Optimization problem (MOP).

3.1 Constraints

Temperature. The impact of temperature on the hydrothermal gasification process is very obvious. It can be seen in Table 1 that the TOC transformation rate is promoted greatly with increasing temperature. At the same time, the content of CH$_4$ increases but the content of CO$_2$ decreases. It shows that more carbon in the OMW transform into flammable gases. When the temperature continues to rise more than 550°C, the TOC transformation rate continues to increase but the content of CH$_4$ decreases and the content of CO$_2$ increases. It's not conducive to the gasification gas. The effect of gasification will be better if the temperature is around 550°C [8]. Therefore, the range of temperature was selected as 500~600°C in this paper.

Residence time. The major impacts of residence time on hydrothermal gasification process are TOC transformation rate and the gas composition under

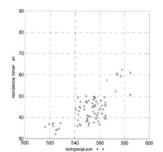

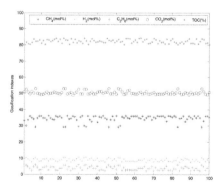

Figure 2. Pareto-optimal solutions.

the same temperature. It can be seen in Table 1 that the TOC transformation rate is relatively low when t = 30 s and the temperature is same. It shows that the carbon in OMW is not transformed fully. Extending the residence time, the TOC

Table 3. Optimization results of OMW hydrothermal gasification.

Temperature (°C)	Residence time (s)	CH_4 (mol%)	H_2 (mol%)	C_2H_6 (mol%)	CO_2 (mol%)	TOC transformation rate (%)
557.6190	36.9841	34.1107	9.6915	4.7145	49.8476	80.8400
559.0476	35.7143	33.7738	9.7655	4.8170	49.9931	81.1291
556.1905	46.8254	34.1783	9.5562	4.6250	50.2777	82.9015
562.5715	47.4603	33.7040	9.6765	4.7914	50.5251	84.2340

transformation rate increases and the content of C_2H_6 decreases which is decomposed into smaller molecules such as CH_4 and so on. However, when the residence time is over 60 seconds, the content of CH_4 decreases but the content of CO_2 increases. Therefore, the range of residence time was selected as 30 s~90 s in this paper.

3.2 Building of the multi-objective optimization function

The main purpose of OMW hydrothermal gasification is to transform the OMW into flammable gases with high quality. Therefore, it is expected that the TOC transformation rate will be higher and the content of H_2, CH_4 will be as high as possible and the content of CO_2 will be as low as possible, which ensures the mixed gases a higher heating value. To this end, the multi-objective optimization function was built as follow:

$$\begin{cases} \max Y = (Y_1(X), Y_2(X), 1/Y_3(X), 1/Y_4(X), \eta_{TOC}) \\ s.t. \ 500°C \leq x_1 \leq 600°C \\ 30s \leq x_2 \leq 90s \end{cases} \quad (3)$$

where $X = (x_1, x_2)^T$, x_1 = temperature and x_2 = residence time, $Y_1(X), Y_2(X), Y_3(X), Y_4(X)$ are the content of CH_4, H_2, C_2H_6, CO_2 respectively (mol%), η_{TOC} = TOC transformation rate (%).

3.3 Optimization for the process of OMW hydrothermal gasification

The parallel selection genetic algorithm [9–10] was used in this paper for the optimization. In the process of the parallel selection genetic algorithm for OMW hydrothermal gasification, based on the LSSVM model, the content of CH_4, H_2, C_2H_6, CO_2 and the TOC transformation rate were selected as five sub-targets for selection. The optimization result was shown in Figure 2. Every solution in Figure 2 meets the target functions of multi-objective optimization as a non-inferior solution. They are also called Pareto-optimal solutions.

3.4 Results analysis

The technology of biomass hydrothermal gasification could transform the organic wastes such as municipal solid wastes, industrial waste and residues etc which have high organic content but hard to treat into clean and efficient gases. The gases could be used for municipal gas system, it provides a new way for municipal gas [11]. According to the national standards for the heating value of biomass gasification gases, it's should higher than 4.6 MJ/Nm^3. To ensure a higher heating value, four solutions with higher CH_4 content from the Pareto-optimal solutions were selected as the optimal subset of OMW hydrothermal gasification. The results are shown in Table 3. It can be seen that the selected optimal subset not only has higher heating value, but also its TOC transformation rate is higher than 80%.

In the process of OMW hydrothermal gasification, the content of CH_4 is raising with the increasing temperature. The maximum content of CH_4 appears around 550°C. Some experimental studies verify the conclusions as well [3, 12]. However, with the temperature continues to rise and the residence time increases, the content of CH_4 decreases instead. The main reason is that the generated CH_4 make further reactions with H_2O to generate CO_2 under a condition of higher temperature and longer residence time. Therefore, the content of CO_2 increases significantly. It is similar to the change of the content of H_2 in the gasification process. The optimization result in Table 3 is in line with the above analysis. It can be seen that the optimal target values of temperature and residence time are around 550°C and 40 seconds respectively. The four groups of optimization results in Table 3 show that both the content of CH_4, H_2 and the TOC transformation rate are close to the maximum experimental values.

4 CONCLUSION

1. The LSSVM model of OMW hydrothermal gasification process was established in

this paper. The results of the model validation show that the maximum relative error is 7.82% between the predictive value and experimental value. The model can simulate the process of OMW hydrothermal gasification in supercritical water effectively.

2. Based on the model built, a multi-objective optimization function was put forward, by which the Pareto-optimal solution set was gained. The optimization results show that the effective gases of, in gasification gas and the TOC transformation rate are close to the maximum experimental values. Thus, it verifies the rationality and effectiveness of the methods in this paper.

REFERENCES

[1] A.D. Taylor, G.J. DiLeo, K. Sun. 2009. Hydrogen production and performance of nickel based catalysts synthesized using supercritical fluids for the gasification of biomass, Applied Catalysis B: Environmental 93: 126–133.

[2] H. Schmieder, J. Abeln, N. Boukis, et al. 2000. Hydrothermal gasification of biomass and organic wastes. Journal of Supercritical Fluids, 17: 145–153.

[3] L. Kong, G. Li, B. Zhang, et al. 2008. Hydrogen production from biomass wastes by hydrothermal gasification. Energy Sources, 30:1166–1178.

[4] Lv Youjun, Ji Chengmeng, Guo Liejin. 2005. Experimental investigation on hydrogen production by agricultural biomass gasification in supercritical water. Journal of Xi'an Jiaotong University, 39(3):238–242.

[5] Ekin Kpcak, Onur O. Sogut, Mesut Akgun. 2011. Hydrothermal gasification of olive mill wastewater as a biomass source in supercritical water. J. of Supercritical Fluids, 57:50–57.

[6] Li Dazhong, Wang Zhen. 2009. Modeling of biomass gasification process based on Least Squares SVM. Journal of System Simulation, 21(3):629–633.

[7] Wang Keqi, Yang Shaochun, Dai Tianhong, Bai Xuebing. 2009. Method of optimizing parameter of Least Squares Support Vector Machines by genetic algorithm. Computer Applications and Software, 26(7).

[8] Yan Qiuhui, Zhao Liang, Lv Youjun. 2009. Effects of operating parameters on performance of product gases from cellulose gasified in supercritical water. Journal of Basic Science and Engineering, 17(2).

[9] Julio Ortega; Javier Fernández; Antonio Díaz. 2002. PSFGA: A parallel genetic algorithm for multi objective optimization. IEEE Computer Society.

[10] Renan Hilbert, Gábor Janiga, Romain Baron, Dominique Thévenin. 2006. Multi-objective shape optimization of a heat exchanger using parallel genetic algorithms. International Journal of Heat and Mass Transfer, 49(15):2567–2577.

[11] J.A. Onwudili, P.T. Williams. 2008. Hydrothermal gasification and oxidation as effective flameless conversion technologies for organic wastes. Journal of the Energy Institute, 81(2): 102–109.

[12] Hao XH, Guo L J, MaoX, et al. 2003. Hydrogen production from glucose used as a model compound of biomass gasified in supercritical water [J]. International Journal of Hydrogen Energy, 28:55–64.

Frontiers of Energy and Environmental Engineering – Sung, Kao & Chen (eds)
© 2013 Taylor & Francis Group, London, ISBN 978-0-415-66159-1

Prediction of hot deformation behavior of biomedical materials TiNiNb SMA

Q. Liu, G. Wu, B. Duan & H.B. Zhang
Tubular Goods Research Institute of China National Petroleum Corporation, Xi'an, China

ABSTRACT: The hot deformation behavior of biomedical material TiNiNb Shape Memory Alloy (SMA) was investigated by isothermal single-pass compression on Gleeble-3500 thermal simulator with the deformation temperature from 800°C to 1050°C and with the strain rate from 0.01 s^{-1} to 10 s^{-1}. The results showed that the true stress-strain curves of TiNiNb SMA increased with decreasing deformation temperature and increasing strain rate, which indicated that the hot deformation behavior of biomedical material TiNiNb SMA was dynamic recrystallization. The hot compression deformation of TiNiNb SMA can be represented by Arrhenius model. The constitutive equation of TiNiNb SMA under hot compression deformation was calculated by a linear regression analysis. The activation energy for hot deformation of the experimental steel is 126.92 kJ/mol.

Keywords: TiNi shape memory alloy; hot deformation; flow stress; activation energy; constitutive equation

1 INTRODUCTION

TiNi Shape Memory Alloy (SMA) has become a new biomedical material due to excellent shape memory effect, biocompatibility, corrosion resistance, wear resistance, non-magnetic and good Super-Elastic (SE).[1-3] Currently, TiNi SMA is widely used in oral orthodontics, artificial heart pacemakers, blood filtration devices, orthopedic surgery and neurosurgery clinical implant materials.[4,5] Scholars all over the world have made lots of research on relationship of TiNi SMA properties and processing technology. S. Wang et al.[6] studied the deformation mechanism and the deformed microstructure of TiNi shape memory alloy. Carl P. Frick's[7] studies have shown that microstructure and texture of TiNi SMA changed significantly after casting, hot extrusion and cold drawing, and found that processing had a great impact on the recovery properties of TiNi SMA. Z.-r. He[8] investigated shape memory effect and its engineering applications of TiNi SMA. However, there is few investigation on deformation process, especially for hot deformation behavior of TiNiNb SMA.

The understanding of TiNiNb SMA behavior at hot deformation condition has great importance because of its effective role on metal flow pattern as well as the kinetics of metallurgical transformation. The size of flow stress is not only an important measure criterion of material plasticity quality, but also related to thermal processing equipment selection and design benchmark.[9]

In the metal forming process, constitutive equation is an important model to describes the relationship of the thermodynamic parameters during TiNiNb SMA deformation, and it is also an important precondition to predict and optimize TiNiNb SMA deformation process parameters with finite element numerical simulation method. In this paper, a comprehensive model describing the relationship of the flow stress, strain rate and temperature of TiNiNb SMA at elevated temperatures was proposed by a series of hot compression tests performed on Gleeble-3500 machine. Then, constitutive equation relating Z parameter and hot deformation activation energy Q were derived for TiNiNb SMA. In addition, the constitutive relation between peak strain and Z parameter was also built.

2 EXPERIMENT

The experimental Ti$_{47}$ Ni$_{44}$ Nb$_9$ (in a.t.%) alloy was melted by vacuum induction melting furnace in argon atmosphere. After cogging, forging and hot swaging, the alloy was machined to cylindrical specimen with a diameter of 10 mm and a height of 15 mm.

Compression tests are carried out on a Gleeble 3500 thermal simulation at temperature range 800–1050°C and with the strain rate from 0.01 s^{-1} to 10 s^{-1}. The samples were heated in vacuum to deformation temperature at heating rate of 1°C/s by thermocoupled feedback-controlled AC current and homogenized at test temperature for 120 s before deformation was initiated. Then, some specimens were compressed at the selected constant temperature and rapidly quenched with water. The reduction in height were 70% at the end of the compression tests.

3 RESULT AND DISCUSSION

3.1 Flow stress behavior

Typical true stress–true strain curves obtained during hot compression are shown in Figure 1. The results showed that the true stress-strain curves of TiNiNb alloy increased with decreasing deformation temperature and increasing strain rate. It can be seen that at the onset of deformation, flow stress increased rapidly at a decreasing rate with increasing strain. After flow stress peak appeared on total true stress-strain curves, with the increase of strain, flow stress decreased. Figure 1 also revealed that at a given strain rate, the higher the deformation temperature, the lower the peak stress.

The deformation at elevated temperature was a competing process of the dynamic softening and the work hardening. At the beginning of the deformation, dislocation density increased dramatically, and the work hardening exceeds the dynamic softening, leading to the rapid increase of stress.[10] As strain increases, dislocation density increased continuously, in addition, dynamic softening such as dynamic recovery and dynamic recrystallization occured, which can offset or partially offset the effect of work hardening. The larger recovery and recrystallization driving force, the stronger dynamic softening effect because of edge dislocation climb, screw dislocation cross-slip, dislocations of opposite signs cancelling and dislocation annihilation.[11,12] The dynamic softening rate was equal to the work hardening rate as flow stress peak appeared. With the true strain increasing, softening effect caused by the dynamic recrystallization was dominant, and the flow stress continued to decline. In view of that, the dynamic recrystallization was the mainly mechanism of dynamic softening for TiNiNb alloy.

3.2 Constitutive equations

Considering hot deformation similar to the creep phenomenon, but occurring at high strain rates and stresses, various constitutive relationships have been developed to model the high temperature deformation behavior of materials. The strain rate ($\dot{\varepsilon}$) is related to the temperature and activation energy for deformation by the Arrhenius equation expressed as:[13–16]

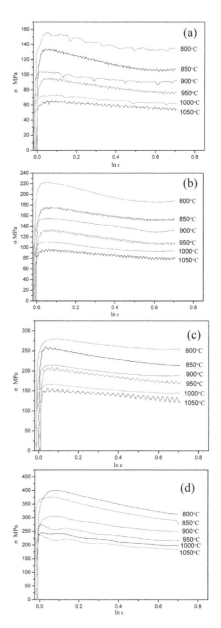

Figure 1. True stress-strain curves of TiNiNb Alloy at different deformation temperatures and strain rates. (a) $\dot{\varepsilon} = 0.01$ s^{-1} (b) $\dot{\varepsilon} = 0.1$ s^{-1} (c) $\dot{\varepsilon} = 1$ s^{-1} (d) $\dot{\varepsilon} = 10$ s^{-1}.

$$\dot{\varepsilon} = Af(\sigma)\exp\left(-\frac{Q_{HW}}{RT}\right) \qquad (1)$$

where Q_{HW} is the activation energy for deformation (J/mol), R is the universal gas constant (= 8.314 J/mol K), T is the temperature (K), A is a constant related with the materials and $f(\sigma)$ is the stress function which can be expressed by any of the following equations:[17-19]

$$f(\sigma) = \sigma^{n_1} \qquad (2)$$

$$f(\sigma) = \exp(\beta\sigma) \qquad (3)$$

$$f(\sigma) = [\sinh(\alpha\sigma)]^n \qquad (4)$$

The term α is the stress multiplier. n_1, β, n, and A are material constants. In the above equations, the peak flow stress σ_p is often taken as σ 13,20, but at few instances, the steady state flow stress (σ_s) is also used. Taking Eqs. (2) and (3) into Eq. (1), we can get the power law equation and the exponential equation respectively. The power law equation breaks down at high stress condition where as the exponential equation breaks down at low stress condition.[20] Over a wide range of stresses, the hyperbolic-sine law is found to be most suitable for explaining the hot deformation behavior of metals and alloys. Combining Eq. (1) with Eq. (4), the following constitutive equation can therefore be obtained:

$$\dot{\varepsilon} = A[\sinh(\alpha\sigma)]^n \exp\left(-Q_{HW}\big/_{RT}\right) \qquad (5)$$

The high temperature deformation behavior of metals and alloys is represented by the Zener–Hollomon parameter (Z), which correlates the strain rate, deformation temperature, and activation energy by the expression:

$$Z = \dot{\varepsilon}\exp(Q_{HW}/RT) \qquad (6)$$

By combining Eqs. (5) and (6), the expression for Z is obtained as:

$$Z = \dot{\varepsilon}\exp(Q_{HW}/RT) = A[\sinh(\alpha\sigma)]^n \qquad (7)$$

For the present study, the peak flow stress (σ_s) is taken for the σ term in the above expressions. Determination of the value of stress multiplier α to obtain a best fit is very important in the analysis of results. The value of α can also be defined as $\alpha \approx \beta/n_1$,[17,19] where β and n_1 are taken as the average values of the slopes of the $\ln(\dot{\varepsilon})$ vs. σ plots and $\ln(\dot{\varepsilon})$ vs. $\ln(\sigma)$ plots, respectively, for the range of temperatures studied. From Figure 2, the values of β was evaluated as 0.028 and n_1 was 5.511, the values of α was determined as 0.0052 for TiNiNb alloy.

3.3 Activation energy for deformation

Taking natural logarithm on the both sides of Eq. (5), we have:

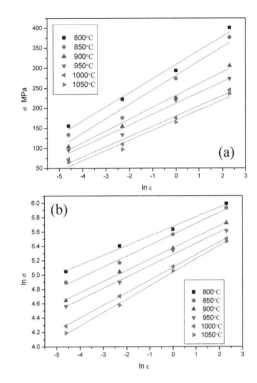

Figure 2. Relationship between strain rate and peak stress at different deformation temperatures. (a) $\sigma - \ln\dot{\varepsilon}$, (b) $\ln\sigma - \ln\dot{\varepsilon}$.

$$Q_{HW}/RT = \ln A - \ln\dot{\varepsilon} + n\ln[\sinh(\alpha\sigma)] \qquad (8)$$

The activation energy (Q_{HW}) for high temperature deformation is determined from the following relationship which calculated partial derivatives of Eq. (8):

$$Q_{HW} = R\left\{\frac{\partial\ln\dot{\varepsilon}}{\partial\ln[\sinh(\alpha\sigma)]}\right\}_T \left\{\frac{\partial\ln[\sinh(\alpha\sigma)]}{\partial\left(\frac{1}{T}\right)}\right\}_{\dot{\varepsilon}} = RnS \qquad (9)$$

where n is the mean slope of $\ln(\dot{\varepsilon})$ vs. $\ln[\sinh(\alpha\sigma)]$ plots at different temperatures and S is the mean slope of the $\ln[\sinh(\alpha\sigma)]$ vs. (1000/T) plots at various strain rates. These plots for TiNiNb alloy studied were shown in Figures 3 and 4. The activation energy Q_{HW} of TiNiNb alloy was calculated as 126.92 kJ/mol. Generally, the hot deformation activation energy Q_{HW} which related with the main chemical composition of materials will increase as the content of alloying elements increases.

It can be also noted that α is important only for the convenience of mathematical fitting for

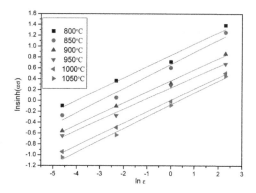

Figure 3. Hyperbolic Sine relationship between peak stress and natural logarithm of strain rate for TiNiNb alloy.

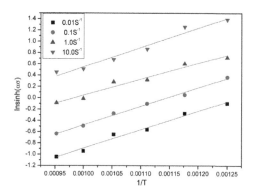

Figure 4. Relationship between peak stress and inverse temperature for TiNiNb alloy.

obtaining linear and parallel lines for the $\ln(\dot{\varepsilon})$ vs. $\ln[\sinh(\alpha\sigma)]$ plots. The value of $\alpha = 0.0052$ gives above 0.99 of coefficient of determination (R^2) values calculated for the six lines corresponding to six temperatures.

3.4 Zener-hollomon parameter (Z)

Variation of flow stress with deformation conditions can well be illustrated by the Zener–Hollomon parameter (Z) of Eq. (7). This expression can also be written as:

$$\ln Z = \ln A + n \ln[\sinh(\alpha\sigma)] \qquad (10)$$

where lnA and n are determined from the ln(Z) vs. $\ln[\sinh(\alpha\sigma)]$ plots. Figure 5 shows the ln(Z) vs. $\ln[\sinh(\alpha\sigma)]$ plots, indicating a good linear fit for the alloys. The coefficient of determination (R^2) values between ln(Z) and $\ln[\sinh(\alpha\sigma)]$ is above 0.989. The values of n and lnA are determine as 4.756 and 15.0 by means of linear regression analysis.

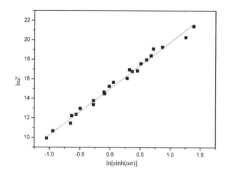

Figure 5. Relationship between peak stress and natural logarithm of Z parameter.

Thus, the constitutive equation of TiNiNb alloy under hot compression deformation can be expressed as the following:

$$\dot{\varepsilon} = e^{15}\left[\sinh(0.00523\sigma)\right]^{4.756}\exp\left(-\frac{126.92\times10^3}{8.31T}\right)$$

$$(11)$$

From Eq. (6) the Zener–Hollomon parameter (Z) can be described as: $Z = \dot{\varepsilon}\exp(126.92\times10^3/8.31T)$. According to the definition of hyperbolic sine function, its inverse function expressed as

$$\sinh^{-1}x = \ln\left(x + \sqrt{x^2+1}\right).$$

Thus the flow stress can also be described by the Zener–Hollomon parameter Z as:

$$\sigma = 191.205\ln\left\{\left(\frac{Z}{3.28\times10^6}\right)^{1/4.756}\right.$$

$$\left.+\left[\left(\frac{Z}{3.28\times10^6}\right)^{2/4.756}+1\right]^{1/2}\right\} \qquad (12)$$

4 CONCLUSIONS

1. The hot deformation behavior of TiNiNb alloy was investigated by isothermal single-pass compression. The flow stress of TiNiNb alloy increased with increasing strain rate and with decreasing deformation temperature. There were obvious flow stress peak observed in true stress-strain curves. After flow stress peak appeared, with the increase of strain, flow stress decreased, indicating that the hot deformations of these conditions are dynamic recrystallization.

2. The flow stress of TiNiNb alloy can be described by constitutive equation in hyperbolic sine function with the hot deformation activation energy Q_{HW} = 126.92 kJ/mol, and can also be described by the Zener–Hollomon parameter (Z) as:

$$\sigma = 191.205 \ln \left\{ \left(\frac{Z}{3.28 \times 10^6} \right)^{1/4.756} + \left[\left(\frac{Z}{3.28 \times 10^6} \right)^{2/4.756} + 1 \right]^{1/2} \right\}$$

where the Zener–Hollomon parameter (Z) can be described as:

$$Z = \dot{\varepsilon} \exp(126.92 \times 10^3 / 8.31T)$$

REFERENCES

1. Otsuka K, Wayman C.M. Cambridge: Cambridge University Press, 1998, 220–224.
2. H.-x. Guo, C.-h. Liang, Q. Mu. The Chinese Journal of Nonferrous Metals, 2001, 11(S2): 272–276M.
3. M. Xue, W.-t. Jia, Journal of Biomedical Engineering, 1987, 4(2): 130–134.
4. Fatiha El Feninat, Gaetan Laroche et al. Advanced Engineering Materials, 2002, 4(3): 91–104.
5. H.-q. Gu, G.-f. Xu, Tianjin: Tianjin Science and Technology Translation Press, 1993: 373–376.
6. S. Wang, Koichi Tsuchiya, L. Wang et al. J. Mater. Sci. Technol., 2010, 26(10), 936–940.
7. Carl P. Frick, Alicia M. Ortega et al. Metallurgical and Materials Transactions A. 2004, 35: 2013.
8. Z.-r. He, F. Wang, J.-e. Zhou, Transaction of materials and heat treatment, 2005, 26:21–26.
9. J.-t. Niu. Beijing: National defense industry press, 2007: 145–146.
10. J.-h. Chen. Dislocation and hardening. Shenyang: Liaoning Education Press, 1991:537.
11. Z.-x. Wang, X.-f. Liu, J.-x. Xie, Acta Metallurgica Sinica, 2008, 44(11):1378.
12. J.-s. Pan, J.-m. Tong, M.-b. Tian, Journal of Materials Science & Technology, 2005:515.
13. W.-j. Liang, Q.-l. Pan, Y.-b. He, Y.-c. Li, X.-g. Zhang, Journal of Central South University of Technology, 15 (2008): 289–294.
14. X.-y. Liu, Q.-l. Pan, Y.-b. He et al. The Chinese Journal of Nonferrous Metals, 2009, 19(2): 201–207.
15. T.-q. Zhang, Y.-j. Wang, Y. Zhou et al. Rare Metal Materials and Engineering, 2005, 34(3): 385–388.
16. W.-b. Li, Q.-l. Pan, W.-j. Liang et al. The Chinese Journal of Nonferrous Metals, 2008, 18(5):777–782.
17. X.-y. Liu, Q.L. Pan, Y.B. He, W.B. Li, W.J. Liang, Z.M. Yin, Materials Science and Engineering A 500 (2009) 150–154.
18. H. Zhang, L. Li, D. Yuan, D. Peng, Material Characterization 58 (2007) 168–173.
19. Z.-y. Chen, S.-q. Xu, X.H. Dong, Acta Metallurgical Sinica 21 (No.6) (2008) 451–458.
20. H.J. McQueen, N.D. Ryan, Materials Science and Engineering A 322 (2002) 43–6.

Frontiers of Energy and Environmental Engineering – Sung, Kao & Chen (eds)
© 2013 Taylor & Francis Group, London, ISBN 978-0-415-66159-1

Analysis on the development strategy China regional new energy industry

J. Chen, Y.N. Wu & X. Ba
Business Management Department of North China Electronic University, China

ABSTRACT: This paper first has a retrospect on China's new energy industry development during the "11th five-year plan" period, on the foundation of this, it analyzes the key factors affecting the industry development, and puts forward some guaranteed measure suggestions to the future new energy industry development. Besides, local governments and communities also need to take a series of safeguard measures to ensure smooth upgrade of local new energy industry by unremittingly deepening systematic construction, cultivating regional innovation network and the innovation system, strengthening the professional talents cultivation etc.

Keywords: regional new energy; industry upgrade; affecting factors; guarantee measures

FOREWORD

Reviewing the global developed countries' energy strategy in recent years, the development of new energy becomes an important composition of the world developed countries' new energy development. After the financial crisis, new energy industry becomes a strategic industry of developed countries to help them out of the crisis. It is already a global consensus to develop new energy industry and vigorously promote low-carbon economy. Unlike traditional industries, China's energy industry almost started with the world, driven by domestic and international markets and relying on abundant domestic resources and manufacturing cost advantages, it grows into a new energy industrial superpower in just a few years. In this process, private enterprises become the new energy industry development subject, industrial parks become an important carrier, and the local governments an essential strength. Under the corporation of the government, enterprises and parks, new energy industry grows into an outstanding contribution in some parts of China. Through the past years of development, great changes have taken place in the policy environment, technology as well as the market environment.

1 CHINA'S ENERGY INDUSTRY HAS DEVELOPED RAPIDLY AND CONSTANTLY UPGRADED

During the "11th five-year plan" period, China's new energy industry has expanded rapidly. Industries like wind power and photovoltaic maintained high growth. Among them, installed capacity of wind power has grown from 2599 MW in 2006 to 26276 MW by 2009, with an over 100% annual compound rate, and the wind power generation capacity being the world leader; moreover China's photovoltaic industry grows as fast as wind power. From 2006 to 2009, driven by international market, the scale of China photovoltaic industry ascends from 438 MW to 3460 MW, its annual compound rate reached 94% on average, ever since 2007, China has become the No. 1 solar battery producer worldwide. China's solar industry scale has reached the top of the world in 2008, being the biggest country in solar heater production and usage. By the end of 2009, China's energy holds more than 9% in total energy production. What's more, biomass, nuclear, geothermal, hydrogen and most of the potential new energy all have got rapid development.

When industrial innovation ascends, so does the sustainable development. With China's energy

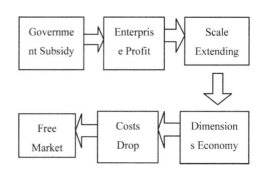

Figure 1. Path of new energy industry development.

industrial robust development, the driving force and innovation of China's energy industry also grew as well as prominent achievements on technology. During the "11th five-year plan" period, China has achieved a series of innovation progress, fields like photovoltaic battery manufacturing, solar battery packaging materials, wind power equipment manufacturing and wind power components manufacturing have emerged large techniques with independent intellectual property rights of the new energy manufacturing technology. The input for new energy industrial technology increases gradually. In order to improve the new energy industrial technology, the state has passed several fund supports like 863 etc. Meanwhile, new energy enterprise's research input improved continuously.

China's new energy industry has preliminarily established an independent innovational system as market-oriented, enterprise-subject, the combination of production and research. The main application of new energy is power generation. Due to the high cost of new energy application currently, it is mainly through government policy guidance and direct financial subsidies, the cultivation of market and guiding residents and other types of users to enlarge the input for new energy generation, to guide the entire industry's development, and ensure enterprise's profit. And since both companies and users have found much confidence in the benefit and development, enterprises begin to enlarge the production scale. On the basis of large scale production, new energy industry has reached a economic effect, which will directly lead to certain competition and thus the whole new energy power generation systems prices will drop, and finally form a free market incorporating both the new energy and traditional energy when they have own certain competitiveness.

Path of new energy industry development.

2 FACTORS TO DEVELOPMENT OF CHINA REGIONAL NEW ENERGY INDUSTRY

New energy, as a burgeoning industry, holds some characteristics as big investment and high technical, high cost of application. Its development is effected by not only the internal and external resources restriction, but also by market demand's push, due to its high investment and cost in application, its development also restricted by government policy factors. In addition, new energy industry realizes international integration development step by step, domestic new energy industry on one hand relies on foreign technology, and on the other hand it must participate in international competition. Therefore, both the domestic

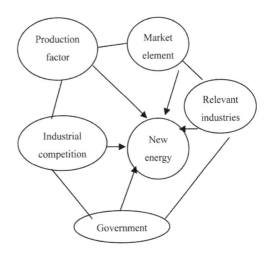

Figure 2. Factors to the development of new energy industry.

and foreign competition and relevant industries supporting ability should be taken into consideration in the position and development of domestic regional new energy industry. Through the analysis above, we conclude that the development of China's energy industry is affected by various factors, such as resource elements, the market demand, the construction of industry chain, industrial competition environment and government, etc.

Among them, the production factors including human resource, material, technical, traffic, new energy natural resources, etc.; market factor refers to the digestion, capacity ability of regional market to new energy application, including the digestive ability in the coming years, as well as specific industries market radius; Industrial competition means the similar product competition in the local market or a same target market of the new energy industry, Government elements refer to the government's attention and support including policy and capital support to the new energy industry. Relevant industry refers to the supporting ability of the relevant industries, including new energy equipment's production capacity, raw material matching production ability, etc.

Construction of regional new energy industry is to choose appropriate industry chain link as a starting point on the condition of the lack of new energy industrial construction, make it a new growth point in the future economic development. The main task of the construction is to determine the key industries direction, and choose an appropriate point. And the construction focus on production factors, industrial competition and the government. Among them, the industrial competition is of the top priority; its analysis consists of different new energy industry subdivision industries competition, supply and demand, future competition and the trend of supply and demand. Factors like resources endowment, traffic, human resources of the production factors are also important, which should be taken into consideration. The government's power in the new energy industry import stage is particularly important, for that to make new energy industry development planning, and perfect the new energy industry carrier construction, will become essential factors whether the new energy industry could root in the local or not.

3 SUGGESTION ON THE NSURING MEASURES OF CHINA REGIONAL NEW ENERGY INDUSTRY DEVELOPMENT

3.1 Strengthen system construction and supporting ability

China's new energy industry system construction and supporting capacity building will be the key content in the next years. Whether it is in a stable new energy industry development region, or a starting region, what it needs is to develop professionalization as a breakthrough, strengthen industrial correlativeness, and improve the overall industry competitiveness. The industrial system construction covers a lot of aspects as new energy equipment manufacturing, new energy industry product manufacture, project financing, project management, project operation and maintenance and matching industry. It requires to start from strengthening overall planning ability of new energy industry to achieve the strengthen of the system construction, to analyze the development needs of subdivide chain, and to deepen regional and international cooperation ability; moreover to cultivate special industries and improve industrial environment.

3.2 Cultivating regional innovation network, and strengthening the grounding and radio activeness of the enterprise

Although the new energy industry has developed into a global industry, the importance of fostering new energy industry and make it rooted in the local society, culture, economy and politics is on the rise as well as improving the innovation ability. The regional new energy industry first needs to realize the root in the local and network agglomeration, and then cooperate with the relevant industry enterprises within the region to establish labor division, and network system, so that it will play its radiation impetus function. An efficient way to accomplish regional innovation network system is

the construction of the industrial park carrier, by this to establish the system of organization, interactivity, innovation, etc. so that to realize the whole innovation system construction.

3.3 Speed up the construction of innovation system; improve the level and stage of industrial cluster

New energy industry construction and upgrade is the first target of regional new energy industry to realize its leaping development, and the second target is realizing the development of new energy industry agglomeration, on this basis, the promotion of new energy industry agglomeration level and stage is an important guarantee to keep regional new energy competitive and sustainable. The construction of innovation system and new energy industry system are the same, they both own a complete industrial chain construction, including research and development, flexible specialized production and marketing. The construction of innovation system is the core competitiveness of the new energy enterprise to keep sustainable development and realize the industry cluster construction.

3.4 Increasing the incentive and cultivation of professional talent, providing personnel security

New energy industry is an industry with relatively high technical barriers; it's still in great need of professional technology, manufacture and management talents to support the industry development. The talents security of regional new energy industry development is closely related to local economy, education and culture. In the talents construction and training, it needs to take more measures to improve talent attraction, by the plans of introducing leading talent, guaranteeing the highly educated and the logistics of staff to perfect the talents attraction system. Only by the cooperation with government, enterprises and schools can the talents training carry out. To strengthen the talents security for the new energy development, it needs to take the combination external absorption and internal training.

4 CLOSING

New energy industry has been identified as one of China's key strategic new industries, the progress in technology and increasing enterprises lead to a more fierce competition environment, and the pulling force from the growing-mature market is also increasing. During the period of industrial challenge and opportunity, both the regions which have hold the foundation of new energy industry and the regions which are entering this industry are facing problems like how to upgrade the local enterprises and to achieve a leap in this industrial development. Facing the increasingly fierce market competition and the severe challenge from domestic and foreign large enterprises, regional new energy industry must accelerate the construction of innovation system, and constantly promote technology, system and organization innovation so that it can attract more excellent enterprises into local place, ascend industry development level and expand the conglomeration effect.

REFERENCES

[1] Schmalensee Richard. Evaluating Policies to Increase Electricity Generation from Renewable Energy. Review of environmental economics and policy, 2012, 6: 1.

[2] Liu Guozhong Effect of Emission Right and Renewable Energy Development to Power Market Power System Automation 2010, 07.

[3] Capik Mehmet; Yilmaz Ali Osman; Cavusoglu Ibrahim. Present situation and potential role of renewable energy in Turkey. Renewable energy, 2012.02.031.

[4] Feng yuanqi China's 6 Key Points of Energy Planning in the "Twelfth-five" Period [J] Sulphur Phosphorus & Bulk Materials Handling Related Engineering 2010,(02).

[5] China's Establishment on the Energy Planning in the "Twelfth-five" Period [J] Chemical Fertilizer Design 2010, (01).

[6] Qi Kai the Position of City Energy Planning in the City Planning and its Establishment Idea [J] Beijing Planning Review 2010, (02).

[7] Kautto N.; Peck P. Regional biomass planning—Helping to realise national renewable energy goals? Renewable energy, 2012.03.024.

[8] Lin Boqiang Necessary Consideration on New Energy in the "Twelfth-five" Period [J] China Petrochem 2010, (12).

[9] Pasicko Robert; Brankovic Cedo; Simic Zdenko. Assessment of climate change impacts on energy generation from renewable sources in Croatia. Renewable energy, 2012, 3(46) 224–231.

[10] Wang Dengyun, Xu Wenfa, Low Carbon City and New Energy Planning of Constructional Region [J] Construction technology 2010,(13).

Frontiers of Energy and Environmental Engineering – Sung, Kao & Chen (eds)
© 2013 Taylor & Francis Group, London, ISBN 978-0-415-66159-1

The feasibility study of new energy technology used for military vehicles

Y. Wang & L. Jin
School of Automobile Engineering, Wuhan University of Technology, Wuhan, P.R. China

H.G. Jia & Z.H. Li
General Logistics Department of Chinese 71897 Armed Forces, P.R. China

ABSTRACT: This paper studies the new energy technology in the military vehicle area in order to solve the shortcomings of traditional automobiles, such as serious fuel consumption, high temperature, and heavy pollution etc. Firstly this paper expounds the characteristics of new energy vehicles; And then as for the application of new technology in the military vehicle area, the paper makes the feasibility study separately from 4 aspects: technical feasibility, security feasibility, economic feasibility, and operating environmental feasibility; Finally, the paper proposes some measures, such as perfecting the infrastructure, reinforcing the development of army and local enterprises, perfecting the mechanism to promote the application of new energy technology on military vehicles.

Keywords: new energy technology; military vehicle; feasibility

1 INTRODUCTION

In recent years, with the rapid development of auto industry, our demand for oil is always very high, and international crude oil price is rising constantly. Taking energy security and environmental protection into account, countries around the world are increasingly focusing on the development and promotion of new energy vehicles. Military vehicles, which assume special tasks, also take up a certain proportion in the total number of vehicles in China. Therefore, promoting new energy technology has a profound meaning for military vehicles.

2 THE FEATURES OF NEW ENERGY VEHICLES

New energy vehicles mean all energy cars except those whose power comes from the gasoline or diesel engines. Researches of new energy vehicles in China mainly focus on electric vehicles. Electric vehicles are divided into four categories: Hybrid Electric Vehicles (HEV), Plug-in Hybrid Electric Vehicles (PHEV), Pure Electric Vehicles (PEV), and Fuel Cell Electric Vehicles (FCEV). Characteristics and application of new energy vehicles are shown in Table 1.

Hybrid electric vehicle is driven by two or more than two kinds of power source, which is a transition model from traditional fuel car to pure electric vehicle; Compared to traditional HEV, plug-in hybrid

electric vehicle can be directly charged by power grid, moreover electric driving often takes up a higher proportion in PHEV, while the power dependence on engine is less than common HEV; Pure electric vehicle is mainly driven by electric power, consisting of battery, motor and control system etc.

3 FEASIBILITY ANALYSIS OF THE NEW ENERGY TECHNOLOGY'S USE IN MILITARY VEHICLES

From the Anti-Japanese War period, the vehicles equipped in CCP army have experienced 3 stages. It is more than half a century from the end of 1937, when leading authorities of the central military commission formed the 1st motor transport troops in order to meet the needs of the new situation of the war, to the present. Chinese military vehicles started from the first generation, and now the products of second generation have fully equipped the army while the products of third generation are developing and researching, and have been finished the trial assembly in army. Large-scale auto companies in China are producing all kinds of military vehicles, but energy is mainly still fuel.

Under the condition of modern war, the battle style, battlefield environment and target characteristics are changing constantly. This requires that military vehicles should have higher mobility, higher passing ability, lower fuel consumption and higher energy efficiency. Traditional internal-combustion

Table 1. Comparison of characteristics and application of various new energy vehicles.

Energy form	New energy vehicle features	Application situation
HEV	Advantages: promoting fuel utilization rate; long driving range Disadvantages: passive fuel efficiency when driving at high speed for long distance; more complex structures; price is relatively expensive	We can realize industrialization more easily; it is the mainstream route of new energy car currently, widely promoted in city buses, encouraged by the State, and a transition model for new energy vehicles
PHEV	Advantages: charged by power grid directly; much lower fuel consumption per 100 km and carbon dioxide emissions volume Disadvantages: relatively higher cost of battery's using, maintenance and replacement	It is considered the most promising new energy automotive technology in energy conservation and environmental protection; it is also a transition model and has been a hot research of national auto enterprises and related organizations
PEV	Advantages: relatively more mature and simple technology; positive energy conservation and environmental protection Disadvantages: lower continuous mileage, life and speed while higher vehicle price	In China pure electric bus has been put into operation, and the State encourages the development It is the development trend of new energy automobile in the future
FCEV	Advantages: higher energy conversion rate, no pollution Disadvantages: very high price without mature technique	It is still in the initial stage in China, and the State encourages the development It is considered to be the ultimate direction of electric vehicle development

engine cars have already reached mature period, whose potential of improving fuel consumption is limited; these cars can hardly realize flexible decoration in overall design, and already can't keep up with military vehicles' demand in the 21st century. With the continuous development of new energy technology, new energy vehicles have not only obviously improved dynamic property and passing ability compared to traditional cars, but also they have the abilities of being controlled more flexibly, saving fuel positively and having the function of concealed driving, which make new energy technology extremely attractive in military vehicle application. Now this paper will make feasibility application analysis for new energy technology in the military vehicles.

3.1 Technical feasibility

The technology of Chinese new energy vehicles will be able to keep up with the developed countries', so that the application of new energy technology in the military vehicle area has certain technical feasibility. Hybrid technology is the current relatively mature technology, and the most suitable technology for the application in the military vehicles. Hybrid car's tandem arrangement scheme can make full use of the excellent electromotor characteristics such as constant torque in low speed and constant power output in high speed. This will improve the mobility of the vehicles, let military vehicles possess the covert combat capability of "mute" driving and large power capacity of power station moving outdoor, conduct the regenerative braking energy recovery, and improve the energy utilization ratio for the

vehicle. Parallel type and hybrid layout scheme can realize three driving modes: the mixed mode, electricity storage mode and the engine model. So that these models can greatly improve the potential of the vehicle's cross-country mobility and concealment in order to improve battlefield survivability.

3.2 Security feasibility

Taking the use particularity into account, military vehicles require great security. In the process of the new energy vehicles research, the engineers must make deep analysis in every detail of the total process from the manufacture, use, services to recovery, so as to provide security solutions for military new energy vehicles. For example, in normal driving process of electric car, a full range advanced monitoring system can monitor each battery to ensure that battery can maintain the correct voltage value and cooling system can provide the best battery operating temperature; In order to alleviate the impact of collision, the battery can be properly placed between the rear wheels, and transmission shaft tunnel area between two wheels can protect battery from the influence of the rear crash. When crash happened, the crash sensor linked to battery will sent the crash information to CPU, that will cut off the power supply automatically in order to avoid the risk of short circuit. For another example, in emergency road conditions, increased battery weight, assembled in plug-in hybrid cars, maybe affect the car's motion characteristics, but the existing platform and brake system have the capacity to deal with the increased weight, and

DSTC system and traction control system can help drivers control the situation to ensure safety. Therefore, in the current technology conditions, new energy vehicles, which can be used in military purpose, have the security feasibility.

3.3 Economic feasibility

New energy vehicles have better economic feasibility in energy use than the traditional fuel cars. We make the following analysis in economic efficiency for existing micro city hybrid coaches, pure electric coaches (all used as city buses), taking the new energy vehicles used in Wuhan city for example (shown in Tables 2 and 3). The current operating hybrid buses in Wuhan are mainly EQ6110HEV manufactured by Dongfeng Electric Vehicle Company Limited, and we use Ankai car as pure electric bus in this paper.

The Table 3 analysis shows that in the same oil price and the same trip mileage, new energy buses show their economic advantage in fuel cost. But because the hybrid buses' fuel proportion is still very high, fuel consumption still take the lead place in the process of driving, so that the energy use cost will still be influenced by the rising oil prices. As power-driven proportion going up in the hybrid electric cars, the fuel economical efficiency will be more apparent,

Table 2. Fuel and new energy bus' calculating parameter under the same condition.

Name	Fuel bus	Hybrid buses	Pure electric buses
Car name	Yangtze	Dongfeng	Ankai
Energy consumption/ km	32 L diesel oil	30 L diesel oil	30 kWh power
Battery purchase expenses	0	40.32 thousand	405 thousand

Table 3. Energy costs compare for three kinds of buses under the condition of different electricity and oil prices (Yuan).

Pure electric buses	Electricity prices (Yuan/kWh)	0.5	0.8	1.0	1.5
	Electricity costs (Yuan)	15	24	30	45
Fuel bus	Oil prices (Yuan/L)	6	6.5	7	8
	Oil costs/km (Yuan)	192	208	224	256
Hybrid buses	Oil prices (Yuan/L)	6	6.5	7	8
	Oil costs/km (Yuan)	180	195	210	240

and pure electric bus will be the most prominent one. Therefore, there is economic feasibility for new energy technology's application in the troops.

3.4 Operating environment feasibility

New energy vehicles have the characteristics of environmental protection, non-pollution, short running mileage, economic energy use and so on, which are applicable for area using. Chinese troops have the relative concentrate location, and the new energy vehicles can be well used in the army hospital, scientific research institutions, schools, military representative offices, major organs and residential places etc. The fact, that the new energy vehicles are use in troops, can make it convenient, short-cut and easy to carry out for the car's research and development, procurement, management, operation, and maintenance. We can energetically promote the new energy military vehicles in residential areas, such as the patrol car, passenger car etc, duly promote heavy SUVs, special operation vehicles and so on. And this has the following advantages.

3.4.1 The clean, comfortable and safety feature can create a quiet and neat work, training, studying and living environment for troops

Traditional vehicles use gasoline or diesel fuel and result in serious pollution and undesired sound, especially those heavy SUVs let out a mass of emissions in using process, which will affect our normal working, training, studying and life. New energy vehicles have the advantages of "mute" driving along with less pollution or even zero emissions, and they can provide clean, comfortable and safe environment.

3.4.2 It's convenient to build infrastructure

In order to promote the new energy vehicles in military vehicle market, there must be a lot of charging station. In the army, where related units and personnel distribution are relatively concentrate as well as military vehicles are relatively abundant and frequently applied, it is suitable for the new energy vehicles. It can be not only special for troops, but also appropriately open to society, so that the infrastructure can be taken full advantage and the use of resources and realize optimization.

3.4.3 It is convenient to management and maintenance

Troops also have the advantage of unified management; it can train related personnel and repair equipments together, which is beneficial for the popularization of new energy vehicles. Therefore, new energy technology can be applied to military vehicles in the current policy environment of domestic rapid development of new energy vehicle technology.

4 ADVICE FOR THE NEW ENERGY TECHNOLOGY PROMOTION IN MILITARY VEHICLE AREA

4.1 *Pilot successively, and then promote*

In view of the particularity of military vehicles, when new energy technology is extend to military cars, we can take measures of "pilot successively, and then promote", which will promote new energy vehicle firstly in a few troops area, then promote in large area after gaining a certain experience of managing the new energy vehicle.

4.2 *Unified management, and establish the new energy vehicles supporting facilities*

A full equipped supporting facility is essential for new energy vehicle's going to scale in army. In order to promote the application of new energy military vehicle, troops should establish a new large amount of facilities. Comprehensively considering the power production, transport, construction of the charging station, battery sale & repair and other issues, corresponding strategies are formulated. For example, troops can build a certain number of public charging stations, charging pile, special cable and socket, so as to extend the new energy vehicle mileage. These utilities can be unified planned and managed by the troops. The establishment of charging station can be combined with the existing parking lot, by setting charging facilities in the parking lot. This method can adopt the intelligent billing and management that there is no person on duty in order to strengthen the new energy vehicle's convenience.

4.3 *Focus on strengthening the building between the military and civilian*

There is something in common in auto technology between army equipments and civil products; therefore, after drawing on the experience of the military equipment development in developed countries, the development of Chinese military vehicle equipment should be army-civilian combination. The development should rely on the domestic auto industry, and take the road of synchronous development. Only in this way can we make full use of the domestic auto industry strength that is in the continuous growth and development, shorten the development cycle of the military vehicle equipment and reduce initial cost. Especially the new energy vehicle belongs to the high-tech projects, and with slight modifications, the existing local models can be transformed into military vehicles. There is a very large market space in the development, production and management of civil-military dual-use new energy vehicle. At the same time, troops can import advanced experiences of local governments and enterprises when constructing the supporting infrastructure in the period of developing new energy vehicle. After the completion, troops can open it to the society appropriately to achieve maximum resource utilization, which will stimulate local economic development in a certain extent.

4.4 *Perfecting the use mechanism of new energy vehicle*

It was important to draw on local development experience to apply new energy technology in military vehicle, while firstly fulfill requirements from the leadership of the Central Military Commission based on the actual situation, and insist on the principles of standardization, serialization and universal. Troops should not only focus on these principles among vehicle models and assembly parts, but also between military and civilian vehicles. By these demands, on one hand, we can lay the foundation for the new energy vehicle promotion in future; on the other hand, we can provide conveniences in the use and maintenance of new energy military vehicles.

5 CONCLUSION

In short, applying the new energy technology to Chinese military vehicles is feasible, which not only has the social benefits of energy conservation and environmental protection, but also has economic benefits of reducing the energy consumption cost. At the same time the new energy vehicle's characteristics, which are energy saving, environmental protection and low temperature in engine, also help improve the military vehicles' property of concealment and dynamic etc. This is also very important for Chinese military vehicles to enhance long-term competitiveness and the ability to make future advances.

REFERENCES

[1] Wang Hongzhi, Huang Jian. Analysis of the hybrid electric drive application on military vehicles. Friend of Science Amateurs, 2008.
[2] Wu Xuelei. Summary of history, current situation and trend in domestic and foreign military vehicle development. Missiles and Space Vehicles, 2001.
[3] Li Dayuan. Countermeasure research in Chinese new energy auto industry development under the background of low carbon economy. Economic Review, 2011.2.
[4] Ma Lige, Wang Jun-xi. Electric vehicle is the development direction for Chinese auto industry. Auto Industry Research, 2011.1.
[5] Zhang Min. Promoting the policy environment for Chinese new energy auto development. Anhui Science & Technology, 2011.1.

Frontiers of Energy and Environmental Engineering – Sung, Kao & Chen (eds)
© 2013 Taylor & Francis Group, London, ISBN 978-0-415-66159-1

Impact of substitution of M (M = Cu, Co) for Ni on hydriding and dehydriding kinetics of as-spun nanocrystalline and amorphous Mg_2Ni alloys

Y.H. Zhang, Y.Y. Xu, T. Yang, C. Zhao & H.W. Shang
Key Laboratory of Integrated Exploitation of Baiyun Obo Multi-Metal Resources, Inner Mongolia University of Science and Technology, Baotou, China

D.L. Zhao
Department of Functional Material Research, Central Iron and Steel Research Institute, Beijing, China

ABSTRACT: The Ni was partially substituted by M (M = Co, Cu) in order to ameliorate the hydriding and dehydriding kinetics of Mg_2Ni-type alloy. The melt spinning technology was used to fabricate the $Mg_{20}Ni_{10-x}M_x$ (M = Co, Cu; $x = 0, 1, 2, 3, 4$) alloys. The structures of the as-spun alloys were characterized by XRD and TEM. The hydriding and dehydriding kinetics of the alloys were measured by an automatically controlled Sieverts apparatus. The results show that the as-spun (M = Co) alloys hold a nanocrystalline and amorphous structure, whereas the as-spun (M = Cu) alloys display an entire nanocrystalline structure, indicating that the substitution of Co for Ni facilitates the glass formation in the Mg_2 Ni-type alloy. Additionally, Co substitution results in the formation of the secondary phase $MgCo_2$ instead of altering the Mg_2 Ni major phase. The substitution of M (M = Co, Cu) for Ni not only engenders an insignificant effect on the hydriding kinetics of the Mg_2 Ni alloy, but also enhances dehydriding kinetics of the alloy dramatically.

1 INTRODUCTION

The Mg_2 Ni-type metallic hydrides have been considered to be quite adequate for commercial applications as a on-board hydrogen sources of the fuel cell vehicles on account of their high hydrogen storage capacity, e.g. 3.6 wt.% for Mg_2NiH_4, 4.5 wt.% for Mg_2CoH_5 and 5.4 wt.% for Mg_2FeH_6 (Ebrahimi-Purkani 2008; Jain 2010). However, the practical application of the alloys is deeply frustrated by their relatively high H-desorption temperatures and sluggish hydriding/dehydriding kinetics. Alloying and microstructure modification are deemed to be the main approaches to improving the hydriding/dehydriding properties. It was documented that Mg and Mg-based alloys with a nanocrystalline and amorphous structure exhibit higher H-absorption capacity and faster kinetics of hydriding/dehydriding than crystalline Mg_2 Ni (Spassov 1999; Orimo 1996). Mechanical alloying (Hong 2002) and melt-spinning (Spassov 2002) are proved to be extremely appropriate techniques to produce amorphous and nanocrystalline Mg-based alloys. It was reported that an amorphous single phase can be obtained by adding a small amount of La, Y or Nd to the Mg_2 Ni alloy (Huang 2006).

In the present study, the investigation emphasis is mainly focused on the hydriding and dehydriding kinetics of the nanocrystalline and amorphous $Mg_{20}Ni_{10-x}M_x$ (M = Co, Cu; $x = 0$–4) alloys prepared by melt spinning. Furthermore, the effects of substituting Ni with M (M = Co, Cu) on the structures of the Mg_2Ni-type alloys have been investigated.

2 EXPERIMENTAL

The experimental alloys, with the chemical compositions of $Mg_{20}Ni_{10-x}M_x$ (M = Co, Cu; $x = 0, 1, 2, 3, 4$), were prepared by using a vacuum induction furnace in a helium atmosphere at a pressure of 0.04 MPa. A part of the as-cast alloys was remelted and spun by melt spinning with a rotating copper roller cooled by water. The spinning rate used in the experiment, approximately expressed by the linear velocity of the copper roller, is 15 m/s.

The phase structures of the as-spun (15 m/s) alloys were determined by XRD (D/max/2400). The thin film samples of the as-spun alloys were prepared by ion etching for observing the morphology with HRTEM (JEM-2100F).

The hydriding and dehydriding kinetics of the alloys were measured by an automatically controlled Sieverts apparatus. The hydrogen absorption was conducted at 1.5 MPa and 200°C, and the hydrogen desorption was carried out at a pressure of 1×10^{-4} MPa and 200°C, too.

3   RESULTS AND DISCUSSION

3.1   *Characterization of the structures*

The XRD profiles of the as-spun (15 m/s) $Mg_{20}Ni_{10-x}M_x$ (M = Co, Cu; x = 0–4) alloys are illustrated in Figure 1. It can be seen from Figure 1 (a) that the structures of the as-spun (M = Co) alloys have an obvious change with the growing of the amount of Co substitution. As the Co content increases to x = 4, the as-spun (M = Co) alloy displays an amorphous structure evidently, whereas it is visible from Figure 1 (b) that all the as-spun (M = Cu) alloys hold an entire crystalline structure, suggesting that the substitution of Co for Ni facilitates the glass formation in the Mg_2Ni-type alloy. Based on the FWHM data of the major diffraction peak (203) in Figure 1, the crystalline sizes $D_{<hkl>}$ (nm) of the as-spun alloys

are calculated by Scherer's equation (Williamson 1953), being in a range of 30 to 80 nm with the rising of the spinning rate. In addition, the substitution of Cu for Ni and the melt spinning do not change the phase structure of the Mg_2Ni alloy, but the substitution of Co for Ni, instead of changing the major phase of Mg_2Ni in the alloys, results in the formation of secondary phase $MgCo_2$.

The TEM micrographs of the as-spun (15 m/s) $Mg_{20}Ni_{10}$, $Mg_{20}Ni_6Co_4$ and $Mg_{20}Ni_6Cu_4$ alloys are illustrated in Figure 2. It can be seen from the amplified morphologies of Figure 2 (a) and (c)

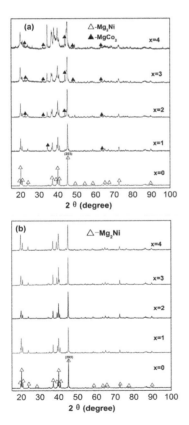

Figure 1.   XRD patterns of the as-spun (15 m/s) $Mg_{20}Ni_{10-x}M_x$ (x = 0–4) alloys: (a) M = Co, (b) M = Cu.

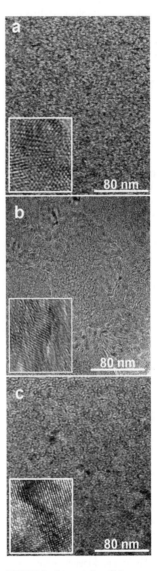

Figure 2.   HRTEM micrographs of the as-spun (15 m/s) alloys: (a) $Mg_{20}Ni_{10}$, (b) $Mg_{20}Ni_6Co_4$, (c) $Mg_{20}Ni_6Cu_4$.

that the as-spun $Mg_{20}Ni_{10}$ and $Mg_{20}Ni_6Cu_4$ alloys display an entire nanocrystalline structure. An amorphous phase is visible on the grain boundaries of the $Mg_{20}Ni_6Co_4$ alloy, which conforms well to the XRD observations depicted in Figure 1.

3.2 Hydriding and dehydriding kinetics

The hydriding kinetic curves of the as-spun (15 m/s) $Mg_{20}Ni_6M_4$ (M = Co, Cu) alloys are presented in Figure 3. From which it is found that all the as-spun alloys display a quite fast initial hydrogen absorption stage after which the hydrogen content is saturated at longer hydrogenation time. In order to establish a direct relationship between the hydriding kinetics and the amount of M (M = Co, Cu) substitution, the hydriding kinetics is symbolized by a hydrogen absorption saturation ratio (R_t^a), the ratio of the hydrogen absorption capacity for a fixed time to the saturated hydrogen absorption capacity of the alloy, which is defined as $R_t^a = C_t^a / C_{100}^a \times 100\%$, where C_{100}^a and C_t^a are hydrogen absorption capacities at 100 min and t min, respectively. It is viewable from Figure 3 that, for all the experimental alloys, the C_{100}^a values are more than 98% of their saturated hydrogen absorption capacities. Therefore, it is justifiable to take the C_{100}^a value as the saturated hydrogen absorption capacity of the alloy. Taking hydrogen absorption time of 5 min as a benchmark, and the relationship between the R_5^a ($t = 5$) values of the as-spun (15 m/s) alloys and the amounts of the M (M = Co, Cu) substitution are also inset in Figure 3. It reveals that the R_5^a ($t = 5$) values first augment and then fall with the growing of M (M = Co, Cu) contents.

The dehydriding kinetic curves of the as-spun (15 m/s) $Mg_{20}Ni_{10-x}M_x$ (M = Co, Cu; x = 0–4) alloys are depicted in Figure 4. It shows an important feature in the dehydriding process of the alloys, very fast initial hydrogen desorption and followed by slack increase of the amount of hydrogen desorbed. To establish the direct relationship between the dehydriding kinetics and the M (M = Co, Cu) content, the dehydriding kinetics of the alloy is signified by hydrogen desorption ratio (R_t^d), the ratio of the hydrogen desorption capacity for a fixed time to the saturated hydrogen absorption capacity, which is defined as $R_t^d = C_t^d / C_{100}^a \times 100\%$, where C_{100}^a is the hydrogen absorption capacity at 100 min and C_t^d is the hydrogen desorption capacity at the time of t min, respectively. Taking hydrogen desorption time of 20 min as a benchmark, the R_{20}^d ($t = 20$) values of the as-spun (15 m/s) alloys as a function of the M (M = Co, Cu) content are presented in Figure 5.

It indicates that the substitution of M (M = Co, Cu) for Ni enhances the dehydriding kinetics of the as-spun (15 m/s) alloys dramatically. The R_{20}^d values

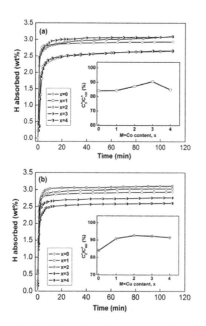

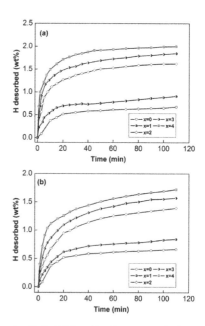

Figure 3. Hydriding kinetic curves of the as-spun (15 m/s) $Mg_{20}Ni_{10-x}M_x$ (M = Co, Cu; x = 0–4) alloys: (a) M = Co, (b) M = Cu.

Figure 4. Dehydriding kinetic curves of the as-spun (15 m/s) $Mg_{20}Ni_{10-x}M_x$ (M = Co, Cu; x = 0–4) alloys: (a) M = Co, (b) M = Cu.

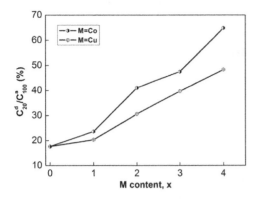

Figure 5. Evolution of the R^d_{20} values of the as-spun (15 m/s) $Mg_{20}Ni_{10-x}M_x$ (M = Co, Cu; x = 0–4) alloys with M (M = Co, Cu) content.

grows from 17.66% to 64.79% for the (M = Co) alloy, and from 17.66% to 48.32% for the (M = Cu) alloy with the rising of the M (M = Co, Cu) content from 0 to 4. The ameliorated dehydriding kinetics by M (M = Co, Cu) substitution is ascribed to the fact that the partial substitution of M (M = Co, Cu) for Ni in Mg_2 Ni compound decreases the stability of the hydride and makes the desorption reaction easier (Woo 1999). It is evident that the Co substitution exhibits a much distinguished impact on the dehydriding kinetics of the Mg_2 Ni alloy as compared to the Cu substitution, which is ascribed to the glass forming ability enhanced by Co substitution.

4 CONCLUSIONS

The Mg_2 Ni-type $Mg_{20}Ni_{10-x}M_x$ (M = Co, Cu; x = 0, 1, 2, 3, 4) alloys with a nanocrystalline and amorphous structure were successfully synthesized by melt spinning technology. The investigation on the structures of the alloys reveals that the as-spun (M = Co) alloys exhibit a nanocrystalline and amorphous structure, but no amorphous phase is detected in the as-spun (M = Cu) alloys. The substitution of M (M = Co, Cu) for Ni engenders a notable impact on the hydriding and dehydriding kinetics of the as-spun (15 m/s) alloy. The hydrogen absorption saturation ratio (R^a_5) first mounts up and then falls, while hydrogen desorption ratio (R^d_{20}) grows dramatically with the growing of the amount of M (M = Co, Cu) substitution.

ACKNOWLEDGEMENTS

This work is supported by National Natural Science Foundations of China (51161015 and 50961009), Natural Science Foundation of Inner Mongolia, China (2011ZD10 and 2010ZD05).

REFERENCES

Ebrahimi-Purkani, A. & Kashani-Bozorg, S.F. 2008. Nanocrystalline Mg_2Ni-based powders produced by high-energy ball milling and subsequent annealing. *Journal of Alloys and Compounds* 456: 211–215.

Hong, T.W. & Kim, Y.J. 2002. Synthesis and hydrogenation behavior of Mg–Ti–Ni–H systems by hydrogen-induced mechanical alloying. *Journal of Alloys and Compounds* 330–332: 584–589.

Huang, L.J. Liang, G.Y. Sun, Z.B. & Wu, D.C. 2006. Electrode properties of melt-spun Mg–Ni–Nd amorphous alloys. *Journal of Power Sources* 160: 684–687.

Jain, I.P. Lal, C. & Jain, A. 2010. Hydrogen storage in Mg: A most promising material. *International Journal of Hydrogen Energy* 35: 5133–5144.

Orimo, S. & Fujii, H. 1996. Hydriding properties of the Mg_2 Ni-H system synthesized by reactive mechanical grinding. *Journal of Alloys and Compounds* 232: L16–L19.

Spassov, T. & Köster, U. 1999. Hydrogenation of amorphous and nanocrystalline Mg-based alloys. *Journal of Alloys and Compounds* 287: 243–250.

Spassov, T. Solsona, P. Suriñach, S. & Baró, M.D. 2002. Nanocrystallization in $Mg_{83}Ni_{17-x}Y_x$ (x = 0, 7.5) amorphous alloys. *Journal of Alloys and Compounds* 345: 123–129.

Williamson, G.K. & Hall, W.H. 1953. X-ray line broadening from filed aluminium and wolfram. *Acta Metallurgica* 1: 22–31.

Woo, J.H. & Lee, K.S. 1999. Electrode characteristics of nanostructured Mg_2 Ni-type alloys prepared by mechanical alloying. *Journal of The Electrochemical Society* 146: 819–823.

Frontiers of Energy and Environmental Engineering – Sung, Kao & Chen (eds)
© 2013 Taylor & Francis Group, London, ISBN 978-0-415-66159-1

The study on membrane fouling mechanism of model EPS in different cations

L. Wu & F. Li, B. Yang & L.L. Fu
College of Environmental Science and Engineering, Donghua University, Shanghai, China

ABSTRACT: In this paper, Microfiltration (MF) and Ultrafiltration (UF) membranes were tested with the solutions model EPS, which contained Sodium Alginate (SA), Bovine Serum Albumin (BSA) and Humic Acid (HA), mixed with NaCl, $CaCl_2$, $AlCl_3$ solution in an unstirred dead-end ultrafiltration cell. The results showed that calcium ion and aluminum ion can improved permeate flux and reduced the resistances, while sodium ion has no significant influence on that. The dominant fouling mechanism are cake formation for the model EPS with the presence of Na^+, Ca^{2+}, Al^{3+} in ultrafiltration, and as for microfiltration process the best models was much complex for different solutions. It was also found that with increasing concentration of calcium and alum ions, the decline of permeate flux become more slowly and the specific cake resistance is gradually being small in ultrafiltration.

Keywords: cation; model EPS; membrane fouling; Hermia law

1 INTRODUCTION

Microfiltration and ultrafiltration recently received considerable attention in drinking water treatment as well as in domestic and industrial wastewater treatment.[1] It was evidenced that the Extracellular Polymeric Substances (EPS) was the main causes of membrane fouling.[2,3] These macromolecular components could easily cause membrane fouling in operation of filtration, lead to deterioration in permeate quality and quantity or an increase of transmembrane pressure. Some studies revealed that the polysaccharides seem to simulate EPS behavior during filtration better than proteins and some kind of Natural Organic Materials (NOMs).[4]

Recently, many studies have focused on the effect of presence of cations such as calcium and aluminum on membrane fouling performance.[5,6] The interaction between EPS and cations has significant effect on membrane fouling, which makes the mechanism and resistances more complex than single EPS.[6] Some literature indicates that the ionic environment in which alginate macromolecules are dissolved has significant influence on their properties. Alginates can form aggregates, and even gels with added bivalent cations.[7,8] DeKerkhove reports on the effect of monovalent and divalent ions on the structural growth and viscoelastic properties of alginate layers.[9,10] The interaction between EPS and cations has great effect on membrane fouling, making the mechanism and resistances more complex than single EPS. Nowadays, there is still lack of knowledge on the fouling mechanism and its contribution to the resistances when EPS consists of three components of sodium alginate, BSA and HA added sodion, calcium and alum respectively in the system. In our study, the mixture Solution of Alginate (SA), Bovine Serum Albumin (BSA) and Humic Acid (HA) is initially used to model EPS and investigate the influence by different cations.

This paper mainly aims to research the complex fouling mechanism of model EPS solutions with the different cation (Na^+, Ca^{2+}, Al^{3+}) by using PES ultralfiltration membranes with MWCO 10k and microfiltration membrane with mean pore size of 0.02 um. These experiments were operated under 0.15 MPa and 0.05 MPa TMP, respectively. The experimental data for the flux decline is examined by Hemia's laws in order to analyze the membrane fouling mechanisms. The specific cake resistance (r) is also analyzed in this study. We systematically investigated the performance membrane fouling by model EPS composed of SA, BSA and HA in the absence and presence of these three ions. Different concentration of these cations impact on membrane performance was also investigated in this study.

2 MATERIAL AND METHODS

2.1 Experimental equipment and operating conditions

Dead-end constant pressure filtration tests were conducted in an unstirred Millipore UF cell (model 8400, Amicon Corp) with a volume of 380 ml and filtration membrane area of $45.36 \, cm^2$, the reservoir with a volume of 1000 ml. Constant-pressure filtration was maintained by gas pressure regulated from a nitrogen cylinder. The flux data are measured by electronic balance (model MP 51001, Yushun (shanghai) Co, Ltd) and logged by computer.

2.2 Membranes and solutions

The UF membrane with Molecular Weight Cut Off (MWCO) of 10 K and MF membrane with pore size of 0.2 μm supplied by Synder (USA) were used in the experiments.

In this study, The Sodium Alginate (SA), Bovine Serum Albumin (BSA) and Humic Acid (HA) mixture solution were used as model EPS with a concentration of 60 mg/L + 20 mg/L + 20 mg/L. Sodium chloride (NaCl), calcium chloride ($CaCl_2$) and alum chloride ($AlCl_3$) were used analytical grade obtained from China National Medicines Co, Ltd. Three kinds of cations (Na^+, Ca^{2+}, and Al^{3+}) were respectively added in the model solutions and the concentration of all ions varied from 0 to 30 mg/L. All feed solutions for the experiment were prepared by using deionized water (Millipore-MilliQ) and added the buffer solutions which is preparation by Na_2HPO_4 (0.2 mg/L) and 16 ml NaH_2PO_4 (0.2 mg/L) to ensure the stability of pH and preserve at room temperature (25°C).

3 RESULT AND DISCUSSION

3.1 Zeta potential and particle size distribution of model EPS solutions

The particle size distribution and Zeta potential of model EPS and mixture solutions were estimated by the Particle size & Zeta Potential Analyzer (Nano ZS, Malvern, UK measuring range was from 0.6 nm to 6000 nm.) All the samples were freshly prepared and filtrated through 0.4 μm filter prior to measurement. The value of PH was maintained at 7.0 by using buffer solutions and the temperature keeping at 25°C. The hydrodynamic diameter distribution and characteristics of the particles were shown in Table 1.

From Table 1, the size of the particles in the mixture solution was much larger than that in the single solution of model EPS. It could be the specific

binding interactions occurred between cations and model EPS, forming large flocs when Al^{3+} or Ca^{2+} was added. The increasing of zeta potential indicated that addition of cations make the solution unstable.

3.2 Contact angle of membrane

The value of contact angle of new membrane and fouling membranes are shown in Table 2. The contact angle value of new ultrafiltration membrane is larger than that of microfiltration membrane, and the reason may be the different membrane surface has identified chemical composition and roughness. After filtration of model EPS solution, we could observe that the value of contact angles of all fouling membranes increased. However, in ultrafiltration experiments, the value of single model EPS was larger than that of mixture solutions added with cations, while the opposite results were found in the microfiltration process. When three cations added in these solutions, the value of contact angle keep to the following order: $Na^+ < Ca^{2+} < Al^{3+}$ in ultafiltration and microfiltration.

3.3 Analysis of membrane fouling mechanism

The permeate flux declined trends of the four investigated systems (model EPS; model EPS and sodion; model EPS and calcium; model EPS and alum) were plotted under the same pressures of 0.15 MPa in ultrafiltration and 0.05 MPa in microfiltration. From Figure 1(a), we could find that the most severe flux decline occurs in system containing only model EPS, whereas the least severe occurs when alum was present in the system and Na^+ had little influence on permeate flux decline. It might be explained that the cations could neutralize the repulsive negatively charged alginate, protein chain and form bridges between adjacent molecules and hence promoted the aggregation of model EPS. The average particle size and the strength of cations could be the main reasons leading to the improvement of permeate flux.

The permeate flux trends of microfiltration of feed solution were shown in Figure 1(b). We had

Table 1. Zeta potential and particle size distribution of the model EPS and Mixture solutions.

Solution	Zeta potential (mV)	Mean particle size (nm)
Model EPS	−0.0196	107.35
Model EPS + NaCl	−0.143	120.83
Model EPS + $CaCl_2$	−0.210	997.02
Model EPS + $AlCl_3$	0.367	1824.23

Table 2. The contact angles of the virgin membrane and the fouled membrane with different solutions.

		Membrane	
	Solution	MF	UF
Virgin membrane	Milli-Q water	64.47°	33.61°
Fouled membrane	Model EPS	76.89°	73.56°
	Model EPS with Na^+	80.36°	59.65°
	Model EPS with Ca^{2+}	90.56°	63.05°
	Model EPS with Al^{3+}	92.65°	66.37°

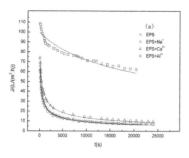

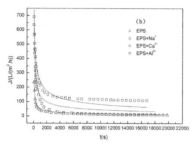

Figure 1. Permeate flux as a function of time for (a) model EPS solution with different ions in ultrafiltration (b) model EPS solution with different ions in microfiltration.

observed that the initial flux was much higher than that of UF membrane. But the flux decline value of steady-state was about 8–10 L/(m²·h) for model EPS and model EPS with sodion, calcium, which was similar to that of UF membrane. The steady flux was about 100 L/(m²·h) in microfiltration, much higher than that of ultrafiltration. We also found that the rapid of flux decline was quickly in microfiltration than that of ultrafiltration when alum was added in the solutions. The pore size of microfiltration was larger than that of ultrafiltration, at the presence of alum, the particle size of solutions was about 1800 nm of diameter, leading to great influence in flux decline.

From Table 3, we observed that the cake filtration model was the best model to fit the values of regression coefficient (R^2) of the experiments of UF, but in microfiltration process the best models was much complex for different solutions. The experimental data of single model EPS and solution with Ca^{2+} fitted the intermediate blocking model, while the solution added Na^+, it more suit for standard blocking model and at presence of Al^{3+}, the best model was also cake filtration. The reason could be the different size of molecular and membrane pore and the different repulsive force.

3.4 Analysis of specific cake resistance

The values of specific cake resistance (r) for two kinds of membrane with different concentration of

cations were shown in Table 4. In this experiment, under the single modle EPS solution, the values of specific cake resistance were 9.7761E+15 and 3.98764E+15 in microfiltration and ultrafiltration respectively. At the same concentration of cations, when Na^+, Ca^{2+} and Al^{3+} respectively added in this solutions, the value of r keep to the following order: $Na^+ < Ca^{2+} > Al^{3+}$ in ultafiltration and microfiltration. The specific cake resistance of the mixture solution with Al^{3+} is found to be the lowest in all solution. When Na^+ added in the solution, the value of r changed a little compared to single model EPS. From Table 4, with increasing concentration of Al^{3+} and Ca^{2+}, the value of specific cake resistance gradually become small, ranging from 4.36856E+13 to 2.253E+14 and 8.35165E+15 to 2.59798E+15 respectively in ultrafiltration. However, in microfiltration process, the value of r did not follow this regular. Instead, it reaches to the lowest value of 3.35811E+13 and 1.32646E+15 respectively when Al^{3+} and Ca^{2+} at the concentration of 20 mg/L.

But according to the Carman-Kozeny equation as in Eq. (1), the specific cake resistance should be inversely proportional to the diameter of the molecule and related to the porosity of the cake:

$$\alpha_c = \frac{180(1-\varepsilon)}{\rho d_p^2 \varepsilon^2} \tag{1}$$

where ρ: the density of the wet cake; ε: the porosity of the cake; d_p: the hydrodynamic diameter of particles.

When the system containing calcium or alum has the relatively lower specific cake resistance, implies that there is a synergistic effect between calcium or alum and model EPS, which results in more permeable cake layer. When the concentration of Al^{3+} and Ca^{2+} varied from 10 mg/L to 30 mg/L, the hydrodynamic diameter of particles increased, leading to the value of r lower. It could be explained that the large and loose flocs formed when present of alum which captured and bound the macromolecular components together during coagulation/flocculation.

The addition of calcium also resulted in larger particle size. The resulting incompressible cake suggested that the flocs formed are more compact/rigid compared to that formed in the presence of alum, however, the different situation was found in microfiltration. This main reason could be the different MWCO and the membrane surface morphology. It should be noted that the values of r were lower for the membrane with larger MWCO. The mean particle size distribution was much similar to the pore size of microfiltration when Al^{3+} and Ca^{2+} at the concentration of 0.03 ml/L. The porosity of the cake layer which formed by model EPS were different when different cations added in

Table 3. Values of R2 of fitting to the experimental data obtained for blocking models.

Model	Menbrane	Mixture solution	Mixture with Na^+	Mixture with Ca^{2+}	Mixture with Al^{3+}
Complete blocking	UF	0.81	0.82	0.84	0.82
	MF	0.98	0.96	0.89	0.63
Standard blocking	UF	0.89	0.89	0.90	0.84
	MF	0.99	0.99	0.94	0.73
Intermediate blocking	UF	0.95	0.95	0.96	0.85
	MF	0.99	0.98	0.97	0.83
Cake filtration	UF	0.99	0.99	0.99	0.89
	MF	0.93	0.93	0.96	0.96

Table 4. Specific cake resistance of model EPS and mixture solution measured with different membranes (m/kg).

Cation	Membrane	Concentration (mg/L)		
		0.01	0.02	0.03
Na^+	MF	4.06E+15	3.99E+15	3.98E+15
	UF	9.86E+15	9.68E+15	9.78E+15
Ca^{2+}	MF	2.89E+15	1.33E+15	3.89E+15
	UF	8.35E+15	5.25E+15	2.59E+15
Al^{3+}	MF	6.74E+13	3.36E+13	4.79E+13
	UF	2.25E+14	7.76E+13	4.37E+13

the solutions. Compared to model EPS, when Al^{3+} and Ca^{2+} were present, large and loose flocs were formed, increasing the porosity of the cake layer and reducing the packing density, while in the case of Na^+, little influence was caused.

4 CONCLUSIONS

Model EPS caused the largest flux decline and resistances. When added calcium and alum, the permeate flux became larger, especially at the presence alum. However, the addition of Na^+ had little influence in flux decline and resistances.

Through the fitting equation of Hermia's laws, the dominant fouling mechanism was found to be cake formation when the solution of model EPS and added sodium or calcium ions, but the fouling mechanism was mainly much complex at the presence of aulm.

With increasing concentration of Al^{3+} and Ca^{2+}, the value of specific cake resistance gradually became small in ultrafiltration and the situation did not follow this regular in microfiltration process.

ACKNOWLEDGEMENTS

This work was supported by the National Natural Science Foundation of China for the Youth (Grant No. 20906011), and the Natural Science Foundation of Shanghai (Grant No. 102R1401200).

REFERENCES

[1] Ahmed, Z, Cho, J., B.-R., Song, K.-G., Ahn, K.-H., 2007. Effects of sludge retention time on membrane fouling and microbial community structure in a membrane bioreactor. Journal of Membrane Science 287(2), 211–218.
[2] I.S. Chang, P. Le Clech, B. Jefferson, S.J. Judd, 2002. Membrane fouling in membrane bioreactors for wastewater treatment, J. Environ. Eng. 128 1018–1029.
[3] M.A. Shannon, P.W. Bohn, M. Elimelech, J.G. Georgiades, B.J. Marinas, A.M. Mayes. 2008. Science and technology for water purification in the coming decades, Nature 452 (132) 301–310.
[4] B.P. Frank, G. Belfort. 2003. Polysaccharides and sticky membrane surfaces: critical ionic effects, J. Membr. Sci. 212 (15) 205–212.
[5] G.T. Grant, E.R. Morris, D.A. Rees, P.J.C. Smith, D. Thom. 1973. Biological interactions between polysaccharides and divalent cations: the egg-boxmodel, FEBS Letters 32, 195–198.
[6] Karin, Listiarini, Wei Chun, Darren D. Sun, James O. Leckie. 2009. Fouling mechanism and resistance analyses of systems containing sodium alginate, calcium, alum and their combination in dead-end fouling of nanofiltration membranes. Journal of Membrane Science 344 (12) 244–251.
[7] A.J. De Kerchove, M. Elimelech. 2006. Structural growth and viscoelastic properties of adsorbed alginate layers in monovalent and divalent salts, Macromolecules 39 (19) 6558.
[8] Y. Ye, V. Chen, A.G. Fane. 2006. Modeling long-term subcritical filtration of model EPS solutions, Desalination 191(23) 318–327.
[9] H. Susanto, M. Ulbricht. 2005. Mechanisms for polysaccharide fouling of ultrafiltration membranes, Extended Abstracts of the International Congress on Membranes and Membrane Processes, Seoul, Korea, pp. 341–342.
[10] A.J. De Kerchove, M. Elimelech, 2006. Structural growth and viscoelastic properties of adsorbed alginate layers in monovalent and divalent salts, Macromolecules 39 (19) 6558.

Effect of detachment on nitrification process in biofilm system

J. Yin, H.J. Xu & C.Y. Dong

School of Environmental Science & Engineering, Zhejiang Gongshang University, China

ABSTRACT: The aim of this study is to evaluate the effect of detachment on nitrification process in a mixed-population biofilm. The biofilm was grown under constant flow condition and at low C/N ratio in a tube biofilm reactor. Detachment experiment was performed by temporarily increasing the influent flow rate into the tube reactor Changes in the composition of biofilm were examined by Denaturing Gradient Gel Electrophoresis (DGGE). Results showed that ammonium oxidation rate reached above 99% after three weeks. Then, nitrate was detected. However, during the experiment nitrite in the effluent was observed all the time. It was found that biofilm sloughing could cause the increase of nitrite concentration in the effluent. Sloughing events happened mainly under constant hydrodynamic condition, even though flow rate in the tube was increased to 4000 mL/min, it did not cause a significant biofilm sloughing. DGGE fingerprints revealed that there was no significant shift in the composition of bacterial community throughout biofilm, but some differences in each lane were noted in band intensities in DGGE pattern. Therefore, biofim sloughing could induce the change of bacterial abundance in biofilms to influence the nitrification process.

Keywords: biofilm; detachment; nitrification

1 INTRODUCTION

Microbial nitrification is the key process in the removal of ammonium from wastewaters and is becoming more important due to strict regulations on nitrogen discharge. Biological nitrogen removal is done by nitrification and denitrification. The nitrification process is primarily accomplished by two groups of autotrophic nitrifiers. Biofilm reactors are used in wastewater treatment to removal organic carbon and to oxidize ammonium. Long solids retention time makes biofilm reactors suitable for the retention of slow-growing organisms such as nitrifiers. Compared with suspended nitrification reactors, biofilm systems can provide high sludge age and avoid washout of nitrifiers from the system.

To design biofilm reactors, more information about the growth and detachment of the mixed-culture biofilms must be available. Detachment from biofilms is caused by a combination of processes. Detachment occurs when external forces are larger than the internal strength of the matrix that is holding the biofilm together. Thus, in principle there are two mechanisms that can lead to detachment: i) increase of the external shear forces, or ii) decrease of the internal strength (Horn *et al.*, 2003). The mechanism of detachment is still discussion (Garny *et al.*, 2009; Coufort *et al.*, 2007; Derlon *et al.*, 2008). The specific effects of detachment on biofilm reactor performance are not well understood. Previous researches mainly focus on the influence of detachment on the characteristics of biofilm itself, such as biofilm thickness, density and composition, rather than the performance of reactor. Therefore, the objectives of this study were to induce detachment of biofilm by varying external shear forces and to evaluate the influence of detachment on nitrification process.

2 MATERIALS AND METHODS

2.1 Biofilm reactor

The experiment was done using similar biofilm system (2.0 cm in diameter, 1.41 m in length), as described by Horn et al. (2003). The tube was made of Plexiglas and was equally divided into three segments (S1, S2 and S3) which could be weighed separately. Medium from the mixing tank were recycled through the external loop of a pipe reactor at a flow rate of 495 mL/min, so that the system can be regarded as a continuous-flow stirred-tank reactor (Turakhia *et al.*, 1983). Only during detachment experiment the reactor was operated without recirculation. The initial total volume of the reactor (mixing tank, tubular part and tubing) was 1.1 L.

In order to avoid the influence of suspended solid on nitrification and for some practical reasons, the system was operated in a sequencing batch mode where the entire water was exchanged daily. Dissolved oxygen concentration was kept in a range of 5 to 6 mgO_2/L. Aeration in the mixing tank provided a constant oxygen concentration. The reactor was operated with an ammonium load of 1.91 $gN/m^2 \cdot d$. Sodium acetate was added as the organic substrate at C/N ratio of 0.5 and 0.25 during phases I (day 0~15) and II (days 16~86), respectively. Temperature and pH were controlled for optimum nitrification.

2.2 Detachment experiment

Detachment experiment was performed by temporarily increasing the influent flow rate into the tube reactor (Telgmann et al., 2004). During the experiment, the recirculation through the gear pump was disconnected to avoid breakup of the detached biomass in the pump and potential reattachment in the tubular reactor. The feed solution during the detachment experiment was tap water without acetate. During the detachment experiment the flow rate was gradually increased and samples were collected directly at the outlet. For the detachment experiment, flow rates of 1000 mL/min (4 min), 2000 mL/min (4 min) and 4000 mL/min (2 min) were chosen in the bulk phase. These flow velocities were equivalent to Reynolds numbers of 1062, 2123 and 4246, respectively, in the tube reactor without biofilm. The drained reactor was weighed before and after each experiment and the biofilm thickness was determined.

2.3 Chemical analyses

Samples from the effluent were treated as follows. All of effluent samples were prefiltered with 0.45 μm membrane filters before chemical analysis. The concentration of ammonium was measured using the Nessler method. Nitrite and nitrate were analyzed by spectrophotometric methods. In this study, detached biomass included particles resulted from erosion and large pieces of biofilm caused by sloughing, which were not collected separately. Detached biomass was measured by passing all of the effluent 1.1 L through a pre-dried (105°C for 2 h) and pre-weighed filter (average pore size, 0.45 μm). After filtration, the filters were dried for 2 h at 105°C and reweighed. The average biofilm thickness was determined following the method described by Horn and Hempel (1997).

2.4 Denaturing gradient gel electrophoresis (DGGE)

The genomic DNA was extracted by using DNA kit for environmental samples (Shanghai Biocolor,

Bioscience & Technology Company, China). PCR was performed with about 50 ng of template DNA with the Eubacterial primers 338f and 518r with the addition of a 5' GC-clamp (Muyzer et al., 1993) to assess bacterial community diversity. DGGE was performed using the Bio-Rad D Gene System (Bio-Rad, Hercules, CA, USA). DNA was visualized after silver staining by UV transillumination, and gel images were stored using the Gel Doc 2000 System (Bio-Rad Laboratories).

3 RESULTS

The reactor was started by inoculation with a mixed culture of nitrifying bacteria for about 1 week to provide initial colonization and accumulation. Subsequently, the reactor was operated under the defined hydrodynamic conditions. The NH_4^+-N removal rate (R_{NH4-N}), NO_2-N and NO_3-N production rate (i.e. P_{NO2-N} and P_{NO3-N}) were determined. When the system reached steady state, on day 79 detachment experiment was carried out.

3.1 System performance

Performance of the tubular biofilm reactor in terms of various nitrogen concentrations is shown in Figure 1. During the first phase (days 0–15), at the C/N ratio of 0.5, nitrite was the only nitrification product in the effluent, which can be explained by Ammonium-Oxidizing Bacteria (AOB), Nitrite-Oxidizing Bacteria (NOB) and heterotrophic bacteria competing for oxygen and space within the biofilm. The acetate oxidation hindered oxygen flow to NO_2^-–N oxidation, but not to NH_4^+-N oxidation. High ammonium concentration and pH (8.0) against NOB and promoted the growth of AOB (Chung et al., 2007). The occurrence of nitrite indicated that autotrophic biofilm slowly developed. At the end of this phase, the R_{NH4-N} and P_{NO2-N} were 1.56 g $N/m^2 \cdot d$ and 1.36 g $N/m^2 \cdot d$, respectively.

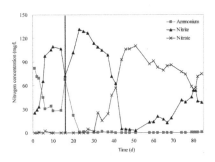

Figure 1. Concentration changes of ammonium, nitrite and nitrate in the effluent.

In order to achieve more nitrifying biofilm, on day 16, the C/N ratio in the influent was decreased from 0.5 to 0.25. As shown in Figure 1, there is a slight increase in ammonium concentration in the effluent due to the change of C/N ratio. From day 23 on, 99% of the ammonium was completely removed, resulting in R_{NH4-N} of 1.90 g N/m$^2 \cdot$d, hereafter almost no ammonium was detected in the effluent. Simultaneously, on day 23 the P_{NO2-N} reached the maximum, i.e. 1.63 g N/m$^2 \cdot$d, and then nitrite concentration in the effluent took on a decline. With the acclimation of NOB, from day 27 on, nitrate was detected and accumulated with the time. Different result has been reported for other biofilm system (Elenter *et al.*, 2007). Long HRTs of 24 h resulted in high nitrate concentrations, providing an advantage for NOB. Meanwhile nitrite concentration in the effluent continued to decrease and dropped to 4.97 mg/L on day 44, which indicated that nitrite was further oxidized to nitrate by NOB. However, from day 58 on, in the effluent nitrite concentration went up again and nitrate declined. A possible explanation could probably be the decrease of activity of NOB by the sloughing event of biofilm. On day 79, detachment experiment was carried out in order to investigate if a sloughing event can be triggered by rapidly increasing shear and its effect on nitrification process. It can be seen that detachment experiment had no influence on ammonium removal; however, there was a slight fluctuation in nitrite concentration.

According to the mass balance for nitrogen and the degradation product proportions, the difference of nitrogen between the influent and the effluent can to some extent be attributed to ammonium used for heterotrophic and autotrophic cell synthesis or potentially to heterotrophic denitrification. The biofilm thickness increased slowly and the biofilm was 227 ± 53 μm thick at the end of phase I (day 15). The biofilm finally reached a thickness of 465 ± 54 μm by the end of phase II before the detachment experiment. During phase II, small sloughing event could cause the reduction of the average biofilm thickness, but it had little effect on the nitrification process. After the detachment experiment, the biofilm thickness decreased from 465 to 406 μm.

3.2 Detachment experiment

In Figure 2, the overall detachment throughout the experiment is shown. As shown in Figure 2, the detachment experiment did not cause a significant sloughing event, even though flow rate in the tube was increased to 4000 mL/min. Significant detached biomass was lost even though constant hydrodynamic condition was maintained. Thus, under the conditions of this study, it was not possible to

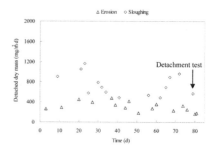

Figure 2. Biofilm erosion and sloughing during the experiment.

control the occurrence of sloughing events by controlling shear stress. During detachment experiment, each increase in flow rate corresponded to a limited sloughing event. This indicates that there is no single level of shear above, which all biofilm will detach. Moreover, no single shear could result in continuous detachment, which showing bioflim detachment depends not only on local shear but also on the cohesiveness of the biofilm.

3.3 Bacterial community composition by denaturing gradient gel electrophoresis

Analysis of the bacterial community composition by PCR-DGGE was performed on biofilm samples collected during the experiment. Banding patterns for the 16S rDNA DGGE-PCR amplicons is presented in Figure 3. Individual lanes contain 16S rDNA PCR products from total DNA extracts at different operating times. Results obtained from the DGGE analysis were very reproducible across the sampling points throughout the experiment (data not shown). As shown in Figure 3 (a), DGGE fingerprints of attached biofilms were similar to detached biofilms, indicating that there was no significant shift in the composition of bacterial community throughout biofilm. Some differences in each lane were noted in band intensity in the total bacterial DGGE patterns. Nevertheless, DGGE band pattern had an obvious change from day 0 to 20, which just corresponds to the lag phase at the beginning stage.

DGGE gel profiles were also analyzed statistically using the Shannon index H. Diversity indices are useful as a first approach to estimate the diversity of microbial communities, i.e., the higher H, the greater the diversity of the microbial community. As shown in Figure 3 (b), Attached biofilms and detached biofilms exhibited different trends. There was an increase followed by a slight decrease over time for the Shannon index of attached biofilms; in contrast, the Shannon index of detached biofilms gradually increased with the time. It was

(a)

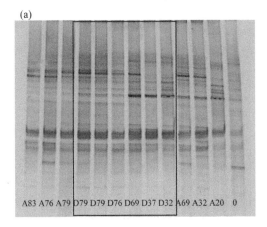

A83 A76 A79 D79 D79 D76 D69 D37 D32 A69 A32 A20   0

(b)

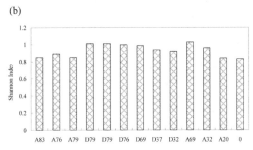

Figure 3.   DGGE analysis of 16S rRNA fragments of total bacterial population from (a) biofilms; (b) Numerical analysis of the above DGGE samples with the Shannon index of diversity.
Note: A means attached biofilm; D means detached biofilm; the number means the sampling time.

supposed that biofilm detachment could affect the bacterial abundance in biofilms.

4   CONCLUSIONS

During the experiment, ammonium loading rate was kept at 1.91 $gN/m^2 \cdot d$. Ammonium oxidation rate reached above 99%. From day 27 on, nitrate was detected, which is indicating the complete nitrification occurred. However, during the experiment, nitrite in the effluent was observed all the time. It was found that biofilm sloughing could cause the increase of nitrite concentration in the effluent.

Detachment experiment showed that sloughing events happened mainly under constant hydrodynamic condition, even though flow rate in the tube was increased to 4000 mL/min which did not cause a significant biofilm sloughing. Thus, under the conditions of this study, it was not possible to control the occurrence of sloughing events by controlling shear stress.

DGGE analysis of biofilter samples revealed that no significant shifts in the community structure of biofim were observed, some differences in each lane were noted in band intensity. From the view of DGGE pattern and Shannon Index, it was supposed that biofim detachment could affect bacterial abundance in biofilms.

ACKNOWLEDGEMENTS

The authors would like to express our appreciation to National Natural Science Foundation of China (No. 50908209) and Science Technology Department of Zhejiang Province (No. 2011R10081) for funding support for this project.

REFERENCES

Chung J, Bae W, Lee YW, Rittmann BE, 2007. Shortcut biological nitrogen removal in hybrid biofilm/suspended growth reactors. *Process Biochemistry*, 42(3): 320–328.

Coufort C, Derlon N, Ochoa-Chaves J, Line A, Paul E, 2007. Cohesion and detachment in biofilm systems for different electron acceptor and donors. *Water Science and Technology*, 55(8–9): 421–428.

Derlon N, Masse A, Escudie R, Bernet N, Paul E, 2008. Stratification in the cohesion of biofilms grown under various environmental conditions. *Water Research*, 42(8–9): 2102–2110.

Elenter D, Milferstedt K, Zhang W, Hausner M, Morgenroth E, 2007. Influence of detachment on substrate removal and microbial ecology in a heterotrophic/autotrophic biofilm. *Water Research*, 41(20): 4657–4671.

Garny K, Neu TR, Horn H, 2009. Sloughing and limited substrate conditions trigger filamentous growth in heterotrophic biofilms—Measurements in flow-through tube reactor. *Chemical Engineering Science*, 64(11): 2723–2732.

Horn H, Hempel DC, 1997. Growth and decay in an auto/heterotrophic biofilm. *Water Research*, 31(9): 2243–2252.

Horn H, Reiff H, Morgenroth E, 2003. Simulation of growth and detachment in biofilm systems under defined hydrodynamic conditions. *Biotechnology and Bioengineering*, 81(5): 607–617.

Muyzer G, Ellen CW, Andre GU, 1993. Profiling of complex microbial populations by denaturing gradient gel electrophores is analysis of polymerase chain reaction genes coding for 16S rRNA. *Applied and Environmental Microbiology*, 59(3): 695–700.

Telgmann U, Horn H, Morgenroth E, 2004. Influence of growth history on sloughing and erosion from biofilms. *Water Research*, 38(17): 3671–3684.

Turakhia MH, Cooksey KE, Characklis WG, 1983. Influence of a calcium-specific chelant on biofilm removal. *Applied and Environmental Microbiology*, 46(5): 1236–1238.

Frontiers of Energy and Environmental Engineering – Sung, Kao & Chen (eds)
© 2013 Taylor & Francis Group, London, ISBN 978-0-415-66159-1

The hazard study of toxic gas in the fire smoke

F. Xie
Tianjin General Fire Brigade, Tianjin, China

F. Xie, W.H. Song, L.Y. Lv & Z. Chen
Tianjin Polytechnical Univeisity, Tianjin, China

ABSTRACT: The mechanism of fire accident in the chemical company is very complex with many risk factors, and the consequence of the accident is usually very serious. A large number of study data suggests that, the asphyxiating or toxic flue gas formed under the conditions of the fire burning has become the main reason leading to the death of the presence of personal. This paper takes the dichloropropane tank area in a chemical company for an example, and adopts the fire simulation software FDS to make numerical simulation for the pool fire in the dichloropropane tank. The variation of the flue gas in the combustion process of pool fire is obtained from the simulation. And then the effect and harm of the main toxic and harmful gas carbon monoxide, carbon dioxide and hydrogen chloride on the human health and environment in the scene of pool fire is analyzed. For the people's health, the toxic and harmful gas can make great impact, such as dizziness, weakness of four limbs, more seriously it can cause some diseases, even it brings death. For the surrounding environment and even the whole environment of the area, it also has a great effect.

Keywords: FDS; the hazard of the flue gas; carbon monoxide; carbon dioxide; hydrogen chloride

1 INTRODUCTION

The components of fire smoke are complex, which not only depends on the ventilation conditions of the fire scene and the building structure, but also depends on the physical and chemical properties of combustible, the flow conditions of flue gas and other so on.[1] At present, there are dozens of known compositions in the toxic fumes or toxic flue gas of the pool fire, including inorganic toxic and harmful gas (CO, CO_2, NOx, HCL, H_2S, NH_3, HCN, P_2O_5, HF, SO_2, etc.) and organic toxic and harmful gas (phosgene, aldehydes gas and hydrogen cyanide, etc.), and some scholars also classify fine particulate matter (aerosols), smoke and heavy metal powder produced from the fire as the toxic and hazardous substances.[2] In the catastrophic fire in Luoyang in 2000, more than 300 people died from severe poisoning suffocation due to the inhalation of toxic fumes. There are many similar examples abroad. The serious fire fact makes people become more aware of the damage of fire and smoke to the personals, which causes even more attention to the fire smoke problem.[3] FPRF, NIST and other research institutions abroad are actively carrying out research in this area, and hope to establish the safety standards related to these work. At present, our country has also gradually begun such study, but remains gaps compared with the level of foreign research.[4] This paper takes the dichloropropane tank area in a chemical company for an example, and adopts the fire simulation software FDS to make numerical simulation, obtaining the smoke conditions in the pool fire, and then analyze the consequences of the toxic and harmful gas on the personals and the environment. Through the contrast between the simulation and analysis results in this paper and the existing actual situation as well as related research, it shows that the impact on the environment of the flue gas in the pool fire is great, causing the deterioration of the air temperature, precipitation, soil erosion and other environmental factors, which is worthy of further attention and taking the relevant measures for prevention and remediation. In this paper, the great danger of the flue gas in the pool fire bringing to the personal and property safety, as well as the surrounding environment, obtained by the simulation analysis, could alert the relevant personnel to strengthen the monitoring to avoid accident and reduce losses, which also provides a reference to the further fire and smoke research.

2 THE FLUE GAS COMPUTER SIMULATION IN THE FIRE

In recent years, computer simulation models of fire and smoke movement has been widely developed and applied. The computational fluid dynamics simulation software CFD begun from mid-1980s in the 20th century is calculation technique used to analyze the fluid flow properties. FDS, as one kind of CFD model, is a practical fluid engineering analysis tool, mainly used to simulate the fluid flow, heat transfer, multiphase flow, the chemical reactions, combustion problems in the fire state.[5]

2.1 The basic fire simulation principle of software FDS

The simulation software FDS is developed by the American National Standards Institute (NIST) Building and Fire Research Laboratory (BFRL). The model of FDS is three-dimensional Computational Fluid Dynamics software (CFD) based on Large Eddy Simulation (LES), and can simulate the flow process of fire turbulent.[3] The software uses numerical method to solve NS equation with flows of low Mach number driven by fire buoyancy, and focus on the calculation of smoke and heat transfer process.[6]

The fire simulation with simulation software FDS mainly includes four basic process of the simplification of the actual combustion process, the set of solving parameters, the establishment of the physical model and fire simulation.[7]

2.2 The model establishment of FDS simulation for pool fire inside the tank in the dichloropropane storage tank area

According to the basic fire simulation principle of the simulation software FDS, combined with the actual situation of the dichloropropane storage tank area, the model established with the FDS software to simulate the pool fire in a single tank is as follows:

2.2.1 The setting of fire source
Select 4 # tank in the dichloropropane tank area to be target tank, and the fire form is whole-area pool fire on the top of the tank. The situation is outdoor with open environment and windless.

2.2.2 The simplification of actual combustion process
Assuming in windless condition, burning parameters of the pool fire in the tank is symmetrically distributed. Choose the radial section taking off the tank central axis as the research object. Ignore the radiative transfer heat between the sun and the tank; ignore the internal heat produced by the physical and chemical factors of dichloropropane itself; ignore the latent vaporization heat of the dichloropropane on the edge and bottom of the tank; The wall and the bottom of the tank is single steel plate sheet of uniformly thick.

2.2.3 The set of solving parameters
The environment temperature is set to be 25°C. The combustion model is established by setting fixed heat release rate, and the fixed heat release rate is calculated to set to be 3000 kJ/m²:[8] Determine the simulation time of initial fire is 60 seconds, and the time of the whole fire simulation process is set to be 600 seconds. The simulation space is selected for the cube. In the cube's six sides, four sides and top is set to be "OPEN" state, the bottom of the tank affixed to the ground is set to be "INERT" state, and the top surface of the tank is set to be flame.

2.2.4 The establishment of physical model
The fire form is whole-area pool fire on the top of the tank, and the heat release rate changes in accordance with the law of steady-state fire source. Therefore, the form of plane grid is designed to be 10 equilateral rectangle, which is plane symmetric distributed., and whose the of for The longest side is 6 meters, which is also the length of axis diameter, and the short side is 0.6 meters. The simulate space is set to be 60 m * 60 m * 20 m.

2.3 The flue gas results and analysis from FDS simulation

The schematic diagram of the flue gas in the burning process of pool fire is obtained by running simulation software FDS, and then according to the screenshots of the set observation point. The diagram can show the shape and scope changes of the pool fire in the combustion process dynamically and visually. Because the schematic diagram is so much, we select a representative diagram example and analyze.

After the pool fire happens 2 seconds, the flue gas is small, and only in a small portion of space of the tank upper part it could be seen, shown in Figure 1; after the pool fire happens 4 seconds, the flue gas begins to expand upwards, and the scope also expands, shown in Figure 2; after the pool fire happens 6 seconds, the scope of flame is further expanding, and it can reach the highest point of the set simulation space, like mushroom cloud, shown in Figure 3; after the pool fire happens 7 seconds, the flue gas is like cylindrical, and there is flue gas like mushroom cloud on the top, shown in Figure 4; after the pool fire happens 8 seconds, the height range of the flue gas

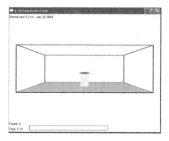

Figure 1.   Flue gas simulation diagram in 2 s.

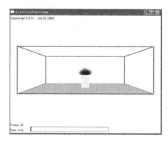

Figure 2.   Flue gas simulation diagram in 4 s.

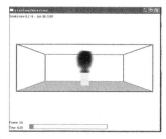

Figure 3.   Flue gas simulation diagram in 6 s.

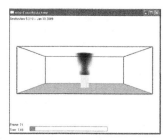

Figure 4.   Flue gas simulation diagram in 7 s.

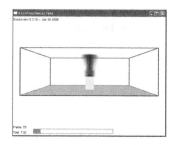

Figure 5.   Flue gas simulation diagram in 8 s.

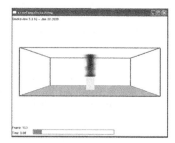

Figure 6.   Flue gas simulation diagram in 9 s.

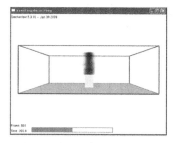

Figure 7.   Flue gas simulation diagram in 300 s.

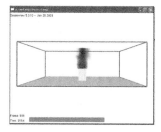

Figure 8.   Flue gas simulation diagram in 600 s.

still remains intact, but the flame in the upper part becomes from like mushroom cloud into like a round pie, as shown in Figure 5; after the pool fire happens 9 seconds, the circular ring flue gas on the top basically disappears, and the flue gas becomes like the cylindrical, as shown in Figure 6; after the pool fire happens 10 seconds, the range, height and shape of flue gas maintains in a stable state, as shown in Figures 6 and 7 as well as shown in Figure 8.

490

3 THE RISK ANALYSIS OF TOXIC GAS IN THE FIRE SMOKE

For the most of fire, the largest proportion of the damage caused by the fire is flue gas. The statistical results show that more than 85 percent of the fire deaths are due to the impact of the flue gas, most of which die from the inhalation of smoke and toxic gas and coma. Various toxic and harmful ingredients in the flue gas, corrosive ingredients, particulate matter, and the fire environment of high temperature, hypoxia, caused great damage to the life, property and the environment.[9] In this article, the main toxic and harmful gas formed from the pool fire in the dichloropropane tank is carbon monoxide, carbon dioxide, hydrogen chloride.

3.1 The risk analysis of carbon monoxide

Carbon monoxide is the product of incomplete combustion of carbonaceous materials. The follow-up investigation of the fire accident proves that the number of death from carbon monoxide in the fire accounted for more than 40% in the total number of death. The toxic gas composition which people understand more and which is only confirmed reason causing a large number of casualties is carbon monoxide, and therefore people have also given sufficient attention. Although carbon monoxide is only a component in the flue gas, the volume fraction of it is always higher than the other components in the flue gas, and most of the flue gas poisoning accident is caused by the role of carbon monoxide

The danger of carbon monoxide on the human body is to hinder the transmission of oxygen in the body, so the body is lack of oxygen. The inhaled carbon monoxide into the blood circulation can rapidly combine with the hemoglobin to generate carbonyl hemoglobin, influencing the oxygen-carrying and solution of blood to make human tissues lack of oxygen, because the affinity capacity of carbon monoxide and hemoglobin is 200–300 times bigger than that of oxygen with hemoglobin, and it can crowd out the oxygen from the oxygen hemoglobin. Even a small amount of carbon monoxide can form carbonyl hemoglobin, resulting in the lack of oxygen in the human tissues, so the content of carbonyl hemoglobin in the blood is an important basis for the diagnosis of carbon monoxide poisoning. Long-term inhalation of low concentrations of carbon monoxide, there may be headache, dizziness, insomnia, weakness of four limbs, fast heart rate.[10]

Hazardous content of CO is: the lethal volume fraction human exposed 30 minutes given by British Naval Engineering Standard (NES) is 0.4% (ie 4000 ppm), the reference volume fraction leading to half of the animals' death 30 minutes after the exposure given by the International Organization for Standardization (ISO) is 0.57%.

Some studies suggest that cardiovascular disease is related to carbon monoxide pollution. High concentration of carbon monoxide in the environment, may affect the body's the function of heart, brain and skeletal muscle. Low concentration of carbon monoxide can also enable the exercise tolerance of patients with atherosclerosis reduce. If exposed in the atmosphere polluted by carbon monoxide, it can promote the development of atherosclerosis, increase the myocardial infarction, or death. In addition, carbon monoxide can also make patients with angina pain prolong the pain time. For the patients with cerebrovascular disease and insufficient oxygen supply in the center of the brain are more sensitive to the exposure of carbon monoxide, and even the several carbonyl hemoglobin formed in the blood can generate serious harm.

3.2 The risk analysis of carbon dioxide

Carbon dioxide is greenhouse gas the content of which the biggest and the role of which is the most. The presence of carbon dioxide makes the outgoing radiation in the earth-atmosphere system reduce, and the temperature on the surface of the ground and in the lower atmosphere maintain in a high standard, which is so-called "greenhouse effect". It is estimated that the carbon dioxide content in the entire atmosphere is less than 0.03%, while in the absorption of thermal radiation it accounts for about 20% of the entire atmosphere, showing its impact on the exchange in the earth-atmosphere system is not small.[11] In addition, in the greenhouse effect, carbon dioxide plays a catalytic role on the water in the atmosphere. When carbon dioxide increases, which will lead to the temperature rise on the surface of the ground and in the atmosphere, prompting the sea evaporation, the water content in the atmosphere increases, make the results of the greenhouse effect increase. This is a positive feedback mechanism with great effect in the numerical simulation, which is much greater than expected rising temperature by simply increasing carbon dioxide.

The concentration increase of carbon dioxide would make the temperature on the ground surface increase, and the winter warming is more evident than in summer, warming in high latitudes is more significant than in low latitudes. In addition, the concentration increase of carbon dioxide will also affect the precipitation and drought. Changes in temperature and precipitation will change the heat and moisture conditions of crop growth, which will produce the effect on the agriculture in many areas and even the development of the economy that can not be ignored.

On a global scale, the agriculture irrigation water of many areas comes from the snow ablation in the mountain. In the conditions of carbon dioxide concentrations doubling, the water resources in these areas will be influenced appropriately.

The increase of carbon dioxide concentrations leads that, in precipitation increasing areas, it is possible to cause the increase of soil erosion, reducing the soil nutrients. In the precipitation reducing region caused by the increase of carbon dioxide concentrations, the frequency of drought may increase, which causes serious soil erosion.

3.3 *The risk analysis of hydrogen chloride*

The hydrogen chloride may be generated from all chlorinated organic compounds by thermal decomposition. The lethal concentration is: the allowance concentration in 10 minutes is 5 ppm.[12]

Hydrogen chloride gas has a strong irritant on the human body, corrosive to the skin and mucosal, causing nasal mucosa ulcer and cornea opacity, in severe cases, pulmonary edema and even death may exist, which also seriously pollute the environment strongly corrosive to the equipment and buildings. The hydrogen chloride gas is easy to valotile, water-soluble strongly, not easy to be adsorbed by particles, thus the capacity of diffusion is strong, which can be arbitrarily mixed with the air. Its hazard range is wide. Hydrogen chloride belongs to the stimulate pollutant gas and the harmful chemical which can directly irritate the airways, and damage the human respiratory tract after people inhaled.

4 CONCLUSION

The flue gas in the fire is the main damage factor for the life. The flue gas in the process of pool fire is simulated through the computer simulation software FDS, and the variation of flue gas in the combustion process can be broadly obtained. The main toxic and harmful gas in the flue gas from the pool fire in the dichloropropane tank carbon monoxide, carbon dioxide and hydrogen chloride is analyzed about the impact on the human health and the environment. Through discovering and mastering the damage action mechanism of smoke in the fire on the human body, appropriate evaluation index should be established to reduce the fire loss, protect the human life and the environment and maintain ecological balance.

REFERENCES

[1] Riehardson R. 200l. What fire statistics tell us about our fire and building codes for housing and small buildings and fire risk for occupants of those structures. Fire and Material, 25:255–271.

[2] Rohr D. 2001. An update to what's burning in home fires. Fire and Materials, 25: 43–45.

[3] Klingsch W, Wittbeeker FW. 2000. Burning characteristics of building products in the light of European harmonization. Fire and Materials, 24:253–258.

[4] ZhaoQiang, YueHailing. 2008. The computer simulation model review of the smoke in the fire. Security Technology, 1:9–11.

[5] Alarie Y. Comparison 1992. Comparison of the upitt, the SwRI/NIST and the upitt Πtest methods for smoke toxicity. Joumal of Fire Science, 10:458–468.

[6] Stephen M. Olenick, Douglas J. Carpenter. 2003. An updated international survey for computer models for fire and smoke. Journal of fire protection engineering, Vol. 13.

[7] WangFujum. 2004 Computational fluid dynamics analysis—CFD software principle and application Beijing: Tsinghua University Press.

[8] ChengYuanping, ChenLiang, ZhangMengjun. 2002. The heat release rate model of fire source and its experimental test method in the fire course. Fire Science, 11(2):70–74.

[9] HuangRui, YangLizhong, FangWeifeng, FanWeicheng. 2002. The risk study and its progress of smoke in the fire. China Engineering Science, 4(7):80–85.

[10] QiuRong, FanWeicheng. 2001. The bio-toxicology of common hazardous combustion products in the fire (I)—carbon monoxide, hydrogen cyanide. Fire Science, 10(3):154–203.

[11] LiQiang, LiuQinghui, ZhangHui, LianChenzhou, ZhanLiping, LvZian. 2003. The volume fraction distribution and damage of the toxic gas in the fire smoke. Natural disaster Journal, 12(3):69–74.

[12] HakkoinenT, MikkolaE, Jan Laperre, et al. 2000. Smoke gas analysis by Fourier transform infrared spectroscopy-summarry of the SAFIR project results. Fire and Materials, 24:101–112.

Frontiers of Energy and Environmental Engineering – Sung, Kao & Chen (eds)
© 2013 Taylor & Francis Group, London, ISBN 978-0-415-66159-1

Study on impacts of net-cage culture on marine benthos in Xiangshan Bay East China sea

G.H. Zhu
Key Laboratory of Marine Ecosystem and Biogeochemistry, State Oceanic Administration, Second Institute of Oceanography, SOA, Hangzhou, China

M. Jin
Key of Laboratory of Engineering Oceanography, Second Institute of Oceanography, SOA, Hangzhou, China

Y.B. Liao
Key Laboratory of Marine Ecosystem and Biogeochemistry, State Oceanic Administration, Second Institute of Oceanography, SOA, Hangzhou, China

J. Zhang & Q.S. Shi
Key of Laboratory of Engineering Oceanography, Second Institute of Oceanography, SOA, Hangzhou, China

Q.Z. Chen
Key Laboratory of Marine Ecosystem and Biogeochemistry, State Oceanic Administration, Second Institute of Oceanography, SOA, Hangzhou, China

ABSTRACT: The seasonal variations of the density of benthic microalgae, the concentration of chlorophyll *a*, the biomass, individual abundance and dominant species of benthos, TN, TP, OC and sulfides between January 2001 and May 2012 have been investigated in this study. The results showed that the abundance of benthic microalgae was 108–812 cell/g, the benthic biomass was 6.68 25.82 g/m^2, the density of benthos was 31–124 ind./m^2. The residual feed in net-cage cultural areas caused nitrogen, phosphorus and organics contamination and the community structure was destroyed by the rapid deterioration of sediment environment. Benthos in non-cultural areas and inner bay were more than that in cultural areas and bay mouth respectively. The concentration of TN, TP, OC and sulfides in cultural areas and inner bay was significantly higher than that in non-cultural areas and bay mouth respectively. With the increasing of cultural time, the concentration of TN, TP and sulfides were increased, while the density of benthos and benthic microalgae were decreased at the surface sediments in cultural areas.

Keywords: the East China Sea; Xiangshan Bay; net-cage culture; marine benthos; impacts

1 INTRODUCTION

Xiangshan Bay water is located in the northeast coast of Zhejiang Province, with Zhoushan archipelago on the East, Hangzhou Bay on the North and Sanmen Bay on the South. It's a northeast-southwest oriented, long, narrow and half-closed bay and the total area is about 563 km^2 with about 70% waters and about 30% beach area. It's one of the most important marine fishery depots in China for its good water quality and rich fishery resources. It's also the important basement for ocean fishery in Zhejiang province. The sea-beds of Xiangshan Harbor is mainly composed of silt mud, then is muddy silt and small amounts of shell sand and gravel, which is the eligible place for various kinds of fishes, prawn, shellfish and algae. So it is suitable for developing net-cage culture. Now the total net-cases are about 70,000 and commercial fishes cultured are large yellow croaker (*Pseudosciaena crocea*), common sea bass (*Lateolabrax japonicus*), (*Sciaenops ocellatus*).

Benthos is one of the most important parts of marine organisms. The species composition, regional distribution and the quantity variation with the seasons are closely related to the living environment. The population composition of marine benthos and their quantitative relation may change with the living environment. Recently, some researchers have reported the analysis of the bottom environment in Xiangshan Bay.[1–6] An investigation on parameters of marine benthos

Figure 1.　The survey stations in Xiangshan Harbor.

and geologic environment was conducted in net-cage cultural areas of Xiangshan Harbor between January 2001 and May 2012. The aim is to provide scientific basis for developing reasonable aquaculture model and protecting marine ecology environment.

2　MATERIALS AND METHODS

2.1　Survey stations and sampling

23 sampling sites were settled in the waters of Xiangshan Harbor between January 2001 and May 2012, as shown in Figure 1. The cultural areas were mainly distributed near apex of the bay and the middle of the bay, and the natural areas covered nearly all the waters of the bay. Benthic sampling included collecting benthos and benthic algae, detecting chlorophyll a and chemical elements including Total Nitrogen (TN), Total Phosphorus (TP), Organic Carbons (OC), Sulphide (Sul), etc.

2.2　Methods

Samples of benthos were collected with 0.1 m^2 box sampler. Every sample was repeated three times and the average value was gained. The treatment of the samples and analysis of chemical elements in sediments (TN, TP, OC, Sul) were according to the Specifications of Oceanographic Survey (GB17378-2007) and the standard of Marine Sediment Quality (GB18668-2002).[3,4]

3　RESULTS AND DISCUSSION

3.1　Seasonal variation of the individual abundance of benthic microalgae and density of chlorophyll a

The individual density of benthic microalgae ranged from 108 to 264 cell/g with the variety of season in cultural areas, averaging being 180 cell/g, while it ranged from 316 to 812 in non-cultural areas,

averaging being 627 cell/g. The density of chlorophyll α varied between 1.036 and 5.024 g/g with the seasons, averaging being 1.740 g/g in cultural areas and 4.113 g/g in contrasting areas respectively. The change trend of benthic microalgae density was completely consistent with the seasonal variation of chlorophyll α. And the trend was as follows: spring > summer > autumn > winter (Table 1).

3.2　Seasonal variation of biomass and individual density of benthos

Benthic biomass in surface sediments varied between 6.68 and 16.42 g/m^2 with the seasons in cultural areas, averaging being 12.67 g/m^2, while it ranged between 12.01 g/m^2 and 25.82 g/m^2 in contrasting areas, averaging being 18.78 g/m^2. The density range of zoobenthos was 31 ~ 80 g/m^2 with the seasons and the average value was 64 ind./m^2 at the sediments in cultural areas, while the density of zoobenthos ranged from 82 ind./m^2 to 124 ind./m^2 in natural areas, averaging being 105 ind./m^2. That's to say, the biomass and density of benthos in surface sediments in non-cultural areas were significantly higher than that in cultural areas. The annual trend of benthic density was summer > spring > autumn > winter (Table 1).

The kinds of benthic composition, distribution and their quantity variation with the seasons were closely related to their living environment. The community structure was changed and some species were decreased even disappeared due to the change of environment. Due to the rapid development of tidal flat aquaculture in Xiangshan Bay, farm scale exceeded the affordability of the water ecosystem. The residual feed and excrements of fishes cultivated had caused severe effects on Xiangshan Harbor ecosystem. So the diversity index of benthos was low and community structure was unstable. Furthermore, the ecological balance of local water environment and sediment environment had been disturbed to some degree by the marine breeding and other production activities. And the imbalance of the ecosystem had interfered with marine organisms multiply and survival, decreased the quantity of marine benthos and the biological diversity index in cultural areas. However, the biological diversity index was well-distributed in non-cultural areas and the value fluctuated between 1.792 and 1.971, averaging being 1.871.

3.3　The composition of main dominant species

The main dominant species of benthos in Xiangshan Harbor are as follows: Lumbrineris heteropoda, Nephtys polybranchia, Nassarius variciferus, Nassarius siquinjorensis, Amphiura vadicola, Amphioplus praetans, Protankyra bidentata,

Table 1. The seasonal variation of benthos of Xiangshan Harbor.

Season	Areas	Zoobenthos biomass/g·m²	Zoobenthos density/ind·m²	Benthic algae/cells·g⁻¹	Chl a/g·g⁻¹
Spring	Culltured	12.26	71	264	2.361
	Non-cultured	17.61	116	812	5.024
Summer	Culltured	16.42	80	182	1.905
	Non-cultured	25.82	124	762	4.532
Autumn	Culltured	15.32	74	166	1.658
	Non-cultured	19.68	98	618	4.024
Winter	Culltured	6.68	31	108	1.036
	Non-cultured	12.01	82	316	2.872
Annual	Culltured	12.67	64	180	1.740
	Non-cultured	18.78	105	627	4.113

Glycera chirori, Sternaspis sculata, Laonice cirrata, Diapatra neapolitana, Capitella capitata etc. The main dominant species of benthic microalgae in Xiangshan Harbor are *Coscinodiscus jonesianus, Coscinodiscus radiatus, Coscinodiscus oculusiridis* etc.

Sediment environment were getting worse gradually after several years' cage culture. The original dominant species of zoobenthos and the community structure was changed during the transformation from low-oxygen sediment conditions to anaerobic environment. As the result, zoobenthos migrated to other places. The dominant species of macro benthos were distributed unevenly in natural areas of Xiangshan Harbor, which exhibited significantly regional distribution and the species in bay were much more than those at bay mouth. Large quantities of benthos such as *Nassarius* spp., *Protankyra* spp., *Sternaspis sculata* etc. were distributed in cultural areas, and *Nassarius* spp. were especially abundant. *Nassarius* spp. take other animal carcass for food and they are considered to be scavengers in bottom water environment. In the seriously polluted underwater environment, some benthos were gradually adapt to the environment and acquired great tolerance to the pollution during the long-term environment evolution.[1]

3.4 Seasonal variation of Total Nitrogen (TN) and Total Phosphorus (TP)

The content of TN in surface sediments of cultural areas varied between 0.075% and 0.092% with the variation of seasons, average being 0.084%; while it ranged from 0.046% to 0.052% in contrasting areas with the average value 0.049%. So TN in cage culture areas was much higher than that in non-culture areas and it was higher in bay than that at the bay mouth. The annual change trend on TN in non-cultural areas was as follows: spring > summer > autumn > winter (Table 2).

Table 2. The seasonal variation of surface sediments of Xiangshan Harbor.

Season	Areas	TN/%	TP/%	OC/%	Sul/mg/kg
Spring	Culltured	0.089	0.041	0.866	56.026
	Non-cultured	0.052	0.022	0.070	2.127
Summer	Culltured	0.081	0.043	0.822	27.423
	Non-cultured	0.050	0.020	0.066	1.052
Autumn	Culltured	0.075	0.052	0.912	39.459
	Non-cultured	0.048	0.023	0.065	5.601
Winter	Culltured	0.092	0.049	0.904	32.222
	Non-cultured	0.046	0.019	0.063	1.568
Annual	Culltured	0.084	0.046	0.876	38.033
	Non cultured	0.049	0.021	0.066	2.587

The content of TP ranged from 0.041% to 0.052% in surface sediments of culture areas in four seasons, averaging being 0.046%; the content of TP ranged from 0.019% to 0.023% in surface sediments of non-culture areas in four seasons, averaging being 0.021%. TP in culture areas was much higher than that in non-culture areas and it was higher in bay than that at the bay mouth. The annual change trend on TP in non-culture areas was as follows: autumn > spring > summer > winter (Table 2).

The main sources of Nitrogen and Phosphorus in cultural areas were pollution of culture itsel in the net-cage culture areas of Xiangshan Harbor. Nitrogen and Phosphorus in surface sediments were used up by marine benthos in seasons with high benthic biomass. According to Kapsar' research on feed, 3/4 TN and TP of feed was drained into waters, of which 65% TN and 10% TP sank to the submarine. Others reported that 20%–30% of residual feed sank to the submarine.[2,7] The residual feed caused Nitrogen, Phosphorus and organic compound contamination to the water environment and sediment environment and the structure of biotic community around would change with the environment. Only 15%–30% of Phosphorus in feed was used by fishes, 16%–26% was dissolved in water and over 50% was existed as particles.[5] Organic matters in sediment of Xiangshan Harbor mainly came from the runoffs of the Yangtze River and the Qiantang River. And the particles with little organic matters sank to the bottom firstly at the bayhead of Xiangshan Harbor, while particles with more organic matters moved on to inner harbor. The content of organic compounds in sediment at the mouth of Xiangshan Harbor was relatively low, but it was higher in cage culture areas because of feeding pollution. The main causes for high content of Sulphide (Sul) in the bottom sediment of Xiangshan Harbor were that the surplus of feed and excrements of aquaculture organisms sunk to the submarine, then they were degraded to release hydrogen sulfides and other matters. With the rapid development of the cage culture in Xiangshan Harbor, large quantities of residual feed and excrements were deposited to the submarine at the rate of over ten centimeters per year, which caused serious deterioration of the environment under the net-cases. Furthermore, during the degradation of high content of organic matters and nutrients in the sediments, they were released to the water again, causing "secondary pollution", serious deterioration of waters in the culture areas, aggravation of eutrophication and red tide sometime. The content of Nitrogen and Phosphorus in the sediment increased significantly due to the large scale of bait casting during the breeding season and the deposition of excrements to the submarine. Secondly, the content of Nitrogen and Phosphorus in the sediments increased because of the continental input in rainy seasons. In Xiangshan Harbor, the content of TP in sediments in culture area wass higher than that in contrasting areas, and it was higher at the bottom of harbor than that at bay mouth. Cage culture had important impacts on phosphoric substances accumulation at the submarine and increase of organic materials at the sediments.

3.5 Seasonal variation of organic carbon (OC) and sulfides (Sul)

The content of OC ranged from 0.822% to 0.912% in surface sediments of culture areas in four seasons, averaging being 0.876%; the content of OC ranged from 0.063% to 0.070% in surface sediments of non-culture areas in four seasons, averaging being 0.066%. The content of OC in culture areas was much higher than that in non-culture areas and it was higher in bay than that at the bay mouth. The annual change trend on OC in non-culture

areas was as follows: autumn > winter > spring > summer (Table 2).

In Xiangshan Harbor, the content of organic matters in cage culture areas was high, which is higher than the average value of the whole harbor by 0.05% and by 0.14% of that in non-culture areas. In Huangdun Harbor, the concentration of organic compounds gradually increased from bay mouth to bay bottom. While in Tie Harbor, the concentration was high at the bay mouth, densely net-cage culture areas.

The content of sulfides ranged from 27.423 to 56.026 in surface sediments of culture areas in four seasons, averaging being 38.033; the content of sulfides ranged from 1.052 to 5.601 in contrasting areas in four seasons, averaging being. 587. The content of sulfides in culture areas was much higher than that in non-culture areas and it was higher in bay than that at the bay mouth. The annual change trend on sulfides in non-culture areas was as follows: spring > autumn > winter > summer (Table 2).The biomass and inhabit density at the station with lower sulfides were much higher than that at other stations. In summer, the feeding was reduced and a little residual feed or fish metabolites could meet the needs for benthos growth.The concentration of sulfides and organic compounds were low and the biomass was high in summer.[2,3]

All the major factors were higher in the sediments around cages in culture areas than that in non-culture areas, among them the content of sulfides was four fold higher in culture areas than that in contrasting areas. The benthic biomass was high at the stations with low concentration of sulfides around the net-cage cultural areas. The sediment was piled up at the rate of 25 cm/year in the center of net-cage cultural areas and it was black and smelly, no benthos were found there. After 3 years' culture, the sediment was black and smelly silt with the thickness about 1 m, the bottom environment was already deteriorated and the benthos were nearly disappeared. The content of sulfides in the sediments of bay mouth was higher than that at the bottom of the bay. The residual feed and excrements produced during culture accumulated at submarine in net-cage cultural areas and the content of sulfides was higher at the stations in cultural areas than other stations due to the release of sulfides during the degradation of accumulations. The content of sulfides increased from bay mouth to the bottom of bay. The content of sulfides in cultural areas was higher than the contrasting areas by over 100 folds because after plenty of residual feed and excrements sank to the submarine, dissolved oxygen was consumed and sulfides were released during the degradation of accumulations.[5]

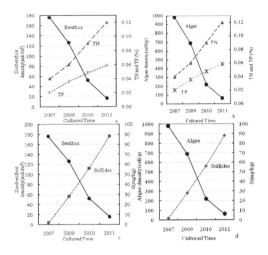

Figure 2. The relationship between zoobenthos density, algae density and TN, TP, Sulfides in cage cultural area.

3.6 The relationship between benthos and TN, TP, sulfides in net-cage cultural areas

The longer the cultural time was, the higher the content of TN, TP and sulfides was and the lower the density of benthos and benthic microalgae was in cultural areas (Fig. 2a–d).Negative correlation were observed between density of benthos and benthic microalgae and the content of TN and TP (Fig. 2a & 2b); and the negative correlation were also observed between the density of benthos and benthic microalgae and the content of sulfides (Fig. 2c & 2d).

4 CONCLUSIONS

In the surface sediments of Xiangshan Bay China East Sea, the annual average density of microalgae and chlorophyll α were 404 cell/g and 2.927 g/g respectively; the annual average benthic biomass and density were 29.45 g/m² and 85 ind./m² respectively; the annual average TN, TP,

OC and sulfides were 0.67%, 0.034%, 0.471% and 20.31 mg/kg respectively. The content of TN, TP, OC and sulfides were significantly higher in cultural areas than that in non-cultural areas and in bay was higher than that at the bay mouth. The concentration of TN, TP and sulfides increases with the cultural time, while the density of benthos and benthic microalgae decreases with the cultural time of the surface sediments in cultural areas.

ACKNOWLEDGEMENTS

This work was financially supported by the National Natural Science Foundation of China (Grant No. 41176142).

REFERENCES

[1] Gao Ai-gen, Yang Jun-yi, Chen Quan-zheng et al., Comparative studies on macrobenthos between cultured and non-cultured areas in Xiangshan Bay. *Jour. Fish. China*, 2003,27(1):25–31.

[2] Gao Ai-gen, Chen Quan-zheng, Hu Xi-gang, Ecological characteristics on macrobenthos of net-cage-cultural areas in the Xiangshan Bay. *Acta Oceanol. Sinica*, 2005,27(4):108–113.

[3] Ning Xiu-ren, Hu xi-gang, et al., The research and evaluation on culture ecology and pot pisciculture breeding capacity at Xiangshan Harbor, Ocean Press, Beijing, 2003, pp.25–32.

[4] Zeng Jiang-ning, Pan Jian-ming, Liang Chu-jin, et al., the ecological environmental comprehensive investigation report in the key harbor of Zhejiang Province, Ocean Press, Beijing, 2011, pp.39–73.

[5] Zhang Jian, Wu Ao-yu, Shi Qing-song, Mariculture and its environment effect in the Xiangshangang Bay, *Donghai Mar. Sci.*, 2003,21(4):54–59.

[6] Zhang Hai-bo, Cai Yan-hong, Ye Hui-ming et al, The assessment on the environment of surface sediments and the type of sediments in Xiangshangang Bay. *Jour. Mar. Sci.*, 2007,25(4):51–58.

[7] Kapsar H.F, Hall G.H., Holl A.J., Effects of sea cage salmon farming on sediment nitrification and dissimilatory nitrate reduction [J]. *Aquaculture*, 1988,70:333–334.

Frontiers of Energy and Environmental Engineering – Sung, Kao & Chen (eds)
© *2013 Taylor & Francis Group, London, ISBN 978-0-415-66159-1*

Analysis on the upgrade processes of urban wastewater treatment plant and the aeration energy-saving technologies

Z.L. Zhu
School of Municipal and Environmental Engineering, Shandong Jianzhu University, Jinan, China

C.J. Zhang
School of Thermal Energy Engineering, Shandong Jianzhu University, Jinan, China

H.L. Weng
Shandong Construction Project Bidding Center, Jinan, China

Q. Li
Jinan Water Group Co., Ltd., Jinan, China

ABSTRACT: Urban wastewater treatment plant faces the task of improving water quality by upgrading process. A simple series of traditional treatment processes, based on the secondary treatment, result in a long process, the unreasonable unit, higher investment and operating costs. With the current advanced nutrient removal process, using the concept of wastewater reclamation, this article describes the three urban sewage recycling process, as the recommended case of urban sewage treatment plant transformation. While the article discusses the municipal wastewater treatment plant aeration energy-saving technologies.

Keywords: urban wastewater treatment plant; process upgrade; wastewater recycling process; aeration energy-saving

At present, China put forward the urban sewage treatment plants effluent to gradually achieve the «GB18918-2002 urban sewage treatment plant comprehensive discharge standard» level of A requirements, and this standard close to many users' recycled water standard. For example, $TN \leq 15\,mg/L$, $TP \leq 0.5\,mg/L$. Water quality engineers are faced with transforming and upgrading the wastewater treatment plant secondary sewage treatment technology, improving the sewage effluent standards, and realizing the depth purification of the sewage, while facing other problems, for example, wastewater treatment plant's high operating costs and complicated management. So the research and implementation of energy-saving technology is also imminent. Combining with the whole process of wastewater reclamation concept, the upgrade of the wastewater treatment plant and aeration energy-saving technologies are discussed in the paper.

1 THE FULL FLOW TECHNICAL PROCESS OF THE WASTEWATER-REUSE

Based the contradiction in the simultaneous nitrogen and phosphorus removal, used the theory of the optimal complete flow as the guidance, combined the new technology and traditional technology of the biological phosphorus and nitrogen removal, introduced 3 kinds of wastewater-reuse technical process of urban sewage which were upgrading and reconstruction technology in the urban sewage treatment plant.

(1) A/O biological phosphorus removal- wastewater reclamation flow of the biological nitrogen removal by anaerobic ammonium oxidation.

The city sewage was used as the treatment object, combined the anaerobic/aerobic activated sludge biological phosphorus removal and anaerobic ammonium oxidation biological autotrophic nitrogen removal process with anaerobic aerobic, produced a wastewater-reuse technical process of urban sewage, as shown in Figure 1.

The characteristics of this process didn't change the operation parameters of the secondary treatment, only changed the front of the biochemical reaction tank into the anaerobic stage, on this premise that the capital expenditure wasn't increased, the nutritive matter-phosphorus was removed, and owing to the existing of the anaerobic stage, the growth of the filamentous bacteria was inhibit, the activated sludge bulking was avoided,

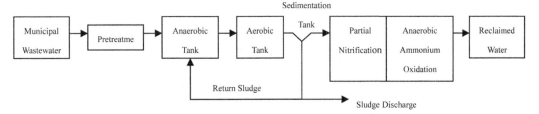

Figure 1. Wastewater reclamation flow including A/O and anammox.

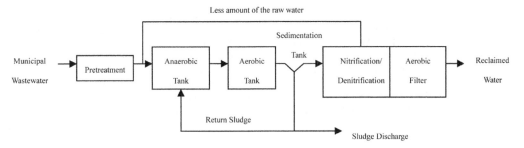

Figure 2. Reclamation flow including A/O and shortcut nitrification and denitrification.

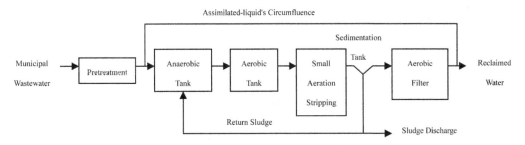

Figure 3. Wastewater reclamation flow including dephosphorization and aerated biofilter.

which made the running more stable. Secondly, let the phosphorus removal occurred in the secondary treatment, the nitrogen removal occurred in the advanced treatment, applied the process of the partial nitrification/anaerobic ammonium oxidation to the nitrogen removal, which was an economic and efficient biological autotrophic nitrogen removal technology, lowly amount of aeration, without additional carbon sources, low sludge production, avoided the production of the secondary pollution.

(2) Reclamation flow of the A/O phosphorus removal—shortcut nitrification/denitrification.

When the raw sewage through the anaerobic/aerobic tank, the biological phosphorus removal was occurred, meanwhile, the organic compounds was degraded, the effluent was took into the system of shortcut nitrification/denitrification to realize the biological nitrogen removal, then it was took into

the terminal aerobic filter to further remove the organic matter, nitrogen and the suspended matter, et al. The process was showed in Figure 2.

(3) Wastewater reclamation flow including dephosphorization and aerated biofilter.

The sewage was contact with the return sludge which was rich in DPAOs in the anaerobic tank, after DPAOs were fully released the phosphorus, the mixed liquid entered into the anoxic tank, then it were fully mixed with the nitrification liquid which came from the aerobic biofilter tank, DPAOs uptake the phosphorus through the NO3- which was regarded as the electron acceptor, at this time, the process of the denitrifying phosphorus-uptake was finished. The purpose of the small aeration stripping is to stripping the bubbles of nitrogen which adhere the activated sludge particles. Meanwhile, the refractory biodegradation organic matter was further degraded by the aerobic filter, the quality of the effluent was also improved. The process is shown in Figure 3.

Combined the dephosphorization and aerated biofilter to produce a full regeneration process, fundamentally, it could resolve the contradiction of the conventional technical process that contend for carbon source was existed between phosphate accumulating organisms and the denitrifying bacteria. Meanwhile, owing to the majority of the organic matter was degraded in the anaerobic stage, so it also reduced the air consumption.

2 THE ANALYSIS OF THE ENERGY CONSUMPTION OF SEWAGE PLANT AND THE TECHNOLOGY OF AERATION ENERGY CONSERVATION

Energy conservation in the operation of sewage treatment plant contains two factors, one is to improve the motor performance, the other is aeration energy conservation. Jacobus [1] proposed that using efficient motor and variable speed drive can increase power factor (pf). Lamers and Li Yafeng [2] etc. proposed that the energy conservation of treatment plant pump system focuses on the selection and tie-in, driving way selection, system maintenance of pump; Improving the aeration system efficiency and excessive aeration is the main issue in energy conservation of aeration system.

A large number of studies have demonstrated that microporous aeration can achieve the effective mixing and oxygen spread fully, but the head of spread often block and difficult to clean. There is no same issue with surface aeration equipment. Ognean [3] studied the relationship of surface aeration equipment oxygen transfer efficiency, the oxygen rate and energy consumption etc., set up the relationship of ideal aeration machine and oxygen transfer rate, the diameter and rotate speed which related to the actual production.

Aeration intelligence control systems provide a new idea for energy conservation of aeration and biological treatment system. Olsson [4] etc. studied aeration pool DO control at earliest, this is also the most direct way to control air-supply. Charpentier [5] etc. studied REDOX potentials ORP control. They indicated that ORP control

is easier than DO control to achieve. Several productive experiment in the sewage treatment plant in France shows that ORP control can optimize energy using, can improve removal rate of carbon sources, nitrogen sources pollutant specially. In addition, the uses of fuzzy logic control also have more significant effect in energy conservation.

3 CONCLUSIONS

Through to the research of urban sewage the regeneration process, research and develop the unit and key technology that can remove COD, phosphorus and nitrogen pollutants effectively. Based on this, optimize the processing unit to establish a set of relatively economic and efficiency urban sewage regeneration process. In order to cut short the treatment process of recycling water, save infrastructure investment and running costs and improve the benefit of the recycling water factory. Meanwhile, if these can be done, that can reduce the urban sewage discharge, reduce the pollution of water sewage, decrease the extraction to the amount of the city natural water, and ease the contradiction between the supply and demand of urban water. The process of the treatment and recycling of sewage will be improved. The recovery of water environment and sustainable use of water resources will be true.

REFERENCES

[1] Jacobs A. Managing Energy at Water Pollution-Facilities [J]. Water Sewage Works, 1980, 1(4): 28–31.
[2] Li yafeng, Ma xuewen. On the energy of urban sewage treatment plant [J]. Energy saving, 1998, (1):35–37.
[3] Ognean T. Relationship between oxygen mass transfer rate and power consumption by vertical shaft aerators [J]. Wat. Res., 1997, 31(6):1325–1332.
[4] Olsson G, Andrews JF. Dissolved oxygen control in the activated sludge process [J]. Wat. Sci. Tech., 1981, 13(10); 341–347.
[5] Charpentier J, Florentz M, David G. Oxidation-Reduction Potential (ORP) regulation: a way to optimize pollution removal and energy savings in the low load activated sludge process [J]. Wat. Sci. Tech., 1987, 19 (Rio); 645–655.

Frontiers of Energy and Environmental Engineering – Sung, Kao & Chen (eds)
© *2013 Taylor & Francis Group, London, ISBN 978-0-415-66159-1*

The application of industry comprehensive water quota in water-saving management

J.X. Fu, H.B. Liu & Q. Liu
School of Municipal and Environmental Engineering, Shenyang Jianzhu University,
Liaoning Shenyang, China

ABSTRACT: Planning design and projecting management are the two complementary components of water-saving management. The rationality can be ensured, by researching the importance of trade comprehensive water quota in planning management. Through the research of water quota management system, the necessity of water-saving management can be defined. According to the water date of 15 hospitals in Shenyang, the water saving level of these hospitals can be judged by using two kinds of water quota. The results show that there are 10 hospitals meet the requirements by using comprehensive water quota and only 1 hospital fully meet the requirements by detailed water quota. The comprehensive water quota is more suitable for assessment of water as a fixed quota to determine the water saving level. Public water waste is serious, so it has a lot of water saving space. Combined with comprehensive water quota to determine whether the industry meet norm water requirements, the industry which fails to reach the requirements of section water requirements can timely take corresponding measures, make it as soon as possible to meet the requirements.

1 WATER QUOTA MANAGEMENT

In order to ensure the rational allocation principle, water-saving sector use water quota as the fundamental basis of quota management to project planning and policy of water price, which improves water use. The management process including setting macroscopic water quotas, determining the plan of water allocation principle, designing the right policy of water price, making scientific management detailed rules for the implementation of the scheme. Carry out water programs and water management must make water quota in basis. So making rational water quota can fully promote in quota management and effective implementation.

The water industry quota for water saving can be divided into two aspects. First from planning and design aspects, in order to establish water-saving society, need to build water saving building. In order to meet the requirements of the water saving building, it need to clear the several important problems, such as the design of water saving water, water saving system and fixed water-saving equipment.

The norm of "Standard for water design in civil building" designs these problems, which determines the water quota range of public industries and clear classification. It also provides clear guidance for making water quota in view of the different industries. Besides from the planned management into consideration, waste is serious in various industries in the process of water use. The public water consumption and the residents water consumption are the same, so the public water consumption also is affected by the regional influence, not content with use uniform quota standards. The country had been 26 cities began to aim at each its own conditions to develop local water quota standards, so as to avoid the problem of waste water.

2 COMPREHENSIVE WATER QUOTA

In the past, when making industrial water quota must determine the overall water unit, the overall water unit, such water quota generally called detailed parts water quota. Because some parts of water is hard to separate the industry statistics, and in statistical or report data will have waste phenomenon. Practice shows that this kind of practice is not suitable for practical operation process. According to the requirements of this kind of situation, it need of water assessment index quota. When implementing water quota management, the important problem is appraisal on water quota, which water quota can be not overmuch, improve the quota management operation, avoid waste happens.

Comprehensive water quota that is made for public industry, is the ratio between the total water consumption and the water unit. It not only simplify the refinement of too much water data collection, still can the actual situation of water should be unit, which can be used as evaluation index of the water industry. Meanwhile, combined with the regulation effects of price lever and use a combination of water as the basis for the water price fixed ladder, to promote the construction of water-saving city.

In some cities formulated in local standard, the comprehensive assessment of water use fixed as fixed value, but our country has no unified and request, to inconvenience water-saving management, which make not directly between the provinces and provincial water for comparison, intuitive analysis of water-saving management of provinces and cities.

Don't reach the assessment requirements of the industry unit, combined with the requirements of the fine water quota, need to take corresponding measures of water saving, find out how quota management meet requirements.

3 RESEARCH BASED ON COMPREHENSIVE WATER QUOTA OF SHENYANG WATER LEVEL OF HOSPITAL

3.1 *Water consumption constitute of hospital*

With the continuous improvement of living standards, the government's commitment to medical enterprise, medical and health institutions had the very big development. The hospital is used for a high proportion of public life in the water. Hospitals are the highly liquidity unit, so will the hospital water management into an important content of water-saving management. In order to better determine water-saving management measures, whether meet the requirements assessment hospital water is very necessary.

According to the administrative department of health care has been the hospital strict rating, divided into three grades. According to "the hospital classification management standard" for sure a, b, and c, which level 3 hospital add special grade, so the hospital is divided into ten part. According to different grade hospital, the change trend of water is clear. The hospital total water consumption constitute with outpatient service water and other water composition in hospital. According to the hospital in recent years of water balance test results, the hospital water has demographics: outpatient 30%, hospital 50%, other water by 20%.

3.2 *The quota on hospital water standard*

The level in "Standard for water design in civil building" is based on the water between outpatient and hospital, as Table 1. And according to oneself circumstance, Liaoning province each formulate local standard "The Liaoning province water quota". In the industry standard of classification standard request, the hospital is divided into following level 2, level 2b, level 2a and level 3a. And determine comprehensive water quota of hospital of water consumption quota appraisal value, as shown in Table 2. In order to study integrated water Standard in the inspection of water-saving management level, respectively for two water standard inspection Shenyang 15 hospitals whether meet the water requirements.

3.3 *The analysis of the water water consumption of hospital*

In Table 3, there are 15 hospitals water data of 2010, which are provided from water supply company in Shenyang. And at the same time further investigate the basic situation of the hospital to get classification, medical number and the bed number in hospital.

Through analyzing water consumption of 15 hospitals, calculate the hospital comprehensive water consumption and water consumption and water consumption in hospital outpatient service.

3.4 *The judgment of water-saving level*

As the water data of 15 hospitals shown in Table 2, the comprehensive water consumption of level 3 a hospitals is about 610–1050 L/d; the comprehensive water consumption of level 2 a hospitals is about 245–935 L/d; the comprehensive water consumption of level 2 b hospitals is about 233–205 L/d; the comprehensive water consumption of following level 2 hospitals is about 120–185 L/d. Contrasting

Table 1. Standard for water design in civil building for hospitals' water quota.

Classification	Water quota
Hospital	130–200 L (b·d)–1
Outpatient	6–12 L (cap·t)–1

Table 2. Liaoning province water quota for hospitals' water quota.

Classification	Water quota/L (b·d)–1
Level 3	900
Level 2a	650
Level 2b	250
Following level 2	150

Table 3. Water consumption and relevance information of hospitals in Shenyang.

Numbers	Classification	Year water consumption m³·a⁻¹	Bed number	Medical number
1	Level 3a	302400	800	600
2	Level 3a	204228	930	958
3	Level 3a	247680	800	1500
4	Level 3a	362880	800	1700
5	Level 3a	215784	888	3000
6	Level 3b	198720	600	676
7	Level 2a	161568	480	245
8	Level 2a	47988	310	340
9	Level 2a	112320	600	200
10	Level 2a	5733	65	188
11	Level 2b	5032	60	148
12	Level 2b	5166	70	74
13	Level 1a	4320	100	148
14	Level 1a	2556	50	38
15	Level 1a	3330	50	450

Table 4. Calculation on water data of hospitals.

Numbers	Comprehensive water consumption L (b·d)⁻1	Outpatient L(cap·t)⁻¹	Hospital L(b·d)⁻¹
1	1050	420	525
2	610	178	305
3	860	138	430
4	1260	178	630
5	675	60	338
6	920	245	460
7	935	550	468
8	430	118	215
9	520	468	260
10	245	26	122
11	233	28	116
12	205	58	103
13	120	24	60
14	142	56	71
15	185	6	93

the water data with Liaoning Standard, there are ten hospitals meeting the water-saving requirement, others do not meet the requirement.

Contrast the water data with the "Standard for water design in civil building", the result is very different. There is only one hospital meeting the outpatient requirement and six hospitals meeting the hospital requirement. Others are all not meeting requirement. Because of the liquidity, it is difficult to get the accurate water data of outpatient. And the classification makes an effect on water consumption, the more higher the classification is, the more water consumption using. Invariable quota don't adapt on developing quota management.

4 CONCLUSIONS

Through contrasting the water-saving level between comprehensive water quota and part water quota, the comprehensive water quota is fitter to be the evaluation index, which can evaluate whether the public industry meet the water-saving requirement. The industry that water consumption is excess have to make measure to meet the requirement.

According to the water-saving level of 15 hospitals, water waste in public industry is serious, which has a pace to make measure to decrease the water consumption.

REFERENCES

2005. Quota Management of the Hotel's Water Use in Beijing. Resources Science 27(5):107–112.
Bernhardi, L. & G.E.G. Beroggi & M.R. Moens. 2000. Sustainable water management through flexible method management. Water Resources Management 14:473–495.
Hao. G.Z & Liu. J.L & Liu. Z.L. 2001. Urban water resource and water using management. Journal of Hebei Institute of Architectural Engineering (4):88–89.
Hong.Y & Wang G.H & Sun. X.F. 2008. Measures of Saving Power and Saving Water in Building Water Supply and Drainage Design. Building Energy Efficiency 6(36):40–49.
Huang. Y. 2007. Urban hospital water rule and countermeasures of Kunming. Zhejiang university.
Jiang. Y.L & Chen. Y.S. 2007. The exploration of university water quota management in Beijing. Water and wastewater engineering 6(33):68–72.
Liu. J.L. 2003. Theory and Technology of Urban Water Conservation Planning. Beijing: Chemical Industry Press.

Luo. T.L & Zhang. S.J & Guo. X.H, et al. 2010. Establishment of indicator system on water consumption quota. China water resources 9:40–24.

Mohamed, A.S, & H.H.G. Savenije. 2000. Water demand management: Positive incentives, negative incentives or quota regulation? Physics and Chemistry of the Earth 25: 251–258.

Weng. J.W & Jiang. Y.L & Chen. Y.S. 2007. Present Status, Problem s and Countermeasures of Public Domestic Water Consumption in Beijing. China Water & Wastewater 23(14):77–82.

Zang. J.H & Yang. Y.S, et al. 2001. Planning life water quota and index system. China Water & Wastewater (1): 66–67.

Frontiers of Energy and Environmental Engineering – Sung, Kao & Chen (eds)
© 2013 Taylor & Francis Group, London, ISBN 978-0-415-66159-1

Research on traffic operation of Fekai expressway and surrounding road network

K.M. Wu & L.D. Zhong
Research Institute of Highway, Ministry of Transportation, Beijing, China

ABSTRACT: A traffic operation plan of Highway reconstruction and widening is presented, in which computer simulation technique is employed, in combination with actual traffic data and traveling OD matrix gathered from the project of reconstruction and widening of Fekai Expressway. VISSIM supported modeling technique of large scale road network and traffic operation plan of Highway reconstruction and widening are illustrated, that is, with the help of OD matrix, the best efficiency of the whole road network is realized through the control of queue length of congested intersections, using dynamic allocation method.

Keywords: traffic simulation; dynamic allocation; traffic operation; OD matrix

1 TRAFFIC OPERATION PROJECT OF HIGHWAY RECONSTRUCTION AND WIDENING

The existing Expressways or main national and provincial Highways are generally main shipping routes in their region, with high daily traffic volumes. Once they are closed, huge pressure will be brought to the corresponding regional road networks. To ensure safe operation of Highway is the prerequisite of reconstruction and widening projects and is also the basic principle to be considered during planning. To comprehensive analyze the influence of Highway reconstruction and widening projects on Highway traffic safety, to pay great attention to traffic operation during the design stage of Highway reconstruction and widening, and to operate and guide traffic initiatively based on a method of system theory will make a significant sense.

The construction characteristic of Highways being planned, and background information such as the traffic characteristic of surrounding road network etc., are the base to determine a rational traffic operation project. Through analyzing the existing contradictory problems, design principles and methods of Highway reconstruction and widening are work out, and traffic operation projects from macroscopic regional road network to microscopic roads and key notes are formed. Through setting of traffic simulation model, traffic operation research during Highway reconstruction and widening based on actual traffic survey is developed. Process of traffic operation design is shown in Figure 1.

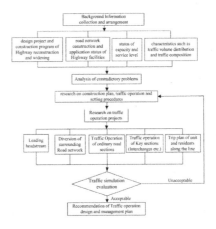

Figure 1. Design process of traffic operation.

2 SUMMARY OF FEKAI EXPRESSWAY

Fekai Expressway opened to traffic on December, 1996, having 4 lanes now with a design speed of 120 km/h. With the rapid development of regional economic, Fekai Expressway has been gradually linked network, which causes a huge increase in traffic volume. The widening of Fekai Expressway will raise the capacity of national main trunk line and give full play to overall effectiveness of road network.

During the widening, Fekai Expressway will limited only part time from current two way four lanes to demi two way two lanes, and most of the time the whole Fekai Expressway will open with limited speed without closing lanes.

3 PATHS FOR DIVERTING TRAFFIC

During the time part of lanes of Fekai Expressway are closed, the two-way-two-lane traffic can't satisfy the requirement of peak hour volume. Vehicles need to be diverted to other paths. The following 5 diverting paths are planned.

3.1 Scheme 1

As shown in Figure 2, in the direction from north to south the diverting point is set at Shayong Interchange, vehicles are diverted to Feshanyihuan highway, at national highway 325 Interchange, enter into national highway 325, and then via Longshan Interchange go back to Fekai Expressway. From south to north, the diverting point is set at Longshan Interchange, the diverting path is the same as the above with opposite direction.

3.2 Scheme 2

From north to south, the diverting point is set at Shabei Interchange, vehicles are diverted to west ring section of Guangzhou Huancheng Expressway, at Haibalu Interchange enter into Feishanyihuan Highway, at national Highway 325 Interchange into national Highway 325, and then via Chenshan Interchange go back to Fekai Expressway. Form south to north the diverging point is set at Chenshan Interchange, diverting path is the same as the above with opposite direction, as shown in Figure 3.

3.3 Scheme 3

From north to south, the diverting point is set at Yayao Interchange, vehicles are diverted to Guangsan Expressway, via Xiaotang Interchange to south section of west ring, via Jiujiang Interchange back to Feikai Expressway, then via Longshan

Figure 2. Diverted path one.

Figure 3. Diverted path two.

Figure 4. Diverted path three.

Interchange diverted to G325, along G325 via Chenshan Interchange back to Fekai Expressway again; from south to north, the diverting point is set at Chenshan Interchange, diverting path is the same as the above with opposite direction, as shown in Figure 4.

3.4 Scheme 4

From north to south, the diverting point is set at Xiebian Interchange, vehicles are diverted to G325, along G325 via Longshan Interchange back to Fekai Expressway; from south to north, the diverting path is the same as the above with opposite direction, as shown in Figure 5.

506

Figure 5.   Diverted path four.

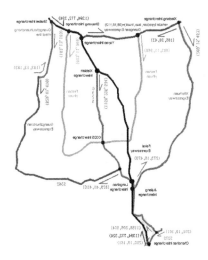

Figure 6.   Schematic diagram of OD Matrix.

3.5   *Scheme 5*

A combined diverting plan is provided, that is, all the diverting plan (diverting points from path one to four) are available, and, according to the practical diverting effect, the diverting points are used as required.

4   SIMULATION METHOD

The project analyzes traffic diverting mechanism and method, uses traffic flow simulation technique, targets the best operating efficiency of road net, presents a traffic diverting model for reconstruction and widening project of Fekai Expressway, and provides reference for other traffic diverting plan of expressway reconstruction and widening.

The difficulty of the Project is to set up a dynamic automatic assignment road net through OD Matrix, to adjust fees of different roads of the road net and expense coefficients of different vehicle types to make vehicles choose the most economical route, and to optimize the operation efficiency of road net throughout the whole process by controlling the queen length on the traffic bottleneck sections.

Dynamic traffic assignment is one of the core modules of traffic flow simulation softwares. The Model of dynamic traffic assignment plays an important role in Advanced Traveler Information System (ATIS) and Advanced Traffic Management System (ATMS). Advanced simulation software such as VISSIM is needed to make dynamic assignment simulation for large scale road net. VISSIM microscopic traffic flow simulation software, produced by PTV Company in Germany, which was researched and developed by Karlsruhe University in Germany and then purchased by PTV limited Company, is one of the most commonly used mainstream traffic flow

simulation softwares in the world today. The software can achieve traffic flow simulation of dynamic assignment of large scale road net. For research purpose, the project uses the software.

5   OD MATRIX OF ROAD NET

The OD Matrix data of Fekai Expressway and surrounding road net is derived based on actual measured data and observer theory.

VISSIM software has two kinds of modes for traffic volume assignment, which are static and dynamic assignments. The road net model of the present project is builded on the basis of the mode of dynamic assignment. The so-called static assignment means entering artificially traffic volume, traffic composition and desired speed etc. for every road of the road net. This mode is not suitable for large scale road net. The so-called dynamic assignment means dividing the road net into a plurality of zones e.g. N zones, the traffic volume between zones can be indicated by a N*N traveling OD Matrix. At the starting point and end point of roads of the road net it is needed to set parking lots, which are also zone connectors, from which the begin and end points of vehicles are generated.

A schematic diagram of the OD Matrix is shown in Figure 6.

6   PRINCIPLE OF ASSIGNMENT

Simply speaking, the principle of VISSIM dynamic assignment is to choose a route according to the total cost of each route. The total cost includes travel time, distance and fee of the roads.

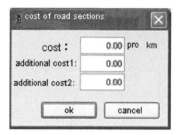

Figure 7. Cost of road sections.

The cost of Roads is made of 3 parts, as shown in Figure 7, which are: cost (pro km), additional cost1 and additional cost2.

Different cost coefficient can be given to travel time, travel distance and economic cost for different vehicle types. The total cost is calculated by the following formulas:

$$\text{Total cost} = \alpha \cdot \text{travel time} + \beta \cdot \text{travel distance} + \gamma \cdot \text{economic cost} + \sum \text{additional cost 2} \tag{1}$$

$$\text{Economic cost} = \text{Distance} \times \text{Cost} + \text{Additional cost 1} \tag{2}$$

Through the above formulas we can artificially adjust the result of dynamic assignment. The method used in the project is that queuing counters are set at the desired cross-sections of roads and cost of different roads is adjusted to make the queue lengths of the desired cross-sections reach a reasonable level. With the increase of the total cost of a route, its traffic attraction reduces, and thus the queue length at some congested cross-sections can also be reduced.

Choosing a route from a series of optional routes is a special case of general "dispersed routing modeling". If a series of optional routes and its total cost are known, the proportions of different types of the vehicles which run on all the chosen routes can be calculated. Logit Model is the most commonly used model to solve these problems. VISSIM uses an improved Logit Model to solve routing problems.

The process of iterative simulation keeps running until traffic volume and travel time obtained at every evaluation time interval makes no obvious changes, which means a state of convergence is reached.

7 OTHER PARAMETERS

The distribution of speed has a great effect on the road capacity and traveling speed, so it is an important parameter for any vehicle type. Suppose that

Table 1. Desired speed.

Highway	Car	Bus	Truck
Fekai expressway (2-way 4-lane)	(55, 60)	(50, 60)	(40, 60)
Feshanyihuan (2-way 8-lane)	(60, 100)	(40, 80)	(40, 70)
G325 (2-way 6-lane)	(40, 100)	(40, 90)	(40, 80)
Guangzhou west ring expressway (2-way 4-lane)	(60, 120)	(60, 110)	(60, 100)
Xi'erhuan expressway (2-way 6-lane)	(60, 120)	(60, 110)	(60, 100)
S272 (2-way 4-lane)	(20, 90)	(20, 80)	(20, 60)
Guangsan expressway (2-way 4-lane)	(60, 120)	(60, 110)	(60, 100)
S113 (2-way 4-lane)	(20, 90)	(20, 80)	(20, 60)
S362 (2-way 4-lane)	(40, 100)	(40, 90)	(40, 80)
Guangzhu west line expressway (2-way 6-lane)	(60, 120)	(60, 110)	(60, 100)

vehicles run without the interference from other vehicles, drivers will drive their vehicles with their desired speed (random changes slightly). The more vehicles there are, which are driven with different desired speeds of their own, the greater the interference among the vehicles is. If conditions allow, so long as a vehicle's desired speed exceeds the current speed of the one before it, it will take chance to overtake if the overtaking will not pose dangers to other vehicles. Choosing proper desired speed has great effect on the simulation authenticity. Desired speed for various kinds of vehicle types is different on different roads in view of different highway grade and limited speed. The desired speed of different vehicle types on all of the highways is shown in Table 1.

8 DIVERTING RESULTS

The vehicles in the project are diverted through the method of dynamic assignment, which can be carried out through 3 plans as follows. The first plan is to change the way of driving from 2-way-4-lane into 2-way-2-lane without intervention, with vehicles automatically diverted; the second one is also to change the way of driving from 2-way-4-lane into 2-way-2-lane, but with trucks prohibited and other vehicles also automatically diverted; the third is to remain 2-way-4-lane driving way, with vehicles automatically diverted. All of the 3 plans need speed limit along the whole Fekai Expressway, and the outcome of simulation includes queue length, speed and travel time of the vehicles before the confluence points of work zones prior to and after vehicles being diverted, as well as traffic volume of all the diverting highways.

The routes of dynamic assignment are shown in Table 2, and the routes from south to north and those from north to south are different.

During the diversion, the queue length at each diverting intersection is controlled below 750 m. The percentage of vehicles on the diverting routes by the method of dynamic assignment is shown in Tables 3 and 4. The percentage obtained using the first and the second plans is the same. When one lane is closed, almost half of the total traffic volume needs to be diverted; when lanes are not closed, almost 40% of the total traffic volume needs to be diverted because of the speed limit along the whole Fekai Expressway. The routes 2, 3 and 4 in both directions have taken a majority of diverted vehicles. In the direction from south to north, when a part of lanes are closed, route 4 has taken the most percentage of diverted vehicles; when no lane is closed, route 3 has taken the most percentage of diverted vehicles. All of routes 2, 3 and 5 pass through the road section from Longshan along G325 to the intersection of Feshanyihuan and G325. Besides the original traffic volume on G325, 30% of original traffic volume of Fekai Expressway will also be taken by the road section. Such a road section with too much traffic may be called traffic bottleneck. In the direction from north to south, when a part of lanes are closed, route 3 has taken the most percentages of diverted vehicles; when no lane is closed, route 2 has taken the most percentage of diverted vehicles. When a part of lanes are closed, all of routes 2, 3 and 4 pass through a road section from Longshan along G325 to the intersection of S362 and G325. Besides the original traffic volume on G325, 42% of original traffic volume on Fekai Expressway will also be taken by the road section. This road section taking too much traffic is also called traffic bottleneck.

When the method of automatic assignment is used, the intersection with the longest queue length 745 m is the one between S362 and G325 in the direction from north to south; When the method of the prohibition of Trucks is used, the intersection with the longest queue length 723 m is the one between Fekai and Xi'erhuan Expressway in the direction from south to north; compared to the method of closing part of lanes, the method without closing lanes has reduced the burden of diversion obviously. The queue length of each main intersection is not more than 500 m and the intersection with the longest queue length 493 m is the one between G325 and Fekai Expressway.

Table 2. Diverted routes.

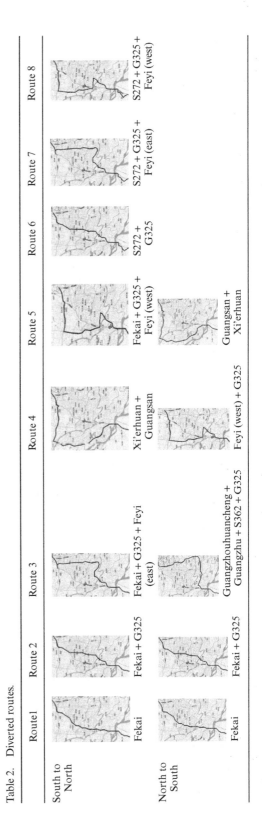

	Route1	Route 2	Route 3	Route 4	Route 5	Route 6	Route 7	Route 8
South to North	Fekai	Fekai + G325	Fekai + G325 + Feyi (east)	Xi'erhuan + Guangsan	Fekai + G325 + Feyi (west)	S272 + G325	S272 + G325 + Feyi (east)	S272 + G325 + Feyi (west)
North to South	Fekai	Fekai + G325	Guangzhouhuancheng + Guangzhu + S362 + G325	Feyi (west) + G325	Guangsan + Xi'erhuan			

509

Table 3. South-North percentage of assignment of traffic volume on routes.

	Automatic assignment	Prohibition of trucks	No lane is closed
Route 1	52%	52%	62%
Route 2	12%	12%	7%
Route 3	11%	11%	16%
Route 4	17%	17%	15%
Route 5	7%	7%	0
Route 6	0.9%	0.9%	0.04%
Route 7	0.5%	0.5%	0
Route 8	0	0	0.12%

Table 4. North-South percentage of assignment of traffic volume on routes.

	Automatic assignment	Prohibition of trucks	No lane is closed
Route 1	51%	51%	67%
Route 2	15%	15%	17%
Route 3	18%	18%	4%
Route 4	9%	9%	12%
Route 5	7%	7%	

Table 5 gives the queue length of vehicles on the confluence points before the work zone before and after diversion when parts of lanes are closed. It is composed of an average queue length and a longest queue length. The average queue length is to record a current queue length on every simulation step and arithmetically average all queue lengths recorded in the test time interval; the longest queue length is to record the current queue length on every simulation step and find the longest queue length in the test time interval. It can be seen from the following 2 tables that the two methods can both reduce queue length obviously.

Table 6 gives travel time of vehicles in 4 plans. The travel time in VISSIM is average travel time (includes stopping or waiting time), which means the time interval from vehicles run through the start point of tested road section to they leave the end point of it. Here includes all the routes from starting point to end point, namely throughout the whole diverting routes. It can be seen that although vehicles have chosen to detour when diversion is applied, the average travel time is shorter than the one when vehicles are not diverted. So, diversion has improved, rather than reduced, the operational efficiency of road network.

Figures 8 and 9 show the changes in traffic volume on Fekai Expressway and diverting road sections in 3 plans. It can be seen that although diversion has reduced the traffic pressure on

Fekai Expressway obviously, it increases that on the diverting routes. G325 is the highway that has taken the most diverted vehicles. But from the results of simulation, its capacity has not been reached. So, the road net can take the traffic pressure caused by reconstruction and widening of Fekai Expressway.

Figures 10–13 show the changes in speed of vehicles at confluence points and middle portion of the work zone before and after diversion. Because part of Fekai Expressway has changed from the original 2-way-4-lane to 2-way-2-lane during the reconstruction, a portion of vehicles need to converge and change lane before the road becomes narrow. Changes of speed make the confluence points become traffic bottlenecks. From it can be seen that, speeds of vehicles at middle portion of the work zone are obviously higher than those at confluence points; the average speeds of vehicles on fast lane and kerb lane after diversion are higher than those before diversion, because reduced queue time makes speeds increase; when the method of prohibition of trucks is used, trucks are all diverted to other road sections, so the speeds of vehicles on Fekai Expressway improved obviously.

When the method of the dynamic automatic assignment with no lane being closed is used, the following Figures 14 and 15 show speeds of vehicles on Fekai Expressway at the points, which is consistent with confluence points while the methods of parts of lanes being closed are used. The speeds of vehicles at the fast lane and kerb lane don't change obviously. Compared with the circumstance closing parts of lanes, speed of vehicles increases obviously and vehicles run more stably.

From the above, through the traffic flow simulation on Fekai Expressway and its surrounding road network, it is not difficult to see that the problem of traffic congestion because of reconstruction and widening of Fekai Expressway can be solved through diverting vehicles to surrounding road sections. The preferred plan that this project recommends is to try not to close lanes during the reconstruction and to remain 2-way-4-lane with speed limit along the whole Expressway. Furthermore, traffic flow should be appropriately guided. If necessary, vehicles should be orderly diverted to G325, Feshanyihuan, Guangsan Expressway, Xi'erhuan Expressway, Guangzhouhuancheng Expressway, Guangzhuxixian and S362, and the percentage of diverted vehicles can refer to the results of simulation, so as to ensure the queue length on each intersection is less than 500 m. When parts of lanes must be closed because of reconstruction, the method of prohibition of trucks is recommended considering the controllability of traffic guidance. The percentage of

Table 5. The queue length of vehicles on the confluence points before the work zone before and after diversion.

| Queue length | Before diversion | | Automatic assignment | | Prohibition of trucks | |
	South-North	North-South	South-North	North-South	South-North	North-South
Average	1772	3335	215	115	6	41
Longest	1872	5012	696	531	196	432

Table 6. Travel time for each plan.

| | Before diversion | | Automatic assignment | | Prohibition of trucks | | No lane is closed | |
	South-North	North-South	South-North	North-South	South-North	North-South	South-North	North-South
Travel time	4686	4439	4311	3924	4335	3826	4161	3991

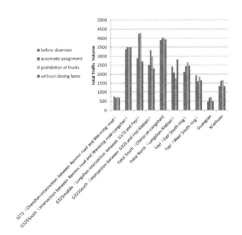

Figure 8. Traffic volume for each road section from south to north (0–7200s).

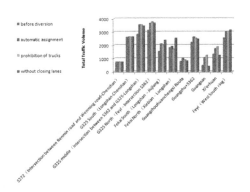

Figure 9. Traffic volume for each road section from south to north (0–7200s).

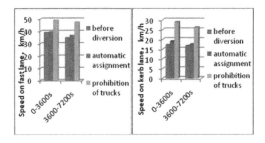

Figure 10. Speed on confluence points from south to north.

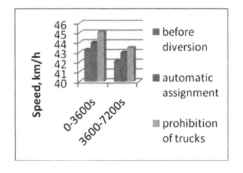

Figure 11. Speed on middle parts of work zone from south to north.

diverted vehicles can refer to the above results of simulation. The main objective is to ensure the queue length at each intersection less than 800 m. The recommended diverting route is G325, namely scheme 4, mainly considering the shorter detour and less expense. During traffic guidance, the latest

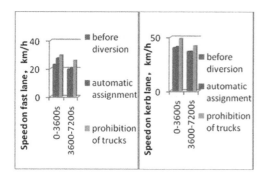

Figure 12. Speed on confluence points from north to south.

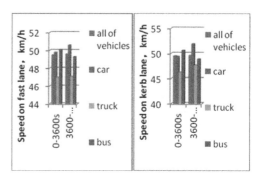

Figure 14. Speeds on Fekai Expressway from South to North without closing lanes.

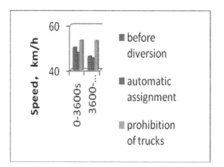

Figure 13. Speed on middle parts of work zone from north to south.

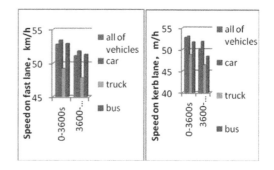

Figure 15. Speeds on Fekai Expressway from North to South without closing lanes.

information of traffic congestion should be monitored and updated through a rapid and efficient way. When traffic congestion happens on G325, secondary diversion of vehicles to other highways depending on specific traffic congestion points should be used to ensure traffic of the whole road network smooth.

REFERENCE

[1] VISSIM5.20 User Manual, PTV traffic technology (Shanghai) limited company (Chinese Copyright). (2009).

Frontiers of Energy and Environmental Engineering – Sung, Kao & Chen (eds)
© 2013 Taylor & Francis Group, London, ISBN 978-0-415-66159-1

The analysis of energy-saving technology and efficiency in operational period of highway

C.Z. Zhu, H.J. Cui & L. Wang
School of Highway, Chang'an University, Xi'an, Shanxi, China
Research Institute of Highway, Ministry of Transport, Beijing, China

ABSTRACT: Through the study of energy consumption indexes of highway in the operation period, based on theoretical or experimental data analyses, the energy-saving effects of three measures of highway were deeply analyzed. The electricity saving rate of the new energy appliance, new energy-saving facilities and energy-saving control technology and measures appliance is about 30%, 40% and 5% respectively.

Keywords: energy-saving technology; energy-saving efficiency; operational period of highway

1 INTRODUCTION

Nowadays, highway promotes the economic development of local area substantially, but meanwhile its huge energy consumption becomes a big problem of highway operation management. As the highway network gradually improved, a large number of facilities along the highway, tunnel, toll station, service areas, play a more and more important role in ensuring efficient and safe operation of highway. Energy-saving measures of highway in operation period mainly consist of the new energy, new energy-saving facilities and energy-saving control technology and measures appliance. These energy-saving measures are applicable to different highway facilities, and their energy-saving effects vary remarkably. Through the study of energy consumption indexes in the operation period, based on theoretical or experimental data analyses, the contribution of different amounts of energy-saving measures were obtained finally. The conclusion is that the electricity saving rate of the new energy appliance, new energy-saving facilities and energy-saving control technology and measures appliance is about 30%, 40% and 5% respectively.

2 ANALYSIS INDEX AND CALCULATION FUNCTION

The energy consumption index E is a synthesized index which takes all the actual energy consumption in highway operation into account, in accord with the national standard of the range of synthesized energy consumption, including energy use in the safe operation of tunnel, toll stations, service area, assistant facilities and raw materials etc. The synthesized energy consumption can be calculated as Equation 1.

$$E = \sum_{i=1}^{n} (e_i \times p_i) \tag{1}$$

where E is synthesized energy consumption; n is the number of the energy kinds consumed in operation enterprise; e_i is the consumption amount of energy i in operation; p_i is the conversion coefficient of energy i. The primary energy consumption in highway operation is electricity, and also includes gasoline, diesel, solar, coal and natural gas etc. Therefore, the analysis of the economic benefits to promote the application of operational energy-saving technology should be based on the synthesized energy saving of various energy-saving technologies.

Electricity is the primary energy consumption within the synthesized energy consumption calculation of highway operating energy-saving technologies. The amount of the electricity saving induced by the implementation of the energy-saving technologies is calculated in Equation 2 in kWh,

$$\Delta A_c = a_q - a_h \tag{2}$$

where ΔA_c is electricity saving in kWh, usually calculated in accordance with years, also calculated according to a defined period as needed; a_q is the actual electricity consumption before the implementation of the energy-saving technologies; a_h is the actual electricity consumption after the implementation of the energy-saving technologies;

Electricity saving rate of the energy-saving technology is:

$$a_c = (a_q - a_h)/a_q \times 100\% \qquad (3)$$

where a_c is the electricity saving rate.

3 THE EFFICIENCY ANALYSIS OF ENERGY-SAVING TECHNOLOGIES

3.1 The energy-saving efficiency analysis of new energy technology appliance

According to the basic data analysis of solar energy application, average electricity consumption of each toll lane is about 3 kW, which contains the electricity for equipment charging and lighting of toll station ceiling. After the appliance of solar as power supply, the annual supply capacity can be deployed to meet the requirement of toll station. According to the conditions of geography, environment, traffic and other factors, if the number of toll stations which have been transformed to be solar-powered, accounts for about 1/3 of the total number of toll stations, that

$$a_q = 3\,kW \times 24h \times 365 \times N$$

$$a_h = 3\,kW \times 24h \times 365 \times \frac{2}{3}N$$

$$a_c = (a_q - a_h)/a_q \times 100\% = 33\%$$

N is the number of toll lanes contained in all the toll stations. For example, the number of toll stations which have been transformed to be solar-powered is 6 and the number of toll lanes contained in the 6 toll stations is 30, after the appliance of solar supply without the consumption of public grid energy, that

$$a_h = 0$$

$$a_q = 3\,kW \times 24\,h \times 365 \times 30 = 788400\,kWh$$

$$\Delta A_c = a_q - a_h = 788400\,kWh$$

The synthesized energy consumption savings according to equivalent standard coal can be calculated as

$$E = 0.1229 \times \Delta A_c = 96894\,kgce$$

Based on the above analysis, the energy-saving efficiency of new energy appliance in highway operations is very impressive. Take constrains of

conditions and adaptability factors into account, the electricity saving rate is about 30% in practical experience.

3.2 The energy-saving efficiency analysis of new energy-saving facilities appliance

According to the basic data analysis of new kind of LED lighting lamp appliance in tunnel, the power use of traditional high pressure sodium lamps is about 115 kW/km per kilometer in tunnel (two-way, medium length). At the same level of security, the power use of LED lighting lamps is about 65 kW/km per kilometer, thence the energy-saving per kilometer in tunnel is

$$a_q = 115\,kW \times 24\,h \times 365 = 1007400\,kWh$$

$$a_h = 65\,kW \times 24\,h \times 365 = 569400\,kWh$$

$$a_c = (a_q - a_h)/a_q \times 100\% = 43\%$$

$$\Delta A_c = a_q - a_h = 438000\,kWh$$

If this new kind of energy-saving facilities are implemented in 5 tunnels of testing section within engineering project, the length of which is about 9.2 km, thence

$$\Delta A_{cc} = 438000 \times 9.2 = 4029600\,kWh$$

The synthesized energy consumption savings according to equivalent standard coal can be calculated as

$$E = 0.1229 \times \Delta A_{cc} = 495238\,kgce$$

Based on the above analysis, the energy-saving rate of new energy-saving facilities appliance in highway operation is about 40%.

3.3 The energy-saving efficiency analysis of energy-saving control technology

According to the basic data analysis of reactive power compensation and harmonic technology appliance in tunnel power supply system, the quality of power supply can be improved significantly after the appliance of this technology; as a result, the power factor can exceed 97% after power compensation, which generally keeps at 90% to 95% (average 92%) of former power supply system. Thence the energy-saving after the appliance of this technology is

$$a_q = P/92\%$$

$$a_h = P / 97\%$$

$$a_c = (a_q - a_h) / a_q \times 100\% = 5\%$$

P is the active power of project.

If reactive power compensation and harmonic technology are applied in 3 tunnels of testing section within engineering project, the length of which is about 6.3 km, and only traditional high pressure sodium lamps are implemented in the tunnels as lighting facilities, thence,

$$a_q = \frac{1007400 \ kWh \times 6.3}{92\%} = 6898500 \ kWh$$

$$a_h = \frac{1007400 \ kWh \times 6.3}{97\%} = 6542907 \ kWh$$

$$\Delta A_c = a_q - a_h = 355593 \ kWh$$

The synthesized energy consumption savings according to equivalent standard coal can be calculated as

$$E = 0.1229 \times \Delta A_{cc} = 43702 \ kgce$$

Based on the above analysis, the energy-saving rate of energy-saving control technology and measures appliance in highway operations is about 5%.

4 THE ENERGY-SAVING EFFICIENCY ANALYSIS OF INTEGRATED TECHNOLOGY APPLIANCE IN HIGHWAY OPERATION

Based on the above analysis, considering the promoting and optimizing effects, the energy-saving efficiency can be deduced as below. New energy appliance is the most effective measure. Yet it may have some constraints in adaptation in certain conditions. According to actual condition of testing section within engineering project, the electricity saving rate may reach 30% as estimate. The energy-saving efficiency of new facilities appliance is quite impressive, the electricity saving rate may reach 40% approximately after applied in tunnel lighting. The energy-saving efficiency of energy-saving control technology and measures appliance is indirect, and the electricity saving rate may reach 5% approximately.

REFERENCES

Dehong Cao&Guowen Li. 2010. Analysis of Energy-Saving Potential of Expressway Tunnel Illuminance. *Highway* 09:248–252.

GB/T 13471–2008. Methods for calculating and evaluating the economic value electricity saving measures (in Chinese).

Jianguo Shen & Huawei Tong & Shan Shi & Shanzeng Jin. 2011. Current Specifications and Trends of Highway Tunnel Lighting: A Discussion of Energy-saving Techniques. *Modern Tunnelling Technology* 48 (04):49–54.

Murat K, Mehmet S, Yunus B, et al. 2004. Determining Optimum Tilt Angles and Orientations of Photovoltaic Panels in Sanlinifa, Turkey. *Renewable Energy:*29: 1265–1275.

Zhongping Kuang. 2012. Research and Application of Integrated Energy-Saving Technologies in Expressway Tunnel Lighting. *Transportation Standardization* 17:117–120.

Frontiers of Energy and Environmental Engineering – Sung, Kao & Chen (eds)
© 2013 Taylor & Francis Group, London, ISBN 978-0-415-66159-1

Discrete element simulation of flow pattern of burden in CDQ shaft

A.Q. Zhang, Y.H. Feng, J. Liu, X.X. Zhang & E. Dianyu
School of Mechanical Engineering, University of Science and Technology Beijing, Beijing, China

P.L. Guo
National Center for Materials Service Safety, University of Science and Technology Beijing, Beijing, China

ABSTRACT: A mathematical model based on the discrete element method is built to describe coke descending behavior in the 1/7-scaled-down experimental setup of an actual 75 t/h cooling shaft. It turns out that cokes flow in the cooling shaft is laminar and no mixing occurs among different coke layers. Due to the resistances of the blast cap and the shaft wall to cokes, the streamline is evolved from "-" to "W" from upper shaft to lower shaft. The change law of time lines is consistent with streamline basically. The velocity distribution of cokes is approximate to "M" shape in contrast to the streamlines and time lines. Compared with potential flow and viscous flow model, the results of DEM are better coincided with the experimental data and can improve the simulation of the behavior of cokes.

Keywords: coke dry quenching; coke descending; discrete element method; streamline; time line

1 INSTRUCTIONS

Cokes play functions of heat generating compound, reducing material and skeleton of stock column in the blast furnace, so it has great influences on coke quality. Coke Dry Quenching (CDQ) technology is currently considered as a technology greatly saving energy in the coke planting, owing to its many advantages such as energy saving, pollution reducing and quality improving. In CDQ, cokes flow from the top to bottom in the CDQ cooling shaft, and descending velocity and distribution of cokes would greatly influence production capacity and quality of quenched coke. Apparently, understanding the fundamentals of coke descending is quite important for improving coke quality and worthwhile to large-scale CDQ system design and operation (Pang, L.H. et al. 2005 & Xu, L. et al. 2009).

Many efforts have been made to develop the models and methods to describe fluid flow, heat transfer and descending behavior of cokes in CDQ shaft. Teplitskil, M.G. et al. (1981) & Starovoit, A.G. et al. (1988) investigated the empirical correlation equations of gas pressure loss and coke quenching time in a CDQ unit experimentally. Grishchenko, A.I. et al. (1984), Sugano, N. et al. (1994) & Matsuhisa, H. et al. (1995) proposed a one-dimensional plug flow model to evaluate the temperature distribution of gas and coke, assuming all solids descended in the same velocity. Yuto, K. et al. (1982) & Kataoka, S. et al. (1990) developed a two-dimensional mathematical model of the flow

and heat transfer in CDQ shaft according to porous medium theory. Liu, H.F. et al. (2003, 2005) investigated the coke cooling process experimentally and proposed a two-dimensional mathematical model based on non-Darcian flow and non-thermal equilibrium methods for fluid flow and heat transfer in the cooling shaft of CDQ unit. Zhang, X.X. et al. (2004) & Feng, Y.H. et al. (2007) studied the local and average heat transfer coefficients of coke bed in a coke dry quenching chamber. Based on the fluid dynamics theory and the premise of continuity, Yu, Q. et al. (2005) established the viscous flow model to describe the descending behavior of cokes. Liu, Z.C. et al. (2006, 2008) studied the coke size distribution in bell-type charging in CDQ shaft experimentally. Xu, M.C. et al. (2006) established a one-dimension heat transfer mathematical model with combustion in the cooling chamber of CDQ. Song, B. et al. (2007) proposed a mathematical model of the pressure drop of coke bed in CDQ shaft. Feng, Y.H. et al. (2007, 2008) studied the law of coke descent for three different blast-cups under five conditions of coke charging.

Above all, these models for fluid flow and heat transfer were mostly assumed to be one- or two-dimensional and the pseudo-fluid models were established to describe the descending behavior of cokes. However, these models based on the continuum assumption were very limited because of particles discreteness in CDQ shaft. It was an attempt to investigate the descending behavior of cokes based on the discrete element method.

From its original development by Cundall, P.A. & Strack, O.D.L. (1979), the Discrete Element Method (DEM) has become a feasible numerical method for analyzing discontinuous media. The technique has already been extensively applied to simulate different granular flows in the industries, including process units such as drum mixers (Kano, J. et al. 2008 & Stewart, R.L. et al. 2001), fluidized beds (Kaneko, Y. et al. 1999 & Maio, F.P.D. et al. 2009) and hopper charging and discharging flows (Chou, C. et al. 2009 & Nguyen, V.D. et al. 2009).

In this paper, the mathematical model based on the discrete element method was proposed to investigate the behaviors of solid particulate motion from particle-scale view. The accuracy and applicability of discrete element method were primarily discussed and validated in details.

2 DISCRETE ELEMENT MODELING

Moving particles in a granular system undergo translational and rotational motions which can be described by Newton's second law of motion. In the DEM, the inter-particle contact model, as illustrated in Figure 1, is composed of spring and dashpot, which correspond to the elastic and plastic nature of particles in the normal direction, respectively. In the tangential direction, the model consists of slider, spring and dashpot. The governing equations for a particle (i) interacting with another particle (j) can be written as (Zhou et al., 2008).

$$m_i \frac{du_i}{dt} = m_i g + \sum_{j=1}^{N} \left(F_{c,ij} + F_{d,ij} \right) \tag{1}$$

$$I_i \frac{d\omega_i}{dt} = \sum_{j=1}^{N} \left(T_{t,ij} + T_{r,ij} \right) \tag{2}$$

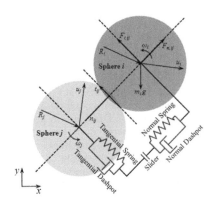

Figure 1. Model of forces acting on particle i from contacting particle j.

where m_i = mass, u_i = translational velocity; I_i = inertia velocity; and w_i = angular velocity of particle i. The forces involved are the gravitational force $m_i\,g$ and inter-particle forces between the particles, which include the contact force $F_{c,ij}$ and damping force $F_{d,ij}$. These inter-particle forces can be resolved into the normal and tangential components at a contact point. The inter-particle forces are summed over the N particles in contact with particle i and depend on the normal and tangential deformation, δ_n and δ_t. The torque acting on particle i includes two components. One arises from tangential force $T_{t,ij}$, and another one is the rolling friction $T_{r,ij}$. All of forces and torques are given in Equation (3–6).

$$F_{c,ij} = F_{cn,ij} + F_{ct,ij} = -k_n \delta_n^{3/2} n - k_t \delta_t \tag{3}$$

$$F_{d,ij} = F_{dn,ij} + F_{dt,ij} = -\eta_n u_{n,ij} - \eta_t u_{t,ij} \tag{4}$$

$$T_{t,ij} = R_i \times \left(F_{ct,ij} + F_{dt,ij} \right) \tag{5}$$

$$T_{r,ij} = -\mu_r \left| F_{cn,ij} \right| \omega_i / \left| \omega_i \right| \tag{6}$$

where k = stiffness; and η = damping coefficient.

$$k_n = \frac{4}{3} \frac{\sqrt{R_i R_j / R_i + R_j}}{\left(1 - v_i^2 / E_i \right) + \left(1 - v_j^2 / E_j \right)} \tag{7}$$

$$k_t = 8 \sqrt{\frac{G_i G_j}{G_i + G_j}} \sqrt{\frac{R_i R_j}{R_i + R_j}} \delta_{n,ij} \tag{8}$$

$$\eta_n = 2\sqrt{\frac{5}{6}} \frac{\ln e}{\sqrt{\ln^2 e + \pi^2}} \delta_{n,ij} \frac{m_i m_j}{m_i + m_j} u_{n,ij}$$
$$\sqrt{2 \left(\frac{1-v_i^2}{E_i} + \frac{1-v_j^2}{E_j} \right) \sqrt{\frac{R_i R_j}{R_i + R_j}}} \tag{9}$$

$$\eta_t = 2\sqrt{\frac{5}{6}} \frac{\ln e}{\sqrt{\ln^2 e + \pi^2}} \delta_{n,ij} \frac{m_i m_j}{m_i + m_j} u_{t,ij}$$
$$\sqrt{2 \left(\frac{1-v_i^2}{E_i} + \frac{1-v_j^2}{E_j} \right) \sqrt{\frac{R_i R_j}{R_i + R_j}}} \tag{10}$$

$$F_{ct,ij} + F_{dt,ij} \le \mu_s F_{cn,ij} \tag{11}$$

$$n = \delta_{n,ij} / \left| \delta_{n,ij} \right| \tag{12}$$

$$G = E / (2 + 2v) \tag{13}$$

where E = Young modulus; v = Poisson ratio of the particles; R = radius modulus of each particles; G = shear modulus of each particles; e = coefficient of restitution, μ_s = sliding friction; and μ_r = rolling friction (including inter-particles and particles and wall).

In DEM, the motion of each particle is tracked, and interaction with other particles or boundaries is considered in the simulation. The hardness of the particles and the dashpot are related to the Young's modulus and coefficient of restitution, respectively. The friction between entities is defined with a Coulomb-type of friction law, limited below by the maximum static friction (Eq. (11)) and implemented with a friction factor based on the physical properties. The particle forces and torques are calculated which leads to new particle movement (Eqs. (1) and (2)) and the particles' new positions after each time step are determined by an explicit method.

3 SYSTEM STUDIED

In order to verify the model, the geometry of the CDQ model used in this work is the one-seventh scale model of an actual 75 t/h cooling shaft, as shown in Figure 2. All models were solved numerically using the EDEM (DEM Solutions Limited) software. The materials and conditions used in the simulations are presented in Table 1.

Since the effects of gas flowing through the shaft on coke descending are neglectable, gas was not introduced into the apparatus in the simulation. The simulation is started with the random generation of certain number of uniform spheres without overlaps, followed by a gravitational settling process. Then the particles are discharged at a preset rate from the bottom outlet and are no longer charged from the top inlet in the process of calculation.

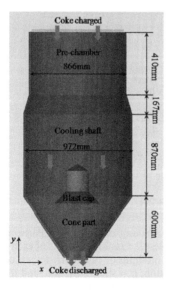

Figure 2. Model of particle movement in CDQ shaft.

Table 1. Materials and conditions used in simulation.

Parameters	Value
Number of particles, N	90000
Particle diameter (mm), d_p	19.4
Particle density ($kg \cdot m^{-3}$), ρ_p	1073
Particle poisson ratio, v_p	0.25
Particle young modulus ($N \cdot m^{-2}$), E_p	1.00E + 07
Particle restitution coeff., e_p	0.20
Inter-particle sliding friction coeff., μ_{sp}	0.50
Inter-particle rolling friction coeff., μ_{rp}	0.05
Wall poisson ratio, v_w	0.30
Wall young modulus ($N \cdot m^{-2}$), E_w	1.00E + 10
Particle-wall restitution coeff., e_w	0.30
Particle-wall rolling friction coeff., μ_{rw}	0.05
Time step (s)	10^{-4}

4 NUMERICAL RESULTS AND DISCUSSION

4.1 Streamline

Some coke particles at the same level were traced in simulation. The streamline of cokes was shown in Figure 3. The simulated results display that coke flow in the cooling shaft is laminar and no mixing occurs among different coke layers. At the upper shaft, bulk cokes descend almost uniformly and streamline is smooth since gravity contributes more than the wall friction. At the lower shaft, with the increasing of the normal wall pressure, the wall friction increases and even surpasses the gravity, so the coke flow near the wall turns slow gradually. Subsequently, the lag spreads to cokes far from the wall via the internal friction between coke particles. The streamline is evolved from "–" to "W" and the upper streamline is much smoother than the lowers due to the resistances of the blast cap and the shaft wall to cokes, which also accounts for that. Obviously, both the inter-particle friction and particle-wall friction play important roles in the coke descending behavior.

4.2 Time lines

In Figure 4, the time lines of cokes are illustrated and compared with experimental data. It can be seen that the results of DEM are in good agreement with the experimental data (Feng, Y.H. et al. 2008) and the change law of time lines is consistent with streamline (in Fig. 3) basically.

Aiming to study the accuracy of DEM, the numerical results are compared with the potential flow model (Liu, H.F. et al. 2005) and the viscous flow model (Yu, Q. et al. 2005), as shown in Figure 5. Both of the DEM and the viscous flow model con-

Figure 3. Particles burden layers motion in CDQ shaft.

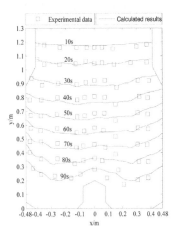

Figure 4. Comparison between simulated timeline and experimental data in CDQ shaft.

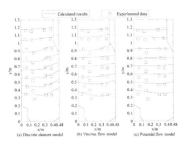

Figure 5. Comparison of simulated results by the DEM, viscous flow, and potential flow models.

sidering the effects of the normal wall stress could well describe the coke flow in a CDQ cooling shaft while the potential flow model couldn't. But compared with the viscous flow model, the results of the DEM fit better with the experimental data and can improve the simulation of the behavior of cokes.

4.3 Axial velocity

In Figure 6, the dimensionless average axial velocity $U^*(u/u_0)$ is plotted versus the dimensionless

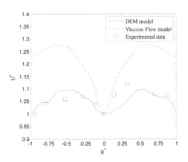

Figure 6. Dimensionless average velocity of Particles moving in CDQ shaft.

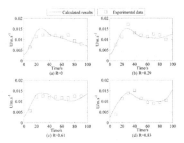

Figure 7. The comparison of tracer particles vertical velocity between simulation and experiment.

$R^*(r/R_0)$. It can be seen that the velocity distribution of cokes is approximate to "M" shape in contrast to the streamlines and time lines (in Fig. 4). The measured value basically agrees well with calculated value, while the viscous flow model has a larger error.

We also got the axial velocity evolutions of different particles in CDQ shaft, as shown in Figure 6. At the very beginning, the axial velocity of all particles is small because it is still in a startup. The axial velocity increases along with the time and reaches the maximum at 30s because of the sudden expansion of furnace. And then it decreases slowly and stabilizes in the later phase. It can be seen that the simulated velocity basically agrees with the measurements data. This validates the proposed DEM method.

5 CONCLUSION

The mathematical model based on the discrete element method was proposed to investigate the behaviors of solid particulate motion from particle-scale view. It turned out that the burden shape, time line and velocity of coke were in good accordance with experimental results.

The whole descending behavior of cokes was uniform and the burden shape near the blast cup like "W" due to the wall friction and axial forces of blast cup. Both the inter-particle friction and particle-wall friction play important roles in the coke descending behavior.

The axial velocity increased along with the time and reached the maximum at 30 s because of the sudden expansion of furnace. And then it decreased slowly and stabilized in the later phase.

The dimensionless average axial velocity distribution of cokes was approximate to "M" shape in contrast to the streamlines and time lines.

REFERENCES

Chou, C., Lee, A. & Yeh, C. 2009. Placement of a non-isosceles-triangle insert in an asymmetrical two-dimensional bin-hopper. Advanced Powder Technology 20, 80–88.

Cundall, P.A. & Strack, O.D.L. 1979. A discrete numerical model for granular assemblies. Geotechnique 29(1): 47–65.

Feng, Y.H., Zhang, X.X. & Wu, M.L. 2007. Experimental Study of Coke Descent in CDQ Cooling Chamber. Iron & Steel 42(07): 15–17.

Feng, Y.H., Zhang, X.X., Liu, Z.C. & Wu, M.L. 2007. Local and average heat transfer coefficients of coke bed in a coke dry quenching chamber. Journal of University of Science and Technology Beijing 29(12): 1268–1272.

Feng, Y.H., Zhang, X.X., Yu, Q., Shi, Z.Y., Liu, Z.C., Zhang, H. & Liu, H.F. 2008. Experimental and numerical investigations of coke descending behavior in a coke dry quenching cooling shaft. Applied Thermal Engineering 28(11–12): 1485–1490.

Grishchenko, A.I., Ereskovskii, O.S. & Kukhar, N.P. 1984. Investigating the thermal systems of a chamber belonging to a dry coke-quenching unit. Coke & Chemistry (USSR) 8: 23–27.

Kaneko, Y., Shiojima, T. & Horio, M. 1999. DEM simulation of fluidized beds for gas-phase olefin polymerization. Chemical Engineering Science 54, 5809–5821.

Kano, J., Kasai, E., Saito, F. & Kawaguch, T. 2008. Numerical simulation model for granulation kinetics of iron ores. In: Ariyama, T. (Ed.), Recent Progresson Mathematical Modelingin Ironmaking 2008, 8. The Iron and Steel Institute of Japan, Tokyoin Japan.

Kataoka, S., Otsuka, J., Yasukouchi, N. & Katahira, H. 1990. Establishment of coke dry quenching with a coke throughput of 200t/h. Proc. 6th Int. Iron & Steel Congress (Nagoya ISIJ): 337–344.

Liu, H.F., Zhang, X.X. & Wu, M.L. 2003. Investigation on the Transient Fluid Flow and Heat Transfer in Coke Packed Bed. Journal of Engineering Thermophysics 24(6): 1022–1024.

Liu, H.F., Zhang, X.X., Feng, Y.H., Zhang, H., Wu, M.L., Xu, L. & Zheng, W.H. 2005. Potential Flow Model of Coke Descending Movement and Experimental Verification. Fuel & Chemical Processes 36(1): 22–24.

Liu, Z.C, Feng, Y.H., Zhang, X.X., Shi, Z.Y., Xu, M.C., Xu, L. & Yu, Z.D. 2006. Study on coke size distribution in bell-type charging in CDQ shaft. Journal of Thermal Science and Technology 5(3): 251–256.

Liu, Z.C., Feng, Y.H., Zhang, X.X., Xu, L. & Yu, Z.D. 2008. Numerical and experimental study on coke size distribution in bell-type charging in the CDQ shaft. Journal of University of Science and Technology Beijing 15(03): 236–240.

Maio, F.P.D., Renzo, A.D. & Trevisan, D. 2009. Comparison of heat transfer models in DEM-CFD simulations of fluidized beds with an immersed probe. Powder Technology 193, 257–265.

Matsuhisa, H., Youjin, W., Toyoda, N., Matsubara, A., Honda, Y. & Sato, S. 1995. Study on efficiency improvement of CDQ waste heat recovery system (2nd report, operating method for obtaining maximum waste heat recovery). Trans. of the Japan Society of Mechanical Engineers, Part C, 61(587): 2916–2921.

Nguyen, V.D., Cogne, C., Guessasma, M., Bellenger, E. & Fortin, J. 2009. Discrete modeling of granular flow with thermal transfer: application to the discharge of silos. Applied Thermal Engineering 29, 1846–1853.

Pang, L.H. & Wei, S.B. 2005. Coke Dry Quenching Technology. Beijing: Metallurgical Industry Press.

Song, B., Feng, Y.H., Zhang, X.X., Xu, L. & Yu, Z.D. 2007. Numerical Study and Experimental Verification of the Pressure Drop of Coke Bed in CDQ Shaft. Industrial Furnace 29(06): 1–4.

Starovoit, A.G., Goncharov, V.F. & Anisimov, V.A. 1988. Coke Quenching Time in a Coke Dry Quenching unit. Coke & chemistry (11): 23–26.

Stewart, R.L., Bridgwater, J., Zhou, Y.C. & Yu, A.B. 2001. Simulated and measured flow of granules in a bladed mixer—a detailed comparison. Chemical Engineering Science 56, 5457–5474.

Sugano, N., Tanigaki, T., Kawai, S., Itou, M., Harakawa, T., Tahara, T. & Katahira, H. 1994. Study on the efficiency improvement of cdq waste heat recovery system (1st report, dynamic characteristics of CDQ cooling chamber). Trans. of the Japan Society of Mechanical Engineers, Part C, 60(577): 3076–3080.

Teplitskil, M.G., Gordon, I.Z. & Kudryavaya, N.A. 1981. Coke Dry Quenching. Beijing: Metallurgical Industry Press.

Xu, L., Zhang, Q.Q., Dong, X.H. & Shao, F. 2009. Present Situation Technical Analysis on Coke Dry Quenching in China. 2009 CSM Annual Meeting Proceedings: 92–95.

Xu, M.C., Feng, Y.H., Zhang, X.X., Xu, L. & Yu, Z.D. 2006. One-dimension heat transfer simulation with combustion in the cooling chamber of CDQ. Energy for Metallurgical Industry 25(04): 16–19.

Yu, Q., Zhang, X.X., Feng, Y.H., Jiang, Z.Y., Xu, L., Zheng, W.H. & Dong, X.H. 2005. Viscous Flow Model of Coke Descending Behavior and Comparison Study in CDQ Cooling Chamber. Industrial Heating 34(01): 11–13.

Yuto, K., Nishihara, N., Kimura, M., Ishida, Y., Terai, Y., Kobayashi, M. & Yamamoto, H. 1982. Application of techniques for packed bed analysis to studies on the construction of large coke dry quenching plants. Nippon Steel Technical Report 20: 95–104.

Zhang, X.X., Si, J.L., Wang, D.N., Liu, H.F., Wu, M.L., Xu, L. & Zheng, W.H. 2004. Study on Local Heat Exchanger Factor and Inverse Problem of Heat Conduction in Coke. Fuel & Chemical Processes 35(06): 8–11.

Frontiers of Energy and Environmental Engineering – Sung, Kao & Chen (eds)
© 2013 Taylor & Francis Group, London, ISBN 978-0-415-66159-1

The analysis of concrete intensity based on waste scallop shell aggregate

J.H. Sui & J.D. Yang
College of Navigation and Shipbuilding engineering, Dalian Ocean University, Liaoning Province, China

ABSTRACT: A simple and practical solution was put forward to the waste scallop shell mountain from Dalian ZhangZiDao fishery Co., Ltd. and also a new method was obtained to make concrete. According to the national standards, the concrete briquette was made by mixing the waste scallop shell, sand and cement in proportion. The intensity test result was shown that the concrete intensity completely meet national standard, which was made in accordance with the appropriate mix their proportion, the feasibility of waste scallop shell aggregate concrete be proved, and the new better way is found to deal with the waste scallop shell mountain.

1 GENERAL INSTRUCTIONS

The scallop is one of the backbone products in North fishery. Since Patinopecten yessoensis taste as good, nutritious, high economic value characteristics, its bottom sowing culture very rapidly in recent years. Continue to increase production of Patinopecten yessoensis, but the rapid development of scallop breeding, while the number of shells waste after processing has also increased year by year, with natural shellfish shells, shell will have a lot of abandoned each year. The abandoned shells as garbage can not be timely and effective treatment, piles up the mountain, not only occupy valuable land, while also seriously pollute the marine environment around them. They have become a public nuisance.

The main ingredient of shells is calcium carbonate (limestone), accounting for 95% of the total ingredients, other ingredients are magnesium, iron, calcium phosphate, calcium sulfate, and silicate and other inorganic substances, as well as a small amount of organic matter such as shell elements. Its shell is qualitative hard, not easy processing, reuse high treatment cost, energy consumption, and a peculiar smell. Often to be thrown as waste, thus not only wasted resources, and pollution to the environment also more and more serious. Abandoned scallop shell free-pollution disposal is always a problem. At present, the domestic and international the treatment measures also has a lot of, such as microwave high-temperature thermal decomposition of Shell produce the calcium gluconate [1], Activity of calcium oxide, taken from the scallop shell, works well for antimicrobial, making disinfection liquid production

preservatives [2], feed additives, soil amendments [3], green decoration materials and so on. Such treatment methods although effective, but the process requires complicated and need to consume energy, the higher the cost of processing, with less shell and can not quickly and effectively solve the problem of abandoned scallop Shell Mountain.

This paper presents a simple and practical solution is to abandon the scallop shell as aggregate to produce a new type of concrete. There are a lot of methods of making concrete, conventional concrete is generally by sand, cement and gravel mixture. In concrete gravel more than 70% of the total, sand and gravel excavation can directly affect the sustainable development of concrete, some people also use abandoned clay brick as recycled aggregate made of concrete, but the intensity and durability has been questioned. [4] Concrete intensity is its main index, mainly by the slurry intensity, cement and aggregate interface bonding intensity, aggregates particle intensity decision. During the 1950s China for 15 MPa designs intensity of concrete, in the 1970s, for 20 MPa average intensity after 80 times rose to 25 MPa above average intensity. [5] So the intensity of concrete test and analysis is very important segment. A test method was built to test and analysis concrete intensity containing abandoned scallop shell in this paper.

2 EXPERIMENTAL

2.1 Materials and equipment

Materials: abandoned scallop shell: the abandoned scallop shell hill of Dalian Zhangzidao Fishery Co., Ltd;

Cement: Tangshan Jidong Cement production of "The Shield"PO32.5 Cement;
Sand: A construction site in Dalian city.
Equipment: Wuxi building materials equipment factory production NYL—200 types of 200 tons of pressure test machine: Meter stick; Shelf; Crusher; Adhesive etc.

2.2 *The method of making concrete block*

Taking the crushed of abandoned shells rough sorting according to size of screen diameter of 7 mm, 15 mm, 40 mm three levels, and were placed respectively. According to cement, sand and shells with different proportion respectively into six test-components Z1–Z6, the specific practices are as follows:

1. Mark Numbers for Z_1: Taking the size of 40 mm shells in accordance with the 2:1:1 ratio (cement: sand: Shell) complexes, even after mixing with water made of 150 mm × 150 mm × 150 mm size of the test block.
2. Mark Numbers for Z_2: Taking the size of 40 mm shells in accordance with the 1:1:1.5 ratio (cement:sand:Shell) complexes, even after mixing with water made of 150 mm × 150 mm × 150 mm size of the test block.
3. Mark Numbers for Z_3: Taking the size of 15 mm shells in accordance with the 2:1:1 ratio (cement: sand: Shell) complexes, even after mixing with water made of 150 mm × 150 mm × 150 mm size of the test block.
4. Mark Numbers for Z_4: Taking the size of 15 mm shells in accordance with the 1:2:3 ratio (cement: sand: Shell) complexes, even after mixing with water made of 150 mm × 150 mm × 150 mm size of the test block.
5. Mark Numbers for Z_5: Taking the size of 7 mm shells in accordance with the 2:1:1 ratio (cement: sand: Shell) complexes, even after mixing with water made of 150 mm × 150 mm × 150 mm size of the test block.
6. Mark Numbers for Z_6: Taking the size of 7 mm shells in accordance with the 1:2:3 ratio (cement: sand: Shell) complexes, even after mixing with water made of 150 mm × 150 mm × 150 mm size of the test block.

According to the national standard making good test block placed in temperature is 20 ± 2°C, relative humidity for 95% of the environment maintenance, maintenance time for 11 days.

2.3 *Anticipated intensity calculation on test cube*

Intensity refers to the ability of material resists the external force damage. Stress will be generated when the materials withstand external force. With the increasing external force, the internal stress also

increases. The material damage will be occurred until the force between particles can't withstand any longer. At this time, the ultimate stress value is the intensity of the material [6].

Formula for calculating the intensity of the material is as follows:

$$f = \frac{P}{F} \qquad (1)$$

where f is the material intensity, P is the maximum load when destruction occurred, F for stress section area.

3 INTENSITY TEST RESULTS AND DISCUSSION

3.1 *Intensity test results*

Taking the maintenance concrete test blocks carry on the intensity test. It should measure the effective pressure area of concrete test block on test machine. Make a record as shown in Table 1 Before testing. The third column contains the diameter of abandoned scallop shell; the fourth column is the measurement of the area of concrete block. The test block is taken flat on the platform of the NYL-200 Type 200 tons pressure testing machine to carry on the intensity test. Clearly read data from the dial.

3.2 *Analysis of testing intensity data*

According to the intensity data in table 1, we take the intensity analysis, from horizontal to vertical two aspects, to the concrete mixed with abandoned scallop shell. Mainly includes the following: Under precondition of same proportion of the abandoned scallop shell, relationships between the testing intensity of concrete blocks and the diameter of abandoned scallop shell; Under precondition of same size of shell, relationships between the testing intensity of concrete blocks and the proportion of the abandoned scallop shell.

3.2.1 *Relationships between the intensity and the diameter with the same proportion of shell*

Select two kinds of schemes with abandoned scallop shell proportion in 25% and 50%, by comparative analysis, results as shown in graph 1 below. From the graph we can clearly see the scheme, same 25% proportion of shell, the testing intensity results of concrete blocks shows the downtrend with the increase in the number of the shells. While, in the scheme of 50% proportion, with the increase in the number of the shells, the testing intensity results of concrete blocks shows going from weakness to intensity and then weakness.

Table 1. Different test blocks intensity test data.

TB	c: sa: sh	Dia.	Compression area	Damaging loading	Test intensity	Shell ratio
Z_1						
1	2:1:1	40	0.0214	532	24.86	25%
2	2:1:1	40	0.02025	448	22.12	
3	2:1:1	40	0.019125	537	28.09	
Avg.		40	0.020258	505.7	25.023	
Z_2						
1	1:1:1.5	40	0.01175	260	22.13	33%
2	1:1:1.5	40	0.01575	321	20.38	
3	1:1:1.5	40	0.0135	286	21.19	
Avg.		40	0.01377	289	21.23	
Z_3						
1	1:2:3	15	0.0085	290.7	27.98	50%
2	1:2:3	15	0.0090	304.9	27.7	
3	1:2:3	15	0.0082	306.4	30.6	
Avg.		15	0.00857	300.6	27.8	
Z_4						
1	2:1:1	15	0.0096	295.4	25.7	25%
2	2:1:1	15	0.0088	298.6	27.8	
3	2:1:1	15	0.0094	330	28.7	
Avg.		15	0.0093	308	27.1	
Z_5						
1	1:2:3	7	0.0090	183.9	16.7	50%
2	1:2:3	7	0.0078	168.1	17.6	
3	1:2:3	7	0.0084	216.9	21.1	
Avg.		7	0.0084	189.7	18.5	
Z_6						
1	2:1:1	7	0.0085	297	28.6	25%
2	2:1:1	7	0.0082	246.8	30.09	
3	2:1:1	7	0.0099	245.5	24.8	
Avg.		7	0.00887	245.2	27.64	

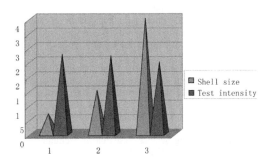

Figure 1. The concrete cube strength contrast diagram of shell proportion 25%.

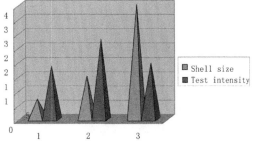

Figure 2. The concrete cube strength contrast diagram of shell proportion 50%.

3.2.2 Relationships between the intensity and the proportion with the same size

Selecting the diameter of abandoned scallop shell as 40 mm, 15 mm and 7 mm is being the different scheme to take comparative analysis, as shown in graph 3, 4, 5. From the graph we can clearly see, in

the scheme with diameter 40 mm, with the increase of the proportion of the shell, the testing intensity results of concrete blocks showing the downtrend. In the scheme with diameter 15 mm, with the increase of the proportion of the shell, the testing intensity results of concrete blocks shows distinctly growing.

In the scheme with diameter 7 mm, with the increase of the proportion of the shell, the testing intensity results of concrete blocks shows relative downtrend.

3.3 *Trend analysis of testing intensity*

According to the data in table 1 and 3.2 section's analysis, draw the relationship trend between the testing intensity of concrete blocks and the

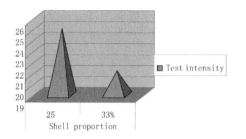

Figure 3. The concrete cube strength contrast diagram of 40 mm diameter Shell.

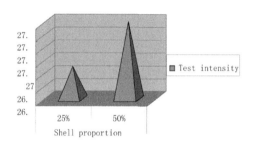

Figure 4. The concrete cube strength contrast diagram of 15 mm diameter Shell.

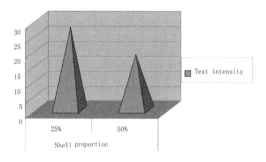

Figure 5. The concrete cube strength contrast diagram of 7 mm diameter Shell.

diameter of containing abandoned scallop shell, as figure 6 shows. Results show that the intensity of the concrete blocks, with the diameter of scallop shell increases, changes following curves, but 15~40 millimeter for the best diameter; The relationship trend between the testing intensity of concrete blocks and the containing proportion of the shell shows as figure 7. Results show that the intensity of the concrete blocks, with the containing proportion of scallop shell increases, changes following curves, But 30%~39% for the best proportion.

4 CONCLUSION

A new method of making concrete was put forward by smashing scallop shell as concrete aggregate in order to deal with the abandoned scallop shell mountain in ZhangZi island of dalian fishery group Co., Ltd in this paper. The concrete intensity was figured out by the special test. The conclusion was shown: The scheme of mixing concrete buildings aggregate with a suitable size and proportion of the scallop shell fragments is feasible. It not only effectively resolves the coastal scallop shell mountain problem, but also provides a new scheme for the sustainable development of concrete.

REFERENCES

[1] ZHAO Guifeng, YANG Lin, CHEN Lei and WU Lin, "Production of calcium gluconate from shell by microwave high-temperature refining", Journal of Dalian Institute of Light Industry, 2009, vol. 28, no. 6, pp:438–440.
[2] WEI Wei, HE Yma- cai "Peparation of Activie CaO from-Seashell and -its Bacterium Inhibition Experiment", Biotechnology, 2005, vol. 15, no. 5, pp: 77–78.
[3] Wang yunchun, Xu li, "the Green Decorate Material—Shell · Rice Hull Paint", Journal of Construction industry information, 2001, vol.6, pp: 12–16.
[4] Yan Handong, Chen Xiufeng, "Study on effects of waste clay brick recycled aggregate on concrete properties" Building Science Research of Sichuan, 2009, vol. 35, no. 5, pp: 179–182.
[5] Su Dagen, "Civil Engineering Materials", Advanced Education Press, 2003. 8.
[6] LIU Jian, "Types of concrete strength and comparison of detection methods", Shanxi Architecture, 2012, vol. 38, no. 10, pp: 139–140.

Frontiers of Energy and Environmental Engineering – Sung, Kao & Chen (eds)
© *2013 Taylor & Francis Group, London, ISBN 978-0-415-66159-1*

The FRP chimney design and construction technology for coal-fired power plant FGD system

D.H. Zhang
Wuhan University of Technology, China
The Central Research Institute of Building and Construction Co. Ltd., MCC Group, China

J.H. Wang
Wuhan University of Technology, China

ABSTRACT: Based on the coal-fired power plant desulphurization chimney of the serious corrosion environment, on the introduction of international FRP chimney based on the application, analysis of China's current domestic FRP chimney application, design, construction technology, presents FRP chimney design, processing and installation of technical matters needing attention.

Keywords: FRP chimney; coal fire power plant; FGD system; design; construction

1 INTRODUCTION

With the coal-fired power plant flue gas desulfurization project progress, a large number of China's domestic new and modified desulphurization chimney emerging severe corrosion, leakage accident, bring huge economic loss to the power plant, but also a serious threat to the safe operation of power plants. It turns out not to be the coal-fired power plants, electric power design institute favors FRP chimney, gradually incorporated into the relevant units in the field.

It is based on glass fiber reinforced plastic materials used as desulphurization chimney has a high performance price ratio, the America Department of Energy in May 1999, take the FRP chimney as the America's desulphurization chimney only recommend material [2]. The IEA clean coal centre also published one book "*SOx Emission and Control*" in 2006, recommend to use FRP chimney [3]. And power company, industry organizations and government for FRP chimney recognition should be relative, American Society for Testing and Materials (ASTM) and International Committee for Industrial Chimneys (CICIND) enacted the FRP chimney design standard [4] [5].

In China, the Central Research Institute of Building & Construction of MCC in the early 1970's, has built several blocks over 100 m high FRP chimney for discharging strong corrosive gas for metallurgy factory, effect of application is well [6].

However, because the FRP material belonging to the anisotropic material, and used to power construction professionals involved in concrete, steel

Table 1. Cost comparison of absorber construction materials. (Cost obtained from a major power engineering firm in 2003) [1].

Construction material	Installed cost ($)	Cost ratio
Epoxy vinyl ester resin-coated carbon steel	1735000	0.70
FRP made with epoxy vinyl ester resin	2475000	1.00
Acid brick or rubber-lined carbon steel	3025000	1.22
C-276 clad carbon steel	4500000	1.82

Table 2. Epoxy vinyl ester resin chemical resistance compared to metal. [1].

Materials	Sulphuric acid	Hydrochloric acid	Acid chloride salts
317 L stainless steel	25°C < 5%	NR	NR
FRP made with bis A epoxy vinyl ester resin	100°C to 30%	80°C to 15%	100°C all conc.
Alloy C-276	100°C to 30%	80°C to 15%	65°C to 20 M ppm at low pH

and other materials are entirely different in nature, and related design, decision making units failed to account for this difference, as well as China's FRP industry processing equipment, technology and foreign products of the same kind of difference, blindly copying steel design, it may cause new project quality accident.

The purpose of this paper, introduction to FRP materials and steel performance differences between, for the design, decision making units desulfurization chimney design selection, decision with respect to the reference.

2 THE BASIC PROPERTIES OF STEEL AND THE STEEL CHIMNEY DESIGN

The basic properties of steel for isotropic, i.e. steel in each direction, with its basic performance are consistent, will not change the direction and the differences. Therefore, in the design of steel structure, only need to consider the basic properties of steel can, without the need for different stress direction of the stress, strain and so on, are analyzed, design calculation.

For example, in desulphurization chimney design, for the domestic large number of brick chimney anticorrosion of inner cylinder desulfurization transformation, more reliable corrosion prevention methods: adding new inner steel cylinder, and then the inner steel cylinder corrosion can be.

The new chimney, at the beginning of the design, use the sleeve type inner steel cylinder scheme, then the inner steel cylinder for anticorrosion construction. If the owners wish to reduce the amount of steel, the self-supporting steel inner cylinder can be changed to suspensory inner steel cylinder, can significantly reduce the steel consumption, the load transfer to the low cost of the concrete shell wall.

At present, China has a large number of new sleeve type inner steel cylinder has adopted this design philosophy.

3 FRP MATERIAL PROPERTIES AND DESIGN MATTERS NEEDING ATTENTION

FRP means glass fiber reinforced plastics, it consisting of two main materials: glass fiber, VE resin.

Wherein, the glass fiber has very good tensile strength, good corrosion resistance, good processing performance; while the resin part according to material selection can be different, having an excellent corrosion resistance, good mechanical properties; the glass fiber, resin was a regular design, a combination of processing, can be the

FRP material having excellent design performance, processing performance, corrosion resistance and heat resistance, so as to meet the needs of corrosion protection for desulphurization chimney.

It is because the FRP's composition characteristics, also causes the FRP belonging to the anisotropic material: along the direction of glass fiber, FRP having excellent tensile strength; and in the direction perpendicular to the glass fiber, FRP reinforced low strength, only the resin provide tensile strength, generally below 80 MPa; if the FRP shear effect, FRP bearing capacity is lower, only resin cohesive strength, is in commonly 10~13 MPa. FRP member is connected between the glass fiber and resin, mainly used for bonding, rather than as steel's free welding operation; if the FRP piece between the bonding operations is not up to the force transmitted smoothly, will be in the bonding position caused by shearing action, leading to joint bearing capacity decreased dramatically.

Therefore, FRP chimney design work, must pay attention to the following matters:

a. FRP chimney must by have the FRP material design and technical knowledge of professionals, and the power of civil engineering professional and technical personnel are matched, common design;

b. Based on the part of China generating units also exist bypass operation reality, the resin must have sufficient heat resistance, whereby the resin heat resistance is not less than the bypass gas temperature; reinforcing material must be made of corrosion resistant performance and excellent mechanical properties of ECR glass fiber;

c. Must give full consideration to the FRP chimney of intensity, rigidity, heat stress release, electrostatic discharge, creep problem;

d. Consideration must be given to the FRP chimney process repeatability, installation feasibility.

4 CHINA'S FRP CHIMNEY PROCESSING LEVEL

China's domestic and other emerging industries development law, FRP technology as a new type of materials and processing technology, in the domestic development may also experience by low to tall development process. At present, most of the domestic FRP processing factory is still in the preliminary stages of mechanical processing, manufacturing and bid quote is still in low price competition stage, mainly in personnel quality is poorer, equipment is pallet. In product quality, mainly in the processing size deviation, poor dimensional stability.

Figure 1. Reinforcing internal voids reduced by enhancing effect.

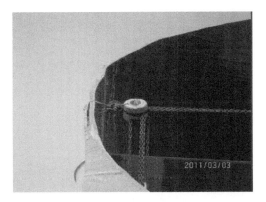

Figure 2. Mold size deviation resulted in FRP chimney can end corrugation.

The following pictures come from a domestic cooling tower and chimney unification project FRP duct processing site:

For the FRP chimney section, if the suspended installation, apparently, in the segment joints of 2 section FRP chimney, stress cannot be smooth transmission, will produce shear force at the connecting position. This makes the suspension type FRP joints become weak point, influence of suspension type FRP chimney of safe use.

And if use the free-standing installation, these parts of the role is to FRP chimney section of the fixing and positioning function, and does not generate shear force, so it is safe. The German FRP chimney construction case in 2001 [7], is still in use for self-supporting FRP chimney.

On the other hand, hanging FRP chimney while in a certain extent can reduce resin, glass fiber use, reduce the cost of raw materials, however, hanging FRP chimney need to increase a lot of installation parts of the special location, the force transmission device, can also add some cost. The above 2 aspects of balance of increase and decrease, not on the cost to produce obvious difference.

5 SUMMARY AND RECOMMENDATIONS

a. From the above on steel, FRP, the basic properties of domestic state of development and foreign related cases, can be clearly seen, for the design of FRP chimney, should not simply copy the steel chimney design ideas, we cannot simply copy foreign case information, but should be based on the domestic FRP processing technology development actual level, the reasonable structure design;

b. From the domestic FRP chimney of the actual situation, in addition to the Central Research Institute of Building & Construction of MCC in this respect detailed research and engineering practice, other units on the FRP chimney design, research, is also limited to the theory discussion stage, there is no practical experience for reference of FRP material in suspension; type structure with respect to the specific problems, is the lack of basic technology research, suggest not rashly use, otherwise the resulting negative effect, will further hindered the domestic FRP chimney of development.

c. Suspension FRP chimney, is the development trend of the future, we must through the test, through smaller project inspection, gained enough FRP hanging type mounting technology experience, and with a free-standing FRP chimney overall cost are compared, have construction cost advantage after, can carry out stage by stage.

d. We must speed up the fiber winding processing equipment renewal pace, rely on existing simple winding machine, it is not possible to obtain a high quality, reliable FRP chimney engineering.

AUTHOR

Zhang Dahou, Professor level senior engineer in Anticorrosive Materials of the Central Research Institute of Building & Construction Corp. Ltd, MCC Group. He is a Executive Councilor of China composites Industry Association, Executive Councilor of China Industry Anticorrosion Technology Association (CIATA), Vice-director of The Expert Committee of CIATA, and the Chief Editor of CECS 'Technical Specification for Steel Structures with Painting Anticorrosion'.

He is actively involved in promoting the use of FRP for corrosion resistant applications in China. He has design & construction the first FRP Chimney in China.

REFERENCES

[1] Don Kelley. The use of FRP in FGD applications. Reinforced Plastics [J]. 2007.01.

[2] U.S. Department of Energy, Office of Fossil Energy. Market-based Advanced Coal Power Systems. Final Report. May 1999.

[3] IEA Clean Coal Centre. SOx Emissions and Control. 2006.

[4] ASTM D 5364-2002 Standard Guide for Design, Fabrication, and Erection of Fiberglass Reinforced Plastic Chimney Liners with Coal-Fired Units.

[5] CICIND. Model Code for FRP Liners in Chimneys.

[6] Zhang Dahou. The Wet Chimney Inner Wall Anti-corrosion Material Properties Analysis. Engineering Journal of Wuhan University [J]. 2005.10.

[7] Reinhard Lux. Beheizbare GFK-Schornsteinröhre Ø4800 mm zur Lösung der Aerosol-Emissionsproblematik im Kraftwerk Simmering. Erfahrungen mit GFK im Rohrleitungs- und Anlagenbau 3. Tagung. Oktober 2001 in München.

Frontiers of Energy and Environmental Engineering – Sung, Kao & Chen (eds)
© 2013 Taylor & Francis Group, London, ISBN 978-0-415-66159-1

The used and developing of FRP in the FGD Stack of coal fire power plant in China

D.H. Zhang
Wuhan University of Technology, China
The Central Research Institute of Building and Construction Co., Ltd., MCC Group, China

J.H. Wang
Wuhan University of Technology, China

ABSTRACT: In this paper, the history and the actual state of FRP application in the FGD stack of coal fire power plant, and the application experience of FRP in America, Japan and Europe are briefly introduced. Then combined with the raw material supply, the development trend of FRP in the FGD stack of coal fire power plant is presented.

Keywords: composite; coal fire power plant; FGD stack; actual state; develop

1 SO_2 DAMAGE AND FLUE GAS DESULFURIZATION PROGRESS

SO_2 is atmospheric pollution caused by one of the main pollutants, effectively control in industrial flue gas SO_2 is the current environmental protection project to brook no delay.

According to the United Nations Environment Program announced in 1988 statistical data shows, SO_2 has become the world's first major pollutants emissions into the atmosphere each year, the human SO_2 up to 1.18×10^{10} t.

China's 2005 national high sulfur dioxide emissions of 2.549×10^7 tons, has become the world's first power SO_2 emission. The resulting economic losses more than 5×10^{11} Yuan [1]! According to the State Environmental Protection Department *2010 China Environmental Status Bulletin* statistics, in 2010 the national emissions of sulfur dioxide is still high amounts to 2.2944×10^7 tons, thermal power plant FGD installed capacity of more than 5.78×10^8 kilowatts, accounting for about 82.6% of total installed capacity of coal.

In the SO_2 emission, industrial sources emissions account for 83% of all emissions. One of our current first energy consumption, coal accounted for 76%, in the next few years the trend is increasing. Our country discharged into the atmosphere each year 87% SO_2 derived from the direct combustion of coals. Of which about half from the thermal power plant, as our country industrialized process accelerate ceaselessly, SO_2 emissions are also increasing.

Due to high sulfur content coal, heavy oil and mineral raw material itself sulfur, calcium fluoride, flue gas containing sulfur dioxide, HF and other toxic and harmful gas, bring serious pollution to atmosphere, is the main cause of acid rain. At present, China's land area of 40% has been turned into acid rain, acid rain and sulfur dioxide pollution caused by crops, forests and human health aspects of the economic loss serious, restricting China's economic and social development of important factors.

To curb acid rain pollution further development, in January 21, 1998, the State Council to country letter [1998] No. 5 sign language approved the State Environmental Protection Agency to develop a *"Control Area of Acid Rain and Sulfur Dioxide Pollution Control Areas Division Scheme"*. In a thermal power plant as an example, the new, alteration of coal-fired power plant sulfur content greater than 1%, must construction of desulphurization facilities; existing coal-fired power plant sulfur content greater than 1% before 2000, to take measures to reduce emissions, in 2010 before the batch built desulphurization facilities or take other considerable effect measure. 2008, the State Environmental Protection Administration also compiled a "National *Acid Rain and Sulfur Dioxide pollution control the Eleventh Five-Year Plan*", ensure to 2010 countrywide total sulfur dioxide emissions reduced by 10% compared with 2005, control within 2.2944×10^7 tons; thermal power industry emissions of sulfur dioxide control in 1.0×10^7 tons within, unit power output SO_2 emission intensity is lower than in 2005

Table 1. Since 2000, sulfur dioxide emissions changes over the years.

Annual	2000	2001	2002	2003	2004	2005	2006	2007	2008	2009	2010
SO$_2$ emissions (k tons)	19951	19478	19266	21585	22549	25494	25888	24681	23212	22144	22944
Rate of change %		−2.4	−1.1	+12.0	+4.5	+13.1	+1.6	−0.5	−0.6	−0.05	+0.04

Note: The data from *China Environment Yearbook 2010* and *The state of environment in China 2010*.

50%. By 2020, the total amount of sulfur dioxide emissions in 2010 based on the decrease.

According to the plan, in 2009 the newly installed capacity of thermal power industry desulfurization of 5000 MW, according to the current desulfurization market, need to invest 8.0×10^9 yuans RMB. It is expected that by 2020, the total installed capacity will be increased to 7.23×10^5 MW desulfurization.

In other countries, since 1860, coal-fired power plants for fuel containing sulfur impurities generated SO$_2$ combustion flue gas, begin to use water or slurry to remove device [2]. At present, the United States of America flue gas desulfurization system of unit capacity has more than half of the world capacity, that is more than 7.2×10^4 MW generating capacity, desulfurization process with limestone/gypsum slurry processing using the most widely used [3].

2 ANTICORROSION TECHNOLOGY OF DESULPHURIZATION CHIMNEY

2.1 China's desulfurization chimney anticorrosion status

In order to reduce the harm of SO$_2$ in the coal to the environment, China began to coal-fired power plant flue gas desulfurization test as early as 1992 [4], [5], high desulfurization efficiency more than 95%. Since 2002, the state environmental protection departments in the implementation nationwide large-scale flue gas desulfurization, asks 2010, large power and other coal-fired flue gas desulfurization device must be completed.

According to the US coal-fired power plant for wet flue gas desulfurization project experience and research results, with wet flue gas desulfurization device, corrosion of flue gas from the original weak corrosion to strong corrosion [6].

However, due to various reasons, China's power industry in wet flue gas desulfurization process for flue gas corrosion, lack of recognition, is simply to imitate the foam glass brick chimney anticorrosion system since 2003, use scrap glass foam glass brick and silicone rubber adhesive anticorrosion system, and a large number of used, which resulted in the installation of wet flue gas desulfurization device after a relatively short time, the desulphurization

chimney appeared relatively serious corrosion: some power plant has been made of foam glass brick anti corrosion construction for desulphurization of steel chimney, only about 2 months, appeared in diameter 210 mm perforation [7]; the power plant desulphurization chimney with steel inner cylinder, domestic waste glass foam glass tiles do within the corrosion, results in less than 1 years, is found throughout the inner steel cylinder have been corroded badly, must be replaced immediately [8]!

There are also some power plants using polyurea as within the corrosion resistance layer, the less than half that of large-area cracking, deciduous; some power plants using cement internal anti-corrosion layer, most quickly put to use 28 days only, desulfurizing condensate water penetrated brick inner cylinder; the author involved in the handling of the severe corrosion case with desulphurization chimney, is in production for only 1 months, i.e. in the chimney above 100 m site, the emergence of a large number of leakage, 6 months after the detection of concrete structure, the chimney has 42 mm corrosion, reinforced diameter decreases more than 2 mm but had to withdraw from the operation, the structure reinforcement and anti-corrosion construction, it is expected that the entire renovation work will last for 1 years, huge economic loss!

2.2 Foreign desulphurization chimney anti-corrosion technology introduction

In other countries, along with 1970's of flue gas desulphurization technology for flue gas desulphurization and large-scale popularization and application, the anti corrosion technology research has also started.

In the flue gas corrosion and anticorrosion technology research of various international organizations, such as American Electric Power Research Institute (referred to as EPRI) which created in 1973 and the IEA clean coal Centre (Clean Coal Centre, International Energy Agency) which founded in 1976 in the flue gas desulphurization anticorrosion technology research comprehensive and authoritative.

Through 30 years of the United States of America flue gas corrosion case research and summarize the experience, EPRI wrote *WET STACKS*

DESIGN GUIDE in 1996, on the United States once used in coal-fired power plant desulphurization chimney wall of various types of anti-corrosion materials for a long-term follow-up investigation, and the material cost of the construction, used in the process of maintenance cost and life cycle cost undertook comparative analysis and evaluation. American Society for Testing and Materials (ASTM) is since 1970 time is developed, promulgated the FRP chimney specification [9], used to guide the FRP chimney design, manufacture, installation and use, and timely new FRP technology through the revision of code reflected. The Office of Fossil Energy of United States Department of Energy which was launched *MARKET—BASED ADVANCED COAL POWER SYSTEMS FINAL REPORT MAY 1999* (DOE/FE-0400) [10],"Reinforced Concrete Shell + FRP Chimney" as the only recommended scheme for coal-fired power plant desulfurization chimney in the US.

Because the FRP material for desulfurizing chimney liner has so many technical, economic advantage, in recent years, "Reinforced Concrete Shell + FRP Chimney" desulfurization chimney form, has become the United States of America coal-fired power plant desulphurization chimney main structural form, see below Table 2.

In Europe, the IEA clean coal centre in 2006 published *SOx Emissions and Control*, their Twelfth chapters devoted to the presentation of the institution of coal-fired power plant desulfurization system corrosion materials research. The book on the desulfurization system corrosion and anticorrosive material's resistance performance evaluation results are shown in Table 3, different materials used as desulphurization chimney anti-corrosion liner of high stress fatigue performance test results are shown in Table 4, and if by these materials to build the desulphurization chimney anti-corrosion liner, the cost of such as are shown in Table 5.

From the above comparison data visible, FRP is used in coal-fired power plant desulphurization chimney anti-corrosion liner performance than the other anticorrosive materials.

2.3 China's FRP chimney research and application status

In China, use FRP as corrosive gas emission chimney with the international synchronization, began in the 1970s. According to the Central Research Institute of Building and Construction (CRIBC) technology archives records, as early as in December 1965 in cold rolling plant of Anshan Iron and Steel Corp., CRIBC design and built the pickling exhaust steel chimney with FRP liner, 40 m height, safe use to 1976 because of the cold rolling plant expansion requires the removal of the chimney, FRP lining remains intact; CRIBC also built the first whole winding FRP chimney in Shenyang smelting factory in 1975, height 102 m, diameter 2.5 m. The chimney is safe to use for nearly 20 years, until around 1996 the need for capacity expansion was dismantled. These cases fully shows the FRP's excellent performance used for gas emission chimney.

Based on the China's large-scale coal-fired power plant wet flue gas desulphurization to chimney brought serious corrosion problems need to be solved urgently, CRIBC began in 2005, jointly with the relevant Electric Power Design Institute of glass fiber chimney in coal-fired power plant desulfurization chimney design and application technology research, design, and in 2006 the domestic construction of the first power plant in glass steel chimney height, diameter, 180 m 6.6 m, maximum use temperature of 180°C. However, at the end of the installation phase, due to the installation of worker's exercise of violate the rules and regulations, causing the chimney was burned.

Table 2. In recent years, the United States of America FRP chimney construction volume [11].

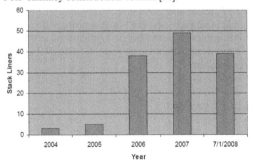

Table 3. Different materials in desulphurization system in corrosion resistance [12].

Materials	Sulphuric acid	Hydrochloric acid	Acid chloride salts
317L stainless steel	25°C below 5%	Not recommended	Not recommended
FRP based on novolac epoxy vinyl ester resin	95°C to 30%	82°C to 37%	100°C all concentrations
Alloy C-276	95°C to 30%	82°C to 5%	65°C to 20,000 ppm

Table 4. Different materials used as desulphurization chimney anti-corrosion lining of high stress fatigue [12].

Materials	Cycles to failure at 30% ultimate stress	Cycles to failure at 10% ultimate stress
317L stainless steel	5,000	>100,000
FRP based on novolac epoxy vinyl ester resin	200	4,000
Alloy C-276	200	5,000

Table 5. Different materials for the construction of desulphurization chimney anti-corrosion liner cost contrast [12].

Material	Installation	cost
FRP based on epoxy vinyl ester resin	$700,000	1.0
316 stainless steel	$800,000	1.15
317LM stainless steel	$1,100,000	1.57
Alloy C-276 clad steel	$1,200,000	1.71
Alloy C-276	$2,300,000	3.28

Note: Cost comparison for a 70 × 3 meters stack liner made from various materials

In addition, the unit also launched a prefabricated FRP plate assembling type FRP chimney, i.e. by titanium alloy nail prefabricated FRP plate fixed on the existing chimney wall, then using hand paste sealing seam and titanium alloy fixing hole. This design has certain novelty, but failed to take into account the chimney of the actual operating condition, leading to put into operation after 2 months is prefabricated shedding, chimney repair after a leak, still can't solve the problem, the owners had to use other programs [13].

3 FRP CHIMNEY'S APPLICATION PROSPECTS IN CHINA

a. From the domestic coal-fired power plant desulphurization chimney corrosion situation, due to corrosion protection scheme and construction quality control and other reasons, have to power industry brought billion yuans RMB of economic losses, to the extent that must solve as soon as possible brook no delay;

b. The whole winding FRP chimney, in China's metallurgical industry has a long history of safe use performance, although in the electric power industry 's first performance as with the types of reasons there was an accident, but from the developed history of more than 30 years and the recent construction of the case, will solve domestic coal-fired power plant desulphurization chimney corrosion problems the only way out.

c. In addition, the Hushan power plant newly build 2 × 600 MW coal-fired units desulfurization chimney has been using the whole winding FRP chimney, Hunan 2 × 600 MW coal-fired units desulfurization chimney has preliminary design for the whole winding FRP chimney, there are a number of other power plants are using similar chimney design.

AUTHOR

Zhang Dahou, Professor level senior engineer in Anticorrosive Materials of the Central Research Institute of Building & Construction Corp. Ltd, MCC Group. He is a Executive Councilor of China composites Industry Association, Executive Councilor of China Industry Anticorrosion Technology Association (CIATA), Vice-director of The Expert Committee of CIATA, and the Chief Editor of CECS 'Technical Specification for Steel Structures with Painting Anticorrosion'.

He is actively involved in promoting the use of FRP for corrosion resistant applications in China. He has design & construction the first FRP Chimney in China.

REFERENCES

[1] 柯伟, 中国工业与自然环境腐蚀调查, 全面腐蚀控制 [J], 2003年第1期, P1–10.
[2] A History of Fuel Gas Desulfurization System Since 1850, Journal of Air Pollution Control Association, Oct. 1977.
[3] "FGD Installations on Coal-fired plants", 1989, IEA Coal Research Report.
[4] 馨喆. 珞璜电厂首次引进大型排烟脱硫装置. 中国电力报[P]. 2006年6月6日.第007版.
[5] 张可钜. 珞璜电厂4 × 360 MW机组烟气脱硫工程评述.电力环境保护[J]. 2000年第4期, P1–11.
[6] Electric Power Research Institute (USA), WET STACKS DESIGN GUIDE（1996.11）, TR-107099 9017.

[7] 沈宝中. 华能上海石洞口第二电厂#2 烟囱钢内筒水平烟道入口处局部腐蚀及修复情况. 火力发电厂脱硫烟囱防腐技术研讨会议. 2009. 08，上海.

[8] 杨清发. 某电厂烟囱钢内筒腐蚀破坏的案例介绍. 火力发电厂脱硫烟囱防腐技术研讨会议. 2009. 08，上海.

[9] ASTM D 5364–93 (Reapproved 2002) Standard Guide for Design, Fabrication, and Erection of Fiberglass Reinforced Plastic Chimney Liners with Coal-Fired Units.

[10] Office of Fossil Energy, U.S. Department of Energy, MARKET-BASED ADVANCED COAL POWER SYSTEMS FINAL REPORT, DOE/FE-0400. MAY 1999.

[11] Thom Johnson, Don Kelley, Mike Stevens. The Rapid Growth of Composites in Air Pollution Control Processes. 8th Annual COAL-GEN Conference, August 13–15, 2008, Kentucky International Convention Center.

[12] Deborah M B Adams, Anne M Carpenter, Lee B Clarke, Robert M Davidson, Rohan Fernando, Kazunori Fukasawa, David H Scott, Irene M Smith, Lesley L Sloss, Herminé Nalbandian Soud, Mitsuru Takeshita, Zhangfa Wu. SOx emissions and control. 2006, IEA Clean Coal Centre.

[13] 王洪斌. 一期烟囱防腐整改技术研究. 火力发电厂脱硫烟囱防腐技术研讨会议. 2009. 08，上海.

Frontiers of Energy and Environmental Engineering – Sung, Kao & Chen (eds)
© 2013 Taylor & Francis Group, London, ISBN 978-0-415-66159-1

The research on accelerator drive systems deal with nuclear waste

Z.S. Hou, T. Zhou, F. Luo, J. Chen & L. Liu
North China Electric Power University Beijing, China

ABSTRACT: The minimum of nuclear waste is the key problem to nuclear fission to Safety and efficiency and sustainable development. On the basis of 1 million kW PWR power station and fuel consumption for 300 days, calculating fuel burned, times actinides elements, plutonium and several long lived fission products through the Dragon, and according to the goal of China's nuclear power development, obtaining the forecasted data of cumulative amount. At the same time analysing transmutation results of times actinides elements, plutonium and several long lived fission products through Accelerator Driven time critical System (ADS), and obtaining the transmutation results of accelerator driven time critical system time is than 4.5~6.5 times of the same power of critical reactor, and nuclear waste can be processed well, but also meets the sustainable development of China's nuclear energy decision-making.

Keywords: spent nuclear fuel; Dragon; ADS; transmutation

1 INTRODUCTION

The nuclear crisis is caused by Japan Fukushima nuclear power accident, making the public pay high attention to the nuclear power security. Japan East Electric Company displacements high concentration of radioactive wastewater into the sea, causing the people of the world panic and governments criticized severely. The radioactive waste safety management regulations is passed by the State Council 183th executive meeting on November 30, 2011, and it is executed from March 1, 2012, to ensure the safety of nuclear energy efficient sustainable development. Thus nuclear power plant spent fuel reprocessing and nuclear waste disposal has become the key problems for the development of nuclear energy. Spent fuel management mainly have three kinds of fuel cycleway[1] at present: (1) "once-through operation" approach, the spent fuel is disposed directly; (2) "Fuel after treatment and thermal reactor cycle" approach, recycling U and Pu in the spent fuel and resued in thermal reactor; (3) "advanced nuclear fuel cycle" approach, recycling the U and Pu in spent fuel and resued in the fast reactor or the Accelerator Driven-System, making Minor Actinide elements (MA) transmutation. Among them, the Accelerator Driven System has high evolution rate, can be deal with nuclear waste well, so it is necessary to ADS processing efficiency of the nuclear waste in depth research.

2 CALCULATION PROCEDURE AND RESEARCH OBJECT

2.1 Calculation procedure and formula

Dragon calculation procedure and MA transmutation rate calculation formula for calculation is used in the study. Dragon calculation procedure[2-4] is developed by Ecole Polytechnique DE Montreal in Canada, it can simulate reactor fuel assembly and neutron characteristic of lattice cell, it not only can be used to transport-transport and transport-diffusion equivalent calculation, also can be used for Neutron characteristic homogenization treatment and nuclide burn up calculation. In this study, the Dragon burn up module (EVO), neutron flux module (FLU) and Geometric (GEO) module are used mainly. One million kW of Pressurized Water Reactor (PWR) power station and burn up time 300 days as benchmark are introduced in this paper, through the ADS for MA evolution calculation, it can be used to indicate transmutation consumption rate:

$$R_{MA} = \frac{M_{MA,BOL} - M_{MA,EOL}}{M_{MA,BOL}} \tag{1}$$

where $M_{MA,BOL}$ is MA quality in initial life; $M_{MA,EOL}$ is MA quality in final life, the unit are grams.

2.2 The research object

Pressurized water reactor fuel assembly and the Accelerator Driven System as the research object. Fuel assembly section as shown in Figure 1.

Where 1 and 2 is uranium fuel, using zoning device, inner area 1 is low concentration fuel, outside area 2 is high concentration fuel; 3 is cladding; 4 is moderator (H_2O); equivalent side length of lattice cell is 1.26209 cm. The Accelerator Driven System is one of be used for radioactive nuclear waste transmutation, and uses nuclear resources effectively and output nuclear energy, separate nuclear waste, and has high security of new nuclear power system[5]. Since 2000, Accelerator Driven System related research has become increasingly active. The United States, Japan, Germany, France and Russia lead the world in the Accelerator Driven System research, India as emerging research state, has produced a certain influence[6] in the Accelerator Driven System research in recent years. In our country the Accelerator Driven System has been included in the "National Key Infrastructure Project Research and Development Planning Project", the device system as shown in Figure 2.

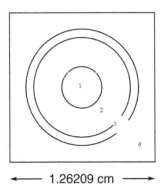

← 1.26209 cm →

Figure 1.   PWR bundle.

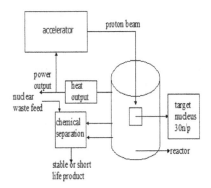

Figure 2.   Diagram Accelerator Driven System.

From graph 2 it is known that, the Accelerator Driven System includes proton accelerator, external neutron generator, subcritical reactor, transmutation system and power output system. The accelerator produces high energy proton beam to bombard heavy metal target parts (such as liquid Pb or Pb-Bi alloy) of subcritical reactor to cause spallation reaction, an energy proton for 1 Gev produces about 30 neutrons. In addition to maintain the chain reaction, the rest of the "neutron balance" can be used to "burn" authigenic or additional long life nuclide in spentfuel. Because of the subcritical operation condition and accelerator fast response, making the ADS has higher flexibility and security.

3   CACULATION RESULTS AND ANALYSIS

3.1   The main nuclide calculation results and analysis

The fuel density is about 10.4375 g/cm^3, and one fuel rod total volume is about 185.3582 cm^3, obtaining a fuel rod quality is about 1934.6766 g. So a 1 million kW and 17 × 17 component pressurized water Reactor power plant (contain 41448 fuel rods) that loading UO_2 amount is about 80.1885t. Through the Dragon calculation to get daya bay nuclear power station with the average fuel for nuclear fuel consumption, loss and nuclear manufacturing value, the result as show in Table 1, and obtain several highly toxic long life fission products, the result as show in Tables 2–5.

From Table 1, it is known that spent fuel quantity of a nuclear power plant that the installed capacity is 1 Gw is about 25t, containing recycled uranium is about 23.75t, Pu is about 200 kg, short life of Fission Products (FPs) is about 1000 kg, minor Actinides Nuclide (MAs) is about 18 kg, Long Life Fission Products (LLFPs) is about 30 kg. These nuclear waste have long life and radiation toxicity, that constitute longterm harm on human environment.

From Table 2, it is known that the main nuclides is the long life and high toxicity Np and Am in minor actinide nuclides, therefore the key of spent fuel processing lies in Np and Am processing.

From Table 3, it is known that the main nuclides is the long life and high toxicity ^{239}Pu and ^{240}Pu in Pu isotopes, therefore the key of Pu processing lies in ^{239}Pu and ^{240}Pu processing.

From Table 4, it is known that the main nuclides is ^{129}I, ^{99}tc, ^{135}cs and ^{93}zr in long life fission products, and ^{129}I and ^{99}tc have bad influence on human body, therefore the key of longlife fission products processing lies in ^{29}I and ^{99}tc processing.

Table 1. The daya bay nuclear power station with the average fuel for nuclear fuel consumption, loss and nuclear manufacturing value.

Unit	Component	UO$_2$ quality/t	U quality/t	Average burnup/ MWD/T	U consumption/t	Pu increment/t
1	157	81.65	71.96	12141	1.29	0.38
2	157	81.86	72.15	13794	1.46	0.42

Table 2. Minor actinides nuclide proportion of spent nuclear fuel (%).

Nuclide	T$_{1/2/a}$	Kg·(GW·a)$^{-1}$	Proportion
^{237}Np	2.14×10^6	11.89450	0.04758
^{241}Am	433	1.733	0.00693
^{242}Am	–	0.03	0.00012
^{243}Am	7370	3.466	0.01386
^{243}Cm	0.446	0.015	0.00006
^{244}Cm	18.11	0.899	0.00360
^{245}Cm	8.53	0.04	0.00016

Table 4. Several long life fission products proportion of spent nuclear fuel (%).

Nuclide	T$_{1/2/a}$	Kg·(GW·a)$^{-1}$	Proportion
^{79}Se	6.5×10^4	0.21	0.00084
^{93}Zr	1.5×10^6	21.16	0.08464
^{99}Tc	2.14×10^6	27.41	0.10964
^{107}Pd	6.5×10^6	6.59	0.02636
^{126}Sn	10^5	1.25	0.00500
^{129}I	1.6×10^7	6.32	0.02528
^{135}Cs	3×10^6	10.88	0.04352

Table 3. Plutonium isotopes proportion of spent nuclear fuel (%).

Nuclide	Pu isotopes quantity (Kg)	Spent fuel	T$_{1/2/a}$
^{238}Pu	3.45	0.0345	87.74
^{239}Pu	54.3	0.543	2.41×10^4
^{240}Pu	24.1	0.241	6570
^{241}Pu	11.9	0.119	14.4
^{242}Pu	7.777	0.07777	3.76×10^6

Table 5. Change of MA stock.

Nuclide	Amount in run time/kg	Amount in 300 days/kg	Changing amount
^{237}Np	745.00	633.11	−111.89
^{241}Am	447.80	390.68	−57.12
^{242}Am	1.23	6.8	5.45
^{243}Am	235.93	233.87	−2.06
^{243}Cm	0.00	16.73	16.73
^{244}Cm	75.30	36.68	38.62
^{245}Cm	3.99	13.37	9.38

From Tables 2–4, and 40 Gwe (confirmed to be working) and 18 Gwe (under construction) of nuclear power installed capacity in 2020, it is estimated that spent fuel accumulated stock will reach 6000t to 10000t in our country. If 80 Gwe to 100 Gwe of the nuclear power installed capacity in 2030, then the spent fuel accumulated stock will reach 20000t to 25000t, Pu will reach 160t to 200t, MAs will reach 16t to 20t, LLFPs (long lived fission products) will reach 24t to 30t. Therefore, how to deal with properly high level radioactive waste in nuclear power plant, has become the key point to ensure the sustainable development of China's nuclear energy.

3.2 Calculation results and analysis of transmutation of MA

By code Dragon and calculation formula of RMA the consumption rate of MA, we can get the results of the transmutation of MA of the spent fuel from Accelerator Driven System, as shown in Table 5.

From Table 5, Am and Np, which account for the major part of MA reduce quickly, the quality of transmutation in 300 days reaches to 100.89 kg, the consumption rate of transmutation is 6.69%, the average quality of transmutation annual is 111.00 kg. The minor actinide wastes generated in pressurized water reactor of thermal power of 1 Gwe each year is about 34 kg. So the ratio of transmutation for 800 MW ADS is 12.96, which is 4.5 to 6.5 times of the ratio of transmutation for the same power of critical reactor. Therefore ADS has good effect of transmutation, and can process the spent fuel better.

4 CONCLUSION

Through fuel assembly calculation research of the Dragon and transmutation calculation analysis research of the Accelerator Driven System, can come to the following conclusions:

1. The transmutation ratio and support ratio of the Accelerator Driven System better than that of the critical reactor, and the transmutation support ratio of the Accelerator Driven System is 4.5 to 6.5 times than the the critical reactor, the Np and Am quantity are greatly reduced in actinide elements.
2. The high transmutation ratio and support ratio of the Accelerator Driven System, can deal with nuclear waste well, therefore the Accelerator Driven System accords with the sustainable development of China's nuclear energy decision.

ACKNOWLEDGEMENTS

The research was supported by Chinese Academy of Science ADS plan and the state key laboratory of reactor system design of National Pesticide Information Center and the special fund of North China Electric Power University.

REFERENCES

[1] Hyo-Jik Lee; Byung-Suk P-ark; Sung-Hyun Kim; Hee-Sung Park; Ji-Sup Yoon; Spent Nuclear Fuel Disposal and Recycling Process and Its Operational Experience [D]. International Conference on Smart Manufacturing Application. 2008 April. 9–11, in KINTEX, Gyeonggi-do, Korea.

[2] G.Marleau and A.H´ebert, A New Driver for Collision Probability Transport Codes [D], Int. Top. Mtg. on Advances in Nuclear Engineering Computation and Radiation Shielding, Santa Fe, New Mexico, 1989 April 9–13.

[3] G.Marleau, R.Roy and A.H´ebert, "DRAGON: A Collision Probability Transport Code for Cell and Supercell Calculations", Report IGE-157, Ecole Polytechnique de Montr´eal 1993.

[4] I.R.Suslov, "An Algebraic Collapsing Acceleration Method for Acceleration of the Inner Scattering Iterations in Long Characteristics Transport Theory", Int. Conf. on Supercomputing in Nuclear Applications, Paris, France, 2003 September 22–24.

[5] Xia Haihong, Luo Zhanglin, Zhao Zhixiang. Strengthen the ADS technology research-Promote nuclear energy sustainable development [J], Modern Physics, 2011, 4.

[6] Li Zexia, Liu Xiaoping, Zhu Xiangli, Huang Longguang. Development Trend Analysis of the Accelerator Driven System [J], Science Focus, 2011, 3.

Frontiers of Energy and Environmental Engineering – Sung, Kao & Chen (eds)
© 2013 Taylor & Francis Group, London, ISBN 978-0-415-66159-1

Study of natural ventilation numerical simulation on the effect of energy conservation in typical building

J.P. Zhao & P.P. Niu

Institute of Occupational Health and Safety, Xi'an University of Architecture and Technology Xi'an, Shanxi, China

ABSTRACT: This paper establish the appropriate model of typical residential building by using computational fluid dynamics software airpak. The energy utilization factor indoor under natural ventilation is greater than 1 in variety of conditions and higher than mechanical ventilation. The inlet air temperature has a direct impact on energy utilization factor; energy utilization factor is lower when the inlet air temperature is in line with the human thermal comfort index, when the inlet air temperature does not comply with the human thermal comfort index higher energy utilization factor. It can be seen that the cross-flow ventilation is better; indoor air short-circuit does not form. The overall air flow is relatively smooth.

Keywords: natural ventilation; building energy conservation; numerical simulation; energy utilization factor

1 INTRODUCTION

Since the United Nations Climate Change Conference held in Copenhagen, Denmark, on December 7, 2009. Energy conservation, the path of low-carbon economy has increasingly become a national consensus; vigorously developed energy conservation technology has become an irreversible trend. In our country, about 90% buildings are not energy conservation buildings; the proportion of building energy consumption in the total social is increasing at the rate of one percentage point a year. It is a huge challenge for our country. For this reason, people gradually pay attention to building energy conservation, a variety of building energy conservation technologies and methods have emerged. Natural ventilation can reduce the indoor temperature, take away wet gas, exclude polluted air and achieve human thermal comfort in the case of consumption of non-renewable energy by reasonable architectural design. Another benefit is it can reduce the dependence of people on the air-conditioning system, thereby save energy, reduce pollution and prevent air conditioning disease[WEI]. Office buildings Castine BRE (Building Research Establishment) in the United Kingdom is a typical instance, which is built by hot natural ventilation pressure differential tissue to achieve the energy saving[LI].

In this paper, starting from the point of energy utilization factor of building indoor, numerical simulate a typical residential building by using Air-pak software. It sets point's indoor typical location to measure the air temperature. By changes of air temperature import and export, it analyses the changes of energy utilization factor. From the energy utilization factor can be seen, the natural ventilation of the building energy conservation has enormous energy-saving potential.

2 THE INDICATOR OF BUILDING ENERGY CONSERVATION EVALUATION

Energy utilization factor η, also known as Temperature Efficiency, is an indicator of the system energy economic evaluation[3].

It is defined as follows:

$$\eta = \frac{t_0 - t_p}{t_n - t_p} \tag{1}$$

Type of t_p—supply air temperature (°C)
T_0—exhaust temperature (°C)
T_n—Area of the average air temperature indoor work (°C)

In general, the mechanical ventilation system, the energy utilization coefficient η = 1.0; under the air or the displacement ventilation system, the energy utilization coefficient η > 1.0; in the airflow short circuit condition, the energy utilization coefficient η < 1.0. Coefficient of utilization of energy actually reflects the indoor temperature gradient, which shows the indoor thermal stratification characteristics.

For fresh air or pure ventilation system, the higher energy utilization factor, the greater energy saving potential of the system. But it also cannot only overemphasize the importance of increasing utilization factor of energy, because the higher energy utilization factors, the larger indoor temperature gradient, which may affect the thermal comfort. In addition, when indoor air specific heat cannot be considered as a constant, it should be used enthalpy to replace temperature, still known as the factor of utilization of energy or Thermal Efficiency:

$$\eta_k = \frac{i_p - i_0}{i_n - i_0} \qquad (2)$$

3 RESEARCH METHOD

3.1 Introduction to physical model

The dimension of physical model is 14 m × 8 m × 3.3 m, internal structure of building is shown in Figure 1.

This residential building contains three bedrooms, two living rooms, a kitchen and a bathroom. The master bedroom is next to the living room and dining room but separates with the second bedroom, near the door is bathroom, and behind the kitchen is dining room. This layout is in line with the typical residential building.-X positive direction north.

Main indoor objects shown in Figure 1: cabinet.1, cabinet.2, TV, table, couch, person.1, person.2, person.3, closet, bed. The specific sizes are shown in Table 1.

3.2 The determination of the boundary condition

1. Supply air temperature and exhaust temperature. Energy utilization factor η directly reflect

Table 1. Object size.

Object name	Size/m (L × W × H)	Temperature
Cabinet.1	1.5 × 0.6 × 1.2	20°C
Cabinet.2	1.5 × 0.6 × 0.5	20°C
TV	0.6 × 0.4 × 0.5	20°C
Table	1.8 × 1.2 × 1	20°C
Couch	2.2 × 1 × 0.8	20°C
Person.1	H:1.73	36.6°C
Person.2	H:1.68	36.6°C
Person.3	H:1.2	36.6°C
Closet	2 × 0.8 × 2.2	20°C
Bed	2 × 1.8 × 0.5	20°C

indoor degree of utilization of the energy, and it associates with the blowing air temperature and exhaust temperature. The initial indoor temperature is set to 24°C, body temperature is set to 36.6 °C, natural ventilation inlet air temperature are set to 0 °C, 5 °C, 10 °C, 15 °C, 20 °C, 25 °C, 30°C. By setting the measuring point in the interior of the window, it measures the temperature of the exhaust port of the wind.

2. Wind speed. In reality, wind speed is subject to change, to simplify the calculations, we assume that the outside blowing indoor wind speed to a fixed speed within a particular time, taking into account the wind speed to the impact of human comfort, assuming wind speed 0.2 m/s into the interior (summer indoor wind speed does not exceed 0.3 m/s, and winter indoor wind speed does not exceed 0.2 m/s, while the human body will feel comfort[4]).

3. Room opening's size. The size of room opening directly impacts on wind speed and the amount of air inlet. When the opening is large, the flow rate is large; when the opening is small, the flow rate increases relatively, but the airflow field is narrow, and the proportional relationship does not exist between the size of the opening and ventilation efficiency. According to the determination, when the opening width is 1/3 to 2/3 of the bay width, and the opening area is 15% to 25% of the floor area, the ventilation efficiency is the best. When the air inlet is larger than the air outlet, the discharging outdoor wind speed will increase. The relative position of the opening, whether it is the flat position, or the level of the cross-sectional view, the opening and closing of the room opening will directly affects airflow route[5]. The indoor window position and size are shown in Table 2.

4. Form of indoor ventilation. The form of indoor ventilation has a very important effect on energy efficiency, References[6] from the room in

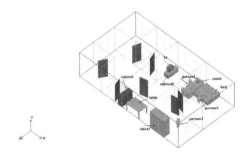

Figure 1. Building inside view.

Table 2. The indoor window position and size.

Name	Position X	Y	Z	Size/m L × W
Window.1	(1, 3.2)	(1.2, 2.4)	(−5, 5)	(2.2 × 1.2)
Window.2	(−6, −6)	(1.2, 2.4)	(0.4, −1.2)	(1.6 × 1.2)
Window.3	(−3.9, −5.1)	(1.2, 2.1)	(3, 3)	(1.2 × 0.9)
Window.4	(−3.9, −5.1)	(1.2, 2.1)	(−5, −5)	(1.2 × 0.9)
Window.5	(−0.9, −2.1)	(1.2, 2.1)	(−5, 5)	(1.2 × 0.9)
Window.6	(8, 8)	(1.2, 2.1)	(−1.9, 0.7)	(1.2 × 0.9)
Window.7	(8, 8)	(1.2, 2.4)	(−1.8, −3.6)	(1.8 × 1.2)

the depth (d) and height (h) of the relationship to consider that when d = 2 h, a single outlet unilateral ventilation is better; Two outlet unilateral ventilation is better, when d = 2.5 h; When d = 5 h, the cross-flow ventilation is better. References[7] suggested that when the room section width is less than or equal to 2.5 times the height of the room and the window area of approximately 5% of the gross floor area of the room, it could use unilateral ventilated; When room section width is equal to five times the height, it can use wind-driven cross-flow ventilation. Under normal circumstances, the cross-flow ventilation is better than unilateral ventilation effect[DUAN]. In order to simplify the model, paper takes better ventilation of the cross-flow ventilation, all the windows indoor and doors are in the open state, the natural wind flowing into the indoor outside is from a particular direction.

4 THE RESULTS ANALYSIS OF NUMERICAL SIMULATION

4.1 The analysis of indoor energy utilization coefficient

It is impossible to measure the temperature of all the points in the large indoor area, so after natural ventilation, the temperature in the room is replaced by the indoor average temperature of a few typical measuring points. The indoor activities of human body range height 1 m to 2 m, so it arranges eight measuring points from the ground height of 1.5 m, after obtaining these eight measuring points temperature, it can compute the average temperature indoor approximately. Similarly, six points were set up at a slight distance of the openings, in order to measure the temperature of the air vents. The exhaust temperature can be obtained by calculate the average.

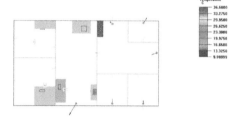

Figure 2. Position of measuring point for the air vents.

Figure 3. Position of measuring point indoor.

Because the height of indoor window is 1.2 m from the ground, so the height of points is chose at the 1.5 m.

Calculating the average indoor temperature and the average exhaust air temperature, and then taking into the formula (3–1), it can be drawn from the interior of the energy utilization factor η. Figures 2 and 3 show each measurement point position and the temperature of the measuring point cloud images at 10°C of the supply air temperature:

After seven calculations, each measuring point temperatures are shown in Tables 3 and 4.

The temperature, is brought into the above formula, can be obtained at different temperatures

Table 3. Measurement point temperature indoor.

Inlet air temperature	Measurement point temperature indoor (°C)								Average temperature
	1	2	3	4	5	6	7	8	
0°C	13.96	13.39	11.33	9.77	7.15	10.38	13.38	13.50	11.60
5°C	16.26	15.80	11.56	11.94	10.22	11.25	15.08	15.86	13.50
10°C	17.40	18.72	15.07	13.82	13.29	14.04	16.86	17.19	15.75
15°C	18.98	19.94	17.71	17.85	16.88	17.74	20.19	19.39	18.59
20°C	22.14	21.82	21.19	21.39	20.58	21.50	22.24	22.04	21.61
25°C	24.46	24.55	24.97	25.24	25.07	24.04	24.40	24.42	24.64
30°C	25.49	26.28	28.65	28.12	29.35	29.00	25.20	25.47	27.20

Table 4. Exhaust temperature of the measuring point.

Inlet air temperature	Exhaust temperature of the measuring point (°C)						Average temperature of exhaust
	9	10	11	12	13	14	
0°C	18.91	14.12	19.26	9.32	15.80	18.76	16.10
5°C	20.73	16.09	19.16	11.92	17.14	22.16	17.87
10°C	21.24	17.71	21.83	14.53	17.65	22.13	19.18
15°C	22.16	19.77	21.65	17.91	20.69	22.59	20.80
20°C	21.98	21.95	21.70	21.12	22.51	23.64	22.15
25°C	24.85	24.90	24.01	25.13	24.08	24.01	24.50
30°C	26.08	25.65	24.03	28.16	24.17	24.04	25.36

Table 5. Energy utilization factor.

Inlet air temperature	Energy utilization factor
0°C	1.39
5°C	1.50
10°C	1.59
15°C	1.61
20°C	1.34
25°C	1.38
30°C	1.66

indoor energy utilization factor, expressed in Table 5.

From the above data and charts, the following conclusions can be drawn:

1. As the blowing air temperature changes, energy utilization coefficient changes turbulently, from0 degrees to 15 degrees between the inlet temperatures, energy utilization coefficient increased gradually; from 20 degrees to 25 degrees decreased, from 25 degrees to 30 degrees began to rise. According to the passage, we know that energy utilization coefficient is bigger, the indoor temperature gradient is larger, and when the indoor temperature gradient is over large, it can influence human thermal comfort. As can be seen from the 4–4, the blowing air temperature of 20 degrees Celsius, the lowest energy utilization coefficient, and the blowing air temperature is more comfortable for the human body. When the blowing air temperature is low or high, the energy utilization coefficient is higher, and the air temperature is not suitable for the human body. Therefore, in consideration of building energy saving at the same time, but also from the thermal comfort of the human body's point of view, comprehensive analysis, to find a balance between the two, so as to achieve the target of energy saving and comfortable.

2. According to the passage, in considering human comfort conditions, energy utilization coefficient is bigger that the ventilation energy saving potential is greater. No matter the building is in various conditions, the energy utilization coefficient is greater than 1 and the energy saving potential is greater than the mechanical ventilation (mechanical ventilation energy utilization coefficient is equal to 1). Therefore, we should do our best to use natural ventilation to improve the energy utilization rate when taking the building design.

4.2 Indoor airflow form analysis

Indoor airflow pattern has a very important impact on interior energy-saving effect; good air

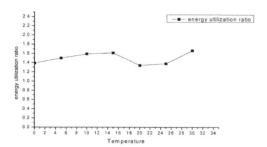

Figure 4. Different supply air temperatures on energy utilization factor.

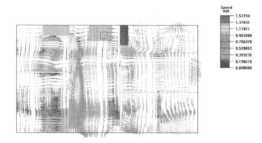

Figure 5. Wind speed at the height of 1.5 m.

distribution mode not only can achieve the effect of saving energy, but also can maintain the health of indoor air environment. Since the range of human activities is about 1.5 m, we choose the room height of 1.5 meters to observe indoor wind speed conditions, as shown in Figure 5.

From the speed of the arrow diagram, we can be see that at the beginning of the natural wind from the window into the room, the change of wind speed is not big, because the body temperature influence wind speed around human body to form a vortex, the limited capacity of body temperature field does not play a major influence factors on wind speed. The convection of the window and the door played a major role on wind speed, which makes wind speed become bigger in the process of the wind through the door into the living room. Natural wind relatively gently through by the living room, when it entered the gate and was near the kitchen, the velocity in the kitchen is increased quickly and forms the vortex as the convection of the door and the window, ultimately excluded from the outside. If buildings take the through-flow ventilation, the indoor air distribution is more reasonable, the overall velocity is smooth and uniform,

organizational form is good. At the same time, we also can see that in some corner of the room, such as a corner, corner of the door, wind speed is very small, almost zero, they easily become pollutants accumulation point.

5 CONCLUSION AND PROSPECT

With the numerical simulation of natural ventilation, we can see that energy utilization coefficient is greater than 1 while buildings use natural ventilation buildings, and that has the quite big energy saving potential. Interior energy utilization coefficient has a direct relationship with the blowing air temperature, however in the practical application can not only from the energy point of view; consideration should also be given to the relationship between temperature and human comfort. The numerical simulations of natural ventilation use the constant boundary conditions, and it is different from the actual situation. Study on the unsteady boundary conditions will be the focus of the analysis in the future, and the numerical simulation of wind speed and direction change with time will be fit with the actual conclusion. This will give the new guidance at the design of building energy saving.

REFERENCES

BRE. Natural ventilation in non 2 domestic buildings. Building Research Establishment, Garston, Watford, UK.

DUAN Shuang-ping, ZhANG Guo-qiang, ZhOUJun-li. Natural ventilation technology research progress [J]. HV & AC, 2004, 34, (3).

Levermore Geoff J. Simulation of a naturally ventilated building at different locations. In: ASHRAE Trans. 2000, 106. 402–407.

LI Min, YANG Zugui. Study on the design of natural ventilation in ecological building design [J]. Sichuan Construction, 2006, 26 (4):29–30.

Ministry of Construction of the People's Republic of China, Administration of Quality Supervision, Inspection and Quarantine, GB 50019–2003. Heating, ventilation and air conditioning design [M]. Beijing: China plans Press, 2003.

NKBANSAL, RAJESH MAHUR, MSBHANDARL Solar Chimney for Enhanced Stack Ventilation, Building and Environment, 1993, Vol28 (3): 373 ~ 377.

PENG Xiaoyun. Natural ventilation and building energy efficiency [J]. Industrial Construction, 2007, 37(3).

WEI Feng. Building natural ventilation applies on design [J]. Journal of Henan University of Science and Technology: Natural Science Edition, 2004, 25(5).

Frontiers of Energy and Environmental Engineering – Sung, Kao & Chen (eds)
© *2013 Taylor & Francis Group, London, ISBN 978-0-415-66159-1*

The prediction of comprehensive pollution indexes of Taihu Lake based on BP network

Z.H. Ma, T. Wang & J.R. Zhou
Changzhou University, Changzhou, China

ABSTRACT: In order to provide decision support for management platform of Taihu Lake, this paper studied BP(Back Propagation) neural network model and used it into predicting of the comprehensive pollution index of inter rivers and out rivers of Taihu Lake. This paper established a three-layer BP neural network prediction model, based on Taihu routine environmental monitoring data, predicted the quality of water, studied the relationship between the variation of water quality of Taihu and inflow and outflow rivers of Taihu. The prediction of 2005 year Taihu's water quality showed that, in 2005 the water quality pollution is more seriously than before, generally V water quality. The result is in line with the pollution situation development trend of Taihu Lake. The reason why BP neural network was selected are as follows. Compared with the former methods, BP network method have good adaptability, higher precision, better response indexes of water indexes' internal change rules. What this paper studied can provide scientific basis for limiting water environmental pollution.

Keywords: BP network; Taihu Lake; comprehensive pollution index; prediction

1 INTRODUCTION

With the continuous development of the social-economic, population was increasing, and large amounts of pollutants were discharged into Taihu Lake, lead water quality and the surroundings of Taihu Lake increasingly serious. This would do great harm to our society and lives.[1] In order to have a clearly understanding of water quality of Taihu Lake, many scientists have done studies on the analysis of water quality. The neural network proved to be an effective way to predict water quality. BP neural network is one kind error propagation neural network, which can effectively solve the nonlinear relation problems, and also one of the most widely used neural network model.[2]

2 THE ESTABLISHMENT OF BP NEURAL NETWORK

2.1 *BP neural network theory*

BP neural network is a multilayer forward neural network, the neuronal transfer function is S-shaped function, the output is a continuous quantity between 0–1, and can reach the input to the output of any non-linear mapping. Three-layer BP neural network consists of one input layer, one output layer and one hidden layer.[3] In order to let the output may take any value, this paper choose tansig function between input layer and hidden layer, purelin function of linear type between hidden layer and output layer.[4] Three-layer BP neural network topology structure is shown in Figure 1.

The guiding principle of the BP network learning is: the correction network weights and threshold have to go along with the negative gradient direction which the performance function drops the fastest.

$$x_{k+1} = x_{k+1} - a_k g_k \qquad (1)$$

Among them, x_k means the current weight and threshold matrix, g_k means the current performance function, and means the learning rate.[5]

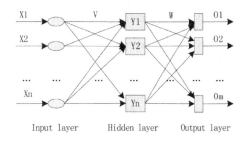

Figure 1. Three-layer BP neural network topology structure.

In the three-layer BP neural network, x_i is the input nodes, y_j is the output nodes, the weight between the input nodes and hidden nodes is w_{ji}, the weight between the hidden nodes and output nodes is v_{lj}. When t_l is assumed to be the expectations of the output node, the calculation formula of model is as follows.

The output of hidden node:

$$y_i = f\left(\sum_i w_{ji} x_i - \theta_j\right) \tag{2}$$

The calculation formula of output node:

$$z_l = f\left(\sum_j v_{lj} y_j - \theta_j\right) \tag{3}$$

The error of the output node:

$$E = \frac{1}{2}\sum_l \left(t_l - z_l\right)^2$$

$$= \frac{1}{2}\sum_l \left(t_l - f\left(\sum_j v_{li} f\left(\sum_j v_{lj} y_j - \theta_j\right) - \theta_j\right)\right)^2 \tag{4}$$

2.2 Data source

The data of 1997 to 2004 per month survey were selected as evaluation objects. The evaluation is shown in Table 1. According to the data of the environmental inspection monitoring department of Jiangsu Province, there are 20 sections as the routine monitoring sections in our province, contains 13 inflow rivers, 7 outflow rivers. Figure 2 shows the distribution of the cross sections.

2.3 The topology of BP network

In order to avoid too larger BP network structure, too long computing time, and fall into local minima, this paper choose the compact structure neural network model, which consists three layers. There are many advantages of three-layer network model, for one thing, it can approximate any rational function with arbitrary precision, for another, the training time is short. BP neural network was put into use for prediction and the network design is as follows: the output layer contains 1neruons which means the comprehensive index prediction. The input layer has 6 neurons which are the routine monitoring data [6]. According to

Table 1. Margin settings for A4 size paper and letter size paper.

Water quality standard	Pollution level	Grading standard
<0.2	Clean	Most items were not detected, individual items also belowstandard
0.2–0.4	Minor clean	Average value is below the standard, the individual item value near the standard
0.4–0.7	Light pollution	He average of one individual items all exceeds the standard
0.7–1.0	Medium pollution	Two items exceeds the standard
1.0–2.0	Heavy pollution	A considerable portion of items exceed the standard
>2.0	Serious pollution	A considerable portion of items exceed the standard several times

Figure 2. Distribution of inflow and outflow rivers of Taihu Lake.

the actual training results, 8 neurons were found to be the best.

S-shaped function and pruelin function are often used in BP neural network as the neuron transfer function. BP neural network training take the momentum gradient descent training function into use which is one batch mode of training function. Traingdm function not only have faster convergence rate, but also can avoid the local minimum problem in the training processing. The so-called accession of the momentum term added the previous variables to the next change.

From the above analysis, the BP neural network model structure defined as follows:[7] the network structure: 6-8-1; transfer function combination: tansing function—purelin function.

2.4 *The selection of evaluation and standards*

This paper choose comprehensive pollution index as evaluation index.

At present the surface water evaluation using "Surface water quality standard" (GB3838-2002). The water evaluation based on the worst individual pollutants. The pollution index is classified. Table 1 is the water pollution grading standard.

3 RESULT AND DISCUSSION

3.1 *BP neural network training*

The data of 1997 to 2002 is the inputs, the data of 2003 is the desired output, than trained the network.[8]

Parameters of network are set as follows:

1. network structure: 6-7-1;
2. net.trainparan.epochs = 1000;
3. net.trainparan.mc = 0.9;
4. net.trainparam.lr = 0.05;
5. net.trainparan.goal = 1e-5;
7. net.trainParam.show = 50;

3.2 *Precision test of network*

The value of comprehensive pollution index of 2004 year can be predicted by the trained network. Results of the comparison showed that the predicted value and the true value are very close. The comparison of the predicted value and the true value are showed in Figure 3.

From the simulation results, BP neural network have good learning performance. The network output and target output has achieved pre-requirement. The forecasting results showed that BP neural network have a higher fitting precision to the data of Taihu Lake's historical data (1997–2002)

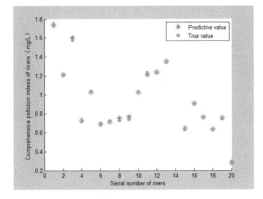

Figure 3. The contrast of real and predictive value of comprehensive pollution indexes.

and the relative error is small. At the same time, it can predict the trend of comprehensive pollution indexes of 2005. The model can meet the needs of daily forecast work.

3.3 *The prediction analysis*

Use the already trained neural network model to predict the comprehensive pollution index of 2005. The result can be seen in Table 2. According to the Table 1, the forecast results are classified.

The predictions of 2005 are in line with the development of the river quality of Taihu Lake. From the analyst of data from 1997 to 2004, pollutions from inflow rivers brought too much load to Taihu Lake. All this made the water quality more serious. At the year of 2001, WuxiCity and Wujin City take effective measures to change the situation, as a result, the pollution got controlled, and the quality of water changed a lot. It is obvious that the comprehensive pollution situation still need great attention in 2005. The quality of water of inflow rivers is still better than outflow rivers. In the inflow rivers, the most seriously polluted rivers are Zhihu Harbour, Xiaoxi Harbour and the Wujin Harbour. There is no clear water of grade II in the year of 2005. The results showed that we have to put more attention to the water quality of Taihu Lake.

Table 2. Predictions of comprehensive pollution indexes of Taihu in 2005.

| River's name | Cross-section name | |
	mg/L	Letter size paper
Xiaoxi harbour	1.7422	V
Zhihu harbour	1.3488	V
Liangxi river	1.0805	V
Hongxiang riveryin	0.9693	IV
Yincun harbour	0.9250	IV
Wuxi harbour	0.8916	IV
Taipu harbour	0.8587	IV
Chendong harbour	0.8416	IV
Guangdu harbour	0.8475	IV
Shedu harbour	1.0957	V
Wujin harbour	1.2432	V
Taige river	1.2391	V
Sudong river	1.2300	V
Xu river	1.2099	V
Xuguang river	1.0605	V
Muguang river	0.8486	IV
Wangyu river	0.7037	IV
Wusong jiang	0.6448	III
Taipu river	0.6225	III
He river	0.6019	III

4 CONCLUSION

Using the nonlinear dynamical characteristics and self-learning features, this study established one three-layer BP neural network to predict the comprehensive pollution indexes of 2005. This study took six sets of data which form 1997 to 2002 as training samples. Then the network is trained by fix the neural network weights and thresholds. This paper predicted the comprehensive indexes of inflow and outflow rivers of Taihu Lake. The results showed that pollution is more serious than before, in line with the water quality. The water quality is almost V grade. BP neural network has strong nonlinear mapping ability, flexible network architecture. Compared with the traditional statistical modeling methods, BP neural network have high prediction accuracy, can reflect the inherent variation of water quality indicators. It can provide scientific support for control water pollution.

Foundation item: The part of science and technology support program of Jiangsu Provincial (BE2011061).

Project name: Project name: Research on the application technology of water monitoring and analysis system of Taihu Lake based on sensor network.

Biography:

1. *Ma Zhenghua (1962–) male, Jiangsu Changzhou, Professor, bachelor degree, main research direction: The development and application of embedded system;*
2. *Wang Teng (1986–) female, Jiangsu Xuzhou, postgraduate, research direction: The development and application of embedded system E-mail: 15961292936@163.com.*
3. *Zhou Jiongru (1962–) female, Jiangsu Changzhou, Senior Engineer, bachelor degree, main research direction: The development and application of embedded system.*

After two months, this paper finally finished. Thanks to my classmates and friends who helped me. Thanks to the scholars referred to this paper, without their research, I will be very difficult to complete the writing. Particularly tanks to my teacher Ma Zhenghua and Zhou Jiongru, their selfless guidance and help benefit me a lot. My thesis may have deficiencies due to my limited academic standards, welcome teachers and students' criticism and correction, I want to learn together with you.

PREFERENCES, SYMBOLS AND UNITS

Qiao Junfei, Huang Xiaoqi & Han Honggui. 2012. Recurrent neural network-based control for wastewater treatment process. Lecture Notes in Computer Science 7638:495–506.

Tan Ying & Ruan Guangchen. 2010. A three-layer back-propagation neural network for span detection using artificial immune concentration. Soft Computing—A Fusion of Foundations, Methodologies and Applications 14(2):139–150.

Tian Yong, Sun Yue, Su Yugang, Wang Zhihui, Tang Chunsen. 2012. Neural network-based constant current control of dynamic wireless power supply system for electric vehicles. Information Technology Journal 11(7): 876–883.

Wang Lili, Chen Deyun, Xu Liyuan, Zhang Zhen and Yu Xiaoyang. 2012. An image reconstruction method based on simulated annealing and back algorithm for electrical capacitance tomography. Information Technology Journal 11(7):936–940.

Wang Yong, Wang Gang, Shengjing Feng. 2012. The dynamic model prediction study of the forest disease, insect pest and rat based on BP neural networks. Journal of Agricultural Science 4 (3) pp. 291–297.

Wang Yuhong, Xu Jun & Wang Guan. 2006. Assessment for production and operation ability of medium and small-sized enterprises based on neural network. Joural of China University of Mining and Techology 16(3):336–380.

Zang Wenke, Liu Xiyu, Xia Ruifang. 2012. Spiking DNA computing with applications to BP neural Networks classification. Research Journal of Applied Sciences, Engineering and Technology 4(16):2765–2772.

Zeng Shuiping, Cui Lin, Li Jinhong. 2012. Diagnosis system for alumina reduction based on BP neural network. Journal of Computers Yea 7(4): 929–933.

Zhao Jun, Cheng Xinlu, He Bi & Yang Xiangdong. 2006. Neural network study on the correlation between impact sensitivity and molecular structures for nitramine explosives. Structural Chemistry 17(5):501–507.

Zhuang Baoyu, Yu Jiaying, Sun Jingmei, Shan Jinlin. 2011. Research and application of non-line early warning system for source water quality of reclaimed water plant. Chinese Journal of Environment Engineering 5(6):1232–1236.

Frontiers of Energy and Environmental Engineering – Sung, Kao & Chen (eds)
© 2013 Taylor & Francis Group, London, ISBN 978-0-415-66159-1

Design and research of the fresh corn peeling machine's roller

C.H. Zhao, B.L. Hu & Z.Y. Cheng
Engineering college, Gansu Agricultural University, Lanzhou, China

ABSTRACT: The peeling roller which is the key components of the corn peeling machine was given an innovative design. Each pair of peeling rollers is composed by a metal roller and a rubber roller. There is spiral metal band on the surface of the metal peeling roller. On the recessed portion, many wood stoppers on which fixed "person" shaped pierce leaves teeth on the metal peeling roller. Rubber peeling roller's work area is divided into two parts. Spirally arranged pattern rod peeling tooth are on the feed side of the peeling roller. On the discharge side of the peeling roller are inclined ladder shaped rubber bulges. This kind of peeling device is suitable for the high and bract tight fresh edible corn in humidity.

Keywords: corn peeling machine; peeling mechanism; peeling roller; peeling teeth

1 INTRODUCTION

Corn peeling roller is the most important working parts of the corn peeling machine. Generally, it uses the bulge on the surface of the corn peeling to shred the bract. Two rollers acted a relative rotation and at the same time, the bract is clamped by the rollers and stripped. Peeling machine's working quality depends on the corn peeling roller's structure parameters and movement parameters, such as peeling net rate, breakage rate, seed expulsion rate, productivity and so on[1,2]. Fresh edible corn peeling machine designed in this topic is consisted by two groups of peeling rollers to meet the corn peeling machine's size and economy.

2 THE WORKING PRINCIPLE OF THE PEELING ROLLER

Generally, Corn along the inclined plate installed at the feed inlet slides into the peeling roller which acts a relative rotation and has an inclination angle with the horizontal direction. Rollers use surface structure and cooperate with each other to capture and peel off the bracts of ear. For example, some turning rollers are provided with rigidity peeling teeth on the roller's recessed portion. Tooth end higher than the outer surface of the roller. In the peeling process, first of all, peeling tooth tore ear bract. Then, the relative rotation of the two pair of rollers would clamp the bracts which had been torn. The two rollers used rotating force to pull the bracts. When the friction between the peeling rollers and the bracts was bigger than the binding force between the bracts

and the spike stalks, bracts would be peeling off from the spike stalks. When the peeling roller rotating, because the two pairs of roller's materials are different, the tangential friction force they produced was not equal. By that reason, ear rotated around its axis and the peeling process became more fully. In the process of the peeling roller's rotation, ear glided down along the leaning peeling rollers and sent out of the machine. At the same time, the mixture of the bracts and the stems were discharged from the gap between the two rollers[3].

2.1 Overall design and structure composition

As shown in Figure 1, Corn peeling machine's overall structure is mainly composed of peeling

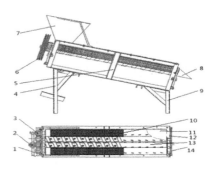

Figure 1. Corn peeling machine structure figure.
1. gear 2. pulley 3. gear join key 4. base1 5. chassis 6. pulley 7. hoppe 8. guide groove join key 9. base2 10. rubber peeling roller 11. cover 12. bearing seat 13. iron peeling roller 14. bearing.

mechanism, material inlet, material outlet and transmission system. This peeling mechanism use a combination of segmented metal roller which with spiral bulge, septa and "person" shaped peeling tooth structure on the surface and sectional rubber roller which with rectangular peeling tooth and trapezoidal convex structure on the surface.

The combination structure can effectively tear corn bract, increase the contact area between corn bract and metal roller, rubber roller and take the corn bract segmentation peeling. Especially the rotary friction that rubber roller to bract increased. So it can improve the humidity and bract compact corn's peeling rate. At the same time, due to projection which been staggered arrangement and ladder shaped is arranged on the outer surface of the rubber roller, it also has rip work ability. Therefore the combination structure can hold firmly ground corn bract to improve stripping ratio. The peeling mechanism has been the national patent. The patent number is 201120335781.1.

Peeling roller is an important working part of the corn peeling device. Axis's height of each pair of peeling rolls is different. They were V-shaped or grove. The angle between the axis and the horizontal direction is 8°–12°, which is conducive to ear along the axis to the decline. Currently, metal roller and rubber roller are not segmented designed. In the peeling operation this structure cannot take the corn bract segmentation peeling, and the rotary friction between corn bract and metal roller, rubber roller is low [4]. On humidity and bract compact corn, peeling rate is low. This study focused on a new type of high efficient maize peeling mechanism which on humidity and bract tight fresh corn (fruit corn, glutinous corn, sweet corn) having a good peeling effect.

2.2 Corn peeling machine's working principle equations

Sending the corn to the middle of peeling roller and pressing conveying plate which with an angle to the horizontal and rotate in the opposite direction, the range of H should focus the center of gravity of the ear corn on the level of two-peeling roller as shown in Figure 2, corn which under the action of pressing conveying plate and gravity moves along the surface

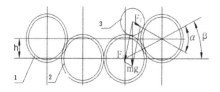

Figure 2. The layout of the peeling roller.
1. rubber roller 2. cast iron roller 3. corn.

of peeling rollers. Since the two peeling roller's height and material are different, so that they have different level of friction to the corn. Leading to ears act rotation along axis. During the pivoting and sliding, pull leaf device which on the iron roller lacerates ear bract. And then the pressing pair of rollers tears up the bracts. The bracts discharged out of the peeling mechanism. At last, ear glide down along the peeling rollers and send out of the machine. The whole peeling process is completed.

The peeling roller's structure as shown below: each pair of rollers is in one high and one low-layout. There is a height difference H between the two rollers. So that corn ear's weight Q have a different pressure to the two peeling rollers. They are Q1 and Q2. Calculation formula respectively:

$$Q_1 = \frac{Q\cos(\alpha-\gamma)}{\sin 2\beta} : Q_2 = \frac{\cos(\alpha+\gamma)}{\sin 2\alpha} \qquad (1)$$

Because $\cos(\alpha-\gamma) > \cos(\alpha+\gamma)$ so $Q_1 > Q$.

Due to the difference pressure, the two peeling rollers' friction is also different. Thus urging the ear acting rotation along its axis, and then improving the peeling rate.

According to the above analysis, peeling effect related to height difference H's size. The smaller H values and the lower the peeling net rate. But when the H value is greater than the radius of the peeling roller, center of gravity of the ear corn near the axis of the peeling roller in an unstable state, easily rolls to the other side of the stripping roll and c rawl out of peeling roller, thereby reducing the stripping net rate. Analysis and testing shows that: center distance near the midpoint. The formation of relatively stable along its own axis of rotation couple choose this point allows ear of corn, thereby enhancing the stripping net rate. Select that point allows the ear corn to the formation of relatively stable along its own axis of rotation couple, thereby reducing the peeling net rate[5].

3 STUDY AND DESIGN OF THE PEELING ROLLER'S STRUCTURE

3.1 Analysis of the influence factors

As can be seen from the working process of corn peeling roller stripping the bracts, that the shape of the surface structure of the corn peeling roller directly affect on the peeling. The peeling effect related to turning roller's surface hardness, the connected curved surface between the turning roller's bulge part and the base surface, the machining accuracy of the equipment, and so on. The affect results of the different factors of the turning roller to the peeling effect as shown in Table 1.

Table 1. The influence of skinning quality caused by factors of peeling roller.

Number	Factor	Peeling net rate	Breakage rate	Production rate
1	Roller hardness	Exist the best value	The bigger the value is, the higher it is	A little influence
2	Roller speed	Exist the best value	The bigger the value is, the higher it is	Direct ratio
3	Roller diameter	Exist the best value	The bigger the value is, the higher it is	A little influence
4	Roller section	The more the corner angle is, the higher it is	The smoother the corner angle is, the lower it is	A little influence
5	Roller material	Directly proportional to bracts adhesion rate	Directly proportional to bracts adhesion rate	A little influence

Therefore, through improved corn peeling roller surface structures and materials, as well as to optimize the structural parameters and motion parameters of the peeling roller could enhance the peeling net capacity, improve peeling net rate, reduce the damage to the corn and reduce the breakage rate and seed shedding rate.

3.2 *Peeling roller design*

In the working process of corn peeling machine, peeling roller's shape, speed and tilt angle are the main factors affecting its performance[6,7]. Metal husking roll structure as shown in Figure 3. A peeling roller with spiral around the metal belt, the recessed portion with a cork, cork top fixed herringbone tooth row bract, row tooth top is slightly higher than that of spiral surface. Metal husking roll on the grasping ability of corn bract was improved in this structure. The draw leaves teeth arranged in cork for wear after adjustment and replacement.

Rubber peeling roller structure as shown in Figure 4. The work area is divided into two parts in a debarking roller feed side is spirally arranged on the threaded rod type debarking teeth, the discharging side is an inclined ladder shaped rubber

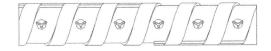

Figure 3. Metal peeling roller structure.

Figure 4. Rubber peeling roller structure.

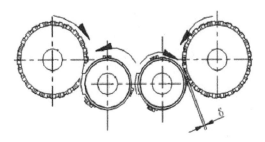

Figure 5. The arrangement of the peeling roller.

bulges. Trapezoidal tooth from short rubber peeling with inclined trapezoid convex rubber ring spirally arranged closely, in order to facilitate replacement after abrasion. The anterior part of the rotation of corn bract cut, after part and a metal roller with corn bracts tearing, peeling efficiency significantly improve.

The arrangement of the peeling roller is shown in Figure 5. The two pairs of peeling rollers arranged in a groove shape, the one is high and the other is low for each pair of peeling rollers arranged. Metal peeling rollers are fixed in the middle of the machine, and their centers below the centers of the rubber rollers. Rubber peeling rollers in floating state are connected on both sides. According to the specific situation of the corns, researchers adjusted the gap δ between the two metal rollers. During the work, adjusting the rubber peeling rollers made the gravity center of the corncob near the midpoint of two peeling rollers horizontal center distance. Choosing the point could make corncob form more stable and revolved along its axis couple and improved peeling rate.

4 SUMMARY

The corncob peeling net rate, broken rate, and fallen seed rate were all affected by the length of peeling roller, especially to peeling net rate. If peeling roller is too short, it can shortening the skinning bractea time and reducing peeling net rate. As a result, it won't reach the corn peeling requirements. If skinning roller is too long, it can increase their working time and give rise to the result of which the bractea is neatly peeled in the first part of peeling roller and bit wound in the second half and causes to vast broken and fallen seed. It was measured by test that the peeling net rate of corncob is about 60% in 250 mm skinning roller, 82% in 500 mm, 95% in 750 mm, and 98% was in 1000 mm. Test results show that longer peeling roller length and peeling work time can slight increase the skinning net rate, while broken rate and fallen seed rate can be added obviously. In conclusion, the effective length of peeling roller is determined 1000 mm.

ACKNOWLEDGEMENTS

The mission has been funded by The Master tutor project of Gansu province Educate hall (1002-08); National Natural Science Foundation of China (50965001); 2011 innovation talent plan in Gansu province and The Gansu province agricultural accomplishment of science and technology transforms (1105 NCNA0).

REFERENCES

[1] B Bell. Farm Machinery. 2nded. England: Farming Press, 1983.
[2] Culpin cloude. Farm Machinery. 12thed.oxford: Blackwell Scientific PVmublication, 1992.
[3] Wenqi Shang, SCH-4 type fresh corn peeling machine inspection report [R], Beijing: Department of Agriculture Agricultural Machinery Testing Centre, 2009.
[4] Yuqiang Zhao, Xiaopeng He, Design and Test of fresh corn peeling machine [J], Agricultural Engineering Newspaper, 2011, 2, 114–118.
[5] Yuan Li, Chunhong Mu, Mingzhi Huang, 5BY–7. 0 type corn peeling machine research and design [J], Mechanization of rural pasturing area, 2007, 3, 12.
[6] Taizhu Wang, Min Li, 6YBJ-2 type high efficiency wind clear corn peeling machine Research and design [J], Design and manufacturing, 2007, 6B, 85–87.
[7] Huiquan Zhang, Design and Study on the corn sheller [J], Agricultural science technology and equipment, 2009, 12, 45–47.

Frontiers of Energy and Environmental Engineering – Sung, Kao & Chen (eds)
© 2013 Taylor & Francis Group, London, ISBN 978-0-415-66159-1

Analysis of energy demand and CO$_2$ emissions of Guangdong electricity sector until 2020

P. Wang, B.B. Cheng, L. Li, P.G. Wen & D.Q. Zhao
Guangzhou Institute of Energy Conversion, Chinese Academy of Science, Guangzhou, Guangdong, China

ABSTRACT: In this paper, the electricity sector of Guangdong is selected to analyze the energy demand and CO$_2$ emissions as it consumed the largest share of primary energy and had the highest level of CO$_2$ emission. In 2007, more than 40% of primary energy and about 81% of coal were consumed by the electricity sector. These caused about 50% of the total CO$_2$ emission in Guangdong. For limiting the CO$_2$ emissions, a techno-economic model was built to simulate the development of the Guangdong electricity supply system, in which the objective function of minimizing the total cost of the electricity supply system during the investigation. Results show that considering costs, an optimal electricity development path is obtained through the model. To achieve this target, several important actions are proposed: increasing the capacity of nuclear power plants to 30 GW; decreasing the capacity of low-efficiency coal-fired power plants by 9 GW; exploiting up to 18 GW of renewable energy, mainly wind power.

Keywords: low carbon scenario; techno-economic model; guangdong; installed capacity

1 INTRODUCTION

As one of the most developed and one of the largest energy consumption areas in China, Guangdong has been chosen by the Chinese government as the country's low-carbon pilot provinces. In recent years, Guangdong has achieved the highest GDP among all the provinces of China, and contributed more than 11% to the total GDP of the country. It also accounts for about 8% of total energy consumption, and coal still dominates as the primary energy supply. For example, Guangdong consumed 19.6 million tons of coal equivalents in 2007, and coal accounted for 49.1% of this consumption[1].

In this paper, the electricity sector of Guangdong was selected to analyze the CO$_2$ mitigation potential as it consumed the largest share of primary energy and had the highest level of CO$_2$ emission. In 2007, more than 40% of primary energy and about 81% of coal were consumed by the electricity sector. These caused about 50% of the total CO$_2$ emission in Guangdong. So the electricity sector of Guangdong should afford the greater carbon emission targets for the decomposition of the task. Recent years, many studies on CO$_2$ mitigation in the electric power sectors of China have been conducted. Li et al. assessed the resource, technology status, and CO$_2$ mitigation potential of renewable energy in China [2]. Vuuren et al. developed a set of energy and emission scenarios to analyze available options including measures in the electricity sectors to mitigate carbon emissions[3].

Hu and Jiang (2001) and Zhou et al. (2003), several major emission-intensive sectors in China were chosen and the electricity sector was included. Hu and Jiang (2001) used the Assessment Integrated Model (AIM) and analyzed the current development and future trend of electric technologies[4]. However, Zhou et al.'s (2003) study use the Long-range Energy Alternatives Planning (LEAP) system model to analyze the technology cost of different technology. Cai et al. analyzed the projected CO$_2$ reduction potential of China's electricity sector in 2030 using the long-range energy alternative planning system model to simulate the different development paths in electricity sector[5]. Yi investigated the pathways and costs of a transition to low-carbon electricity by 2020 in Guangdong using a techno-economic model[6].

However, most of these studies deal with this research in China as a whole; few focused on a special region of China, least of all Guangdong. To identify optimal electricity development pathways for Guangdong, we built a modified techno-economic model based on Yi's model. It is still an optimal model with the objective function of minimizing the total cost of electricity supply system during the evaluation period. It simulates the development process of the electricity sector by building new generation capacity and phasing out old ones. In this modified model, the CO$_2$ emission was limited, and the optimal electricity development path can be obtained. The model includes results on the development of the electricity supply system, the investments required for this development, the cost of electricity supply.

2 METHODOLOGY

A techno-economic model was built to simulate the development of the Guangdong electricity supply system. Figure 1 shows the basic structure of the model.

2.1 Electricity demand and supply

In the model, electricity demand is exogenously projected, and electricity supply is supposed to equal electricity demand. Electricity supply is described in terms of the production output from every electric power generation technology, such as Supercritical (SC) steam plants, nuclear power plants, and hydropower plants.

2.2 Objective function

The objective of the model is to minimize the total system cost during the evaluated period. Thus, the total system cost includes investment, operation, and maintenance costs, as well as termination costs from power plants phased out in the middle of their technical life time. Therefore, the objective function of the model is

$$\min C_{total} = \sum_{y=start}^{y=end} (1+d)^{start-y} \times C_{y,an} \quad (1)$$

$$C_{y,an} = \sum_{k} (Fx_{y,k} \times EQ_{y,k} + Vr_{y,k} \times ELE_{y,k}) + \sum_{f} EN_{y,f} \times P_{y,f} + \sum_{k} QEQ_{y,k} \times QC_{y,k} \quad (2)$$

where C_{total} is the total cost during the entire investigated period; d is the discount rate; $C_{y,an}$ denotes

Table 1. Projections on electricity demand of Guangdong.

	2009	2010	2015*	2020*
Electricity demand (GWh)	361000	406000	63000	780000

*The number of 2007, 2009, 2010 year is actual value, others are projected[7-12].

the total cost in year y; $Fx_{y,k}$ represents the fixed cost, including the annualized investment through the assumed lifetime for the installed power plants in technology k in year y; $Vr_{y,k}$ is the variable cost of active capacity in technology k in year y; $ELE_{y,k}$ denotes the sum of electricity generated by technology k in year y; $P_{y,f}$ is the price of fuel f in year y; $QEQ_{y,k}$ is the phased-out capacity in technology k in year y; and $QC_{y,k}$ is the cost of the phased-out capacity in technology k in year y. The cost mainly considers wasted investment from phased-out power plants, which no longer run to the end of their designed technical life. The model equations are solved using the General Algebraic Modeling System, a high level programming syntax suitable for mathematical programming including optimization[6].

3 ASSUMPTIONS AND INPUT PARAMETERS

The projections on electricity demand are based on the study of the South China power grid planning in the 12th five-year plan and medium long term scheme, in which the electricity demand level up to 2020 is analyzed. The projections on electricity demand (Table 1) consider GDP development, population, industry structure, and so on. These are currently the most reliable projections. The core of the model is electric power generation technologies. More details are described in Yi's paper[6].

4 RESULTS AND DISCUSSION

4.1 Scenarios of the electricity supply system

The electricity capacity is based on the state of the Guangdong electricity supply system in 2010. The total local installed capacity was 68 GW in 2010. The capacity of coal-fired power plants was 42 GW. The capacity of natural gas power plants was 6 GW. The capacity of oil power plants was 6 GW. The Nuclear Power Station had a capacity of 5 GW. The capacity of hydro power plants was 8 GW, nearing their upper limit. The rest, with a capacity of 1 GW, include wind power plants,

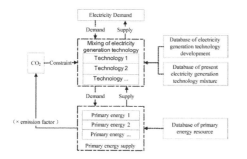

Figure 1. Basic structure of the Guangdong electricity system model.

municipal waste power plants, and others. Aside from the local installed capacity, the electricity supply system also imported electricity of 90.7 TWh from the west in 2010.

For the first one (Business as usual scenario), it is assumed that expansion of the electricity supply system is based only on coal-fired power plants. For the second scenario (Energy-saving scenario), it is supposed that the decision makers have already envisaged the use of natural gas Combined Cycle Power Plants (CCPP) in addition to middle nuclear power plants. The third scenario mixes natural gas CCPP, large nuclear power plants and more renewable energy power plants. In the LCS scenario (Figure 4), more efforts are implemented to achieve the target of reducing CO_2 emission per unit electricity supply in 2020. Although coal-fired power plants have been developed less intensively, they continue to dominate the electric power generation system in Guangdong with a share of 45% of local installed capacity. Oil power plants are initially phased out in 2020. New natural gas power plants need to be constructed during China's 13th five-year plan (2015–2020). Renewable energy undergoes intense development. Hydro power, biomass, and PV are all developed to their maximum limits because of their lower potential. The capacity of wind power increases to about 12 GW by 2020. Nuclear power is exploited to their maximum capacity of 30 GW in the LCS scenario.

4.2 Scenarios of energy consumption and CO_2 emissions

Figures 5 and 6 show the progress of total energy consumption of electricity sector and CO_2 emission from 2010 to 2020 in the three scenarios. For realizing the energy-saving and low-carbon scenarios, some technical and policy measures should be arranged to the electricity sector of Guangdong province until 2020. In the BAU case, as the share

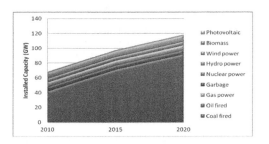

Figure 2. Development of installed capacity until 2020 (BAU scenario).

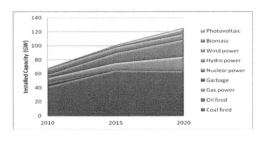

Figure 3. Development of installed capacity until 2020 (Energy-saving scenario).

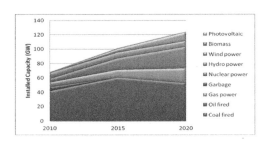

Figure 4. Development of installed capacity until 2020 (Low-carbon scenario).

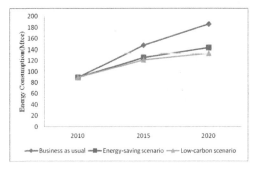

Figure 5. Energy consumption from electricity supply in BAU, ESC, and LCS up to 2020.

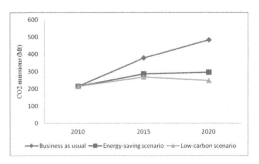

Figure 6. CO_2 emission from electricity supply in BAU, ESC, and LCS up to 2020.

of coal-fired capacity increase, the average CO_2 emissions per kWh of electricity will increase and the total energy consumption and CO_2 emissions will be a significant increase trend with growth rate of 110%. In the ESC and LCS case, the use of Demand Side Management (DSM), optimization of power installed capacity structure, energy efficiency enhancement and other energy saving measures are deployed. So the trend of energy consumption and emissions has been improved in Figures 5 and 6. Compared with the BAU scenario, the electricity sector of Guangdong province appears the 23%-29% of the energy saving potential and 39%-49% of the emission reduction potential. According to the preferred measures for emission reduction potential contribution, the biggest contribution depends on the development of the clean power. Developing nuclear power gives a big reduction contribution, followed by natural gas power generation and wind power. Demand side management, power plant system transformation, photovoltaic, biomass and other measures to reduce emissions is not large, but the emission reduction potential of these measures can not be ignored.

5 CONCLUSION

In this study, several measures will contribute to achieving CO_2 reduction and realizing low carbon electricity development in three scenarios. The most important is developing nuclear power to the planned upper limit, which contributes the most to CO_2 reduction. Second, improving the efficiency of coal-fired power plants also contributes to a great extent. The third is expanding renewable energy capacity by about 18 GW, mainly the wind power. More researches in other sectors need to be further conducted.

ACKNOWLEDGEMENTS

This work was co-funded by Guangdong Natural Science Foundation (GNSF) under the Grant No. S2011040002839, Director Innovation Fund (y007r81001), National Key Technology R&D Program (2011BAG07B00).

REFERENCES

1. National Bureau of Statistics of China, National Energy Administration of China. China energy statistics yearbook 2008. Beijing, China Statistics Press; 2009 [in Chinese].
2. Li J, Wan Y, Ohi J. Renewable energy development in China: Resource assessment, technology status, and greenhouse gas mitigation potential. Applied Energy 1997; 56: 381–394.
3. Vuuren D, Zhou FQ, Vries B, et al. Energy and emission scenarios for China in the 21st century—exploration of baseline development and mitigation options. Energy Policy 2003; 31: 369–387.
4. Hu X, Jiang K. Evaluation of Technology and Countermeasure for Greenhouse Gas Mitigation in China. Beijing, China Environmental Science Press; 2001 [in Chinese].
5. Cai W, Wang C, Wang k, et al. Scenario analysis on CO_2 emissions reduction potential in China's electricity sector. Energy Policy 2007; 35:6445–6456.
6. Yi Jingwei, Zhao Daiqing, Hu Xiulian, Cai Guotian. Transition to Low-Carbon Electricity by 2020 in Guangdong, China: Pathways and Costs. Strategic Planning of Energy and the Environment (in published).
7. Guangdong Bureau of Statistics, Guangdong Investigation Team of National Bureau. Guangdong Statistical Yearbook 2008. Beijing: China Statistics Press, 2008.
8. Guangdong Bureau of Statistics, Guangdong Investigation Team of National Bureau. Guangdong Statistical Yearbook 2009. Beijing: China Statistics Press, 2009.
9. Guangdong Bureau of Statistics, Guangdong Investigation Team of National Bureau. Guangdong Statistical Yearbook 2010. Beijing: China Statistics Press, 2010.
10. Guangdong Bureau of Statistics, Guangdong Investigation Team of National Bureau. Guangdong Statistical Yearbook 2011. Beijing: China Statistics Press, 2011.
11. National Bureau of Statistics of China. Main Data of the Fifth National Population Census. http://www.stats.gov.cn/tjsj/ndsj/renkoupucha/2000pucha/pucha.htm.
12. Guangdong Bureau of Statistics. Guangdong Statistical Yearbook 2006. Beijing: China Statistics Press, 2006.

Frontiers of Energy and Environmental Engineering – Sung, Kao & Chen (eds)
© 2013 Taylor & Francis Group, London, ISBN 978-0-415-66159-1

Removal of MC-LR in photoelectrocatalysis process

M.G. Wang, Q.K. Qi, Y.J. Jin, Y. Zhang, X.Y. Du, Y.B. Zuo & Y. Zhang
School of Environmental Science and Engineering, Zhejiang Gongshang University, Hangzhou, Zhejiang, China

Q. Xin
College of Electronic Information, Hangzhou Dianzi University, Hangzhou, China

ABSTRACT: The removal of MC-LR by photoelectrocatalysis method with different TiO_2 nanotube films was studied in this study. Different kind of electrolytes was used to prepare TiO_2 films by electrochemical oxidation method. The surface topography of TiO_2 films with different electrolyte was also detected. There are alveolate TiO_2 nanotubes formed with Citric Acid + NaF electrolyte. While with Ethylene Glycol + NaF electrolyte, the TiO_2 nanotubes had several floccus. And with Oxalic Acid + NH_4F electrolyte, TiO_2 nanotubes were arranged uniformly and orderly. The removal of MC-LR was also investigated with different electrolyte in photoelectrocatalysis system. The highest of removal efficiency was obtained with Citric Acid + NaF electrolyte. In addition, the kinetic model of MC-LR degradation was also analyzed, which was fit to the first-order kinetic model.

Keywords: MC-LR; photoelectrocatalysis; TiO_2 film; kinetic model

1 INSTRUCTION

Microsystems (MCs) were the toxins produced by cyanobacteria in the lake eutrophication, which is a class of hepatotoxic monocyclic heptapeptides. During large bloom events microcystins can reach concentrations in the ppm range which polluted the drinking water sources (He et al. 2012). Microcystin-LR (MC-LR) is one of the most toxic pollutants among the MCs having set $1 \ mg \cdot L^{-1}$ of MC-LR as a guideline value in drinking water which was set by WHO (WHO, 1998).

Conventional water treatment methods, such as flocculation, adsorption, chloride oxidation, were used for MC-LR removal (Acero et al., 2008, Lee & Walker, 2006), but these methods were not removal the MC-LR completely and produced the second pollutants probably. It was important to find a useful method to removal MC-LR in the water. Advanced Oxidation Technologies (AOTs) are a serial of methods to generate active radicals such as hydroxyl radicals to removal such toxic organic chemicals, in which the photoelectrocatalysis is an efficient method (Antoniou et al., 2008).

Photoelectrocatalysis is a method for combination of the TiO_2 photocatalysis and electrocatalysis. TiO_2 photocatalyst was induced by UV radiation (< 387 nm) to separate an electron (e⁻) and vacancy hole (h⁺), which could generate ·OH, as shown in Equations (1)–(3). However, there was a high recombination of electron-hole pairs on the surface of TiO_2, which was disadvantage to photocatalytic reaction (Triantis et al., 2012). Electrocatalysis is one of method to enhance the separation of electron and vacancy hole. In additional, electrocatalysis could produce direct and indirect oxidation of pollutants as shown in Equation (4) (Feng et al. 2005). Therefore, combination of photocatalysis and electrocatalysis could enhance the MC-LR degradation.

$$TiO_2 + hv \rightarrow h^+ + e^- \tag{1}$$

$$h^+ + H_2O \rightarrow 2H^+ + \cdot OH \tag{2}$$

$$h^+ + OH^- \rightarrow \cdot OH \tag{3}$$

$$2H_2O \rightarrow 2H^+ + 2 \cdot OH + 2e^- \tag{4}$$

In this study, the kinds of TiO_2 film was prepared and analyzed by SEM image. And then the removal of MC-LR during the different types of TiO_2 film was also compared. Finally, the kinetic model of MC-LR removal was established to analyze the photoelectrocatalysis process for MC-LR degradation.

2 EXPERIMENTS

2.1 *Materials*

Titanium plate and nickel plate were supplied by the Baoji Deli Titanium Industry Co., Ltd, Shanxi, China. MC-LR was supplied by Wuhan Pulan technology Co., Ltd. $C_2H_2O_4$, $C_6H_8O_7$, $C_2H_6O_2$, NaF, NH_4F were analytical grade. 78-1 magnetic stirrer, KQ2200DE ultrasonic cleaner, WYL-302S power supply and PHS-25CW microprocessor pH/mV meter were used in this study.

2.2 *Preparation of the TiO_2 films*

The TiO_2 nanotubes (NTs) film is prepared by electrochemical oxidation (Lei et al., 2007). At first, high purity titanium plate was cleaned by acetone and then by deionized water. A DC power supply was utilized for electrochemical experiment. Titanium plate was an anodic electrode and nickel plate was a cathode electrode. The electrolytes were prepared as $C_2H_2O_4$ + NaF or $C_6H_8O_7$ + NaF or $C_2H_6O_2$ + NH_4F. Then the electrochemical experiment reacted for 2 h at voltage of 20 V. Finally, the simple was calcined for 2 h at 400 °C.

2.3 *Setup*

The experimental apparatus was a beaker with rotor mixed. TiO_2 film was the anode electrode and stainless steel was the cathode electrode. The electrode distance and input voltage were both adjustable. The external high pressure mercury lamp was placed at one side of beaker with power of 250 W. The initial MC-LR concentration was 1 ppm and treated volume was 25 ml. The electrolyte concentration of Na_2SO_4 was about 0.1 mol·L^{-1}.

2.4 *Analysis methods*

Agilent HPLC with a C18 column was used to analysis the concentration of MC-LR with ultraviolet detector setting wavelength of 238 nm. The volume ratio of mobile phase of 0.05 mol·L^{-1} K_2HPO_4 solution and methanol was 60:40.

3 RESULTS AND DISCUSSION

3.1 *SEM of TiO_2 film with different electrolyte types*

Figure 1 shows the SEM images of TiO_2 films with Citric Acid + NaF electrolyte. It was clearly that it was alveolate TiO_2 nanotubes were arranged uniformly and orderly on the surface of titanium plate. But the wall thickness was large. As shown in Figure 1, the pipe diameter was about 60–80 nm, while the wall thickness was about 60 nm.

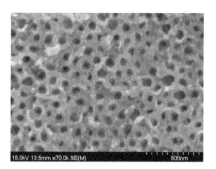

Figure 1. SEM image of TiO_2 film with Citric Acid+NaF electrolyte.

Figure 2 shows the SEM images of TiO_2 films with Ethylene Glycol + NaF electrolyte. The TiO_2 nanotubes were formed but not very clear and irregularity. Among the nanotubes there were several floccus formed. It was probably due to the electrochemical oxidation was not very uniform during the titanium plate surface and also because the titanium was not clean clearly.

Figure 3 shows the SEM images of TiO_2 films with Oxalic Acid + NH_4F electrolyte. TiO_2 nanotubes were arranged uniformly and orderly on the surface of titanium plate with the pipe diameter was about 60–90 nm, and the wall thickness was only about 12 nm, which was much less than the pipe diameter.

3.2 *Removal of MC-LR in photoelectrocatalysis process*

Figure 4 presents MC-LR removal under different TiO_2 film types with Citric Acid + NaF or Ethylene Glycol + NaF or Oxalic Acid + NH_4F electrolyte. During the photoelectrocatalysis system, the TiO_2 film was the anode electrode and stainless steel was the cathode electrode. The electrode distance was about 10 mm and the voltage was about 20 V. The initial MC-LR concentration was 1 ppm. As shown in Figure 4, MC-LR was graded completely under each TiO_2 film. At the TiO_2 film with Citric Acid + NaF electrolyte, the degradation efficiency of MC-LR was the highest that 99% of MC-LR was degraded after 5 min reaction. At the TiO_2 film with Ethylene Glycol + NaF electrolyte, the removal efficiency of MC-LR was about 99% after 10 min degradation. And at the TiO_2 film with Oxalic Acid + NH_4F electrolyte, the MC-LR was degraded completely after 20 min reaction.

3.3 *Reaction kinetic model analysis*

In order to investigate the kinetics of MC-LR degradation at photoelectrocatalysis, the experimental were assumed to the first-order kinetic:

Figure 2. SEM image of TiO_2 film with Ethylene Glycol + NaF electrolyte.

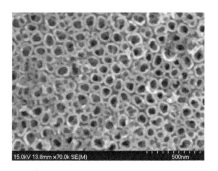

Figure 3. SEM image of TiO_2 film with Oxalic Acid + NH_4F electrolyte.

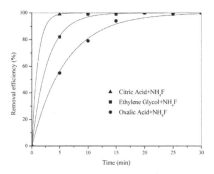

Figure 4. Effect of TiO_2 films on MC-LR degradation.

$$\frac{dc}{dt} = -kc \qquad (5)$$

where, c is the MC-LR concentration (mg·L^{-1}), t is the reaction time (min), k is the apparent rate constant (min^{-1}).

The apparent rate constant and correlation coefficient was fitted by MC-LR degradation in Figure 4 and list in Table 1. As shown in Table 1, the correlation coefficients R^2 were all larger than

Table 1. The apparent rate constant and correlation coefficient of phenol degradation.

Discharge mode	Rate constant (min^{-1})	Correlation coefficient R^2
Oxalic Acid + NH_4F	0.17	0.997
Ethylene Glycol + NaF	0.35	0.999
Citric Acid + NaF	0.92	0.999

0.99, which shows that the assumption for first-order kinetic model was fit with the data well. It was shown that apparent rate constant of MC-LR degradation with Citric Acid + NaF electrolyte was 0.92 min^{-1}, which was the largest, and then was that with Ethylene Glycol + NaF electrolyte of 0.35 min^{-1}, and the lowest of all was that with Oxalic Acid +NH_4F electrolyte of 0.17 min^{-1}.

4 CONCLUSIONS

The removal of MC-LR by photoelectrocatalysis system was investigated. The types of TiO_2 film used as the anode electrode were also studied with three different electrolytes. The SEM of TiO_2 films was also detected. With Citric Acid + NaF electrolyte, there are alveolate TiO_2 nanotubes were formed with pipe diameter was about 60–80 nm, but the wall thickness was about 60 nm. With Ethylene Glycol + NaF electrolyte, the TiO_2 nanotubes were not very clear and had several floccus. With Oxalic Acid + NH_4F electrolyte, TiO_2 nanotubes were arranged uniformly and orderly. During the photoelectrocatalysis, the removal of MC-LR was the highest with Citric Acid + NaF electrolyte and then was with Ethylene Glycol + NaF electrolyte and finally was with Oxalic Acid + NH_4F electrolyte. The MC-LR degradation by photoelectrocatalysis with different TiO_2 films was fitted to first-order kinetic model.

ACKNOWLEDGEMENT

This work is financially supported by National Natural Science Foundation of China (No.20906079, No. 20976162, No. 21006017), and National Natural Science Foundation of Zhejiang Province (Y5100356).

REFERENCES

Acero, J.L., Rodriguez, E., Majado, M.E., Sordo, A. & Meriluoto, J. 2008. Oxidation of microcystin-LR with chlorine and permanganate during drinking water treatment. Journal of Water Supply: Research and Technology-AQUA 57 (6): 371–380.

Antoniou, M.G., Shoemaker, J.A., de la Cruz, A.A. & Dionysiou, D.D. 2008. Unveiling new degradation intermediates/pathways from the photocatalytic degradation of microcystin-LR. *Environmental Science and Technology* 42 (23): 8877–8883.

Feng, C.P., Sugiura, N., Masaoka, Y. & Maekawa, T. 2005. Electrochemical Degradation of Microcystin-LR. *Journal of Environmental Science and Health, Part A: Toxic/Hazardous Substances and Environmental Engineering* 40(2): 453–465.

He, X.X., Pelaez, M., Westrick, J.A., O'Shea, K.E., Hiskia, A., Triantis, T., Kaloudis, T., Stefan, M.I., de la Cruz, A.A. & Dionysiou, D.D. 2012. Efficient removal of microcystin-LR by UV-C/H$_2$O$_2$ in synthetic and natural water samples. *Water Research* 46(5): 1501–1510.

Lee, J. & Walker, H.W. 2006. Effect of process variables and natural organic matter on removal of microcystin-LR by PAC-UF. *Environmental Science & Technology* 40(23): 7336–7342.

Lei, L., Su, Y., Zhou, M., Zhang, X. & Chen, X. 2007. Fabrication of multi-non-metal-doped TiO$_2$ nanotubes by anodization in mixed acid electrolyte. *Material Research Bulletin* 42(12): 2230–2236.

Triantis, T.M., Fotiou, T., Kaloudis, T., Kontos, A.G., Falaras, P., Dionysiou, D.D., Pelaea, M. & Hiskia, A. 2012. Photocatalytic degradation and mineralization of microcystin-LR under UV-A, solar and visible light using nanostructured nitrogen doped TiO$_2$. *Journal of Hazardous Materials* 211–212: 196–202.

WHO, 1998, Guidelines for *Drinking-water Quality*, second ed. Addendum to vol. 2, Health Criteria and Other Supporting Information. World Health Organization, Geneva.

Frontiers of Energy and Environmental Engineering – Sung, Kao & Chen (eds)
© *2013 Taylor & Francis Group, London, ISBN 978-0-415-66159-1*

Study on specific grinding energy of fine ELID cross grinding

K.L. Xu, Z.W. Tan & S.Y. Hu
Hunan Radio & TV University, Changsha, Hunan, China

ABSTRACT: Tungsten carbide is widely used as the most important substrate material in integrated circuit and micro electronic devices field. Electrolytic in-process dressing (ELID) grinding technique is an effective grinding process especially for machining hard and brittle material. In this paper, the mechanism of super high grinding, the specific grinding energy in grinding was introduced. The effect factors of specific grinding energy with ultra-high speed grinding were analyzed. The inverse relationship between specific grinding energy and processes parameter was established. In general, specific grinding energy is found to decrease with the increase of table speed, grinding speed, actual grinding depth and the rate of material removal.

Keywords: ELID grinding; tungsten carbide; specific grinding energy; super fine abrasive wheel

1 GENERAL INSTRUCTIONS

In recent years, Tungsten carbide (WC) is considered as an important substrate material, and largely demanded [1]. Current manufacturing processes for WC include slicing, lapping, etching, polishing and grinding [2]. The high brittleness and hardness make it difficult to be ground, and result in some cracks and surface damages in conventional grinding. Grinding is considered as an important and effective method for obtaining high quality surface in high efficiency. However, WC is difficult to be ground in conventional grinding, which resulted in grinding marks, subsurface cracks and damages on the ground surface [3,4]. Furthermore, grinding wheel couldn't maintain self-dressing ability. Electrolytic in-process dressing (ELID) technique has great potential as an effective dressing process especially for grinding hard or brittle materials and obtaining better surface roughness and less subsurface damages [5–6].

In this paper, the influences of various conditions on specific grinding energy of WC were studied using super fine abrasive wheel. A set of cross grinding experiments have been conducted through changing various grinding conditions including grain sizes, rotation speeds of grinding wheel and work piece, ELID conditions on specific grinding energy.

2 CALCULATION PRINCIPLE OF THE GRINDING FORCE IN GRINDING PROCESS

The grinding force measured by the dynamometer represents the sum of the forces between the active grains and the work piece. The tangential component of the grinding force can be used to compute the specific grinding energy, u, which is defined as the energy required removing a unit volume of material. This parameter is widely used in metal grinding research.

$$u = \frac{F_H V}{Vbd} \tag{1}$$

For surface grinding with wheel speeds much greater than table speeds, the horizontal force, F_H, multiplied by the wheel velocity, v, is the total power input. With a high G-ratio, the metal removal rate is the product of the table speed, v, the down feed or wheel depth of cut, d, and the width of cut, b. The specific grinding energy is given by the ratio of the power input to metal removal rate.

3 EXPERIMENTAL SET-UP AND PROCEDURE

3.1 *Experimental set-up for ELID cross grinding*

Schematic illustration of the experimental set-up for ELID cross grinding process is shown in Figure 1. The grinding wheel is connected to the positive terminal of a supply power and a negative electrode is installed near the wheel surface. The grinding wheel and the wafer rotate about their own rotation axes simultaneously, and the wheel is fed towards the wafer. A single crystal silicon wafer is installed on a vacuum chuck. The silicon wafer

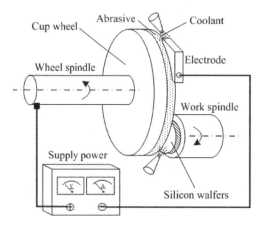

Figure 1. Schematic illustration of ELID cross grinding system.

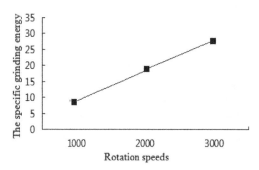

Figure 2. Rotation speeds.

or the work spindle rotates at a lower speed, and the wheel rotates at a higher speed. The rotation axis for the grinding wheel is offset by a distance of the wheel radius relative to the rotation axis for the silicon wafer. The electrolytic fluid is supplied into the small gap between the negative electrode and the grinding wheel. Because of anodic electrolysis process, bond material on wheel surface is dissolved and new sharp abrasive is exposed. At the same time, an oxide film is formed on the wheel surface, and then resists excessive electrolysis. This ELID cycle continues and makes wheel always keep best state during ELID grinding process [5–9]. Figure 2 shows the view of ELID cross grinding.

3.2 *Experimental conditions*

Grinding experiment was carried out using cast iron bonded diamond wheels and resin-metal bonded wheel on an ultra-precision cross grinder. The shape of grinding wheel was cup; the diameter

and width was 5 mm; grit sizes of wheel included #4000. The peak currents were set at 6 A; the open voltages were set at 80V. The feed rates were 0.5, 1, 2, 3 and 4 μm/min; the total of grinding depth was 14 μm. The wheel rotation speeds were maintained at 1000, 2000 and 3000 rpm; the workpiece rotation speeds were 150 and 250 rpm. The gap between the wheel and the electrode was kept at 0.2 mm. An electric current in the form of square pulse wave was supplied from the ELID power supply and the pulse time was 2/2 μs. It noted that an ELID pre-dressing process was required for maintaining the protrusion of diamond grains on the wheel surface before grinding. Five points of ground surface were respectively chosen and measured by contact surface roughness measurement.

4 RESULTS AND DISCUSSIONS

4.1 *The influence of rotation speeds*

In general, the specific grinding energy will increase to a particular value with an increase of wheel speed. Figure 2 shows the influence of wheel rotation speeds on the specific grinding energy using #4000 wheel. When the rotation speed is 1000 rpm, the specific grinding energy is lower. But if the rotation speed increases to 3000 rpm, the roughness will be higher.

4.2 *The influence of wheel depth of cut*

Figure 3 shows the influence of different wheel depth of cut and work piece on the specific grinding energy using #4000 wheels. For #4000 wheels, the grinding mode is just brittle-to-ductile transition mode. With an increase of cut depth of wheel, the specific grinding energy will be smaller. The change of the former (i.e. from 0.5 to 2) is obvious, but the smaller cut depth (i.e. from 2 to 4), the change of specific grinding energy is smaller.

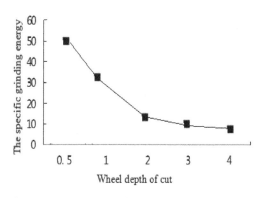

Figure 3. Wheel depth of cut.

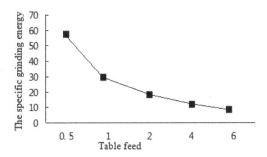

Figure 4. Table feed.

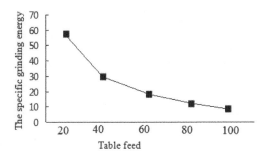

Figure 5. The rate o f material removal.

4.3 *The influence of table feed*

Figure 4 shows the influence of different table feed and work piece on the specific grinding energy using #4000 wheels. With an increase of table feed, the specific grinding energy will be smaller. The change of the former (i.e. from 0.5 to 2) is obvious, but the smaller cut depth (i.e. from 2 to 6), the change of specific grinding energy is smaller.

4.4 *The influence of the rate o f material removal*

Figure 5 shows the influence of different rate o f material removal and work piece on the specific grinding energy using #4000 wheels. With an increase of the rate of material removal, the specific grinding energy will be smaller. The change of the former (i.e. from 20 to 60) is obvious, but the smaller cut depth (i.e. from 60 to 100), the change of specific grinding energy is smaller.

5 CONCLUSIONS

The ELID cross grinding for WC was performed to study on the influences of different grinding conditions on the specific grinding energy when the grain size was very small. The effects of table speed, grinding speed, actual grinding depth and the rate of material removal on the specific grind-ing energy were examined, respectively. It appears that the relative high rotation speed and low feed rate, table speed, low actual grinding depth and the low rate of material removal are the appropriate experimental conditions for ELID grinding WC while using super fine abrasive grain. The experimental results can give good references for further researches on effectively grinding high quality surface of WC.

REFERENCES

[1] Yong Chul Lee, Tae Soo Kwak, H.Ohmori, et al. A Study on Nano-precision Combined Process of ELID Grinding and Magnetic Assisted Polishing on Optics Glass Material. Proceedings of the 41th ELID Seminar, 2005, 41:158–166.
[2] Hong Chen, James C.M. Li. Anodic metal matrix removal rate in electrolytic in-process dressing II; Protrusion effect and three-dimensional modelling. Journal of Applied Physics, 2000, 87(6):3159–3164.
[3] Jianyun Shen, Weimin Lin. Surfaces Formation of Precision ELID Grinding Rock Minerals. The 11th International Conference on Precision Engineering (ICPE), 2006, 8:16–18.
[4] Shaohui Yin, S. Morita, H.Ohmori, et al. ELID precision grinding of large special Schmidt plate for fibre multi-object spectrograph for 8.2 m Subaru telescope. International Journal of Machine Tools and Manufacture, 2005, 45 (14):1598–1604.
[5] Feihu Zhang, Wei Li, Zhongjun Qiu et al. Application of ELID grinding technique to precision machining of optics. Proceedings of SPIE, Advanced Optical Manufacturing and Testing Technology 2000, 2000, 4231:218–223.
[6] Tong Fuqiang, Zhang Yong, Zhang Feihu, et al. Experimental Study on Carbide Concave Spherical ELID Ultra-precision Grind. The First International ELID-Grinding Conference, 2008, 6: 45–51.

Energy conservation and environmental protection

Frontiers of Energy and Environmental Engineering – Sung, Kao & Chen (eds)
© 2013 Taylor & Francis Group, London, ISBN 978-0-415-66159-1

Immobilization of alkali protease with magnetic nanoparticles modified by amino-silane

S.N. Wang, L.Z. Jiang, Y. Li & D.D. Li

Food Collage, Northeast Agriculture University, Harbin, China

ABSTRACT: Fe_3O_4 magnetic nanoparticles were modified by 3-aminopropyl triethoxysilane (APTES) in the supporting of ultrasound, with glutaraldehyde as the crosslinking agent in inmobilization. Using X-ray photoelectron spectroscopy, infrared spectroscopy and vibrating sample magnetometer to characterize the prepared particles. The results show that the magnetic nanoparticles before and after immobilization with superparamagnetic, and easily recycled from the system under the conditions of the external magnetic field, while the thermodynamic stability of the immobilized enzyme prepared better than the free enzyme.

Keywords: magnetic nanoparticles, amino-silane, immobilization, alkali protease

1 INTRODUCTION

At present, the enzyme immobilization technology has been widely used in the food industry, pharmaceutical, chemical and other natural polymer carrier, with mining and exploration, modification, the use of supercritical technology, nanotechnology and membrane technology to immobilized enzyme became a development trend of immobilized enzyme technology[1]. Alkaline proteases are a class of suitable enzymes under alkaline conditions to hydrolyze protein peptide bonds, with the deeper understanding of the alkaline protease, it is widely used in hydrolyzed vegetable protein, but the enzyme is less stable under the influence of external factors such as temperature, pH and inorganic ions, easily denatured and mixed with substrate, also difficult to recycle[2-5] Enzyme mix with solution, no doubt bring some difficulties to further separation and purification of product. Immobilized enzyme compared to free enzyme stability has improved greatly, stability of the heat, pH, etc., reduced sensitivity to inhibitors, been separated easily. Immobilized system suitable for continuous, automated production, enzymatic process control, has improved the utilization efficiency of the enzyme and reduced the cost of production.

Magnetic nanoparticles are novel functional nanomaterials with article size between 1–100 nm, with huge surface area, high surface activity and adsorption capacity[6-8].

Magnetic nanoparticles can be wrapped by the polymer material, making the particle surface with specific groups, such as: -OH, $-NH_2$, -COOH, -CHO, etc., can be used as a vector carrying for protein and drug[9-13]. Magnetic nanoparticles as excellent carrier can be easily separated from the system and recycling because of its magnetic properties; the magnetic orientation of the particles can make the particles change with the changes of the external magnetic field[14-18].

2 PREPARATION OF FE_3O_4 MAGNETIC NANOPARTICLES AND SURFACE MODIFICATION

Fe_3O_4 magnetic nanoparticles were prepared with ultrasonic-assisted chemical coprecipitation method. $FeSO_4 \cdot 7H_2O$ and $FeCl_3 \cdot 6H_2O$ were mixed into a flask in a certain ratio, with continuous nitrogen accessed. During the reaction, ammonia was added into the reactor to maintain pH value of mixture is 11 under the action of the ultrasonic oscillation. The mixture was placed at room temperature for 30 min. After centrifugation, washing, ultrasonic cleaning machine to remove impurities on the surface and vacuum freeze-drying, Fe_3O_4 magnetic nanoparticles were prepared.

1 g of Fe_3O_4 magnetic nanoparticles dispersed in ethanol-water (volume ratio 50:1), then a certain amount of APTES was added. After ultrasound for 30 min, precipitation was recycled with a magnet and washed several times with ethanol and ultrapure water, pouring the supernatant body, dried by vacuum freeze-drying. Nanoparticles adsorbed -OH in the water, the following reacted with the amino-silane modified particle with NH_2 (Fig. 1).

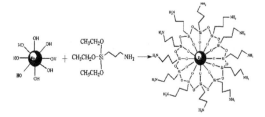

Figure 1. Schematic diagram of Fe_3O_4 encircled by amino-silane.

3 ANALYSIS METHOD

Sample was taken into agate mortar and grounded into small and uniform particles, adding KBr solution as diluent. Fourier transform infrared spectroscopy was used to test, resolution was set to 4 cm⁻¹, the scans number was 40 times.

Hysteresis loop of Fe_3O_4 nanoparticles was tested by Vibrating Sample Magnetometer (VSM) under 300 K and magnetic field strength from −7000 Oe to 7000 Oe.

4 EXPERIMENTAL RESULT

From Figure 2 and Table 1 we can see that particles without modification does not contain Si, while the Si element amount of particles modified by aminosilane is 1.21%, which proves the aminosilane was modified on the surface of the particle. Due to the sample was test with carbonmembrane method, so the carbon element can not directly explain the existence of the aminosilane[19].

The results shown in Figure 3 that diffraction peak of a located at 2θ = 30.390, 35.649, 43.571, 53.831, 57.468, 62.987, the characteristic peaks corresponding to the inverse spinel fcc phase of Fe_3O_4, (220), (311), (400), (422), (511), (440). Correspondence of crystal surface location and characteristic diffraction peaks of Fe_3O_4, indicating that the particles with high purity Fe_3O_4. According to the Debye-Sherrer formula: Dhkl = k λ/ βcos(2θ/2)$_{hkl}$ (D_{hkl} is particle size of crystal grain vertical to the (hkl) direction, k is a constant, λ is X-ray wavelength, βis crystal diffraction peak width at half maximum of (hkl), hkl is the Bragg diffraction angle of (hkl) crystal face), average particle size of a, b, c can be calculated as 15.8, 17.1, 18.7. Modification does not have of effect on polymorphs, has little influence on particle size, with an average particle size less than 30 nm, the description of the particles remain superparamagnetic.

From Figure 4 we can see that 3417.52 cm⁻¹ of a and 3418.51 cm⁻¹ of b were created by stretching vibrationmay of -OH, because the moisture exist

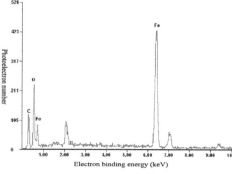

(a) Unmodified

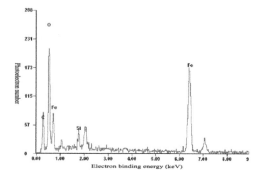

(b) Modified

Figure 2. Elemental analysis of particles.

Table 1. Elemental analysis of particles.

| Sample | Mass fraction of element | | | |
	C	O	Fe	Si
Unmodified	0.368	17.82	78.5	–
Modified	12.76	15.25	70.77	1.21

in the sample or in the air, while the corresponding -OH bending vibration appear at 625.52 cm⁻¹ of a and 11618.17 cm⁻¹ of b. 565.24 cm⁻¹ of a and 582.21 cm⁻¹ of b are typical stretching vibration of Fe-O-Fe. The results prove that a is pure Fe_3O_4 magnetic nanoparticles, 992.49 cm⁻¹ of b is stretching vibration of Si-O, and further certificate the existence of the aminosilane on magnetic nanoparticles surface. c show clear emergence of stretching vibration peak at 3417.45 cm⁻¹ of OH, stretching vibration peak at 583.71 cm⁻¹ and 434.36 cm⁻¹ of Fe-O-Fe, stretching vibration peak at 1058.75 cm⁻¹ and 919.74 cm⁻¹ of Si-O, also appeared stretching

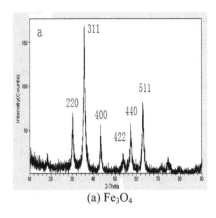

(a) Fe₃O₄

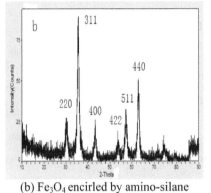

(b) Fe₃O₄ encirled by amino-silane

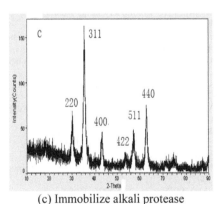

(c) Immobilize alkali protease

Figure 3. Image of X-Ray.

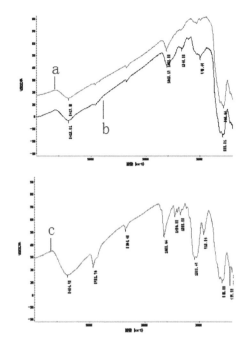

Figure 4. FTIR image. a) Fe_3O_4; b: Fe_3O_4 encirled by amino-silane; c: immobilize alkali protease.

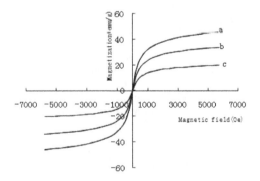

Figure 5. VSM image. a) Fe_3O_4; b: Fe_3O_4 encirled by amino-silane; c: immobilize alkali protease.

vibration peak at 2926.43 cm⁻¹ of CH_2, stretching vibration peak at 1411.83 cm⁻¹ and 1352.96 cm⁻¹ of N-O, stretching vibration peak at 1652.60 cm⁻¹ of C = O, indicating that the compounds containing N-O, C = O and CH_2, can be an indirect proof of the existence of enzyme.

It can be seen from the Figure 5, the saturation magnetization of a (45.753 emu/g)> b (33.617 emu/g) > c (20.249 emu/g), amino-silane

modified and immobilized enzyme with Fe_3O_4 particles also have higher saturation magnetization and the magnetic coercive force and residual magnetic intensity are extremely low, near zero, indicating that the modification and immobilization of magnetic nanoparticles with superparamagnetic and can be recovered with a magnet. Compared with pure Fe_3O_4 nanoparticles, saturation magnetization of the Fe_3O_4 nanoparticles in aminosilane-modified particles and immobilized enzyme decline slightly, the main reason may be the saturation

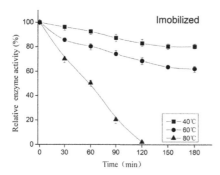

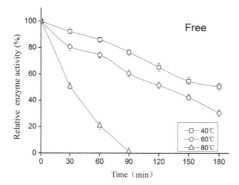

Figure 6. Thermal stability of immobilized alkaline protease and free.

magnetization relate to material content, with the accession of aminosilane and enzyme, content of magnetic Fe_3O_4 reduced, resulting in the decrease of saturation magnetization[20].

It can be seen from Figure 6, immobilized enzyme compared to free enzyme has strong thermodynamic stability. Activity of Immobilized enzyme maintain 80% at 40°C for 180 min, and activity of free enzyme was lost almost 50%. Free enzyme stability is lower than immobilized enzyme at 60°C. Activity of free enzyme loss at 80°C for 90 min, while the immobilized enzyme loss its activity until 120 min. This phenomenon may be due to the micro-environment plays a protective role in the structural stability of the immobilized enzyme that carrier provides.

5 CONCLUSIONS

This method use magnetic nanoparticles modified by 3-aminopropyl triethoxysilane as support for imobilization. Magnetic nanoparticles have huge surface area, high surface activity and adsorption capacity, superparamagnetic, it can be recycled easily from the system. X-ray photoelectron spectroscopy, infrared spectroscopy and vibrating sample magnetometer were used to characterize the prepared particles, results indicate that magnetic nanoparticles with superparamagnetic is a good support to inmobilize alkali protease. Furthermore, it has been demonstrated the thermodynamic stability of the immobilized enzyme prepared better than the free enzyme.

Acknowledgments: This work was supported in part by China's Ministry of Agriculture under Grant No. nycytx-004.

REFERENCES

1. A. Tanriseven, Z. Olcer, Bio. Eng. J. 39 (2008).
2. A.M. Girelli, E. Mattei, J. Chromatogr. B 819 (2005).
3. H.A.S. Amin, H. A. El-Menoufy, A. Mostafa, J. Mol. Catal B: Enzym. 69 (2011).
4. F.L. Hu, C.H. Deng, X.M. Zhang, J. Chromatogr. B 71 (2008).
5. L. Yang, C.Y. Chen, Y.F. Chen, Anal. Chim. Acta 683 (2010).
6. M.D. Busto, K.E. García-Tramontín, N. Ortega, Bioresour. Technol. 97 (2006).
7. C. Rocha, M.P. Goncalves, J.A. Teixeira, Process Biochem. 46 (2011).
8. J.F. Ma, L.H. Zhang, Z. Liang, Trends Analy Chem. 30 (2011).
9. E. Quiroga, C. OIllanes, N.A. Ochoa, Process Biochem. 46 (2011).
10. M.C. Yen, W.H. Hsu, S.C. Lin, Process Biochem. 45 (2010).
11. K.B. Hartman, L.J. Wilson, M.G. Rosenblum, Mol. Diagnosis & Therapy 12 (2008).
12. A.H. Latham, M.E. Williams, Acc. Chem. Research, 41 (2008).
13. W. Ying, J. Nanosci. Nanote. 8 (2008).
14. L. An-hui, E.L. Salabas, F. Schiith, Angew. Chem. Int. Ed. 46 (2007).
15. P. Wang, X. Wang, T. Xu, Thin Solid Films, 515 (2007).
16. Uhlen M. Nature, 340, 6236, (1989).
17. M. Simone, R.R. Carlo, P. Pierluigi, Trends Mol. Med, 9, 5 (2003).
18. C. Poeckler-Schoeniger, J. Koepke, F. Gueckel, J. Sturm, M. Georgi. Magn. Reson. Imaging 17 (1999).
19. M. Chen, J.P. Liu, S.H. Sun, J. Amer. Chem. Soc. 126 (2004).
20. M.A. Neouze, U. Schubert, Monatsh. Chem. 139 (2008).

Frontiers of Energy and Environmental Engineering – Sung, Kao & Chen (eds)
© 2013 Taylor & Francis Group, London, ISBN 978-0-415-66159-1

Enzymological properties of magnetic chitosan microspere-immobilized alkaline protease

Y. Li, S.N. Wang, L.Z. Jiang & D.D. Li
Food Collage, Northeast Agriculture University, Harbin, China

ABSTRACT: Alkaline protease was immobilized with Magnetic chitosan microspheres compound with Fe_3O_4 and chitosan, glutaraldehyde as cross-linking agent. Watching and analysising its appearance and structural properties, the results showed that: it had good Spherical appearance, its size increasing from 15 to 20 nm after immobilized; Fe_3O_4 was surround by chitosan well analyzed with FTIR; it had full crystal structure and good ability of magnetic responsivenes and super magnetism before and after immobilization. The optimum pH of immobilized enzyme is 10, optimum temperature is 60°C, operational stability of the immobilized enzyme is better. Furthermore, the enzyme activity of immobilized enzyme is only reduced by 40% after used five cycles.

Keywords: magnrtism; chitosan; immobilize; alkaline protease

1 INTRODUCTION

Chitosan is the only basic polysaccharide rich in nature, safe and nontoxic, it has a unique molecular structure and easy to modify, usually as carrier to immobilize enzyme and cell.[1] Magnetic chitosan microspheres have overcome the shortcomings that loose structure and large aperture of the chitosan microspheres; have stronger mechanical properties; easy to recovery from the system; move with changes in magnetic field; has a huge specific surface area that provide more binding sites for other active groups.[2,3] In addition, it can be recovered and re-used, improving the economic efficiency.[4] These advantages make magnetic nano-chitosan microspheres have broad application prospects in emerging green materials (Yen et al. 2010).

Alkaline protease is a class of suitable enzymes can hydrolyze protein peptide bonds under alkaline conditions, which has strong capacity for hydrolysis, resistance to acid and heat.[5,6] With the deeper understanding, alkaline protease is widely used in hydrolysis vegetable protein, such as: soy protein, corn protein, rice protein (Latham et al. 2008). Enzyme often reacts in the liquid phase, it is difficult recovery from the system, resulting in unnecessary waste,[7-12] thus the alkaline protease immobilization may be the solution to this problem.

2 PREPARATION OF CHITOSAN MAGNETIC NANOPARTICLES AND IMMOBILIZED ALKALINE PROTEASE

Take 0.3 g chitosan into 20 mL 1.5% (v/v) acetic acid solution, add 0.5 g Fe_3O_4, dispersed uniformly. Take 80 mL liquid paraffin, 6 mL Tween 80 and 0.05 g magnesium stearate into 250 mL three neck flask, stir and mix in water bath at 40°C, and then add Fe_3O_4-chitosan to the three neck flask. Add 10 mL 5% glutaraldehyde solution after stirring about 1 h, continue the reaction for 1 h, make pH 10 with 0.5 mol/L NaOH, than heating to 60°C, continue to react for 2 h. Than filtration, and wash a few times by ether, acetone, ethanol, and ultrapure water.

Place the above-prepared magnetic microspheres 0.50 g in the flask, add a certain amount of pH 9.5 borax—sodium hydroxide buffer solution to swell, add 10 mL 1% the alkaline protease buffer to absorb a certain period of time, and then add 5 mL 5% glutaraldehyde solution, oscillation at room temperature overnight. Use a magnet to separate magnetic microspheres after the reaction, and pour out of the upper liquid, use ultrapure water to wash until the supernatant have no UV absorption. The microspheres were vacuum freeze drying, stored at −20°C.

3 ANALYSIS METHOD

Particle size and external morphology were observed by transmission electron microscopy. Powder sample was diluted in ethanol solution, ultrasonic dispersion for 30 min, scatter a good liquid drop in the copper, and then grid into the vacuum oven drying for 48 h.

Crystal structure of the particles was tested by XRD, copper target was used, ray wavelength was 0.154 nm, the scanning angle was range of 10°~90°, scanning speed was 5°/min, step was 0.02°, tube voltage was 40 kV, tube current was 30 mA.

Sample was taken into agate mortar and grounded into small and uniform particles, adding KBr solution as diluent. Fourier transform infrared spectroscopy was used to test, resolution was set to 4 cm⁻¹, scan for 40 times.

Hysteresis loop of nanoparticles was tested by Vibrating sample magnetometer under 300 K and agnetic field strength from −7000 Oe to 7000 Oe.

4 EXPERIMENTAL RESULT

From Figure 1A and B, we can see that the size of particles is between 10–20 nm, average size is about 15 nm. Particles size of B is slightly larger than A, the morphology of the basic is spherical, slight agglomeration, this is because these particles with strong magnetic and enhance the intermolecular interactions.[13]

The results shown in Figure 2 that diffraction peak of A located at $2\theta = 30.073, 35.649, 43.212, 53.577, 57.299, 62.847$, B located at $2\theta = 30.390, 35.649, 43.571, 53.831, 57.468, 62.987$, the characteristic peaks corresponding to the inverse spinel fcc phase of Fe_3O_4, (220), (311), (400), (422), (511), (440). Correspondence of crystal surface

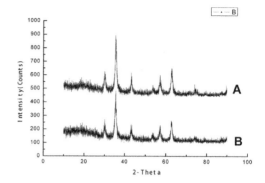

Figure 2. Image of XRD. A: Fe_3O_4 magnetic microspheres; B: chitosan magnetic microspheres.

location and characteristic diffraction peaks of Fe_3O_4, indicating that chitosan magnetic microspheres have no effect to crystal. According to the Debye-Sherrer formula: $Dhkl = k\ \lambda/\beta\cos(2\theta/2)$ hkl (Dhkl is particle size of crystal grain vertical to the (hkl) direction, k is a constant, λ is X-ray wavelength, β is crystal diffraction peak width at half maximum of (hkl), hkl is the Bragg diffraction angle of (hkl) crystal face), average particle size of A, B can be calculated as 16.53, 22.04. Chitosan magnetic microspheres immobilized enzymes have a certain influence on size,[14] this is because the role of cross-linking of chitosan on the particle surface increases the particle size.

Fe_3O_4 peaks in A at 565.64 cm⁻¹ and 3417.52 cm⁻¹ are OH stretching vibration. 2922.87 cm⁻¹ and 2872.29 cm⁻¹ is aliphatic C–H vibration peak of B. 1599.98 cm⁻¹ and 1651.99 cm⁻¹ are N–H and N–C stretching vibration, 1378.56 cm⁻¹ is C–H deformation vibration peak of CH_3, due to chitosan is not completely deacety, 1077.90 cm⁻¹ and 1032.63 cm⁻¹ are C–O vibration peak of primary alcohols and secondary alcohols, 896.77 cm⁻¹ is D–pyran glycosides peaks.[15] Characteristic peaks 2942.22 cm⁻¹ and 2584.16 cm⁻¹ of B not only appear, but also increase and offset. In addition, C have characteristic peaks of A, indicate chitosan and Fe_3O_4 have a adsorption, have also appar 1641.42 cm⁻¹ characteristic absorption peak of the schiff base, which is caused by the stretching of C=N groups, indicating reaction between E dialdehyde and NH_2. 1077.90 cm⁻¹ of B after the formation of the complexes shift to 1067.46 cm⁻¹ of C, indicating the −OH of chitosan play a role in the coordination with Fe, explain the parcel is successful.

Saturation magnetic intensity of Fe_3O_4 is 33.8274 emu/g, chitosan magnetic microspheres immobilized enzyme is 24.0095 emu/g. Saturation magnetic intensity of B compared to A is decline, this is because saturation magnetic intensity related to

A. Fe_3O_4 magnetic microspheres

B. chitosan magnetic microspheres

Figure 1. Image of TEM.

substances amount, decrease of Fe_3O_4 content and adding chitosan, resulting in decrease of saturation magnetic intensity;[16] chitosan wrapped in surface of Fe_3O_4 hindered orientation effect of magnetic field to particle, resulting in the decrease. The magnetic coercivity and residual magnetic intensity of A and B are very low, near zero, indicating its have been provided with superparamagnetic.

Enzyme activity of free and immobilized enzyme was measured at different pH boric acid—sodium hydroxide buffer, and relative enzymatic activities were calculated, the results shown in Figure 5. It can be seen from the figure, the most suitable pH of immobilized enzyme is 10, to move a unit to the alkaline direction than free enzyme that the optimum pH is 9. Relative enzymatic activities of immobilized enzyme are higher than the free at the limit pH. The reason may be that the presence of carrier in immobilized enzyme affect the structure of the enzyme, hinder ions.[17]

Enzyme activity of immobilized and free enzyme was determined at different temperatures, the results shown in Figure 6. From the figure

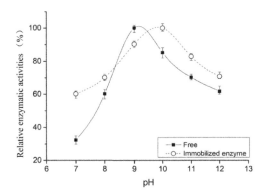

Figure 5. Relative enzymatic activities of free and immobilized alkaline protease under different pH value conditions.

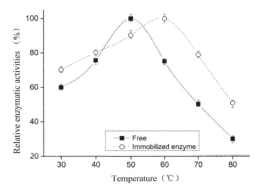

Figure 6. Relative enzymatic activities of free and immobilized alkaline protease at different temperatures.

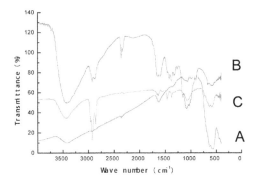

Figure 3. Image of FT-IR. A: Fe_3O_4; B: chitosan; C: chitosan magnetic microspheres.

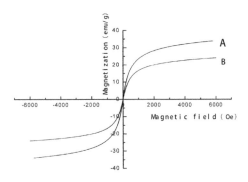

Figure 4. Image of VSM. A: Fe_3O_4 magnetic microspheres; B: chitosan magnetic microspheres.

we can see that the optimum temperature of free enzyme is 50°C, immobilized enzyme is 60°C, due to overheating, part of the enzyme are inactivation, activity was significantly decreased when temperature higher than 60°C. The results show that the optimal temperature of immobilized enzyme has improved, may be due to a change in the structure of the enzyme in immobilization process, the carrier has some protective effect on the enzyme also.

The results can be seen from Figure 7, with the increase of the cycle, the relative enzyme activity was reduced, resulting in loss of enzyme activity may be due to the carrier by the wear and tear, enzyme is detached from the carrier, or sites of action of the enzyme gradually exposed and easily damaged with the ongoing of hydrolyzed casein. The test results show that after recycle five times, enzyme activity of immobilized enzyme can still maintain 60% of the original.

571

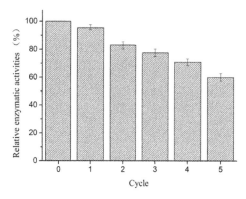

Figure 7. Operation stability of immobilize alkali protease.

5 CONCLUSIONS

Magnetic nano-chitosan microspheres have good spherical appearance, its size is about 15 nm, size of immobilized particle is approximately 20 nm; change before and after immobilization, the reason is that the coupling effect of the chitosan and Fe_3O_4 increases the size. Infrared spectroscopy confirmed Fe_3O_4 has been wrapped by chitosan by the apparent of Schiff base; particle crystal form is complete before and after immobilization, with good magnetic responsiveness and superparamagnetic. Compared with free enzyme, the optimum temperature of immobilized enzyme increase from 50°C to 60°C, the optimum pH move from 9 to 10, significantly enhance the alkaline environment adaptability. Furthermore, after recycle five times, enzyme activity of immobilized enzyme can still maintain 60% of the original.

ACKNOWLEDGMENTS

This work was supported in part by China's Ministry of Agriculture under Grant No. nycytx-004.

REFERENCES

1. M.C. Yen, W.H. Hsu, S.C. Lin, Process Biochem. 45 (2010).
2. A.H. Latham, M.E. Williams, Acc. Chem. Research, 41 (2008).
3. W. Ying, J. Nanosci. Nanote. 8 (2008).
4. L. An-hui, E.L. Salabas, F. Schiith, Angew. Chem. Int. Ed. 46 (2007).
5. L. Yang, C.Y. Chen, Y.F. Chen, Anal. Chim. Acta 683 (2010).
6. M.D. Busto, K.E. García-Tramontín, N. Ortega, Bioresour. Technol. 97 (2006).
7. C. Rocha, M.P. Goncalves, J.A. Teixeira, Process Biochem. 46 (2011).
8. J.F. Ma, L.H. Zhang, Z. Liang, Trends Analy Chem. 30 (2011).
9. E. Quiroga, C. OIllanes, N.A. Ochoa, Process Biochem. 46 (2011).
10. A. Tanriseven, Z. Olcer, Bio. Eng. J. 39 (2008).
11. K.B. Hartman, L.J. Wilson, M.G. Rosenblum, Mol. Diagnosis & Therapy 12 (2008).
12. A.M. Girelli, E. Mattei, J. Chromatogr. B 819 (2005).
13. H.A.S. Amin, H.A. El-Menoufy, A. Mostafa, J. Mol. Catal B: Enzym. 69 (2011).
14. F.L. Hu, C.H. Deng, X.M. Zhang, J. Chromatogr. B 71 (2008).
15. P. Wang, X. Wang, T. Xu, Thin Solid Films, 515 (2007).
16. Uhlen M. Nature, 340, 6236, (1989).
17. M. Simone, R.R. Carlo, P. Pierluigi, Trends Mol. Med, 9, 5 (2003).

Frontiers of Energy and Environmental Engineering – Sung, Kao & Chen (eds)
© *2013 Taylor & Francis Group, London, ISBN 978-0-415-66159-1*

Research of real-time monitoring technology for surface water's oil pollution

Y. Yao & Y.C. Yao
School of Automation and Electronic Information, Sichuan University of Science and Engineering, Zigong, China

ABSTRACT: This paper introduces the application of the microwave technology in surface water oil pollution monitoring, analyzes the microwave detection principle, the application characteristics of monitoring system and the basic microwave detection model; moreover, it presents an application example of real-time monitoring system for surface water oil's pollution.

Keywords: microwave; surface water; oil pollution; real-time monitoring

1 INSTRUCTIONS

The surface water pollution mainly includes four types: chemical pollution, heavy metal pollution, life pollution, biological pollution; the oil of surface water is caused by the chemical pollution and life pollution, which will do great harm to the river, lake, reservoir pollution. However, the petroleum substances can be dissolved in carbon tetrachloride which cannot be magnesium silicate adsorption of a class of organic compounds, including hexane, benzene, polycyclic aromatic hydrocarbons, fluorescent agent and so on; these organic molecules are some large molecules, which combined with the suspended substances in the water; for the oil pollution of life, which is mixed together with the other food and formed the more large molecules with the other organic molecules, or formed a large area of oil pollution combined with the suspended substances, and if there are more oil substances floating on the water surface, it will influence the oxygen exchange between the air and the water interface, make the oil which is dispersed in the water, absorbed on the suspended particulates or exists in emulsified state in the water to be oxidized and decomposed by the organisms, which will consume the dissolved oxygen in the water, so it will make the water quality be deteriorated and do great harm to the water ecological balance. In addition, the oil substances are also extremely serious for the soil pollution.

1.1 Measuring system of oil concentration

Oil content monitoring in the water is essential for the control of oil pollutions, the detection of water quality changes, the protection of water resources.

The degree of the oil pollutants is one of the important indicators of the water quality, so real-time monitoring of oil pollutants in the water has important practical significance.

With regard to the oil content detection, the traditional techniques include chromatography, turbidity, ultrasonic, gravimetric, ultraviolet spectrophotometry, fluorescence spectrophotometry, infrared spectrophotometry and non dispersive infrared photometric method and so on; all of these methods are indirect measurement, which need to make the monitoring samples and convert the measurement results so that obtain the concentration for the measured substances; but the operation processes are very complex, easy to introduce errors, and can not be operated field, furthermore, from extracting samples to outputting the detecting results, it ordinarily takes three to five days, which the main reason is that it needs a longer time to cultivate monitoring samples, that is using chemical method to release the oil from the conjugates of the oil and the suspended substances.

1.2 Research objectives

The research objective is to develop a real-time monitoring system of the oil pollution of surface water. Using this technology can be developed the real-time monitoring system for oil pollution in surface water, which is composed of a number of electromagnetic wave transmitting and receiving devices, data processing interface, communication circuit, data acquisition model, upper computer data processing, analysis software expert system and so on. During the monitoring process, a number of electromagnetic wave transmitting and receiving devices are respectively arranged at different monitoring points, which will preprocess signals through data processing interface and then transfer these signals to upper computer for analyzing data by communication circuit.

2 MICROWAVE DETECTION METHOD FOR OIL CONTENT

Compared with the traditional low frequency detection methods, such as capacitance and resistance methods, microwave oil detection is different, which is high frequency detection method. Microwaves' wavelength are the most short for electromagnetic waves in the radio band, according to the wave band, they can be divided into decimeter, centimeter and millimeter waves. Considering the oil molecules themselves are some big molecules, they and other molecules are combined together and formed bigger molecules, of course, their absorbing frequencies of the resonances with electromagnetic waves will not in the optical band, also may be not in the infrared band, which is the reason that it need to make water samples for using the optical spectroscopy and chromatography. Therefore, in this topic, the absorbing frequencies of the resonances for the oil combined molecules will be researched in the microwave, short wave and even the medium-frequency wave's bands, moreover, the real-time monitoring method for the oil pollution in interface water will be studied and realized. Using the power, amplitude, phase or frequency changes information when microwaves act on the oil surface compute the oil contents in the water. Essentially, it is to use microwave reflection, transmission, scattering and cavity perturbation and other physical properties' change, through detecting microwave signal parameters (such as amplitude, phase, frequency, etc.) changes to realize detection.

The macroscopic effect for the oil molecules affecting the microwave field is that it will make the microwave electric field energy change, so that this change extracted by microwave receiver will reflect the oil molecule number, namely, the oil content in the water. Microwave transmission method based on oil molecules in microwave field polarization loss of the microwave electric field strength according to the material dielectric constant associated with power-law decay law principle, using the oil molecules in a resonant frequency excitation than pure water obviously absorb more energy, which can not only detect the system launched change of energy, and can also detect the corresponding microwave parameter changes, again through the measurement of a mathematical model derived from the water oil. Microwave reflection method is also based on the microwave energy on the measured water quality effect, but the record was water reflection electromagnetic wave energy.

3 OIL POLLUTION REAL-TIME DETECTION SYSTEM DESIGN

3.1 *Oil content detection principle and model*

A transmission microwave detection system composition diagram is shown in Figure 1. The microwave power and the oscillator compose the microwave generator, which selects the small-sized microwave oscillator, namely, the body effect tube solid state signal generator. This type of microwave generator has some advantages, such as high quality factor, low temperature, small volume, light weight, etc. The non-reciprocal microwave device's field displacement isolator make the positive transferring wave pass without or with little attenuation through, while the reverse transmission wave will has a greater attenuation. In the microwave detection system, the purpose of using the isolator is to make the signal source work stably. Absorptive attenuator is used to limit or control the power level in the system, which is a waveguide absorbing sheet. They can be divided into fixed attenuator and tunable attenuator, that the attenuator indicators include the inserting loss, the maximum attenuation, standing-wave ratio and the band. Microwave sensor means the transmitting, receiving microwave antennas. Traditionally, it is using the horn antenna, but should pay attention to the influence of the edge effect on measurement, which performance indexes include the transmission performance and the anti-interference ability.

The detector and the amplifier are essentially an amplifying and filtering circuit. Because the microwave signal source is logarithm amplified after it needs squared modulation, demodulation, amplifying by the band pass filter so that it can ensure the measurement accuracy. In addition, the display is used for displaying the values of the detection signal.

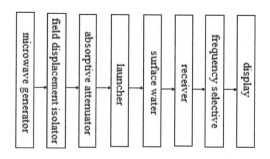

Figure 1. The transmission type microwave detection system principle.

3.2 *Experiments and testing system*

Using experiments study the attenuation of the electromagnetic wave in the pollution-free river in various seasons to define the distance for the transmitting and receiving antennas. The specific method is that prepares different pollution-free river water according to the sediment concentration for various seasons' river water, installs the receiving antenna in the 10 meters in the pool placed the river water samples from the transmitting antenna, which the transmitting antenna will sweep frequency emission and measure the attenuation of the electromagnetic wave and the frequency characteristics of the attenuation.

Adding a single measured pollutant in the oil content into the pollution-free river water make them form the river water samples having the standard content, at the same time, keep the receiving antenna from the transmitting antenna of 2 m, and then gradually increase the frequency of the transmitting antenna from 300 MHZ, moreover, find out the maximum absorbing frequency (that is the sensitive frequency) in this river water sample and plot the absorbing frequency curve. Finally, determine the attenuation values for the Ki1, ki2 ... Kij ... Kin when the electromagnetic wave propagation distance is unit. If the absorbing frequency is not obvious, decrease the frequency to 1 MHZ to 300 MHZ and then carry on the sweep frequency experiment.

For the sensitive frequency of the measured pollutant in oil, using experiment study the relation of the pollutants' affecting on the electromagnetic wave absorption and content, plot the absorption-content curve. If in the sensitive frequency f of the measured pollutant i, gradually increase the content of the measured pollutant i, measure the change of the absorption following the content. At the same time, respectively measure the changes in the other measured pollutant sensitive frequencies f.

Establish a database for basic experimental data, which includes the sensitive frequencies of the various measured substances; in the various sensitive frequencies, the attenuation parameter Kii when the electromagnetic wave propagation distance is unit and the attenuation Kij for the other pollutants in this frequency; the relation of the absorption—content for the every pollutants in every sensitive frequency.

Develop the algorithm for monitoring results.

Assuming the electromagnetic wave attenuation in the every sensitive frequencies respectively s_1, s_2 ... s_n, yields that the unit distance attenuation are $s_1 = s_1/d$, $s_2 = s_2/d$, ... $s_n = s_n/d$, where d is the distance of the transmitting and the receiving antennas, which can be expressed as follows.

$$\begin{pmatrix} x_1 \\ \vdots \\ x_n \end{pmatrix} = \begin{pmatrix} k_{11} & \cdots & k_{1n} \\ \vdots & & \vdots \\ k_{n1} & \cdots & k_{nn} \end{pmatrix}^{-1} \begin{pmatrix} s_1 \\ \vdots \\ s_n \end{pmatrix} \qquad (1)$$

Finally, according to these calculation results, using look-up table method looks up the absorption-content relation table and obtains the content of the oil pollutants in river water.

4 CONCLUSIONS

The microwave detecting method for the oil content in the surface water has high detection precision, fast speed, good repeatability, stable performance, anti-interference and strong shock resistance ability. Applying this method can provide the whole oil control for the surface water pollution processing. This microwave monitoring system for the surface water pollution is a kind of new high performance real-time monitoring system in the environmental monitoring, which has great practical value.

REFERENCES

Cheng Wenfang, Sheng Baizhen. Microwave semiconductor power devices and its applications. Electronics and packaging, 2003, Vol. 3, No. 1: 26–33.

Hashem M. A. Al-Mattarneh, Deepad K. Ghodgaoankar, Hanizah Ab-dul Hamdetal. 2002. "Microwave Reflectometer System for Continuous Monitoring of WaterQuality," Conference on Research and 2002 Student Development Proceedings. Shah Alam, Malaysia.

Kesheng Chen, Chuqiang Lei. 2001. Trophic State Evaluation Using Landsat Reservior TMData. Journal of the American Water Resource Association, 2001, 37 (5): 30–33.

Zhang Zhichao, Huang Shi-liang, Wu Haiyun. 2004. antenna communication equipment and system design. Beijing: People's post and Telecommunications Publishing House.

Frontiers of Energy and Environmental Engineering – Sung, Kao & Chen (eds)
© *2013 Taylor & Francis Group, London, ISBN 978-0-415-66159-1*

Rural small hydropower management system in Jiangxi, China

W. Cai, Y.H. Guo, Y. Wang & J.B. Xin
Jiangxi Electric Power Research Institute, Nanchang, Jiangxi, China

Y.J. Yu & J.F. Ling
Automation Department of Nanchang University, Honggutan New District, Nanchang, Jiangxi, China

ABSTRACT: The rural small hydropower is a very important part of Jiangxi Grid Corp, and how to adapt to the requirement of 'Five Big' Planning of State Grid is an urgent task. This paper presents the status of rural small hydropower and analyzes the problems of rural small hydropower management system in Jiangxi, China from the following three aspects: (1) The imperfect of national regulation and management system; (2) The problems of small hydropower plant operation; (3) The uncoordinated of grid merge. Finally, the paper gives some suggestions about how to adapt to the requirement of 'Five Big' system of China State Grid.

Keywords: rural small hydropower; 'Five Big' system; management system

1 INTRODUCTION

The technical, economic and environmental benefits of hydropower make it an important contributor to the future world energy mix, particularly in the developing countries (Ibrahim Yuksel, 2010). The small hydropower resources in China are abundant and distribute widely (XiaoLin Chang, 2010). Developing small hydropower in rural areas can not only benefit the local economy, support the modernization of China's rural areas, but also can assess huge environmental benefits, social benefits and prevent the transfer of environment pollution from urban to rural areas (LI Jia-kun, 2012). These provide hydropower construction with unprecedented advantages and opportunities.

China State Grid Corp released "On promoting the 'Three Ji Five Big' system" at January 2012. And Jiangxi Grid Corp is the first batch of 'Five Big' system construction promotion unit. The rural small hydropower is a very important part of Jiangxi Grid Corp, and how to adapt to the requirement of 'Five Big' Planning of State Grid is an urgent task. But what is the 'Five Big' system? The 'Five Big' system include 'Big Planning' system, 'Big construction' system, 'Big production' system, 'Big operation' system and 'Big marketing' system.

"Big Planning" system is to establish the integrated planning organization system, formulate unified planning which cover all levels of the organization, all the business field and all voltage levels of the grid, and ensure the scientific, integrity and consistency of the planning program.

"Big construction" system is to establish the intensive, specialization, flattening construction organization system, unified management, technical specifications and construction standards, and improve project quality and benefits.

"Big production" system is to establish the production organization system and the equipment management is the core of this system. It is to implement the life cycle management of asset and improve equipment utilization efficiency and reliability.

"Big operation" system is to implement the integration of various regulation including state grid integration and province grid integration, enhance the control ability of the power system and big range of the optimal allocation of resources, and ensure electricity network safety, economic, high-quality, efficient operation.

"Big marketing" system is to establish customer demand-oriented, efficient synergistic integration of marketing organization system which includes unified service platform, business mode and management standard to improve the service ability.

2 THE STATUS OF RURAL SMALL HYDROPOWER IN JIANGXI PROVINCE, CHINA

Small hydropower resources are very abundant in China, and the national developable small hydropower installed capacity is approximately 1.2 billion kW. It is widely distributed and fit the vast rural areas and remote mountainous areas to develop and utilize the small hydropower according to local conditions, which could add the local electricity supply and develop the local economy or bring considerable benefits return to investors.

There are many rivers in Jiangxi including Gan River, Fu River, Xin River, Rao River and Xiu River. These five rivers run through the region and import to the Poyang Lake, then through the Hukou into the Yangtze River. So hydropower resources are abundant in Jiangxi which water reserves are 6,845,600 kW, of which can be developed volume is 6,330,000 kW. Small hydropower resources (referring to a single station installed capacity of less than 50 000 kW hydropower station) at the amount of 4,228,500 kW, accounting for 67% of the province's hydro technology can be developed. By the end of 2009, the province has been built nearly 4000 small hydropower generation stations, installed capacity of 2,588,500 kW, the annual generation capacity of 6.309 billion kWh (Xiaowen Li, 2010), the province's total electricity consumption in 2009 was 60.922 billion kWh (National Bureau of Statistics, 2010), the small hydropower power generation accounts for 10.35% of the consumption of electricity in the province.

In recent years, the development of small hydropower stations in Jiangxi Province is very fast, especially the development of small hydropower in rural areas. It makes a strong complement to the lack of electricity, and plays an important role in promoting the province's rapid economic development especially in rural economic and social development. Take the Yongfeng country in Jiangxi for example. From January to June 2009, capacity generated by the small hydropower has reached 75.76 million kWh, which is an increase of 113.17% over the same period in previous year, exceeds the highest value of the calendar year and makes a new record. At present, the total installed capacity of the Yongfeng small hydropower has more than 45,000 kW.[6] Rural small hydropower stations provide a large amount of electrical energy to the county's power grid, especially the peaking power generation in small hydropower stations greatly eased the tensions of power supply.

3 THE PROBLEMS OF RURAL SMALL HYDROPOWER MANAGEMENT SYSTEM IN JIANGXI, CHINA

China's small hydropower development has made great achievements, but there are many problems in the development inevitably. Specific influences go as follows.

3.1 The imperfect of national regulation and management system

1. The regulation system is imperfect and national tax policies have not been finalized limiting the development of small hydropower. Rural small hydropower system engineering is designed to coordinating poverty alleviation, natural resources, ecology and water conservancy construction, so the government issued a series of policies to support the small hydropower. However it is lack of policies support in some localities and relevant departments, manifested in: (1) Lack of specific supporting policies and operation; (2) The large grid take a high degree of monopoly to rural electricity market; (3) In many parts of the electricity sector refuses to implement national tax policy in order to pursue their own benefit.

2. Lacking of investment from the government and no preferential policies support its development. Compared with the important role small hydropower playing in the coordinated economic development between regions, the national investment in small hydropower is very limited. It is mainly in: There is not standard construction to bring small hydropower projects into the budgets at all levels and programs for a long time, lack of capital investment and other necessary financial support on the development of small hydropower from the government. National Energy "12th Five-Year" Plan required that to 2015, proportion of electricity of other non-fossil energy including natural gas, hydropower, nuclear power and wind power increase from the current 3.9%, 7.5%, 0.8% to 8.3%, 9%, 2.6%. (Xiaowen Li, 2010).

Small hydropower is a clean and renewable energy. It plays an important role in the aspects of improving the energy structure, alleviating the tensions around the power supply, saving resources, reducing emissions and promoting regional economic and social development. The state encourages and supports the use of renewable energy and clean energy generation, which takes development and utilization of renewable energy as the priority areas of energy development. Countries issued a series of supportive policies for energy which is clean, renewable energy source while none for small hydropower. Such as, wind power equipment manufacturers in accordance with the prescribed conditions will have 500 yuan per kW by the standard as subsidy, and machine manufacturing enterprises and key components suppliers each get 50%. Another example, National Development and Reform Commission, State Electricity Regulatory Commission makes detailed provisions to additional deployment and subsidies on wind power, photovoltaic, biomass electricity, and other related matters in "notification of renewable energy price subsidies and quota trading scheme at January 2009 to June 2009", with an average subsidy of 0.22 yuan/kWh. (Xiaowen Li, 2010).

3. The electricity price of small hydropower is low while the cost of power generation is high. Currently, the small hydropower tariff average is about 0.2 yuan/kWh, while the tariff of thermal power plants is about 0.48 yuan/kWh (Shuangfenf Dai, 2011) in most regions. It is substantially lower compared with thermal power. Small hydropower is a clean, renewable energy, which would not emit harmful gases and could keep the coordinated development between power plant operations and the natural environment during the operation of power plant. Wastewater, exhaust residue discharges from thermal power plants lead to serious environmental pollution. Highly polluting thermal power electricity use at high prices while green small hydropower is cheap, which is obviously unfair and unreasonable. Tariff revenue is the main income of small hydropower stations, low tariff and the high cost of power generation lead most small hydropower stations at a loss, which restricts the development of small hydropower.

3.2 The problems of small hydropower plant operation

1. Low degree of modernization result in less competitive in the market. A limited electricity production and production cost rising along with the project cost and construction cost rising year by year directly impact the competitiveness of small rural hydropower enterprises.

2. Decentralized management lead to lack of cohesion. Small hydropower is a vulnerable industry and the future development largely depends on whether its social and environmental benefits as well as its public status could be recognized by society. Many small rural hydropower enterprises run in a extensive way. The small hydropower industry cohesion is low without a national trade organizations safeguarding the interests and guiding for their development.

3. The size of the rural hydropower stations is small and the mode of operation is single. Beside power generation, flood control, irrigation, soil conservation, aquaculture, shipping, fruit production, water recreation, tourism and other business project can generate profits too. But most of the small hydropower plants are lack of the diversified hematopoietic function, which is another dilemma leading to low economic and social benefits.

4. The limitations of the small hydropower industry is a big issue, Most are run-of-river power station resulting in limited power production and high production cost. In addition, small hydropower are also affected with seasonal factors, power output is difficult. There is a certain gap between our technical level and international advanced level.

5. The small hydropower plants pay much attention to construction while underestimate management. System and security and technical management of small hydropower is lagging behind due to the limit of size, capital, technology and other factors for a long time. In recent years, with further diversification of the main body of investment, a large number of contracting business and private power plants make industry management even more confusing and take very large risks to the development of small hydropower. The enterprises lack long-term awareness and enthusiasm to application of advanced technology but focus only on immediate interests.

3.3 The uncoordinated of grid merge

1. There are several manage office. As everyone knows, all levels water conservancy government departments are the competent department of small hydropower and in charge of formulation, resource planning, project review, and production safety. With the institutional change of state investment, the relevant ministries and energy companies implement various management on small hydropower which leading to overlapping functions, the responsibilities and ownership are unknown. Rights and economic interests of the investment subject are confused.

2. The property right is uncertain. The state-owned small hydropower stations accounted for a large proportion of the small hydropower stations. As the uncertainty of property rights, this part of the state-owned enterprises still enjoy subsidies and other traces of the planned economy, so it is difficult to improve the incentive and restraint mechanisms. This part of the state-owned enterprises' economic benefits is worst and all losses.

3. The coordination of the relationship between the big and small grid is a big problem. At initial stage small hydropower are generally issued for an independent run, mainly to supply electricity for the local county, township and rural areas, agricultural processing and agricultural production. The small hydropower form a network (small network) of the local power grid with economic development, which become the main power supply source to local urban and rural residents living and production, local

industry and township enterprises. National grid continuously extends around based on large and medium-sized cities, and builds up the backbone of the national inter-regional power supply network. The connections of big net with small net is inevitable under the correct guidance of the principle of central and local "walking on two legs" and economic development, which may produce a series issues about social, economic and interest. The key is how the state properly to guide and take timely measures to address these issues to promote small hydropower.

4 HOW TO ADAPT TO THE REQUIREMENT OF 'FIVE BIG' SYSTEM OF CHINA STATE GRID

The unprecedented rate of expansion of area and quantity of hydropower plants have led to significant changes in management and operation of hydropower systems which have become one of the significant factors in constraining the security and economic operation of power grid in China (Chun-Tian Cheng, 2012).

For 'Big Planning' system, governments are drawn into the hydropower department as societies increasingly demand public services. As life styles change and gain more complexity, the hierarchization services of the spatial organization tends to reach farther limits of the state, Such as using Geographic Information System (GIS) to select feasible sites father for small run-of-river hydropower projects (Pannathat Rojanamon, 2009).

For 'Big construction' system, it is necessary to make construction designs of small hydropower more economically viable.

For 'Big operation' system and 'Big production' system, We can enhance small hydropower system from the following aspects (Juan I, 2010):

1. Modeling for the complicated hydropower systems
   Reservoir hydropower systems involve power generation, flood control, water supply, irrigation, ecology and other comprehensive purposes.
2. Parallel computation of optimal operation of hydropower systems
   The computation efficiency not only affects the feasibility, but also determines the utilization of the method to optimal operation of hydropower systems. For hydropower system in region-oriented or provincial power grid, the attribute of natural spatial distribution and coordination management of intra-regional and inter-regional power system are suitable for the solution by the parallel computation.
3. Knowledge management of optimal operation of hydropower system
   For operation problem of hydropower systems with strong engineering attributes, simplifications are necessary and useful to enhance the solution efficiency and feasibility. During the past decades, it is feasible to introduce intelligent techniques, such as knowledge management for optimal operation.
4. Uncertainty analysis of optimal operation of hydropower system
   The uncertainty of runoff is the key factor affecting the feasibility of hydropower system scheduling, in particular for middle to long term optimal operation. The occurrences of extreme climate in recent years also aggravate the difficulties of operation and management of hydropower system. To enhance prediction accuracy, forecast period based on uncertainty is very important.

For 'Big marketing' system, electricity is a metered commodity and output is registered with notable accuracy; hence its pricing is calibrated by accountants. In a world with multiple energy sources, competition for market share cannot be neglected. Project costs, electricity markets and electricity prices are all variables, so these have to anticipate what the electricity market will be for the coming two or three decades. Assessments of projects are based on cost, and it is equally important to assess the multiplier effect of hydropower for the economies.

REFERENCES

Chun-Tian Cheng, Jian-Jian Shen, Xin-Yu Wu, Kwok-wing Chau. 2012. Operation challenges for fast-growing China's hydropower systems and respondence to energy saving and emission reduction. *Renewable and Sustainable Energy Reviews* 16: 2386–2393.

Ibrahim Yuksel. 2010. Hydropower for sustainable water and energy development. *Renewable and Sustainable Energy Reviews* 14: 462–469.

Juan I. Pérez-Díaz, José R. Wilhelmi, Luis A. Arévalo. 2010. Optimal short-term operation schedule of a hydropower plant in a competitive electricity market. *Energy Conversion and Management* 51: 2955–2966.

Jia-kun LI. 2010. Research on Prospect and Problem for Hydropower Development of China. *2012 International Conference on Modern Hydraulic Engineering. Procedia Engineering* 28: 667–682.

National Bureau of Statistics. 2010. *China Statistical Year book 2010.*

Pannathat Rojanamon, Taweep Chaisomphob, Thawilwadee Bureekul. 2009. Application of geographical information system to site selection of small run-of-river hydropower project by considering engineering/economic/environmental criteria and social impact. *Renewable and Sustainable Energy Reviews* 13: 2336–2348.

Shuangfenf Dai. 2011. Dilemma and Outlet of small hydropower stations in rural areas (In Chinese). *Electricity and social* 8: 5–7.

The record of small rural hydropower generating capacity in YongFeng in Jiangxi Province (In Chinese). *Chinese hydropower and electrification* 7: 71–72.

XiaoLin Chang, Xinghong Liu, Wei Zhou. 2010. Hydropower in China at present and its further development. *Energy* 35: 4400–4406.

Xiaowen Li, Kezhao Wu, Chen Luo. 2010. Profiles and main problem and countermeasures of Small hydropower in Jiangxi Province (In Chinese). *Jiangxi Water Science and Technology* 36 (4): 285–287.

Frontiers of Energy and Environmental Engineering – Sung, Kao & Chen (eds)
© 2013 Taylor & Francis Group, London, ISBN 978-0-415-66159-1

Analysis of UV detection in Seergu substation

Z.J. Jia, S.B. Liu, D.G. Gan & F. Liu
Sichuan Electric Power Research Institute, Chengdu, P.R. China

ABSTRACT: Audible noise caused by corona discharge of equipments with 500 kV voltage levels in Seergu substation is higher than that of other 500 kV substations. The main purpose of UV detection technology is to analyze and decide defects and faults of electrical apparatus by monitoring corona discharge around electrical apparatus. UV detection in Seergu substation and electric field calculation of CVT and arrester grading ring have been elaborated in this paper. The result shows that abnormal phenomenon of UV detection is mainly caused by colored paint stripping from equipments' grading ring. Colored paint stripping is brought about by high altitude, strong wind force and high day-night temperature difference. Finally, improvement measures are proposed.

Keywords: corona discharge; UV detection; high altitude; wind force; day-night temperature difference; colored paint

1 INTRODUCTION

The elevation of 500 kV Seergu substation is 1850 m. The substation was put into operation in December 2011. Until now operating staff have found that Audible noise caused by corona discharge of equipments with 500 kV voltage levels is higher than that of other 500 kV substations. The abnormal phenomenon of corona discharge is ought to be analyzed by both field detection and electric field calculation.

With regard to external insulation of equipments in air, corona discharge and surface partial discharge can't be completely eliminated. With the reduction of insulation performance, appearance of structural defects and the increase of surface contamination, the possibility and intensity of discharge increase. Thus the generation and enhancement of corona discharge and surface partial discharge can be utilized to indirectly evaluate running equipment's insulation condition and discover the equipment's defects in time [1].

Using the characteristics that corona discharge in air can produce ultraviolet, research on corona discharge by UV detection technology have been gradually developed home and abroad. The EPRI combined with several power companies have researched on UV detection technology aiming at corona discharge of substation and overhead line for three years. Finally, *Guide to Corona and Arcing Inspection of Substations* and *Guide to Corona & Arcing Inspection of Overhead Transmission Lines* are written [2,3]. In China, the first standard of UV detection, *charged device technology application guidelines for UV diagnostics*, has

been established [4], but the guideline centers on qualitative analysis more than quantitative analysis. Therefore, our country recently still in the primary stage of technology import and absorption.

Considering results of the previous studies on UV detection and evaluation of insulation defect and contamination of high voltage apparatus, this paper addresses field UV detection in Seergu substation and electric field calculation of CVT and arrester grading ring. Meanwhile, improvement measures to reduce corona discharge intensity are proposed.

2 THE PROCESS OF UV DETECTION

While UV detection was carried out in 500 kV Seergu substation, the meteorological condition was as

Position- CVT grading ring of No. 1 transformer 500 kV side (C phase).
Gain-120; Measuring distance-19.09 m; Photon counting-9010.

Figure 1. The first image of UV detection.

follows. The temperature was 20°C, relatively humidity was 40% (RH), and wind speed was 6 grade (10.8–13.8 m/s). UV detector is produced by Israel OFIL Company, the type of which is DayCor SUPERB. The range finder is produced by Switzerland LEICA Company, the type of which is D5.

UV detection of high voltage apparatus in Seergu substation have discovered such positions with intense corona discharge as is shown in Figures 1 to 4. Positions with intense corona

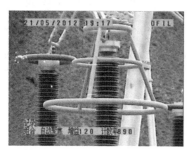

Position- Arrester grading ring of Maose line (A phase).
Gain-120; Measuring distance-35.8 m; Photon counting-890.

Position- Arrester grading ring of Maose line (B phase).
Gain-120; Measuring distance-35.85 m; Photon counting-9340.

Position- Arrester grading ring of Maose line (C phase).
Gain-120; Measuring distance-35.89m; Photon counting-2010.

Figure 2. The second image of UV detection.

Position- Arrester grading ring of No. 1 transformer 500kV side (A phase).
Gain-120; Measuring distance-41.87 m; Photon counting-7820.

Figure 3. The third image of UV detection.

Position- CVT grading ring of the second SeMao line (C phase).
Gain-120; Measuring distance-22.29 m; Photon counting-10710.

Figure 4. The fourth image of UV detection.

discharge concentrate on arrester/CVT grading rings of outgoing/incoming line.

3 ANALYSIS OF DETECTION RESULTS

3.1 Influence of geographical and meteorological factors

From the view of geographical and meteorological factors, corona discharge in Seergu substation is mainly influenced by elevation and wind force.

With the elevation getting higher, gas pressure reduced which induces corona inception voltage to decrease [5]. For example, if the elevation increases by 1000 m, corona inception discharge reduces by about 7%. The elevation of 500 kV Seergu substation is 1850 m, so this substation locates in high elevation area. Electric field concentration of high voltage charged equipments is prone to corona discharge.

In comparison with ac field of Deyang inverter station the elevation of which is only 550 m, the result of UV detection by the same detector shows that corona discharge intensity of the whole ac field is obviously lower than that of Seergu substation. The image of UV detection with intense corona discharge is as shown in Figure 5. Photon counting in Figure 5 is almost the same as that in Figures 1 and 2 (b), but with regard to the influence of detector's gain, photon counting in Figure 5 is obviously lower than that in Seergu substation under the condition of the same gain.

With regard to ac corona discharge, if there is wind, spacer charge generated in negative half-period can be easily blown off. Thus positive polarity streamer becomes more active. If there is no wind, spacer charge can alleviate surface voltage gradient of corona source. Thus formation of positive polarity streamer is checked [6]. Accordingly wind makes corona discharge more active. According to climatic condition in Seergu, gale with wind speed of 7 to 8 grade is common in the whole year which induces corona discharge to be intense.

3.2 *Influence of equipment itself*

The phenomenon of colored paint shedding is evident in Seergu substation. From Figure 2, corona discharge of B phase is more active than A and C phase. In the investigation, construction technique is to paint with brush instead of baking or spray painting. Thus adhesive force of colored paint is small. In addition, if surface treatment of substrate is weak before paint or crosslink density of paint's base material isn't high, adhesive force of colored paint is weakened obviously.

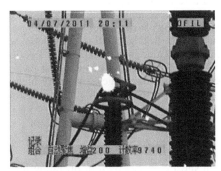

Position- Post insulator's grading ring of station transformer 500 kV side.
Gain-200; Measuring distance-21 m; Photon counting-9740.

Figure 5.   The image of UV detection in 500 kV ac field of Deyang inverter station.

The difference of thermal expansion coefficient between paint film and aluminum alloy substrate has effect on adhesive force of paint film. Once paint film is brushed on the surface of grading ring substrate, boned spots between paint and substrate surface can be damaged due to thermal expansion and contraction. Because thermal expansion coefficient of colored paint ($6.8 \times 10^{-5}/°C$) is obvious greater than that of aluminum alloy substrate ($2.2 \times 10^{-5}/°C$), the extent of expansion or contraction of paint film is greater than that of aluminum alloy of grading ring. Thus adhesive force of paint film is reduced by its distortion or rumple. Day-night temperature difference of Seergu substation is great in the whole year. Daily range of temperature keeps from 15°C to 30°C. Therefore, the difference of thermal expansion between colored paint and aluminum alloy is great which is caused by big day-night temperature difference. Hence paint can't adhere to grading ring well.

Gale with wind speed of 7 to 8 grade is common in Seergu substation in the whole year. Paint which is newly brushed may easily be blown off, turned over, stripped or cracked by gale. Thus burr appears on grading ring to form local non-uniform electric field. 500 kV ac field of Seergu substation has been in operation for less than one year, and the phenomenon of colored paint shedding of CVT and arrester is serious. However, the phenomenon of colored paint shedding of the same kinds of equipments isn't evident in Maoxian substation which has been in operation for more than five years and the elevation is about 1630 m. Wind speed of 2 to 4 grade is common in Maoxian substation in the whole year.

Because of weak adhesive force of brush painting, great day-night temperature difference and strong wind, colored paint shedding from CVT and arrester grading ring is serious in the capital construction stage. It forms local non-uniform electric field which aggravates corona discharge. Corona discharge associates with acoustic-electro-optic phenomenon, especially strong oxidant such as O_3 and NO_2 which can easily combines with water to form nitric acid. Decomposition and shedding of Colored paint is accelerated by substance above with strong corrosiveness.

4   ELECTRIC FIELD CALCULATION OF CVT AND ARRESTER GRADING RING

The phenomenon of intense corona discharge of CVT and arrester grading ring can be analyzed by electric field calculation software ANSOFT. The first step is to build model of fittings and assembly for verifying design rationality.

The second step is to simulate surface paint shedding using simplified model for understanding the influence of paint shedding on fittings' field distribution.

4.1 The model of CVT grading ring and calculation result of electric field distribution

According to the size of CVT grading ring provided by substation and producer (outer diameter is 1000 mm and pipe diameter is 100 m), simplified model of CVT is built by ANSOFT to calculate electric field distribution, as is shown in Figures 6 and 7.

From Figures 6 and 7, most of field strength along surface of CVT grading ring is higher than design criteria for 500 kV fittings (15 kV/cm). Electric field concentrates on the upper side of lateral surface of grading ring, and the maximum field strength is 27.5 kV/cm which is quite near breakdown strength in air (30 kV/cm). The image of UV detection in Figure 1 can demonstrate that. Therefore, the size of CVT grading ring causes practical field distribution to approach breakdown strength in air. If the surface of grading ring exists tip or burr, corona discharge can easily occur.

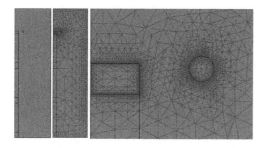

Figure 6. The axisymmetric model and partition of CVT grading ring.

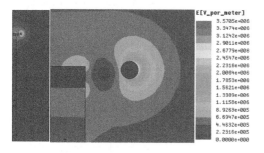

Figure 7. Field distribution of CVT grading ring.

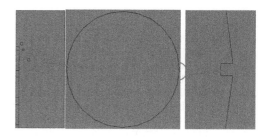

Figure 8. Simulation of surface paint off of arrester grading ring.

Figure 9. Surface field distribution of the third grading ring of arrester.

4.2 Simplified model of paint shedding of arrester grading ring and calculation result of electric field distribution

According to the size of arrester grading ring provided by substation and producer, axisymmetric simplified model of arrester with surface paint off is built by ANSOFT to calculate electric field distribution, as is shown in Figures 8 and 9. The surface of grading ring is painted the coating with 1 mm in height. Then a cavity with 1 mm in depth is opened at the lateral surface of the third grading ring of arrester from up to bottom.

From Figures 8 and 9, electric field concentrates on grading ring's surface with paint shedding, and the maximum field strength is 18.8 kV/cm which is higher than design criteria for 500 kV fittings (15 kV/cm). Therefore, the possibility of the occurrence of corona discharge is relatively high. The image of UV detection in Figure 2 can demonstrate that.

5 CONCLUSIONS

Seergu substation has been in operation for only half years, so rough surface or burr may exist on new equipments. Moreover, colored paint shedding, high elevation, strong wind and imperfect preventive measures of corona all cause corona discharge intensity more active than that of other 500 kV substations.

Most of field strength along surface of CVT grading ring is higher than design criteria for 500 kV fittings, and field strength of some part is quite near breakdown strength in air. Therefore, the surface along CVT grading ring is ought to be readjusted.

Electric field concentrates on arrester grading ring's surface with paint shedding, the field strength of which is higher than design criteria for 500 kV fittings. Thus the possibility of the occurrence of corona discharge is relatively high.

In order to alleviate the hazard induced by wind and corona, colored paint is ought to be daubed on relatively low position such as uncharged support.

REFERENCES

[1] Degang Gan, Hongbo Chen, Ping Liu, etc, "Application and outlook of ultraviolet detection technique in electric power system," Electrotechnical Application, vol. 7, pp. 77–81, 2012 (in Chinese).

[2] 1001792, "Guide to Corona and Arcing Inspection of Substations," US: Electric Power Research Institute, 2002.

[3] 1001910, "Guide to Corona & Arcing Inspection of Overhead Transmission Lines [S]," US: Electric Power Research Institute, 2003.

[4] DL/T 345-2010, "Charged device technology application guidelines for UV diagnostics," Beijing: China Electric Power Press, 2011 (in Chinese).

[5] Zhicheng Guan, Fenglin Chen, Xingming Bian, etc, "Analysis on onset voltage of positive corona on stranded conductors in high-altitude condition," High Voltage Engineering, vol. 37(4), pp. 809–816, 2011 (in Chinese).

[6] Huibin Wang, "Research on corona onset characteristics of conductors at different altitudes based on UV imaging technology," Baoding: North China Electric Power University, 2009 (in Chinese).

Frontiers of Energy and Environmental Engineering – Sung, Kao & Chen (eds)
© *2013 Taylor & Francis Group, London, ISBN 978-0-415-66159-1*

Characterization and source identification of PM₁₀-bound polycyclic aromatic hydrocarbons in urban air of Daqing, China

X.Y. Han
College of Architecture and Engineering, Kunming University of Science and Technology, Kunming, Yunnan, China

J.W. Shi
College of Environmental Science and Engineering, Kunming University of Science and Technology, Kunming, Yunna, China

ABSTRACT: The concentrations of 17 selected PAHs in PM₁₀ were quantified at three sites in city center of Daqing from April 2008 to January 2009. Total concentration of 17 selected PAHs was 67.09 ng/m³ in average, and the dominant PAHs were PA, FLu, Ant, FL, BaA, CHR and Pyr accounting for above 80% of 17 selected PAHs. Spatial variations were predominantly due to the different strengths of source emission. Higher PAHs concentrations during heating period and lower concentrations during no-heating period were observed at the three sampling sites, which may be caused by the stronger emissions from stationary combustion sources in heating period and the quicker air dispersion, washout effects, photo-degradation and higher percentage in the air in vapor phase in no-heating period. The contributions from potential sources to PAHs in PM₁₀ were estimated by the Principal Component Analysis (PCA). In whole sampling period, oil refinery and coal combustion were found to the predominant contributor of PM₁₀-bound PAHs, followed by vehicles emission and wood combustion.

Keywords: PAHs; PM₁₀; PCA; sources

1 INTRODUCTION

PAHs are ubiquitous constituents on particulate matter that mainly originate from incomplete combustion of organic matter such as petroleum and coal (Liu et al., 2009; Guo et al., 2003; Khalili et al., 1995). In urban ambient air, PAHs are almost entirely emitted from anthropogenic sources (Shi et al., 2010; Mantis et al., 2005; Caricchia et al., 1999). In this study we present the characterization of PM₁₀-bound PAHs in urban area of Daqing. The possible sources of PM₁₀-bound PAHs in Daqing are also discussed based on PCA.

2 MATERIALS AND METHODS

Daqing (45°46′–46°55′N, 124°19′–125°12′E), located about 150 km southeast of Harbin, is the biggest petroleum industry base of China, with an urban area of 5107 km² and a population of 1.2 million. Compared with other cities, Daqing has vast territory and less population density. Daqing has the largest oil field, and petrochemical is the supporting industry.

Three sites were chosen as the sampling sites according to their different function in the city: Sa ertu (SET) was chosen as a typical commercial site; Longfeng area (LFQ) was chosen as a typical industrial/residential site; Honggang area (HGQ) was chosen as a typical site in industrial area. In winter, there are large amount of coal combustion boilers for domestic heating around the three sites.

Ambient PM₁₀ samples of 20 days were collected from April 2008 to January 2009. To study the seasonal variation, the samples of 5 days were collected for each season, using a standard medium-volume PM₁₀ sampler at a flow rate of 100 L/min. A gas chromatography coupled to mass spectrometry (trace 2000GC-MS, Thermo Finnigan, USA) was used for determining PAHs with Selected Ion Monitoring (SIM), and the analysis process and quality control descriptions on this study are in publications by Shi et al. (2010). In this study the 17 PAHs were analyzed including naphthalene (NaP), acenaphthylene (AcPy), acenaphthene (Acp), fluorene (FLu), phenanthrene (PA), anthracene (Ant), fluoranthene (FL), pyrene (Pyr), benzo[a] anthracene (BaA), chrysene (CHR), benzo[b] fluoranthene (BbF), benzo[k]fluoranthene (BkF), benzo[a]pyrene (BaP), indeno[1, 2, 3-cd]pyrene (IND), dibenz[a,h]anthracene (DBA), benzo[ghi] perylene (BghiP) and Coronene (COR).

3 RESULTS AND DISCUSSION

3.1 *Distribution of PAHs in different rings*

Average total concentration of 17 selected PAHs was 67.09 ng/m³, and the dominant PAHs were PA, FLu, Ant, FL, BaA, CHR and Pyr accounting for above 80% of 17 selected PAHs. Average concentration of individual PAHs in the whole sampling time varied from 0.29 (AcPy) to 14.22 (PA) ng/m³.

Examined PAHs could be classified according to their number of aromatic rings as follows: 2-ring including Nap; 3-ring including Acpy, Acp, Flu, PA and Ant; 4-ring including FL, Pyr, BaA and CHR; 5-ring including BbF, BkF and BaP, 6-rings including IND, DBA, and BghiP; 7-rings including COR. They can be further classified into lower molecular weight (LMW, 2-and 3-rings PAHs), middle molecular weight (MMW, 4-rings PAHs), and higher molecular weight (HMW, 5-,6-and 7-rings PAHs). LMW PAHs can be tracers for wood combustion (Khalili et al., 1995) or industrial combustion of oil (Park et al., 2002). MMW PAHs such as FL, Pyr, BaA and CHR are usually associated with coal combustion and can be identified from this source (Khalili et al., 1995). HMW PAHs such as BbF, BkF, BghiP, IND, and COR may be associated with vehicles emission, and can be regards as tracers for this source (Marr et al., 1999).

Figure 1 shows ring number distribution of PAHs for PM₁₀ in Daqing in spring, summer, autumn, and winter. LMW PAHs were found to be abundant in all seasons, reflecting the strong contribution from oil refinery emission, while HMW PAHs in PM₁₀ had lowest concentrations in all seasons. Additionally, MMW PAHs are semi-volatile organic compounds and distribute higher in particle phase when air temperature is low. LMW PAHs were the most abundant in winter, which may be caused by wood and coal combustion. Therefore, the concentration proportion variation of LMW, MMW and HMW PAHs in different seasons can be used to reflect the variation in the categories of PAHs sources in these seasons.

3.2 *Variations of PAH profiles*

The year was divided in two periods, heating period (H, 15 November-31 March) and no-heating period (NH, 1 April-14 November). In Daqing, there is very significant seasonal variation in PAHs concentrations.

Figure 2 shows the mean PAH profiles (percent contribution of each PAH compound to ∑PAH) for the three sampling sites in Daqing. A relatively similarity was existed in the PAH profiles in both heating and no-heating period, which reveals the sources of PAHs around these sampling sites were extremely similar. PA, Ant, Flu and FL were all abundant compounds at these sampling sites in heating and no-heating period, indicating coal combustion emission for house heating was not the major source of these PAHs. This phenomenon revealed the importance of vehicular emissions and oil refinery emission as PAH sources in the whole year.

3.3 *Principal component analysis (PCA)*

We used PCA to identify the emission sources of PAHs in no-heating period. The principle of PCA is to transform an original set of variables into a smaller set of linear combinations that account for most of the variance of the original set. The factor loadings which are obtained for each variable

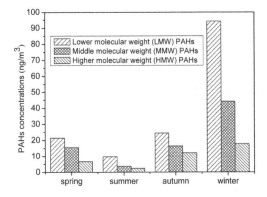

Figure 1. Ring number distribution of PAHs for PM₁₀ in Daqing.

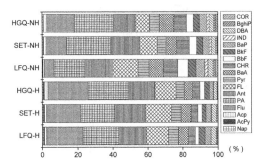

Figure 2. Mean PAH profiles of ambient PM₁₀ in the central urban of Daqing.

within the factors generated by the PCA, are a type of correlation coefficient, and higher values are therefore associated with greater significance. Due to non-detectable or too low concentrations of NaP, AcPy and AcP in PM_{10}, only 14 PAH species measured in PM_{10} in no-heating period were subjected to PCA with varimax rotation. Table 1 provides the results of PCA for 14 PAHs in no-heating.

Factor 1: it explains 32.24% of the variance with a high load for Flu, PA and Ant–the three lightest PAHs studied. Oil refinery emissions were found to be enriched by more volatile PAH species (Harrison, et al., 1996). Biomass burning is another potential source of this factor, because ANT and PHE were sometimes used as indicators for wood combustion sources (Li and Kamens, 1993). In the rural areas, burning of straw and firewood is a common practice for cooking and heating, which contributes to more or less PAH pollution in urban atmosphere. The two sources could not be distinguished in this apportionment.

Factor 2: it explains 25.59% of the variance with a high load for FL, Pyr and BaA, which were identified as fingerprints for coal combustion (Zuo et al., 2007).

Factor 3: it explains 22.17% of the variance with a high load for IND, DBA, BghiP and COR. BghiP and COR were typical tracers for gasoline exhaust (Li and Kamens, 1993; Rogge, et al., 1993). IND was also found in both diesel and gasoline engine emissions.

Factor 4: it explains 17.96% of the variance with a high load for BbF and BkF, which were found in

large amount in coke oven emissions (Zhu et al., 2002). Coking is an important industrial sector as well as PAH source in the northeast region.

According to the PCA results, Oil refinery, coal combustion, gasoline emissions, wood combustion, coke oven emission and diesel emissions are the major sources of PAHs in Daqing in no-heating period.

4 CONCLUSIONS

PAHs in PM_{10} in Daqing had been determined at three sampling sites for one year. PAH levels were the highest at heavy industrial area in heating period and the lowest at a commercial area in no-heating period. Obviously, the PAHs levels in PM_{10} were closely associated with source strength and meteorological conditions. Higher heating period PAHs concentrations and lower no-heating period concentrations were observed in all sampling sites. Based on the PCA analysis, oil refinery, coal combustion and gasoline engine emissions were found to be the predominant source to PM_{10}-bound PAHs in Daqing in no-heating period. Other sources such as emission from wood combustion and diesel engine emission also play an important role in PAHs production. In heating period, coal combustion was regards as the more predominant source to PM_{10}-bound PAHs than that in no-heating period. Additionally, meteorological conditions also promote the PAHs pollution in heating period, such as lower dispersion of air pollutants, less photo-degradation and higher percentage in the air in solid phase.

Table 1. PCA analysis of PAHs in PM_{10} in central urban of Daqing.

PAHs variance (%)	Factor 1 32.24	Factor 2 25.59	Factor 3 22.17	Factor 4 17.96
Flu	0.95			
PA	0.68			
Ant	0.73			
FL		0.82		
Pyr		0.80		
BaA		0.80		
CHR				
BbF				0.86
BkF				0.90
BaP				
IND			0.63	
DBA			0.63	
BghiP			0.67	
COR			0.82	

Extraction Method: Principal Component Analysis.
Rotation Method: Varimax with Kaiser Normalization.
Number of factors: 4. Factor loading ≥ 0.6 listed.

ACKNOWLEDGEMENTS

This study was funded by the China National Natural Science Foundation program (Grants 21207055) and the Special Environmental Research Fund for Public Welfare, No. 200709013). Shi Jianwu is the corresponding author of this paper, his e-mail address: shijianwu2000@sina.com.

REFERENCES

Caricchia, A.M., Chiavarini, S., Pezza, M. (1999). Polycyclic aromatic hydrocarbons in the urban atmospheric particulate matter in the city of Naples (Italy). Atmospheric Environment. 33: 3731–3738.

Guo, H., Lee, S.C., Ho, K.F. (2003). Particle-associated polycyclic aromatic hydrocarbons in urban air of Hong Kong. Atmospheric Environment. 37: 5307–5317.

Harrison, R.M., Smith, D.J.T., Luhana, L. (1996). Source apportionment of atmospheric polycyclic aromatic hydrocarbons collected from an urban location in Birmingham U.K. Environmental Science and Technology. 30: 825–832.

Khalili, N.R., Scheff, P.A., Holsen, T.M. (1995). PAH source fingerprints for coke ovens, diesel and gasoline engines, highway tunnels, and wood combustion emissions. Atmospheric Environment. 4: 533–542.

Li, C. K., Kamens, R. M., 1993. The use of polycyclic aromatic hydrocarbons as sources signatures in receptor modeling. Atmos. Environ. 27A, 523–532.

Liu, W. X., Dou, H., Wei, Z. C., Chang, B., Qiu, W. X., Liu, Y., Tao, S. (2009). Emission characteristics of polycyclic aromatic hydrocarbons from combustion of different residential coals in North China. Science of the Total Environment 407(4): 1436–1446.

Mantis, J., Chaloulakou, A., Samara, C. (2005). PM_{10}-bound Polycyclic Aromatic Hydrocarbons (PAHs) in the Greater Area of Athens. Greece. Chemosphere. 59: 593–604.

Marr, L.C., Kirchstetter, T.W., Harley, R.A. (1999). Characterization of polycyclic aromatic hydrocarbons in motor vehicle fuels and exhaust emissions. Environmental Science and Technology. 33: 3091–3099.

Rogge, W.F., Hildemann, L.M., Mazurek, M.A., Cass, G.R., Simoneit, B.R.T., 1993. Sources of fine organic aerosol. 2. Non-catalyst 479 and catalyst-equipped automobiles and heavy duty diesel trucks. Environ. Sci. Technol. 27, 636–651.

Shi Jianwu, PengYue, Li Weifang, et al (2010). Characterization and Source Identification of PM_{10}-bound Polycyclic Aromatic Hydrocarbons in Urban Air of Tianjin, China. Aerosol and Air Quality Research. Vol. 10, No. 5, 507–518.

Zhu, X., Liu, W., Lu, Y., Zhu, T., 2002. A comparison of PAHs source profiles of domestic coal combustion, coke plant and petroleum asphalt industry. Acta Scientiae Circumstantiae, 22, 199–203. (in Chinese).

Zuo, Q., Duan, Y., Yang, Y., Wang, X., Tao, S., 2007. Source apportionment of polycyclic aromatic hydrocarbons in surface soil in Tianjin, China. Environ. Pollut. 147, 303–310.

Frontiers of Energy and Environmental Engineering – Sung, Kao & Chen (eds)
© *2013 Taylor & Francis Group, London, ISBN 978-0-415-66159-1*

Research status and development of oxygen-enriched combustion technology

L. Pan

Energy and Power Engineering Departement of Changchun Institute of Technology, Changchun, China

ABSTRACT: Currently, the Oxygen-enriched combustion technology of boiler is gradually developing and its application dimensions and area are continuously expanding. The paper discussed the background and significance of oxygen-enriched technology, and described its theatrical basis. Introducing the developing course and instance of membrane method for oxygen making and oxygen-enriched combustion technology, the paper prospecting the technology, pointed out that oxygen-enriched combustion technology would has wide foreground at energy saving and environment protection.

Keywords: oxygen-enriched combustion; research; application status; development

1 BACKGROUND AND SIGNIFICANCE OF OXYGEN-ENRICHED COMBUSTION TECHNOLOGY

Combustion is the most main method for human to get energy. The energy through mineral burning is over 90 per of the total energy wasting in the world. Weather the combustion process is reasonable or not may affect the energy using degree and energy consumption deduce. When human use mineral materials to burn, much greenhouse gas and acid gas are producing which are the main effect factors of global environmental deterioration. The biggest problem human facing in the 21st century is the energy and environment. Especially, in the developing country which is developing industrialization as our country, more people, less average resource, the problem of resources and environment are more outstanding.

The energy in China is not rich, so energy saving is very important. Energy saving has became the basic national policy as energy saving strategic position, building economical society to realize sustainable development. As the reality in our country, the fuel consumption is very big, so it's urgent to strengthen the combustion theory research, improve fuel organization technology. So using heat equipment to improve the combustion efficiency is a very important subject.

Oxygen enriched combustion is one of the technologies for energy saving in modern times. Oxygen-enriched combustion technology can reduce fuel ignition point, accelerate burning speed, promote burning completely, improve flame temperature, deduce burning smoke, improve heat using efficiency, and reduce excess air coefficient. It is called "resource produce technology" by western countries.

2 THEORY BASIS OF OXYGEN-ENRICHED TECHNOLOGY

Combustion is because the combustible molecular in fuel and oxygen molecular highly energy collide, so the oxygen supply determines the combustion process full or not. The combustion using oxygen enriched air with high per oxygen (21% oxygen) than normal air is called oxygen enriched combustion, abbr. OEC. It is one combustion technology with high energy saving efficiency and it's used widely in glass industry, metallurgical industry, thermal engineering, etc. oxygen enriched combustion has the advantages as below than normal air combustion.

2.1 *High flame temperature and blackness*

Combustion process is the process with oxygen and fuel oxide to spread light and heat. Heat transfer usually through radiation, conduction and convection. Radiation heat transfer is one main method for boiler heat transfer. According air radiation character, only triatomic and polyatomic air have radiation power, atom air almost has no radiation ability. So under normal air assisted, because the N_2 has high percent with no radiation ability, the smoke blackness is very low. And this affects the

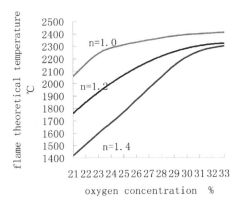

Figure 1. Relation graph city gasoxygen concentration and oxygen concentration.

Table 1. Combustion speed of several gas fuel.

Fuel	Combustion speed in air (cm/s)	Combustion speed pure O_2 (cm/s)
H2	250–360	890–1190
Natural gas	33–34	325–480
Propane	40–47	360–400
Butane	37–46	335–390
acetylene	110–180	950–1280

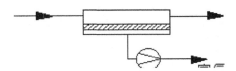

Figure 2. Figure of m embrace separation process of oxygen-enriched making.

transferring of smoke to boiler radiation transfer surface. In oxygen-enriched combustion, because the N_2 reduces, air and smoke reduce significantly. So the flame temperature and blackness increase significantly as O_2 increasing, then improving flame radiation intensity and strengthening radiation heat transfer. From Figure 1, we can see that with O_2 is increasing, theoretical flame temperature improving decreasing. But oxygen-enriched air couldn't concentration be very high, usually 26%–31%. Because flame temperature increasing decrease, when oxygen-enriched concentration goes higher. But the oxygen making investment goes faster, comprehensive benefit decreases.[1]

2.2 Speed up combustion speed, promote combustion completely

Combustion speed differs in air and pure O_2, for example, the combustion speed of H_2 in pure O_2 is 4.2 times than in normal air, furthermore 10.7 times in natural gas. So using oxygen-enriched combustion aided can shorten and improve combustion intensity, and speed up combustion speed, obtain better heat conduction. At the same time, because temperature is improved, it is beneficial to combustion completely. Comparison of several gas fuel combustion speed in air and pure O_2 can be seen in Table 1.[2]

2.3 Decrease point temperature, reduce burning-out time

Ignition point temperature of fuel is changing with combustion condition. Ignition point temperature of fuel isn't a constant, for example, the temperature of CO is 609°C in air, while 388°C in pure oxygen. So using oxygen-enriched for combustion aided can improve flame intensity, increase heat release. Figure 2 is the ignition point temperature of several

fuels in air and oxygen.[3] Visibly, adding oxygen is helpful to decrease ignition point temperature. For example, municipal refuse is unease to burn with normal air for combustion aided because of high ignition point. Mitsubishi applied oxygen-enriched combustion aided technology in municipal refuse combustion, and obtained considerable economic and environment benefit.

2.4 Decrease excess air coefficient, reduce flue gas[4,5]

Using oxygen-enriched replacing air for combustion aided can decrease excess air coefficient, reduce flue gas. Using back balance method to calculate boiler efficiency, we can see that percentage of flue gas loss is very big in heat loss, especially with normal air for combustion aided, nitrogen, which is 4/5 percentage of air for combustion aided, doesn't participate combustion. Furthermore, in combustion process, it is heated, taking more heat.

If using normal air with oxygen of 21% concentration, according to flue gas is 1 in theatrical combustion, with oxygen improving, flue gas has the tendency to decrease. Comparison with oxygen-enriched air with 27% oxygen and normal air with 21% oxygen, when excess air coefficient a = 1, flue gas volume decrease 20%, heat loss of gas exhaust decrease corresponding, energy is saving.

3 COMPARISON OF OXYGEN MAKING METHOD

People have some understanding of oxygen-enriched combustion aided mechanism home and abroad, and have researched in practical projects. The key factor in oxygen-enriched combustion aided practical process is oxygen-enriched air making. This is the main subject in research home and abroad.

At present, oxygen making methods are liquefied air distillation (deep cooling method), using kinds of absorbents for Pressure Swing Adsorption (PSA) and different membrane permeability of gas (membrane method), etc. Usually, oxygen concentration under 40%, oxygen-enriched air flow less than 6000 Nm^3/h, membrane method is more economical; PSA is more economical while oxygen concentration is between 60%~93%, in medium oxygen concentration or medium scale range; deep cooling method is used in high oxygen concentration or larger scale. Membrane method has the advantages as simple device, convenient operation, safe, quick operation, no environmental pollution, less investment, widely used, etc. which oxygen concentration is about 30%, scale less than 15000 Nm^3/h, investment, maintenance and operation spent of membrane method is only 2/3 and 3/4 of deep cooling method and PSA. Furthermore, for membrane method less scales, more economical. In recent years, membrane method of oxygen making has attracting widely attention home and abroad.

Membrane method of oxygen-enriched on oxygen making application is fast growing, and is replacing other separation technology for high cost and inconvenient operation. Its basic process is as Figure 2.

Air Membrane device Nitrogen-enriched air.

4 RESEARCH AND APPLICATION ON MEMBRANE METHOD OF OXYGEN-ENRICHED COMBUSTION TECHNOLOGY

4.1 *Abroad instance*

Many developed countries has input much human and material resources to research membrane method of oxygen-enriched combustion technology since early 80 s in last century. Especially Japan's MITI financial aided and organized 7 companies and institutes to compose "membrane method of oxygen-enriched combustion technology study group". Because of energy shortage, there were nearly 20 companies pushed out membrane method of oxygen-enriched combustion technology devices successively. They used gas oil and coal for combustion on oxygen-enriched application experiment, and the conclusion was: using 23% oxygen-enriched air for combustion supporting could save 10%~25% energy; using 25% oxygen-enriched air for combustion

supporting could save 20%~40% energy; using 27% oxygen-enriched air for combustion supporting could save reached 10%~25% energy. Germany used 27% oxygen-enriched air for experiment on horseshoe regenerative furnace, and this method improved 56.2% melting rate, decreased 20% energy, improved 100°C melting temperature. Sweden, Britain and Germany used 25%~27% oxygen-enriched air on roll and aluminum smelting furnace to save 12%~28% fuel, improve original equipment productivity 17%~39%. Wolverine used 29% oxygen-enriched air and saved fuel over 30%. Furthermore, there were reports of membrane method of oxygen-enriched combustion aided in former USSR, Britain, France and Czech.

4.2 *Domestic instance*

Our country has studied this technology since middle 80 s in last century and gain favorable results. There are more than ten units to study this technology, such as Tsinghua University, Northeastern University. they have actively explored and apply on membrane method of oxygen making and oxygen-enriched combustion technology.

Dalian Institute of Chemical Physics has engaged in national "75", "85" project since 1986, such as spiral-wound oxygen-enriched membrane and device application and development., and successfully developed LTV-PS oxygen-enriched membrane which obtained First Prize of Science and Technology Progress of Chinese Academy of Sciences.

The results as "Polymer Membrane Oxygen-Enriched Device and Combustion Technology for Glass Furnace" passed the appraisal of Chinese Academy of Sciences and Beijing Municipal Government in 1990, which was determined as National "85" New Technology Mainly Extension Project.

Liaoning Province boiler technology Institute has widely studied and investigated in home and abroad related field. They introduced oxygen making technology, and designed and made oxygen-enriched combustion aided device system. They cooperated with some glasswork, successfully transformed two glass kiln and through relative department acceptance.

Daqin is in the head of developing and extending boiler oxygen-enriched combustion aided device and obtained fund assistance of national technology-based small and medium-sized enterprises in 2005. The company configured oxygen-enriched combustion aided device on six spreader stoker steam boilers for Heilongjiang hu Da run alcohol company. When oxygen-enriched concentration was 29%, energy decreased 6.37%, and gas blackness decreased from Lindeman 5 to lindeman 2, which met environmental requirement in this area.

Jilin University completed the project as "experimental and theatrical research on oxygen increasing aided combustion of boiler" which was fund assistance of national technology-based small and medium-sized enterprises. In this case, they designed and made experiment platform of oxygen-enriched aided combustion of boiler. The experiment showed that the technology of oxygen-enriched aided combustion could accelerate burning speed, improve flame temperature, decrease burning gas, save energy. Jilin University built mathematical model of oxygen-enriched aided combustion experiment, using computer to theatrical calculation and simulation analysis, laying a foundation of oxygen-enriched aided combustion research. The experimental and theatrical research had significance for the designing and improvement of oxygen-enriched aided combustion device.

REFERENCES

[1] Masaharu Kira. Development of New Stoker Incinerator for Municipal Solid Wastes Using Oxygen Enrichment. Mitsubishi Heavy Industries, Ltd. Technical Review 2001. No. 2:78–81.
[2] Masao Takuma. Contribution of Waste to Energy Technology to Global Warming. Mitsubishi Heavy Industries, Ltd. Technical Review 2004. No. 4:1–4.
[3] Kiga T, Takano S, Kimura N, et al. Characteristics of pulverized-coal combustion in the system of oxygen-recycled flue gas combustion [J]. Energy Conversion and Management, 1997, (38), Supplement 1:129–134.
[4] Kimura, Browall W.R. Membrane Oxygen Enrichment Demonstration of Membrane Oxygen Enrichment for Natural Gas Combustion [J]. Membrane Sei, 1986:62.
[5] Till marc Numerical simulation of oxygen-enriched combustion in industrial processes. Computational Fluid Dynamics. 2003. 3:42–52.

Frontiers of Energy and Environmental Engineering – Sung, Kao & Chen (eds)
© 2013 Taylor & Francis Group, London, ISBN 978-0-415-66159-1

Study on nitrogen removal in A²O reactor with fluidized carriers

B. Wang, S. Han & Y.F. Li
Municipal & Environmental Engineering College, Shenyang Jianzhu University, Shenyang, China

ABSTRACT: A²O process with fluidized carriers was used to treat municipal wastewater. Put fluidized carriers sharing 40% volume into aerobic zone to enhance nitrifying capacity of the system. The average of NH_4^+-N removel rate was about 83.7%, and the tiptop could reach 95.16%. The average of TN removal rate was 60.3%, and the tiptop could reach 73.7%. The TN removal was beyond the bacteria assimilation. TN removal rate could be as high as 16.3% of the whole reactor. The system showed a significant phenomenon of simultaneous nitrification and denitrification.

Keywords: A²O process; fluidized carrier; nitrogen removal; municipal wastewater

1 GENERAL INSTRUCTIONS

1.1 *Type area*

According to traditional theory, nitrification occurred in aerobic environment and denitrification in anaerobic environment[1–3]. When the non-assimilation reduction of total nitrogen was found by researchers, Simultaneous Nitrification and Denitrification (SND) was investigated[4–5]. This process could reduce the relatively large reactor volumes and energy costs for recirculation required for separated aerobic and anoxic system [6–8].

A²O process was currently the most widely used in urban wastewater treatment[9,10]. On the basis of A²O process, the pollutants of urban sewage was detailedly studied. The A²O process was combined with biofilm by adding fluidized carrier into the aerobic tank to enhance the nitrogen and phosphorus removal efficiency ratio. A pilot scale study was conducted to determine whether SND was feasible using A²O process with fluidized carrier. In this study that anoxic zones would form in the deeper layers of the biofilm. The formation of dense flocs disengaged from the fluidized carrier also played an important role in the realization of SND.

2 MATERIAL AND METHOD

2.1 *System description*

As shown in Figure 1, the reactor was made by organic glass. It was composed of the anaerobic-anoxic-aerobic tank and secondary settling tank. The effective volume was 1.8 m³. the reactor was divided into four rooms. The first two

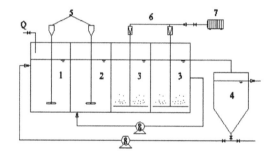

Figure 1. Experimental device.
1. anaerobic zone 2. anoxic zone 3. Aerobic zone 4. the secondary settling tank 5. stirrer 6. gas flow meter 7. air pump 8. nitrification liquid reflux pump 9. sludge return pump 10. aeration tube.

rooms were anaerobic zone and anoxic zone, aerobic zone for the last two to the volume ratio of 1:1:2. Anaerobic zone and anoxic zones were separated by partition to ensure the volume ratio was adjustable. Stirrers were equipped in these two rooms. Perforated pipes were located at the bottom of aerobic zone. The fluidized carriers could flow throughout the reactor when oxygen is provided. The effective volume of the secondary settling tank was 0.45 m³.

The influent water was supplied in the form of gravity flow. Return sludge and return nitrification were controlled by peristaltic pump. The water inflow was 208.3 L/h, and the HRT was 8h.

2.2 *Water quality and seed sludge*

The effluent from vortex sand-basin of sewage plant was used for influent of A²O reactor. The water quality was volatile. The average COD was 245 mg/L, BOD was 95 mg/L, PH was 7.5, NH_4^+-N was 29.9 mg/L, TN was 32.7 mg/L, TP was 2.36 mg/L, and SS was 145 mg/L. The seed sludge was from secondary sedimentation tank.

2.3 Abstract methods

Standard methods were performed to analyze the following parameters: NH_4^+-N, NO_2-N, NO_3-N, TP and COD. The concentration of Mixed Liquor Suspended Solids (MLSS) was measured inaccordance with the standard method after drying for 2 h at 105°C. The MLSS were separated from the carrier material by sieving. The temperature and dissolved oxygen concentration were measured with a DO meter and probe. The pH was determined with a pH meter. Biofacies was observed with a microscope.

2.4 Fluidized carrier

The fluidized carrier was polyurethane. The diameter was 20 mm. The height was 20 mm. The specific surface area was 510 m^2/m^3, and the density was in the range of 0.96–0.98 kg/L.

3 RESULTS AND DISCUSSION

3.1 Start-up of the reactor

During the early time, sewage, seed sludge and the fluidized carriers were put into the reactor. The 40% volume of the reactor was filled with fluidized carriers. The reflux pump was run without water inflow. After 48 h the Raw water pump was opened. Then the sludge incubation period started. The influent quantity was 208 L/h. The sludge age was controled in 13 d. The sludge return ratio was 100%, and the nitrification liquid reflux ratio was 200%. DO was aboult 2 mg/L. After ten days, the thickness of the biomembrane on the fluidized carrier was aboult 1 mm. As shown in Figure 2, the Amoebae, infusorian, infusorian and other protozoa were found by microscope.

3.2 Nitrogen removal effectiveness of the reactor

During the formal operation period, the influent NH_4^+-N was in the range of 14.8–47.2 mg/L, and the average was 30 mg/L. As shown in Figure 3, the Influent NH_4^+-N concentration fluctuated greatly, while the effluent NH_4^+-N concentration was quite stable. The effluent NH_4^+-N concentration was in the range of 1.5–6.9 mg/L, and the average was 4.6 mg/L. The NH_4^+-N removel

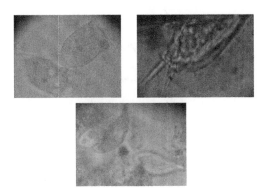

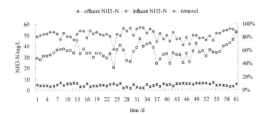

Figure 2.   Microorganism photo.

Figure 3. NH_4^+-N removel curve.

efficiency was high by biological fluidized bed reactor. The average of NH_4^+-N removel rate was about 83.7%, and the tiptop could reach 95.16%. 65% of NH_4^+-N was removed in the aerobic tank. It showed that nitrifying bacteria could attach on the fluidized carriers in abundance. The contact area of nitrifying bacteria and fluidized carrier was enlarged and the nitrification characteristics was strengthened.

During the formal operation period, the influent TN was in the range of 21.6–49.0 mg/L, and the average was 32.7 mg/L. As shown in Figure 3, the Influent TN concentration fluctuated greatly, so the removal efficiency changed obviously. The average of effluent TN was 12.8 mg/L, and the valley value was 9.1 mg/L. The average of removal rate was 60.3%, and the tiptop could reach 73.7%. Compared with traditional A^2O process, the TN removal efficiency improved greatly. Partial anaerobic micro-environment was formed between the carrier and the biomembrane on it. In the presence of dissolved oxygen, oxygen was used as electron-acceptor by denitrifying bacterium. When there was no dissolved oxygen in the water or in the micro-oxygen state, NO^{4+}-N was used as electron-acceptor. TN could be removed in this two conditions.

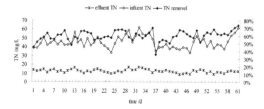

Figure 4. TN removel curve.

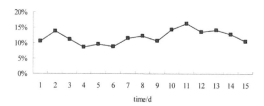

Figure 5. TN removal rate in aerobic.

3.3 *Phenomenon of SND in the reactor*

3.3.1 *TN removal in aerobic area*

In order to investigate SND phenomenon, TN removal rate in aerobic area was measured every two days during the last month. The minimum was 8.7%, and the maximum was 16.3%. It was beyond the bacteria assimilation of the TN removal. The loss of the excess part of the TN showed that denitrification happened. With the reactor running, NH_4^+-N was transformed into NO_2-N and NO_3-N by the nitrifying bacteria attached to the carrier. Then denitrification occurred by the help of denitrifying bacteria. Oxidation of organic matter, nitrification and denitrification occurred simultaneously. This also showed that aerobic nitrification and aerobic denitrification happened at the same time. A good TN removal rate was achieved.

3.3.2 *SND phenomenon analysis*

The fluidized carriers in A^2O process were attached by biofilm. The emergence of anaerobic micro-environment was affected by the change of substrate concentration and film thickness. DO gradient existed in microorganisms. DO concentration was high on the surface carrier. Aerobic bacteria, nitrifying bacteria concentrated on the carrier surface. As oxygen transfer delayed and the DO was consumed a lot by microorganisms in the outer carrier, anaerobic environment was formed insade the biofilm. It was conducive to the breeding of denitrifying bacteria. SND phenomenon emerged in the system.

4 CONCLUSIONS

A^2O process with fluidized carriers was used to treat municipal wastewater. nitrogen removal rate was better than traditional process. The effluent NH_4^+-N concentration was in the range of 1.5–6.9 mg/L, and the average was 4.6 mg/L. The average of TN removal rate was 60.3%, and the tip-top could reach 73.7%.

Added fillers, even the influent water quality was highly fluctuant, the system removal efficiency remained very high. The effluent water quality was stable. The impact resistance of system was greatly enhanced.

The fluidized carriers were attached by biofilm. Because of DO gradient, anaerobic-anoxic microenvironment emerged. The experimental datas showed that the SND phenomenon occurred obviously in the aerobic area. TN removal rate could be as high as 16.3% of the whole reactor.

Water quality could be improved by using A^2O process with fluidized carriers. This method was easy to operate. This process can be used for reform of sewage plant.

REFERENCES

[1] Trine Rolighed Thomsen, Yunhong Kong, and Per Halkjær Nielsen, "Ecophysiology of abundant denitrifying bacteria in activated sludge," FEMS Microbiology Ecology, vol. 60, pp. 370–382, June 2007.

[2] Masahito Hayatsu, Kanako Tago, and Masanori Saito, "Various players in the nitrogen cycle: Diversity and functions of the microorganisms involved in nitrification and denitrification," Soil Science & Plant Nutrition, vol. 54, pp. 33–45, February 2008.

[3] Sujay S. Kaushal, Peter M. Groffman, Paul M. Mayer, et al., "Effects of stream restoration on denitrification in an urbanizing watershed," Ecological Applications, vol. 18, pp. 789—804, April 2008.

[4] Baek, Seung H, Pagilla, et al., "Simultaneous Nitrification and Denitrification of Municipal Wastewater in Aerobic Membrane Bioreactors," Water Environment Research, vol. 80, pp. 109–117, February 2008.

[5] Gulsum Yilmaz, Romain Lemaire, Jurg Keller, et al., "Simultaneous nitrification, denitrification, and phosphorus removal from nutrient-rich industrial wastewater using granular sludge," Biotechnology and Bioengineering, vol. 100, pp. 529–541, June 2008.

[6] Li Bo, "Research on Simultaneous Nitrification and Denitrification in an Integrated Reactor with Liquid Circulation," Environmental Science and Management, vol. 9, pp. 97–99, 172 2008.

[7] Li YZ, He YL, Ohandja DG, et al., "Simultaneous nitrification-denitrification achieved by an innovative internal-loop airlift MBR: Comparative study," Bioresource Technology, vol. 99, pp. 58–67, 2008.

[8] Evalyn Walters, Andrea Hille, Mei He, et al., "Simultaneous nitrification/denitrification in a biofilm airlift suspension (BAS) reactor with biodegradable carrier material," Water Research, vol. 43, pp. 4461–4468, 2009.

[9] Wu CY, Peng YZ, Peng Y, et al., "Influence of carbon source on biological nutrient removal in A²O process," Huan Jing Ke Xue, vol.30, pp. 798–802, Mar 2009.

[10] J. Rajesh Banu, Do Khac Uan, Ick-Tae Yeom, "Nutrient removal in an A²O-MBR reactor with sludge reduction," Bioresource Technology, vol. 100, pp. 3820–3824, 2009.

Frontiers of Energy and Environmental Engineering – Sung, Kao & Chen (eds)
© 2013 Taylor & Francis Group, London, ISBN 978-0-415-66159-1

Statistic of the energy output and collectable by city car

G.J. Wang, S.S. Zhu, Y. Zhu & A.T. Xu
Department of Automobile Engineering, Academy of Military Transportation, China

ABSTRACT: The energy output and absorbed by a city car were got from tests. The tested car was a typical Chinese city car. It was tested in 5 Chinese cities, including Beijing, Guangzhou, Shanghai, Shantou, Wuhan. The torque and rotating speed of driving shaft was measured. Statistic of the exhausted energy and collectable energy were based on these data. The results show that collectable energy was 15% of total exhausted energy on average. The energy exhausted by the car was about 15 kWh/100 km on average.

Keywords: city car; exhausted energy; collectable energy

1 INTRODUCTION

The energy exhausted by city car became a hot point with more and more cars appeared in cities. The models of oil exhausted by city cars were studied extensively in order to invent simple and high efficiency device to rebuild the energy. Although problem above was studied, the amount of the energy which was able to rebuild for city cars was discussed little. The actual energy exhausted and brake energy for city cars were measured. A car with engines of 1.6 liters was used in the test. The test was performed on typical roads in Beijing, Guangzhou, Shanghai, Shantou, and Wuhan. The energy exhausted and brake energy were calculated and the difference was compared for different cities. The results were important data support for brake energy rebuild device study and application.

2 TEST CITIES, ROADS AND TIME

2.1 Choose the typical road type

Road network is similar in different cities in china, including expressway, arterial road and secondary trunk road. Expressways and arterial roads are framework of city road net, which was called main road system. Secondary trunk road connected main roads which was the most common road in city. Three road lines were selected in different cities, including one express road, arterial road and secondary trunk road, which were taken as samples.

2.2 Choose the typical cities

There are nearly 700 cities in China. The cities in which to perform the test should be typical most of the cities in China. So the cities including Beijing, Guangzhou, Shanghai, Shantou and Wuhan were taken as the samples. The cities were typical from view of geometry and traffic. From view of geometry, Beijing is in the north of china; Shanghai is in the east of China; Wuhan is at the center of China; Guangzhou is in the south of China; Chongqing is in the west south of China. There are more cloverleaf junction in Beijing; there are more viaduct in Shanghai; there are ramp in Chongqing. These cities represent the traffic of most cities in China.

2.3 Choose the typical time

In consideration traffic influenced by time, the test was made at two conditions including traffic busy time and free time. Traffic busy time included 8AM–10AM and 14PM–18PM. Traffic free time included 11AM–14PM. The test was made on Tuesday or Thursday. Data acquisition was made on the appointed road out and home.

In consideration the operation method of different driver, indigene driver was engaged to drive the test car.

3 TEST AND STATISTICAL METHOD

3.1 Install torque sensor on the driving shaft

Variation of the driving shaft Torque and velocity of vehicle with time were measured. The torque sensor include Strain gauge, signal amplifier, wireless signal transmitting device. The wireless signal receiving device was installed on eDAQ. The strain gauge was bonding on the half shaft which was shown in Figure 1. In order to avoid the strain gauge being destroyed by outer impact, wet, dust,

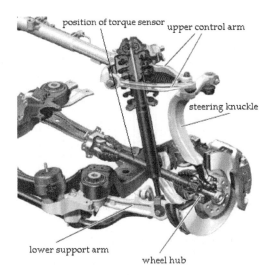

Figure 1.   Position of torque sensor (Strain gauge).

Figure 2.   Torque sensor on the driving shaft was protected by plastic adhesive tape.

etc, plastic adhesive tape was used to protect them which was shown in Figure 2.

3.2   *The principal of torque measurement*

The principal of torque measurement was shown in Figure 3 M_e was torque from engine and M_b was torque form brake. M_g was the torque acted on wheel tire from ground. The torque measured on the driving shaft was M_m. Relationships among torques were shown in Equation 1 when the velocity of car in steady velocity.

$$M_e = M_b + M_g \qquad (1)$$

If the car at a velocity steady condition, $M_m = M_e$, $M_m > 0$.

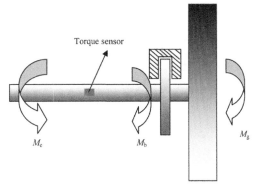

Figure 3.   Torque measurement principal.

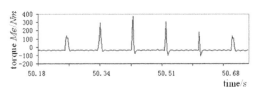

Figure 4.   Torque-time history measured on the driving shaft.

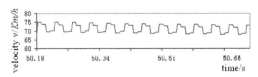

Figure 5.   Velocity-time history of test car.

If the car at a acceleration condition, $M_e = M_b + M_g$, $M_m > 0$.

If the car at a decelerations condition, $M_e = M_b + M_g$, $M_m > 0$.

The variation of torque measured on the driving shaft with time was shown in Figure 4 at steady velocity. The variation of velocity measured on the testing car was shown in Figure 5.

3.3   *Calculation principal of energy output*

Torque-time history and velocity-time history were simplified, which were shown in Figures 6 and 7. When the measurement of torque $M_m > 0$, the power was transmitted from engine to driving shaft. When the measurement of torque $M_m > 0$, the power was transmitted from wheel to engine and absorbed by it.

The energy was absorbed by engine, which was defined as A_1.

$$A_1 = \int_{t_1}^{t_2} T(t) \cdot v(t) dt \qquad (2)$$

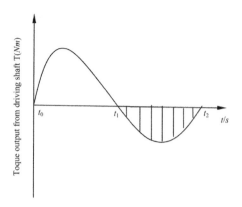

Figure 6. Simplification of torque-time history.

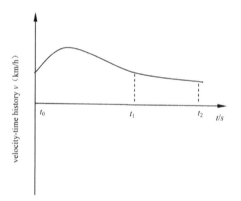

Figure 7. Velocity-time history.

The power output by engine was defined by A_0. It was able to be recollected, which was called recollectable energy.

$$A_0 = \int_{t0}^{t_1} T(t)v(t)dt \qquad (3)$$

4 STATISTIC OF TEST RESULTS

4.1 Energy exhausted and energy recllectable at different time

Energy exhausted by the car on different city per hundred miles road, time and road was calculated through Equation 2 and Equation 3. The results were shown in Figure 8.

The symbols meaning on t x axle was shown below.

1—expressway, traffic free time
1F—expressway, traffic busy time
2 —arterial road, traffic free time

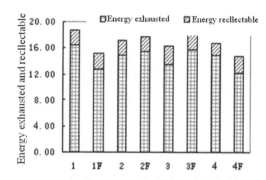

Figure 8. Energy exhausted and recollectable at different time and road type in Beijing.

Table 1. Statistic of energy exhausted and recollectable for the car in different city.

City	Energy exhausted kWh/ 100 km	Energy recollectable kWh/ 100 km	Recollectable/ exhausted%
Guangzhou	14.82	3.13	21.10%
Shantou	14.49	2.49	17.16%
Beijing	16.89	2.38	14.07%
Wuhan	13.38	1.60	11.96%
Shanghai	16.16	1.74	10.77%
Chongqing	16.73	0.83	4.98%
Average	15.41	2.03	13.15%

2F—arterial road, traffic busy time
3—secondary trunk road, traffic free time
3F—secondary trunk road, busy time
4—fourth circle road, traffic free time
4F—fourth circle road, traffic busy time.

Figure 8 shows that there was small difference energy exhausted and recollectable between traffic busy time and traffic free time. For the energy exhausted at the idle state of the car was not measured.

4.2 Statistic of energy exhausted and energy recollectable in different city

Statistic of energy exhausted and recollectable for the car in different city were made, which was shown in Table 1.

The car exhausted more energy in Chongqing than any other cities. The reason was that there were so many mountains in Chongqing. The car in Wuhan exhausted the least energy of the 6 typical cities. The reason was that the road and traffic condition was better than other 5 cities.

5 SUMMARY

There was little difference of recollectable energy between time in consideration of traffic different. The recollectable energy was more than 13% of the energy exhausted, so the benefit of recollect this part of energy. For the mountain city, taking chongqing as example, the recollecatble energy was small, for most of the energy was absorbed by brake hub and tire.

REFERENCES

[1] Xiang Qiaojun. Research on Vehicle Fuel Consumption in the Urban Transportation System. Nanjing:south west University, 2000. 1 (in Chinese).

[2] Gao Lei. Research on Vehicle Fuel Consumption Model Based on Urban Road Operations. Changchun: Jilin Univercity, 2007.09.

[3] Wu Guangqiang; Wang Huiyi; Ju Lijuan, etc. Theoretical Approach to the Reuse of Retrieved Energy of the City Bus with an Accumulator of Flywheel. Chian Journal of Highway and Transport. 1996,9(1).

[4] Zhang Tiezhu. Study on the new type of energy saving system of power transmission for city vehicles. Journal of QingDao University Engineering & Technology edition. 1997 (2).

Frontiers of Energy and Environmental Engineering – Sung, Kao & Chen (eds)
© *2013 Taylor & Francis Group, London, ISBN 978-0-415-66159-1*

Salt inhibition kinetics in Anaerobic Sequencing Batch Biofilm (ASBBR) treating mustard tuber wastewater at low temperature

H.X. Chai & W. Chen

Ministry of Education Chongqing University, Key Laboratory of Three Gorges Reservoir Region's Eco-Environment, Chongqing, China

ABSTRACT: Mustard tuber wastewater is of high organic content (COD = 4000 ± 100 L^{-1}), high salinity ([Cl^{-1}] = 18~23 g L^{-1}) and biodegradability (BOD$_5$/COD $\approx$ 0.5). The Anaerobic Sequencing Batch Biofilm Reactor (ASBBR) pretreatment was employed to reduce much of the organics in mustard tuber wastewater. In this study the substrate removal performance and salt inhibition kinetics at low temperature and saline environment were investigated by ASBBR. The result indicated that at low temperature, an aerobic polishing step is necessary to degrade organics in hypersaline wastewater. The maximum specific substrate utilization rate (d^{-1}) k and substrate saturation constant (mg L^{-1}) K_s could be expressed as a function of salinity, K_s = 24.19 salinity + 691.21 and k = −0.0004 salinity + 0.0251, respectively. Thus effluent COD concentration could be predicted, this may have some applications for ASBBR design and selection.

Keywords: ASBBR; saline wastewater; salt inhibition; optimization

1 INTRODUCTION

Anaerobic Sequencing Batch Reactors (ASBR) was applied to wastewater with large amount of organic matter, like dairy (Dugba 1999), brewery (Shao et al. 2008) and slaughterhouse wastewater (Masse et al. 2001). The operational advantage of this system was mainly attributed to the metabolism of microorganisms (Lefebvre & Moletta 2006). The Anaerobic Sequencing Bath Biofilm Reactor (ASBBR), proposed by Ratusznei et al. (Ratuszne et al. 2000) achieved high biomass concentration by immobilizing biomass as its inner support. Besides, mechanical stirring accelerated the hydrolysis process and reduces the size of particulate organic matters, both of which exerted a crucial influence on the performance of anaerobic system (Pinho et al. 2005; Christian et al. 2010).

Many manufacturing processes generate saline wastewater, such as sea-food processing, tanneries and chemical industry. The metabolism and biodegradation in microorganisms could be remarkably deteriorated for the presence of sodium and other cations (Doudoroff 1940), then low biomass concentration would cause sludge bulking (Kargi & Uygur 2005) and low nutrient removal efficiency (Tsuneda et al. 2005; Kartal et al. 2006). The utilization of salt-tolerant bacteria could provide a solution to biological wastewater treatment system. Most biological saline wastewater treatment employs aerobic halophilic bacteria and technologies (Dincer & Kargi 2001; Uygur Kargi 2004), while anaerobic treatment is rather a new approach for salt containing wastewater (Omil et al. 1995).

The treatment system could be greatly optimized by designing desirability function (Cuetos et al. 2007). Shen et al. (Shen et al. 2009) developed an empirical model for COD removal in Biological Aerated Filter (BAF) which could predict COD removal profiles at different water depths along the reactor height. Villasenor et al. (Villasenor et al. 2011) studied the kinetics of COD removal by Constructed Wetlands (CW) at different CW depths and different plant species. The experimental data could help the design of CW, especially in temperate periods.

Few researches focused on the kinetics of saline wastewater COD removal by ASBBR, particularly at low temperature. In this study, salt inhibition kinetics at low temperature and saline environment were investigated by ASBBR, which had some practical applications for ASBBR design and selection.

2 MATERIALS AND METHODS

2.1 *Experimental set-up*

A schematic diagram of the experimental reactor was proposed in Figure 1. The reactor was made of plastic and semi-soft fiber was used as its filler material. The working volume of the reactor was

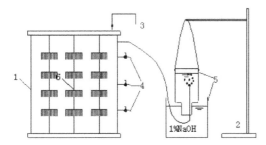

Figure 1. A schematic diagram of ASBBR.
1 ASBBR Reactor, 2 Bracket, 3 Influent, 4 Sampling and effluent, 5 Biogas collection device, 6 Biological packing.

Table 1. The characteristics of influent.

COD (mg L^{-1})	640~8000
BOD$_5$ (mg L^{-1})	1400~5600
Total nitrogen (mg L^{-1})	560~1100
Total phosphours (mg L^{-1})	8~19
Cl$^-$ (g L^{-1})	18~23
Total dissolved solids (g L^{-1})	38~47
Total suspended solids (g L^{-1})	2.2~4.6
Volatile suspended solids (g L^{-1})	0.8~1.4
pH	4.5~6.2

2.4 L (length, 30.0 cm; breadth, 16.0 cm; height 50.0 cm). Methane was measured by displacement of 1% NaOH in a 2 L serum bottle. The biofilm density of this ASBBR was 50%.

2 d batch cycles were used. At the beginning of a cycle, the ASBBR was fed in 0.5 h and reaction phase was 47 h. The effluent was discharged in 0.5 h at the end of a cycle. The excess sludge at the bottom was withdrawn by a pump regularly.

All experiments were conducted at low temperature (around 10°C) and the volumetric exchange ratio was controlled 1/4 throughout the study. The Hydraulic Retention Time (HRT) was 2 d.

2.2 Water and sludge

The experimental wastewater was generated by Fuling Mustard Tuber Group Co., Ltd, Chongqing, China. Wastewater quality was listed in Table 1. Our previous study found that the effect of pH value on the performance of ASBBR was quite significant (Chai 2005), thus we adjusted the pH value to 7.0 ± 0.2 at the beginning of the test.

Halophilic bacteria selected from the hypersaline wastewater were inoculated into conventional anaerobic sludge. The acclimatization was achieved by gradually increasing the organic load and salinity at room temperature (25°C). Before our experiment, the influent salinity of each acclimatization step was controlled to be 1000, 5000, 10000, 15000, 20000 mg L^{-1}. If COD removal efficiency attained 80% at a specific salinity, this step was considered finished then proceeded the next level of salinity until 20000 mg L^{-1}. The whole acclimatization lasted 8 months.

2.3 Analytical methods

The effluent COD was chosen to evaluate the inhibition of salinity because of its simple measurement and reliable accuracy. The test methods employed in this study based on Standard Methods for the Examination of Water and Wastewater (APHA 2005).

2.4 Kinetic study

In operational phase of ASBBR, the various substrates in influent were homogenized by stirring. The fed and discharge phase were not taken into consideration. Substrate balance in the ASBBR resulted in the following equation:

$$(1-\lambda)VC_e + \lambda VC_0 + V\frac{\mathrm{d}C}{\mathrm{d}t} = VC_e \tag{1}$$

where, λ is volumetric exchange ratio; C_0 and C_e are influent and effluent COD concentration (mg L^{-1}), respectively; $\mathrm{d}C/\mathrm{d}t$ is the substrate degradation rates, V is the volume of the ASBBR (L).

Rearranging Eq. (1) yields:

$$\frac{\mathrm{d}C}{\mathrm{d}t} = \lambda(C_0 - C_e) \tag{2}$$

In reaction phase, the degradation kinetics could also be presented by Monod equation:

$$-\frac{\mathrm{d}C}{\mathrm{d}t} = \frac{kXC}{k_s + C} \tag{3}$$

$$-\frac{\mathrm{d}t}{\mathrm{d}c} = \frac{K_s + C}{kXC} \tag{4}$$

where, k is the maximum specific substrate utilization rate (t^{-1}); X is microbial concentration (mg L^{-1}); K_s is substrate saturation constant (mg L^{-1}).

Substitution of Eq. (3) into Eq. (1), leads to:

$$C_e = C_0 - \frac{K_s + C}{\lambda kXC} \tag{5}$$

Integrating Eq. (5) yields:

$$-t = \frac{K_s}{kX} \cdot \ln C + \frac{1}{kX} \cdot C + Z \tag{6}$$

where, t is the contact time; Z is a constant.

During reaction phase at t = 0, the substrate concentration could be described as:

$$C = (1-\lambda)C_e + \lambda V C_0 \qquad (7)$$

Substitution of Eq. (7) into Eq. (6), Z value could be obtained:

$$Z = -\frac{K_s}{kX} \cdot \ln[(1-\lambda)C_e + \lambda V C_0) \qquad (8)$$

Substitution of Eq. (8) into Eq. (6) yields:

$$-t = \frac{K_s}{kX} \cdot \ln C + \frac{K_s}{kX} \cdot C - \frac{K_s}{kX} \cdot \ln[(1-\lambda)C_e + \lambda C_0]$$
$$-\frac{1}{kX} \cdot [(1-\lambda)C_e + \lambda C_0] \qquad (9)$$

when $t = $ HRT, $C = C_e$, thus Eq. (9) could be written as:

$$\mathrm{HRT} = \frac{Ks}{kX} \cdot \ln[(1-\lambda)C_e + \lambda C_0] - \frac{K_s}{kX} \cdot \ln C_e$$
$$+\frac{1}{kX} \cdot [(1-\lambda)C_e + \lambda C_0 - C_e) \qquad (10)$$

Rearranging Eq. (10) leads to:

$$\lambda(C_0 - C_e) = -K_s[\ln(C_e - \lambda C_e + \lambda C_0) - \ln C_e]$$
$$+ \mathrm{HRT} \cdot kX \qquad (11)$$

By plotting $\lambda(C_0-C_e)$ against $[\ln(C_e-\lambda C_e + \lambda C_0) - \ln Ce]$, the slope $-K_s$ can be obtained for different value of C_0 and C_e. The intercept is HRT $\cdot k \cdot X$.

3 RESULTS AND DISCUSSIONS

3.1 ASBBR performance

To make experimental data more accurate, four identical reactors were employed on every trial. Salinity from 4000–20000 mg L^{-1} was tested. The effect of salinity on effluent COD value was presented in Figure 2.

The final effluent COD value was obtained by calculating mean and standard deviation of COD concentration in the four reactors. The higher salinity caused increasing in effluent COD. The maximum COD removal efficiency (82 ± 0.9%) was observed at 4000 mg L^{-1} salt content and 640 mg L^{-1} organic load. The minimum removal efficiency was determined as 23.1% at influent COD concentration of 8000 mg L^{-1} and salinity of 20000 mg L^{-1}. The removal efficiency was far below the results reported by other anaerobic

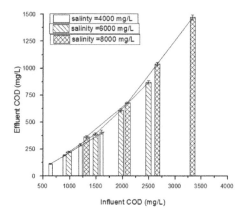

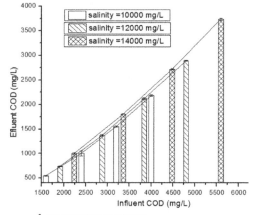

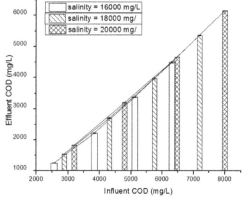

Figure 2. Organic matter concentration of the effluent at different salinity.

treatment (Gebauer 2004; Jeison et al. 2008). And this could also proved that low temperature exerted a strongly adverse effect on biological treatment as well as ASBBR (Lettinga et al. 1999; Chai 2010). At low temperature, an aerobic polishing step is necessary to degrade organics in hypersaline wastewater.

3.2 Salt inhibition kinetics at low temperature

As shown in Figure 3, $\lambda(C_0 - C_e)$ varies linearly with $[\ln(C_e - \lambda C_e + \lambda C_0) - \ln Ce]$, and $-K_s$ is the slope [Eq. (10)], $HRT \cdot k \cdot X$ is the intercept, with R^2 values between 0.986 and 0.999. Thus the values of maximum specific substrate utilization rate k and substrate saturation constant K_s could be obtained, as shown in Table 2. The relationship between salinity and K_s or k was presented in Eq. (11) and Eq. (12), which may be applied to predicting the effluent COD concentration at different salinity.

3.3 Model simulation

Validation of simulated model was carried out with wastewater containing 8000 mg L^{-1},

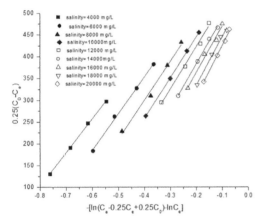

Figure 3. The linear fitting of 0.25 (C_0-C_e) as a function of $[\ln(C_e - \lambda C_e + \lambda C_0) - \ln C_e]$.

Table 2. Values of K_s and k as a function of salinity

Salinity [mg L^{-1}]	k [d^{-1}]	K_s [mg L^{-1}]
4000	0.02412	777.2
6000	0.02306	845.3
8000	0.02193	894.2
10000	0.02108	935.3
12000	0.02058	970.3
14000	0.01973	1041
16000	0.01924	1065
18000	0.01861	1120
20000	0.01815	1185

$K_s = 24.19$ salinity + 691.21. $R^2 = 0.993$ (12).

$k = -0.0004$ salinity + 0.0251. $R^2 = 0.977$ (13).

Table 3. Comparison of predicated COD with experimental data.

Salinity (mg L^{-1})	8000	10,000	15,000
Influent COD (mg L^{-1})	3200	4000	6000
Experimental COD (mg L^{-1})	1470	2070	3620
Ks	884.73	933.11	1078.25
k	0.022	0.021	0.019
Predicted COD (mg L^{-1})	1370	1920	3810
Relative error (%)	7.3	7.8	5

10000 mg L^{-1} and 15000 mg L^{-1} salinity, respectively. The experiment was also conducted at 10°C at volumetric exchange ratio of 1/4. Sludge concentration was 15000 mg L^{-1}. The results was presented in Table 3.

As described by Table 3, experimental results were in agreement with simulated values, which in turn proved the validity of the simulation.

4 CONCLUSION

An empirical model for treatment of hypersaline industrial wastewater was developed based on the influent, effluent COD content and salinity in the ASBBR. The maximum specific substrate utilization rate (d^{-1}) k and substrate saturation constant (mg L^{-1}) K_s could be expressed as a function of salinity, $K_s = 24.19$ salinity +691.21 and $k = -0.0004$ salinity + 0.0251, respectively. This model may be applied to predicting the effluent COD concentration and it has some practical use for design and selection of ASBBR.

ACKNOWLEDGEMENT

This project was supported by the Ministry of Education Doctoral Fund Issues (New Teacher Category, People's Republic of China, Grant No. 2009019 1120036).

REFERENCES

American Public Health Association (APHA) 2005. Standard methods for the examination of water and wastewater, 21th ed Washington, DC.

Chai, H.X., 2005. Research on the efficiency of ASBBR reactor for pickle wastewater treatment, MA Thesis, Chongqing University, Chongqing, China.

Chai, H.X., Li, X.P., Zhou, J., Chen, Y., Long, T.R. 2010. Treatment of mustard tuber wastewater by anaerobic sequencing batch biofilm reactor-two-stage sequencing batch biofilm reactor-chemical-dephosphorization process. *Chinese Journal of Environmental Engineering*, 4, 785–789, in Chinese.

Christian, R., Sheng F., Frank U., Axel B., Ingo S., Michael S. 2010. Stirring and biomass starter influences the anaerobic digestion of different substrates for biogas production. *Journal Engineering in Life Science*, 10 (4), 339–347.

Cuetos, M.J., G'omez, X., Escapa, A., Moran, A. 2007. Evaluation and simultaneous optimization of biohydrogen production using 3(2) factorial design and the desirability function. *J. Power Sources*, 169 (1), 131–139.

Dincer, A.R., Kargi, F. 2001. Performance of rotating biological disc system treating saline wastewater. *Process Biochem*, 36 (8–9), 901–906.

Doudoroff M. 1940. Experiments on the adaptation of Escherichia coli to sodium chloride. *J. Gen. Physiol.*, 23 (5), 585–611.

Dugba, P.N., Zhang, R. 1999. Treatment of dairy wastewater with two-stage anaerobic sequencing batch reactor systems—thermophilic versus mesophilic operations. *Biores. Technol.*, 68, 225–33.

Gebauer, R. 2004. Mesophilic anaerobic treatment of sludge from saline fish farm effluents with biogas production *Biores. Technol.*, 93 (2), 155–167.

Jeison, D., Kremer, B., van Lier, J.B. 2008. Application of membrane enhanced biomass retention to the anaerobic treatment of acidified wastewaters under extreme saline conditions. *Separation and Purification Technology*, 64 (2), 198–205.

Kargi, F., Uygur, A. 2005. Improved nutrient removal from saline wastewater in an SBR by Halobacter supplemented activated sludge. *Environmental Engineering Science*, 22 (2), 170–176.

Kartal, B., Koleva, M., Arsov, R., van der Star, W., Jetten M.S.M., Strous M. 2006. Adaptation of a freshwater anammox population to high salinity wastewater. *Journal of Biotechnology*, 126 (4), 546–553.

Lefebvre, O., Moletta, R. 2006. Treatment of organic pollution in industrial saline wastewater: A literature review. *Water Research*, 40 (20), 3671–3682.

Lettinga, G., Rebac, S., Parshina, S., Nozhevnikova, A., van Lier, J.B., Stams, A.J.M. 1999. High-rate anaerobic treatment of wastewater at low temperatures. *Applied and Environmental Microbiology*, 65 (4), 1696–1702.

Masse, D.I., Masse, L. 2001. The effect of temperature on slaughterhouse wastewater treatment in anaerobic sequencing batch reactors. *Biores. Technol.*, 76 (2), 91–98.

Omil, F., Mendez, R., Lema, J.M. 1995. Anaerobic treatment of saline wastewaters under high sulphide and ammonia content. *Bioresour. Technol.*, 54 (3), 269–278.

Pinho, S.C., Ratusznei, S.M., Rodrigues, J.A.D., Foresti, E., Zaiat, M. 2005. Influence of bioparticle size on the degradation of partially soluble wastewater in an anaerobic sequencing batch biofilm reactor (ASBBR). *Process Biochemistry*, 40 (10), 3206–3212.

Ratusznei, S.M., Rodrigues, J.A.D., Camargo, E.F.M., Zaiat, M., Borzani, W. 2000. Feasibility of a stirred anaerobic sequencing batch reactor containing immobilized biomass for wastewater treatment. *Biores. Technol.*, 75 (2), 127–33.

Shao, X.W., Peng, D.C., Teng, Z.H., Ju, X.H. 2008. Treatment of brewery wastewater using anaerobic sequencing batch reactor (ASBR). *Biores. Technol.*, 99, 3182–3186.

Shen, J.Y., He, R., Wang, L.J., Han, W.Q., Sun, X.Y., Li, J.S. 2009. Kinetics of COD removal in a biological aerated filter in the presence of 2,4,6-trinitrophenol (picric acid). *Chinese Journal of Chemical Engineering*, 17 (6), 1021–1026, in Chinese.

Tsuneda, S., Mikami, M., Kimohi, Y., Hirata, A. 2005. Effect of salinity on nitrous oxide emission in the biological nitrogen removal process for industrial wastewater. *Journal of Hazardous Materials*, 119 (1–3) 93–98.

Uygur, A., Kargi, F. 2004. Salt inhibition on biological nutrient removal from saline wastewater in a sequencing batch reactor. *Enzyme and Microbial Technology*, 34, 313–318.

Villasenor, J., Mena, J., Fernandez, F., Gomez, R., de Lucas, A. 2011. Kinetics of domestic wastewater COD removal by subsurface flow constructed wetlands using different plant species in temperate period. *International Journal of Environmental Analytical Chemistry*, 91 (7–8) 693–707.

Frontiers of Energy and Environmental Engineering – Sung, Kao & Chen (eds)
© 2013 Taylor & Francis Group, London, ISBN 978-0-415-66159-1

The Design of the automatic monitoring system of water regimen based on GPRS

H.S. Yu, S.H. Zhang, X.J. Liu & X.K. Ji
School of Physical Science and Information Technology, Liaocheng University, Liaocheng, Shandong, China

ABSTRACT: A new type monitoring system of water regimen is designed by using current popular ARM9 processor and GPRS technology to achieve hardware of poor scalability, Single function and data transmission. The system can effectively overcome the many shortcomings to the use of traditional wired transmission, such as Lines of difficult, overcome the shortcomings of not strong, secondary development and upgrading of inconvenient by through the ARM9 processor instead of microcontroller. The preliminary result shows that the system owns such advantages as high reliability, scalability, testing accuracy, being used easily and so on. The technology and methods adopted in the system are practical.

Keywords: AT91SAM9260; GPRS; ARM9

1 INTRODUCTION

At present, the typical hydrologic telemetry system generally have sensors, remote sensing stand and data center (PC) constitutes. Hydrologic telemetry system complex and huge and how to realize the remote sensing data center with the communications of the station and hydrologic telemetry system should be resolved. This paper put forward based on GPRS hydrologic telemetry system, by the use of the GPRS wireless communication technology means the telemetry data center of standing with between transparent transmission, We through the general packet Radio service GPRS (Genelal packet Radio Serviee), to realize to provide users with GSM group forms of data services by practical application, the system has high reliability, can be expanded, testing accurate, convenient maintenance, etc.

GPRS technology greatly improves the network resource utilization, is the wireless data transmission in the application of the most extensive one of technology.

2 THE WHOLE SYSTEM DESIGN

Field monitoring terminal of the sensor parameters of hydrologic information collection, and embedded monitoring terminal by for the collected signal analysis, data through the GPRS module real-time transmission to the data center, data center will receive the data to be saved and Inquires the convenient. Data center can also through the GPRS

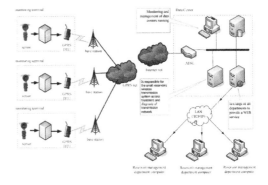

Figure 1. The whole system structure.

module to realize the control of site monitoring terminal. A major system including four parts the scene monitoring terminal, wireless transmission network, data centers and remote users.

3 FIELD MONITORING TERMINAL OF THE HARDWARE DESIGN

Field monitoring terminal mainly has water level sensor, embedded control module and GPRS module. AT91SAM9260ARM processor by the A/D channel acquisition MH-GA the ultrasonic thing location instrument signal, after analysis save, again through the RS-232 serial will level the data sent to GPRS wireless module, and through the GPRS network will the data sent to the data center. Field monitoring terminal hardware general structure shown in Figure 2.

3.1 Acquisition module

The system USES the ARM9 processor AT91SAM9260 internal 10 A/D realize the transformation of the data, Because MH-GA the ultrasonic thing location instrument is 4–20 output current of mA, and A/D receive must be that of the voltage signal, Therefore need to current sampling, in MH-GA and A/D with A high accuracy between resistance, let electricity pass 150 Ω sampling resistance after conversion, through the RC filter circuit for filter, and finally leads sampling voltage signals connected to A/D converter. Sensor output and A/D converter the connection will shown in Figure 3.

3.2 GPRS transmission module

GPRS communication module is monitoring terminal access wireless network interface, and at the same time also is the realization monitoring

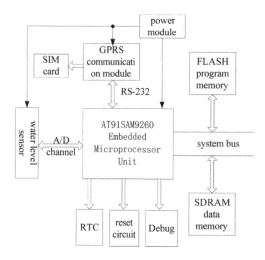

Figure 2. The general structure of the system hardware.

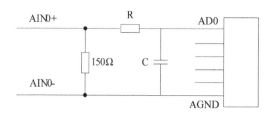

Figure 3. Sensor output and A/D converter of connection.

terminal and wireless two-way communication data center of important component, the performance of the regime is directly related to whether the measurement system can normal and stable and accurate operation. GPRS MODEM chooses is XIAMEN four letter communication technology Co., Ltd. Of the production of the F2103 GPRS DTU, to provide users with high speed, stable and reliable, data terminal always online, a variety of protocol conversion of virtual private network. F2103 GPRS DTU with AT91SAM9260 connected through a serial port.

4 SOFTWARE DESIGN

4.1 GPRS data transmission

The system continuously detects transmitted command of GPRS module and alarm detection module is through the AT91SAM9260. If there is the occurrence of an alarm or the data sending instructions from the data center, regardless of whether the time data arrive, the data will directly packaged and sent to the data center; if not, then start the A/D data collection timely, the system time set for one hour, after the completion of the acquisition data is stored, and the stored data according to a transmission protocol packaged is sent to the GPRS communication module, through the serial port GPRS communication module sends the data to the Internet network, the data can be transmitted to a data center. However in practice, there will often be dropped, therefore, before sending data to the GPRS there will be on-line monitoring. GPRS data transmission process is shown in Figure 4.

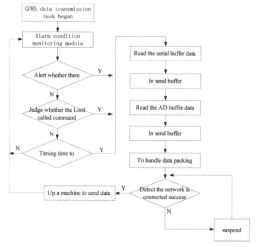

Figure 4. Data transmission flow chart.

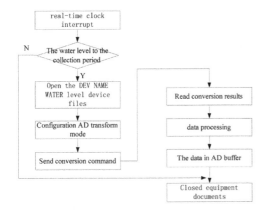

Figure 5.   Data collection part of the flow chart.

4.2   *The water level sensor acquisition software design*

Monitoring terminal achieves the level of data collection using ultrasonic Position-meter, and makes full use of the water level signal itself relatively stable characteristics, and selects timing acquisition ways to complete the level signal acquisition software design. Through the monitoring terminal real time clock interrupt, the water level acquisition time is reached or not firstly. If the water level acquisition time reaches the standard value level set by the acquisition function, it will open DEV_NAME_WATER level equipment files and configuration the conversion mode, and sends the conversion commands. Through the A/D conversion complete analog to the actual value of the voltage conversion, it can process the corresponding data to be the actual value of water level according to the water Gao Cheng and water level. After that the data is sent into a AD buffer. Water level data flow diagram as shown in Figure 5.

5   SUMMARY

Using GPRS technology to construct a remote wireless data acquisition and transmission system has been widely used in various fields, and satisfactory results were obtained. Therefore, in this paper, the use of GPRS technology achieves regimen measuring and reporting system in field telemetry station and data center between the data signal transmission. The paper designs a new type of water regime measurement system giving full play to advantages on t the GPRS network technology he basis and at the same time with the aid of modern computer technology and embedded technology. The system successful implements the water regime of real-time wireless measurement report. The application results show that, the system is stable and reliable, and is able to meet the disaster prevention and satisfies water regime forecast requirements. It has a strong promotion of the value, and the system is low cost, high reliability, stable performance and other advantages, for hydrological measuring and reporting system construction has an important role in promoting.

REFERENCES

[1] Cui yixin. Based on the CDMA network of water regime measurement system research [D] Shanxi. taiyuan University of Technology.

[2] Liu youzhu, Li shuliang, Zhu jiebin. GPRS based low voltage power distribution network management system [J] Electric power automation equipment 2009, (4): 131–134.

[3] Li changsheng, Jin ou. GPRS based vending machine monitor system design and Realization [J] Computer measurement & Control. 2008, 16(3):327–329.

[4] Wang dianhong, Liang juan, Xiong yuehuaetc. Power monitoring terminal based on the MC55 and LPC2136 GPRS [J]. Data acquisition and processing, 2006, 21:258–261.

Frontiers of Energy and Environmental Engineering – Sung, Kao & Chen (eds)
© *2013 Taylor & Francis Group, London, ISBN 978-0-415-66159-1*

The study on course construction of environmental engineering principles

C.N. Liu, P. Cao & W.H. Xia

College of Chemical and Environmental Engineering, Shanghai Institute of Technology, Shanghai, China

ABSTRACT: This paper has discussed the construction of course *Environmental Engineering Principles* from the practice of environmental engineering undergraduate major in Shanghai Institute of Technology, it has studied the course set up, the building of teaching resources database, the application of simulation technology in teaching, bilingual teaching, and the understanding of practical application of the basic principles. Results have shown that the teaching effectiveness has been greatly improved, the useful experiences have been gained, and the references for future teaching reform have been provided.

Keywords: course construction; building resources database; application of simulation; bilingual teaching; practical application

1 INTRODUCTION

In rencent years, the environmental pollution problems have become more and more complicated and shown composite characteristics, time characteristics, and geographical features. To solve these complicated environmental issues, the technical experts in environmental engineering should have strong comprehensive abilities and systemic perception, while the solid theoretical foundation is the cornerstone of these capabilities. In view of these facts, we have carried out the construction of course *Environmental Engineering Principles* in order to improve the professional basic education of environmental engineering undergraduate major.

2 THE COURSE SET UP OF ENVIRONMENTAL ENGINEERING PRINCIPLE

Since the environmental engineering major had been set in our school in 1998, the course *Chemical Engineering Principles* was offered as the professional foundation requisition to the undergraduate majoring in environmental engineering. After years of teaching practice, curriculum has been already mature and complete, a wealth of teaching experiences have been accumulated, and good teaching accomplishments have achieved. However, we also realized that the examples, exercises, experiments, and project examples in the teaching are chemical background. These make the environmental

engineering students lacked the intuitive and in-depth understanding of the relationship between the basic knowledge learned and pollution control technology; and the connection of the basic course and follow-up of professional teaching is not tight enough and smooth. Therefore, in order to improve the teaching quality of the undergraduate majoring in environmental engineering, we had to reform the teaching and practice of the course.

At 2006, our school set up the course *Environmental Engineering Principles* to replace the course *Chemical Engineering Principles*. In this course, the basic principles of environmental engineering are divided into three parts: the environmental separation engineering (to separate the pollutants from the environment, does not change its chemical properties), environmental chemical engineering (to chemically convert pollutants to harmless), and environmental biological engineering (to use biotechnology to achieve the pollutants harmless). Three basic principles can be applied to treat various pollutants in various environmental media. Therefore, this course is more favorable to the environmental engineering student; it has universal significance and fully embodies the characteristics of environmental engineering.

With the principals of ensuring the teaching needs of the professional knowledge, after seriously comparison, the "15" national planning textbook *Environmental Engineering Principles* (Hong-Ying Hu, ed.) was selection as the textbook of this course. This book systematically presents the basic concepts of the common technologies involved in the environmental engineering. Compared with other textbooks available in the market, this textbook has more integrated, theoretical and comprehensive contents. For the contents which are duplicated with other courses, we reduced them in this textbook. For example, the content of fluid flow in chapter 3 in this textbook repeats the related content in the course *Fluid Dynamics*; the content of heat transfer in chapter 4 repeats the related content in the course *Thermal Engineering*, etc. Then we began the construction of course *Environmental Engineering Principles* at 2008.

3 BUILDING THE TEACHING RESOURCES DATABASE

In the teaching process, we have proposed the "Teacher-led, student-centered" teaching philosophy which ensure that the students are the teaching subject while the teachers are the dominant; both are very important roles. Teachers should create the learning environment and provide effective means to enable students to actively participate into teaching process. In order to excite students' curiosity and creativity, we have adopted a lot of interactions between teaching and learning, which achieved satisfactory results. Furthermore, we have designed examples and exercises according to the environmental engineering context, and build the exercises and test questions databases which cover more comprehensive contents of environmental engineering principles. The databases include various forms and diverse kinds of questions, by which the students have strengthened their understanding to the basic concepts, better grasped and applied the basic concepts.

Besides, we have made the illustrated, information-rich multimedia courseware to improve the teaching effectiveness. Based on the actual needs from the environmental engineering, we have strengthened the training for professional technical ability via the practice example of environmental engineering, and focused on analyzing and solving problems skills to train knowledge and capable undergraduate.

Numerous cases of integrating theory with practice, the development trends of environmental engineering, and the latest research results have been complemented in the teaching process. All of these have excited students' thirst for knowledge, and improved their ability to solve practical problems. These also help the development of environmental engineering.

After several years of teaching practice, we have achieved very good results. As the basic course and main course in environmental engineering, the course *Environmental Engineering Principles* plays an important role and has formed the teaching model with certain characteristics.

4 APPLICATION OF SIMULATION IN TEACHING

In order to deepen students' understanding of the theoretical knowledge, to improve students' innovative spirit and practical ability, and especially to stress on training of applied talents according to our school characteristics, we have greatly developed the experiment teaching section of the course. The teaching of course *Environmental Engineering Principles* should enable students to establish the complete concept of systematic purification and pollution control; the experiment teaching should make students to have a comprehensive, in-depth understanding of environmental purification and pollution control system components, and lay a solid foundation for the follow-up of environmental engineering courses. Thus we have implemented a multi-level experiment teaching model, Including the basic experiment, comprehensive experiment, design experiments, etc. In the experiment teaching process, some experiment devices for important operation units are so expensive that they are not yet fully equipped in our school, but they are essential for students to master the basic principles and grasp the device performance. Therefore, we have carried out the experiments via the interactive simulation teaching software which includes filtration experiment, absorption experiment, extraction experiment, precipitation experiment, centrifugal pump performance curve determination, and so on. Through the practices of measuring and calculating parameters based on experiment simulation, the students have understood the principle of operations and are familiar with the use of equipment. Good teaching results have been achieved.

Meanwhile in the theory teaching process, we also use a lot of vivid three-dimensional animations to demonstrate the working principle of cyclone, bridging in filtering process, activated carbon adsorption, adsorption motion and penetration, ion exchange process, etc. Through these interesting animated interpretations, the boring and unintelligible theories become easier to understand and more conducive to students' grasping and understanding of each knowledge point. At the same time, the classroom atmosphere is lively and easy to inspire students' thinking.

Student' average scores of the course in recent years are given in Table 1; it shows that students

Table 1. Students' average scores in recent years.

Academic year	2006	2007	2008	2009	2010	2011
Average scores	69.2	68.5	75.5	78.8	80.3	80.1

can improve their academic performance in a relaxed state by these ways.

5 BILINGUAL TEACHING

Bilingual teaching is taught in two languages: the Chinese and English. Developing bilingual teaching of this course is one of key means which can bring our teaching in line with international practices, accelerate the learning of foreign scientific and technological achievements in environmental engineering, and short the gap with foreign. The bilingual teaching is very popular and welcome by the students.

At present, some important chapters in the course such as mechanical separation, mass transfer, absorption, etc. have adopted the bilingual teaching in which the foreign advanced teaching philosophy has been learned and the teaching contents and teaching methods have reformed. The teaching by discussion method has been applied to improve teaching effectiveness. This has evoked students' interest in learning.

The materials used in the bilingual teaching are the appropriate choices from original textbooks in English including Environmental Engineering Science edited by Willianm W. Nazaroff and Lisa Alvarez-Cohen, Unit Operations of Chemical Engineering edited by McCabe and Warren L, and so on. We have adopted several teaching modes: teaching in English with reference to the English materials, teaching in English with reference to the Chinese materials, teaching in Chinese with reference to the English materials, or the combination of the above-mentioned ways, for example, the combination of teaching in Chinese and English or the combination of teaching materials in English and Chinese. The English glossary of technical terms in the course has been compiled and offered to students. It provides definitions of key terms used throughout the guide to facilitate students' learning and mastering.

Recent years, the bilingual teaching practices have received initial results and have also showed its significant role. Students have been familiar with the English expression of technical term and the latest development abroad in environmental engineering, their professional English proficiency has been greatly improved.

Some reforms to the course evaluation which reflects the bilingual feature in this course have been carried out. For example, answering the questions in English by oral or in essay is implemented; and in the final test or quizzes, there is a certain portion of questions in English. The students need to answer these questions in English. By doing in this way, we can evaluate the students' ability to

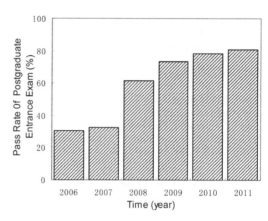

Figure 1. The percent of pass postgraduate entrance examination in recent years.

answer these questions correctly and completely in English. The answers of students are scored and revised accord to academic and English standard, and then feedbacked to the students. In this way, the students generally think that they are benefit a lot from it.

Our teaching practices have shown that the bilingual education in this course has expanded students' horizons, improved their technical ability in English, and enhanced their self-confidence in learning.

Since 2008 the bilingual teaching has been implemented in this course, there are remarkable teaching accomplishments has been achieved. Upon the survey of the feedbacks from graduates majoring in environmental engineering, Environmental Engineering Principles course is one of the most practical courses. It has greatly cultivated students' study interests and innovative ability. Students' initiative and creativity have been to give full play; the students' ability of competition for jobs has improved greatly.

The bilingual teaching has conducive to get better grades for students in the entrance exam of the graduate school than the conventional teaching method. Figure 1 shows changes on the percent of pass postgraduate entrance examination in recent years since the bilingual teaching has been performed. As can be seen from Figure 1, bilingual teaching has the great advantage for students to pass this examination and become postgraduates.

6 UNDERSTANDING THE PRACTICAL APPLICATION OF BASIC PRINCIPALS

This course discusses the fundamental principles of solving problems. Therefore, one of the most important focuses in our teaching is to develop students' abilities of applying the theoretical knowledge into practice to solve practical problems. We have used pictures, videos and other examples, such as applications of environmental engineering principles in sewage treatment plants, chemical plants, to enhance students' theoretical knowledge and its application in realistic situations. Through the perceptual cognition from the internship and realistic practice, students can understand the theoretical knowledge deeper, and furthermore to apply it into practice. Students' comprehensive abilities are greatly improved that solid their foundation capabilities in solving practical problems in the future. Moreover, students feel much more confident and adaptable in their practical work after graduation; while the employers also give satisfied feedback of students' performance in work.

7 CONCLUSIONS

Along with the social development, the improvement of training goals and training models will eventually be implemented on the reform of the course. Through years of teaching practice in the course construction, we have greatly optimized the teaching results and achieved the remarkable teaching accomplishment. At the same time, we have recognize that the teaching activities of *Environmental Engineering Principles* course still need us to explore and develop; we will continuously accumulate experience in practice to gradually improve and perfect the curriculum development.

ACKNOWLEDGEMENTS

This work were financial supported by the key discipline fund of Shanghai education committee (J51502).

REFERENCES

Hongying Hu, Jining Cheng, and Gang Yu. 2005. Exploration and Practice of High-Quality Talent Training System in Environmental Disciplines. *The 2th Course Forum proceedings of Environmental undergraduate*: 536–540, Beijing, Higher Education Press.
Pei Shen & Shuiping Feng. 2005. Exploration and practice of promoting bilingual education. *China university edication* 8(2): 36.
Shiyu Li. 2003. Research and Practice in Chemical Simulation Practice Teaching. *Chemical Higher Education* 6(2): 49–52.
Zhenhuan Zhang, Xiangnan Luo, and Bimgyuan Zhao. 2003. The Analysis of Education Development and Talent Demand of Chinese Environmental Engineering Undergraduate. *Environmental Protection* 15 (2):59–64.

Frontiers of Energy and Environmental Engineering – Sung, Kao & Chen (eds)
© 2013 Taylor & Francis Group, London, ISBN 978-0-415-66159-1

Numerical simulation of COD concentration distribution in Yanming 2# lake based on MIKE 21

Y. Wang & Y.M. Yu
Key Laboratory of Northwest Water Resources and Environmental of Education Ministry, Xi'an University of Technology, Xi'an, China

L. Lu
Jining City Investment Liability Company, Shandong, China

ABSTRACT: With the development of industry and urban population, the river water is rapidly deteriorating. The water environment problem solving is imminent. In the paper, the MIKE21 is used to simulate COD concentration distribution of Yanming 2# lake, based on Hydrodynamics module (HD) and Advection-Diffusion module (AD). According to the contrast results with the filed actual measured data in 2008, the relative error is low, so simulate result is reasonable and reliable. Through the simulation result, flow velocity is large in the inward and outward, COD concentration is closely related with the flux and concentration in the inward area, and it is affected by seasonal influence, which is the highest in winter and lowest in spring.

Keywords: Yanming 2# lake; MIKE21 model; COD concentration distribution; numerical simulation

1 GENERAL INSTRUCTIONS

In order to improve the urban worsening environment, more and more city landscape lakes arise at the historic moment. As the symbol of green city, the city landscape lakes can not only adjust the regional climate, but also play an important role in the improvement of ecological environment system for the city. However, due to the influence of many factors, especially the unreasonable utilization of humans, ecological system structure and function of the lakes degenerate, and water shortage due to water quality increase seriously, which lead to huge losses[1].

This article selects MIKE21, the MIKE series software of the Danish water resources and water environment institute (DHI). The software is of power in a two-dimensional free surface flow simulation, which is widely used in lakes, estuaries, the gulf and coastal areas to simulate hydraulics and the related phenomenon. Qin Qiaoli[2] used MIKE21 to simulate flood routing in Bailanghe River and its flood control planning. He Wenhua[3] researched on the effect of urbanization on storm flood and flood simulation at Jinan City. Rohit and Goyal[4] used the software to model hydrodynamics for water quality of the Dwarka Region (Gujarat). Xiao-kang[5] did research about reservoir operation schemes for water pollution accidents in

Yangtze River. All these prove the practicability of the software.

This paper build 2d water hydrodynamic model of Yanming 2# lake, depending on related hydrodynamic theory combined with MIKE21[6]. According to the actual measured data of environmental science research institute of Xi'an university of technology from June 2007 to March 2009. The change trend of COD concentration distribution in the lake is simulated based on the COD measurement data of 2008, which can provide technical basis for further study of comprehensive governance about lake and environment optimal allocation and management.

2 SIMULATE CONDITIONS

2.1 *Calculation area and grid figure*

The simulation research is based on Yanming 2# lake. According to the measurement data for several years, in the paper the lake is generalized to be an artificial shallow water lake, which is 1240 m long from south to north and 170 m wide from east to west with the average depth of 1.5 m. The lake area is set to be calculation area and then model grids are established, with the grids size for 5 m × 2 m. In order to facilitate calculation and analysis, water inflow in the south and outflow in the north are set at the left and right of the calculated regional figure, which can be seen in Figure 1.

2.2 Initial conditions

Before the simulation, initial conditions such as velocity and water level must be set. Initial velocity is set to be 0. And average water level of these years is taken to be the initial water level, which is 1.5 m. According to the actual measured data, the average COD concentration in the lake of December of 2007 is chosen to be the initial concentration, which is 15 mg/L.

2.3 Boundary conditions

While simulating practical problems, boundary conditions, as part of Definite solution conditions which is important to solve problems, must be given properly. The model boundary consists of two parts: inflow boundary and outflow boundary. The value of the two boundary is given in the

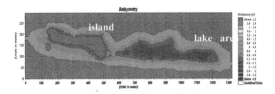

Figure 1. Yanming 2nd Lake topographic map generalization.

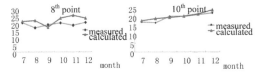

Figure 2. The comparison chart of model calculated values and measured values.

form of two dimensional data file, which is given by day.

2.4 Calculation method

ADI method in rectangular staggered grids is used in the numerical simulation model. Through differing the mass equation and momentum equation in time and space, each physical quantity such as z, h, u, v stays at different points on the staggered grid. The specific calculation process is shown in reference[7].

2.5 Model diffusion coefficient

In MIKE21, model diffusion coefficient is normally confirmed from empirical formula[8]. The diffusion coefficient in AD module is selected through two methods, one is based on velocity and the other is based on flux. And no matter what methods we choose, the diffusion coefficient is always depends on the grid space step, time step and flow velocity. According to the situation of the actual lakes,and combined with space step and time step, According to the equation $K = k\Delta x \times u$, Δx is grid step and u is flow velocity.we choose model diffusion coefficient $K_{COD} = 1.1 \times 10^{-7} S^{-1}$.

2.6 Model validation

The Model validation is based on the data from July to December, and two measuring points, the 8th point located at the east bank of the lake and the 10th point located near the outflow area, are selected for error analysis. The analysis result is listed below as Figure 2.

Error analysis between the simulation value and measured value is listed in Table 1.

It can be easily seen that the error rate between simulation value and measured value is not big, which means that the simulation value can reflect the measured value accurately. So MIKE21 is proved to be suitable for the simulation of Yanming 2# lake.

Table 1. Comparison table of the simulated and measured values.

Month	Simulation material	8th point			10th point		
		Measured value	Simulation value	Relative error (%)	Measured value	Simulation value	Relative error (%)
Jul.	COD	20.29	21.26	16.71	16.9	17.79	16.32
Apr.		16.75	21.8	1.07	16.57	18.9	13.30
Sep.		18.79	17.1	−7.13	20.13	19.53	14.21
Oct.		20.16	23.6	0.40	20.08	20.36	13.73
Nov.		18.33	25.43	15.28	21.13	21.48	15.53
Dec.		20.51	23.45	−7.70	22.09	23.76	1.32

3 SIMULATION RESULT ANALYSIS

3.1 Model validation

Through simulation, we can get flow velocity contour line map and velocity vector map, as is shown in Figures 3 and 4. From Figure 3 we can find than annual average flow velocity is bigger in inflow and outflow area, between 0.04 to 0.07 m/s, and the area is small, about 5% of the lake area. Flow velocity in main area of the lake is between 0.001 to 0.003 m/s, especially after the water flowing through the island into the body area, the velocity remains in 0.001 to 0.002 m/s stably. In the east and west of the body area, the velocity is below 0.001 m/s, which is about 15% of the lake. In Figure 4, direction of arrow represents for flow direction and the length means the value of the velocity. According to Figure 4, influenced mainly by topographic condition, water flows from south to north, which is mainly accordant to the inflow direction.

3.2 Simulation result analysis of COD concentration distribution

Through simulation, we can get the mean value of COD concentration of every month, then the migration change rule can be found out based on hydrodynamic conditions. The simulation results are shown in Figures 5–16.

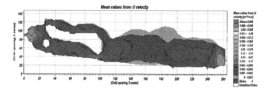

Figure 3. Flow field flow velocity contour map.

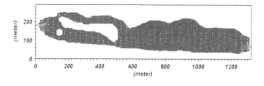

Figure 4. Lake flow field velocity vector of the Lake.

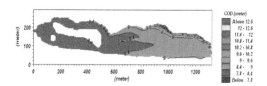

Figure 5. Lake COD distribution in January.

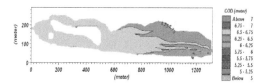

Figure 6. Lake COD distribution in February.

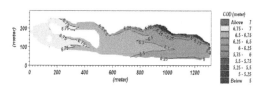

Figure 7. Lake COD distribution in March.

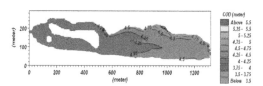

Figure 8. Lake COD distribution in April.

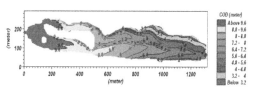

Figure 9. Lake COD distribution in May.

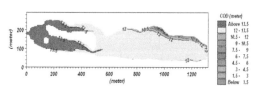

Figure 10. Lake COD distribution in June.

As is shown in Figures 5–16, the COD concentration comes to the following rules:

1. COD concentration distribution is closely linked with the inflow flux. For example, with the minimum flux (0.047 m³/s) in May and the maximum flux (0.295 m³/s) in November, COD concentration spread area can reach 15% and 75% of the lake.

2. COD concentration distribution changes mainly through water replacement, And obviously presents stratification phenomenon.

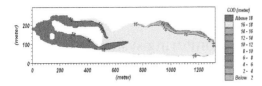

Figure 11.　Lake COD distribution in July.

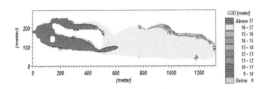

Figure 12.　Lake COD distribution in August.

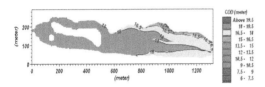

Figure 13.　Lake COD distribution in September.

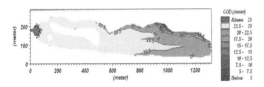

Figure 14.　Lake COD distribution in October.

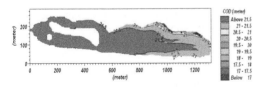

Figure 15.　Lake COD distribution in November.

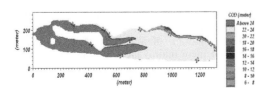

Figure 16.　Lake COD distribution in December.

3. COD concentration is low in spring, and the minimum value appears in April, about 4–5 mg/l; COD concentration is high in winter, and the maximum value appears in December, about 20–24 mg/l. In addition, COD concentration is closely related with the inflow COD concentration.

4　CONCLUSION

In the paper, the MIKE21 is used to simulate COD concentration distribution of Yanming 2# lake, based on Hydrodynamics module (HD) and Advection-Diffusion module (AD). The simulate result is reasonable and reliable. It comes to the following conclusion:

1. Annual average flow velocity is bigger in inflow and outflow area, and the flow direction is from south to north.
2. COD concentration distribution is closely linked with the inflow flux.
3. COD concentration distribution changes according to season, which is low in spring and high in winter.
4. COD concentration distribution changes mainly through water replacement, And obviously presents stratification phenomenon.

REFERENCES

[1] Jian Pan, Junli Ma.Existence Crisis and Protection Measures of Urban Lakes [J]. Northern Environment, 2011(6):115,131.
[2] Qin Qiaoli. Reaserch on Numerical Simulation of Flood Routing in Bailanghe River and Its Flood Control Planning [D]. Shandong university, 2009.
[3] He Wenhua. Study on Effect of Urbanization on Storm Flood and Flood Simulation at Jinan City. South China University of Technology, 2010.
[4] Rohit,Goyal. Modelling Hydrodynamics for Water Quality of the Dwarka Region (Gujarat). Proceedings of International Conference on Environmental Science and Development (ICESD 2011) Mumbai, India, 2011.
[5] Xiao-kang, Xin Weiyin, Meng Wang. Reservoir operation schemes for water pollution accidents in Yangtze River. Water Sciences and Engineering. 2012,05(1) 10.3882/j.issn.
[6] Ting Xu. MIKE21 Model Summary and Application Example.Water Conservancy Technology and Economy, 2010(8).
[7] Danish Hydraulic Institute(DHI).MIKE21 FLOW MODEL:Hydrodynamic Module Scientific Documentation [M]. DHI, 2007.
[8] Luo L-C, Qin B-Q. Numerical simulation based on a three-dimensional shallow-water hydrodynamic model in Lake Taihu current circulations in Lake Taihu with \\\prevailingwind-forcing [J]. Hyd-Dyn, 2003, 18(6):686–691.

Frontiers of Energy and Environmental Engineering – Sung, Kao & Chen (eds)
© *2013 Taylor & Francis Group, London, ISBN 978-0-415-66159-1*

Technical study on building gas real-time monitoring system based on Ajax and ArcIMS

S.H. Liang
Department of Electronic and Information Engineering, Jiangsu Institute of Architectural Technology, Xuzhou, China
School of Environment and Spatial Informatics, Chinese University of Mining and Technology, Xuzhou, China

Y.J. Wang
School of Environment and Spatial Informatics, Chinese University of Mining and Technology, Xuzhou, China

S.S. Zhu
Jiangsu Key Laboratory of Resources and Environmental Information Engineering, Xuzhou, China
Airforce Service Colledge, Xuzhou, China

ABSTRACT: In view of the problems of heavy burden of the server and presence of a huge amount of repeated data in the returned pages due to adoption of synchronous interaction technology for web services in traditional WebGIS system, in this paper a concept of incorporating AJAX (Asynchronous JavaScript + XML) technology into WebGIS system is proposed, and the coupling model and realization method for building the gas real-time monitoring system through integration of AJAX technology and ArcIMS Java Connector are described, and finally data asynchronous transfer at browser and server under B/S structure is achieved and the service performance is improved.

Keywords: AJAX; ArcIMS Java Connector; WebGIS; gas real-time monitoring

1 INTRODUCTION

Gas outburst is the largest threat to the safety of coal mine, while the mining gas is a kind of dynamically changing information closely related to the spatial location, how to transfer such information accurately and in real time to the remote control center and relevant production management departments (production section and ventilation section, etc.) and the superior department is of great significance for guiding the production safety of coal mine[1], The traditional practice is: each browser automatically updates the pages at a certain time interval, the client dynamically requests them, then the server transfers the processed new results to the client to display the dynamic change. However, this also brings about a series of problems. First, since the client and the server adopt synchronous interaction technology, when the client frequently sends request to the server, the network transfer load will be increased, and this easily results in "crash" phenomenon at the client due to too heavy load at the server; second, since the browser updates the whole page, the page frequently flickers, which will bring difficulty to users

in operation. because ArcIMS has already had Ajax concept from 3.X edition, in other words, it has been able to transfer XML format based data and dynamically refresh part of pages, it is only short of asynchronous transfer[2].

2 CONSTRUCTION OF NETWORKING SYSTEM FOR REAL-TIME MONITORING OF COAL MINING GAS

The system is mainly divided into three subsystems: real-time monitoring and management subsystem, graphic data management subsystem and data management subsystem. Functional structure of the system is shown in Figure 1.

1. Real-time monitoring and management subsystem
   The functions of real-time monitoring and management mainly include connection, monitoring, alarm and statistics etc. of the system as follows:
   1. The connection status of the present monitoring system and each intelligent substation;

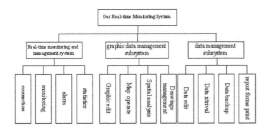

Figure 1. Functional structure of the system.

2. Real-time field monitoring display, inquiry and statistics of monitoring data of every mine;
3. Dynamic curve display of change trend of various monitoring data;
4. Real-time dynamic alarm prompt of various monitoring data;
5. Alarm statistics in the specified time range;
6. Display of dynamic theme map of data
2. Graphic management subsystem

The production mines have accumulated a lot of gas geologic data in the course from the geological exploration, well construction to the normal production, such huge amount of information is scattered in geological reports, technical literature, drawings and various working records, they are separate and are difficult to share. This module has such functions as of plotting, management, spatial data analysis and printout, etc. of professional graphic drawings:

1. Can turn various professional drawings into vector digital drawings for storage and editing;
2. Can carry out basic GIS operations (zoom, move and modify etc.) for various professional drawings;
3. Spatial analysis of various professional drawings (analysis of theme drawing and buffer area);
4. Printout and output of various professional drawings.
3. Data management subsystem

This module is mainly used for management of basic data as of ventilation, gas and geological data etc., preview and print of various reports, data backup and recovery:

1. Maintenance of basic attribute data such as gas information;
2. Maintenance of such real-time monitoring information as concentration of gas, wind speed and CO;
3. Maintenance of inspection information of various sensors and production equipment;
4. Printing of data sheet.

3 ESTABLISHMENT OF DEVELOPMENT PRINCIPLE AND TECHNICAL SCHEME OF GAS MONITORING SYSTEM

The system builds development platform based on ArcIMS software, since the basic functions for customized publishing provided by Manager in ArcIMS is very limited, it is required to make further customization and development for ArcIMS so as to further satisfy the requirements of system performance. ArcIMS provides two development modes, i.e. ArcXML language and connectors (Servlet connector, ActiveX connector, Could-Fusion connector and Java connector), among them ArcXML is the basic method for realizing more flexible and complex system functions and map display, while the connector is a package of ArcXML language, which enables the developer to avoid dealing with ArcXML and lowering the development difficulty.

This paper adopts the development mode of combining the secondary development language ArcXML and ArcIMS Java Connector of ArcIMS to realize expansion of more complex functions. The technical composition principle of the system is shown in Figure 2.

1. Secondary development method of combining ArcXML and Java Connector[3,4]

The basic functions and presentation of gas system, they can be realized through JAVA Viewer of ArcIMS customized by the author, besides, through the four sub-tags CONFIG, REQUEST, RESPONSE and MARKUP of ArcXML, it is very convenient to modify the source of data and display mode of map, add symbol, and display pattern of theme drawings, etc. In the system, the basic map service

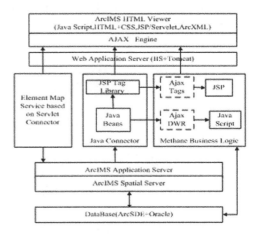

Figure 2. Principle of system development technology.

619

is provided by the default Servlet Connector set by ArcIMS. For the relatively complex logic request of gas service, for example, the real-time field monitoring and display of gas data, generation of dynamic theme drawing, etc.

2. Integration of Ajax and ArcIMS

AJAX framework DWR (Direct Web Remoting) is a remote process framework [5]. Another AJAX framework AJAX Tag is an open-resource JSP tag base, used for simplifying the use of AJAX technology in JSP page, it provides tags of some common functions such as drop-down cascade selection, when the user enters characters in the text box, it automatically matches the characters entered by the user from the designated data. With the aid of powerful function of objects transmission of DWR between the browser and server and the high efficiency advantage in developing JSP pages containing Ajax application using AJAX tag, the system realizes the integration of Ajax technology, ArcIMS Java Connector and JSP Viewer.

4 REALIZATION OF GAS REAL-TIME MONITORING SYSTEM

The running environment for gas real-time monitoring system is IIS5.0 + Tomcat5.0 + ArcIMS9.2 + ArcSDE9.0 + Oracle9.0, it edits and processes relevant space data by ArcMap, including the gas geologic map, plan of mining works, drawing of mine ventilation system, topographic map, etc., then store the space data into the Oracle database using Arc-Catalog through the database engine ArcSDE. After developing the application function program of gas real-time monitoring WebGIS system using such languages as HTML, JavaScript, JSP, Java, XML and ArcXML, the development personnel create and initiate map service using Administrator in ArcIMS software to establish site publishing data. The running interface of system is shown in Figure 3.

5 CONCLUSION

In this paper the gas real-time monitoring WebGIS system is built based on AJAX technology and ArcIMS secondary development, a detailed introduction is given to the key technologies related to the realization and optimization of the system scheme, and finally the Internet publishing of real-time monitoring information of coal mines is realized. Compared with the traditional WebGIS development platform, this gas WebGIS real-time monitoring system with asynchronous request/response mechanism which adopts AJAX technology has faster response speed, higher service performance and better customer experience.

ACKNOWLEDGEMENT

This research was supported by the National Natural Science Fund of China (NO: 40971275).

REFERENCES

ArcIMS—Using the Java Connector [EB/OL], http://www.esri.com. ArcIMS9 Customizing.

Guo Xing-hua, Cuihu-ping, Shan Guo-hui.: The design and practice of Web Map service based on Ajax. J. Science of Surveying and Mapping, 33(2), 205–206 (2008).

Liu Hai-xin, Yang Qin.: Based on Com GIS Research and Application Of Coal Mine Gas Manage information System. J. Coal Engineering 7, 103–104 (2007).

Nathanic. T. Schutta, Ryan Asteson.: Pro Ajax and Java Frameworks [M]. Post & telecom Press. BeiJing (2007).

Song Jian-cheng, Mao Shan-jun.: Research and application on Navigation system for mine gas prevention and treatment [M]. China Coal Industry Publishing House. BeiJing (2007).

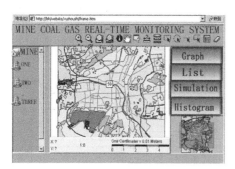

Figure 3. Sectional view of gas real-time monitoring system.

Frontiers of Energy and Environmental Engineering – Sung, Kao & Chen (eds)
© 2013 Taylor & Francis Group, London, ISBN 978-0-415-66159-1

Analysis on characteristics of pollution layer on the North Coast of Liaodong Bay

Y.H. Zhang, H. Ye & Y.J. Ma
Institute of Atmospheric Environment, CMA, Shenyang, China

ABSTRACT: By using meteorological data in the pollution boundary layer which was observed in two ground observation sites: coast and land on the north coast of Liaodong bay during January–February in 2007, the daily change characteristics of pollute boundary layer in winter in the area were discussed. The results showed that the pollute boundary layer on the north coast of Liaodong bay was affected by the sea and land. In the certain weather condition, maybe the sea-land breeze appeared in the low altitude which was below 200 m in the coastal zone. The stability change in the different height in the coastal zone was more stable than in the land zone, and the wind field change in the area was mainly in 300 m low altitude. At night, the temperature inversion often appears in the area, and the thickness of temperature inversion layer is stably during 200–300 m. The thermal internal boundary layer penetrated deeply into the land about 10 km, and the height could reach 800 m. The atmospheric diffusion ability in the coastal area was weaker and stronger in the land area.

Keywords: The north coast of Liaodong bay; pollution characteristics; boundary layer; atmospheric diffusion; observational research

1 INTRODUCTION

In the border area between the sea and the land, for thermal force difference of different under laying surfaces in the sea and the land, the meso-scale circulation variation which was caused appeared all the year round. The variation of meso-scale circulation not only could affect the regional weather change but also had the close relationship with the transmitting and diffusion of atmospheric pollutant. For the influence of sea, land regional circulation on the pollution diffusion, the pollution was induced, and the coastal fumigation pollution was generated when the thermal boundary layer existed. It had the practical significance to study the regional weather, the zone environment protection and so on[1].

In recent years, as "five points one line" strategy in Liaoning coastal areas was carried out, the industry in Liaoning coastal areas developed quickly, and the coastal development zone gradually expanded. It made that the pollution boundary layer condition which affected the atmospheric environment had the great change. The north coast of Liaodong bay is one part of Liaoning coastal industrial development areas. It is in the south of old city proper in Yingkou and the east bank of Liao River. For it is in the special position of coast, suburb and plain estuary, the atmospheric diffusion capability and the boundary layer characteristic have the sharp region features. According to the boundary layer observational research in the area in winter, we analyzed and studied the basic characteristic of pollution boundary layer and wanted to provide the reference for the prophylaxis and treatment of atmospheric pollution in the area.

2 METHODS

The observation research sites were selected to be in 2 representative stations which were in Yingkou industry development zone in the east bank of river outlet in Grand Liao River. They were the seaside station which was 200 m away from the coast and big pond station in the south suburb of Yingkou which was 6.4 km away from the seaside station in the northeast direction. They attempted to stand for the coast, the flat ground of river outlet, the pollution boundary layer surrounding the city. The air sounding height was 1500 m, and the probe items had the wind field, the temperature field in the boundary layer. The probe time was from January 24 to February 7, 2007. We observed 9 times every day and respectively at 06:00, 08:00, 10:00, 12:00, 14:00, 16:00, 18:00, 20:00 and 22:00. On February 3 and 4, we done the dense observation research and increased 2 time levels which were at 02:00 and 04:00.

The observation instruments adopted TK-II electronic air sounding instrument receptor machine of Beijing University and XJE50 automatic transit instrument of Shenyang 3301 Factory in Zhongche group. An air sounding temperature data was received every 5s. The air sounding launch plate was the thermistance, and the precision was 0.1°C. The wind field situation in different height was probed by the transit instrument, and a group of data was gained every 10s. The air sounding balloon which carried the instruments adopted 50 g standard meteorological air sounding balloon. Via the pullback weight, 100 m/min lifting speed was maintained. In this research, every probe site both gained 103 effective observation values which included the temperature, the wind direction and speed and so on.

3 RESULTS AND DISCUSSION

3.1 Boundary layer test

Weather during the test period: The north coast of Liaodong bay is the extremely north coastal site in China. The climate characteristics mainly presented as the continental climate. It was in the relative low-pressure zone between the polar region high pressure and the subtropical high pressure. The westerly circulation was prevalent in the whole year. The activity in the trough ridge of low-pressure zone was frequent, and the cold air was active. In winter, it was mainly controlled by the continental cold high pressure, and the cold air activity was frequent. The cold air in the polar region went down south frequently. The strong north gale appeared, and the temperature sharply declined. After the cold air passed, the weather was stable. The temperature was low, and the wind force was steady[2].

During the observation research period, a strong cold front transit weather process appeared in the north coast of Liaodong bay. In 500 hPa high altitude on January 26, 2007, there was a cold vortex center in the west of Lake Baikal. The center was near 54°N, 100°E, and the circulation was straight.

The ground was affected by the cold front, and the northerly was comparatively strong. The cold vortex shifted from the northwest to the southeast on 27th. The area was controlled by the front part of cold air which was in the front of ground high pressure, and the weather had no change. In 500 hPa high altitude on 29th, the cold vortex was packed and formed the high-altitude trough. The ground cold air shifted to the south direction, and the ground air pressure gradient was dense in the north coast of Liaodong bay. The northerly increased, and the cloudy weather appeared. The temperature reduced, but there had no precipitation. After the cold air passed, the ground in the north coast of Liaodong bay area was controlled by the front of Mongolia high pressure. The wind force wasn't great, and the weather was sunny. During February 1–2, there was the small group of cold air which affected the area. The weather was sunny interval, and the change of wind speed, wind direction, temperature, etc. wasn't great. During February 5–7, the high-altitude frontal zone advanced southward, and the ground weather in the north coast of Liaodong bay was good. The wind, temperature had no obvious change, and the weather situation was seen from Table 1.

Characteristics of wind field: Power law distribution of wind speed profile in the tower layer. The wind speed profiles of stratification stability in the atmospheric near the ground layer could be expressed by the power law. To understand the situation near the ground layer, the power law of wind speed was applied to the height below 300 m. If the wind speed in z height was u, the relational expression was as below:

$$u = u_1(z/z_1)^m \tag{1}$$

Table 2 was the power law value m of different stabilities in 2 observation sites: seaside and big pond. The level of stability was the ground stability which was classified by the correctional Pasquill stability method. Seen from Table 2, for the ground roughness in two observation sites was different, m value in the big pond was bigger than in the seaside. When the atmosphere was in the stable state (E–F class), m value was bigger. When the

Table 1. Weather situation in the low-altitude probe period during.

Item	1/24–26	1/27–28	1/29–31	2/1–3	2/4–7
Weather system	Ground cold high pressure	Cold front transit	Mongolia high pressure	Mongolia high pressure	Ground cold high pressure
Actual situation of weather	Low temperature, northerly	Cloudy, north gale, temperature reduction	North gale	Cloudy, stable wind temperature reduction	Good weather

Table 2. Power law values of different stabilities.

Stability	A	B	C	D	E–F
Seaside		0.09	0.13	0.19	0.28
Big pond		0.16	0.22	0.28	0.34

atmosphere was unstable, m value was smaller. In the stable condition, the atmospheric momentum transformation was weaker than in the unstable condition. It caused that the wind speed enlarged with the increase rate of height, and m value also increased. It was clear that the stability in the seaside in different height was more stable than in the land.

Frequency of wind direction: The analysis showed that the difference of wind direction wasn't great in 2 observation sites, but the change of wind direction as the height in every observation site was obvious. The low-layer wind direction in the seaside was mainly the south and north directions, and the frequencies of other wind directions were lower. However, it was even in the big pond, and the distribution of different wind direction frequencies was even. Seen from Tables 3 and 4, the wind direction distribution of seaside and big pond was closed, and it was mainly the southerly below 200 m. The frequencies of prevalent wind direction in the seaside in different height were respectively 25 m, ESE: 13.33%; 75 m, S: 14.42%; 125 m, S: 13.73%; 175 m, S: 13.86%. In the big pond, the frequencies of prevalent wind direction in different height were respectively 25 m, ESE: 13%; 75 m, SSW: 12.12%; 125 m, SSW: 13.13%; 175 m, SSE: 12.63%. In addition, the frequency of southeaster in the low layer in the seaside area was bigger than in the big pond. The wind directions in the two observation sites above 300 m gradually inclined to be consistent, and the northeaster and the southwester were prevalent.

Land-sea breeze: According to the position of the north coast of Liaodong bay, the wind directions of sea wind were defined as S, SW and SW. The wind directions of land wind were N, NNE and NE. During the observation period, one times land-sea breeze process appeared during February 3–4 in the seaside observation site, and the duration was about 8 h. At 22:00 on 3th, the ground wind direction turned from NNE (wind speed was 1.2 m/s) to S (the wind speed was 1.3 m/s). At 02:00 and 04:00 on 4th, the wind direction the wind speeds were respectively 1.3 and 1.4 m/s. At 06:00, the wind direction was NNE, and the wind was 0.8 m/s. From the ground to 175 m height, the transformation process of wind direction appeared during 22:00 on 3th to early hours on 4th. Above 175 m, the wind directions in the seaside and big

pond gradually inclined to be consistent and were NNE or N. It was clear that the land-sea breeze maybe appeared in the north coast of Liaodong bay in the certain condition. Moreover, it mainly happened in the low layer which was below 200 m and the area which was near the coast. As the height increased, and it penetrated deeply the inland, the land-sea breeze phenomenon gradually disappeared[3–5].

Thermal internal boundary layer: For the difference of temperature in the water-land boundary in the coastal zone, it made that the atmospheric temperature changed in the structure, and it started to form the thermal force internal boundary layer from the boundary surface of sea and land to the inland. For the existence of thermal internal boundary layer, the influence on the regional atmospheric stability was great. The hypatmism phenomenon often happens, which isn't favorable to the atmospheric diffusion. The height and the change grade rate of thermal internal boundary layer were determined according to the air sounding data of temperature. In the temperature profiles which were in different observation sites, they all had the flex points. In generally, the height of flex point heightens as the distance from the bank increases. The flex point heights of temperature profiles in different observation sites were drew in the same chart. The corresponding distance which was from the observation site to the bank was found by the terrain map, and the change grade rate in 6 km of thermal force boundary layer in the land was during 1/20–1/10. By using the observation research data and referring other observation data and the relative research results in Liaodong Bay[6,7], the effect factors in the thermal internal boundary layer were analyzed and selected. Finally, the below formula was selected to be the fitting formula of height in the thermal internal boundary layer:

$$h = A \cdot X n_1 \cdot \Delta T n_2 \cdot u n_3 \cdot r n_4 \qquad (2)$$

h was the height in the thermal internal boundary layer (m), and A was the coefficient. X was the distance from the bank (m), and ΔT was the temperature difference between the sea and the land (°C). u was the average wind speed in the boundary layer (m/s), and r was the temperature gradient of sea incoming flow (°C/100 m).12 groups of data were selected to do the multivariate linear regression. Via the calculation, the corresponding results were gained as below: A = 2.934, n_1 = 0.411, n_2 = 0.703, n_3 = −0.249 and n_4 = −0.51.

The actual observation height was compared with the calculation height by using the above fitting formula. The results showed that the calculation height in the thermal internal boundary layer fit closely with the actual observation height.

Table 3. Change of wind frequency in the seaside site in different height.

Height (m)	N	NNE	NE	ENE	E	ESE	SE	SSE	S	SSW	SW	WSW	W	WNW	NW	NNW
25	13.33	0.95	0	2.86	5.71	13.33	8.57	12.38	5.71	2.86	7.62	0.95	2.86	4.76	7.62	10.48
75	14.42	1.92	0.96	1.92	3.85	8.65	8.65	8.65	14.42	3.85	5.77	6.73	0.96	1.92	8.65	8.65
125	11.76	3.92	1.96	2.94	4.9	6.86	9.8	6.86	13.73	6.86	4.9	3.92	1.96	0.98	8.82	9.8
175	7.92	5.94	0	4.95	3.96	7.92	6.93	8.91	13.86	6.93	3.96	3.96	3.96	1.98	7.92	10.89
225	8.91	2.97	1.98	6.93	3.96	4.95	6.93	8.91	14.85	7.92	1.98	6.93	0.99	3.96	6.93	10.89
275	7.14	4.08	2.04	3.06	8.16	4.08	7.14	8.16	11.22	11.22	3.06	5.1	3.06	5.1	5.1	12.24
325	8.16	2.04	2.04	6.12	4.08	4.08	5.1	13.27	11.22	9.18	3.06	3.06	3.06	5.1	7.14	13.27
375	10.42	2.08	2.08	2.08	5.21	4.17	6.25	9.38	13.54	11.46	3.13	3.13	2.08	3.13	8.33	13.54
425	7.29	4.17	1.04	3.13	5.21	3.13	9.38	6.25	10.42	9.38	6.25	4.17	2.08	6.25	8.33	13.54
475	6.45	4.3	2.15	2.15	2.15	3.23	7.53	6.45	15.05	8.6	4.3	5.38	3.23	6.25	10.75	15.05
550	5.88	0	2.35	0	1.18	5.88	7.06	9.41	8.24	10.59	4.71	5.88	2.35	3.23	10.59	18.82
650	6.25	1.25	2.5	2.5	0	6.25	6.25	6.25	7.5	6.25	10	6.25	2.5	10	10	16.25
750	6.94	0	1.39	1.39	2.78	6.94	4.17	6.94	5.56	5.56	11.11	5.56	11.11	6.94	11.11	12.5
850	4.55	1.52	0	4.55	4.55	3.03	6.06	4.55	7.58	4.55	9.09	9.09	6.06	9.09	10.61	15.15
950	3.39	5.08	1.69	1.69	3.39	5.08	1.69	3.39	6.78	3.39	10.17	10.17	10.17	15.25	5.08	13.56
1050	2.04	4.08	2.04	2.04	2.04	6.12	2.04	2.04	6.12	6.12	12.24	6.12	10.2	12.24	8.16	16.33
1150	0	2.56	2.56	2.56	0	7.69	0	2.56	5.13	12.82	7.69	15.38	7.69	12.82	12.82	7.69
1250	0	3.03	0	3.03	3.03	9.09	0	0	3.03	9.09	18.18	9.09	6.06	18.18	9.09	9.09

Table 4. Change of wind frequency in the big pond site in different height.

Height (m)	N	NNE	NE	ENE	E	ESE	SE	SSE	S	SSW	SW	WSW	W	WNW	NW	NNW
25	6	11	6	2	5	13	9	8	2	6	9	2	7	4	4	4
75	9.09	8.08	6.06	1.01	0	10.1	8.08	11.11	6.06	12.12	6.06	4.04	7.07	5.05	0	6.06
125	7.07	10.1	7.07	1.01	0	9.09	6.06	12.12	4.04	13.13	8.08	4.04	8.08	2.02	1.01	7.07
175	10.53	9.47	7.37	1.05	1.05	6.32	4.21	12.63	5.26	7.37	10.53	5.26	7.37	3.16	2.11	6.32
225	7.37	9.47	10.53	0	1.05	4.21	9.47	7.37	7.37	8.42	9.47	5.26	8.42	1.05	1.05	9.47
275	6.32	15.79	3.16	0	4.21	3.16	6.32	10.53	6.32	6.32	9.47	9.47	4.21	3.16	1.05	10.53
325	5.49	17.58	3.3	0	3.3	2.2	8.79	8.79	7.69	8.79	8.79	12.09	1.1	2.2	2.2	7.69
375	4.4	17.58	4.4	1.1	2.2	3.3	4.4	9.89	5.49	13.19	8.79	10.99	3.3	2.2	1.1	7.69
425	6.82	18.18	2.27	2.27	3.41	1.14	3.41	11.36	6.82	9.09	13.64	6.82	5.68	1.14	2.27	5.68
475	8.24	15.29	3.53	0	3.53	1.18	1.18	14.12	7.06	8.24	14.12	8.24	3.53	3.53	2.35	5.88
550	4.76	17.86	1.19	1.19	2.38	0	1.19	11.9	9.52	13.1	7.14	9.52	2.38	4.76	5.95	7.14
650	9.21	13.16	2.63	2.63	2.63	0	1.32	10.53	2.63	7.89	7.89	13.16	5.26	5.26	10.53	10.53
750	8.96	13.43	2.99	1.49	2.99	1.49	0	5.97	4.48	7.46	10.45	10.45	7.46	2.99	4.48	14.93
850	11.86	6.78	1.69	3.39	5.08	0	0	5.08	3.39	8.47	10.17	5.08	16.95	6.78	5.08	10.17
950	13.46	7.69	0	3.85	1.92	3.85	0	3.85	1.92	7.69	11.54	3.85	23.08	3.85	0	13.46
1050	10.64	8.51	6.38	0	0	4.26	0	4.26	2.13	4.26	6.38	12.77	21.28	2.13	6.38	10.64
1150	14.29	7.14	2.38	0	0	4.76	0	2.38	2.38	7.14	2.38	14.29	11.9	11.9	9.52	9.52
1250	18.42	0	0	0	0	5.26	0	0	0	0	10.53	5.26	18.42	15.79	18.42	7.89

The thermal internal boundary layer in the area could be expressed by the above values (Figure 1).

3.2 Analysis on the influence of atmospheric diffusion

During the observation period, the appearance frequency of thermal internal boundary layer in the north coast of Liaodong bay was 10%. The development of height wasn't sufficient and was basically below 450 m. Even if the occurrence frequency of thermal internal boundary layer wasn't high, when the discharge gases entered the thermal internal boundary layer from the stable atmospheric layer, the gases quickly went downward and formed the hypatmism for the exchange of up and down turbulence was quick. It caused the influence on the atmospheric diffusion capability in the area. In addition, the wind speed in the whole low altitude below 300 m was very small during the transformation period of land-sea breeze, which also had the certain influence on the atmospheric diffusion.

Statistical distribution of temperature inversion: The temperature inversion layer had the important significance on the atmospheric diffusion in the boundary layer. The temperature inversion was the strong stable index and had the strong repression effect on the atmospheric turbulence. The intensity,

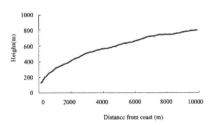

Figure 1. Change curve of height in the thermal internal boundary layer as the distance.

thickness and position of temperature inversion layer had the certain control effect on the local or regional atmospheric diffusion. The observation research divided the temperature inversion into three classes: a class was the earth connection temperature inversion, and the temperature inversion started from the ground, b class was the non-earth connection temperature inversion, and the temperature inversion was formed after leaved the ground, c class was the temperature inversion with two layers. The temperature inversion still existed in the above layers of a and b classes temperature inversion, which was called two layers temperature inversion.

According to the above definition, the temperature inversion in the seaside and big pond was done the statistics and analysis. The results were seen from Tables 5 and 6.

Seen from Tables 5 and 6, the temperature inversion appeared in two observation sites. The occurrence frequency of temperature inversion in the seaside was 69.9% and was 67% in the big pond. It was clear that the occurrence frequency of temperature inversion in the big bond where was far away from the coast was comparatively small. The thickness of earth connection (temperature inversion was above 200 m, and the intensity was above 1.4°C/100 m. The occurrence frequencies of multi-layer temperature inversion were respectively 14.5% in the seaside and 25.3% in the big pond.

Analysis on the height of atmospheric mixture layer: The height of mixture layer represented the atmospheric self-purification capacity. The mixture layer was higher, which favored the atmospheric diffusion. When the temperature curve which was probed in the low altitude was selected, a line which was from the ground temperature to the cross point of thermal insulation rise line and air sounding curve was the mixture layer height at that time. Seen from Table 7, the average height of mixture layer in the seaside was lower than in the big pond. The average height of mixture layer in the seaside was 355.5 m and was 405.3 m in the big

Table 5. Characteristic statistics of temperature inversion in the seaside.

Seaside	Times	Intensity (°C/100 m)	Layer thickness (m)	Bottom height (m)	Top height (m)	Frequency (%)
A	39	1.6	242.3	0	242.3	35.9
B	18	0.9	575.0	677.8	1252.8	19.4
C						
One layer	2	1.5	350.0	75.0	425.0	1.9
Two layer		1.3	350.0	1450.0	1800.0	
D						
One layer	13	2.1	242.3	0	242.3	12.6
Two layer		1.4	196.1	915.4	1111.5	
Occurrence times	72					69.9

Table 6. Characteristic statistics of temperature inversion in the big pond.

Big pond	Times	Intensity (°C/100 m)	Layer thickness (m)	Bottom height (m)	Top height (m)	Frequency (%)
A	24	1.4	204.2	0	204.2	23.3
B	19	0.8	342.0	574.7	916.7	18.4
C						
One layer	12	1.4	187.5	295.8	483.3	11.7
Two layer		1.2	320.8	1029.2	1350.0	
D						
One layer	14	1.8	178.6	0	178.6	13.6
Two layer		0.9	350.0	1203.6	1553.6	
Occurrence times	69					67.0

Table 7. Heights of average mixture layers in the seaside and big pond (m).

Time	Seaside	Big pond
08	191	219
10	262	287
14	598	678
16	371	437
Mean	355.5	405.3

bond. The heights of mixture layers in two observation sites both reached the highest at 14:00.

4 CONCLUSIONS

The observational research showed that the change of pollution boundary layer from the seaside to the land near the sea wasn't very obvious in the north coast of Liaodong bay. In the certain weather condition, the land-sea breeze maybe appeared in the low altitude below 200 m in the zone which was near the coast. The stability change in different heights in the seaside was more stable than in the land zone, and the wind field change was mainly in the low altitude in 300 m. At night, the temperature inversion often appears in the area, and the thickness of temperature inversion layer is during 200–300 m. The thermal boundary layer penetrated deeply 10 km into the land, and the height could reach 800 m. In the stable weather condition, the land-sea breeze phenomenon sometimes appears in the low altitude, and the inland zone has no land-sea breeze phenomenon. For the existence of thermal internal boundary layer, the influence on the regional atmospheric stability was greater. The hypatmism phenomenon sometimes happens, which isn't favorable to the atmospheric diffusion. The height of atmospheric mixture layer in the land was higher than in the seaside, and the atmospheric diffusion in the land area was strong. For it was only the short-term air sounding observational research in winter and only had 2 observation sites, the research couldn't totally represent the total characteristics in the area. For the change and characteristic of pollution boundary layer in the north coast of Liaodong bay, it needed the further observation and analysis.

ACKNOWLEDGEMENTS

The financial support of Scientific Research Special Fund for Public Service (GYHY201106033), Key Technology Integration and Application Project CMA (CAMGJ2012M14), Institute of atmospheric environment, CMA Fund (2012IAE-CMA02) are gratefully acknowledged.

REFERENCES

[1] Cai RS, Yan B, Huang RH. A numerical model and its simulation of the land and sea breeze over the Taiwan Strait [J]. Chinese Journal of Atmospheric Sciences, 27(1): 86–95.(2007).
[2] Chen LQ, Yang HB. Synoptic patterns of regional air pollution in Liaoning Province [J]. Environmental Pollution and Control, (6): 435–436. (2006).
[3] Mu YC, Yang S. Characteristics of sea-land breeze in Jinzhou area of Dalian [J]. Journal of Meteorology and Environment, 23(2): 13–16. (2007).
[4] Zheng XI, Zhang SL, Chen DH, et al. Analysis on climatic change of surface wind on the west coast 01 Taiwan Strait[J]. Journal of Oceanography in Taiwan Strait, 28(4): 569–576. (2009).
[5] UY, Wang YQ, Liu Q. Influences of weather and climatic conditions on the distribution of air pollutants [J]. Journal of Anhui Agricultural Sciences, 37(4): 1781–1782. (2009).
[6] Wang YG, Wu ZM, Chang ZQ. Statistic characteristics of sea-land breeze in west coast of Liaodong Bay [J]. Marine Forecasts, 21 (3): 57–564. (2004).
[7] Mitsumoto S, Ueda H, Ozeo H. A laboratory experiment the dynamics of the land and sea breeze [J]. Journal of the Atmospheric Sciences, 40:1228–1245. (1983).

Frontiers of Energy and Environmental Engineering – Sung, Kao & Chen (eds)
© 2013 Taylor & Francis Group, London, ISBN 978-0-415-66159-1

An analysis of building long-term safeguard mechanism on ideological and political education of college students

Z.G. Xu
Hubei University of Technology Engineering and Technology College, Wuhan, Hubei, China

ABSTRACT: Building long-term safeguard mechanism on ideological and political education of college students will be realized through making policies and rules, making ideological belief true, paying more attention to details around college students, linking the powers of society, college, family, and individuals, which will provide strong bases of establishing ideological and political education.

Keywords: college students; ideological and political education; long-term safeguard mechanism

The ideological and political education on college students is a complicated systematic program, which needs the participation and mutual help of colleges, families, and society to produce resultant force. College students are in the important stages of forming views of life and values, and whether their views are right or not will produce vital impacts on the development of the states and society. The ideological and political education involves how to be a good person, and everyone will change with the outer change of environment and regions, so we can not guide the students well just through a successful lecture, a wonderful class meeting, a chat with the students. The character of the ideological and political education determines it a long-term, repeated, finding-problem and solving-problem process.

1 BUILDING LONG-TERM SAFEGUARD MECHANISM ON IDEOLOGICAL AND POLITICAL EDUCATION OF COLLEGE STUDENTS SHOULD BE REALIZED FROM THE POLICY AND SYSTEM

The system is the base of establishing the mechanism, which involves establishing, perfecting, and realizing every system. Deng Xiaoping said that: The system problem is more fundamental and basic. The system is normative, stable, and conditional, which can be depended on to develop ideological and political education, and through which the ideological and political education can be operated smoothly.

1.1 *Study and follow out the spirit of party central committee documents, and strengthen the system construction*

Central Government has regulated the position, effects, tasks, policies, and principles of ideological and political education of the universities, which shows the authority and stability of the education. However, there should be relevant regulations and laws to guarantee the following of the educators, students, and relevant social institutions. With the rapid development of society, the contents of ideological and political education have changed continuously with the improvement of requirements on the ideological and political education, which makes us produce the corresponding guide and solution of the obvious problems according to the ideological situation and current reality, to ensure the ideological and political education in system level.

1.2 *Set up working staff and put the staff training system into practice*

The ideological and political education needs a working staff with powerful politics, skillful operation, upright style to guarantee the practice. The university should hire a group of high educated graduates to join in the instructor staff, and select some lecturers and administrators to be part-time instructors, who should enhance their operating and working ability on ideological and political education through pre-post training, on-the-job training, assigning the instructors study out to realize the professionalization, specialization, and expertise.

1.3 Establish the linking system of colleges, families, society, and individuals

The colleges should use the power of society, families, and individuals to set up a three-dimensional and networking educational structure, to develop the multiple forces and ensure college students to accept ideological and political totally. In this linking system, the most important task is to guide and cultivate self-education consciousness of the colleges, because the college students are not only the objects of the ideological and political education, but also the subjects. Only should the contents of ideological and political education be internalized into self-conscious behavior that it can really be put into practice.

2 BUILDING LONG-TERM SAFEGUARD MECHANISM ON IDEOLOGICAL AND POLITICAL EDUCATION OF COLLEGE STUDENTS SHOULD BE PUT INTO PRACTICE FROM IDEAL AND FAITH EDUCATION

A. The education should be around the ideal and faith education and involve right views of the world, life and values. The colleges should make all of the students know that Party and people hope them deeply to construct the Well-off Society and realize the socialist modernization, to resurge Chinese nation, to make their life colorful through devoting themselves to the national construction, to make their work involve in the state and nationality.

B. The college should focus on the patriotism education and develop national spirit education, guiding the college students to promote the national self-respect, self-confidence and sense of pride, encouraging the students to love our country and devote themselves to constructing the socialist country as the utmost honor, making them know that every behavior involving the damages of motherland interests, dignity or honor will be the deepest disgrace.

C. The college should recognize the fundamental ethnic regulations as the base of the educational system, develop ethnic education deeply, guide the college students to follow the fundamental ethnic regulation, such as being a patriot, abiding by the law, being honest, being friendly, being diligent, being dedicated, etc, and develop the good ethnic traits and behavior.

3 BUILDING LONG-TERM SAFEGUARD MECHANISM ON IDEOLOGICAL AND POLITICAL EDUCATION OF COLLEGE STUDENTS SHOULD BE REALIZED FROM THE FIELD AND REGIONS

A. Occupy the internet field. The colleges should set up the leading group of internet on the ideological and political education to regulate and manage the college websites, to guide the students to pay attention to country affairs and current affairs, to analyze the highlight of modern society, to publicize the right direction.

B. Occupy the living region. The colleges should build up some bulletin boards in the public areas such as dormitory, dining halls, and sports grounds, assign relevant teachers to manage these regions, develop ideological and political education in time, and let the students who need the help contact the teachers at any time, which will make the ideological and political education around every student.

C. Occupy the corporate communities. The Communist Youth League should enforce the guide of the student union and other unions, develop the flexibility of the student unions in the context of plenty of time for them. The instructors should try some new methods to build up party organizations and CY organizations actively, foster a group of powerful and strong star communities carefully, and strengthen the self-education effectively.

4 BUILDING LONG-TERM SAFEGUARD MECHANISM ON IDEOLOGICAL AND POLITICAL EDUCATION OF COLLEGE STUDENTS SHOULD BE REALIZED THROUGH REAL WORK

4.1 Grasp the leading function of classroom teaching on the ideological and political education of college students

Classroom teaching is the main approach and area for the ideological and political education, which is the vital way to help college students to judge what is right and what is wrong, to set up the right views of life, world, and values. The colleges should strengthen the subject construction, educating construction, and textbook construction of the ideological and political education, reform the contents, methods, styles of education, combine the society with ideological reality, make the vital theories and real problems which attract the students as the emphasis of classroom teaching, develop the activeness of the students totally, guide the students to analyze and solve problems, and encourage the students to develop entirely and harmoniously. The colleges should develop the educating function of the campus culture, cultivate the tastes, purify the souls, and enrich the cultural life of the college students.

4.2 *Grasp the vital function of Party and CY organizations on the ideological and political education*

The Party should develop their advantages of politics and organizations to attract excellent college students to join in, realizing the targets of "there being Party members in the first year of the college students, there being Party group in the second year of the college students, there being Party branch in the latter year of the college students", setting up selected courses on Party knowledge, carrying out Party school training strictly, regulating development procedures, improving the quality of the Party members, strengthening the construction of basic Party organization, making the Party branch into the strong barriers of ideological and political education. CY organization should develop the advantages on educating, helping, and contacting college students, and serve the college students to grow up.

4.3 *Combine the aid work with educating work*

The colleges should carry out the aid work according to the aims of "People first" and "Service education" in the principles of selecting openly, checking carefully, supervising totally, allocating in time, follow-up survey. The colleges should produce the requirements of positive selection rules, reward the students who study hard and need financial aids, guide students to study hard and develop their interests widely, form the good study environment of "competition, study, follow, help" and positive attitudes of life, and cultivate graduates with high quality of "Specialist in one field while possessing all-around knowledge and ability".

4.4 *Focus on the psychological education for the students*

Nowadays, mental heath is an important problem annoying and preventing college students. The colleges should help students have some self-adjusted methods to enhance their mental ability to endure troubles through setting up special mental health columns and psychological consultation, establishing mental health union, organizing mental health group in the class level, and holding on psychological lectures, psychological essays, psychological movie show.

4.5 *Strengthen social practice activities of college students*

Social practice is the significant procedure of ideological and political education. Social practice can help college students to know about society, experience folk life, strengthen the mind, improve the ability, and train character, which can make the theory internalize own consciousness and behavior, enhance the feeling of social responsibility and service, and realize the targets of ideological and political education.

REFERENCES

[1] Liu Xiangwu, Zheng Yanping. "Dissection on building long-term mechanism on ideological and political education of college students" [J]. Journal of Hunan University of Science and Engineering, 2005, 12:101–102.

[2] Wen Gu. "Challenges and suggestions on ideological and political education of current college students" [J]. Journal of Southwest. Agricultural University: Social Sciences Edition, 2006(3):222–225.

[3] Ministry of propaganda department of Central Committee of Communist Party of China, Ministry of Education, "Proposals on further strengthening lecturer group of ideological and political education for institution of higher learning". Teach Social Science [2008] the fifth, 2008-09-23.

Frontiers of Energy and Environmental Engineering – Sung, Kao & Chen (eds)
© 2013 Taylor & Francis Group, London, ISBN 978-0-415-66159-1

Development of risk-based management strategies at a Brownfield site

W.Y. Huang
China Petrochemical Development Corp., Kaohsiung, Taiwan

H.Y. Chiu, Z.H. Yang, P.J. Lien & C.M. Kao
Institute of Environmental Engineering, National Sun Yat-Sen University, Kaohsiung, Taiwan

W.P. Sung
Department of Landscape Architecture, National Chin-Yi University of Technology, Taichung, Taiwan

ABSTRACT: A 40-year old petroleum tank farm site was closed in 2005. The Triad approach was performed to streamline the Environmental Site Assessment (ESA) and potential Risk-Based Corrective Actions (RBCA) at this facility. Contaminated site conceptual models (SCM) were established by in situ real-time Membrane Interface Probe (MIP) detection system, direct-push sampling, chemical laboratory analysis, and hydrogeological testing. A noticeable contaminated area of 1,200 m² were found with maximum soil Total Petroleum Hydrocarbon (TPH) concentration of 90,000 mg/kg and groundwater naphthalene concentration of 3,500 mg/L. Soil samples were collected from selected locations with depths of 10 m below ground surface, and were screened by MIP and off-site Gas Chromatograph (GC). Results of tiered Health Risk Assessment (HRA) show that the contaminants in soil and groundwater at this site pose a remarkable risk to human health via inhalation [carcinogenic risk was 3.08E-3 and chronic hazard quotient (HZ) was 1.1E+1], exceeding the acceptable target risk-based levels (cancer risk of 1E-6 and HZ of one). Thus, several engineering controls, redevelopment options, and remediation for the site were proposed in this study. Results provide the site owner a view of financial, legal, and environmental considerations in decision-making.

1 INTRODUCTION

A seven-hectare petroleum fuel oil storage and transport tank farm facility in Taiwan with more than 40 years operation history was closed in early 2005. In very urbanized and industrialized regions, water resources, and particularly groundwater, are subject to many pollution pressures related to different kinds of socio-economic activities and contaminants. Vulnerability and risk assessments are becoming a standard approach in groundwater management when dealing with water quality and contamination issues (Ellen Milnes 2012). Moreover, groundwater contamination is an imperceptible and irreversible process, and prohibitive costs and time requirements may limit efforts to improve the groundwater condition (Causape et al., 2006; Yu et al., 2010; Wang et al., 2012). Several projects have been dedicated recently to the development of methodologies for contaminated site management (Jamin et al., 2012; Li et al., 2012). Figure 1 presents the site map showing the suspected contaminant source area, groundwater flow direction, the monitoring system, and surrounding facilities. All the oil storage and transport equipments, e.g. oil storage

tanks, pipelines, and pumps, were torn down as well. Combined residential and commercial areas of concern are located hydraulically downgradient of the groundwater beneath the north side of the facility and are also located in the east nearby areas. Future reuse and redevelopment plans for

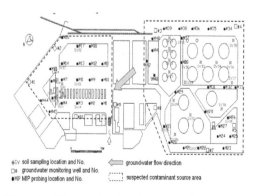

Figure 1. Site map showing the suspected contaminant source area, groundwater flow direction, the monitoring system, and surrounding facilities.

the property are still uncertain. However, it might be for the residential, commercial, combination thereof, cultural and educational, or entertaining use instead of industrial or agricultural property in order to heighten the property value. The proprietor (land owner and property manager) has completed a series of actions, i.e. environmental site assessment, subsurface environmental investigation, HRA, engineering control and remediation alternatives assessment, with US$170,000 control budget in 6 months so as to identify the subsurface environmental risks and proper risk controls.

2 FRAMEWORK AND METHODS

This project was initiated with an aim of managing uncertain factors and risk-based control strategies depending on rapid, cost-effective, and efficient identifications of site environmental risks to facilitate the decision-making for risk-based control. The proposed study minimizes site variable factors effectively, quickly, and efficiently by integrating various ASTM technical guides (ASTM 2004a, b & 2005c). with the Triad approach (Crumbling 2004; ASTM 2010d). The framework and process consist of four steps, i.e. subsurface environment and contaminants characterization delineation, SCM development, HRA, and RBCA and financial assessment. So as to characterize the site geology, hydrogeology, and contaminant and to establish SCM, the study excludes the conventional investigation measures and procedures and adopts the Triad Approach incorporating systematic planning, dynamic work strategies, and real-time measurement systems together with a Membrane Interface Probe (MIP) detection system, a real-time in situ screening technology (Tsai et al., 2011; Baciocchi et al., 2010) to screen the areas of concern (AOCs), soil geology and the chemicals of concern (COCs) consisting of Benzene, Toluene, Ethyl-benzene, Xylenes (BTEX), naphthalene, and Total Petroleum Hydrocarbon (TPH) for gasoline and diesel oil, which is followed by a number of concise off-site chemical laboratory analysis tests of groundwater and soil samples. Based on the SCM, HRA have been evaluated in conformity with ASTM Standard Guidance for Risk-Based Corrective Action Applied at Petroleum (RBCA) and the Health Risk Assessment and Analysis Standard Guideline for Contaminated Sites newly developed and implemented by Taiwan Environmental Protection Administration (TEPA) for the sake of assessing the health risks of exposure pathways to human receptors. Depending on the risk levels of exposure pathways, several corrective action options and their financial cost evaluations have been proposed, which meet current and future possible

redevelopments and could serve as references of risk-based control discussion and decision-making for the responsible Brownfield managers or decision makers on accounts of the sustained control, the improvement of environmental quality for the affected property, and the property reuse and redevelopment while achieving sustainable economic development and environmental protection.

3 RESULTS AND DISCUSSION

Contaminated Site Conceptual Model. The study commences a series of in situ and on-site investigations for the subsurface environmental characterizations, i.e. groundwater hydrogeology, geology, and contamination, with direct-push technology. Figure 2 is visualized subsurface contaminated SCM including hydrogeology and contamination together with the EC and soil stratification distributions, illustrating geological

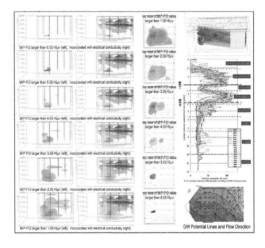

Figure 2. Site conceptual model showing visualized subsurface contamination.

Table 1. Maximum soil and groundwater contaminant concentrations at this studied site.

COCs/mediums	Groundwater, mg/L	Soil, mg/kg
Benzene	24.8	199
Toluene	14.1	760
Ethyl-benzene	22.3	611
Xylenes	38	1,478
TPH-g	–	20,016
TPH-d	–	69,474
Naphthalene	3,500	–

631

Table 2. Site-specific carcinogenic and non-carcinogenic risk.

Exposure Pathway	Target level	Benzene	Ethyl-benzene	Toluene
Soil				
Soil volatilization to outdoor air, mg/kg	Cancer risk	1.76E-03	–	–
	Chronic HQ	–	9.39E-02	7.73E-01
Soil-vapor intrusion from soil to buildings, mg/kg	Cancer risk	1.10E-03	–	–
	Chronic HQ	–	1.43E-01	3.73E+00
Surficial soil (0–1.0 m) ingestion/dermal/ inhalation, mg/kg	Cancer risk	7.86E-05	–	–
	Chronic HQ	–	3.23E-01	1.01E+00
Soil-leachate to protect groundwater ingestion target level, mg/kg	Cancer risk	2.30E-10	–	–
	Chronic HQ	–	2.03E-15	5.66E-45
Soil risk subtotal	Cancer risk	2.94E-03	–	–
	Chronic HQ	–	5.60E-01	5.52E+00

Exposure pathway	Target level	Xylenes (Mixed)	Naphthalene	Soil risk sub-total
Soil				
Soil volatilization to outdoor air, mg/kg	Cancer risk	–	–	1.76E-03
	Chronic HQ	6.22E-02	4.65E-01	1.39E+00
Soil-vapor intrusion from soil to buildings, mg/kg	Cancer risk	–	–	1.10E-03
	Chronic HQ	1.66E-01	7.08E-01	4.75E+00
Surficial soil (0–1.0 m) ingestion/dermal/ inhalation, mg/kg	Cancer risk	–	–	7.86E-05
	Chronic HQ	7.62E-02	3.25E+00	4.66E+00
Soil-leachate to protect ground water ingestion target level, mg/kg	Cancer risk	–	–	2.30E-10
	Chronic HQ	1.69E-12	1.92E-13	1.89E-12
Soil risk subtotal	Cancer risk	–	–	2.94E-03
	Chronic HQ	3.05E-01	4.42E+00	1.08E+01

Exposure Pathway	Target Level	Xylenes (Mixed)	Naphthalene	GW Risk sub-total
Ground water				
Groundwater volatilization to outdoor air, mg/L	Cancer risk	3.36E-05	–	–
	Chronic HQ	–	1.18E-02	1.88E-02
Groundwater ingestion, mg/L	Cancer risk	2.55E-10	–	–
	Chronic HQ	–	2.72E-45	2.72E-45
Groundwater vapor intrusion from groundwater to buildings, mg/L	Cancer risk	1.03E-04	–	–
	Chronic HQ	–	3.79E-02	5.92E-02
GW risk subtotal	Cancer risk	1.36E-04	–	–
	Chronic HQ	–	4.97E-02	7.80E-02
Soil & GW total risk	Cancer risk	3.08E-03	–	–
	Chronic HQ	1.10E+01	–	–

Exposure pathway	Target level	Xylenes (Mixed)	Naphthalene	GW risk sub-total
Ground water				
Groundwater volatilization to outdoor air, mg/L	Cancer risk	–	–	3.36E-05
	Chronic HQ	2.60E-03	1.00E-02	4.32E-02
Groundwater ingestion, mg/L	Cancer risk	–	–	2.55E-10
	Chronic HQ	2.72E-45	2.72E-45	1.09E-44
Groundwater vapor intrusion from groundwater to buildings, mg/L	Cancer risk	–	–	1.03E-04
	Chronic HQ	8.23E-03	2.28E-02	1.28E-01
GW Risk Subtotal	Cancer risk	–	–	1.36E-04
	Chronic HQ	1.08E-02	3.28E-02	1.71E-01

distributions in side view and groundwater fluctuation varying 3–5 meters BGS. Groundwater flow directions, mainly from south to north or northwest with hydraulic conductivity (K-value) 1.04E-3 to 9.28E-5 cm/sec, groundwater gradient around 0.00225, groundwater apparent velocity 0.07 to 0.74 m/year. With a former soil gas investigation at 700 points of the property, one major contaminated site has been screened out of 30 potentially contaminated sites to further proceed with MIP at 20 points of the site and the results are as Figure 2 with an average probing depth 10 meters under the ground surface. Figure 2 also shows the probing results employing 20 points of MIP (equipped with dipole electrical conductivity EC sensors and linked with above ground flame ionization detector-FID), discovering the depth of affected site up to 10 meters underground surface and the effected area of 1,200 square meters. The direct-push technology is also applied to sample soil and groundwater, which are subsequently sent for laboratory tests, in MIP-FID high concentration response areas, resulting in the findings that the most contaminated soils and some apparent petroleum oil plumes are found at the depth of 7 to 10 meters below groundwater surface and the principal contaminants including naphthalene, and BTEX with THP concentrations around 90,000 mg/kg are identified, as shown in Table 1.

4 HUMAN HEALTH RISK ASSESSMENT

The clarifications of potential risks to human health posed by various exposure pathways lead to the proper decisions of corrective actions so as to continuingly decrease the risks to acceptable levels and secure the safety of property reuse. Table 2 is the assessment results of potential carcinogenic and non-carcinogenic risks to human health resulted from different exposure pathways and various COCs based on the conservative development plan of residential use. The total cancer risk for soil and groundwater in Table 2 is 3.08E-03, which exceeds the acceptable risk-based carcinogenic risk 1.0E-6; the total chronic hazardous quotient is 1.10E+01, which also exceeds the risk-based non-carcinogenic risk 1.0. The exposure pathways and routes include soil volatilization to outdoor air inhalation, soil-vapor intrusion from soil to buildings inhalation, soil ingestion, soil dermal absorption, ground water ingestion, ground water volatilization to outdoor air inhalation, groundwater-vapor intrusion from ground water to buildings inhalation, and ground water dermal absorption. For the exposure pathways and routes of soil, the soil-vapor intrusion

from soil to indoor buildings or outdoor air inhalation ranks the highest risk; in regard to groundwater, the groundwater-vapor intrusion from ground water to buildings or outdoor air inhalation rates the highest risk as the property has been prohibited from installing any pumping well and from pumping the groundwater as drinking or domestic water supply. Among the soil and groundwater, the long-term inhalation of vapor intrusion to buildings or outdoor air possibly poses the highest risk.

5 CONCLUSIONS

This study applied the Triad approach and in situ and real-time site assessment tools to effectively and rapidly characterize the studied petroleum Brownfield site. Health and risk assessment was also performed to determine the site-specific carcinogenic and non-carcinogenic risks caused by the detected contaminants. Diverse packages of engineering controls and/or cleanup or remediation are established based on the human exposure risks, the property redevelopment plans, the expected consuming timetable, and the initial cost evaluation for all redevelopment scenarios is accomplished as well. In the event of making decisions, the property managers and/or decision makers in charge could refer to these implementing framework, process, and outcomes intended for minimizing indefinite risks.

REFERENCES

ASTM International (American Society for Testing and Materials). 2004a. Standard Guide for Accelerated Site Characterization for Confirmed or Suspected Petroleum Releases. E1912-98, West Conshohocken, PA, USA.

ASTM International (American Society for Testing and Materials). 2004b. Standard Guide for Risk Based Corrective Action. E2081-00, West Conshohocken, PA, USA.

ASTM International (American Society for Testing and Materials). 2005c. Standard Guide for Use of Activity and Use Limitations, Including Institutional and Engineering Controls. E2091-05, West Conshohocken, PA, USA.

ASTM International (American Society for Testing and Materials). 2010d. Standard Guide for Application of Engineering Controls to Facilitate Use or Redevelopment of Chemical-Affected Properties, Including Institutional and Engineering Controls. E2435-05, West Conshohocken, PA, USA.

Baciocchi, R., Berardi, S., & Verginell, I. 2010. Human health risk assessment: Models for predicting the effective exposure duration of on-site receptors exposed to contaminated groundwater. *J. Hazard. Mater.,* 181(1–3): 226–233.

Crumbling, D.M. 2004. Summary of the Triad approach, U.S. Environmental Protection Agency, Washington, DC, USA.

Causape´, J., Quı´lez, D., & Arague´s, R. 2006. Groundwater quality in CR-V irrigation district (Bardenas I, Spain): Alternative scenarios to reduce off-site salt and nitrate contamination. *Agric. Water Manage.,* 84: 281–289.

Ellen Milnes. 2012. Process-based groundwater salinisation risk assessment methodology: Application to the Akrotiri aquifer (Southern Cyprus). *J. Hydrol.,* 399: 29–47.

Jamin, P., Dollé, F., Chisala, B., Orban, P., Popescu, I.C., Hérivaux, C., Dassargues, A., & Brouyère, S. 2012. A regional flux-based risk assessment approach for multiple contaminated sites on groundwater bodies. *J. Contam. Hydrol.,* 127: 65–75.

Li, Y., Li, J., Chen, S., & Diao, W. 2012. Establishing indices for groundwater contamination risk assessment in the vicinity of hazardous waste landfills in China. *Environ. Pollut.,* 165: 77–90.

Tsai, T.T., Kao, C.M., Surampalli, R., Huang, W.Y. & Rao, J.P. 2011. Sensitivity analysis of risk assessment at a petroleum-hydrocarbon contaminated site. *J. Hazard. Toxic Radioactive Waste Manage.,* 15: 89–98.

Wang, J., He, J., & Chen, H. 2012. Assessment of groundwater contamination risk using hazard quantification, a modified DRASTIC model and groundwater value, Beijing Plain, China. *Sci. Total Environ.,* 432: 216–226.

Yu, C., Yao, Y., Hayes, G., Zhang, B., & Zheng, C. 2010. Quantitative assessment of groundwater vulnerability using index system and transport simulation, Huangshuihe catchment, China. *Sci. Total Environ.,* 408: 6108–6116.

Frontiers of Energy and Environmental Engineering – Sung, Kao & Chen (eds)
© *2013 Taylor & Francis Group, London, ISBN 978-0-415-66159-1*

Simulation design of field kitchen kits based on principles of material mechanics

X.H. Zhang, H. Yang, X.Q. Liu & Y.J. Li
The Quartermaster Institute of General Logistics Department of the PLA, Beijing, China

ABSTRACT: This paper has made a finite element mechanical optimization analysis towards the rotationally molded box and the aluminum alloy adopted through software modeling to ensure the required performance standards of the field kitchen kit are satisfied. The comprehensive analysis showed that the rotationally molding material is more suitable for box-type military equipments.

Keywords: field kitchen kit, rotational molding, finite element analysis

1 INTRODUCTION

Field kitchen kit is a kind of sustenance support equipment used as military temporary kitchen under field conditions[1–3]. With a box pallet for carrying primary kitchen wares, it can be carried by four people or a forklift truck. The loading capacity of single kit should reach 100 kg and four of such kits can be stacked together to meet the military needs.

The four considerations for selection materials of box-type military equipments are stiffness, intensity, weight and cost, among which stiffness and intensity are the most important[4]. Overweight should be avoided for easy carrying and handling. Therefore the densities of different materials should be compared and the one with low density should be selected, with the premise that it is stiff and intense enough. At the same time cost-saving is another important factor.

2 MATERIALS AND METHODS

2.1 *Modeling method*

For convenience of calculation, the box is simplified into a mechanical model as shown in Figure 1, and we suppose that the boxboards (whose thickness is δ) bear all the loads.

The total mass of the box (M) is 50 kg or 100 kg, the length (L) is 1200 mm, the width (H) is 800 mm, and the thickness of the board (δ) is 1.2 mm or 3 mm. Let the acceleration of the dynamic load exerted on the box be 3 g, and P1 = P2 = 750 N, the intensity of the box is checked.

The maximum normal stress σ max generated from longitudinal bending moment is:

$$\sigma \, max = Mb/W = (P1 \, L)/(2 \, W) \tag{1}$$

$$W = (BH3 - bh3)/6H \tag{2}$$

The maximum tangential stress τ_{max} generated from the pure toque within the transverse section is:

$$\tau_{max} = (P_2 H)/2\delta(H - \delta)^2 \tag{3}$$

Then the strength after the combined deformation due to torsion and compression is calculated:

$$\sigma \, x = (\sigma \, max2 + 4 \, \tau \, max2)1/2 \tag{4}$$

The calculated intensity parameters of materials of different shapes and sizes are shown in Table 1. The minimum allowable stress of cold-rolled steel sheets (σ) ranges from 235 MPa to 650 MPa, that of LF21 aluminum plates is 111.0 MPa, and that of rotationally molded boxboards is 10.0 MPa. Obviously, σx<[σ], which indicates that the safety of the box is ensured.

2.2 *The calculation of the bending rigidity of the box structure*

Suppose the maximum bending deflection of the middle part is Y$_{max}$, then

$$Y_{max} = (P1 \, L3)/(48EJX) \tag{5}$$

Generally speaking, for the simply supported beam which needs more rigidity, the condition [Y$_{max}$] ≤ 0.0002 L = 0.12 mm should be satisfied. The calculated rigidity parameters of materials of different shapes and sizes are shown in Table 1.

It is obvious that Y$_{max}$ < [Y$_{max}$], which proves the flexural rigidity of the box is satisfying.

The maximum loading capacity of the box that can be carried by 4 people is 100 kg, and the load-bearing capacity should be 300 kg to meet

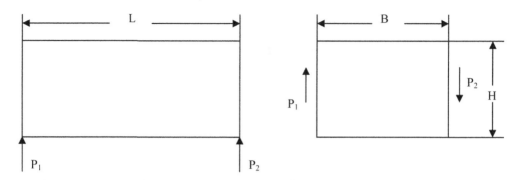

Figure 1.   Mechanical model of the box.

Table 1.   Performance parameters of materials of the box.

Material	Thickness (mm)	Inside dimension (mm)	Density (kg/m³)	Net weight (kg)	Intensity (MPa)	Rigidity (mm)
Aluminum	1.2	1197.6 × 997.6 × 797.6	2700	19.84	13.950	7.035×10^{-4}
RM material	3(2)	1194 × 994 × 794	930	16.42	6.975	1.695×10^{-3}
Steel	1.2	1197.6 × 997.6 × 797.6	7800	55.28	13.950	3.165×10^{-4}

Notes: (1) Suppose the contour dimension of the box is 1200 mm × 1000 mm × 800 mm; (2) The bulge of the extended reinforcing material (5 mm) is excluded from the result.

the requirements of a 4-layer stack, excluding the weight of hinges, handles and attachments. Reinforcing ribs are formed on both the boxes made of aluminum and rotationally-molding materials through compression molding. After calculation and comparison we know that under the same loading conditions, the thickness of the rotationally-molded boxboard is 3 mm, excluding the height of the rib (5 mm). Its overall density is 930 kg/m³, which is smaller than the density of the aluminum board. Due to a lower rigidity, the thickness of the board should be increased to meet the load-bearing capacity for stacking (300 kg), and consequently the weight of the box will be increased.

The finite element analysis models of the box whose contour dimension is 1200 mm × 1000 mm × 800 mm established through COSMOSWorks function of Solidworks[5] is shown in Figure 2.

2.3 Finite element analysis

In order to test the loading capacity of boxes made of different materials, we analyzed the carrying capacity of steel box, aluminum box and rotationally-molded box with the same contour dimension (1200 mm × 1000 mm × 800 mm) under the same load cases.

1. *Analysis on the bearing capacity of aluminum box*
   Suppose the box is made of Al6061 aluminum plate whose thickness is 1.2 mm, elastic modulus is 7310 N/mm², Poisson's ratio is 0.33, density is 2700 kg/m³, and coefficient of thermal expansion is 9.63×10^{-5}/°C, and the model is simplified. After establishing the analyzing task and fixing the bottom, we analyzed the bearing capacity of the model when a pressure of 1000 N was exerted on its top and got the stress and deformation pattern, Figure 3.
   From the calculated result we can see that the maximum stress of the cover under the pressure of 1000 N is 186 MPa which is higher than the yield strength of the same area (62.05 MPa). The maximal deflection appears at the center of the cover and reaches 4.385 mm.
   Similarly, the maximum stress of the box bottom is 148.7 MPa when a pressure of 750 N is exerted to the bottom from the inside, higher than the yield strength (62.05 MPa). The maximal deflection is 3.582 mm.

2. *Analysis on bearing capacity of rotationally-molded box*
   At first the material performance parameters required for model analysis was set: the box is made of LLDPE suitable for rotational molding with an average thickness of 4.5 mm. The elastic

Figure 2.   Finite element models of the cover and the bottom of the box.

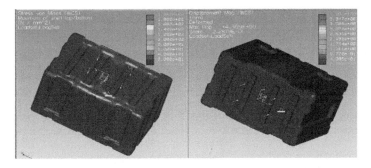

Figure 3.   Analysis on the bearing capacity of aluminum box.

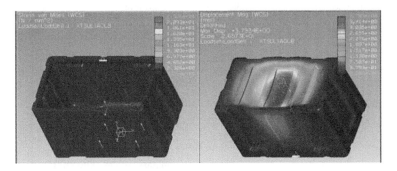

Figure 4.   Analysis on the bearing capacity of rotationally-molded box.

modulus, Poisson's ratio, density and coefficient of thermal expansion of the material are 2500 N/mm², 0.4, 934 kg/m³, and 9×10^{-5}/°C respectively. Then the lower surface of the box was constrained, and a surface load of 1000 N was exerted on the upper surface. After establishing the task and analyzing, we got the stress nephogram and deformation pattern as shown in Figure 4.

From the calculated result we can see that the maximum stress of the cover under the pressure of 1000 N is 23.26 MPa which is lower than the yield strength of the same area (27.44 MPa). The maximal deflection appears at the center of the cover, with a value of 3.793 mm. Similarly, the maximum stress of the rotationally-molded bottom is 17.07 MPa when a pressure of 750 N is exerted to the bottom from the inside, lower than the yield strength (27.44 MPa). The maximal deflection is 3.012 mm.

3. *Analysis on bearing capacity of steel box*
We could also draw a conclusion from Figure 5, the maximum stress of the rotationally-molded bottom is 186.6 MPa when a pressure of 750 N is exerted to the bottom from the inside, lower than the yield strength (282.69 MPa). The maximal deflection is 1.255 mm.

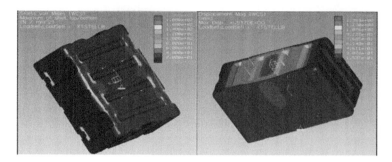

Figure 5.　Analysis on the bearing capacity of steel box.

Table 2.　Performance parameters of steel, aluminum and rotationally-molded boxes.

Material	Aluminum	Rotationally-molding material	Steel
Model	AL6061	LLDPE	1023
Thickness	1.2 mm	4.5 mm	1.2 mm
Elastic modulus	7310 N/mm²	2500 N/mm²	205000 N/mm²
Yield strength	62.05 MPa	27.44 MPa	282.69 MPa
Poisson's rate	0.33	0.40	0.29
Density	2700 kg/m³	934 kg/m³	7858 kg/m³
Net weight of the box	13.416 kg	17.17 kg	38.756 kg
Bearing capacity of the cover			
Distributed load	1000 N	1000 N	1000 N
Maximum stress	186 MPa	186.6 MPa	23.26 MPa
Maximum deflection	4.385 mm	1.537 mm	3.793 mm
Bearing capacity of the bottom			
Distributed load	750 N	750 N	750 N
Maximum stress	148.7 MPa	144.5 MPa	17.07 MPa
Maximum deflection	3.582 mm	1.255 mm	3.012 mm

Table 2 shows the performance parameters and analysis results of boxes made of different materials.

3 CONCLUSION

Table 2 shows that when exposed to the same conditions, the rotationally-molded boxes will not suffer strength failure as the aluminum ones. For the aluminum boxes, the stress coming from the pressure on both the cover (outside) and the bottom (inside) has exceeded the yield strength, although they are lighter than the rotationally-molded ones. The loading capacity can meet the requirements for box-type military equipments (100 kg for each box) well.

REFERENCES

[1] Zhang Yanyang, Zhou Bin. The military rolling kitchen. Commercial vehicle, 1997(2):67–70.
[2] Zhang Xianghong. The brief introduction of military rolling kitchen (1). Commercial vehicle, 1999(2):32–35.
[3] Zhang Xianghong. The brief introduction of military rolling kitchen (2). Commercial vehicle, 1999(3):47–51.
[4] Wang Shuqin. Floating military kitchen. Auto application, 1999(5):69–72.
[5] DS SolidWorks. The basic tutorial of SolidWorks Simulation. Mechanical Industry Press, 2010.
[6] Li Bing. Engineering Application of ANSYS Software. Thsinghua University Press, 2010.

Frontiers of Energy and Environmental Engineering – Sung, Kao & Chen (eds)
© *2013 Taylor & Francis Group, London, ISBN 978-0-415-66159-1*

Product development and innovation in the lifestyle model

Y. Ren
College of Fine Arts, Hubei Normal University Huang Shi, China

ABSTRACT: As the concept of product development and innovation is grabbing more attention from the public, many countries are taking it as a method of enriching the nation and strengthening the people, and corporations think it as the most powerful way of enhancing their core competency. Today with the market competition getting fiercer than ever, those roles, possess an insight for market's demands instantly, seize first market opportunities and take accurate orientation of design, have an incredible importance for the enterprise to win market share and obtain excess profits. Therefore, in order to lead the market trends and stimulate market consumption it's important to conduct a study on the lifestyle of customers, know the specific and substantial needs for target consumers group, and develop products meeting their needs. The concept of lifestyle has been gradually applied to the development and innovative design of products, and in the long run, the research on the lifestyle model will be helpful to find a new direction for product innovation. This article offers an analysis of the new trends in product development and innovation through study on the lifestyle model of consumers, and discusses the guiding meaning of lifestyle provided in product developing and innovating process.

Keywords: lifestyle; product development; innovation

1 INTRODUCTION

(This article can be used for ministry of education humanities and social sciences research projects of department of education of HuBei province, and also can be used for planning and development of creative industry in Wuhan metropolitan area .NO.2010q110). Innovation is a constant theme for products, and is the only applicable and effective way to keep the survival of a product. For corporations, innovation is what keeps them vitalized and full of energy. It's the rule of surviving, and only by conducting continuous innovation can a corporation remain invincible in the fierce market competition. The purpose of product innovation is to increase the competitive edge. However, the key to competition lies not only in what products a corporation produces, but also in whether the products can really satisfy consumers' desire and cater to their needs, both at material and spiritual level. An American scholar Theodore, Levitt pointed out: "the new competition is not occurring at all what the company's factory products, but took place in what their products can provide additional benefits (such as packaging, service, advertising, customer advice, financing, delivery, warehousing, as well as with the other values of the form)." Therefore within a lifestyle model, the product innovation isn't just about the products innovation in various types; more importantly,

it is the innovation in the methods of increasing products' additional benefits.

Study on target consumers and make breakthroughs in innovation.

Lifestyle symbolizes different ways of living. It not only reveals the developing process of culture, but also records human activities in daily life. With society developing and fashion trends changing, the consumer demand is leaning more on culture and civilization. Designing will surely become the dominator in the formation and transformation of lifestyles, which will strongly lead to development and innovation of products.

The lifestyle design concept is to know about the tendencies among possible consumer groups, design new products according to their needs and spot new niches in the market. Lifestyle has made a new interpretation to and an application of product development and innovation, helping corporations make breakthroughs. Meanwhile designers will take customer's living background and characteristics as well as social trends, culture shock, users' requirement and usage environment, as key factors in product development and innovation. What's more, during the developing and innovating process a practical and effective market positioning can be conducted through the analysis of the lifestyle model, which will enable the products to conform to the leading tendencies in the consumer market and meet consumers' requirements.

In this way, a bridge of effective communication between products and consumers will be built and the products designed will have real values and practical use to better satisfy the diversified needs of consumers.

However, corporations may also encounter many challenges if they wanted to take the lion's share in the market of a certain product. These challenges may force them to establish a healthy internal structure which can generate profits, encourage innovation and keep the continuity of a brand at the same time. On one hand, a corporation needs to gain more profits; on the other hand, it needs to handle well the development and innovation of its new products, spot new niches in the market and find the key to breakthroughs in innovation. In this way the corporations can apply the result obtained from the research on the lifestyle theory to product development and innovation, decreasing products recognition difference among consumers, so as to build up and maintain good corporation image by products, leading market trends and maximizing profits.

2 FIND COMMON FEATURES AND DIFFERENCES IN VARIOUS CONSUMER CLANS

With social changes going on through the times, the consumer market has been diversified and science and technology have been greatly improved. Therefore before the development and innovation of product, the corporations must consider the diversified needs and consumers' different habits of using product. The establishment of living clans brings up various lifestyle and consumer trends Different consumer clans have their own needs at material or spiritual dimensions in the same or similar or different from each other. These people live in different environment, belong to different social classes and live with different lifestyles and characteristics. As a result, corporations need to develop and innovate their products accordingly and assimilate themselves into the life of their consumers, experiencing their lifestyles while looking for common features and differences among them. Only in this way can they find valuable opportunities for their products and make them more innovative, flexible and suitable for the market. Meanwhile, corporations themselves can expand their influence, set up brands and earn profits by selling more consumer-oriented products.

How to define and distinguish the "clans" is the key to product innovation within a lifestyle model. People all live in different environment, and their habits in life also vary. But just as the famous Chinese saying goes: "one takes on the color of one's company," so if they live in the same or a similar environment, they may have similar habits, too. Therefore, product innovation in accordance with the needs of different clans is the highlight in a lifestyle model. In China there is an idiom saying "if you don't know me, how will you know what I'm thinking about?" It is the same with product innovation. In a lifestyle model, product innovation should blend into consumers' lives, and it requires the producers to learn about features of the clans and listen to their voices. Take the innovative design of toys as an example. Though the consumer clans are kids, the designers and producers are adults. So how can they design and produce toys popular among kids if they don't have the state of mind as a kid to play? As a result, an analysis on a certain consumer clan is a must. Only by applying the result of the analysis to the product development and innovation can suitable products be produced to better meet the needs of its users.

For corporations, their analysis and research on the lifestyle model have made up for their lack of knowledge on consumers' needs, greatly improved their innovation efficiency and lessened the time for developing products, thus Creating great value to clients and consumers, and building up contributions to product innovation in China.

3 CULTURAL CONTENT

Culture is the basis of designing. As consumers today are no longer easily satisfied with products having simple functions only, the design of products are bringing more emotional, aesthetics, spiritual, psychological and cultural connotations. The "pure material" elements are leaving less effect decreasing while the "non-material" elements are gaining momentum, leading to the co-existence between spiritual and functional demands rather than opposition and conflicts. Therefore exhumation of products' deep culture intension is the inevitable requirement of product innovations, the only way to improve design quality and the fundamental method to strengthen the core corporation competency.

With people increasingly close in contact and more cross-cultural communication, Multi-culture is formed nowadays. The Theory of Evolution raised by Charles Darwin states that existence originated from evolution, therefore the diversified culture is the inevitable result of social evolution and development interacting with trends of living Different groups have their own cultural backgrounds, and the formation of their lifestyle is closely related to social and cultural changes and the characteristics of their lives. So the cultural factor can not be neglected when conducting product development and innovation in the lifestyle model. The research and study on lifestyle can help corporations get to know consumers' values under a cultural shock, their perspectives and attitudes toward life as well as the features of cultural codes, facilitating development of products and the decision of innovation planning.

When an object was first made to meet the preliminary needs of human survival, it came into being because it had certain functions, and the decoration was not always necessary. However, when a person had more time for deliberation, he would think of making the object better, adding his signs to make it have personal characteristics. "Consumer groups' nowadays have higher level and more diversified values. They pursue individuality, tastes and cultural values, and they want the products to bring out their unique styles and satisfy their psychological needs as well. Various lifestyles show different personal life values and lead to discrepancy among personal culture attainment, taste of life, and consuming capacity. Therefore, when looking at the changes of trends in life, culture is the accumulated knowledge about life, and life is the reflection of culture, while consumption, a small part of the mixture of life and culture, acts as a motivator to push the trends forward. The three elements circulate and interact with each other, and it is in this circulation and interaction process that the innovation of products is realized.

For an example as design a ball arm of golf, you must know the consumption custom and make a market research for consumption custom of the consumer first. Golf is popular entertainment project in the USA. Below Table 1 is the users' questionnaire of the golf in the USA.

The data provides a basic reference value of the users who playing golf in the USA. The product made position in the "often played golf enthusiasts more than 12 times in a year" as the ball arm is made by titanium alloy and the elasticity is better than others. Then we can clearly know the consumer groups of the ball arm by the market research. After accurate positioning, we should make more market research. The basic design direction is based on the reason for the user's life style formation, such as what colors users like, how the feel is having handle, what kind of pack is felt better by users.

4 ADAPT ONESELF TO CHANGES AND DEMANDS OF THE TIMES

The application of the lifestyle model is a major part in product design and development as well as an important mean to achieve products innovation and differentiation. Devin Moore, an American industrial designer said, in one of his articles "Design + Innovation = Business", that in any preliminary stages, the design practice and development of new products should always be consumer-oriented. It is very important to know what people need and give a quick response even before they speak. Only corporations like this can perform effective innovation.

People have accepted and experienced all kinds of knowledge, cultures, customs and activities throughout the evolution of human society. In this process they have developed the ability to adapt themselves to social life, and at the mean time they seek for personal satisfaction from it. They want their work or jobs done or improved by using a

Table 1.

| | The total population of the USA | | The people often play golf (once or more than once a month) | | |
	Thousands of people	Percentage	Thousands of people	Thousands of people	Index
Adult	197 462	100.0%	13 097 (6.6%)	100.0%	100
Male	94 827	48.0%	9 882	75.6%	157
Female	102 635	52.0%	3 198	24.4%	47
Age 18–24	64 961	32.9%	3 584	27.4%	83
Age 35–64	100 241	50.8%	7 410	56.7%	112
Age over 65	32 260	16.3%	2 086	15.9%	98
University education	43 406	22.0%	4 693	35.9%	163
Manager	18 969	9.6%	2 330	17.8%	185
Annual income more than 50 thousand	22 865	11.6%	3 380	25.8%	223

Data source: Mediamak company in spring of 1999.

certain product; they hope the product can enrich their life experience and relate it to some dreams of their own. Seen from this, a product should be able to help to realize one's certain emotional value and satisfy an emotional need at a higher level. The old concept of "form should conform to the functions" is not applicable here any more, while a new concept of "the common dream for harmony of form and functions" is taking place instead. Nowadays people want the products to offer something more than just practical functions-through the process of buying a product, they also seek for psychological satisfaction, and further more the realization of their personal value. That is to say that people want a way to show their "style of living" more than just a pleasant product.

Today people are having more "spiritual consumption" than the simple "material consumption" in their consuming activities in daily life, since they have developed a different material and spiritual demand on products. Therefore, the lifestyle model will be playing an increasingly important role in product development and innovation. There is no doubt that product development and innovation will get greatest improvement in the lifestyle model and reap the harvest of success in the future.

(This article can be used for ministry of education humanities and social sciences research projects of department of education of Hubei province, and also can be used for planning and development of creative industry in Wuhan metropolitan area. NO.2010q110.)

REFERENCES

[1] Du Ruize, Life Styling Design, Taipei: Asiapac Books, 2004.
[2] Haishi, eds., Yang Huiming, trans., Realize: Design Means Business, Beijing: Jinghua Press, 2008.
[3] Craig M. Vogel, Jonathan Cagan, Creating breakthrough products, China MachinePress, October 2003.
[4] Walter Miles, Handbook for Designers, Beijing: Jinghua Press, 1995.
[5] Chen Hanqing & Wang Chunxia, Innovation and Harmony, 2006.

Frontiers of Energy and Environmental Engineering – Sung, Kao & Chen (eds)
© 2013 Taylor & Francis Group, London, ISBN 978-0-415-66159-1

Research on U-shape arm length range for directly buried heating supply pipeline

P. Zhang & F. Wang
College of Environmental Science and Engineering, Taiyuan University of Technology, Taiyuan Shanxi, China

Y.L. Chen
College of Resources and Environmental Sciences, Chongqing University, Chongqing, China

ABSTRACT: Aimed DN800, DN1000 and DN1200 directly buried U-shape elbow for heating supply, use ANSYS finite element application to do stress analysis for the elbow which is based on the same stress and curvature radius. And by doing this, it is clear that the effect to the stress by using different pipe diameter, length of overhang arm and displacement load. The conclusion shows us that the reasonable overhang arm length range of the U-shape pipeline is between 1.5DN to 3DN.

Keywords: directly buried heating supply pipeline; U-shape pipeline; finite element analysis; load

1 GENERAL INSTRUCTIONS

As the fast development of central heating in city, more and more U-shape pipeline is adopted extensively in parts of directly buried ductwork. It is helpful to avoid barrier and make up the extending of straight pipeline, in a further step, it makes a replacement of compensator and effectively reduces the cost of the construction. Compared with trench and aerial laying, there is an obvious difference in the stress characteristic of directly buried U-shape pipeline for it supports the stress from soil. Applying the theory of trench and aerial laying to calculate the stress, there would must be a big error. By doing the wrong calculation, it makes the stress of the elbow bigger than the rated stress of the material and then may even lead to an elbow break and affect the safely operation of the ductwork. Presently, the design range about the U-shape elbow in the code of directly buried pipe is still only limited to suit for the less or equal to DN500 pipe which is far away to match the need for the construction of reality.

Recently, some investigators made some experiments and studies to the regularity of the U-shape pipe stress (Guo Ruiping, 1998). But few of them focused on its overhang arm length range. Aiming the problems which is pressing to be solved in the engineering design, this thesis shows us a analysis finite element method by using ANSYS finite element software to do a data analysis to the stress changing situation of DN800, DN1000 and DN1200 U-shape elbows for directly buried heating supply ductwork. And give a reasonable overhang arm length suggestion.

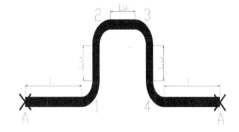

Figure 1. Structure of U-shape elbow.

2 U-SHAPE PIPELINE

2.1 Structure of U-shape elbow

As shown in Figure 1. Where, L-compensated straight pipeline; Lu-overhang arm of U-shape pipeline; 1, 2, 3, 4-No. series of elbow for U-shape pipeline; A-natural point and stagnation point.

2.2 Specifications and dimension of pipe (Table 1)

2.3 The physical parameters of the computation module (Table 2)

3 ESTABLISHMENT OF MODULE

3.1 Assumptions for calculation

The calculation of stress is based on the assumptions that the tubular product is stretch, seriate

Table 1. Specifications and dimension of pipe.

Pipe diameter	External diameter × wall thickness (mm)	External thermal insulation thickness (mm)	Material (steel)
DN800	820 × 10	960	Q235
DN1000	1020 × 13	1155	Q235
DN1200	1220 × 14	1370	Q235

Table 2. Physical parameters of the computation module.

Steel elasticity modulus/MPa	Steel pipeline swell factor m/(m · °C)	Reaction force coefficient of compressed soil/(N/m³)	Soil density/ (kg/m³)	Poisson's ratio
19.6×10^4	12.6×10^{-6}	4×10^6	1800	0.3

(pipes are a seriate whole), well-proportioned (pipes have the same elasticity) and isotropy. In addition, unevenness and ellipticity of pipe wall thickness should be ignored (Liu Shimin, 2006).

3.2 Establishment of module

According to the symmetry of the boundary conditions and module, we select U-shape elbow and half part of the both sides' pipe as a module. The plane of symmetry is the central cross-section of outside bend arm of the U-shape elbow.

3.3 Dividing mesh

In order to analysis the stress distribution of U-shape pipeline better and increase the calculation accuracy when the ductwork is in operation, we need to choice appropriate mesh to be divided after the U-shape elbow is established. The SOLID95 solid element which has 20 nodes is used here for the module of elbow and its section of arm. In the other side, 8 nodes SOLID45 element is adopted for the both sides' pipe of elbow. COMBIN14 element spring-damper is selected for soil action. The mesh mode divided after scan is used for all the parts of the module. The calculation model is shown in Figure 2.

3.4 Exerting load

The load be exerted to U-shape elbow mainly includes displacement load, stress load, temperature load and soil load. The acting force between soil and pipes is simulated by changing the parameter which is due to the spring action to pipes. The following figure is the module of the DN1000 U-shape pipeline whose arm length type is 1DN.

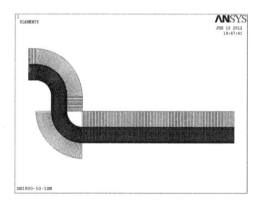

Figure 2. Calculation model.

4 CALCULATION RESULTS AND DATA ANALYSIS OF THE MODULE

1.6 MPa stress is acted to the DN800, DN1000 and DN1200 U-shape elbow and both sides' pipes separately. The design temperature is 130°C, the environment temperature is 10°C, length of overhanging arm is 2 m, depth is 1.5 m, and elbow's radius of curvature is 1.5DN. We do the analysis for different displacement load. The results based on the fourth strength theory are shown in Figure 3.

The results indicate: Based on fix design stress, circular temperature differences, depth of burial and length of the overhang arm, the elbow stress increase with the raise of the displacement load. Comparing with different pipe diameter pipes, the stress increase with the decrease of the pipe diameter, and when the displacement is bigger than 60 mm, it becomes to a linear trend.

Based on the assumptions of DN1000, 1.6 MPa stress, 130°C design temperature, 10°C environment temperature, 1.5 m depth of burial and 1.5DN

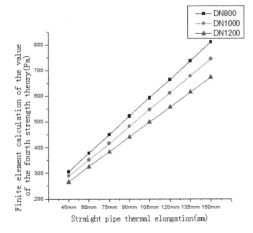

Figure 3. Results (DN800, DN1000 and DN1200 U-shape elbow) under different displacement load based on the fourth strength theory.

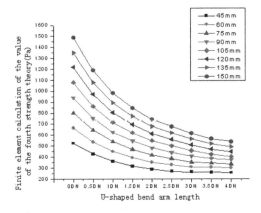

Figure 4. Results of DN1000 under different displacement load based on the fourth strength theory.

as the elbow curve radius, we do the analysis under the actions in different displacement load based on the fourth strength theory. The results based on the fourth strength theory are shown in Figure 4.

The results state: Based on fix design stress, circular temperature differences, depth of burial and pipe diameter, the elbow stress increase with the raise of the displacement load, the increase range fall with the swell of the bend arm length, and in a further step, the decrease range become less and less.

5 CONCLUSIONS

We adopt the analysis theory for the elbow tiredness from Eurocode, the reasonable arm length range for U-shape elbow should be 1.5DN to 3DN. When the straight pipe's displacement is less than 100 mm or its pipe diameter is less than DN1000, the bigger arm length should be selected, oppositely, the smaller one should be adopted.

The compensation ability of the U-shape elbow is great, we can use in the practical projects. It can be a replacement of compensator, further reduce the investment and bring down the hazard rate of the ductwork.

REFERENCES

CHEN Gang, ZHANG Chuanyong, LIU Yinghua. 2005. Finite element analysi of plastic limit loads of locally thinned elbows under internal pressure and in-plane bending moment. *Engineering Mechanics*, (22) 43.

CHEN Gang, ZHANG Chuanyong, LIU Yinghua. 2005. Finite element analysi of plastic limit loads of locally thinned elbows under internal pressure and in-plane bending moment. *Engineering Mechanics*, (22) 43.

(CJJ/T81-98) Technical specification for directly buried heating pipeline engineering in city. 1998.

(CJJ/T81-98) Technical specification for directly buried heating pipeline engineering in city. 1998.

(CJJ/T81-98) Technical specification for directly buried heating pipeline engineering in city. 1998.

(CJJ/T81-98) Technical specification for directly buried heating pipeline engineering in city. 1998.

Grove, A.T. 1980. Geomorphic evolution of the Sahara and the Nile. In M.A.J. Williams & H. Faure (eds), *The Sahara and the Nile*: 21–35. Rotterdam: Balkema.

Grove, A.T. 1980. Geomorphic evolution of the Sahara and the Nile. In M.A.J. Williams & H. Faure (eds), *The Sahara and the Nile*: 21–35. Rotterdam: Balkema.

Guo Ruiping, Li Guangxin.1998. Study of mechanical law of "Ω" type section of buried heating pipe. Journal of Tsinghua University (Sci & Tech). (38) 23.

Jappelli, R. & Marconi, N. 1997. Recommendations and prejudices in the realm of foundation engineering in Italy: A historical review. In Carlo Viggiani (ed.), *Geotechnical engineering for the preservation of monuments and historical sites*; Proc. intern. symp., Napoli, 3–4 October 1996. Rotterdam: Balkema.

Jappelli, R. & Marconi, N. 1997. Recommendations and prejudices in the realm of foundation engineering in Italy: A historical review. In Carlo Viggiani (ed.), *Geotechnical engineering for the preservation of monuments and historical sites*; Proc. intern. symp., Napoli, 3–4 October 1996. Rotterdam: Balkema.

Johnson, H.L. 1965. Artistic development in autistic children. *Child Development* 65(1): 13–16.

Johnson, H.L. 1965. Artistic development in autistic children. *Child Development* 65(1): 13–16.

Liu Shimin. 2006. Analysis of the Thermal Stress of the Elbow of Heat-Serve Pipe by ANSYS Software. Journal of Yancheng Institute of Technology (Natural Science), (19) 28.

Polhill, R.M. 1982. *Crotalaria in Africa and Madagascar*. Rotterdam: Balkema.

Polhill, R.M. 1982. *Crotalaria in Africa and Madagascar*. Rotterdam: Balkema.

Wang Guowei. 2010. Large-diameter directly buried heating pipe 90° elbow fatigue life of the finite element analysis. Shanxi: Taiyuan University of Technology.

Wang Guowei. 2010. Large-diameter directly buried heating pipe 90° elbow fatigue life of the finite element analysis. Shanxi: Taiyuan University of Technology.

Wang Guowei. 2010. Large-diameter directly buried heating pipe 90° elbow fatigue life of the finite element analysis. Shanxi: Taiyuan University of Technology.

Yaze YU, Xiaogong LI. 2009. Foundation of finite element model for directly buried hot water heat-supply pipeline. *Gas & Heat*, (29) 8.

Yaze YU, Xiaogong LI. 2009. Foundation of finite element model for directly buried hot water heat-supply pipeline. *Gas & Heat*, (29) 8.

Frontiers of Energy and Environmental Engineering – Sung, Kao & Chen (eds)
© 2013 Taylor & Francis Group, London, ISBN 978-0-415-66159-1

Numerical analysis of the directly buried heating supply horizontal elbow stress under different angles

G. Du & F. Wang
College of Environmental Science and Engineering, Taiyuan University of Technology, Taiyuan Shanxi, China

Y.L. Chen
College of Resource and Environmental Sciences, Chongqing University, Chongqing, China

ABSTRACT: In studying DN1000 directly buried heating horizontal elbow, numerical simulation method was used to analyze the law of stress change of the angle elbow (40°~110°) under the conditions of same pressure load, temperature load, and curvature radius. The result showed neutral lines parts in both top and bottom of the elbow had the largest stress after the pipe was exerted displacement load, pressure load, and temperature load. Besides, with the increase of the adjacent supplementary angle, the stress gradually decreased; for large-diameter buried elbow, when using angle for natural compensation, the degree of angle can range from 70° to 110°. The result of elastic flexure resistance hinge analysis was conservative and the potential of the elbow cannot be tapped.

Keywords: directly buried heating elbow, angle, finite element, stress

1 GENERAL INSTRUCTIONS

The implemented Technical specification for directly buried heating pipeline engineering in city (hereinafter called "Technical specification") set standards for the design of directly buried heating pipeline. However, the Technical specification was limited to the design and application of pipes smaller than DN500. The reason is that the experimental data which the Technical specification were based on only concerned specification less than DN500. When verifying the elbow's fatigue strength, the Technical specification used the elastic flexure resistance hinge analysis method and deals with load when doing the pipe calculation. The dimension and geometrical shape of elbow were neglected because these elements are not important when dealing with small-diameter pipes. However, currently the diameter of the pipe used in engineering design has expanded to DN1400. Whether those two elements have effect on the large-diameter ones or not is not clear. Therefore, it is necessary to analyze the effect of size and geometrical shape on the angle change of large-diameter elbows and the change of relevant stress.

Ansys finite element analysis software was used in this thesis to make numerical analysis of the angle's effect on the stress of horizontal elbow of directly buried heating pipeline and make contrastive analysis with the elbow's analytic calculation given by the Technical specification; The analysis determined the maximum critical stress that can be sustained by the elbow according to the elbow action circle index requirement from Euro code. Also the L-shape syphon angle used to do nature compensation was given. Both these two elements can be the references for design.

2 MODEL BUILDING

To make a better simulation about the real performance of elbow, the effect of loading method of loading end was taken into account and 10 m straight pipes were added to both ends of the elbow.

In the finite element model of the elbow, the loads include temperature load, pressure load, displacement load and soil load. According to Winkler's Foundation Theory, the deformation of certain point in the foundation is irrelevant to the pressure acting on other points. Therefore, soil is seen as formulated by numerous earth columns with frictionless sides. It is further assumed that soil becomes a system of disconnected springs after using springs to replace earth columns. Based on this theory, soil is seen as the equivalent of soil springs. The effects of soil on pipe were dispersed to element nodes of the pipe. Soil spring is simulated by spring element COMBIN14 and the reaction force of side soil on pipe was simulated by adjusting elasticity coefficient of spring. The finite element model is seen as Figure 1.

2.1 Properties and dimension parameters of piping material

Table 1. Properties of piping material (Steel grade Q235).

Modulus of elasticity/MPa	19.6×10^4
Coefficient of thermal expansion α/K^{-1}	12.6×10^{-6}
Poisson's ratio	0.3
Allowable stress/MPa	125
Yield stress/MPa	235

Table 2. Dimension parameters of elbow.

Diameter	Outer diameter/mm	Inner diameter/mm
DN1000	1020	994

Table 3. Synthesized bedding value (kN/m²).

Standard	DN1000
Expansion cushion 40 mm	3433
Expansion cushion 60 mm	2785
Expansion cushion 100 mm	2019
Expansion cushion 120 mm	1774
Expansion cushion 180 mm	1298

2.2 Element type

Proper elements were needed to build the finite element model. In this thesis, elbows and bent arms were 3D 20-node solid element Solid95. Soild95, higher degree form of solid45, was exerted to irregular shape and cannot damage precision. Besides, this element has the shape which is coordinated by displacement and can be applied to simulate curved boundary. This element was defined by 20 nodes. Every node has three degrees of freedom: X, Y and Z. Moreover, this element has all spatial directions, with qualities like plasticity, creep, swelling, stress strengthening, large deformation, and large strain. Soil spring was simulated by spring element COMBIN14.

3 EXERTING LOAD

The designed water temperature was 130°C. The minimum cycle temperature was 10°C. 1.6 MPa pressure was exerted on elbow and inner surfaces of both arms. It was assumed that the lengths of elbow arms were 10, 30, 60, 90, 120 m respectively and relative displacement load was exerted too. The results were obtained after using the order Solution > Solve > Current LS. The results applied with the fourth strength theory were shown in Figure 2.

4 LOW CYCLE FATIGUE FAILURE

Low cycle fatigue failure is the major form of failures for elbows. Therefore, the determination of the maximum temperature cycle index is crucial for verifying the fatigue life. Features of low

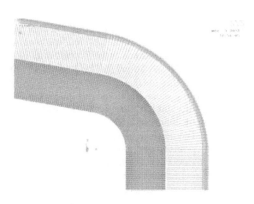

Figure 1. Model of 80° directly buried heating elbow.

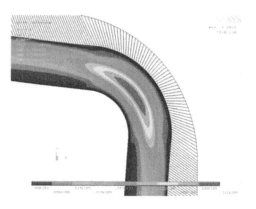

Figure 2. Stress cloud picture of the fourth equivalence of DN1000 80°elbow.

Table 4. Items of safety coefficients for fatigue.

Project class	γ_{fat}
A	5
B	6.67
C	10

Table 5. Maximum action cycle index.

Major pipelines	100
Main pipelines	250
House service connections	1000

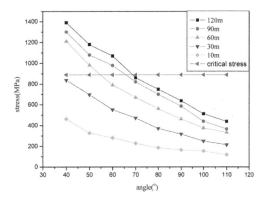

Figure 3. Stress theory resolutions with the fourth strength theory.

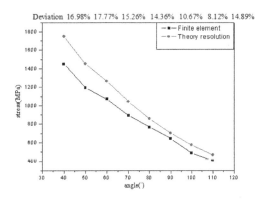

Figure 4. The third strength theory finite element values and the theory resolutions.

cycle fatigue were that cyclic stress amplitude was relatively high, stress cycle index of fatigue failure was relatively low. Ultimate state of low fatigue plays an important role in elbow, tee joint and pipe reducer. The straight pipe part which bears high axial stress should be verified too. It is specified by the Technical Specification that the range of stress should be less than three times of allowable stress. In fact, it simplified fatigue analysis when doing the verification. A brief introduction on European safety verification method on low cycle fatigue was given in the following part. According to Table 4, when doing calculation, the maximum cycle index should not be less than the given figures in Table 5.

Verification of sufficient safety against fatigue fracture is made using Palmgren-Miner's formula:

$$\sum_i \frac{n_i}{N_i} \le \frac{1}{\gamma_{fat}} \qquad (1)$$

where n_i is the number of cycles with stress range, during the required design life; N_i is the number of cycles of stress range, to cause failure; γ_{fat} is the safety coefficient of fatigue fracture, as shown in Table 4; i is the serial number of different stress ranges.

$$N_i = \left(\frac{k}{S_i}\right)^m = \left(\frac{5000}{S_i}\right)^4 \qquad (2)$$

where S_i the design stress range, MPa.

For the directly buried heating pipe with large diameters, project class was determined as C. According to formulas (1) and (2), the fourth strength critical stress was calculated as 889 MPa.

5 RESULT OF THE SIMULATION

The lengths of DN1000 bent arms are 10, 30, 60, 90, 120 m respectively. The numerical simulation of stress theory resolutions with the fourth strength theory were shown in Figure 3. It shows the changes of the fourth strength theory finite element theory resolutions in neutral line part of elbows under the conditions of pressure 1.6 MPa, curvature radius 1.5 DN, angle ranges from 40° to 110°; the numerical simulation of the third strength theory finite element values and the theory resolutions based on Technical Specification under the conditions of arm length 90 m, angle DN1000, and angle ranges from 40° to 110° were shown in Figure 4. The figure shows the changes of both numerical simulation of the third strength theory finite element values and the theory resolutions based on Technical Specification in neutral lines part of elbows under the conditions of pressure 1.6 MPa, curvature radius 1.5DN, angle ranges from 40° to 110°.

6 CONCLUSIONS

1. For elbow angle ranged from 40°~110°, neutral lines in both top and bottom of the elbow have the largest stress, and further, with the increase of the adjacent supplementary angle, the stress gradually decreases.
2. When using angle for natural compensation for directly buried heating pipeline, when the elbow angle is less than 70°, stress is larger than critical stress; for large-diameter directly buried heating pipe, when using angle for natural compensation, the degree of angel can range from 70° to 110° without taking security measures.
3. The result of elastic flexure resistance hinge analysis was conservative and the potential of the elbow cannot be tapped.

REFERENCES

Bolt, S.E. Greenstreet, W.L., "Experimental Determinations of Plastic Collapse Loads for Pipe Elbows" American Society of Mechanical Engineers, Pressure Vessels & Piping Conference, 1971, P71-PVP-37.

CHEN Gang, ZHANG Chuanyong, LIU Yinghua. 2005. Finite element analysi of plastic limit loads of locally thinned elbows under internal pressure and in-plane bending moment. Engineering Mechanics, (22)43.

Cross, N., "Experiments in Short-Radius Pipe Bends", Proceedings of Institution of Mechanical Engineers, (B), Vol. 1B, 1952–1953, P465.

(CJJ/T81-98) Technical specification for directly buried heating pipeline engineering in city. 1998.

District heating pipes-Preinsulated bonded pipe systems for directly buried hot water networks-Pipe assembly of steel service pipe, polyurethane thermal insulation and outer casing of polyethylene, BS EN253: 2009.

Grove, A.T. 1980. Geomorphic evolution of the Sahara and the Nile. In M.A.J. Williams & H. Faure (eds), The Sahara and the Nile: 21–35. Rotterdam: Balkema.

Guo Ruiping, Li Guangxin. 1998. Study of mechanical law of "Ω" type section of buried heating pipe. Journal of Tsinghua University (Sci & Tech). (38) 23.

Jappelli, R. & Marconi, N. 1997. Recommendations and prejudices in the realm of foundation engineering in Italy: A historical review. In Carlo Viggiani (ed.), Geotechnical engineering for the preservation of monuments and historical sites; Proc. intern. symp., Napoli, 3–4 October 1996. Rotterdam: Balkema.

Johnson, H.L. 1965. Artistic development in autistic children. Child Development 65(1): 13–16.

Liu Shimin. 2006. Analysis of the Thermal Stress of the Elbow of Heat-Serve Pipe by ANSYS Software. Journal of Yancheng Institute of Technology (Natural Science), (19) 28.

Marcal, P.V., "Elatic-Plastic Behaviour of Pipe. Bends With In-Plane Bending", Journal of Strain Analysis Vol. 2, Nol., 1967. P. 84.

Polhill, R.M. 1982. Crotalaria in Africa and Madagascar. Rotterdam: Balkema.

Spence, J., Findlay, G.E., "Limit Loads for Pipe Bends Under In-Plane Bending", Second International Conference On Pressure Vessel Technelogy. Part1, Design and Analysis, 1973, P. 393.

The M.W. Kellogg Company. Design of Piping Systems. 1952. 52~60.

Wang Guowei. 2010. Large-diameter directly buried heating pipe 90⊕ elbow fatigue life of the finite element analysis. Shanxi: Taiyuan University of Technology.

Yu Yaze, Li Xiaogong. 2009. Foundation of finite element model for directly buried hot water heat-supply pipeline. Gas & Heat, (29) 8.

Frontiers of Energy and Environmental Engineering – Sung, Kao & Chen (eds)
© 2013 Taylor & Francis Group, London, ISBN 978-0-415-66159-1

Effect of Fe^0-PRB process degradation of 2, 4-dichlorophenol in groundwater

R. Wang, Y.X. Zhang, J.M. Gao & J.K. Liang
Beijing Key Lab of Water Quality Science and Water Environmental Restoration Engineering, College of Architecture and Civil Engineering, Beijing University of Technology, Beijing, China

ABSTRACT: 2, 4-Dichlorophenol (2, 4-DCP) is a typical groundwater phenolic pollutants. In this paper, the zero-valent iron-Permeable Reactive Barrier (Fe^0-PRB) is being used on the degradation of 2, 4-DCP in groundwater. Monitoring data include pH, redox potential and the concentration of 2, 4-DCP. Finally, the analysis result shows that Fe^0-PRB technology has the high-value in removal 2, 4-DCP in groundwater, which support for engineering applications.

The phenolic compounds are common organic pollutants in soil and groundwater. They are widely concerned because its higher toxicity. 2, 4-dichlorophenol (2, 4-DCP) is a typical phenolic pollutants used extensively for agricultural herbicides 2, 4-D, phosphonothioic acid, P-phenyl-, O-(2, 4-dichlorophenyl) O-ethyl ester and bithionol production of the synthesis. Ministry of Environmental Protection of the People's Republic of China and United States Environmental Protection Agency all listed it as priority control of pollutants[1,2].

Compared with the traditional groundwater ectopic treatment method, in situ Permeable Reaction Barrier (PRB) technology, which do not need extra motive power, has a better effect and handle a variety of pollutants. Besides, long-time processing and low operation fee also show the advantages of PRB. Since 1994, Gilliam and Ohannesin first field trials investigated the effect of zero-valent iron filing reductive DE chlorination of chlorinated organics. Zero-valent iron filing as a reaction medium PRB technology is famous for low cost and good treatment effect[3,4]. This paper studies the influencing factors, effect and the reaction mechanism about Fe^0-PRB degradation 2, 4-DCP in groundwater.

1 EXPERIMENTAL

1.1 *Experimental materials*

Experiment materials and medicines, including cast iron filing, slag, river sand, 2, 4-dichlorophenol, 1, 4-dichlorobenzene, phenol, 2-chloro-phenol, 4-chloro-phenol, hydrochloric acid and sodium hydroxide, they are all chemical analytical. Sewage for experiment is synthesized and pH is 7.5.

1.2 *Instruments and test methods*

The experimental instruments including the Agilent GC6890 N gas chromatograph (ECD detector), solid phase micro extraction device (with 85 µm thickness Poly-acrylate PA-SPME, Supelco Inc.), precision acidity meter (Leici Instrument PHSJ-3F), Constant temperature magnetic blender (Jintan medical instrument factory HJ-3, control the temperature of ±1°C) and shaking incubator (Suzhou Wier experiment Goods Co., Ltd. BS-2FD) Headspace vial 20 ml, with a sealed aluminiferous cap. At the very first time use it, to soak it with 1:1 nitric acid and then clean it by distilled water boiled for 40 minutes. At last, to bake it in an oven for 2 hours. For further using, clean it in the same way.

Using the internal standard method to test the concentration of 2, 4-DCP, the internal standard is 1, 4-dichlorobenzene and water is Wahaha purified water. Sample extraction 30 min, adsorption 10 min, analytical 3 min; Chromatographic column for HP-5 capillary column (30 m × 0.32 mm × 0.25 µm); Gasification chamber temperature 260°C; Detector temperature 280°C; The column temperature (program warming): 60°C (keep 5 min), with 15°C/min to 130°C, with 30°C/min to 280°C (keep 2 min); The carrier (high purity nitrogen) velocity 2 ml/min; Tail gas blowing 60 ml/min. Determination of six times the peak area (close to the blank of low concentrations), results to calculate the relative standard deviation, to take unilateral 99% confidence level (t = 3.365). 2, 4-DCP detection limit is 6.54 µg /L and the recovery was 90%-103%[5].

1.3 *Experimental device*

Laboratory PRB installations using inert PTFE material, which diameter is 8 cm, high 60 cm, is made of 5 reaction column, labeled as columns 1, 2, 3, 4 and 5. Columns 1 and 2 are filled the iron filing, slag and river sand. Column 3 is filled with river sand and running the column in parallel with the 1, 2. The column 4 is a releasing oxygen-aerobic biological column and the 5 is the comparison column which filled with river sand. They all set up five sampling ports. Sewage goes through the column 1, and then enters the series on the column 4. This test mainly to study the effect of degradation after sewage through the columns 1 and 2. Figure 1 shows the schematic diagram of PRB.

1.4 *Test methods*

Pretreatment process of the test material: put iron filings into excess of 0.5 mol/L of NaOH soak for 1 d to remove grease and water washing, and then put those iron filings into the excess of 0.5 mol/L HCL soak for 30 min to remove the iron oxide layer of the surface, Then use water to wash until acid-free residue, using water sealing technology to save stand, iron filings, slag and river sand were selected 2 to 5 mm, 0.3 to 0.6 mm and 1 to 2 mm particle size[5]. Through pre-test to determine the best ratio of the reaction medium for packed columns in Table 1.

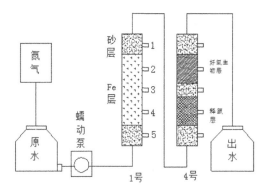

Figure 1. Schematic diagram of PRB.

2 RESULTS AND ANALYSIS

2.1 *Degradation effect*

Column 3 is the comparison (control column) of column 1 & 2, which mainly filled with river sand. Those three test columns were running individually. The original concentration of 2, 4-DCP in sewage is about 18.6 mg/L. Compared the effect of degradation in three columns to determine the removal efficiency about Fe^0 for 2, 4-DCP. The effect of degradation of three columns during 400 days to run in Figures 2–4.

Accordingly, three columns' removal efficiency for 2, 4-DCP is shown in Figures 5–7.

Figures 5 and 6 show during the PRB continuous running for 400 days, the removal of 2, 4-DCP of each reaction layer in column1 and 2 is basically stable and the mainly reaction area is between sampling ports 2, 3, 4. The effluent concentrations about top sampling ports have fluctuate slightly over time, following 1 mg/L. Column 1 and 2 filled basically the same and the different between them is effective porosity (1 higher than 2). Thus, the effluent concentration and removal efficiency trend is basically the same, slightly different.

2.2 *pH changes*

Groundwater composed of the sewage directly through the peristaltic pumps into the reaction columns. This experiment investigated the conditions of this process in practical engineering applications. Therefore, not change the pH value of the raw water and five pH probe installed in each column to monitor the pH value. Figures 8–10 shown columns' pH changes during the running days.

Raw water pH is about 7.5. Figures 8 and 9 show that bottom of the columns 1and 2 sampling ports almost have no change and with the flow direction, the pH value becoming high trend. The obvious area of change in pH is the reaction layer between the sampling port 3 and 4, because the reductive DE chlorination reaction mainly occurs in this

Table 1. Experimental column fill data.

Composition	Proportion-on	Column volume (L)	Effective volume (dm³)	Porosity (%)	Effective porosity (%)
1. River sand, Fe⁰, slag	10:31:9	3.0144	0.66	40.50	21.90
2. River sand, Fe⁰, slag	10:31:9	3.0144	0.55	40.10	18.20
3. River sand,		3.0144	0.36	28.53	11.94
4. River sand, and oxygen release material, sawdust	And oxygen release material: river sand = 1:16 sawdust: river sand = 1:4	3.0144	0.35	23.22	11.61
5. River sand		3.0144	0.32	23.22	10.62

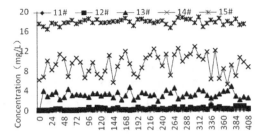

Figure 2.. The curve of 2, 4-DCP of different sampling ports with the time in column 1.

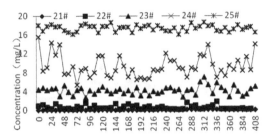

Figure 3. The curve of 2, 4-DCP of different sampling ports with the time in column 2.

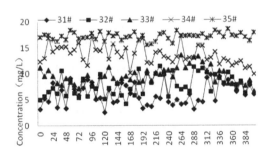

Figure 4. The curve of 2, 4-DCP of different sampling ports with the time in column 3.

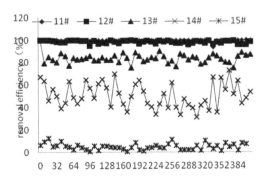

Figure 5. The curve of removal efficiency of 2, 4-DCP of different sampling ports with the time in column 1.

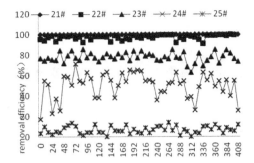

Figure 6. The curve of removal efficiency of 2, 4-DCP of different sampling ports with the time in column 2.

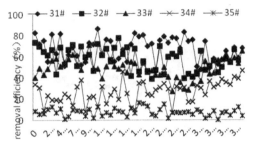

Figure 7. The curve of removal efficiency of 2, 4-DCP of different sampling ports with the time in column 3.

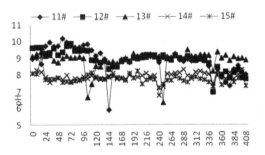

Figure 8. The curve of pH in different sampling ports in column 1.

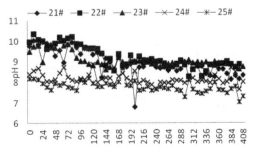

Figure 9. The curve of pH in different sampling ports in column 2.

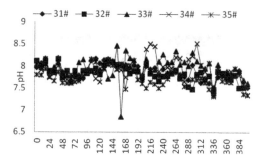

Figure 10. The curve of pH in different sampling ports in column 2.

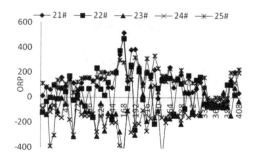

Figure 12. The curve of ORP in different sampling ports in column 2.

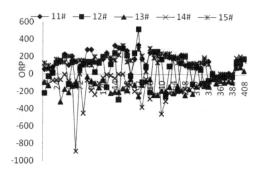

Figure 11. The curve of ORP in different sampling ports in column 1.

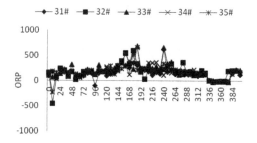

Figure 13. The curve of ORP in different sampling ports in column 3.

area and the pH changes more apparent. Monitor the top sampling ports in columns 1 and 2 indicated the pH up to 10.18 and 10.03 and the pH values can be maintained at following nine at the end of 400 days. Figure 10 shows that the sampling ports' pH value did not change significantly along the flow direction in column 3. The reason is Column 3 mainly filled with river sand which only has a small amount of adsorption to 2, 4-DCP and can't happen reductive DE chlorination and effective primary batteries reaction.

2.3 Oxidation—reduction potential change

Fe^0 has a strong reduction (E^0 (Fe^{2+}/Fe^0) = −0.44V). Under hypoxic conditions, low redox potential in the reaction column will cause anaerobic corrosion. That is, water is decomposed to generate OH^- and H^+, Fe^0 oxidized to Fe^{2+} ($Fe^0 + 2H_2O \rightarrow Fe^{2+} + 2OH^- + H_2$). Test using the ORP electrode to monitor the ORP for three columns. Figures 11–13 show columns ORP changes during the running days.

Figures 11 and 12 show that sampling ports 1 and 5 in columns 1 and 2 ORP are basically positive, in an oxidizing environment; sampling ports 2, 3 and 4 ORP basically negative, in a reducing

environment, because this region is the main area of the degradation reaction occurred which is match with the reaction principle. Sampling ports in column 3 (comparison column)'s ORP value are basically positive, because the filler in column 3 is river sand which does not have a reduction and the column is oxidation environment.

2.4 Reaction mechanism

The main principles by Fe^0 as the filler in PRB on the 2, 4-DCP degradation of following 2 points:

1. The original battery reaction. The Fe^0 is generally iron-carbon alloys. When immersed in aqueous solution to form a complete micro-battery circuit, it can form numerous tiny original batteries. Iron as the anode and carbonized iron as cathode.

The electrode reaction is as follows:

Anodic reaction:

$$Fe \rightarrow Fe^{2+} + 2e^- \ E \ (Fe^{2+}/Fe^0) = -0.44v$$

Cathode reaction:

$$2H^+ + 2e^- \rightarrow 2H \rightarrow H_2 \ E \ (H^+/H_2) = 0.00 \ v$$

With O_2:

$$O_2 + 4H^+ + 4e^- \rightarrow H_2O \; E(O_2) = 1.23 \, v$$

$$O_2 + 2H_2O + 4e^- \rightarrow 4OH^- \; E(O_2/OH^-) = 0.40 \, v$$

2. Reduction of the iron. The Fe^0 has strong reducing ability. 2, 4-DCP first get an electron on the cathode surface that is reduced to chlorophenol (2-chloro-phenol and 4-chlorophenol); chlorophenol continue to receive an electron restored to phenol. Iron filing reduction reaction equation is as follow:

$$Fe^0 + 2H_2O + 2RCl_x \rightarrow 2RCl_{x-1} + 2OH^- + Fe^2 + 2Cl^-$$

In addition, there is iron combined effect of physical adsorption, electrochemical enrichment and iron coagulation, etc.

3 CONCLUSION

Fe^0-PRB has sustained good results on the degradation of 2, 4-DCP during the running 400 days. Lower the price of zero-valent iron systems has and this process has a higher cost performance.

However, due to the reaction column sealed, the steps of the conversion between the various reaction steps and reactants and the environment column can't monitor exactly.

Recommending use bioreactor column after the chemical reaction column in practical applications and put into and oxygen release agent to extend the PRB service life.

ACKNOWLEDGEMENTS

Funding for this research was provided through one of the Major Projects of National Science and Technology Projects: South-to-North Water Diversion Project.

REFERENCES

[1] LI Tai-you, Liu Qiong-yu, Xiao Fei. Experimental Study on Catalytic Reductive DE chlorination of 2, 4-Dichlorophennol Wastewater by Zero-valent Iron Systems [J]. Journal of Jianghan University (Natural Sciences), 2002, 19(3):27–30.

[2] Luo Qi-shi, Zhang Xi-hui, Wang Hui, Qian Yi. The migration and its mechanism of phenolic contaminants in soil by electro kinetics [j]. China Environmental Science, 2004, 24(2):134–138.

[3] Qiu, Jin-an, Zhang Cheng-bo, Li Hong-yi, etc. Application and research progress of PRB in remediation of polluted groundwater [J]. Guangdong Agricultural Sciences, 2011, 13:144–152.

[4] Yang Wei, Wang Li-dong, YANG Jun-feng, etc.. Experimental Study on Using PRB to Treat Groundwater Polluted by PCBs and Heavy Metal [J]. Environmental Sciences, 2007, 33(2):15–18.

[5] Liang Jian-kui, Zhang Yong-xiang, Wang Ran, etc. Experimental Study on influence factors DE chlorination of 2, 4-Dichlorophennol wastewater by Zero-valent Iron Systems [J]. Journal of Ecology and Rural Environment. 2011, 27(6):64–67.

Frontiers of Energy and Environmental Engineering – Sung, Kao & Chen (eds)
© 2013 Taylor & Francis Group, London, ISBN 978-0-415-66159-1

Distribution and source identification of PM_{10}-bound polycyclic aromatic hydrocarbons in urban air of Anshan, China

L.N. Pu
College of Environmental Science and Engineering, Kunming University of Science and Technology, Kunming, Yunnan, China

X.Y. Han
College of Architecture and Engineering, Kunming University of Science and Technology, Kunming, Yunnan, China

J.W. Shi & P. Ning
College of Environmental Science and Engineering, Kunming University of Science and Technology, Kunming, Yunnan, China

ABSTRACT: The concentrations of 17 selected PAHs in PM_{10} were quantified at three sites in city center of Anshan in 2008 year. Total concentration of 17 selected PAHs was 117.10 ng/m^3 in average, and the dominant PAHs were FL, PA, Ant, Pyr, IND, BaA, CHR and BbF accounting for 80.1% of 17 selected PAHs. Higher PAHs concentrations during heating period and lower concentrations during no-heating period were observed at the three sampling sites, which may be caused by the stronger emissions from stationary combustion sources in heating period and the quicker air dispersion, washout effects, photo-degradation and higher percentage in the air in vapor phase in no-heating period. The contributions from potential sources to PAHs in PM_{10} were estimated by the Principal Component Analysis (PCA). In whole sampling period, vehicles emission, oil refinery, iron and steel, and coal combustion were found to the predominant contributors of PM_{10}-bound PAHs, followed by coke oven and wood combustion.

Keywords: PAHs; PM_{10}; PCA; sources

1 INTRODUCTION

PAHs are ubiquitous constituents on particulate matter that mainly originate from incomplete combustion of organic matter such as petroleum and coal (Liu et al., 2009; Guo et al., 2003; Khalili et al., 1995). In urban ambient air, PAHs are almost entirely emitted from anthropogenic sources (Shi et al., 2010; Mantis et al., 2005; Caricchia et al., 1999). In this study we present the characterization of PM_{10}-bound PAHs in urban area of Anshan. The possible sources of PM_{10}-bound PAHs in Anshan are also discussed based on PCA.

2 MATERIALS AND METHODS

Anshan (40°27'–41°34'N, 122°10'–123°13'E), a typical iron and steel city, is located 90 km south of Shenyang, with an urban area of 624.3 km^2 and a population of 1.3 million. Anshan is the country's vital reserve area of three mineral resources–iron ore,

magnesite and alum mine. Metallurgy and mining are the leading industries.

Three sites were chosen as the sampling sites according to their different function in the city: Tai Yang Castle area (TYC) was chosen as a typical commercial/residential site; ShenGou Shrine area (SGS) was chosen as a typical residential site; Tiexi station area (TXZ) was chosen as a typical traffic site. In winter, there are large amount of coal combustion boilers for domestic heating around the three sites. Being one of the largest iron and steel companies, Anshan steel owns one third area of the whole Anshan city, and is a major emission source of PM_{10}.

Ambient PM_{10} samples of 20 days were collected in 2008. To study the seasonal variation, the samples of 5 days were collected for each season, using a standard medium-volume PM_{10} sampler at a flow rate of 100 L/min. A gas chromatography coupled to mass spectrometry (trace 2000GC-MS, Thermo Finnigan, USA) was used for determining PAHs with Selected Ion Monitoring (SIM), and the analysis process and quality control descriptions on this study are in publications by Shi et al. (2010). In this study the 17 PAHs were analyzed including naphthalene (NaP), acenaphthylene (AcPy), acenaphthene (Acp), fluorene (FLu), phenanthrene (PA), anthracene (Ant), fluoranthene (FL), pyrene (Pyr), benzo[a]anthracene (BaA), chrysene (CHR), benzo[b]fluoranthene (BbF), benzo[k]fluoranthene (BkF), benzo[a]pyrene (BaP), indeno[1,2,3-cd]pyrene (IND), dibenz[a,h]anthracene (DBA), benzo[ghi]perylene (BghiP) and Coronene (COR).

3 RESULTS AND DISCUSSION

3.1 Distribution of PAHs in different rings

Average total concentration of 17 selected PAHs was 117.10 ng/m³, and the dominant PAHs were FL, PA, Ant, Pyr, IND, BaA, CHR and BbF accounting for 80.1% of 17 selected PAHs. Average concentration of individual PAHs in the whole sampling time varied from 0.32 (AcP) to 19.29 (FL) ng/m³.

Examined PAHs could be classified according to their number of aromatic rings as follows: 2-ring including Nap; 3-ring including Acpy, Acp, Flu, PA and Ant; 4-ring including FL, Pyr, BaA and CHR; 5-ring including BbF, BkF and BaP, 6-rings including IND, DBA, and BghiP; 7-rings including COR. They can be further classified into lower molecular weight (LMW, 2-and 3-rings PAHs), middle molecular weight (MMW, 4-rings PAHs), and higher molecular weight (HMW, 5-,6-and 7-rings PAHs). LMW PAHs can be tracers for wood combustion (Khalili et al., 1995) or industrial combustion of oil (Park et al., 2002). MMW PAHs such as FL, Pyr, BaA and CHR are usually associated with coal combustion and can be identified from this source (Khalili et al., 1995). HMW PAHs such as BbF, BkF, BghiP, IND, and COR may be associated with vehicles emission, and can be regards as tracers for this source (Marr et al., 1999).

Figure 1 shows ring number distribution of PAHs for PM₁₀ in Anshan in spring, summer, autumn, and winter. HMW PAHs were found to be abundant in summer and autumn, reflecting the strong contribution from vehicles emission, coke and iron and steel industries emission, while MMW PAHs in PM₁₀ had higher concentrations in winter than other seasons, owing to the higher emission from coal combustion sources for house heating. Additionally, MMW PAHs are semi-volatile organic compounds and distribute higher in particle phase when air temperature is low. LMW PAHs were the relatively abundant in winter, which may be caused by wood combustion for domestic

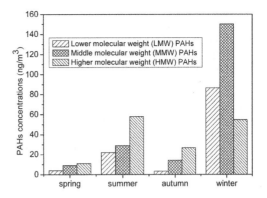

Figure 1. Ring number distribution of PAHs for PM₁₀ in Anshan.

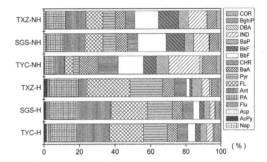

Figure 2. Mean PAH profiles of ambient PM₁₀ in the central urban of Anshan.

cooking or heating. Therefore, the concentration proportion variation of LMW, MMW and HMW PAHs in different seasons can be used to reflect the variation in the categories of PAHs sources in these seasons.

3.2 Variations of PAH profiles

The sampling year was divided in two periods, heating period (H, 15 November–31 March) and no-heating period (NH, 1 April–14 November). In Anshan, there is very significant seasonal variation in PAHs concentrations. Higher PAHs concentrations during heating period and lower concentrations during no-heating period were observed at the three sampling sites, which may be caused by the stronger emissions from stationary combustion sources in heating period and the quicker air dispersion, washout effects, photo-degradation and higher percentage in the air in vapor phase in no-heating period.

Figure 2 shows the mean PAH profiles (percent contribution of each PAH compound to ΣPAH) for the three sampling sites in Anshan.

A relatively similarity was existed in the PAH profiles in both heating and no-heating period, which reveals the sources of PAHs around these sampling sites may be similar. PA, Ant, Pyr and FL were abundant compounds at these sampling sites in heating, indicating coal combustion emission for house heating, coke and iron and steel industries were the major source of these PAHs. In no-heating period PA, Ant, Pyr and FL also were abundant, and BbF, BkF were high contribution to ∑PAH. This phenomenon revealed the importance of iron and steel industries and coke emission as PAH sources in the whole year, and vehicular emissions as a major PAH source in no-heating period.

3.3 *Principal component analysis (PCA)*

We used PCA to identify the emission sources of PAHs in no-heating period. The principle of PCA is to transform an original set of variables into a smaller set of linear combinations that account for most of the variance of the original set. The factor loadings which are obtained for each variable within the factors generated by the PCA, are a type of correlation coefficient, and higher values are therefore associated with greater significance. Due to non-detectable or too low concentrations of NaP, AcPy and AcP in PM_{10}, only 14 PAH species measured in PM_{10} in no-heating period were subjected to PCA with varimax rotation. Table 1 provides the results of PCA for 14 PAHs in no-heating.

Table 1. PCA analysis of PAHs in PM_{10} in central urban of Anshan.

PAHs variance (%)	Factor 1 31.23	Factor 2 22.47	Factor 3 19.48	Factor 4 16.31
Flu		0.94		
PA		0.93		
Ant		0.93		
FL			0.84	
Pyr			0.84	
BaA	0.82			
CHR	0.91			
BbF				0.80
BkF				0.91
BaP				
IND	0.82			
DBA				
BghiP	0.82			
COR	0.92			

Extraction Method: Principal Component Analysis.
Rotation Method: Varimax with Kaiser Normalization.
Number of factors: 4.
Factor loading ≥ 0.6 listed.

Factor 1: It explains 31.23% of the variance with a high load for BaA, CHR, IND, BghiP and COR BghiP and COR were typical tracers for gasoline exhaust (Li and Kamens, 1993; Rogge, et al., 1993). IND was also found in both diesel and gasoline engine emissions.

Factor 2: It explains 22.47% of the variance with a high load for FLu, PA and Ant. Oil refinery emissions were found to be enriched by more volatile PAH species (Harrison, et al., 1996). Biomass burning is another potential source of this factor, because ANT and PA were sometimes used as indicators for wood combustion sources (Li and Kamens, 1993). In the rural areas, burning of straw and firewood is a common practice for cooking and heating, which contributes to more or less PAH pollution in urban atmosphere. The two sources could not be distinguished in this apportionment.

Factor 3: It explains 19.48% of the variance with a high load for FL and Pyr, which were identified as fingerprints for coal combustion (Zuo et al., 2007).

Factor 4: It explains 16.31% of the variance with a high load for BbF and BkF, which were found in large amount in coke oven emissions (Zhu et al., 2002). Coking is an important industrial sector as well as PAH source in the northeast region.

According to the PCA results, gasoline/diesel emissions, oil refinery, wood combustion, coal combustion, and coke oven emission are the major sources of PAHs in Anshan in no-heating period.

4 CONCLUSIONS

PAHs in PM_{10} in Anshan had been determined at three sampling sites for one year. PAH levels were the highest at heavy industrial area in heating period and the lowest at a residential area in no-heating period. Obviously, the PAHs levels in PM_{10} were closely associated with source strength and meteorological conditions. Higher heating period PAHs concentrations and lower no-heating period concentrations were observed in all sampling sites. Based on the PCA analysis, gasoline/diesel emissions, oil refinery, wood combustion, coal combustion, and coke oven emission were found to be the predominant source to PM_{10}-bound PAHs in Anshan in no-heating period. Other sources such as emission from wood combustion also play an important role in PAHs production. In heating period, coal combustion was regards as the more predominant source to PM_{10}-bound PAHs than that in no-heating period.

ACKNOWLEDGEMENTS

This study was funded by the China National Natural Science Foundation program (Grants 21207055) and the Special Environmental Research Fund for Public Welfare, No. 200709013). Shi Jianwu is the corresponding author of this paper, his e-mail address: shijianwu2000@sina.com.

REFERENCES

Caricchia, A.M., Chiavarini, S., Pezza, M. (1999). Polycyclic aromatic hydrocarbons in the urban atmospheric particulate matter in the city of Naples (Italy). Atmospheric Environment. 33: 3731–3738.

Guo, H., Lee, S.C., Ho, K.F. (2003). Particle-associated polycyclic aromatic hydrocarbons in urban air of Hong Kong. Atmospheric Environment. 37: 5307–5317.

Harrison, R.M., Smith, D.J.T., Luhana, L. (1996). Source apportionment of atmospheric polycyclic aromatic hydrocarbons collected from an urban location in Birmingham U.K. Environmental Science and Technology. 30: 825–832.

Khalili, N.R., Scheff, P.A., Holsen, T.M. (1995). PAH source fingerprints for coke ovens, diesel and gasoline engines, highway tunnels, and wood combustion emissions. Atmospheric Environment. 4: 533–542.

Li, C.K., Kamens, R.M., 1993. The use of polycyclic aromatic hydrocarbons as sources signatures in receptor modeling. Atmos. Environ. 27 A, 523–532.

Liu, W.X., Dou, H., Wei, Z.C., Chang, B., Qiu, W.X., Liu, Y., Tao, S. (2009). Emission characteristics of polycyclic aromatic hydrocarbons from combustion of different residential coals in North China. Science of the Total Environment 407(4): 1436–1446.

Mantis, J., Chaloulakou, A., Samara, C. (2005). PM_{10}-bound polycyclic aromatic hydrocarbons (PAHs) in the Greater Area of Athens. Greece. Chemosphere. 59: 593–604.

Marr, L.C., Kirchstetter, T.W., Harley, R.A. (1999). Characterization of polycyclic aromatic hydrocarbons in motor vehicle fuels and exhaust emissions. Environmental Science and Technology. 33: 3091–3099.

Rogge, W.F., Hildemann, L.M., Mazurek, M.A., Cass, G.R., Simoneit, B.R.T., 1993. Sources of fine organic aerosol. 2. Non-catalyst 479 and catalyst-equipped automobiles and heavy duty diesel trucks. Environ. Sci. Technol. 27, 636–651.

Shi Jianwu, Peng Yue, Li Weifang, et al (2010). Characterization and Source Identification of PM_{10}-bound Polycyclic Aromatic Hydrocarbons in Urban Air of Tianjin, China. Aerosol and Air Quality Research. Vol.10, No.5, 507–518.

Zhu, X., Liu, W., Lu, Y., Zhu, T., 2002. A comparison of PAHs source profiles of domestic coal combustion, coke plant and petroleum asphalt industry. Acta Scientiae Circumstantiae, 22, 199–203. (in Chinese).

Zuo, Q., Duan, Y., Yang, Y., Wang, X., Tao, S., 2007. Source apportionment of polycyclic aromatic hydrocarbons in surface soil in Tianjin, China. Environ. Pollut. 147, 303–310.

Frontiers of Energy and Environmental Engineering – Sung, Kao & Chen (eds)
© 2013 Taylor & Francis Group, London, ISBN 978-0-415-66159-1

Research on control strategies in solar aided coal-fired power plant

R.R. Zhai, Y. Zhu, K.Y. Tan, D.G. Chen & Y.P. Yang
North China Electric Power University, Changping District, Beijing, China

ABSTRACT: Solar aided coal-fired power generating system couples solar thermal power with coal-fired power generating system for substituting part of coal consumption. This paper deals with the quantitative relationship between light and heat during the process of solar radiation and transmission of energy by using the improved mathematical model of solar integrated system. After analyzing the solar integrated system and the coal-fired power generating thermodynamic system, the analytical modes of coal-fired unit and solar aided power generating system have been explored with data validation of typical units. Considered with the constraints of external factors, the performances of various solar aided coal-fired power generating systems have been presented on the system and unit devices level.

Keywords: solar power; coal-fired power plant; aided power generation; control strategy

1 INTRODUCTION

China is a major producer and a great consumer of coal which accounts for about 70% of primary energy consumption. And over half of the coal consumption was taken by coal-fired power plants (Yang, Y.P. et al. 2010). For power generation industry, measures of energy saving can be classified into two ways: structural and technological energy conservation. So far China has established plenty of units with higher parameters and larger capacity. Solar energy can be used as the source to supply extra energy to the system (Cui, Y.H. 2009).

In the process of solar aided coal-fired power generating, the control strategy of heat flow is a significant problem which is still unsolved. Energy gain maximization through mass flow rate control in stand-alone solar thermal power generation system have been reported by Kovarik and Horel (Kovarik, M. et al. 1976, Horel, J.D. et al. 1978). Hollands dealt with water flow rate optimization for a closed loop system (Hollands. K.G.T. et al. 1992). Based on the exergetic analysis, control strategies of various flow amounts were proposed by Bejan (Bejan, A. 1982). Simulation and analysis on improving control of the thermal energy storage system outlet temperature was explored by Kody (Powell, K.M. et al. 2012). Techniques of controlling the solar collector outlet temperature by varying the heat transfer fluid flow rate through the collector field were concluded by Camacho (Camacho, E.F. et al. 2007). Moreover, there are lots of researches on control strategies of traditional coal-fired power plants. Two different ways of regulating superheated and reheated steam

temperatures were reported by Sanchez-Lopez (Sanchez-Lopez, A. et al. 2004).

The approaches of heat flow control can be classified into two kinds: outlet temperature control and mass flow rate control. This paper refers to heat flow control strategies of solar subsystem in solar aided coal-fired power generation system. And the specific operation strategy can be confirmed by analyzing the thermal power of working medium.

2 SYSTEM DESCRIPTIONS

The basic flow of solar aided coal-fired power plant is similar to the traditional coal-fired power plant, while the only difference is the additional parabolic trough collector with thermal storage system. The heat transfer fluid of solar thermal collector is heat transfer oil which heated boiler feedwater by new oil-water heat exchanger. Besides, HP FWH1, the most powerful device, need to be shut down, while the substituted steam will continue working in the turbine. Therefore, coal consumption will be reduced and the indirect solar thermal power generation will be realized. The solar contribution of the system lies in heating feedwater. The heated feedwater enters in to the boiler and then generates the high temperature and high pressure steam after going through the steam-water system. The outlet superheated steam will be transported to the turbine for power production, and the solar thermal energy will finally transform into electrical energy. According to the principle of system flow, the system can be divided

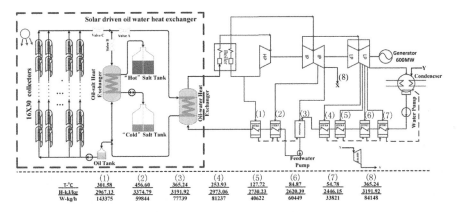

T-°C	(1) 301.58	(2) 456.60	(3) 365.24	(4) 253.93	(5) 127.72	(6) 84.87	(7) 54.78	(8) 365.24
H-kJ/kg	2967.13	3374.79	3191.92	2973.06	2730.23	2620.39	2446.15	3191.92
W-kg/h	143375	59844	77739	81237	40622	60449	33821	84148

Figure 1.   The layout of the solar aided coal-fired power plant.

into two parts: solar part and coal-fired power generation part. The solar part is shown in dotted box of Figure 1 and the outside is the coal-fired power generation part. The structure of oil-water heat exchanger system driven by solar energy is shown in dotted box of Figure 1. This system includes three parts: thermal concentration circulating system, thermal storage circulating system and heat exchanger circulating system. Thermal concentration circulating system is constituted by 16 × 30 collectors. Thermal storage system contains "Hot" Salt Tank, "Cold" Salt Tank and oil-salt heat exchanger (SDOSHE). Heat exchanger system is composed of oil-water heat exchanger (SDOWHE). The energy of heat transfer oil, molten salt and feedwater will be exchanged in the corresponding circulating system.

3   CONTROL STRATEGY OF SOLAR AIDED COAL-FIRED POWER PLANTS WITH THERMAL STORAGE

Modeling of the system is needed for exploring the control strategy of the system showed in Figure 1. Firstly, the models of solar part and coal-fired power generation part will be built. And the control strategy modeling will be carried out.

Since the influences of weather, season and other natural conditions, solar radiation intensity is extremely unsteady. For the normal operation of solar aided coal-fired power plants, the thermal storage system is quite necessary. According to the various radiation intensity, the operation strategy can be classified in three kinds based on the working conditions: radiation intensity is higher than the critical value, radiation intensity is lower than the critical value, radiation intensity equals to zero (Stuetzle, T. et al. 2009).

3.1   Radiation intensity is higher than the critical value

When sunshine is adequate, which means solar radiation intensity is higher than the critical value of direct insolation intensity ($G > G_m$), the transferred heat by absorber with heat transfer fluid is fairly enough for the demand of oil-water heat exchange after collector absorbed heat from solar radiation ($Q_1 > Q_2$). The convection transferred heat between absorber and fluid can be divided into two kinds: one part for the demand of heat exchanger, the other for thermal storage tank.

Flux of valve A is given by:

$$q_{m1} = \frac{Q_2}{c\,(T_{HTF} - T_1)} \tag{1}$$

The flow direction is from collector to heat exchanger.
Flux of valve B is given by:

$$q_{m2} = q_{m3} - q_{m1} \tag{2}$$

The flow direction is from collector to thermal storage tank.

3.2   Radiation intensity is lower than the critical value

When solar radiation is little weak, which means radiation intensity is lower than the critical value of direct insolation intensity ($G < G_m$), the heat absorbed by collector is not enough for the heat demand of oil-water heat exchanger ($Q_1 < Q_2$).

When there is $Q_{sto} + Q_1 > Q_2$, thermal storage tank and collector will provide heat for heat exchanger at the same time. At this time, the thermal storage tank is in "heat release" state.

Flux of valve B is given by:

$$q_{m2} = \frac{Q_2 - Q_1}{c(T_{sto} - T_1)} \tag{3}$$

The flow direction is from thermal storage tank to heat exchanger.

Flux of valve A is given by:

$$q_{m1} = q_{m2} + q_{m3} \tag{4}$$

The flow direction is from thermal storage tank to heat exchanger.

When there is $Q_{sto} + Q_1 < Q_2$, the summation of the heat stored by thermal storage tank and the heat absorbed by collector will not be enough for the heat demand of oil-water heat exchanger. At this moment, we need to shut down valve A and open valve B. Then the heat absorbed by collector from solar radiation will store in thermal storage tank. And the state of heat storage tank, at this time, can be called "heat storage".

Flux of valve B is given by:

$$q_{m2} = q_{m3} \tag{5}$$

The flow direction is from collector to thermal storage tank.

Flux of valve A equals to zero.

3.3 *Radiation intensity equals to zero*

At this moment, collector is unable to provide heat for heat exchanger.

If $Q_{sto} > Q_2$, it's needed to turn off valve C and keep the flux of valve A and valve B equal, which means:

$$q_{m1} = q_{m2} = \frac{Q_4 - Q_3}{c(T_{sto} - T_6)} \tag{6}$$

If $Q_{sto} < Q_2$, it's needed to turn off all the valves and cut off the oil-water heat exchanger driven by solar energy. Then the steam extraction should be recovered.

4 CASE STUDY

4.1 *System parameters*

A 600 MW supercritical coal-fired power generation unit has been taken as the research case in this part. After establishing the simulation model of solar aided coal-fired power generation system with thermal storage system, the control strategy can be applied in this system. The major param-

eters of this 600 MW power generation unit are as follows: Feedwater flux of 1645.15 t/h; Condenser pressure of 4.9 kPa; Feedwater temperature of 272.3 °C; Coal consumption rate of 257.4 g/kWh.

The major parameters of solar part in the system are as follows: Input solar radiation of 925 W/m2; Thermal concentration area of 235 m2 Inlet temperature of heat transfer oil of 250 °C; Outlet temperature of heat transfer oil of 328 °C. It's assumed that the storage capacity of thermal storage tank is large enough.

The solar radiation intensity of a typical day has been considered as reference data for analysis. And the distribution of radiation intensity is shown in Figure 2. According to Figure 2, it's found that the radiation intensity increases after 5:30 while there are some slight fluctuations from 7:30 to 9:30, and then it continue increasing until reaching the maximum at 15:30.

4.2 *Result of calculation and process of control*

After simulation, the variation of flux of each valve with hours is shown in Figure 3. Assume that the values of flux flowing through valve B will be

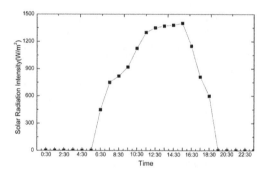

Figure 2. Variation of solar radiation intensity of a typical day.

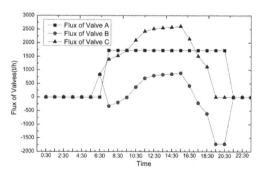

Figure 3. Variation of each flux of electric valve with hours.

positive and others will be negative. Moreover, the losses of process and heat transfer will be ignored in the calculation.

Depending on Figure 3, it's found that

1. Before the sun rises (0:30–5:30), all the valves should be turned off and coal-fired part maintains previous operation.
2. During the weak radiation morning (5:30–6:30), valve A should be turned off and valve B and valve C should be opened. At this moment, the absorbed heat by collector is totally stored in thermal tank and coal-fired part still maintains previous operation.
3. Since radiation intensity increasing gradually (6:30–9:30), the summation of heat absorbed and stored is enough for the demand of oil-water heat exchanger while the heat absorbed by collector is not enough. Thus all the valves should be opened and the heat absorbed and stored start to provide for heating feedwater. And the value of valveB flux will be negative, which means releasing heat.
4. With the radiation intensity increasing more and more (9:30–16:30), the heat absorbed by collector is fairly enough for the demand of oil-water heat exchanger. So, all the valves should keep open. The heat absorbed will provide for heat exchanger and storage.
5. On the late afternoon (16:30–19:30), the radiation intensity starts to decrease. At this time, storage tank and collector provide heat for oil-heat exchanger together. Therefore, all the valves will still be open while the flux value of valve B turns to be negative, which means heat releasing.
6. After the sun went down (19:30–20:30), thermal storage tank will provide heat for oil-water heat exchanger alone. For this reason, valve A and valve B will keep open while valve C should be turned off, which means heat releasing.
7. When the heat stored in thermal storage tank is insufficient (20:30–23:30), all the valves should be shut down and the coal-fired power generation part will switch back to the previous operation.

Figure 3 shows that the flux of valve C and the heat provided for feed water from solar part maintain stability since the addition of thermal storage tank and the application of reasonable control strategy. In this way, the operation of power generation system will keep steady and the service lives of devices will also be extended.

5 CONCLUSIONS

According to the data of operating coal-fired power plant, the control strategy of solar aided coal-fired power generation system with thermal storage tank has been explored in this paper. Considering the particularity of solar energy, the basic research of critical issues on the solar aided coal-fired power generation system field has been taken as the major point. Therefore, the control strategy of the integration of coal-fired power generation system with solar energy has been presented. At the system level, the research method combined with theory study, numerical simulation and computer simulation has been applied for exploring control strategy and thermodynamic performance.

ACKNOWLEDGEMENT

The research work is supported by China National Natural Science Foundation (No. 51106048) and National High-tech R&D Program of China (863 Program) (2012AA050604).

NOMENCLATURE

G	direct solar radiation intensity
G_m	the minimum solar radiation intensity
Q_1	convection heat transfer of collector tube with heat transfer fluid
Q_2	heat needed by the boiler feedwater
Q_{sto}	the heat storage by thermal storage tank
q_{m1}	the flow of Valve A
q_{m2}	the flow of Valve B
q_{m3}	the flow of Valve C
T_{HTF}	temperature of heat transfer fluid
T_1	export oil temperature of heat exchanger
T_{sto}	the temperature in thermal storage tank
c	the specific heat capacity of heat transfer fluid.

REFERENCES

Bejan, A. 1982. Extraction of exergy from solar collectors under time varying conditions. Int J Heat Fluid Flow 3:67–72.

Cui, Y.H. 2009. Research on Coupling Mechanism and Thermal Character of the Solar Supported Coal-Fired Electric Generation System [D]. PhD Thesis. Beijin:North China Electric Power University.

Camacho, E.F., Rubio, F.R., Berenguel, M., Valenzuela, L. 2007. A survey on control schemes for distributed solar collector fields. Part II: Advanced Control Approaches. Solar Energy 81:1252–1272.

Horel, J.D., Winter, F. De. 1978. Investigations of methods to transfer heat from solar liquid-heating collectors to heat storage tanks. Final report on US Department of Energy Contract E(04)1238, Altas Corporation, Santa Cruz CA; April 1978.

Hollands, K.G.T., Brunger, A.P. 1992. Optimum flow rates in solar water heating systems with a counter flow exchanger. Solar Energy 48:15–19.

Kovarik, M., Lesse, P.F. 1976. Optimal control of flow in low temperature solar heat collectors. Solar Energy 17:431–435.

Powell, K.M., Edgar, T.F. 2012. Modeling and control of a solar thermal power plant with thermal energy storage. Chemical Engineering Science 71:138–145.

Sanchez-Lopez, A., Arroyo-Figueroa, G., Villavicencio-Ramirez, A. 2004. Advanced control algorithms for steam temperature regulation of thermal power plants. International Journal of Electrical power & Energy Systems 26:779–785.

Stuetzle, T., Blair, N., Mitchell, J.W., Beckman, W.A. 2004. Automatic control of a 30MWe SEGS VI parabolic trough plant. Solar Energy 76:187–193.

Yang, Y.P., Guo, X.Y., Wang, N.L. 2010. Power generation from pulverized coal in China. Energy 35(11): 4336–4348.

Frontiers of Energy and Environmental Engineering – Sung, Kao & Chen (eds)
© 2013 Taylor & Francis Group, London, ISBN 978-0-415-66159-1

Experimental study on the relationship between shear strength of red strata sandstones undergoing different times of dry-wet-cycle procedures and their quality

Z.H. Zhang & Q.C. Sun
Key Laboratory of Geological Hazards on Three Gorges Reservoir Area, China Three Gorges University, Ministry of Education, Yichang, P.R. China

J.J. Xue
Hubei Provincial Key Laboratory of Hazards Prevention and Alleviation, China Three Gorges University, Yichang, P.R. China

ABSTRACT: Some rock slides have taken place in Three Gorges Reservoir Area, China. Many Researches on the causes of the slides shows that the more sensitive of rock shear strength to the reservoir water level fluctuation, the easier to fail of the rock slopes. In order to study the relationship between shear strength parameters of the red strata sandstones undergoing different times of dry-wet-cycle procedures and their quality, triaxial compression tests were taken to obtain the cohesions and friction angles of the sandstones undergoing different times of dry-wet-cycle procedures, and longitudinal wave velocity test were adopted to test the quality of the sandstones undergoing different times of dry-wet-cycle procedures. The research results can give references to judge the variation of the shear strength parameters of the red strata sandstones undergoing different times of dry-wet-cycle procedures by longitudinal wave velocity test.

Keywords: red strata sandstone; dry-wet-cycle procedures; shear strength; quality

1 INSTRUCTIONS

The total length of bank slopes in Three Gorges Reservoir Area is about 5300 km. In order to utilize the Three Gorges Dam safety, much research work has been done to study the stability of bank slope in Three Gorges Reservoir (Yin et al, 2004). After the impoundment of Three Gorges Reservoir, one large scale rock slide has taken place in Three Gorges Reservoir Area, China (Chen et al, 2003). Many Researches on the causes of the slides shows that the more sensitive of rock shear strength to the reservoir water level fluctuation, the easier to fail of the rock slopes (Li et al, 2003). For evaluate the stabilities of rock bank slopes under the conditions of water level fluctuation, it is necessary to distinguish the rock shear strength difference when the rock undergo different times of dry-wet-cycle procedures. In order to study the relationship between shear strength parameters of the red strata sandstones undergoing different times of dry-wet-cycle procedures and their quality, triaxial compression tests were taken to obtain the cohesions and friction angles of the sandstones undergoing different times of dry-wet-cycle procedures, and

longitudinal wave velocity test were adopted to test the quality of the sandstones undergoing different times of dry-wet-cycle procedures.

2 PROPERTIES OF SANDSTONE SAMPLES

2.1 *Mineral composition analysis of the samples*

The test sandstone samples of red strata sandstones is shown as Figure 1. The geometric shape of each test sandstone sample is cylinder with 50 mm diameter and 100 mm height.

The mineral identification results of the red strata sandstones were shown in Table 1.

2.2 *Micro-structural characteristics of the sandstone samples*

The micro-structural pictures of the red strata sandstones are obtained by microscope. The Orthogonal polarized pictures of the red strata sandstone is shown in Figure 2.

The mineral signed by number 1 in picture (a) of Figure 2 is feldspar, mineral signed by number 2 is

Figure 1. Rock samples of the red strata sandstones.

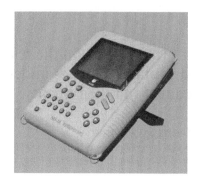

Figure 3. NM-4A nonmetal ultrasonic testing analyzer.

Table 1. Percentage of mineral compositions of the red strata sandstones.

Type	Quartz	Feldspar	Muscovite	Calcite
Percentage	73%	15%	2%	10%

Table 2. Longitudinal wave velocities of the sandstone samples undergoing different times of dry-wet-cycle procedures (m/s).

Times	0	1	2	4	6	8	10
Velocities	1971	1641	1541	1469	1484	1521	1533

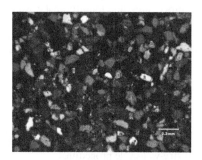

(a) crossed polarizer

(b) mono polarizer

Figure 2. Micro-structural pictures of the samples.

component of sample is quartz with small granule and the minor is feldspar.

3 LONGITUDINAL WAVE VELOCITY TESTS OF THE SANDSTONE SAMPLES

After every time of dry-wet-cycle procedure, the sandstone samples were taken to test longitudinal wave velocity by NM-4A Nonmetal Ultrasonic Testing Analyzer (shown in Fig. 3). The test results are shown in Table 2.

The data in Table 2 shows that the red strata sandstone samples in dry state have high longitudinal wave velocity. After once dry-wet-cycle procedure, the longitudinal wave velocity of the sandstone had a great attenuation. The rate of descent is about 16.7%. After two times dry-wet-cycle procedures, the longitudinal wave velocity declined slightly about 6.05%. It decreased 4.7% smoothly after four times dry-wet-cycle procedures, and rebounded about 1.1% after six times dry-wet-cycle procedures. After ten times dry-wet-cycle procedures, the longitudinal wave velocity continued to fall with the descent rate of 0.7%.

4 TRIAXIAL COMPRESSIVE TEST OF THE SAMPLES

In order to obtain the shear strength parameters (cohesion and friction angle) of the red strata sandstone s undergoing different times of dry-wet-cycle

quartz, mineral signed by number 3 is zircon, and mineral signed by number 4 is muscovite. The mineral signed by number 1 in picture (b) of Figure 2 is grain crumbs and mineral signed by number 2 is clay cement. The Figure 2 demonstrates that main procedures, the triaxial compressive test was carried out by RMT-150C Rock Mechanical Test Machine (shown in Fig. 4).

The sandstone samples were put into dryer (shown in Fig. 5) to dry for 24 hours at constant temperature of 45°C, and then were taken out from the dryer and were put into the pure water to saturate at room temperature for 48 hours, which were called one time of dry-wet-cycle procedure. After the dry-wet-cycle procedure, the sandstone samples were taken to triaxial compression tests.

The shear strength parameters cohesion and friction angle (c and φ) of the sandstone samples were calculated by the following formulas, the dates are shown in Table 3 (Xu et al, 2007).

$$c = \frac{\sigma_c}{2\sqrt{K}} \tag{1}$$

$$\varphi = \tan^{-1}\left(\frac{K-1}{2\sqrt{K}}\right) \tag{2}$$

where, c = cohesion of sandstone; φ = friction angle of sandstone; φ_c = vertical intercept of the

Table 3. Shear strength parameters of the sandstone samples undergoing different times of dry-wet-cycle procedures under different confining pressures.

Times of dry-wet-cycle procedures	Confining pressure σ_3 (MPa)	Peak strength σ_1 (MPa)	Cohesion c (Mpa)	Friction angle φ (°)
0 time	0	37.087	7.31	43.77
	5	67.309		
	10	93.435		
	15	111.903		
1 time	0	22.450	4.84	41.91
	5	52.519		
	10	81.333		
	15	108.358		
4 times	0	22.450	4.66	41.40
	5	52.519		
	10	72.595		
	15	89.575		
8 times	0	15.146	4.73	40.76
	5	50.043		
	10	73.461		
	15	86.692		

Figure 4. RMT-150C rock mechanical test machine.

Figure 5. Dryer.

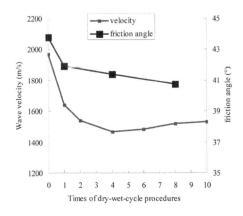

Figure 6. The curve of friction angle and longitudinal wave velocity vs. times of dry-wet-cycle procedures.

linear relationship between the first principal stress σ_{1f} and the third principal stress σ_{3f} and K = slope of the linear relationship between the first principal stress σ_{1f} and the third principal stress σ_{3f}.

The friction angles of the sandstone samples were shown in Figure 6. For Figure 6, the friction angles decreased gradually with the increase of the times of dry-wet-cycle procedures. After the first dry-wet-cycle procedure, the descent rate of friction angles is about 4.24%. It got to 1.21% after four times of dry-wet-cycle procedures, and to1.54% after eight times.

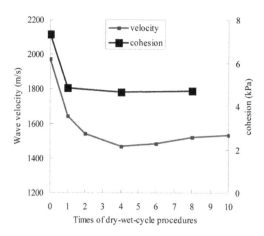

Figure 7. The curve of cohesion and longitudinal wave velocity vs. times of dry-wet-cycle procedures.

The cohesions of the sandstone samples were shown in Figure 7. From Figure 7, the cohesions of the sandstone samples decreased substantially after undergoing once dry-wet-cycle procedure, the descent rate of cohesions is about 33.78%, and after four times of dry-wet-cycle procedures, the cohesions nearly kept constant.

Obviously, the cohesions and the friction angles of the sandstone samples decreased with the increase of times of dry-wet-cycle procedures.

5 THE RELATIONSHIP BETWEEN SHEAR STRENGTH AND WAVE VELOCITY

Ultrasonic is one of the material damage evaluation indexes (You et al, 2008). Longitudinal wave velocity test can make an initial judgment for the rock mechanics characteristics. The curves (Figs. 5 and 6) demonstrated that with the increase of the times of dry-wet-cycle procedures, the shear strength parameters of the red strata sandstone had an overall trend of decline as well as the longitudinal wave velocity of the samples.

6 CONCLUSION

In this paper, triaxial compression tests and wave velocity tests were taken to obtain shear strength parameters and longitudinal wave velocity of the red strata sandstone samples undergoing different times of dry-wet-cycle procedures. Based on the test results, the following conclusions can be drawn.

1. The cohesion and friction angle of the sandstones decreased with the increase of times of dry-wet-cycle procedures, which indicates that the shear strength of the sandstones undergoing different times of dry-wet-cycle procedures has a downward trend.
2. The longitudinal wave velocity of the sandstones decreased with the increase of times of dry-wet-cycle procedures as well, thus we can make a preliminary judgment that dry-wet-cycle procedures deteriorate the quality of the sandstones.
3. Longitudinal ave velocity and triaxial shear strength of red strata sandstones undergoing different times of dry-wet-cycle procedures are well relevant, by which, the variable tendency of the shear strength parameters of the red strata sandstones undergoing different times of dry-wet-cycle procedures can be estimated by longitudinal wave velocity test.

ACKNOWLEDGEMENTS

This work was financially supported by the National Natural Science Foundation of China (No. 50909053) and the Third Stage Foundation of Geological Hazards Control in Three Gorges Reservoir Area (No.SXKY3-2-2-200903).

REFERENCES

B.X. Li, T.D. Miao. 2003. Strength controlling forecast method of critical landslide along red-soft-mudstone layer. *Chinese Journal of Rock Mechanics and Engineering.* 22(s2):2703–2706.

M.Q You, C.D. Su, X.S. Li. 2008.Study on relationship between mechanical properties and longitudinal wave velocities for damaged rock samples. *Chinese Jounal of Rock Mechanics and Engineering.* 27(3):258–267.

S. Qin, C.Q. Wang. 2007. *The Basis of Mineralogy.* Beijing: Peking University Press.

Y.B. Chen, C.H. Wang, X.Y. Fan. 2003.The characteristics and genetic analysis of Qianjiang Ping landslide in Hubei province. *Journal of Mountain Science.* 21(5):633–634.

Y.P. Yin, R.L. Hu. 2004. The disintegration features of the red mud in the Three Gorges. *Journal of Engineering Geology.* 12(2):124–135.

Y.T, Xu C, Wang BX, Zhang L and Liao GH. 2007. The cohesion strength and the friction angle in rock-soil triaxial tests. *China Mining Magazine.* 16: 104–107.

Frontiers of Energy and Environmental Engineering – Sung, Kao & Chen (eds)
© *2013 Taylor & Francis Group, London, ISBN 978-0-415-66159-1*

Research on the characteristics of exhaust gas energy flow for piston engine

F.Z. Ji
School of Transportation Science and Engineering, Beihang University, Beijing, China

F.R. Du
School of Jet Propulsion, Beihang University, Beijing, China

X.B. Zhang
College of Aeronautical Engineering, Civil Aviation University of China, Tianjin, China

ABSTRACT: A GT-power model was founded and calibrated using the cylinder data experimented aiming at a two-stroke piston engine. The exhaust parameters and energy was calculated utilizing the calibrated model. The exhaust energy was classified into three kinds of pressure energy, kinetic energy and thermal energy. The change rules of exhaust energy were studied. The percentage of different form of exhaust energy was compared. The calculated results show that energy increases with the raising of speed and load, the impact of speed is greater than one of load; thermal energy is the major form, it occupies more than 90% at full speed range, 95% above at high speed operating mode; the pressure energy takes second place, while the kinetic energy is least; the exhaust energy has a greater recovery potential at high speed and heavy load running conditions. These results provide a data reference for recovering and utilizing the exhaust energy.

Keywords: piston engine; two stroke; exhaust gas; energy flow; energy recovery

1 INTRODUCTION

The main problems are energy loss and exhaust pollution for piston engine now (Liu, C.Z. et al. 2012). The energy flow distribution was studied by Taymaz I et al from the perspective of energy flow by testing. The research results indicate that, about 33% energy released from burning fuel was taken away by engine exhaust; 29% was carried off by cooling water and thermal radiation; and the lacking 40% energy was translated into the engine effective power (Taymaz, I. 2006, Kauranen, P. et al. 2010 & Xiong, Y. et al. 2006). The researching work mainly focuses on two aspects: energy recover methods and recovering efficiency. The research on the characteristic of exhaust energy flow was seldom reported (Vaja, I. & Gambarotta, A. 2010, Huang, K.D. et al. 2009 & Shi, J.M. 2012). Prof. Liu was studied the exhaust energy flow characteristic based on a vehicle four-stroke piston engine, and the results provide the reference data for recycling the exhaust energy (Liu, J.P. et al. 2011).

The model of exhaust energy flow was established by combining experiment and simulation based on some two-stroke piston engine. The characteristics of energy flow were simulated.

The calculation results could lay a foundation for recovering exhaust energy further.

2 THE THEORETICAL BASIS OF ENERGY FLOW ANALYSIS

We know that by Bernoulli total energy equation, the total exhaust energy flow rate can be expressed as:

$$\dot{Q}_{ex} = \dot{Q}_k + \dot{Q}_p + \dot{Q}_t \qquad (1)$$

where, $\dot{Q}_{ex}$ stands for the total exhaust energy flow rate, J/s; $\dot{Q}_k$ is exhaust kinetic energy flow rate, J/s; $\dot{Q}_p$ is exhaust pressure energy flow rate, J/s; $\dot{Q}_t$ is exhaust thermal energy flow rate, J/s.

The exhaust energy is closely related to mass flow rate. The mass flow rate is describe as follows,

$$\dot{m}_{ex} = Au\rho \qquad (2)$$

where, $\dot{m}_{ex}$ is the exhaust mass flow rate, kg/s; μ stands for the flow coefficient; A is the vent-pipe section, m²; u is exhaust velocity, m/s; ρ is the exhaust density, kg/m³.

The approximate calculation formula of exhaust pressure flow rate is described as follows:

$$\dot{Q}_P = \dot{m}_{ex} \frac{k}{k-1} R_g T_{ex} \left[1 - \left(\frac{p_0}{p_{ex}} \right)^{\frac{k-1}{k}} \right] \qquad (3)$$

where, k is the specific heat capacity; R_g stands for the exhaust gas constant, J/kg·K; T_{ex} is the exhaust temperature, K; p_{ex} is exhaust pressure, Pa; p_0 is the atmospheric pressure at standard condition, Pa.

The exhaust process can be simplified a constant volume process. So the exhaust thermal energy flow rate is shown as follows,

$$\dot{Q}_t = \dot{m}_{ex} \int_{T_0}^{T_{ex}} C_{v,ex} dT \qquad (4)$$

where, T_0 is the environmental temperature, K; $C_{v,ex}$ stands for the constant-volume exhaust specific heat capacity, J/kg · K. The gas specific heat capacity is a complex function of component and temperature. The exhaust specific heat capacity could be attained by piecewise linear interpolation (Fu, Q.S. 2012).

The thermal energy is described as follows:

$$\dot{Q}_k = \frac{1}{2} \dot{m}_{ex} u^2 \qquad (5)$$

The calculation results are the theoretical recoverable energy at the ground conditions while the standard atmospheric pressure and temperature was taken as the reference point (Liu, J.P. et al. 2011). But the atmospheric pressure and temperature vary with the height changes. Therefore, the recoverable pressure energy and thermal energy could change also. But the detailed calculation couldn't be presented because of the limited length in this paper.

3 SIMULATION MODEL ESTABLISH AND DEMARCATE

3.1 The GT-power simulation model

The exhaust pressure, temperature and velocity vary with the engine speed and load. And the transient pulse characteristics belong to them. Therefore, the energy carried by exhaust presents a characteristic of irregular wave with the change of working condition. The accurate measurement will not only cost more but also harder to achieve (Liu, Z.M. et al. 2007). The GT-power simulation models were established for air intake system, fuel injection system, cylinder, crank case and exhaust

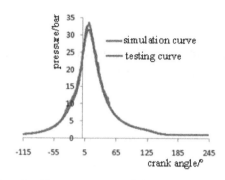

Figure 1. The pressure curve of simulation and testing.

system based on some two-stroke piston engine. GT-power model simulated and calibrated

The pressure and temperature of cylinder gas were calculated according to established model above and the main technical parameters. The calculated results were compared with testing data for checking the model accuracy and calibrating the model. The contrast of pressure was presented in Figure 1. It can be seen that the calculated results are in good agreement with the measured ones. The relative error is less than 5%. This result supports that the accuracy of model could meet the needs of engineering calculation.

4 THE CALCULATION RESULTS AND ANALYSIS

4.1 The change rule of exhaust mass flow rate

The characteristic of exhaust energy flow is connected with exhaust mass flow rate from (3)~(5). The exhaust mass flow rate varies with the engine speed as the load is fixed. While the load is full, the exhaust mass flow rate at different speed was shown in Figure 2. It can be seen that the mass flow rate increased with the speed rise. There had obvious fluctuations in a cycle. The speed is lower, the greater the fluctuation. It emerged a negative value as speed is 800 rpm. The inverse differential pressure was appeared in the exhaust system at this time. The phenomenon of waste gas flowing back was occurred. It is bad for low speed conditions.

4.2 The change rules of total energy flow rate

The total energy flow rate is connected with engine speed and load. The change rules of exhaust total energy flow rate at different speed and load were presented in Figure 3. It can be seen that the exhaust total energy flow rate increases as adding of speed and load. But the influence of speed is more obvious than load. The maximum is 92 kW,

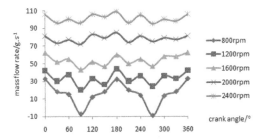

Figure 2. The exhaust mass flow rate at different speed.

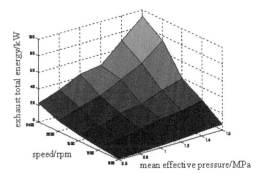

Figure 3. The change rule of exhaust energy with speed and load.

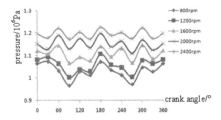

(a)The variation of exhaust pressure with crank angle

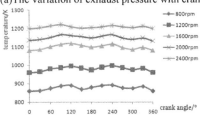

(b) Variation of exhaust temperature with crank angle

Figure 4. The variations of exhaust pressure and temperature with crank angle.

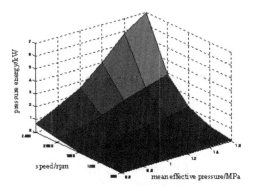

Figure 5. Variation of pressure energy with speed and load.

and appeared at working condition n = 2400 rpm, pe = 1.6 MPa. The maximum of exhaust total energy is more than engine power (75 kW) in this working condition. It declared that the exhaust energy has larger recycling potential.

4.3 The change rule of exhaust pressure energy

At full working condition, the change rules of pressure and temperature with crank angle at different speed were shown in Figure 4. It can be seen that the exhaust pressure gradually increases as rising of speed from Figure 4(a); the higher the speed the smaller pressure fluctuations; the pressure fluctuations is more large at low speed and appears a vacuum. From Figure 4(b), we can also find that change trend is the same of temperature and pressure; the temperature curve is relatively stable. It demonstrated that the exhaust temperature is relatively stable. The heat transfer closes to prospective steady and useful to recovery of the waste heat.

The change rule of pressure energy with speed and load was shown in Figure 5. We can find that the pressure energy increases with the rising of speed and load. But the affection of speed is more obvious than the one of load. The pressure energy

is less at the whole working condition of engine, and it mainly focus on area of high speed and high load. The maximum is 6.97 kW at full load condition and appears at the speed of 2400 rpm.

4.4 The change rule of kinetic energy

The kinetic energy is connected with the exhaust mass flow rate and velocity from formula (5). We know that exhaust velocity relates to the vent-pipe section based on theory of the flow inside the canal. The greater the cross-sectional area, the smaller the flow velocity is. The change rule of exhaust kinetic energy with speed and load was shown in Figure 6. We can easily find that the variation trend is the

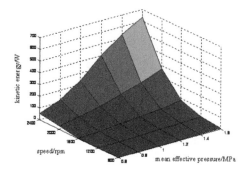

Figure 6. The change of kinetic energy with speed and load.

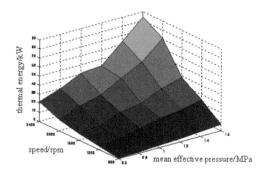

Figure 7. The change of thermal energy with speed and load.

same as pressure energy, that is, it increases as the speed and load rising, and the affection of speed is larger than that of the load. The exhaust kinetic energy is little and the maximum is 0.631 kW only. The kinetic energy is far less than engine effective power or exhaust total energy. So the exhaust kinetic energy usually was ignored in order to simplify the recycling equipment.

4.5 *The change rule of exhaust thermal energy*

The change rule of exhaust thermal energy with speed and load was shown in Figure 7. The rule is the same as pressure energy and kinetic energy, that is to say, the exhaust thermal energy increases gradually with rising of speed and load, and it concentrate on the area of high speed and high load. The value of exhaust thermal energy is far more than that of pressure energy and kinetic energy. And it is the main form of exhaust energy. The maximum of thermal energy reaches 84.4 kW at the calculated condition of high speed. This data is greater than the effective power at the same condition. Therefore, the thermal energy is one of the most recycle potential exhaust energy.

4.6 *The distribution rule of various exhaust energy*

We know easily from analysis above, the proportion occupied by various energy from total energy is different. The proportional relationship of various energy forms at full working condition was shown in Figure 8. We can find that the pressure energy and kinetic energy is tiny, the thermal energy occupies more than 95% of exhaust total energy. The scale of pressure energy and kinetic energy increases with rising of speed, but the extent of variation is less; the thermal energy still is the main form when speed reaches the rating speed, and it accounting for more than 90%. The exhaust kinetic energy is very small at the whole speed range, and

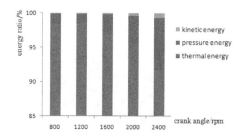

Figure 8. The energy ratio of various energy.

it could be dismissal. The exhaust pressure energy is little at low speed, thus it could be ignored. The pressure energy scale is 2%~8% at middle and high speed, so it can be considered or ignored according to the specific conditions.

5 CONCLUSION

1. There are many energy forms in the engine exhaust. It can be divided into pressure energy, thermal energy and kinetic energy. Thermal energy is the main form of exhaust energy, and the proportion is more than 90% at the whole engine speed range. The recovery potential is tremendous. The kinetic energy is little at the whole work range, so it could be dismissal. The pressure energy is between thermal energy and kinetic energy, and it could be selected according to the specific circumstances.
2. All kinds of exhaust parameters various with the engine speed at full load condition. There has been a large fluctuation of exhaust mass flow rate, pressure, temperature and velocity at low speed. The mass flow rate and velocity become negative and the waste gas flow bake at low speed. These parameters are stable relatively at high speed. The temperature is little

fluctuation at high and low speed, and is advantage for thermal recycle.
3. All kinds of exhaust energy MAP calculated can reflect the change rules of exhaust energy with speed and load. We can quantitatively search out the energy value corresponding to the different working conditions. The calculation results could provide an important basis for analyzing the potential and choosing recycle method.

REFERENCES

Fu, Q.S. 2012. Engineering thermodynamics. Beijing: China Machine PRESS, 2012.

Huang, K.D., Quang, K.V. & Tseng, K.T. 2009. Study of recycling exhaust gas energy of hybrid pneumatic power system with CFD. Energy Conversion and Management 50(5): 1271–1278.

Kauranen, P., Elonen, T. & Wikstrom, L. et al. 2010. Temperature optimisation of a diesel engine using exhaust gas heat recovery and thermal energy storage (diesel engine with thermal energy storage). Applied Thermal Engineering, 30: 631–638.

Liu, C.Z., Liu, G.F. & Hao, Y.G. et al. 2012. Method Discussion on Performance Simulation of Opposed Piston and Opposed Cylinder Two-stroke Engine. Diesel Engine 34(1):31–34.

Liu, J.P., Fu, J.Q. & Feng, K. et al. 2011. Characteristics of engine exhaust gas energy flow. Journal of Central South University (Science and Technology) 42(11): 3370–3376.

Liu, Z.M., Yu, X.L. & Shen, Y.M. 2007. Engine exhaust quantity of heat measuring method. Transactions of the Chinese Society for Agricultural Machinery 38(7): 193–195.

Shi, J.M. 2012. A Prospect of The Two-stroke Engine Development. Small Internal Combustion Engine and Motorcycle 41(2): 88–90.

Taymaz, I. 2006. An experimental study of energy balance in low heat rejection diesel engine. Energy 31(2): 364–371.

Vaja, I. & Gambarotta, A. 2010. Internal combustion engine (ICE) bottoming with organic Rankine cycles (ORCs). Energy 35(2): 1084–1093.

Xiong, Y., Xu, L.H. & Zhong, Y.L. 2006. Automobile energy conservation technology principle and Application. Beijing: China Petrochemical Press, 2006:9–13.

Frontiers of Energy and Environmental Engineering – Sung, Kao & Chen (eds)
© 2013 Taylor & Francis Group, London, ISBN 978-0-415-66159-1

Kinetic modelling of the hydrolysis of carbon disulfide catalyzed by titania based CT6-8

R.H. Zhu, H.G. Chang, J.L. He & C.R. Wen
Research Institute of Natural Gas Technology, PetroChina Southwest Oil and Gas Field Company, Chengdu, Sichuan, P.R. China

R.H. Zhu & J.H. Li
State Key Joint Laboratory of Environment Simulation and Pollution Control, School of Environment, Tsinghua University, Beijing, P.R. China

ABSTRACT: The carbon disulfide hydrolysis kinetic over titania based sulfur recovery catalyst CT6-8 was studied in a continuous flow fixed bed reactor. Without both internal and external diffusion limitation, kinetic data were obtained by orthogonal experimental design under the conditions of 280–350 °C, contact time 3.5×10^{-6}–10^{-5} h, CS_2 concentration 0.1–1%, H_2O concentration 5–30%, H_2S concentration 2–5%, SO_2 concentration 1–2.5%. The reaction rate was calculated through the data fitting method. The relationship between the reaction rate and the experimental conditions was fitted by power function model. The CS_2 hydrolysis reaction kinetic equation was obtained.

Keywords: kinetic; carbon disulfide; hydrolysis; titania; CT6-8

1 GENERAL INSTRUCTIONS

The Claus process was developed by Chance and Claus in 1894 and works well for gas streams containing greater than 20% (by volume) H_2S and less than 5% hydrocarbons. The first stage of the Claus process is the thermal oxidation of one-third of the initial H_2S concentration in a high temperature. Formation of COS and CS_2 results from hydrocarbons present in the flue gas, reacting with many of the other sulfur species present in the combustion step. Attention has been recently focused on the hydrolysis of COS and CS_2 to achieve the very high levels of sulfur depletion now considered necessary. In the first reactor, containing an alumina catalyst, the conversion of COS and CS_2 is limited to 75% for COS and 50% for CS_2 at the outlet temperature of 340 °C while titania-based catalysts, such as CRS-31 & CT6-8, are regarded as a more efficient way for COS and CS_2 hydrolysis (Rhodes et al., 2000).

CS_2 and COS are two problem compounds that often appear together in the first Claus converter. The kinetic of COS hydrolysis have been studied in numerous research papers while research of CS_2 hydrolysis kinetic was much less (Svoronos and Bruno, 2002, Rhodes et al., 2000, Tong et al., 1993, Kerr and Paskall, 1976). The kinetic of CS_2 hydrolysis is considered to be more complicated than COS hydrolysis. This is due to the interaction between CS_2 and SO_2, which has high concentration in the first Claus converter, as well as COS being a proposed intermediate in the CS_2 hydrolysis process (Clark et al., 2001, George Z, 1974). Tong studied the hydrolysis of CS_2 for a titania catalyst and found that the hydrolysis of CS_2 can be best described by an Eley-Rideal model over the ranges of temperature from 270 to 330 °C. In this study, the feed gas contains only CS_2, H_2O and N_2, except SO_2 (Tong et al., 1995).

In this research, the kinetic measurements for the hydrolysis of CS_2 using titania-based CT6-8 catalyst was obtained under controlled condition using a continuous flow fixed bed reactor. The reaction temperature varied from 280 to 350 °C. The feed gas contains CS_2, H_2O, H_2S, SO_2 and N_2 to simulate the gases encountered within the first Claus converter.

2 EXPERIMENTAL SECTION

2.1 *Catalyst*

CT6-8 catalyst was produced by Research Institute of Natural Gas Technology, PetroChina Southwest Oil and Gas Field Company.

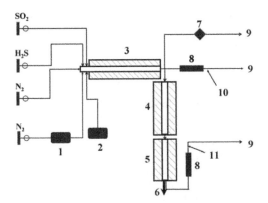

Figure 1. Flow chart of catalysis activity evaluation. (1) liquid CS_2, (2) water (3) preheater, (4) reactor, (5) sulfur condenser, (6) liquid sulfur vent, (7) pressure display, (8) desiccant, (9) vent (10) analysis for inlet gas (11) analysis for outlet gas.

2.2 Experimental apparatus

A schematic flow diagram of the catalysis activity evaluation is shown in Figure 1. A fixed-bed reactor used for this study was constructed with a stainless steel tube 12 mm ID and 300 mm in length. There have also preheater and sulfur condenser before and after the reactor. The concentration of water in the feed was controlled by a microscale water pump. The flow rates of H_2S, SO_2 and N_2 were controlled by mass flow controller. The CS_2 gas was obtained by passing certain flow rate of N_2 through a thermostated liquid CS_2 bubbler kept at 25 °C. In each experiment, 1.0 g catalyst was in use. The inlet and outlet gas were measured by Agilent 7890 gas chromatograph with thermal conductivity cell detector. The conversion rate of CS_2 was calculated as Equation (1):

$$\eta_{CS2} = (1 - Kv \times \varphi/\varphi_0) \times 100\% \qquad (1)$$

where $\varphi_0 = CS_2$ concentration without water in the inlet gas, $\varphi = CS_2$ concentration without water in the outlet gas, Kv = volume adjusting factor, calculated as Equation (2):

$$Kv = [100 - (\varphi_{H_2S} + \varphi_{SO_2})]/[100 - (\varphi'_{H_2S} + \varphi'_{SO_2})] \qquad (2)$$

3 RESULTS AND DISCUSSION

3.1 Blank test

Prior to the measurement of the reaction test, blank runs using an empty reactor or reactor filled with ceramic balls showed no detectable conversion of CS_2.

3.2 The effect of internal and external diffusion

For the purpose of kinetic study, it is important to ensure that the rate data obtained are under the kinetic regime.

To eliminate the external diffusion effect, two group CS_2 conversion rate test with catalyst loading amount of 0.8 g and 1.0 g were carried out. As shown in Figure 2, when the contact time less than 7×10^{-5} h, the CS_2 conversion rate almost the same, which can shown that the conversion rate is uncorrelated with the gas flow rate and it can be supposed that the external diffusion effect has been eliminated.

The catalysts were crushed to different size to test the effect of particle size on internal diffusion effect. As shown in Figure 3, when the catalyst size was smaller than 30–40 mesh, the CS_2 conversion rate remain unchanged with different size, which showed that the reaction was free of internal diffusion with particles within or below the range: 30–40 mesh.

3.3 Kinetic data results

To detailed study each of the reaction conditions and to reduce the experiments amount, orthogonal table ($L_{16}(4^5)$) was applied to arrange the experiments. Considered the actual situation in the first Claus converter, the reaction conditions were selected as follows: temperature range 280–350 °C, CS_2 concentration 0.1–1%, H_2O concentration 5–30%, H_2S concentration 2–5% and SO_2 concentration 1–2.5%.

Kinetic data were collected after the reaction system reached a stable condition. Parallel samples

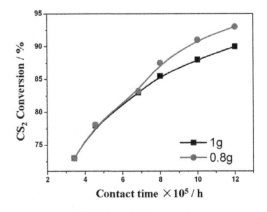

Figure 2. The relationship between the contact time and CS_2 conversion rate (reaction condition: 5.5% H_2S, 2.5% SO_2, 1% CS_2, 30% H_2O, N_2 as balance gas, the reaction temperature was 350 °C and the size of the catalyst was 30–40 mesh).

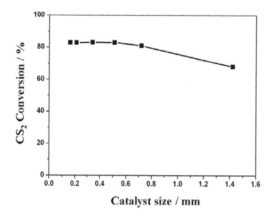

Figure 3. The relationship between the catalyst size and CS_2 conversion rate (reaction condition: 5.5% H_2S, 2.5% SO_2, 1% CS_2, 30% H_2O, N_2 as balance gas, the reaction temperature was 350 °C and GHSV was 10,000 h^{-1}).

Table 1. CS_2 kinetic data.

T/°C	H_2S/ kPa	SO_2/ kPa	CS_2/ kPa	H_2O/ kPa	$-r_{exp}$/ mol/h·L
350	1.88	1.72	0.98	13.30	11.94
350	2.04	2.45	0.69	4.75	6.77
350	3.79	0.86	0.37	28.50	6.10
350	3.98	1.25	0.11	19.95	0.81
320	2.93	1.09	0.93	19.95	7.32
320	1.86	1.53	0.67	28.50	3.32
320	4.69	1.99	0.39	4.75	1.89
320	3.84	2.12	0.11	13.30	0.27
300	3.97	1.67	1.17	4.75	5.70
300	4.25	1.31	0.65	13.30	2.53
300	1.88	2.40	0.38	19.95	1.73
300	2.93	1.80	0.09	28.50	0.11
280	0.93	4.32	2.26	28.50	1.62
280	3.75	1.88	0.64	19.95	0.96
280	2.86	1.31	0.41	13.30	0.57
280	1.90	1.17	0.11	4.75	0.29

were taken at least 3 times for each experimental point. H_2S, SO_2 and CS_2 partial pressure were calculated according to the results of chromatographic data, atmospheric pressure and internal pressure of the kinetic experiment apparatus. The partial pressure of H_2O was calculated by the flow rate of micro-metering pump and the total pressure. The kinetic data were shown in Table 1.

3.4 Kinetic model and parameter analysis

Many types of model have been used to studied COS and CS_2 kinetic. In this study, we selected

power function model as shown in Equation (3) for the complexity of the experimental conditions, easier data processing and parameter estimation.

$$-r_{CS_2} = k_2 e^{E/RT} P_{H_2S}^a P_{SO_2}^b P_{CS_2}^c P_{H_2O}^d \qquad (3)$$

The proposed model for the hydrolysis of CS_2 is non-linear functions of the parameters. The software of Matlab was used to search for best values of the model parameters. The best fitting values of the parameters were shown in Table 2. The value of k_2 and E were all greater than zero, which compliance with the physics-chemistry rule.

The reaction rate comparison of the experimental value ($-r_{exp}$, as shown in Table 1) and the calculated value ($-r_{cal}$, calculated by Equation (3) and the values in Table 2) was shown in Figure 4. The points evenly distributed on both sides of the diagonal, indicating that the deviation between the calculated values and the experimental values is smaller to meet kinetic experiments requirement.

3.5 Comparison of activation energy

Table 3 compared the activation energies of present result with other published data of CS_2 hydroly-

Table 2. Best fitting values of the parameters in Equation (3).

Parameters	Values
k_2	8.86×10^7
E/J/mol	58404
a	−0.195
b	−0.983
c	1.220
d	−0.282

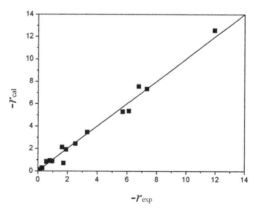

Figure 4. The reaction rate comparison of the experimental value and the calculated value.

Table 3. Comparison of activation energies.

Catalysts	T/°C	E/kJ/mol
CRS-31[*]	270–330	56.73
Kaiser201[*]	270–330	40.41
LS-951[**]	180–280	51.78
CT6-8[***]	280–320	58.40

[*](Tong et al., 1995); [**](Zhang et al., 2006); [***](Present study).

sis. The activation en ergy of CT6-8 compares well with values of CRS-31. The titania catalysts, CRS-31 and CT6-8, showed a larger activation energy than the alumina catalyst Kaiser 201 and alumina based catalyst LS-951. Part of the reasons for this phenomenon is that titania and alumina has different hydrophilic surface property.

4 CONCLUSIONS

Carbon disulfide hydrolysis kinetic have been studied in the simulated first Claus converter condition and the kinetic equation has been obtained as:

$$-r_{CS_2} = 8.86 \times 10^7 e^{-58404/RT} P_{H_2S}^{-0.195} P_{SO_2}^{-0.983} P_{CS_2}^{1.22} P_{H_2O}^{-0.282}.$$

This study lays the foundation for optimization of the Claus reactor design and titania-based catalysts improvements.

ACKNOWLEDGEMENTS

We are thankful for the financial support from post-doctoral research center of PetroChina Southwest Oil and Gas Field Company (No. 20100308-13).

REFERENCES

Clark, P.D., Dowling, N.I. & Huang, M. 2001. Conversion of CS_2 and COS over alumina and titania under Claus process conditions: reaction with H_2O and SO_2. *Applied Catalysis B: Environmental* 31(2): 107–112.

George, Z.M. 1974. Effect of catalyst basicity for COS-SO_2 and COS hydrolysis reaction. *Journal of Catalysis,* 35(2) 218–224.

Kerr, R.K. & Paskall, H.G. 1976. Claus process: catalytic kinetics-2. COS and CS_2 hydrolysis. *Energy Processing/Canada,* 69(2) 38–44.

Rhodes, C., Riddel, S.A., West, J., Williams, B.P. & Hutchings, G.J. 2000. The low-temperature hydrolysis of carbonyl sulfide and carbon disulfide: a review. *Catalysis Today,* 59(3–4): 443–464.

Svornos, P.D.N. & Bruno, T.J. 2002. Carbonyl sulfide: a review of its chemistry and properties. *Industrial & Engineering Chemistry Research,* 41(22): 5321–5336.

Tong, S., Dalla Lana, I.G. & Chuang, K.T. 1993. Kinetic modelling of the hydrolysis of carbonyl sulfide catalyzed by either titania or alumina. *The Canadian Journal of Chemical Engineering,* 71(3): 392–400.

Tong, S., Dalla Lana, I.G. & Chuang, K.T 1995. Kinetic modeling of the hydrolysis of carbon disulfide catalyzed by either titania or alumina. *The Canadian Journal of Chemical Engineering,* 73(2) 220–227.

Zhang, K., Yu, C. & Dai, X. 2006. Kinetic study of LS-951 hydrogenation catalyst for treating tail gas from Claus process. *Petroleum Processing and Petrochiemicals,* 37(1), 38–42.

Frontiers of Energy and Environmental Engineering – Sung, Kao & Chen (eds)
© *2013 Taylor & Francis Group, London, ISBN 978-0-415-66159-1*

Control of deice robot for its straight transmission line followed navigation based on electromagnetic transducer

G.P. Wu, Y. Yan, Z.Y. Yang & T. Zheng
WuHan University, WuHan, HuBei, China

ABSTRACT: Aiming at deice robot, a theory of electromagnetic navigational controlling arithmetic was proposed now. This arithmetic firstly adopts the principle of relative value detection to identify the obstacles along the straight line, then classify the environment where deice robot has located, and design controlling strategy for each classification accordingly. Based on the known information, dedicated to realize the situation transfer of robot. At last, the feasibility of this arithmetic is approved by lines in laboratory.

Keywords: deice robot; electromagnetic navigation; situation transfer

1 INTRODUCTION

High voltage line is the main carrier for power transmission, the stabilized power supply relays mostly on normal work of high voltage line. In recent years, extreme weather like snow storm calamity appears globally. In 2008, disastrous effects were brought to south China by unusual snow storm, and also great damage to the national wealth. The researches of power transmission deice robot poses great meanings for safe and stable running of the state grid.

Katrasnik, Pernus, and Likar (2010) did some research on navigational control of line inspectional robot, and they have provided great facilities for the research of navigational control of deice robot. During three-arm line inspectional robot's obstacle clearing operation, Tang, Fang and Wang (2004) uses optical fibre sensor to judge the relative position between his hands and the overhead earth wire. and meanwhile, controls the process of grasping line. Zhang, Liang, and Tan (2007) designed a program of visional control for obstacles clearing based on image, and adopted Fourier descriptor to build the vallate eigenveetor with features of parallel move, whirl, zoom of size and unchanged starting point, then made the recognition of driving wheel as true. Hu, Wu, and Cao (2008) proposed a method for obstacles inspection and detection based on vision sensor, then improved the arithmetic operator of Canny with Otsu arithmetic to extract the image edge, so as to reducing the effects brought by light variation [4].

Aiming at the electromagnetic working environment that deice robot faced, this article depicts the independent design of a small and exquisite electromagnetic transducer. Taking advantage of this transducer, the electromagnetic transducer array of deice robot was constructed, and also, obstacles indentifying method based on electromagnetic transducer was proposed. The basic thought can be summarized as, utilizing filtered historical electromagnetic transducer and current information of transducer to describe the environment, basing on the principle of relative value detecting, finally make the movement of deice robot along its line. While in different circumstances, deice robot can take correspondent strategies to control its movement.

2 ENVIRONMENTAL PERCEPTION BASED ON ELECTROMAGNETIC TRANSDUCER

2.1 Introduction for the working principle of electromagnetic transducer

The production principles of Magnetic induction electromotance tell us: when a coil is put near lead wire with exchange current flowing, magnetic induction electromotance will appear in the coil.

2.2 Relative value testing principles

According to the Biot Savart Law and Faraday law on electromagnetic induction, if a coil of N turns with a cross-sectional area of S is put near the indefinitely long lead wire with exchange current flowing, magnetic induction electromotance will appear in the ends of the coil circuit.

$$\varepsilon = -\frac{d\varphi}{dt} = -NS\cos\theta * \mu_0 \frac{dI}{2\pi r dt} \qquad (1)$$

In the (1), ε is magnetic induction electromotance in the coil; φ is magnetic flux flowing through the coil; θ is the angle between the normals of the coil's cross-section and magnetic direction; S is the cross-sectional area of the coil; N is the turns of the coil; r is the distance between the coil and the lead wire; dI/dt is the change rate of lead wire's electric current. From (1) it can be seen magnetic induction electromotance is inversely proportional to the distance between the coil and the lead wire. The EHV Transmission Line can be seen as a indefinitely long lead wire with exchange current flowing. If dI/dt, N, S, θ are fixed, ε is only related to r. That's to say, in the measuring heads of a sensor, the measuring head of the sensor with the maxε is closest to the EHV Transmission Line. Chen, Xiao, and Wu (2006) designed a reasonable array of the measuring heads of sensors which can be designed to make certain of the location situation through relative value testing principles.

2.3 Realization way of environment awareness

Environment awareness in this paper is based on the concept of active windows proposed by Borenstein and Keron (1991), as is shown in the Figure 1. A movable, round and robot-centered window is created and this window is attached to the robot. The ice-removing robot use the present and past data of electromagnetism sensors to sense the obstacles in the movable window. Concrete process is as follows. Supposing in some certain coordinate system XOY, at the k moment the ice-removing robot is at $(x_r(k), y_r(k))$, the obstacle information can be described by a point set of N elements in the window. That's $P(k) = \{(xj, yj, tj)|j = 1, 2, ..., N\}$ and (x_p, y_p) is the location

of the obstacle detected by some electromagnetism sensor. t_j stands for the time trust value of the obstacle. As time passes by, t_j gradually decreases. When t_j is zero, the obstacle is removed from the point set. At the $(k + 1)$ moment, the ice-removing robot is at $(x_r(k + 1), y_r(k + 1))$, the description point set of the obstacle, $P(k + 1)$, is made up of 2 parts. The first part is the point set gained by filtering the point set at k moment $P(k)$; the second part is the point set of the new obstacle directly gained from the electromagnetism sensor at $k+1$ moment. In this formula, is a preset value.

3 ELECTROMAGNETIC NAVIGATION CONTROL ALGORITHM

The straight line of the EHV Transmission Line is generally equipped with the support and function accessories such as shockproof hammers and Suspension clamps. They therefore become the main obstacles the ice-removing robot meets when walking on the line. so the robot's working environment can be divided into 3 categories:

The straight-line part; in this part the distribution principles of the electromagnetic field can be nearly seen as several Concentric circles on their axes round the lead wire which are shaped by magnetic lines of force. The sketch map of magnetic lines of force is shown in Figure 2a.
The shockproof hammer part; the shockproof hammers are iron implements which have a high magnetic permeability and can effectively assemble the magnetic lines of force. Because of the shockproof hammers' assembling use, the distribution of magnetic lines of force in the section of the lead wire changes. In the area between the lead wire and the shockproof hammer, magnetic field strength intensifies obviously and the shape of magnetic lines of force changes from circular arc to parabola. The sketch map of magnetic lines of force is shown in Figure 2b.
The Suspension clamp part; similar to the shockproof hammer, both the Suspension clamp and the connecting plate are iron implements. The sketch map of magnetic lines of force is shown in Figure 2c.

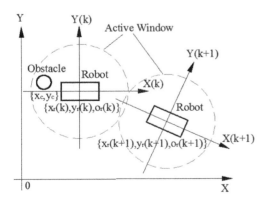

Figure 1. Active windows.

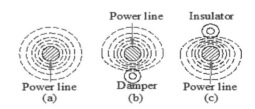

Figure 2. The sketch map of magnetic lines of force.

According to analysis above, the distribution of magnetic field around different obstacles have its own relatively obvious particulars. If the ice-removing robot is equipped with 3 different electromagnetic sensors (Fig. 3), the ice-removing robot can reach the goal of detecting obstacles.

Supposing a plane right angle coordinate system is established on a plane which is through the lead wire and vertical the ground, the direction of the lead wire is the direction of X-axis, and the robot's moving direction is the positive direction of X-axis, and the point of contact of the robot's front wheel and the lead wheel is the origin. It can be seen in Figure 3. There are 2 areas in the robot's moving direction at this moment: One is the x > 0 and y > 0 area where the obstacles of the Suspension clamps lie ac cording to Figure 2c, another is x > 0 and y < 0 area where the obstacles of the shockproof hammers lie according to Figure 2b.

This paper divides electromagnetic navigation control algorithm into 2 parts. One is the ice-removing robot judges the categories of obstacles from the environmental information detected by the electromagnetic sensor array and changes its own status according to the informed overall information, Another is particular measures are taken to avoid these obstacles according to the categories of obstacles.

3.1 *Status change of the ice-removing robot*

The ice-removing robot's status refers to the categories of obstacles around it. When the robot walks in straight line and removes ice, it is in the status of no obstacle. Similarly, the robot can be in other status such as Status Damper. As the robot moves on the line, its status changes correspondently. In order to adapt to the robot's need of navigation, this paper divided its situation into 3 categories: No_Obstacle, Status_Damper, Status_Insulator.

Obstacles-detecting algorithm of the electromagnetic sensor is as follows:

IF ((Value_A—Value_C)) > Threshold_Insulaor), the robot detects the obstacle of an insulator in front;
ELSE IF ((Value_B—Value_C)) > Threshold_Damper), the robot detects the obstacle of a damper in front;
ELSE There's no obstacle in front.
Introduction: Value_A, Value_B, Value_C are the value timely detected by the probe A, probe B, probe C of the electromagnetic sensor in Figure 3. Because probe C is farther from the obstacle, Value_C here is just a reference. Threshold_Insulator and Threshold_Damper are threshold levels dynamically decided by the current flowing through the line. If the threshold levels are smaller, obstacles are detected.

According to the above algorithm, status change of the ice-removing robot has certain conditions. The conditions are shown in Figure 4. Introduction: insulators are not next to one another on the EHV.

Transmission Line, therefore when in Status Insulator, the robot wouldn't go next into Status_Insulator again.

3.2 *Motion strategy of the ice-removing robot*

Corresponding to every status of the robot, the robot should take different motion strategies accordingly to avoid obstacles.

No_Obstacle: The motion strategy taken by the robot is to switch on barrier-detecting thread and move forward on the EHV Transmission Line removing ice until the barrier-detecting thread checks obstacles.
Status_Damper: Here the process in which the robot crosses the damper is divided into 4 phases. Analysis of barriers-crossing process is as follow (Fig. 5).

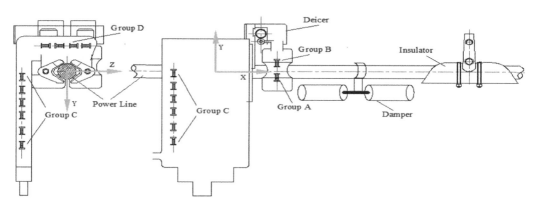

Figure 3.   The sketch map of environment-detecting.

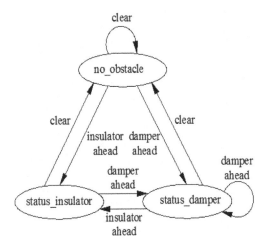

Figure 4. The sketch map of status change of the ice-removing robot.

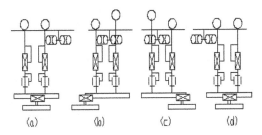

Figure 5. The entire process of the ice-removing robot's crossing the damper.

The preparation phrase: When detecting the damper, the robot stops moving and prepares to cross the damper.

The front arm cross phrase: The clamping device and the hold-down mechanism of the behind arm act to fix the behind arm. Backward shifting the center of gravity and stretching the front arm, the robot switches on the ice-removing equipment in front of the front wheel. The front arm moves forward to remove most of ice on the line and then get backward to make sure that the center of gravity of the robot is right below the point of support of the behind wheel. The clamping device of the behind arm loosens, and the behind arm moves with a slow speed in only one wheel. When the behind wheel gets close to the damper, the robot stops moving, and the clamping device of the upper arm clamps. The front arm moves forward again until it completely crosses the damper.

The upper arm cross phrase: The expanding and contracting equipments of the front arm contract. After the expanding and contracting equipment

goes off the line, the clamping device of the front arm clamps and the hold-down mechanism of the front arm compact. The center of gravity of the robot is adjusted below the point of support of the front wheel. The clamping device of the behind arm loosens, and the hold-down mechanism of the behind arm declines and the expanding and contracting equipments extends. The ice-removing equipment is switched on. Only the front wheel moves forward until the behind arm completely crosses the damper. The ice-removing equipment is switched off.

Reversion phase: The expanding and contracting equipments of the behind arm contract. The hold-down mechanism of the behind arm compacts. The center of gravity of the robot is adjusted below the robot.

Status_Insulator: Here the process in which the robot crosses the insulator is also divided into 4 phases. Analysis of barriers-crossing process is as follow (Fig. 6).

The preparation phrase: When detecting the insulator, the robot stops moving and prepares to cross the insulator.

The front arm cross phrase: The expanding and contracting equipments of the front arm extend, and the swinging gear turned away. The center of gravity is shifted backward through adjustment. The upper arm moves with a slow speed in only one wheel and stops when getting close to the insulator. The clamping devices of the behind arm clamp. The front arm move forward again until they completely cross the insulator. The swinging gears of the front arm turn back. The expanding and contracting equipments of the front arm contract. At the same time, the ice-removing equipment is switched on to remove the ice layer on the line.

The upper arm cross phrase: The clamping devices of the front arm clamp. The center of gravity of the robot is adjusted below the point of support of the front wheel. The clamping devices of the behind arm loosen, and the hold-down mechanisms of the behind arm decline and the expanding and

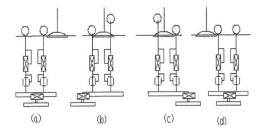

Figure 6. The entire process of the ice-removing robot's crossing the insulator.

contracting equipments extend. The ice-removing equipment is switched on. Only the front wheel moves forward until the behind arm completely cross the damper. The ice-removing equipment is switched off.

Reversion phase: The swinging gears of the behind arm turn back. The expanding and contracting equipments of the behind arm contract and the hold-down mechanisms of the behind arm compact. The center of gravity of the robot is adjusted below the robot.

4 EXPERIMENT FOR ALGORITHM

The above control algorithm is realized in the ice-removing robot based on EVC. Navigation test has been done on the line without ice in the lab. In this test 110 A AC is flowing through the EHV Transmission Line. Parts of the process of the experiment are indicated in Figure 7. The experiment results show the ice-removing robot sample can work normally in Ultrahigh Voltage (UHV) environment where the robot can automatically walk and automatically or semi-automatically crosses obstacles. The robot sample has main functions of removing ice, standing the test of electromagnetic interference.

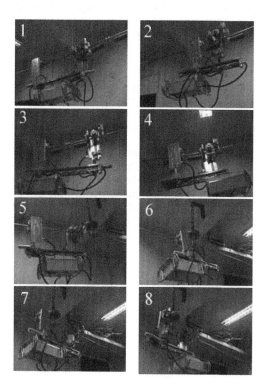

Figure 7. The process of deicing robot crossing obstacle in the Lab.

5 CONCLUSION

This article mainly researched the navigational control of deice robot moving straight along the power transmission line, and proposed a controlling arithmetic based on electromagnetic navigation. This arithmetic filtered the information gained by historical electromagnetic transducer so as to exclude the interferences of hash coming from external environment, reaching the aim of effective identifying the obstacles along straight line with the relative value detection principle. Meanwhile, classifying the condition that deice robot moving along line, and adopting relevant controlling strategy for different classification to perfect satisfy robot's requirement of adapting to the environment. As built-in counter of deice robot made certain errors when counting different speed of each electric motor, causing some uncertain factors while the single rear wheel moving toward the obstacles. All this required that deice robot can obtain more information, for the protection of failed check which caused by false counting, but this will be the next research.

REFERENCES

Borenstein, J. & Keron, Y. 1991. The vector filed histogram-fast obstacle avoidance formobile robots. *IEEE Transactions on Robotics and Automation* 7:278–288.

Chen, Z.W., Xiao, H. & Wu, G.P. 2006. Navigation ways of the detecting line robot on the EHV Transmission Line by Electromagnetic sensors. *Sensors and Microsystems* 25:33–39.

Hu, C.S., Wu, G.P., Cao, H. et al. 2008. Study of the detecting line robot on the EHV Transmission Line visually testing and recognizing obstacles. *Chinese Journal of Sensors and Actuators* 21:2092–2096.

Katrasnik, J., Pernus, F. & Likar, B. 2010. A survey of mobile robots for distribution power line inspection. *IEEE Transactions on Power Delivery* 25:485–493.

Tang, L., Fang, L.J. & Wang, H.G. 2004. Obstacle-navigation control for a mobile robot suspended on overhead ground wires, *2004 8th International Conference on Control Automation, Robotics andVision*: 2082–2087.

Zhang, Y.C., Liang, Z.Z., Tan, M. et al. 2007. The detecting line robot on the overhead Transmission Line overcoming obstacles by visual servo control. *Robot* 29:111–116.

Frontiers of Energy and Environmental Engineering – Sung, Kao & Chen (eds)
© 2013 Taylor & Francis Group, London, ISBN 978-0-415-66159-1

Preparation and photocatalytic activity of continuous Ag-TiO$_2$ fibers

S.Y. Zhang, X.C. Zhang, D.F. Xu, Y. You & Y.L. Li
Institute of New Materials, Department of Biological and Environmental Science, Changsha University, Changsha, P.R. China

ABSTRACT: The continuous Ag-TiO$_2$ fibers were synthesized by the sol–gel method. The as-synthesized samples were characterized using XRD, SEM and XPS. The photocatalytic properties for degrading formaldehyde was primarily investigated. The results demonstrated that the as-synthesized Ag-TiO$_2$ fibers all were crystal of anatase with about 20–40 μm in diameter. The continuous Ag-TiO$_2$ fiber photocatalytic degradation rate of formaldehyde 98.4%, the degradation of formaldehyde was about 75% for resusing 10 times. It would provide promising applications in cleaning indoor air.

Keywords: photocatalytic; continuous fibers; Ag-TiO$_2$

1 INSTRUCTIONS

TiO$_2$ nanomaterials are playing and will continue to play an important role in environmental protection and in the search for renewable and clean energy technologies due to its non-toxic, chemical stability, and high photoactivity[1,2]. However, large band gap (3.2 ev) and low recovery efficiency significantly limits the broad application of TiO$_2$.[3-4] The development of photocatalysts under visible light irradiation, which is one of the challenging tasks in the field of photocatalysis, is one of the major goals of enhancing the efficient utilization of solar energy.

Considerable effort has been made to enhance the photogenerated charge separation and photoresponse range of TiO$_2$ nanomaterials through phase and morphological controls, doping, surface sensitization, and so on[5-7]. Feng et al[8] deposits Ag on the TiO$_2$ nanotubes by photochemical deposition process, and the experiment results showed that the light absorption region red shifted, and the photocatalytic performance improved greatly. Moreover, the catalyst immobilization can avoid particles aggregation and enhance the separation efficiency of the catalyst from the solution[9-11]. Due to the very large aspect ratios (length to diameter) of fiber, photocatalyst in the form of fiber is superior to particles as far as the recycling and aggregation are concerned. Since 1934 Formahals patented the first invention of electrostatic spinning of fibers[12]. Several preparation methods of TiO$_2$ fibers have been reported[13-15], such as the KDC (kneading–drying–calcination) method, sol–gel process, hydrothermal method, and electrospinning.

In this work, the continuous Ag-TiO$_2$ fibers have been synthesized by the sol–gel method using the polymer of titanate as the precursor solutions. The effects of doping Ag on the microstructure of the as-synthesized samples were characterized using Scanning Electron Microscope (SEM), X-Ray diffraction (XRD) and X-ray Photoelectron spectroscopy (XPS) analysis methods. The photocatalytic properties for degrading formaldehyde was primarily investigated. It would provide promising applications in cleaning indoor air.

2 EXPERIMENTAL

2.1 *Preparation of continuous Ag-TiO$_2$ fibers*

The continuous Ag-TiO$_2$ fibers were synthesized by sol–gel method, using titanium tetrabutyloxide (Ti (OC$_4$H$_9$)$_4$, TTBO), anhydrous alcohol (C$_2$H$_5$OH, EtOH), polyvinylpyrrolidone ((C$_6$H$_9$NO)$_n$, molecular weight 1000, PVP), silver nitrate (AgNO$_3$) and hydrochloric acid (HCl, 6 mol/l, HCl) as starting materials. All reagents were analytical grade and used without further purification. In typical synthesis process, 105.10 ml TTBO was mixed with 25.66 ml EtOH and magneticly stirred for 60 min at room temperature and nominated as "solution A". Solution B was prepared by mixing 10.00 g (0.01 mol) PVP, 2.20 g AgNO$_3$ and 5.60 g HCl (6 mol/L) dissolving in 76.98 ml EtOH. Solution B was added to solution A drop by drop keeping vigorous stirring for 120 min, during the stage of stirring process the TTBO was hydrolyzed with the H$_2$O in the HCl (6 mol/L). On the other hand, HCl serves as a stabilizer to hinder the hydrolysis of TTBO. The improvement of viscosity of precursor collosol of polymer titanate is achieved by adding PVP. The chemical composition of the starting alkoxide solution was $AgNO_3$: PVP: TTBO: EtOH: $H_2O = 0.05{:}0.04{:}1{:}8{:}1{:}1$ in molar ratio. The sol was evaporated by a rotary evaporator in oil bath at 110–150°C to prepare the spinnable solution, corresponding to the viscosity of 5 Pa·s. Then, the long continuous precursor fibers were obtained by homemade spinning apparatus. After calcined at 300–800°C with steam activation, the continuous Ag-TiO_2 fibers have been prepared.

2.2 *Characterization*

The continuous Ag-TiO_2 fiber morphology was characterized by scanning electron microscopy (SEM, JSM-6700F) at an operating accelerating voltage of 20 kV. Crystal structure of samples were identified by X-ray diffraction (XRD) using a Rigaku D/max 2550 diffractometer with graphite monochromatized Cu Kα radiation. The X-ray Photoelectron Spectroscopy (XPS) analysis was performed on a VG ESCALAB MK-II spectrometer equipped with an Mg Kα. monochromator X-ray source with a power of 240 W. The test chamber pressure was maintained below 5×10^{-7} Torr during spectral acquisition. The XPS Binding Energy (BE) was internally referenced to the C 1 s peak (BE = 284.6 eV).

2.3 *Photocatalytic property testing*

The photocatalytic properties for degrading formaldehyde were based on Chinese standard of HJ601–2011 (water quality—dertermination of formaldehyde—acetylacetone spectrophotometric method). The concentrations of formaldehyde solution were determined by UV—vis spectroscopy using UV–1601. 100 mL of formaldehyde solution (10×10^{-6} mol/L) was mixed with 0.05 g as-prepared catalysts in a quartz beaker under solar light irradiation with constant mechanical stirring. Before the irradiation, the solution was stirred for 20 min in dark to allow the system to reach adsorption equilibrium. The determined absorbance was converted to concentration through the standard curve method of formaldehyde. The degradation efficiency of formaldehyde was calculated by $R = (1 - C/C_0) \times 100\%$, where C_0 and C were the concentration of formaldehyde when reaction time was 0 and t, respectively.

3 RESULTS AND DISCUSSION

3.1 *SEM analysis*

Digital photo and SEM images of the continuous Ag-TiO_2 fibers were shown in Figure 1. From the

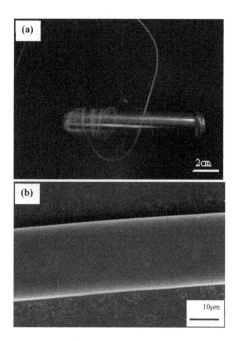

Figure 1. (a) Digital photo and (b) SEM images of continuous Ag-TiO_2 fiber.

SEM image, it can be seen that the as-prepared fibers have smooth surfaces and are composed of cylindrical nanoparticles, and the crystalline size gradually increased with increasing the calcination temperature. The average diameters of the fibers are in the range of 20–30 μm and the length could reach tens of centimeters, even continuous fibers (>1 m).

3.2 *XRD analysis*

XRD was used to characterize the phase structure of the asprepared samples. Figure 2 shows XRD pattern of the continuous TiO_2 and Ag-TiO_2 fiber. As shown in Figure 2, the presence of peaks can be readily indexed to diffraction peak of anatase phase TiO_2 (JCPDS 21-1272) for the continuous TiO_2 fiber. Besides the peak of anatase TiO_2, the new peaks associated with metallic of Ag were observed in the pattern of the sample continuous Ag-TiO_2 fibers. By doping Ag, the peaks became sharper and the intensity increased. The diameter of Ag^+ (about 0.126 nm) is much larger than Ti^{4+} (about 0.068 nm) and it is difficult for Ag^+ to enter the crystal lattice of TiO_2. The deposited Ag^+ was deoxidized to form metallic nanoparticles, dispersing on TiO_2 surface, which restricted the aggregation of TiO_2 nanocrystals and reduced the particle size. It was proved by the peak at 2θ position of 43.4°, referring to (200) crystal plane of Ag.

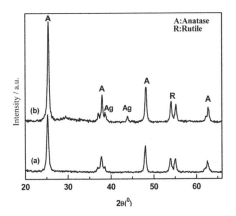

Figure 2. XRD patterns of (a) TiO_2 and (b) Ag-TiO_2 continuous fibers.

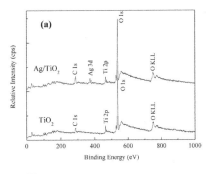

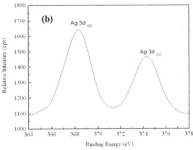

Figure 3. (a) Full survey XPS spectra of samples and (b) high-resolution XPS regional spectra of Ag.

3.3 *XPS analysis*

XPS measurements were performed to elucidate the surface chemical composition and the oxidation state for the as-synthesized samples. Figure 3 (a) shows the XPS full survey spectra in the binding energy (BE) range of 0–1000 eV for continuous Ag-TiO_2 fibers and continuous TiO_2 fibers, respectively. Figure 3 (b) shows the high-resolution XPS regional spectra of Ag.

The XPS results indicated that four elements, namely, Ti, O, Y and C were presented on the continuous Ag-TiO_2 fibers surface, while the continuous TiO_2 fibers surface contained Ti, O and C elements. The trace amounts of carbon originate from the residual carbon in the fibers and the adventitious hydrocarbon in the XPS instrument itself. The peaks observed at 368.4 and 374.4 eV (Figure 3b) can be ascribed to Ag $3d_{3/2}$ and Ag $3d_{5/2}$ of the metallic silver[16]. The result indicated that the chemical composition of the continuous Ag-TiO_2 fiber existed Ag, which were the results from XRD analysis.

3.4 *Photocatalytic activity*

The photocatalytic activity of the catalysts was studied by using formaldehyde degradation experiments as explained in the experimental section. Figure 4 shows the photocatalytic degradation profiles of formaldehyde over the continuous TiO_2 and Ag-TiO_2 fibers, respectively.

Form the figure 4, the highest decomposition rate of formaldehyde for the continuous Ag-TiO_2 fiber was about 98.4% for 90 min, while that for the continuous TiO_2 fiber was only 57%. The results indicated that the photocatalytic activity of formaldehyde degradation was evidently improved for the continuous Ag-TiO_2 fiber than that of the

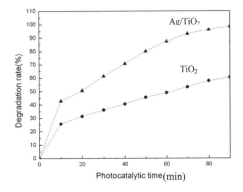

Figure 4. Photocatalytic degradation profiles of formaldehyde over the continuous TiO_2 and Ag-TiO_2 fibers, respectively.

continuous TiO_2 fiber. Research on the reason and the mechanism for enhancing the photocatalytic activity are in progress.

The photocatalytic stability of the continuous Ag-TiO_2 fiber sample under ultraviolet irradiation was also evaluated by reusing experiments, and the result was shown in Figure 5. The experimental parameters and processes were the same as above. After experiments, the filter screen with a pore size of 0.42 mm was used to separate the continuous Ag-TiO_2 fiber. Then, the fibers were washed using

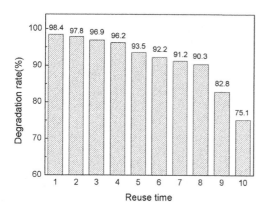

Figure 5. Degradation rate of formaldehyde after 90 min of UV irradiation for different reusing.

deionized water for 3 times and dried at 100 °C for 4 h. After 10 times reuse, the degradation rate of formaldehyde was still above 75%. The continuous Ag-TiO$_2$ fibers exhibited potential value in industrial application because of its easy separation and high photocatalytic activity. It would provide promising applications in cleaning indoor air.

4 CONCLUSIONS

The continuous TiO$_2$ and Ag-TiO$_2$ fibers have been synthesized by the sol–gel method using the polymer of titanate as the precursor solutions. The results demonstrated that the as-synthesized Ag-TiO$_2$ fibers all were crystal of anatase with the uniform size about 20–40 μm in diameter. The continuous Ag-TiO$_2$ fiber photocatalytic degradation rate of formaldehyde 98.4%, the degradation of formaldehyde was about 75% for resusing 10 times. The results would be useful to resolve some problems, such as easy to gather and difficult recovery and low energy efficiency, of the application of nano-TiO$_2$.

ACKNOWLEDGEMENTS

This work was supported by the Program for the National Natural Science Foundation of China (50872014, 51102026) and Aid program for Science and Technology Innovative Research Team in Higher Educational Instituions of Hunan Province.

REFERENCES

[1] A. Kubacka, M.F. Garcia, G. Colon. Advanced nanoarchitectures for solar photocatalytic applications, *Chem. Rev.* 112, 1555, 2012.

[2] H. Tong, S.X. Ouyang, Y.P. Bi, N. Umezawa, M. Oshikiri, J.H. Ye. Nano-photocatalytic materials: possibilities and challenges, *Adv. Mater.* 24, 229, 2012.

[3] X. Chen, S.M. Samuel. Titanium dioxide nanomaterials: synthesis, properties, modifications and applications, *Chemical Reviews*, 107, 2891, 2007.

[4] A. Fujishima, X. Zhang, D.A. Tryk. TiO$_2$ photocatalysis and related surface phenomena, *Sur. Sci. Rep.* 63, 515, 2008.

[5] S.P. Xu, D.D. Sun. Significant improvement of photocatalytic hydrogen generation rate over TiO$_2$ with deposited CuO, *Int. J. Hydrogen Energ.* 34, 6096, 2009.

[6] T. Guo, L.S. Wang, D.G. Evans, W.S. Yang. Synthesis and photocatalytic properties of a polyaniline-intercalated layered protonic titanate nanocomposite with a p-n heterojunction structure, *J. Chem. Phy. C*, 114, 4765, 2010.

[7] O. Akhavan, M. Abdolahad, A. Esfandiar, M. Mohatashamifar. Photodegradation of graphene oxide sheets by TiO$_2$ nanoparticles after a photo-catalytic reduction, *J. Chem. Phy. C*, 114, 12955 2010.

[8] C.X. Feng, J.W. Zhang, R. Lang. Unusual photo-induced adsorption-desorption behavior of propylene on Ag/TiO$_2$ nanotube under visible ligth irradiation. *Applied Sur*, 257:1864, 2011.

[9] J.Y. Hong, X.Y. Hao. TiO$_2$-g-C$_3$N$_4$ composite materials for photocatalytic H$_2$ evolution under visible light irradiation. *J. Alloy. Compound.* 509, 26, 2011.

[10] D. Huang, Y. Miyamoto, J. Ding. A new method to prepare high-surface-area N-TiO$_2$/activated carbon. *Mater. Lett,* 65, 326, 2010.

[11] R. Sasikala, A.R. Shirole, V. Sudarsan. Role of support on the photocatalytic activity of titanium oxide. *Applied Cata A*, 390, 245, 2010.

[12] A. Formalas, Process and apparatus for preparing artificial threads, *US Patent*, no. 1975504, 1934.

[13] L.Y. Zhu, G. Yu, X.Q. Wang, D. Xu. Preparation and characterization of TiO$_2$ fiber with a facile polyorganotitanium precursor method, *J. of Col. Int. Sci.*, 336, 438, 2009.

[14] R.S. Yuan, X.Z. Fu, X.C. Wang, P. Liu, L. Wu, Y.M. Xu, X.X. Wang, Z.Y. Wang. Template synthesis of hollow metal oxide fibers with hierarchical architecture, *Chem. Mater.* 18, 4700, 2006.

[15] Z.Y. Cai, J.S. Li, Y.G. Wang. Fabrication of zinc titanate nanofibers by electrospinning technique, *J. Alloys. Compound.*, 489, 167, 2010.

[16] J. Du, J. Zhang, Z. Liu, B. Han, T. Jiang, Ying Huang. Controlled synthesis of Ag/TiO$_2$ core-shell nanowires with smooth and bristled surfaces via a one-step solution route. *Langmuir, 22,* 1307, 2006.

Frontiers of Energy and Environmental Engineering – Sung, Kao & Chen (eds)
© *2013 Taylor & Francis Group, London, ISBN 978-0-415-66159-1*

Analysis on the energy consumption and usage characteristics of convenience stores

P.Y. Kuo
Department of Architecture, Chaoyang University of Technology, Taiwan

J.C. Fu & C.H. Cho
Graduate School of Architecture and Urban Design, Chaoyang University of Technology, Taiwan

ABSTRACT: According to the statistics, as of the end of 2011, there were a total of 9,871 convenience stores in Taiwan. On average, every 2,352 people owned a convenience store. The density of convenience stores in Taiwan is higher than that in other countries, such as Japan and Korea, and is the highest around the world. However, according to the review on relevant studies, the existing studies on the energy consumption of convenience stores in Taiwan only focused on overall energy consumption and did not probe in to the causes of energy consumption. Therefore, this study intended to investigate the compositions and characteristics of energy use of convenience stores from various aspects. Based on the investigation on a large amount of fundamental data, this study divided the architectural types and locations of convenience stores into four major architectural types and then analyzed the factors affecting the energy consumption of convenience stores, such as total electricity consumption, area, number of customers, categories of equipment, and Energy Use Intensity (EUI), to understand the current energy consumption and characteristics of convenience stores in Taiwan. Moreover, this study also used power monitoring system to measure and to record the electrical equipment of convenience stores, and probed into total electricity consumption month by month. The research results showed that among the four major architectural types of convenience stores, although the EUI of detached bungalow-typed convenience stores was only 1,368 [kWh/(m² · yr)], their average annual electricity consumption was as high as 175,475 kWh when the area was not taken into account. Therefore, detached bungalow-typed convenience stores remained the highest energy-consuming category among all the architectural types. The orientation and number of windows did not significantly affect energy consumption. However, the determination of orientation and installation of windows in future convenience stores are still advised to avoid the 180° scope from the south to the west and the north.

1 BACKGROUND AND PURPOSES

As of the end of 2011, the total population in Taiwan was approximately 23,220,000, and every 2,353 owned a convenience store. The density of convenience store in Taiwan is the highest in the world. However, owing to the 24/7 business hours, the total annual electricity consumption of convenience stores of various architectural types and where both heating and cooling electrical equipment are used is as high as 1.4 billion kWh. The studies concerning energy saving of convenience stores in Taiwan mainly focused on the energy saving technology of single equipment. Moreover, the studies on overall energy consumption of convenience stores only discussed the EUI of few samples. There was a lack of studies on the energy consumption compositions of different types of convenience stores. Because many factors, including equipment, area of the store, number of customers, period of peak hours, external design of architecture, climate, environment, etc., will affect the final energy consumption of convenience stores, it is necessary to analyze the study the compositions of energy use from various aspects. Consequently, to comprehensively probe into the energy use situation and usage characteristics of energy-consuming equipment, this study intended to use statistics and power monitoring to achieve the following purposes:

1. To investigate the fundamental data on the factors affecting energy consumption of convenience stores, such as orientation, building area, total annual electricity consumption, and

number of customers, to find out the critical factors affecting energy consumption of convenience stores.

2. To use 24-hour power monitoring system to understand the usage and characteristics of electrical equipment in convenience stores.

2 LITERATURE REVIEW AND RESEARCH METHOD

In recent years, with the rise of the awareness of the concept of energy saving and carbon reduction, both the operators of convenience stores and governmental departments have aggressively promoted the energy use efficiency of convenience stores, and the research issues of energy saving of convenience stores have become more and more popular. However, there is still a lack of studies on the usage characteristics of electrical equipment which is most significantly correlated with the energy use of convenience stores. Therefore, this study intended to use the following methods to conduct a preliminary investigation and have a better understand.

1. Analyses on fundamental data of convenience stores: This study investigated the fundamental data of 4,319 convenience stores in Taiwan. The data included architectural orientation, direction of window, address, area, total annual electricity consumption (kWh), and total annual number of customers. Moreover, because the architectural type and location of convenience stores also have significant effects on the annual number of customers and total annual electricity consumption (kWh), this study divided the buildings into four major categories, "convenience store along the street," "convenience store with unilateral arcade," "convenience store with bilateral arcade," and "detached bungalow-typed convenience store."would affect the EUI of convenience stores.

2. Power monitoring system for electrical equipment: This study used digital electricity meter to monitor various electrical equipment in convenience stores The digital electricity meter was installed in the completed power system for long-term observation. The monitored data would be immediately uploaded onto the cloud database via wireless internet for analyzing the electricity composition and usage characteristics of electrical equipment in an attempt to seeking the methods that can reduce energy use and improve equipment use efficiency.

3 RESEARCH PROCEDURES AND RESULTS

3.1.1 *Statistical analysis on the architectural types and EUI of convenience stores*

According to the statistical result of architectural types and EUI of convenience stores (Table 1), the four sides of the building of detached bungalow-typed convenience stores were exposed to outdoor air. Therefore, the indoor electricity consumption would be increased owing to the influence of external climate and environment. The annual average electricity consumption of a single detached bungalow-typed convenience store was as high as more than 170,000 kWh. However, detached bungalow-typed convenience stores were mainly established in suburbs where the cost of land and rent was lower. Consequently, the single store area was larger than that of other types of convenience stores, and the EUI was only 1,368 [kWh/(m² · yr)]. The average single store area of conveniences stores with unilateral and bilateral arcade which are commonly seen in downtown is only 113 m² because the rent is higher. To meet customers' needs in such a narrow space, the annual average EUI was 1,534 [kWh/(m² · yr)] and 1,577 [kWh/(m² · yr)], respectively. The convenience stores with unilateral/bilateral arcade were the highest energy-consuming architectural types.

3.1.2 *Statistical analysis on architectural orientation, number of windows, and EUI of convenience stores*

This study divided the orientation of convenience stores into eight directions (according to the position of automatic doors) and divided the installation of windows into three categories. Among the three categories of windows, a single-sided window was mainly observed in convenience stores along the street, while other categories were more frequently observed in detached bungalow-typed convenience stores, convenience stores with unilateral arcade and bilateral arcade. The statistical table (Table 2) of the annual EUI of convenience stores with various orienation and number of windows showed that, because the convenience stores with windows on three sides used more glass materials, the heat load entering the indoor was relatively increased. Therefore, the annual average electricity consumption was the highest (1520 [kWh/(m² · yr)]) among all types of windows. The result is consistent with the awareness of general public. Moreover, in terms of orientation, as shown in the radar chart, the annual average EUI of convenience stores whore orientation was south or northwest was significantly higher. The cause for such a phenomenon could be generally attributed

Table 1. Statistical table of EUI of convenience stores of various architectural types.

Items	Photographs	Average area	Annual average electricity consumption	Average EUI
Detached bungalow-typed convenience store		135.4	175,475	1,368
Convenience store along the street		120.2	163,242	1,469
Convenience store with unilateral arcade		113.5	162,285	1,534
Convenience store with bilateral arcade		113	166,536	1,577

Table 2. Statistical table of annual EUI of Convenience stores with various orientation and number of windows.

Orientation	Window on single side	Window on two sides	Window on three sides
North	1569	1569	1488
Northeast	1460	1462	1449
East	1394	1446	1436
Southeast	1391	1461	1592
South	1471	1525	1485
Southwest	1512	1441	1575
West	1379	1545	1547
Northwest	1543	1562	1539
Average	1468	1502	1520

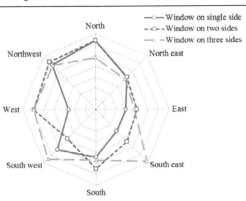

to the large amount of insolation caused by western exposure to the sun. Such convenience stores needed to use air conditioner more frequently to maintain the comfortable indoor shopping environment and further affect the energy consumption of convenience stores.

3.2 Monitoring of electrical equipment of convenience stores

This study further selected two convenience stores with similar number of customers but a significant difference in total annual electricity consumption from convenience stores (Table 3) to use power monitoring system to upload the real-time electricity consumption information onto cloud platform via wireless internet for long-term power monitoring (2010.10.25~2011.10.24), in order to understand the usage of electrical equipment.

3.2.1 Characteristics of electricity usage of annual and single-day equipment

The electricity consumption data of convenience stores sample A and B at traffic artery area obtained from the power monitoring system showed that (Figs. 1 and 2), the monthly average electricity consumption of the two samples was 19,953 (kWh/month) and 12,190 (kWh/month), respectively. Although the total electricity consumption from June to September during the electricity peak season (summer) accounted for 37~38% of the total

Table 3. Table of the sample fundamental data of annual monitoring of convenience stores.

	Sample A	Sample B
Number of customers	295,073	295,076
Total electricity consumption	202,160	155,200
Total area	138.5	135.5
EUI	1,459	1,145
Difference in number of customers	3	
Difference in electricity consumption	46,960	

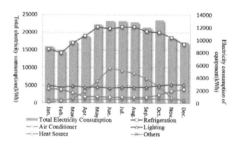

Figure 1. Statistics of total electricity consumption and monthly electricity consumption of various electrical equipment of sample A.

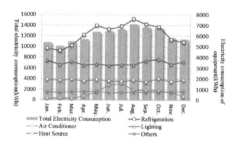

Figure 2. Statistics of total electricity consumption and monthly electricity consumption of various electrical equipment of sample B.

annual electricity consumption, the total electricity consumption [93,158 (kWh/year)] of sample A was still larger than that of sample B under the situation where the number of customers was not significantly increased. Such a fact suggested that it is necessary to probe into the use of various electricity equipment of sample A to reduce unnecessary energy consumption. Moreover, the electrical equipment in convenience stores were divided into five major categories (including "refrigeration system," "air conditioner," "lighting equipment," "heat source equipment," and "others") for further analysis. The electricity

consumption of the refrigeration system which created the largest operating income for convenience stores accounted for 50.4% and 52.9% of the electricity consumption of the two samples, respectively. However, the critical factor leading to the difference in total electricity consumption between two convenience stores was air conditioner. During the electricity peak season (summer), the electricity consumption of air conditioners of sample A and sample B accounted for 5.8% and 12.5% of the total annual electricity consumption, respectively. However, the total electricity consumption of air conditioner in sample A was approximately 4 times of that of air conditioner in sample B, which indirectly led to the significant difference in total electricity consumption between sample A and sample B. Such a phenomenon is worthy of investigation for convenience stores where air conditioner is turned on all day.

The further analysis (Fig. 3 Average daily electricity consumption of various equipment in convenience stores was approximately 510 kWh) on the average daily electricity consumption of electrical equipment of convenience stores showed that the electricity consumption of refrigeration equipment consisting of open refrigerator, combination refrigerator, freezer, horizontal refrigerator, cabinet refrigerator, and smoothie machine was the highest. Under the situation that there was a need to constantly produce ice and maintain the coolness and fresh-ness of products, the average daily electricity consumption was refrigeration equipment was as high as 56.9% (290.2 kWh) of the total electricity consumption, while that of the air conditioner maintaining indoor comfortable shopping environment accounted for 11.7% (59.9 kWh) of the total daily electricity consumption. The electricity consumption of indoor lighting equipment which was turned on all day and that of lighting equipment for signs and arcade which was turned on for as many as 12 hours also accounted for 19.3% (98.6 kWh) of the total electricity consumption because many lamps were used. The electricity

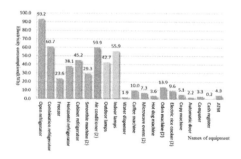

Figure 3. Average daily electricity consumption of various equipment in convenience stores.

consumption of lighting equipment was only second to that of refrigeration equipment.

3.2.2 Hourly electricity usage characteristic of single equipment

The usage characteristics of all the electrical equipment in convenience stores varied with the period of time. If the electricity usage characteristic of electrical equipment was further investigated, the electricity peak hours or irrational electricity usage characteristics could be found and relevant improvement countermeasures could be proposed to improve the irrational characteristics. Owing to the restriction on the length of this study, this study only analyzed the 24-hour electricity usage characteristics of two electrical equipment of sample A and sample B, in order to understand the differences between them.

As shown in Figure 4 Hourly electricity consumption of refrigeration equipment, the electricity consumption of sample A and B was not significantly affected by different periods of time. In addition, the daily electricity consumption was 14.5 kWh and 8.4 kWh, respectively. The cause for the difference in electricity consumption (nearly 6 kWh daily) between these two convenience stores that were both located at traffic artery areas with similar equipment where the difference in number of customers was merely 3 people was the time of usage of equipment. On average, the equipment in sample A was used for more than 7 years, while that in sample B was used for 3 years only. The aging of equipment decreased the operation efficiency and led to the significant increase in electricity consumption. The similar phenomenon was also observed in Figure 5 air condition. The equipment in sample A was used 4 years longer than that used in sample B. Moreover, the air conditioner used in sample A was general variable-frequency air conditioner, while that used in sample B was new variable-frequency air conditioner. The comparison between them showed that the daily average electricity consumption of sample A was approximately 3.4 kWh, which was significantly

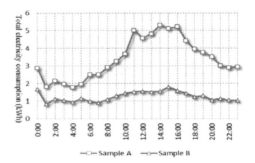

Figure 5. Hourly electricity consumption of air conditioner.

higher than that of sample B (1.3 kWh) by more than more than double, particularly in the period from 10:00 to 18:00. The change in the electricity consumption curve of sample A was significantly higher, and such usage characteristic deserved the attention from operators of convenience stores.

4 CONCLUSIONS AND SUGGESTIONS

After performing various analyses, this study reached the following conclusions based on the aforementioned results: 1) among the four major architectural types of convenience stores, although the average EUI of detached bungalow-typed convenience stores was merely 1368 [kWh/(m^2·yr)], the annual average electricity consumption was as high as 175,475 kWh, suggesting that such architectural type was the highest energy-consuming one; 2) the orientation and number of windows did not have a significant effect on energy consumption. However, the establishment of future convenience stores was advised to avoid the 180° scope from the south to the west and the north; 3) years of use of equipment and whether the air conditioners were variable-frequency one had a significant effect on energy consumption of equipment, as indicated by the figures of electricity consumption of refrigeration equipment, air conditioner, and microwave ovens. Operators of convenience stores were advised to gradually replace the old equipment by new one to reduce energy expenditure.

REFERENCES

1. B.Y. Kuo, 2010, "Green Convenience Store Grading Certification and Incentive Improvement Plan", Architecture and Building Research Institute, Ministry of The Interior.
2. K.P. Li, 2003, "An Analysis on the Effect of Refrigeration System on the Energy Consumption of Buildings of General Merchandise–Taking Convenience Stores for Example", Architectural Institute of the ROC.

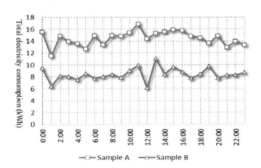

Figure 4. Hourly electricity consumption of refrigeration equipment.

Frontiers of Energy and Environmental Engineering – Sung, Kao & Chen (eds)
© 2013 Taylor & Francis Group, London, ISBN 978-0-415-66159-1

Performance evaluation of solar aided coal fired power plant using structural theory of thermoeconomic

R.R. Zhai, P. Peng, Y.P. Yang & D.G. Chen
North China Electric Power University, Changping District, Beijing, China

ABSTRACT: The structural theory of thermoeconomic is applied to a hybrid system named Solar Aided Coal Fired Power Plant (SACFPP) where solar energy is combined to a 600 MW coal-fired system using the way of taking place the high pressure bled steam feed heater. Based on Fuel-Product concept of structural theory, the performance of components and the interaction between them were evaluated. The performance of components at design condition of solar aided coal fired power plant are evaluated by the way of system simulation. The results prove that the Structural Theory is a useful tool of complex energy system and the solar aided coal-fired power plant can use the solar energy efficiently.

Keywords: unit energy cost, solar aided coal-fired system, structural theory, performance analysis, consists of energy cost

1 INTRODUCTION

The situation of inadequate resources of the fossil fuels and global climate changes as well as environmental deterioration has made rational use of energy increasingly important. Solar aided coal fired power plant, to some extent, overcome the weak points of traditional solar thermal power plant, and now it become a feasible way to use solar energy more effectively (Eric Hu. et al. 2010). The past thermodynamic analysis of solar aided coal fired power plant is based on the first and second law of thermodynamics. In these study, the quantity and quality of energy can be studied well, but they still have some defects: Firstly, it didn't take the equivalence of the energy destruction into account (Lin. W.C. et al. 1994). Secondly, when the thermal performance of an single equipment changed, it failed to reflect the complex impact caused to the system (Lozano M.A, et al. 1993). Moreover, as for multi-energies hybrid power systems, it could not distinguish the accurate thermodynamics difference of these energy resources (Uche. j. et al. 2000). Based on Structural Theory of Thermo-economics (Valero A Serra. et al. 1993), this paper present a novel thermo-economic mathematical model and an exhaustive analysis of every components of this solar aided coal fired power plant. By this way, unit exergy consumption and the composition of unit exergy consumption was researched. In the last, the distinction of different energy resource in hybrid power plant was researched.

2 SOLAR AIDED COAL FIRED POWER PLANT SPECIFICATION

The energy system reviewed in this research is a hybrid power plant consisting of fossil resource and solar resource. There are many schemes about the integration of solar heat with coal fired system (Qin, Yan. 2010). The system studied in this paper is using solar energy to replace the high pressure stream extraction. The integration enables the solar energy take into the energy system at a higher level of temperature, thus, with higher system efficiency (Cui, Y. H. 2009). The detailed working process is as follows. Synthetic oils are used in parabolic trough collectors as a heat transfer medium between collectors and Rankin cycle. It flow across the collectors and are heated to high temperature, and then flow through the heat exchanger to improve the temperature of the feed water. In order to decrease the influence of the boiler, the temperature of the feed water temperature is constant by regulating the flow rate of the synthetic oils (Jing, W 2010). The schematic diagram of the system is shown in Figure 1.

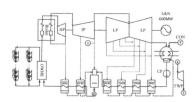

Figure 1. Schematic diagram of the solar aided coal-fired power plant.

3 EXERGTIC MODEL

3.1 *Productive structure*

The productive purpose of a process device measured in terms of energy is named as "product"; and the consumed energy flow to create the "product" is "fuel". Thus, a set of higher-level relationships derived from the productive purpose of each device could be defined (Uche J, 2000). Using this concept of Fuel-Product, the flows of each device in this system can be classified into "product" or "fuel" based on the functionality of each device. Thus the system can be converted into the productive structure. On Structural theory of thermoeconomics, the "product" of the condenser is named as entropy. It is dispatched to every unit, so every component consumes "fuel" and "entropy" two kinds of exergy resource (Valero A. et al. 2006).

The productive structure diagram for the power plant is presented in Figure 3. Beside the components match with the Physical Structure of the system, there are some fictitious components represented by rhombus and circles. Rhombus and circles represent collector and distribution respectively. Exergy is collected together at rhombus and then dispatched to other components form circle (Valero A. et al. 2006). In Figure 3, full line and dotted line represent energy "fuel" and entropy respectively.

3.2 *Characteristic equations*

The thermoeconomic model is formed by a set of "characteristic equations", which relate each inlet flow to outlet flow and internal parameters that depend only on the behaviors of relevant subsystems (Chao. Z. 2005). The characteristic equations of the pant can be rewritten in the form as:

$$R_i = f_i(x_k, R_j) \qquad (1)$$

In this characteristic equation, k represents the sequence number of the components. x_k represents the inherent parameters of unit k. R_j symbolizes the outlet exergy. Considering the inherent parameters is assumed constant, after Euler transformation, this function can be expressed as:

$$R_i = \sum \left(\frac{\delta f_i}{\delta R_i} \right) R_i = k_{ij} p_j \qquad (2)$$

k_{ij} is named as technical production coefficients (as is show in function 2), it represents how much "product" of unit j is cost when unit "product" of unit i is obtained. The technical production coefficients of each unit can be calculated by the Productive structure of the system (see reference 10 for details).

Fictitious units also have their own characteristic equations, they are expressed as:

$$F_i = \sum P_j, F_i = r_{ij} P_j \qquad (3)$$

3.3 *Unit energy cost functions*

Partial Derivative of the characteristic equations is expressed as:

$$\frac{\delta R_0}{\delta R_i} = \sum \frac{\delta R_0}{\delta R_j} \frac{\delta f_j}{\delta x_i} \qquad (4)$$

It is named as unit energy consumptions (k_p^*). It represents the external resource needed for getting unit "product" of unit i. With the above theory of technical production coefficient, function 3 can be translated into:

$$k_{p,i}^* = k_{0,i} + \sum k_{ji} k_{p,j}^* \qquad (5)$$

By solving the equations systems consist of technical production coefficients equation (2), (3) and characteristic equations (5), unit energy cost will be obtained easily. Unit exergy cost represents the consumptions of external exergy when unit "product" of certain components was obtained (Rosen. Ma. et al. 2003). The external consumption of the system this paper descripted is solar energy and coal energy. As the Productive Structure of system shown in Figure 2, the "fuel" of solar energy collector (component 22) and the oil water heat exchanger are all come from solar energy. Other components, including the boiler, their "fuel" is come from the "product" of the virtual components (component B2). Base on the exergy cost analysis, this paper proposes two new parameters to distinguish the solar energy part and coal energy part of "fuel" of every component.

$$k_{P,i,sol}^* = k_{P,i}^* \frac{k_{P,8}^* \ast p_8}{k_{P,9}^* \ast p_9 + k_{P,10}^* \ast p_{10}} \qquad (6)$$

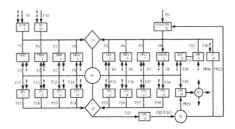

Figure 2.   Productive structure of the solar aided coal-fired power plant.

$$k^*_{P,i,coal} \quad \frac{k^*_{P,9} * p_9 + k^*_{P,10} * p_{10} - k^*_{P,8} * p_8}{k^*_{P,9} * p_9 + k^*_{P,10} * p_{10}} \qquad (7)$$

where p_9, p_{10}, p_{22} represent the "product" of the boiler and the respectively. The $k^*_{P,i,sol}$ was named as Unit Solar Exergy Consumption (USEC) and $k^*_{P,i,coal}$ was named as Unit Coal Exergy Consumption (UCEC).

4 CASE STUDY

4.1 The consist of unit exergy consumption of each unit

Based on a 600 MW generator system, the various parameters of design conditions were calculated. The important nominal parameters of the system are listed below: the direct solar radiation is 925 W/m², in the solar thermal collector, inlet synthetic oils temperature and outlet synthetic oils temperature of OWHE is 250 °C and 328 °C respectively. The outlet temperature of OWHE is kept constant by regulating the flow rate of the synthetic oils.

According to the analysis above, unit exergy consumptions can be described as:

$$k^*_{P,i} = k^*_{FB,i} + KIk^*_{FB,i} + KS_i k^*_{FS,i} \qquad (8)$$

In this function the first part represent the minimum limit theoretical external exergy consumption of unit "product" (TEC), the second part is the external consumption caused by irreversibility (IEC). The last part is the external consumption caused by negative entropy. Through this equation (NEC), it is easy to find the reasons for the growth of the unit exergy cost.

The consist of unit exergy consumption for the solar aided coal fired plant are shown in Figure 3. On the design condition, the kp^* of the feed water heater is decrease progressively form high pressure heater to low pressure heater. the variation law of kp^* and KB (the ratio of "fuel" and "product") was not Coincide. The kp^* of upstream component of the system are higher than that of the downstream components. However there was no obviously trend of KB like this. The kp^* of the units are connected with not only the performance of this unit, but also the construction of the overall system. It reflected the influence to the whole system caused by certain units. On the other hand KB just reflected the thermal performance of single unit. The unit cost of low pressure feed water heater working ineffectively on part load of the system, the necessity to improve the performance of these devices.

The kp^* of water oil heat exchanger (unit 8) is 5.12, the maximum of the system, the reason is that the TEC of it is highest of all. The kp^* of the solar energy collector (unit 22) is the second highest of the system, 3.10, it is caused by the high value of IEC. By the means of thermoecnomic analysis, the reason of the increase of the system can be found easily.

4.2 The energy cost contrast between SACFPP and conventional power plant

The unit exergy cost results for the conventional power plant and the SACFP (USEC and UCEC) are shown in Table 2. It is easy to get the conclusion that almost all the components of SACFP are higher than conventional plant.

The unit energy cost of the generator (unit 23) is 2.352. the USEC of the generator is 0.16, the proportion of solar energy is 6.80% of the unit exergy cost. This parameter is of great important to define the price of electricity price also has the profound significance.

The reciprocal of it is the energy efficiency of the system. The unit exergy consumption increased

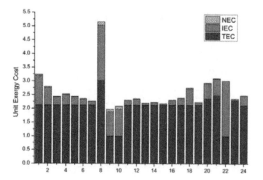

Figure 3. The consist of unit exergy consumption of each unit.

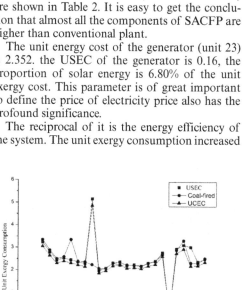

Figure 4. The comparison USEC, UCEC and unit exergy consumption of coal fired power plant.

from 2.300 to 2.352. It was caused by the low-level efficiency of the solar energy collector. However, it is easily to notice that unit coal exergy consumption decrease from 2.300 to 2.192. In other words, the coal we consumed when we get unit "product" of the generator is reduced. The coal consumption of the conventional system is 281.78 g/kWh, the coal consumption of SACFP is 275.82 g/kWh, the ability of solar aided coal fired power plant to save coal resource is obvious.

5 CONCLUSION

Based on thermoeconomic theory, a detailed unit exergy cost analysis on a 600 MW solar aided power plant and conventional power plant are proposed.

By means of Thermoeconomic theory, the value of unit exergy consumption can reflect the necessity to improve the performance of each units. The large the unit exergy consumption is, the necessity to improve the performance of this unit is.

In this analysis, unit exergy cost was divided into three part: the minimum limit theoretical external exergy consumption of unit "product" (TEC), the external consumption caused by irreversibility (IEC). The external consumption caused by negative entropy (NEC). By this means, it is easy to find the reasons for the growth of the unit exergy cost.

Using thermoeconomic theory, the accurate solar energy part (USEC) and coal energy part (UCEC) were obtained, considering the unequivalence between solar energy and coal energy. In this way the proportion of this two energy resource in the product of the plant is obtained, and does this, to define the price of electricity price also has the profound significance.

ACKNOWLEDGEMENT

The research work is supported by China National Natural Science Foundation (No. 51106048) and National High-tech R&D Program of China (863 Program) (2012AA050604).

REFERENCES

Cheng Weiliang, Chen Danghui, Xu Shouchen, Economic Analysis of Thermal Power Plant [J] Power Energy, 2005, 25(24):108–113.

Cui Yinghong, Chen Juan, Efficiency anylasis on different models of cogeneration integrated with solar heating system, 2009. 36(1);69–75.

Hu Eric et al. Solar thermal aided power generation [J]. Applied Energy. 2010, 87(9):2881–2885.

Lozano MA, Valero A. Theory of the exergeticcost [J]. Energy. 1993. 18(9);939–960.

Mills D. Advances in solar thermal electricity technology [J]. Solar Energy, 2004. 76;19–31.

MingShan Zhu. Exergy Analysis of Energy System [M]. Beijing: Tsinghua University, 1988: 229–313.

Rosen, Ma, Dincei, I Thermoeconomic analysis of power plants; An application to a coal fired electrical generating station. Energy Conversion and Management, 2003, 44(17);2743–2761.

Song Zhiping. Principles of Energy-saving [M]. Beijing: Water Conservancy and Hydropower Press. 1985:123–312 (in Chinese).

Song Zhiping. Principles of Energy-saving [M]. Beijing: Water Conservancy and Hydropower Press. 1985:123–312 (in Chinese).

Uche J. Thermoconmic Analysis and simulation of a combined Power and Desalination Plant [D]. Department of Mechanical Engineering; University of Zaragoza; 2000.

Uche J. Thermoconmic Analysis and simulation of a combined Power and Desalination Plant [D]. Department of Mechanical Engineering; University of Zaragoza; 2000.

Valero A, Correas L, Zaleta A et al. On the thermoeconomic approach to the diagnosis of energy system malfunction.

Valero A, Serra L, Lozano MA. Structural Theory of Theory of Thermo economics, international symposium on Thermodynamics and the Design, Analysis ande Improvement of Energy Systems. ASME Book NO. H00874, New Orleans, 1993, 189–198.

Valero A, Serra L, Uche J. Fundamentals of Exergy Cost Accounting and Thermoecomics. Part I Theory. J. Energy Resour. techno, 2006, 128, 128:1–8.

Wanchao Lin, Energy saving theory of Thermal Power Plant [M], Xi' An, Xi' An Jiao Tonguniersity, 1994.

Wu Jing, Wang Xiuyan, Study on Integrated Modes of Solar 2 Coal Hybrid Power Generation Systems [J] Journal of Chinese Society Of Power Engineering, 2010, 30(8);639–643.

Yan Qin. Thermodynamic Characteristic Research on Solar Aided Coal-fired Power Generation System [D], North China Electric Power university, 2010.

Zhang Chao, Thermoeconomic Analysis and Optimization of complex Energy Systems [D], HuBei: Huazhong University of Science and Technology, 2006.

Frontiers of Energy and Environmental Engineering – Sung, Kao & Chen (eds)
© 2013 Taylor & Francis Group, London, ISBN 978-0-415-66159-1

Research on slightly polluted river water treatment of Xinkai river in Changchun city

Z.G. Zhao, Y.X. Zhang, X. Hu, S.J. Peng & Z.Y. Li
Water Resources and Water Engineering Research Institute, Beijing University of Technology, Beijing, China

ABSTRACT: The series combination process of oxidation pond and constructed wetlands was studied on slightly polluted river water treatment of Xinkai River in Changchun city. Based on the study, the series of water treatment process could be used to treat slightly polluted river water, because of the effluent of sewage treatment plant and channel point-source pollution. The main indicators of treated water could achieve or exceed the surface water class IV standard. It can be used for agricultural irrigation, fisheries and wetlands complement; Constructed wetland system plays a main role in the series process, so it should be adopted preferentially in practical engineering applications; In the Spring, the quality of river water was so poor that treated result was bad, but it can improve operation effect by strengthening the oxidation pond aeration.

Keywords: river ecology treatment; oxidation pond; constructed wetlands

1 INSTRUCTION

In recent years, society and economy and construction and other aspects of Changchun city sustain rapid development. Conversely, these result in Changchun northwest water resources shortage and serious pollution, changed its former clean and beautiful ecological environment.

Xinkai River was western the main drainage system and the reclaimed water supplement channel of agricultural irrigation and wetland water. It also supplies water for natural runoff. Due to the river coast of point source pollution, the water quality was unable to reach the river ecological restoration, wetland water and landscape water requirements. In order to make full use of Xinkai River water resources, alleviate water resources shortage in the northwest of the Changchun City, improve ecological environment of Boluohu wetland group, promote agricultural production conditions, the government of Changchun city implemented the reclaimed water utilization project, namely "lead Xinkai River water into the Taipingchi reservoir" engineering. The main target of project was to reduce water pollution load and improve water quality function.

With the implement of the project, it construct test field on the right bank of Xinkai River in Shunshanpu. The series combination process of oxidation pond and constructed wetlands was tested, providing the reliable design parameters for the practical engineering design and construction.

2 EXPERIMENTAL DESIGN

2.1 Experimental design process

Test field covers an area of about 40000 m², size of 635 m × 60 m. Test time was from May to October. Test field was composed of Regulation pool, Oxidation pond, Constructed wetland, Ecological river, Hybrid plant pond and Reservoir. The series combination process was showed in Figure 1.

2.2 Test field profiles

The available volume of Regulation pool was 3000 m³, size of pool body was 55 m × 37 m × 2.5 m, depth of water was 2.0 m.

The oxidation pond for strip corridors, was around with trapezoidal cross section. The size of pool body was 82 m × 15 m × 2.5 m, depth of water was 2.0 m. The available volume was 1058 m³. The oxidation pond range surface placed type Aquamats ecological base, a total of 92 pieces, either pond layout 46 pieces. The ponds both use EPDM pipe to aerate.

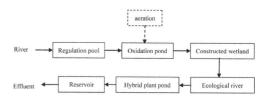

Figure 1. The series combination process figure.

The size of constructed wetland was 30 m × 59 m × 1 m. with 4 units in parallel. Wetland bed body adopts the double-layer structure. A and C wetland unit cover 60 cm original soil in the upper and 40 cm gravel layer in the lower. The gravel gradation was 50 mm, 20 mm, 10 mm and 5 mm. B and D wetland unit cover 40 cm original soil in the upper and 60 cm sand layer in the lower. The sand gradation was 2 mm. The surface layers of constructed wetland plant with reed, cattail and calamus. The planting depth was 30 cm, planting density was 30 strains/m², planting spacing was 0.20 m × 0.17 m. The vegetation porosity was 0.75.

The size of ecological river was 60 m × 59 m, consit of 4 units in parallel. Design water depth was 0.2 m, planting reeds. The planting depth was 30 cm, planting density was 30 strains/m², planting spacing was 0.20 m × 0.17 m. The vegetation porosity was 0.75. Reed planting mode was the same as constructed wetland unit.

The size of hybrid plant pond was 177 m × 59 m, consist of 2 units in series. The former unit of length, width, were respectively 78 m and 59 m; the other later unit of length, width, were respectively 97 m and 59 m. The former unit of the hybrid plant wetland pond plant some reeds, mixed with the local original plants. The latter unit keep local original plants.

The size of reservoir was 210 m × 59 m, water depth was about 2 m. The effluent of treatment water flows into the Xinkai River.

2.3 *Experimental water quality and analytical method*

Experimental water was taken from the actual river water of Xinkai River. The Flow was from 730.1 m³/d to 1407.62 m³/d. The quality of the experimental water indicate: COD was from 49.72 mg/L to 110.70 mg/L, TP was from 0.60 mg/L to 1.18 mg/L, NH_4^+-N was from 4.86 mg/L to 12.02 mg/L, NO_2-N was from 0.74 mg/L to 1.23 mg/L, NO_3-N was from 6.18 mg/L to 32.25 mg/L, DO was from 49.72 mg/L to 110.70 mg/L. Experimental various indicators of monitoring methods accord to the National Protection Environmental Agency "water and wastewater monitoring analysis method".

3 RESULTS AND DISCUSSION

3.1 *COD removal effect analysis*

From the Figure 2, it showed that COD inlet concentration was higher on May, the effluent concentration decreased less, COD removal rate was not high. With the passage of time, the effluent COD concentrations gradually decreased, the removal

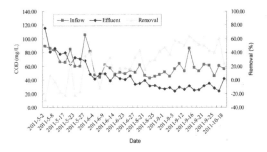

Figure 2. Inlet and outlet water changes and removal of COD.

rate increased gradually. By the middle of August, the effluent concentrations was less than 30 mg/L, achieved the surface water standards of grade IV, removal rate reached about 50%. What were the reasons? I think there were several reasons. In the spring, river flow was less, so water quality was relatively more serious pollution. At the same time, because of lower temperatures, poor plant and microbial activity on the pollutant removal capacity were weak. With temperature increasing and plant growing, plants began to absorb the organic matter, so removal efficiency was significantly improved.

3.2 *Nitrogen and phosphorus removal effect analysis*

From the Figures 3–5, it show that TP, NH_4^+-N and TN inlet concentration was higher on May like COD, the effluent concentration decreased less, TP, NH_4^+-N and TN removal rate was not high. But after June, the effluent concentrations of TP, NH_4^+-N and TN gradually decreased, the removal rate increased gradually. By the middle of August, the effluent concentrations TP, NH_4^+-N and TN separately were less than 0.2 mg/L, 0.78 mg/L and 1.46 mg/L, exceeded the surface water standards of grade IV, removal rate separately reached about 84%, 92% and 91%. The reasons were similar with COD. In the spring, river flow was less, so water quality was relatively more serious pollution. At the same time, because of lower temperatures, poor plant and microbial activity on the pollutant removal capacity were weak. With temperature increasing and plant growing, plants began to absorb the organic matter, so removal efficiency was significantly improved.

3.3 *COD removal efficiency analysis of every process units*

From the Figure 6, it showed that COD removal efficiency of the overall process was not high. In general, the trend for the COD removal efficiency

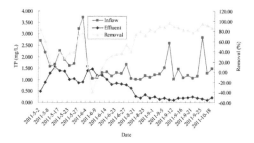

Figure 3. Inlet and outlet water changes and removal of TP.

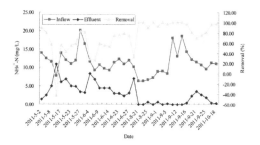

Figure 4. Inlet and outlet water changes and removal of NH_4^+-N.

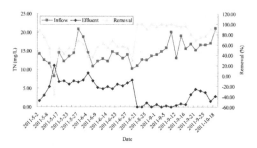

Figure 5. Inlet and outlet water changes and removal of TN.

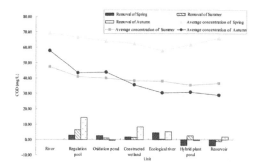

Figure 6. Inlet and outlet water changes and removal of COD.

was, Regulation pool > Constructed wetland > Ecological river > Oxidation pond > Hybrid plant pond > Reservoir, however. But the trend has differences in every season. In spring, removal abilities of ecological river system was the highest, the removal ability of regulation pool, Constructed wetland and Oxidation pond were similar, slightly lower than the ecological river. All the worse, the effluent value of COD increased after Oxidation pond and reservoir because of the decay of plant of last year. With the vigorous growth of the plant in summer and autumn, the removal abilities of the Constructed wetland system and Ecological river system with reeds were significantly higher than other units. So when the water quality was poor in spring, it may adjust the oxidation pond to improve treatment effect. But in summer and autumn, it give full play to the plants' removal ability.

3.4 Nitrogen and phosphorus removal efficiency Analysis of every process units

From the Figures 7–9, it showed that the removal ability of hybrid plant pond and reservoir were significantly higher than the other units in spring. Why? The main reason was that these two units have wide variety of plants. Some plants can quickly

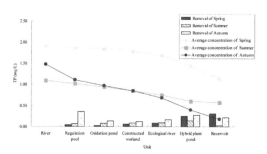

Figure 7. Inlet and outlet water changes and removal of TP.

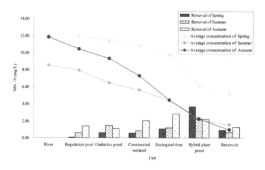

Figure 8. Inlet and outlet water changes and removal of NH_4^+-N.

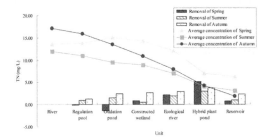

Figure 9. Inlet and outlet water changes and removal of TN.

grow in the spring and absorb a lot of nitrogen and phosphorus. However the reeds of other units grow slowly and its removal efficiency was low. Entering the summer, reeds and other high removal plants had better grow and play a major role.

4 CONCLUSION

The series combination process of oxidation pond and constructed wetlands could be used to treat slightly polluted river water of Xinkai River in Changchun city. The main indicators of treated water could achieve or exceed the surface water class IV standard. It can be used for agricultural irrigation, fisheries and wetlands complement. COD, TP, NH_4^+-N and TN separately removal rate reached about 50%, 84%, 92% and 91%.

Constructed wetland system plays a main role in the series process, so it should be adopted preferentially in practical engineering applications.

In the spring, the quality of river water was so poor that treated result was bad, but it can improve operation effect by strengthening the oxidation pond aeration.

REFERENCES

Chinese Environmental Protection Agency. 2002. Methods of Water and Wastewater Determination. Beijing: China Environmental Science Press.

Neralla S, et al. 2000. Improvement of domestic wastewater quality by subsurface flow constructed wetlands. Bioresource Technology 75(1): 19–25.

Shao-lin He & Qi Zhou. 2004. Application of Constructed Wetland in Non-point Source Pollution Control. Sichuan Environment 23: 71–74, 97.

USA Environmental Protection Agency. 2000. Constructed Wetlands Treatment of Municipal Wastewaters, M. Cincinnati. EPA/625/R-99/010, http://www.epa.gov/ORD/NRMRL.

Wei-gang Yuan & Zhi-yi Fan. 2007. Application of Aquamats technology in lake harness and maintenance. China Water & Wastewater 23: J1–J4, August 2007.

Xiao-dong Chen et al. 2007. Application Study and Engineering Demonstration of Northern Constructed Wetlands Wastewater Dwasposal Process. Environmental Protection Science 33: 25–28, 61.

Ye-chun Li, et al. 2010. Aquamats ecological base in northern China demonstration of wastewater treatment. Science & Technology information: 501–502, March 2010.

Zhi-gang Zhao, et al. 2011. Research on reclaimed water reuse in northwest ecological economic zone of changchun city. 2011 International Conference on Electric Technology and Civil Engineering, ICETCE 2011—Proceedings: 4380–4383.

Zhi-gang Zhao, et al. 2011. Treatment of micro-polluted river water by using integrated horizontal-flow constructed wetland. 2011 International Conference on Remote Sensing, Environment and Transportation Engineering, RSETE 2011—Proceedings: 1856–1858.

Frontiers of Energy and Environmental Engineering – Sung, Kao & Chen (eds)
© 2013 Taylor & Francis Group, London, ISBN 978-0-415-66159-1

A study on evaluation of environmental impacts of shore protection facilities at Toban Coast

Q. Jiang & C.K. Zhang
College of Harbour, Coastal and Offshore Engineering, Hohai University, Nanjing, P.R. China

M. Fukuhama
Water Resources Environment Technology Center, Tokyo, Japan

ABSTRACT: Evaluation of the environmental impacts of shore protection facilities is the basis of protecting coastal environment and ensuring sustainable coastal development. In this study, to estimate the influences of coastal defense structures, such as breakwaters and headlines, on the physical and habitat environment at Toban coast of Japan, field surveys on the habitat environment and inhabitation of coastal life and plants were firstly carried out both in nearshore waters and on land; then the representative species for evaluation were selected on the basis of accumulated field surveying data and basic knowledge of coastal ecosystem; considering the bionomics and inhabitation characteristics of coastal life and plants, the dominant environment impact factors and suitability index models for the selected representative species were examined; finally, the habitat suitability for clams in areas around breakwaters and headland at Toban coast was evaluated by using an integrated coastal environment impact evaluation model that combines coastal hydrodynamic models and HEP technique. Comparisons between the simulated results and measured data indicate that the model can give reasonable habitat impact assessment.

Keywords: coastal ecosystem; coastal life and plants; shore protection facilities; habitat evaluation procedure; environment impact evaluation

1 INTRODUCTION

For sustainable coastal zone development, protection and conservation of physical and habitat environment for natural life and plants attracts more and more attention to not only policy makers and professionals but also ordinary people. The construction of shore protection facilities like headlands, breakwaters, and seawalls results in visible reduction in natural beaches, deterioration in coastal life and plants and influences on nearshore fisheries. From the point of conservation and mitigation of coastal environment, environmental impact evaluation is necessary before and after building coastal protection structures. However, few studies can be found on how to evaluate the impacts of shore protection structures on coastal environment and ecosystem.

In this study, aiming at establishing a method to evaluate habitat changes caused by the construction of coastal protection structures, Toban coast of Japan is chosen as a prototype coast for environmental impact evaluation. To clarify the present habitat environment under the influences of coastal engineering works at Toban coast,

field surveys both in nearshore waters and on land are conducted. Then, the representative species for environment impact evaluation at Toban coast are selected based upon the knowledge on coastal ecosystem and available field surveying data. Considering the inhabitation characteristics and life cycle of the selected representative species, dominant environment impact factors for inhabitations of the selected representative species are determined. Moreover, the SI (Suitability Index) models representing the relationship between dominant impact factors and inhabitation for the representative species are established by using the available measured data. Finally, an integrated coastal environment impact evaluation model is developed by combing hydrodynamic models and Habitat Evaluation Procedure. The HSI (Habitat Suitability Index) for clams is calculated in areas where headlands and breakwaters are built. Comparison of the calculated results and measured data indicates that it is possible to apply the proposed assessment model to real engineering problems if the Suitability Index model for the representative species can be properly built up.

2 HABITAT ENVIRONMENT AT TOBAN COAST

Toban coast located at Hyogo prefecture, Japan. It has a long history of beautiful beaches with white fine sand and pine trees and its total length is about 26 km. However, protection works like breakwaters, headlands and beach nourishment are recently built to against beach erosion and storm surges. To understand the habitat environment changes after construction of these coastal protection structures, field surveys and investigations on the inhabitations of 6 representative species (to be explained in section 3) including clams, eelgrass are carried out at 4 typical coastal districts (Matsue, Fujie, Taniyagi and Yagi) where protection facilities are different in structures as shown in Figure 1.

In the on land survey, the configuration of structures, geographic features (coastal sand, foreshore slope, backshore width, coastal line shapes), and coastal plant inhabitations are investigated. For surveys in nearshore waters, bed materials and inhabitation of sea life and plants (eelgrass, benthos, and fishes) are examined through sampling. Together with the existing field surveying data at Toban coasts, environmental information concerning the constructed coastal defense structures and inhabitation of coastal life and plants are obtained.

More than 10 species of coastal life and plants, including benthos (lancelets), sea grass or seaweed (eelgrass), fishes (flounder), birds (herons and little egret), Sea turtles (loggerhead turtle) are found at the concerned coastal districts. Figure 1 is an example of the observed distribution of inhabited coastal life and plants at Fujie districts.

3 REPRESENTATIVE SPECIES AND SUITABILITY INDEX MODELS

3.1 Representative species

Selection of representative species or biological groups for a coastal ecosystem is crucial for environmental impact assessment. In this study, it is proposed that the representative species or biological groups for evaluation are chosen in terms of the following selecting criteria. The selected representative species or biological groups should be the ones that are: *(a) most sensitive to the human activities, such as the construction of coastal defense structures; (b) most attractive to the public or society in terms of scarcity, or in extinction and economically usefulness; (c) the characters of or of important values in a local ecosystem. Meanwhile, (d) enough data is available for building its SI (suitability Index) model and quantitate evaluation models are applicable.*

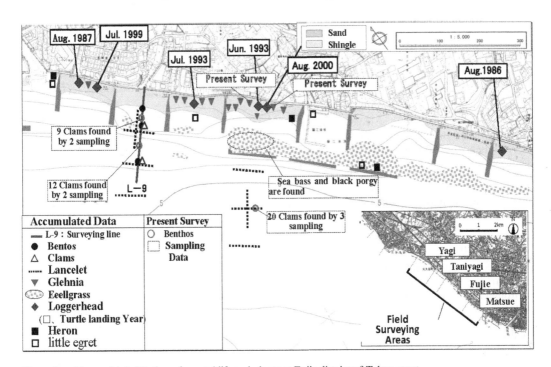

Figure 1.   Observed inhabitation of coastal life and plants at Fujie district of Toban coast.

With these selecting principals, the representative species for Toban coast are examined by analyzing the impact-response relations in case of different type of coastal structures and the characteristics of inhabited coastal life and plants (scarcity, preciousness, usefulness, dominant species, living habits (in sand or on seaweed bed), etc.). It is found that clams and lancelets (benthos), eelgrass (sea grass or seaweed), flounder (fishes), herons and little egret (birds), crinum (coastal plants), and loggerhead turtle (sea turtles) can be chosen as the representative species at Toban coast.

3.2 Dominant environment impact factors

Life and plants populated in coastal area are influenced directly or indirectly by various environmental factors such as water waves, currents, bottom sand, topography, and water quality, etc. The inhabitation of coastal life and plants are usually related to various physical and chemical conditions. Effective evaluation or prediction of the engineering impacts on a coastal ecosystem depends on successful determination of the dominant impact factors to the habitat of the selected representative species.

In this study, we proposed the following criteria to determine the dominant environment impact factors for the representative species: as a dominant impact factor, *(a) it is a direct reason for the changes in inhabitation of coastal life or plants; and meanwhile this factor is also varied due to the construction of coastal protection structures; and (b) it is closely related to the habitat and ecological conditions of the selected coastal life and plants.*

Through analyses of the impact-response relations between inhabitation of coastal life and plants and different type of coastal structures, as well as inhabitation characteristics and life cycles, the dominant environment impact factors for the selected representative species at Toban coast are examined. Table 1 gives an example of the dominant environment impact factors for clams selected at Toban coast.

3.3 Suitability index models

Formularization of the relationship between habitat environment (habitat suitability) and dominant environment impact factors for the selected representative species plays a key role in the quantitative evaluation or prediction of shore protection facilities' impacts on a coastal ecosystem. According to the accumulated environmental surveying data and knowledge on the relationship between suitable habitation condition and dominant

Table 1. Dominant environment impact factors and SI model for clams.

environment impact factors, the quantitative SI (Suitability Index) models or qualitative indexes for each selected representative species at Toban cost are examined. Table 1 gives an example of the obtained SI model or qualitative findings for clams.

702

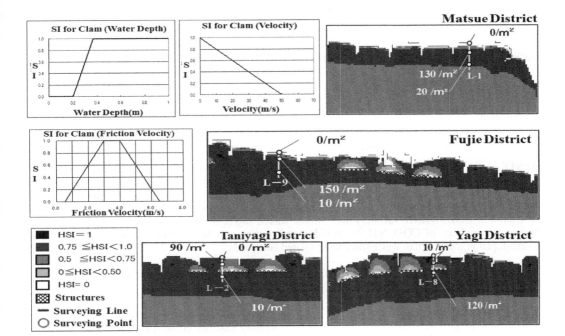

Figure 2. Calculated HSI for clams and measured data.

4 CASE STUDY ON ENVIRONMENT IMPACT EVALUAION AT TOBAN COAST

An integrated environment impact assessment model is constructed by combing coastal hydrodynamic models with Habitat Evaluation Procedure, in order to evaluate the impacts of coastal structures on the physical and habitat environment of a coastal ecosystem. The existing hydrodynamic models (simulation models for waves, currents, topographical changes and water qualities, etc.) can be used to simulate the physical environment changes and then determine the variations in dominant impact factors such as waves, currents, sedimentations and water depths; on the other hand, the Habitat Evaluation Procedure can be used to predict habitat environment changes or HSI (Habitat Suitability Index) by using SI (Suitability Index) models for the representative species.

The developed model is used to evaluate the impacts of coastal structures on the habitat suitability of clams and eelgrass at Toban Coast. Figure 2 gives an example of comparison of the simulated Habitat Suitability Index and field measurement for clams. In the calculation, water depth, nearshore current velocity and bottom friction velocity are chosen as the dominant impact factors and the SI (Suitability Index) models for these dominant impact factors are built upon field surveying data as shown in Figure 2. It shows that

although the simulated HSI for clams near foreshore behind breakwaters in Taniyagi and Yagi districts are not well agree with the measurements, the obtained overall tendency of inhabitation density of clams by the numerical model agrees well with the measurements.

5 CONCLUSION

Environment impact evaluation method for coasts under the influences of shore protection facilities is discussed through case study for Toban coast, Japan. It indicated that the proposed methods for selecting the representative species and dominant environment impact factors are reasonable and effective. Meanwhile, quantitative evaluation of engineering impact on the inhabitation of life and plants in coastal areas, for instance, by using the combined hydrodynamic and HEP method, is possible if habitat SI (Suitability Index) models are properly determined.

REFERENCES

Anton Mc Lachlan, A.C. Brown, 2006. The Ecology of Sandy Shores.
U.S. Fish and Wildlife Service, Division of Ecological Services, 1980. Habitat Evaluation Procedures, http://www.fws.gov/policy/esmindex.html.

Frontiers of Energy and Environmental Engineering – Sung, Kao & Chen (eds)
© *2013 Taylor & Francis Group, London, ISBN 978-0-415-66159-1*

Research on treating the secondary clarifier effluent with BAF

K. You, J.X. Fu, J. Liu, Y.S. Yuan & J.R. Han
Municipal and Environmental Engineering College, Shenyang Jianzhu University, Shenyang, Liaoning, China

ABSTRACT: BAF was used to treat the secondary clarifier effluent. After star-up of BAF was success, discussing the advanced wastewater treatment feasibility by comparing the removal efficiency of COD, NH_3-N, turbidity and chroma of BAF effluent under the conditions of filter velocity 6 m/h, gas-water ratio 1:1 and room temperature. The result showed that the removal efficiency was best. The average concentration of COD, NH_3-N, turbidity and chroma was 32 mg/L, 0.1 mg/L, 3.4 NTU and 37 degree. The average removal rate of COD, NH_3-N, turbidity and chroma was 38.7%, 99.4%, 35.4% and 33.5%. The effluent quality meets urban wastewater reuse related standards.

Keywords: Biological Aerated Filter; COD; advanced wastewater treatment; back-washing

1 INSTRUCTIONS

Biological Aerated Filter (BAF) is wastewater treatment process according to filter process principle of water supply on the basis of ordinary bio-filter in the 80s to 90s. BAF possess many characteristics, such as high load capacity, large hydraulic load, short hydraulic retention time and high-quality water (Yang et al. 2007). Not only can be used in the secondary and tertiary wastewater treatment, but also can be used for the Micro pollution source water pretreatment. BAF technology possesses biological treatment and deep filter in a body. This fully embodies the characteristics of modern water treatment technology (Chang et al. 2002). Now, BAF has developed new wasterwater treatment process which posses denitrification and dephosphorization function (Li et al. 1999; Zheng et al. 2001). It has a wide range of application and promotion in many countries and regions.

The experiment regarded the secondary clarifier effluent from shenyang northern sewage treatment plant as research object. By dynamic continuous test, discussing effluent treatment effect and the advanced wastewater treatment feasibility.

2 MATERIALS AND METHODS

2.1 *The experiment device*

The test taked two up-flow biological aerated filters. BAF was composed of two Φ 120 mm and 3.5 meters high organic glass columns that was filled with biological ceramsite. The first filter column used 4–6 mm ceramsite, the second filter column used 3–5 mm ceramsite. Firstly, the secondary clarifier effluent of wastewater treatment plant was influent of the first filter column, and then effluent of it stored in balance tank. Finally, the water pump ascended into second filter column. Compressed air pumps provided aeration. Rotor flow-meter measured water and gas. Filter columns installed of water inlet, supporting plate, supporting layer, filler, water outlet from top to bottom. Perforated pipe installed under supporting plate, meanwhile, pressure port of filter column side at intervals of 200 mm connected with pressure measuring plate in order to determine pressure change of each point.

2.2 *Test water and test items*

The test was took in the north wastewater treatment plant of Shenyang. The secondary clarifier effluent was as raw water. Water quality was that COD was 30–100 mg/L, NH_3-N was 1–28 mg/L, turbidity was 4–15 NTU, SS was 3–25 mg/L, turbidity was 4–15 NTU, Chroma was 35–100 degree, T was 9–27°C.

Test items adopted method respectively was COD by potassium dichromate method, NH_3-N by sodium reagent photometric method, chroma by colorimetric method. turbidity by photoelectric method.

2.3 *Star-up of BAF*

The star-up of BAF adopted Compound Inoculation. The compound inoculation was carried out in 2 stages. Firstly, Filters was filled with activated sludge, and oxygen was Continuously supplied by the aircompressor. Secondly, after the secondary clarifier effluent flowed into BAF at low rate for a period of time, inlet velocity was increased until filter material surface were covered with stable biofilm (Fu et al. 2006; Fu et al. 2007; Fu et al. 2008;).

After star-up of BAF met with success, filter material surface had transparent and sticky substance. The phenomenon can't be see obvious change in the first week. After two weeks eddish-brown flocking substance had been found in filter materials surface below filter columns. Biofilm grew in number after four weeks. This suggests that biofilm was basic mature. The rich biological phase can be detect. Now, BAF gained start-up success. It can be directly put into operation and deal with the secondary clarifier effluent.

2.4 Operation and control

The test was carried out by continuous aeration and continuous effluent way under the conditions of filter velocity 6 m/h, gas-water ratio1:1 and room temperature. After star-up success, filter velocity adjusted gradually from 4 m/h to 6 m/h. At 6 m/h, the system ran continuously 10d.

3 RESULTS AND DISCUSSION

3.1 Removal efficiency of COD

From Figure 1, we can see that inlet COD was the highest for 64.9 mg/L, the lowest was 43.3 mg/L. In the case of little change of inlet COD, trend of effluent COD was similar with the change trend of inlet COD. In the condition of higher inlet, the effluent was higher. COD of effluent was highest for 40.6 mg/L, and lowest was 25.4 mg/L. COD of effluent was only more than 40 mg/L in the first day during operation time. During the rest of the time, COD of effluent was less than 36.5 mg/L. The average concentration of COD was 32 mg/L. This was mainly two reasons. On the one hand, filter velocity was ajusted from 4 m/h to 6 m/h just which caused the system instability, COD of effluent was too high. On the other hand, COD of inlet was the highest in the first running day during operation time which would cause effluent too high. From the third day, COD of effluent was relatively stable. The average COD was 32.17 mg/L. The effluent quality met urban wastewater reuse related standards. The result showed that organic matter of raw water was degraded by the strong oxidation degradation ability of high concentration biofilm in the surface of filter materials.

COD removal efficiency of the system was not very ideal. The highest was 47.3%, the lowest was 32.79%. The average removal rate was 38.7%. It had larger gap that COD removal rate was more over 85% in the sewage processing (Jiang & Hu. 2002). The main reason was that organic matter of the secondary clarifier effluent was not essyily to be degraded by microbes. That was to say it was not as the carbon sources of microbial metabolism.

3.2 Removal efficiency of NH₃-N

From Figure 2, we can see that inlet NH_3-N was the highest for 18.3 mg/L, the lowest was 14.9 mg/L. In the case of little change of inlet NH_3-N, the effluent quality was Very stable. The average NH_3-N of effluent was 0.1 mg/L or so. The effluent was the highest for 0.12 mg/L, the lowest was 0.08 mg/L. The removal efficiency of NH_3-N in system was very good. The highest removal rate was 99.45%, the lowest was 99.19%. The average removal rate was 99.39%.

Trend of effluent NH_3-N removal rate was similar with the change trend of inlet NH_3-N. From the second day, removal rate appeared slightly from up to down process. It was mainly because of

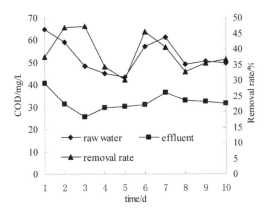

Figure 1. Caption of variation of COD.

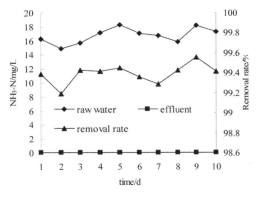

Figure 2. Caption of variation of NH₃-N.

the original water quality change caused. NH$_3$-N removal efficiency was very good because the surface of filter materials grew a lot of nitrobacterium in the condition of enough DO after biological membrane was stability. When wastewater flowed through it, nitrobacterium used its ability of nitrification to purify the sewage in order to degrade NH$_3$-N. The result showed that NH$_3$-N of effluent was far less urban miscellaneous water requirements.

3.3 Removal efficiency of turbidity

CH-161-F From Figure 3, we can see that inlet turbidity was the highest for 6 NTU, the lowest was 4.7 NTU, and the average was 5.27 NTU. In the case of little change of inlet turbidity, turbidity trend of effluent was similar with the change turbidity trend of inlet. In the condition of higher inlet, the effluent was higher. It was highest for 3.7 NTU, and lowest was 3.1 NTU. The average turbidity of effluent was 3.39 NTU. The effluent quality met urban miscellaneous water requirements. Turbidity was removed because that BAF was filled with many smaller particle size biological ceramsite. When wastewater flowed through it, biological ceramsites were compacted which intercepted the numerous suspended matter of wastewater by the role of biological flocculation and the smaller particle size characteristics of ceramsite that lead to effluent turbidity decreased.

Although effluent quality was stable, but removal rate appeared the trend of the lower with the extension of time that dropped from 45.45% to 21.28%. In the first day, the removal rate was only 33.33% because the system operation was not stable. From the second day, The turbidity removal rate began to rise. From the third day to the eighth day the system was in a relatively stable. Since then the removal rate declined greatly. The main reason for

the removal rate reducing may be that the numerous suspended matter and fell off biological film would skip out with water which lead to effluent turbidity went up and removal rate decreased.

3.4 Removal efficiency of chroma

From Figure 4, we can see that chroma of raw water was basically stable between 50–70 degree during the test. The average was 56.8 degree. The chroma of effluent was basically stable that average was 37 degree. It was e between 36–39 degree in the most of the time. The lowest was 34 degree. In the case of relatively larger change of inlet chroma, The chroma of effluent was relatively stable. Chroma removal is similar to the turbidity removal. It was because that the role of biological flocculation and the smaller particle size ceramsites interception effect would result in removal rate reduction.

Although effluent quality was stable, but removal rate appeared the trend of first increased then decreased with the extension of time that dropped from 46.27% to 21.28%. In the first day, the removal rate was only 29.09% because the system operation was not stable. From the second day, The removal rate began to rise. From the fifth day, it declined significantly. The main reason for the removal rate reducing may be that the numerous suspended matter and fell off biological film would skip out with water which lead to chroma effluent went up and removal rate decreased.

3.5 Back-washing of BAF

After BAF ran a period of time, biofilm increased thickness lead to oxygen transfer rate descent, biofilm activity reduction and degradation ability of organic matter decreased. Furthermore, filter material intercepted SS which can jam the porosity of it in order to effluent water quality deteriora-

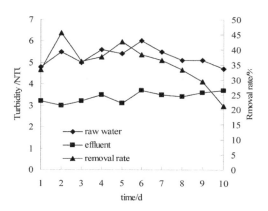

Figure 3. Caption of variation of turbidity.

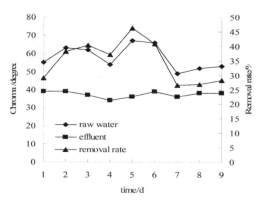

Figure 4. Caption of variation of chroma.

tion. For this reason, BAF need the regular back-washing to restore the handling ability.

At present there were many a back-washing way. the experiment adopted way was carried out 3 stages. Firstly, gas washed. Secondly, gas-water washed. Finally, the way of water rinse washed BAF. During the process gas-wash can separate the interception impurities from filter material which played a role of filter materials loose. Gas-water of back-washing made that most of the dirt would be out of the water. Finally, the aim of water residue was to wash out biofilm which had fallen down in system.

The Back-washing cycle directly related to effluent quality and production water yield. The long cycle would lead to the effluent quality deterioration, but too frequent back-washing can also lead to produce water yield drop. The experiment was performed by controlling head loss means to reverse wash. It was critical that a reasonable head loss was chose at this time. At first, it could ensure effluent quality met requirement when head loss reached it. Secondly, it could make filter layer get sufficient use and prevented frequent recoil to reduce water yield.

The growth trend of head loss was relatively slow, because the concentration of the pollutants of raw water was in lower level. From Figure 5, we can see that the head loss of system was increasing gradually along with the growth of the running time. From the eighth day, head loss sharply ascended and the effluent turbidity became worse at this point. This showed that the system should start the reverse wash. The results indicated that the system can basically guarantee effluent quality when head loss was 50–60 cm. The Back-washing cycle was 8 days when Influent quantity was 70 L/h. Duing the test, effluent quality did not appear deterioration phenomenon.

4 CONCLUSIONS

1. The removal efficiency of BAF was better. The average concentration of COD, NH3-N, turbidity and chroma was 32 mg/L, 0.1 mg/L, 3.4 NTU and 37 degree. The average removal rate of COD, NH3-N, turbidity and chroma was 38.7%, 99.4%, 35.4% and 33.5%.
2. The effluent quality was relatively steady, at the same time, it met urban wastewater reuse related standards.
3. Though biological ceramsites had better effect on the turbidity, chroma removal, considering the activated carbon not only had interception effect, but also had very good adsorption effect, the author suggested the research was carried out that double-deck filter material which filter material replaced a single filter material.
4. The system can basically guarantee effluent quality when head loss was 50–60 cm. The Back-washing cycle was 8 days.

REFERENCES

Chang, W.S et al. 2002. effect of zeolite media for the treatment of textile wastewater in a biological aerated filter. Process Biochemistry. 37(7):693–698.

Fu, J.X et al. 2007. Biological aerated filter for compound inoculation. Journal of Shenyang Jianzhu University (Natural Science). 23(3):478–481.

Fu, J.X et al. 2008. Research on aerobic-anoxic BAF for compound inoculations. Journal of Shenyang Jianzhu University (Natural Science). 24(5): 828–831.

Fu, J.X et al. 2006. Research on start-up of biological aerated filter with different raw water. China Water & Wastewater. 22(11):90–92.

Jiang, P. & Hu, J.C. 2002. Kinetics research on the treatment of domestic sewerage by biological aerated filter. Journal of Nanchang University (Engineering & Technology). 24 (1):62–67.

Li, R.Q. Kong, B. & Qian, Y. 1999. The removal performance of the sewage treatment with the biological aerated filter (BAF). 20(5):69–71.

Yang, Q et al. 2007. Desing parametesr of two-stage biongical aeratel filted treating sewage. Journal of Shenyang Jianzhu University (Natural Science). 23(1):130–133.

Zheng, J et al. 2001. Research on domestic sewage treatment with up-flow biological aerated filter process. China Water & Wastewater. 17(1):51–53.

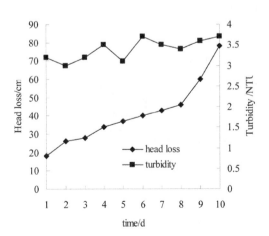

Figure 5. Caption of variation of head loss.

Frontiers of Energy and Environmental Engineering – Sung, Kao & Chen (eds)
© 2013 Taylor & Francis Group, London, ISBN 978-0-415-66159-1

Research of landfill structure impact on percolate water quality

D.H. Li
Shenyang Academy of Environmental Sciences, Shenyang, China

W.S. Dai
Environmental protection Bureau of Kangping County of Shenyang, China

J.D. Huang
Ecological Environment Research Institute, Liaoning University, China

X.H. Fan
Environmental protection Bureau of Kangping County of Shenyang, China

ABSTRACT: Water quality change in different landfill structure was studied. Experiments show that backflow of the percolate is conducive to decline COD_{Cr} and NH_3-N concentration of percolate; Standard aerobic landfill is more useful in improving percolate water quality; The principle of "bioreactor" was further validated.

Keywords: landfill; percolate; backflow; water quality

1 INTRODUCTION

With the increase of urban population, urban expansion and the improvement of living standards, China's MSW production increased dramatically. Currently, the landfill of MSW is a major disposal way. In the MSW landfill process, there will be inevitably the release of CH4, CO2 and other greenhouse gases and percolate contamination, of which the massive percolate produced by landfill has characteristics of high organic content, low carbon and nitrogen, complex compositionand unstable water quality and quantity. The processing of percolate is difficulty, the disposal cost of percolate is high, It has become difficult problem puzzling environmental scientists and the project constructor.

In the course of the percolate study, European and American scholars have proposed a "bioreactor" Principle, namely using landfill layer as a degradative bioreactor of organic matter to dispose highly concentrated organic matter percolate produced by landfill, so that the organic waste can be rapidly degraded. It had already obtained good confirmation and application in Americas, Europe, East Asia and other developed regions, for example, this technology had been applied at 200 landfill in US. 50% of the UK landfill proceed the percolate backflow, Canada, Australia and Japan also proceed field research and engineering practice of the bioreactor. The related bioreactor landfill technology research was started from 1995 in China, but mostly research concentrated in the laboratory simulation and mechanism, the field—scale experimental data and project application is lacking. By means of pilot scale test in real landfill site and comparing two landfill structures difference, this study aimed to analysis influence of percolate backflow from different landfill structure on water quality of percolate produced by landfill trash, to confirms the principle of "the bioreactor", and to provide the powerful parameter for reasonable choosing landfill pattern as well as percolate backflow recirculation in the future garbage field construction.

2 EXPERIMENTAL METHODS

Using a block area of Shenyang Daxin garbage disposal field as tests place, which was uniformly divided into standard aerobic, anaerobic, standard aerobic backflow and anaerobic backflow processing sites. HDPE membrane was laid down at the bottom of standard aerobic landfill structure, 0.1 m diameter corrugated pipes was laid on membrane as percolate main collection pipe line, gravel was laid down around main collecting pipe to protect covers. Two 0.1 m diameter percolate collection branch pipe and a 0.1 m diameter vertical airway tube was laid down in the middle of collection tube. Airway tube is 4 m high and protected

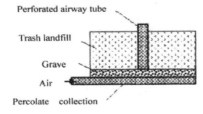

Standard aerobic landfill

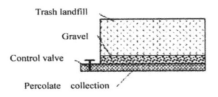

Anaerobic landfill

Figure 1. Experimental device of landfill.

Table 1. Analysis indicators and analysis methods.

Analysis indicators	Analysis methods
COD	Potassium dichromate method
NH3-N	Nessler's reagent colorimetric method

Table 2. Experimental control conditions of each landfill devices.

Device number	Landfill structure	Backflow	Backflow volume	Backflow frequency
A	Standard aerobic	Rinsing	0.4	1
B	Standard aerobic	Percolate	0.4	1
C	Anaerobic	Rinsing	0.4	1
D	Anaerobic	Percolate	0.4	1

with gabion; HDPE membrane, percolate main collection pipe and branch pipe was also laid down in anaerobic landfill structure, but there is no vertical airway tube, and ball valve was arranged in percolate main collection exit to control percolate and keep garbage layer in the anaerobic state.

In loading garbage process, compaction and lay down of the sampling points was conducted when trash was dumped to 0.8 m, 1.6 m and 2.4 m. Ten gas production and five gather percolate device was deployed at each level of each disposal area. Finally, simulated rainfall pipeline was deployed when trash was dumped to 3.2 m, covered with the

HDPE membrane to seal trash layer. Experimental device is seen in Figure 1.

One month after capping landfill devices, COD_{Cr}, NH_3-N concentration of percolate was conducted once a week. Analysis methods are shown in Table 1. Artificial simulation rainfall was conducted once a week, the amount of simulated rainfall was 0.4 m³/week, backflow experiment was proceed in this period, the amount of backflow was 0.4 m³/week. The concrete test control conditions are shown in Table 2.

3 RESULT AND DISCUSSION

Backflow percolate add moisture and microorganism to the landfill layer and accelerate the stabilization of the landfill, simultaneously, during the downward penetration process of the percolate through rubbish heap along the pore, pollutants in the percolate was cleared away due to the physical, chemistry and biological function.

3.1 Percolate water quality changes in the anaerobic landfill structure

The percolate COD_{Cr} concentration of D# landfill site in previous 5 week remained unchanged, on the contrary, the percolate COD_{Cr} concentration of C# landfill site decreased more rapidly, The reasons for this phenomenonmay is that the return line can not guarantee percolate and water to contact sufficiently with the garbage layer, microorganisms in the layer can not decompose organic matter of percolate well, so percolate water quality after backflow in D# landfill site changed little, while the backflow rinsing dilution effect on percolate in C# landfill site is obvious and bring about rapid decline of the COD_{Cr} concentration in C# landfill site. The return pipe was improved at five weeks, the residence time of percolate in the garbage layer increase to some extent, the COD_{Cr} concentration of D# landfill site rose first slightly and then decline sharply, while the COD_{Cr} concentration change of C# landfill site is relatively gentle. On the one hand, backflow percolate has drip washing function on dissolved organic matter in the garbage laye, causing the massive organic matters enter the percolate, this mD_{Cr} concentration decline in C# landfill site is 61.9%.

Ammonia nitrogen concentration change situation of backflow percolate in anaerobic landfill stray result in rising of COD_{Cr} value; On the other hand, backflow percolate contains massive hydrolysate, which has the inhibitory action on microorganism's liveness. After passing through period of time, the microorganism began to adapt to the changing circumstances, its liveness began to strengthen, hereafter It began to decline rapidly, the rate of decline reaches as high as 121.8 mg/L of every day. After end of entire backflow process, the rate of COD_{Cr} concentration decline in $D^{\#}$ landfill site is 80.6%, while the rate of COucture, NH_3-N concentration of percolate in $C^{\#}$ landfill site maintained a high level from beginning to end, Its change fluctuated greatly and declined somewhat slightly, NH_3-N concentration of percolate in $D^{\#}$ landfill site also maintained a high level, Its change fluctuated less and appeared the tendency of rising, It was not obvious in previous four week, Its rising rate was 4.2 percent, and that latter rising rate increased by 20%. NH_3-N concentration decrease in $C^{\#}$ landfill site is due to drip washing and dilution by simulated rainfall on the ammonia nitrogen of garbage layer. After the long-term percolate backflow, landfill layer adsorption vanish as NH_3-N accumulated to a certain extent, simultaneously the ammonia nitrogen produced by organic matter in anaerobic decomposition process enter percolate, thus leading to percolate NH_3-N concentration rising rapidly later in $D^{\#}$ landfill site.

3.2 Backflow percolate water quality changes in the standard aerobic landfill structure

The percolate COD_{Cr} concentration of $B^{\#}$ landfill site in previous 5 week drop slowly, the descending rate is only 19.94%, which is lower than descending rate (47.13%) of $A^{\#}$ landfill site. The percolate COD_{Cr} concentration of B $^{\#}$ landfill site drop significantly after the return pipe was improved, the descending rate achieves 81.6% between fifth week and ninth week, this shows COD_{Cr} eliminated by percolate backflow is obvious.

NH_3-N concentration of percolate in $A^{\#}$ and $B^{\#}$ landfill site maintained a lower level and continued to decline. Becausen $A^{\#}$ and $B^{\#}$ landfill site are standard aerobic landfill structure, most landfill layer are aerobic environment, therefore it is good for nitrification process, simultaneously parts of anaerobic environment is also good for denitrification process, making most ammonia nitrogen produced by organic matter decomposition enter into the air in the nitrogen form, thus resulting in NH_3-N concentration of percolate maintained lower levels and continued to decline. Not only NH_3-N concentration value of percolate in $B^{\#}$ landfill site is lower than the value of $A^{\#}$ landfill site, but also the descending rate in $B^{\#}$ landfill site is also higher than that in $A^{\#}$ landfill site.

4 CONCLUSION

- Backflow of the percolate in anaerobic and standard aerobic landfill is conducive to decline of percolate COD_{Cr} concentration.

- Backflow of the percolate in anaerobic landfill can cause cumulative increase of NH_3-N.
- Backflow of the percolate in standard aerobic landfill has big effect on reducing NH_3-N of percolate.
- Comprehensive evaluation, backflow of the percolate in different landfill is good for improvement of percolate water quality.
- From the final outcome, the standard aerobic landfill structure is more useful in improving percolate water quality.

ACKNOWLEDGMENT

corresponding author: Jindong Huang.

REFERENCES

[1] Huang Jianping and Bao Jianglin, Jan. 2008, "Landfill Percolate Treatment" Environmental science and management. vol. 33, No. 1 pp. 93–98.
[2] Pinel-Raffaitin P,Amouroux D,LeHecho I. et al., May 2008, Occurrence and distribution of organotin compounds in leachates and biogases from municipal landfills [J]. Wat.Res, 42 (4–5): 987–996.
[3] Liu Yigui,Cheng Yingxiang, Li Jin, May 2008, "Study on the treatment technology of urban landfill leachate" Industrial Water Treatment. China, vol. 28 No. 5, pp. 28–30.
[4] Hwang I H, Ouchi Y, Matsuto T., July 2007, Characterisics of leachate from pyrolysis residue of sewage sludge [J].Chemosphere, 68:1913–1919.
[5] J. Wiszniowski, D. Robert, J. Surmacz-Gorska, K. Miksch, J.V. Weber., Jan. 2006, Landfill leachate treatment methods: A review [J]. Environmental Chemistry Letters, 4(1).
[6] Sanphoti N, Towprayoon S, Chaiprasert P, Nopharatana A., Jan. 2006, The effects of leachate recirculation with supplemental water addition on methane production and waste decomposition in a simulated tropical landfill. J Environ Manage, 81 (1): 27–35.
[7] N. Sanphoti, S. Towprayoon, P. Chaiprasert, A. Nopharatana. May 2006, The effects of leachate recirculation with supplemental water addition on methane productionand waste decomposition in a simulated tropical landfill. Journal of Environmental Management (81):27~35.
[8] Fikret K, Yunus MP., June 2004, Adsorbent supplemented biological treatment of pretreated landfill leachate by fed—batch operation [J]. Bioresource Technology, 94(5): 285~291.
[9] Butt T.E., Oduyemik O.K.A, July 2003, holistic approach to concentration assessment of hazards in the risk assessment of landfill leachate. Environment International, 28:597~608D.
[10] Ferhan Çeçen, Didem Çakıroğlu, May 2001, Impact of landfill leachate on the co-treatment of domestic wastewater [J]. Biotechnology Letters, 23(10).

Research of fermentation, separation and purification in production of extracellular glucan from *Sclerotium rolfsii*

B.Q. Wang
Key Laboratory for Food Safety of Binzhou, Binzhou University, Binzhou, Shandong, P.R. China

Z.P. Xu
The Technological Centre of Shandong Chambroad Holding Co., Ltd. Binzhou, Shandong, P.R. China

N.N. Liu
Department of Life Science, Binzhou University, Binzhou, Shandong, P.R. China

ABSTRACT: The glucan from *Sclerotium* spp. fermentation process can be used as oil-displacing agent applied in oil production. It is defined that by single factor test and orthogonal test the optimal conditions were: glucose 5%, sodium nitrate 0.4%, initial pH 5.0 and rotation speed 140 r/min. The rotation speed was the factor influenced the glucan production mostly in all factors. Raising rotation speed, the polysaccharides of mycelium and extracellular was increased in fermentation broth. The yield of crude polysaccharide get by extracting of alkali method was higher than that by extracting of water method. The polysaccharide get by water extraction was more suitable to be used as oil-displacing agent. The deproteinization effects of enzymolysis and ultrafiltration were more outstanding than that of Sevag method.

Keywords: glucan; scleroglucan; *Sclerotium rolfsii*; liquid fermentation; membrane separation and purification; enzymolysis and ultrafiltration deproteinizationto

1   GENERAL INSTRUCTIONS

The glucan produced by *Sclerotium rolfsii* and *S.glucanium*, alternate name Scleroglucan, were the neutral extracellular polysaccharides. The structure of which is β-1,3-D-glucose main chain with β-1,6-D-glucose side chain in every three glucose of main chain. The conformation of the glucan is triply helix structure in water, rigid and clubbed, with molecular weight of 1.4~5.4 × 10⁶ Da and the degree of polymerization of 110~1600. Scleroglucan possessed excellent physicochemical properties of thermal stability, viscosity stability under high temperature and high salinity and shearing tolerance. The main properties of sleroglucan were finer than that of xanthan gum used as thickening agent in oil-displacing liquid, not only the effects of oil-displacing but also adapt the odious rockbottom conditions, for example high temperature and high salinity.

There were few reports on the scleroglucan production. This paper was studied on the glucan fermented production, separation and purification from *Sclerotium rolfsii*.

2   MATERIALS AND METHODS

2.1   Strain and cultivated medium

The tested strain was *Sclerotium rolfsii*, cultivated on PDA medium. The liquid medium consisted of glucose 30 g/L, cubic niter 3 g/L, Yeast Extract 1 g/L, dipotassium hydrogen phosphate 1 g/L, magnesium sulfate 0.5 g/L, kalium chloratum 0.5 g/L and pH 4.5.

2.2   Instruments and reagents

THZ-82 air bath oscillator (Jiangsu Jintan), HJY-125 bottle rocker (Ningbo Xinzhi), GI54D automatic autoclave (shanghai Zhiwei), FD-1 vacuum freeze drier (Beijing Boyikong), DS-1 high speed tissue crusher (Shanghai model factory), Fa1004 balance (shanghai Jingke), 90-2 constant temperature and magnetic force stirrer (shanghai Yarong), GL-21M high rate centrifuge (Hubei Xiangyi), T6 new century ultraviolet-uisible spectrophotometer (Beijing PUxi), Re-301 rotatory evaporator (Gongyi Yingyu yuhua).

The other Agents were AR grade. Yeast extract, basicity protease and agar were biological reagent.

2.3 Defining of liquid fermented conditions

Single factor experimental: the influent factors of the production of mycelium and extracellular polysaccharides were carbon and nitrogen source, initial pH value and rate speed. These factors were defined by Single Factor Experimental. Sucrose, glucose, soluble starch and acetic acid were used as carbon sources and cubic niter, ammonia sulfate, ammonia chloride and peptone as nitrogen source with feed coefficient of 30% in triangular flask and inoculum size of 5%. Initial pH values were arranged with 3, 4, 5, 6, 7 and rate speed were 95, 110, 125, 140 and 145 r/min. All of the treatments were cultured on the Bottle rocker at 25 °C for 5~7 d, triplicated.

Orthogonal experimental: According to experimental purpose, experimental level of the factors were decided as Table 1 and were arranged as L9(3)4 orthogonal layout.

The yield of mycelium and glucan were observe and study indices in experimental. The optimal level of factors was defined by range analysis and ANOVA and the fermented conditions were optimized.

2.4 Extraction and separation of polysaccharide

Alkaline extraction: Preparing 200 mL fermented broth, added 200 mL 4% NaOH, mixed and extracted at 60 °C water bath for 1h, centrifuged at 4000 r/min and the precipitation was gathered and extracted at 60 °C water bath for 1h again, centrifuged at 4000 r/min and precipitate was discard. The tow supernatants were merged and the polysaccharide content was measured, acetic acid was added until to pH 7.0, centrifuged and precipitate was discard. The supernatant was concentrated at rotatory evaporator and was precipitated by adding 2.4 times volume of alcohol. The precipitate was separated and lyophilized.

Water extraction: Preparing 200 mL fermented broth, neutralized with 2% NaOH, added 200 mL

purity water, extracted at 80 °C water bath for 30 min, cooled to room temperature, crashed by tissue crusher at 4 °C. Centrifuged at 15000 r/min for 40 min, the supernatant was gathered and added 2 mL kalium chloratum of 1.0% and 2.4 times volume of alcohol. The precipitate was separated and lyophilized. The other treatment was that, after kalium chloratum was added, 1 times volume of isopropyl alcohol was added instead of alcohol, and then the precipitate was separated and lyophilized.

2.5 Purification of polysaccharide

Sevag method: Preparing 2.00 g crude polysaccharide, added 200 mL water to dissolve it, 1/4 volumes of chloroform—n butyl alcohol (5:2, v/v) were added and mixed, centrifuged at 1000 r/min for 20 min, supernatant was gathered, and chloroform—n butyl alcohol was added once more, until there was no white membrane between two phases. The supernatants were merged and precipitated by 2.5 times volume of alcohol, the precipitate was separated and lyophilized.

Enzymolysis and ultrafiltration method: Preparing 2.0135 g crude polysaccharide, added 200 mL water to dissolve it, 0.1% basicity protease was added and mixed, kept temperature of 50 °C for 1h, microfiltrated by 0.22 μm membrane, ultrafiltrated with membrane of 10000 Dalton MWCO, precipitated by 2.5 times volume of alcohol and lyophilized.

2.6 Determination of mycelium biomass

The fermented broth was filtrated by Buchner funnel and the mycelium was separated, lyophilized and weighed up. The biomass $(g/L) = (W_2 - W_1)/V$, in which, W_2 was the total mass of mycelium and filter paper; W_1 was the mass of filter paper; V was the volume of fermented broth.

2.7 Measurement of polysaccharide content

The method used for measurement of polysaccharide content was phenol—sulfuric acid method.

2.7.1 Construction of standard curve
Preparation of glucose standard solution: Appropriate amount of glucose was dried at 105 °C to constant weight, 25 mg was weighed accurately, dissolved in small amount of water and calibrated to 250 mL measuring flask.

Preparing 11 test tubes added glucose standard solution 0, 0.2, 0.4, 0.6, 0.8, 1.0, 1.2, 1.4, 1.6, 1.8 and 2.0 mL in sequence, and added additionally water to 2 mL, and then 2 mL phenol and 5 mL concentrated sulfuric acid, mixed. The test tube were heated at water bath of 100 °C for 10 min to coloration. Referencing by number 0 test tube,

Table 1. Factor and its levels in orthogonal experimental.

| Level | Factor | | | |
	Carbone source	Nitrogen source	Initial PH value	Rate speed (r/min)
1	4.0%	0.2%	4	110
2	5.0%	0.3%	5	125
3	6.0%	0.4%	6	140

the absorbance value were measured at 485 nm, and the standard curve was drew with horizontal ordinate axis of glucose content and ordinates axis of absorbance value. The regression equation was then calculated.

2.7.2 Measurement of polysaccharide content

Preparing 2 mL sample, diluted with 100 mL water, taken out 2 mL dilution to measure its absorbance value at 485 nm according to the method described as in construction of standard curve. The absorbance value was substituted into the regression equation and the content was calculated.

3 RESULTS AND DISCUSSION

3.1 Defining of liquid fermented conditions

Under the same conditions of the concentration of yeast extract and mineral salt (described in 2.1), carbon or nitrogen source matter concentration (5%), initial pH (5.0), rate speed (110 r/min), the optimal carbon source matter chosen was soluble starch and the nitrogen source matter was peptone (Tables 2 and 3). As slow released carbon source, the soluble starch not resolved in the broth was precipitated by alcohol increasing difficult to purify the polysaccharide. The yield of extracellular polysaccharide of sodium nitrate was higher than that of peptone. So that glucose and sodium nitrate were chosen as carbon source and nitrogen source matter. The optimal initial pH value and rate speed chosen by single factor experimental were 5.0 and 125 r/min respectively.

It had been seen from Table 4 that the factor influence the mycelium product was mainly rate

Table 2. Influences effects of different carbon souce (x ± SD).

Carbon source	Biomass (g/L)	EPSg (g/L)
Acetic acid	1.6729 ± 0.7629	0.1146 ± 0.0188
Sucrose	2.7503 ± 0.2397	0.2256 ± 0.0346
Glucose	2.3259 ± 0.4163	0.2906 ± 0.0732
Soluble starch	4.6950 ± 1.2399	0.3354 ± 0.2389

Table 3. Influences effects of different nitrogen souce (x ± SD).

Nitrogen source	Biomass (g/L)	EPSs (g/L)
Ammonia chloride	2.9681 ± 0.3128	0.2838 ± 0.0325
Ammonia sulfate	3.2081 ± 0.5282	0.2906 ± 0.0058
Sodium nitrate	2.7094 ± 0.7535	0.3475 ± 0.0621
Peptone	3.4121 ± 0.5409	0.3375 ± 0.0502

Table 4. Results of $L_9(3)^4$ orthogonal experiment.

Row factor	A Gluc.	B Nano$_3$	C Ini. pH	D Rate spd	Mycel. Bioms. (g/L)	Yield of EPSs* (g/L)
1	1	1	1	1	2.348	0.482
2	1	2	2	2	2.248	0.688
3	1	3	3	3	4.059	1.061
4	2	1	2	3	3.856	1.187
5	2	2	3	1	2.032	0.632
6	2	3	1	2	2.665	0.688
7	3	1	3	2	1.997	0.723
8	3	2	1	3	3.602	1.147
9	3	3	2	1	1.664	0.794
K_a1	2.885	2.734	2.871	2.014		
K_a2	2.851	2.627	2.589	2.303		
K_a3	2.421	2.796	2.696	3.839		
Range	0.464	0.169	0.282	1.825		
K_b1	0.744	0.797	0.772	0.636		
K_b2	0.836	0.822	0.890	0.699		
K_b3	0.888	0.848	0.805	1.132		
Range	0.144	0.051	0.118	0.496		

*EPBs, extracellular polysaccharides; a, average value of mycelium biomass; b, average value of extracellular polysaccharides.

speed, then initial pH, and then glucose and NaNO$_3$ and the factor influence extracellular polysaccharides yield were rate speed > glucose > initial pH > NaNO$_3$. The results of range analysis were agreed with the analysis of variance (ANOVA). Considered mainly of the yield of extracellular polysaccharides, the following conditions were chosen: set the concentration of glucose and NaNO$_3$ was 5% and 0.4% respectively with initial pH 5.0 and rate speed of 140r/min.

3.2 Construction of standard curve and detection of polysaccharide content

The glucose standard curve (Fig. 1) was drew according to the method given in item 2.7.

It had been seen from Figure 1 that the glucose content was positive correlated with the absorbance value in the range of 0~0.25 mg. Glucose concentration and the correlation had a fine linear relationship. The regression equation calculated by statistical methods was:

$$Y = 3.1382 x + 0.0305 \qquad (1)$$

Correlation coefficient of glucan content and absorbance value was $r^2 = 0.9961$.

The phenol-sulphuric acid method was used for detection of polysaccharide. Before the test, different concentration gradient were diluted and the absorbance value were tested to define the optimal

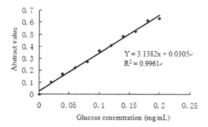

Figure 1. Standard curve of glucose conetent.

Table 5. Polysaccharide yield of different extract method.

Extract method	Yield of polysaccharide (g/L)	Content of polysaccharide (%)
Alkaline extraction	8.7869	42.58
Water extraction	6.4325	38.17
Isopropyl alcohol precipitate	6.1730	54.26

dilution multiple because of the sealed polysaccharide content. The hydrolysis degree or equivalent value of polysaccharide and the oxidized degree of phenol were the important influence factors to the content measurement.

3.3 Extraction and separation of polysaccharide

Three methods were used to extract and separate polysaccharides, namely, alkaline extraction alcohol precipitate, water extraction alcohol precipitate and water extraction isopropyl alcohol precipitate. The yields of polysaccharide with different extract method were tabulated in Table 5. It had been seen that the yield of polysaccharide using alkaline extraction and alcohol precipitate was 8.7869 g/L, higher obviously than that of the other two. The yield of polysaccharide using hot water extraction and isopropyl alcohol precipitate was 6.1730 g/L. The content of polysaccharide precipitated by isopropyl alcohol was 54.26%, higher than that precipitated by alcohol.

The polysaccharide extracted with hot water was almost extracellular polysaccharide. Hardly any polysaccharides in mycelium were dissolving out to extracellular. Because of increasing of the cellular permeability and releasing of intracellular polysaccharides induced by alkaline, and that could be precipitated by alcohol together with extracellular polysaccharide, the yield of polysaccharide extracted with alkaline was higher than that with hot water.

The yield of crude polysaccharide precipitated by ethanol method was higher than that by isopropyl alcohol and the color of ethanol method polysaccharide was deeper than that by isopropyl alcohol. The polysaccharide precipitated by isopropyl alcohol had fewer impurity substances than that by ethanol method.

The extraction effects of different extract method could be also seen on the polysaccharide content indices in Table 5.

There was a solubility difference between the polysaccharides extracted by alkaline and hot water. The polysaccharides extracted by hot water had a

good solubility and high viscosity, while the other with a weak solubility and lower viscosity. So that alkaline extraction method was inappropriate for extraction of gulcan from Sclerotium rolfsii.

3.4 Purification of polysaccharide

In the methods of purification of polysaccharide, Sevag method could dislodge only the free protein, and to the combinative protein and the other small molecule, Sevag method could hardly play a part in. The protease completed the hydrolysis of combinative protein into polypeptide and amino acids, and the polypeptide and amino acids were ultrafiltrated with other small molecule. Though it had a lower yield of polysaccharide, the purification effects of enzymolysis and ultra-filtration was more advantageous than Sevag method.

In this experiment, 1.3761 g refined polysaccharide was get from 2.0135 g crude polysaccharide. The yield of polysaccharide was 68.34% using en-zymolysis and ultra-filtration method, and the glucan content in the crude polysaccharide was 80.45%.

4 CONCLUSION

The fungi Sclerotium rolfsii could be sued to produce extracellular polysaccharide with the mainly composition of glucan, namely, scleroglucan. The optimal mode of production was liquid fermentation and the fermentation conditions were that: The concentration of glucose and $NaNO_3$ was 5% and 0.4% respectively with initial pH 5.0 and rate speed of 140 r/min. The content of polysaccharide in fermented broth and in extract could be measured by the phenol—sulfuric acid method. In which, the glucose content was positive correlated with the absorbance value in the range of 0~0.25 mg. The correlation co-efficient was $r^2 = 0.9961$. The regression equation was: $Y = 3.1382 x + 0.0305$. The suitable extraction method could be used for extracellular polysaccharide extraction and separation was hot water extraction and isopropyl alcohol precipitate method with a high glucan content, good solubility and high viscosity, which was suit to be oil-displacing agent applied in oil production. The crude polysaccharide could be purified by protease enzymolysis and ultra-filtration farther. It had a potential to be industrialization.

ACKNOWLEDGEMENT

This work was financially supported by Shandong Provincial Natural Science Foundation, China ZR2009DL002 and Y2008D39.

REFERENCES

Davison, P. & Mentzer, E. 1982. Polymer flooding in North Sea reservoirs. Society of Petroleum Engineers Journal, 22(3): 353–362.

Li, B. Zhang, J.F. & Jiang, P.J. 2003. The production of fungal Scleroglucan and its application in oil field, Microbiology, 30(5):99–102.

Liu, R.L. Heng, B. & Zhao, D.J. et al. 1990. Study on the production of Sclerotium glucan using staich. Academic journal of Nankai university (natural science version), (2):84–91.

Liu, Y.J. Liu, J.S. & Huang, Zh. Q. et al. 2001. Polymer and carbon dioxide displacement technology of reservoir oil. Beijing: China petrochemical processing press. p. 1–2.

Ren, Y.E. Liu, H.C. & Xu, X.Ch. et al. 1993. Fermentation method production polysaccharide from *Sclerotium rolfsii*. Microbiology, 19(3): 142–145.

Xu, T.T. Zhao, Zh.J. & Yuan, Ch. 2004. New progress on drilling fluid and completion fluid technology abroad. drilling fluid & completion fluid, 21(2): 1–10.

Frontiers of Energy and Environmental Engineering – Sung, Kao & Chen (eds)
© 2013 Taylor & Francis Group, London, ISBN 978-0-415-66159-1

Study on soil infiltration law of unsaturated soil slope

P.M. Jiang
College of Urban Construction and Safety Engineering, Nanjing University of Technology, Nanjing, Jiangsu, China
School of Architecture and Engineering, Jiangsu University of Science and Technology, Zhenjiang, Jiangsu, China

Z.L. Yan, L. Mei & P. Li
School of Architecture and Engineering, Jiangsu University of Science and Technology, Zhenjiang, Jiangsu, China

ABSTRACT: Rainfall infiltration is the main evoked factor of unsaturated soil slope failure, by using the MIDAS software, the paper systematically analyses the parameters which can influence seepage field (Including rainfall intensity, rainfall duration, permeability coefficient and underground water level), summed up the soil infiltration law of unsaturated soil slope. In addition, by analyzing it finds that soil saturation is not just related to the permeability coefficient or rainfall intensity unilaterally, but also has the close relation with their proportion. Through this we obtain a method about the ratio of the permeability coefficient and rainfall intensity.

Keywords: rainfall infiltration; unsaturated soil slope; matrix suction; the ratio of permeability coefficient and rainfall intensity

1 INTRODUCTION

Previous slope seepage field researches usually adopt stable seepage models, most analysts are ignored the movement of the rainfall in the unsaturated zone. To solve this problem, the paper using the MIDAS software systematically analyses the parameters which can influence seepage field (Including rainfall intensity, rainfall duration, permeability coefficient and underground water level), Expecting to find soil infiltration law of unsaturated soil slope. This law can be used for landslide prevention and control under rainfall conditions.

2 UNSATURATED SEEPAGE FLOW MATHEMATIC MODEL OF SOIL SLOPE

2.1 Saturated-unsaturated seepage continuity equation [1]

$$\frac{\partial \theta_w}{\partial t} = -\frac{1}{\rho_w g}\left(\frac{\partial k_{wx}}{\partial x}\frac{\partial \bar{u}_w}{\partial x} + \frac{\partial k_{wy}}{\partial y}\frac{\partial \bar{u}_w}{\partial y} + \frac{\partial k_{wz}}{\partial z}\frac{\partial \bar{u}_w}{\partial z}\right.$$
$$\left. + k_{wx}\frac{\partial^2 \bar{u}_w}{\partial x^2} + k_{wy}\frac{\partial^2 \bar{u}_w}{\partial y^2} + k_{wz}\frac{\partial^2 \bar{u}_w}{\partial z^2}\right) \qquad (2)$$

where, θ_w is volumetric water content; k_{wx}, k_{wy} and k_{wz} are permeability coefficients in three direction which changes with water content; $\bar{u}_w$ is the total head which conclude pore water pressure and the position of head ($\bar{u}_w = u_w + \rho_w g z$); ρ_w and g are the density of water and gravity acceleration respectively, ρ_w is a constant when ignore the compressibility of water.

2.2 The mathematical model of soil-water characteristic curve

Soil-water characteristic function is based on Van Genuchten model [2, 3]:

$$\frac{\theta - \theta_r}{\theta_s - \theta_r} = F(\Psi) = \frac{1}{\left[1 + \left(\dfrac{\Psi}{a}\right)^n\right]^m} \qquad (2)$$

which

$$m = 1 - \frac{1}{b}$$

where, a, n and m are fitting parameters, Ψ is matrix suction; θ is volume moisture content; θ_s is Saturated volume moisture content; θ_r is residual volume moisture content.

3 NUMERICAL SIMULATION OF RAINFALL INFILTRATION

3.1 The finite element model of unsaturated soil slope

According to the classification method of finite element model [4], the size of the model is shown in Figure 1. Elastic modulus of soil is 85.2 MPa, poisson ratio is 0.3, bulk density is 17 kN/m³, saturated density is 20 kN/m³, cohesive force is 25 kPa, friction angle is 15°, volume water content is 30%. The following are the related parameters of soil-water characteristic curve, $\theta_r = 0.05$, $\theta_s = 0.4243$, a = 0.6977, n = 1.525, m = 0.3443.

3.2 Slope rainfall infiltration analysis

In the rainfall analysis, we have to consider not just daily mean rainfall but also the influence of rainfall intensity and underground water level. According to rainfall intensity level division standard [5], the rainfall intensity is divided into weak rainfall and strong rainfall, respectively divided each grade into four grades in this paper; Permeability coefficient will be divided into four penetrate grade according to the degree of difficulty of infiltration; groundwater level is divided into four grades from scratch to 4 m. Grade selection criteria l will be used in the analysis are listed in Table 1.

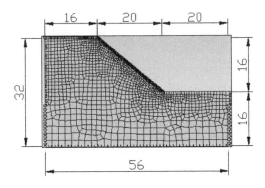

Figure 1. Lope finite element mode.

4 SLOPE RAINFALL INFILTRATION ANALYSIS IN THE STATE OF DIFFERENT RAINFALL INTENSITY

All the data in Table 1 was permutated and combined, and rainfall infiltration was analysed about the combined model.

4.1 Slope rainfall infiltration analysis in the state of long-term weak rainfall

In the state of long-term weak rainfall, Rainwater can infiltrate to the internal slope, but stability time is different. Matrix suction dissipated time is both related to the ratio of the permeability coefficient and rainfall intensity and the position of underground water level. The relation between matrix suction dissipated time and the ratio of permeability coefficient and rainfall intensity in different water level is shown in Figure 2.

Rainfall infiltration would over time with the increase in the ratio of the permeability coefficient and rainfall intensity, as can be seen from Figure 2-a and 2-b; Rainfall infiltration time will correspondingly shorten with the rise of the groundwater level, and the slope of the slash composition is basically the same and when the ratio of both is the same, the slope are basically the same no matter how much the underground water level is. It also has the same rules in Figure 2-c, but stable time has increased dramatically when the ratio of the permeability coefficient and rainfall intensity is about 10. No obvious rules in Figure 2-d, the ratio of the permeability coefficient and rainfall intensity lies between 12.5 to 50, the ratio is too large, rainwater can run free to all the direction and the soil can't reach saturation, so little effect on the slope stability.

4.2 Slope rainfall infiltration analysis in the state of short-term heavy rainfall

If your In the state of short-term heavy rainfall, the ratio of the permeability coefficient and rainfall intensity listed in Table 2. Part of pore water pressure diagram after 24 hours under no water level of different conditions during short-term heavy rainfall as shown in Figure 3.

Table 1. Permeability coefficient, rainfall intensity and underground water level elevation selection.

Permeability coefficient	Long-term weak rainfall	Short-term heavy rainfall	Groundwater level
0.0001 m/hr (I)	0.0001 m/hr (a)	0.001 m/hr (A)	No groundwater (1)
0.0005 m/hr (II)	0.0002 m/hr (b)	0.004 m/hr (B)	4 m (2)
0.001 m/hr (III)	0.0003 m/hr (c)	0.007 m/hr (C)	8 m (3)
0.005 m/hr (IV)	0.0004 m/hr (d)	0.01 m/hr (D)	12 m (4)

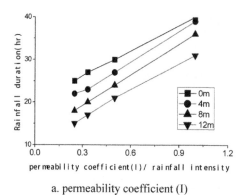

a. permeability coefficient (I)

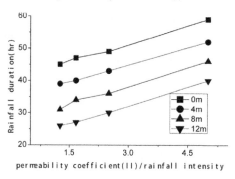

b. permeability coefficient (II)

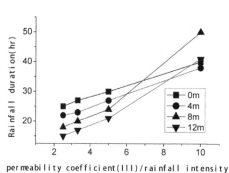

c. permeability coefficient (III)

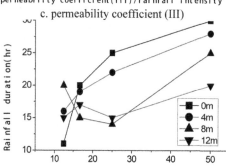

Figure 2. The relation between matrix suction dissipated time and the ratio of permeability coefficient and rainfall intensity in different water level.

Table 2. The ratio of permeability coefficient and rainfall intensity under various operating conditions during heavy rainfall intensity conditions.

Condition	Ratio	Condition	Ratio
I-A	0.1	II-A	0.5
III-A	1	IV-A	5
I-B	0.025	II-B	0.125
III-B	0.25	IV-B	1.25
I-C	0.014	II-C	0.071
III-C	0.143	IV-C	0.714
I-D	0.01	II-D	0.05
III-D	0.1	IV-D	0.5

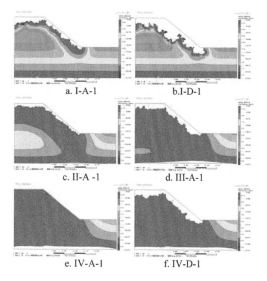

Figure 3. Part of pore water pressure diagram after 24 hours under different conditions during short-term heavy rainfall.

Like the weak rainfall state, when the rainwater infiltrate to the soil, groundwater level will increase, the higher the underground water level, the earlier the water level rising, and the increased range is related to the rainfall intensity, the greater the rainfall intensity, the higher the increased range. In these models, the phenomenon of accumulating water all happens in addition to the IV-A model, and the smaller the ratio of the permeability coefficient and rainfall intensity, the more obvious of hydrops. I think this phenomenon is related to both the permeability coefficient and rainfall intensity, rather than related to a single variable. The IV-A model don't show hydrops, that's because the ratio is too large and it is out of Range of certain limits, the rain can dissipate rapidly throughout the entire process of rainfall infiltration.

Through rainfall infiltration analysis, the following conclusions can be got obtained:

1. Soil saturation is not just related to the permeability coefficient or rainfall intensity unilaterally, but also has the close relation with their proportion.
   1. When the ratio of the permeability coefficient and rainfall intensity is less than 1, the soil surface is not saturated, only the basic loss of surface matrix suction In the long-term weak rainfall condition; In the state of short-term heavy rainfall, the soil surface saturated, total loss of matrix suction, the rainwater into the soil within a certain depth, and can not completely infiltrate, internal soil matrix suction is reduced.
   2. When the ratio of the permeability coefficient and rainfall intensity lies between 1 to 5, The soil surface still has hydrops and the soil matrix suction has almost completely vanished in the range of precipitation area.
   3. When the ratio of the permeability coefficient and rainfall intensity lies between 5 to 10, throughout the process of rainfall infiltration, rainwater can be infiltrate regularly and matrix suction eventually disappeared with the passage of time.
   4. When the ratio of the permeability coefficient and rainfall intensity is greater than 10, rainwater can run free to all the direction and the soil can't reach saturation, appear irregular phenomenon in the long-term weak rainfall condition.
2. With the matrix suction in the slope gradually dissipated, when the ratio of the permeability coefficient and rainfall intensity is greater than 1, groundwater level will slowly increase with the growth of precipitation time; when the ratio of the permeability coefficient and rainfall intensity is less than 1, the rain can't infiltrate to the bottom of the slope, so the groundwater level without any change.

5 CONCLUSIONS

Based on these studies, the following conclusions can be obtained:

The ratio of the permeability coefficient and rainfall intensity is 1, this ratio can be considered to be a critical point in rainfall infiltration. When the ratio is less than 1, it can form surface runoff, that is to say quantity of rainwater is relatively large, it can be considered to be heavy rainfall, this will make the soil surface saturated and the slope surface prone to damage; when the ratio is greater than 1, topsoil can't reach saturation. In this case it can be considered to be weak rainfall, only makes the soil matrix suction reduced and water content changed. The change of the water content will change the strength of the soil, but the change is weak. According to the relationship of moisture content and soil strength, we know it has little influence on the soil strength. Therefore, in the practical engineering, we can first judge the ratio of the permeability coefficient and rainfall intensity, if the ratio is around 1, is suit analysis rainfall infiltration on slope stability, otherwise the analysis of rainfall will not have much practical significance.

ACKNOWLEDGEMENTS

This work was financially supported by Social Development Project of Science and Technology Department of Jiangsu Province (BE2011746), and Natural Science Foundation of Resources Department of Jiangsu of Province (2009015).

REFERENCES

[1] Junping Yuan & Zongze Yin. Rock and soil mechanics. Vol. 25 (2004), p. 1252.
[2] Fredlund et al. Canadian Geotechnical Journal. Vol. 31 (1994), p. 533.
[3] Leong E C & Rahardjo H. Journal of Geotechnical and Geoenvironmental Engineering. Vol. 123 (1997), p. 1106.
[4] Jun Zhu. Journal of Wuhan University (engineering edition), Vol. 34 (2001), p. 5.
[5] Shouren Zheng. Flood control and emergency rescue knowledge manual. (Yangtze River press, Wuhan 2010).

Frontiers of Energy and Environmental Engineering – Sung, Kao & Chen (eds)
© 2013 Taylor & Francis Group, London, ISBN 978-0-415-66159-1

Catalysts pore structure and their performance in low-temperature Claus tail gas treating processes

J.J. Li, L.M. Huang, J.L. He & C.R. Wen
Research Institute of Natural Gas Technology, Southwest Oil and Gas field Co., PetroChina,
Chengdu, Sichuan, P.R. China

ABSTRACT: The effect of pore structures on the catalytic activities was studied in the course of the Claus tail gas treating processes at sub-dewpoint conditions. The distribution of pore size and its effect on the textural properties of the materials were determined by using nitrogen adsorption and mercury porosimetry. The results show that the sulfur capacity ability of the catalysts can be improved by increasing the surface area, the volume and diameter of macropores. The best catalyst had very high macroporosity combining with high microporosity. High volume and smooth channels are contributed to its excellent activity.

Keywords: Claus reaction; hydrogen sulphide; sulphur dioxide; alumina

1 INTRODUCTION

Recovery of elemental sulfur from industrial waste gases is an important problem both from environmental and economic points of view (Goar 1994; Goddin 1974). As is well known, the equilibrium for the Claus reaction:

$$2\ H_2S + SO_2 \rightarrow 3/n\ S_n + 2\ H_2O$$

Though tail gas cleanup processes provide high recovery degree, the major part of sulfur is nevertheless produced in Claus furnace and main Claus catalytic stage. Looking at the history of process development, one may state that during the more than 50 years' experience the basic technological principles of the main catalytic stage of the Claus process remained unchanged. All significant improvements have been made only in relation to the "head" of the process or to its "tail", not touching the "body"-main catalytic stage. The progress here was practically restricted by the development of new catalysts (Pineda 1996).

The Claus reaction is exothermic and reversible, so the minimum reaction temperature is required to provide favorable thermodynamic conditions for maximum conversion. Thus, a sub-dewpoint process is cyclic in nature (Kunkel 1977). Multiple catalyst beds are designed. A bed is firstly fed tail gas from the Claus portion of the plant and then a wave of deactivation moves through it as sulfur is formed and adsorbed on the catalyst surface. When the bed is mostly deactivated, another hot gas stream is introduced to vaporize the adsorbed sulfur and reactivates the catalyst. After regeneration,

the bed is cooled and then fed tail gas again. Thus, in order to achieve high conversion, the process is usually performed in two or three consecutive catalytic reactors with intermediate removal of sulfur in condensers. Examples of sub-dew point sulfur recovery processes include CBA, Sulfreen, and MCRC (Knudtson 1977; Tsbulevski 1996; Alvarez 1996). The conventional two-stage system provides the degree of sulfur recovery of up to 96%, and the three-stage one up to 98%.

It has been well established that the performance of a sub-dewpoint tail gas unit is dependent on the porosity of the activated alumina catalyst employed in the process (Pearson 1977). Specifically, macroporosity has been identified as an important variable in the catalyst's performance. Sub-dew catalysts can be tested by observing their activity for H_2S/SO_2 conversion as a function of time. The catalyst adsorbs the sulfur that is formed. The performance of the catalyst is indicated by the time it continues to deliver good conversion, and this quantity is related to the sulfur it is able to adsorb.

In this paper, the activities of three catalysts with different pore structures are reported. The impact of this improved performance on existing and future plants is discussed.

2 EXPERIMETAL SECTION

2.1 *Materials*

Three catalysts, designated A, B and C, were prepared from similar alumina trihydrate as starting material, but treated under different conditions to obtain different pore structures. General properties and pore volume data of the three catalysts are given as follows.

2.2 Characterization

Surface areas were measured by nitrogen adsorption-desorption isotherms at 77 K in a Micromeritics ASAP 2010. Pore-size distributions were obtained by mercury intrusion porosimetry in a Micromeritics Autopore IV 9500 apparatus.

2.3 Catalytic activity method

The Claus reaction was carried out in a fixed bed reactor. Reactant gas H_2S and SO_2 were fed from cylinders with mass flow meter, and water was pumped into a pre-heater. The temperature of the reactor was set at 127 °C and the reaction temperatures were moniteded by three thermocouples immersed in the catalytic bed top, center and bottom respectively. The observed temperature variations were less than 5 °C with reference to the set value. The inlet and outlet gases from the reactor were analyzed by gas chromatography. The experimental were undertaken at the following conditions: atmospheric pressure, total flow rate of inlet gases 400 ml/min, inlet gas composition C_{H2S} = 4 vol%, C_{SO2} = 2 vol%, C_{H2O} = 30 vol%, N_2 balance and the volume of catalyst 20 ml of 8–12 mesh.

3 RESULTS AND DISCUSSION

Results for the three catalyst samples are given in Table 1. In order to generate comparative test data among the three samples, the end of the loading phase was fixed at which the sulfur conversion declined to 60%. To a good approximation this allows the length of time to achieve this drop to be taken as an indication of the relative sulfur loading capacity of the different catalysts.

Sample C showed a substantially longer life than A and B. This result is also reflected in a higher sulfur loading for catalyst C (1.02 grams of

Table 1. Results of low-temperature Claus catalyst testing.

Catalyst	Reaction time/h	Sample weight/g	Sulfur collected/g	Sulfur loading/g/g
A	6.7	15.8	9.9	0.62
B	8.2	13.1	11.4	0.87
C	8.4	12.5	12.7	1.02

Reaction conditions: Reaction temperature, 127 °C; reaction time, when the sulfur conversion dropped to 60%; Sulfur density, 1.8 g/ml.

sulfur per gram of catalyst) than catalyst A and B (0.62 and 0.87 grams of sulfur per gram of catalyst, respectively).

We also research the rate of decline at the end of each curve, which can be related to the size of the mass transfer zone. Sample A over sample B and C in the decline rate, the larger sulfur adsorption capacity of sample C overshadows the advantage of catalyst B.

Several catalyst samples basic performance data are given in Table 2. These properties will affect the performance of the catalyst in industrial applications to varying degrees. As shown in Table 2, the strength of the catalyst can reach to 150 N/granule per granule. It also can be seen from the wear data, the more homogeneous distribution and higher intensity, the wear is lower. Therefore, strength and distribution are the important two factors that affect the wear and catalyst applications chalking phenomenon.

However, for these three samples, there is no big difference in intensity distribution and the average strength, their differences are reflected in the pore structure.

Table 3 gives several samples of pore volume data. From the table, C and B have larger total pore volume and macropore volume, were 0.591 ml/g, 0.595 ml/g and 0.15 ml/g, 0.17 ml/g. The total pore volume of A was 0.461 ml/g, large pore volume was 0.07 ml/g. The sulfur capacities of these three catalysts were 1.02 g/g, 0.87 g/g and 0.62 g/g, respectively. There is not obviously difference in micropore and mesopore volume.

It can be inferred that, macropore in low temperature sulfur recovery plays an important role in improving the quality of catalyst sulfur capacity. The three catalyst macropore pore volume to total pore volume percentage was 15.2%, 28.6% and 25.4%. As macropore volume percentage changes, the pore structure of catalyst and the sulfur capacity change as well. As macropore volume percentage increases, the sulfur content is increased, C and B catalysts are more than A sulfur capacity.

Figure 1 gives the relationship between porous volume and pore diameter of these catalysts. Two line are positioned the boundary of micropore and macropore, 12 nm and 75 nm. The diagram shows clearly that C and B have similar macropore, and both are much higher than A. The biggest difference between C and B occurred in the range of 150 nm–1620 nm. The components of both of pore volume are 0.12 mL/g, but the contribution to this part of the pore volume comes from different size pore. The pore volume of sample C mainly comes from more lager pore than sample B, indicating that sample C has greater pore porosity than sample B. In low-temperature sulfur recovery process, sample with macropore and ultra-macropore

Table 2. Catalyst properties.

Catalyst	Crush strength/ N/granule	Attrition rate/%	Bulk density/ ml·g⁻¹	Content of Na₂O/%	Surface area/ m²·g⁻¹
A	155.1	0.52	0.84	0.19	263
B	173.2	0.41	0.77	0.20	344
C	150.5	0.45	0.74	0.16	341

Table 3. Catalyst pore volume determined by merury porosimetry and N_2 adsorption-desorption.

| Sample | Pore Volume | | | | | | |
| | Macropore (>75 nm) | | Mesopore (12~75 nm) | | Micropore (12 nm) | | |
	V/ml//g	ratio/%	V/ml/g	ratio/%	V/ml/g	ratio/%	Total/ml/g
A	0.07	15.2	0.047	10.2	0.344	74.6	0.461
B	0.17	28.6	0.046	7.7	0.379	63.7	0.595
C	0.15	25.4	0.053	8.9	0.388	65.7	0.591

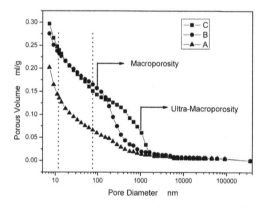

Figure 1. Relationship of sample A, B and C about porous volume to pore diameter by mercury intrusion.

Table 4. Property of macropores.

Sample	Average macropore diameter/nm
A	230
B	180
C	1050

can exhibit excellent properties, with a longer time of adsorption, can accommodate more sulfur, and desorption are more easily.

It is also consistent with the activity test results. As shown in Table 4, sample C sulfur capacity is 1.02 g/g, higher than that of sample B 0.87 g/g and sample A 0.62 g/g. Therefore, macroporosity plays

an important role in low temperature sulfur recovery process.

In B and C, the measured data of macropore volume, accounted for the total pore volume percentage, B is lager than C, but in its activity results in the sulfur content, a lower activity of B than C was obtained about 14.7%.

Investigate its reason, according to the macropore characterization as well as the results given in Table 3 and Figure 2, macroporous average pore diameter of the catalyst, sample C's macropore mainly concentrated in 1000 nm. In sample B, macropore distribution is concentrated mainly at 180 nm, far less than the C pore diameter. The optimized macropore of C plays a more important role in the liquid sulfur capacity, the larger of the pore diameter and the small on the internal diffusion resistance. It can provide relatively smooth response of the channel and environment, not easily blocked in the orifice. During the reaction procession, the sulfur of sample A and B with the smaller pore diameters gradually accumulated, leading to block in orifice, thus affecting their activities. Therefore in the low temperature sulfur recovery catalyst, ensure the catalyst has certain strength, the macropore volume and pore size as large as possible.

Based on McBain catalyst ink bottle model, as shown in Figure 3, the mouth and body of the bottle radius is r_a and r_b. According to Kelvin equation, the capillary condensation steam pressure of mouth and body can be described as follows respectively:

According to capillary condensation theory, $r_b > r_a$, so the occurrence of capillary condensation pressure $P_b > P_a$. That can be interpreted as, with the reaction progresses, sulfur gradually adsorb in

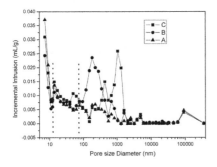

Figure 2. Differencital pore volume distrubution.

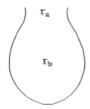

Figure 3. Ink bottle model diagram.

the pore wall, resulting in the changes of the steam pressure. In this case, the orifice is more prone to occur capillary condensation, thereby causing the plug hole. Therefore, the size of the orifice has an important influcncc on rcaction. The orifice is bigger, the less prone to capillary condensation phenomenon. It is also explained that in considerable of macropore volume, sample C has higher sulfur capacity than other samples.

The influence of mesoporous on the reaction can be found from Table 3. Among these catalysts, mesoporous accounted for only a small percentage of total volume, and the proportion of mesopore reduced accordingly with the macropore increases. On the response contributions to reaction, macroporous is better than mesoporous. Furthermore, mesoporous do not provide larger specific surface and also can not provid more pore volume, therefore should try to compress the amount of mesopore in catalyst samples.

Although the micropore diameter is narrow, it also plays an important role in the reaction. Approximate 80% of the surface area is mainly contributed by micropore. The larger specific surface area indicates that the more reaction active center, hence the reaction efficiency can be improved. In addition, micropore is an important factor to ensure the strength of catalyst. In the previous table, the micropore volume accounted for the total pore volume of more than 60%, 74.6% 63.7%, and 65.7%, respectively. The specific surface area is 263, 344, 341 m^2/g, respectively.

Consequently, the impact of micropore and macropore on the reaction efficiency is more than mesopore, and the following pore structure of catalyst distribution form is more recommended: increasing macropore porosity, maintaining micropore porosity, decreasing in mesopore porosity, and making the diameter as large as possible.

4 CONCLUSIONS

From the study of the effect of alumina structure on the Claus reaction in a fixed bed reactor at sub-dewpoint, it can be concluded that, during the reaction proceeds, mass transfer limitations is appear and transport phenomena became the determining step. All the reaction parameters affectthe diffusivity, and the most important factors are surface area, pore radius distribution (include micropores, mesoporoes and macropores), and porosity. By optimization of the structure textural properties, the activities of alumina based catalysts for the Claus reaction at low temperature can be improved efficiently.

ACKNOWLEDGEMENTS

We are thankful for the financial support from postdoctoral research center of PetroChina Southwest Oil and Gas Field Company (No. 20110308-05).

REFERENCES

Alvarez, E., Mendioroz, S., Munoz, V. & Palacios, M.J. 1996. Sulphur recovery from sour gas by using a modified low-temperature Claus process on sepiolite. *Applied Catalysis B:Enviromental* 9: 179–199.
Goar, B.G. & Nasato, E. 1994. Large-plant sulfur recovery processes stress efficiency. *Oil and Gas Journal* 92:61.
Goddin, C.S., Hunt, E.B. & Palm, J.W. 1974. CBA process ups Claus recovery. *Hydrocarbon Processing* 53(10): 122–124.
Knudtson, D.K.1977. Amoco CBA process in commercial operation, 27th Can. Chem. Eng. Conf., 23–27 Oct. Calgary.
Kunkel, L.V., Palm, J.W., Petty, L.E. & Grekel, H. 1977. CBA for Claus tail gas cleanup. *US. Patent* 4035474.
Pearson, M.J. 1977. Alumina Catalysts in Low-Temperature Claus Process. *Industry Engineering Chemistry Product Research and Development* 16(2): 154–158.
Pineda, M. & Palacios. 1996. The performance of a γ-Al2O3 catalyst for the Claus reaction at low temperature in a fixed bed reactor. *Applied Catalysis A: General* 13:681–96.
Tsbulevski, A.M., Morgun, L.V., Sharp, M. & Pearson, M. 1996. Catalysts macroporosity and their efficiency in sulphur sub-dew point claus tail gas treating processes. *Applied Catalysis A: General* 145(1–2), 85–94.

Frontiers of Energy and Environmental Engineering – Sung, Kao & Chen (eds)
© 2013 Taylor & Francis Group, London, ISBN 978-0-415-66159-1

Experimental study of titanium-bearing slag load activated carbon fiber adsorption of formaldehyde

X.G. Ma, C.P. Xu, J.R. Han & X.N. Liu

Municipal and Environmental Engineering College, Shenyang Jianzhu University, Shenyang, China

ABSTRACT: Against the shortcomings of titania-bearing blast furnace slag's application in the photo-catalytic oxidation areas and its difficultion to recycling, in this study, TBBFS is loaded to the Activated Carbon Fiber (ACF) to form complex catalyst, and use the complex catalyst to remove the formaldehyde in the air. In the static experiment, the relationship between the initial concentration of formaldehyde and the removal rate of formaldehyde are studied. In the dynamic experiment, We study four factors as follow: effect of TBBFS/ACF with or without UV light on the removal rate of formaldehyde; effect of the different temperature on the removal rate of formaldehyde; effect of the position of TBBFS/ACF in the reactor on the removal rate of formaldehyde; TBBFS/ACF's inactivation. Conclusion of static experiment as follow: photocatalytic oxidation has good effect on the low concentration of formaldehyde. Conclusion of dynamic experiment as follow: the removal rate of formaldehyde with the UV light is better than that without UV light; when the temperature of formaldehyde is 35°C, the removal rate of formaldehyde can reach 70%; complex catalyst which is placed parallel or perpendicular in the reactor has less effect on the removal rate of formaldehyde; Complex catalyst's best using time is three.

Keywords: Ttitania-Bearing Blast Furnace Slag (TBBFS); activated carbon fiber; complex catalyst; formaldehyde

Titania-Bearing Blast Furnace Slag (TBBFS) from Panzhihua is a unique resource. Its content of TiO_2 is about 22~25%. The problem of comprehensive utilization is still not solved. Therefore, finding new ways to make TBBFS as the available resources have a very important significance for the society.[1,2] At present, TiO_2 as a photocatalyst which is applied in the field of indoor air-purification has matured.[3,4] Because of containing TiO_2 in TBBFS, it is also feasible to some extent which is applied to indoor air-purification.[5,6] In the indoor environment, the concentration of formaldehyde is very high and noxious, which has very bad effect on people's life and make people keep working on it.[6–10] In this study, TBBFS is loaded to the Activated Carbon Fiber (ACF). In the UV light, ACF's adsorption and TBBFS's photocatalytic oxidation are used to remove the formaldehyde in the air, in order to achieve the purpose of comprehensive using TBBFS.

1 EXPERIMENTAL DRUGS AND LABORATORY INSTRUMENTS

1.1 *Experimental drugs*

Titania-Bearing Blast Furnace Slag (TBBFS): retrieved from Panzhihua Iron and Steel Group ironworks.

Activated carbon fiber: Liaoning Anke activated carbon fiber applications technology development company.

Rormaldehyde solution: purity of 35% to 40%, Shenyang Chemical Reagent Factory.

1.2 *Laboratory instruments*

Air pump, Rotameter, Temperature and humidity control instrument, Electronic Libra, UV lamp, Magnetic stirrer, Electric thermostat Blast Oven, Formaldehyde tester, Scanning electron microscopy and Glass Instrument.

2 PREPARATION OF TITANIUM-BEARING SLAG/ACF

Specific preparation process are as follows:

1. Pretreatment of the carrier: the activated carbon fibers are cut into a ring with the outer diameter of 5 cm, and an inner diameter of 1.5 cm, Immersed into 500 mL of distilled water, Soak 30 min, then the ACF was placed in an oven at 120°C to dry 1h, then cooled 1h.
2. Containing titanium slag solution preparation: a certain amount of titanium-bearing slag mixed with 100 mL distilled water, then fast stirred for 10 min under magnetic stirrer.

3. Preparation of supported catalysts: pretreated ACF were put in suspension of titanium-bearing slag, fully mixed 10 min, pulled out at constant rate, dried in an oven at 180°C for 2 h, cooled 1 h.

3   STATIC EXPERIMENTS

Different initial concentrations have a big impact on their removal efficiency, When formaldehyde gas occur the photocatalytic oxidation reaction. Therefore, there is a need for further study of the relationship between different initial concentrations the of formaldehyde with formaldehyde removal rate. Because it is different to occur different concentrations and standard gas source in the present experimental conditions, so it is not easy to study Initial concentration of formaldehyde on the impact of formaldehyde photocatalytic degradation rate. Therefore, I will study this factor in depth in the the static experimental part.

3.1   *Experimental methods*

The experiments occur formaldehyde outside the reactor. Injecting 10 mL formaldehyde solution (analytical grade) into jar, stuff with rubber stoppers, and smear Vaseline in the seams of the rubber stopper and jars, then wrap Teflon tape around the seams, ensure the formaldehyde solution in a sealed state. Under the laboratory temperature conditions, the formaldehyde solution naturally volatiliz from the jar. using it when occur photocatalytic reaction.

In the static experiment, use 10 µL syringe to draw different volume of formaldehyde gas from the wide-mouthed bottle of the storage source of formaldehyde gas, Injecting into the reactor, form the reaction conditions of the different initial concentration of formaldehyde. To ensure the formaldehyde gas a uniform distribution in the reactor, openning the UV lamp after formaldehyde gas wasinjected into the reactor for 10 min in this study, and used temperature an d humidity control device to maintain a temperature of the reactor at 35 °C, then started photocatalytic oxidation reaction for 30 min.

3.2   *Experimental process and results*

Extracting respectively 2 µL (3.83 ppm), 3 µL (5.75 ppm), 4 µL (7.67 ppm), 5 µL (9.58 ppm), 6 µL (11.5 ppm) formaldehyde gas into the reactor, detecting and recording the remaining concentration of formaldehyde gas within the reactor after the reaction was carried out for 30 minutes. Making diagrams between the initial concentration of formaldehyde with formaldehyde removal rate, shown as Figure 1.

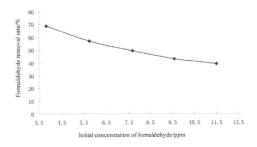

Figure 1.   Effect of concentration on the photocatalysis of formaldehyde.

We can see from Figure 1, initial concentration of formaldehyde is in inverse proportion to its degradation efficiency, promptly the initial concentration of formaldehyde is higher, the degradation efficiency of the composite catalyst is lower.

4   DYNAMIC EXPERIMENTS

4.1   *The impacts of titanium-bearing slag/ACF with or without UV irradiation formaldehyde removal rate*

Under the experimental conditions that experimental temperature was about 25°C, humidity was about 60%, titanium-bearing slag/ACF composite catalyst was placed in photocatalytic reactor, using formaldehyde tester determine formaldehyde gas concentration in the photocatalytic reactor outlet every 5 min, respectively, in the openning violet light, and closing violet lamp case. Comparing formaldehyde gas concentration at the outlet of the photocatalytic reactor in two cases, studying formaldehyde gas concentration value's declining contribution.

We can see from the graph 2, the titanium-bearing slag/ACF composite catalyst in the absence of UV lamp irradiation conditions, relying on ACF adsorption of formaldehyde pollution gas removal rate is about 40%, In an ultraviolet lamp irradiation conditions, adsorption and photocatalytic combined effects of formaldehyde pollution gas removal rate is about 55%. Analysis shows that, under UV radiation degradation efficiency of formaldehyde over that part is just the composite catalyst has photocatalytic oxidation results.

4.2   *Reaction temperature on the effect of formaldehyde removal rate*

This experiment selects the titanium-bearing slag/ACF composite catalysts for photocatalytic degradation of formaldehyde in the experimental study. During the whole experiment, reaction

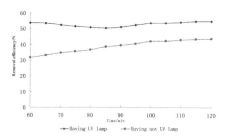

Figure 2. Contrast curve of adsorption and adsorption-photocatalytic oxidation.

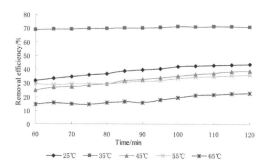

Figure 3. Effect of different temperature on removal rate of formaldehyde.

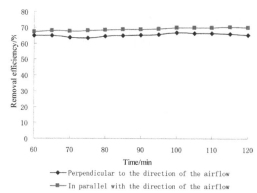

Figure 4. Effect of different position of TBBFS/ACF on removal rate of formaldehyde.

temperature range is 25~65°C, 10°C as the temperature range. In the photocatalytic reactor outlet, detect formaldehyde concentration every 5 min, produce formaldehyde removal rate vs time curve, as shown in Figure 3.

Figure 3 shows, in the process of experiment, formaldehyde removal rates in the range is 14.3% to 71.2% in different temperature condition, in which the formaldehyde removal rate maximum value is under 35°C reaction conditions; formaldehyde removal rate minimum value appeared at 65°C. When the reaction temperature is 35°C, the average removal rate of formaldehyde is about 70%; When the reaction temperature is 65°C, the average removal rate of formaldehyde is about 20%.

4.3 The titanium-bearing slag/ACF's placement on the influence of formaldehyde removal rate

The titanium-bearing slag/ACF composite catalysts in the reactor is placed parallel to the direction of the airflow can make full use of vector purification area, so that the ultraviolet light can be carrier uniform distribution on the activated carbon fiber; the titanium-bearing slag/ACF composite catalysts in the reactor is placed perpendicular to the direction of air flow pattern better can intercept, adsorp formaldehyde gas, provided high concentration environment for

photocatalytic oxidation. Two different placement way have an advantage each, this experiment will study different placements of the titanium-bearing slag/ACF composite catalysts within the reactor on the effect of formaldehyde removal rate, and the two the removal rate of formaldehyde were compared.

Four pieces of circular titanium-bearing slag/ACF composite catalysts respectively parallel and perpendicular to the direction of gas flow is placed in the reaction device, the temperature was 35°C, detected formaldehyde concentrations in the photocatalytic reactor outlet every 5 min, and compute the removal efficiency of formaldehyde in two cases, produced titanium-bearing slag/ACF composite catalysts placement mode curve, as shown in Figure 4.

We can see from the graph 4, the removal effect of titanium-bearing slag/ACF composite catalyst to formaldehyde has little difference when they are placed parallel to the direction of airflow and and perpendicular to the airflow direction within the reactor.

4.4 Titanium-bearing slag/ACF composite catalyst deactivation

Repeating three times to the same experiment as the one using time of titanium-bearing slag/ACF composite catalyst, the removal rate of formaldehyde in the last experiment as the removal rate of a titanium-bearing slag/ACF composite catalyst when using the time to formaldehyde. This experiment in determining the titanium-bearing slag/ACF best frequency of use, used continuously titanium-bearing slag/ACF composite catalyst for four time, studied removal rate changes of formaldehyde. The reaction temperature is 35°C, the reaction humidity is about 60%. The curve of titanium-bearing slag/ACF composite catalyst's using frequency and formaldehyde removal rate was shown in Figure 5.

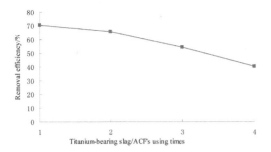

Figure 5. Effect of using times on removal rate of formaldehyde.

Figure 5 shows, as titanium-bearing slag/ACF composite catalyst using times increasing, the removal rate of formaldehyde decreased. titanium-bearing slag/ACF composite catalyst used the first, two or three, four time, its formaldehyde removal efficiency were 70%, 65%, 55%, 40%. Figure 5 shows, the removal rate of only ACF adsorp formaldehyde is about 40%, it is same with titanium-bearing slag/ACF composite catalyst fourth to the formaldehyde removal efficiency, titanium-bearing slag/ACF composite catalyst has deactivated, it needs to be replaced or regenerated.

5 CONCLUSIONS

1. Initial concentration of formaldehyde is in inverse proportion to its degradation efficiency, promptly the initial concentration of formaldehyde is higher, the degradation efficiency of the composite catalyst is lower.
2. Removal rate of titanium-bearing slag/ACF composite catalyst to formaldehyde is higher under UV irradiation than it is in the absence of UV irradiation
3. When the formaldehyde concentration is stable, the formaldehyde removal rate first increases with increasing temperature, reaches a certain value and began to decrease after. When the reaction temperature is 35°C, the average removal rate of formaldehyde was up to 70%; When the reaction temperature is 65°C, the average removal rate of formaldehyde was up to 20%.
4. The removal effect of titanium-bearing slag/ACF composite catalyst to formaldehyde has little difference when they are placed parallel to the direction of airflow and and perpendicular to the airflow direction within the reactor.
5. Formaldehyde removal rate has downward trend as the titanium-bearing slag/ACF composite catalyst increase of the frequency of use.

Author brief introduction: Ma Xingguan,(1972–), Male, Engaged in water pollution control technology research.

REFERENCES

[1] Zou XL, Lu XG. Preparation of titanium alloy by direct reduction of Ti-bearing blast furnace slag [J]. The Chinese Journal of Nonferrous Metals. 2010, 20(9):1829–1835.
[2] Zhou X, Li LJ, Luo CL. A research on high Ti blast furnace slag crushed stone applied for concrete aggregate [J]. Iron Steel Vanadium Titanium. 2001, 22(4):43–46.
[3] Egon Matievie. Preparation and mechanism of formation of titanium dioxide hydrosols of narrow size distribution [J]. J Colloid & Interface Sci. 1977, 61(2):302–311.
[4] Mario Visca. Preparation of uniform colloidal dispersion by chemical reaction in aerosols [J]. J Colloid & Interface Sci. 1979, 68(2):308–319.
[5] Einaga H, Futamura S, Ibusuki T. Heterogeneous photocatalytic oxidation of benzene, toluene, cyclohexene and cyclohexane in humidified air: comparison of decomposition behavior on photoir radiated TiO_2 catalyst [J]. Appl. Catal. B: Environ. 2002, 38(3):215–225.
[6] Vorontsov AV, Dubovitskaya VP. Selectivity of photocatalytic oxidation of gaseous ethanol over pure and modified TiO_2 [J]. J. Catal. 2004, 221(1):102–109.
[7] Kim SB, Hong SC. Kinetic study of photocatalytic degradation of volatile organic compounds in air using thin film TiO_2 photocatalyst [J]. Appl. Catal. B: Environ. 2002, 35(4):305–315.
[8] Sano T, Negishi N, Uchino K, Tanaka J, Mat suzaw a S, Takeuchi K. Photocatalytic degradation of gaseous acetaldehyde on TiO_2 with photodeposited metals and metal oxides [J]. J.Photochem. Photobiol. A: Chem. 2003, 160(1–2):93–98.
[9] Amama PB, Itoh K, Murabayashi M. Photocatalytic degradation of trichloethylene in dry and humid atmospheres: role of gas phase reactions [J]. J. Mol. Cat al. A: Chem. 2004, 217(1–2):109–115.
[10] Shang J, Du Y, Xu Z. Photocatalytic oxidation of heptane in the gas phase over TiO_2 [J]. Chemosphere, 2002, 46(1):93–99.

Frontiers of Energy and Environmental Engineering – Sung, Kao & Chen (eds)
© 2013 Taylor & Francis Group, London, ISBN 978-0-415-66159-1

The application of improved FUZZY-AHP method in groundwater quality evaluation

J. Hao, Y.X. Zhang, Z.Y. Ren & X. Xue
Beijing University of Technology, Beijing, China

ABSTRACT: Groundwater pollution seriously threaten the environment if controlling steps are not implemented immediately, groundwater quality evaluation is one of the significant means to solve these problems. As impacting factors that are evaluated their influences on groundwater pollution, weights of indicators and attributes are determined by the Analytic Hierarchy Process (AHP) method which is performed by matrix calculations. Considering defects of the method for sorting, Improvement on the sorting method will be made. Then, combines the weights that respectively determined by the AHP method and by the improved AHP method with the fuzzy membership degree, to evaluate the level of groundwater quality. The method not only overcomes deviations of personal opinion for experts, but also avoids errors brought by fuzzy uncertainty; Contrast with these two kinds of evaluation methods, the improved AHP method can be more successful to sort the factors, example of application testified the reliability of above method.

Keywords: AHP; sorting; fuzzy; evaluation

1 INSTRUCTION

Groundwater quality evaluation is useful and valuable means for controlling and ensuring groundwater pollution, it can reflect the quality of ground water and its processing of change accurately and effectively, it is placed as the priority to prevent and to control the groundwater environment pollution. It's also an important means of decision-making, supervision and management of water environmental (Guo et al. 2006). The main purpose of the process is to consider of major pollution indicators that influenced the groundwater quality. By using analytic hierarchy process and improved analytic hierarchy process to solve the scheduling problem of index's system. Valuate on the groundwater quality level, provide a scientific basis for the control and management of groundwater water quality pollution.

In view of the above, an AHP-statistics model for the synthetic and dynamic evaluation of groundwater quality was proposed. Analytic hierarchy process (referred to as AHP) is proposed by a operations research experts T.L. Saaty whom is a professor in the United States of America at the beginning of nineteen seventies, it is a statistic method which is used to eliminate the influences arising from the differences in dimension and magnitude of indicators. It is a simple and practical method for multiple criteria decision making (Saaty 1977;

Saaty 1990). On these grounds, an AHP-statistics model is provided for regional groundwater quality assessment. The study case of this paper is a problem that including many kinds of factors by subordinate relationship. By the method of AHP, the problem will be a problem of multi-layered structure model. At last, it should be a matter to determine relative importance weights of the lowest layer to the top layer or a matter to sort by the relative merits.

2 DETERMINE THE WEIGHT OF INDEX BY AHP METHOD

2.1 *The construction of the hierarchical structure model*

The process of construct the hierarchy as follow: first of all, define the problem and determine its goal. Then, a hierarchical model should be established for groundwater quality assessment and a pollution evaluation indexes system is needed to be built. Next, select and determine parameters for different types (Ali et al. 2007). Combine factors according to the subordination relation between them and construct the hierarchical structure model. Structure the hierarchy from the top through the intermediate levels to the lowest level (Kong 2009).

In this paper, according to the "drinking water of source water quality standard (CJ3020-93)" and the "groundwater quality standard (GB T14848-93)", groundwater quality evaluation hierarchy model was established. The level of the groundwater quality is the top level (A), middle layer also known as a constraint layer, including five aspects indexes: sensory index (B1), chemical index (B2), heavy metal index (B3), inorganic toxic substance index (B4) and bacteriological index (B5) (As is shown in Fig. 1).

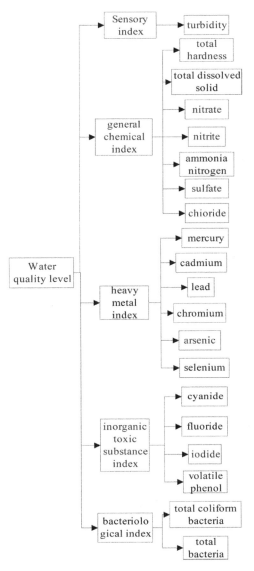

Figure 1. The hierarchy of groundwater quality evaluation.

2.2 The construction of judgment matrix

AHP method give their weights by adopt opinions that are consulted from experienced and knowledgeable experts. The evaluation standard is based on the 1–9 scale method (Saaty 1977) (As listed in Table 1).

According to the contribution of each factor to the groundwater quality evaluation process, a quantitative weight is designated to every factor. Construct a set of pairwise comparison matrices for each of the lower level (formula 1). As to determine which factor is dominated.

For evaluation factors in the system, take factors a_{ij} (i = 1,2 ... n; j = 1,2 ... n) for pair comparison, to determine their relative importance, which is composed of a_{ij} (n × n) matrix that is the judgment matrix.

$$A = \begin{bmatrix} a_{11} & a_{12} & ... & a_{1n} \\ a_{21} & a_{22} & ... & a_{2n} \\ ... & ... & ... & ... \\ a_{n1} & a_{n2} & ... & a_{nn} \end{bmatrix} \qquad (1)$$

Element of a_{ij} (i, j = 1, 2, ..., n) is the ratio of factor i' value with and factors j's. Obviously $a_{ji} = 1/a_{ij}$, $ai_i = 1$. The hierarchical structure of the evaluation is the most important step of building the judgment matrix.

According to the final scores and by comparing the total score of every factor, the judgment matrix of AHP is built (As is shown in Table 2).

Due to multiple factors were included in groundwater quality, including qualitative factors and quantitative factors, this kind of quantitative method is of great advantage to solve different metrics problems of water quality indexes (Liu 2003).

The limit value in the drinking water standard should be considered, when we assigning scales to constraint layers. On the other hand, relative important and collaborative antagonistic action of the water environment must be considered also.

Take matrix of heavy metal index for a example:

1. Arsenic and selenium showed antagonistic effects to each other;

Table 1. Scale of analytic hierarchy process and its description.

Scale a_{ij}	Meaning
1	I compare with j equally important
3	I compare with j slightly important
5	I compare with j significantly important
7	I compare with j strongly important
9	I compare with j top importance
2, 4, 6, 8	Between that of the former two.

Table 2. A–B comparison and judgment matrix.

A	B1	B2	B3	B4	B5	Weight (bi)
Sensory indexe (B1)	1	1/3	1/4	1/4	1/2	0.066
Chemical index (B2)	3	1	1/2	1/2	2	0.197
Heavy metal index (B3)	4	2	1	1	3	0.31
Inorganic toxic substance index (B4)	4	2	1	1	3	0.31
Bacteriological index (B5)	2	1/2	1/3	1/3	1	0.117

Table 3. Average random consistency index (Saaty 1977).

Rank (n)	1	2	3	4	5	6	7	8	9	10
RI	0	0	0.52	0.89	1.12	1.26	1.36	1.41	1.46	1.49

2. Compounds of selenium and mercury showed antagonistic effects;
3. Mercury and lead show a synergistic effect.

Take these for guiding ideology, twenty factors relative importance judgment matrix of drinking water quality standard's could be gained.

2.3 Check consistency

Generally, judgment matrix A is not consistent, therefore, check consistency to deviation degree is indispens able (Ali et al. 2012; Zou et al. 2012).

If it does not consistent, redo the pair wise comparisons. Formulas for calculation are as follows:

$$\lambda_{max} = \sum_{i=1}^{n} \frac{(AW)_i}{nW_i} \qquad (2)$$

where n is the rank of the matrix; λ_{max} is the maximum eigenvalue of the matrix.

$$CI = (\lambda_{max} - n)/(n - 1) \qquad (3)$$

For different matrixes, their CI values are changed, The closer the consistency ratio is to zero, the better the consistency.

$$CR = CI/RI \qquad (4)$$

where RI stands for random index, values of RI are (see Table 3 for correct listings). CR is consistency rate, When CR < 0.10, it is considered that the judgment matrix is consistency; or the judgment matrix is needed to be adjusted, until it is satisfied.

For example: the matrix of A–B comparison and judgment (Table 2): $\lambda_{max} = 5.049$, CI = 0.011 < 0.1. Therefore, the A–B judgment matrix is complete consistency.

3 HIERARCHICAL RANKING ANALYSIS OF A CASE

3.1 Comprehensive ranking based on the traditional AHP

The exact value weight determination and consistencyvalidation were conducted using the method proposed in AHP. The eigenvalues of the judgment matrices were calculated. The process of comprehensive ranking caculation and their checking of consistency are presented (Tables 4–8).

Calculate from the top layer to the bottom layer according to the sorting calculation, For the second floor, the single sort of hierarchical is the total sort (Li 2007; R. et al. 2008.). By the total sequencing result of each programme layer, schemes sort can be gained. Consistency check, formulas of Hierarchy ordering:

$$CR = \sum_{j=1}^{m} a_j CI_j \Big/ \sum_{j=1}^{m} a_j RI_j \qquad (5)$$

if CR < 0.10, the hierarchy ordering is consistency. $\lambda_{max} = 4.066$, CR = 0.022 < 0.1. The judgment matrix of B4-C is fully consistency.

3.2 Determine indexes weights by the improved AHP method

3.2.1 Analysis of the defects of AHP comprehensive sorted algorithm

Due to each involvement index is independent, so no matter a certain index' value is increased or decreased, the importance order of other indicator should not be affected. For example, total sort of hierarchical analysis for importance of indexes are: C1 > C2 > C3 > C4 > C5, suppose index C2 is not participate in the evaluation, importance ranking

Table 4. Comparison and judgment matrix of chemical index (B2-C).

Bi	Bi							
	Total hardness	Total dissolved solid	Nitrate	Nitrite	Ammonia nitrogen	Sulfate	Chloride	Weight (c_{ij})
Total hardness	1	1/4	2	1/2	1	2	3	0.122
Total dissolved solid	4	1	5	3	4	5	6	0.351
Nitrate	1/5	1/2	1	1/4	1/2	1	2	0.068
Nitrite	2	1/3	4	1	2	4	5	0.23
Ammonia nitrogen	1	1/4	2	1/2	1	2	3	0.122
Sulfate	1/2	1/5	1	1/4	1/2	1	2	0.068
Chloride	1/3	1/6	1/2	1/5	1/3	1/2	1	0.038

Table 5. Comparison and judgment matrix of heavy metal index (B3-C).

Bi	Bi						Weight (c_{ij})	Check
	Hg	Cd	Pb	Cr	As	Se		
Hg	1	3	7	5	4	3	0.352	λmax = 6.276
Cd	1/3	1	5	4	3	1	0.219	CI = 0.055,
Pb	1/7	1/5	1	1/3	1/4	1/5	0.033	CR = 0.049
Cr	1/5	1/4	3	1	1	1/4	0.087	
As	1/4	1/3	4	1	1	1/3	0.106	
Se	1/3	1	5	3	3	1	0.204	

Table 6. Comparison and judgment matrix of inorganic toxic substance index (B4-C).

Bi	Bi				Weight (c_{ij})	Check
	Cyanide	Fluoride	Iodide	Volatile phenol		
Cyanide	1	3	4	5	0.518	λmax = 4.066 CI = 0.022
Fluoride	1/3	1	2	3	0.252	CR = 0.024
Iodide	1/4	1/2	1	2	0.149	CR < 0.1
Volatile phenol	1/5	1/3	1/2	1	0.081	

Table 7. B5-C bacteriological index.

Bi	Bi		Weight (c_{ij})	Consistency check
	Total coliform bacteria	Total bacteria		
Total coliform bacteria	1	1	0.5	λmax = 2 CI = 0
Total bacteria	1	1	0.5	CR = 0 CR < 0.1

of index results is: C1> C4 > C3 > C5, because of changes of other indexes'relative order, this sort is illogical (Li 2012).

According to the traditional AHP level sorting meth od, suppose remove nitrite index and chro-mium indictort separately, whether the importance of other ions will change or not will be a question to study.

The calculation results are shown in Table 9, form the table we can see the relative importance

Table 8. The sort results of total level.

Ci	B1 0.066	B2 0.197	B3 0.31	B4 0.31	B5 0.117	W	Level of total order	Consistency check
Turbidity	1	0	0	0	0	0.066	6	CI = 0.044
Total hardness	0	0.122	0	0	0	0.024	15	RI = 1.12
Total dissolved solid	0	0.351	0	0	0	0.069	4	CR = 0.040 < 0.1
Nitrate	0	0.068	0	0	0	0.013	17	
Nitrite	0	0.23	0	0	0	0.045	11	
Ammonia nitrogen	0	0.122	0	0	0	0.024	16	
Sulfate	0	0.068	0	0	0	0.013	18	
Chloride	0	0.038	0	0	0	0.007	20	
Hg	0	0	0.352	0	0	0.109	2	
Cd	0	0	0.219	0	0	0.068	5	
Pb	0	0	0.033	0	0	0.01	19	
Cr	0	0	0.087	0	0	0.027	13	
As	0	0	0.106	0	0	0.033	12	
Se	0	0	0.204	0	0	0.063	7	
Cyanide	0	0	0	0.518	0	0.16	1	
Fluoride	0	0	0	0.252	0	0.078	3	
Iodide	0	0	0	0.149	0	0.046	10	
Volatile phenol	0	0	0	0.081	0	0.025	14	
Total coliform bacteria	0	0	0	0	0.5	0.059	8	
Total bacteria	0	0	0	0	0.5	0.059	9	

Table 9. Comprehensive ranking results's comparison of the AHP method.

Importance	The total sorts	The sort after removed nitrite	The sort after removed cadmium
1	Cyanide	Cyanide	Cyanide
2	Hg	Hg	Hg
3	Fluoride	Total dissolved solid	Se
4	Total dissolved solid	Fluoride	fluoride
5	Cd	Cd	Total dissolved solid
6	Turbidity	Turbidity	Turbidity
7	Se	Se	Total coliform bacteria
8	Total coliform bacteria	Total coliform bacteria	Total bacteria
9	Total bacteria	Total bacteria	Iodide
10	Iodide	Iodide	Nitrite
11	Nitrite	As	As
12	As	Total hardness	Cr
13	Cr	Ammonia nitrogen	Volatile phenol
14	Volatile phenol	Cr	Total hardness
15	Total hardness	Volatile phenol	Ammonia nitrogen
16	Ammonia nitrogen	Nitrate	Nitrate
17	Nitrate	Sulfate	Sulfate
18	Sulfate	Pb	Pb
19	Pb	Chloride	Chloride
20	Chloride		

sorting of some remaining ions vary greatly. We can draw the conclusion that this method is not consistent.

Investigate its reason: Saaty'AHP synthetical arrangement give weights by the linear algorithm. The values be weighted reflect the relative importance degree of indexes weights, but weights are the weights of attribute layer, can not indicate relative importance of weights. This mismatch is the root of the inconsistency integrated sort (Li 2012; Sule et al. 2010.).

3.2.2 The improved AHP method and its application in hierarchical ranking

This method that based on the ideal index' weight make improve on AHP method. Suppose "C*" be ideal index; "b *" as the weight of number i of constraint layer; "c_{ij}" refer as the weight of property i of the index j: the total value of comprehensive weight W (C *) is "1". The calculation formula is as follows (shown in Table 10).

$$W(C_j) = \sum_{i=1}^{m} b_i \cdot \frac{c_{ij}}{c_i^*} \quad (6)$$

It is assumed that nitrite and cadmium ion don't participate in the evaluation, from the remaining indexes sequencing results (shown in Table 11) we can see relative orders of the others are invariable.

After some indexes were deleted, contrast with these two kinds of evaluation methods, by means of case analysis, the overall results of remaining indicators that calculated by the improved AHP algorithm consistent with the original relative order. It can be more successful to sort the factors.

4 FUZZY COMPREHENSIVE EVALUATION

4.1 Determination of membership matrix

As the factor set (the degree of water polluted) and the evaluation set (Water quality evaluation standard) both with some characteristics of fuzzy, so they should be differentiated by membership function, membership functions are as follows (Bojan et al. 2008):

$$r_{ij} = \begin{cases} 1 & \left(c_i \leq a_{i1} \, or \, c_i \leq a_{im} \right) \\ \dfrac{c_i - a_{i(j-1)}}{a_{ij} - a_{i(j-1)}} & \left(a_{i(j-1)} \leq c_i \leq a_{ij} \right) \\ \dfrac{a_{i(j-1)} - c_i}{a_{i(j-1)} - a_{ij}} & \left(a_{ij} \leq c_i \leq a_{i(j-1)} \right) \\ 0 & \left(c_i \leq a_{i(j-1)} \, or \, c_i \geq a_{i(j+1)} \right) \end{cases} \quad (7)$$

Table 10. The hierarchy ordering results of improved AHP.

Indexes	B1 0.066	B2 0.197	B3 0.31	B4 0.31	B5 0.117	W (Cj)	Sort
c_i^*	1	0.351	0.352	0.518	0.5		
Turbidity	1	0	0	0	0	0.028	15
Total hardness	0	0.122	0	0	0	0.029	13
Total dissolved solid	0	0.351	0	0	0	0.084	3
Nitrate	0	0.068	0	0	0	0.016	17
Nitrite	0	0.23	0	0	0	0.055	7
Ammonia nitrogen	0	0.122	0	0	0	0.029	14
Sulfate	0	0.068	0	0	0	0.016	18
Chloride	0	0.038	0	0	0	0.009	20
Hg	0	0	0.352	0	0	0.132	1
Cd	0	0	0.219	0	0	0.082	4
Pb	0	0	0.033	0	0	0.012	19
Cr	0	0	0.087	0	0	0.033	12
As	0	0	0.106	0	0	0.04	10
Se	0	0	0.204	0	0	0.077	5
Cyanide	0	0	0	0.518	0	0.132	2
Fluoride	0	0	0	0.252	0	0.064	6
Iodide	0	0	0	0.149	0	0.038	11
Volatile phenol	0	0	0	0.081	0	0.021	16
Total coliform bacteria	0	0	0	0	0.5	0.05	8
Total bacteria	0	0	0	0	0.5	0.05	9

Table 11. Comprehensive ranking results's comparison of the improved AHP method.

Importance	The total sorts	The sort after removed nitrite	The sort after removed cadmium
1	Hg	Hg	Hg
2	Cyanide	Cyanide	Cyanide
3	Total dissolved solid	Total dissolved solid	Total dissolved solid
4	Cd	Cd	Se
5	Se	Se	Fluoride
6	Fluoride	Fluoride	Nitrite
7	Nitrite	Total coliform bacteria	Total coliform bacteria
8	Total coliform bacteria	Total bacteria	Total bacteria
9	Total bacteria	As	As
10	As	Iodide	Iodide
11	Iodide	Cr	Cr
12	Cr	Total hardness	Total hardness
13	Total hardness	Ammonia nitrogen	Ammonia nitrogen
14	Ammonia nitrogen	Turbidity	Turbidity
15	Turbidity	Volatile phenol	Volatile phenol
16	Volatile phenol	Nitrate	Nitrate
17	Nitrate	Sulfate	Sulfate
18	Sulfate	Pb	Pb
19	Pb	Chloride	Chloride
20	Chloride		

Table 12. Groundwater quality of a well.

Indexes	Turbidity	Total hardness	Total dissolved solid	Nitrate	Nitrite	Ammonia nitrogen	Sulfate	Chloride	Hg	Cd
Measured value	4.5	532	727	8.47	0.0015	0.08	133	125	<0.00005	0.0022
Measured value	<0.0008	0.006	0.006	<0.0001	0.002	0.28	<0.02	<0.001	<3	<1

R_i—index i on level j membership degree of evaluation factor (i = 1,2,3, ... n; j = 1,2,3, ... m);
C_i—measured value of index i;
S_{ij}—standard value of index i on level j.

From the above equations, the fuzzy relation matrix that between evaluation factors and the set of evaluation can be established.

$$R = \begin{bmatrix} r_{11} & r_{12} & \cdots & r_{1n} \\ r_{21} & r_{22} & \cdots & r_{2n} \\ \cdots & \cdots & \cdots & \cdots \\ r_{m1} & r_{m2} & \cdots & r_{mn} \end{bmatrix} \qquad (8)$$

4.2 Comprehensive evaluation

All influences of factors have been considered in fuzzy comprehensive evaluation, build the weight vector 'W' and single factor fuzzy evaluation matrix "R", each fuzzy comprehensive evaluation vector is available.

$$B = WR \qquad (9)$$

4.3 Case study

Take water quality analysis results of a well in Chaoyang district Beijing city for example (as is shown in Table 12). Calculate the membership degree of each index to groundwater quality classification standard of five grades, combine them with the influence weights of the factors to ground water quality pollution in the above hierarchy analysis, then comprehensively evaluate the groundwater quality of the well.

On the principle of membership degree (Table 13), According to the comprehensive evaluation of the formula 9, compare results that calculated by traditional AHP with results obtained from improved AHP method, calculate and

Table 13. Membership function.

	I	II	III	IV	V
Turbidity	0	0	0.786	0.214	0
Total hardness	0	0	0.18	0.82	0
Total dissolved solid	0	0.546	0.454	0	0
Nitrate	0	0.769	0.231	0	0
Nitrite	0.944	0.056	0	0	0
Ammonia nitrogen	0	0.667	0.333	0	0
Sulfate	0.17	0.83	0	0	0
Chloride	0.25	0.75	0	0	0
Hg	1	0	0	0	0
Cd	0	0.867	0.133	0	0
Pb	1	0	0	0	0
Cr	0.8	0.2	0	0	0
As	0.8	0.2	0	0	0
Se	1	0	0	0	0
Cyanide	0.889	0.111	0	0	0
Fluoride	1	0	0	0	0
Iodide	1	0	0	0	0
Volatile phenol	1	0	0	0	0
Total coliform bacteria	1	0	0	0	0
Total bacteria	1	0	0	0	0

Table 14. Comparison result of two assessment methods.

Assessment methods	AHP-fuzzy assessment	IAHP-fuzzy assessment
Assessment value	0.686, 0.172, 0.108 , 0.034 , 0	0.678, 0.202, 0.090, 0.030, 0
The level	I	I

normalize the vector, the maximum weight is the vaule of level one (Table 14), so the well's water quality Be classified as grade one. The result of Fuzzy-AHP is acceptable. The overall evaluation turned out to be scientific and reliability.

5 CONCLUSION

By the potential hazard' degree analysis of groundwater pollution indexes to the human body, indexes have been given importance scales after multiple comparison, the hierarchical structure model of the groundwater quality have been established .

Comprehensive sort results of the water pollution have been calculated, by the analysis of the method' defects and causes which result to their defects, that the AHP algorithm can't keep indexes' independence is the key origin. Finally, the AHP algorithm have been improved and the rationality of the improved scheme have been verified.

By the method of combining with fuzzy mathematics with analytic hierarchy process and improved analytic hierarchy process, we have evaluated and verified groundwater quality of a well.

This method not only overcomes the subjective deviation created by man, but also avoid the errors brought by fuzzy uncertainty of fuzzy mathematics. By its application, the evaluation results are scientific and reliability.

REFERENCES

Ali, Aalianvari et al. 2007. A decision support system using Analytical Hierarchy Process (AHP) for the optimal environmental reclamation of an open-pit mine. *Environ Geol* 2007 (52):663–672.
Ali, Aalianvari et al. 2012 Application of fuzzy Delphi AHP method for the estimation and classification of Ghomrud tunnel from groundwater flow hazard. *Arab J Geosci*. 2012 (5):275–284. Bojan, Srdjevic et al. 2008. Fuzzy AHP Assessment of Water Management Plans. *Water Resour Manage*. 2008 (22):877–894.
Guo, Yanying et al. 2006. Application of Analytic Hierarchy Process (AHP) Method in Comprehensive Assessment of Surface Water Quality, *Journal of Lanzhou Jiaotong University (Natural Sciences)*. 25 (3):70–73.
Kong, Feng, & Liu Hongyan. 2009. Analysis and improvement on final ranking of AHP algorithm. *Journal of Harbin institute of technology*. 41 (4):260–263.

Li, Fengwei et al. 2012. Application of Improved AHP in Risk Identification During Open-cut Construction of a Subway Station. *Journal of Beijing university of technology*. 38 (2):167–172.

Li,.Ruzhong. 2007. Dynamic Assessment on Regional Eco-environmental Quality. Using AHP-Statistics Model—A Case Study of Chaohu Lake Basin. *Chinese Geographical Science* 17 (4): 341–348.

Liu, Jun. 2003. Application of AHP in Comprehensive Assessment of Water Environment Quality. *Journal of Chongqing Jianzhu University*. 25 (1):77–81.

Sinha R. et al. 2008. Flood Risk Analysis in the Kosi River Basin, North Bihar using 10 Multi-Parametric Approach of Analytical Hierarchy Process (AHP). *J. Indian Soc*. 2008 (36): 335–349.

Saaty TL. 1977. A scaling method for priorities in hierarchical structures. J Math Psychol 15:234–281. doi:10.1016/0022-2496 (77) 90033-5.

Saaty TL & Vargas LG. 1990. The analytic hierarchy process series. University of Pittsburg, USA.

Sule, Tudes et al. 2010. Preparation of land use planning model using GIS based on AHP: case study Adana-Turkey. *Bull Eng Geol Environ* 2010 (69):235–245.

Zou, Qiang et al. 2012. Comprehensive flood risk assessment based on set pair analysis-variable fuzzy sets model and fuzzy AHP. *Stoch Environ Res Risk Assess*. DOI 10.1007/s00477-012-0598-5.

Frontiers of Energy and Environmental Engineering – Sung, Kao & Chen (eds)
© 2013 Taylor & Francis Group, London, ISBN 978-0-415-66159-1

Remediation of DDVP-contaminated soil by surfactant

X.J. Yu, F. Zhang, J. Zhang, A.M. Zhang, L.Z. Huang & L.Y. Tang
School of Science, Xi'an University of Technology, Xi'an, Shanxi, China

ABSTRACT: Sodium dodecyl sulfate is selected to eliminate contaminated soil which is polluted by diethyl dichlorovinyl phosphate. This article discusses the effects of temperature, contact time and liquid-soil ratio for the efficiency of elution by means of experiment and aim at determining the optimum experimental conditions. The results indicate that the elution efficiency of diethyl dichlorovinyl phosphate increases with the increasing of sodium dodecyl sulfate concentration, liquid soil ratio, temperature, and extraction time within a certain range. The elution efficiency could achieves the maximum 36% while the concentration of surfactant is 15000 mg/L, temperature is 30°, contact time is 10 hours and liquid-soil ratio is 30:1.

Keywords: diethyl dichlorovinyl phosphate; surfactant; contaminated soil; enhanced remediation

1 INTRODUCTION

With the acceleration of industrialization and the development of human activities, the area of contamination is extending and the degree of contamination is deepening. More and more accidents caused by chemical substances have been heard these years. Pernicious substances contaminate water and plants in the soil, eventually enter into human body through food chains or potable water, resulting in body pathological change[1]. As the detriment of chemical substances is more and more extended, since 1980s, a large quantity of scientists focuses on the remediation of contaminated soil and advance a series of technologies on remediation of contaminated soil[2]. There are two methods dividing into physical technique and chemical technique. Physical technique contains steam extraction technique, solidification/stabilization remediation technique, thermodynamics of repair technique, thermal desorption remediation technique, electro-kinetic remediation technique and so on. Electro-kinetic remediation technique can apply to reconditioning organic and metal contaminated soil, but high costs will be the biggest constraint for practical application. Chemical technique mainly contains chemical leaching technology, solvent extraction technology, chemical oxidation remediation technology and so on[3-8].

Surfactant enhanced remediation technology is a technique of chemistry, which is one of the most effective methods. Surfactant has many properties such as wetting, emulsification, dispersion, solubilization, foaming, and detergency. These properties make surfactant to be potential for the remediation of contaminated soil in the future[9]. In recent years, surfactant enhanced remediation technology is applied successfully to soil flushing and organic pollution of water. For example, Dr. Foutain and his members remove TCE in soils successfully with complex surfactants; Eckenfelder Company in Germany has proposed a method to deal with the waste surfactants recycling[10]. In order to solve the problem of DDVP pollution better, this article analyzes the factors of influencing the efficiency of remediation, such as temperature, concentration, contact time and liquid-soil ratio, by means of surfactant enhanced remediation on Diethyl Dichlorovinyl Phosphate (DDVP) contaminated soil.

2 MATERIALS AND METHODS

2.1 *Soil and pretreatment*

There are 500 g soil samples mixed with acetone, DDVP of known concentrations and 500 mL distilled water all in a 1 L conical flask. The soil is put on the ground for air drying and litter from the soil with 100 mesh sieve after 48 hours oscillating under 25 degrees when it is balanced, then taken into a brown bottle. The DDVP content of the soil sample is 0.5 mg/g.

2.2 *Experimental procedure*

There is 1 g contaminated soil mixed 20 mL Sodium Dodecyl Sulfate (SDS) which is the specified concentration in a 50 mL conical flask. After compounding several same samples, the experiment will be taken through different temperature, contact time and liquid-soil ratio. Subsequently, the eluent added in a 50 mL centrifuge tube is centrifuged at 5000 r/min for 30 min; standing for 10 min, there are 2 mL supernatant liquid and 4 mL n-hexane oscillating in a 20 mL bottle for 15 min, then centrifuge the mixture again and stand for 15 min.

2.3 Analysis method

Supernatant liquid is extracted from mixture. After filtering by means of anhydrous sodium sulfate and adding potassium persulfate, high temperature digestion will take place. Absorbance is determined by phosphorus molybdenum blue photometric method at the wavelength of 690 nm, thus calculating the phosphorus content, then the DDVP content is shown by phosphorus content.

3 RESULTS AND DISCUSSION

3.1 Effect of concentration on remediation

Concentration is an essential factor for reactions. This experiment studies the effect of SDS concentration on remediation of DDVP-contaminated soil. Several SDS solution of different concentrations (from 2000 mg/L to 20000 mg/L) are added into soil samples. The liquid-soil ratio is 30:1 and the temperature is 30 degrees. After 10-hours reaction, the efficiency of remediation of DDVP-contaminated soil shows in Figure 1.

Figure 1 shows that SDS is effective for enhanced remediation. When the concentration of SDS is below the Critical Micelle Concentration (CMC), the elution efficiency is quite low. However, it is increasing sharply when the concentration of SDS is beyond the CMC. It demonstrates that the

concentration of SDS plays an important role and makes a great promotion for eluting DDVP. It is studied that the curve in the coordinates keeps rising whenever the concentration is below or beyond CMC. There are some reasons. Initially, surfactant molecules arrange on the surface of solution until there is no room for more surfactant molecules. During this period, there is slight solubilization and desorption, resulting in a slow rise of the curve. Subsequently, micelles begin to form and more DDVP molecules dissolve in micelles, and then removed from the surface of soil particles. Ultimately, there is also desorption in the reaction. Due to the surface tension of liquid-soil declining, DDVP molecules which are blocked are capable of getting through the interstices between soil particles more easily and eluted by solution. With the concentration increasing, the elution efficiency is raising, meanwhile, the costs will be higher. Therefore, keeping concentration at 15000 mg/L is optimum in this experiment.

3.2 Effect of temperature on remediation

Temperature is one of important factors in the experiment. There are three kinds of SDS samples: 4000 mg/L, 7500 mg/L and 15000 mg/L, on the condition of liquid-soil ratio is 30:1 and contact time is 10 hours. In the experiment, make sure that three samples are always on the same conditions when temperature changes. The result is shown in Figure 2.

It shows that the elution efficiency of DDVP is increasing rapidly with variation of temperature when the concentration is 15000 mg/L in Figure 2, while temperature is barely effective at low concentration. We can analyze the data in the experiment and acquire some information, that is, when the concentration of surfactant is below the CMC, temperature just affects the elution efficiency

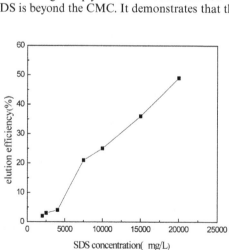

Figure 1.   Effect of concentration on remediation.

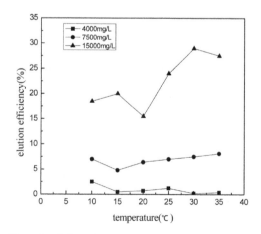

Figure 2.   Effect of temperature on remediation.

slightly. However, if the concentration of surfactant is beyond the CMC, there is a great effect of temperature on elution efficiency. The efficient cause of the results above is that temperature can mainly enhance the solubilization of micelle and improve solubility of surfactant, which does not exist at low concentration. As a result, elution efficiency is developing with temperature rising. Considering about the reality, therefore, the experiment will attain the best elution efficiency at 30°.

3.3 *Effect of contact time on remediation*

The contact time of the reaction between surfactant and contaminated soil is significant to elution efficiency. There are two kinds of SDS concentration: 7500 mg/L and 15000 mg/L, on the same condition of liquid-soil ratio are 30:1 and temperature is 30 degrees, and then elute the samples within different contact time. The result of the experiment is shown in Figure 3.

Figure 3 indicates that during the 50 hours' observation, there are only slight changes at 7500 mg/L while the elution efficiency increases constantly at 15000 mg/L. Experiment data shows reaction tend to keep balance gradually for about 10 hours, no matter when the concentration is below or beyond the CMC. In the following period, the elution efficiency increases slowly or even decreases. Because there are a series of complex reactions internally, and most of the reactions are rapid after interaction of surfactant and soil. They just react and elute most pollution in incipient 10 hours. If the complex compound of surfactant and pollution is inadequate stabilization, there may be a slight decline of the elution efficiency with the reaction after 10 hours. So there are not obvious experimental phenomena after 10 hours. Therefore, it is

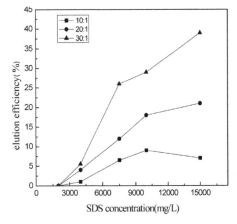

Figure 4. Effect of liquid-soil ratio on remediation.

necessary to stipulate the contact time in practice. In this experiment, the contact time of reaction is selected to be 10 hours.

3.4 *Effect of liquid-soil ratio on remediation*

The experiment studies the effect of liquid-soil ratio on remediation of contaminated soil. Compounds three kinds of liquid-soil ratio samples, which is 10:1, 20:1 and 30:1. Then on the same condition of temperature is 30 degrees and the contact time is 10 hours. The result is shown in Figure 4.

Figure 4 indicates that elution efficiency of DDVP rise with liquid-soil ratio increasing. When liquid-soil ratio is 10:1, the effect of surfactant is not obvious, because there is too little surfactant in the sample so that surfactant cannot contact adequately with soil, consequently, it will not attain apex. When liquid-soil ratio is 20:1 or 30:1, elution efficiency of DDVP leaps obviously. That is, the higher liquid-soil ratio, the more surfactant in the reaction. As a result, in order to attain optimum result, selecting appropriate liquid-soil ratio is quite important.

4 CONCLUSIONS

We can obtain some conclusions below by means of experiments: (1) SDS actually improves the elution efficiency of DDVP. It indicates that surfactant concentration is key for remediation of DDVP-contaminated soil, and elution efficiency increases substantially with surfactant concentration increasing. (2) Various factors, such as temperature, contact time, liquid-soil ratio and so on, more or less affect the elution efficiency of

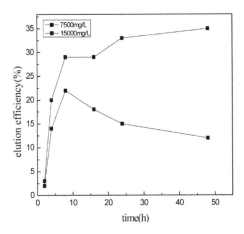

Figure 3. Effect of contact time on remediation.

DDVP. (3) The optimum condition of remediation by SDS is that the concentration of surfactant is 15000 mg/L, while temperature is 30 degrees and liquid-soil ratio is 30:1. After 10 hours reaction, we will attain the best result, and the elution efficiency reaches to 36%.

Nowadays, surfactant is playing a more and more important role on remediation. The surfactant enhanced remediation may show a quite immense potential and the combination between surfactant enhanced remediation and other technologies will be a hotspot on remediation of contaminated soil.

ACKNOWLEDGEMENTS

This work was financially supported by National Nature Science Foundation of China (No. 21276208); The S&T Plan Projects of Shanxi Provincial Science and Technology Department (2012 K07-11) and the S&T Plan Projects of Shanxi Provincial Education Department (No.12JK0635).

REFERENCES

Andreottola G., Foladori P., Ziglio G. 2009. Biological treatment of winery wastewater: an overview. *Water Science & Technology*, vol. 60 (5): 1117–1125.

Dianyou Li, Yinong Lu. 2005. On the harm of arsenic pollution in soil and its countermeasures. *Journal of Xinjiang Normal University*, vol. 24: 89–91.

Donato R. O. Nichols, and H. Possingham. 2007. A critical review of the effects of gold cyanide-bearing tailings solutions on wildlife. *Environment International*, vol. 33: 974–984.

Fang Ma, Yujie Feng, Nanqi Ren. 2003. Environmental biotechnology. *Chemical Industry Press*.

Gene A., Chase G., Foos A. 2009. Electrokinetic removal of manganese from river sediment. *Water Air and Soil Pollution*, vol. 197: 131–141.

Liqin Yang, Sijin Lu, Hongqi Wang. 2008. Study on physical and chemical remediation of contaminated soil. *Environmental Protection Science*, vol. 34 (5): 42–45.

Mei Zhu, Jialin Xu, Honghai Tian. 1996. Study on surfactant remediation of petroleum pollution in aeration zone. *Environmental Science*, vol. 17 (4): 21–24.

U.S.EPA. 2000. Treatment technologies for site cleanup: annual status report. *Eleventh Edition* (EPA 542-R-01-004).

Xia Jiang, Xuesheng Gao, Peifeng Ying. 2003 Compatibilization and motion of surfactant in soil. *Chinese Journal of Applied Ecology*, vol. 14 (2): 2072–2076.

Zuming Mei, Pingfan Yuan, Ting Yin, Jun Lu, Ge Zhu. 2010. Discussion on remediation of contaminated soil. *ShangHai Geology*, vol. B11 (5): 128–132.

Frontiers of Energy and Environmental Engineering – Sung, Kao & Chen (eds)
© *2013 Taylor & Francis Group, London, ISBN 978-0-415-66159-1*

Research on the mode of the development and marketing approaches of cultural products in universities

L.F. Wang, J.N. Zhu, S.W. Xie, C. Zhang & X. Guo
Northwestern Polytechnical University, Xi'an, P.R. China

ABSTRACT: With the constant achievements we have been making on the implementation of the reform and opening-up policy, Chinese culture industry has been hold in increasingly greater account, and especially the university culture, which plays a crucial role in the domestic culture industry. University culture-based products will provide the school culture itself an extremely great platform to increase its influence. This paper analyzes the problems existing in the present development of culture products and also causes of them in scientific ways systematically, and finally raises the idea of applying "cultural product development model" to develop university cultural products. In the meantime, this paper presents some marketing strategies from different angles, aiming at promoting the standardization of University cultural products with the help of the strength provided by both cultural and commercial patterns.

Keywords: culture industry; university culture industry; university influence; cultural product development model; marketing strategies

1 GENERAL INSTRUCTIONS

1.1 *Raise of the problem*

At the beginning of the reform and opening up, the government put forward the development of culture industry and accelerating the construction of the cultural industry base. University culture industry as a part of it is a combination of university culture, local culture and art design which provides a perfect platform for the school to expand its influence, which has drawn great attention. Many famous universities have opened gift stores, selling some small objects. However, many got stuck after a short period for different reasons, such as the product diversity and people's weak awareness of university culture. In a nationwide online survey, 61.1% of the students have access to cultural products of their schools, but in fact only about 10 universities have their own cultural products stores, showing that the access is rather restricted. Apart from that, there are 72 stores selling university cultural products on Taobao (the biggest e-commercial platform in China), but quite a number of them operating status are pale. Shanghai Jiaotong University's official taobao shop only has 9 orders per month, turnover is only 78.4 yuan. It didn't get everyone's favor and attention. However, in the questionnaires to ZheJiang University students, in which the data showed that 89.1% of students would buy the cultural products, especially the postcards, which suggested the tremendous potential of the market.

This paper attempts to answer the following two questions: first, how to design a products grounding the campus culture; Second, how to establish complete marketing system to promote product sales in the cultural marketing angle, and achieve the win-win purpose.

2 ANALYSIS OF THE PROBLEMS EMERGED DURING THE PROCESS OF CULTURAL PRODUCT DEVELOPMENT

2.1 *Introduction to the method—system science method*

System science method means using the theory and point of system science, putting the object into the system, from an overall perspective, to inspect, analyze and study it considering the relationship between system and element, element and element, structure and function, system and environment so as to solve problems in the best way.[2]

In this paper, we put university cultural products in the system of Chinese university culture to inspect, analyze and study the university cultural products considering the controversial but united relationship between the university cultural system and products, university cultural products themselves, university cultural products and the market environment in order to find the best marketing strategies.

2.2 The problems existed in the development of university cultural products and the reasons for them

2.2.1 The relationship between the system and the products of university culture

In recent years, with the further reform in China, the campus cultural establishment also gains good opportunities to develop. However, because of the influence of traditional educational ideas and theories, as well as the educational system's and the campus management's imperfection, and the negative influence of the media on the campus culture, all these make university cultural establishment still has the problems as follows:

The political culture is almost dominant, which makes the university culture too simple, rigid, abnormal, it hides the way which can make university culture has further development, think highly of the establishment of material culture, rather than spiritual culture, it makes university culture lack cohesion, influence and attraction; adores entertainment and popularity, leaving some room for vulgar culture, keeping valuing the modern attraction, ignoring the history and tradition during the developing progress of campus culture.

The weaknesses of university culture establishment make the university culture does not have its special, we should use something like university cultural products to encourage universities as an existence of material culture, and more of spiritual culture.

2.2.2 The relationship between different university cultural products

From the surveys of university cultural products which have already appeared on the market, we can find that most of these have lower quality, they don't perfectly mix with the idea of university special culture, they lack the combination between innovation, sense of times and historic, and they haven't make them like a system. For example, the products just like key chains, bookmarks, ties, scarfs, clothes, the appearance and styles just like other products sold in the supermarket, they just make the sign of universities on them, the style is simple and lack of innovation.

Secondly, because they have to make the sign of universities on the normal products, they have to spend more time on making them and the cost is higher, and the dealers hold the point that everything is based on the interests, all these make the prices of university cultural products are higher than others, so it is difficult for them to meet the needs of consumers and the recognition from the market.

3 THE DEVELOPMENT SCHEME OF UNIVERSITY CULTURAL PRODUCTS

3.1 The combination of product concept and university culture

Culture-oriented products on a school basis have been playing a critical role in effective communication between schools and consumers, because they not only provide tangible products and service to consumers, but in the meantime convey school culture potentially to them to a certain degree. Consequently, enhancing both the image and the recognition of schools profoundly lies in how far we are able to go about combining the existed school culture with products and service to be provided, and eventually, reaching to the consumers.

3.2 The development of university cultural products

3.2.1 Review

Based on recent years' development in this field, design has gradually become the part which plays a decisive role in cultural innovation field. This is because it is the most direct and effective approach offered to us when connecting products to the market closely. It is possible for products to satisfy the various and changeable needs of market by the means of design and as a result of that, to push the sales volume to grow constantly. Also, they can direct and stimulate new demands potentially.

3.2.2 The Model for developing university cultural products

3.2.2.1 The relation map of the model

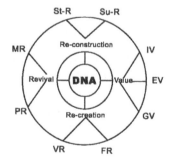

Chart 1. The cultural product model.

3.2.2.2 The introduction to the culture DNA concept

Culture DNA is an abstract concept equivalent to the biological DNA, used to describe the cultural factors. Cultural DNA is the most central essence in diverse cultural forms, which can be presented by other means, and then to be communicated and inherited.

3.2.2.3 *The development process*
- The First Stage: Put the selected cultural DNA at the center of the model and classify 4 types of cultural product cluster, that is Revival, Re-creation, Reconstruction, and Valuing.
- The Second Stage: Cultural product clusters_further classified into 9 types of commercialization methods according to the classified clusters.

    *Nine types of commercialization methods: Meaning Revival, Physical Revival, Visual Re-creation, Functional Re-creation, Story Reconstruction, Surreal Construction, Integrated Value, Experience Value and Global Value.
- The third stage: Cultural product commercialization methods_ Set items to be developed according to 9 types of commercialization methods.

Concrete description of 4 types of product development clusters and 9 types of item development clusters is as follows:

1. Revival: Cultural products that make the archetype and meaning of traditional culture revive.
    ① Meaning Revival: Covers metaphysical revival of intangible thoughts and concepts such as consciousness, norm, way of thinking, etc. as well as the revival of ceremonies through thorough historical research.
    ② Physical Revival: Covers metaphysical revival of proportional revival of tangible prototype of traditional culture in shape, material and size as well as process restoration including fabrication of the actual object.
2. Re-creation: Cultural products re-created based on cultural heritage.

③ Visible Re-creation: Covers the design products that apply the entire or some element of the visual characteristics of tangible cultural assets
④ Functional Re-creation: Covers design products that reflect the original functions of tangible cultural assets or assigning their functionality of modern meaning.
3. Re-construction: Cultural products which include the intricacies and stories inherent in traditional culture.
    ⑤ Story Reconstruction: Covers cultural products that reconstruct the origin of cultural assets and related stories, and reflect tangible and intangible content
    ⑥ Surrealistic Reconstruction: Covers formative or artistic reconstruction through visual re-creation, surrealistic expansion or exaggeration, and utilization of other media.
4. Valuing: Cultural products where general meaning, shape or image of cultural heritage is sublimated into cultural value
    ⑦ Integrated Value: Covers brands, symbols and characters designed based on integrated image.
    ⑧ Experience Value: Covers experience on cultural values through the reappearance of experience or process.
    ⑨ Global Value: Covers global products with values for which people around the world can feel sympathy, over the representative cultural product of an original culture.

3.2.3 *Apply the model into the development of university cultural products*
1. Conduct the selection about university culture in a broad sense, and focus on the most typical and sustainable university culture DNA

Chart 2. Example of the product developing process.

Culture DNA		School Motto: *Loyalty, Integrity, Courage and Perseverance*
Revival	MR	
	PR	
Recreation	VR	Apply the motto into the development of NPU's VI system such as writing paper, markers and hand bags, etc.
	FR	Demonstrate the school motto from different aspects such as students, teachers and school, and illustrate new needs by relevant products.
Re-construction	St-C	
	Su-C	Invite or employ talented artists to translate the school motto into striking artworks like sculpture and pottery to create new attractive landscapes for the school.
Value	IV	For test, design 4 types of fonts specially for authorities' use, embodying the nature of the motto and integrate it with all kinds of school affairs.
	EV	
	GV	

2. Analyze the DNA specifically and closely, and following that polish and filter the extended 9 aspects during the developing process expecting the result of the stretch of highly-related and developable aspects in particular

The following is the example of NPU aiming to illustrate the entire development process of university cultural products (the blank in the chart means that aspect is not suitable for further development).

4 MARKETING STRATEGIES FOR UNIVERSITY CULTURAL PRODUCTS

Cultural marketing is a new effective way of marketing which integrates corporation culture consciously into the ordinary marketing process and furthermore, promotes the marketing ability as a whole with the corporation acting as the motivate, and finally targets on enhancing corporation's capacity to compete with other power. Compared with traditional marketing strategies (tangible-product-oriented), cultural marketing emphasizes on stimulating the cultural attributes of the products, building the attraction during the whole process. As a result, the recognition, acceptance and support coming from consumers and even the public are more accessible.

4.1 Basic strategies

4.1.1 Market positioning

Chart 3. Market positioning of the university cultural products.

Consumer group	Price position	Style
Students	More low and medium standard, less high standard	Trendy and casual
Teachers	More medium and high standard	Serious and grave
External group	A wide range of standards	Combining different styles to express the morale of universities

4.1.2 The width and depth of publicity
4.1.2.1 Advertisement
Publicity is of great importance especially for those products which are ready to come out. The existed university souvenirs adopt old-fashioned approaches to publicize or just sell the products online. That can no longer keep up with the increasing demands given by the market, which forces us to focus on the quality of the product.

Considering about the specific campus features and consumers, together with cost and publicity effects, there are a range of approaches offered to us:

1. Students: use "in" ways such as post and internet to give information on those products with a certain theme like the Olympics and the World Cup to meet student group's demands.
2. Faculty: communicate with relevant school departments and give them an overall introduction to the culture products, striving to bond closely with each department so as to reach different activities timely and precisely.
3. External group: Depending on the experience from former marketing, make a selection of several typical products of highly pratical use and then have tentative sales in some retail stores and supermarkets. Expand the market based on successive analysis of data collected from the test sales.

4.2 Cultural strategies

4.2.1 Link other universities to strengthen the cultural influence—university league
Since NPU has a relatively weak influence, we can make up for that by means of university league-starting from a small group of universities who have enjoyed higher influence domestically and regarded as a whole to practice the strategies-in order to build the fundamental atmosphere for cultural products.

The Ivy League have authorized a brand to design and sell clothes and other accessories with those universities' logo printed and even with some typical cultural factors planted during the design process. The clothes have received plenty of attention and consumers'phraise due to the universities' fame and also the brilliant design. The brand not only attains a high profit, but enables the universities' image to come closer to the public and publicize the school culture.

4.2.2 Independent way encourage the transmission of culture—students'participation
To cater to the market is the key to university cultural products marketing. Few of the universities which have cultural products face the problem: low quantity, few products. The main reason is that they gpt the wrong information of consumers'favour and needs besidesthe relationship with them is not enough.

5 ENDING

The research of University cultural product project on product design and marketing is based on analyzing the market environment, coming up with a fully judgment through the scientific systematic method, it can efficiently react the problems which exist in the production of university culture. Based on these problems, the development of production has introduced the concept of cultural DNA, establishing the developed model of cultural production, from the view of marketing strategy, it has introduced multi-angle ways to develop the market, and it can provide valued views to the demonstration and establishment of university cultural products, it has certain promotion value.

From the progress of the establishment of product developed model and culture marketing strategy, we can see that the quality of products and the development of market have great influence to the production. In order to grasp the market of cultural products, we should pay more attention to how can we combine products and culture to develop, gain the market position and do better market sales, rather than simple manufacture.

REFERENCES

[1] Jinjun Fu. University Campus Culture [M]. Shanghai: Shanghai Jiaotong University Press, 2001, 1.

[2] Xiaonian Liu. The Discussion Of Contradictory Analysis Method and System Analysis Methods [J]. Journal of Systems Science, 2011, 03.

[3] Lixin Ma. The Construction of Campus Culture In Institutions of Higher Learning [D]. The Institute of Foreign Languages and Social Sciences of Tianjin University, 2008, 5.

[4] Banghu Wang. The Theory of Campus Culture [M]. Beijing: People's Education Press, 2000, 16.

[5] Dingxin Yang, Baoling Shi. Several Basic Problems of Cross-cultural Marketing Study [A]. Economics and Management College of Gansu Lianhe University, 2003, 5.

[6] The gift is too homely Chinese university cultural products market is in the doldrums[EB/OL.]China Educational Equipment Procurement Network, 2011, 08.

[7] Yan Li. Of Enterprise's High-end marketing— culture marketing [J]. Heilongjiang Institute of Commerce, 2012, 3.

[8] Chunling Shi. On Cultural Marketing [J]. Market Modernization, 2006, 31.

[9] Fenggang Shi. On Campus Culture and Its Optimization [J]. Higher Education Research, 1998, 1.

[10] Australian Journal of Emerging Technologies and Society, Trading Card Games As A Social Learning Tool [M] Vol. 3, No. 2, 2005, pp: 64–76.

[11] Zhang Xu. Cultural force in the corporate DNA [J]. Dalian: Dalian University of Technology, College of management, Liaoning, 2005, 26 (4).

[12] The Floor Study of The Construction of Enterprise Culture. Modern Enterprise Culture, Academic Edition, First Period in 2012.

[13] HongBing Ni. The Explore of How The Corporate Combine Culture and Products and Services To Customers [J]. Standards and Science. Fifth Period in 2009.

Frontiers of Energy and Environmental Engineering – Sung, Kao & Chen (eds)
© 2013 Taylor & Francis Group, London, ISBN 978-0-415-66159-1

Chemical methods to improve the foam stability for foam flooding

G.Q. Jian
State Key Laboratory of Enhanced Oil Recovery, Research Institute of Petroleum Exploration and Development, CNPC, Beijing, China

S.Y. Chen
School of Chemistry and Biology Engineering, University of Science and Technology Beijing, Beijing, China

S.L. Gao
E&D (Exploration and Development) Research Institute of Daqing Oil Field Company

Y.Y. Zhu, Y.S. Luo & Q.F. Hou
State Key Laboratory of Enhanced Oil Recovery, Research Institute of Petroleum Exploration and Development, CNPC, Beijing, China

ABSTRACT: In this paper, evaluation experiments concerning surfactant foams, polymer enhanced foams and clay nanoparticle enhanced foams were carried out. The results under crude oil free conditions indicated that, polymer enhanced foams could improve foam stability by 10 to 50 times compared with surfactant foams. For nanoparticle enhanced foams, the foams stability could be further enhanced by more than one order of magnitude compared with polymer enhanced foams. With Daqing oil introduced, the stability of foams for different foam formula differed remarkably. For the experiment investigated, foam stability could be improved by certain degree when Daqing crude oil was introduced.

1 INSTRUCTION

The average enhanced oil recovery for main reservoirs in Daqing oil field has reached 53%. Further development of such reservoir became a big challenge for the reservoir heterogeneity and highly dispersed residual oil. Foam flooding can increase the oil displacing efficiency and swept volume simultaneously and now are recognized as a promising EOR technology after polymer flooding for Daqing oil filed. According to recent experience of foam flooding in Daqing oil field, the stability of foam is recognized as a main factor influencing the performance of foam flooding. Hence, high stable foam formula research tailed for foam flooding is of great importance.

Surfactant foam is the simplest foam and has been applied widely in different areas. The surfactant commonly used for foam flooding including two classes. One is inion surfactants and the other is amphoteric surfactants. The former includes sodium dodecyl benzene sulfonate, petroleum sulfonate, sodium dodecyl sulfonate, sodium dodecyl sulfate, α olefins sulfonic acid salt, sodium oleate. The later includes fatty alcohol polyoxyethylene ether sulfate, fatty alcohol glyceryl sulfonate, dodecyl alkylphenol polyoxyethylene ether. With respect to oil displacing surfactant, foamers are usually low carbon chain length surfactants. The hydrophobic chain often includes less than or about 20 carbons. In this paper, C_{14-16} AOS and FC are applied as foamers and their foaming ability and foam stability were investigated.

Polymer enhanced foams [1][2][3] are the most commonly used foam formula for certain foam flooding pilot tests in China. With certain amount of polymers introduced, the foam stability can be improved dramatically. Several representative polymers are chosen including linear partially hydrolyzed polyacrylamide (MO4000 and HPAM2500), salt tolerance polymer (HJKY), temperature resistance polymer (YH1096), hydrophobic associating polymer (SSDH45), star polymer (LWL8036), and surface active polyacrylamide (MPAM). The surface active polyacrylamide MPAM was at last selected from all abovementioned polymers to be applied as a foam stabilizer. HPAM2500 enhanced foam experiment was conducted as a reference.

Nanoparticle foams or nanoparticle enhanced foams are at the forefront of foam research area. B. P. Binks[4][5][6] and Urs T. Gonzenbach[7][8] have conducted several experiments about nanoparticle foams. They have confirmed that special coated nanoparticles can improve the stability of nanoparticles by orders of magnitude. Recent research focused more on silica nanoparticle enhanced foams. Research on clay particle enhanced foams, however, was not so much. In this paper, a novel special coated clay nanoparticle was applied as foam stabilizer. The foaming and foam stability experiments were conducted to investigate the influence of nanoparticle on foam systems.

This paper introduced some main recent progress of foam formulas of SKL-EOR lab in RIPED. These foam formula includes surfactant foams, polymer enhanced foams and nanoparticle foams. We hope our research can facilitate deeper understanding of stable foams for related researchers.

2 EXPERIMENTAL

2.1 *Materials*

The surfactant applied included fluorinated surfactant FC, Alkyl Glucoside and α-sodium olefin sulfonate $C_{14-16}AOS$. The effective content of these surfactant are 25%, 50% and 90% respectively.

The polymers applied are HPAM2500, modified poly-acrylamide MPAM, MO4000, salt tolerance polymer HJKY, hydrophobic associating polymer SSDH45 and star shape polymer LWL8036, temperature resistance polymer YH1096.

The particle was a novel coated clay nanoparticle. The water is synthetic brine with 3652.5 mg/L salinity. The constitution of the water was as shown in Table 1. The oil is Daqing crude oil with density 0.8521 g/cm³ and viscosity 9.6 mPa·s at 45 degrees.

2.2 *Methods*

2.2.1 *Foaming experiment*
The foaming process was carried out in Waring Blender. About 200 ml foaming agent which was heated at 45°C for two hours was initially introduced into the blender and mixed by the propeller in the blender for one minute. After intensively blended, foams were generated and they were then decanted

into the 1000 ml graduated cylinder. The decanting span last for 30 seconds to make sure that most of the foams can be poured out from the blender. The volume of the foams was recorded. Then the cylinder was sealed with a plastic plug and was put into the oven with temperature set at 45°C. After the liquid in the cylinder reached 100 ml, the time was recorder and was defined as the liquid drainage half-life.

2.2.2 *Viscosity experiment*
The viscosity experiment was conducted on Brookfield DV-∏ viscosity equipment. The test temperature was 45°C. The 0 type rotor was applied. The shear rate was 6 r/min.

2.2.3 *Oil tolerance experiment*
The oil applied was Daqing crude oil. 100 ml crude oil and 200 ml foaming agent was simultaneously introduced into Waring blender. The following experiment process was the same with foaming experiment.

3 RESULTS AND DISCUSSIONS

3.1 *Surfactant foams*

As shown in Table 2, the stability of FC surfactant foams was longer than that of $C_{14-16}AOS$ foams. The reason was that the FC surfactant could generate homogenously distributed size foams. And the arrangement of FC surfactant in the lamella of foams was more compact than that of $C_{14-16}AOS$ foams.

The FC surfactant was more tolerant with oil. This was caused by its favorable water and oil repellency properties of its hydrophobic carbon chain. While for hydrocarbon surfactants such as $C_{14-16}AOS$, the light composition in the crude oil would make the surfactant arrangement in the lamella of foams dramatically changed significantly and the foams is thus become instable.

Table 1. Constitution of Daqing synthetic brine.

Inion type	Na+	K+	Ca2+	Mg2+	Cl-	SO4²⁻	CO₃²⁻	HCO₃⁻
Conc. mg/L	1120.4	0.1	26.6	1.5	786.5	57.4	1585.0	75.0

Table 2. Initial foaming volume (V_0), half-life of liquid drainage ($t_{1/2}$) for AOS and FC surfactant foam (oil free).

Surfactant	V_0/mL	$t_{1/2}$/s
C_{14-16} AOS	1000	450
FC	900	702

Table 3. Initial foaming volume (V_0), half life of liquid drainage ($t_{1/2}$) for AOS and FC amphoteric surfactant foam (with 100 ml oil).

Surfactant	V_0/mL	$t_{1/2}$/s
$C_{14-16}AOS$	1000	450
FC	800	690

Table 4. Polymer enhanced foam (oil free).

Foam formula	V_0/mL	$t_{1/2}$/s
AOS 0.4% + HPAM 0.12%	870	2750
AOS 0.4% + MPAM 0.12%	850	1900
FC 0.4% + HPAM 0.12%	810	4715
FC 0.4% + MPAM 0.12%	700	14623

Table 5. Polymer enhanced foam (with 100 ml Daqing crude oil).

Foam formula	V_0/mL	$t_{1/2}$/s
AOS 0.4% + HPAM 0.12%	800	4234
AOS 0.4% + MPAM 0.12%	860	2842
FC 0.4% + HPAM 0.12%	840	7855
FC 0.4% + MPAM 0.12%	710	10827

3.2 Polymer enhanced foams

The polymers applied are commonly linear partially hydrolyzed polyacrylamide HPAM, MPAM, MO4000, salt tolerance polymer HJKY, hydrophobic associating polymer SSDH45, star shape polymer LWL8036 and temperature resistance polymer YH1096. The polymer MPAM with best viscosity enhancement shown in Figure 1 was chosen as a foam stabilizer in our experiment. The HPAM2500 enhanced foams were set as a reference.

The result showed that foam formula FC+HPAM can improve the foam stability dramatically compared with FC+HPAM, C_{14-16}AOS+ HPAM and C_{14-16}AOS+MPAM foam formula systems. This was due to the strong synergistic effect of FC surfactant and polymer MPAM. The deep reason in a molecule scale should be further investigated.

3.3 Nanoparticle enhanced foams

As shown in Figures 3 and 4, the particle can be dispersed in deionized water homogeneously.

From Table 6, we can see that the foam stability can be improved by one order of magnitude when 3 g to 4 g nanoparticles was introduced. As shown in Table 7, the foam stability can be significantly improved while 100 ml Daqing crude oil was introduced. In all, the nanoparticle enhance foams can improve the foam stability sharply with favorable foam volume. Thus, nanoparticle enhanced foams is a promising foam formula which can maintain ultra-stable requirements.

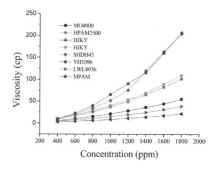

Figure 1. Molecule structure of MPAM.

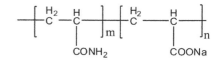

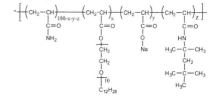

Figure 2. Molecule structure of HPAM.

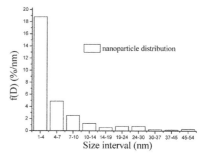

Figure 3. Particle size distribution by SANS.

Figure 4. TEM of nanoparticle in deionized water.

Table 6. Nano-particle enhanced foams (oil free).

Foam formula	V_0/mL	$t_{1/2}$/s
AOS	1000	450
AOS + 0.5 g particle	1000	768
AOS + 1.0 g particle	1000	1350
AOS + 2.0 g particle	850	7590
AOS + 3.0 g particle	780	35600
AOS + 4.0 g particle	570	144000

Table 7. Nano-particle enhanced foams (with 100 ml Daqing crude oil).

Foam formula	V_0/mL	$t_{1/2}$/s
AOS + 4.0 g particle	520	720000

4 CONCLUSIONS

In this paper, evaluation experiments concerning surfactant foams, polymer enhanced foams and clay nanoparticle enhanced foams were investigated. The results under crude oil free conditions indicated that, polymer enhanced foams could improve foam stability by 10 to 50 times compared with surfactant foams. For nanoparticle enhanced foams, the foams stability could be further enhanced by more than one order of magnitude compared with polymer enhanced foams. With Daqing oil introduced, the stability of foams for different foam formula differed. For the experiment investigated,

foam stability could be improved by certain degree when Daqing crude oil was introduced.

REFERENCES

[1] Laurier L. Schramm, Susan M. Kutay, Randy J. Mikula, Vicente A. Munoz. The morphology of non-equilibrium foam and gelled foam lamellae in porous media. J. Pet Sci Engineering, 23 (1999) 117–132.

[2] T. Zhu, D.O. Ogbe, S. Khataniar. Improving the foam performance for mobility control and improved sweep efficiency in gas flooding. Ind. Eng. Chem. Res, 43 (2004) 4413–4421.

[3] S. Kutay, L. Schramm. Foams in Enhancing Petroleum Recovery. J. Can Petrol Technol., 43 (2004) 2.

[4] Bernard P. Binks, Tommy S. Horozov. Aqueous Foams Stabilized Solely by Silica Nanoparticles. Angew. Chem., 117 (2005) 3788–3791.

[5] A. Cervantes Martinez, E. Rio, G. Delon, A. Saint-Jalmes, D. Langevin, B.P. Binks. On the origin of the remarkable stability of aqueous foams stabilised by nanoparticles: link with microscopic surface properties. Soft Matter, 2008, 4, 1531–1535, 1531.

[6] Bernard P. Binks, Mark Kirklandb, Jhonny A. Rodrigues. Origin of stabilisation of aqueous foams in nanoparticle–surfactant mixtures. Soft Matter, 4 (2008) 2373–2382.

[7] Urs T. Gonzenbach, Andr R. Studart, Elena Tervoort, and Ludwig J. Gauckler. Ultrastable Particle-Stabilized Foams. Angew. Chem. Int. Ed., 45 (2006) 3526–3530.

[8] Andre´ R. Studart, Urs T. Gonzenbach, Ilke Akartuna, Elena Tervoort, Ludwig J. Gauckler. Materials from foams and emulsions stabilized by colloidal particles. J. Mater. Chem., 17 (2007) 3283–3289.

Frontiers of Energy and Environmental Engineering – Sung, Kao & Chen (eds)
© 2013 Taylor & Francis Group, London, ISBN 978-0-415-66159-1

The full cost analysis of reclaimed water reuse: Qingdao city as an example

S.L. Yang
Computer Department of Basic Courses, Shandong Institute of Business and Technology, Yantai, China

Z.P. Duan
College of Economics and Management, Shandong University of Science and Technology, Qingdao, China

ABSTRACT: With the ever-increasing urban population and economic activities, water usage and demand are continuously increasing. Hence, finding adequate water supply and fully utilizing wastewater become important issues in sustainable urban development and environmental benign aspect. Developing wastewater reclamation and reuse system is of specific significance to exploit new water resource and save natural fresh water supplied. In view of water reuse cost is the main factor to promote the use of, Qingdao city as an example the full cost analysis of reclaimed water reuse is carried out combination of external theory, this preliminary studies can help to understand and increase knowledge in utilizing reclaimed water, to foresee the feasibility of developing new water resource, to estimate the cost-effectiveness of reclaimed water in metropolitan city.

1 INTRODUCTIONS

Water is the source of life, is the foundation of natural resources and economic resources in coastal area of our country, water resources is serious shortage, especially the Yangtze River to the north of the coastal city is more serious, water shortage has become the important factors of constraining china's coastal city and regional economic and social development.

2 INFLUENCE FACTORS OF RECLAIMED WATER REUSE COST

Design capacity utilization rate is the main affecting factors cost of Reclaimed water reuse cost. There are a variety of other factors, such as raw water quality and treatment difficulty, the selected process and the power consumption of main equipments, depreciation fee which is directly affected by the life of equipment and investment, the day-to-day maintenance repair costs of equipments, the automation degree and artificial cost etc.

2.1 Facilities processing scale and its utilization rate of reclaimed water

Design capacity utilization rate is the greatest indicate associated with the reclaimed water project investment benefit. it is inversely proportional relationship between reclaimed water scale and operation cost. Analysis showed that the bigger the water facilities, the processing capacity utilization is more sufficient, the water facilities operation cost is lower. But if the design capacity is too large, it is easy to reduces the equipment utilization rate. For the different utilization rate, the biggest difference of processing cost is 2 yuan. This illustrate that the effect of utilization rate is bigger than other factors, it is the main factors influencing on the treatment cost, and it should be used as a main respect to reduce operating costs of reclaimed water project in the future.

2.2 The effect of process

In the biological treatment process, the operation cost of biological contact oxidation and biological activated carbon process is more economical, physical method and chemical method, processing costs are higher, and especially the operation cost of chemical method is the highest. To the Beijing MaHe Building as an example, by using ozonation method in reclaimed water treatment, the treatment cost is as high as 4.03 yuan. Therefore, the biological treatment process is the prefered method, careful with chemical method.

2.3 The raw water quality and the applications of reclaimed water

Reclaimed water treatment technological process mainly depends on the water source and water use, water source not only affects the selection of treatment process, but also affects the treatment cost. Therefore, the raw water quality is a very important factor. When the raw water concentration is high, such as domestic sewage as raw water, the organic matter content is 3–4 times of bathing wastewater, Organics bio-oxidation need to be explosive gas aeration. Aeration energy consumption in wastewater biological treatment is usually significant cost, under normal circumstances should account for one third of the processing cost.

Accordance with its purposes, water quality requirements for reclaimed water vary. For flushing, the water quality requirement is lower, for the city virescence, fire control or vehicle cleaning, the water quality requirement is higher than the flushing water, for landscape, water quality is more demanding. Therefore, even for the same reclaimed water system, processing cost of sewage treatment plant varies with the different reclaimed water quality.

2.4 Depreciation of equipment and maintenance and repair costs

Depreciation is mainly affected by the equipment investment and the life.The higher the investment, the higher the depreciation cost is the shorter the life, the higher the depreciation cost is domestic equipment or imported equipment also impact the depreciation cost greatly. Maintenance and repair cost mainly refers to the expenses of the wearing parts and materials consumables. Such as the use of activated carbon adsorption, due to carbon saturation, activated carbon regeneration and renewal fee is larger.

2.5 Auxiliary materials and power cost

Mainly including pharmacy, electricity and other expenses.

For general biological treatment only use a little medicament in the disinfection, the cost is small. But for the chemical method, the effection of the medicament costs are considerably to the water treatment cost. Power consumption cost account for a large proportion in the total operating cost. To reduce the power consumption is the key in cutting down the running cost of reclaimed water.

2.6 Artificial cost

Artificial cost mainly depends on the quality of workers, the convenience and reliability of equipment use, and automation level.

3 PROJECT COST ANALYSIS

Table 1 is the sewage treatment plant cost of Qingdao city in 2010.

To estimate the average cost for sewage treatment in Qingdao city is:1.791yuan/m³.

4 EXTERNALITY ANALYSIS OF RECLAIMED WATER REUSE

Reclaimed water reuse negative externality mainly include sewage sludge and waste gas emission of the sewage treatment plant.

During the reclaimed water treatment, it is inevitably to produce sludge, sludge is the end product of the reclaimed water treatment in city. The sludge is solid, semi-solid waste produced by the sewage treatment process. Sludge contains a lot of organic matter, rich in nutrients such as nitrogen and phosphorus, but it also contains a certain amount of heavy metals as well as a variety of pathogenic bacteria, if it is untreated indiscriminate discharges, after rain erosion and leakage, can easily cause secondary pollution to groundwater

Table 1. The cost of Qingdao four sewage treatment plant in 2010 (unit is ten thousand yuan if there are not give).

Sewage treatment plant	Tuandao	Licun	Haiberhe	Maidao
The annual handling capacity	2047	3320	2965	2899
Material				
Flocculant				
Total (ton)	41	74		
Cost	287	518		
Electric charge	510.08	2102.3	1397.06	1807.02
Salary and welfare	302.35		242.59	657.12
Maintenance cost	98.2		116.63	392.72
Other cost	221.37	547.8	228.91	481.35
Total cost fee	1419	7234	5046	6385
Total cost per unit (yuan/ton)	0.693	2.179	1.702	2.202

anf soil, directly endanger the health of the human body, and bring a serious Hidden danger to the local ecological environment.

The sludge first produced in the primary stage of the sewage treatment plant, followed by the sludge still produced in two processing stages and tertiary treatment stage. A variety of large and small sewage treatment plant will produce large amounts sludge a day. Since in 1857 the world's first sewage treatment plant established in London, sludge disposal problem has been one of the important issues in the municipal administration. With the improvement of sewage treatment capacity and processing deepening, the amount of sludge generated is bound to have a larger increase, how to reasonable disposal the sewage sludge has become a very urgent task to solve.

Sewage treatment plant sewage sludge treatment process will inevitably produce large amounts of malodorous gases. With the economic development of the human society, the improvement of people's living standards and the growing public environmental awareness, malodorous gases problems of sewage treatment plant has attracted more and more attention in the society. so deodorization problem of the sewage treatment plant has inevitably mentioned on the agenda, some have reached the point of urgent needs to be addressed. Currently, activated carbon adsorption method, a thermal oxidation method, the oxygen ionic groups deodorizing method, chemical washing method and biological filtration etc are the primary deodorizing technology in domestic sewage treatment at home and abroad.

5 EXTERNAL COST OF RECLAIMED WATER REUSE

It is sometimes quite difficult to assess the environmental costs directly with the environmental losses caused by the use of reuse water, and it is realistic to assess the environmental costs in accordance with our country's sewage charges method. In the third of sewage charge standard regulation clearly states: sewage charges according to the type and the number of the pollutants.

Charged standard of each pollution equivalent is 0.7 yuan. The number of pollutants types of Sewage Charges for each outfall, don't exceed three in decreasing order of the number of pollution equivalent.

Pollution equivalent number N of some pollutant is:

$$N = \frac{\text{Emission of pollutants}}{\text{Pollution equivalent of pollutants}}$$

Table 2. Pollution equivalent value.

Pollutant	Pollution equivalent value/kg
SS	4
COD	1
Ammonia nitrogen (NH_N)	0.8
Total phosphorus	0.25

The environmental costs of reuse water recycling are:

$$G = 0.7 \times \sum_{i=1}^{3} \frac{\left(w_i - w_i' \times Q_d \times 365 \right)}{1000 D_i}$$

G—the environmental costs of reuse water, ten thousand yuan

W_i—the ith pollutant thinkness in sewage produced by using reuse water, mg/L

W_i'—the ith pollutant thinkness in raw water of reuse water, mg/L

Q_d—Daily processing capacity of water recycling plant, 10^4 m^3/d

D_i—Pollution equivalent value of the ith pollutant, kg.

Qingdao city plan to build reclaimed water demonstration project of 2×10^4 m^3/d water production capacity on the basis of the the Haiberhe sewage treatment plant, using conventional processing techniques (secondary effluent—coagulation—sedimentation—filtration—disinfection). All projects with its own funds, the construction period is one year, and at the same year put into operation and reached the production capacity.

External cost calculate of reuse water by using equivalent calculation method:

The pollutant is COD, SS and NH_N, for the input water of reclaimed water plant COD = 60 mg/L, SS = 15 mg/L, NH_N = 3 mg/L, TP = 1 mg/L, for the raw water COD = 250 mg/L, SS = 280 mg/L, NH_N = 30 mg/L, TP = 5 mg/L, above data into formula (1) can obtain:

G = 0.7 × (0.19 + 0.066 + 0.016) × 2 × 365 = 138.99 (ten thousand yuan)

Then the unit external cost is 0.190 yuan/m^3, the full cost of reclaimed water reuse of Qingdao city is 1.79 + 0.19 = 1.98 yuan/m^3.

6 CONCLUSION

The full cost analysis of reclaimed water reuse is carried out combination of external theory in Qingdao city as an example, this preliminary attempts can help to understand and increase knowledge in utilizing reclaimed water, to foresee the feasibility of developing new water resource, to estimate the cost-effectiveness of reclaimed water in metropolitan city.

REFERENCES

[1] Chang Dun-hu, Ma Zhong. Wastewater reclamation and reuse in Beijing: Influence factors and policy implications [J]. Desalination. Jul 2012, Vol. 297, p 72–78.

[2] Pasqualino, Jorgelina C.; Meneses, Montse; Castells, Francesc. Life Cycle Assessment of Urban Wastewater Reclamation and Reuse Alternatives [J]. Journal of Industrial Ecology. Feb 2011, Vol. 15 Issue 1, p 49–63.

[3] Chiou RJ, Chang TC, Ouyang CF. Aspects of municipal wastewater reclamation and reuse for future water resource shortages in Taiwan [J]. Water Science And Technology: 2007; Vol. 55 (1–2), pp. 397–405.

[4] Takashi Asano, Audrey D. Levine. Wastewater reclamation, recycling and reuse: past, present, and future [J]. Water Science and Technology, Volume 33, Issues 10–11, 1996, Pages 1–14.

[5] Drewes, Jörg E.; Fox, Peter. Effect of Drinking Water Sources on Reclaimed Water Quality in Water Reuse [J]. Systems Water Environment Research, Volume 72, Number 3 May/June 2000, pp. 353–362(10).

[6] Lu W, Leung AY. A preliminary study on potential of developing shower/laundry wastewater reclamation and reuse system [J]. Chemosphere 2003 Sep; Vol. 52 (9), pp. 1451–1459.

Frontiers of Energy and Environmental Engineering – Sung, Kao & Chen (eds)
© *2013 Taylor & Francis Group, London, ISBN 978-0-415-66159-1*

Characterization of PM$_{10}$-bound polycyclic aromatic hydrocarbons in urban air of Haerbin, China

X.Y. Han

College of Architecture and Engineering, Kunming University of Science and Technology, Kunming, Yunnan, China

J.W. Shi

College of Environmental Science and Engineering, Kunming University of Science and Technology, Kunming, Yunna, China

ABSTRACT: PM$_{10}$ samples were collected at five sampling sites in city center of Haerbin from April 2008 to January 2009. The ambient concentrations of 17 selected polycyclic aromatic hydrocarbons (PAHs) in PM$_{10}$ were quantified. Spatial and seasonal variations of PAHs were characterized. Average total concentration of 17 selected PAHs at the five sites was 161.26 ng/m^3, and the dominant PAHs were PA, FL, Pyr, Ant, BaA, CHR, IND and BaP accounting for above 80% of 17 selected PAHs. Average concentration of individual PAHs in the whole sampling time varied from 0.45 (NaP) to 33.17 (PA) ng/m^3. Higher PAHs concentrations in heating period and lower concentrations in no-heating period were observed at the five sampling sites, which may be caused by the stronger emissions from stationary combustion sources in heating period, and the weak source emissions, the quicker air dispersion, washout effects, photo-degradation and higher percentage in the air in vapor phase in no-heating period.

Keywords: PAHs; PM$_{10}$; ambient concentrations; temporal and spatial distributions

1 INTRODUCTION

PAHs are ubiquitous constituents on particulate matter that mainly originate from incomplete combustion of organic matter such as petroleum and coal (Liu et al., 2009; Guo et al., 2003; Khalili et al., 1995). In urban ambient air, PAHs are almost entirely emitted from anthropogenic sources (Shi et al., 2010; Mantis et al., 2005; Caricchia et al., 1999). With the rapid urbanization, the consumption of petroleum and coal has grown considerably in Haerbin, and the air pollution from particulate matter becomes gradually serious. Therefore, it is utmost necessary to investigate the abundance, speciation and distributions of PAHs in aerosols to control the PAHs pollution in Haerbin. In this study we present the characterization of PM$_{10}$-bound PAHs in urban area of Haerbin.

2 MATERIALS AND METHODS

Haerbin (44°04′–46°40′N, 125°42′–130°10′E), the capital of Heilongjiang province, is the largest city in the northeast region of China, with an urban area of 7086 km^2 and a population of 4.8 million.

Foodstuff, medicine, equip manufacturing and petrochemical are major industries of the city.

Five sites were chosen as the sampling sites according to their different function in the city. The field descriptions were given as follows and the locations of the sites are shown in Figure 1.

Xuefulu (XFL) and Jianguo street (JGJ) site were chosen as typical traffic density/residential area; Chinese Medicine university (ZYY) site was chosen as a typical commercial/cultural area; Muli park (MLY) site was chosen as a commercial area; University of Commerce (SYDX) site was chosen as a typical cultural area.

In winter, there are large amount of coal combustion boilers for domestic heating around the five sites.

Ambient PM$_{10}$ samples of 20 days were collected from April 2008 to January 2009. To study the seasonal variation, the samples of 5 days were collected for each season, using a standard medium-volume PM$_{10}$ sampler at a flow rate of 100 L/min. A gas chromatography coupled to mass spectrometry (trace 2000GC-MS, Thermo Finnigan, USA) was used for determining PAHs with Selected Ion Monitoring (SIM), and the analysis process and quality control descriptions on this study are in

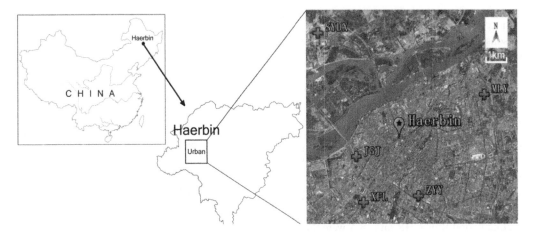

Figure 1. Location map of the sampling sites in Haerbin.

publications by Shi et al. (2010). In this study the 17 PAHs were analyzed including Naphthalene (NaP), Acenaphthylene (AcPy), Acenaphthene(Acp), Fluorene (FLu), phenanthrene (PA), Anthracene (Ant), fluoranthene (FL), pyrene (Pyr), Benzo[a] anthracene (BaA), Chrysene (CHR), Benzo[b] fluoranthene (BbF), benzo[k]fluoranthene (BkF), Benzo[a]pyrene (BaP), Indeno[1,2,3-cd]pyrene (IND), Dibenz[a,h]anthracene (DBA), Benzo[ghi] perylene (BghiP) and Coronene (COR).

3 RESULTS AND DISCUSSION

3.1 Distribution of PAHs in different rings

Average total concentration of 17 selected PAHs was 161.26 ng/m³, and the dominant PAHs were PA, FL, Pyr, Ant, BaA, CHR, IND and BaP accounting for above 80% of 17 selected PAHs. Average concentration of individual PAHs in the whole sampling time varied from 0.45 (NaP) to 33.17 (PA) ng/m³.

Examined PAHs could be classified according to their number of aromatic rings as follows: 2-ring including Nap; 3-ring including Acpy, Acp, Flu, PA and Ant; 4-ring including FL, Pyr, BaA and CHR; 5-ring including BbF, BkF and BaP; 6-rings including IND, DBA, and BghiP; 7-rings including COR. They can be further classified into lower molecular weight (LMW, 2-and 3-rings PAHs), middle molecular weight (MMW, 4-rings PAHs), and higher molecular weight (HMW, 5-, 6-and 7-rings PAHs). LMW PAHs can be tracers for wood combustion (Khalili et al., 1995) or industrial combustion of oil (Parka et al., 2002). MMW PAHs such as FL, Pyr, BaA and CHR are usually associated

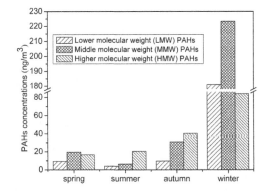

Figure 2. Ring number distribution of PAHs for PM₁₀ in Haerbin.

with coal combustion and can be identified from this source (Khalili et al., 1995). HMW PAHs such as BbF, BkF, BghiP, IND, and COR may be associated with vehicles emission, and can be regards as tracers for this source(Marr et al., 1999).

Figure 2 shows ring number distribution of PAHs for PM₁₀ in Haerbin in spring, summer, autumn, and winter. HMW PAHs were found to be abundant in summer and autumn, reflecting the strong contribution from vehicles emission, while MMW PAHs in PM₁₀ had higher concentrations in winter than other seasons, owing to the higher emission from coal combustion sources. Additionally, MMW PAHs are semi-volatile organic compounds and distribute higher in particle phase when air temperature is low. LMW PAHs were the relatively abundant in winter, which may be caused by wood combustion for domestic cooking or heating.

Therefore, the concentration proportion variation of LMW, MMW and HMW PAHs in different seasons can be used to reflect the variation in the categories of PAHs sources in these seasons.

3.2 Seasonal and spatial variations of PAHs in PM_{10}

The year was divided in two periods, heating period (H, 15 November–31 March) and no-heating period (NH, 1 April–14 November). In Haerbin, there is very significant seasonal variation in PAHs concentrations.

Figure 3 shows the mean PAH profiles (percent contribution of each PAH compound to ∑PAH) for the five sampling sites in Haerbin. During no-heating period BbF, CHR and BaA were the most abundant compounds at all sampling sites, while during heating period PA, FL, Pyr, and Ant were the most abundant compounds. This phenomenon revealed the importance of vehicular emissions as PAH sources in no-heating period and coal combustion as PAH sources in heating period, because the compounds BbF, CHR and BaA were identified as vehicles emission and the compounds PA, FL, Pyr, and Ant were abundant in coal combustion emission (Khalili et al., 1995). A significant similarity was existed in the PAH profiles in both heating and no-heating period, which reveals the sources of PAHs around these sampling sites were extremely similar.

The seasonal and spatial variations of PAHs in PM_{10} in Haerbin are presented in Figures 4 and 5. Higher PAHs concentrations in heating period and lower concentrations in no-heating period were observed at the five sampling sites. The ratio of heating period to no-heating period total PAHs was 9.3 at JGJ site, 13.8 at XFL site, 7.4 at ZYY site, 7.6 at MLY site, and 9.2 at SYDX site, respectively. That phenomenon confirms that the coal combustion emission was an important source to PAHs in heating period since so many coal

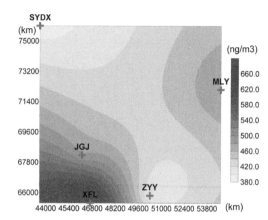

Figure 4. Distribution of PAHs concentrations in Haerbin during heating period.

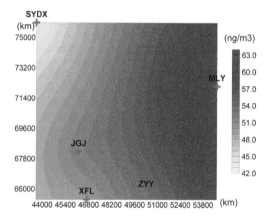

Figure 5. Distribution of PAHs concentrations in Haerbin during no-heating period.

combustion boilers were used for house heating in Haerbin.

The total PAHs concentration at MLY site was higher than that at other four sites in no-heating period, and it was higher at XFL and MLY sites in heating period. However, the PAHs concentration at JGJ, SYDX and ZYY sites was lower in whole sampling campaign. The major cause was the different surrounding of sampling sites, and no significant emission sources were observed at JGJ, SYDX and ZYY sites.

4 CONCLUSIONS

PAHs in PM_{10} in Haerbin had been determined at five sampling sites for one year. PAHs levels were the highest at traffic and resident density

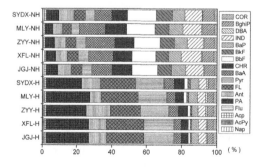

Figure 3. Mean PAH profiles of ambient PM_{10} in the central urban of Haerbin.

area in heating period and the lowest at the cultural area no-heating period. Obviously, the PAHs levels in PM_{10} were closely associated with source strength and meteorological conditions. Higher heating period PAHs concentrations and lower no-heating period concentrations were observed in all sampling sites. In heating period, coal combustion could be regards as the more predominant source to PM_{10}-bound PAHs than that in no-heating period. Additionally, meteorological conditions also promote the PAHs pollution in heating period, such as lower dispersion of air pollutants, less photo-degradation and higher percentage in the air in solid phase.

ACKNOWLEDGEMENTS

This study was funded by the China National Natural Science Foundation program (Grants 21207055) and the Special Environmental Research Fund for Public Welfare, No. 200709013). Shi Jianwu is the corresponding author of this paper, his e-mail address: shijianwu2000@sina.com.

REFERENCES

Caricchia, A.M., Chiavarini, S., Pezza, M. (1999). Polycyclic aromatic hydrocarbons in the urban atmospheric particulate matter in the city of Naples (Italy). Atmospheric Environment. 33: 3731–3738.

Guo, H., Lee, S.C., Ho, K.F. (2003). Particle-associated polycyclic aromatic hydrocarbons in urban air of Hong Kong. Atmospheric Environment. 37: 5307–5317.

Khalili, N.R., Scheff, P.A., Holsen, T.M. (1995). PAH source fingerprints for coke ovens, diesel and gasoline engines, highway tunnels, and wood combustion emissions. Atmospheric Environment. 4: 533–542.

Liu, W.X., Dou, H., Wei, Z.C., Chang, B., Qiu, W.X., Liu, Y., Tao, S. (2009). Emission characteristics of polycyclic aromatic hydrocarbons from combustion of different residential coals in North China. Science of the Total Environment 407(4): 1436–1446.

Mantis, J., Chaloulakou, A., Samara, C. (2005). PM_{10}-bound polycyclic aromatic hydrocarbons (PAHs) in the Greater Area of Athens. Greece. Chemosphere. 59: 593–604.

Marr, L.C., Kirchstetter, T.W., Harley, R.A. (1999). Characterization of polycyclic aromatic hydrocarbons in motor vehicle fuels and exhaust emissions. Environmental Science and Technology. 33: 3091–3099.

Parka, S.S., Kima, Y.J., Kang, C.H. (2002). Atmospheric polycyclic aromatic hydrocarbons in Seoul, Korea. Atmospheric Environment. 36: 2917–2924.

Shi Jianwu, Peng Yue, Li Weifang, et al(2010). Characterization and Source Identification of PM_{10}-bound Polycyclic Aromatic Hydrocarbons in Urban Air of Tianjin, China. Aerosol and Air Quality Research. Vol. 10, No. 5, 507–518.

757

Frontiers of Energy and Environmental Engineering – Sung, Kao & Chen (eds)
© *2013 Taylor & Francis Group, London, ISBN 978-0-415-66159-1*

The comparison of the Wedge and Hotspot radioactive aerosol diffusion mechanisms

F. Liu
School of Nuclear Science and Engineering, North China Electric Power University, Beijing, China
Northwest Institute of Nuclear Technology, Xi'an, China

J.X. Cheng, R.T. Niu, X.K. Zhang & W. Sun
School of Nuclear Science and Engineering, North China Electric Power University, Beijing, China

ABSTRACT: The ground deposition activity and dose equivalent parameters are analyzed using the Wedge and Hotspot model. The computer simulation is performed and results show that the Wedge model is more security than Hot spot model. The results are of great importance to the rescue of nuclear accident, delimitation of cordon zones for rescuer, decision of effective measures of residents of the moving around, aid decision in the scene of the accident for nuclear radioactive protection and the defense and decontamination of radioactive aerosol.

Keywords: Wedge model; Hotspot model; aerosol diffusion

1 INSTRUCTIONS

As China's nuclear power industry gets into a high stage of development, environmental radiation safety situation is becoming increasingly grim. In order to better protection of the environment and public, it is necessary to monitor the radioactivity around the nuclear power plants. There are three forms of radiation in nuclear power plant, neutron radiation, gamma radiation and radioactive aerosol. Among them, radioactive aerosols is easy to diffuse, and it is the nuclear power plant area as well as the internal and external residents around. Thus the study of the radioactive aerosol diffusion mechanism is of great significance for nuclear power plant safety.

Although the researches on radioactivity aerosol have been carried on internationally, there are still many problems to be solved. When the Nuclear accident happened, the radioactive aerosols will be released and diffused into the atmosphere, leading to harm caused by irradiation after personnel inhaled or eating in. At present, China nuclear plant constructions spread to the inland areas from the coastal gradually and deeply. Once nuclear accident occurred, inland area's special geographical environment and the dense city population were detrimental to the radioactive nuclide transport and diffusion; it is easy to cause serious radioactive pollution accidents [1]. Therefore, the research of radioactive nuclide dispersion in the atmosphere is very important to evaluate the construction of

nuclear power plant, normal working conditions and accident conditions to protect people from radioactive material of radiation damage.

However, radioactive nuclide in the atmosphere is dispersion in a very complicated way. We should consider not only meteorological conditions such as wind speed, atmospheric stability, the dry deposition, suspension, rain again wash effect, and the underlying surface conditions such as terrain features, buildings, surface roughness, also the nuclide itself size distribution characteristics, gravity settlement, the nuclide concentration spread area changes caused by radioactive and half-life. So we can use different mode of study for the spread of radioactive nuclides different range area. This paper focused on the research of contrasting the two kinds of model and the influence of the human. This is significant for the protection and accident treatment after the nuclear accident.

2 THE SYSTEM MODEL

2.1 The wedge model

The Gaussian plume model or Lagrange model is used generally for radioactive aerosol large scale spread, and for smaller scale spread, experience is often more accurate formula. We introduced the wedge mode to forecasting and analysis of the aerosol diffusion [2].

In nuclear accident, the atmospheric is mixed with radioactive aerosols rapidly after the explosion, and produced gas mixture in the height of about 300~2500 meters [3]. If we assume that T is diffusion time, r is the diffusion radius, the aerosol content of Smoke plume is Q(r). At the same time, the amount of aerosol is proportional to the amount of the plume, the Q(r) is expressed as:

$$Q(r) = Q_0 e^{\frac{-r}{L}} \tag{1}$$

where, Q_0 is the gross of aerosol released from nuclear accident, L is the average diffusion distance and can be obtained as:

$$L = \frac{Hu}{V} \tag{2}$$

where, V is the deposition velocity.

Based on the geometrical relationship in Figure 1, the deposition amount of radioactive aerosol in unit area can be expressed as:

$$\sigma(r) = \frac{-1}{r\theta} \times \frac{dQ}{dr} = \frac{1}{rL\theta} Q_0 e^{\frac{-r}{L}} \tag{3}$$

where, $Hrd(r)\theta$ is volume of plume and the concentration is:

$$\rho(r,\theta,z) = \frac{1}{rH\theta d(r)} Q_0 e^{\frac{-r}{L}} \tag{4}$$

To estimate the inhaled irradiation, we assumed that radioactive plume thickness is d (r), the residence time in the respiratory tract is $\tau = d(r)/u$, and the respiratory rate is b (m^3/s). So, the individual total inhaled aerosol (mg) is:

$$m(r) = \frac{Q(r)}{Hrd(r)\theta} \times \frac{bd(r)}{u} = \frac{Q_0 b}{Hru\theta} e^{\frac{-r}{L}} \tag{5}$$

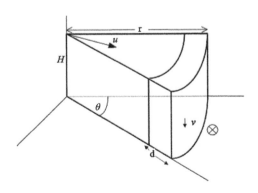

Figure 1. The wedge model.

2.2 The hotspot model

Hotspot is a model system which used in the radioactive materials spread in nuclear accident mainly for the personnel emergency response and contingency plans provide rapid assessment and portable service, and also can be used for nuclear facility safety analysis. Hotspot model is the first approximation for health effects caused in the atmosphere of radioactive material release and diffusion, also the mixed application of gauss smoke plume model.

In the Hotspot model, atmospheric concentration of radioactive aerosol following Gaussian diffusion mode:

$$C(x,y,z,H) =$$
$$\frac{Q}{2\pi\sigma_y\sigma_z u}\exp\left[-\frac{1}{2}\left(\frac{y}{\sigma_y}\right)^2\right]\left\{\exp\left[-\frac{1}{2}\left(\frac{z-H}{\sigma_z}\right)^2\right]\right.$$
$$\left.+\exp\left[-\frac{1}{2}\left(\frac{z+H}{\sigma_z}\right)^2\right]\right\} \times \exp\left[-\frac{\lambda x}{u}\right] DF(x)$$

$$\tag{6}$$

where, C is time related aerosol concentration ($Ci \cdot sec/m^3$); Q is the source term (Ci); H is effective release (m); λ is radioactive decay constant (m); X is the direction of the underdog distance (m); The wind direction for y distance (m); Z is vertical axis distance (m); σ_y is the wind direction on the comprehensive concentration distribution of the standard deviation (m); σ_z is the vertical axis direction comprehensive concentration distribution of the standard deviation (m); U is effective height (H) release on average wind speed (m/s); DF (x) is the loss factor smoke plume.

If the temperature inversion layers is chose effectively, and σ_z is more than temperature inversion layer height L, the equation (6) can be simplified as:

$$C(x,y,z,H) = \frac{Q}{\sqrt{2\pi}\sigma_y Lu}\exp\left[-\frac{1}{2}\left(\frac{y}{\sigma_y}\right)^2\right]$$
$$\times \exp\left[-\frac{\lambda x}{u}\right] DF(x) \tag{7}$$

where, L is the height of temperature inversion layer.

3 SIMULATION RESULTS

In this section, we assumed that the aerosol deposition speed $v = 0.003$ m/s, nuclear accident U source term $Q_0 = 1$ kg, wind speed $u = 1$ m/s (10 m place), respiratory rate $b = 3.33 \times 10^{-4}$ m^3 [2].

3.1 The comparison of ground deposition activity

In simulation, the radioactive aerosol ground deposition activity under the wedge model is according to the formula (3) [4] [5]. The corresponding calculation of Hotspot is drawing for comparisons in Figure 2. It is clear that the two model calculation result is more similar. It can provide decision support for nuclear radioactive decontamination and recovery.

3.2 Doses of comparison

The equivalent dose under the wedge model is according to the formula (5) and application of Hotspot calculation. As shown in Figure 3, if we are more concerned with equivalent dose, the calculation result of Wedge model is more security than Hot spot model. Therefore, it can be applied in the cancer risk assessment caused by radioactive aerosol inhalation [6].

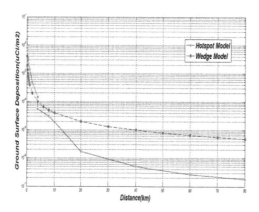

Figuer 2. The comparison of ground deposition activity.

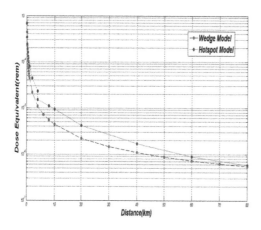

Figure 3. The comparison of equivalent dose.

4 CONCLUSIONS

Under given nuclear accident cases, the Wedge model is practical and can achieve rapid response, which is more security than Hot spot. So it can offer aid decision in the scene of the accident for nuclear radioactive protection, and can be applied in the cancer risk assessment caused by radioactive aerosol inhalation.

ACKNOWLEDGMENTS

This work is supported by the Fundamental Research Funds for the Central University (No. 11QG14).

REFERENCES

[1] Theodor E. Liolios. Broken Arrows: Radio lo gical hazards from nuclear war head accidents. 2008. 01.
[2] Wei Dong, Dong Fa·jun, Dong Xi·lin: Prediction of Diffusion Concentration of Radioactive Nuclides during Nuclear Accident; China Safety Science Journal. Mar. 2006.
[3] Zia Mian, M.V. Ramana, R. Rajaraman. Risks and consequences of nuclear weapons accidents in south Asia. Prince to n. USA.
[4] Homann, S.G., HOT SPOT Health Physics Codes for the PC, Lawrence Livermore National Laboratory, Livermore, CA, UCRLMA106315. 1994.
[5] U.S. Environmental Protection Agency, Washington, DC, Federal Guidance Report No. 11. 1988.
[6] Cheng Jinxing, The Spread Forecast and Consequences Assessment of Radioactive Aerosol from Nuclear Weapon Accidents. Nuclear Electronics & Detection Technology. Vol. 30 No. 1 Jan. 2010.

Frontiers of Energy and Environmental Engineering – Sung, Kao & Chen (eds)
© 2013 Taylor & Francis Group, London, ISBN 978-0-415-66159-1

The theoretical prediction and simulation of radioactive aerosol diffusion based on the revised Gaussian plume model

F. Liu
School of Nuclear Science and Engineering, North China Electric Power University, Beijing, China
Radiation Detection Research Center, Northwest Institute of Nuclear Technology, Xi'an, China

R. Zhao, Y. Liu, X.L. Cheng, R.T. Niu, X.K. Zhang & W. Sun
School of Nuclear Science and Engineering, North China Electric Power University, Beijing, China

ABSTRACT: A radioactive aerosol diffusion Gaussian plume model is revised in the vicinity of nuclear power plant considering effects of the emission heights, ground reflection, dry deposition, wet deposition, radioactive decay and other factors based on the open field diffusion model. Numerical simulations are performed to calculate the concentration distribution of the radioactive aerosol and to determine the evacuation zone. Simulation results show that the atmospheric stability tends to be the key factor of the radioactive aerosol distribution and the major diffusion zone prove to be in the downwind direction. With atmospheric stability degrees ranging from A to F, the area of the evacuation zone increases, whose maximum span in downwind direction expand to 28 km compared with the 0.8 km in crosswind direction.

Keywords: radioactive aerosol; gaussian plume model; atmospheric stability

1 INTRODUCTION

When the uncontrollable radioactive aerosol emission happened in the nuclear power plant the radioactive material diffused by atmosphere has a strong impact on the environment. confirming the accurate boundary of the radionuclide-polluted region is the prerequisite of all protection measures [1]. Hence, it is crucial to predict the concentration distribution of the radionuclide in nuclear leak process which is also the foundation and precondition of emergency response. [2].

Gaussian plume model is widely used in the calculation of atmospheric pollution movement and dispersion. The simulation results are also reasonable in most weather condition making it the classic method dealing with the pollution dispersion problem [3]. The Gaussian plume model was modified by Wei, D. to predict the dispersion pattern in radioactive cloud in nuclear accident. An oblique particle dispersion model was adopted to propose a fast method [4] to estimate the radionuclide concentration in atmosphere considering the gravity deposition, wet deposition and radioactive decay. Hu, E. B. et al. [5] built a relatively complete numerical model of mixed atmospheric diffusion based on Gaussian plume model with the environmental data from Qinshan Stage III and the data from domestic and foreign articles

in consideration of the influences of calm wind, mixed layer, wet deposition, radioactive decay and buildings. Raze et al. [6] simulated the scenario that radionuclide diffuses when severe accident happens to the Unit I of research nuclear power plant in Pakistan at the power of 10 MW and predicted the effective dose in the calculated region based on Gaussian plume model. The results point out that the reference man at the position 500 meters downwind from the leak source suffers an overdose. Venkatesan et al. [7] revised the Gaussian plume model adding the influence of sea wind discovering that the monitored radioactivity at the position 6 kilometers downwind from nuclear power plant is twice the value of the calculated result of Gaussian plume model.

In this paper, an iodine-containing radioactive aerosol dispersion Gaussian model in the vicinity of nuclear power plant is established with appropriate correlative parameters on the basis of the open field diffusion model. And modification has been made to the Gaussian model considering the effects of the emission heights, ground reflection, dry deposition, wet deposition, radioactive decay and other factors. Meanwhile, a computer simulation is applied to get the distribution of the radioactive aerosol concentration and the emergency zone with different atmospheric stability degrees.

2 A GAUSSIAN PLUME MODEL FOR A SUSTAINED POINT SOURCE

Concerning the ground holophote shown in the Figure 1, the revised Gaussian model [8] is given as:

$$C_{(x,y,z,H)} = \frac{Q}{2\pi \bar{u}\sigma_y\sigma_z}\exp\left(-\frac{y^2}{2\sigma_y^2}\right)$$
$$\times\left\{\exp\left[-\frac{(z-H)^2}{2\sigma_z^2}\right]+\exp\left[-\frac{(z+H)^2}{2\sigma_z^2}\right]\right\} \quad (1)$$

where, $C_{(x,x\,y,\,z)}$ is the radionuclide concentration of a certain point (x, y, z) in downwind direction, Q is the (Bq/s) the release rate, H is the release height, $\bar{u}$ (m/s) is the average wind speed, δ_y, δ_z are respectively the cross-wind and vertical standard deviation at the downwind distance x which both increase as the x grows.

3 THE REVISED GAUSSIAN MODEL

There have been many researches aiming at the atmospheric radionuclide dispersion calculation based on the conventional Gaussian model, however, in which the dry deposition, wet deposition and radioactive decay are ignored especially in practical diffusion process. Hence, revisions need to be taking to get the more precise radionuclide dispersion pattern reflecting the fact.

3.1 The influence of dry deposition

Concerning the effect of dry deposition in continuous releasing point model, the particles with a diameter larger than 10 um have an evident gravity deposition effect, whose velocity largely depends on the balance of air resistance and the gravity. The velocity can be expressed by Stokes Equation given as:

$$V_s = \frac{\rho g D^2}{18\mu} \quad (2)$$

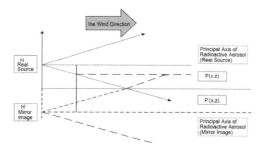

Figure 1. Ground holophote Gaussian diffusion model.

where ρ—density of the particle (kg/m³);
G—acceleration of gravity (9.8065 m/s²);
D—t he diameter of the particle (m);
m—the dynamic viscosity coefficient of the air (kg/m³).

Due to the incline of plume central line by gravity, all particles disperse in the direction of an oblique central line. For this process combined of dispersion and deposition the actual source moves downward with a speed of V_s, so the height can be expressed as $V_s t = V_s x/\mu$ at position x, in other words, the height of source moves down to the $H - V_s/\mu$.

3.2 The influence of wet deposition

The raindrops can wash the particles in the plume, moreover the soluble gas and stream can also dissolve in it. So, the rainfall is another important factor of the radionuclide deposition to the ground. The removal rate representing the radionuclide washing effect of rains can be described using a coefficient Λ (s⁻¹). The relationship between Λ and rainfall is given as:

$$\Lambda = aI^b \quad (3)$$

where I is the rainfall (mm/h), a and b are empirical coefficients with different values according to distinct release substances. To be more specific, for the material contain iodine, $a = 8 \times 10^{-5}$ and $b = 0.6$, otherwise, $a = 1.2 \times 10^{-4}$ and $b = 0.5$.

The total amount of material remaining in the plume Q discharged by wet deposition can be calculated as:

$$Q(x) = Q\exp\left(-\frac{\Lambda x}{\bar{u}}\right) \quad (4)$$

3.3 The influence of radioactive decay

Besides the dry and wet deposition, Radioactivity is also accounted for the atmospheric radionuclide dispersion which subjects to the law of radioactive decay namely decay equation as follows:

$$c = c_0 e^{-\lambda t} \quad (5)$$

where c_0—the initial concentration; λ—decay constant; t—decay time h.

Decay removal factor is proposed to revise the source Q. With theoretical inference, we attain the formula:

$$Q(x) = Q\exp\left(-\frac{0.693x}{3600T_{0.5}\bar{u}}\right) \quad (6)$$

Considering the wet deposition and the radioactive decay, the source Q are supposed to be modified by multiplying the equation (6) and equation (8):

$$Q(x) = Q \exp\left(-\frac{0.693x}{3600T_{0.5}\bar{u}}\right)\exp\left(-\frac{\Lambda x}{\bar{u}}\right) \qquad (7)$$

4 THE REVISED CONTINUOUS POINT SOURCE GAUSSIAN MODEL WITH WIND

The ground does not absorb the entire aerosol owing to the atmospheric turbulence and other driving force so the reflection of the ground has to be taken into account. However, the deposition of particles, rather than total reflection, by all means plays a dominant role in the process, hence, the reflection term needs to be multiplied with a reflection coefficient α ($\alpha < 1$), meanwhile, the effective height in reflection term changes into $H - V_s/\mu$, so the corresponding concentration formula lies as follows:

$$C_{(x,y,z,H)} = \frac{Q}{2\pi\bar{u}\sigma_y\sigma_z}\exp\left(-\frac{y^2}{2\sigma_y^2}\right)$$
$$\times \left\{\exp\left[-\frac{\left(z-H+\frac{V_s x}{\bar{u}}\right)^2}{2\upsilon_z^2}\right] + \alpha\exp\left[-\frac{\left(z+H-\frac{V_s x}{\bar{u}}\right)^2}{2\sigma_z^2}\right]\right\}$$
$$(8)$$

The results of concentration distribution with dry deposition can be obtained by equation (10), where α ($0 < \alpha < 1$) is the reflection coefficient, the experiential value is 0.5 for radionuclide.

5 CALCULATION OF THE PARAMETERS

5.1 The calculations of diffusion parameters with wind

According to the Chinese National Standard (GB/T 13201-1991), Atmospheric stability is categorized into 6 degrees ranging from A to F of which A is the most instable while F is the most stable.

In the open field the connection between diffusion parameters δ_y, δ_z and the atmospheric stability is shown in Table 1.

5.2 The effective release height

If the temperature of the nuclear leakage source is different from that of the environment, the buoyancy force has to be taken into account. Compared

Table 1. The relationship between δ_y, δ_z and atmospheric stability.

Atmospheric stability	δ_y	δ_z
A	$0.22/(1+0.0001x)^{0.5}$	$0.2x$
B	$0.16/(1+0.0001x)^{0.5}$	$0.12x$
C	$0.11/(1+0.0001x)^{0.5}$	$0.08/(1+0.0002x)^{0.5}$
D	$0.08/(1+0.0001x)^{0.5}$	$0.06/(1+0.0015x)^{0.5}$
E	$0.06/(1+0.0001x)^{0.5}$	$0.03/(1+0.0003x)^{0.5}$
F	$0.04/(1+0.0001x)^{0.5}$	$0.016/(1+0.0003x)^{0.5}$

with the particles elevation formula by Wei et al. [2] the method we finally adopt is as follows:

1. In wind condition, namely, $\mu > 0.1$ m/s, when the temperature difference is larger than 35 °C, $T_s - T_a \geq 35$ K, the elevated height can be expressed as:

$$\Delta H = \frac{1}{\mu_s}\left(0.92V_s D + 0.792Q_{hkj}^{0.4}H_s^{0.6}\right) \qquad (9)$$

where H_s—the geometry height of stack, (m); D—the diameter of the stack, (m); μ_s—the average wind speed of the stack exit, deriving from formula 16; V_s—the velocity of the plume, (m/s); Q_{hkj}—heat emission ratio, kW, the expression is as follows.

$$Q_{hkj} = 0.275PD^2V_s\frac{\Delta T}{T_s} \qquad (10)$$

where $\Delta T = T_s - T_a$, T_s—the plume temperature, (K); T—environmental temperature, (K); P—the barometric pressure, (kPa).

2. In calm wind, $\mu < 0.1$ m/s, when the temperature difference is larger than 35 °C, $T_s - T_a \geq 35$ K, the elevated height is:

$$\Delta H = 5.5Q_{hkj}^{0.25}\left[\frac{dT_a}{dZ} + 0.0098\right]^{-0.375} \qquad (11)$$

where, dT_a/d_z is the atmospheric temperature gradient above the geometry height of stack, usually less than 0.01 K/m.

3. When cold discharge happens, namely, when the temperature difference is less than 35 °C, $T_s - T_a < 35$ K, the elevated height is:

$$\Delta H = 2\left(1.5V_s D + 0.01Q_{hkj}\right)\big/\mu_s \qquad (12)$$

When the wind speed is less than 1.0 m/s, μ values 1.0 m/s.

Table 2. The wind profile coefficient m.

Underlying surface	The Degree of stability					
	A	B	C	D	E	F
Open plain	0.10	0.15	0.20	0.25	0.35	0.40

Now we calculate the effective height of release source. The height of releasing source in diffusion formula refers to the effective one. It is the sum of geometry stack height and the elevated plume height:

$$H = H_s + \Delta H \tag{13}$$

where, H_s—the geometry height of stack, (m); ΔH—the elevated height of plume, (m); H—the effective height of release source, (m).

5.3 The average wind speed at the effective releasing height

Calculating with above mentioned formulas, the wind profile has to be applied, and we choose the wind speed on the ground μ_{10} into calculation. Wind profile is a mathematical power function [2] describing the relationship between wind speed and height.

$$\bar{u} = \bar{u}_{10} \left(\frac{H}{10} \right)^m \tag{14}$$

where $\bar{u}$—the average wind speed at the effective release height; $\bar{u}_{10}$—the average wind speed at the position 10 meters above the ground; H—the effective height of releasing source derived from equation 13; m—the coefficient of wind profile, dimensionless, associated with the terrain roughness and the air stratification. The relationship between m and atmospheric stability is given in Table 2.

6 THE PROGRAM SIMULATION AND NUMERICAL CALCULATION

Radioactive aerosol concentration at a certain point (x, y, z) in a nuclear accident can be obtained. The distribution of aerosol diffusion concentration and the boundary of evacuation region can be calculated using MATLAB simulation. This paper mainly discusses the diffusion concentration distribution of continuous releasing point in wind and we focus on the concentration of the ground, $(z = 0)$. According to the typical parameters in

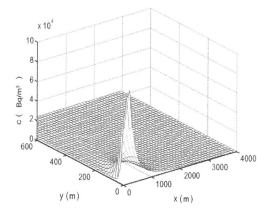

Figure 2. The distribution of the radionuclide aerosol concentration atmospheric stability with degree A.

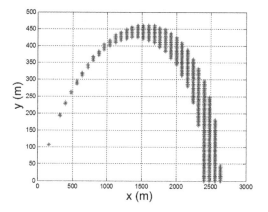

Figure 3. The evacuation region of atmospheric stability with degree A.

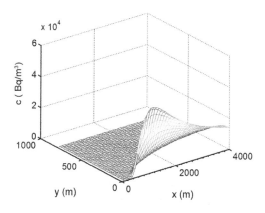

Figure 4. The distribution of the radionuclide aerosol concentration of atmospheric stability with degree E.

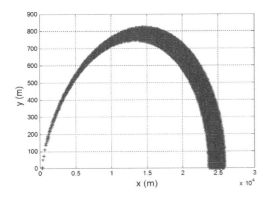

Figure 5. The evacuation region of atmospheric stability with degree E.

references assuming that the source intensity is 2.14×10^{-9} Bq/s, environmental temperature is 293 K, plume temperature is 375 K, the average wind speed is 1.6 m/s at the position 10 meters above the ground and diameter of stack is 1 meter, the results are shown below categorized by different atmospheric range.

1. If the atmospheric stability is degree A, the distribution of the radioactive aerosol concentration are given in Figure 2. The maximum radioactivity concentration one can bear is 10^3 orders. So residents should evacuate the region. The simulated picture of the evacuation region by MATLAB is given in Figure 3.

   Figure 2 shows that the influence of radioactive aerosol on X axis is more distinct than that on Y axis so the span of the region in X direction should be larger. In Figure 3, the external boundary of shaded part represents the demarcation of the evacuation region and the evacuation region in negative direction of Y axis is symmetric to the positive part. In this situation, the evacuation region in X direction spans 2500 m while that in Y direction spans 450 m.

2. If the atmospheric stability is Degree E, the distribution of the concentration is given in Figure 4.

In this condition, the evacuation region in X direction spans 28000 m while that in Y direction spans 800 m. Compared with stability of degree D, the span of Degree E in X direction increase sharply, however, the changes seem to be tiny in the Y direction.

7 CONCLUSIONS

The diffusion model of radioactive aerosol containing iodine is revised. The radionuclide concentration distribution in different atmospheric stability degrees and the maps of evacuation region are analyzed and simulated. Simulation results show that atmospheric stability has a significant influence on the radioactive aerosol, especially its diffusion in downwind direction. The evacuation region increases when atmospheric stability degrees changing from Degree A to F and the span of region downwind should be larger than that crosswind. When the atmospheric stability is Degree F the maximum evacuation region spans 2.8 kilometers downwind.

ACKNOWLEDGMENTS

This work is supported by the Fundamental Research Funds for the Central University (No. 11QG14).

REFERENCES

[1] Chi, B. et al. 2006, "Development and application of random walk model of atmospheric diffusion in emergency response of nuclear accidents", in *Chinese Journal of Nuclear Science and Engineering*, Vol. 26, No. 1, 40–45.
[2] Yang, X. S. et al. 2005, "Current Status of Nuclear Security and Radiation Safety in China and Their Countermeasures", in *China Safety Science Journal*, Vol. 15, No. 7, 48–51.
[3] Liu, A.H., Kuai, L. 2011, "A review on radionuclides atmospheric dispersion modes", *Journal of Meteorology and Environment*, Vol. 27, No. 4, 59–65.
[4] Wei, D. et al. 2006, "Prediction of Diffusion Concentration of Radioactive Nuclides during Nuclear Accident", in *China Safety Science Journal*, Vol.16, No.3, 107–113.
[5] Hu, E.B., Gao, Z. 1998, "Model, Parameter and Code of Environmental Dispersion of Gaseous Effluent under Normal Operation from Nuclear Power Plant with 600 MW", *China Nuclear Science and Technology Report*, S6, 1–17.
[6] Raza, S.S., Iqbal, M. 2005, "Atmospheric Dispersion Modeling for an Accidental Release From the Pakistan Research Reactor-1 (Parr-1)" [J], *Annals of Nuclear Energy*, 32(11):1157–1166.
[7] Venkatesan, R. et al. 2002, "Study of Atmospheric Dispersion of Radionuclides at a Coastal Site Using a Modified Gaussian Model and a Mesoscale Sea Breeze Model" [J], *Atmospheric Environment*, Vol. 36, No. 18, 2933–2942.
[8] Zhiquan Tong, Evaluation of atmospheric environment, China environment science press, 1988.
[9] Qing Gu, Yunsheng Li. "The calculation method of atmospheric environment mode". Meteorology press, 2002.

Frontiers of Energy and Environmental Engineering – Sung, Kao & Chen (eds)
© 2013 Taylor & Francis Group, London, ISBN 978-0-415-66159-1

Optimization and simulation of groups discussing based on immune genetic algorithm

S. Lou, H.M. Wang & W.J. Niu

*State Key Laboratory of Hydrology Water Resource and Hydraulic Engineering, HOHAI University,
Nanjing, China*
Management Science Institute of HOHAI University, Nanjing, China
Business School of HOHAI University, Nanjing, China

ABSTRACT: Researching the characteristics of decision-making for the water resource system under hwme and build a group decision making model of water resource allocation. Analyze the main particularity of water resource allocation. Getting the optimum configuration mode program and simulation calculation the process of multi-stage group decision-making by combined with the WAA operators and immune genetic algorithm.

Keywords: water resources management; immune genetic algorithm; group decision making; HWME

1 INTRODUCTION

Water resources management HWME is a "place" for the activities of water resource management under the effective form of the application of HWME. HWME is a "man-in-loop" system which is more prominent role of expert[1][2]. But water resources management system as a complex giant system, expert group covering many fields; expert individuals have different views on the complex issues of water resource management[3]. So HWME has the drawbacks of decentralized thinking in the group activities of the complexity problems.

This paper adopts the immune genetic algorithm and presents a group decision making optimization model, and gets the satisfactory result to solve the problem by processing the different opinions of various experts.

2 PROBLEM DESCRIPTION

2.1 *The mode water resources allocation*

The sustainable water resources optimal allocation model is a complex system of multi-objective which is considered the aspects of economic, social, ecological environment and resources have to be thought over[4]. Because of the scarcity of water resource we can't achieve the optimal of all the goals, so according to the difference and local situation of the basin water resource, water resource allocation model could highlighting single target while taking into account the multi-target, making multi-objective water resource allocation decision-making problem transformed into a selection problem in different configuration models[5][6]. According to the characteristics of multi-objective water resource systems, water resources allocation model is divided into the economic efficiency mode, social efficiency mode, eco-environmental benefits mode and efficiency of resource use mode. The decision-making of water resources allocation is also a multi-attribute decision-making, so we can do group decision-making through the attributes of the river basin water resource system, and the final decision-maker select the basin water resource allocation mode reference every mode's tendency value G_m obtained from the result of group decision-making.

2.2 *Optimization of group decision-making for water resource allocation*

Water resource allocation decision is a multi-attribute decision-making, in order to avoid the deviation that due to the limitations of knowledge and every stakeholder as an individual expert in the problem when expert make qualitative evaluation[8], introducing the adjustment factor $(\varsigma^s, s = 1, 2 \cdots n)$ on the basis of expert evaluation. In the range of adjustment factor elements, the experts scoring matrix can be adjusted appropriately. The problem can be described as:

For one water resource allocation decision, the expert group is $E = \{e_1, e_2 \cdots e_n\}$; $\omega = \{\omega_1, \omega_2 \cdots \omega_n\}$ represents the expert weights $(\sum_{i=1}^{n} \omega_i = 1, 0 \le \omega_i \le 1)$.

First, setting of attribute collection ($R = \{r_1, r_2 \ldots r_k\}$) of water resource allocation decision-making by experts under the guidance of the host of HWME after the preliminary seminar, using $\lambda = \{\lambda_1, \lambda_2 \cdots \lambda_k\}$ represents the expert's weights ($\sum_{j=1}^{k} \lambda_j = 1, 0 \leq \lambda_j \leq$); then the host give the adjustable range of every attribute adjustment factor $-\xi_j < \xi_j < \xi_j, 1 \leq j \leq k$.

Expert group E scoring evaluation of the decision-making attributes, obtain evaluation matrix $A_i = (a_{ij}), 1 \leq i \leq n, 1 \leq j \leq k$, a_{ij} express the evaluation value of attribute r_j made by expert e_i. Comprehensive evaluation value of some attribute may obtain from a_{ij} and ω_i though WAA operator[7][9], expressed by y_j;

$$y_j = \sum_{i=1}^{n} a_{ij}\omega_i \qquad (1)$$

So get the result of group decision-making $Q = (Q_1, Q_2 \cdots Q_k)$,

$$Q_j = y_j = \sum_{i=1}^{n} a_{ij}\omega_i, 1 \leq i \leq n, 1 \leq j \leq k \qquad (2)$$

Introducing the adjustment factor, getting the expert evaluation matrix after adjustment:

$$A_i' = \left(a_{ij}'\right), 1 \leq i \leq n, 1 \leq j \leq k, a_{ij}' = a_{ij} + \xi_j \qquad (3)$$

Then get the results of group decision-making after adjustment $Q' = (Q_1', Q_2' \cdots Q_k'$

$$Q_j' = \sum_{i=1}^{n} a_{ij}'\omega_i = y_j + \sum_{i=1}^{n} \xi_j\omega_i \qquad (4)$$

Measure the consensus of the expert groups use the strong consistency between the evaluation vectors and the result vector of group decision-making. Individual strong consistency index AGE_i^t is the consistent level between the results (E_i^t) of the genetic variation of generation t of individual i, the result ($Q'' = (Q_1^{t'}, Q_2^{t'} \cdots Q_k^{t'})$) of group decision-making of generation t and other members' opinions ($E_1^t, E_2^t \cdots E_n^t$) in the group, expressed by the cosine value between the vectors; vectors represent the closeness between the vector. After adjusted the best individual provided by experts is $E^t = \{E_1^t, E_2^t \cdots E_n^t\}, i = 1, 2 \cdots n$; Individual strong consistency index can be defined as

$$AGE_i^t = \frac{(E_i^t \cdot Q'')}{(\| E_i^t \| \times \| Q'' \|)} \qquad (5)$$

After considering the weight of every expert, the group consistency index can be defined as

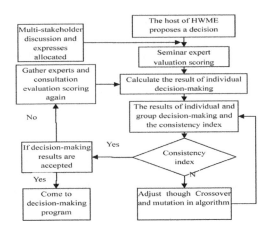

Figure 1. Group decision-making process based on HWME.

$$IAI^t = \frac{\sum_{i=1}^{n} \omega_i AGE_i^t}{n} \quad 1 \leq i \leq n \qquad (6)$$

IAI^t is the group consistency index and action as the fitness value in the algorithm. The smaller IAI^t the stronger the group consistency, group convergence is the better. Figure 1 is the group decision-making process in the environment of HWME.

3 IMMUNE GENETIC ALGORITHM BASED ON ADJUSTMENT FACTOR

Immune Genetic Algorithm (IGA) is an improved genetic algorithm which is proposed based on biological immune mechanism[10]. The study of complex decision problems in water resources management HWME realizable as the following scene: due to the experts group in HWME represent different stakeholders and the limitations of qualitative knowledge will bring subjectivity bias. So the adjustment factor can be added to adjust the results of single expert decision in order to avoid the impact on the effectiveness and consistency of the results of group decision-making. When optimize the expert group decision-making, we can change the adjustment factor to adjust single expert decision-making, in order to better achieve convergence. Then we can see the decision-making goal as an antigen, the adjustment factor for each expert as the antibody. In order to maintain the value and diversity of single expert decision only crossover and mutation on the adjustment factor, then generate the next generation of antibody population though update group strategy, until meet the termination condition.

3.1 Chromosome coding

This paper adopts a floating-point encoding format which is based on the scoring matrix of decision-making groups and combined with the adjustment factor [11]. The length of coding is equal to the number of decision variables. Shown in Table 1 is the coding of the decision vector of expert s.

3.2 Antibody fitness

In the expert group decisions based on HWME, The consensus of expert opinions can be achieved though the degree of difference of the antibody formed by the evaluation of expert group, so we can define the degree of difference as the objective function:

$$\min f = \min IAI^t = \frac{\sum_{i=1}^{n} \omega_i AGE_i^t}{n} \quad (7)$$

3.3 Population initialization

The algorithm of generating the initial individual as follows:

To getting the decision matrix of every attribute of the decision-making program evaluation by experts $A_s = (a_{ij}^s), 1 \le i \le m, 1 \le j \le k$, randomly n generated adjustment factor according to the number of experts as the initial antibody $\xi^s, s = 1,2 \cdots n, -\zeta' \le \zeta^s \le \zeta$; range of adjustment of expert evaluation which is discussed by experts and settled by the host of HWME; Combined with the use of formula and adjustment factor calculated expert decision vector based on adjustment factor $Y_s' = \{y_1^{s'}, y_2^{s'} \cdots y_m^{s'}\}, \quad s = 1,2 \cdots n$.

3.4 Groups demonstrated method based on immune genetic algorithm

The steps of groups demonstrated based on immune genetic algorithm[12]:

Step 1. The host put forward the problem of water resources allocation, and gives adjustment factors of decision attribute and indexes, as the antigen;

Step 2. Experts give the evaluation value of indicator parameters, and with adjustment factor calculated evaluation vector of expert;

Step 3. Encode the initial antibody, and in accordance with the selection probability to select antibodies carry out crossover and

mutation operations, produce a new generation program;

Step 4. Select strong antibody populations which have high fitness, crossover and mutation to produce a new generation program in accordance with the memory strategy of ensuring quality;

Step 5. After iterative calculation to determine maybe a satisfactory program or the evolutionary whether get the specified generation of stop, if is, go to Step 6, else go to Step 3;

Step 6. Reached a consensus of opinion, generate the demonstration plan.

4 ALGORITHM SIMULATION

Assume there are 10 experts in the field of water resources participated in the demonstrated in the HWME. First, with the help of HWME, make the first round of seminars by the expert group to consensus determine decision-making indicators of water resources allocation (show in Fig. 2). Weight of index $\lambda = \{0.1, 0.07, 0.09, 0.11, 0.06, 0.08, 0.06, 0.05, 0.05, 0.06, 0.04, 0.03, 0.08, 0.07, 0.05\}$, weight of every expert is $\omega = \{0.07, 0.1, 0.13, 0.08, 0.09, 0.1, 0.11, 0.13, 0.12, 0.07\}$.

The host of HWME gives the range of adjustment factors to the subjective judgment of experts

$$\zeta = \begin{bmatrix} -5 & -10 & -5 & -5 & -10 & -5 & -10 & -5 & -5 & -5 & -10 & -10 & -10 & -10 \\ 5 & 10 & 5 & 5 & 10 & 5 & 10 & 5 & 5 & 5 & 10 & 10 & 10 & 10 \end{bmatrix}$$

And set the crossover probability is 0.65; the mutation probability is 0.1; the maximum generation of iteration is 100.

The consistency value of individual initial of the experts before the introduction of adjustment

Table 1. Chromosome encoding format.

ξ_{i1}^t *	ξ_{i2}^t	ξ_{i3}^t	ξ_{i4}^s	...	ξ_{ik}^t

*ξ_i^t is the adjustment factor of a_{ij}^s.

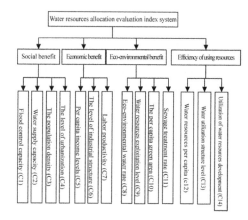

Figure 2. Index system of water resources allocation evaluation.

Table 2. The value of expert evaluation.*

Attribute	Social benefit				Economic benefit			Eco-environmental benefit				Efficiency of using resources		
Expert	C1	C2	C3	C4	C5	C6	C7	C8	C9	C10	C11	C12	C13	C14
1	85	65	75	95	65	85	55	85	95	75	65	85	95	75
2	65	95	75	65	95	90	65	70	75	80	65	90	80	90
3	70	80	65	90	75	90	75	80	80	65	90	85	90	85
4	65	70	65	90	65	75	75	90	85	90	85	85	70	80
5	65	60	65	70	80	75	65	85	65	70	75	80	70	65
6	90	80	80	70	65	70	70	85	90	80	75	90	85	80
7	70	75	70	80	90	85	85	80	65	70	85	85	90	80
8	70	75	65	75	85	80	85	90	65	75	70	75	80	75
9	85	85	90	75	65	70	65	75	80	85	85	70	75	80
10	70	75	75	80	80	85	75	80	65	80	75	80	80	75

* The value given by experts.

Table 3. The consistency value of individual initial of experts.

Expert i	1	2	3	4	5
AGE_i^0	0.141	0.133	0.084	0.109	0.080

Expert i	6	7	8	9	10
AGE_i^0	0.103	0.084	0.085	0.119	0.051

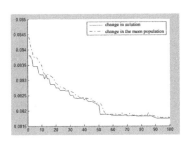

Figure 3. Curve of optimal solution of group decision making.

Table 4. Consistency value of the optimal expert individual.

Expert i	1	2	3	4	5
AGE_i^{92}	0.120	0.090	0.068	0.085	0.077

Expert i	6	7	8	9	10
AGE_i^{92}	0.083	0.080	0.072	0.097	0.052

factors can be calculated according to the fitness formula (show in Table 3). And obtained the initial consistency value of groups is $IAI^0 = 0.0984$.

We can see from simulation (Fig. 3), with the iterative calculation the results tend to a minimum, When $t = 9$, $IAI^{92} = 0.0817$ is the minimum, the value of individual consistency in Table 4.

The results of group decision-making after adjust in Table 5.

The tend value of group decision-making result is:

$$G_1 = \frac{Q_1' + Q_2' + Q_3' + Q_4'}{\sum_{i=1}^{i=14} Q_i'} = 0.277,$$

$$G_2 = \frac{Q_5' + Q_6' + Q_7'}{\sum_{i=1}^{i=14} Q_i'} = 0.213,$$

$$G_3 = \frac{Q_8' + Q_9' + Q_{10}' + Q_{11}'}{\sum_{i=1}^{i=14} Q_i'} = 0.287,$$

$$G_4 = \frac{Q_{12}' + Q_{13}' + Q_{14}'}{\sum_{i=1}^{i=14} Q_i'} = 0.223;$$

Obtain the result $G_3 > G_1 > G_4 > G_2$. Based on the above data, the eco-environmental benefits mode is more conducive to the sustainable development of the basin during the water resources allocation.

5 CONCLUSIONS

This paper using HWME as the implement platform of group decision-making, using optimization method based immune genetic algorithm for the consistency optimization of group decision-making has obvious advantages. It carefully handled by the computer after experts give the initial judgment, and then come to the optimization program of decision-making convergence. Save time for discussion, avoided can't get the end of discussion due to wide divergence of views and avoid the herd behavior due to the influence of the group members' role, authority and power to

Table 5. The group decision-making results after optimized.

Configuration mode attribute	Social benefit G1				Economic benefit G2		
	C1	C2	C3	C4	C5	C6	C7
Q'	75.93	80.9	71.3	78.35	81.05	80.55	74.4
	Eco-environmental benefit G3				Efficiency of using resources G4		
	C8	C9	C10	C11	C12	C13	C14
Q'	83.75	79.4	76.77	77.75	84.6	82.55	80.45

decision-making behavior during the decision-making process.

ACKNOWLEDGEMENT

This research is supported in part by: (i) the National Natural Science Foundation of China (NSFC) under Grant 50979024; (ii) the Fund for 2011 Graduate Students research and innovation projects of Jiangsu Province, China. CXLX11_0463.

REFERENCES

[1] Jeffery R. Seaton R. 1997. Evaluation methods for the design of adaptive water supply systems in urban environments [J]. *Water Science and Technology*. 35(9): 45–51.

[2] Wang Danli, Dai Ruwei. 2001. Behavior criterion for expert group in hall for workshop of meta-synthetic engineering [J]. *OURNAL OF MANAGEMENT CIENCES IN CHINA*. 4 (2):1–7.

[3] Srinivasan B, Palanki S, Bonvin D. 2003. Dynamic optimization of batch processes I: Characterization of the nominal solution [J]. *Computers & Chemical Engineering*. 27(1): 1–26.

[4] Frank Messner, Oliver Zwirner, Matthias Karkuschke. 2006. Participation in multi-criteria decision support for the resolution of a water allocation problem in the Spree River basin [J]. *Land Use Policy*. 23(1): 63–75.

[5] S. Yu. Schreider. 2005. Integrative modeling for sustainable water allocation: Editorial notes on the special issue [J]. Journal of Environmental Management. 77 (4): 267–268.

[6] Matsatsinis N, Grigoroudis E, Samaras A. 2005. Aggregation and disaggregation of preferences for collective decision-making [J]. *Group Decision and Negotiation*. 14: 217–32.

[7] Xu Z. 2005. Deviation measures of linguistic preference relations in group decision-making [J]. *Omega*. 33: 249–54.

[8] Mikhailov L. 2002. Fuzzy analytical approach to partnership selection in formation of virtual enterprises [J]. *Omega the International Journal of Management Science*. 30: 393–401.

[9] Mikhailov L. 2004. A Fuzzy Approach to Deriving Priorities from Interval Pair wise Comparison Judgments [J]. *European Journal of Operational Research*. 159.

[10] Sung-BaeCho, Joo-Young Lee. 2002. A human- oriented image retrieval system using interactive genetic algorithm [J]. *Systems, Man and Cybernetics, IEEE*. 32(3): 452–458.

[11] Hong J, Lim W, Lee S. An Efficient Production Algorithm for Multihead Surface Mounting Machines Using Biological Immune Algorithm [J]. *International J of Fuzzy Systems*. 2000, 2(1): 45253.

[12] Alisantoso D, Khoo L.P, Jiang P.Y. 2003. An immune algorithm approach to the scheduling of a flexible PCB flow shop [J]. *Advanced Manufuring Technology*. 22: 819–827.

Frontiers of Energy and Environmental Engineering – Sung, Kao & Chen (eds)
© 2013 Taylor & Francis Group, London, ISBN 978-0-415-66159-1

Discussion and analysis of landscape design for land art style commercial plaza in the city

N. Song
College of Horticulture, Huazhong Agricultural University, Wuhan City, Hubei Province, China

ABSTRACT: With the rapid development of domestic economic level and mass consumption level, landscape design of commercial plaza has attracted more attention. The landscape design of land art style commercial plaza can get everything under control with ease during construction of commercial plaza space by virtue of the features of its humanity, ecological nature and diversification. We can learn some experiences from domestic design projects while drawing lessons from excellent design cases abroad, so as to discuss the landscape design of commercial plaza which is more suitable to current social and economic development.

Keywords: commercial landscape; land art style; landscape of the plaza

FOREWORD

With the rapid development of domestic urbanization process and continuous rising of the people's living standard, the demand for commercial environment has become higher increasingly. In particular, the landscape of the plaza in some large-scale commercial space plays a pivotal role in connecting commercial buildings and arterial streets; therefore, the demand for landscape design has become increasingly higher from the angle of function and form. Although there is a certain gap between domestic landscape design of commercial plaza and that of developed countries in terms of spatial form and design method, we can still learn some experiences from domestic design projects while studying, so as to discuss the landscape design of commercial plaza which is more suitable to current social and economic development.

1 THE CONCEPT OF COMMERCIAL LANDSCAPE AND LAND ART STYLE COMMERCIAL LANDSCAPE

Commercial landscape is a kind of business scenario that links the three basic elements of commercial activities (commodity, consumer and space) together reasonably and effectively based jointly on architecture landscape and garden landscape. In comparison with general landscape design, the commercial landscape has more explicit sense of purpose and economic nature, including the plaza, courtyard, pedestrian street, colonnade, terrace,

sculptures etc that match the commercial building in order to provide the shoppers with a place for leisure and gathering. Meanwhile, it can also create a space for merchants that are convenient to carry out various sales promotion activities, enterprise image shows or other public social activities. It is a transitional space that falls in between shopping districts in commercial buildings and the urban roads, the particularity of which is that it is not only part of private sector within the scope of merchants, but also has a certain degree of openness. Therefore, it requires a certain comfortable and safe environment and also meets the functional requirements of broad vision and convenient transportation at the same time.

Land art style commercial landscape means literally the commercial space landscape that is either open or surrounded and half-surrounded, being built on the basis of the ground. In addition to reflecting the commercial culture and commercial value, the commercial space landscape also shoulders various quests of people in the commercial environment. It involves not only the functional requirements of practicability in business transaction, but also requirements of entertaining activities.

2 THE CONNOTATION OF LANDSCAPE FOR LAND ART STYLE COMMERCIAL PLAZA

2.1 *Landscape design of commercial plaza*

For the purpose of presenting a convenient, comfortable, safe and beautiful shopping place, the

771

landscape design of commercial plaza will take behavior and activities of human being as the basic principles, and build with independent forms and form factors within the scope of specific space —by means of reasonable configuration of commercial buildings, vehicle operating area and area of pedestrian circulations. It is a kind of creative activity with certain cultural connotation and aesthetic value.

2.2 Characteristics of land art style commercial landscape

1. The integration of culture & art and commercial atmosphere

   Culture and art is the direct reflection of richness of human life. It is also the image measure to evaluate the health status of the commercial activities. The land art style commercial landscape is more in need of implantation and participation of art and culture on the premise that the requirements of commercial activities have been met. The land art style commercial landscape built on the ground will not be vacuous or impoverishing any more with the blending of culture and art—as if the material life of human being has its own thoughts and soul. People in this area will experience more humanities and will place more spiritual connotation on it. The commercial landscape formed simply by means of demonstrating a wide variety of commodities will only present a kind of dull and jerky perceptual knowledge, and will only bring very little commercial appeal. To fulfill more emotional experiences during the shopping and promote the development of commercial activities, a high level, humanized and artistic commercial environment has to be formed with the modeling of commercial landscape by means of the culture and art (acting as the carrier).

2. Integration of ecological benefit and economic benefit

   Comfortable and pleasant shopping environment should be created for commercial space landscape and the "comfortable and pleasant" here has to be understood in two perspectives. On the one hand, the commercial environment must facilitate customers to complete shopping behavior in this place, and help the merchants to popularize their commodities and improve their sales of commodity; on the other hand, the commercial space landscape has to create an entertaining environment for the people exposed to commercial space, namely it ought to meet people's aesthetic needs and also beneficial to their physical and psychological health. As known to all, in case one person has exposed to beautiful green surroundings, all the functions of the human body will get relaxed and rested, which is beneficial to relieve tension in the fast-paced commercial activities; appropriate commercial landscape design will be more helpful in promoting economic benefit to get twice the result. Therefore, landscape design can create distinctive commercial landscape in the commercial space by use of seasonal aspects and ecological features of the plants. Luxuriant arbor trees can keep us shady and cool in the hot summer, while a gleam of flickering sunshine will come through deciduous trees in cold winter.

3. The diversification of texture of material for landscape

   The so called texture of material means the material properties determined by its physical characteristics and chemical characteristics. The designer is capable of delivering distinct regional culture and all sorts of commercial air by combination of different materials in the commercial landscape design. Famous American landscape designer—Professor Martha Schwartz fully deserves the outstanding representative in terms of configuration of texture of materials for commercial landscape design. The landscape works of Martha Schwartz are collection of daily necessities and materials that can be found in most hardware stores or catalogue of garden supplies—for instance, the clay pot, color crushed stone, bright yellow paint, plastic plant, artificial turf, garden ornament, lime and cord etc. Her selection of source materials derived from her interesting in Pop-Art, while it also reflects the postmodernism features of colloquialisms and slang.

3 ANALYSIS OF LANDSCAPE DESIGN OF THE MANCHESTER CITY STOCK EXCHANGE PLAZA

Manchester City is an important industrial city in the United Kingdom. The orientation of EDAW regarding Stock Exchange Plaza in urban planning: "an important node in public domain" that needs "an amazing iconic design". In face of problems such as complicated site condition and irregular quality of construction, Martha Schwartz studied several factors locally including the natural aspects, cultural aspects and social aspects, and then found out everything available. Finally she completed the design successfully. Martha Schwartz hoped that the plaza would "act like a living room where people can play freely" and also regard it as an important part of street furniture. There is a curved stream at the lowest part of northern plaza being filled up with stones internally that complies with the curve of the street, which suggests that there used to be a canal here. The curved road shaped by parapets and stool has made the elevation difference of the plaza to be unified in one surface. These curved roads possess the linking function just like the bridge; in the meantime, it can also be served as the theatre. There is a blue plate supported by train wheel placed on the "railway" that are in linear arrangement at the southern plaza, which is a mobile and also can be served as a sliding seat on "railway". Martha Schwartz has also arranged eight man-made palm trees of 10 meters high with steel cast trunk and colored leaves in southern plaza, and these palm trees will provide "an intervention of heterogeneous medium and implies that everything could happen in the world." But in reality, a lot of design contents made by Martha Schwartz were not fulfilled since Party A terminated the contract of design in advance. Consequently, the blue plate supported by train wheel placed in plaza had been removed and the man-made palm tree has never showed up.

4 ANALYSIS OF LANDSCAPE DESIGN FOR WUHAN NORTHERN HANKOU INTERNATIONAL BUSINESS CENTER PLAZA

4.1 *Project background*

Northern Hankou International Commodity Exchange Centre is located in hinderland of air-transportation related economic zone in Wuhan City—Wuhan Panlongcheng Economic Development Zone that located in the intersection of center surrounded by Wuhan Tianhe Airport, marshalling station in northern Wuhan, Yangluo hundred million tons class deepwater port as well as seven highways. The orientation of Northern Hankou International Commodity Exchange Centre is the fourth-generation specialized wholesale market which has sufficiently drawn lessons from advanced planning ideal and style of architecture design of the international and domestic large-scale commerce market. The international commodity exchange centre has centralized specialized wholesale market specialized in shoes, small commodities, leather articles and cases, hotel supplies, hardware and mechanical & electrical products, clothing, home textile and appliance, daily necessities, supplies for children and automobiles, which is supported by international brand anchor stores and also accompanied by large scale commercial plaza, large scale logistics center and e-commerce trading platform.

4.2 *Design concept*

1. *Incorporation of pattern and business function in land art design*

The landscape design of this project has discarded traditional commercial plaza modeling method for pure aesthetic appreciation, and combines the business function of the place sufficiently. From the practical perspective of comprehensive business center plaza, it utilizes relatively strong transportation orientation function of land art and characteristics of visual metaphor nature, which differentiates and strengthens effectively the function of different areas in large-scale commercial plaza. People coming here would not lose their shopping direction owing to the openness of the place, and they can enter into the commercial shopping environment conveniently while enjoying beautiful scenery. It will also play a preferable role of propaganda and popularization in business and trade per se. For example, as for the land art treatment of the large-scale waterscape and sculpture at the entrance of plaza, on the one hand, from the functional perspective, it will play a role of guiding the streams of pedestrian and cohesion of popularity; on the other hand, applying the logo of the enterprise into waterscape facilities will generate preferable marketing and popularization effects for enterprise brand.

2. *Humanized design of spatial scale in land art design*

A lot of landscape designs for commercial plaza have overlooked "comfortableness" owing to pursuing aggressively the "spacious feature" of the "plaza", thus the people in the place would look particularly tiny in comparison with excessive broad pedestrian walkway. Particularly, as for some commercial plazas, greening design is rejected in order to satisfy permeability of sight. In this way, the plaza has become a real "bare place". The landscape design for this project takes comfortable, safe, amiable, convenient mental feelings into account from appropriate and moderate perspectives, in consideration of several aspects such as spatial scale, plant configuration, street furniture etc. For example, as for the tree array design in sunken plaza, on the one hand it provides entertaining space for shoppers; on the other hand, it can bring people and the spatial scale of commercial environment closer. It makes people to perceive the characteristics of commodity vividly, and it also strengthened the commercial atmosphere further. In addition, the organic integration of waterscape design and tree pool design in sunken plaza has been realized. It takes circles as fundamental modeling elements, with the addition of round pool that float downstream along with cascade. Therefore, a well-proportioned space modeling has been formed in vertical position and a varying flat pattern has been formed in horizontal position. In this way, the esthetic perception of people towards different spaces has been satisfied. "Circles" in the water are the carriers for the pool, and they would be turned into carriers for tree pool on the ground, which allow the pedestrian to have a rest. It also embodies the organic and vivid features of landscape design elements.

3. *Harmony and unity between pavement and greening in land art design*

Although the landscape design of commercial plaza is hard landscape design oriented and supplemented by soft landscape design, the openness of commercial plaza would be lost in case the hard landscape and soft landscape are designed respectively. To this end, hard pavement design has been combined w3ith peripheral greening design during the landscape design of this project after comprehensive consideration of overall environment around the place. It is capable of not only create a comfortable, convenient and relaxed shopping environment, but also enhance the overall commercial atmosphere. For example, as for the forms of pavement and plane layout of greening in this project, integral unity has been paid great attention to by the designer. The overall geometrical pavement layout and concise green belts are interspersed with each other with the changing and unified forms

of "point—line—surface". As a result, the concise and generous principle for large scale commercial environment can be met.

5 CONCLUSION

Commercial plaza, acting as the portrait of the city and also the principal element constituting the urban space, has drawn public attention increasingly owing to people's emphasis on commercial space. One-sided pursuing landscape design of "big" and "foreign flavor" will not satisfy the current needs of commercial plaza. Comfortable, harmonious and efficient commercial plaza space could be created only by means of overall design, form combined with function, humanized and ecological landscape design.

REFERENCES

[1] Huang Lei & Zhang Bo. *Commercial Landscape Design: A Compelling Frontier* [J]. Shanghai Business (2009, 05): P36–39.
[2] Yang Xuelan Environmental Landscape Art Design of Commercial Plaza [J] Arts Exploration (2011, 02).
[3] Chen Dongchang & Zhang Jianhua The Future of Modern Commercial Landscape–Land Art Style Landscape and Imaginary space [J] Shanghai Business (2010, 11): P52–55.
[4] Lv Guiju Distinctive Landscape Design of Commercial Space in Modern City [J] Sichuan Architecture 2010. 08: P40–41.
[5] Su Xiaogeng A deviant—Interpretation of Martha Schwartz's Works [J] Chinese Garden (2000, 04).

Study on operation parameters of capacitive deionization to reclaimed water treatment

Z.L. Zhang
Department of Environmental Engineering, Tianjin Polytechnic University, Tianjin, China

Z.H. Zhang & X.F. Guo
Department of Environmental Engineering, Tianjin Polytechnic University, Tianjin, China
State Key Laboratory of Hollow Fiber Membrane Materials and Processes, Tianjin Polytechnic University, Tianjin, China

Q. Zhang & J.C. Wang
Department of Environmental Engineering, Tianjin Polytechnic University, Tianjin, China

ABSTRACT: Active carbon fiber clothes are used as electrode materials. The paper has studied the main operating parameters which affect Capacitive Deionization (CDI) process, such as operating voltage, flow velocity of feed water, Hydraulic Retention Time (HRT) and regenerative method. Raw water is secondary treatment effluent of municipal sewage treatment plant. And the water quality has reached the discharge standard. The results show that it will achieve the best desalination effect under the operating conditions: operating voltage 1.7 V, velocity 0.04 m/min, HRT 120s, the regenerative way of short circuit connection. At the same time, the stability of the electrode materials and capacitive deionization performance are considerate by continuous "capacitive deionization-desorption regeneration" experiment and the test of cycle current-voltage curve, The experiment results show that capacitive deionization has great stability in dealing with reclaimed water desalination.

Keywords: reclaimed water; capacitive deionization; Activated Carbon Fiber (ACF); conductivity

1 INSTRUCTION

The volume of reclaimed water is large, and the water quality is stable. It is widely applicated in irrigation, botanical garden afforestation, landscape water and industrial water etc. Due to the large consumption of industrial water, it is important to improve the integrated water resources utilization of reclaimed water, which is applied in industrial water after desalination processing. Besides, It can relieve the shortage of present situation of urban water resources effectively. The reclaimed water is widely used as the main circulating cooling water in industry. To make reclaimed water more suitable for industrial cooling water circulation, it is necessary to have the secondary effluent water of municipal sewage treatment plant desalted (Seo et al. 2010).

The common desalination methods contain distillation, membrane method (electrodialysis method, reverse osmosis), ion exchange method, etc. Distillation has large energy consumption, high investment, low volume, which limits its application. The desalting efficiency of electrodialysis, reverse osmosis desalting and ion exchange is very high, but they have some common problems Feed water quality requirements is very hard, so it is necessary to form a complex pretreatment device to guarantee the operation of the desalination system working smoothly (Oren 2008, Anderson & Cudero 2010). Capacitive deionization technology is a new type of desalination technology (Huang & Su 2010, Broséus et al. 2010). Compared with other desalination technologies, capacitive deionization technology operates under atmospheric pressure, low voltage operation (1.2~1.5 V), operating energy consumption is very low. If taking energy recovery and system design into consideration, the operation energy consumption can be as low as 0.1 kW·h/t (Welgemoed & Schutte 2005). Another significant advantage of capacitive deionization system is that the requirments of feed water is very low, especially for organic matter and the requirements of suspended solids (organic content of feed water could be up to 100 mg/L, SS up to 5 mg/L), so there is no need to prepare strict and complex pretreatment process device (Ryu et al. 2010). Because the water anionic and cationic are enriched on both different sides electrode plate, it will not exist scaling problems even though feed water with high hardness (Chen et al. 2011). The electrode produced by nonvalent material can defend the oil pollution. In these respects membrane separation and electrodialysis technologies can not be comparable with capacitive deionization.

Because capacitive deionization technology has advantage that other desalination technology could incomparable with it. It is very suitable for reclaimed water desalination. However, at present the research about capacitive deionization technology applying on reclaimed water deep treatment is lack. This paper is to research operation parameters about the capacitive deionization technology desalting reclaimed water.

2 THE EXPERIMENT DEVICE AND METHOD

2.1 Raw water quality

Raw water is taken from the wastewater treatment plant, and the raw water quality is shown in Table 1.

2.2 Electrode preparation

Take active carbon fiber (ACF) cloth (Anhui J.L.Q Carbon Fiber Co, Ltd China) as electric adsorption electrode materials. Besmear a thin layer of preparative conductive glue on the graphite plates. Then, stick on the preparative ACF, press the plates tightly. Cure 3 hours at temperature of 105 °C, then make the plates cool down to normal.

Table 1. Raw water quality.

Parameter	Value	Unit
COD$_{Cr}$	95–118	[mg/L]
Turbidity	2–3	[NTU]
pH	6.5–8.5	
SS	20–28	[mg/l]
Total hardness	320–415	[mg/L(CaCO$_3$)]
Conductivity	1490–1550	[μs/cm]

Table 2. Performance parameters.

Parameter	Value	Unit
Specific surface area	1257	[m²/g]
Pore volume	0.5242	[cc/g]
Pore diameter	1.7	[nm]
Contact angle	110.782	[°]
Adsorption capacity	8.837	[mg/g]

The main property parameters of the selected ACF electrode materials are shown in Table 2. A pore size distribution is measured by specific surface area and pore size distribution analyzer (Quanta chrome Co., Ltd, America). The contact angle is analyzed by Contact Angle tester (Beijing J.S.X Testing Machine Co, Ltd. China).

2.3 The experiment device

The experiment device is shown in Figure 1. 12 plates (100 mm · 100 mm · 5 mm) are arranged in electric adsorption module, separation distance is 5 mm between the every two plates. Use nylon nets to separate the flow between electrodes, so as to prevent electrode short circuit. Electric adsorption module is made up with organic glass, 200 mm long, 100 mm wide and 90 mm high.

The raw water is pumped into the precision filter (5 micron filtration precision) by peristaltic pump, filtered into the electric adsorption modules to proceed desalination processing. The electric adsorption module's voltage is supplied by regulated dc power. The raw water is pumped into the precision filter (5 micron filtration precision) by peristaltic pump, filtered into the electric adsorption modules to proceed desalination processing. The electric adsorption module's voltage is supplied by regulated dc power.

2.4 The experimental method

The electric adsorption device's desalination effect is reflected by raw water's conductivity changes. Conductivity is measured by an electrical conductivity analyzer (DDS-307, China). Use electrochemical workstation testing (LK98B, China) to detect electrochemical performance changes of the electrode material. Activated carbon fiber electrode is taken as working electrode, platinum electrode as counter electrode Saturated Calomel Electrode (SCE) as reference electrode; 0.5 mol/L NaCl solution as electrolyte. To prevent water electrolyte electrolyzing, the cyclic

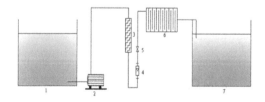

Figure 1. Schematic diagram of the experimental CDI system: 1 influent water tank; 2 peristaltic pump; 3 refined filtration. 4 rotor flow meter; 5 valve; 6 CDI device; 7 effluent water tank.

voltammetry curves' operating voltage ranges from 0 to 4V. The scan rate is 5 mV/s.

3 RESULTS AND DISCUSSION

3.1 Operating voltage's influence on capacitive deionization

Figure 2 shows the effect of operating voltage on the conductivity of water samples. At a flow velocity of 0.02 m/min, the operating voltage was varied from 1.0 to 2.0 V. With the rise of voltage, the time of the device achieving the biggest and stable desalination rate is shortened, which indicates the device desalination rate increasing gradually. At the operating voltage of 2 V, desalination rate is fastest. For the further study of the operating voltage's influence on capacitive deionization, take the data on Figure 2 further analysis, as shown in Figure 3. It indicated that the higher the operation voltage, the higher the device's desalination rate can be obtained. Under the operating voltage ranges from 1.0 to 1.7V, the desalination rate increased

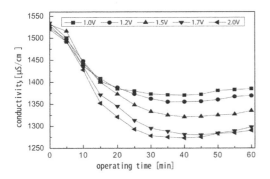

Figure 2. Water conductivity changes at applied voltages of 1.0, 1.2, 1.5, 1.7, and 2.0 V, respectively.

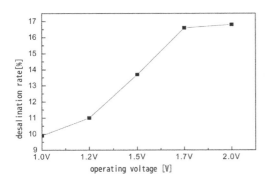

Figure 3. Desalination rate changes at applied voltages of 1.0, 1.2, 1.5, 1.7, and 2.0 V, respectively.

obviously with the rise of the operation voltage. The desalination rate no longer increases significantly at the voltage beyond 1.7 V. Desalination rate merely increased 0.2% under the operating voltage ranging from 1.7 to 2.0 V. When operating voltage rises to 2.0 V, water electrolytic reaction will happen, this reduces the energy utilization ratio. Beside, the gas electrolytic produced will be adsorbed on the electrode surface of activated carbon fiber, the lower active carbon fiber and solution of the contact area, further make electrode electric adsorption effect reduces. Which will lessen the contact area between the activated carbon fiber and solution. And then, it will reduce the electrode's desalination ratio. Although the higher operating voltage has advantages on raising desalination ratio, considering energy consumption and desalination ratio, optimized voltage is 1.7 V.

3.2 Flow velocity and HRT's influence on capacitive deionization

With the speeding up of the flow velocity, the desalination rate of the device also reduces gradually. The thickness of the electric double layer on the surface of electrode gets thinner, which reduces the ability of electrode adsorption. At the same time, speeding up the water flow velocity increases the level of the water turbulence, and transfer effect is intense. It can reduce the processing cycle time.

Another important parameter that affect the desalination rate is Hydraulic Retention Time (HRT), which is determined by both flow velocity and height of the plate. Figure 5 shows that the HRT varied from 48 to 40 s affect the conductivity of the solution. Under the HRT ranges from 40 to 120s, the desalination rate increased obviously. The desalination rate no longer increased significantly at the HRT beyond 120s. Desalination rate merely

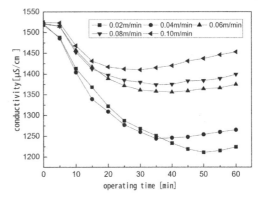

Figure 4. Water conductivity changes under different flow velocity.

increased 2% under the HRT ranging from 120 to 240s.

3.3 Different ways of regeneration's influence on capacitive deionization

In the process of capacitive deionization, after the electro-adsorption get saturated, the device need to take off the attached to restore the desalination efficiency. Thus, the effect of regeneration will directly influence the quality of capacitive deionization. Different ways of regeneration can bring about different regeneration effect, the regeneration methods of reverse connection and short circuit connection are investigated in this experiment. Figure 6 indicates that the conductivity changes in the "capacitive deionization—desorption regeneration" process of the two methods in a complete operation cycle.

As seen in Figure 6, at the beginning of the regeneration, the desorption rate of reverse connection is obviously faster than that of short circuit connection. But as the process goes on, the regeneration effect of short circuit connection is much better than that of reverse connection. The cause of the

phenomenon is that in the process of reverse connection, the driving force is stronger than short circuit connection, thus, at the beginning of the desorption rate is faster. At the same time, because of the electrode upside down the released ions get adsorbed by the electrode of opposite charges before they outflow from the flow channel, as a result, the regeneration effect of reverse connection is worse than that of short circuit connection. Thus, in the actual use of reclaimed water desalination, the regeneration method of short circuit connection can be used to save energy and get better regeneration effect.

3.4 The stability of continuous operation cycle

Figure 7(a) and Figure 7(b) reflects continuous operation cycle under the condition of the desalination efficiency and electrode properties. Figure 7(a) reflects that under the conditions of operating voltage 1.7V, flow velocity 0.04 m/min, take complete "capacitive deionization—desorption regeneration" for a cycle, and operating five cycle continuously.

It can be seen from the Figure 7(a), in the five continuous periods operations, the device desalination

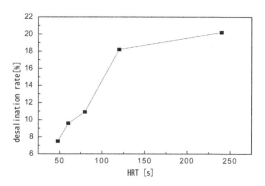

Figure 5. Desalination rate changes under different HRT.

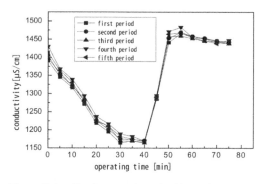

Figure 7(a). Continuous adsorption/desorption curve.

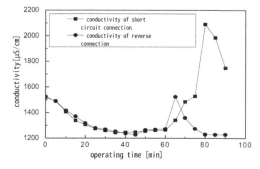

Figure 6. Comparison different regeneration effect Along with different ways of regeneration.

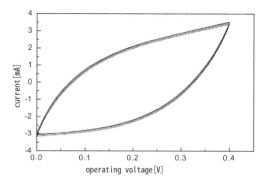

Figure 7(b). Circulation current-voltage curve.

rate is stable and the largest stable desalination rate can reach 22% in 40 min. Desalination effect has not decayed obviously, it can realize stable regeneration and desorption, electric adsorption is reversible well in the 30 min. At the same time, test the electrode materials in three electrodes system test cycle current-voltage curve after the end of each period. The results is shown in Figure 7(b). It shows the test of five cycle current-voltage curve, the coincidence degree of the curve is high, the area around the curves have no great changes to be seen. It shows that the electrode capacitance does not decrease, which has verified that it is reversible well to desalt reclaimed water with technology of capacitive deionization.

4 CONCLUSIONS

The higher the operation voltage, the better the device's desalination effect can be obtained. However, when operating voltage rises beyond 2.0 V, water electrolytic reaction will happen, which reduces the energy utilization ratio. So, taking energy consumption and raising desalination rate into consideration, the optimized voltage is 1.7V. The slower the flow velocity and the longer HRT are, the better the device's desalination effect can be obtained. However, reclaimed water treatment capacity will decrease if the flow velocity is too slow and HRT is too long. Then, determine the flow velocity of 0.04 m/min HRT for 120s as the optimized flow velocity and HRT. Though the desorption rate of reverse connection is obviously faster than that of short circuit connection at the beginning of the regeneration, from the whole regeneration stage the regeneration effect of reverse connection is worse than that of short circuit connection. The device desalination rate is stable and has not decayed obviously. Besides, the electrode capacitance hardly decreased, through continuous "capacitive deionization- desorption regeneration" periods. Those all show that capacitive deionization has good operation stability in reclaimed water desalination treatment.

ACKNOWLEDGEMENTS

This work was financially supported by the National Natural Science Fund of China (51108315), the Tianjin Technology Support Program of China (11ZCGYSF01500) and Tianjin Science and Technology Innovation Special Funds of China (08FDZDSF03200).

REFERENCES

Anderson, M.A. & Cudero, A.L. 2010. Capacitive deionization as an electrochemical means of saving energy and delivering clean water. Comparison to present desalination practices: Will it compete? *Electrochimica Acta* 55: 3845–3856.

Broséus, R. et al 2009. Removal of total dissolved solids, nitrates and ammonium ions from drinking water using charge-barrier capacitive deionization. *Desalination* 249: 217–223.

Chen, ZL. et al 2011. Kinetic and isotherm studies on the electrosorption of NaCl from aqueous solutions by activated carbon electrodes. *Desalination* 267:239–243.

Huang, C.C. & Su, Y.J 2010. Removal of copper ions from wastewater by adsorption/electrosorption on modified activated carbon cloths. *Journal of Hazardous Materials* 175: 477–483.

Oren, Y. 2008. Capacitive deionization (CDI) for desalination and water treatment—past, present and future (a review) [J]. *Desalination* 228:10–29.

Ryu, J.H. et al 2010. A study on modeling and simulation of capacitive deionization process for wastewater treatment. *Journal of the Taiwan Institute of Chemical Engineers* 41:. 506–511.

Seo, S.J. et al 2010. Investigation on removal of hardness ions by capacitive deionization (CDI) for water softening applications. *Water Research* 44: 2267–2275.

Welgemoed, T.J. & Schutte, C.F. 2005. Capacitive Deionization Technology™: An alternative desalination solution. *Desalination* 183: 327–340.

Frontiers of Energy and Environmental Engineering – Sung, Kao & Chen (eds)
© 2013 Taylor & Francis Group, London, ISBN 978-0-415-66159-1

Research on preservation performance adjustment of Changbai Mountain Abies

X.Y. Niu & J.H. Wu
Key Laboratory of Wood Material Science and Engineering of Jilin Province, Jilin City, China
Beihua University, Jilin City, China

Y.G. Wang
Petrochina Jilin Petrochemical Company Refinery, Jilin City, China

Y. Zhang
Beihua University, Jilin City, China

ABSTRACT: This paper improves the preservation performance of Changbai Mountains Abies by impregnating its specimens with preservative, and then gets the optimum preservation parameters through comparative analysis of the specimens' weight loss rate after preservative treatment by different concentration, which improves the practicality of Changbai Mountain Abies, to achieve the purpose of best use of inferior materials.

Keywords: Changbai mountain abies; xylan; wood preservation

Because of the lack of wood resource in our country and the deficiency of wood quality, some people advocate substitute wood material with steel, aluminum products and cement. This restricts the development of wood products and furniture industry. For a long-term plan, wood material, which is a renewable material, has not been brought into full play. Wood preservative technology can overcome its partial deficiency, and improve its physics property, thus expand the applying range of woods and improve the utilization efficiency. In addition, it can save wood resource in the end.

The Changbai Mountain abies is a kind of loose material, it has a low density, and is easy to perish and be moth-eaten, these defects restricted it from widely using. Nevertheless, we can improve the Changbai Mountain abies's preserve performance through wood preservative technology, and extent its service life.

1 MATERIAL AND METHOD

1.1 Material

Our experiment uses 36 abies lumbers (from Changbai Mountain area) as specimen, size 20 mm × 20 mm × 10 mm, water content 12%.

Xylan. outsourcing, molecular weight 200~20000, white or slightly yellow powder.

Culture medium: dry sand, masson pine sapwood sawdust, corn flour, brown sugar, potatoes, feeding wood, etc.

1.2 Instruments

Steam autoclave, bacteria culture room or electric incubator, drying oven.

1.3 Method

Specimens amount: It needs 36 Abies specimens (20 mm × 20 mm × 10 mm) to determine the preservation performance of a preservative against the testing bacteria. The blank specimen is treated with solvent or formula not including an effective preservative ingredient, thus preservation performance of non-active ingredients in solvent or formula against the testing bacteria is determined; another group of 9 specimens without any treatment is used to determine the activity of the testing bacteria.

Preparation of preservative solution: preservative was confected to 4 concentration gradient, which is 9%, 6%, 3% and 0% respectively, each concentration was used in 9 specimens.

Preservative impregnation and preservative treatment of specimens:

Before impregnation, specimens are drying to constant firstly on the blast drying oven. It was weighed with the accuracy of 0.01 g. And its quality is represented with T1, this is the quality before treatment.

The specimens treated with preservative of the same concentration, is impregnated for 30 min in a vacuum desiccator and then taken out. After liquid on surface is absorbed with filter paper, each specimen is weighed immediately with the accuracy of 0.01 g, this quality is represented with T2.

The specimens are put into a blast oven and dried to constant at 40°C and then weighed (with the accuracy of 0.01 g). This quality before decaying is represented with T3.

Specimens of the same group are wrapped in muti-layer gauze, put into steam sterilizer and been steamed for 30 min at 100°C ± 2°C. After cooled and under sterile condition, the specimens are put on feeding wood covered with mycelium in cultivating bottle, with the broad surface in touch with feeding wood. Each specimen is placed on different feeding wood. The cultivating bottle is wrapped in cotton and waterproof paper, and placed in an incubator, keeping temperature of 28°C ± 2°C and relative humidity of 75% ~ 85%. The specimens are cultured for 6 weeks incubator. After clearing out mycelium on surface, specimens are weighed (with

the accuracy of 0.01 g). The quality of decayed specimens is represented with T4.

Maintain amount of preservative of specimen is calculated as follows:

$$R = \frac{(T_2 - T_1) \times c}{V} \times 10 \qquad (1)$$

In this formula:

R—maintain amount of preservative of specimen, kg/m³;
V—specimen volume, cm3;
C—concentration of preservative fluid (mass fraction), %.

Percentage of mass loss of specimen is calculated as follows:

$$L = \frac{T_3 - T_4}{T_3} \times 100 \qquad (2)$$

In this formula:

L—percentage of mass loss of specimen, %;
T3—the constant weight quality of specimen before decaying, g

Table 1. Comparison of weight loss rate and preservative impregnated amount of different concentrations.

ID	Xylan concentration (%)	Preservative impregnated amount (Kg/m²)	Weight loss rate (%)
1	0	0	35.16
2	3	0.024975	18.44
3	6	0.079950	11.55
4	9	0.203175	21.42

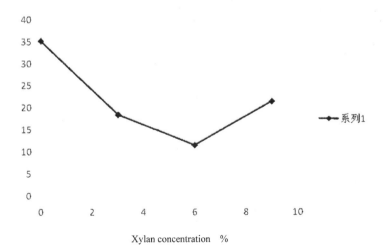

Xylan concentration   %

Figure 1.   Relationship between xylan concentration and weight loss rate.

T4—the constant weight quality of specimen after decaying, g.

2 EXPERIMENT RESULT AND ANALYSIS

Under appropriate condition, we use xylanase preservative of different concentration to impregnate and degenerate the 36 Changbai Mountain Abies specimens in cultivating box for 6 weeks. Then we test and analyze the quality of the specimens, and get the relationship between weight loss rate and preservative impregnated amount of different concentrations (See Table 1 and Fig. 1).

After analyzing each group experiment data, we can draw conclusion that xylan preservative has certain preservative effect on Changbai mountain abies, and the impregnation amount of wood preservative will increase with the raise of density.

Basing on analysis of comparing experiment group with the xylan concentration of 3%, 6%, 9%, and consideration of practical situation, we can determine the optimum process parameters: when preservative has the concentration of 6%, it will get the best preservation performance, and the least average of weight loss rate(decaying rate) is 11.54%.

3 CONCLUSION

By preservative impregnation treatment to Changbai mountain abies specimens with xylan preservative of different concentration, and spend 2 months to prepare Culture medium, preserve, dry, process experiment data and analyze result, we can draw the following conclusion: preservative of 6% concentration correspond to the best preservation performance, when the impregnation amount is 0.07995 Kg/m^2, the least average of weight loss rate (decaying rate) is 11.54%. And under this condition, wood material has the best performance adjustment by xylan preservative.

ACKNOWLEDGEMENT

Research work of this paper has been supported by Science and Technology Department Project of Jilin Province (No. 20120256).

REFERENCES

[1] Li Jian. 2009. Monitoring of indoor wood preservation experiments and wood properties, Wood Research, 415–423.
[2] Morris PIMJJ, Ruddick JNR. 1994. A review of incising as a means of improving treatment of sawnwood. 25.
[3] Torgovnikov G, Vinden P. 2002a. Microwave modification of wood properties_improvements in wood permeability. 31stannual meeting of international researchgroup on wood preservation. Kona, Hawaii, U.S.A., Document No: IRGWP 00-40181.
[4] Sanders MG, Amburgey TL, Barnes HM. 2000. Innovations in the treatment of southern pine heartwood. 31stannual meeting of international research group onwood preservation. Kona, Hawaii, USA, Document No: IRGWP00-40172.
[5] Torgovnikov G, Vinden P. 2002a. Microwave modification of wood properties_improvements in wood permeability. 31st annual meeting of international researchgroup on wood preservation. Kona, Hawaii, U.S.A., Document No: IRGWP 00-40181.
[6] Zhou Huiming. Wood Protection [M], Beijing: China Forestry Publishing House, 1991:1–3.
[7] Jiang Mingliang. Discussion on the registration and quality supervision system of the wood preservatives [J], Wood Industry, 2005, 19(2): 43–45.
[8] Shao Peilan, Xu Ming, Li Haifeng, Zhu Xiaohong. Research on Alkali extraction of the corn cob xylan [J], Journal of Ningxia Agricultural College, 2000.

Frontiers of Energy and Environmental Engineering – Sung, Kao & Chen (eds)
© 2013 Taylor & Francis Group, London, ISBN 978-0-415-66159-1

Community structure of methanotrophs under wetland evolution of WuLiangSuHai lake by T-RFLP analysis

J.Y. Li
College of Life Sciences, Inner Mongolia University, Huhhot, Inner Mongolia, China

Y. Zhou, Y. Zhuang & J. Zhao
College of Environment and Resources, Inner Mongolia University, Huhhot, Inner Mongolia, China

ABSTRACT: The aim of this study was to investigate community composition and diversity of Methanotrophs and their relationship with soil environmental variables under wetland evolution of WuLiangSuHai lake. Terminal Restriction Fragment Length Polymorphism was explored. As a result, the diversity and evenness index fluctuated from the sample S1 to S5 with high→low→high→low→high trend, and pH value was most closely related to the shift of Methanotrophs community structure (correlation coefficient, 0. 7052); the combination of total phosphorus and pH value was closely related to the shift of Methanotrophs community structure (correlation coefficient, 0.6565). Methanotrophs community structure changed along with gradient under wetland evolution, Methanotrophs adapted the terrestrial environment from the aquatic environment.

Keywords: wetland evolution; methanotrophs; community structure; T-RFLP

1 INTRODUCTION

Methane (CH_4) is a greenhouse gas, which preceded only by carbon dioxide (CO_2) in the atmosphere. The CH_4 content is 1/27 of the carbon dioxide, but the greenhouse effect is 20~30 times more than that of the CO_2 under same quality. Some reports indicated that the contribution to greenhouse gas emission from wetland ecosystem is more than half of that from water, accounting for 15%~20% of total CH_4 emission. The wetlands have significant impact on global climate change, and are considered as the major source of CH_4 emissions. The only biological sink for CH_4 is the oxidation of methane by Methanotrophs in soil, which is responsible for 10% atmospheric CH_4 of the sinks. In methanotrophs cells, Methane Monooxygenase (MMO) is a first key enzyme, which oxidise methane to methanol under molecular oxygen exist. pMMO is catalytic oxidation methane enzyme which exist in all methanotrophs except *Methylocella silvestris*. The conservative subunit α-subunit gene of *pmoA* is most commonly used for exploring methane oxidizing bacteria diversity as functional genes. Based on the consistent phylogenetic analysis from *pmoA* and 16S rRNA sequences, *pmoA* gene has been widely used for detecting the diversity of methane oxidizing bacteria as a marker in wetlands.

WuLiangSuHai wetland is a typical eutrophic lake, which was officially listed in the "Ramsar List of Wetlands of International Importance" by the Ramsar Convention on Wetlands in January 2002. In this study, Terminal Restriction Fragment Length Polymorphism (T-RFLP) was used to characterize the composition and diversity of methanotrophs community and their relationship with soil environmental variables, and to find out their distribution features. Finally, we can provide scientific data for controlling greenhouse from natural wetlands.

2 MATERIALS AND METHODS

2.1 Soil sampling

WuLiangSuHai (40°47′–41°03′N, 108°43′–108°57′E), an oxbow lake, is formed by the Yellow River diversion. It is a typical shallow weedy lake of west arid region in Inner Mongolia Plateau, and has multifunctions, such as biodiversity protection and environmental protection. It is also the largest natural wetland in the same latitude on the earth. The lake is located in Urad Front Banner, BaYan Nur City, Inner Mongolia Autonomous Region. The water area is 333.48 km². Farmland drainage is the mainly supplementary water for the lake. Therefore, the eutrophication has been intensified in a recently decade. Now, WuLiangSuHai has been a serious eutrophic weedy lake, which can be characterized by excessively growth of macrophytes, and accumulating the decay grasses on the bottom under the lake at a speed of 9~13 mm per year. It has become one of the fastest swamp lakes, and the evolution direction is as follow: *Phragmites autralis* marsh→*Suaeda glauca* salinized meadow land→*Nitraria tangutorum Bobr* desert.

Sediment (or soil) samples were collected according to the evolution gradient mentioned above: the sample S1 was collected from the growth zone of *Phragmites australis* in WuLiangSuHai; the sample S2 collected from the growth zone of *Scirpus triqueter*; the sample S3 collected from the marsh zone of *Phragmites autralis*; the sample S4 collected from the zone of *Suaeda glauca* salinized meadow land; the sample S5 collected from the desert zone of *Nitraria tangutorum Bobr*. Five point sampling swas used to collect the sample with a depth of 0~10 cm. The soil sample should be kept fresh for molecular biology analysis, and the soil physical, chemical properties and soil respiration are analyzed in our lab (Table1).

2.2 *DNA extraction*

According to the glass bead/calcium chloride/SDS method, total DNA was extracted from the soil sample.

2.3 *pmoA gene nested PCR*

The *pmoA* gene of methanotrophs was amplified using nested PCR. The first round of the nested PCR amplification from 1 µL of extracted soil DNA template was conducted in a total volume of 25 µL by using 2.0 µL of 2.5×10^{-3} M dNTP, 1.0 µL of 1.0×10^{-5} M A189, 1.0 µL of 1.0×10^{-5} M A682, 2.5 µL of $10 \times$ buffer, and 0.2 µL of 5U/µL Taq under the following conditions: 3 min at 95°C,

Table 1. Properties of soil sample used in DNA extraction.

Basal physical and chemical properties	S1	S2	S3	S4	S5
Organic carbon (g/kg)	11.82	33.47	7.99	3.73	12.93
Total nitrogen (g/kg)	1.07	2.64	0.69	0.32	1.33
Total phosphor (g/kg)	0.78	1.08	0.57	0.47	0.66
Total water soluble salt (g/kg)	9.11	14.87	7.11	13.65	20.46
pH	8.04	7.98	8.61	9.25	8.69

35 cycles of 60s at 95°C, 60s at 62°C, and 60s at 72°C, and an additional 10-min cycle at 72°C. The second round of the nested PCR amplification from 1 µL of extracted soil DNA template was conducted in a total volume of 25 µL using 2.0 µL of 2.5×10^{-3} M dNTP, 1.0 µL of 1.0×10^{-5} M A189, 1.0 µL of 1.0×10^{-5} M mb661, 2.5 µL of $10 \times$ buffer, and 0.2 µL of 5U/µL Taq under the following conditions: 3 min at 95°C, 30 cycles of 60 s at 95°C, 60s at 55°C, and 40s at 72°C, and an additional 10-min cycle at 72°C.

2.4 *The purification of pmoA gene amplification products, restriction and purification of the restriction enzyme products*

The products are purified by TransGen according to the manufacture's instruction, and the purified products are digested by *Alu*| and *Hha*|. Double digestion reaction system *of Alu| and Hha|* is as follow: 15 µL purified PCR products, 2 µL $10 \times$ H Buffer, 0.5 µL *Alu*| (10 U/µL), 0.5 µL *Hha*| (10 U/µL), add ddH$_2$O to 20 µL, the temperature and time of restriction are 37°C and 4 h, respectively.

The products of restriction enzyme follows the purified steps: (1) add 1 µL glycogen (20 mg/mL, Beckman Coulter, USA) and 2 µL Na$_2$Ac (3 M) into the products, 2.5 times volume of 75% ice ethanol, centrifuge (14000 rpm, 10 min), the supernatant was decanted; (2) add 100 µL 70% ice ethanol, centrifuge (14000 rpm, 5 min), the supernatant was decanted; (3) add 100 µL 70% ice ethanol, centrifuge (14000 rpm, 5 min), the supernatant was decanted; (4) avoid light and air-dried, then add 40 µL formamide (Beckman Coulter, USA) mix.

2.5 *T-RFLP capillary electrophoresis*

39.5 µL of the prepared sample and 0.5 µL Size Standard-600 (Beckman Coulter, USA) were mixed and transferred to the 96-wells plate, using GenomeLab GeXP genetic analysis system (Beckman Coulter, USA) for capillary electrophoresis, the voltage is 4.8 kV, the time is 60 min.

2.6 *Community structure and diversity analysis*

The results of capillary electrophoresis were analyzed by GeXP Fragment Analyse, every peak was considered as one Operational Taxonomic Unit roughly. Diversity indexs are as follows:

Richness index: the number of S = T-RFs (1)

Shannon-Weiner (H') index: H = $-\sum$(pi)(log$_2$pi) (2)

Evenness index: E = H/H$_{max}$ (3)

where H$_{max}$ = log$_2$(S)

Simpson index: $D = 1 - \sum pi^2$ (4)

Where pi is the ratio of one peak area divided by total peak area of all samples.

Bray-Curtis dissimilarity coefficient was calculated by R language vegan package, environmental data were processed by Z-score using SPSS 16.0, and the relationship between environmental factors and community of Methanotrophs were analyzed, and Canonical correspondence analysis (CCA, Canoco for Windows 4.5) was used to find out the influence of environmental factors on T-RFs distribution.

3 RESULT AND ANALYSIS

3.1 Community and diversity of methanotrophs

T-RFLP results of Methanotrophs community structure are presented in Figure 1. 56 bp, 128 bp, 511 bp are the common prevailing T-RFs in S1 and S2, their relative abundances of S1 are 34.1%, 33.1%, 15.1%, respectively, and they are account for 82.3%

of the total community. Their relative abundances of S2 are 25.3%, 35.2%, 34.5%, respectively, and they are 95.0% of the total community. 128 bp exists only in S1, S2 and S3; while 63 bp exists in S3, S4 and S5, and its relative abundances are 0.6%, 11.8% and 3.7%, respectively. It don't exist in S1 and S2; the relative abundances of 129 bp, 139 bp, 148 bp, 166 bp, 186 bp and 578 bp are 11.3%, 12.3%, 8.0%, 6.6%, 6.3% and 18.0%, respectively; the relative abundances of 178 bp and 231 bp are 24.8% and 58.1%, respectively; the relative abundances of 89 bp, 287 bp, 511 bp are 32.4%, 10.6% and 5.1% in S5, respectively. The diversity index of Methanotrophs are presented in Table 2. The diversity and evenness indices indicate a high→low→high→low trend. S3 has high diversity, good distribution, and complicated structure; the dominant species of S1 and S2 are the same, they have high relative abundance; the dominant species of S3 is different from S1 and S2, it has high diversity, increased evenness. There are two new dominant species 178 bp and 231 bp from S4 with simple structure and the lowest diversity and evenness.

3.2 Comparative analysis of community structure of methanotrophs during wetland evolution process

Community structure of Methanotrophs during evolution process was analyzed by Bray-Curtis dissimilarity Coefficient matrix, in which 0 represents two samples have the same community structure, 1 represents two samples have different community structure. The results are presented in Table 3. Community structure of S1 and S2 are

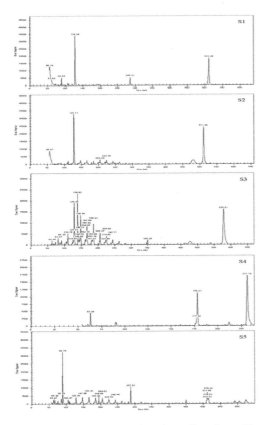

Figure 1. Capillary electrophoresis profiles of *pmoA* by T-RFLP.

Table 2. Diversity indexes of the methanotrophs.

Samples	Richness (S)	Shannon-Weiner (H)	Simpson (D)	Evenness (E)
S1	6	2.19	0.74	0.85
S2	5	1.82	0.69	0.79
S3	27	4.10	0.92	0.86
S4	4	1.54	0.58	0.77
S5	18	3.58	0.86	0.86

Table 3. Bray-Curtis dissimilarity matrix of the methanotrophs bacteria between different samples.

Samples	S1	S2	S3	S4
S2	0.264655			
S3	0.963235	0.950327		
S4	1	1	0.976066	
S5	0.905265	0.956837	0.863195	0.962515

similar; S1, S2 and S3, S5 are not similar, and they are different from S4; community structure of S3 is not similar with S4, S5, and S4 is not similar with S5. S1 and S2 have similar community structure due to the same sediments from WuLiangSuHai lake, although different vegetation type and no evolution gradient. Community structure of other samples are not similar, caused by the changed microhabitat during wetland evolution. By further analyzing the correlation between different combination of environmental factors and dissimilarity of community structure of Methanotrophs, we find that the pH are closely related to changed community structure of Methanotrophs with a correlation coefficient of 0.7052. Additionaly, total P and pH combination is related to changed community structure of Methanotrophs with a correlation coefficient of 0.6565. Results are showed in Table 4.

3.3 Environmental factors effect on community structure

Effects of environmental factors on community structure of Methanotrophs were analyzed by

Table 4. The relationship between different combination of environmental variables and methanotrophs community dissimilarities.

Different combination of environmental factors	Correlation
pH	0.7052
Total phosphor + pH	0.6565
Total nitrogen + total phosphor + pH	0.3708
Organic carbon + total nitrogen + total phosphor + pH	0.2736
Organic carbon + total nitrogen + total phosphor + total water soluble salt + pH	−0.0547

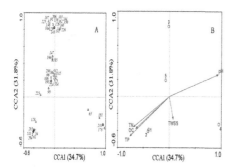

Figure 2. Canonical correspondence analysis of the methanotrophs community in relation to the environmental factors. TN, Total nitrogen; OC, Organic carbon; TP, Total phosphor; TWSS, Total water soluble salt.

CCA, the results are shown in Figure 2. pH, total P, total N and TOC are highly correlated with first axle, and correlation coefficients are 0.9377, −0.7618, −0.6917 and −0.6482, respectively. The total water soluble salt is highly correlated with the third axle, and correlation coefficient is 0.8616.

4 DISCUSSION

Methanotrophs are the key bacteria which play important roles in reducing the CH_4 emission. As a biofilter, methanotrophic bacteria can oxide CH_4 released from soil and sediment. Thus, they are considered to be biological weapon to deal with global climate change. Their diversity and distribution were explored in order to find out the influence of their diversity and community structure on atmospheric greenhouse effect, and controlling the atmospheric greenhouse. In this study, community structure of Methanotrophic bacteria in WuLiangSuHai lake were characterized by T-RFLP, and we find that their community structure and dominant species are quite different in different stages, especially dominant species. There are two types of Methanotrophs existing in soils, type I and type II. They have different niches in wetlands: type I grow under high oxygen, low concentration of methane and has high affinity for CH_4. Type II grow under low oxygen, high concentration of methane and has low affinity for CH_4, and use high-concentration methane from soils as substrate. Among the five samples, S1 and S2 come from similar sediments in the lake and there is no evolution gradient between them. The aquatic plant of S1 is *Phragmites australis* and the aquatic plant of S2 *Scirpus triqueter*. According to the results of T-RFLP analysis, we can find that different plants have no influence on community structure of Methanotrophs in the lake sediments. S1 and S2 have the same community structure of Methanotrophs. S1, S2 have evolution gradient with S3, S4, S5, with gradient *Phragmites autralis* marsh→*Suaeda glauca* salinized meadow land→*Nitraria tangutorum Bobr* desert. Community structure of Methanotrophs has changed by evolution gradient, and Methanotrophs begin to adapt the terrestrial environment from the aquatic environment.

In summary, the T-RFLP technology is helpful for us to screen and cultivate Methanotrophic bacteria. Some scientific question should be further solved: ①the dynamic changes of Methanotrophic bacteria and the influence of microhabitat and changed substrates on them ②the relationship between spatial heterogeneity of Methanotrophic bacteria distribution and methane emission ③how to screen efficient Methanotrophic bacteria and how to find microbial resources which can control global greenhouse gas emission?

This work was financially supported by the National Natural Science Foundation of China (Grant No. 31160129) and National Basic Research Programs (973) of China (Grant No. 2009CB125909).

REFERENCES

Bender M, Conrad R. Kinetics of CH4 oxidation in oxic soils. Chemosphere, 1993, 26:687–696.

Blake DR, Rowlands FS. Continuing worldwide increase in tropospheric methane, 1978 to 1987. Science, 1988, 239: 1129–1131.

Costello AM, Lidstrom ME.Molecular Characterization of Functional and Phylogenetic Genes from Natural Populations of Methanotrophs in Lake Sediments. Appl Environ Microbiol, 1999, 65(11):5066–5074.

Duxbury JM, Mosier AR. Status and issues concerning agricultural emissions of greenhouse gases. In: Drennen TE, eds. Agricultual Dimensions of Global Climate Change. Delray Beach, Fla: St. Lucie Press, 1993, 229–258.

Edwards C, Hales BA, Hall GH, et al. Microbiological processes in the terrestrial carbon cycle: methane cycling in peat. Atmos Environ, 1998, 32(19): 3247–3255.

Fisk MC, Ruethe KF, Yavitt JB. Microbial activity and functional composition among northern peatland ecosystems. Soil Biol Biochem, 2003, 35(4): 591–602.

Hanson AL, Swanson D, Ewing G, et al. Wetland Ecologial Functions Assessment: An Overview of Approches. Atlantic Region, 2008, Canadian Wildlife Service Technical Report Series No. 497.59pp.

Holmes AJ, Costello AM, Lidstrom ME, et al. Evidence that particulate methane monooxygenase and ammonia monooxygenase may be evolutionarily related. FEMS Microbiol Lett, 1995, 132(3): 203–208.

Li JY, Li B, Zhou Y, et al. A rapid DNA extraction method for PCR amplification from wetland soils. Lett App Microbiol, 2011, 52:626–633.

LI Jun, TONG Xiao-Juan, YU Qiang. Methane uptake and oxidation by unsaturated soil, Acta Ecologica Sinica 2005, 25(1):141–147.

Matthews E, Fung I. Methane emission from natural wetlands: Global distribution, area, and environmental characteristics of sources. Global Biogeochem Cycles, 1987, 1: 61–86.

McDonald IR, Bodrossy L, Chen Y, et al. Molecular ecology techniques for the study of aerobic methanotrophs. Appl Environ Microbiol, 2008, 74(5): 1305–1315.

Rodhe H. A comparison of the contribution of various gases to the greenhouse effect. Science, 1990, 248: 1217–1219.

Sun Hui-min, HE Jiang, GAO Xing-dong, et al. Distribution of Total Phosphorus in Sediments of Wuliangsuhai Lake. Acta Sedimentological Sinica, 2006, 24(4): 579–584.

SUN Hui-min, HE Jiang, Lv Chang-wei, et al. Nitrogen pollution and sptial distribution pattern of Wuliangsuhai Lake. Geographical Research, 2006, 25(6): 1003–1012.

Theisen AR, Ali MH, Radajewski S, et al. Regulation of methane oxidation in the facultative methanotroph Methylocella silvestris BL2. Mol Microbiol, 2005, 58(3): 682–692.

Zhao Ji, Li Jingyu, Zhou Yu, et al. Methane-and Ammonia-Oxidation microorganisms and their coupling functions. Advances in Earth Science, 2012, 27(6):651–659.

Frontiers of Energy and Environmental Engineering – Sung, Kao & Chen (eds)
© 2013 Taylor & Francis Group, London, ISBN 978-0-415-66159-1

Study on the regularity of fault water inrush in Yuzhou mine area

C.H. Huang
Institute of Resources and Environment, Henan Polytechnic University, Jiaozuo Henan, China

J.J. Huang
School of Mechanical and Power Engineering, Henan Polytechnic University, Jiaozuo Henan, China

ABSTRACT: Fault is one of the important reasons causing floor water invasion of coal seam. It shows that the water irruption quantity of fault is large, the water bursting source is the limestone water of the cambrian mainly, and the types of fault water inrush is diversity through the characteristic analysis of all previous fault water inrush in Yuzhou mine area. It also can be found that water inrush quantities of faults present positive correlation with the water inrush channel area of faults, the porosity of fault zones and the water pressure value of aquifer by the water inrush quantity calculated of main influence faults. The fault type, water inrush quantity, cutting the cambrain limestone aquifer or not and the water pressure value of aquifer are chosen as discrimination factors, through which the risk of fault water inrush can be divided into four kinds. With this method, four strips of main influence faults in Yuzhou mine area are classified, where the number of the first risk fault is four and other faults can be distinguished as the lower grades.

Keywords: fault; water inrush; the cambrian limestone; the classification of water inrush risk

1 INTRODUCTION

Fault is the main reason to cause the water inrush of coal seam floor. According to statistics, about eighty percents of floor water inrush accidents relate to the faults directly or indirectly [1–3]. The several water inrush accidents of Yuzhou mine area proved it. For example, the water inrush of the cambrian limestone happened in Tongshuzhang fault in 1985, the mine was flooded because the water inruption quantity exceeded the drainage capacity of the mine facillities. Also in 2005, the delayed water inrush happened in Pingyu first mine, the reason was that the buried fault between the coal seam and the cambrian limestone guided the aquifer and the maximum water inrush quantity reached 38056 cubic meter per hour. Therefore the regularity study of fault water inrush in Yuzhou mine area is very important to ensure the safe production of the mine. Many scholars research the prediction of fault water inrush with mechanics theory, numerical simulation method and similarity simulation experiment et al. [4–8], but the results are not ideal in engineering practice. Based on the analysis of the hydrogeology features and the fault water inrush characteristics in Yuzhou mine area, the main faults' water inrush quantities are calculated with fluid mechanics theory, the risk of faults' water inrush is classified by their geological features and water inrush quantities.

2 THE ANALYSIS OF THE FAULT WATER INRUSH CHARACTERISTICS IN MINE

2.1 The water inrush quantity is large

In the three faults water inrush accidents, the minimum water inrush is the one happened in Baimiao mine, and its water inrush quantity is 1400 cubic meter per hour. The maximum water inrush is the one happened in the east channel of Pingyu first mine, and its water inrush quantity reaches 38056 cubic meter per hour. Also the water inrush quantities of all the three faults water inrush accidents are over 1000 cubic meter per hour.

2.2 The water inrush sources are from cambrian limestone aquifer

In three faults water inrush accidents, the the sources of water inrush are all from the cambrian limestone aquifer. Under the condition of mining, high fall fault can cut the cambrian limestone aquifer, mining fissures hold through fault's fissures, and lead the lower cambrian limestone water to mining face or tunnel. Furthermore, the thickness of the cambrian limestone is big in Yuzhou mine area, and the groudwater recharge source is sufficient, so the water inrush quantities of previous fault water inrush accidents are large.

2.3 *The fault water inrush types are complex*

The types of fault water inrush in coal seam floor can be divided into four ones, including passage type, concealed fault type, delayed type and fault water filling one. In Yuzhou mine area, the fault water inrush of west return airway in Pingyu first mine and Baimiao mine belong to the passage type. And the fault water inrush of east tunnel in Pingyu first mine is not only concealed fault type, but also delayed one. Therefore, the fault water inrush types in Yuzhou mine area shows complexity and diversity.

3 THE CALCULATING OF WATER INRUSH

3.1 *The calculation principle*

In fact, the process of water inrush in coal seam floor is a kind of seepage process when the groundwater flows out from a certain aquifer along the mining fractures. Obviously if the fault water inrush happens, there are three different stages of seepage process including the pore flow (or Darcy flow), fracture flow and pipe flow. So based on seepage mechanism, the initial value of fault water inrush quantity in Yuzhou mine area can be calculated. The calculating formula is

$$Q_0 = \frac{a^2 b^2 \gamma n^2 p_W}{8\pi\mu L}. \tag{1}$$

where, Q_0 is the initial value of fault water inrush quantity, its unit is cubic meter per hour (m³/h). And a or b is the length and width of fault water inrush channel respectively, which unit is meter. γ represents the bulk density of groundwater, its unit is Newton per cubic meter. μ is the coefficient of dynamic viscosity of groundwater. And n expresses the porosity of fault zone rock. L is the distance between the fault water inrush point and confined aquifer, its unit is meter. Lastly, p_W is the water pressure, and its unit is Million Pascals (MPa).

3.2 *Calculatin results*

To study the effects of faults on the coal seam floor in Yuzhou mine area, Dianchili normal fault, Liantang normal fault, Yungaishan normal and Xiabaiyu normal fault are chosed as main effect faults. The faults' water inrush quantities are calculated with the seepage formula introduced above considering all

five factors, such as the length and width of fault water inrush channel, the distance between the fault water inrush point and confined aquifer, the porosity of fault zone rock and the water pressure. The parameters are chosed from the real ones and calculating results are shown in Figures 1 and 2.

The effect laws of four faults on the coal seam floor in Yuzhou mine area are summarized as follows according to the calculating results. Firstly, under the coupling action of mining and aquifer water pressure, the bigger the area of fault water inrush channel (the length of fault water inrush channel multiplied by the width of fault water inrush channel), the larger the initial value of fault water inrush quantity will be. For example, the area of Xiabaiyu normal fault is the biggest, and its water inrush quantity is the largest. Secondly, the porosity of fault zone is an important effect factor on the water inrush of coal seam floor, because the bigger the porosity of fault is, the larger the initial value of fault water inrush quantity is. From the Figures 1 and 2, we can see that when the porosity of fault is 0.05, the fault water inrush quantity can reach over 4×10^3 m³/h. Thirdly, the initial value of

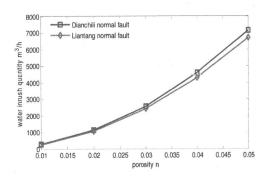

Figure 1. The relationship between fault water inrush and porosity in Pingyu first mine.

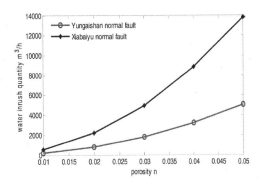

Figure 2. The relationship between fault water inrush and porosity in Baimiao mine.

fault water inrush quantity increases linearly with the value of aquifer water pressure. Such as the water pressure of Dianchili normal fault and Liantang normal fault is the biggest (about 4.5 MPa), and their calculating results are too large.

4 THE RISK CLASSIFICAITON OF FAULT WATER INRUSH IN RUZHOU MINE AREA

4.1 The calssification standards

The risk of fault water inrush is affect by natural factors and human factors. The former includes the fault features and the water pressure value of every aquifer. The research results show that the water inrush risk of the normal fault is greater than that of the pressure fault [10,11]. And if the fault cuts the strong aquifer such as the cambrian limestone aquifer, the water inrush quantity will be large. In addition, the water inrush risk of high water pressure is bigger than that of low water pressure in aquifer. The human factors are the effects of the mining methods, which can cause the increasing or decreasing of the fault water inrush quantity. According to the analysis of the fault water inrush characteristics and the calculating results of fault water inrush quantities in Yuzhou mine area, the risk of fault water inrush is classified by above factors.

The classification grades can be divided into four ones with four effect factors.

The classified conditions of the I grade risk is that the fault is normal fault, it cuts the cambrian limestone aquifer, the water pressure value is more than 2 MPa and the fault water inrush quantity is greater than 120 m³/h.

The classified conditions of the II grade risk is that the fault is normal fault, it cuts the cambrian limestone aquifer, the water pressure value is more than 2 MPa and the fault water inrush quantity is greater than 60 m³/h and less than 120 m³/h.

The classified conditions of the III grade risk is that the fault is normal fault, it cuts the cambrian limestone aquifer, the water pressure value is more than 1 MPa and less than 2 MPa, the fault water inrush quantity is greater than 5 m³/h and less than 60 m³/h.

The classified conditions of the IV grade risk is that the fault is pressure fault, it doesn't cut the cambrian limestone aquifer, the water pressure value is less than 1 MPa, and the fault water inrush quantity is less than 5 m³/h.

4.2 The classification results

On the basis of above classification standards, the water inrush risk classificaitons of main faults in

Yuzhou mine area are distinguished. The grades of all faults are the I grade. Especially, though Dianchicli fault doesn't cut the cambrian limestone aquifer, because its water inrush quantity reaches 286.1 m³/h and the water pressure is the biggest (4.5 MPa), its risk grade can be discriminanted as the I grade risk to ensure production safety. Therefore the water inrush risk grades of the main faults are all I grades. Other faults not introduced in the paper can be distinguished as the lower grades, such as the II grade, or the III grade, even the IV grade.

5 CONCLUSIONS

Based on the methods of theory analysis, seepage calculation and risk classification, the regularity of fault water inrush in Yuzhou mine area is studied, and the main conclusions are as follows.

1. The fault water inrush features are that the water inrush quantity is large, the sources of water inrush are from the cambrian limestone aquifer and the types of fault water inrush are complex.
2. The calculation results of the fault water inrush quantity show that the fault water inrush quantity increases linearly with the channel area of the fault water inrush, the porosity of fault zone and the value of aquifer water pressure.
3. The risk classification grades of fault water inrush can be divided into four ones with four effect factors, including the normal fault or not, cutting the cambrian limestone or not, the water inrush quantity and the value of aquifer water pressure. Therefore the water inrush risk grades of the main faults in Yuzhou mine area are all I grades. Other faults not introduced in the paper can be distinguished as the lower grades, such as the II grade, or the III grade, even the IV grade.

ACKNOWLEDGEMENTS

This work was supported in part by the Important Special Subject of National Technology under Grant 2011ZX05060-006 and the Doctoral Fund of Henan Polytechnic University under Grant B2011-019.

REFERENCES

[1] SHI Longqing, HAN Jin. Floor water inrush mechanism and forecasting [M]. Xuzhou: China University of Mining and Technology Press, 2004:35–70.
[2] WANG Enying. Lithologic structure analysis in coal-seam fault formation [J]. Journal of China Coal Society, 2005, 30 (3):319–321.

[3] HU Weiyue. Controlling theory and method of mining water disaster [M]. Beijing: China coal industry publishing house, 2005: 40–110.

[4] TAN Zhixiang. Preliminary Analysis of Mechanical property of Water Inrush from Faulits [J]. Mining safety and environmental protection, 1999,3:21–23.

[5] LI Qingfeng, WANG Weijun, ZHU Chuanqu, "Analysis of Fault Water-Inrush Mechanism Based on the Principle of Water-Resistant Key Strata," Journal of Mining & Safety Engineering, Vol. 26, No. 1, pp. 87–90, Mar. 2009.

[6] LI Lian-chong, TANG Chun-an, LIANG Zheng-zhao, et al., "Numerical analysis of pathway formation of groundwater inrush from faults in coal seam floor," *Journal of Rock Mechanics and Engineering*, Vol. 28, No. 2, pp. 290–297, Feb. 2009.

[7] BU Wan-kui, MAO Xian-biao, "Research on effect of fault dip on fault activation and water inrush of coal floor," Journal of Rock Mechanics and Engineering, Vol. 28, No. 2, pp. 386–394, Feb. 2009.

[8] LIU Shucai, LIU Xinming, JIANG Zhihai, et al. Research on electrical prediction for evaluating water conducting fracture zones in coal seam floor [J]. Journal of Rock Mechanics and Engineering, 2009, 28(2):348–356.

[9] WU Yuhua, ZHANG Wenquan, ZHAO Kaiquan, et al. Comprehensive prevention technology research of mine water disaster [M]. Xuzhou: China University of Mining and Technology Press, 2009:200–202.

[10] PENG Wen-qing, WANG Wei-jun, LI Qing-feng. Reasonable width of waterproof coal pillar under the condition of different fault dip angles [J]. Journal of Mining & Safety Engineering, 2009, 26 (2): 179–184.

[11] HUANG Cun-han, FENG Tao, WANG Wei-jun, et al. Research on the Failure Mechanism of Water-Resisting Floor Affected by Fault [J]. Journal of Mining and Safety Engineering, 2010, 2 (27): 219–222, 227.

Frontiers of Energy and Environmental Engineering – Sung, Kao & Chen (eds)
© 2013 Taylor & Francis Group, London, ISBN 978-0-415-66159-1

Study on hydrodynamics of the vertical-axis turbine with variable radius

X.H. Su, X.Y. Gao, H.Y. Zhang & Y. Li
School of Energy and Power Engineering, Dalian Univeristy of Technology, Dalian, China

J.T. Zhang
Institute of Water Resource & Hydropower, Dalian University of Technology, Dalian, China

ABSTRACT: In this paper, a novel conception of Variable Radius Vertical-Axis Turbine (VRVAT) has been proposed for the design of tidal current power generator turbine. Different from conventional Vertical-Axis Turbine (VAT), the radius of VRVAT varies along with azimuthal angle. The investigation on hydrodynamic performance of the proposed turbine has been made in details by using numerical program developed based on multiple streamtube theory. Firstly, mathematical formulas have been derived for the new type turbine. Then hydrodynamic coefficients, including lift, drag, torque and thrust have been analyzed and compared with conventional VAT. The numerical investigations release that the new type tidal power turbine really generates much more torque. For instance, the maximum value of Torque for VRVAT (oval trajectory with a = 1.6b when the tip speed ratio is 3.0) is nearly 2.6 times the value of conventional VAT. Although more work should be done to fulfill the method of variable radius, the bright future of VRVAT can be predicted.

Keywords: VRVAT; structure optimization design; tidal current; hydrodynamic performance

1 INTRODUCTION

Since electrical power generation is highly concerned in recent years, tidal current power which is clean and renewable is getting interests (Fraenkl.P). There are many researches working on numerical simulation and optimum theory all around the world (Kopeika.O.V&Tereshchenko.A.V). Currently, most researchers focus on the airfoil optimization, AOA control in order to improve efficiency, but rare attention is paid to the improvement of the structure of the turbine. In this paper, a new structure design method is put forward. Figure 1 shows torque distribution along with the change of azimuthal angle. The results are reached for single blade conventional VAT with the help of the computing software based on multiple stream model. It can be figured out that the torque reaches positive and maximum at point A, and negative and minimum at point B. Suppose the radius of blade around point A is lengthened, while at point B shortened, then the torque for the full cycle will clearly be improved, and then power coefficient will be enhanced.

In this study, a novel tidal current turbine has been proposed and the common circle blade orbit is replaced by oval orbit in order to demonstrate the new idea, actually the blade trajectory could also be other shapes. In the following sections,

Figure 1. Torque varies with azimuthal angle for single NACA 0018 blade VAT at the tip speed ratio of 1.73.

investigations on its performance of the new turbine will be done.

2 MATHEMATIC FORMULA

2.1 Derivation of induced velocity using in the hydrodynamic model

The derivation process of hydrodynamics model for VRVAT is based on conventional multiple stream model (Strickland.J.H) with proper modifications for describing the situation of various arm length. For the integrity of hydrodynamics mode, the detail derivation has been written in this section.

In the model, there is one disk as shown in Figure 1. At current stage only one straight blade is considered in the turbine and no installation angle is concerned in this thesis for the comparing purpose.

The stream-wise component of the induced flow at this disk is given by $-uV_\infty$, where u is an induction factor. Thus, the velocity of the water passing through disk is

$$V = (1 - u)V_\infty \qquad (1)$$

Applying momentum equation in the stream-tube, the force exerting on the stream disk can be expressed:

$$(p^+ - p^-)A_d = (V_\infty - V_e)\rho A_d V = (V_\infty - V_e)\rho A_d(1 - u)V_\infty \qquad (2)$$

Thus, the force on the disk in unit time can be given by

$$F_{disk} = (p^+ - p^-)A_d = 2\rho A_d V_\infty^2 u(1 - u) \qquad (3)$$

2.2 Calculation of the angle of attack

As shown in Figure 2, in the orbit of oval, the trajectory could be described as below, where θ is the azimuthal angle:

$$R = [a^2b^2(1 + \tan^2\theta)/(a^2\tan^2\theta + b^2)]^{1/2} \qquad (4)$$

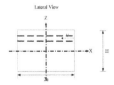

Figure 2(a). Definition of rotor geometry for the turbine.

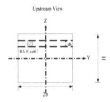

Figure 2(b). Definition of rotor geometry for the turbine.

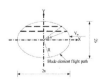

Figure 2(c). Definition of rotor geometry for the turbine.

Define β as the angle composed by vectors $\overline{V}$ and $\omega\overline{R}$. As shown in Figure 3, $\cos\beta = \sin\theta$ ($-\pi/2 \leq \theta \leq 3\pi/2$). Using cosine theorem in the triangle created by the velocity vectors $\overline{V}, \omega\overline{R}$ and $\overline{W}$, the following expressions can be given:

$$\cos\beta = \sin\theta = [V^2 + (\omega R)^2 - W^2]/2V\omega R \qquad (5)$$

Let X denote the local tip speed ratio $X = \omega R/V$, Eq. (5) can be rewritten:

$$W^2 = V^2[\cos^2\theta + (X-\sin\theta)^2] \qquad (6)$$

Using sine theorem in the triangle created by the velocity vectors $\overline{V}, \omega\overline{R}$ and $\overline{W}$, the below expressions can be obtained:

$$\cos\alpha = (X-\sin\theta)/[\cos^2\theta + (X-\sin\theta)^2]^{1/2} \qquad (7)$$

2.3 Derivation of hydrodynamic model

If C_N is defined as the normal force coefficient and C_T as tangential force coefficient, then C_N and C_T could be described as follow,

$$C_N = C_L\cos\alpha + C_D\sin\alpha \qquad (8)$$

$$C_T = C_L\sin\alpha - C_D\cos\alpha \qquad (9)$$

where C_L is lift coefficient and C_D is drag coefficient, which is obtained from Ref (Sandia Laboratory Report).

Thus, the force on the disk in unit time can be given by

$$F_{disk} = 0.5\rho c(C_N\cos\theta - C_T\sin\theta)W^2 \qquad (10)$$

where c is the foil chord.

Based on Eq. (3), a new equation can be obtained:

$$2\rho A_d V_\infty^2 u(1 - u) = 0.5\rho c(C_N\cos\theta - C_T\sin\theta)W^2 \qquad (11)$$

where, $A_d = R\Delta\theta |\cos\theta|$. Substituting A_d into Eq. (11),

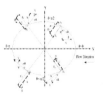

Figure 3. The sketch of force and velocity vectors in the hydrodynamic model.

$$\Delta\theta V_\infty^2 u(1-u)\backslash V^2 = \Delta\theta u/(1-u) = 0.25$$

$$\frac{c}{R}\left(C_N\cos\theta - C_T\sin\theta\right)\frac{W}{(V)^2}/|\cos\theta| \tag{12}$$

Define F as a non-dimensional force coefficient,

$$F = \int_{-\pi/2}^{3\pi/2}\frac{u}{1-u}d\theta = 2\pi u/(1-u)$$

$$F(1-u) = 2\pi u \tag{13}$$

The probability for the blade to appear in the position of $\Delta\theta$ is $\Delta\theta/2\pi$, so the non-dimensional force coefficient on the disk in time that the rotor sweeps the angle $\Delta\theta$ is

$$f = 0.25\frac{c}{R}\left(C_N\cos\theta|\cos\theta| - C_T\sin\theta/|\cos\theta|\right)$$

$$\frac{W}{(V)^2}\frac{\Delta\theta}{2\pi} \tag{14}$$

$$F = \frac{c}{8\pi}\int_{-\pi/2}^{3\pi/2}\frac{1}{R}\left(C_N\cos\theta - C_T\sin\theta\right)$$

$$\frac{W}{(V)^2}/|\cos\theta|d\theta \tag{15}$$

Defining the blade Reynolds number as Re_b, for local conditions, Re_b is given by the expression:

$$Re_b = Wc/V_\infty \tag{16}$$

Considering the time term $\Delta\theta/2\pi$ appearing in Eq. (14), the expression can be given:

$$C_p = 0.5\rho cH\omega\frac{1}{2\pi}\int_{-\pi/2}^{3\pi/2}C_T RW^2 d\theta/\left(0.5\rho SV_\infty^3\right)$$

$$= \int_{-\pi/2}^{3\pi/2}\omega R\backslash V_\infty)\frac{cH}{2\pi S}\left(W/V_\infty\right)^2 C_T d\theta \tag{17}$$

$$C_p = \int_{-\pi/2}^{3\pi/2}X_\infty C_Q d\theta \tag{18}$$

where $C_Q = cH/2\pi S\left(W/V_\infty\right)^2 C_T$ is non-dimensional torque coefficient and X_∞ is the local tip speed ratio when the incoming speed is V_∞.

For a given rotor geometry and a given rotational speed ω and a local velocity, a value of an induced tip speed ratio is chosen by assuming that the induction factor u is unity.

3 NUMERICAL INVESTIGATION

Based on the above mathematic formulas, a real-time computing software for simulating and analyzing the hydrodynamics of the turbine is developed. With the help of the developed software, a series of theoretical statistics for the oval orbit is calculated and compared with the traditional round orbit. The simulated VRVAT states that there is only one blade rounding with oval orbit of which the minor radius is fixed as 0.6 m and major radius is various from 0.6 m to 0.96 m, one time to 1.6 times of minor radius, respectively. The height of the VRVAT is 0.66 m, and rotating speed is 55 rpm, incoming velocity differs according to the tip speed ratio. The parameters concerned in the three cases are shown in Table 1. Since there is only one blade thought in current VRVAT, induced velocity factor u is considered as one in the hydrodynamic model derived above, which makes the theory in this paper same as BEM theory.

3.1 Resultant velocity and AOA for VRVAT

Figures 4(1)–(2) show AOA and resultant velocity profiles along with azimuthal angle. Figure 4(1) releases that the absolute values of AOA decrease and values of W increase with the increase of a/b. The max AOA values in cases a = 1.6b, a = 1.4b, a = 1.2b are 27.5°, 28.9°, 31.2°, which are 0.78, 0.82, 0.88 times of 35.3° in case a = b, respectively. At the same time, the max values of W are bigger than that in case a = b, respectively. The azimuthal angle where the max AOA value is generated is smaller and smaller with the increase of a/b. The significant change of AOA happens as azimuthal angle is from 0°–60°, 120°–220° as well as 320°–360° while the profiles of AOA are almost same in other azimuthal angle. In the other two cases, the similar profiles can be achieved. Moreover, the max values of W and AOA decrease with the increase of $\omega b/V_\infty$ and the increase of rotating speed.

Table 1. the descriptions of parameters in cases for single blade (NACA0018) VRVAT.

Case	$\omega b/V_\infty$	a/b	b(m)	Rotating speed(rpm)	Incoming velocity(m/s)
11	1.73	1	0.6	55	2
12	1.73	1.2	0.6	55	2
13	1.73	1.4	0.6	55	2
14	1.73	1.6	0.6	55	2
21	3	1	0.6	95.5	2
22	3	1.2	0.6	95.5	2
23	3	1.4	0.6	95.5	2
24	3	1.6	0.6	95.5	2

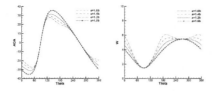

Figure 4(1). Comparisons of AOA and resultant velocity along with azimuthal angle for Cases 11–14. (solid line with square: a = b; solid line: a = 1.2b; dashed line: a = 1.4b and dashdot line: a = 1.6b).

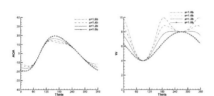

Figure 4(2). Comparisons of AOA and resultant velocity along with azimuthal angle for Cases 21–24. (solid line with square: a = b; solid line: a = 1.2b; dashed line: a = 1.4b and dashdot line: a = 1.6b).

3.2 FT and FN

Figures 5(1)–(2) show the profiles of F_T and F_N along with azimuthal angle. F_T is force in the tangent direction while F_N is the force in radius direction (See in Fig. 3). These two forces mentioned here are originally obtained from lift force and drag force through projection method. Figure 5(2) shows that the profile of F_T changes significantly in case 34. The region of positive F_T becomes larger at several theta segments, 0°–60°, 120°–220° as well as 320°–360°. And the profile of F_T at other theta angle left does not change too much. Moreover, the theta value where maximum F_T happened shifts back, from 210° in case a = b to 182°. It is believed that both higher F_T maximum value and larger positive F_T region will play an important role in increasing effective torque and show a great enhancement of the performance of the turbine. The conception of various radius is designed by increasing R as there exists positive F_T. Since torque is obtained by F_T *R, the value of torque then increase accordingly. So an unexpected improvement is achieved in the conception of various radius. However, one attention has to be paid that although the various radius concept indeed generates larger F_T, the force F_N will increase at the same time. Those F_N will be an important factor during the mechanical design and control design. In the other two cases, the similar conclusions can be achieved.

3.3 Torque and thrust

The profiles of torque and thrust coefficient along with theta have been plotted in Figures 6(1)–(2). Torque is indeed improved dramatically. Figure 6(2) releases that the max values of torque in cases a = 1.6b, a = 1.4b, a = 1.2b are bigger than that

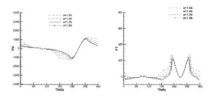

Figure 5(1). Comparisons of F_T and F_N along with azimuthal angle for Cases 11–14. (solid line with square: a = b; solid line: a = 1.2b; dashed line: a = 1.4b and dashdot line: a = 1.6b).

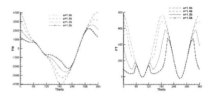

Figure 5(2). Comparisons of F_T and F_N along with azimuthal angle for Cases 21–24. (a. F_N; b. F_T. solid line with square: a = b; solid line: a = 1.2b, dashed line: a = 1.4b and dashdot line: a = 1.6b).

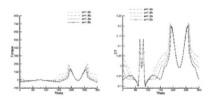

Figure 6(1). Comparisons of torque and thrust coefficient along with azimuthal angle for Cases 11–14. (solid line with square: a = b; solid line: a = 1.2b; dashed line: a = 1.4b and dashdot line: a = 1.6b).

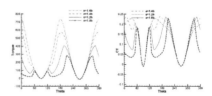

Figure 6(2). Comparisons of torque and thrust coefficient along with azimuthal angle for Cases 21–24 (solid line with square: a = b; solid line: a = 1.2b; dashed line: a = 1.4b and dashdot line: a = 1.6b).

in case a = b, respectively, which indicates a great success of the novel concept. Even though the maximum and minimum values of C_T are nearly constant, the figure still shows a great profile improvement. In the other two cases, the similar conclusions can be achieved, while the change tendencies are much greater, which shows a better enhancement.

4 CONCLUSION

In this paper, a novel conception of VRVAT has been proposed for the design of tidal current power generator turbine. Conventional multiple stream-tube theory has been revised in order to investigate hydrodynamic performance of the proposed turbine. In fact, the modified multiple streamtube theory has the more general characteristics of VAT because the equal-radius VAT is only one typical instance VRVAT at a = b. Numerical program is developed based on modified multiple streamtube theory. To clearly explain hydrodynamic performance, a VRVAT with only one NACA0018 blade is investigated in this paper. Numerical investigations release that the proposed novel conception of VRVAT can improve the performance of the turbine dramatically. Although more work should be done to fulfill the method of variable radius, the bright future of VRVAT can be predicted.

REFERENCES

Aerodynamic Characteristics of Seven Symmetrical Airfoil Sections Through 180-Degree Angle of Attack for Use in Aerodynamic Analysis of Vertical Axis Wind Turbines, Sandia Laboratory Report SAND 80-2114.

Fraenkl.P, Tial Current Energy Technologies. Marine Current Turbines Ltd, Ibis, 148(sl): 145–151.

Kopeika.O.V&Tereshchenko.A.V, Wind power transforming systems. Journal of Mathematical Sciences, pp. 1631–1634.

Strickland.J.H, The Darrieus Turbine: A Performance Prediction Model Using Multiple St reamtubes, Sandia Laboratory Report SAND 75-0431.

Frontiers of Energy and Environmental Engineering – Sung, Kao & Chen (eds)
© 2013 Taylor & Francis Group, London, ISBN 978-0-415-66159-1

Design of hybrid energy storage in solar LED streetlight

X.J. Yan, G.H. Li & Y. Geng
Electronic Engineering Department, North China Institute of Aerospace Engineering, China

ABSTRACT: The shortage of battery for solar LED streetlight and advantage of ultra capacitor were analyzed in this paper, the hybrid system based on ultra capacitor/battery was put forward. The matching with ultra capacitor and battery via DC/DC convertor could been able to collectted the energy effectually, and protected the battery fall under the striking of high current, and could still be storaged little energy which emerged by PV on cloudy and rainy days. Experiments and simulation shown that the hybrid storage module could be optimized storage battery charging and discharging process, and improved the shortage of battery.

Keywords: ultra capacitor; storage battery; DC/DC convertor

1 INTRODUCTION

Solar LED streetlight is a strong proponent of green energy products in today's society[1]. But the solar LED lights must focus on solving the scientific and technical issues: electrical energy storage efficiency and reliable power supply problem. Involved in key aspects of this problem is: in the case of certain photoelectric conversion rate, how to sufficient store the electricity energy, and increasing the stored energyto meet the needs of LED lighting in different weather conditions, and significant enhance the LED lighting time and improve battery life.

In response to these scientific issues, we explored ultracapacitor/battery hybrid energy storage for solar LED lights, to take full advantage of ultra capacitor solve the shortcomings of traditional solar LED lights, as low current uncharging, charging and discharging efficiency lower and battery life short, and provide reliable support for the application of solar LED lights.

2 ULTRA CAPACITOR AND BATTERY PERFORMANCE COMPARISON

2.1 The lack of battery

The biggest advantage of batteries relative ultra capacitor is high energy density. However, the biggest shortcomings of battery are low temperature properties and high-rate or deep discharge will greatly shorten the life[2]. During discharge process, battery stored chemical energy into usable electrical energy. Seen from Figure 1, the terminal voltage of battery decreases with increasing discharge time and discharge rate. High current charging,

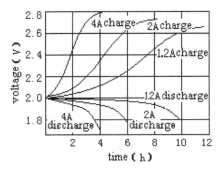

Figure 1. A battery charge-discharge curves at different current.

the voltage rises rapidly, and ultimately achieve a higher voltage. From the battery principle, still can not fundamentally change the battery life, the best solution to improve the battery life is reduce the rate of discharge current.

Another weakness of battery is low temperature properties. In low temperature environment, there will be appeared serious problems as can not be started or underpowered at first, this is also determined by the battery itself. As shown in Figure 2, the lower electrolyte temperature lead to the average voltage lower, while the charging voltage higher in the discharge. Conversely, the higher temperature of electrolyte, the discharge average voltage higher and the charging voltage lower.

2.2 Advantage of ultra capacitor

Ultra capacitor is a new charge storage element, which realized the charge and discharge process

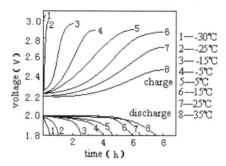

Figure 2. Temperature and charge-discharge curve relationship.

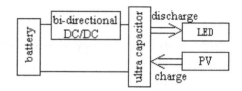

Figure 3. System block diagram.

based on the principle of capacitor energy storage. Charge-discharge process of ultra capacitor does not exist electrochemical reaction, and thus does not exist problems based on the electrochemical principle of energy storage. Ultra capacitor can charge and discharge of high-rate current and have good low temperature properties[3].

Ultra capacitor in –40 °C, the electrical properties is basically the same as room temperature, ensured the excellent performance of ultra capacitor in low temperature state. The energy density of ultra capacitor can reach 6 Wh/kg, the energy density of lithium batteries can be achieved 80 Wh/kg, so the energy density of ultra capacitor was significantly lower than the battery. In need of continuous power, the ultra capacitor appeared to be inadequate[4]. In the normal conditions, the life of ultra capacitor more than 10 years, fully charge and discharge cycle life more than 100 000 times.

2.3 Combination of ultra capacitor and battery complementary advantages

Through the above analysis, if ultra capacitor and battery combination to form composite power system, can improve the store energy of power system and short time high power output capability, but also have long-lasting power supply performance. Through the combination of ultra capacitors and batteries, can give full play to the advantage of ultra capacitor and batteries, supply continuous power and short-term high-power to meet the solar LED lights requirements, also be collected energy in lower current and rapid charge and discharge.

3 ULTRA CAPACITOR-BATTERY HYBRID ENERGY STORAGE MODULE

Parallel ultra capacitors and batteries can generally be summarized as three: direct parallel, through the inductor in parallel, and through power converter in parallel[5]. The first two ways discussed in detail

in many literature, this is not in the repeat. In this design ultra capacitors and batteries used power converters (DC/DC) in parallel. The terminal voltage of battery and ultra capacitor banks can be different as introduction of power converter, and thus have greater flexibility in the design. According to the actual situation of the system, the power converter designed for the buck-boost to match the voltage of battery and ultra capacitor banks.

The structure of hybrid energy storage system shown in Figure 3. This system can bring more advantages. First of all, system can achieve short-term high-current energy efficient collection, to avoid the impact of battery by high-current and appear energy harvesting incomplete as difficult to accept short-term high charging current. Second, you can achieve the efficient collection of small current. Rainy days, photovoltaic cells can only output a lower voltage and current, the traditional battery charger can not charging when voltage is lower, or current is too small. Ultra capacitor makes the hybird system will still be able to storage energy on rainy days. Finally, when the battery's energy is not enough, the ultra capacitor energy can also be discharged to the load, especially in low temperature environment, the battery charge/discharge capacity will be greatly weakened, while the temperature features of ultra capacitor far superior to the battery.

4 DC/DC CONVERTER DESIGN

4.1 The main control loop design

Ultra capacitor voltage may be higher or lower than battery voltage, so DC/DC converter is a buck-boost converter, the main circuit topology shown in Figure 4. DC/DC converter master chip used switching regulator controller LTC3780, which seamlessly switch buck-boost switch mode, high efficiency, output voltage automatically and stabilize.

4.2 Switch control mode

In order to achieve the design requirements, the DC/DC conveter working on three modes:

1. Buck mode ($V_{IN} > V_O$). In this mode, switch Q_4 is always turned on and switch Q_2 is always turned off. The starting point in each clock cycle, controlling

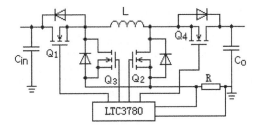

Figure 4. Buck-Boost DC/DC converter based on LTC3780.

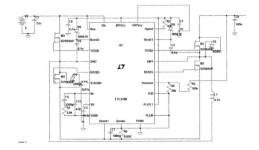

Figure 5. Simulation circuit of DC/DC converter.

switch Q_3 turns on, and at the same time to detect the inductor current. Switch Q_3 shutdown, and switch Q_1 turns on in the remaining time of the cycle when the inductor voltage lower than the reference voltage. Q_1 and Q_3 as synchronous buck regulator as alternately conduction.

2. Buck-boost mode ($V_{IN} \approx V_O$). When V_{IN} is close to V_O, the controller is in Buck-Boost mode. Every cycle, if the controller starts with Switches Q_3 and Q_4 turned on, Switches Q_1 and Q_2 are then turned on. Finally, Switches Q_1 and Q_4 are turned on for the remainder of the time. If the controller starts with Switches Q_1 and Q_2 turned on, Switches Q_3 and Q_4 are then turned on. Finally, Switches Q_1 and Q_4 are turned on for the remainder of the time.

3. Boost mode ($V_{IN} < V_O$) Switch Q_1 is always on and synchronous switch B is always off in boost mode. Every cycle, switch Q_2 is turned on first. Inductor current is sensed when synchronous switch Q_2 is turned on. After the sensed inductor current exceeds the reference voltage, switch Q_2 is turned off and synchronous switch Q_4 is turned on for the remainder of the cycle. Switches Q_2 and Q_4 will alternate, ehaving like a typical synchronous boost regulator.

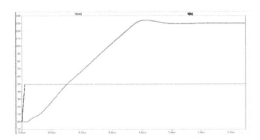

Figure 6. Output simulation results when VIN = 5 V.

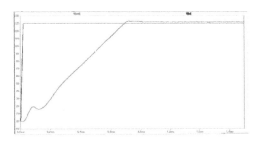

Figure 7. Output simulation results when VIN = 13 V.

5 SIMULATION AND EXPERIMENT

5.1 Simulation and analysis of DC/DC converter

Design parameters: 5–28 V input voltage and output voltage of 13.2 V, output current of 5 A, the switching frequency of 400 kHz. The converter simulation circuit shown in Figure 5. The simulation results are shown in Figures 6–8. According to the simulation results whereous input voltage above, below or close to the output voltage, DC/DC converter can output a stable voltage of 13.2 V.

5.2 Experiments and analysis of hybrid energy storage

The authors designed a buck-boost DC/DC converter based on LTC3780, and anti-parallel two

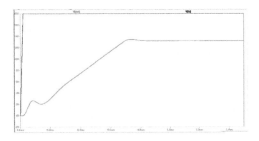

Figure 8. Output simulation results when VIN = 18 V.

DC/DC converter constitute a bi-directional DC/DC converter for connecting ultracapacitor and battery. Hybrid energy storage include eight 2.7 V/1000 F ultra capacitor in series, 12 V/150 Ah battery and 50 W-bright white LED for load.

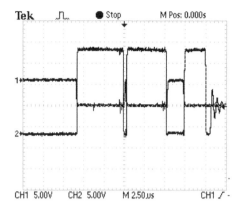

Figure 9. Q1, Q3 gate drive waveform when VIN = 15.36 V.

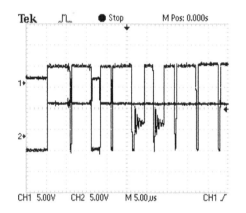

Figure 10. Q1, Q3 gate drive waveform when VIN = 24 V.

DC/DC converter input voltage V_{IN} = 5~28 V, output voltage V_O = 0~30 V, output current I_O = 0~8 A, efficiency up to 96%, and standby power is 0.2 W to 1 W at no load. When battery voltage is below 11.1 V, battery stop discharge. When ultra capacitor voltage higher than 13.2 V, battery started charging.

When input voltage is 15.36V and full load, battery voltage across the Vo = 13.204 V, switch Q_1 and Q_3 gate drive waveform shown in Figure 9 curves 1 and 2. When input voltage is 24 V, battery voltage across the Vo = 13.206 V, switch Q_1 and Q_3 gate drive waveform shown in Figure 10 curves 1 and 2.

By the perfect protection function of LTC3780, system can upgrade the application performance of battery more reliable and secure. Ultracapacitor could instantly absorb and instantaneous output current, those characteristics could extend battery life, and significant improvement the supply capacity of battery, and allows the user to get a better effect than direct use of battery.

6 SUMMARY

Hybrid energy storage system is able to collect tiny electric power to ultra capacitors and batteries. In addition to the extreme case, battery is always in a mild discharge. The other, charge and discharge current of hybrid energy storage system is stable, make battery charge and discharge curve in ideal. The system extended the life of battery and considerable economic benefits.

ACKNOWLEDGEMENTS

This work was financially supported by the 2009 Hebei Province Department of Education Science and Technology Research Project (ID: 2009401) and Langfang City Technology Support Program project (No: 2009113501-2).

REFERENCES

[1] WANG Bin, SHI Zhang-rong, ZHU Tuo. 2007. Design of controller with ultracapacitor-battery hybrid energy on a stand-alone PV system. Energy Engineering. 5:37–41.
[2] Gao L, Dougal RA, Liu S. 2005. Power enhancement of an actively controlled battery/ultracapacitor hybrid [J]. IEEE Trans on Power Electronics. 20(1): 236–243.
[3] CHEN Yong-zhen. 2005. Capacitor and its applications. Beijing: Science Press.
[4] DOUGAL RA, LIU Sheng-yi, WHITE RE. Power and life extension of battery-ultracapacitor hybrids [J]. IEEE Trans on Components and Packaging Technologies, 2002, 25(1): 120–131.
[5] TANG Xi-sheng, QI Zhi-ping. 2006. Study on an actively controlled battery/ultracapacitor hybrid in stand-alone PV system. Advanced Technology of Electrical Engineering and Energy. 25(3):37–41.

Frontiers of Energy and Environmental Engineering – Sung, Kao & Chen (eds)
© 2013 Taylor & Francis Group, London, ISBN 978-0-415-66159-1

Fundamental model of hazard prevention system with remote monitoring and health diagnosis for aged dam structure

H.C. Lin
Department of Health Risk Management, China Medical University, Taichung, Taiwan

Y.M. Hong
Department of Design for Sustainable Environment, MingDao University, Changhua, Taiwan

W.P. Sung
Department of Landscape Architecture, National Chin-Yi University of Technology, Taichung, Taiwan

Y.C. Kan
Department of Communications Engineering, Yuan Ze University, Taoyuan, Taiwan

ABSTRACT: Climate change frequently caused severe natural disasters in the last decade in Taiwan. Risk assessment of structure safety for the aged dam becomes a quite important topic to avoid hazard. This study establishes a prototype of hazard prevention system with remote monitoring and health diagnosis for aged dam structure by creating the risk criteria upon failure simulation module associated with measurement data for evaluating dam body behaviors. Development of system prototype employs the open-source program to simulate the failure modes of a practical dam due to critical geotechnical and geometrical conditions. The engineering data warehouse required to support the expert knowledge bank at system backbone is built up to collect monitoring data relative to the dam structure and circumambient information. Risk indexes feedback by expert opinions are further formed as assessment criteria for the hazard prevention model. As the result, the model provides web-based interface with data integration and management for hazard alert as well as founds the base of the remote monitoring and risk assessment center.

Keywords: risk assessment; failure simulation; remote monitoring; hazard prevention

1   INTRODUCTION

Nature hazards attacked construction safety and people life more frequency in the past years due to unexpected climate change. Risk assessment of aged dam structure safety hence becomes important beyond the prevention of hazards. In Taiwan, Sun Moon Lake is one of the major reservoir lakes located in the mid-western mountain area. Several aged dams have been located around the lake for over 70 years. It is helpful and logical for engineers to efficiently evaluate and plan the maintenance procedure if there are online criteria of risk indexes to be followed. The risk indexes relative to health diagnosis of aged dam structure or real time prediction of the web-based hazard alert system can be established by integrating the techniques of machine learning and data mining.

Internet data transportation and computation should be considered for system requirement of the remote monitoring platform to satisfy the load balance of resources [1]. Therefore the database management system for heavy native data and the unified schema for light decision data can be functionally allocated in the database and web servers for daemon process and online analysis, respectively. The concept above is workable for developing an online monitoring system of structure safety. That is, the monitored field data can be delivered to the laboratory through internet for planning engineering data warehouse and employing analysis modules to determine risk indexes of structure safety behind feeding back the decision support criterion to the expert system [2,3]. Meanwhile, the instant data transaction interface required by online monitoring can be functioned with Online Analytical Process (OLAP) based on Web technology [4]. Furthermore, the remote backup schedule can be practiced by connecting database server according to user privilege administration for data backup and restore [5].

Hazard prevention is the best solution for the remedy of damage. Computer assistant evaluation with failure simulation can provide a risk assessment model on health diagnosis and collapse analysis of the large structure. Based on the energy conservation principle, the method of Discontinuous Deformation Analysis (DDA) with Numerical Manifold Method (NMM) can calculate block kinematics and elastic deformation of separated blocks in a system due to arbitrary polygon element [6]. It enables evaluation of structure stability analysis for practical problem. Hence the analytical results can be reference for data mining in the engineering data warehouse to found the health diagnosis system. In order to integrate multidisciplinary modules within web-based system for hazard prevention, the well known Model-View-Controller (MVC) design pattern [7] allows independent components for support one another as well as to enhances system efficiency. For online risk assessment requirement, the presentation interface of platform should involve uniform web services schema and scheme parser module beyond the prediction model [9]. It implies the decision support functionality that retrieves instantaneous monitoring data by matching the risk criteria after discovering historical database. Thus, the risk map can be designed to demonstrate the potential hazard area for making alert and decision.

In this study, we explored the potential risk factors within engineering data warehouse to collect historical and monitoring data for the Shui-She Dam in Sun Moon Lake area. A prototype of expert model for risk assessment of dam was finally created as the monitoring and diagnosis system for hazard prevention.

2 SYSTEM DEVELOPMENT

The proposed remote monitoring and diagnose system is based on a client-server architecture to construct the framework of engineering data warehouse for acquiring measurement data. A failure simulation module accompanies with environmental parameters relative to structure safety and carries out the risk indexes for yielding assessment criteria of expert knowledge bank.

2.1 Design of system architecture

The hazard prevention network of the model for collaborating distributed digital surveillance data is illustrated in Figure 1. The model is charged with three major challenges. (1) The monitoring station at the aged dam offers a data acquisition server to collect and classify diverse measured data that can be automatically evaluated by risk assessment

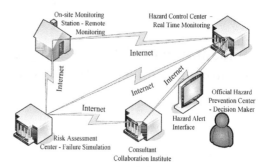

Figure 1. Hazard prevention model upon client-server architecture and surveillance environment.

and decision support modules due to criteria. All data are delivered to backend server for advanced analysis beyond feedback to expert knowledge bank. (2) With remote measurement data control and backup, it involves the analytical modules for risk estimation as well as hazard prediction due to instantaneous records such as crack, groundwater level, rainfall, and so on. (3) The failure simulation module is applied for simulation at the backend of system by combing analytical model and practical data stored in the database. The failure criteria can be expected after computation under critical condition to feedback necessary risk indexes for the threshold of assessment criteria.

2.2 Model of risk assessment

Data computation procedure is designed in four layers for the risk assessment model as shown in Figure 2. They are described below to support automatic functionality of data input, control, allotment, computation, output and presentation. (1) Data integration—provides heterogeneous integration for diverse measurement and analysis data under user privilege administration for generating data warehouse. In which, the engineering data may primarily require deformation of dam structure, landslide of hillslope, damage of machine, etc. for efficient analysis. (2) Data analysis—imports and exports engineering data converted in uniform format for advanced calculation and simulation based on potential risk factors. Herein, the efficient analysis tools such as DDA + NMM are employed to feedback the expert knowledge bank and define risk indexes for the assessment criteria. (3) Risk assessment—formulates risk assessment criteria for possible hazard information to establish the risk pattern library. Thresholds of potential risk factors are hence formulated for critical deformation, dam crack, machine failure, groundwater flow channel, hillslope collapse, structure stability, and so on. (4) Risk information presentation—present

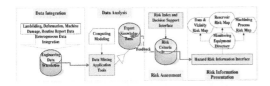

Figure 2. Four-layer architecture of system data integration.

hazard risk information interface and create the risk map to visually demonstrate the relationship between risk indexes and environmental hazards. The interface incorporates risk information with online statistical diagram upon the risk map to support necessary hazard prediction and alert.

The designed layers above considers various requirements for acquiring, computing, processing, and importing native data as well as exporting risk indexes behind practicing automatic control flow.

2.3 Scheme of data warehouse

The proposed engineering data warehouse for monitored data related to dam structure, machine, and hydraulic computation within the reservoir area. The schemas are created as follows. (a) Inspection overview: records deployment of monitoring equipments and measurement items such as groundwater level wells, inclinometers, settlement gauges, piezometers. (b) Measurement data: trace monitored water or groundwater levels, displacement, rainfall, etc. (c) Possible max precipitation: estimate possible maximum precipitation data during rainfall period. (d) Rainfall history: record daily rainfall information within the reservoir area. (e) Safety Evaluation—store risk indexes with respect to failure criteria of the dam. (f) Failure mode and effects analysis: estimate potential failure modes of the dam for damage management and hazard alert.

The scheme above defines the fields of tables within the data warehouse. The XML-based schema is employed to parse Web data transmission files with convertibility and expandability for developing the system interface of Risk Assessment Of Reservoir Safety (RAORS).

3 RESULTS AND DISCUSSIONS

3.1 Risk assessment for structure failure

A macro concept upon large scale analysis is applied for the risk assessment criteria of dam structure failure modes by simulating interaction of discontinuous blocks due to the possible joint distribution within the system. Theoretically, various output sets of structure deformation could result from the corresponding loading and joint conditions in many combinations. However, practically, the simulation is experienced that the geotechnical coefficients and geometrical properties are the primary factors to lead failure modes under critical loads. Therefore, we consider the extreme water pressure and uplift forces in potential for evaluating possible structural failures under critical conditions with tremendous parameters. According to simulation, four failure modes can be estimated by geometrical and geotechnical properties: (A) high water level flow lines, (B) low water level flow lines, (C) both of high and low flow lines, and (D) high, medium, and low water level flow lines; in which, a flow line means one possible joint to separate blocks under the critical condition. Therefore, a variety of assessment combinations including block stiffness (soft, normal, hard), motion status, and deformation can lead the three-level risk index of criteria as shown in Table 1.

3.2 Risk management of hazard prevention

The web-based risk management interface assists the departments responsible for supervision and hazard prevention. The monitoring data are collaborated with the risk assessment criteria above to provide information required for routine maintenance and damage control while the data flow is illustrated in Figure 3. The developed interface provides management of risk assessment criteria and diagram-based statistics of monitoring data.

Table 1. Failure criteria and risk assessment upon DDA + NMM simulation for Shui-She Dam.

Type	Stiff.	Deform.	Status	Risk
A	Soft	Large	Static	High
	Hard	None	Dynamic	Low
	Normal	None	Static	Low
	Normal	Large	Dynamic	High
B	Soft	Large	Static	High
	Hard	None	Dynamic	Low
	Normal	None	Static	Low
	Normal	Moderate	Dynamic	High
C	Soft	Large	Static	High
	Hard	None	Static	Low
	Hard	Small	Dynamic	Med.
	Normal	None	Static	Low
	Normal	Moderate	Dynamic	High
D	Soft	Large	Static	High
	Hard	None	Static	Low
	Hard	Small	Dynamic	Med.
	Normal	None	Static	Low
	Normal	Large	Dynamic	High

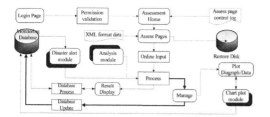

Figure 3. Data transportation flowchart of web-based monitoring and hazard prevention system.

Herein, convertible XML-based files are designed for web data transportation in the system. The data parsing module is available to flexibly retrieve and backup records within the database. In this prototype, essential models of data coalescence are established for online and remote data input, monitoring image library, routine report upload, and data maintenance while privilege administration is required for data management. Furthermore, visualized statistical diagrams interface the user and enormous monitoring data with the real time online analytical mechanism. It enables graphical information presentation to support risk assessment graphically. As soon as the thread of risk criteria is reached for some critical status, the supervisor can immediately activate the emergent procedure for hazard prevention for early warning.

4 CONCLUSION REMARKS

This study developed a prototype of web-based aged dam management and expert system with remote monitoring and health diagnosis functionality for hazard prevention. The Shui-She Dam at Sun Moon Lake area in Taiwan is exampled by practically measurement records with respect to the dam and circumambient data for evaluating the system development. The risk assessment model is based on the failure criteria through the four-layer data computation procedure. The potential hazard factors are resulted by numerical simulation and compared with the scheme of engineering data warehouse. The flexible internet transportable schema and friendly diagram-based interface can enhance integration of monitoring data as well as support information management and decision making.

ACKNOWLEDGEMENT

The authors would like to appreciate the research support from National Science Council of the Republic of China with the Grant numbered NSC 100-2625-M-039-001 and 101-2625-M-039-001, and Sinotech Engineering Consultants, Inc., Taiwan.

REFERENCES

[1] Ni, L.M. and Hwang, K., "Optimal Load Balancing in a Multiple Processor System with Many Job Classes," Software Engineering, Vol. SE-11, Issue 5, pp. 491–496 (1985).
[2] Darryll, J.P. and Lovell, P.A., "Conceptual framework of a remote wireless health monitoring system for large civil structures," Smart Material Structure Vol. 7 627–636 (1998).
[3] Straser, E.G., Kiremidjin, A.S., and Meng, T.H., "Modular, Damage Monitoring System for Structures," United States Patent, No. 6292108B1 (2001).
[4] Chaudhuri, S. and Dayal, U., "An Overview of Data Warehousing and OLAP Technology," Volume 26, Issue 1, pp. 65–74 (1997).
[5] King, R.P., Halim, N., Garcia-Molina, H., and Polyzois, C.A., "Management of a remote backup copy for disaster recovery," ACM Trans. on Database Systems, vol. 16, no. 2, pp. 338–368 (1991).
[6] Crawford, C.M., "Internet Online Backup System Provides Storage for Customers Using IDs and Passwords Which Were Interactively Established When Sign up for Backup Services," United States Patent, No. 5771354 (1998).
[7] Jing, L., "A review of techniques, advances and outstanding issues in numerical modelling for rock mechanics and rock engineering," Rock Mech. & Mining Sci., vol. 40, pp. 283–353 (2003).
[8] Gamma, E., Helm, R., Johnson, R., and Vlissides, J., "Design Patterns: Elements of Reusable Object-Oriented Software," Addison-Wesley (1994).
[9] Lin, H.-C., Hong, Y.-M., and Kan, Y.C., "The back-end design of an environmental monitoring system upon real-time prediction of groundwater level fluctuation under the hillslope," Environmental Monitoring and Assessment, vol. 184, no. 1 (2012), 381–395.

Frontiers of Energy and Environmental Engineering – Sung, Kao & Chen (eds)
© 2013 Taylor & Francis Group, London, ISBN 978-0-415-66159-1

The analysis of China's regional environmental protection input efficiency based on DEA

Y.C. Li & M. Su

Department of Economics and Management of North China Electric Power University, Baoding Hebei, China

ABSTRACT: Environmental Protection (EP) is the important problem in the world. Most countries pay more attention to environmental protection, and they should consider the output effect while making the input to EP. Environmental Protection Input Efficiency (EPIE) refers to using the effects of pollution control to measure effectiveness of environmental investment. By using the Data Envelope Analysis (DEA) method, we studies China's 7 regional waste water treatment and air pollution situation. So we can draw the conclusion that regional environmental protection input efficiency in China is generally good, but the efficiency is low in a few of regions. If the government wants to improve the effect of environment protection, it cannot only rely on the inputs of capital and equipment, and it also need to seek the new methods to pollution governance.

Keywords: DEA; regional; EPIE

1 INTRODUCTION

With the rapid development of China's economy, the contradiction between economic growth and environmental protection are increasingly highlights, while the environmental efficiency of resource allocation problem is gaining more attention. Environmental governance investments are important means of promoting economic growth and harmonious development of environmental governance. Environmental Protection Input Efficiency (EPIE) can be used to evaluate the effects on the environment of investment. EPIE refers to using the effects of pollution control to measure effectiveness of environmental investment. The indicator can appraise the environmental protection government effect which can achieve to the environment government investment. Meanwhile it can get up the effective survey to the sustainable development, and will make the contribution to different departments to make policy.

Data Envelopment Analysis (DEA) method can assess the efficiency of different decision-making unit in multi-input and multi-output case. It can overcome the subjective factors of influence when set weight in the traditional performance evaluation method. So in this paper, we analyze regional input efficiency of environmental protection in China, especially research the efficiency of waste water and waste gas treatment efficiency, by using the data of China's environmental input and output in 2009 and the DEA method.

2 DEA MODEL

2.1 Introduce of DEA method

The Data Envelopment Analysis (DEA) is one of efficiency assessment method which is proposed by American famous operation scientist A.Chames and W.W.Cooper et al. It based on the concept of relative efficiency, and extended the engineering efficiency concept of single-input and single-output to the effective evaluation of similar Decision Making Unit (DMU) of multi-input and multi-output. DEA method enriches microeconomic production function theory and application, and at the same time it avoids subjective factors, simplifies algorithm and reduces errors.

Using linear programming, DEA method reflects the input and output values of DMU to the efficiency space, and finds the efficient frontier. The combination of input and output in the efficient frontier has Pareto optimal efficiency, and its efficiency value is 1. Since 1978, DEA method developed rapidly, and had many advances in the theory and applications. The main mathematical model of DEA method is shown in Table 1.

2.2 Model implementation steps

DEA method steps can be divided into the following four steps in general: a) Define questions and determine the target; b) Establish indicator system and select DMU, input variables and output vari-

Table 1. The main mathematical model of DEA.

Parameter setting	Suppose there are n DMU, using m kinds of inputs $x_i (i = 1, 2, ..., m)$, producing s kinds of output $y_r (r = 1, 2, ..., s)$. θ_k represents the equal proportion reduction potential of all input items of DMU_k; the weight $\lambda = (\lambda_1, \lambda_2, ..., \lambda_n)$ represents a polyhedron vector linking of all data.
Fractional programming	$\max h_k = \sum_{r=1}^{s} U_r y_{rk} \Big/ \sum_{i=1}^{m} V_i x_{ik}$
	s.t. $\sum_{r=1}^{s} U_r y_{rk} \Big/ \sum_{i=1}^{m} V_i x_{ik} \leq 1$
	$U_r \geq \varepsilon > 0 \quad V_i \geq \varepsilon > 0$
Linear programming	$\max h_k = \sum_{r=1}^{s} U_r y_{rk}$
	s.t. $\sum_{i=1}^{m} V_i x_{ik} = 1,$
	$\sum_{r=1}^{s} U_r y_{rj} - \sum_{i=1}^{m} V_i x_{ij} \leq 0$
	$U_r \geq \varepsilon > 0 \quad V_i \geq \varepsilon > 0$ (making $U_r = t u_r$, $V_i = t v_i$ Charnes – Cooper transform)
Dual programming	$\max[\theta - \varepsilon(\hat{e}^T s^- + e^T s^+)]$
	s.t. $\sum_{j=1}^{n} x_j \lambda_j + s^- = \theta x_{j0},$
	$\sum_{j=1}^{n} y_j \lambda_j + s^+ = y_{j0}$
	$\lambda_j \geq 0 \ (j = 1, 2, ..., n); \ s^-$ is slack variable; s^+ is surplus variable. $\hat{e}$ and e are respectively the component of an m-dimensional and s-dimensional column vector; ε is non-Archimedean infinitesimal (smaller than any amount greater than zero)

ables; c) collect data and select model; d) analyze the model and draw the conclusion.

Taking into account the current effect of environmental investment and governance, we study industrial pollution control inputs and the control effect outputs of China's region. The implementation steps are as follows: a) Determine environmental protection input efficiency of China's regional as the evaluation objective; b) Select input variables (e.g. industrial pollution management investment and the number of waste water and waste gas treatment facility etc.) and output variables (e.g. industrial waste emission standard rate and industrial sulfur dioxide removal rate etc.); c) Collects the data of input variables and the output variables in 2009 from China Statistics Yearbook, and select CCR model to analyze our regional EPIE; d) Use DEAP software to analyze the data and draw the conclusion.

2.3 Sample data

China has vast territory. It can usually be divided into the North China area, the East China area, the Central China area, the South China area, the Northeast area, the Southwest area, the Northwest area as well as the Hong-Kong, Macao and Taiwan area according to the geographical position. We do not analyze the data of Tibet and the Hong-Kong, Macao and Taiwan area temporarily due to it's hard to obtain. So we consider China as 7 areas and use the data from China Statistics Yearbook 2009, which include (1) industrial pollution management investment, (2) the number of waste water treatment facility and (3) the number of waste gas treatment facility as input variables, (4) industrial waste emission standard rate and (5) industrial sulfur dioxide removal rate as output variables. The industrial waste emission standard rate is equal to the amount of industrial wastewater discharge standards of compliance divided by the total amount of industrial wastewater discharge. The industrial sulfur dioxide removal rate is equal to the volume of industrial sulfur dioxide removed divided by the sum of volume of industrial sulfur dioxide emissions and removal of industrial sulfur dioxide. The variables value of DMU is shown in Table 2.

Table 2. The input variables and output variables of China's regional EPIE.

DMU	Input			Output	
	(1)	(2)	(3)	(4)	(5)
The North China	911716	8678	31761	93.56	57.32
The East China	1285291	27569	53411	97.54	66.38
The Central China	608920	10299	23021	94.51	61.69
The South China	348145	12632	19414	93.61	60.07
The Northeast	375135	3901	17417	86.17	42.77
The Southwest	351293	10153	18109	93.13	62.84
The Northwest	545709	3770	13305	83.97	59.21

*From China Statistics Yearbook 2009.

3 EMPIRICAL RESULTS AND ANALYSIS

We calculate and analyze the EPIE of China's 7 areas by using CCR model of DEA. The result is shown in Table 3. Crste is the technical efficiency which does not include returns to scale. Vrste is the technical efficiency which includes returns to scale. Scale is the scale efficiency which includes returns to scale. The relationship between them is crste = vrste × scale.

We can see from Table 3 that there are 4 DMUs on the frontier of China's regional EPIE: the South China, the Northeast, the Southwest and the Northwest. Their EPIE are relatively optimal. These four samples accounted for 57% of the total number of DMU, and the number of samples which is not on the frontier of EPIE is 43%. So we can find China's regional EPIE is generally good.

From Table 3, we can discover the vrste of China's 7 areas all is 1, which means the environmental protection input in China can achieve good returns to scale. The average of crste of environmental protection input is 0.805, so that we can note the overall regional environmental governance in China is relatively good. The value of EPIE is up to 1 in the South China, the Northeast, the Southwest and the Northwest, so the governance efficiency of industrial waste water and gas is relatively high in these 4 areas. However the value of EPIE is only 0.339 in the East China, which is the lowest. The values of EPIE are 0.547 and 0.746 respectively in the North China and the Central China, so the environmental protection input efficiency needs to improve in these two areas.

From the last row of Table 3, it is obvious that the returns to scale are unchanged in the South China, the Northeast, the Southwest and the Northwest. The proportion of environmental protection input and industrial scale is appropriate in these four areas. While we can realize the returns to scale are decreasing in the North China, the East China and the Central China. The governance effect is not satisfactory in these three areas. If the government wants to improve the EPIE in the North China, the East China and the Central China, it cannot only rely on increasing the investment of industrial pollution management and the number of waste water and waste gas treatment facility, and it can seek some new methods to improve the regional environmental protection input efficiency, such as enhancing internal management of industrial pollution treatment or increasing the precise degree of treatment equipment etc.

4 CONCLUSIONS AND IMPLICATIONS

With the implement of the sustainable development strategy in China, it is obvious that the government pays more attention to environmental protection. But it is no effect only investing capitals and equipments while ignoring the efficiency of pollution treatment. We use DEA model to study the input and output of regional environmental protection in China. By analyzing the efficiency of China's regional environmental protection investment and returns to scale, we can notice that the average of EPIE in China is 0.805, which tells us the environmental protection input efficiency is still in high level. From the data of regional EPIE, we find the value of EPIE is up to 1 in the South China, the Northeast, the Southwest and the Northwest, so the environmental protection input efficiency is relatively high in these 4 areas. While the value of EPIE is less than 1 in the North China, the East China and the Central China, which shows the EPIE is poor in these three areas. The returns to scale are unchanged in the South China, the Northeast, the Southwest and the Northwest, so that the proportion of environmental protection input and industrial scale is appropriate in these four areas. While we can realize the returns to scale are decreasing in the North China, the East China and the Central China, so they cannot only rely on the inputs of capital and equipment, but also need to seek the new methods to solve pollution problem.

According to the analysis above, the government can use some new methods, such as updating pollution control equipment, use of new energy, changing the structure of pollution investment, to improve environmental protection input efficiency. Although the EPIE is optimal in the South China, the Northeast, the Southwest and the Northwest, the industrial waste water emission standard rate is lower than the national average 91.78% in the Northeast and the Northwest, meanwhile the industrial sulfur dioxide removal rate is much less than the national average 58.61% in the Northeast. In this model the environmental efficiency is optimal in these four areas, but maybe the efficiency is not optimal if the input or output variables changed. Therefore we should make a great effort to regional environmental protection in order to improve our environment quality.

REFERENCES

Chames A, Cooper W.W, Rhodes E. Measuring the efficiency of decision making units [J]. European Journal of Operational Research, 1978,2(6):429–444

China Statistics Yearbook 2009. Chinese Statistics Bureau. Beijing: China statistics press, 2010.

Ping-lin He, Ya-dong Shi & Tao Li. Data Envelopment Analysis Method of Environmental Performance—A study based on the case of China's thermal power plant [J]. Accounting Research, 2012(2):11–17.

Shan-shan Song, Feng-ping Wu. Analysis of Regional Environmental Investment Efficiency in DEA [J]. Value Engineering, 2012(4):78–79.

Tim Coelli. A Guide to DEAP Version 2.1: A Data Envelopment Analysis (Computer) Program. Centre for Efficiency and Productivity Analysis Department of Econometrics University of New England Armidale, NSW, 2351. Australia.

Zhan-xin Ma (2010). Data Envelopment Analysis Model and Method [M]. Beijing: Science press.

Frontiers of Energy and Environmental Engineering – Sung, Kao & Chen (eds)
© *2013 Taylor & Francis Group, London, ISBN 978-0-415-66159-1*

Evaluation on the efficiency of regional carbon emissions based on non-radial DEA model

Y.C. Li, Q. Zhang & H.X. He
Department of Economics and management of North China Electric Power University, Baoding Hebei, China

ABSTRACT: Global warming, which is affecting the economic and social development profoundly, is increasingly becoming the focus of our attention. Reducing greenhouse gas emissions has been a problem to be solved immediately. This article made an analysis of carbon emission efficiency and the factors by way of related models. It aims to improve the efficiency of carbon emissions and provide effective recommendations on Low-carbon economy and its trading system. First of all, it measured the static efficiency of carbon emissions in China from 1995 to 2009 through on-radial DEA model, and make the analysis of the dynamic level.

Keywords: DEA; carbon emissions; low-carbon economy

1 INTRODUCTION

In recent years, along with our country of green GDP and low carbon economy advocate, theory circle and practice circle pay more and more attention to the influence of economic efficiency of carbon emissions. Many empirical studies have examined variables include environmental pollution, changes in conditions of economic efficiency. Such as Li Jing (2009) empirical studies have shown that the introduction of environmental variables significantly reduces the average efficiency of Chinese regional level [1]. But at present, at home and abroad of carbon emissions on economic efficiency effects of little research literature.

In this paper, DEA-SBM model to deal with non-desired outputs (undesirable output), empirical research to consider carbon emissions in China after the economic efficiency of state. The analysis of carbon emissions on China's regional economic efficiency and the effect of regional differences. On this basis, for the development of low carbon economy and construction of carbon emissions trading system to put forward effective suggestions.

2 RESEARCH METHODS

Assumed production system has n decision unit which has three input and output vector: input, output and the expected output expectation, The three vector representation: $x \in R^m, y^g \in R^{s_1}, y^b \in R^{s_2}$, You can define a matrix X, Y^g, Y^b $X = [x_1, ..., x_n] \in R^{m \times n}$, $Y^g = [y_1^g, ..., y_n^g] \in R^{s_1 \times n}$ As well as, In which

$X > 0, Y^g > 0$ and $Y^b > 0$, Constant returns to scale of production possibility set P can be defined as:

$$P = \left\{ (x, y^g, y^b) \mid x \geq X\lambda, y^g \leq Y^g\lambda, y^b \geq Y^b\lambda, \lambda \geq 0 \right\} \quad (1)$$

In accordance with the Tone [12] proposed SBM model approach, a desired outputs of the SBM model:

$$\rho^* = \min \frac{1 - \frac{1}{m} \sum_{i=1}^{m} \frac{s_i^-}{x_{i0}}}{1 + \frac{1}{s_1 + s_2} \left(\sum_{r=1}^{s_1} \frac{s_r^g}{y_{r0}^g} + \sum_{r=1}^{s_2} \frac{s_r^b}{y_{r0}^b} \right)} \quad (2)$$

S. t.

$$x_0 = X\lambda + s^-$$
$$y_0^g = Y^g\lambda - s^g$$
$$y_0^b = Y^b\lambda + s^b$$
$$s^- \geq 0, s^g \geq 0, s^b \geq 0, \lambda \geq 0$$

Among them, λ is the weight vector, s said input, output amount of slack.

Further evaluation unit, if and only if $\rho^* = 1$, $s^- = 0, s^g = 0, s^b = 0$ is efficient, $\rho^* < 1$, described being evaluated unit is not valid, the existence of input and output on the necessity of improvement. On the other hand, also solved Compared with the C2R model and BC2 model, SBM model can solve the presence of undesirable output efficiency evaluation problem is also solved the input-output relaxation problem, make the efficiency evaluation of unbiased.

3 CHINA'S REGIONAL DIFFERENCES IN THE EFFICIENCY OF CARBON EMISSIONS

3.1 Data sources and processing

The main purpose of this paper is to study the existence of carbon emissions under the condition of regional economic efficiency in China, Therefore draw the relevant literature, in this paper from 1995 to 2009 a total of 30 provinces in Chinese mainland 15 years of panel data as samples, Because the data is from Chongqing began keeping statistics in 1997, so to be removed, also, the paper smoothed the data of Sichuan. Relevant data are shown in Table 1.

According to the theory of modern economics, the main production elements for labor, land and capital. In the past years labor index employees number instead of labor quota; Because land Is basically constant indicators, directly as input indicators do not effectively distinguish between

Table 1. Carbon emissions under the condition of regional economic efficiency of each provinces.

Province	1995	2000	2005	2009
Beijing	1.000	1.000	1.000	1.000
Tianjin	1.000	1.000	1.000	1.000
Hebei	0.482	0.435	0.322	0.447
Shanxi	0.302	0.262	0.234	0.297
Mongolia	0.408	0.346	0.370	0.400
Liaoning	0.559	0.484	0.476	0.557
Jilin	0.526	0.553	0.486	0.622
Heilongjiang	0.500	0.481	0.591	0.695
Shanghai	1.000	1.000	1.000	1.000
Jiangsu	0.811	0.935	0.885	1.000
Zhejiang	1.000	0.829	0.753	0.768
Anhui	0.681	0.661	0.688	0.768
Fujian	0.910	1.000	0.688	0.913
Jiangxi	0.665	1.000	0.616	0.646
Shandong	1.000	0.660	0.542	0.833
Henan	0.549	0.460	0.373	0.523
Hubei	0.623	0.649	0.524	0.649
Hunan	0.530	0.637	0.505	0.621
Guangdong	1.000	1.000	1.000	1.000
Guangxi	0.522	0.480	0.390	0.522
Sichuan	0.479	0.553	0.573	0.572
Guizhou	0.357	0.268	0.304	0.413
Yunnan	0.522	0.443	0.320	0.500
Hainan	0.521	0.500	0.493	0.527
Shanxi	0.480	0.533	0.383	0.500

each province area annual differences, and carbon emission efficiency of little relevance, therefore the use and production process and carbon emissions is closely related to energy consumption alternatives; With the constant capital stock instead of capital. Over the years the employees data from China Statistical Yearbook of the calendar year and the provincial statistical yearbook; Energy consumption data derived from the new China 55 years compilation of statistics, provincial statistical yearbook, lack in part by the development and Reform Commission issued the provinces of GDP energy consumption [12] is multiplied by the corresponding year and GDP transformation; Capital stock processing is more complex, We follow Zhang Jun (2004) [13] approach, As the Perpetual inventory method (Perpetual Inventory Method) of the capital stock is very sensitive to initial value selection, The base year data should be preceded by a longer history investment series, This paper in 1952 for the initial years, in the provinces of fixed capital formation as the capital stock of the calculation basis, namely in the provinces of fixed capital formation 10% as the initial capital stock. When investment index is the use of gross fixed capital formation, investment in fixed capital formation as an index of price index calculation basis, and then united with fixed price was calculated 1995, 9.5% of the depreciation rate, according to the perpetual inventory method to simulate the Chinese provincial capital, relevant data are derived from the statistical yearbook.

According to the foregoing description we will output into desired output (Y^g) and undesirable outputs (Y^b), we choose the provinces GDP representation, to the comparability of results, we are in constant 1995 prices processing. Undesirable outputs (Y^b), this paper refers to the amount of carbon emissions, with coal, oil and natural gas consumption is multiplied by the corresponding discharge coefficient is derived. Coal, oil and natural gas emissions over the China Statistical Yearbook, discharge coefficients were 0.7559, 0.5042 and 0.4483.

3.2 Study on the efficiency of Chinese regional carbon emissions

The use of SBM model respectively in 30 provinces in China from 1995 to 2009 were calculated the carbon emission efficiency, major provinces carbon emission efficiency values are shown in Table 2.

Following on the carbon emission efficiency values for data analysis:

1. Average efficiency analysis
   Figure 1 visibly indication from 1995 to 2009 in carbon emission efficiency trends. From 1995 to 2005, carbon emission efficiency of the overall

Table 2. The main year of carbon emission efficiency of the provinces.

Gansu	0.343	0.301	0.301	0.418
Qinghai	0.419	0.293	0.240	0.294
Ningxia	0.317	0.276	0.174	0.256
Xinjiang	0.540	0.434	0.347	0.465
Tibet	0.409	0.400	0.391	0.413

Figure 2. The trend of the major regional differences in the efficiency of carbon emissions.

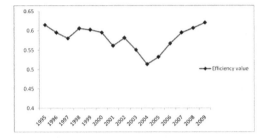

Figure 1. The Chinese calendar year is average efficiency trend graph.

downward trend, From 1998 to 2004 is obvious, 2005 is a significant turning point, The reason is mainly from the government, In 2005 the "eleven five" planning" compendium" put forward clearly the unit GDP energy consumption to be reduced by about 20% 5 years. This is the first time for China to be specific and clear energy saving targets included 5 year planning. The national level of positive action to drive a nationwide energy-saving emission reduction work, thereby reversing the downward trend in average efficiency. In 2008 2007, the efficiency level to accelerate growth, The article thinks, the main reason is that after 2007, domestic and foreign government departments and related organizations to advocate low carbon economy, advocate energetically to reduce carbon dioxide emissions.

2. Regional difference analysis

In order to compare Chinese mainland regional carbon emission efficiency of regional differences, We follow the traditional method will China provinces division is divided into East, West, in three regions. Figure 2 clearly shows in three regions of China carbon emission efficiency differences of status and trend. As you can see from Figure 2, area of the east part carbon emission efficiency is the highest, in middle, the lowest in the west. Western region and the eastern region difference, Area of the east part efficiency values maintained at about 0.78, mid area is in 0.52 the left and right sides, western the area is only about 0.4. People often

have such errors, The higher levels of economic development, carbon emissions are higher, the lower the efficiency of, Through the data analysis, we can see, a developed economy, efficiency is often higher carbon emissions. Main reason is, carbon emission efficiency research has considered the region economy development the input-output ratio, At the same time, economically developed areas focus on cleaner energy and capital intensive industry development.

3. Regional differences in dynamic analysis

In order to further study the development trend of carbon emission efficiency, this paper selects three regions of our country after 2000, carbon emission efficiency to the time dummy variable regression, three area efficiency on time elasticity respectively: East efficiency of $y_1 = 0.003x_1 - 5.9$, mid area efficiency $y_2 = 0.006x_2 - 12.48$, $y_3 = 0.002x_3 - 5.459$ western area efficiency. Visible, after 2000, three regions of our country carbon emission efficiency rise, mainly because of, enter twenty-first Century, China's social market economic system ceaseless and perfect, domestic and foreign to the energy saving and emission reduction, green GDP and development of low carbon economy advocate energetically, our country in maintaining social economy rapid development at the same time, more and more attention to environmental protection and GDP energy consumption reduction. At the same time, also can be seen, the central region and the eastern region difference of efficiency decreases slowly with time, the western region and the central region of the efficiency differences tended to increase, western regions on carbon emission efficiency.

4 CONCLUSION

This paper first reviews the treatment of desired outputs of the DEA model and the development of low carbon economy and related literature, through the use of DEA-SBM model to study the 1995–2009 of our country year 30 provinces of carbon emissions on the influence of economic efficiency (carbon emission efficiency) and regional differences. Main conclusions and suggestions are as follows.

Study finds, carbon emissions in China Eastern region has the highest efficiency, then the middle region, the lowest in the west, and the eastern and central regions difference narrows gradually, the western area and mid area was increased. The use of 2000–2009 data for dynamic analysis, enter twenty-first Century, especially after 2005, China's carbon emission efficiency is gradually improved, the efficiency level is rising.

REFERENCES

Charnes A, W.W. Cooper, E. Rhodes. Measuring the efficiency of decision making units [J]. European Journal of Operational Research, 1978, 2 (6):429–444.

Fang Gang, Wang Xiaolu, Zhu Hengpeng. China market index [M]. Beijing: Economic Science Press, 2010.

Fu Yun Development patterns of low carbon economy [J]. China Population, Resources and Environment, 2008, 18 (3).

Gou Heng, Sun Lei. The development of low carbon economy and strategic choice of macroscopically economy [J], 2010, Phase I.

Hailu, Atakelty, Terrence S, et al. Non-Parametric Productivity Analysis with Undesirable Outputs: An Appliction to the Canadian Pulp and Paper Industry [J]. American Journal of Agricultural Economics, 2001, 83(3):805–816.

Haorui. Economic efficiency and regional equity: empirical analysis on Chinese provincial economic growth and the gap [J]. World Economic Forum, 2006, (2):11–29.

Hu Dan, Xu Kaipeng, Yang Jianxin and so on. Economic development on environmental quality—the influence of environment of domestic and foreign research on Kuznets curve [J]. Journal of Ecology, 2004, 24 (6): 1259–1266.

Lijing. Chinese environmental efficiency differences and influencing factors study [J]. South China Journal of Economics, 2009, 12.

Liu Shijin. The current development of low carbon economy key and policy proposal [J]. China Investment of Science and Technology, 2010.

Scheel, H. Undesirable outputs in efficiency valuations [J]. European Journal of Operational Research, 2001, (132):400–410.

Tone, K. A slacks-based measure of efficiency in data envelopment analysis [J]. European Journal of Operational Research, 2001, (130):498–509.

Yuan Nanyou. Low carbon economy concept [J]. City Environment and Ecology, 2010, (1).

Zhu, J. Quantitative Models for Performance Evaluation and Benchmarking: Data Envelopment Analysis with Spreadsheets and DEA Excel Solver [M]. Netherland: Kluwer Academic Publishers, 2003.

Frontiers of Energy and Environmental Engineering – Sung, Kao & Chen (eds)
© 2013 Taylor & Francis Group, London, ISBN 978-0-415-66159-1

An empirical study on total factor energy efficiency of China's thermal power generation

Y.C. Li, Q. Liu & Q. Zhang
Department of Economics and Management of North China Electric Power University, Baoding Hebei, China

ABSTRACT: China's coal-fired power conversion efficiency is about 30%, less than the one of 40% in developed countries. Therefore promoting the energy efficiency of China's power industry, especially thermal power industry, will be undoubtedly significant to promoting the country's energy efficiency and energy saving. Traditional measure methods of total factor energy efficiency often ignore the undesirable output and only consider the desirable output; therefore, the value of total factor energy efficiency measured has a lager deviation from actual one. The paper using Malmquist-Luenberger productivity index measures the growth of total factor energy efficiency of 30 provinces' thermal power generation of China over 2003 to 2010. The results indicate that the growth rate of total factor energy efficiency of Thermal Power Generation considering environmental constraint is much lower than the one not.

Keywords: energy efficiency; power generation; Malmquist-Luenberger Index; empirical study

1 INTRODUCTION

Coal is China's main energy, accounting for more than 70% of primary energy, while the coal to power generation accounts for more than half of the consumption of raw coal; China's power generation is main thermal power, and coal accounts for more than 95% of the energy of thermal power.

At present, China's coal-fired power conversion efficiency is about 30%, less than the one of 40% in developed countries. On the other hand, the thermal power generation of which fuel is main coal not only consumes a large number of non-renewable resources, but also generates a large number of polluting gases, which serious pollute our living environment and even global climate. Therefore promoting the energy efficiency of China's power industry, especially thermal power industry, will be undoubtedly significant to promoting the country's energy efficiency and energy saving. Thus, the research on thermal power industry's total factor energy efficiency, particularly the one under environment constraint, not only help to identify the reasons for low energy efficiency of China's thermal power generation, so as to improve it, but also are of great practical significance to improve the situation of environmental resources and achieve our targets of energy saving and emission reduction.

The empirical studies of energy efficiency mostly focused on industrial areas or the entire economy, however, rarely involve in electricity industry, energy efficiency, especially the one under environmental constraint. Even if the existing studies of the thermal power industry are mainly focused on the aspect of technical efficiency, few concerns to energy efficiency. For the deficiencies of the studies above, this paper adopts the non-parametric Malmquist-Luenberger index method to measure the growth of total factor energy efficiency and its components of China's Thermal Power Generation over the period 2003 to 2010 under the environmental constraint.

2 RESEARCH METHODS

This paper regards pollution variable as undesirable output, and makes use of the directional distance function to construct non-parametric Malmquist index to measure the total factor energy efficiency of China's thermal power generation under environmental constraints.

2.1 Environmental technology

In order to integrate environmental considerations into the framework of total factor energy efficiency, we need to first construct a production possibility set contained both desirable output and undesirable output, that is, environmental technology. we assume that each province or region uses N kinds of inputs $x = (x_1, x_2 \cdots, x_N) \in R_+^N$ to produce M kinds of desirable outputs $y = (y_1, y_2, \cdots, y_M) \in R_+^M$ and I kinds of undesirable outputs $b = (b_1, b_2 \cdots, b_I) \in R_+^I$.

The relationship between input and output is represented by the following output set:

$$P(x) = \{(y,b) : x \text{ can produce}(y,b)\}, \quad x \in R_+^N \quad (a)$$

If the output set represent environmental technology, the output set will has the following four properties.

1. "Null-jointness":

$$(y,b) \in p(x); b = 0 \Rightarrow y = 0 \quad (b)$$

It means if there is no undesirable output there will be no desirable output, or if there is desirable output, there must be undesirable output.

2. Desirable and undesirable outputs are jointly weakly disposable:

$$(y,b) \in p(x) \text{ and } 0 \leq \theta \leq 1, \quad \text{then } q(y,b) \in P(x) \quad (c)$$

This implies that the reduction in undesirable output is always along with the reduction in desirable output.

3. Strong disposability of desirable output:

$$(y,b) \in P(x) \text{ and } y^0 \leq y, \quad \text{then } (y^0,b) \in P(x) \quad (d)$$

This implies that desirable output can also be reduced without reducing the undesirable one. The amount of desirable output shows the level of technical efficiency under environmental constraint.

4. Inputs can be freely disposed:

$$\text{If } x^0 \geq x, \quad \text{then } P(x^0) \supseteq P(x) \quad (e)$$

Now, assume period $t = 1,2, ..., T$, and J decision making unit (DMU), $j = 1,2, ..., J$, input-output vector is $(x_{(J \times N)}^t, y_{(J \times M)}^t, b_{(J \times I)}^t)$, Using data of inputs, outputs and pollution, the DEA models for measuring the environmental technology model, meeting those above characteristics, can be represented by the following model

$$P^t(x^t) = \left\{ \begin{array}{l} \sum_{j=1}^{J} y_{j,m}^t \lambda_j \geq y_{j,m}^t, \quad m = 1, ..., M; \sum_{j=1}^{J} b_{j,i}^t \lambda_j = b_{j,i}^t, \quad i = 1, ..., I; \\ \sum_{j=1}^{J} x_{j,n}^t \lambda_j \leq x_{j,n}^t, \quad n = 1, ..., N; \lambda_j \geq 0, \quad j = 1, ..., J \end{array} \right\} \quad (1)$$

2.2 Directional distance function

Environmental technology gives the Frontier of possibility of production under environmental constraints, which is the set maximum output and minimum pollution under a given input. But the environmental output set that meet the above properties can not be calculated by the traditional Shephard distance function. According to Luenberger's (1992, 1995) shortage function, Chung et al (1997) build a directional environment distance function:

$$D(x,y,b,g) = \sup\{\beta : (y,b) + \beta g \in P(x)\} \quad (f)$$

$g = (g_y, g_b)$ is the direction vector of the expansion of desirable output and contraction on of undesirable output. β implies that for a given input x, when output y and b expand and contract by same proportion, the maximum possible reduction in desirable output y and undesirable output b. Therefore, the directional distance function measure the level of producer's inefficiency relative to the level of environmental technology. The firm, which operates on the frontier, has technical efficiency, when the value of directional distance function, β is zero. We can solve the following mathematical linear programming to calculateβ, the value of directional distance function.

Maximize β

$$\text{s.t.} \sum_{j=1}^{J} \lambda_j^t y_{jm}^t \geq (1+\beta) y_{jm}^t, \quad m = 1, ..., M; \quad (i)$$

$$\sum_{j=1}^{J} \lambda_j^t b_{ji}^t = (1-\beta) b_{ji}^t, \quad i = 1, ..., I; \quad (ii)$$

$$\sum_{j=1}^{J} \lambda_j^t x_{jn}^t \leq x_{jn}^t, \quad n = 1, ..., N; \quad (iii)$$

$$\sum_{j} \lambda_j = 1; \lambda_j \geq 0; \quad j = 1, ..., J \quad (v) \quad (2)$$

2.3 Malmquist-Luenberger Index

According to Chung et al. (1997), the output-oriented Malmquist-Luenberger Index (ML) between t period and the period t+1 is represented as follows:

$$ML_t^{t+1} = \left\{ \left[\frac{\left[1 + \vec{D}_o^t(x^t, y^t, b^t; g^t) \right]}{\left[1 + \vec{D}_o^t(x^{t+1}, y^{t+1}, b^{t+1}; g^{t+1}) \right]} \right] \times \left[\frac{\left[1 + \vec{D}_o^{t+1}(x^t, y^t, b^t; g^t) \right]}{\left[1 + \vec{D}_o^{t+1}(x^{t+1}, y^{t+1}, b^{t+1}; g^{t+1}) \right]} \right] \right\}^{\frac{1}{2}} \quad (3)$$

ML index can be decomposed into Efficiency Change (EFFCH) and Technological progress (TECH):

$$ML = EFFCH \times TECH$$

$$EFFCH_t^{t+1} = \frac{1 + \vec{D}_o^t\left(x^t, y^t, b^t; g^t\right)}{1 + \vec{D}_o^{t+1}\left(x^{t+1}, y^{t+1}, b^{t+1}; g^{t+1}\right)} \qquad (4)$$

$$TECH_t^{t+1} = \left\{ \left[\frac{\left[1 + \vec{D}_o^{t+1}\left(x^t, y^t, b^t; g^t\right)\right]}{\left[1 + \vec{D}_o^t\left(x^t, y^t, b^t; g^t\right)\right]} \right] \times \left[\frac{\left[1 + \vec{D}_o^{t+1}\left(x^{t+1}, y^{t+1}, b^{t+1}; g^{t+1}\right)\right]}{\left[1 + \vec{D}_o^t\left(x^{t+1}, y^{t+1}, b^{t+1}; g^{t+1}\right)\right]} \right] \right\}^{\frac{1}{2}} \qquad (5)$$

That the value of ML, EFFCH and TECH is greater than 1 indicates productivity growth, efficiency improvements and technological advances, less than 1 indicates productivity decline, efficiency deterioration and technological regress, equal to 1 indicates that the productivity, efficiency and technological progress have no change. Based on above method, this paper measures the China's 30 provinces, productivity index, efficiency change index and the index of technological progress over the period 2003 to 2010.

3 INDICATOR SELECTION AND DATA CONSOLIDATION

In accordance with the above theoretical approach, combined with the theme of this paper, and considering data's availability and integrity, this paper selected thermal power industry input and output data of China's 30 provinces (except Tibet and Taiwan) from the year of 2003 to 2010. The Inputs include fuel (10,000 tons), installed capacity (MW), labor (people), and outputs are power generation (100 million kwh) and sulfur dioxide emissions (tons). Power is the desirable output, and undesirable output is measured by sulfur dioxide (SO_2). For the absence of statistics of employment and sulfur dioxide emissions of the thermal power generation, so the variables of labor is replaced by the number of employees that is highly correlated with the industry of thermal power generation, electricity, heat production and supply, and sulfur dioxide is replaced by sulfur dioxide emissions

from industry. Using installed capacity to replace capital investment. Fuels are converted into standard coal

For lack of the direct fuel data, it is calculated by making power generating multiply by the standard coal consumption in the technical and economic indicators. The data of sulfur dioxide emissions and labor force comes from "China Statistical Yearbook", the rest were from the "Power Book" for the indicate years.

4 EMPIRICAL RESULTS

According to non-parametric DEA model, this paper measures, with the software of DEAP2.1, the total factor energy efficiency's average change and its decomposition results of China's thermal power generation that both has consideration of environmental constraints and hasn't environmental constraints from the year of 2003 to 2010, and the results are presented in Table 1 below. The total factor energy efficiency can be decomposed into technical efficiency, technological progress, and technical efficiency includes pure technical efficiency and scale efficiency. Effch stands for change of technical efficiency, techch for technological change, pech for pure technical efficiency change, sech for scale efficiency change, tfpch for the total factor energy efficiency change.

4.1 The average change and decomposition in total factor energy efficiency of thermal power industry considering environmental effect

We can find from Table 1 that the average total factor energy efficiency of thermal power industry considering environmental constraint over the period 2003 to 2010 is 1.01, and the average growth rate is 1%, however, the one is 1.018 not considering environmental constraint, and its average growth rate is 1.8%, significantly higher than the former case, nearly twice the former. To further analyze the source of changes in total factor energy efficiency, the changes in total factor energy efficiency is decomposed into technological progress and technical efficiency change, and technical efficiency is further decomposed into pure technical efficiency and scale efficiency in this paper.

From the results of decomposition, when considering the environmental pollution, the contribution of technical progress is slightly higher than the one of technical efficiency, the average growth rate of technical efficiency being 0.5%, the average growth rate of technical efficiency being 0.4%. Viewing from the decomposition of technical

Table 1. The average change and decomposition of total factor energy efficiency of thermal power industry from 2003 to 2010.

Year		effch	techch	pech	ech	tfpch
2003	E	0.998	1.011	0.996	1.002	1.009
	NE	1.009	1.024	1.004	1.004	1.033
2004	E	1.001	1.014	1.003	0.998	1.015
	NE	1.012	1.013	1.012	1.000	1.026
2005	E	1.011	1.028	1.000	1.011	1.039
	NE	1.005	1.020	0.999	1.005	1.025
2006	E	1.011	0.998	1.010	1.001	1.009
	NE	1.009	1.003	1.007	1.002	1.012
2007	E	0.999	1.014	1.000	0.999	1.012
	NE	1.002	1.001	0.999	1.003	1.003
2008	E	0.997	1.004	1.008	0.989	1.000
	NE	0.995	1.010	1.007	0.988	1.005
2009	E	1.019	0.986	1.009	1.011	1.005
	NE	1.009	1.024	1.005	1.005	1.033
2010	E	1.000	0.990	0.996	1.004	0.991
	NE	0.996	1.011	0.993	1.003	1.007
Mean	E	1.004	1.005	1.003	1.002	1.010
	NE	1.005	1.013	1.003	1.001	1.018

E—Considering environmental constraint.
NE—No considering environmental constraint.

efficiency, the average growth rate of pure technical efficiency and scale efficiency is 0.3% and 0.2%. Without considering the environmental pollution, the contribution of technical progress to the growth of total factor energy efficiency of thermal power industry is particularly prominent, with the average growth rate being 1.3%, while the average growth rate of technical efficiency is only 0.5%, and the one of pure technical efficiency is 0.3%, the one of the scale efficiency is 0.1%. Thus, the total factor energy efficiency of thermal power industry considering environmental constraint is significantly lower than the one that doesn't. And considering environmental constraint has greatly reduced the growth level of technical progress, while almost no effect on technical efficiency.

4.2 *The average change and decomposition of the total factor energy efficiency of each province's thermal power industry considering environmental effect*

As can be seen in Table 2, the annual average growth of total factor energy efficiency of China's thermal power industry is 1%, with technical efficiency increasing by 0.4%, technological progress increasing by 0.5%. The provinces of which total factor energy efficiency increases by more than 2% have Shanxi, Inner Mongolia, Shanghai, Jiangsu, Zhejiang, Hainan, Gansu and Ningxia.

Table 2. Total factor energy efficiency and its decomposition of each provinces.

Province	effch	techch	pech	sech	tfpch
Beijing	1.000	1.014	1.000	1.000	1.014
Tianjin	0.999	1.010	1.000	0.999	1.008
Hebei	1.000	1.004	1.000	1.000	1.003
Shanxi	1.010	1.027	1.006	1.004	1.038
Inner Mongolia	1.009	1.041	1.007	1.002	1.050
Liaoning	1.009	1.001	1.005	1.004	1.010
Jilin	1.007	1.002	1.007	1.000	1.010
Heilongjiang	1.000	1.008	1.002	0.998	1.008
Shanghai	0.998	1.023	0.999	0.999	1.021
Jiangsu	1.000	1.061	1.000	1.000	1.061
Zhejiang	1.005	1.017	1.003	1.002	1.022
Anhui	1.001	1.011	1.000	1.001	1.013
Fujian	1.002	1.011	1.002	1.000	1.012
Jiangxi	1.005	1.001	1.005	1.000	1.006
Shandong	0.999	1.011	1.000	0.999	1.010
Henan	1.011	1.009	1.011	1.000	1.019
Hubei	0.997	1.012	0.999	0.998	1.009
Hunan	0.994	1.012	0.996	0.998	1.006
Guangdong	1.003	1.011	0.999	1.004	1.014
Guangxi	1.000	0.956	1.000	1.000	0.956
Hainan	1.006	1.016	1.000	1.006	1.022
Chongqing	1.000	0.958	1.000	1.000	0.958
Sichuan	0.998	0.995	1.000	0.998	0.993
Guizhou	1.000	0.973	1.000	1.000	0.973
Yunnan	1.002	1.006	1.000	1.002	1.008
Tibet	1.004	1.003	1.003	1.001	1.007
Shaanxi	1.002	1.000	1.002	1.000	1.002
Gansu	1.041	0.988	1.000	1.041	1.028
Qinghai	1.000	0.998	1.000	1.000	0.998
Ningxia	1.034	0.992	1.034	1.000	1.026
Average	1.004	1.005	1.003	1.002	1.010

The provinces whose total factor energy efficiency increases by between 1% and 2% have Beijing, Liaoning, Anhui, Fujian, Shandong, Henan, Guangdong, and others provinces' is less than 1%. Even in Guangxi, Chongqing, Sichuan, Guizhou and Qinghai, the growth of total factor energy efficiency is negative. From the above analysis we can conclude that the provinces with higher growth rate of total factor energy efficiency are all either developed coastal provinces or coal-rich provinces (Shanxi, Inner Mongolia, etc.). The provinces with negative growth are developing areas, and the reason of causing its negative growth was mainly due to technical regress, while the efficiency of these provinces almost has no changes.

In Table 2, we also can find the growth of total factor energy efficiency of thermal power industry in most provinces is mainly credited to technical progress, while the contribution of efficiency improvement only accounts for a small proportion, and only a few provinces' is due to technical efficiency improvement.

5 CONCLUDING REMARKS

Traditional measure methods of total factor energy efficiency often ignore the undesirable output and only consider the desirable output; therefore, the value of total factor energy efficiency measured has a lager deviation from actual one. After taking environmental constraint into account, the paper using Malmquist-Luenberger productivity index measures the growth of total factor energy efficiency of 30 provinces' thermal power generation of China over 2003 to 2010, and test its convergence, the results show that:

1. From a national perspective, the growth rate of total factor energy efficiency in developed or coal-rich provinces is generally higher than other provinces; from the sub-regional perspective, the growth of total factor energy efficiency of thermal power generation in western region is the fastest, followed by the eastern region, central region slowest, this is mainly because of the most abundant coal resources in the western region, the most Developed economies in eastern region, while central region doesn't possess these advantages above, This indicates that resource endowment and the level of economic development have a great impact on the growth of total factor energy efficiency of the thermal power generation.
2. From the national calculation results, the growth of total factor energy efficiency of national thermal power generation is result of the common effect of technological progress and efficiency improvement; from the sub-region measured results, the one of eastern region is also result of the common effect of technological progress and efficiency improvement, while the one of central and western region is mainly caused by efficiency improvement, and technical progress play little role. This shows that by increasing the technical progress, the total factor energy efficiency of thermal power generation in China's central and western region is still much potential to promote. In future development, we should more focus on the improvement of technological progress.

REFERENCES

Charnes A, W.W. Cooper, E. Rhodes. Measuring the efficiency of decision making units [J]. European Journal of Operational Research, 1978, 2 (6): 429–444.

Fang Gang, Wang Xiaolu, Zhu Hengpeng. China market index [M]. Beijing: Economic Science Press, 2010.

Hu Dan, Xu Kaipeng, Yang Jianxin and so on. Economic development on environmental quality—the influence of environment of domestic and foreign research on Kuznets curve [J]. Journal of Ecology, 2004, 24 (6): 1259–1266.

Lijing. Chinese environmental efficiency differences and influencing factors study [J]. South China Journal of Economics, 2009, (12).

Shi Dan. The improvement of energy consumption efficiency in China's economic growth. Economic Research Journal [J], 49–56.

State Statistical Bureau Secretary General. The provinces, autonomous region, municipality directly under the central government the unit GDP energy consumption indicators communiqué in 2010:http://www.stats.gov. cn/tjgb/qttjgb/qgqttjgb/t20100715_402657560.htm.

Yuan Nanyou. Low carbon economy concept [J]. City Environment and Ecology, 2010, (1).

Zhang Jun, Wu Guiying, Zhang Jipeng. Chinese provincial capital stock estimation: 19522000 [J]. Economic Research Journal, 2004, (10): 35–44.

Frontiers of Energy and Environmental Engineering – Sung, Kao & Chen (eds)
© 2013 Taylor & Francis Group, London, ISBN 978-0-415-66159-1

Investigation on surface related multiple suppression by predictive deconvolution in linear Radon domain

Y. Shi, Y. Li & H.L. Jing

School of Earth Sciences, Northeast Petroleum University, Daqing, Heilongjiang, China

ABSTRACT: Surface related multiple suppression by predictive deconvolution method in $t - x$ domain usually suppose that multiples have strict periodicity, which limits its application greatly. Linear Radon transform in the paper improves multiples periodicity effectively, and investigate multiples suppression by predictive deconvolution in linear Radon domain. Firstly, the seismic data with surface related multiple is transformed into linear Radon domain, then predictive deconvolution method is used to suppress surface related multiple. This method can determine the predictive deconvolution operator by auto-correlation operation. It discusses the influence of predictive operator length on multiples suppression effect by theoretical models. This method can overcome periodicity limit that predictive deconvolution demand multiple effectively, and has practical value and bright future application.

1 INTRODUCTION

Multiples often exist in seismic data, and surface related multiples is very strong in ocean seismic data, also it will decrease migration effect, influence seismic data interpretation, unknow earth structure if we cope with multiples badly. Currently, there are two solutions about multiple problems, that is multiple migration and multiple suppression. We investigate multiple by viewing it as coherent noise to suppress in the paper. Multiple suppression method can be classified into two types, one is prediction subtraction method based on wave equation prediction, and the other is filter method based on signal processing. The former mainly includes wave field extrapolation, inverse scattering series and feedback iteration method, and the latter mainly predictive deconvolution (Peacock, 1969), F-K filter (Backus, 1959), beamforming and Radon transform filter and so on. It belongs to filter method to apply predictive deconvolution in linear Radon domain to suppress multiples in the paper.

Predictive deconvolution can suppress multiples effectively in seismic data. However, some assumption conditions related to the method limit its application. Conventional predictive deconvolution (Lokshtanov, 1995) method demands multiple with strict periodicity, but the periodicity of multiple often occur near zero offset in horizontal layer media, and multiples do not appear periodicity for far offset. Linear Radon transform, i.e. $\tau - p$ transform can improve periodicity of multiples. Multiple has strict periodicity in $\tau - p$ domain, and it does not vary with nonzero offset, also it is

the difficulty to suppress long-period multiple in $t - x$ domain by predictive deconvolution method exactly.

So it is feasible in theory to suppress surface related multiple by predictive deconvolution in $\tau - p$ domain (Lokshtanov, 1999).

The proposed method is not only simple and efficient, but also applicable, which can process the seismic data with surface related multiple. The theoretical model tests proposed algorithm in the paper.

2 BASIC PRINCIPLE

One weak part of the predictive deconvolution is that its processing effect relies on strict periodicity of multiples. This is the only case around the zero offset for horizontal layer media. For far offsets, the periodicity of multiples breaks down (Lokshtanov, 2000). In other words, for the given non-zero offset, multiples do not have periodicity in time. This is because each higher order multiple arrives at a different angle, and the difference in travel-time between each order of multiples becomes smaller, as shown in Figure 1 (a). When the data is transformed into the linear Radon domain, the multiples become periodic again if one horizontal ray parameter is considered, as shown in Figure 1 (b). Therefore, predictive deconvolution method can be applied per ray parameter independently, and the second-order multiple is shown in Figure 1 (b) in dotted line.

Wu (2003) proved that the delay time between the nth reflect multiple and the $(n - 1)th$ reflect

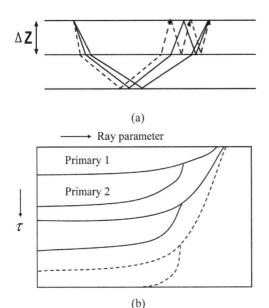

(a)

⟶ Ray parameter

Primary 1

Primary 2

(b)

Figure 1. The water layer multiple and reverberations become periodic per horizontal ray parameter in linear Radon domain.

multiple can be expressed for a certain offset or a certain trace in the $x - t$ domain as follows,

$$\Delta t = t_n - t_{n-1} = \left(\frac{x_n^2}{v^2} + n^2 t_0^2 \right)^{\frac{1}{2}} - \left(\frac{x_{n-1}^2}{v^2} + (n-1)^2 t_0^2 \right)^{\frac{1}{2}} \quad (1)$$

where t_n and t_{n-1} represent travel time of the nth and the $(n - 1)th$ reflect multiple respectively; x_1 to x_n are offset; v is velocity in homogeneous media.

While in $\tau - p$ domain, we can get

$$\Delta \tau = \tau_n - \tau_{n-1} = (1 - p^2 v^2) \tau_0 \quad (2)$$

where p is ray parameter, and

$$\tau_0 = t_0 = 2 \Delta z / v \quad (3)$$

Compare equation (1) and (2), in $x - t$ domain, we can see that the repeat periodic or delay time Δt of multiples have strict periodicity only at non-zero offset, i.e. $x = 0$. Non-zero offset are related to n, and it is a complex function relation. The delay time of multiples Δt has no relationship to x or n, only relates to τ_0. This shows that, to each p, all whole-trip multiples produced from horizontal

reflection interface increases one relative travel time when one reflection is increased.

It needs to select different delay time for different ray parameter when predictive deconvolution is computed. The delay time at zero ray parameter equals to the vertical two-way travel time in the first layer, that i

$$\Delta \tau(0) = 2 \Delta z / v \quad (4)$$

From Equation (2), the delay time for other horizontal ray parameters p_x is given as follows:

$$\Delta \tau(p_x) = \Delta \tau(0) \sqrt{1 - v^2 p_x^2} \quad (5)$$

We can also see that $\Delta \tau(0)$ is the predictive distance when p equals zero. We can adjust predictive distance according to p value when the operator length is constant.

Predictive deconvolution in linear Radon domain can be defined as follows:

$$p_0 (p_x, \tau) = p(p_x, \tau) + f(\tau) * p(p_x, \tau - \Delta \tau(p_x)) \quad (6)$$

Suppressing multiples mainly use the accurate periodicity $\Delta \tau = \tau_n - \tau_{n-1}$ in $\tau - p$ domain, and the easy prediction properties of energy amplitude between the adjacent order multiples of $\tau - p$ domain. Through autocorrelation process, we can determine periodicity of multiples and provide predictive deconvolution operator. Conventional predictive deconvolution is used in $\tau - p$ domain to suppress multiples, then the data in $\tau - p$ domain is returned to $x - t$ domain to do other processing.

3 EXAMPLE ANALYSIS

The proposed method is tested by theoretical synthetic data in the paper.

The shot record contains primary and multiple, as shown in Figure 2 (a), and its auto-correlation result for each trace is shown in Figure 2 (c), also multiple periodicity can be seen at near offset, and multiples have no periodicity at far offset. Frequency dispersion properties can be observed from auto-correlation profile, so it is not desirable for shot gather record to suppress multiples by predictive deconvolution.

The $t - x$ domain data shown in Figure 2 (b) is from the data in Figure 2 (a) by linear Radon transform. The data in $\tau - p$ domain improve multiple periodicity obviously, and multiple periodicity can be seen in $\tau - p$ domain for most ray parameter, which can be observed from auto-correlation record in $\tau - p$ domain, and the auto-correlation record in $\tau - p$ domain can show

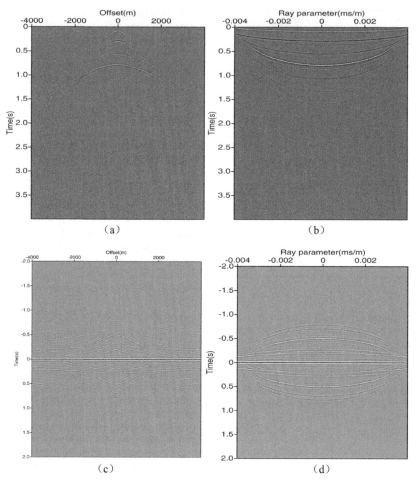

Figure 2.   Linear Radon transform of the shot record and its auto-correlation.

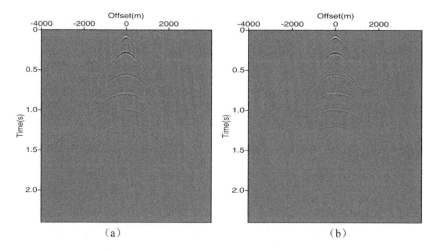

Figure 3.   Surface related multiples suppression by predictive deconvolution in $\tau - p$ domain.

multiple periodicity well, as shown in Figure 2 (d). However, we can also observe that multiple period is vary with different ray parameter, as is described at Ellipse Equation (4). The periodicity in $\tau - p$ domain provides a good chance for applying predictive deconvolution.

In order to observe the suppression result better, we use predictive deconvolution to suppress multiples in the shot record shown in Figure 2, and the result is shown in Figure 3, where only partial profile is cut to display. The longer predictive operator, the more multiple remains. While the predictive operator is shorter, false event can be seen. So it is very important to define reasonable and effective parameters for this method.

4 CONCLUSIONS

Using predictive deconvolution method to suppress surface related multiples in linear Radon domain is put forward in the paper, which overcomes conventional periodicity limit of multiples by predictive deconvolution in $t - x$ domain. Linear Radon transform can improve periodicity of multiples effectively, and predictive deconvolution operator can be determined by auto-correlation operation. Theoretical model test result show that the method can suppress surface multiple effectively.

ACKNOWLEDGMENTS

This research was supported by the Project of National Natural Science Foundation of China. (grant No. 41004057, 41104088), the Project of Science Technology Investigation of Heilongjiang Province Education office. (grant No. 12511025), the Open Project of Earth Information Detection Technology and Instrument Education Ministry Key Lab. (grant No. GDL0903).

REFERENCES

Backus, M.M., 1959, Water reverberations – Their nature and elimination: Geophysics, Soc. of Expl. GEOPHYS., 24, 233–261.

Lokshtanov, D., 1995, Multiple suppression by single channel and multichannel deconvolution in the tau-p domain, 65th Ann. Internat. Mtg:Soc. of Exel. Geophys., 1482–1485.

Lokshtanov, D., 1999, Multiple suppression by data-consistent deconvolution: The Leading Edge, 18, no. 1, 115–119.

Lokshtanov, D., 2000, Suppression of water-layer multiples from deconvolution to wave-equation approach: 70th Annual Internation Meeting, SEG, Expanded Abstracts, 1981–1984.

Peacock, K.L., and S. Treitel, 1969, Predictive deconvolution: Theory and practice: Geophysics, Soc. of Exel. Geophys., 34, 155–169.

Frontiers of Energy and Environmental Engineering – Sung, Kao & Chen (eds)
© 2013 Taylor & Francis Group, London, ISBN 978-0-415-66159-1

Relational study on the public library alliance based on the public cultural services system

L. Li

The Propaganda Department, Jiangxi Science & Technology Normal University, Nanchang, Jiangxi, China

H. Xiao

Library of Jiangxi Agricultural University, Nanchang, Jiangxi, China

ABSTRACT: Cultural gradually become an important aspect to measure the comprehensive strength of a country or region and the protection of citizens' cultural rights and the increase of cultural soft power are an important goal of the public culture system. Public libraries are not only an important part of our public cultural service system, but also an important development objects for national and regional culture-building strategies. Library Union, as the organizational form of mutually beneficial cooperation and literature resource sharing between libraries, is gradually developed along with the development of information technology, network technology and database technology. The paper analyzed the background of the public library alliance based on the public cultural services system and dissected the development stage and research status of public library's alliance. Finally, the paper focused on the distinction and future development of the public library's alliance.

Keywords: public library; alliance; public cultural services system

1 INTRODUCTION

With the growing trend of China's socialist market economic development and the increasingly evident trend of mutual integration of the cultural and economic, cultural gradually become an important aspect to measure the comprehensive strength of a country or region. Therefore, it is an important goal of our public culture system for both the protection of citizens' cultural rights and the increase of "cultural soft power". It is the officially launched of "National Cultural Information Resources Sharing Project" in 2002 and the release of "Number of views on strengthening the construction of public cultural service system" in 2007 by China government as well as all over the country's cultural construction practices that is conducive to the realization of cultural information resources sharing, and also pointed out the direction for public library construction and development.

In network environment, the majority of single public libraries face with the lack of resources, shortage of funds and obsolete equipment, and other problems and it is difficult to meet all the cultural information needs. Therefore, it is the development direction for public library construction with the realization of cultural information

resources common building and sharing, the joint establishment of the sound document information security system, and the providing with the one-stop cultural information service for users. Library Alliance is an important way for information resources sharing between individual library, and it is with the way of resource sharing and mutually beneficial cooperation to meet the growing and complex needs of readers. However, the domestic and foreign scholars' study focused primarily mainly on the Library Union Alliance, the Library Union Building, the Regional Library Alliance, University Library Alliance and other aspects. While the study on Public Library Union is so less, which posed a challenge to the development of public libraries, created opportunities. Present, China has a large number of regional or small-and medium-sized public library service network, such as Shanghai Document Resource sharing and collaboration network, the network of public libraries of the Pearl River Delta Region, Chongqing Fuling District's Digital Library Federation, Zhejiang Province Public Library Information Services Union, and so on. The public library service network to a certain extent achieve the purpose of sharing resources and mutual benefit, but still not really play the effectiveness of the Library Consortium. And also it lacks planning of cultural information resources, business processes, and services content of the public libraries based on the Union as a whole.

2 DEVELOPMENT STAGE AND RESEARCH STATUS OF PUBLIC LIBRARY'S ALLIANCE

Library Union, as the organizational form of mutually beneficial cooperation and literature resource sharing between libraries, is gradually developed along with the development of information technology, network technology and database technology. At present, domestic and foreign scholars made a summary to the library union's theoretical research and the construction of practical experience and have got some successes. Foreign scholars' survey direction on the Library Union is mainly contain the concept of Library Alliance (Carlson & Scott, 2003), the formation of elements, alliance management, impact factors (Kirlidog & Melih, 2007), alliance-building practice experience (Matthews, 2009) and other aspects. Despite Chinese scholars' study on Library Consortium have a late start, but in recent years the research is expanded and research efforts is deepening which mainly focus in the overview study, the introduction of foreign alliances research, coalition-building and management, regional Library consortium research, digital Library Federation research and other aspects.

United States, Britain, Japan and other developed countries shows the characteristics of the foreign public library service system and alliance-building in the Union research and construction of public libraries. The management mode, the resources set, the geographical distribution and other aspects are worthy of our public library construction the cause carried out for reference. Therefore, we have chosen the United States, Britain, Japan, the representative countries to introduce foreign public library alliance-building and Research status.

The United States, the world's power, has established a comprehensive public library system. Its service contents, organizational models and management mechanisms are worthy of reference of the Public Libraries in the world. The American public library is the people's community center. The public can borrow books, use of computer resources, access to financial information and get policy advice in the public library. The scope of services of the public library service system covering all aspects of civic life and work and its service contents are variety too. American Public Library Service System services include basic services and extension services in two parts. And the basic services is including lending service, information

inquiry service, interlibrary interactive service, Regardless of which library you can borrow and return service, multi-cultural lending services and so on and the extension services is including the special services for vulnerable groups, the object-centered services to children, organizing various cultural activities and other manners. The heart of the system of public library services is the community library (Yu, 2009). Lots of community libraries which provide loan services, news and information, such as diversification, personalized service for the residents are situated in place where traffic convenient. American Public Library System mainly uses "Center Museum-Branch Museums" organizational model to provide citizens with universal and equal with the shape of the network to maximize the effectiveness of service. According to statistics, there are a total of 9221 public libraries in the United States in 2008, of which 1559 libraries have build one or more branch libraries. And the total number of branch libraries is 7629 (Smith, 2007). The U.S. public library system is a very large social service system, but the government did not establish a single management agencies. The supervision and management of public libraries is by the laws, regulations and industry standards.

The United Kingdom is one of the oldest countries which built the public library system. As early as 1985, Britain promulgated the first Public Libraries Act and revised and published the "Public Libraries and Museums Act" in 1964. Currently, it has been forming a better library legal system. These laws and regulations of the Public Library played in a very good effective oversight and policy guidance for public library development. At present, the UK public library service system, provided book lending services, document delivery services, Reference Services and so on for more than sixty million UK residents, has been around the British urban and rural areas. In recent years, depending on the cultural needs of the urban and rural residents, the British public library system expands the various forms of library services (Parker, 2006). Developed network of public libraries is the main development path of the Public Library of the British cities. The City of London, for example, over 390 public libraries in the London area which having a total of 16 million paper books and 1.5 million sets of CDs, audio books, tapes, CD-ROM, was formed the service network and the London Library Development Bureau is responsible for the operation and management. UK City Public Library which build with Commercial Chain Center, not only can promote the library's own social value can also be in order to mutually beneficial cooperation with the operators. For the rural residents of the United Kingdom, the British public libraries mainly provide mobile library services by through books mobile van.

The development of public libraries in Japan is in the world leading level, especially that the library has a perfect legal system. At present, Japan's national legislation of the library career has a total of three: "National Diet Library Law", "Library Law" and "School Library", in which the "Library Law" is a specialized norms of public libraries law (Rockman, 2005). Japan's public library system includes four levels of the main library, branch libraries, books docked and mobile library. And the public library service network is set mostly in accordance with administrative divisions. Of Tokyo, for example, public library service network uses the Main-Branch library system which made the area as the geographical units. Each district has set up a library for the main library and the rest of the library as the branch. The region where the Main Library and branch libraries cannot be covered is complemented through the mobile service car. The Japanese government at all levels is the main body of the public library service system construction and the public library services system carried out capital unified management, personnel unified deployment, resources unified procurement, unified services policies and other systems.

By exploring and analyzing the United States, Britain, Japan and other countries' construction and management experience of public library service system, it will benefit to the construction and research of the Public Libraries Union in the new era.

3 CONSTRUCTION OF PUBLIC LIBRARY UNION IN CHINA

Public Library Union (Public Library Service System) is an important part of our public cultural service system. At present, Chinese scholars' study on Public Library Union mainly include three areas: comprehensive theoretical study, the topic of theoretical research and practical experiences.

3.1 *Comprehensive theoretical study*

Comprehensive theoretical research is means that around the theme of the Public Library Union or the public library service system to carried out allround elaboration. Public Library Union and the Public Library Service System are the organization forms of public libraries in the system interlibrary cooperative. And the theoretical studies and construction practices of both of them have commonality. So the related study of the system of public library services for public libraries alliance building and research provides an important theoretical and practical basis.

Making the Public Library Service System as the theme and Droved in the universal and equal services target, Yu-liang Chi et al introduces the main content of public library service system building which including the construction of grass-roots library, branch library building and regional services network construction. Xing Jie, Li Lingjie described the construction status and strategies of the two public library service system models, which are the Regional Service Network based on the big cities which made extend service as characteristics and the Main-Branch library model based on small and medium-sized cities. Ruan Shengli's analyzed of the level problem of public library services system, the positioning problem of the urban and rural community libraries and the "universal and equal" principle problem of public library services, aims to provide a reference for the building of grass-roots public library services system. In addition, there are many scholars of public library services system as the theme for a comprehensive discussion and research, aimed at provide the underlying theoretical basis to the public library service system. Although the public Library Union research yielded few results, but also reflects the concern of the research scholars of the Union of the Public Library. He Yan Public analyzed Library Union Development and the constraints in "Thinking on Public Library Union development", and combined with examples which discuss the prospects for the development of the Public Library Union, then proposed the development of countermeasures. Based on the description of the alliance development status of our public libraries, Zheng Bangkun analyzed the constraints of the Union the development of small and medium-sized public libraries. And Sichuan Province Public Library, for example, he putted forward construction strategy of the provincial small and medium-sized public library alliance.

3.2 *Theoretical monograph*

Compared with comprehensive theoretical study, Theoretical Monograph pay more attention to theoretical issues of a particular aspect of the alliance-building of public libraries, and in-depth exploration, such as construction mode, governance model, performance evaluation, and close analysis.

In the construction mode, Liang Xin proposed Main-Branch library and Rural Community Book Museum building mode aimed at China's actual conditions on the basic of in-depth study of our public library service system factors in the mode of construction in the book of "China's public library service system construction mode study". Li Mingsheng analyzed the close degree of the public library service system, and according to the close degree divided public library services system into six modes. By analyzing the Zhejiang's part of the county, town, village three library construction surveys and visiting the library in Jiangsu, Guangdong and other places, Li Hongxiang et al. carried out the construction model which based on County Public Library Three Service System. The study on the construction mode is the main aspects of the topic of Union of Public Library. There are also many scholars study the governance model of the Public Library Service System. Li Mingsheng have analyzed the pure Main-Branch library system, close Main-Branch library system, semi-compact Main-Branch library system and other public library services system of six kinds' board governance model and governance structure from the property rights perspective. In addition, The performance evaluation for the subjects of the study is also one of the topics of Public Library Alliance.

3.3 *The practical experiences*

Since trying to construct the library consortia in the 1990s, China has built a number of Universities Library Alliance, Regional Library Union, Digital Library Federation, scientific research Library Consortium and the Public Library Alliance. At present, Chinese public library alliance which has built or are building mainly include: Shanghai Literature resources sharing Collaboration, the Beijing Municipal Public Library Information Service Network, collaboration Library of Changchun City, the Pearl River Delta Region Public Library Union and other large public libraries Union, as well as small and medium-sized public library alliance. In order to further promote the development of our public libraries, a lot of research scholars were summarized and analyzed the practical experience of the public library alliance-building. And then can it help to provide science building strategy for the future public library alliance-building.

At present, the pace of development of the alliance-building of public libraries in China's developed eastern regions is significantly faster and it has established a number of successful public library alliances. Wang Xuexi in the article "Current Situation and Countermeasures of the public library service system" focuses on the public library alliance-building in the eastern region, such as "mobile library" model in Guangdong, "City of Libraries" mode in Shenzhen, Tianjin Book Museum's "Community Branch Library, Industry Branch Library mode", Shanghai's "Central Library mode", Jiangsu's diversified modes. In addition, on the basis of the analysis of

domestic and foreign public library service system status, and with the Jiaxing branch library system for example, Li Chaoping uses empirical research way to examine the design of the system, operation mechanism and operation effect of Jiaxing branch library system. Ye Yanping analyzed the Hangzhou City Public Library's four network system, and proposed a number of measures on the public library network system improving. Research Scholar made an in-depth exploration and study on Jilin Province Public Library Service System, the Pearl River Delta Public Library Service System and so on.

4 DISTINCTIONS AND FUTURE DEVELOPMENT OF THE PUBLIC LIBRARY'S ALLIANCE

The foreign experience in the construction shows that Main-Branch library system is an effective system model of a co-ordinate the resources of public library services in the aspect of management model. Therefore, China should be on the basis of foreign advanced experience and successful experience in practice and in accordance with the actual situation of the region, to explore a management mode which truly suitable for our public library alliance-building. In resource allocation, although the main library of the public library system in different countries which involved in the branch libraries business's scope and depth is different, but they are all stress the Main Library's unified management and scheduling on the resources, personnel, funds of branch libraries. And then it can achieve the goal that made the allocation of resources and shared to maximize efficiency. While there have many serious management compartmentalization issues in the building of the Public Library Network in China and have constraints in fund management and resource scheduling of the public library system, it makes the entire public library system cannot achieve maximum benefit. So we can learn from the advanced experience of foreign countries and combined with China's actual situation and to solve these problems. In terms of layout planning, foreign public library network follows ideas of the overall planning, rational distribution and makes the convenience of the reader as the first criteria. All these were not based entirely on the administrative divisions. While public library network construction in China are mostly based on the administrative divisions. Although it is reasonable, but it should be combined with the actual situation of population distribution, economic development and social needs to do the overall planning and rational distribution.

REFERENCES

Carlson Scott (2003). Libraries' Consortium Conundrum. *Chronicle of Higher Education*, 50(7): A30–A31.

Ilene F. Rockman (2005). Distinct and expanded roles for reference librarians, *Reference Services Review*, 33(3):257–258.

Kirlidog, Melih, Bayir, Didar (2007). The effects of electronic access to scientific literature in the consortium of Turkish university libraries. *Electronic Library*, 25(1):102–113.

London Libraries. http://www.londonlibraries.org/, 2012-01-13.

MacKenzie Smith (2004). Libraries in the lead: the institutional repository phenomenon, *paper presented at Breaking Boundaries: Integration and Interoperability*: VALA, 12th Biennial Conference and Exhibition.

Sandra Parker (2006). The performance measurement of public libraries in Japan and the UK. *Performance Measurement and Metrics*, 7(1):29–36.

Tansy E. Matthews (2009). Improving Usage Statistics Processing for a Library Consortium: The Virtual Library of Virginia's Experience. *Journal of Electronic Resources Librarianship*, 21(1):37–47.

YuDajin (2009). Research on the SuPer-Network Model for Eco-industrial Convergence-A case of Poyang Lake Eco-economic Zone. *Proceedings of 2009 International Conference on Management and Service Science*, IEEE, 9.

Frontiers of Energy and Environmental Engineering – Sung, Kao & Chen (eds)
© 2013 Taylor & Francis Group, London, ISBN 978-0-415-66159-1

Characteristics of global cruise tourism consumption and their inspirations to Shanghai

G.D. Yan
School of Management, Shanghai University of Engineering Science, China

J.C. Kang, G.D. Wang & Q.C. Han
Urban Ecology and Environment Research Center, Shanghai Normal University, China

ABSTRACT: According to "12th Five-year Program of Shanghai Cruise Industry", by the end of the 12th five-year program, Shanghai will have become a cruise hub port in East Asia and one of the three int'l cruise centers in Asian-Pacific Region after Singapore and HK. How to fulfill the abovementioned development goals, conform to the development trend of global cruise tourism consumption, and facilitate development of cruise market with success has become a critical issue in urgent need of solution. Therefore, the paper attempts to probe into characteristics of global cruise tourism consumption focusing on Consumption Level, Consumption Will, Leisure Time, Route Demand, Features of Information Demand based on related research findings, and raises countermeasures based on the current situation of Shanghai cruise tourism consumption, in the hope of promoting sustainable development of cruise tourism and providing foundations for decision making of the government's related management policies. We got that we should improve the consumption level of cruise tourism, bridge this gap between cruise tourists and non-cruise tourists and fully meet the consumption will of tourists, cruise companies and travel agencies should focus on the launch of routes within 6 days for route design and product promotion. To strive for the affiliation of int'l cruise routes from Europe, America and Asian-Pacific Region boost rapid growth of travel agencies and enhance information publicizing about cruise tourism through travel agencies.

Keywords: cruise tourism; consumption characteristics; Shanghai

1 INSTRUCTION

Since 1980, the growth speed of global cruise tourism has been about 8% on average annually, and the total number of tourists has risen from 6.8 million in 1996 to 17 million in 2008, with an average growth rate of 9.3% between 1996 and 2006, much higher than the overall growth rate of int'l tourism. All these showed the rapid development of cruise tourism market, in which the North American Market takes up an absolute share. In 2009, the number of cruise tourists from North America seized 75.8% among the world cruise tourist gross. Despite the impact of world financial crisis, the cruise tourist source market in North America fell by 1.5% year-on-year, but the world cruise tourist gross still rose by 4% year-on-year. Meanwhile, the cruise tourism market in Europe witnessed a boom. From 1999 to 2009, the European cruise market demand maintained a growth rate of over 10%, and in 2009 the market size grew by 166%, 13.3 times of the growth scale of land tourism market at the same period of year (China Communications and Transportation

Association—Cruise & Yacht Industry Association, 2010–2011).

The Asian cruise market bears a great development potential. In the last 8 years, the number of cruise tourists in Asia has tripled, and the Asian-Pacific Region has become an important engine for the future growth of global cruise market. In this region, China is deemed the major power zone for the resurgence of Asian cruise tourism market. In the past few years, the number of Chinese cruise tourists showed an obvious growth, and it's forecasted that by 2020, China will have become the largest tourist destination in the world and attracted more int'l cruises to berth alongside (China Communications and Transportation Association—Cruise & Yacht Industry Association, 2010–2011). The Yangtze River Delta City Agglomeration centered at Shanghai has become the most energetic area of economic development in China, and the leaping growth of economy drove the rapid increase of resident's income and consumer spending in the city, thus leading to rapid growth of cruise tourism demands. According to "12th Five-year Program of Shanghai Cruise Industry", by the end of the 12th five-year program, Shanghai will have become a cruise hub port in East Asia and one of the three int'l cruise centers in Asian-Pacific Region after Singapore and HK; it was also estimated that the number of inbound and outbound cruises in Shanghai would reach 500 ship times, the number of cruise tourists would achieve 1 to 1.2 million, and 5 to 8 cruises would have their home port in Shanghai.

How to fulfill the abovementioned development goals, conform to the development trend of global cruise tourism consumption, and facilitate development of cruise market with success has become a critical issue in urgent need of solution. Therefore, the paper attempts to probe into characteristics of global cruise tourism consumption based on related research findings, and raises countermeasures based on the current situation of Shanghai cruise tourism consumption, in the hope of promoting sustainable development of cruise tourism and providing foundations for decision making of the government's related management policies.

2 CHARACTERISTICS OF GLOBAL CRUISE TOURISM CONSUMPTION

It's well known that tourism consumption behavior is usually impacted by factors like income, price, motive, attitude and leisure time (Lin Zengxue, 1999). And as a special type of tourism product, the consumption behaviors of it must bear some uniqueness. Based on extant researches, this paper tries to make analysis of characteristics of cruise tourism consumption by focusing on consumption level, consumption will, leisure time, route demand and information demand.

2.1 Consumption level

The average consumption level of cruise tourists is about 50% higher than that of non-cruise tourists. A region possesses the basic condition of developing cruise tourism economy when the GDP per capita is above USD 6,000. In 2009, cruise economy brought about direct economic contributions worth USD 17.2 billion and EUR 14.1 billion respectively to the world's top 2 cruise market, namely, North America and Europe, as well as an overall economic impact of USD 35.1 billion and EUR 34.1 billion respectively. From 2002 to 2010, the British tourist source market size was doubled, and the cruise tourism market penetration rate was enhanced year by year. Thus, the market share that cruise tourism took up in Britain's package tourism rose from 8.33% in 2008 to 10.6% in 2009, making Britain the world's second largest cruise tourist

source country (CLIA, 2009). As the world's third largest inbound tourism receiving country and the fourth largest outbound tourism consumption country, China's domestic tourism income rose by 12% year-on-year in 2011, leading to obvious increase of consumption needs of cruise tourism and the feature of low price leading.

2.2 Consumption will

Consumption will has a direct impact on decision making about cruise tourism consumption. Between 1980 and 2009, the total tourists joining in over 2 days of cruise tourism in North America amounted to 176 million, among whom 68% chose cruise tourism in 10 years, and 40% in 5 years. It's expected that in the following 3 years, the number of tourists willing to take part in the tourism will exceed 50 million, owing to an obvious increase of consumption will in cruise tourism. In the US, about 20% US citizens have taken part in cruise tourism, among whom tourists who did it for the first time are mainly concerned about the experience of cruise tourism, and they take up 70% of the total. Currently, more than 68 million US citizens wish to join in cruise tourism, and 69 million are hoping to achieve this in the following 5 years. More than 43 million are certain that they will put it into practice. That means the potential cruise tourism consumption amount will reach USD 57 billion to 85 billion. All these show obvious cruise tourism consumption will. The outbound tourism that Chinese citizens take is mainly directed to short-distant destinations, like North America, Africa, Hong Kong, Macao, Taiwan, Japan, Korea and ASEAN countries. Lately, destinations of medium- and long-distant tourism enjoy greater popularity, which drove the increase of the consumption will of cruise tourism.

2.3 Leisure time

Leisure time is a necessary factor for realizing cruise tourism consumption behavior. Now regional tours lasting 6 to 7 days take up the largest proportion, staying at about 60%; and the proportion of offshore tours lasting 2 to 3 days keeps rising (Marti, B.E, 2004; Zhang Yanqing, 2010). Besides, cruise tourism in different parts varies greatly in the lasting time. In 2009, European and US cruise tourism generally lasts 5 to 12 days, among which 7 is the mainstream; in North American market, 6 to 8 days' tours take up the largest proportion, 49.1%, and tourists joining in 2 to 8 days' tours take up 79.1% among the total tourists; the average time of cruise holidays in North America increased from 6.6 days in 1999 to 7.2 days in 2009, and routes in North American cruise market went up slowly

(CLIA, 2009). Between 2006 and 2012, the cruise routes in China mainly lasted from 5 to 7 days, and a tendency of short-distant routes at cruise home ports became extremely evident.

2.4 Route demand

Now Caribbean and Europe/Mediterranean are two regions most concentrated with cruise routes in the world, witnessing cruise visit rates of 47.2% and 24% respectively. Tourists from different regions show obviously different demands of cruise destinations. In 2009, cruise tourists from North America favored Caribbean, Mediterranean, continent of Europe, Bahamas and Alaska, and the total days that North American cruise tourists spent in Mediterranean seized 18.2% of the total bed days of North American cruise market. Mediterranean, Baltic Sea and other European regions are port areas that received the largest number of cruises in Europe, and the routes of Mediterranean contributed 2,823,000 bed days to European cruise market, which was about 57.1% of the total European bed days. It was forecasted that in the following 3 years, cruise tourism in South America and Antarctica, Mediterranean, Norway, Baltic Sea and Asian-Pacific Region would enjoy the greatest development potential, and Europe, Asia and North America would witness the fastest growing demands of cruise tourism. France, Italy, Germany and Spain are the most popular countries in Europe and Asia, and the favored destinations of Chinese tourists include Mediterranean and the Far East, followed by Caribbean and North Europe. From 2009 to 2011, Chinese mainland saw 106 more int'l cruise voyages, among which 44 more for coastal cities. The cruise route demand showed an evident growth.

2.5 Features of information demand

Judging from the planning time of cruise tourism, from 2006 to 2008, the advance planning time for North American cruise tourists extended from 4.3 months to 5.6 months, among which 4 to 6 months took up the largest proportion in 2006 and 6 to 12 months took up the largest in 2008 (CLIA, 2009; Zhang Yanqing, 2010). This indicated that tourists' advance booking amount of cruise tourism further increased, and tourists spent a longer time in acquiring cruise tourism information. As for information channels, in 2006, cruise tourists acquired information in the sequence of verbal publicity, spouse or travel partner, websites of destination and cruise, and travel magazines; compared with non-cruise tourists, the most evident difference in their source of information was the website of cruise, in which

non-cruise tourists were 28% less than cruise ones; then it comes to recommendation by travel agency, in which non-cruise tourists were 7% less; but in items of "verbal publicity and spouse or travel partner", non-cruise tourists were 5% more than cruise tourists (CLIA, 2009). All these show that cruise tourists are more willing to acquire information via cruise website and travel agency than non-cruise tourists do, while non-cruise tourists are more willing to acquire information through verbal publicity and spouse or travel partner than cruise ones do. Therefore, publicity through these channels should be strengthened. It's been not long since Chinese tourists became acquainted with cruise tourism, and introduction from relatives and friends, TV and broadcast and newspapers, magazines and books capture the top 3 among all channels of information about cruise tourism; the functions of the Internet and related tourist agencies in publicizing cruise tourism are yet to be explored (Zheng Hui, 2009).

3 CURRENT SITUATION OF SHANGHAI CRUISE TOURISM CONSUMPTION

3.1 Consumption level

In 2005, the per capita GDP in Shanghai amounted to USD 6414, and the figure exceeded 10,000 in 2008, largely helping rapid development of cruise tourism consumption in Shanghai. Between 1997 and 2005, the number of cruises inbound and outbound at Shanghai Port totaled 25, with cruise tourists adding up to 75,000. This indicated the very beginning of cruise consumption. From 2007 to 2010, the number of tourists in inbound and outbound cruises at Shanghai Port rose from 100,000 to 340,000, a very evident increase. On Feb. 16th, 2010, "Queen Mary 2" Cruise brought about 3700 tourists and sailors in total, making it the largest luxury cruise since Shanghai Port was opened for business. During World Expo 2010, solely Pujiang Frontier Inspection Station checked 141 cruises, including 102 Sino-Japan passenger cruises, which brought nearly 260,000 tourists and sailors, witnessing a passenger flow volume rise by 176% year-on-year. Shanghai Expo led to tourist peak in cruise tourism consumption. In 2012, the number of cruise tourists amounted to 300,000 based on the current information of cruise routes already confirmed. And by the end of the "12th Five-year Program", the figure was forecasted to achieve 1 million to 1.2 million; the direct economic contributions that the cruise industry makes to Shanghai will reach 5 billion to 8 billion yuan, and its overall economic contribution will get to 15 billion to 20 billion yuan. All these are symbols of rapid growth of cruise tourism consumption.

3.2 Consumption will

Consumption will directly impacts decision making about cruise tourism consumption. In 2010, among all Shanghai tourists, 57.77% are willing to go traveling with their families; 20.39% are willing to choose the time for honeymoon; 16.99% are willing to go with their friends. It could thus be seen that cruise tourists going with families bear an obvious consumption will, and there's a great potential in the consumption market of tourists traveling with friends. In addition, non-cruise tourists and cruise tourists in Shanghai show certain difference in their demand preference. In 2010, non-cruise tourists in Shanghai believed that the top 3 major attraction factors about cruise tourism were respectively the exotic flavor, facilities and activities of the port berthed, as they deemed them symbols of their identities; at the same year, a survey was carried out targeted at tourists of "Romantica" from Costa Cruise Lines, the result showed that the top 3 attractions included rich activities, shore sightseeing and hearty catering. It could thus be concluded that non-cruise tourists favored a comfortable and romantic atmosphere, while cruise ones paid greater attention to cruise experience. The difference of demand preference was quite obvious between the two types of tourists. Therefore, how to bridge the gap between them in their consumption will is an important premise for popularization of cruise tourism consumption.

3.3 Leisure time

In 2010, 71.84% Shanghai tourists accepted cruise tourism products lasting over 5 days, and were willing to choose May for traveling. The top three lasting periods they accepted were 6 to 8 days, 2 to 5 days and 9 to 15 days respectively, which indicated that the characteristic of Shanghai tourists' time demand of cruise tourism products complied with int'l trend while tended to be shorter (China Communications and Transportation Association—Cruise & Yacht Industry Association, 2010–2011). The leisure time of Chinese residents mainly lies in May Day Holiday, National Day Holiday and Spring Festival, but a 5-day cruise line requires at least 6 days. Such a tight schedule drove away many tourists. Besides, Shanghai possesses a complete three-dimensional traffic network, with over 60 foreign airline companies launching regular flights in Shanghai and an average daily handling capacity of 300 int'l flights; also the high-speed railway attracted some tourists. Therefore, more importance should be attached to routes lasting not more than 6 days.

3.4 Route demand

Shanghai is the center of China's offshore routes and Asian cruise routes, having a natural advantage in developing cruise routes. Centered at Shanghai, cruises can reach South Korea, Japan, Singapore, HK and Taiwan within 48 hours, thus obviously helping with the launch of new int'l cruise routes. Currently, the world's top 3 cruise groups have set up offices in Shanghai and launched regional cruise tourism routes with the home port at Shanghai. In 2010, the three cruises "Costa Classica", "Costa Romantica" and "Legend of the Seas", launched "Shanghai-Japan and South Korea" marine outbound routes, the amount and scale both record in Shanghai Port, and formed cruise route demands mainly oriented to white-collar middle-aged and young people. On July 7th, 2012, the two home port cruises "Voyager of the Seas" and "Costa Victoria" berthed at Shanghai Wusongkou Int'l Cruise Port at the same day, bringing about a passenger flow of 16,000. Both of the two cruises launched Asian routes based on Shanghai as the home port for the first time, and the total tonnage and passenger capacity were both record in the operation of China's cruise ports. To continue the expansion of cruise route demands in Shanghai, it's necessary to strive for attracting the affiliation of int'l cruise routes from Europe, America and Asian-Pacific Region, while developing key products based on seasonal characteristics of Shanghai routes. For instance, March to October is the sailing season, and focus should be on developing routes to north coastal region, Japan, South Korea and Russia; Nov. to Feb. the next year should witness more routes to south coastal region and southeast Asia, so that the cycle of slack season and boom season can be minimized, while the economic effect of cruises can be maximized.

3.5 Characteristic of information demand

In 2009, in Shanghai Port International Technical Center, a survey was carried out targeted at tourists of "Costa Allegra" and it was found out that the top 4 channels for tourists to obtain information about cruise were respectively travel agency, recommendation by friends, newspapers and magazines and the Internet; TV, broadcast and pamphlets were of low ranks; and for tourists of "Costa Romantica" starting from Shanghai in 2010, the top 4 channels to obtain information about cruise were the Internet, travel agency, recommendation by relatives and friends, and newspaper commercials; TV, broadcast and travel exhibition were of low rank. These showed that travel agency, relatives and friends, newspapers and magazines and the Internet had become major information sources about cruise for Shanghai cruise tourists. In 2010, based on surveys, the top 3 channels for non-cruise tourists in Shanghai to obtain information about cruise were respectively the Internet, TV and broadcast, and recommendation by relatives and friends.

By comparing the two types of tourists, it can be seen that the difference in travel agency was the most evident, where cruise tourists are 40.07% higher than non-cruise tourists; the second place was the recommendation by relatives and friends and the third was TV and broadcast. All these showed that cruise tourists are more willing to obtain information through travel agency and the Internet, while non-cruise tourists are more willing to do this via the Internet and TV and broadcast. Therefore, publicity about cruise tourism through travel agency, the Internet, and TV and broadcast should be further stressed. In 2011, there were altogether 1175 travel agencies in Shanghai, 46 of which were permitted to operate outbound travel businesses for Chinese citizens and serve on behalf of cruise companies to organize travel groups; as the cruise reception capacity reached 60 to 100 voyages annually on average, it's urgent to boost rapid increase of travel agencies that are able to serve cruise business and enhance information publicizing about cruise tourism through travel agencies.

4 INSPIRATIONS TO SHANGHAI

Firstly, In terms of the consumption level, cruise tourism consumption is developing rapidly in Shanghai but mainly oriented to low price still, so it's necessary to speed up market promotion to attract more consumers to accept high-end cruise tourism products and improve the consumption level of cruise tourism. Second, in terms of the consumption will, there's an evident cruise tourism consumption will in the form of family traveling in Shanghai; non-cruise tourists favored a comfortable and romantic atmosphere, while cruise tourists pay greater attention to cruise experience, therefore, it's urgent to bridge this gap between cruise tourists and non-cruise tourists and fully meet the consumption will of tourists, so as to promote the popularization of cruise tourism consumption. The third, in terms of leisure time, Shanghai tourists' time demand of cruise tourism products complied with international trend while tended to be shorter. Therefore, cruise companies and travel agencies should focus on the launch of routes within 6 days for route design and product promotion. Fourth, In terms of route demand, Chinese tourists give priority to Mediterranean and Far East, so it's good to strive for the affiliation of int'l cruise routes from Europe, America and Asian-Pacific Region; and during March to October, attention should be paid to developing routes to north coastal region, Japan, South Korea and Russia, and Nov. to Feb. the next year should witness more routes to south coastal region and southeast Asia, so that the cycle of slack season and boom season can be minimized, while the economic effect of cruises can be maximized. At last, in terms of characteristics of information demand, Shanghai cruise tourists are more willing to obtain information via travel agency and the Internet, while non-cruise ones favor the Internet and TV and broadcast for information, so it's urgent to boost rapid growth of travel agencies that are able to serve cruise business and enhance information publicizing about cruise tourism through travel agencies.

ACKNOWLEDGEMENTS

This work was financially supported by the Connotation Construction Project of the Twelfth Five-Year Guideline for the Shanghai local undergraduate universities: Construction of Research Bases for Intelligent Management Engineering of Modern traffic and Key Discipline Construction of Decision Support System for modern public transportation and the Public Decision Consulting research (0852011XKZY15), the Key Scientific Research Project for the Ministry of Education (206051/05ZZ13), Key Scientific Research Project of Science and Technology Commission of Shanghai Municipality (062412049) and Geography and Urban Environment, the leading Academic Discipline Project of Shanghai Municipal Education Committee (J50402). We are grateful for their valuable comments.

REFERENCES

China Communications and Transportation Association. Cruise & Yacht Industry Association, 2012. *2010–2011 China Cruise Industry Development Report*, Shanghai.

CLIA. 2009 *Cruise Market Profile Study.* Cruise Lines International Association, New York.

Lin Zengxue 1999. *An Analysis on Tourists' Consumption Behavior Model.* Journal of Guilin Institute of Tourism.

Marti, B.E. 2004. Trends in world and extended-length cruising (1985–2002). Marine Policy, 28(1):199~211.

Zhang Yanqing, Ma Bo, Liu Tao. 2010. Characteristics of International Cruise Tourism Market and Chinese Prospect. Tourism Forum, 3(4) 468–472.

Zheng Hui. 2009. Research on the Development Strategy of Cruise Tourism Products Based on the Demand of Domestic tourist Master Degree Thesis of China Ocean University.

Evaluation of trihalomethanes risk in water system: A case study

C.C. Chien
Industrial Technology Research Institute, Tainan, Taiwan

Y.T. Tu, H.Y. Chiu & C.M. Kao
Institute of Environmental Engineering, National Sun Yat-Sen University, Kaohsiung, Taiwan

C.R. Jou
Department of Safety, Health and Environmental Engineering, National Kaohsiung First University of Science and Technology, Kaohsiung, Taiwan

W.P. Sung
Department of Landscape Architecture, National Chin-Yi University of Technology, Taiwan

ABSTRACT: The main objective of this study was to undertake the multipathway exposure assessment for different water borne disinfection byproduct Trihalomethanes (THM) species contained in the chlorinated drinking water samples collected from different district areas of Kaohsiung City, Taiwan. Tap water samples from 10 locations were collect and analyzed quarterly during a two-year investigation period. Results show that the lifetime cancer risks for Bromodichloromethane (BDCM) contributed the highest percentage (55%) to the total risks, followed by chloroform (9%), and Chlorodibromomethane (CDBM) (36%). The lifetime cancer risks of total THMs in all administration districts were lower than 10^{-5}. Results show that the District 7 had the highest cancer risk mainly due to the exposure to chloroform and BDCM through multipath ways. Intermediate chlorination station located inside District 7 caused high concentrations of free residual chlorine in drinking water, which resulted in the formation of THMs. Compared to other THMs, chloroform posed a higher cancer risk to city residents through dermal exposure. This was due to the fact that tap water would be further treated (e.g., boiling) by city residents before it was used for drinking. Thus, oral ingestion was not the main exposure route. Results from this study would be helpful in developing a drinking water treatment and management plan for the city water treatment system.

1 INSTRUCTIONS

Chlorine is the most widely used disinfectant in drinking water. However, a link between the chlorine disinfectants (e.g., chlorine, chloramines, chlorine dioxide) and Disinfection Byproducts (DBPs) has been found (Alejandro et al. 2012, Bull et al. 2001). These chemical disinfectants react with Naturally Occurring Organic Material (NOM) to create a series of compounds that have been identified as potential carcinogens and the cause of reproductive and development defects in laboratory animals and human (Dell'Erba et al. 2007, Ashley & Julian 2012). Varying water quality and treatment characteristics of the processing of tap water and different exposure route would cause varying cancer risks. Since 1990, scientists proposed that inhalation and dermal absorption be considered in the risk assessment of drinking water (Sadiqa & Rodriguez 2004, Krasuera & Wright 2005, Wright & Murphy 2006).

In southern Taiwan, fresh water comes directly from the Kaoping River, by gravity or via pumps, and goes through large-diameter pipelines to the treatment works. In Kaohsiung City, drinking water is always disinfected with chlorine to kill microorganisms, such as bacteria, viruses, and protozoa that can cause serious illnesses and deaths. This kind of disinfection process is the most commonly employed chemical disinfectant in drinking water treatment nowadays (Chien et al. 2007).

Drinking water is used not only for drinking but also for cooking, washing, bathing, showering, laundering, and cleaning. Thus, humans are exposed to DBPs [e.g., Trihalomethanes (THMs) and Haloacetic Acids (HAAs)] through multiple routes. Ingestion is one of the exposure pathways, such as inhalation and dermal contact during showers, baths, swimming, dish washing, and clothes washing. Traditional risk assessments for water are restricted to ingestion exposure to toxic chemicals. However, studies on the relative importance of these pathways indicate that exposure to toxic chemicals through routes other than direct ingestion may be larger than the exposure from ingestion alone (Tokmark et al. 2004, Wright et al. 2006). This indicates that there is a need to develop exposure models for the entire range of chemical contaminants found in drinking water and for the three primary exposure pathways. The main objective of this study was to undertake multi-pathway exposure assessment of the drinking water from different areas on the concentrations of THMs in water samples collected from Kaohsiung City, Taiwan.

2 MATERIALS AND METHODS

The risk assessment of DBPs in drinking water in Kaohsiung City was analyzed. Triplicate tap water samples were collected and analyzed from 50 locations, representing different districts in Kaohsiung City, for THMs concentration measurement. THMs considered in this study were chloroform, bromoform, BDCM (bromodichloromethane), and CDBM (chlorodibromomethane). Collected samples were analyzed for THMs by GC/MS (APHA 2001). In this study, the approach to human health risk assessment used for exposure to THMs from drinking water in Kaohsiung City followed the guidelines developed by US Environmental Protection Agency (EPA) (US EPA 2002). The risk assessment process included the following four components: data collection and evaluation, exposure assessment, toxicity assessment, and risk characterization.

The source of exposure, exposure pathways (e.g., ingestion route, dermal absorption, inhalation exposure), potentially exposed population, magnitude, duration, and frequency of exposure to site contaminants for each receptor group were identified based on the lifestyle of city residents and the behavior of contaminated chemicals in drinking water. The Reference Doses (RfDs) and reference concentrations were calculated to evaluate non-carcinogenic and developed effects on humans. Cancer slope factors and unit risk estimates were calculated to evaluate the carcinogenic effects. Results were then integrated and compared with the estimates of intake with appropriate toxicological values to determine the likelihood of adverse effects on potentially exposed populations (Chen et al. 2003, Chen 2006).

3 RESULTS AND DISCUSSION

3.1 *THMs concentrations in the distribution system*

The analytical results are shown in Table 1 and are compared with the World Health Organization

Table 1. Summary of THMs levels in tap water of different districts in Kaohsiung City.

Dist.	BDCM	DBCM	THMs
1	3.67E + 00	1.21E + 00	1.34E + 01
2	1.82E + 00	9.03E – 01	6.74E + 00
3	1.40E + 00	8.56E – 01	5.47E + 00
4	7.87E – 01	6.21E – 01	2.84E + 00
5	2.68E + 00	1.35E + 00	8.30E + 00
6	1.05E + 00	8.81E – 01	3.29E + 00
7	9.74E + 00	4.67E + 00	2.75E + 01
8	5.31E + 00	2.52E + 00	1.63E + 01
9	6.09E + 00	2.86E + 00	1.86E + 01

*Unit = $\mu g\ L^{-1}$; ND, not detectable: BDCM < 0.048 $\mu g\ L^{-1}$, DBCM 0.13 $\mu g\ L^{-1}$.

(WHO) (WHO 2006) guideline values (300, 60, 100, and 100 mg L^{-1} for chloroform, BDCM, DBCM, and bromoform, respectively) (WHO 2006). Results show that THMs concentrations were lower than EU (European Union), US EPA, and TEPA (Taiwan) guideline values of 100, 80, and 80 $\mu g\ L^{-1}$, respectively. The total concentrations of THMs in tap water samples were in the range from 2.84 to 27.5 $\mu g\ L^{-1}$. Results show that the THMs concentrations were within the following ranges: chloroform, 1.24 to 12.8 $\mu g\ L^{-1}$; BDCM, 0.787 to 9.74 $\mu g\ L^{-1}$; DBCM, 0.621 to 4.67 $\mu g\ L^{-1}$, and bromoform, 0.039 to 0.428 $\mu g\ L^{-1}$. Chloroform was the major species of THMs. The bromo-THMs concentrations were lower than the concentrations of chloroform. Concentrations of bromoform were undetectable and concentrations of DBCM were also low in most samples. The observations were consistent with other studies (Ashley & Julian 2012, Sadiqa & Rodriguez 2004, Nallanthigal & Amadeo 2012).

3.2 *Evaluations of multi-pathway cancer risk caused by THMs*

The results of cancer risk through oral ingestion are shown in Figure 1 for all districts. The lifetime cancer risks for chloroform, BDCM, and DBCM from tap water of all districts were lower than 10^{-6}, the de minimum or negligible risk level defined by US EPA. For chloroform, the highest lifetime cancer risk of 6.54×10^{-7} was observed in District 7. Among the four THMs, the highest lifetime cancer risk of 5.07×10^{-6} was detected for BDCM in District 7. For bromoform, the concentrations were below detection limit for most samples. Results show that BDCM caused the highest average lifetime cancer risk followed by chloroform, DBCM, and bromoform.

Because exposure to multiple toxicants may result in additive and/or interactive effects, interactive

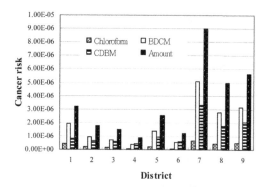

Figure 1. Cancer risk of THMs through oral route in tap water.

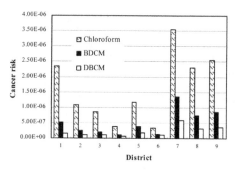

Figure 2. Cancer risk for males from THMs through dermal contact (during bathing and showering) in tap water.

effects may be synergistic or antagonistic (Ashley & Julian 2012). Furthermore, the highest lifetime cancer risk for total THMs was 9.02×10^{-6} in District 7 while the lowest lifetime cancer risk for total THMs was 9.02×10^{-7} in District 4. The lifetime cancer risks for total THMs in all districts were lower than 10^{-5}. Higher THMs concentrations were due to local accumulation at locations close to pipe-ends in the distribution system (Chien et al. 2008). District 7 was located near the pipe-end of network and there were higher concentrations of organic pollutants. The intermediate chlorination process caused the increase in THMs values and cancer risks in this district. District 4 was located in the pipe-front network region, and thus, lower THMs concentrations and lower cancer risks were observed.

The cancer risk of THMs through dermal absorption exposure for both males and females are shown in Figures 2 and 3, respectively. The values of potency factors for oral route were used to calculate the cancer risks of THMs through dermal contact. The lifetime cancer risks of chloroform through dermal contact with tap water for males in most districts were higher than 10^{-7}, which was in the lower end of the acceptable risk range determined by US EPA (Fig. 3). The values for bromoform were either undetectable or lower than the acceptable risk level by a factor of 100 to 1,000. Among the four THMs, chloroform had the highest lifetime cancer risk, followed by BDCM, DBCM, and bromoform. Comparing the cancer risk in different districts, for chloroform, District 7 had the highest cancer risk value of 3.54×10^{-6} while the lowest cancer risk of 8.55×10^{-7} appeared in the District 3. For BDCM, the highest and the lowest cancer risks were in Districts 7 and 4, respectively.

Inhalation exposure occurred when the air breathed contained compounds volatilized during water usage, such as bathing, showering, washing, and cooking. Showering has been identified as the

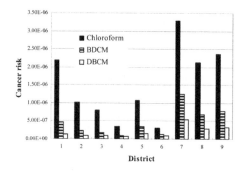

Figure 3. Cancer risk for females from THMs through dermal contact (during bathing and showering) in tap water.

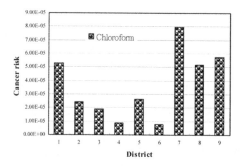

Figure 4. Cancer risk from THMs through inhalation route (during bathing) in tap water.

activity contributing the greatest amount to inhalation exposure to volatile compounds (Chien et al. 2008, Glauner et al. 2005). Due to its property of a lower boiling point, chloroform is assumed to be the major compound to which people are exposed during showering and bathing. As a result, the calculation of cancer risk of THMs through inhalation was only carried out for chloroform compound. The results are shown in Figure 4. The chloroform from tap water in the districts was at or above the 10^{-6} risk level. Residents in District 7 had the highest cancer risk through inhalation of chloroform among the different districts, and the residents in District 6 had the least cancer risk.

4 CONCLUSIONS

The association between THMs exposure through three different pathways and lifetime cancer risks was evaluated. The results show that people had a higher risk of cancer through inhalation route. This was different from the results reported by other researches. This is due to the fact that most residents in Taiwan do not drink tap water directly without further treatment (e.g., boiling). Thus, oral ingestion was not the main exposure route. The lifetime cancer risks through oral ingestion of BDCM and DBCM from tap water of all districts were higher than US EPA suggested limiting value (10^{-6}).

The highest lifetime cancer risk for total THMs was observed in District 7 and the lowest lifetime cancer risk for total THMs was detected in District 4. District 7 was located near the pipe-end of network and there were higher concentrations of organic pollutants. The intermediate chlorination process caused the increase in THMs values and cancer risks in this district. District 4 was located in the pipe-front network region, and thus, lower THMs concentrations and lower cancer risks were observed. Thus, proper water system management should be applied in districts in the pipe-end regions. Results from this study would be helpful in developing a drinking water treatment and management plan for the city water treatment system.

ACKNOWLEDGEMENTS

This study was funded in part by Taiwan National Science Council. Additional thanks to the personnel of Taiwan Environmental Protection Administration and Environmental Protection Bureau of Kaohsiung County, and National Sun Yat-Sen University for their laboratory and field studies cooperation and assistance with this research project.

REFERENCES

Alejandro, M.L. Rincon, A.G. César, P. Norberto, B. 2012. Significant decrease of THMs generated during chlorination of river water by previous photo-Fenton treatment at near neutral pH. *Journal of Photochemistry and Photobiology A: Chemistry* 229: 46–52.

American Public Health Association (APHA). 2001. Standard Methods for the Examination of Water and Wastewater, 20th ed. American Public Health Association/American Water Works Association/Water Environment Federation, Washington DC, USA.

Ashley, D.P. & Julian, L.F. 2012. Improving on SUVA254 using fluorescence-PARAFAC analysis and asymmetric flow-field flow fractionation for assessing disinfection byproduct formation and control, *Water Research* In press.

Bull, R.J. Krasner, S.W. Daniel, P.A. Bull, R.D. 2001. Health effects and occurrence of disinfection by-products. American Water Works Association Research Foundation and American Waterworks Association, Denver, CO, USA.

Chao, W.S. 2006. Evaluation of the Drinking Water Quality and Risk Assessment of TTHM of Kaohsiung City. Thesis of master, Taiwan.

Chen, M.J. Wu, K.Y. Chang, L. 2003. A New Approach to Estimate the Volatilization Rates of Volatile Organic Compounds during Showering. *Atmospheric Environment* 37: 4325–4333.

Chien, C.C., Kao, C.M., Dong, C.D., Chen, T.Y., Chen, J.Y. 2007. Effectiveness of AOC removal by advanced water treatment systems: a case study. *Desalination* 202: 318–325.

Chien, C.C. Kao, C.M. Chen, C.W. Dong, C.D. Chien, H.Y. 2008. Evaluation of biological stability and corrosion potential in drinking water distribution systems: a case study. *Environ Monit Assess* 153: 127–138.

Dell'Erba, A. Falsanisi, D. Liberti, L. Notarnicola, M. Santoro, D. 2007. Disinfection by-products formation during wastewater disinfection with peracetic acid. *Desalination* 215: 177–186.

Glauner, T. Waldmann, P. Frimmel, F.H Zwiener, C. 2005. Swimming pool water-fractionation and genotoxicological characterization of organic constituents. *Water Research* 39: 4494–4502.

Krasnera, S.W. Wright, J.M. 2005. The effect of boiling water on disinfection by-product exposure. *Water Research* 39: 855–864.

Nallanthigal, S.C. & Amadeo, R.F. 2012. Determination of volatile organic compounds in drinking and environmental waters. *Trends in Analytical Chemistry* 32: 60–75.

Sadiqa, R. & Rodriguez, M.J. 2004. Fuzzy synthetic evaluation of disinfection of by-products a risk-based indexing system. *Journal of Environmental Management* 73: 1–13.

Tokmark, B. Capar, G. Dilek, F.B. Yetis, U. 2004. The effect of boiling water on disinfection by-product exposure. *Water Research* 39: 855–864.

US EPA. 2002. Guidelines for Carcinogen Risk Assessment. Risk Assessment Forum, U.S. Environmental Protection Agency, *Washington DC*. NCEA-F-0644.

WHO (World Health Organization). 2006. Guidelines for Drinking-Water Quality, 3nd Edition. WHO, Geneva.

Wright, J.M. Murphy, P.A. Nieuwenhuijsen, M.J. Savitz, D.A. 2006. The impact of water consumption, point-of-use filtration and exposure categorization on exposure misclassification of ingested drinking water contaminants. *Science of the Total Environment* 366: 65–73.

Frontiers of Energy and Environmental Engineering – Sung, Kao & Chen (eds)
© *2013 Taylor & Francis Group, London, ISBN 978-0-415-66159-1*

Application of multivariate statistical analysis to evaluate the effects of natural treatment system on water quality improvement

C.Y. Wu, H.Y. Chiu, T.Y. Tu & C.M. Kao
Institute of Environmental Engineering, National Sun Yat-Sen University, Kaohsiung, Taiwan

T.N. Wu
Department of Environmental Engineering, Kun Shan University, Tainan, Taiwan

W.P. Sung
Department of Landscape Architecture, National Chin-Yi University of Technology, Taichung, Taiwan

ABSTRACT: In recent years, many natural treatment systems in Taiwan have been built for the purposes of wastewater treatment, river water purification, and ecology conservation. To evaluate the effectiveness of natural treatment systems on water purification, frequent water quality monitoring is needed. In this study, the multivariate statistical analysis was applied to evaluate the contaminant removal efficiency at a constructed wetland, and the time series method was then used to predict the trend of the indicative pollutant concentration in the wetland. In this study, a constructed wetland locates in the Kaoping River Basin was used as the study site. The statistical software SPSS was used to perform the multivariate statistical analysis to evaluate water quality characteristics and effectiveness of water purification. Results from this study show that the removal efficiencies were 98% for the Total Coliforms (TC), 55% for Biochemical Oxygen Demand (BOD), 53% for Chemical Oxygen Demand (COD), 55% for ammonia nitrogen (NH_3-N), and 39% for Total Nitrogen (TN). Results from the factor analysis show that 17 water-quality items of the study site could obtain four to six principal components, including nitrate nutrition factor, phosphorus nutrition factor, eutrophication factor, organic factor, and environmental background factor, the major influencing components were nutrition factor and eutrophication factor.

1 INTRODUCTION

Applying the natural treatment systems as alternative methods for river water quality improvement/purification and wastewater polishment has become a very popular issue in Taiwan because natural treatment systems have the benefits of operational simplicity and cost efficiency (El-Khatee et al., 2009). Wetlands are one of the most biologically productive natural ecosystems on earth, a large number of physical, chemical and biological processes are involved in these systems influencing each other (Tanveer et al., 2011; Seid et al., 2012). Human activities have already negatively influenced water quality and aquatic ecosystem functions (Dimitriou & Zacharias, 2010). This situation has generated great pressure on these ecosystems, resulting in a decrease of water quality and biodiversity, loss of critical habitats, and an overall decrease in the quality of life of local inhabitants (Herrear & Morales, 2009; Wu et al., 2010a). In this study, a constructed wetland [named Kaoping River Rail Bridge Constructed Wetland

(KRRBCW)] located next to the old Kaoping River Rail Bridge at Dashu District, Kaohsiung, Taiwan was selected as a case study site to evaluate the effectiveness of the natural treatment system on river water quality improvement. The major influents come from the local drainage system containing untreated domestic, agricultural, and industrial wastewater, and effluent from the secondary wastewater treatment plant of a paper mill.

Since the year of 2000, applying the natural treatment systems (e.g., Constructed Wetlands, CWs) as alternative methods for river water quality improvement/purification and wastewater polishment has become a very popular issue because natural treatment systems have the benefits of operational simplicity and cost efficiency (Wu et al., 2010b; María et al., 2011; Wu et al., 2012). Several studies on the historical behavior of real-scale constructed wetlands have been published (Wu et al., 2010a; Zhang et al., 2010; Vera et al., 2011). In this study, the Dashu Township located in the downstream of Kaoping River watershed was selected as the case study site. The major influents come from the local drainage system containing untreated domestic, agricultural, and industrial wastewater, and effluent from the secondary wastewater treatment plant of a paper mill. The KRRBCW commissioned in 2004, is said to be functioning well, and has probably reached final mature performance and at the same time provides an expanded wildlife habitat. In addition, the studied constructed wetland has been designed and is planted in such a way that wastewater treatment can be combined with recreational functions.

In this study, multivariate statistical methods were employed as the diagnosis tool to evaluate the factors affecting constructed wetland water quality and identify the domain of the potential contamination. This study provided a novel thought of applying statistical diagnosis approach on anatomizing constructed wetland monitoring data. Furthermore, the findings and discovered natures would improve the management constructed wetland, and thus, enhance the treatment efficiency.

2 MATERIALS AND METHODS

Figure 1 shows the schematic diagram of the KRRBCW. The wetland has two different systems, Systems A and B. The major influent for System A is the treated industrial wastewater from a paper mill. The major influent for System B is the water from local drainage system containing untreated domestic, agricultural, and industrial wastewaters. Systems A and B contain six (A1 to A6) and seven (B1 to B7) basins, respectively. The proposed statistical diagnosis approach was illustrated in Figure 2. Several steps of data processing were followed by data pre-processing, data reduction, information extraction, data clustering, information interpretation, neighborhood survey and statistical diagnosis (Wu et al., 2010a).

In this study, the SPSS-12.0 software was employed to conduct the regarding statistical analyses. Before subjecting to data processing, data selection and data standardization were the essential steps of data pre-processing. The criterion of

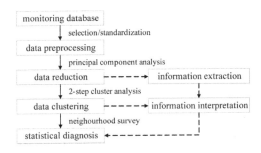

Figure 2. Data processing flowchart.

data selection was to abandon the variables that originally had more than 50% missing values or Not Detected (ND) values. Since 2005 to 2009, only 22 variables of water quality data remained after data selection. The measurement scales and numerical range of the original variables evaluated in this study varied widely, so standardization for all variables were performed. Z-score standardization method was employed that each variable within the original data matrix subtracted the column mean and then was divided by the column standard deviation. Principal Component Analysis (PCA), a method of factor analysis, was applied for data reduction (Matthias et al., 2012). The original p-dimensional standardized data matrix was transformed into m-dimensional Principal Component (PC) matrix with less degree of freedom, and thus the complexity of the data structure were simplified by applying PCA, the organic compounds and geochemical indicators were performed in accordance with APHA Methods (APHA, 2005).

Considering the correlations presented in the original data, PCs can still reserve inherent interdependencies of the data after reducing the data dimension. As a general rule, the first few PCs accounted for the majority of the variance within the original dataset, then the first one explained the most variance and each subsequent PC explained progressively less. The factor loadings were responsible for the correlations between PCs and selected variables, and those with the greatest positive and negative loadings made the largest contribution. In this case, water quality reflected the integrated effects of all underlying processes. Each PC obtained in PCA possibly corresponded to a specific type of water contamination. The loadings of each PC could provide more information to identify the sources that was responsible for the similarities of collected samples in water quality. As a result, each PC could be characterized by a certain mould of water contamination based on the extracted information.

Cluster Analysis (CA) was a method of data classification that classified true groups of dataset

Figure 1. Site map showing the wetland system.

according to their similarities to each other. In general, Euclidean distant was served as a measure of the similarity. A short Euclidean distant implied the high similarity between the measured objects, and vice versa. The dataset could be sorted to the distinct groups based on either the interaction among the variables or the interrelation among the samples. Two common types of CA methods were Hierarchical Cluster Analysis (HCA) and non-hierarchical cluster analysis. The HCA method could be implemented without prior knowing of the cluster numbers, but it was sometimes knotty to distinguish the clusters clearly as using HCA. Hence, the 2-step cluster analysis was employed in this study.

The HCA method was first used to identify the number of clusters by Ward's clustering procedure. Subsequently, the number of clusters was assigned as the number of the identified PCs in K-means's clustering procedure to obtain the correct classified observations. Accordingly, all samples were distributed into the pre-classified clusters through 2-step cluster analysis. The concept of statistical diagnosis was to integrate the information extracted from PCA and the message interpreted by clustering. In such a way, the similarity of the clustering to the PCA conforms to the interpretation of the PCA. The PCA results might discover the potential sources of water contamination, while the clustering results could appoint the monitoring wells to different contaminant sources. Furthermore, neighborhood survey was indispensable to conclude the nature of underlying processes affecting water quality.

3 RESULTS AND DISCUSSION

Investigation results (n = 9) show that the averaged inflow rate measured in the influent of the first basin (A1), HLR, and HRT for System A were 13,454 m^3/day, 0.08 m/day, and 5.5 days, respectively. The measured averaged inflow rate measured in the influent of the first basin (B1), HLR, and HRT for System B were 5,309 m^3/day, 0.04 m/day, and 13.3 day, respectively. Table 1 presents the averaged influent and effluent concentrations of SS, BOD, TN, TP, and TC as well as the Removal Efficiency (RE) for Systems A and B. Table 2 presents the variance rotated factor loading of 17 physic-chemical variables on 5 PCs.

Results indicate that significant variations in RE were observed in those five water quality indicators. Results show that more than 48% and 96% of BOD and TC removal were observed in Systems A and B, respectively. Results also show that significant TN removal were detected (52% for System A and 61% for System B). To successfully

Table 1. The averaged influent and effluent concentrations for major water quality indicators.

Items	System	Influent	Effluent	RE (%)
SS (mg/L)	A	10.2 ± 7.03	34.3 ± 18.2	–
	B	19.2 ± 16.4	50.9 ± 31.2	–
BOD	A	29.3 ± 20.2	12.1 ± 9.58	58
(mg/L)	B	29.3 ± 22.4	13.1 ± 11.0	48
TN (mg/L)	A	4.69 ± 2.73	2.04 ± 0.61	52
	B	5.64 ± 3.04	3.47 ± 1.74	61
TP (mg/L)	A	0.21 ± 0.20	0.42 ± 0.91	40
	B	0.63 ± 0.81	0.27 ± 0.36	66
TC	A	2.3E + 05	9.4E + 02	96
(CFU/ 100 mL)	B	8.0E + 05	5.7E + 03	97

Table 2. Variance rotated factor loading of 17 physic-chemical variables on 5 PCs.

Variables	PC 1	PC 2	PC 3	PC 4	PC 5
NOX	0.971	−0.095	0.022	0.035	−0.055
NO3-N	0.963	−0.090	0.015	0.035	−0.050
TN	0.916	0.321	−0.021	0.060	0.013
NO2-N	0.796	−0.152	0.130	0.016	−0.119
NH3-N	−0.011	0.879	0.021	−0.102	0.141
TKN	−0.171	0.873	−0.092	0.052	0.145
OP	0.019	0.719	0.309	−0.183	−0.176
TC	0.063	0.305	−0.233	−0.218	0.162
EC	−0.410	0.048	−0.772	0.054	0.139
TP	−0.143	0.129	0.716	0.024	0.226
Temp.	0.353	−0.107	0.359	−0.079	−0.155
Chl-a	−0.044	−0.011	0.006	0.799	0.027
DO	0.140	−0.315	−0.118	0.764	−0.056
pH	−0.021	−0.088	−0.438	0.458	−0.105
SS	0.095	0.123	0.131	0.431	−0.422
BOD	−0.003	0.101	0.346	0.016	0.821
COD	−0.165	0.149	−0.130	−0.091	0.810
Variance (%)	23.98	16.98	10.03	8.45	7.96
Cumulative variance (%)	23.98	40.97	51.00	59.46	67.42

reduce the data dimension, the requisite condition of running PCA was the strong linear correlation among the original data variables. The Kaiser-Meyer-Oklin (KMO) test and the Bartlett's sphericity test were commonly used to examine the suitability of applying PCA.

The KSP showcase passed the KMO test that obtained a calculated KMO value 0.611 greater than the acceptable value 0.5. The Bartlett's sphericity test provided a similar result as well showing a calculated χ^2 = 10,106 with probability

P < 0.01 and 136 degrees of freedom. Both the KMO test and the Bartlett's sphericity test had verified that PCA was suitable for the processes of data reduction. The major criterion of becoming a PC was its Eigen value greater than unity, meaning that the contribution of the PC on variance was more significant than the original variable.

As shown in Table 1, the obtained five PCs account for 67.42% of the variance or information contained in the original data set. The highlighted variables had their absolute values of loadings greater than 0.7 for the reason that it was an indicator of the participation of the variables in each PC. PC 1 accounted for 23.98% of the total variance and was characterized by very high loadings of NO_x, NO_3-N, NO_2-N and TN. Nitrogen-nutrients served as an indicator of and PC 1 was defined as the nitrogen factor. PC 2 added 14.54% of the total variance and was mostly participated by NH_3-N, TKN and OP. Nitrite and nitrate are the intermediate of ammonia oxidation, which sometimes serves as the sign of nutrient pollution relating agricultural activities. However, the co-existence of sulfate and nitrate could be resulted from the industrial leakage of acidic wastewater. PC 2 was appointed as nutrient pollution factor. PC 3 represented 10.43% of the total variance and was distinguished by EC and TP. Manganese and iron are the major constituents of the earth shell, and their presence in water is essentially ascribed to mineral dissolution. As a result, PC 3 was identified as the phosphorus factor. PC 4 explained 10.22% of the total variance and was mainly associated with very high loadings of Chl-a and DO. As a general rule, the mineral contents found in water samples were closely related to dissolution processes of geological formation in the studied area. Arsenic is released from the geologic formation in a reduction state. PC 4 was suitably assigned as the eutrophication factor. PC 5 gave 10.20% of the total variance and was contributed by BOD and COD. Accordingly, PC 5 was recognized as the organic matter factor. We observed strong and positive correlations: chloride and TDS (r = 0.999), chloride and conductivity (r = 0.997), chloride and hardness (r = 0.972), chloride and sulfate (r = 0.908), sulfate and iron (r = 0.704). The Kaiser-Meyer-Oklin (KMO) test carried out on the correlation matrix shows a calculated value, KMO = 0.764 greater than the acceptable value 0.5, thus meaning that PCA can successfully reduce the dimensionality of the original data set. The Bartlett's sphericity test provided a similar result as well showing a calculated χ^2 = 1239.9 (P < 0.01 and 78 degrees of freedom).

4 CONCLUSIONS

The studied constructed wetland fulfilled the requirements for wastewater purification and combined this function with the creation of a pleasing environment for public and wildlife. Results from this study show that the overall removal efficiencies were 96% for TC, 48% for BOD, and more than 40% for nutrients (e.g., TN, TP). The factor analysis results reveal that there were 17 water-quality items could obtain four to six principal components, including nitrate nutrition factor, phosphorus nutrition factor, eutrophication factor, organic factor, and environmental background factor, the major influencing components were nutrition factor and eutrophication factor. Results indicate that the wetland system was able to remove pollutants from the influent significantly. Thus, constructed wetlands are useful alternative for river water quality improvement.

ACKNOWLEDGEMENTS

This study was funded in part by Taiwan National Science Council. Additional thanks to the personnel of Taiwan Environmental Protection Administration and Environmental Protection Bureau of Kaohsiung County, and National Sun Yat-Sen University for their laboratory and field studies cooperation and assistance with this research project.

REFERENCES

APHA (American Public Health Association). 2005. Standard Methods for the Examination of Water and Wastewater. 23th Ed, APHA-AWWA-WEF, Washington, DC, USA.

Dimitriou, E. & Zacharias, I. 2010. Identifying microclimatic, hydrologic and land use impacts on a protected wetland area by using statistical models and GIS techniques, *Mathematical and Computer Modelling,* 51(3–4): 200–205.

El-Khatee, M.A., Al-Herrawy, A.Z., Kamel, M.M., & El-Gohary, F.A. 2009. Use of wetlands as posttreatment of anaerobically treated effluent. *Desalination,* 245: 50–59.

Herrera-Silveira & Morales-Ojeda. 2009. Evaluation of the health status of a coastal ecosystem in southeast Mexico: assessment of water quality, phytoplankton and submerged aquatic vegetation, *Marine Pollution Bulletin,* 59: 72–86.

María, H.V., Ricardo, S.C., Javier, M.V., M. Cruz, V.B., Josep, M.B., & Eloy, B. 2012. Statistical modelling of organic matter and emerging pollutants removal in constructed wetlands, *Bioresource Tech.,* 102(8): 4981–4988.

Matthias, R., Paul, A.W., Christopher, J.D., Constanze, T., Peter, D., & Sophie, E.B. 2012. Three-dimensional geological modelling and multivariate statistical analysis of water chemistry data to analyze and visualize aquifer structure and groundwater composition in the Wairau Plain, Marlborough District, New Zealand, *J. Hydrol*, 433–437: 13–34.

Seid, T.M., Pieter, B., Argaw, A.B., Asgdom, M., Zewdu, E., Addisu, S., Hailu, E., Menberu, Y., Amana, J., Luc, D.M., & Peter, L.M.G. 2012. Analysis of environmental factors determining the abundance and diversity of macroinvertebrate taxa in natural wetlands of Southwest Ethiopia, *Ecological Informatics*, 7(1): 52–61.

Tanveer Saeed & Guangzhi Sun. 2011. Kinetic modelling of nitrogen and organics removal in vertical and horizontal flow wetlands, *Water Res.*, 45(10): 3137–3152.

Vera, I., García, J., Sáez, K., Moragas, L., & Vidal, G. 2011. Performance evaluation of eight years experience of constructed wetland systems in Catalonia as alternative treatment for small communities, *Ecol. Eng.*, 37(2): 364–371.

Wu, C.I., Fu, Y.T., Yang, Z.H., Kao, C.M., & Tu, Y.T. 2012. Using multi-function constructed wetland for urban stream restoration. *Applied Mechanics and Materials*, 121–126: 3072–3076.

Wu, C.Y., Kao, C.M., & Lin, C.E. 2010b. Using a constructed wetland for non-point source pollution control and river water quality purification: a case study in Taiwan, *Wat. Sci. Tech.*, 61: 2549–2555.

Wu, M.L., Wang, Y.S., Sun, C.C., Wang, H., Dong, J.D., Yin, J.P., & Han, S.H. 2010a. Denitrification of coastal water quality by statistical analysis methods in Daya Bay, South China Sea, *Marine Pollution Bulletin*, 60(6): 852–860.

Zhang, L., Wang, M.H., Hu, J., & Ho, Y.S., 2010. A review of published wetland research, 1991–2008: Ecological engineering and ecosystem restoration, *Ecol. Eng.*, 36(8): 973–980.

Frontiers of Energy and Environmental Engineering – Sung, Kao & Chen (eds)
© 2013 Taylor & Francis Group, London, ISBN 978-0-415-66159-1

Theoretical study of measuring oil pump efficiency through thermodynamic method

J.G. Wang & Y.B. Ren
School of Mechanical Engineering, Xi'an ShiYou University, Xi'an, China

ABSTRACT: The main index of the fuel pump is the pump efficiency, to direct at traditional pump performance test standards unable to meet the requirements of the field performance test, based on the thermodynamic principle of the measurement methods in ISO5189. Proposed simplified formula applicable to field testing while using thermodynamic method to measure the pump efficiency, defined how to determine the various parameters in the formula. Approximately calculate the specific heat through state—contrast method, reference to the "water equivalent" concept, design experiments to seek specific heat, combine these two methods to determine the specific heat finally.

Keywords: oil pump; efficiency; thermodynamic approach; specific heat

1 INTRODUCTION

The classical method for measuring pump performance is hydraulics law. This method using the power balance principle. The ratios of the product of flow and total head to the input pump axis power is pump efficiency. But the vast majority do not have a straight pipe which is needed in pump stand-alone placed in the flow meter, the complexity of measuring the pump shaft power beyond the bounds of the capacity of scene, therefore, the site basically does not have the test conditions for the hydraulics. The thermodynamic method does not need to measure flow and shaft power, only measuring pump temperature difference and pressure difference between imports and exports will be able to determine the pump efficiency, and have a particularly simple and convenient features in the field, will play an underestimated role in the determination of the high lift pump efficiency.

2 THE PRINCIPLE OF THERMODYNAMIC METHOD FOR MEASURING PUMP EFFICIENCY

2.1 *The original efficiency formula*

The definition of the pump efficiency can be expressed as:

$A^{[1]}$ = energy absorbed by the equivalent entropy flow of fluid Ne

$B^{[1]}$ = the energy of the actual flow pump supply ΔN+ various types of losses $\Delta Em + Ex$ efficiency:

$$\eta = \frac{A}{B} \quad (1)$$

ΔEm is amendment for take into account the balance disc leakage and shaft seal leakage losses; Ex is external loss the shaft has provided but lost for the fluid away, including bearings, shaft seal friction, pump shell heat loss.

2.2 *Derivation formula*

According to the basic theory of the pump:
Effective power:

$$N_e = \rho g V H = mgH$$

In it,

$$H = (h_2 - h_1)_s + \frac{1}{2}(c_2^2 - c_1^2) + g(Z_2 - Z_1)$$

$(h_2 - h_1)_s$ is enthalpy rise in the isentropic process

c_1, c_2 are average speeds at imports and exports of the pump

Z_1, Z_2 are the heights at the inlet and outlet of the pump.

In the isentropic process,

$$(h_2 - h_1)_s = \bar{V}(P_2 - P_1)$$

$\bar{V}$ is the average specific volume of fluid

P_1, P_2 are pressures at inlet and outlet of the pump

In actual measurement, measurement location is usually taken as $c_1 = c_2$, $Z_1 = Z_2$

$$\eta = \frac{\overline{V}(P_2 - P_1)}{\overline{a}(P_2 - P1) + \overline{c}_p(T_2 - T_1) + \Delta Em + Ex} \quad (2)$$

$\overline{c}_p$ is specific heat at constant pressure, a is oil isentropic compression coefficient. It varies in small range, and express as $\overline{a}$, would gain correspond temperature value by checking chemical manual.

2.3 Final computation formula

In actual measurement, measurement location is usually taken as $c_1 = c_2$, $Z_1 = Z_2$

$$\eta = \frac{\overline{V}(P_2 - P_1)}{\overline{a}(P_2 - P_1) + \overline{c}_p(T_2 - T_1) + \Delta Em + Ex} \quad (3)$$

$\overline{c}_p$ is specific heat at constant pressure, a is oil isentropic compression coefficient. It varies in small range, and express as $\overline{a}$, would gain correspond temperature value by checking chemical manual.

3 EXPERIMENTAL METHODS

Because crude oil and its fraction is a system that extremely complex and can't do accurate analysis. its heat capacity under Constant pressure is hard to predict as predicting pure substances or given given the composition of the mixtu with the theoretical or semi- theoretical prediction method. Generally need to rely on the experimentally determined or experience related to calculate. Experimental methods are: (1) Direct method, in orther words, traditional calorimeter. (2) Comparative method, such as DSC. Although the former contains large workload, it is highly accurate, can be used as the standard method. In this article, of measures the specific heat capacity of the fluid in the pump through calorimetric method, in order to lay a solid foundation for the accurate measurement of the efficiency of the pump running.

3.1 Principle of experiment

When an isolated thermal system is in equilibrium, it has an initial temperature T_1, when the system has absorbed some external heat, it is in a new equilibrium, it has an end temperature T_2, if the system does not undergo chemical changes and phase transitions heat that the system aquire is:

$$Q = (m_1 c_1 - m_2 c_2 + \dots)(T_2 - T_1)$$

In the formula, m_1, m_2 ... are masses of the various substances which make up the system, c_1, c_2 ... are the specific heats of corresponding substances. While measuring specific heat of liquid. Usually use the calorimeter, agitator, thermometer or sensor, should also consider the heat exchange between the calorimetric system and the external environment. For simplicity without affecting the measurement results, heat capacity of all material in the calorimetric system except the liquid sample can be converted into heat capacity of considerable water which is z, called them "water equivalent[3]". We can considered like this, as long as the initial temperature and the end temperature of each in the experimental process is basically the same, to set the amount of thermal systems, and each measurement process is basically the same. We can make the "water equivalent" to maintain a constant. Calculate it through data processing. Becasue heating system is a purely resistive load, when given plus the voltage U through current I at both ends of heating plate inside calorie meter, the heat released by the heating wire within a certain time t is

$$Q = IUt$$

If the mass of the liquid sample is m, specific heat is c, initial temperature T_1 and end temperature T_2 of liquid sample before and after is energized, in the measurement without heat dissipation:

$$IUt = (mc + Z)(T_2 - T_1) \quad (4)$$

Calculate from (3)

$$c = \frac{1}{m}\left[\frac{IUt}{T_2 - T_1} - Z\right] \quad (5)$$

As long as the Z remain unchanged during the experiment, then we can calculate the c, also can obtained Z under this condition.

In the calorimeter barrel in the same thermal systems, each holds liquid sample of different mass, do many measurements, and maintain the initial temperature and the end temperature of each system while was measuring is basically the same, and time while was measuring is basically the same, then "water equivalent" z remain unchanged. As long as determined the mass m_i and corresponding total heat capacity of liquid in each measurements,

Table 1. Calculate pure substance specific heat.

Temperature T	Specific heat Calculated value	Specific heat Published values
85	302.31	295.35
105	310.65	304.39
125	319.26	313.33
145	328.64	322.36
165	347.21	331.30
185	336.24	340.32
205	340.26	349.18
225	351.50	358.04
245	361.29	366.92
265	369.97	376.78
285	378.56	384.63
295	381.54	389.05

Units of temperature T, specific heat capacity c_p are °C and $J/mol \cdot K$

Table 2. Calculate some cut.

Temperature T	Specific heat Calculated value
28	1.854
33	1.868
38	1.881
43	1.905
48	1.936
53	1.953
58	1.977
63	2.001
68	2.032
73	2.051
78	2.081
83	2.087

and then you can find specific heat capacity of the liquid which need to be tested through the method of fitting a straight line.

$$c_i = Z + m_i c \qquad (6)$$

shows that the total system heat capacity will vary linearly with mass m_i, slope of the linear is specific heat capacity, the intercept of the line is the "water equivalent" Z.

3.2 *Experimental results and analysis*

Here take ether[4] as example, use this device to measure and calculate the specific heat, verify the accuracy of this experimental method.

To further confirm the reliability of the test method, calculated the specific heat capacity of the ether by the experimental method. In result, error does not exceed 3%, fits the literature values well.

These results indicate that: inaccuracy in measuring the specific heat volume through the heating system is less than before and it comply with the requirements of engineering applications. In this way, we can use this experimental method to measure the the media within the pump.

Specific heat of liquid under 28–83 degrees have been calculated, We can use this data to calculate the efficiency of the pump where the fluide under that temperature state.

4 CONCLUSION

This paper mainly introduced the principle of detecting efficiency of oil pump, have derived the formula, have cleared how to determining every parameter especially the constant pressure specific heat, this laid the necessary theoretical basis for the designing how to test efficiency of oil pump. Compared with the previous research in this area, in this paper, there are several new breakthroughs in the domain of following aspects: have derived formula for calculating efficiency under normal conditions, have determined the ratio between the energy loss and useful work; Corresponding state methods to determine medium heat at constant pressure.

REFERENCES

[1] C.X. Li .The research progress in measuring efficiency of pump through thermodynamic method proceedings of North China Electric Power University, 2000.
[2] Y.T. Zhu, W.H. Qu, P.Y. Yu. Chemical Equipment Design Manual, Chemical Industry Press, 2004.
[3] L. Yu. Measure specific heat of liquid, proceedings of sichuan normal university, 2003.
[4] Y. Sun, determination of specific heat and enthalpy of oil with thermal conductivity and ether, Journal of Petroleum & Chemical, 1986.

Frontiers of Energy and Environmental Engineering – Sung, Kao & Chen (eds)
© 2013 Taylor & Francis Group, London, ISBN 978-0-415-66159-1

Author index

Printed and bound by CPI Group (UK) Ltd, Croydon, CR0 4YY

01/05/2025

01858565-0001